T0342191

E instein Gravity in a Nutshell

Einstein Gravity in a Nutshell

A. Zee

PRINCETON UNIVERSITY PRESS • PRINCETON AND OXFORD

Copyright © 2013 by Princeton University Press

Published by Princeton University Press, 41 William Street,
Princeton, New Jersey 08540
In the United Kingdom: Princeton University Press,
6 Oxford Street, Woodstock, Oxfordshire OX20 1TW

press.princeton.edu

Cover art by Jane Callister

Library of Congress Cataloging-in-Publication Data

Zee, A.
    Einstein gravity in a nutshell / A. Zee.
        pages   cm — (In a nutshell)
    Summary: "This unique textbook provides an accessible introduction to Einstein's general theory of
relativity, a subject of breathtaking beauty and supreme importance in physics. With his trademark blend
of wit and incisiveness, A. Zee guides readers from the fundamentals of Newtonian mechanics to the most
exciting frontiers of research today, including de Sitter and anti–de Sitter spacetimes, Kałuza-Klein theory,
and brane worlds. Unlike other books on Einstein gravity, this book emphasizes the action principle and
group theory as guides in constructing physical theories. Zee treats various topics in a spiral style that is easy
on beginners, and includes anecdotes from the history of physics that will appeal to students and experts
alike. He takes a friendly approach to the required mathematics, yet does not shy away from more advanced
mathematical topics such as differential forms. The extensive discussion of black holes includes rotating and
extremal black holes and Hawking radiation. The ideal textbook for undergraduate and graduate students,
Einstein Gravity in a Nutshell also provides an essential resource for professional physicists and is accessible
to anyone familiar with classical mechanics and electromagnetism. It features numerous exercises as well
as detailed appendices covering a multitude of topics not readily found elsewhere.   Provides an accessible
introduction to Einstein's general theory of relativity   Guides readers from Newtonian mechanics to the
frontiers of modern research   Emphasizes symmetry and the Einstein-Hilbert action   Covers topics not found
in standard textbooks on Einstein gravity   Includes interesting historical asides   Features numerous exercises
and detailed appendices   Ideal for students, physicists, and scientifically minded lay readers   Solutions manual
(available only to teachers) "— Provided by publisher.
    Includes bibliographical references and index.
    ISBN 978-0-691-14558-7 (hardback)
    1. General relativity (Physics)—Textbooks.   I. Title.
QC173.6.Z44   2013
530.11—dc23   2012040613

British Library Cataloging-in-Publication Data is available

This book has been composed in Scala LF with ZzTEX
by Princeton Editorial Associates Inc., Scottsdale, Arizona

Printed on acid-free paper ∞

Printed in the United States of America

10 9 8 7 6 5 4

To WW and Max

Contents

II Part II: Action, Symmetry, and Conservation

III Part III: Space and Time Unified

IV Part IV: Electromagnetism and Gravity

TWO Book Two: From the Happiest Thought to the Universe

V Part V: Equivalence Principle and Curved Spacetime

$\boxed{\text{X}}$ Part X: Gravity Past, Present, and Future

Preface

Not simple, but as simple as possible

> Physics should be made as simple as possible,
> but not any simpler.
> —A. Einstein

Einstein gravity should be made as simple as possible, but not any simpler.

My goal is to make Einstein gravity* as simple as possible. I believe that Einstein's theory should be readily accessible to those who have mastered Newtonian mechanics and a modest amount of classical mathematics. To underline my point, I start with a review of $F = ma$.

Seriously, what do you need to know to read this book? Only some knowledge of classical mechanics and electromagnetism! So I fondly imagine, perhaps unrealistically. More importantly, you need to be possessed of what we theoretical physicists call sense— physical, mathematical, and also common.

I wrote this book in the same spirit as my *Quantum Field Theory in a Nutshell*.[1] In his *Physics Today* review of that book, Zvi Bern wrote this lovely sentence aptly capturing my pedagogical philosophy: "The purpose of Zee's book is not to turn students into experts— it is to make them fall in love with the subject." I might extend that to "fall in love with the subject so that they might desire to become experts." Here I am echoing William Butler Yeats, who said, "Education is not the filling of a pail, but the lighting of a fire."

---

* Also known as general relativity.

A portion of this book can be used for an undergraduate course. I have done it, and I provide a detailed course outline later in this preface.

Accessible is not to be equated with dumbed-down or watered-down. Also, accessible is not necessarily the same as elementary: in the last parts of the book, I include some topics far beyond the usual introductory treatment.

My strategy to make Einstein gravity as simple as possible has two prongs. The first is the emphasis on symmetry. As some readers may know, I have written an entire book[2] on the role of symmetry in physics, and I absolutely love how symmetry guides us in constructing physical theories, a notion that started with Einstein gravity, in fact. The second is the extensive use of the action principle. The action is invariably simpler than the equations of motion and manifests the inherent symmetry much more forcefully. I can hardly believe that some well-known textbooks on Einstein's theory barely mention the Einstein-Hilbert action. Symmetry and the action principle constitute the two great themes of theoretical physics.

To get a flavor of what the book is about, you might want to glance at the recaps first; there is one at the end of each of the ten parts of the book.

How difficult is Einstein gravity?

> Any intelligent student can grasp it without too much trouble.
> —A. Einstein, referring to his theory of gravity

When Arthur Eddington returned from the famous 1919 solar eclipse expedition that observed light from a distant star bending in agreement with Einstein gravity, somebody asked him if it were true that only three people understood Einstein's theory. Eddington replied, "Who is the third?" The story, apocryphal[3] or not, is one of many[4] that gives Einstein's theory its undeserved reputation of being incomprehensible.

I believe that in some cases, people like to persist in believing that Einstein's theory is beyond them. A renowned philosopher who is clearly well above average in intelligence (and who understands things that I find impossible to understand) once told me that he was tired of popular accounts of general relativity and that he would like to finally learn the subject for real. He also emphasized to me that he had taken advanced calculus[5] in college, as if to say that he could handle the math. I replied that, for a small fee, my impecunious graduate student could readily teach him the essence of general relativity in six easy lessons. I never heard from the renowned philosopher again. I was happy and he was happy: he could go on enunciating philosophical profundities about relative truths[6] and physical reality.

The point of the story is that it is not that difficult.

For whom is this book intended

Experience with my field theory textbook suggests that readers of this book will include the following overlapping groups: students enrolled in a course on general relativity, students and others indulging in the admirable practice of self-study, professional physicists in other research specialties who want to brush up, and readers of popular books on Einstein gravity who want to fly beyond the superficial discussions these books (including my own[7]) offer. My comments below apply to some or all of these groups.[8]

Personally, I feel special sympathy for those studying the subject on their own, as I remember struggling[9] one summer during my undergraduate years with a particularly idiosyncratic text on general relativity, the only one I could find in São Paulo back in those antediluvian times. That experience probably contributed to my desire to write a textbook on the subject. From the mail I have received regarding *QFT Nut,* I have been pleasantly surprised, and impressed, by the number of people out there studying quantum field theory on their own. Surely there are even more who are capable of self-studying Einstein gravity. All power to you! I wrote this book partly with you in mind.

Serious students of physics know that one can't get far without doing exercises. Some of the exercises lead to results that I will need later.

Quite naturally, I have also written this book with an eye toward quantum field theory and quantum gravity. While I certainly do not cover quantum gravity, I hope that the reader who works through this book conscientiously will be ready for more specialized monographs[10] and the vast literature out there.

So, I prevaricated a little earlier. In the latter part of the book, occasionally you will need to know more than classical mechanics and electromagnetism. But, to be fair, how do you expect me to talk about Hawking radiation, a quintessentially quantum phenomenon, in chapter VII.3? Indeed, how could we discuss natural units in the introduction if you have never heard of quantum mechanics? For the readers with only a nodding acquaintance with quantum mechanics, the good news is that for the most part, I only ask that you know the uncertainty principle.

I do not doubt that some readers will encounter difficult passages. That's because I have not made the book "any simpler"!

In the preface to the second edition of my quantum field theory book, I mentioned that Steve Weinberg and I, each referring to his own textbook, each said, "I wrote the book that I would have liked to learn from." So this is the book I would have liked as an undergrad* eager to learn Einstein gravity. I would have liked having at least a flavor of what the latest

---

* In a letter to the editors of *Physics Today* in 2005, A. Harvey and E. Schucking wrote that, in view of the "monumental lip service" paid to Einstein in the physics community, "it is a scandal" that Einstein gravity is still not regularly taught to undergraduates. I find it even more of a scandal that many physics professors proudly profess ignorance of Einstein gravity, saying that it is irrelevant to their research. Yes, maybe, but this is akin to being proudly ignorant of Darwinian evolution because it is irrelevant to whatever you are doing.

excitement was all about. In this spirit, I offer chapter X.6 on twistors, for example, trusting the reader to be sophisticated enough to know that all one should expect to get from a single textbook chapter is an entry key to the research literature rather than a complete account of an emerging area.

The importance of feeling amazed, and amused

I am amazed that students are not amazed.

The action principle amazed Feynman when he first heard about it. In learning theoretical physics, I was, and am, constantly amazed. But in teaching, I am amazed that students are often not amazed. Even worse, they are not amused.

Perhaps it is difficult for some students to be amazed and amused when they have to drag themselves through miles of formalism. So this exhortation to be amazed is related to my attempt to keep the formalism to an absolute minimum in my textbooks and to get to the physics.

To paraphrase another of my action heroes, students should be required to gasp and laugh[11] periodically. Why study Einstein gravity unless you have fun doing it?

As much fun as possible

Bern started his review of my quantum field theory textbook thus:

> When writing a book on a subject in which a number of distinguished texts already exist, any would-be author should ask the following key question: What new perspectives can I offer that are not already covered elsewhere? . . . perhaps foremost in A. Zee's mind was how to make *Quantum Field Theory in a Nutshell* as much fun as possible.

Good question! My answer remains the same. I want to make Einstein gravity as much fun as possible.

Sidney Coleman, my professor in graduate school and thesis advisor, once advised me that theoretical physics is a "gentleman's diversion." I was made to understand that I should avoid doing long sweaty calculations. This book reflects some of that spirit. Thus, in chapter VI.1, instead of deriving Einstein's field equation as a true Confucian scholar would, I try to get to it as quickly as possible by a method I dub "winging it southern California style." Similarly, in chapter VI.2, I get to cosmology as quickly as possible.

This invariably brings me to the dreaded topic of drudgery in general relativity. Many theory students in my generation went into particle physics rather than general relativity to avoid the drudgery of spending an entire day calculating the Riemann curvature tensor. I did.[12] But that was the old days. Nowadays, students of general relativity can use ready-made symbolic manipulation programs[13] to do all the tedious work. I strongly urge you, however, to write your own programs, as I did, rather than open a can. It also goes without saying that you should calculate the Riemann curvature tensor from scratch at least a few times to know how all the cogs fit together.

You make the discoveries

My pedagogical philosophy is to let students discover certain things on their own. Some of these lessons evolved into what I call extragalactic fables. For example, in part IV, I let the extragalactic version of you discover electrodynamics and gravity. In chapter IV.3, you discover that gravity affects the flow of time.

I also whet your appetite by anticipating. For example, I mention the Einstein-Rosen bridge already in chapter I.6. In working out the shortest distance between two points in chapter II.2, I mention that you will encounter the same equations when you study motion around black holes. In part II, I note that the peculiar replacement of a simple equation by a more complicated looking equation foreshadows Einstein's deep insight about gravity to be discussed in part V.

The return of Confusio

Readers of *QFT Nut* might be pleased to hear that Confusio makes a return appearance, together with other characters, such as the Smart Experimentalist. Some other friends of mine, for example the Jargon Guy, also show up. Here I am alluding to what Einstein referred[14] to as "more or less dispensable erudition."

An outline of this book

This book appears to start at a rather low level, with a review of Newtonian mechanics in part I. The reason is that I want to treat two topics more thoroughly than usual: rotations and coordinate transformations. A good understanding of these two elementary subjects allows us to jump to the Lorentz group and curved spacetime later. My pedagogical approach is to beat 2-dimensional rotations to death. Depending on how mechanics is taught, students typically miss, or fail to grasp, some of the material in the chapter on tensors. I repeat the discussion of tensors under various guises and in different contexts. One of my students who read the book points to various places where I appear to repeat myself, but I told her that it is better to hear some key point for the third time[15] than not to have understood it at all. A respected senior colleague and pioneer in Einstein gravity said to me that a good teacher is someone who never says anything worth saying only once.

I devote part II to a discussion of the all-important action principle, because I believe that it provides the quickest, and the most fundamental, way to Einstein gravity (and to quantum field theory). Part III is devoted to special relativity but, in contrast to some elementary treatments, the emphasis is on geometry and completion, not on a collection of paradoxes. In part IV, as was mentioned earlier, I let you discover electromagnetism and gravity, and so the treatment is somewhat nonstandard. Thus, even if you feel that you already know special relativity, you might want to take a quick look at part III and part IV.

Many readers probably pick up this book because of a burning desire to learn Einstein gravity. These readers would have already mastered Newtonian mechanics and special relativity, and they could probably cut to the chase and skip directly to part V. To them, the first four parts may appear to be a rather leisurely preparation for Einstein gravity. Still, I would counsel skimming, rather than skipping, the first four parts. At the very least, parts I–IV set down the conventions and notation. More importantly, they offer up the ideology of this text, an ideology that can be simply stated: action!

While I appear to start slow in parts I–III, I am actually setting things up so that we can go fast in parts V and VI. For example, all the discussion about coordinate transformation and curved spaces is to prepare the reader for a quick plunge into curved spacetime in chapter V.1. Similarly, the action principle enables the geodesic equation to be introduced early on, in part II, so that it is "ready to trot" when needed in part V. In considering whether to sign up for my course that grew into this book, some students ask how fast I will be zooming through special relativity to get to the "good stuff." But special relativity is good stuff! In particular, it is essential to understand special relativity as the geometry of spacetime* before moving on to general relativity.

The essence of Einstein gravity is explained in parts V and VI. The rest of the book contains what may be regarded as applications of the theory as developed in part VI. Part X contains extras that some might consider beyond the scope of an introductory text. The title is thus something of a misnomer, but to please my publisher, I am obliged to keep up a running joke I started with my field theory book. A better title might be *Gravity from Newton to the Brane World*.

The role of appendices

As a textbook writer, I am torn between being concise and being complete. One way out is to place numerous topics in appendices to various chapters. Some are fun, such as Einstein's derivation of $E = mc^2$ in his 1946 Haifa lectures (see chapter III.6), which, unfortunately, is in danger of being forgotten and which I much prefer to his 1905 derivation. Another example is Weyl's shortcut to the Schwarzschild solution (see chapter VI.3). Some are results I will need later, but often much later. For example, I talk about the speed of sound in an appendix to chapter III.6, but I won't need it until I get to the cosmic microwave background. Some appendices are peripheral or technical. When possible, I try to give an intuitive and heuristic understanding before launching into a long development, such as

---

* A multitude of books treat special relativity, but while they all get the job done, they differ widely in conceptual clarity. Besides the geometrical view of special relativity, I also want to emphasize the Lorentz action as leading to a unified approach to both massive and massless particles.

the treatment of Fermi normal coordinates. Some are for enrichment. In sum, the use of appendices represents my effort to appeal to a broad range of readers with enormously different levels of knowledge and sophistication. The reader should not feel obliged, upon first reading this book, to study all the appendices. Each should exercise his or her own judgment.

Still, a book this size is inevitably incomplete, and so it comes down to the author's choice (of course). So many beautiful results, so little space and time! I regard certain topics, though important, as better covered in more specialized tomes, such as gravitational lensing, and prefer to include some topics not discussed in several standard textbooks, such as anti de Sitter spacetime, brane worlds, and twistors.

The most incomprehensible thing about some physics textbooks

> The most incomprehensible thing about the physical world is that it is comprehensible.
>
> —A. Einstein

The most incomprehensible thing about some physics textbooks is that they are incomprehensible.

They manage to render the easily comprehensible into the nearly incomprehensible. Some textbook writers are simplifiers, others are what I call complicators. In defiance of Einstein's exhortation, many authors strive to make physics as complicated as possible, or so it seems to me. In the research literature, the cause of obscurity may be unintentional or intentional: either the author has not understood the issues involved completely (often laudably so, when the author is at the cutting edge), or the author wants to impress upon the reader the profundity of his or her paper by resorting to obfuscations. But in a textbook?

My task, and hope, in my textbooks is to make physics as simple as possible, as the "old man" with his toy[16] said. Having written both a textbook and a couple of popular books, I am perhaps qualified to express my opinions here. Popular books attempt to make physics simpler than it really is, thus in some sense deceiving the reader. Textbooks are different: they must make the reader work to master the subject. But making the reader work is not the same as making the reader suffer by rendering simple things obscure.

No bijective maps in this book

I am puzzled by students who profess no trouble with the physics but moan* about the math. All the "grown-ups" would say the opposite. The pros regard Riemannian geometry,

---

* Indeed, many of the postings on the sites of online booksellers regarding general relativity texts lament the difficulty of the math. At the other extreme, a few, by misguided individuals in my opinion, complain about the lack of rigor.

which is after all totally logical and algorithmic, as easy, but continue to lose sleep over Einstein's theory. Regarding the math, I can say, with only slight exaggeration, that mastery of the index notation and the chain rule almost suffices. Indeed, any serious student with a future in theoretical physics should be continually puzzled by the physics but not at all by the math.

Einstein did not say that physics should be made simple. Of course, physics is not simple, and understanding Einstein's theory does require effort. Surely you have heard that Einstein gravity involves curved spacetime, so there is no getting around learning the language needed to describe curvature. My strategy is to introduce math only when necessary, and then to illustrate the key concepts with plenty of examples. I dislike the Red Army[17] approach, and so I do not start by defining bundles on the tangent plane. I bring in the math gently and sneak in curvature early on via the familiar change of coordinates.

As for rigor, I will let yet another of my action heroes speak. "I'll differentiate any function, even the freaking delta function, as many times as I darn well please." So if you have to differentiate, just differentiate until the expression you are differentiating starts bleating for mercy. The trick is to know when it is absolutely necessary to be rigorous (which is seldom—I would never say never).

I respectfully submit that this book is not for those who want rigor.

While I realize the need for and the benefit of precise definition, for the most part I simply plead membership in the Feynman[18] "Shut up and calculate" school of physics.[19] Thus, I won't trouble your sleep with assertions such as "A bijective differentiable map of a manifold, whose inverse is also differentiable, is called a diffeomorphism." Regarding statements like this, I think that another Einstein quote may be apropos: "We should take care not to make the intellect our god; it has, of course, powerful muscles, but no personality."[20] Yet another relevant quote: "The people in Göttingen sometimes strike me, not as if they wanted to help one formulate something clearly, but instead as if they wanted only to show us physicists how much brighter they are than we."[21] Alas, "the people in Göttingen" have now gone off and multiplied,* and some even live in our midst. Precise definitions are indeed necessary occasionally, but by and large, they don't do much good in theoretical physics. Some things are better left undefined. In this connection, also keep in mind the distinction between true clarity and false clarity.[22] For example, I consider the insistence on saying "pseudo-Riemannian manifolds" in a book of this level false clarity at best.

As I was putting the finishing touches on this book, I read about some notes[23] Feynman scribbled to himself before teaching some course: "First figure out why you want the student to learn the subject and what you want them to know, and the method will result more or less by common sense." Well said! As it turned out, that was the method I followed when writing this book.

If you feel that bijection is indispensable for your existential essence, then I also respectfully submit that this book is not for you.

---

* One tribe is known to look at "old fashioned" indices with contempt. Only coordinate-free notations[24] are good enough for them.

But of course I am not against mathematics. For instance, I am all for differential forms (see chapters IX.7 and IX.8). However, when faced with a new formalism, I tend to be practical and ask, "For the time invested in learning it, what is the payoff?" How significant is it for the physics?

Teaching from this book and self-studying

It would be ideal to teach a leisurely year-long course based on this book. But I have also taught Einstein gravity at the University of California, Santa Barbara, as a scandalously short one-quarter undergraduate course consisting of only 29 lectures. The students allegedly knew the action principle and special relativity, but I was appropriately skeptical. Here is the actual course plan.

Lecture 1 gives an overview. Lectures 2–6 cover chapters I.5 and I.6, starting with the notion of a metric and illustrated with numerous examples, including the Poincaré half plane, and ending with locally flat coordinates and a count of the components contained in the curvature tensor. Lectures 7 and 8 cover part II, and lectures 9 and 10 part III. In lectures 11 and 12, I let the students discover electromagnetism and gravity and derive how gravity affects the flow of time. Lectures 13–15 introduce the equivalence principle and cover part V up to chapter V.3, ending with closed, flat, and open universes.

The second half of the course proceeds as follows:

Lecture 16: the geodesic equation reduced to Newton's equation, gravitational redshift, spherically symmetric spacetime with time dependence

Lecture 17: the motion of particles and light in static spherically symmetric spacetime

Lecture 18: covariant differentiation, the geometrical picture

Lecture 19: to Einstein's field equation as quickly as possible

Lecture 20: the Riemann curvature tensor and its symmetry properties

Lecture 21: the Einstein-Hilbert action

Lecture 22: the cosmological constant and the expanding universe

Lecture 23: Schwarzschild metric, with precession of planets and radar echo delay described in words and pictures

Lecture 24: the energy momentum tensor

Lecture 25: general proof of energy momentum conservation

Lecture 26: the Einstein tensor and the Bianchi identity

Lecture 27: black holes in various coordinates

Lecture 28: the causal structure of spacetime

Lecture 29: Hawking radiation and a grand review

So it is entirely possible to cover the bulk of this book in a one-quarter course! I did it. Students were expected to do some reading and to fill in some gaps on their own. Of course, instructors could deviate considerably from this course plan, emphasizing one topic at the expense of another. They might also wish to challenge the better students by assigning the appendices and some later chapters.

Here I come back to those I applauded earlier for self-studying Einstein gravity. Some of you might want to know which chapters to read. The answer is of course that you should read them all, in an ideal world. But if you want to get "there" quickly, I suggest the following. You are on your own regarding the first three parts: it all depends on what you already know. So try starting with part IV and see how often you need to refer back to an earlier chapter. Part V is indispensable, particularly the equivalence principle and the tour of curved spacetimes. You need to understand the covariant derivative, but you could skip the somewhat heavier appendices in chapter V.6. After the covariant derivative, you are ready for the heart of the matter, Einstein's field equation, in chapter VI.1. The rest of part VI forms the core of a traditional course on general relativity, but my emphasis is somewhat less on working out orbits in detail. That's it! You would have then reached a certain level of mastery of Einstein gravity. You could then regard the rest of the book, parts VII–X, as a buffet of topics that you could browse at your leisure. Part X contains more speculative topics, including some that may not be of lasting value. Be warned!

Acknowledgements

I thank Lasma Alberle, Nima Arkani-Hamed, Yoni BenTov, Zvi Bern, Ta-pei Cheng, Karin Dahmen, Stanley Deser, Doug Eardley, Martin Einhorn, Joshua Feinberg, Matthew Fisher, Gary Gibbons, David Gross, John Haberstroh, Christine Hartmann, Gavin Hartnett, Gary Horowitz, Steve Hsu, Ted Jacobson, Shamit Kachru, Joshua Katz, Zoltan Kunszt, Josh Lapan, Ian Low, Slava Mukhanov, Zohar Nussinov, Don Page, Joe Polchinski, Rafael Porto, John Rehr, Subir Sachdev, Richard Scalettar, Nicholas Scherrer, Eva Silverstein, Andy Strominger, Bill Unruh, Daniel Walsh, Richard Woodard, and Tzu-Chiang Yuan for reading one or more chapters of my draft at various points along the way and for their very helpful comments. I am especially grateful to Lasma Alberle, Yoni BenTov, Stanley Deser, Joshua Feinberg, and Christine Hartmann for their careful reading and suggestions. Yoni BenTov convinced me to switch from the $(+ - - -)$ to the $(- + + +)$ signature. See the collection of formulas in the back of the book and the table of sign conventions.

I wrote the bulk of this book in Santa Barbara, California, but some parts were written while visiting the Academia Sinica in Taipei, the Republic of China. I deeply appreciate the generous hospitality of Maw-kuen Wu, the director of the Institute of Physics. Bits and pieces were also written in Munich, Germany, and in Beijing, the People's Republic of China.

My editor at Princeton University Press, Ingrid Gnerlich, has always been a pleasure to talk to and work with. She has listened patiently to my ranting and raving for years. For my copyeditor, I am delighted to have, once again, Cyd Westmoreland, who worked on both editions of my *Quantum Field Theory in a Nutshell*. I am much impressed by the meticulous work of Peter Strupp and Princeton Editorial Associates. I also thank Craig Kunimoto for his indispensable computer help.

I am deeply grateful to my wife Janice for her loving support and encouragement throughout the writing of this book. As this book was nearing completion, she gave birth to our son Max.

Notes

1. Hereafter referred to as *QFT Nut*.
2. A. Zee, *Fearful Symmetry*. Hereafter, *Fearful*.
3. See chapter VI.3.
4. Chaim Weizmann, the first president of Israel and a chemist, once crossed the ocean with Albert Einstein on the same liner, and Einstein tried to explain the theory of relativity to him. When asked about this later, Weizmann said something like "I did not understand his theory, but he certainly convinced me that he did."
5. For the record, I took a philosophy course in college. To further emphasize that I am not totally lacking in "philosophical credentials," I was once invited by a philosophy professor to lecture, thanks to one of my popular books, to an auditorium full of philosophers. I like philosophers.
6. Einstein once said that he should have called his work "invariance theory" and lamented his use of the word "relative."
7. A. Zee, *An Old Man's Toy*. Hereafter, *Toy/Universe*.
8. In my introduction to Feynman's book on quantum electrodynamics, I wrote about three different kinds of readers of that book. Only part 0 of this book will be comprehensible to the first kind. See R. P. Feynman, *QED: The Strange Theory of Light and Matter,* with a new introduction by A. Zee, Princeton Science Library, 2006.
9. An undergrad friend had also deluded me into thinking that it was salutary to read Einstein in the original German!
10. Read J. Polchinski, *String Theory,* for example.
11. *QFT Nut*, p. 473.
12. For the record, I started my research career with John Wheeler, studying gravitational wave emission from neutron stars. For Wheeler's influence on his students, see Charles W. Misner, "John Wheeler and the Recertification of General Relativity as True Physics," in *General Relativity and John Archibald Wheeler,* ed. I. Ciufolini and R. Matzner, Springer, 2010.
13. See my remarks in chapter IX.9, for example.
14. A. Einstein, *Autobiographical Notes*, Open Court, 1999.
15. In any case, if you think that I talk too much about tensors, you could simply feel smugly superior to those poor souls who never get it.
16. See *Toy/Universe*. Also see figure 2b in the prologue to book two.
17. I learned this terminology (which, I should clarify, referred to the Russian, not the Chinese, version) in a conversation with Steve Weinberg about textbooks. It has something to do with lining up all the tanks first.
18. A colleague who got his doctorate at Caltech told me the following story. He was examined by a committee consisting of Feynman and a bunch of lesser lights. One of the lesser lights posed a question to my friend, who proceeded to answer it perfectly, outlining the calculation necessary and explaining the physical significance of the result. The lesser light then opined ominously, "You should have also said . . . " and hereforth issued from his mouth a long string of highfalutin hundred-dollar words. Feynman turned to the lesser light and announced to the rest of the room, "But that's exactly what he said!"

    Here is a totally gratuitous Feynman story that has nothing to do with the discussion at hand. During the exam, Feynman asked a question about quantum mechanics that the student was unable to answer. Feynman exploded, saying something like "Quantum mechanics was invented in the 1920s and it's now 1972; you really should have mastered quantum mechanics by now!" A committee member turned to Feynman and said softly, "Dick, Dick, it's now 1973."
19. A colleague told me his retort to Feynman: "Shut up and contemplate." Of course, Feynman is capable of doing both. Contrary to myth, Feynman won the national Putnam mathematics competition. Here we are talking about people who can only talk and not calculate.

20. The quote is possibly apocryphal.

21. Quoted in C. Reid, *Hilbert*, Springer, 1996, p. 142.

22. As one of my professors, an exceedingly distinguished theoretical physicist, used to say, the main purpose of all the talk about tangent bundles and pullback is to frighten young children. This is not entirely true, but, oh well.

23. R. P. Feynman, R. B. Leighton, and M. Sands, *The Feynman Lectures on Physics*, volume III, Addison Wesley (Commemorative issue 2004), p. xi.

24. I am certainly not against coordinate-free notations. In physics, the only issue is which notation is best suited for the job at hand. Coordinate-free notations are great for proving general theorems but are not so good for calculating. In this connection, I might regale the reader with a story. At a recent Santa Barbara conference on black holes, dS, AdS, gravity dual, and so on—in short, the latest hot stuff—I was chatting at lunch with two leading young researchers, up and coming stars, not some aging curmudgeons with congealed opinions. When I mentioned how some people clamored for index-free notations, one of these two leading lights basically said to please get those people out of her sight. The other told me a more illuminating story. During grad school, to deepen his understanding of Einstein gravity, he enrolled in a course taught by a famous mathematician. As it happened, he was the only student able to do the problems in the final exam involving actual calculations: he did them by first using old fashioned indices and then translating back into the abstract notation used in the course.

    The index-free notation in Einstein gravity is somewhat analogous to using vectors without committing to any specific coordinate choice. For example, one can prove easily that $\vec{L} = \vec{r} \times \vec{p}$ is conserved, but try to do the spinning top on an oscillating inclined plane without setting up coordinates! The difference between the uninitiated and the misinformed is that the uninitiated is not acquainted with a particular formalism, while the misinformed insists that only the particular formalism he or she likes is any good.

Part 0 | Setting the Stage

Prologue

Three Stories

Story 1: The drowning beauty and the scrawny lifeguard

Since I started my quantum field theory text[1] with a story, possibly apocryphal, about Feynman in a quantum mechanics class, I feel compelled to start this text also by telling a story, possibly true,[2] about Feynman. The movie opens on a gorgeous southern California beach. We zoom in on a lifeguard, noticeably scrawnier than the other lifeguards. But on the other hand, we soon discover that he is considerably smarter. Egads, it is Dick Feynman, in the days before Baywatch! Perched on his high chair, he has been watching an attractively curvaceous swimmer with great interest, plotting how he could win the girl's affection, all the while solving a field theory problem in his head. Suddenly, he notices that the girl is splashing about frantically. She is going under! Must be a cramp! An action hero is as an action hero does: Feynman jumps down from his lookout and goes into action.*

The other lifeguards are already proceeding in a straight line (starting from point F, the lifeguard station, in figure 1, going along the dotted line) toward the girl (at point G). That would be the path of least distance. But no, Feynman has already calculated the path that would allow him to reach the girl in the least amount of time. Time counts more than space here: least time trumps least distance. Our hero (like other humans) can run much faster, even on a soft sandy beach, than he can swim. So the rescuer should spend more time running before plunging into the sea. A simple high school level calculation (exercise 1) shows Feynman the best path to take (see the solid line in figure 1). Our hero beats the other guys and gets to the eternally grateful girl first!

---

* "Physics is where the action is." See chapter III.2.

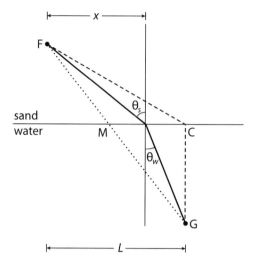

Figure 1 The best possible path for Feynman to follow to get to the drowning girl is along the solid lines from F to G.

But you don't have to calculate to see that there is an optimal path. Only a cretin would follow the third path (the dashed line) shown in the figure!

In the 17th century, Fermat discovered that light, just like Feynman, also follows a least time principle, and as a result "bends" as it enters from one medium (say, air) into another (say, water). To read these very words, you have, or rather your saintly mother has, cleverly positioned in your eyes a blob of watery substance (known to the cognoscenti as a lens) that you squeeze just so, using tiny muscles, to bend light to your advantage and bring the ambient light bouncing off these words on the printed page into focus. Your mother, as the product of eons of evolution, was oh so clever, giving you eyes. As we speak (so to speak), you are using precisely this phenomenon of light bending to save the light entering your eyes some time, a phenomenon known as refraction, and to gain yourself some knowledge about physics and the universe—an activity evolution applauds: reading this book could conceivably boost your reproductive advantage.

We all know that light travels in a straight line, but we also notice easily that when light enters water from air, it bends (as shown in figure 1 with "sand" replaced by "air"). Indeed, that explains why people standing in swimming pools appear to have comically short legs,* a phenomenon you can test by sticking a pencil in a glass of water.

It has also been known ever since Euclid† that the shortest path between two points is a straight line. Ergo, if light is always in a hurry to get from one point to another, it

---

* If you can't explain that, see figure 7.1 in *Fearful*. See also the common mirage shown in figure 7.2: on a hot day, the highway beneath a distant car appears to be wet, but is in fact dry. This mirage shows that light only cares about the local, not the global, minimum in time of transit.

† Babies have no need for Euclid; as soon as they can crawl, they move toward the obscure objects of their desire along a straight line.

wants to move in a straight line. Fermat and others realized that the bending of light could be explained if light moves more slowly in water than in air. Indeed, if light were really stupid, it would move in a straight line through point M to get from F to G, just like the other lifeguards.

Story 2: An ant and her honey

When I was a kid, I was challenged by a puzzle about an ant and a drop of honey. An ant located on the outside of a cylindrical glass of radius R and a vertical distance d below the rim, sees, never mind how, or perhaps smells, a drop of honey directly opposite her, but on the inside of the glass (see figure 2a). The ant wants to get to the honey in the shortest possible time,[3] crawling at some constant speed.

The solution depends on a cute trick. Imagine that the glass is made of paper. Tear out the bottom and cut the cylindrical glass down some vertical line. Lay the paper down flat, as shown in figure 2b. Further, imagine the paper to be double-sheeted, so the side with the drop of honey could be folded out, as shown in figure 2c. Now clearly, the path of shortest distance between the ant and the honey is a straight line, with distance $\sqrt{(\pi R)^2 + (2d)^2}$. The path is also indicated in figure 2b, with the segment inside the glass indicated by a dotted line. A really dumb ant would go up vertically to the rim of the glass, then move along the rim to a point above the honey, and then go down (or along a number of similar paths equal in distance to the one just described).

This puzzle contains two of the themes central to this book: the shortest path between two points and curvature, intrinsic and extrinsic.

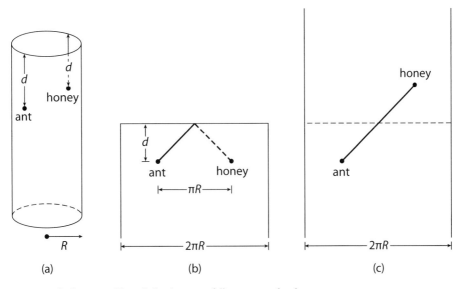

Figure 2 The best possible path for the ant to follow to get to her honey.

Draw circles and triangles on a flat piece of paper. Then roll the paper up into a cylinder. The radius and circumference of a circle maintain the same value as when flat: the paper is neither stretched nor compressed in any way. Similarly, the three angles in the triangle remain the same. A cylinder has extrinsic curvature, but zero intrinsic curvature: it is intrinsically flat. In contrast, the sphere is intrinsically curved: there is no way to construct a sphere from a flat piece of paper without stretching and compressing the paper.

The proverbial guy and gal in the street think that cylinders are curved, but you and the ant* know better. The uninitiated are talking about extrinsic curvature, regarding how the 2-dimensional surface of a cylinder is embedded into an external 3-dimensional Euclidean space.

Imagine a civilization of mites living on some curved surface. The mites are much smaller than the characteristic radius of the curvature of the surface. Once they learn how to measure the distance along any path (by pacing off the steps they have to take, for instance) they are ready for geometry. They could define the straight line between two points P_1 and P_2 as the path of least distance. Eventually, the mite professors of geometry could determine whether the world of mites is curved without getting out of their world to take a look. For example, with enough government funding, the professors could organize teams of mites to draw small circles of any desired radius by finding the set of all points a fixed distance from a given point P. Then they can measure the circumference of the circle and compute

$$R = \lim_{\text{radius} \to 0} \frac{6}{(\text{radius})^2} \left(1 - \frac{\text{circumference}}{2\pi \, \text{radius}} \right) \tag{1}$$

as the circle shrinks to zero. For flat space, R vanishes everywhere. Thus, a nonvanishing value of R gives the mites a measure of the intrinsic curvature at P—of how the geometry of their world differs[†] from Euclid's flat geometry. (The factor of 6 provides a convenient normalization to match another definition of R to be given later.) Another measure would be the extent that the sum of the angles enclosed by a triangle deviates from π.

Our mites are not interested in the extrinsic curvature, since they cannot get off the surface to take a look. Similarly, we are only interested in the intrinsic curvature of our universe, not in the extrinsic curvature, since we cannot get out[‡] of the universe to take a look.

---

* Ants will eventually find the shortest path to food if the starting point is the location of the colony, but you need a whole colony of them to do so. Their trick is to lay down pheromone on the path as they go along and to prefer to follow paths with the stronger pheromone. It is crucial that the pheromone evaporates at some fixed rate and that ants often wander off the beaten paths to try out nearby paths. (Moral: wander off the beaten paths!) We explore this variational principle in chapter II.2. A multitude of physicists may also eventually solve the mystery of quantum gravity. The paths correspond to published papers, the strength of the pheromone to the prestige of the authors and the number of citations received, and so on and so forth. Not a perfect analogy by any means.

† Early in the 20th century, a distinguished professor, Sir Arthur Eddington, did precisely that, defining a straight line by the trajectory of light. See chapter VI.3.

‡ There exist some wild speculations that our universe is embedded in a much larger spacetime, but even in these theories, it does not appear that their proponents can get out of our universe, at least not until after this book is published. See chapter X.2.

Story 3: Dueling thinkers

Professor Vicious and Dr. Nasty have been at each other's throats for decades. Theoretical physicists are forever fighting over who did what when. They are constantly bickering, telling each other (as the joke goes), "Nyah, nyah, what you did is trivial and wrong, and I did it first!"

Of course, the fight for credit goes on in every field, but in theoretical physics, it is almost a way of life, since ideas are by nature ethereal. And the stakes are high: the victor gets to go to Stockholm, while the loser is consigned to the dustbin of history, a history largely written by the victor with the help of an army of idolaters and science writers.

We are finally going to settle matters between Vicious and Nasty once and for all. We place the two of them at two ends of a long hall, Vicious at $x = 0$ and Nasty at $x = L$.

We now tell Vicious and Nasty to solve the basic mystery of why the material world comes in three copies.[4] As soon as they figure it out, they are to push a button in front of them. When the button is pushed, a pulse of light is flashed to the middle of the room where, at $x = L/2$, our experimental colleague, an electronics wiz, has set up a screen. When the screen detects the arrival of a light pulse, all kinds of bells and whistles are rigged to go off. In particular, if, and only if, two light pulses arrive at the screen at precisely the same instant, a huge imperial Chinese gong will be bonged.

"Fair is fair, any and all priority claims will be settled," we tell Vicious and Nasty. "Now go to work and solve the mystery of the family problem: why do quarks and leptons come in three sets?" The dueling duo immediately assume the Rodinesque pose of the deep thinker and lock themselves in a think to the death.

Meanwhile, you are sitting on a train, moving smoothly relative to the dueling thinkers. Denote the time and space coordinates in your rest frame by t' and x'. In the Newtonian universe, time is absolute, and so we have $t' = t$. In your frame, Vicious and Nasty are moving by according to $x' = vt$ and $x' = L + vt$, respectively, but you are sitting at $x' = 0$. Of course, in the duelists' frame, you are the one who appears to be moving, gliding by at $x = -vt$ (see figure 3).

Some time passes, and all of a sudden we all hear a loud bong of the gong. "The best possible outcome, you solved the problem simultaneously!" we exclaim joyously with much relief. "You guys are equally smart and you should go to Stockholm together!"

The arrangement is electronically fool-proof. We won't have either of them gloating, "I did it first!" Peace shall reign on earth. But guess what?

A Swede is sitting next to you. He, too, heard the gong. That's the whole point of the gong: you either heard it or you didn't. It is all admissible in a court of law. Now, not only is the Swede on the Committee, but he also happens to be an intelligent Swede. He reasons as follows.

The two thinkers are gliding by as described by $x' = vt'$. When Professor Vicious pushed the button, she sent forth a multitude of photons surging toward the screen at the speed of light c. But the screen was also moving forward, away from the surging photons. Of

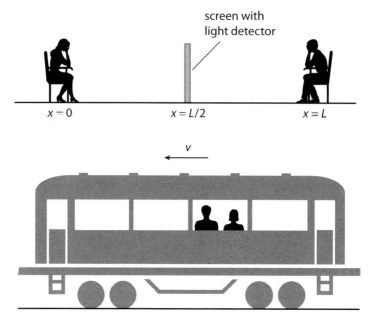

Figure 3 Professor Vicious versus Doctor Nasty.

course, light moves at the maximum allowed speed in the universe, and it soon catches up with the screen. The opposite is true for Dr. Nasty. The screen is moving toward the photons he sent forth. Thus, to reach the screen, his photons have less distance to cover than Vicious's photons.

Hence, reasons the Swede, for the two bunches of photons to reach the screen at the same time and so cause the gong to bong, the photons sent out by Vicious must have gotten going earlier. Thus, Vicious solved the problem first. With malicious glee, the Swede solemnly intones, "After Professor Vicious is awarded the Nobel Prize, she will kindly help us stuff Dr. Nasty into the dustbin of history!"

As Vicious[5] enjoys her fleeting immortality, we bemoan or toast, as our taste might be, the fall of simultaneity. Nasty, trying to climb out of the dustbin, insists that he and Vicious had been sitting still, thinking hard, and it was the Swede who was moving. Since the gong had bonged, Nasty is absolutely sure that he and Vicious hit their buttons at the same instant and so he is entitled to half the prize, while the Swede is equally sure that Vicious hit her button before Nasty hit his.

The very notion of simultaneity depends on the observer!

Meanwhile, another Swede, also on the Committee, is moving by on another train described in the duelists' frame by $x = vt$. You can fill in the rest.

Young Einstein has bent the stately flow of time out of shape. Albert himself thought up this gedanken experiment—I have merely added a few dramatic details—showing that the constancy of the speed of light necessarily has to alter our notion of simultaneity in time.

In theoretical physics, we say, "Mind-boggler in, mind-boggler out!" We feed the mind-boggling fact that the speed of light does not depend on the observer into the wondrous machinery of logic and out pops another mind-boggling fact, namely that simultaneity is in the mind of the beholder. Making up one gedanken experiment after another, Einstein showed that our common sense notion of time must be modified.

Exercises

1   Derive Snell's law: $\sin \theta_w / \sin \theta_a = c_w/c_a < 1$, where c_w and c_a denote the speed of light in water and in air, respectively.

2   Suppose the ant is outside a hemispherical bowl and the drop of honey is inside the bowl directly across from her. Find the shortest distance.

3   What happens if the ant can crawl faster on the outside of the glass than on the inside?

Notes

1. *QFT Nut.*
2. R. P. Feynman, *QED: The Strange Theory of Light and Matter,* with a new introduction by A. Zee, Princeton Science Library, 2006.
3. A colleague told me that this reminded him, at least superficially, of the umveg test (http://www.guidehorse .com/intellig.htm) for assessing intelligence in horses.
4. I am referring to the fact that quarks and leptons come in three families.
5. In his autobiography, Michael Faraday wrote of his conception of scientists: "My desire to escape from trade, which I thought vicious and selfish, and to enter into the service of Science, which I imagined made its pursuers amiable and liberal. . . . " Do I detect in the word "imagined" a trace of cynical disillusion?

Introduction

A Natural System of Units, the Cube of Physics, Being Overweight, and Hawking Radiation

Planck gave us natural units

Max Planck* is properly revered for his profound contribution to quantum mechanics. But he is also much loved for his second greatest contribution to physics: in a far-reaching and insightful paper, he gave us a natural system of units.

Once upon a time, we used some English king's feet to measure lengths.† Einstein recognized that with the universal speed of light c, we no longer need separate units for length and time. Even the proverbial guy and gal in the street understand that henceforth, we could measure length in lightyears.

We and another civilization, be they in some other galaxy, would now be able to agree on a unit of distance, if we could only communicate to them what we mean by one year or one day. Therein lies the rub: our unit for measuring time derives from how fast our home planet spins and revolves around its star. Only homeboys would know. How could we possibly communicate to a distant civilization this period of rotation we call a day, which is merely an accident of how some interstellar debris came together to form the rock we call home?

---

* In his personal life, Planck suffered terribly. He lost his first wife, then his son in action in World War I, then both daughters in childbirth. In World War II, bombs totally demolished his house, while the Gestapo tortured his other son to death for trying to assassinate Hitler.

† Notions we take for granted today still had to be thought up by someone. Maxwell, in his magnum opus on electromagnetism, proposed that the meter be tied to the wavelength of light emitted by some particular substance, adding that such a standard "would be independent of any changes in the dimensions of the earth, and should be adopted by those who expect their writings to be more permanent than that body." The various eminences of our subject could be quite sarcastic.

Newton's discovery of the universal law of gravity brought another constant G into physics. Comparing the kinetic energy $\frac{1}{2}mv^2$ of a particle of mass m in a gravitational potential with its potential energy $-GMm/r$ and canceling off m, we see that the combination* GM/c^2 has dimensions of length. In other words, having two universal constants c and G at hand allows us to measure masses in terms of our unit for length (or equivalently time), or lengths in terms of our unit for mass.

Planck with his constant $\hbar$ made a monumental contribution to physics by noting that the quantum world gives us for free a fundamental set of units that physicists call natural units.

Three big names, three basic principles, three natural units

To see how, note that Heisenberg's uncertainty principle tells us that $\hbar$ divided by the momentum Mc is a length. Equating the two lengths GM/c^2 and $\hbar/Mc$, we see that the combination $\hbar c/G$ has dimensions of mass squared. In other words, the three fundamental constants G, c, and $\hbar$ allow us to define a mass,[1] known rightfully as the Planck mass

$$M_{\mathrm{P}} = \sqrt{\frac{\hbar c}{G}} \tag{1}$$

We can immediately define, with Heisenberg's help, a Planck length

$$l_{\mathrm{P}} = \frac{\hbar}{M_{\mathrm{P}}c} = \sqrt{\frac{\hbar G}{c^3}} \tag{2}$$

and, with Einstein's help, a Planck time

$$t_{\mathrm{P}} = \frac{l_{\mathrm{P}}}{c} = \sqrt{\frac{\hbar G}{c^5}} \tag{3}$$

Einstein, Newton, and Heisenberg—three big names, three basic principles, three natural units to measure space, time, and energy by. We have reduced the MLT system to nothing! We no longer have to invent or find some unit, such as the good king's foot, to measure the universe with. We measure mass in units of M_{P}, length in units of l_{P}, and time in units of t_{P}. Another way of saying this is that in these natural units, $c = 1$, $G = 1$, and $\hbar = 1$. The natural system of units is understood no matter where your travels might take you, within this galaxy or far beyond.

Newton small, so Planck huge, and the Mother of All Headaches

The Planck mass works out to be 10^{19} times the proton mass M_p. That humongous number 10^{19}, as we will see, is responsible for the Mother of All Headaches plaguing fundamental

---

* You will learn shortly what this combination means physically.

physics today.[2] That M_P is so gigantic compared to the known particles can be traced back to the extreme feebleness of gravity: G is tiny, so M_P is enormous.

As the Planck mass is huge, the Planck length and time are teeny. If you insist on contaminating the purity of natural units by manmade ones, t_P comes out to be $\sim 5.4 \times 10^{-44}$ second, the Planck length $l_P \sim 1.6 \times 10^{-33}$ centimeter, and the Planck mass $M_P \sim 2.2 \times 10^{-5}$ gram!

It is important to realize how profound Planck's insight was. Nature herself, far transcending any silly English king or some self-important French revolutionary committee, gives us a set of units to measure her by. We have managed to get rid of all manmade units. We needed three fundamental constants, each associated with a fundamental principle, and we have precisely three!

This suggests that we have discovered all* the fundamental principles that there are. Had we not known about the quantum, then we would have to use one manmade unit to describe the universe, which would be weird. From that fact alone, we would have to go looking for quantum physics.

The cube of physics

Here is a nifty summary of all of physics as a cube (see figure 1). Physics started with Newtonian mechanics at one corner of the cube, and is now desperately trying to get to the opposite corner, where sits the alleged Holy Grail. The three fundamental constants, c^{-1}, $\hbar$, and G, characterizing Einstein, Planck or Heisenberg, and Newton, label the three axes. As we turned on one or the other of three constants (in other words, as each of these constants came into physics), we took off from the home base of Newtonian mechanics.[†] Much of 20th century physics consisted of getting from one corner of the cube to another. Consider the bottom face[3] of the cube. When we turned on c^{-1} we went from Newtonian mechanics to special relativity. When we turned on $\hbar$, we went from Newtonian mechanics to quantum mechanics. When we turned on both c^{-1} and $\hbar$, we arrived at quantum field theory, in my opinion the greatest monument of 20th century physics.

Newton himself had already moved up the vertical axis from Newtonian mechanics to Newtonian gravity by turning on G. Turning on c^{-1}, Einstein took us from that corner to Einstein gravity, the main subject of this book.[‡] All the Stürm und Drang of the past few decades is the attempt to cross from that corner to the Holy Grail of quantum gravity, when (glory glory hallelujah!) all three fundamental constants are turned on.[§]

---

* These days, fundamental principles are posted on the physics archive with abandon. There might be hundreds by now.

† By this I mean the three laws, $F = ma$ and so on, not including the law of universal gravitation.

‡ The corner with $c^{-1} = 0$ but $\hbar \neq 0$ and $G \neq 0$ is relatively unpublicized and generally neglected. It covers phenomena described adequately by nonrelativistic quantum mechanics in the presence of a gravitational field. Two fascinating experiments in this area are: (1) dribbling neutrons like basketballs, and (2) interfering a neutron beam with itself in a gravitational field.[4]

§ This statement carries a slight caveat, which we will come to in chapter VII.3.

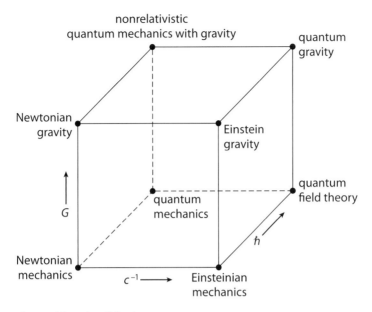

Figure 1 The cube of physics.

In our everyday existence, we are aware of only two corners of this cube, because these three fundamental constants are either absurdly small or absurdly large compared to what humans experience.

The universe's obesity index

As the obesity epidemic sweeps over the developed countries, one government after another has issued some kind of obesity index, basically dividing body weight by size. As we have seen, for an object of mass M, the combination GM/c^2 is a length that can be compared to the characteristic size of the object. So, Nature has her own obesity index for any object, from electron to galaxy. Indeed, as is well known, John Michell in 1783 and the Marquis Pierre-Simon Laplace in 1796 pointed out that even light cannot escape from an object excessively massive for its size.

More precisely, consider an object of mass M and radius R. A particle of mass m at the surface of this object has a gravitational potential energy $-GMm/R$ and kinetic energy $\frac{1}{2}mv^2$. Equating these two energies gives the escape velocity $v_{escape} = \sqrt{2GM/R}$. Setting v_{escape} to c tells us that if $2GM > Rc^2$, not even light can escape, and the object is a black hole.[5] Remarkably, even though the physics behind the argument* is not correct in detail (as we now know, we should not treat light as a Newtonian corpuscle with a tiny mass), this

---

* This often cited Newtonian argument actually does not establish the existence of black hole defined as an object from which nothing could escape. The escape velocity refers to the initial speed with which we attempt to fling something into outer space. In the Newtonian world, we could certainly escape from any massive planet in a rocket with a powerful enough engine.

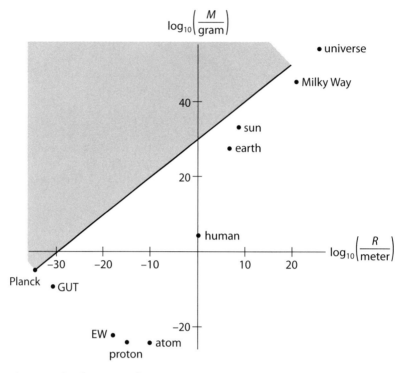

Figure 2 A plot of M versus R for various objects in the universe. EW stands for electroweak and GUT for grand unified theory. The shaded area represents the "black hole" regime with $2GM > R$.

criterion, including the factor of 2, turns out to hold in Einstein's theory. Figure 2 shows a plot of M versus R for various objects in the universe.

Hawking radiation

Unless you have been hiding out in the jungles of New Guinea, you would have heard that in an extremely influential paper, Stephen Hawking, building on the earlier work of Jacob Bekenstein and others, and working in collaboration with Gary Gibbons, pointed out this purely classical argument needs to be amended when quantum effects are included: black holes evaporate and radiate particles.

In fact, the temperature of the radiation, known as the Hawking temperature T_H of the black hole, can be estimated by using dimensional analysis. You may be puzzled,* since there are two masses in the problem, the mass M of the black hole and the Planck mass M_P. With two masses, any function of M/M_P is admissible, and so dimensional analysis appears to be inapplicable. Indeed, we need one more piece of information. The key is that

---

* I was talking to a distinguished condensed matter physicist just the other day, and he was puzzled about precisely this point. So your unspoken question may be widespread.

Newton's constant G is a multiplicative measure of the strength of gravity. In Einstein's theory as well as in Newton's, the gravitational field around an object of mass M can only depend on the combination of GM. Let us now set c and $\hbar$ (but not G) to 1. The combination GM is a length and hence an inverse mass. On the other hand, Boltzmann and the founding fathers of statistical mechanics had long ago revealed to us that temperature, a highly mysterious concept at one time, is merely the average energy[6] of the microscopic constituents of macroscopic matter. Hence temperature has the dimensions of energy, that is, of a mass in units with $c = 1$.

It follows immediately that $T_{\mathrm{H}} \sim \frac{1}{GM}$. This "sophisticated" dimensional analysis captures an essential piece of physics: the radiation is explosive! As the black hole radiates energy, M goes down and T_{H} goes up, and thus the black hole radiates faster. The radiative mass loss accelerates. Certainly not something you want to see in the kitchen: an object that gets hotter as it loses energy.

In chapter VII.3, we will see that the overall numerical constant can be determined in a couple of lines of algebra, so that

$$T_{\mathrm{H}} = \frac{\hbar c^3}{8\pi GM} \tag{4}$$

We have restored c and $\hbar$ by high school dimensional analysis using everyday unnatural units. It is gratifying to see that indeed, with $\hbar = 0$ and quantum effects turned off, $T_{\mathrm{H}} = 0$, and the black hole does not radiate.

Thermodynamics states that entropy S is given by $dE = T dS$. Here E is just the mass of the black hole. Integrating $\frac{dS}{dM} = \frac{1}{T_{\mathrm{H}}} \sim GM$, we obtain

$$S \sim GM^2 \sim \left(\frac{M}{M_{\mathrm{P}}}\right)^2 \tag{5}$$

Note that, as expected, S is dimensionless.

Using the fact that the black hole has radius $R \sim GM$ and hence surface area $A \sim R^2$, we conclude that

$$S \sim \frac{R^2}{G} \sim \frac{A}{l_{\mathrm{P}}^2} \tag{6}$$

You should be shocked, shocked, shocked. Most theoretical physicists were, and are. Not shocked?

Normally, the entropy of a system is extensive, that is, proportional to its volume. Somehow, a black hole has an entropy proportional to its surface area rather than to its volume. This fact has led to the so-called holographic principle. Many fundamental physicists believe that this mysterious property of black holes holds the key to quantum gravity.

All of this merely from dimensional analysis!

Notes

1. Some readers might wonder why we do not use the mass of the electron m_e. In modern particle physics, the electron may not always have had the mass it has now, and in fact it might have been massless in the early universe. The masses of elementary particles depend on quantum field theoretic notions known as spontaneous symmetry breaking and the Higgs mechanism. We should express m_e in terms of M_P, not M_P in terms of m_e. In different areas of physics, different units are used: for example, the size of the hydrogen atom might be used as a length unit.
2. I return to this problem in due course, in chapter X.8, for example.
3. This face, regarded as a square, was discussed in the very first section of the first chapter in *QFT Nut*.
4. See appendix 5 to chapter X.8; for more details, see J. J. Sakurai and J. Napolitano, *Modern Quantum Mechanics*, pp. 110 and 133.
5. Named by John Wheeler almost 200 years later.
6. The Boltzmann constant k, which is merely a conversion factor between energy units and the markings on some tubes containing mercury known as degrees, has been set to 1.

Prelude

Relativity Is an Everyday and Ancient Concept

Butterflies will fly indifferently toward every side

Relativity is all about the notion that you are as good as the next guy, or to put it relatively, the other guy is as good as you.

More seriously, relativity expresses the fact that the laws of physics as deduced by two observers in uniform motion with respect to each other must be the same.

We physicists believe in the fundamental principle that physics should not depend on the physicist, unlike some other academic disciplines we need not name, in which the truth can vary according to the practitioner.

The proverbial guy in the street thinks that relativity started with Albert Einstein (1879–1955), but you know better, of course. Surely, some smart human had an inkling of it as soon as sufficiently smooth transport* became available, perhaps even the proverbial "cave man"† drifting downriver on a log watching his buddies moving by. Galileo Galilei (1564–1642) first[1] explicitly stated the principle of relativity. In *Dialogue Concerning the Two Chief World Systems* (first published in 1632) the character Salviati says:

> Shut yourself up with some friend in the main cabin below decks on some large ship, and have with you there some flies, butterflies, and other small flying animals. Have a large bowl of water with some fish in it; hang up a bottle that empties drop by drop into a wide vessel beneath it. With the ship standing still, observe carefully how the little animals fly with equal speed to all sides of the cabin. The fish swim indifferently in all directions; the drops fall into the

---

* Of course, we are on a spinning rock orbiting a star in a rotating galaxy hurtling toward its neighbor at high speed, but our transport is so smooth that we didn't notice it for the longest time.

† Or a Sung dynasty poet in a boat; see *Fearful*, p. 52.

vessel beneath; and, in throwing something to your friend, you need throw it no more strongly in one direction than another, the distances being equal; jumping with your feet together, you pass equal spaces in every direction. When you have observed all these things carefully (though doubtless when the ship is standing still everything must happen in this way), have the ship proceed with any speed you like, so long as the motion is uniform and not fluctuating this way and that. You will discover not the least change in all the effects named, nor could you tell from any of them whether the ship was moving or standing still. In jumping, you will pass on the floor the same spaces as before, nor will you make larger jumps toward the stern than toward the prow even though the ship is moving quite rapidly, despite the fact that during the time that you are in the air the floor under you will be going in a direction opposite to your jump. In throwing something to your companion, you will need no more force to get it to him whether he is in the direction of the bow or the stern, with yourself situated opposite. The droplets will fall as before into the vessel beneath without dropping toward the stern, although while the drops are in the air the ship runs many spans. The fish in their water will swim toward the front of their bowl with no more effort than toward the back, and will go with equal ease to bait placed anywhere around the edges of the bowl. Finally the butterflies and flies will continue their flights indifferently toward every side, nor will it ever happen that they are concentrated toward the stern, as if tired out from keeping up with the course of the ship, from which they will have been separated during long intervals by keeping themselves in the air. And if smoke is made by burning some incense, it will be seen going up in the form of a little cloud, remaining still and moving no more toward one side than the other. The cause of all these correspondences of effects is the fact that the ship's motion is common* to all the things contained in it, and to the air also. That is why I said you should be below decks; for if this took place above in the open air, which would not follow the course of the ship, more or less noticeable differences would be seen in some of the effects noted.[2]

That[†] is so beautifully stated! Much better than most popular physics books on the market (see figure 1).

Galileo's ship was updated to Einstein's train[‡] and later to rocket ships and other space vehicles. Let's use Einstein's train, moving smoothly along the x-axis with velocity u (see figure 2). Let an event occur at the point (x, y, z) at time t for the observer on the train (call her Ms. Unprime) and at the point (x', y', z') at time t' for the observer on the ground (Mr. Prime). We are of course utilizing the profound and brilliant insight of Galileo's contemporary René Descartes (1596–1650) that geometry can be reduced to algebra by associating three numbers with each point in space. The Galilean transformation states that

$$t' = t \tag{1}$$

---

* The phrase "common to all the things contained in it" will play a starring role when we get to Einstein's equivalence principle, as we will see in part V.

† Galileo intended this passage as a refutation of the argument that the earth could not rotate since otherwise objects would fall toward the west.

‡ The historian Peter Galison has pointed out that in the period leading up to 1905, the year Einstein proposed his theory of special relativity, high speed trains and the telegraph linked the cities of Europe, and an increasingly technological society was preoccupied with clock synchronization among other things.[3]

Figure 1 Galileo's vision: butterflies fly normally in a cabin on a smoothly moving ship.

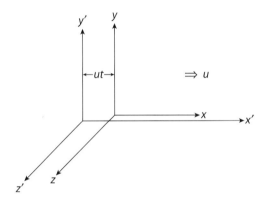

Figure 2 Galilean transformation.

and

$$x' = x + ut, \quad y' = y, \quad \text{and} \quad z' = z \tag{2}$$

with u the constant relative velocity between the two observers.

We simply differentiate: $\frac{dx'}{dt'} = \frac{dx'}{dt} = \frac{dx}{dt} + u$. Thus, if Ms. Unprime tosses a ball forward with speed v, Mr. Prime sees the ball moving forward with speed $v' = v + u$, in accordance with everyday observation, as known to you, me, and Salviati. We have derived the Galilean law* for the addition of velocities:

$$v' = v + u \tag{3}$$

Differentiating again, we obtain the ball's acceleration $a' = \frac{dv'}{dt} = \frac{dv}{dt} = a$. Since Newton's law of motion $F = ma$ involves acceleration, we conclude that Newtonian mechanics is invariant under the Galilean transformation, as Salviati told us.

* Which you can verify these days at any major airport with a moving sidewalk.

Special relativity in one minute

Special relativity can be simply summarized. (Of course, we will be going through it in much greater detail later.) Maxwell's laws of electromagnetism turned out not to be invariant under the Galilean transformation. The speed of light c is determined by how fast an electric field can turn into a magnetic field and vice versa and so does not depend on the observer. In total defiance of (3), Maxwell had

$$c \neq c + u \tag{4}$$

In the high noon showdown between Maxwell and Galileo, Maxwell won. The Galilean transformation had to be replaced by the Lorentzian transformation involving that universal constant of Nature, c for celeritas.* The relations (1) and (2) between space and time were modified.

General relativity in 30 seconds

That was special relativity in 60 seconds. But then we could ask, what would happen if u were not constant, if Salviati's ship encountered a storm, as it were? In deriving $a' = a$, we used $\frac{du}{dt} = 0$, but if that were not so, we would have

$$a' = a + \frac{du}{dt} \tag{5}$$

Multiply this by m, the mass of the ball Ms. Unprime tossed forward, to obtain $ma' = ma + m\frac{du}{dt}$. Mr. Prime, invoking Newton's law, thus sees an additional force $m\frac{du}{dt}$ acting on the ball.

What could that force possibly be? The answer to that question will lead us to curved spacetime and Einstein gravity.[4]

Truth is not relative

Later in life, Einstein moaned that he should have called his work "invariant theory" instead of "relativity theory." Had he been more judicious in his choice of words, you, I, and Einstein would have been spared the spectacle of eminent humanities scholars asserting that "Truth is relative" since "There is no absolute truth: Einstein proved it so." Of course, you know that Einstein said exactly the opposite. Physics must be invariant and true.

Notes

1. Perhaps some historian will track down others before Galileo.
2. Galileo, *Dialogue Concerning the Two Chief World Systems*, trans. S. Drake, University of California Press, 1953, pp. 186–187.
3. See P. Galison, *Einstein's Clocks, Poincaré's Maps: Empires of Time*, W. W. Norton, 2004.
4. Nitpickers, please! It's what I could say in 30 seconds!

———

* Einstein used V in his 1905 paper.

BOOK ONE

From Newton to the Gravitational Redshift

Part I | From Newton to Riemann: Coordinates to Curvature

1.1 Newton's Laws

The foundational equation of our subject

> For in those days I was in the prime of my age for invention and minded Mathematicks & Philosophy more than at any time since.
>
> —Newton describing his youth in his memoirs

Let us start with one of Newton's laws, which curiously enough is spoken as $F = ma$ but written as $ma = F$. For a point particle moving in D-dimensional space with position given by $\vec{x}(t) = (x^1(t), x^2(t), \cdots, x^D(t))$, Mr. Newton taught us that

$$m\frac{d^2x^i}{dt^2} = F^i \tag{1}$$

with the index* $i = 1, \cdots, D$. For $D \leq 3$ the coordinates have traditional "names": for example, for $D = 3$, x^1, x^2, x^3 are often called, with some affection, x, y, z, respectively.

Bad notation alert! In teaching physics, I sometimes feel, with only slight exaggeration, that students are confused by bad notation almost as much as by the concepts. I am using the standard notation of x and t here, but the letter x does double duty, as the position of the particle, which more strictly should be denoted by $x^i(t)$ or $\vec{x}(t)$, and as the space coordinates x^i, which are variables ranging from $-\infty$ to ∞ and which certainly are independent of t.

The different status between x and t in say (1) is particularly glaring if $N > 1$ particles are involved, in which case we write $m\frac{d^2x^i{}_a}{dt^2} = F^i{}_a$ or $m\frac{d^2\vec{x}_a}{dt^2} = \vec{F}_a$ with $x^i{}_a(t)$ for $a = 1, 2, \cdots, N$. But certainly t_a is a meaningless concept in Newtonian physics. In the Newtonian universe, t is the time ticked off by a universal clock, while $\vec{x}_a(t)$ is each particle's private business. We will have plenty more to say about this point. Here $x^i{}_a(t)$ are $3N$ functions of t, but there are still only 3 x^i.

---

* See appendix 2.

Some readers may feel that I am overly pedantic here, but in fact this fundamental inequality of status between x and t will come to a head when we get to the special theory of relativity. (I now drop the arrow on $\vec{x}$.) Perhaps Einstein as a student was bothered by this bad notation. One way to remedy the situation is to use q (or q_a) to denote the position of particles, as in more advanced treatments. But here I bow to tradition and continue to use x.

Have differential equation, will solve

After Newton's great insight, we "merely" have to solve some second order differential equations.

To understand Newton's fabulous equation, it's best to work through a few examples. (I need hardly say that if you do not already know Newtonian mechanics, you are unlikely to be able to learn it here.)

A priori, the force F^i could depend on any number of things, but from experience we know that in many simple cases, it depends only on x and not on t or $\frac{dx}{dt}$. As physicists unravel the mysteries of Nature, it becomes increasingly clear that fundamental forces are derived from an underlying quantum field theory and that they have simple forms. Complicated forces often merely result from some approximations we make in particular situations.

Example A

A particle in 1-dimensional space tied to a spring oscillates back and forth.

The force F is a function of space. Newton's equation

$$m\frac{d^2x}{dt^2} = -kx \qquad (2)$$

is easily solved in terms of two integration constants: $x(t) = a\cos\omega t + b\sin\omega t$, with $\omega = \sqrt{\frac{k}{m}}$. The two constants a and b are determined by the initial position and initial velocity, or alternatively* by the initial position at $t = 0$ and by the final position at some time $t = T$. Energy, but not momentum, is conserved.

Example B

We kick a particle in 1-dimensional space at $t = 0$.

The force F is a function of time. This example allows me to introduce the highly useful Dirac[1] delta function, or simply delta function.[2] By the word "kick" we mean that the time scale τ during which the force acts is much less than the other time scales we are

---

* See part II.

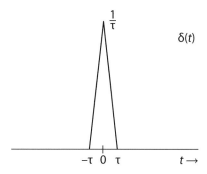

Figure 1 The delta function, which could be thought of as an infinitely sharp spike, is strictly speaking not a function, but the limit of a sequence of functions.

interested in. Thus, take $F(t) = w\delta(t)$, where the function $\delta(t)$ rises sharply just before $t = 0$, rapidly reaches its maximum at $t = 0$, and then sharply drops to 0. Because we included a multiplicative constant w, we could always normalize $\delta(t)$ by

$$\int dt\, \delta(t) = 1 \tag{3}$$

As we will see presently, the precise form of $\delta(t)$ does not matter. For example, we could take $\delta(t)$ to rise linearly from 0 at $t = -\tau$, reach a peak value of $1/\tau$ at $t = 0$, and then fall linearly to 0 at $t = \tau$. For $t < -\tau$ and for $t > \tau$, the function $\delta(t)$ is defined to be zero. Take the limit $\tau \to 0$, in which this function is known as the delta function. In other words the delta function is an infinitely sharp spike. See figure 1.

The δ function is somehow treated as an advanced topic in mathematical physics, but in fact, as you will see, it is an extremely useful function that I will use extensively in this book, for example in chapters II.1 and III.6. More properties of the δ function will be introduced as needed.

Integrating

$$\frac{d^2x}{dt^2} = \frac{w}{m}\delta(t) \tag{4}$$

from some time $t_- < 0$ to some time $t_+ > 0$, we obtain the change in velocity $v \equiv \frac{dx}{dt}$:

$$v(t_+) - v(t_-) = \frac{w}{m} \tag{5}$$

Note that in this example, neither energy nor momentum is conserved. The lack of conservation is easy to understand: (4) does not include the agent administering the kick. In general, a time-dependent force indicates that the description is not dynamically complete.

Example C

A planet approximately described as a point particle of mass m goes around its sun of mass $M \gg m$.

This is of course the celebrated problem Newton solved to unify celestial and terrestrial mechanics, previously thought to be two different areas of physics. His equation now reads

$$m\frac{d^2\vec{r}}{dt^2} = -GMm\frac{\vec{r}}{r^3} \qquad (6)$$

where we use the notation $\vec{r} = (x, y, z)$ and $r = \sqrt{\vec{x} \cdot \vec{x}} = \sqrt{x^2 + y^2 + z^2}$.

John Wheeler has emphasized the interesting point that while Newton's law (1) tells us how a particle moves in space as a function of time, we tend to think of the trajectory of a particle as a curve fixed in space. For example, when we think of the motion of a planet around the sun, we think of an ellipse rather than a spiral around the time axis. Even in Newtonian mechanics, it is often illuminating to think in terms of a spacetime picture rather than a picture in space.[3]

Newton and his two distinct masses

> By thinking on it continually.
> —Newton (reply given when
> asked how he discovered
> the law of gravity)

Conceptually, in (6), m represents two distinct physical notions of mass. On the left hand side, the inertial mass measures the reluctance of the object to move. On the right hand side, the gravitational mass measures how strongly the object responds to a gravitational field. The equality of the inertial and the gravitational mass was what Galileo tried to verify in his famous apocryphal experiment dropping different objects from the Leaning Tower of Pisa. Newton himself experimented with a pendulum consisting of a hollow wooden box, which he proceeded to fill with different substances, such as sand and water. In our own times, this equality has been experimentally verified[4,5] to incredible accuracy.

That the same m appears on both sides of the equation turns out to be one of the greatest mysteries in physics before Einstein came along. His great insight was that this unexplained fact provided the clue to a deeper understanding of gravity. At this point, all we care about this mysterious equality is that m cancels out of (6), so that $\ddot{\vec{r}} = -\kappa \frac{\vec{r}}{r^3}$, with $\kappa \equiv GM$.

Celestial mechanics solved

Since the force is "central," namely it points in the direction of $\vec{r}$, a simple symmetry argument shows that the motion is confined to a plane, which we take to be the $(x$-$y)$ plane. Set $z = 0$ and we are left with

$$\ddot{x} = -\kappa x/r^3 \quad \text{and} \quad \ddot{y} = -\kappa y/r^3 \tag{7}$$

I have already, without warning, switched from Leibniz's notation to Newton's dot notation

$$\dot{x} \equiv \frac{dx}{dt} \quad \text{and} \quad \ddot{x} \equiv \frac{d^2x}{dt^2} \tag{8}$$

Since this is one of the most beautiful problems[6] in theoretical physics, I cannot resist solving it here in all its glory. Think of this as a warm-up before we do the heavy lifting of learning Einstein gravity. Also, later, we can compare the solution here with Einstein's solution.

Evidently, we should change from Cartesian coordinates (x, y) to polar coordinates (r, θ). We will do it by brute force to show, in contrast, the elegance of the formalism we will develop later. Differentiate

$$x = r\cos\theta \quad \text{and} \quad y = r\sin\theta \tag{9}$$

twice to obtain first

$$\dot{x} = \dot{r}\cos\theta - r\sin\theta\,\dot{\theta} \quad \text{and} \quad \dot{y} = \dot{r}\sin\theta + r\cos\theta\,\dot{\theta} \tag{10}$$

and then

$$\ddot{x} = \ddot{r}\cos\theta - 2\dot{r}\sin\theta\,\dot{\theta} - r\cos\theta\,\dot{\theta}^2 - r\sin\theta\,\ddot{\theta}$$
$$\text{and} \quad \ddot{y} = \ddot{r}\sin\theta + 2\dot{r}\cos\theta\,\dot{\theta} - r\sin\theta\,\dot{\theta}^2 + r\cos\theta\,\ddot{\theta} \tag{11}$$

(Note that in each pair of these equations, the second could be obtained from the first by the substitution $\theta \to \theta - \frac{\pi}{2}$, so that $\cos\theta \to \sin\theta$, and $\sin\theta \to -\cos\theta$.)

Multiplying the first equation in (7) by $\cos\theta$ and the second by $\sin\theta$ and adding, we obtain, using (11),

$$\ddot{r} - r\dot{\theta}^2 = -\frac{\kappa}{r^2} \tag{12}$$

On the other hand, multiplying the first equation in (7) by $\sin\theta$ and the second by $\cos\theta$ and subtracting, we have

$$2\dot{r}\dot{\theta} + r\ddot{\theta} = 0 \tag{13}$$

I remind the reader again that we are doing all this in a clumsy brute force way to show the power of the formalism we are going to develop later.

After staring at (13) we recognize that it is equivalent to

$$\frac{d}{dt}(r^2\dot{\theta}) = 0 \tag{14}$$

which implies that

$$\dot{\theta} = \frac{l}{r^2} \tag{15}$$

for some constant l. Inserting this into (12), we have

$$\ddot{r} = \frac{l^2}{r^3} - \frac{\kappa}{r^2} = -\frac{dv(r)}{dr} \tag{16}$$

where we have defined

$$v(r) = \frac{l^2}{2r^2} - \frac{\kappa}{r} \tag{17}$$

Multiplying (16) by $\dot{r}$ and integrating over t, we have

$$\int dt \, \frac{1}{2} \frac{d}{dt} \dot{r}^2 = \int dt \, \dot{r} \ddot{r} = - \int dt \frac{dr}{dt} \frac{dv(r)}{dr} = - \int dr \frac{dv(r)}{dr}$$

so that finally

$$\frac{1}{2} \dot{r}^2 + v(r) = \epsilon \tag{18}$$

with ϵ an integration constant.

This describes a unit mass particle moving in the potential $v(r)$ with energy ϵ. Plot $v(r)$. Clearly, if ϵ is equal to the minimum of the potential $v_{\min} = -\frac{\kappa^2}{2l^2}$, then $\dot{r} = 0$ and r stays constant. The planet follows a circular orbit of radius l^2/κ. If $\epsilon > v_{\min}$ the orbit is elliptical, with r varying between $r_{\min}$ (perihelion) and $r_{\max}$ (aphelion) defined by the solutions to $\epsilon = v(r)$. For $\epsilon > 0$ the planet is not bound and should not even be called a planet.

We have stumbled across two conserved quantities, the angular momentum l and the energy ϵ per unit mass, seemingly by accident. They emerged as integration constants, but surely there should be a more fundamental and satisfying way of understanding conservation laws. We will see in chapter II.4 that there is.

Orbit closes

One fascinating apparent mystery is that the orbit closes. In other words, as the particle goes from $r_{\min}$ to $r_{\max}$ and then back to $r_{\min}$, θ changes by precisely 2π. To verify that this is so, solve (18) for $\dot{r}$ and divide by (15) to obtain $\frac{dr}{d\theta} = \pm(r^2/l)\sqrt{2(\epsilon - v(r))}$. Changing variable from r to $u = 1/r$, we see, using (17), that $2(\epsilon - v(r))$ becomes the quadratic polynomial $2\epsilon - l^2 u^2 + 2\kappa u$, which we can write in terms of its two roots as $l^2(u_{\max} - u)(u - u_{\min})$. Since u varies between $u_{\min}$ and $u_{\max}$, we are led to make another change of variable from $u = u_{\min} + (u_{\max} - u_{\min}) \sin^2 \zeta$ to ζ, so that ζ ranges from 0 to $\frac{\pi}{2}$. Thus, as the particle completes one round trip excursion in r, the polar angle changes by (note that $u_{\min} = 1/r_{\max}$ and $u_{\max} = 1/r_{\min}$)

$$\Delta\theta = 2 \int_{r_{\min}}^{r_{\max}} \frac{l\, dr}{r^2 \sqrt{2(\epsilon - v(r))}} = 2 \int_{u_{\min}}^{u_{\max}} \frac{l\, du}{\sqrt{2\epsilon - l^2 u^2 + 2\kappa u}}$$

$$= 2 \int_{u_{\min}}^{u_{\max}} \frac{du}{\sqrt{(u_{\max} - u)(u - u_{\min})}} = 4 \int_0^{\frac{\pi}{2}} d\zeta = 2\pi \tag{19}$$

That this integral turns out to be exactly 2π is at this stage nothing less than an apparent miracle. Surely, there is something deeper going on, which we will reveal in chapter I.4. Note also that the inverse square law is crucial here. Incidentally, the change of variable here indicates how the Newtonian orbit* (and also the Einsteinian orbit, as we will see in part VI) could be determined. See exercise 2.

Bad notation alert! In (1), the force on the right hand side should be written as $F^i(x(t))$ (in many cases). In C, the gravitational force exists everywhere, namely $F(x)$ exists as a function, and what appears in Newton's equation is just $F(x)$ evaluated at the position of the particle $x(t)$. In contrast, in A, with a mass pulled by a spring, $F(x)$ does not make sense, only $F(x(t))$ does. The force exerted by the spring does not pervade all of space, and hence is defined only at the position of the particle $x(t)$, not at any old x. I can practically hear the reader chuckling, wondering what kind of person I could be addressing here, but believe me, I have encountered plenty of students who confuse these two basic concepts: spatial coordinates and the location of particles. I may sound awfully pedantic, but when we get to curved spacetime, it is often important to be clear that certain quantities are defined only on so-called geodesic curves, while others are defined everywhere in spacetime.

A historical digression on the so-called Newton's constant

> Wouldn't we be better off with the two eyes we now have plus a third that would tell us what is sneaking up behind? . . . With six eyes, we could have precise stereoscopic vision in all directions at once, including straight up. A six-eyed Newton might have dodged that apple and bequeathed us some levity rather than gravity.
>
> —George C. Williams[7]

Physics textbooks by necessity cannot do justice to physics history. As you probably know, in the *Principia*, Newton (1642–1727) converted his calculus-based calculations to geometric arguments,[8] which most modern readers find rather difficult to follow. Here I want to mention another curious point: Newton never did specifically define what we call his constant G. What he did with $ma = GMm/r^2$ was to compare the moon's acceleration with the apple's acceleration: $a_{\text{moon}} R^2_{\text{lunar orbit}} = GM_{\text{earth}} = a_{\text{apple}} R^2_{\text{radius of earth}}$. But to write $GM_{\text{earth}} = a_{\text{apple}} R^2_{\text{radius of earth}}$, he had to prove what is sometimes referred to as the first of Newton's two "superb theorems," namely that with the inverse square law the gravitational force exerted by a spherical mass distribution acts as if the entire mass were concentrated in a point at the center of the distribution. (See exercise 4.) Even with his abilities, Newton had to struggle for almost 20 years, the length of which contributed to the bitter priority fight he had with Hooke on the inverse square law, with Newton claiming that he had the law a long time before publication. You should be able to do it faster by a factor of $\sim10^4$ as an exercise.

---

* On the old one pound note, a portrait of Newton together with his orbits appears on the back. Amusingly, the artist felt compelled to put the sun at the center, rather than one of the foci, of the ellipse.

Knowing the moon's period and $R_{\text{lunar orbit}}$, Newton could calculate a_{moon}. Since $R_{\text{radius of earth}}$ had been known since antiquity, he was thus able to calculate a_{apple} and obtained agreement* with Galileo's measurement of a_{apple}. This of course represents one of the most magnificent advances in physics history, with Newton unifying[9] the previously disparate subjects of celestial and terrestrial mechanics in one stroke. I don't have space to dwell on this here, but I do want to call your attention to the fact that Newton did not need to know G and M_{earth} to perform his feat.

Indeed, G was not measured until 1798 by Henry Cavendish (1731–1810) using equipment built and designed by his friend John Michell (1724–1793), now of black hole fame, who died before he could carry out the experiment.

Needless to say, what I have presented here should only be regarded as a comic book version of history.

Appendix 1: Where is hell?

You will find it in this appendix, sort of.

Curiously, contrary to what some textbooks and popular books stated, Cavendish's goal was not to measure G, but M_{earth} and hence the earth's density. Why this was of more interest to physicists of the time than G is in itself another interesting tidbit in physics history.

I mentioned that Newton had two superb theorems and that the first triggered his feud with Hooke. His second superb theorem states that there is no gravitational force inside a spherical shell.[10] Are you curious why Newton would even attack such a problem? An erroneous calculation had convinced him that the earth was much less dense than the moon, which led his friend Edmond Halley (1656–1742), who by the way published the *Principia* at his expense, to propose the hollow earth theory.[11] Witness the popularity of the idea in science fiction, notably Jules Verne's *Journey to the Center of the Earth* (1864). The idea may seem absurd to us, but at that time, a location for hell had to be found, and leading physicists gave serious thought to this pressing problem. Every epoch in physics has its own top ten problems.

So now we understand Cavendish's interest in M_{earth} and hence in the density of the earth rather than in G. Some textbooks give the impression that people easily obtained M_{earth} by multiplying the density of rock and the volume of the earth. Not so easy if you think that the earth might be hollow! We learn from Newton's second theorem that there is no gravitational force in hell, so the usual portrayal of the leaping flames can't be right!

Appendix 2: Fear of indices

Occasionally, a student or two would profess, unaccountably, a "fear of indices." In fact, there is nothing to fear.[12] At this stage, just stand back and admire how clever the invention of indices is. Instead of giving names to each coordinate axis, such as x, y, and z, we could pass fluidly between different dimensions by writing x^i, with $i = 1, 2, \cdots, D$. The length of the alphabet we use does not limit us, and we could easily go beyond 26 dimensions.

When we get to Einstein's theory, there will be a flood of indices, and we will have to distinguish between upper and lower indices. In Newtonian mechanics, there is no significance to whether we write the index as a superscript or a subscript. Have no fear: we will discuss each of these features of indices when the need arises. At this point, we merely note that a quantity can carry more than one index. In the text, we wrote x^i_a, with $i = 1, 2, \cdots, D$ labeling the different spatial directions, and $a = 1, 2, \cdots, N$ labeling the different particles. We will encounter more examples as we go along.

* Newton's first try did not lead to excellent agreement, because the value for the earth's equatorial radius was off. Just a reminder that physics never progresses as smoothly as textbooks say.

With only slight exaggeration, we could say that the invention of indices represents one of the really clever ideas[13] in the history of physics and mathematics, almost a "magic trick" that enables us to deal with as many particles in as many spatial dimensions as we like with the mere addition of some subscripts and superscripts.

Exercises

1 Show that for some suitably smooth function $f(x)$, the integral $\int_{-\infty}^{\infty} dx \delta(x) f(x) = f(0)$. Then show that $\delta(ax) = \delta(x)/|a|$ by evaluating the integral $\int_{-\infty}^{\infty} dx \delta(ax) f(x)$ for some smooth function $f(x)$.

2 Determine the orbit $r(\theta)$ by changing variable from r to $u = 1/r$. We will need the result of this exercise later.

3 Newton thought that light consists of "corpuscles." Calculate the deflection of light by the sun, applying what you learned in the text to the case $\epsilon > 0$. Note that the mass of these minute "particles of light" drops out in Newtonian theory anyway. We will need this result to compare with Einstein's theory later in chapter VI.3.

4 Prove Newton's first superb theorem: the gravitational force exerted by a spherical mass distribution acts as if the entire mass were concentrated in a point at the center of the distribution.

5 Prove Newton's second superb theorem.

6 Suppose engineers can build a straight tunnel connecting two cities on earth. Then we could have a free unpowered "gravity express"[14] by simply dropping a railroad car into the tunnel, allowing it to fall from one city to the other. Use Newton's two superb theorems to calculate the transit time.

Notes

1. Also introduced by Cauchy, Poisson, Hermite, Kirchoff, Kelvin, Helmholtz, and Heaviside. See J. D. Jackson, *Am. J. Phys.* 76 (2008), pp. 707–709.
2. Rigorous mathematicians go berserk at physicists' use of the word "function" here; they prefer to call it a distribution, defined as the limit of a function. But working physicists do not give a flying barf about such niceties. In any case, I do not personally know a theoretical physicist suffering any harm by calling $\delta(t)$ a function.
3. Consider a game of tennis. Compare a hard drive down the line and a soft lob high over the net. In both cases, we are to solve Newton's law $\frac{d^2 x}{dt^2} = 0$, $\frac{d^2 y}{dt^2} = -g$, with the boundary conditions $x(0) = 0$, $x(T) = L$, and $y(0) = y(T) = 0$. (The problem is so elementary that we won't bother to explain the notation, that y denotes the vertical direction, that $y = 0$ is the ground, that T is the time of flight before the ball hits the ground, that L is the length of the tennis court, and so on and so forth. You might want to draw your own figure.) The solution is $x = Lt/T$, $y = \frac{1}{2}g(T - t)t$. Note that the two types of tennis shots are governed by the same equation and the same L. Hence we obtain the same solution, but keep in mind that T is small in the case of the hard drive and that T is large in the case of the soft lob. Now eliminate t to obtain y as a function of x, namely $y(x) = \frac{1}{2}gT^2(1 - \frac{x}{L})\frac{x}{L}$, a parabola in both cases (of course). But compare the curvature of the two parabolas: we have $\frac{d^2 y}{dx^2} = -g(T/L)^2$, very small in the case of the hard drive (small T) and very large in the case of the lob (large T). The hard drive down the line barely skimming over the net, and the soft lob climbing lazily high up into the sky, look and feel totally different pictured in space. In contrast, consider y as a function of t. We also have two parabolas (of course), namely $y(t) = \frac{1}{2}g(T - t)t$, as given earlier. Now compare the curvature of the two parabolas: we have $\frac{d^2 y}{dt^2} = -g$, the same in both cases. The curvature of the ball's trajectory in spacetime is universal (universal gravity, get it?). But we tend to see in our mind's eye the two parabolas $y(x)$ in space, one for the hard drive and one for the lob, which look quite different, rather than the parabolas $y(t)$ in spacetime, which have the same curvature. I learned this long ago from John Wheeler.
4. Currently to one part in 10^{13}. The modern round of experiments started with Loránd Eötvös in 1885 and continues with the Eöt-Wash experiment led by E. Adelberger in our days.

5. The equality of the gravitational and inertial mass of the neutron has also been verified to good accuracy using neutron interferometry.

6. For Newton's letter to Halley about Hooke on the inverse square, see P. J. Nahin, *Mrs. Perkins's Electric Quilt,* Princeton University Press, 2009.

7. G. C. Williams, *The Pony Fish's Glow,* Basic Books, 1997, p. 128.

8. S. Chandrasekhar, *Newton's Principia for the Common Reader,* Oxford University Press, 2003.

9. *Fearful,* pp. 74–75.

10. For a popular account, see *Toy/Universe.*

11. N. Kollerstrom, "The Hollow World of Edmond Halley," *J. Hist. Astronomy* 23 (1992), p. 185.

12. Surely most readers are familiar with indices. My son the biologist informs me that even biologists use indices routinely; for example, on p. 20 of *Genetics and Analysis of Quantitative Traits* by M. Lynch and B. Walsh, indices appear without explanation or apology.

13. A colleague told me to mention that indices are crucial in computer programming, something that many readers can relate to.

14. *Toy/Universe,* p. xxix.

1.2 | Conservation Is Good

An integrability condition

Conservation has been important to physics from day one.[1] In this chapter, we discuss the origin of various conservation laws in Newtonian mechanics.

The most important case is when the force F^i depends only on x and can be written in the form

$$F^i(x) = -\frac{\partial V(x)}{\partial x^i} \tag{1}$$

for $i = 1, 2, \cdots, D$. As we all learned, $V(x)$ is called the potential.

Suppose such a function $V(x)$ exists; then a clever person might have the insight to multiply each of Newton's equations

$$m\frac{d^2 x^i}{dt^2} = F^i = -\frac{\partial V(x)}{\partial x^i} \tag{2}$$

by $\frac{dx^i}{dt}$ to obtain the D equations

$$m\frac{d^2 x^i}{dt^2}\frac{dx^i}{dt} = -\frac{\partial V(x)}{\partial x^i}\frac{dx^i}{dt}, \quad \text{with } i = 1, \cdots, D \tag{3}$$

He or she would then recognize that the sum of these D equations could be written as

$$\frac{d}{dt}\left[\frac{1}{2}m\sum_i \left(\frac{dx^i}{dt}\right)^2 + V(x)\right] = 0 \tag{4}$$

which we could verify by explicit differentiation. Lo and behold, the total energy, defined by

$$E = \frac{1}{2}m\sum_i \left(\frac{dx^i}{dt}\right)^2 + V(x) \tag{5}$$

is conserved. It does not change in time.

For $D = 1$, (1) holds automatically: $V(x)$ is simply given by $-\int^x dx' F(x')$. For $D > 1$, the D equations in (1), namely $F^i(x) = -\frac{\partial V(x)}{\partial x^i}$, imply the consistency or integrability condition

$$\frac{\partial F^i(x)}{\partial x^j} = \frac{\partial F^j(x)}{\partial x^i} \tag{6}$$

(Since derivatives commute, both sides of (6) are equal to $-\frac{\partial^2 V(x)}{\partial x^i \partial x^j}$.) Thus, given $F^i(x)$, we merely have to check to see whether (6) holds. If not, then V does not exist. If yes, then we could integrate $F^i(x) = -\frac{\partial V(x)}{\partial x^i}$ for each i to determine V.

Apples do not fall down

Suppose $V(r)$ depends only on $r \equiv \left(\sum_{i=1}^{D}(x^i)^2\right)^{\frac{1}{2}}$. In other words, the potential does not pick out any preferred direction. We take this for granted nowadays, but it represents one of the most astonishing insights of physics.[2] Newton realized that the apple did not fall down, but toward the center of the earth.

Differentiating $r^2 = \sum_{i=1}^{D}(x^i)^2$, we obtain $r\,dr = \sum_i x^i dx^i$ (an "identity," which we will use again and again in this text) or $\frac{\partial r}{\partial x^j} = \frac{x^j}{r}$, so that

$$F^i = -\frac{x^i}{r}V'(r) \quad \text{and} \quad \frac{\partial F^i(x)}{\partial x^j} = -\frac{1}{r}[\delta^{ij}V'(r) + \frac{x^i x^j}{r^2}(-V'(r) + rV''(r))]$$

which is manifestly symmetric under $i \leftrightarrow j$.

Here we have introduced the Kronecker delta δ^{ij}, defined by

$$\delta^{kj} = 1 \text{ if } k = j, \qquad \delta^{kj} = 0 \text{ if } k \neq j \tag{7}$$

(which we can think of as an ancestor of the Dirac delta function[3] introduced in chapter I.1).

The important point is not the somewhat involved expression for $\frac{\partial F^i(x)}{\partial x^j}$, but that it is a linear combination of δ^{ij} and $x^i x^j$. We haven't talked about tensors yet (see chapter I.4), but this result could have been anticipated by a "what else can it be?" type of argument. Not having any preferred direction, we could only construct an object with indices i and j out of δ^{ij} and $x^i x^j$. We could have seen immediately that the integrability condition (6) holds.

Note that this discussion holds for any value of D.

Conservation of angular momentum

Suppose the force in (2) points toward the center, so that it has the form $F^i = f(r)x^i$ (with $f(r) = -V'(r)/r$, as we just saw). Then we obtain angular momentum conservation immediately. To see this, multiply Newton's equation (2)

$$m\frac{d^2 x^i}{dt^2} = f(r)x^i \tag{8}$$

by x^j, so that $m \frac{d^2 x^i}{dt^2} x^j = f(r) x^i x^j$. Subtract from this the same equation but with i and j interchanged. Regardless of the function $f(r)$, we find

$$x^j \frac{d^2 x^i}{dt^2} - x^i \frac{d^2 x^j}{dt^2} = 0 \qquad (9)$$

But this is the same as

$$\frac{d}{dt}\left(x^j \frac{dx^i}{dt} - x^i \frac{dx^j}{dt} \right) = 0 \qquad (10)$$

Clever, eh? I am constantly amazed by how brilliant early physicists were.

The quantity $l^{ij} \equiv \left(x^j \frac{dx^i}{dt} - x^i \frac{dx^j}{dt} \right)$, the angular momentum per unit mass, is conserved. Recall that in the preceding chapter, this fact seemingly fell out when we changed to polar coordinates. Note also that the argument given here holds for any $D \geq 2$.

Exercise

1 Let N particles interact according to

$$m_a \frac{d^2 x_a^i}{dt^2} = -\frac{\partial V(x)}{\partial x_a^i} \qquad (11)$$

with $a = 1, \cdots, N$. Suppose $V(x_1, \cdots, x_N)$ depends only on the differences $x_a^i - x_b^i$, with $a, b = 1, \cdots, N$. Show that the total momentum $\sum_a m_a \frac{dx_a^i}{dt}$ is conserved.

Notes

1. *Fearful.*
2. I once explained this point to humanists using Einstein's terminology by saying that "The words up and down have no place in the Mind of the Creator." See A. Zee, *New Lit. Hist.* 23 (1992), pp. 815–838. See also web.physics.ucsb.edu/jatila/supplements/zee lecture.pdf.
3. In the sense that $\delta(x - y)$ is zero for $x \neq y$.

I.3 | Rotation: Invariance and Infinitesimal Transformation

Rotation in the plane

My pedagogical strategy for this chapter is to take something you know extremely* well, namely rotations in the plane, present it in a way possibly unfamiliar to you, and go through it slowly in great detail, "beating the subject to death," so to speak.

I have already mentioned that Monsieur Descartes had the clever idea of reducing geometry to algebra. Put down Cartesian coordinate axes so that a point P is labeled by two real numbers (x, y). Suppose another observer (call him Mr. Prime) puts down coordinate axes rotated by angle θ with respect to the axes put down by the first observer (call her Ms. Unprime) but sharing the same origin O. Elementary trigonometry tells us that the coordinates (x, y) and (x', y') assigned by the two observers to the same point P are related by[†] (see figure 1)

$$x' = \cos\theta\, x + \sin\theta\, y, \qquad y' = -\sin\theta\, x + \cos\theta\, y \tag{1}$$

The distance from P to the origin O of course has to be the same for the two observers. According to Pythagoras, this requires $\sqrt{x'^2 + y'^2} = \sqrt{x^2 + y^2}$, which you can check using (1).

Introduce the column vectors $\vec{r} = \begin{pmatrix} x \\ y \end{pmatrix}$ and $\vec{r}' = \begin{pmatrix} x' \\ y' \end{pmatrix}$ and the rotation matrix

$$R(\theta) = \begin{pmatrix} \cos\theta & \sin\theta \\ -\sin\theta & \cos\theta \end{pmatrix} \tag{2}$$

so that we can write (1) more compactly as $\vec{r}' = R(\theta)\vec{r}$.

---

* If you don't know rotations in the plane extremely well, then perhaps you are not ready for this book. A nodding familiarity with matrices and linear algebra is among the prerequisites.

[†] For example, by comparing similar triangles in the figure, we obtain $x' = (x/\cos\theta) + (y - x\tan\theta)\sin\theta$.

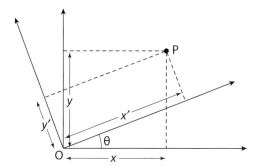

Figure 1 The same point P is labeled by (x, y) and (x', y'), depending on the observer's frame of reference.

As you recall from a course on mechanics, we can either envisage rotating the physical body we are studying or rotating the observer. We will consistently rotate the observer.

We have already used the word "vector." A vector is a physical quantity (for example the velocity of a particle in the plane) consisting of two real numbers, so that if Ms. Unprime represents it by $\vec{p} = \left(\begin{smallmatrix} p^1 \\ p^2 \end{smallmatrix} \right)$, then Mr. Prime will represent it by $\vec{p}' = R(\theta)\vec{p}$. In short, a vector is something that transforms like the coordinates $\left(\begin{smallmatrix} x \\ y \end{smallmatrix} \right)$ under rotation.

Given two vectors $\vec{p} = \left(\begin{smallmatrix} p^1 \\ p^2 \end{smallmatrix} \right)$ and $\vec{q} = \left(\begin{smallmatrix} q^1 \\ q^2 \end{smallmatrix} \right)$, the scalar or dot product is defined by $\vec{p}^T \cdot \vec{q} = p^1 q^1 + p^2 q^2$. Here T stands for transpose and $\vec{p}^T$ the row vector (p^1, p^2). By definition, rotations leave $\vec{p}^2 \equiv \vec{p}^T \cdot \vec{p} = (p^1)^2 + (p^2)^2$ invariant. In other words, if $\vec{p}' = R(\theta)\vec{p}$, then $\vec{p}'^2 = \vec{p}^2$. Since this works for any vector $\vec{p}$, including the case in which $\vec{p}$ happens to be the sum of two vectors $\vec{p} = \vec{u} + \vec{v}$, and since $\vec{p}^2 = (\vec{u} + \vec{v})^2 = \vec{u}^2 + \vec{v}^2 + 2\vec{u}^T \cdot \vec{v}$, rotation also leaves the dot product between two arbitrary vectors invariant: the invariance of $\vec{p}^2$ implies that $\vec{u}'^T \cdot \vec{v}' = \vec{u}^T \cdot \vec{v}$.

Since $\vec{u}' = R\vec{u}$ (to unclutter things, we often suppress the θ dependence in $R(\theta)$) and so $\vec{u}'^T = \vec{u}^T R^T$, we now have $\vec{u}^T \cdot \vec{v} = \vec{u}'^T \cdot \vec{v}' = (\vec{u}^T R^T) \cdot (R\vec{v}) = \vec{u}^T \cdot (R^T R)\vec{v}$. (The transpose M^T of a matrix M is of course obtained by interchanging the rows and columns of M.) As this holds for any two vectors $\vec{u}$ and $\vec{v}$, we must have the matrix equation

$$R^T R = I \tag{3}$$

where, as usual, I denotes the identity or unit matrix: $I = \left(\begin{smallmatrix} 1 & 0 \\ 0 & 1 \end{smallmatrix} \right)$. Indeed, we could verify (3) explicitly:

$$R(\theta)^T R(\theta) = \begin{pmatrix} \cos\theta & -\sin\theta \\ \sin\theta & \cos\theta \end{pmatrix} \begin{pmatrix} \cos\theta & \sin\theta \\ -\sin\theta & \cos\theta \end{pmatrix} = \begin{pmatrix} 1 & 0 \\ 0 & 1 \end{pmatrix} \tag{4}$$

Matrices that satisfy (3) are called orthogonal.

Taking the determinant of (3), we obtain $(\det R)^2 = 1$, that is, $\det R = \pm 1$. The determinant of an orthogonal matrix may be -1 as well as $+1$. In other words, orthogonal matrices also include reflection matrices, such as $\mathcal{P} = \left(\begin{smallmatrix} 1 & 0 \\ 0 & -1 \end{smallmatrix} \right)$, a reflection in the y-axis.

To focus on rotations, let us exclude reflections by imposing the condition (since $\det \mathcal{P} = -1$)

$$\det R = 1 \tag{5}$$

Matrices with unit determinant are called special.

We define a rotation as a matrix that is both orthogonal and special, that is, a matrix that satisfies both (3) and (5). Thus, the rotation group of the plane consists of the set of all special orthogonal 2 by 2 matrices and is known as $SO(2)$.

Note that matrices of the form $\mathcal{P}R$ for any rotation R are also excluded by (5), since $\det(\mathcal{P}R) = \det \mathcal{P} \det R = (-1)(+1) = -1$. In particular, a reflection in the x-axis $\begin{pmatrix} -1 & 0 \\ 0 & 1 \end{pmatrix}$, which is the product of $\mathcal{P}$ and a rotation through $90°$, is also excluded.

Act a little bit at a time

The Norwegian physicist Marius Sophus Lie (1842–1899) had the almost childishly obvious but brilliant idea that to rotate through, say, $29°$, you could just as well rotate through a zillionth of a degree and repeat the process 29 zillion times. To study rotations, it suffices to study rotation through infinitesimal angles. Shades of Newton and Leibniz! A rotation through a finite angle could always be obtained by performing infinitesimal rotations repeatedly. As is typical with many profound statements in physics and mathematics, Lie's idea is astonishingly simple. Replace the proverb "Never put off until tomorrow what you have to do today" by "Do what you have to do a little bit at a time."

When the angle is small enough, the rotation is almost the identity, that is, no rotation at all. Thus, we can write

$$R(\theta) \simeq I + A \tag{6}$$

where A denotes some infinitesimal matrix.

Now suppose we have never seen (2). Indeed, suppose we have never even heard of sine and cosine. Instead, let us define rotations as the set of linear transformations on 2-component objects $\vec{u}' = R\vec{u}$ and $\vec{v}' = R\vec{v}$ that leave $\vec{u}^T \cdot \vec{v}$ invariant. Following Lie, we solve this condition on R, namely (3) $R^T R = I$, by considering an infinitesimal transformation $R(\theta) \simeq I + A$. Since by assumption, A^2 can be neglected relative to A, $R^T R \simeq (I + A^T)(I + A) \simeq (I + A^T + A) = I$. We thus obtain $A^T = -A$, namely that A must be antisymmetric. But there is basically only one 2-by-2 antisymmetric matrix:

$$\mathcal{J} = \begin{pmatrix} 0 & 1 \\ -1 & 0 \end{pmatrix} \tag{7}$$

In other words, the solution of $A^T = -A$ is $A = \theta \mathcal{J}$ for some real number θ. Thus, rotations close to the identity have the form $R = I + \theta \mathcal{J} + O(\theta^2) = \begin{pmatrix} 1 & \theta \\ -\theta & 1 \end{pmatrix} + O(\theta^2)$. The antisymmetric matrix $\mathcal{J}$ is known as the generator of the rotation group.

An equivalent way of saying this is that for infinitesimal θ, the transformation $x' \simeq x + \theta y$ and $y' \simeq y - \theta x$ (you could verify that (1) indeed reduces to this to leading order in

θ) obviously satisfies the Pythagorean condition $x'^2 + y'^2 = x^2 + y^2$ to first order in θ. Or, write $x' = x + \delta x$, $y' = y + \delta y$ and solve $x\delta x + y\delta y = 0$.

Alternatively, simply draw figure 1 for θ infinitesimal. Since we know the transformation is linear, we could determine the matrix R in (6) by looking at the figure to see what happens to the two points $(x = 1, y = 0)$ and $(x = 0, y = 1)$ under an infinitesimal rotation.

Now recall the identity $e^x = \lim_{N \to \infty}(1 + \frac{x}{N})^N$ (which you can easily prove by differentiating both sides). Then, for a finite (that is, not infinitesimal) angle θ, we have

$$R(\theta) = \lim_{N \to \infty} R\left(\frac{\theta}{N}\right)^N = \lim_{N \to \infty}\left(1 + \frac{\theta \mathcal{J}}{N}\right)^N = e^{\theta \mathcal{J}} \tag{8}$$

The first equality represents Lie's profound idea. For the last equality, we use the identity just mentioned, which amounts to the definition of the exponential.

Some readers may not be familiar with the exponential of a matrix. Given a well-behaved function f with a power series expansion, we can define $f(M)$ for an arbitrary matrix M using that power series. For example, define $e^M \equiv \sum_{n=0}^{\infty} M^n/n!$; since we know how to multiply and add matrices, this series makes perfect sense. (Whether or not any given series converges is of course another issue.) We must be careful, however, in using various identities that may or may not generalize. For example, the identity $e^a e^a = e^{2a}$ for a a real number, which we could prove by applying the binomial theorem to the product of two series (square of a series in this case) generalizes immediately. Thus, $e^M e^M = e^{2M}$. But for two matrices M_1 and M_2 that do not commute with each other, $e^{M_1}e^{M_2} \neq e^{M_1 + M_2}$.

This provides an alternative but of course equivalent path to our result. To leading order, we have every right to write $R\left(\frac{\theta}{N}\right) = 1 + \frac{\theta \mathcal{J}}{N} \simeq e^{\frac{\theta \mathcal{J}}{N}}$ and thus $R(\theta) = R\left(\frac{\theta}{N}\right)^N = e^{\theta \mathcal{J}}$.

Finally, we easily check that the formula $R(\theta) = e^{\theta \mathcal{J}}$ reproduces (2) for any value of θ. We simply note that $\mathcal{J}^2 = -I$ and separate the exponential series into even and odd terms. Thus

$$e^{\theta \mathcal{J}} = \sum_{n=0}^{\infty} \theta^n \mathcal{J}^n/n! = \left(\sum_{k=0}^{\infty} (-1)^k \theta^{2k}/(2k)!\right) I + \left(\sum_{k=0}^{\infty} (-1)^k \theta^{2k+1}/(2k+1)!\right) \mathcal{J}$$

$$= \cos\theta \, I + \sin\theta \, \mathcal{J} = \cos\theta \begin{pmatrix} 1 & 0 \\ 0 & 1 \end{pmatrix} + \sin\theta \begin{pmatrix} 0 & 1 \\ -1 & 0 \end{pmatrix} = \begin{pmatrix} \cos\theta & \sin\theta \\ -\sin\theta & \cos\theta \end{pmatrix} \tag{9}$$

which is precisely $R(\theta)$ as given in (2). Note this works because $\mathcal{J}$ plays the same role as i in the identity $e^{i\theta} = \cos\theta + i\sin\theta$.

Poor Lie, he never made it into the 20th century.

Two approaches to rotation

Notice that I actually gave you two different approaches to rotation. Let us summarize the two approaches. In the first approach, applying trigonometry to figure 1, we write down (1) and hence (2). In the second approach, we specify what is to be left invariant by rotations and hence define rotations by the condition (3) that rotations must satisfy. Lie then tells us that it suffices to solve (3) for infinitesimal rotations. We could then build up rotations through finite angles by multiplying infinitesimal rotations together, thus also arriving at (2).

It might seem that the first approach is much more direct. One writes down (2) and that is that. The second approach appears more roundabout and involves some "fancy math." It might even provoke an adherent of the first "more macho" approach to wisecrack, "Why, with a bit of higher education, sine and cosine are not good enough for you any more? You have to go around doing fancy math!" The point is that the second approach generalizes to higher dimensional spaces (and to other situations) much more readily than the first approach does, as we will see presently. Dear reader, in going through life, you would be well advised to always separate fancy but useful math from fancy but useless math.

Before we go on, let us take care of one technical detail. We assumed that Mr. Prime and Ms. Unprime set up their coordinate systems to share the same origin O. We now show that this condition is unnecessary if we consider two points P and Q (rather than one point, as in our discussion above) and study how the vector connecting P to Q transforms.

Let Ms. Unprime assign the coordinates $\vec{r}_P = (x, y)$ and $\vec{r}_Q = (\tilde{x}, \tilde{y})$ to P and Q, respectively. Then Mr. Prime's coordinates $\vec{r}'_P = (x', y')$ for P and $\vec{r}'_Q = (\tilde{x}', \tilde{y}')$ for Q are then given by $\vec{r}'_P = R(\theta)\vec{r}_P$ and $\vec{r}'_Q = R(\theta)\vec{r}_Q$. Subtracting the first equation from the second and defining $\Delta x = \tilde{x} - x$, $\Delta y = \tilde{y} - y$, and the corresponding primed quantities, we obtain

$$\begin{pmatrix} \Delta x' \\ \Delta y' \end{pmatrix} = \begin{pmatrix} \cos\theta & -\sin\theta \\ \sin\theta & \cos\theta \end{pmatrix} \begin{pmatrix} \Delta x \\ \Delta y \end{pmatrix} \tag{10}$$

Rotations leave the distance between the points P and Q unchanged: $(\Delta x')^2 + (\Delta y')^2 = (\Delta x)^2 + (\Delta y)^2$. You recognize of course that this is a lot of tedious verbiage stating the perfectly obvious, but I want to be precise here. Of course, the distance between any two points is left unchanged by rotations. (This also means that the distance between P and the origin is left unchanged by rotations; ditto for the distance between Q and the origin.)

Invariance and geometry

> There is no royal road to geometry.
> —Euclid's advice to a prince

> Let no one unversed in geometry enter here.
> —Plato's motto, carved over the
> entrance to his academy

Let us take the two points P and Q to be infinitesimally close to each other and replace the differences $\Delta x'$, Δx, and so forth by differentials dx', dx, and so forth. Indeed, 2-dimensional Euclidean space is defined by the distance squared between two nearby points:

$$ds^2 = dx^2 + dy^2 \tag{11}$$

Rotations are defined as linear transformations* $(x, y) \rightarrow (x', y')$, such that

$$dx^2 + dy^2 = dx'^2 + dy'^2 \tag{12}$$

The whole point is that this now makes no reference to the origin O (and whether Mr. Prime and Ms. Unprime even share the same origin).

The column $d\vec{x} = \begin{pmatrix} dx^1 \\ dx^2 \end{pmatrix} \equiv \begin{pmatrix} dx \\ dy \end{pmatrix}$ is defined as the basic or ur-vector, the template for all other vectors. To repeat, a vector is defined as something that transforms like $d\vec{x}$ under rotations.

So, a vector is defined by how it transforms. An array of two numbers $\vec{p} = \begin{pmatrix} p^1 \\ p^2 \end{pmatrix}$ is a vector if it transforms according to $\vec{p}' = R(\theta)\vec{p}$.

Sometimes it is very helpful, in order to understand what something is, to be given an example of something that is not. As a simple example, given a $\vec{p}$, then $\begin{pmatrix} ap^1 \\ bp^2 \end{pmatrix}$ is definitely not a vector if $a \neq b$. (You could easily write down more outrageous examples, such as $\begin{pmatrix} (p^1)^2 p^2 \\ (p^1)^3 + (p^2)^3 \end{pmatrix}$. That ain't no vector!) You will work out further examples in exercise 1. An array of numbers is not a vector unless it transforms in the right way.[1]

Oh, about the advice Euclid gave to the prince who wanted to know a quick way of mastering geometry. Mr. E is also telling you that, to master the material covered in this book, there is no way other than to cogitate over the material until you get it and to work through as many exercises as possible.

From the plane to higher dimensional space

The reader who has wrestled with Euler angles in a mechanics course knows that the analog of (2) for 3-dimensional space is already quite a mess. In contrast, Lie's approach allows us, as mentioned above, to immediately jump to D-dimensional Euclidean space, defined by specifying the distance squared between two nearby points (compare this with (11)), as given by the obvious generalization of Pythagoras' theorem:

$$ds^2 = \sum_{i=1}^{D} \left(dx^i\right)^2 = \left(dx^1\right)^2 + \left(dx^2\right)^2 + \cdots + \left(dx^D\right)^2 \tag{13}$$

This is as good a place as any to say a word about indices. As I said in chapter I.1, in my experience teaching, there are always a couple of students confounded by indices. Dear reader, if you are not, you could simply laugh and skip to the next paragraph. Indices provide a marvelous notational device to save us from having to give names to individual elements belonging to a set. (For example, consider all humans h^i now alive, with $i = 1, 2, \cdots, P$ where P denotes the population size.) Take a look at the 19th century physics literature, before the use of indices became widespread. I am always amazed by

---

* Indeed, most, but not all, of the readers[2] of this book are constantly rotating between two coordinate systems.

the fact that, for example, Maxwell could see through the morass of the electromagnetic equations written out component by component.

Rotations are defined as linear transformations $d\vec{x}' = R d\vec{x}$ that leave ds unchanged. The preceding discussion allows us to write this condition as $R^T R = I$. As before, we want to focus on rotations by imposing the additional condition $\det R = 1$. The set of D-by-D matrices R that satisfy these two conditions forms the simple orthogonal group $SO(D)$, which is just a fancy way of saying the rotation group in D-dimensional space.

Lie in higher dimensions

The power of Lie now shines through when we want to work out rotations in higher dimensional spaces. All we have to do is satisfy the two conditions $R^T R = I$ and $\det R = 1$.

So let us follow Lie and write $R \simeq I + A$. Then $R^T R = I$ is solved by requiring $A = -A^T$, namely that A must be antisymmetric. But it is very easy to write down all possible antisymmetric D-by-D matrices! For $D = 2$, there is basically only one: the $\mathcal{J}$ introduced earlier. For $D = 3$, there are basically three of them:

$$
\mathcal{J}_x = \begin{pmatrix} 0 & 0 & 0 \\ 0 & 0 & 1 \\ 0 & -1 & 0 \end{pmatrix}, \quad
\mathcal{J}_y = \begin{pmatrix} 0 & 0 & -1 \\ 0 & 0 & 0 \\ 1 & 0 & 0 \end{pmatrix}, \quad
\mathcal{J}_z = \begin{pmatrix} 0 & 1 & 0 \\ -1 & 0 & 0 \\ 0 & 0 & 0 \end{pmatrix}
\tag{14}
$$

Any 3-by-3 antisymmetric matrix can be written as $A = \theta_x \mathcal{J}_x + \theta_y \mathcal{J}_y + \theta_z \mathcal{J}_z$, with three real numbers θ_x, θ_y, and θ_z. At this point, you can verify that $R \simeq I + A$, with A as given here, satisfies the condition $\det R = 1$.

The three matrices $\mathcal{J}_x$, $\mathcal{J}_y$, $\mathcal{J}_z$ are known as the generators of the 3-dimensional rotation group $SO(3)$. They generate rotations, but are of course not to be confused with rotations, which are by definition 3-by-3 orthogonal matrices with determinant equal to 1.

The upshot of this whole discussion is that any 3-dimensional rotation (not necessarily infinitesimal) can be written as $R(\theta) = e^A$ and is thus characterized by three real numbers. As I said, those readers who have suffered through the rotation of a rigid body in a course on mechanics must appreciate the simplicity of studying the generators of infinitesimal rotations and then simply exponentiating them.

Index notation and rotations

Some readers will find this obvious, but others might find it helpful if we derive the condition $R^T R = I$ explicitly once again using the index notation. I prefer to go slow here, since we will need some of the same formalism later when we get to special relativity. Once the reader feels sure-footed, we could then dispense with indices.

Let me start by reminding the reader that a D-by-D matrix M carries two indices and has entries M^{ij}, with the standard convention that the first index labels the rows, the second the column (for i, $j = 1, 2, \cdots, D$). For example, for $D = 2$, $M = \begin{pmatrix} M^{11} & M^{12} \\ M^{21} & M^{22} \end{pmatrix}$, and M^{12} is the entry in the first row and the second column, whereas M^{21} is the entry in the second row and the first column. Note that the transpose of a matrix M is given by $(M^T)^{ji} \equiv M^{ij}$. Thus, if $\vec{v}$ is a column vector with entries v^j, then the entries of the column vector $\vec{u} = M\vec{v}$ are given by $u^i = \sum_j M^{ij} v^j$. For A and B two D-by-D matrices, the product AB is defined as the matrix with the entries $(AB)^{ij} = \sum_k A^{ik} B^{kj}$. (If everything here is news to you, see the first footnote in this chapter.)

Under a rotation,

$$dx'^i = \sum_j R^{ij} dx^j = R^{i1} dx^1 + R^{i2} dx^2 + \cdots + R^{iD} dx^D \tag{15}$$

(I have written the sum out explicitly for the benefit of the rare reader afflicted by fear of indices.) Also, as was mentioned in chapter I.1, at this stage it is completely arbitrary whether we write upper or lower indices.

Let us pause and recall the Kronecker delta symbol δ^{ij} introduced in (I.2.7), defined to be equal to $+1$ if $i = j$ and 0 otherwise, and which we can also think of as a D-by-D unit matrix. We will be encountering the highly useful Kronecker delta often in this book. For example, $\sum_j A^j B^j = \sum_k \sum_j \delta^{kj} A^k B^j$. Since δ^{kj} vanishes unless k is equal to j, the double sum on the right hand side collapses to the single sum on the left hand side. In other words, the Kronecker delta allows us to write a single sum as a double sum. It seems like a really silly thing to do, but as we will see presently, it is an extremely useful trick that we use quite often in this book.

We now determine how the matrix R must be restricted for it to be a rotation. The statement that $ds^2 = \sum_{i=1}^{D}(dx^i)^2$ as defined in (13) is left unchanged by the rotation implies that (with all indices running over $1, \cdots, D$)

$$\sum_i (dx'^i)^2 = \sum_i \sum_k \sum_j R^{ik} dx^k R^{ij} dx^j = \sum_j (dx^j)^2 = \sum_k \sum_j \delta^{kj} dx^k dx^j \tag{16}$$

In the last step, we used what we just learned.

Since the infinitesimals dx^i can take on arbitrary values, to have the second term equal to the last term in (16), we must equate the coefficients of $dx^k dx^j$ and demand that

$$\sum_i R^{ik} R^{ij} = \delta^{kj} = \sum_i (R^T)^{ki} R^{ij} = (R^T R)^{kj} \tag{17}$$

Indeed, we obtain $R^T R = I$ just as in (3), but now in D-dimensional space for any D.

We end this section with a trivial remark. So far in this chapter, we have written the column vectors as columns. But columns take up so much space, and so for typographical convenience (editors must be placated!) we will henceforth write the entries of a column vector as $d\vec{x} = (dx^1, dx^2, \cdots, dx^D)$, a practice we will indulge in throughout this book. (If we want to be insufferably pedantic, we could put in a T for transpose: the column ur-vector $d\vec{x} = (dx^1, dx^2, \cdots, dx^D)^T$.)

Einstein's repeated index summation

Observe that in all those sums in (16) the indices to be summed over always appear twice, that is, they are repeated. For example, in the second term in (16), $\sum_i \sum_k \sum_j R^{ik} dx^k R^{ij} dx^j$, the indices i, k, and j all appear repeated. Thus, we could adopt the so-called repeated index summation convention proposed by Albert Einstein himself: omit the pesky summation symbol and agree that if an index is repeated, then it is to be summed over. For example, $dx'^i = \sum_j R^{ij} dx^j$ can now be written as $dx'^i = R^{ij} dx^j$: in the expression on the right hand side, the index j appears twice and is thus to be summed over.* In contrast, i is a "free" index and does not appear twice in the same expression. Notice that free indices must match on opposite sides of any equation. It is rightly said that one of Einstein's greatest contribution to physics is the repeated index summation convention.† When we get to Einstein gravity, we will meet lots of indices to be summed over, and it would be silly to keep on writing the summation symbol.

Vector fields

The vectors we encounter may well vary in space. For example, the flow velocity in a fluid in general would depend on where we are. We are then dealing with a vector field $\vec{V}(\vec{x})$. Again, consider two observers studying the same vector field. Mr. Prime would see

$$\vec{V}'(\vec{x}') = R\vec{V}(\vec{x}) \tag{18}$$

with $\vec{x}' = R\vec{x}$ of course. In other words, the two observers are studying the same vector field at the same point P. See figure 2. As another example, the familiar electric $\vec{E}(\vec{x})$ and magnetic fields $\vec{B}(\vec{x})$ are both vector fields.

Physics should not depend on the observer

Let me stress again why physicists constantly talk about vectors. The laws of physics often involve the statement that one vector is equal to another, for example, Newton's law states $m\vec{a} = \vec{F}$. Applying a rotation matrix $R(\theta)$, we obtain $mR(\theta)\vec{a} = R(\theta)\vec{F}$. If $\vec{F}$ transforms like a vector, then $m\vec{a}' = \vec{F}'$. Ms. Unprime and Mr. Prime see the same Newton's law, and more generally, the same laws of physics!

This statement, while self-evident, is profound, and in some sense, it is what makes physics possible. Physics should not depend on the physicist. Ms. Unprime and Mr. Prime

---

* When a pair of repeated indices, such as j here, is summed over, they are often said to be contracted with each other. In a tiny abuse of terminology, people also say that R^{ij} is contracted with dx^j.

† It appeared only in his later work. In 1905, Einstein did not even use vector notation! In one system, the coordinates were denoted by x, y, z, in the other, by ξ, η, ζ; the components of the force acting on the electron were called X, Y, Z. To modern eyes, his notation was a horrific mess.

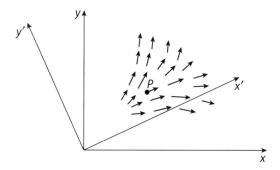

Figure 2 Two observers studying the same vector field.

see different accelerations $\vec{a}$ and $\vec{a}'$, and different forces $\vec{F}$ and $\vec{F}'$, but the same Newton's law. We say that Newton's law is invariant—that is, it does not change—under rotation.*

We should also remind ourselves that mass is an example of a scalar: a physical quantity that does not change under rotation. If it does change, Newton's law would not be invariant under rotation and one observer would be preferred over another, which is unacceptable. Physics rests on the democratic ideal.

Let me remind you that the gravitational force in the planetary problem studied in chapter I.1 is derived from what is sometimes called a central potential, namely one without a preferred direction: $F^i(x) = -\frac{\partial}{\partial x^i} V(r) = -\frac{x^i}{r} V'(r)$. Hence, $\vec{F}$ is proportional to $\vec{x}$ and so a fortiori transforms like a vector.

At this point, it may be worthwhile to be a bit more pedantic and professorial. Some authors give long-winded speeches about covariance versus invariance, and take great pain to distinguish the two. We should too. The equation $m\vec{a} = \vec{F}$ is covariant, that is, the two sides transform the same way under rotations. The physics expressed by Newton's second law is, however, invariant, that is, independent of observers related by a rotation. If physics depends on how you tilt your head, we are in trouble. Physics does not, but the way physics is expressed, in terms of equations, does.

Here is the profound and trivial statement. Under a certain set of transformations, a purportedly fundamental equation is said to be covariant if the two sides of the equation transform in the same way. If so, then that transformation is known as a symmetry of physics.[3] Physics is said to be invariant under that transformation. As we will see, both sides of Einstein's field equation transform in the same way, as tensors, under what are known as general coordinate transformations. I will explain what a tensor is in the next chapter. I will allow myself the luxury of using the words invariance and covariance interchangeably and simply trust you to be discerning.

Since we can always move the quantity on the right hand side of an equation to the left hand side, we can rewrite a physical law of the form $\vec{u} = \vec{v}$ in the form $\vec{w} \equiv \vec{u} - \vec{v} = 0$. Physics students sometimes joke that they could already write down the ultimate

---

* The reader who has already been exposed to the special theory of relativity knows that this notion of invariance represents the essence of Einstein's insight. We will of course have a great deal more to say about that!

equation of physics, namely $\mathcal{X} = 0$, whatever $\mathcal{X}$ is. Thus, the statement of invariance merely expresses the mathematically obvious fact that if $\vec{w} = 0$, then $R(\theta)\vec{w} = 0$. (Strictly speaking, the 0 on the right hand side should be written as $\vec{0}$, but we don't want to be that pedantic!)

Descartes versus Euclid

I remember how excited I was when I learned about analytic geometry. Surely you were excited too. What a genius, that Descartes! Henceforth, we could prove geometric theorems by doing algebra. After Descartes,[4] physics can no longer live without the concept of coordinates,* but he also managed to obscure what was once obvious to Euclid. We now must also insist on invariance. Indeed, the notion of invariance is at the heart of what we mean by geometry.

For example, suppose somebody hands you a formula for the area of a triangle with vertices at (a_1, b_1), (a_2, b_2), (a_3, b_3). You better insist that the formula is invariant under rotation. In fact, this requirement, plus the requirement that the area should scale as the square of the separation between the three vertices, suffices to determine the formula. This simple example rings in the central motif of this book.

Appendix 1: Differential operators rather than matrices

Here I have to divide readers into the haves and the have-nots, but only temporarily. What I will say may sound difficult, but really, it amounts to not much more than a notational triviality.

If you have studied quantum mechanics, you would know that the generators $\mathcal{J}$ of rotation studied here are related to angular momentum operators. You would also know that in quantum mechanics, observables are represented by hermitean operators. However, in our discussion, the $\mathcal{J}$s come out naturally as antisymmetric matrices and are thus antihermitean. To make them hermitean, we multiply them by some multiples of i.

If you have not studied quantum mechanics, then the preceding would sound like gibberish to you, but do not worry. Simply take the attitude that, hey, it is a free country, and we can always invite ourselves to define a new set of physical quantities by multiplying an existing set of physical quantities by some constant. Heck, we could multiply by $\sqrt{17}i$ if we want.

Even though here we are nowhere near quantum mechanics, we will bow to customary usage and define $J_x \equiv -i\mathcal{J}_x$ and so forth. From (14) we see that, for example, J_z acting on the column vector (x, y, z) gives $i(y, -x, 0)$. Thus, instead of using matrices, we could also represent J_z by $i(y\frac{\partial}{\partial x} - x\frac{\partial}{\partial y})$, since $J_z x = i(y\frac{\partial}{\partial x} - x\frac{\partial}{\partial y})x = iy$, $J_z y = i(y\frac{\partial}{\partial x} - x\frac{\partial}{\partial y})y = -ix$, and $J_z y = i(y\frac{\partial}{\partial x} - x\frac{\partial}{\partial y})z = 0$. Note that J_z is precisely the z-component of the angular momentum operators in quantum mechanics. We can naturally pass back and forth between matrices and differential operators. We will not make use of this differential representation until a later chapter.

---

* Regarding the argument (which I mentioned in a footnote in the preface) between those who live with coordinates and those who live coordinate free, I would say that the proof of angular momentum conservation, which I already gave, not once, but twice in the two preceding chapters using coordinates, provides an example in favor of the latter group: $\frac{d}{dt}\vec{l} = \frac{d}{dt}(\vec{r} \times \vec{p}) = m\frac{d}{dt}(\vec{r} \times \frac{d\vec{r}}{dt}) = m\frac{d\vec{r}}{dt} \times \frac{d\vec{r}}{dt} + m\vec{r} \times \frac{d^2\vec{r}}{dt^2} = 0$ for rotationally symmetric potentials. While this indeed looks simpler than the two previous discussions, the former group could also say that this requires learning "considerable formal math," such as the cross product and its various properties.

Appendix 2: Rotations in higher dimensional space

Here we discuss rotations in D-dimensional Euclidean space. As you have no doubt heard, Einstein combined space and time into a 4-dimensional spacetime. Thus, what you will learn here about $SO(4)$ will be put to good use.* If you prefer, you could skip this discussion and come back to it later.

Start with a D-by-D matrix with 0 everywhere. Generalize (14). Stick a 1 into the mth row and nth column, and a (-1) into the nth row and mth column. Call this matrix $\mathcal{J}_{(mn)}$. We put the subscripts (mn) in parentheses to emphasize that (mn) labels the matrix. They are not indices to tell us which element of the matrix we are talking about. As explained before, we define $J_{(mn)} = -i\mathcal{J}_{(mn)}$ so that explicitly

$$J_{(mn)}^{ij} = -i(\delta^{mi}\delta^{nj} - \delta^{mj}\delta^{ni}) \tag{19}$$

To repeat, in the symbol $J_{(mn)}^{ij}$, the indices i and j indicate respectively the row and column of the entry $J_{(mn)}^{ij}$ of the matrix $J_{(mn)}$, while the indices m and n, which I put in parentheses for pedagogical clarity, indicate which matrix we are talking about. The first index m on $J_{(mn)}$ can take on D values, and then the second index n can take on only $(D-1)$ values since, obviously, $J_{(mm)} = 0$. Also, since $J_{(nm)} = -J_{(mn)}$, we require $m > n$ to avoid double counting. Thus, there are only $\frac{1}{2}D(D-1)$ real antisymmetric D-by-D matrices $J_{(mn)}$, and A could be written as a linear combination of them: $A = i\sum_{m>n}\theta_{mn}J_{(mn)}$, where θ_{mn} denote $\frac{1}{2}D(D-1)$ real numbers. (As a check, for $D = 2$ and 3, $\frac{1}{2}D(D-1)$ equals 1 and 3, respectively.) The matrices $J_{(mn)}$ are known as the generators of the group $SO(D)$.

Notice a notational peculiarity: for $SO(3)$, the Js could be labeled with one index rather than two indices. The reason is simple. In this case, the indices m, n take on 3 values, and so we could write $J_x = J_{23}$, $J_y = J_{31}$, and $J_z = J_{12}$. We will, as we do here, often pass freely between the index sets (123) and (xyz). In general, rotations are labeled by the plane they occur in, say the (m-n) plane spanned by the mth and nth axes. In 3-dimensional space, and only in 3-dimensional space, a plane is uniquely specified by the vector perpendicular to it. Thus, a rotation commonly spoken of as a rotation around the z-axis is better thought of as a rotation in the (1-2) plane, that is, the (x-y) plane. (In this connection, note that the $\mathcal{J}$ in (7) appears as the upper left 2-by-2 block in $\mathcal{J}_z$ in (14).) In contrast, for $SO(4)$ it makes no sense to speak of a rotation around, say, the third axis.

The reader who has studied some group theory knows that the essence of the group is captured by the extent to which the multiplication of two group elements does not commute. For rotations, everyday observations show that $R(\theta)R(\theta')$ is in general quite different from $R(\theta')R(\theta)$. See figure 3.

Following Lie, we could try to capture this essence by focusing on infinitesimal rotations. Let $R_1 \simeq I + A$ and $R_2 \simeq I + B$. Then $R_1 R_2 \simeq (I + A)(I + B) \simeq I + A + B + AB + O(A^2, B^2)$ (where rather pedantically we have indicated that to the desired order if we keep AB, we should also keep terms of order $O(A^2, B^2)$, but we will see immediately that they are irrelevant). If we multiply in the other order, we simply interchange A and B, thus $R_2 R_1 \simeq (I + A)(I + B) \simeq I + B + A + BA + O(A^2, B^2)$. Hence, $R_1 R_2$ and $R_2 R_1$ differ by the amount $[A, B] \equiv AB - BA$, a quantity known as the commutator between A and B.

More formally, given two matrices X and Y, to measure how they differ from each other, we could ask how $X^{-1}Y$ differs from the identity. If $X = Y$, then this product is equal to the identity. Now, the inverse of a matrix $I + A$ infinitesimally close to the identity is easy to determine: it is just $I - A$, since $(I - A)(I + A) = I + O(A^2)$. Thus, let us calculate $(R_2 R_1)^{-1}R_1 R_2$:

$$(R_2 R_1)^{-1}R_1 R_2 = [I - (B + A + BA + O(A^2, B^2))][I + A + B + AB + O(A^2, B^2)]$$
$$= I + [A, B] + \cdots \tag{20}$$

For $SO(3)$, for example, A is a linear combination of the J_is, known as the generators of the Lie algebra. Thus, we could write $A = i\sum_i \theta_i J_i$ and similarly $B = i\sum_j \theta_j' J_j$. Hence $[A, B] = i^2\sum_{ij}\theta_i\theta_j'[J_i, J_j]$, and so it suffices to calculate the commutators $[J_i, J_j]$.

Recall that for two matrices M_1 and M_2, $(M_1 M_2)^T = M_2^T M_1^T$. Transpose reverses the order. Thus $([J_i, J_j])^T = -[J_i, J_j]$. In other words, the commutator $[J_i, J_j]$ is itself an antisymmetric 3-by-3 matrix and thus could be written as a linear combination of the J_ks:

$$[J_i, J_j] = ic_{ijk}J_k \tag{21}$$

---

* Higher dimensional rotation groups often pop up in the most unlikely places in theoretical physics. For example, $SO(4)$ is relevant for a deeper understanding of the spectrum of the hydrogen atom.[5]

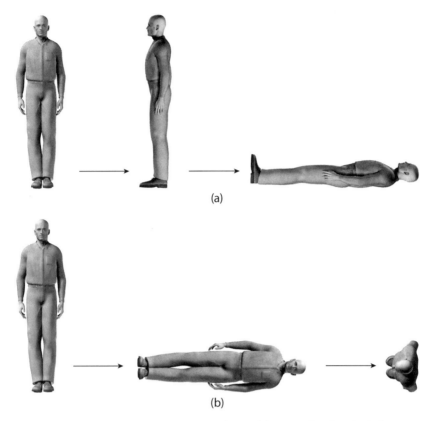

(a)

(b)

Figure 3 A marine recruit in a boot camp is standing and facing north. When the drill sergeant shouts, "Rotate by 90° eastward around the vertical axis" our recruit turns to face east. Suppose the sergeant next shouts, "Rotate by 90° westward around the north-south axis." Our recruit ends up lying down on his back with his head pointing west, his feet pointing east. But what would happen if the sergeant reverses his two commands? You could easily verify that our recruit now ends up lying down on his left elbow, with his head pointing north. The order matters. For this reason, the study of rotations has been a *bête noire* for generations of physics students.

for a set of real (convince yourself of this!) numbers c_{ijk}. The summation over k is implied by the repeated index summation convention.

By explicit computation using (14), we find

$$[J_x, J_y] = i J_z \tag{22}$$

You should work out the other commutators or argue by cyclic substitution $x \to y \to z \to x$. The three commutation relations may be summarized by

$$[J_i, J_j] = i\epsilon_{ijk}J_k \tag{23}$$

We define the totally antisymmetric symbol ϵ_{ijk} by saying that it changes sign upon the interchange of any pair of indices (and hence it vanishes when any two indices are equal) and by specifying that $\epsilon_{123} = 1$. In other words, we found that $c_{ijk} = \epsilon_{ijk}$.

Lie's great insight is that the preceding discussion holds for any group whose elements are labeled by a set of continuous parameters (such as θ_i, $i = 1, 2, 3$ in the case of $SO(3)$), groups now known as Lie groups. Expanding the group elements around the origin, we arrive at (20) and hence the structure (21) for any continuous group. The set of all commutation relations of the form (21) is said to define a Lie algebra, with c_{ijk} referred to as the structure constants of the algebra. The matrices J_i are called the generators of the Lie algebra. The idea is that by studying the Lie algebra, we go a long way toward understanding the group.

You should now work out (exercise 4), starting from (19), the Lie algebra for $SO(D)$:

$$[J_{(mn)}, J_{(pq)}] = i(\delta_{mp}J_{(nq)} + \delta_{nq}J_{(mp)} - \delta_{np}J_{(mq)} - \delta_{mq}J_{(np)}) \tag{24}$$

This may look rather involved to the uninitiated, but in fact it is quite simple. First, the right hand side, a linear combination of the Js, as required by the general argument above, is completely fixed by the first term by noting that the left hand side is antisymmetric under three separate interchanges: $m \leftrightarrow n$, $p \leftrightarrow q$, and $(mn) \leftrightarrow (pq)$. Next, all those Kronecker deltas just say that if the two sets (mn) and (pq) have no integer in common, then the commutator vanishes. If they do have an integer in common, you simply "cross off" that integer. This is best explained by using $SO(4)$ as an example. We have $[J_{(12)}, J_{(34)}] = 0$, $[J_{(12)}, J_{(14)}] = i J_{(24)}$, $[J_{(23)}, J_{(31)}] = -i J_{(21)} = i J_{(12)}$, and so forth. The first of these relations says that rotations in the (1-2) plane and in the (3-4) plane commute, as you might expect. Do write down a few more and you will get it.

Exercises

1   Suppose we are given two vectors $\vec{p}$ and $\vec{q}$ in ordinary 3-dimensional space. Consider this array of three numbers: $\begin{pmatrix} p^2q^3 \\ p^3q^1 \\ p^1q^2 \end{pmatrix}$. Prove that it is not a vector, even though it looks like a vector. (Check how it transforms under rotation!) In contrast, $\begin{pmatrix} p^2q^3 - p^3q^2 \\ p^3q^1 - p^1q^3 \\ p^1q^2 - p^2q^1 \end{pmatrix}$ does transform like a vector. It is in fact the vector cross product $\vec{p} \times \vec{q}$.

2   Show that the product of two delta functions $\delta(x)\delta(y)$ is invariant under rotation around the origin.

3   Using (14) show that a rotation around the x-axis through angle θ_x is given by

$$R_x(\theta_x) = \begin{pmatrix} 1 & 0 & 0 \\ 0 & \cos\theta_x & \sin\theta_x \\ 0 & -\sin\theta_x & \cos\theta_x \end{pmatrix}$$

Write down $R_y(\theta_y)$. Show explicitly that $R_x(\theta_x)R_y(\theta_y) \neq R_y(\theta_y)R_x(\theta_x)$.

4   Calculate $[J_{(mn)}, J_{(pq)}]$.

5   Given a 3-vector $\vec{p}$, show that the quantity $p^i p^j$ when averaged over the direction of $\vec{p}$ is given by $\frac{1}{4\pi} \int d\theta d\varphi \cos\theta\, p^i p^j = \frac{1}{3}\vec{p}^2 \delta^{ij}$.

Notes

1. Outside of physics, people often erroneously call any array of numbers a vector. Of course, people are free to call anything anything, so let's not quibble about the word "erroneously."
2. I say "most, but not all," because it is conceivable that you are a native speaker of Guugu Yimithirr. See G. Deutscher, *Through the Language Glass*, H. Holt and Co., 2010, p. 161.
3. The intellectual precision of our definition of symmetry is necessary lest we make the same mistake as the ancient Greeks. See *Fearful*, pp. 11–12 and figure 2.2.
4. According to one story, take it or leave it, Descartes was lying in bed when he noticed a fly buzzing around the room. He then realized that he could fix the fly's position given how far the fly was from two intersecting walls and the ceiling.
5. For example, J. J. Sakurai and J. Napolitano, *Modern Quantum Mechanics*, pp. 265–268.

1.4 Who Is Afraid of Tensors?

A tensor is something that transforms like a tensor

Long ago, an undergrad who later became a distinguished condensed matter physicist came to me after a class on group theory and asked me, "What exactly is a tensor?" I told him that a tensor is something that transforms like a tensor. When I ran into him many years later, he regaled me with the following story. At his graduation, his father, perhaps still smarting from the hefty sum he had paid to the prestigious private university his son attended, asked him what was the most memorable piece of knowledge he acquired during his four years in college. He replied, "A tensor is something that transforms like a tensor."

But this should not perplex us. A duck is something that quacks like a duck. Mathematical objects could also be defined by their behavior. We already saw in the preceding chapter that a vector is defined by how it transforms: $V'^i = R^{ij}V^j$. Consider a collection of "mathematical entities" T^{ij} with $i, j = 1, 2, \cdots, D$ in D-dimensional space. If they transform under rotations according to

$$T^{ij} \to T'^{ij} = R^{ik}R^{jl}T^{kl} \tag{1}$$

then we say that T transforms like a tensor, and hence is a tensor. (Here we are using the Einstein summation convention introduced in the previous chapter: The right hand side actually means $\sum_{k=1}^{D}\sum_{l=1}^{D} R^{ik}R^{jl}T^{kl}$ and is a sum of D^2 terms.) Indeed, we see that we are just generalizing the transformation law of a vector.

Fear of tensors

In my experience teaching, a couple of students are invariably confused by the notion of tensors. The very word "tensor" apparently make them tense. Dear reader, if you are not one of these unfortunates, so much the better for you! You could zip through this chapter. But to allay the nameless fear of the tensorphobe, I will go slow and be specific.

Think of the tensor T^{ij} as a collection of D^2 mathematical entities that transform into linear combinations of one another. To help the reader focus, I will often specialize to $D = 3$. Compounded and intertwined with their fear of tensors, the unfortunates mentioned above are also unaccountably afraid of indices, as mentioned in chapter I.1. For them, let us list T^{ij} explicitly for $D = 3$. There are $3^2 = 9$ of them: $T^{11}, T^{12}, T^{13}, T^{21}, T^{22}, T^{23}, T^{31}, T^{32}, T^{33}$. That's it, 9 objects that transform into linear combinations of one another. For example, (1) says that $T'^{21} = R^{2k}R^{1l}T^{kl} = R^{21}R^{11}T^{11} + R^{21}R^{12}T^{12} + R^{21}R^{13}T^{13} + R^{22}R^{11}T^{21} + R^{22}R^{12}T^{22} + R^{22}R^{13}T^{23} + R^{23}R^{11}T^{31} + R^{23}R^{12}T^{32} + R^{23}R^{13}T^{33}$. This shows explicitly, as if there were any doubt to begin with, that T'^{21} is given by a particular linear combination of the 9 objects. That's all: the tensor T^{ij} consists of 9 objects that transform into linear combinations of themselves under rotations.

We could generalize further and define* 3-indexed tensors, 4-indexed tensors, and so forth by such transformation laws as $W'^{ijn} = R^{ik}R^{jl}R^{nm}W^{klm}$. Here we will focus on 2-indexed tensors, and if we say tensor without any qualifier, we often, but not always, mean a 2-indexed tensor. With this definition, we might say that a vector is a 1-indexed tensor and a scalar is a 0-indexed tensor, but this usage is not common. A scalar transforms as a tensor with no index at all, namely $S' = S$; in other words, a scalar does not transform.

Tensor field

In the preceding chapter, we introduced the notion of a vector field $V^i(\vec{x})$, nothing more or less than a vector function of position. That it is a vector means that it transforms according to $V'^i(\vec{x}') = R^{ij}V^j(\vec{x})$. Now consider the derivative of this vector field $\frac{\partial V^j(\vec{x})}{\partial x^k}$, which we will call $W^{kj}(\vec{x})$.

Use the fact that $\vec{x}' = R\vec{x}$ implies $\vec{x} = R^{-1}\vec{x}' = R^T\vec{x}'$ and thus $\frac{\partial x^k}{\partial x'^h} = (R^T)^{kh} = R^{hk}$. (The O in the rotation group $SO(D)$ is crucial: the inverse of a rotation is its transpose.) Then

$$\frac{\partial}{\partial x'^h} = \frac{\partial x^k}{\partial x'^h}\frac{\partial}{\partial x^k} = R^{hk}\frac{\partial}{\partial x^k} \tag{2}$$

Thus

$$W'^{hi}(\vec{x}') \equiv \frac{\partial V'^i(\vec{x}')}{\partial x'^h} = R^{hk}\frac{\partial}{\partial x^k}(R^{ij}V^j(\vec{x})) = R^{hk}R^{ij}\frac{\partial V^j(\vec{x})}{\partial x^k} = R^{hk}R^{ij}W^{kj}(\vec{x}) \tag{3}$$

Comparing with (1) we see that $W^{kj}(\vec{x})$ transforms like a tensor and, hence, is a tensor. Indeed, it is a tensor field.

Notice that a tensor T^{ij} transforms as if it were composed of two vectors v^iw^j, that is, T^{ij} and v^iw^j transform in the same way. (Compare $v^iw^j \rightarrow v'^iw'^j = R^{ik}v^kR^{jl}w^l = R^{ik}R^{jl}v^kw^l$ with (1).) It is important to recognize that only in exceptional cases does a tensor T^{ij} happen to be equal to v^iw^j for some v and w. In general, a tensor cannot be

---

* Our friend the Jargon Guy tells us that the number of indices carried by a tensor is known as its rank. (The Jargon Guy is a new friend of the author; he did not appear in *QFT Nut*.)

written in the form $v^i w^j$. Our tensor field $W^{kj}(\vec{x})$ offers a ready example: in general, it is not equal to some vector U^k multiplied by $V^j(\vec{x})$.

Also, note in our example that the differential operator $\frac{\partial}{\partial x^k}$ transforms (2) like a vector. For example, if $\phi'(x') = \phi(x)$ transforms like a scalar, then $\frac{\partial \phi}{\partial x^k}$ transforms like a vector. Indeed, that's why you have encountered the notation $\vec{\nabla}$ for the gradient in an elementary physics course. This remark will be important later when we revisit Newton's inverse square law in chapter II.3. Do exercise 1 now.

Representation theory

Go back to the 9 objects T^{ij} that form a tensor. Mentally arrange them in a column

$$\begin{pmatrix} T^{11} \\ T^{12} \\ \vdots \\ T^{33} \end{pmatrix}$$

The linear transformation on the 9 objects can then be represented by a 9-by-9 matrix $\mathcal{D}(R)$ acting on this column. (Here we are going painfully slowly because of common confusion on this point. Some authors refer to this column as a 9-component "vector," which is a horrible abuse of terminology. We reserve the word "vector" for something that transforms like a vector $V'^i = R^{ij} V^j$. It is not true that any old collection of stuff arranged in a column is a vector. Don't call anything with feathers a duck!)

For every rotation, specified by a 3-by-3 matrix R, we could thus associate a 9-by-9 matrix $\mathcal{D}(R)$ transforming the 9 objects T^{ij} linearly among themselves. We say that the 9-by-9 matrix $\mathcal{D}(R)$ represents the rotation matrix R in the sense that

$$\mathcal{D}(R_1)\mathcal{D}(R_2) = \mathcal{D}(R_1 R_2) \tag{4}$$

Multiplication of $\mathcal{D}(R_1)$ and $\mathcal{D}(R_2)$ mirrors the multiplication of R_1 and R_2, as it were. The tensor T is said to furnish a 9-dimensional representation of the rotation group $SO(3)$. The 9-by-9 matrices $\mathcal{D}(R)$ represent R. Notice that with this jargon, the vector furnishes a 3-dimensional representation of the rotation group, known as the defining or fundamental representation.

Reducible versus irreducible

Let us now pose the central question of representation theory. Given these 9 entities T^{ij} that transform into each other, consider the 9 independent linear combinations that we can form out of them. Is there a subset among them that only transform into each other? A secret in-club, as it were.

A moment's thought reveals that there is indeed an in-club. Consider $A^{ij} \equiv T^{ij} - T^{ji}$. Under a rotation,

$$A^{ij} \to A'^{ij} = T'^{ij} - T'^{ji} = R^{ik}R^{jl}T^{kl} - R^{jk}R^{il}T^{kl}$$
$$= R^{ik}R^{jl}T^{kl} - R^{jl}R^{ik}T^{lk} = R^{ik}R^{jl}(T^{kl} - T^{lk}) = R^{ik}R^{jl}A^{kl} \tag{5}$$

I have again gone painfully slow here, but it is obvious, isn't it? We just verified in (5) that A^{ij} transforms like a tensor and is thus a tensor. Furthermore, this tensor changes sign upon interchange of its two indices ($A^{ij} = -A^{ji}$) and is said to be antisymmetric. The transformation law (1) treats the two indices democratically, without favoring one over the other, and thus preserves the antisymmetric character of a tensor: if $A^{ij} = -A^{ji}$, then $A'^{ij} = -A'^{ji}$ also.

Let us count. The index i in A^{ij} could take on D values; for each of these values, the index j could take on only $D - 1$ values (since the D diagonal elements $A^{ii} = 0$ for $i = 1, 2, \cdots, D$, no Einstein repeated index summation here); but to avoid double counting (since $A^{ij} = -A^{ji}$) we should divide by 2. Hence, the number of independent components in A is equal to $\frac{1}{2}D(D - 1)$. For example, for $D = 3$, we have the 3 objects: A^{12}, A^{23}, and A^{31}. The attentive reader would recall that we did the same counting in the previous chapter.

Obviously, the same goes for the symmetric combination $S^{ij} \equiv T^{ij} + T^{ji}$. You could verify as a trivial exercise that $S'^{ij} = R^{ik}R^{jl}S^{kl}$. A tensor S^{ij} that does not change sign upon interchange of its two indices ($S^{ij} = S^{ji}$) is said to be symmetric. Evidently, the symmetric tensor S has more components than the antisymmetric tensor A. In addition to the components S^{ij} with $i \neq j$, S also has D diagonal components, namely $S^{11}, S^{22}, \cdots, S^{DD}$. Thus, the number of independent components in S is equal to $\frac{1}{2}D(D - 1) + D = \frac{1}{2}D(D + 1)$.

For $D = 3$, the number of components in A and S are $\frac{1}{2} \cdot 3 \cdot 2 = 3$ and $\frac{1}{2} \cdot 3 \cdot 4 = 6$, respectively. (For $D = 4$, the number of components in A and S are 6 and 10, respectively.) Thus, in a suitable basis, the 9-by-9 matrix referred to above actually breaks up into a 3-by-3 block and a 6-by-6 block. We say that the 9-dimensional representation is reducible: it could be reduced to smaller representations.

But we are not done yet. The 6-dimensional representation is also reducible. To see this, note

$$S'^{ii} = R^{ik}R^{il}S^{kl} = (R^T)^{ki}R^{il}S^{kl} = (R^{-1})^{ki}R^{il}S^{kl} = \delta^{kl}S^{kl} = S^{kk} \tag{6}$$

where we have used the O in $SO(D)$. (Here we are using repeated index summation: the indices i and k are both summed over.) In other words, the linear combination $S^{11} + S^{22} + \cdots + S^{DD}$, the trace of S, transforms into itself, that is, does not transform at all. It is a loner forming an in-club of one. The 6-by-6 matrix describing the linear transformation of the 6 objects S^{ij} breaks up into a 1-by-1 block and a 5-by-5 block. See figure 1.

Again, for the sake of the beginning student, let us work out explicitly the 5 objects that furnish the representation 5 of $SO(3)$. First define a traceless symmetric tensor $\tilde{S}$ by

$$\tilde{S}^{ij} = S^{ij} - \delta^{ij}(S^{kk}/D) \tag{7}$$

(The repeated index k is summed over.) Explicitly, $\tilde{S}^{ii} = S^{ii} - D(S^{kk}/D) = 0$, and $\tilde{S}$ is traceless. Specialize to $D = 3$. Now we have only 5 objects, namely $\tilde{S}^{11}, \tilde{S}^{22}, \tilde{S}^{12}, \tilde{S}^{13}, \tilde{S}^{23}$.

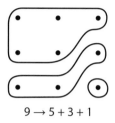

$$9 \rightarrow 5 + 3 + 1$$

Figure 1 How the collection of 9 objects T^{ij} splits up. The figure is meant to be schematic: the dots do not represent the original 9 objects, but linear combinations of them, and the positions of the dots are not meaningful.

We do not count $\tilde{S}^{33}$ separately, since it is equal to $-(\tilde{S}^{11} + \tilde{S}^{22})$. Under an $SO(3)$ rotation, these 5 objects transform into linear combinations of one another, as we just explained.

Let us be specific: the object $\tilde{S}^{13}$, for example, transforms into $\tilde{S}'^{13} = R^{1k}R^{3l}\tilde{S}^{kl} = R^{11}R^{31}\tilde{S}^{11} + R^{11}R^{32}\tilde{S}^{12} + R^{11}R^{33}\tilde{S}^{13} + R^{12}R^{31}\tilde{S}^{21} + R^{12}R^{32}\tilde{S}^{22} + R^{12}R^{33}\tilde{S}^{23} + R^{13}R^{31}\tilde{S}^{31} + R^{13}R^{32}\tilde{S}^{32} + R^{13}R^{33}\tilde{S}^{33} = (R^{11}R^{31} - R^{13}R^{33})\tilde{S}^{11} + (R^{11}R^{32} + R^{12}R^{31})\tilde{S}^{12} + (R^{11}R^{33} + R^{13}R^{31})\tilde{S}^{13} + (R^{12}R^{32} - R^{13}R^{33})\tilde{S}^{22} + (R^{12}R^{33} + R^{13}R^{32})\tilde{S}^{23}$, where in the last equality, we used $\tilde{S}^{ij} = \tilde{S}^{ji}$ and $\tilde{S}^{33} = -(\tilde{S}^{11} + \tilde{S}^{22})$. Indeed, $\tilde{S}^{13}$ transforms into a linear combination of $\tilde{S}^{11}, \tilde{S}^{22}, \tilde{S}^{12}, \tilde{S}^{13}, \tilde{S}^{23}$.

To summarize, what we found is that if, instead of the basis consisting of the 9 entities T^{ij}, we use the basis consisting of the 3 entities A^{ij}, the single entity S^{kk} (remember repeated index summation!), and the 5 entities $\tilde{S}^{ij}$, the 9-by-9 matrix $\mathcal{D}(R)$ (that represents rotation in the sense of (4)) breaks up into a 3-by-3 matrix, a 1-by-1 matrix, and a 5-by-5 matrix "stacked on top of each other." This is represented schematically as

$$\mathcal{D}(R) = \text{(9-by-9 matrix)} \rightarrow \begin{bmatrix} \text{(3-by-3 block)} & 0 & 0 \\ 0 & \text{(1-by-1 block)} & 0 \\ 0 & 0 & \text{(5-by-5 block)} \end{bmatrix} \tag{8}$$

Note that once we chose the new basis, this decomposition holds true for all rotations. (For the readers who know their linear algebra, the technical statement is that there exists a similarity transformation that block-diagonalizes $\mathcal{D}(R)$ for all R. Incidentally, we will encounter plenty of similarity transformations later.)

More generally, the D^2 representation furnished by a general 2-indexed tensor decomposes into a $\frac{1}{2}D(D - 1)$-dimensional representation, a $(\frac{1}{2}D(D + 1) - 1)$-dimensional representation, and a 1-dimensional representation. We say that in $SO(3)$, $9 = 5 + 3 + 1$. (In $SO(4)$, $16 = 9 + 6 + 1$.)

You might have noticed that in this entire discussion we never had to write out R explicitly in terms of the 3 rotation angles and how the 5 objects $\tilde{S}^{11}, \cdots, \tilde{S}^{23}$ transform into one another in terms of these angles. It is only the counting that matters. You might regard that as the difference between mathematics and arithmetic.

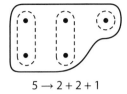

$$5 \rightarrow 2 + 2 + 1$$

Figure 2 Under $SO(3)$, the 5 objects inside the solid line transform into linear combinations of each other, but under the smaller group of transformations $SO(2)$, the objects inside each of the 3 dashed lines transform into linear combinations of each other. The 5 breaks up as $5 \rightarrow 2 + 2 + 1$. As in figure 1, this figure is meant to be schematic.

Restriction to a subgroup

You definitely do not have to master group theory[1] to read this book, but it would be useful for you to learn a few basic concepts and to be able to count. For instance, the notion of a subgroup. Consider the group $SO(2)$ that we studied to exhaustion, consisting of rotations around the z-axis, say. Evidently, $SO(2)$ is a subgroup of $SO(3)$ in that its elements are all elements of $SO(3)$ and form a group all by themselves. The components of the 3-vector V^i could be split into two sets: (V^1, V^2) and V^3. Under a rotation around the z-axis, (V^1, V^2) transform as a 2-vector and V^3 as a scalar. We say that upon restriction to the subgroup $SO(2)$, the irreducible representation 3 breaks up into the representations 2 and 1 of the subgroup, a decomposition we write as $3 \rightarrow 2 + 1$. All the group theoretic results we need in this book could be obtained by explicit listing and simple counting.

Look at the 5 objects, S^{11}, S^{22}, S^{12}, S^{13}, S^{23}, that furnish the representation 5 of $SO(3)$. Now consider a restriction to the subgroup $SO(2)$. In other words, we restrict ourselves to rotations around the z-axis, that is, rotations under which $V^3 \rightarrow V'^3 = V^3$, namely rotations with $R^{33} = 1$ and R^{13}, R^{23}, R^{31}, R^{32} all vanishing. Since $SO(2)$ does not touch the index 3, we conclude immediately that the combination $S^{11} + S^{22} = -S^{33}$ does not transform, or in other words, it transforms as a singlet under $SO(2)$. Similarly, the pair (S^{13}, S^{23}) transforms as a doublet, since the index 3 is "invisible" to $SO(2)$: the group transforms the indices 1 and 2 into each other, while leaving the index 3 alone. Indeed, we see that our earlier expression for S'^{13} collapses to $S'^{13} = R^{11}S^{13} + R^{12}S^{23}$, as expected. Finally, you can verify that the remaining combinations $(S^{12}, S^{11} - S^{22})$ transform like a doublet. These results could be summarized by saying that, upon restriction to the subgroup $SO(2)$, the irreducible representation 5 of the group $SO(3)$ breaks up as $5 \rightarrow 2 + 2 + 1$. See figure 2.

Tensors in Newtonian mechanics

Let us give another example, particularly apt for a book on gravity, of a Newtonian tensor. Consider two nearby particles moving in a potential. Denote their trajectories by $\vec{x}(t)$

and $\vec{y}(t)$, respectively, determined by $\frac{d^2x^i}{dt^2} = -\partial^i V(\vec{x})$ and $\frac{d^2y^i}{dt^2} = -\partial^i V(\vec{y})$. (I am also testing whether there are any readers who do not understand thoroughly the concept of notational freedom.) We want to know how the separation vector $\vec{s} \equiv \vec{y} - \vec{x}$ changes with time, keeping terms to leading order in $\vec{s}$:

$$\frac{d^2s^i}{dt^2} = \frac{d^2y^i}{dt^2} - \frac{d^2x^i}{dt^2} = -\partial^i[V(\vec{y}) - V(\vec{x})] = -\partial^i[V(\vec{x} + \vec{s}) - V(\vec{x})] \simeq -\partial^i\partial^j V(\vec{x})s^j$$

The object $\mathcal{R}^{ij}(\vec{x}) \equiv \partial^i\partial^j V(\vec{x})$ is manifestly a tensor if $V(\vec{x})$ is a scalar. For example, verify that $\mathcal{R}^{ij} = GM(\delta^{ij}r^2 - 3x^ix^j)/r^5$ for the gravitational potential $V(\vec{x}) = -GM/r$. Note that $\mathcal{R}^{ij}$ is a symmetric traceless tensor. Since $\mathcal{R}^{ii} = \partial^i\partial^i V(\vec{x}) = \vec{\nabla}^2 V$, the tracelessness merely reaffirms the fact that the $1/r$ potential satisfies Laplace's equation $\vec{\nabla}^2 V = 0$. Also, $\mathcal{R}^{ij}$ is manifestly not the product of two vectors, but it transforms as if it were.

Let us see how rotational covariance works in the equation

$$\frac{d^2s^i}{dt^2} = -\mathcal{R}^{ij}s^j \tag{9}$$

The right hand side has to be linear in the vector $\vec{s}$. Since the left hand side transforms like a vector, the right hand side must also: indeed, it is given by a tensor $\mathcal{R}$ contracted* with a vector $\vec{s}$. A tensor is needed on the right hand side.

Imagine yourself falling toward a spherical planet or star. With no loss of generality, let your location at some instant be $(0, 0, r)$ along the z-axis. The tensor $\mathcal{R}$ written out as a matrix is then diagonal and is given by (for example, $\mathcal{R}^{33} = GM(\delta^{33}r^2 - 3x^3x^3)/r^5 = GM(1 - 3)/r^3$)

$$\mathcal{R} = \frac{GM}{r^3}\begin{pmatrix} 1 & 0 & 0 \\ 0 & 1 & 0 \\ 0 & 0 & -2 \end{pmatrix} \tag{10}$$

Thus, the sign of $\frac{d^2\vec{s}}{dt^2}$ depends on the orientation of $\vec{s}$.

To see why this is so and to understand what tensors are all about, imagine surrounding yourself with a circular arrangement of balls lying in the $(x\text{-}z)$ plane (see figure 3a) and initially at rest in your frame. Using (9) and (10), we can now write down how the separation between two balls along different directions changes.

Since we are going to specify the direction, we will denote the separation simply by s. Along the z-axis, s grows according to (see (9)) $\frac{d^2s}{dt^2} = -\mathcal{R}^{33}s = +2\frac{GM}{r^3}s$. The plus sign indicates that the two balls move away from each other. In contrast, along the x-axis, s decreases according to $\frac{d^2s}{dt^2} = -\mathcal{R}^{11}s = -\frac{GM}{r^3}s$. The two balls approach each other. (Similarly for two balls aligned along the y-axis.) (Note that acting on $\vec{s}$ on the right hand side of (9) by a tensor makes it possible for $\frac{d^2s}{dt^2}$ to change sign depending on the orientation of $\vec{s}$.)

Inspecting figure 3a, you see why. Look at it as an observer on the planet. In the first case, one of the two balls, being closer to the planet, is falling faster than the other. Thus, they

---

* When a pair of repeated indices, such as j in (9), is summed over, they are often said to be contracted with each other (as mentioned in a footnote in the preceding chapter) in the sense that this index no longer appears in the result, as shown by the left hand side of (9).

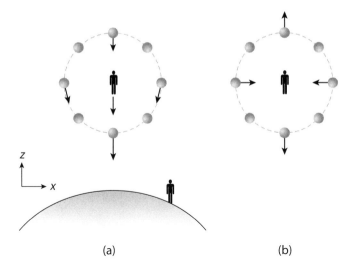

Figure 3 A falling ring of balls as seen by an observer on the planet (a), and as seen by an observer falling with the balls (b).

are moving away from each other. In the second case, the two balls are coming closer due to spherical symmetry: they are both heading toward the center of the planet. As Newton pointed out, objects do not fall down to earth, but toward the center of the earth.

In your rest frame (figure 3b) as you fall along with the balls, however, you see a tidal force acting on the circular ring (or a spherical shell if you prefer) of balls. The force appears to stretch the ring in the z-direction and to squeeze it in the orthogonal direction. When we come to Einstein's prediction of gravitational waves in chapter IX.4, we will see that gravitational waves act on the detector according to equations analogous to (9) and (10). Note also for future reference that the tidal force $\mathcal{R}^{ij}(\vec{x}) \equiv \partial^i \partial^j V(\vec{x})$ involves two derivatives acting on the gravitational potential $V(\vec{x})$.

Invariant tensors

In D-dimensional space, define the antisymmetric symbol $\varepsilon^{ijk\cdots n}$ carrying D indices to have the following properties:

$$\varepsilon^{\cdots l \cdots m \cdots} = -\varepsilon^{\cdots m \cdots l \cdots} \quad \text{and} \quad \varepsilon^{12\cdots D} = 1 \tag{11}$$

In other words, the antisymmetric symbol ε flips sign upon the interchange of any pair of indices. It follows that ε vanishes when two indices are equal. (Note that the second property listed is just normalization.) Since each index can take on only values $1, 2, \cdots, D$, the antisymmetric symbol for D-dimensional space must carry D indices as already noted. For example, for $D = 2$, $\varepsilon^{12} = -\varepsilon^{21} = 1$, with all other components vanishing. For $D = 3$, $\varepsilon^{123} = \varepsilon^{231} = \varepsilon^{312} = -\varepsilon^{213} = -\varepsilon^{132} = -\varepsilon^{321} = 1$, with all other components vanishing (as was already noted in the preceding chapter).

Using the Kronecker delta and the antisymmetric symbol, we can write the defining properties of rotations $R^T R = I$ and $\det R = 1$ as

$$\delta^{ij} R^{ik} R^{jl} = \delta^{kl} \tag{12}$$

and

$$\varepsilon^{ijk\cdots n} R^{ip} R^{jq} R^{kr} \cdots R^{ns} = \varepsilon^{pqr\cdots s} \det R = \varepsilon^{pqr\cdots s} \tag{13}$$

respectively. In (13) we used the definition of $\det R$. (Verify this for $D = 2$ and 3.)

Referring to (1), we see that we can describe δ^{ij} and $\varepsilon^{ijk\cdots n}$ as invariant tensors: they transform into themselves. For the rest of this text, we will often use, implicitly or explicitly, the notion of invariant tensors.

For example, for $SO(3)$, using (13) you can show that $\varepsilon^{ijk} A^i B^j \equiv C^k$ defines a vector $\vec{C} = \vec{A} \times \vec{B}$, the familiar cross product. Various identities follow. Consider, for example,

$$\varepsilon^{ijk} \varepsilon^{lnk} = \delta^{il} \delta^{jn} - \delta^{in} \delta^{jl} \tag{14}$$

To prove this, simply note that both sides transform as invariant tensors with four indices, and the symmetry properties (such as under $i \leftrightarrow j$) of the two sides match. Contracting with A^j, B^l, and C^n, we obtain an identity you might recognize: $\vec{A} \times (\vec{B} \times \vec{C}) = \vec{B}(\vec{A} \cdot \vec{C}) - \vec{C}(\vec{A} \cdot \vec{B})$.

Closing of Newtonian orbits once again

We can now go back to the apparent mystery in chapter I.1, that the Newtonian orbits in a $1/r$ potential close. Out of the conserved angular momentum vector $\vec{l} = \vec{r} \times \vec{p} = \vec{r} \times \dot{\vec{r}}$ (we are using the notation of chapter I.1; we have effectively set the mass to unity and hence the second equality) we can form the Laplace-Runge-Lenz vector $\vec{\mathcal{L}} \equiv \vec{l} \times \dot{\vec{r}} + \frac{\kappa}{r}\vec{r}$. Computing the time derivative $\dot{\vec{\mathcal{L}}}$, you can verify (see exercise 4) that $\vec{\mathcal{L}}$ is conserved for an inverse square central force. When $\dot{\vec{r}}$ is perpendicular to $\vec{r}$, which occurs at perihelion and aphelion, the vector $\vec{\mathcal{L}}$ points in the direction of $\vec{r}$. We could take the constant vector $\vec{\mathcal{L}}$ to point toward the perihelion, and thus the position of the perihelion does not change. Hence the orbit closes.

This result does not hold in Einstein gravity. The precession of the perihelion of Mercury, which we will discuss in chapter VI.3, is of course one of the classic tests of general relativity.

Appendix: Two lemmas for future use

There is a lot more we could say about tensors, but let me mention two simple lemmas that we will happen to need later.

Let S^{ij} and A^{ij} be two arbitrary and unrelated tensors, symmetric and antisymmetric, respectively. Then $S^{ij} A^{ij} = 0$. (See exercise 5.)

Tensors can have all kinds of symmetry properties, which you can explore on your own and in the exercises. For example, a totally antisymmetric 3-indexed tensor T^{ijk} is such that T flips sign under the interchange of any pair of indices (for example, $T^{ijk} = -T^{jik} = +T^{jki}$). A multi-indexed tensor can also have symmetry properties under the interchange of a specific pair, or may have no symmetry at all. Consider, for example, a tensor G^{kij} symmetric under the interchange of the first pair of indices only, that is, $G^{kij} = G^{ikj}$. To be pedantic and absolutely clear, sometimes I like to put a space or a dot between the indices, thus $G^{ki\ j}$ or $G^{ki \cdot j}$ to separate the "special" pair from the other indices. For example, our tensor could happen to be $G^{ki \cdot j} = \partial^k \partial^i W^j(\vec{x})$ for some vector field W^j.

Given $G^{ki \cdot j}$, define $H^{k \cdot ij} \equiv G^{ki \cdot j} + G^{kj \cdot i}$. (Note that $H^{k \cdot ij} = H^{k \cdot ji}$ by definition, but $H^{i \cdot kj}$ is in general not equal to $H^{k \cdot ij}$.) Then we can solve for G in terms of H:

$$G^{ki \cdot j} = \frac{1}{2}(H^{k \cdot ij} + H^{i \cdot jk} - H^{j \cdot ki}) \tag{15}$$

(See exercise 8.)

Exercises

1   Define $\vec{\nabla} \equiv \left(\frac{\partial}{\partial x^1}, \frac{\partial}{\partial x^2}, \cdots, \frac{\partial}{\partial x^D} \right)$. Show that if ϕ is a scalar, then $(\vec{\nabla}\phi)^2 = \vec{\nabla}\phi \cdot \vec{\nabla}\phi = \sum_k \left(\frac{\partial \phi}{\partial x^k} \right)^2$ and $\nabla^2 \phi$ transform like a scalar. The Laplacian is defined by

$$\nabla^2 = \vec{\nabla} \cdot \vec{\nabla} = \frac{\partial^2}{\partial(x^1)^2} + \frac{\partial^2}{\partial(x^2)^2} + \cdots + \frac{\partial^2}{\partial(x^D)^2}$$

2   Show that the symmetric tensor S^{ij} is indeed a tensor.

3   Show that the infinitesimal volume element d^3x is a scalar.

4   Show that the Laplace-Runge-Lenz vector is conserved.

5   Show that $S^{ij}A^{ij} = 0$ if S^{ij} is a symmetric tensor and A^{ij} an antisymmetric tensor.

6   Let T^{ijk} be a totally antisymmetric 3-indexed tensor. Show that T has $\frac{1}{3!}D(D-1)(D-2)$ components. Identify the one component for $D = 3$.

7   Consider for $SO(3)$ the tensor T^{ijk} from exercise 6. Show that it transforms as a scalar.

8   Prove the lemma in (15).

9   Verify (13) for $D = 2$ and 3.

Note

1. For a concise introduction to some of the group theory needed in theoretical physics, see *QFT Nut*, appendix B.

I.5 | From Change of Coordinates to Curved Spaces

Euclidean spaces described with different coordinates

In discussing rotations in chapter I.3, I emphasized that Euclid is defined by Pythagoras. That the square of the distance between two neighboring points in 2-dimensional Euclidean space with coordinates (x, y) and $(x + dx, y + dy)$ is given by $ds^2 = dx^2 + dy^2$ defines what we mean by Euclidean space.

But even the familiar Euclidean space can look unfamiliar. You know well that in many physics problems, one set of coordinates is often much more convenient than another. Indeed, in discussing Newton's planetary orbit problem in chapter I.1, we changed from Cartesian* coordinates (x, y) to polar coordinates (r, θ), with $x = r \cos \theta$ and $y = r \sin \theta$. Differentiating, we have $dx = dr \cos \theta - r \sin \theta \, d\theta$ and $dy = dr \sin \theta + r \cos \theta \, d\theta$, so that

$$ds^2 = dx^2 + dy^2 = (dr \cos \theta - r \sin \theta \, d\theta)^2 + (dr \sin \theta + r \cos \theta \, d\theta)^2 = dr^2 + r^2 d\theta^2 \qquad (1)$$

We are free to make any coordinate transformation we feel like. Consider the most general transformation $x = f(u, v)$, $y = g(u, v)$. Then $dx = f_u(u, v)du + f_v(u, v)dv$ where $f_u \equiv \frac{\partial f}{\partial u}$ and so on, and $dy = g_u(u, v)du + g_v(u, v)dv$. Just plug in to obtain $ds^2 = dx^2 + dy^2 = (f_u^2 + g_u^2)du^2 + (f_v^2 + g_v^2)dv^2 + 2(f_u f_v + g_u g_v)dudv$. With a gunky choice of f and g you will end up with a mess of a coordinate system that would only make your life miserable. (Note that even the innocuous change $x = u + v$ and $y = v$ leads to $ds^2 = du^2 + 2dv^2 + 2dudv$ with the rather unpleasant $dudv$ cross term.) Of course, it was discovered long ago that by choosing $f(u, v) = u \cos v$ and $g(u, v) = u \sin v$, we can get rid of the cross term. By now probably all the nice choices for f and g have already been published by someone.

---

* When I was in high school, I got the erroneous impression that the notion of coordinates originated with Descartes. In fact, by the time of Ptolemy, astronomers in the West certainly had latitudes and longitudes. In China, Chang Heng, roughly a contemporary of Ptolemy, was said to have derived, by watching a woman weaving, a system of coordinates to map heaven and earth with. The Chinese words for latitudes and longitudes, "jing" and "wei," are just the terms for warp and weft in weaving.

I presume that you also know how to go from Cartesian coordinates (x, y, z) in 3-dimensional Euclidean space E^3 to spherical coordinates (r, θ, φ), with $x = r \sin \theta \cos \varphi$, $y = r \sin \theta \sin \varphi$, $z = r \cos \theta$. The more-than-familiar (and who can blame you if you have been in it all your life?) E^3 could be described by either $ds^2 = dx^2 + dy^2 + dz^2$ in Cartesian coordinates or by

$$ds^2 = dr^2 + r^2 d\theta^2 + r^2 \sin^2 \theta d\varphi^2 \tag{2}$$

in spherical coordinates.

From Latin to Greek

We can systematize and generalize this to D-dimensional space easily enough. In previous chapters, I used Latin letters for the index on the coordinates. I now switch, for later convenience, from Latin to Greek and call the coordinates $x^\mu = (x^1, x^2, \cdots, x^D)$. Then, for Euclid's spaces E^D, Pythagoras said that $ds^2 = \sum_{\mu=1}^{D} (dx^\mu)^2$.

We write this in the fancier form $ds^2 = \sum_{\mu=1}^{D} \sum_{\nu=1}^{D} g_{\mu\nu} dx^\mu dx^\nu$ by introducing a D-by-D matrix g whose diagonal elements are all equal to one and whose other elements are all zero, the famous matrix known far and wide as the identity matrix. To repeat, the indices μ, ν run over 1, 2, $\cdots$, D, and $g_{\mu\nu}$ is defined by $g_{\mu\mu} = 1$ and $g_{\mu\nu} = 0$ if $\mu \neq \nu$. (In other words, it is just the Kronecker delta introduced in chapter I.3: $g_{\mu\nu} = \delta_{\mu\nu}$.) Thus, in the double sum for ds^2, the terms with $\mu \neq \nu$ drop out and we are left with $ds^2 = \sum_{\mu=1}^{D} (dx^\mu)^2$.

Now a word on notation. In the chapter on rotation, I have already introduced this expression for ds^2, and furthermore, the repeated index summation convention. Einstein suggested that between us friends we could omit the cumbersome summation symbol and agree that if an index is repeated, then it is to be summed over. Thus, we suppress the double summation $\sum_{\mu=1}^{D} \sum_{\nu=1}^{D}$ and write simply $ds^2 = g_{\mu\nu} dx^\mu dx^\nu$. Here μ and ν are both repeated and hence summed over. Unless there is a risk of confusion, no more summation symbols!

The metric

The matrix $g_{\mu\nu}$ is called the metric, a word meaning measure, as in geometry, the science of measuring the earth. We use the metric to measure space. This step of introducing a metric for Euclidean spaces seems like one of those totally senseless moves that certain academics like and publish. In the discussion just given, the metric is simply the identity matrix.

But as soon as we change coordinates, the metric is no longer so simple. As we have already noted in (1), with polar coordinates, the plane E^2 is described by a metric with $g_{rr} = 1$, $g_{\theta\theta} = r^2$, and $g_{r\theta} = 0 = g_{\theta r}$. With spherical coordinates, E^3 is described by a metric with $g_{rr} = 1$, $g_{\theta\theta} = r^2$, $g_{\varphi\varphi} = r^2 \sin^2 \theta$, with all other entries zero, as in (2).

In both examples, the metric is not given by the identity matrix. Furthermore, the metric $g_{\mu\nu}(x)$ varies from point to point. For example, $g_{\varphi\varphi}$ depends on both r and θ. Note, however, that for these examples, the metric is diagonal. (That is why polar and spherical coordinates are so popular!) In general, the metric $g_{\mu\nu}$ need not be diagonal (as shown in the example $ds^2 = du^2 + 2dv^2 + 2dudv$, for which $g_{uu} = 1$, $g_{vv} = 2$, $g_{uv} = g_{vu} = 1$). However, in this text, for the sake of simplicity, we will mostly stick to metrics that are diagonal. Furthermore, since $dx^\mu dx^\nu = dx^\nu dx^\mu$, the metric is symmetric under interchange of indices: $g_{\mu\nu} = g_{\nu\mu}$. It goes without saying that the reader encountering all this for the first time should verify everything I say.

Lower indices appear

The attentive reader might have noticed that lower indices have sneakily appeared! The metric $g_{\mu\nu}$ carries lower indices, while dx^μ carries an upper index. When I taught Einstein gravity, the appearance of upper and lower indices invariably confused some students. In this text, I will try to motivate the point of introducing upper and lower indices, more from a utilitarian, rather than a profoundly mathematical, point of view. My strategy is to introduce this business of two kinds of indices in stages.

At this stage, the motivation, to put it bluntly, is that we just feel like it. But this caprice immediately leads to a useful rule. In the Einstein repeated index summation, we will insist that when we sum over a pair of repeated indices, one of them must be upstairs, the other downstairs. This is manifestly, and trivially, satisfied by the only example $ds^2 = g_{\mu\nu}(x)dx^\mu dx^\nu$ we have encountered thus far. The whole business of two kinds of indices may seem unnecessary at this point, but later, you will see that the distinction between upper and lower indices becomes essential, or at least highly useful.

A word about terminology: Some authors refer to $ds^2 = g_{\mu\nu}(x)dx^\mu dx^\nu$ as the square of the line element, reserving the term metric for the object $g_{\mu\nu}(x)$ contained in the line element. I find it convenient to abuse terminology and simply refer to both as the metric.

Let me mention one trivial point, but one with the potential for confusing beginners. Some years ago, when I surveyed the students in my class for points of confusion, one student told me that for quite a while he did not realize that $g_{\rho\mu}(x)dx^\rho dx^\mu$, $g_{\zeta\psi}(x)dx^\zeta dx^\psi$, and so on, all denote the same thing! Perhaps this is because the summation symbol has been suppressed: the same student could recognize that $\sum_{\mu=1}^{D}\sum_{\nu=1}^{D} g_{\mu\nu}(x)dx^\mu dx^\nu = \sum_{\rho=1}^{D}\sum_{\mu=1}^{D} g_{\rho\mu}(x)dx^\rho dx^\mu = \sum_{\zeta=1}^{D}\sum_{\psi=1}^{D} g_{\zeta\psi}(x)dx^\zeta dx^\psi$.

Change of coordinates, curved space, and curved spacetime

We all know that in Euclidean 3-space, if we restrict r to be equal to a, we would find ourselves on the surface of a sphere of radius a. In other words, the set of points at a distance a from the origin form a sphere with radius a.

This procedure gives us an easy way to determine the metric on a sphere. Simply take the metric (2) $ds^2 = dr^2 + r^2 d\theta^2 + r^2 \sin^2 \theta d\varphi^2$ and set $r = a$. Then $dr = 0$, so that ds^2 collapses to $a^2(d\theta^2 + \sin^2 \theta d\varphi^2)$. From 3-dimensional flat space we have "lost" the coordinate r and gone to a 2-dimensional curved space with coordinates $x^\mu = (x^1, x^2) = (\theta, \varphi)$. Without loss of generality, we can take a as the unit of distance and set $a = 1$. So, on the unit 2-dimensional sphere S^2

$$ds^2 = d\theta^2 + \sin^2 \theta d\varphi^2 \tag{3}$$

with a metric given by $g_{11} = g_{\theta\theta} = 1$, $g_{22} = g_{\varphi\varphi} = \sin^2 \theta$, and $g_{12} = g_{\theta\varphi} = g_{21} = g_{\varphi\theta} = 0$.

The take-home message here is that curved space is just a skip and a hop away from the familiar change of coordinates. This is fortunate for students of physics: when you learned to change coordinates, you were actually also learning about curved spaces. We are now going to develop a general formalism for changing coordinates. Even though you already know how to change coordinates, it pays to learn this formalism, because we can also use it to study curved space and curved spacetime (which, as you have surely heard, plays a central role in Einstein gravity).

Change of coordinates, curved space, and curved spacetime: basically the same deal, as you will see.

How do we know whether a space is curved or not?

This raises an exceedingly interesting and crucial question: given a space with the metric $g_{\mu\nu}(x)$, how do we know whether it is curved or flat?

A complicated looking metric does not necessarily mean that the space is curved, since somebody could have simply chosen an especially gunky coordinate system. It could be flat space in disguise. To forcefully bring home this point, I invite you to consider $ds^2 = (1 + u^2)du^2 + (1 + 4v^2)dv^2 + 2(2v - u)dudv$ and $ds^2 = (1 + u^2)du^2 + (1 + 2v^2)dv^2 + 2(2v - u)dudv$. One describes flat space, the other a space that at some points is violently curved. Which is which?

Puzzled, you reply: "How could I possibly tell?"

That's in fact the correct answer at this stage of this discussion. The two metrics I just gave you look almost identical except for one single $2 \to 4$. In one of the most famous episodes in mathematics, Carl Friedrich Gauss (1777–1855) solved this problem for 2-dimensional spaces. His work was then generalized by his student Bernhard Riemann (1826–1866). Later, in chapter VI.1, given any metric in any number of dimensions, you will be able to calculate, and even better, to train the computer to calculate, something called the Riemann curvature tensor, which will tell you once and for all if the space is flat or curved. No more thinking involved! Gauss and Riemann did it for you.

But for now, let me ask you to think about two simple examples in good old 2-dimensional space, for which our intuition is allegedly pretty good. We know that $ds^2 = dr^2 + r^2 d\theta^2$ describes flat space. Consider

$$ds^2 = d\rho^2 + \sin^2 \rho \, d\theta^2 \tag{4}$$

Is the space being described flat or curved? Or consider the space described by

$$ds^2 = \cos^2 \rho \, d\rho^2 + \sin^2 \rho \, d\theta^2 \tag{5}$$

Is it flat or curved? You should think about this before reading on. The answers are given in appendix 1.

Remember the civilization of mites in the prologue? You are in the same position as the mite professors of geometry: they can measure the distance between infinitesimally separated points, and from that they have to figure out whether their world is curved. We will face the same problem as the mites when we get to cosmology in parts V and VI.

The logic of differential geometry

Differential geometry, as developed by Gauss and Riemann, tells us that given the metric, we can calculate the curvature. The logic goes as follows. The metric tells you the distance between two nearby points. Integrating, you can obtain the distance along any curve joining two points, not necessarily nearby. Find the curve with the shortest distance. By definition, this curve is the "straight line" between these two points. Once you know how to find the "straight line" between any two points, you can test all of Euclid's theorems to see whether our space is flat. For example, as described in the prologue, the mite geometers could now draw a small circle around any point, measure its circumference, and see if it is equal to 2π times the radius. (See appendix 1.) Thus, the metric can tell us about curvature.

Take an everyday example: given an airline table of distances, you can deduce that the world is curved without ever going outside. If I tell you the three distances between Paris, Berlin, and Barcelona, you can draw a triangle on a flat piece of paper with the three cities at the vertices. But now if I also give you the distances between Rome and each of these three cities, you would find that you can't extend the triangle to a planar quadrangle (figure 1). So the distances between four points suffice to prove that the world is not flat. But the metric tells you the distances between an infinite number of points.

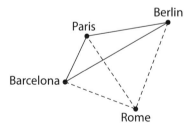

Figure 1 The distances between four cities suffice to prove that the world is not flat.

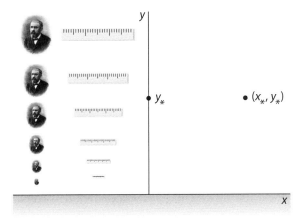

Figure 2 The Poincaré half plane (including pictures of Poincaré).

Poincaré half plane

Let me tell you the interesting example of the Poincaré half plane.* Consider the upper half plane covered by the usual coordinates (x, y) with $y > 0$, but endowed with the peculiar metric

$$ds^2 = \frac{dx^2 + dy^2}{y^2} \tag{6}$$

See figure 2. In other words, $g_{xx} = g_{yy} = \frac{1}{y^2}$. The space is translation invariant in x, that is, $g_{\mu\nu}(x + a, y) = g_{\mu\nu}(x, y)$ for any a, and the space at one value of x looks exactly the same as the space at some other value of x, but it is not translation invariant in y. Evidently, this space has an edge at $y = 0$. Consider a standard ruler with length l, pointing along the x-axis, so that the two ends of the ruler are separated by $\Delta y = 0$. According to (6), $l = \Delta s = \Delta x / y$. Take the ruler closer and closer to the edge. The ruler covers less and less Δx as you approach the edge: indeed $\Delta x = yl \to 0$ as $y \to 0$. It would appear that your ruler is shrinking relative to the milestones the inhabitants of this world have helpfully erected at fixed values of x. (We point the ruler along the x-axis merely for pedagogical clarity; in fact, we would reach the same conclusion regardless of the ruler's orientation. For example, a ruler pointing along the y-axis would cover $\Delta y = yl$, which $\to 0$ as you approach the edge.)

In reality, the edge is infinitely far away, since the actual distance between the points (x, y_*) and $(x, 0^+)$ along the line of constant x is given by

$$\int ds = \int_{0^+}^{y_*} dy/y = \log(y_*/0^+) \to \infty$$

---

* First discovered by the Italian mathematician Eugenio Beltrami (1835–1899) long before Poincaré.

Imagine, as in a sci-fi movie, finding yourself in such a weird space. Before reading on, can you figure out the straight line joining two points, say $(0, y_*)$ and (x_*, y_*)? The phrase "straight line joining two points" is of course, as already mentioned, defined to be the path of shortest distance between the two points.

If you were to go along the line of fixed $y = y_*$, the distance would be $\int ds = (\int_0^{x_*} dx)/y_*$ $= x_*/y_*$. Clearly, you could do better. To economize on ds, you should curve away from the edge at $y = 0$ into the region of larger y and hence smaller ds for a given stretch of (dx, dy). At this point, we can only discuss this curved path qualitatively. In chapter II.2, we will learn how to determine this curve. In this sci-fi movie, you see your favorite person in the distance. You run to him or her, but you sure don't want this person to think you are an idiot, which you would be if you ran along a "straight" line expressed in (x, y) coordinates. You try to be like Feynman the lifeguard in the prologue, so you follow the curve that minimizes $\int ds$.

In fact, this simple example of the Poincaré half plane is behind a recent advance in quantum gravity and string theory, known as AdS/CFT (anti de Sitter/conformal field theories). See chapter IX.11.

A pervasive theme of theoretical physics

The central message here is that coordinates do not have intrinsic geometric significance. If you use the coordinates x^μ, somebody else could perfectly well use coordinates x'^μ, with the two sets of coordinates related by the D functions of D variables $x'^\mu(x)$ and their inverses $x^\mu(x')$.

Again we come to the pervasive theme of theoretical physics already alluded to in chapter I.3: transformations. Given a set of physical laws, we ask: What are the transformations that leave these laws unchanged? How do the relevant physical quantities transform? What combinations of these quantities are left invariant? Here we are concerned with coordinate transformations.

Rotations furnish the prototypical example. Indeed, we see that we can think of rotation $x'^\mu = R^\mu{}_\nu x^\nu$ as a special example of a class of coordinate transformations, namely those that are linear. Notice that, obeying our rule of only summing an upper with a lower index, we moved the column index on the rotation matrix R downstairs. We also graduated from Latin to Greek.

General coordinate transformation

In elementary discussions of change of coordinates, brute force substitution, as in (1), suffices, but as you will see, it pays to formalize the steps involved, so that the formalism applies to more general situations. I have already advertised that many concepts involved in changing coordinates are also relevant for discussing curved spaces.

We will use as our canonical example the transformation between Cartesian and polar co-ordinates, between $(x^1 \equiv x, x^2 \equiv y)$ and $(x'^1 \equiv r, x'^2 \equiv \theta)$. We have the two functions $x^1 \equiv x = r \cos \theta, x^2 \equiv y = r \sin \theta$ and their inverses $x'^1 \equiv r = \sqrt{x^2 + y^2}, x'^2 \equiv \theta = \arctan \frac{y}{x}$. Note that in many cases, x^μ and x'^μ all have common colloquial names, namely x, y and r, θ, respectively, in this example. We will pass freely without any further remark between the "academic" names x^μ and x'^μ and their street names.

A word about notation: I trust the reader not to be confused by unavoidable but trivial notational abuse. For example, here the letter x does double duty: generically, it represents x^μ and, in the special case of the standard Cartesian coordinates, also the first component of x^μ.

In general, coordinate transformations are definitely nonlinear. For example, the polar angle $x'^2 \equiv \theta = \arctan \frac{y}{x}$ is defiantly not a linear function of x and y. However, and this is a crucial point, the infinitesimals dx^μ do transform linearly. Indeed, applying elementary calculus, we have

$$dx'^\mu = \frac{\partial x'^\mu}{\partial x^\nu} dx^\nu \tag{7}$$

For the sake of the reader seeing this for the first time, let's go slow here. Calculate dx'^3, for example (supposing that $D \geq 3$):

$$dx'^3 = \frac{\partial x'^3}{\partial x^1} dx^1 + \frac{\partial x'^3}{\partial x^2} dx^2 + \frac{\partial x'^3}{\partial x^3} dx^3 + \cdots + \frac{\partial x'^3}{\partial x^D} dx^D = \sum_{\nu=1}^{D} \frac{\partial x'^3}{\partial x^\nu} dx^\nu = \frac{\partial x'^3}{\partial x^\nu} dx^\nu \tag{8}$$

where in the last step we invoke the repeated index summation convention and drop the summation sign. This is of course just (7) with $\mu = 3$.

It is useful to define the matrix

$$S^\mu_{\;\nu}(x) \equiv \frac{\partial x'^\mu}{\partial x^\nu} \tag{9}$$

which we could regard as a matrix with a row index μ and column index ν. Then we can write $dx'^\mu = S^\mu_{\;\nu}(x) dx^\nu$. In accordance with the repeated index summation convention, the index ν is summed over. Note that we put the second index on S downstairs to obey the rule of only summing an upper with a lower index.

The infinitesimals dx'^μ and dx^ν are related linearly by the matrix $S^\mu_{\;\nu}(x)$, just as $dx'^\mu = R^\mu_{\;\nu} dx^\nu$ for rotation. The big difference from simple rotation is of course that $S^\mu_{\;\nu}(x)$ depends on x. (This fact leads to all the mathematical complications in Einstein gravity that you may or may not have heard about.)

In the polar example, we have

$$dx'^1 \equiv dr = \frac{x dx + y dy}{\sqrt{x^2 + y^2}} \quad \text{and} \quad dx'^2 \equiv d\theta = \frac{x dy - y dx}{x^2 + y^2}$$

so that

$$S^1_{\;1} = \frac{x}{\sqrt{x^2 + y^2}}, \quad S^1_{\;2} = \frac{y}{\sqrt{x^2 + y^2}}, \quad S^2_{\;1} = \frac{-y}{x^2 + y^2}, \quad \text{and} \quad S^2_{\;2} = \frac{x}{x^2 + y^2}$$

(Note that, if we think of x, y, and r as having dimensions of length, the different components of the matrix $S^\mu_{\ \nu}$ do not even have the same dimension—hardly surprising, since $x'^1 = r$ and $x'^2 = \theta$ have different length dimensions.)

Going from primed to unprimed coordinates, we expect to encounter the inverse of (7): duh, we are going the other way. Again, using elementary calculus, we write (with the index ρ summed over as per the repeated index convention)

$$dx^\mu = \frac{\partial x^\mu}{\partial x'^\rho}dx'^\rho = (S^{-1})^\mu_{\ \rho}(x')dx'^\rho \tag{10}$$

where

$$(S^{-1})^\mu_{\ \rho} \equiv \frac{\partial x^\mu}{\partial x'^\rho} \tag{11}$$

Since we are simply performing the inverse transformation, S^{-1} in (10) has to be, as we just said, the inverse of the matrix S introduced in (7). Nevertheless, let us pause to verify the obvious. Use the chain rule of calculus: $(S^{-1})^\mu_{\ \rho}S^\rho_{\ \nu} = \frac{\partial x^\mu}{\partial x'^\rho}\frac{\partial x'^\rho}{\partial x^\nu} = \frac{\partial x^\mu}{\partial x^\nu} = \delta^\mu_\nu$, where the Kronecker delta is defined in analogy with the Kronecker delta used in (I.3.16) for rotations, namely that $\delta^\mu_\nu = 1$ if $\mu = \nu$ and 0 otherwise. In other words, the Kronecker delta is just a fancy way of describing the identity matrix.* Thus, we have verified that, indeed, $S^{-1}S = I$.

In the simple polar example, (10) just consists of $dx = \cos\theta\, dr - r\sin\theta\, d\theta$ and $dy = \sin\theta\, dr + r\cos\theta\, d\theta$. Thus, $(S^{-1})^1_{\ 1} = \cos\theta$, $(S^{-1})^1_{\ 2} = -r\sin\theta$, $S^2_{\ 1} = \sin\theta$, and $S^2_{\ 2} = r\cos\theta$.

Let me also quickly put another potential source of confusion to rest. I have written S as a function of x and S^{-1} as a function of x', but of course any function of x could be written as a function of x' and vice versa.

A general formalism for changing coordinates

We want to know how the metric $g_{\mu\nu}$ transforms when we change from one set of coordinates x^μ to another set of coordinates x'^μ. We know that the square of the infinitesimal distance between two neighboring points does not depend on our choice of coordinate system:

$$ds^2 = g_{\mu\nu}(x)dx^\mu dx^\nu = g'_{\rho\sigma}(x')dx'^\rho dx'^\sigma \tag{12}$$

This requirement fixes the relation between $g_{\mu\nu}$ and $g'_{\rho\sigma}$, as we will now see.

Let me reassure the reader seeing this for the first time and finding all these indices a bit fearsome. Keep in mind that what we are doing is nothing but a more general and compact packaging of something that is conceptually quite simple, almost trivial, if you have mastered calculus. For this class of readers, I will, in what follows, jump back and forth between the general and the specific. At each step, we will review the familiar change from

---

* Note that the δ here carries one upper and one lower index.

Cartesian (x, y) to polar (r, θ) coordinates, and point out how this illustrates the general point. Indeed, in this example, (12) is just (1), namely $ds^2 = dx^2 + dy^2 = dr^2 + r^2 d\theta^2$ written in index notation.

Eliminate $dx^\mu dx^\nu$ in (12) in favor of $dx'^\rho dx'^\sigma$ using (10), and obtain

$$g'_{\rho\sigma}(x')dx'^\rho dx'^\sigma = g_{\mu\nu}(x)\frac{\partial x^\mu}{\partial x'^\rho}\frac{\partial x^\nu}{\partial x'^\sigma}dx'^\rho dx'^\sigma \tag{13}$$

Note that all repeated indices in (13) are summed over. To help the abecedarian reader, I now restore temporarily the summation symbol and write the right hand side of (13) more slowly and explicitly as

$$\sum_\mu \sum_\nu g_{\mu\nu}(x)dx^\mu dx^\nu = \sum_\mu \sum_\nu g_{\mu\nu}(x)\left(\sum_\rho \frac{\partial x^\mu}{\partial x'^\rho}dx'^\rho\right)\left(\sum_\sigma \frac{\partial x^\nu}{\partial x'^\sigma}dx'^\sigma\right)$$
$$= \sum_\mu \sum_\nu \sum_\rho \sum_\sigma g_{\mu\nu}(x)\frac{\partial x^\mu}{\partial x'^\rho}\frac{\partial x^\nu}{\partial x'^\sigma}dx'^\rho dx'^\sigma$$

I hope you see the advantage of following Einstein and dropping all summation symbols.

Since the infinitesimals dx'^ρ and dx'^σ in (13) are arbitrary, we can identify the coefficient of $dx'^\rho dx'^\sigma$ to find

$$g'_{\rho\sigma}(x') = g_{\mu\nu}(x)\frac{\partial x^\mu}{\partial x'^\rho}\frac{\partial x^\nu}{\partial x'^\sigma} \tag{14}$$

This tells us how the metric $g'_{\rho\sigma}(x')$ in the primed coordinate system is related to the metric $g_{\mu\nu}(x)$ in the unprimed coordinate system. Note that we are relating the two metrics at the same point P: x and x' are merely the different coordinate values corresponding to P.

Given the metric in one coordinate system, we can work out the metric in some other coordinate system using* (14). For example, applied to the polar case, one of the equations in (14) says that $g'_{22}(x') = g_{11}(\frac{\partial x^1}{\partial x'^2})^2 + g_{22}(\frac{\partial x^2}{\partial x'^2})^2 = r^2$. (As remarked earlier, we could write this also as $g'_{\theta\theta}(r, \theta) = g_{xx}(\frac{\partial x}{\partial \theta})^2 + g_{yy}(\frac{\partial y}{\partial \theta})^2 = r^2$.) You should work out the other components of g' in this way. As another exercise, check the formalism here for spherical coordinates.

Upper versus lower

In going from unprimed to primed coordinates, we have (7) $dx'^\mu = S^\mu_{\ \nu}(x)dx^\nu$ and (14) $g'_{\rho\sigma}(x') = g_{\mu\nu}(x)(S^{-1})^\mu_{\ \rho}(S^{-1})^\nu_{\ \sigma}$. Practical calculation aside, we notice that, in going from unprimed to primed coordinates, an upper index μ transforms with $S^\mu_{\ \nu}$ and a lower index ρ transforms with the inverse matrix $(S^{-1})^\mu_{\ \rho}$. An important conceptual point, but clearly this must be the case: for $ds^2 = g_{\mu\nu}(x)dx^\mu dx^\nu$ to remain invariant, the upper and lower

---

* In practice, it is actually somewhat more direct to apply calculus to (12), rather than to use (14): again for the polar example, going the other way from (r, θ) to (x, y) for a change, we have

$$ds^2 = dr^2 + r^2 d\theta^2 = \left(\frac{xdx + ydy}{r}\right)^2 + r^2\left(\frac{xdy - ydx}{r^2}\right)^2 = dx^2 + dy^2$$

indices must transform oppositely, so that the S and the S^{-1} can knock each other out. Fancy people call the upper index contravariant and the lower index covariant—I can never remember which is which. If you like big words, go for it.

From (9), we notice another important point. Look at $S^\mu_\nu(x) = \frac{\partial x'^\mu}{\partial x^\nu}$: the upper index ν on the right hand side emerges as a lower index on the left hand side. This makes a certain amount of "intuitive" sense: on the right hand side we sort of "divide" by dx^ν. We will use the shorthand notation for the differential operator

$$\partial_\nu \equiv \frac{\partial}{\partial x^\nu} \tag{15}$$

To check that it is consistent for ∂_ν to carry a lower index, note that, using the chain rule,

$$\partial'_\mu \equiv \frac{\partial}{\partial x'^\mu} = \frac{\partial x^\nu}{\partial x'^\mu} \frac{\partial}{\partial x^\nu} = (S^{-1})^\nu_{\ \mu} \partial_\nu \tag{16}$$

This is precisely how a lower index should transform, with the inverse matrix S^{-1}. Notice also that this generalizes what we learned in chapter I.3 about how $\frac{\partial}{\partial x^k}$ transforms under a rotation.

In the two preceding paragraphs, we talked about going from unprimed to primed coordinates. Going from primed to unprimed coordinates, we have of course, as already remarked, the inverse transformations: $dx^\nu = (S^{-1})^\nu_{\ \mu}(x')dx'^\mu$ and $g_{\mu\nu}(x) = g'_{\rho\sigma}(x')S^\rho_{\ \mu}S^\sigma_{\ \nu}$. The roles played by S and S^{-1} are interchanged: an upper index transforms with S^{-1} and a lower index with S.

It is also illuminating to write the transformation law of $g_{\mu\nu}$ in terms of matrices. Given a matrix $M^\mu_{\ \rho}$, define the transpose by $(M^T)^{\ \mu}_\rho = M^\mu_{\ \rho}$. Then we can rephrase the transformation law of the metric $g'_{\rho\sigma}(x') = g_{\mu\nu}(x)(S^{-1})^\mu_{\ \rho}(S^{-1})^\nu_{\ \sigma}$ as

$$g'(x') = (S^{-1})^T g(x) S^{-1} \tag{17}$$

Now you can make contact with something you know only all too well, rotations. Let us see how the formalism just developed works for rotations. For a rotation matrix R, $R^T R = RR^T = I$, so that $R^{-1} = R^T$. Thus, if S happens to be a rotation matrix R, that is, if the transformation from unprimed to primed coordinates is a mere rotation, then in this special case (and only in this special case) $(S^{-1})^T = (S^T)^T = S$. The transformation law (17) of the metric collapses to $I = SIS^{-1}$, since in this case the metric is just the identity matrix I in both the unprimed and primed coordinates. In other words, rotations are defined as linear coordinate transformations that leave the Euclidean metric invariant, as we learned in chapter I.3.

Regarding $g_{\mu\nu}$ as a matrix, we naturally think of its inverse, which we denote by $g^{\mu\nu}$. In other words, the matrix $g^{\mu\nu}$ is defined by

$$g^{\mu\nu} g_{\nu\lambda} = \delta^\mu_\lambda \tag{18}$$

Here the index ν is summed over. (Since the inverse of a symmetric matrix is also symmetric, $g^{\mu\nu} = g^{\nu\mu}$.) Note that, in accordance with our rule that we are allowed only to sum over a lower index with an upper index, the inverse of the metric $g^{\mu\nu}$ carries two upper indices. Normally, we denote the inverse of a matrix M by M^{-1}. However, here it is customary to omit the superscript (-1): the placement of the indices suffices to tell us that $g_{\mu\nu}$ is the metric and $g^{\mu\nu}$ is its inverse.

Most of the metrics we deal with are diagonal. If so, then the inverse metric is a cinch to write down. For example, for spherical coordinates $g^{rr} = 1$, $g^{\theta\theta} = 1/r^2$, $g^{\varphi\varphi} = 1/(r^2 \sin^2\theta)$.

Not surprisingly, when we go from unprimed to primed coordinates, while the metric $g_{\mu\nu}$ transforms with S^{-1}, the inverse metric transforms with S:

$$g'^{\mu\nu}(x') = S^{\mu}{}_{\rho} S^{\nu}{}_{\sigma} g^{\rho\sigma}(x) \tag{19}$$

(This sounds entirely plausible, since $g^{\mu\nu}$ is, after all, the inverse of the metric, and so should transform oppositely as the metric, but you can also easily check that the $g'^{\mu\nu}$ given in (19) is indeed the inverse of $g'_{\nu\sigma}$: $g'^{\mu\nu}(x')g'_{\nu\sigma}(x') = (S^{\mu}{}_{\rho} S^{\nu}{}_{\sigma} g^{\rho\sigma}(x))(g_{\eta\kappa}(x)(S^{-1})^{\eta}{}_{\nu}(S^{-1})^{\kappa}{}_{\sigma})$ $= S^{\mu}{}_{\rho} g^{\rho\sigma}(x) g_{\sigma\kappa}(x)(S^{-1})^{\kappa}{}_{\sigma} = S^{\mu}{}_{\rho}(S^{-1})^{\rho}{}_{\sigma} = \delta^{\mu}{}_{\sigma}$.)

A word or two of encouragement: I agree that for the reader seeing this for the first time, all these indices might look overwhelming, but it is actually quite simple, once you get used to how the indices hang together. You might also remind yourself that we are doing nothing more involved than changing coordinates, but instead of doing it one coordinate system at a time as in elementary treatments, here we want to do it in general, for any coordinate system. In any case, don't worry about it if you are having some difficulty. As I said, we will discuss all this in more depth later.

Vectors, scalars, and tensors

I will now tell you what a vector is in curved coordinates (for example, spherical coordinates). Eventually, we will discuss vectors, tensors, and all that good stuff in curved space in great detail, but for now let's do it on the quick. The easiest path is to simply generalize the familiar notions of scalars and vectors under rotation. We say that $W^{\mu}(x)$ is a vector with an upper index if it transforms just like dx^{μ} does. In other words,

$$W'^{\mu}(x') = S^{\mu}{}_{\nu}(x)W^{\nu}(x) \tag{20}$$

when we go from unprimed to primed coordinates. Note that we are simply generalizing (I.3.18), which describes how a vector field transforms under a rotation.

Indeed, we could now take over our discussion of vectors and tensors in chapter I.4 almost in its entirety. The one novelty, as already noted, is that we now have upper and lower indices, or as the Jargon Guy would say, contravariant and covariant indices. In contrast to a vector with an upper index W^{μ}, which transforms like dx^{μ}, a vector with a lower index W_{μ} transforms like ∂_{μ}:

$$W'_{\mu}(x') = W_{\rho}(x)(S^{-1})^{\rho}{}_{\mu}(x) \tag{21}$$

I like to think of dx^{μ} and ∂_{μ} as the two "primeval" vectors.

A scalar field $\phi(x)$ transforms like $\phi'(x') = \phi(x)$, again simply generalizing a basic notion we learned in studying rotation.

From (16), we learn that $\partial'_\mu \phi'(x') = (S^{-1})^\nu_{\ \mu} \partial_\nu \phi(x)$. We say that $\partial_\mu \phi(x)$ transforms like a vector with a lower index. As I said, this subject will be developed in more detail in part V, but at the moment all we need is that $W^\mu(x)$ and $\partial_\mu \phi(x)$ transform oppositely, with S and S^{-1}, respectively. Thus, $W^\mu(x) \partial_\mu \phi(x)$ (with the index μ summed over, of course) transforms like a scalar. It's easy to check:

$$W'^\mu(x') \partial'_\mu \phi'(x') = S^\mu_{\ \nu} W^\nu(x) (S^{-1})^\lambda_{\ \mu} \partial_\lambda \phi(x) = W^\nu(x) \partial_\nu \phi(x)$$

where we used $S^\mu_{\ \nu} (S^{-1})^\lambda_{\ \mu} = \delta^\lambda_\nu$.

Given V_μ and W^μ, you can verify easily that $V'_\mu(x') W'^\mu(x') = V_\mu(x) W^\mu(x)$: the S and S^{-1} in (20) and (21) knock each other off. Colloquially, we say that summed indices disappear. Indeed, the astute reader will notice that we just showed this in the preceding paragraph, since V_μ and $\partial_\mu \phi(x)$ transform in exactly the same way.

The concept of tensors arises naturally, just as in chapter I.4, but now tensors can carry both upper and lower indices. For example, the tensor $T_\omega^{\mu\nu\lambda}$ transforms like

$$T'^{\mu\nu\lambda}_\omega(x') = S^\mu_{\ \rho} S^\nu_{\ \sigma} S^\lambda_{\ \tau} (S^{-1})^\xi_{\ \omega} T^{\rho\sigma\tau}_\xi(x) \tag{22}$$

In other words, $T_\omega^{\mu\nu\lambda}$ transforms as if it were composed of the product of four vectors $W^\mu V^\nu U^\lambda Y_\omega$.

Again, summed indices disappear. For example, setting λ equal to ω in (22) and summing, we obtain $T'^{\mu\nu\lambda}_\lambda(x') = S^\mu_{\ \rho} S^\nu_{\ \sigma} T^{\rho\sigma\tau}_\tau(x)$. In other words, $T^{\mu\nu\lambda}_\lambda$ is actually a tensor with two upper indices.

A basic "theorem" is that we could use the metric to lower indices and its inverse to raise indices. Given a vector with an upper index W^ν, the claim is that the object $V_\mu \equiv g_{\mu\nu} W^\nu$ transforms like a vector with a lower index. Simply plug how $g_{\mu\nu}$ transforms into (20) to obtain $V'_\rho = g'_{\rho\sigma} W'^\sigma = g_{\mu\nu} (S^{-1})^\mu_{\ \rho} (S^{-1})^\nu_{\ \sigma} S^\sigma_{\ \kappa} W^\kappa = g_{\mu\nu} (S^{-1})^\mu_{\ \rho} W^\nu = (S^{-1})^\mu_{\ \rho} V_\mu$, in agreement with (21), as claimed. Indeed, this little exercise shows that there is no need to introduce another letter: it is customary to simply write $W_\mu = g_{\mu\nu} W^\nu$. I will leave it to you to verify that, similarly, given a vector with a lower index U_ν, the object $g^{\mu\nu} U_\nu$ transforms like a vector with an upper index. Since tensors transform as if they were products of vectors, we can use the metric to lower indices and its inverse to raise indices on tensors at will.

Here is a simple practice problem involving upper and lower indices. Solve the equation $g_{\mu\rho} A^\rho = B_\mu$ for A. Multiply by $g^{\sigma\mu}$: we obtain $g^{\sigma\mu} g_{\mu\rho} A^\rho = \delta^\sigma_\rho A^\rho = A^\sigma = g^{\sigma\mu} B_\mu$. In other words, $A^\sigma = g^{\sigma\mu} B_\mu$. We now rename the index σ and call it ρ and write $A^\rho = g^{\rho\mu} B_\mu$. With practice, you could simply omit the intermediate steps and go directly from $g_{\mu\rho} A^\rho = B_\mu$ to $A^\rho = g^{\rho\mu} B_\mu$. We may think of this colloquially as flipping the metric from the left hand side to the right hand side, where it flips over to become its inverse. Some readers might think that I am belaboring the obvious, but then they would not believe how some students flip out if I omit the intermediate steps when I teach. Indeed, it really is clear if we regard g as a matrix, and A, B as vectors: $gA = B$ implies $A = g^{-1}B$.

Regard this as a first brush with tensors. Much more later.

Area and volume

As you will see, in studying relativity, we constantly ask how various quantities transform. The importance of transformation is in fact central to theoretical physics. In elementary physics, this importance is disguised to some extent, as the quantities the student is likely to encounter have already been formulated to transform properly. In more advanced physics, quantum field theory, for example, the requirement of transforming properly often defines the concept in question.

Let me give you an elementary example that beginning students can all relate to: the concept of area and volume. In flat space, the infinitesimal volume element is given by d^3x in Cartesian coordinates, but what is it in spherical coordinates? You probably obtained the answer in a course on integral calculus by drawing a curved infinitesimal volume element bounded by the eight points $(r + \eta_r dr, \theta + \eta_\theta d\theta, \varphi + \eta_\varphi d\varphi)$, where $\eta_r, \eta_\theta, \eta_\varphi = 0$ or 1, and approximating it by a rectilinear volume. Clearly, d^3x cannot be a volume except in Cartesian coordinates. In spherical coordinates, $d^3x = dr d\theta d\varphi$ does not even have dimensions of length cubed. The power of the metric tensor formalism developed here is that it will tell us what the correct volume element is in any coordinates, even those for which $g_{\mu\nu}$ is not diagonal, and in any dimensions (for $D = 1$, we are talking about a length element, for $D = 2$ an area element, for $D = 3$ a volume element, and so on).

The point is that $d^D x$ does not transform properly under a coordinate transformation $x \to x'$. This is just a fancy way of stating the obvious, $dx dy dz$ and $dr d\theta d\varphi$ are surely not equal to each other. (They don't even have the same dimension!) Indeed, you know that a Jacobian J is needed when you change integration variables. You learned in calculus that $d^D x = d^D x' J$, where J is the determinant of the D-by-D matrix $\frac{\partial x^\mu}{\partial x'^\rho}$.

Go back to (14), which shows how the metric transforms, and take the determinant of that equation. Use the fact that the determinant of a product of matrices is the product of their determinants. You obtain $g' = g J^2$, denoting the determinant of $g_{\mu\nu}$ regarded as a D-by-D matrix by* g and similarly for the primed quantities.

Putting these two facts together, we obtain

$$ d^D x \sqrt{g} = d^D x' J \sqrt{g} = d^D x' J \sqrt{\frac{g'}{J^2}} = d^D x' \sqrt{g'} \tag{23} $$

In other words, $d^D x \sqrt{g}$ is invariant under coordinate transformation. We learned that it is not $d^D x$, but the combination $d^D x \sqrt{g}$, that has intrinsic geometric significance as a volume element, intrinsic in the sense that it does not depend on our coordinate choice.[†] I will show you presently that this reproduces the volume element you have known since childhood.

---

* We have also used the letter g to refer to the metric, but the context should remove any potential confusion.
[†] The Jargon Guy would say that (23) defines $\sqrt{g}$ as a scalar density. A tensor density is then defined as a tensor multiplied by $\sqrt{g}$. For me, the less jargon the better.

In spherical coordinates, $ds^2 = dr^2 + r^2 d\theta^2 + r^2 \sin^2 \theta d\varphi^2$ with $(x^1, x^2, x^3) = (r, \theta, \varphi)$. We regard the metric $g_{\mu\nu}$ as the matrix

$$\begin{pmatrix} 1 & 0 & 0 \\ 0 & r^2 & 0 \\ 0 & 0 & r^2 \sin^2 \theta \end{pmatrix}$$

You can calculate the determinant of this matrix by eyeball: $g = r^4 \sin^2 \theta$. In Cartesian coordinates, the metric is the identity matrix, and so you don't even need to exercise your eyeball to calculate the determinant: $g = 1$. Thus, applied to these two coordinate systems, the statement (23) that $d^D x \sqrt{g}$ is an invariant amounts to saying that $dx\,dy\,dz = dr\,d\theta\,d\varphi\, r^2 \sin\theta$, as you have always known.*

You should now reproduce all the area and volume elements in all the coordinate systems you know. The important point to appreciate, as I have already stressed, is the generality of (23). In exercise 11, you will see that this formalism enables you to calculate the area of higher dimensional spheres.

As a preview of exciting things to come, let me tell you that Einstein's field equation contains a "famous" $\frac{1}{2}$ that both Einstein and Hilbert failed to obtain in their separate first tries. At this stage, I will whet your appetite with a cryptic remark that the square root in (23) is responsible for this all-important $\frac{1}{2}$ in the history of physics. (See chapter VI.4.)

Local versus global

A couple of remarks are in order.

Typically, we need more than one set of coordinates to cover an entire space. The sphere provides an example. The (θ, φ) coordinates (aka latitude and longitude) fail at the north pole and the south pole. (Didn't you ask your teacher what the longitude of the north pole was? The correct answer is that it is undefined.) One symptom of this failure is that $g_{\varphi\varphi} = \sin^2 \theta$ vanishes at the poles. Since the coordinate system fails[†] at only two points, most physicists simply ignore this failure as a technicality and happily use the spherical coordinates until they run into trouble,[1] which is almost never. In this example, clearly nothing intrinsically bad happens at the two poles: on the sphere, every point is as good as any other. All we have to do is to set up a coordinate system with coordinates (θ', φ') with some point (other than the two poles) and its antipodal point playing the role of the north and south poles for this primed coordinate system.

In general, we may need several "coordinate patches" to cover the entire space. One patch must overlap with another such that in the overlap region, the two sets of coordinates are related to each other by smooth differentiable functions.

---

* Note that although we are using a formalism appropriate for curved space to derive these expressions, when we apply them to spherical coordinates, we are dealing with flat space, of course.

† Incidentally, this also implies that g vanishes and hence the inverse metric $g^{\mu\nu}$ fails to exist at that point.

Similarly, the polar coordinates are, strictly speaking, not defined at the origin, since $g_{\theta\theta}$ vanishes there.

You might have also noticed that in the metric $ds^2 = dr^2 + r^2 d\theta^2$, and in (4) and (5), I have implicitly assumed, as is customary, that (r, θ) and $(r, \theta + 2\pi)$ describe the same point. In other words, θ is an angular variable. But nothing in the metric itself tells us that. By its very nature, the infinitesimal distance between two nearby points cannot tell us anything about the global character of the space. The identification $\theta \equiv \theta + 2\pi$ is a global statement.

Indeed, we can imagine cutting out a wedge of angle α from the flat piece on which we have mentally drawn the polar coordinates and gluing the two edges together. Evidently, this forms a cone characterized by the angle α. In other words, we now identify (r, θ) and $(r, \theta + 2\pi - \alpha)$. Remember the ant in the prologue?

A quick summary

Since this has been a fairly long trek, it may be helpful to have a brief summary. The metric was introduced, and you learned that the metric enabled you to calculate the curvature of space (only in principle, not yet in practice, for any dimensions). The Poincaré half plane gave an interesting example of how a simple looking metric could describe a strange space. Along the way, we also got acquainted with upper and lower indices, and understood how vectors and tensors could be defined by how they transformed under coordinate transformation. In particular, we figured out how the metric transformed. As an application, we determined the generalized volume element for any curved space.

Appendix 1: Mite geometers draw circles and dream of black holes

Go back to the mite professors of geometry in the prologue, busily drawing circles of radius ϵ around any given point P and calculating the curvature R by evaluating $R = \lim_{\text{radius}\to 0} \frac{6}{(\text{radius})^2}\left(1 - \frac{\text{circumference}}{2\pi\ \text{radius}}\right)$. If the space is flat, R vanishes.

Let us see how this works for the metric (4) with θ and $\theta + 2\pi$ identified. The arithmetic simplifies enormously if we pick P to be the origin. The distance from the origin $(0, 0)$ to the point (ϵ, θ) is given by $\int_0^\epsilon ds = \int_0^\epsilon d\rho = \epsilon$. In other words, the set of points with coordinates (ϵ, θ) form a circle. Then the circumference is given by $\int d\theta \sin \epsilon = 2\pi \sin \epsilon$. Thus, $R = \lim_{\epsilon \to 0} \frac{6}{\epsilon^2}(1 - \frac{\sin \epsilon}{\epsilon}) = 1$ and the space is curved. Perhaps you have already recognized that the metric (4) is just the metric of the unit sphere (3) with θ and φ disguised as ρ and θ, respectively.

As for the metric (5), it turns out that the space it describes is flat. The distance from the origin $(0, 0)$ to the point (ϵ, θ) is now given by $\int_0^\epsilon ds = \int_0^\epsilon \cos r\, dr = \sin \epsilon$. Now $R = 0$ at the origin. Indeed, you might have seen immediately that the metric in (5) is just $ds^2 = dr^2 + r^2 d\theta^2$ with the change of variable $r \to \sin \rho$.

You are now catching on and can tackle any metric. An interesting class of $D = 2$ metric is given by $ds^2 = f(r)dr^2 + r^2 d\theta^2$. The geometrical meaning is clear. We have filled space with circles (consisting of the points (r, θ) with θ varying from 0 to 2π) centered around the origin. For each circle, the radius is equal to $\int_0^r dr'\sqrt{f(r')}$, while the circumference is given by $\int_0^{2\pi} d\theta\, r = 2\pi r$. A simple calculation shows that for $f(r) \simeq 1 + \gamma r^2$ and small r, the curvature at the origin is given by $R = \gamma$.

We could generalize the preceding discussion to $D = 3$ for $ds^2 = f(r)dr^2 + r^2 d\theta^2 + r^2 \sin^2 \theta\, d\varphi^2$. Fill space with spheres, with the surface area of each sphere given by $4\pi r^2$ and radius depending on $f(r)$. We will study metrics of this form when we encounter black holes.

Appendix 2: A peculiar description of flat space

Here, at this early stage in this text, I give you a particularly fun description of flat space, which we will need much later in part VII, when we study rotating black holes. Spherical coordinates are so nice that normally hardly anybody would think of mucking around with them, but let's do precisely that and write $x = f(r) \sin\theta \cos\varphi$, $y = f(r) \sin\theta \sin\varphi$, $z = r \cos\theta$. Start with Pythagoras and describe flat space with $ds^2 = dx^2 + dy^2 + dz^2$. Plug in $dx = f'(r) \sin\theta \cos\varphi \, dr + f(r) \cos\theta \cos\varphi \, d\theta - f(r) \sin\theta \sin\varphi \, d\varphi$, and so on, to obtain

$$
\begin{aligned}
ds^2 &= dx^2 + dy^2 + dz^2 \\
&= (f'^2 \sin^2\theta + \cos^2\theta) dr^2 + (f'^2 \cos^2\theta + r^2 \sin^2\theta) d\theta^2 + f^2 \sin^2\theta \, d\varphi^2 \\
&\quad + 2(ff' - r) \sin\theta \cos\theta \, dr \, d\theta
\end{aligned}
\tag{24}
$$

You might have recognized that we have merely worked out various versions of (12), (13), and (14) for a specific example.

A couple of lessons here. For some arbitrary $f(r)$, (24) is a metric that nobody would love, or even want to calculate with. But by construction, it certainly describes flat space. While we are used to diagonal $g_{\mu\nu}$, we see that an off-diagonal term $dr \, d\theta$ appears easily. To get rid of this diagonal term, set $ff' = r$, which implies $f^2 = r^2 + a^2$. For $a = 0$, we recover the usual spherical coordinates, but perhaps surprisingly, we find that for $a \neq 0$, we could describe flat space with what are known as Boyer-Lindquist coordinates:

$$
ds^2 = \frac{r^2 + a^2 \cos^2\theta}{r^2 + a^2} dr^2 + (r^2 + a^2 \cos^2\theta) d\theta^2 + (r^2 + a^2) \sin^2\theta \, d\varphi^2
\tag{25}
$$

Instead of spheres, the surfaces of fixed r are now ellipsoids with cylindrical symmetry. Strangely, at least for those seeing this for the first time, $r = 0$ is not a point! Instead, it is a disk (a flattened ellipsoid) with radius a. Set $r = 0$ to obtain $x = a \sin\theta \cos\varphi$, $y = a \sin\theta \sin\varphi$, $z = 0$, and we see that as θ and φ vary, this sweeps out a disk. In fact, θ plays the role of a "radial" rather than an "angular" variable: as θ goes from 0 to $\frac{\pi}{2}$, we go from the center of the disk to its edge. In other words, $(r, \theta) = (0, \pi/2)$ describes a ring with radius a.

After examples like this, I hope that you will find black holes a little less strange when you eventually encounter them.

Appendix 3: Divergence, Laplacian, and all the rest

When I was an undergrad, to do the exercises in a course on electromagnetism, for example, I had to know the form of the divergence, the Laplacian, and things like that in various coordinate systems. Since I was taught to derive things rather than to look them up or to memorize them, I must have derived the Laplacian in spherical coordinates about a hundred times. It got to be kind of annoying. No doubt many readers have had the same experience.

I now show you that the high-powered metric formalism you just learned provides an easy way to derive all these objects like the Laplacian directly from the metric $g_{\mu\nu}$. Of course, they could all be derived also by brute force substitution (as I painfully recall).

First, we will determine the divergence of a vector field $W^\mu(x)$. The simple minded divergence that works in Cartesian coordinates $\partial_\mu W^\mu(x) = \sum_{\mu=1}^{D} \frac{\partial W^\mu}{\partial x^\mu}$ does not work in general, because it does not transform like a scalar (as you can verify; also see below). We want a divergence that does not depend on the coordinate system, that is, one that transforms like a scalar.

The clever trick is to invoke the integral $\mathcal{I} \equiv \int d^D x \sqrt{g} \, W^\mu(x) \partial_\mu \phi(x)$. We learned earlier that the integrand $W^\mu(x) \partial_\mu \phi(x)$ is a scalar, and we know from (23) that $d^D x \sqrt{g}$ is also a scalar. Hence $\mathcal{I}$ is invariant under coordinate transformations. Integrate by parts to find that $\mathcal{I} = -\int d^D x \, \partial_\mu(\sqrt{g} \, W^\mu) \phi = -\int d^D x \sqrt{g} \frac{1}{\sqrt{g}} \partial_\mu(\sqrt{g} \, W^\mu) \phi$. We deduce from the coordinate invariance of $\mathcal{I}$ and $d^D x \sqrt{g}$ that the combination

$$
D_\mu W^\mu \equiv \frac{1}{\sqrt{g}} \partial_\mu(\sqrt{g} \, W^\mu) = \partial_\mu W^\mu + \left(\frac{1}{\sqrt{g}} \partial_\mu \sqrt{g} \right) W^\mu
\tag{26}
$$

must be a scalar. At this stage, $D_\mu W^\mu$ is just a convenient symbol; you won't learn what D_μ means until a later chapter. What you have learned here is that the Cartesian divergence $\partial_\mu W^\mu$ must be corrected by adding an extra term. Of course, in Cartesian coordinates, $g = 1$ and the extra term vanishes. But in spherical coordinates, $\sqrt{g} = r^2 \sin\theta$, as noted before, so that the correct expression for the divergence is $\partial_r W^r + \partial_\theta W^\theta + \partial_\varphi W^\varphi + \frac{2}{r} W^r + \frac{\cos\theta}{\sin\theta} W^\theta$, as you may or may not recall. The point is that (26) gives the divergence in any coordinates and any curved space.

Next, to the Laplacian. Here we are after a general result, independent of specific coordinate systems. The trick is to consider the integral $\int d^D x \sqrt{g} g^{\mu\nu} \partial_\mu \phi \partial_\nu \phi$, which is manifestly a scalar, since the integrand is a scalar (since $g^{\mu\nu}$ and $\partial_\mu \phi$ transform with S and S^{-1}, respectively). Integrate by parts to obtain $- \int d^D x \phi \partial_\mu (\sqrt{g} g^{\mu\nu} \partial_\nu \phi) = - \int d^D x \sqrt{g} \phi (\frac{1}{\sqrt{g}}) \partial_\mu (\sqrt{g} g^{\mu\nu} \partial_\nu \phi)$. We thus conclude that the Laplacian is given by

$$D^2 \phi \equiv \frac{1}{\sqrt{g}} \partial_\mu (\sqrt{g} g^{\mu\nu} \partial_\nu \phi) \tag{27}$$

In Cartesian coordinates, this reduces to the usual expression. But, for example, in spherical coordinates, plugging in $\sqrt{g}$ and $g^{\mu\nu}$, we obtain the Laplacian $(\partial_r^2 + \frac{2}{r}\partial_r + \frac{1}{r^2}\partial_\theta^2 + \frac{\cos\theta}{r^2 \sin\theta}\partial_\theta + \frac{1}{r^2 \sin^2\theta}\partial_\varphi^2)\phi$, again, as you may or may not recall.

My pedagogical strategy is to go from change of coordinates to curved spaces, from which it is a short hop over to the curved spacetimes we need for Einstein gravity. You learn here that it pays to learn this general formalism, even if you don't intend to deal with curved spacetimes any time soon.

Exercises

1 Suppose we defined $ds^2 = g_{\mu\nu}(x)dx^\mu dx^\nu$ as the square of the distance from x to $x + dx$. One consistency requirement is that this is the same as the square of the distance from $x + dx$ to x. Show that this requirement is satisfied.

2 A race of Eskimo mites living around the north pole naturally uses the north pole as the origin of their coordinate system and the "walking" distance from the north pole as a distance measure. After years of study, their mite geometers figured out that the metric of their world is given to second order by (set the radius of the sphere to 1)

$$ds^2 = \left(1 - \frac{y^2}{3}\right)dx^2 + \left(1 - \frac{x^2}{3}\right)dy^2 + \frac{2}{3}xy\,dxdy + \cdots$$

For $x, y \ll 1$, the space is flat and as Euclidean as it could be. But note that in second order the metric is not diagonal.

You of course know that their coordinates (x, y) are related to the usual spherical coordinates by $x = \theta \cos\varphi$ and $y = \theta \sin\varphi$. Furthermore, you know, but Eskimo mites don't, that actually $ds^2 = d\theta^2 + \sin^2\theta d\varphi^2$. Given your knowledge, derive the metric above.

3 The familiar Mercator map of the world is obtained by transforming spherical coordinates θ, φ to coordinates x, y given by $x = \frac{W}{2\pi}\varphi$, $y = -\frac{W}{2\pi}\log\tan\frac{\theta}{2}$. (This was first derived by the English mathematician Edward Wright in 1599.) Show that $ds^2 = \Omega^2(x, y)(dx^2 + dy^2)$. Determine Ω.

4 Consider the space described by

$$ds^2 = \frac{\rho^2}{\rho^2 + a^2}dr^2 + \rho^2 d\theta^2 + (r^2 + a^2)\sin^2\theta d\varphi^2, \text{ where } \rho^2 \equiv r^2 + a^2 \cos^2\theta$$

Is the space curved or not? (Hint: $r = 0$ does not represent a point. Also, study lines of fixed θ and φ.)

5 Consider the metric $ds^2 = dr^2 + (rh(r))^2 d\theta^2$, with θ and $\theta + 2\pi$ identified. For $h(r) = 1$, this is flat space. Let $h(0) = 1$. Show that the curvature at the origin is positive or negative according to whether $h(r)$ starts to turn downward or upward. Calculate the curvature for $h(r) = \frac{\sin r}{r}$ and for $h(r) = \frac{\sinh r}{r}$.

6 Consider a "fixed latitude" circle defined by $\theta = \epsilon$ around the north pole of a unit sphere. The radius is θ, while the circumference is $2\pi \sin\theta$. Show that $R = 1$.

7 Show that $d^D x \sqrt{g}$ gives the correct length element on the unit circle.

8 Here is a simple way of understanding that $d^3 x \sqrt{g}$ gives the volume element. Consider the metric $ds^2 = a^2 dx^2 + b^2 dy^2 + c^2 dz^2$. Calculate the volume of an infinitesimal rectilinear container with sides dx, dy, dz.

9 In several areas of theoretical physics, we need to talk about higher dimensional spheres. The d-dimensional unit sphere S^d is embedded into E^{d+1} by the usual Pythagorean relation $(X^1)^2 + (X^2)^2 + \cdots + (X^{d+1})^2 = 1$. (We will discuss embedding more in the next chapter. See also exercise 16 below.) Thus, S^1 is the circle and S^2 the sphere. Indeed, we may even live on S^3. You can readily generalize what you know about polar and spherical coordinates to higher dimensions by defining

$$X^1 = \cos \theta_1, \qquad X^2 = \sin \theta_1 \cos \theta_2, \quad \cdots$$

$$X^d = \sin \theta_1 \cdots \sin \theta_{d-1} \cos \theta_d, \qquad X^{d+1} = \sin \theta_1 \cdots \sin \theta_{d-1} \sin \theta_d$$

where $0 \leq \theta_i < \pi$ for $1 \leq i < d - 1$, but $0 \leq \theta_d < 2\pi$. (For S^2, $\theta_1 = \theta$ and $\theta_2 = \varphi$. Note that the usual Cartesian coordinates are trivially permuted from the coordinates used here: $X^1 = Z$, $X^2 = X$, $X^3 = Y$.) Verify the Pythagorean relation. Show that the metric on S^d works out to be

$$ds_d^2 = \sum_{i=1}^{d+1} (dX^i)^2 = d\theta_1^2 + \sin^2 \theta_1 d\theta_2^2 + \cdots + \sin^2 \theta_1 \cdots \sin^2 \theta_{d-1} d\theta_d^2$$

10 Show that the metrics on the unit spheres satisfy the iterative relation

$$ds_d^2 = d\theta^2 + \sin^2 \theta ds_{d-1}^2$$

Note that the common observation that curves of fixed latitude on the globe are circles is an example of this relation.

11 Use the formalism in the text to calculate the area (used in the generalized sense, of course) of S^d. Verify that you recover what you know for $d = 2$ and 3. Show that the area of S^3 is equal to $2\pi^2$, a result that you will need in quantum field theory.[2]

12 A squashed sphere: consider $ds^2 = (b^2 + a^2 \cos^2 \theta) d\theta^2 + \frac{(b^2 + a^2)^2}{b^2 + a^2 \cos^2 \theta} \sin^2 \theta d\varphi^2$, with $0 \leq \theta \leq \pi$ and $0 \leq \varphi \leq 2\pi$, as usual. Find the length of the equator (that is, the curve defined by $\theta = \pi/2$) and of the circle of fixed longitude going through the two poles (for this, give the numerical result when $b = a$), and the area of this squashed sphere. We will need these results later when we discuss rotating black holes.

13 Stereographic projection of the sphere: start out with $ds^2 = \frac{dr^2}{1 - \frac{r^2}{L^2}} + r^2 d\varphi^2$ for a sphere of radius L. (This is of course the usual spherical coordinates with $L \sin \theta$ disguised as r.) Now imagine the sphere as a transparent globe with the south pole touching a plane, as shown in figure 3, and a light affixed to the north pole. The shadows of various points cast on the plane defines the stereographic map. Show that the point (r, φ) is mapped to the point (ρ, φ) on the plane, with

$$r = \frac{\rho}{1 + \frac{\rho^2}{4L^2}}$$

and by substitution

$$ds^2 = \frac{1}{\left(1 + \frac{\rho^2}{4L^2}\right)^2} (d\rho^2 + \rho^2 d\varphi^2)$$

For S^d the projection generalizes to $ds^2 = \frac{1}{\left(1 + \frac{\rho^2}{4L^2}\right)^2} (d\rho^2 + \rho^2 d\Omega_{d-2}^2)$.

14 Conformally flat: a space described by $ds^2 = \Omega^2(x) ds_{\text{flat}}^2$ is said to be conformally flat. We also say that the metric $g_{\mu\nu}$ related to the flat metric by $g_{\mu\nu}(x) = \Omega^2(x) g_{\mu\nu}^{\text{flat}}$ is conformally flat. The metric in exercise 13 provides an example. Show that lengths are not the same but that angles are the same as calculated with the

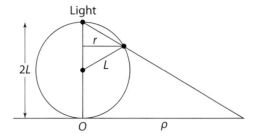

Figure 3 Stereographic projection of the sphere.

two different metrics. This is clear for the stereographic projection of the sphere. Note, for example, that as $\rho \to \infty$, a given $\Delta\rho$ corresponds to an ever-smaller Δs, but as $\rho \to 0$, the two metrics approach each other. Bad terminology alert: A conformally flat metric does not describe a flat space, as the example of the sphere makes clear! More generally, two metrics are said to be conformally related to each other if there exists a function $\Omega(x)$ such that $\tilde{g}_{\mu\nu}(x) = \Omega^2(x)g_{\mu\nu}(x)$. Again, it is worth emphasizing that the two metrics are not related by a coordinate transformation.

15 Verify that the expression given for $D_\mu W^\mu$ in (26) reproduces the usual formula for the divergence in spherical coordinates.

16 Find the metric on the torus.

17 Show that any $d = 2$ space is conformally flat.

18 Show that $ds^2 = e^{2u}(du^2 + dv^2)$ is not only conformally flat but literally flat.

Notes

1. One situation that requires more care is in the discussion of magnetic monopoles. See, for example, *QFT Nut*, chapter IV.4.
2. *QFT Nut*, p. 273. For a rather nifty trick giving the area of S^d for any d, see *QFT Nut*, p. 539.

I.6 Curved Spaces: Gauss and Riemann

Fear of curves

Surely you have heard that Einstein gravity involves curved spacetime. In my experience, the very mention of the Riemann curvature tensor, which we will come to in due time, inspires fear and trembling in many beginning students of general relativity. "I was doing well in the course," students would moan, "until we started doing Riemannian geometry!" I actually have a great deal of sympathy for these students: very little in their background prepares them for the Riemann curvature tensor. Unless they have taken an exceptionally good mechanics course, with studies of particle motion on curved surfaces, they typically have no prior exposure to the massive amount of rather technical material needed to calculate the Riemann curvature tensor. Dear reader, take heart: even Einstein had to struggle to master differential geometry.

Given this frightening reputation of Riemannian geometry, we will take baby steps toward curvature. My pedagogical strategy is to first deal with surfaces that you can practically imagine holding in your hands.

Introducing Professor Flat

As I was encouraging a fearful student, Professor Flat came ambling along. A mild mannered man with a somewhat disheveled look, he inquired gently, "Why all the trembling?" We told him, and he nodded. Then he asked the fearful student, "Do you know why it took humans so long to realize that the world was round?"

FS: "Because the world is locally flat, and humans can't walk very fast."

PF: "Exactly! In everyday life, we have no need to know that the world is actually round, as long as the distance scale of interest is small compared to the earth's radius. If you focus on a small enough region on any surface—or any space in any dimension for that matter—it's going to look flat, and you could simply study the small deviation from flat space. Nothing terribly frightening at all!"

Tangent plane

So indeed, let's follow Professor Flat's advice. To get some feel for curvature, let us keep things simple and be content to take some nice curved space, roundish like the sphere. Imagine approaching it with a flat plane until it touches the surface at some point P. The plane is then known as the tangent plane. To focus our minds, consider a sphere of radius L. Since all points on it are equivalent, we might as well have the plane touch the south pole and be perpendicular to the z-axis joining the north and south poles. The southern hemisphere is defined by $z = -\sqrt{L^2 - x^2 - y^2} + L$, where we added the constant L to the usual definition of z so that for convenience, $z = 0$ at the point P we are interested in (the south pole in this case). Near the south pole, $z \simeq \frac{1}{2L}(x^2 + y^2)$: the sphere is locally a parabolic bowl and is well approximated by the tangent plane. It is in this sense that Professor Flat says that the surface is locally flat.

So, consider the tangent plane touching the roundish surface we are studying at some point P. See figure 1. Let z denote the coordinate perpendicular to the plane, and (x, y) the coordinates in the plane, with the point P coordinatized by $(x, y) = (0, 0)$. Locally, we have a quadratic expansion in the small quantities x and y: $z = \frac{1}{2}ax^2 + cxy + \frac{1}{2}by^2$. Here a, b, c have dimension of inverse length, and thus the local region is defined by x, y small compared to a^{-1}, b^{-1}, c^{-1}. (For the sphere, $c = 0$ and $a = b = 1/L$.) Applying Pythagoras, we obtain the distance squared between two nearby points with coordinates (x, y) and $(x + dx, y + dy)$:

$$ds^2 = dx^2 + dy^2 + dz^2 = dx^2 + dy^2 + [(ax + cy)dx + (by + cx)dy]^2$$
$$\equiv g_{xx}dx^2 + g_{yy}dy^2 + 2g_{xy}dxdy \tag{1}$$

with the metric

$$g_{xx} = 1 + (ax + cy)^2, g_{yy} = 1 + (by + cx)^2, g_{xy} = (ax + cy)(by + cx) \tag{2}$$

Note that even for the sphere, the metric regarded as a 2-by-2 matrix contains an off-diagonal term in this locally flat coordinate system. Of course, for $x, y \to 0$, the off-diagonal term vanishes and the metric approaches the identity matrix.

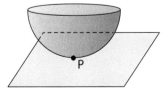

Figure 1 The tangent plane to a curved surface.

We could also write $z \simeq \frac{1}{2}\vec{x}^T M \vec{x}$, with $M = \begin{pmatrix} a & c \\ c & b \end{pmatrix}$ and $\vec{x} = (x, y)$ a column vector which we are writing as a row vector for typographical reasons mentioned in chapter I.3. But we can always rotate the coordinates $\vec{x} = (x, y)$ to some other coordinates $\vec{x}' = (u, v)$ given by $\vec{x}' = R\vec{x}$. We want the curvature to be an invariant geometric concept not dependent on our coordinate choices. In the new coordinates, M is replaced by $M' = R^{-1}MR$. We know that M possesses two invariant quantities, namely its two eigenvalues μ and ν. But this dovetails perfectly with our discussion in the prologue.

Remember the ant going for her honey? We learned from that story that at a given point, a curved space has an intrinsic and an extrinsic curvature. It all makes sense, then: these two curvatures correspond to the two invariant quantities contained in M. As you will see, we could exploit similar reasoning, asking how things transform under a change of coordinates, to determine the curvature of Riemannian spaces of any dimension.

Let us diagonalize M, so that in the new coordinate basis* $z = \frac{1}{2}\mu u^2 + \frac{1}{2}\nu v^2$. Thus, we have the intuitive result that any surface is locally the sum of two parabolas, $z = \frac{1}{2}\mu u^2 + \frac{1}{2}\nu v^2$, with μ^{-1} and ν^{-1} the radius of curvature of the parabola in the u and in the v direction, respectively. (Expand a circle of radius L around some point, in the same way that we expanded a sphere earlier, $y = -\sqrt{L^2 - x^2} + L \simeq \frac{1}{2L}x^2$, and we see that it is locally a parabola with radius of curvature equal to L.)

Intrinsic versus extrinsic

For a 2-by-2 matrix, its determinant and trace, or equivalently, its two eigenvalues, constitute its two basis-independent attributes. Our insistence that the measure of surface curvature does not depend on whether we use the x, y or the u, v coordinates means that we can have two measures of surface curvature:

$$\text{Intrinsic curvature} = \det M = \mu\nu = ab - c^2 \tag{3}$$

$$\text{Extrinsic curvature} = (\tfrac{1}{2}\mathrm{tr}M)^2 = \tfrac{1}{4}(\mu + \nu)^2 = \tfrac{1}{4}(a + b)^2 \tag{4}$$

(Note that I have defined the extrinsic curvature to have the same dimension as the extrinsic curvature. The normalization factor $\frac{1}{4}$ is such that the intrinsic and extrinsic curvatures of the sphere, for which $a = b$, $c = 0$, are equal.)

How do I know which is which, intrinsic versus extrinsic curvature? We appeal to the cylinder in the prologue as an example, knowing that it is intrinsically flat. Indeed, for a cylinder† of radius L, $z = -\sqrt{L^2 - x^2} + L$ independent of y and hence $b = c = 0$. We have intrinsic curvature $= 0$ and extrinsic curvature $= a^2 = 1/L^2$, as we would expect. In

---

* Note that while M is diagonal, the metric is not.

† The cylinder example also underlines the fact that we are always talking about local curvature, as explained in the preceding chapter. Globally, the mites could of course go all the way round and come back to the same place.

Einstein gravity, we are normally only interested in the intrinsic curvature of spacetime, since we can't get out* of spacetime and look at its extrinsic curvature.

Negative curvature

A sphere of radius L has intrinsic and extrinsic curvatures both equal to $1/L^2$. Since every point on the sphere is the same as the south pole, the sphere has constant intrinsic and extrinsic curvatures.

A beginning student named Confusio[1] looked, well, confused. He said: "You mean the curvature at the bottom of a bowl and at the top of a dome are not opposite in sign? Everybody knows that the bowl is curved upward, the dome downward."

Confusio's everyday intuition caused him to think that he was somehow inside the bowl and outside the dome. In fact, the sphere is an infinitely thin surface. There is no conceptual distinction between being on the "inside" and the "outside" of the surface. Of course, we could also turn the bowl upside down. Let's calculate the curvature at "the top of the dome," as Confusio calls it, otherwise known as the north pole. Near the north pole, $z = \sqrt{L^2 - x^2 - y^2} - L \simeq -\frac{1}{2L}(x^2 + y^2)$, so that $a = b = -\frac{1}{L}$, $c = 0$. Thus, the intrinsic curvature is $ab - c^2 = \frac{1}{L^2}$, as expected. Compared to the neighborhood around the south pole, z flips sign, but dz^2 does not, so that the metric, and hence the curvature, have the same value at the north pole and at the south pole, as they should.

Thus, in contrast to everyday perception, a negative curvature surface[2] is one in which the two parabolas bend in opposite directions, $\mu = -\nu$, such as the proverbial saddle. A more contemporary example is the kind of potato chips that come in a cylindrical container. I am surmising that the typical reader of this text is more likely to eat potato chips than to gallop across the steppes with the Golden Horde. An example is a surface that goes like $z = xy$ for $x \simeq 0$, $y \simeq 0$. Then $a = b = 0$, $c = 1$, and $\mu = 1$, $\nu = -1$, and it has negative curvature $R = -1$ at the origin.

Professor Flat: "Confusio, you still look confused. Think of a donut or a bagel, otherwise known as the torus. Do you see that along its hole, the curvature is negative? Along the outer edge, in contrast, the curvature is positive."

Incidentally, it then follows that somewhere on the torus the curvature (we are always talking about intrinsic curvature) vanishes. Do you see where? You could always use the tangent space method here to calculate the curvature and thus check your intuition.

Embedding of curved spaces in higher dimensional flat spaces

In general, we could embed[3] a D-dimensional space in N-dimensional Euclidean space E^N. Write $X^A(x^1, \cdots, x^D)$ for $A = 1, \cdots, N$. For everyday surfaces, $D = 2$ and $N = 3$, but the formalism to be given presently works for arbitrary values of $N > D$. (For the iconic

* Except in some speculative and unproven theories!

sphere, $(x^1, x^2) = (\theta, \varphi)$ and $(X^1, X^2, X^3) = (X, Y, Z) = (\sin\theta\cos\varphi, \sin\theta\sin\varphi, \cos\theta)$.)
We see that x represents coordinates on the surface: as x^μ varies, the points $X^A(x)$ sweep
out a D-dimensional space in E^N. In this age of spectacular computer graphics, you can
easily generate all kinds of interesting surfaces by playing around with three functions
$X^A(x)$ of two variables.

Consider two neighboring points: P described by x^μ and Q described by $x^\mu + dx^\mu$. If
the Euclidean coordinates of P are given by X^A, then the Euclidean coordinates of Q are
given by $X^A + dX^A = X^A + \frac{\partial X^A}{\partial x^\mu}dx^\mu$. Pythagoras gives us the distance squared between
P and Q

$$ds^2 \equiv \sum_A (dX^A)^2 = \sum_A \frac{\partial X^A}{\partial x^\mu}dx^\mu\frac{\partial X^A}{\partial x^\nu}dx^\nu \equiv g_{\mu\nu}dx^\mu dx^\nu \tag{5}$$

with the metric

$$g_{\mu\nu} = \sum_A \frac{\partial X^A}{\partial x^\mu}\frac{\partial X^A}{\partial x^\nu} \tag{6}$$

Here I choose to display the summation over A explicitly. The attentive reader wlll realize
that (1) and (2) are examples of these two equations. The metric $g_{\mu\nu}$ is said to be induced
by the ambient Euclidean metric.

Another common embedding method is to restrict X^A to satisfy certain conditions.
Again, we have the familiar example of the unit sphere defined by $X^2 + Y^2 + Z^2 = 1$.
Indeed, writing (X, Y, Z) in terms of (θ, φ) amounts to solving this equation. Instead,
we could choose to eliminate* Z. In the notation used above, $(x^1, x^2) = (X, Y)$. Then,
$ds^2 = dX^2 + dY^2 + dZ^2 = dX^2 + dY^2 + \frac{(XdX+YdY)^2}{1-X^2-Y^2}$. We can clearly exploit the rotational
invariance in the $(X\text{-}Y)$ plane and write[4] $X = r\cos\varphi$, $Y = r\sin\varphi$, $X^2 + Y^2 = r^2$, and
$XdX + YdY = rdr$, thus obtaining

$$ds^2 = dr^2 + r^2d\varphi^2 + \frac{r^2dr^2}{1-r^2} = \frac{dr^2}{1-r^2} + r^2d\varphi^2 \tag{7}$$

Since the metric in (7) and the usual $ds^2 = d\theta^2 + \sin^2\theta d\varphi^2$ both describe the sphere, they
must be related by a coordinate transformation. Do you see how? Another question for you:
Why does (7) become singular as $r \to 1$? See appendix 1. Further examples of embedding
are given in appendix 2.

Locally flat

The fearful student looks much more relaxed. He asks, "All this is clear enough for actual
2-dimensional surfaces I can visualize, but is it obvious that we can always choose our
neighborhood to be locally flat for any space of any dimension D?"

---

* The two solutions of the quadratic equation for Z define two coordinate patches covering the northern and
southern hemispheres (minus the equator, strictly speaking), as mentioned in the preceding chapter.

Professor Flat: "It is fairly obvious for any sufficiently smooth[5] space. Look at how the metric transforms (I.5.14) when you go to a new set of coordinates:

$$g'_{\lambda\sigma}(x') = g_{\mu\nu}(x)\frac{\partial x^\mu}{\partial x'^\lambda}\frac{\partial x^\nu}{\partial x'^\sigma} \tag{8}$$

Within reason, you could choose any x' you want, and for each choice, you get a new form for the metric. You have a lot of freedom to massage the metric into the form you want. The proof simply amounts to counting how much freedom you have on hand."

So, look at our space around a point P. First, for writing convenience, shift our coordinates so that the point P is labeled by $x = 0$. Expand the given metric around P out to second order: $g_{\mu\nu}(x) = g_{\mu\nu}(0) + A_{\mu\nu,\lambda}x^\lambda + B_{\mu\nu,\lambda\sigma}x^\lambda x^\sigma + \cdots$. (The commas in the subscripts carried by A and B are purely for notational clarity, to separate two sets of indices.)

Again, let me assure the abecedarians that nothing profound is going on. We are merely expanding $g_{\mu\nu}(x)$ in a power series, with the coefficients given names $A_{\mu\nu,\lambda}$ and $B_{\mu\nu,\lambda\sigma}$ (with indices that accord with the rule of repeated summation having to involve an upper and a lower index). Note that the lower indices on the left hand side remain lower indices on the right hand side, and similarly for upper indices.

As always, if you get confused, you should simply refer to the sphere. Thus, let the coordinates of the point P be (θ_*, φ_*), so that $x^1 = (\theta - \theta_*)$, $x^2 = (\varphi - \varphi_*)$. (Of course, in this simple case, nothing depends on φ_*.) What we just wrote down is then simply, for example, $g_{\varphi\varphi} = \sin^2\theta = \sin^2\theta_* + 2\sin\theta_*\cos\theta_* x^1 + \cdots$, so that $A_{\varphi\varphi,1} = 2\sin\theta_*\cos\theta_*$ and $A_{\varphi\varphi,2} = 0$. Nothing profound at all.

Change coordinates according to $x^\mu = K^\mu{}_\nu x'^\nu + L^\mu{}_{\nu\lambda}x'^\nu x'^\lambda + M^\mu{}_{\nu\lambda\sigma}x'^\nu x'^\lambda x'^\sigma + \cdots$. Again, nothing profound: K, L, M, $\cdots$ are just a bunch of coefficients to be determined. At the point P, the new metric is given by (8):

$$g'_{\lambda\sigma}(0) = g_{\mu\nu}(0)K^\mu{}_\lambda K^\nu{}_\sigma \tag{9}$$

Regard this as a matrix equation $g' = K^T g K$, where T denotes transpose. Since $g_{\mu\nu}(0)$ is symmetric and real, there always exists a matrix K that will diagonalize it. After $g_{\mu\nu}(0)$ becomes diagonal (with positive diagonal elements—we will take that as a definition of space), we could scale each coordinate, one by one, by an appropriate factor,* so that the diagonal elements become 1. We end up with the Euclidean metric $g_{\mu\nu}(0) = \delta_{\mu\nu}$, with $\delta_{\mu\nu}$ equal to 1 if $\mu = \nu$ and 0 otherwise. (We shall keep dropping primes as we move along, renaming the new coordinates and metric x^μ and $g_{\mu\nu}$, respectively.)

Let us count

The object $K^\mu{}_\nu$ has D^2 elements to start with. How many are left, now that we have fixed $g_{\mu\nu}(0)$?

---

* $x^\mu \to x^\mu/\sqrt{g_{\mu\mu}(0)}$, no sum over repeated indices here.

Let's count. We are in D-dimensional space. First, note that the number of independent elements in an antisymmetric D-by-D matrix $F_{\mu\nu} = -F_{\nu\mu}$ is equal to $\frac{1}{2}D(D-1)$, since, for each of the D values the first index can take on, the second index can take on only $D-1$ values.* In contrast, a symmetric D-by-D matrix has $\frac{1}{2}D(D-1)$ off-diagonal elements and D diagonal elements, making for a total of $\frac{1}{2}D(D+1)$ elements.[†]

Since $g_{\mu\nu}(0)$ had $\frac{1}{2}D(D+1)$ arbitrary elements to begin with, we had to use this many elements in $K^\mu_{\ \nu}$ to adjust these to $\delta_{\mu\nu}$. Hence, the object $K^\mu_{\ \nu}$ has $D^2 - \frac{1}{2}D(D+1) = \frac{1}{2}D(D-1)$ elements left over. This is exactly the number of independent elements in an antisymmetric D-by-D matrix. Hardly an accident! As discussed in detail in chapters I.3 and I.4, the number of generators in the rotation group $SO(D)$ relevant for D-dimensional space is $\frac{1}{2}D(D-1)$. (For instance, for $D=3$, $\frac{1}{2}D(D-1) = 3$, and we have precisely three rotations that leave the identity matrix $\delta_{\mu\nu}$ invariant.) We have just shown the fairly obvious fact that in D-dimensional space, the freedom we have left in K is precisely the freedom to rotate.

Now onward. We proceed to the next step and claim that the linear terms in $g_{\mu\nu}(x) = \delta_{\mu\nu} + A_{\mu\nu,\lambda}x^\lambda + \cdots$ can be removed by suitable choices of $L^\mu_{\ \nu\lambda}$ in $x^\mu = x'^\mu + L^\mu_{\ \nu\lambda}x'^\nu x'^\lambda + \cdots$. (Evidently, $A_{\mu\nu,\lambda}$ and $L^\mu_{\ \nu\lambda}$ have been modified already by what we have done thus far, but we do not want to introduce more letters.)

I urge the reader to expand (8) to first order in x to see what is going on. Since $A_{\mu\nu,\lambda}$ is symmetric in $\mu\nu$, it contains $\frac{1}{2}D(D+1)D = \frac{1}{2}D^2(D+1)$ elements (18 for $D=3$). But $L^\mu_{\ \nu\lambda}$ also has $\frac{1}{2}D^2(D+1)$ elements: like $A_{\mu\nu,\lambda}$ it has three indices and is symmetric in two of them. Yes! We have enough Ls to knock off[6] the As. Thus, locally around any point P in a Riemannian manifold, the metric can always be chosen to be $g_{\mu\nu}(x) = \delta_{\mu\nu}$ plus second order terms.

Thus, at any point P in a Riemannian manifold, not only can we choose the metric to be Euclidean, but we can also arrange for the first order deviations from Euclidean to vanish. Indeed, in our simple example (2), the deviation from locally Euclidean is quadratic. (See also exercise I.5.2.) That the corrections to the locally flat Euclidean metric are second order, rather than first order, is the mathematical explanation for why humans thought their world was flat for so long.

Curvature

Let's keep going! At this stage, we have $g_{\mu\nu}(x) = \delta_{\mu\nu} + B_{\mu\nu,\lambda\sigma}x^\lambda x^\sigma + \cdots$ and $x^\mu = x'^\mu + M^\mu_{\ \nu\lambda\sigma}x'^\nu x'^\lambda x'^\sigma + \cdots$. How many components of B can we knock off by judicious choices of $M^\mu_{\ \nu\lambda\sigma}$?

---

* Recall that we did this sort of counting in chapter I.4.

[†] Alternatively, knowing that $n(D)$, the number of independent elements in a symmetric D-by-D matrix, can be at most quadratic in D, we write $n(D) = c_0 + c_1 D + c_2 D^2$ and fix the coefficients instantly from $n(0) = 0$, $n(1) = 1$, and $n(2) = 3$. We obtain $n(D) = \frac{1}{2}D(D+1)$. A similar argument gives the number of independent elements in an antisymmetric D-by-D matrix.

Let's count. The number of components in $B_{\mu\nu,\lambda\sigma}$ is easy to count, since B has two pairs of symmetric indices, and so we have $(\frac{1}{2}D(D+1))^2$ components. The number of components in $M^\mu_{\ \nu\lambda\sigma}$ is harder to count. Focus on the symmetric triplets $\nu\lambda\sigma$. We know that the number $f(D)$ of possible choices is cubic in D. Again, the quickest method, efficient though not particularly clever, is to write $f(D)$ as a cubic polynomial in D and then determine the coefficients by "fitting to data." Start with $f(1) = 1$. To find $f(2)$, simply exploit the symmetry to arrange the indices in the order $\nu \geq \lambda \geq \sigma$, and list the possibilities: 222, 221, 211, 111, and so $f(2) = 4$. Similarly $f(3) = 10$. (Also, the process is clearly inductive: $f(D) = f(D-1) + \frac{1}{2}D(D+1)$.) In this way, we obtain $f(D) = \frac{1}{6}D(D+1)(D+2)$, and hence $M^\mu_{\ \nu\lambda\sigma}$ has $\frac{1}{6}D^2(D+1)(D+2)$ components. We use these to knock off some components in B, leaving $\frac{1}{4}D^2(D+1)^2 - \frac{1}{6}D^2(D+1)(D+2) = \frac{1}{12}D^2(D^2-1)$ elements that we can't get rid of.

If you didn't quite get all that, just write it out for $D = 2$, and you will see what's going on. At this stage of the cancellation game, we have 3 coefficients a, b, c in $g_{11} = 1 + a(x^1)^2 + b(x^2)^2 + c(x^1x^2) + \cdots$. Similarly for g_{22} and g_{12}, for a total of $3 \times 3 = 9$ coefficients we want to cancel. On the other side of the ledger, we can adjust 4 parameters p, q, r, s in $x^1 = x'^1 + p(x'^1)^3 + q(x'^1)^2x'^2 + rx'^1(x'^2)^2 + s(x'^2)^3 + \cdots$. Similarly for x^2 for a total of $2 \times 4 = 8$ parameters we can adjust to knock off the 9 coefficients in the metric. So we are left with $9 - 8 = 1 = \frac{1}{12}2^2(2^2-1)$ number we can't get rid of.

As another example, for $D = 4$, B has 100 components, and M 80 components, leaving us with $100 - 80 = 20 = \frac{1}{12}4^2(4^2-1)$ numbers we can't get rid of.

The measure of curvature is what we can't iron flat. We thus conclude that at any given point on a Riemannian manifold, we need Riemann$(D) \equiv \frac{1}{12}D^2(D^2-1)$ numbers to specify the curvature. In particular, Riemann$(1) = 0$ and Riemann$(2) = 1$, confirming[*] what we already know. The number Riemann(D) increases rapidly: Riemann$(3) = 6$, and for $D = 4$, which, I'm sure you've heard is relevant for Einstein gravity, the curvature has Riemann$(4) = 20$ components, a number that sets our student FS to fear and trembling again. It is of course reasonable that it takes a lot of numbers to describe curvature in higher dimensional space, since we have to specify how the space is curving in many different directions.

To make sure that you follow this discussion, I suggest you try this fun exercise. Suppose you were given a space described by the metric $ds^2 = dr^2 + r^2d\theta^2$. This is of course a plane as flat as Kansas, but suppose you didn't know that. Calculate the curvature by first transforming polar coordinates into locally flat coordinates[†] at the point $(r, \theta) = (r_*, 0)$ by going through all the steps here. Then extract the combination of the $B_{\mu\nu,\lambda\sigma}$s giving the intrinsic curvature. By the end of this straightforward exercise, you will probably agree that there ought to be a better way to get at the curvature.

---

[*] Note that a curve has no intrinsic curvature, only extrinsic curvature, as is intuitively clear, while a surface, as described in chapter I.5, is characterized by two numbers specifying the intrinsic and extrinsic curvatures. We are evidently talking about the intrinsic curvature here.

[†] Also known as Riemann normal coordinates. Thanks, Jargon Guy.

It is commonly said in academia that the best way to master a subject is to teach it. In this connection, the computer is a more-than-willing student. Write a program such that, given a metric, the program is able to find the locally flat coordinates at an arbitrary point. If you do this, as I did while writing this chapter, you will truly understand the counting above.

Guessing what the Riemann curvature must look like

One significant by-product of this counting and subsequent understanding of local flatness is that we see what the general expression for curvature must involve. Since the curvature at the point P is described by those components in $B_{\mu\nu,\lambda\sigma}$ that we cannot transform away, we conclude* that the definition of curvature must involve two powers[†] of derivatives acting on the metric $g_{\mu\nu}$. As we have already seen, it takes lots of numbers to describe curvature completely. But we also know that we could change coordinates (from (x, y) to (u, v), for the simple curved surface that started this chapter). Thus, we don't want these numbers to gallop wildly out of our control when we change coordinates. Now you see that the concept of a tensor, as discussed in chapter I.4, is going to play a big role. In fact, the $\frac{1}{12}D^2(D^2 - 1)$ numbers are the components of a tensor, quite naturally called the Riemann curvature tensor. With our intuitive discussion here, we can anticipate that the curvature tensor will have the schematic form $R_{....} \sim \partial\partial g_{..}$; since the number of components grows quartically with D, we can even guess that it will carry 4 indices. Very nice: this is all consistent with the curvature being related to $B_{\mu\nu,\lambda\sigma}$.

What did Riemann want?

The great insight of Carl Friedrich Gauss and other pioneers of differential geometry, not to mention the mite professors of geometry, is that given the metric, we should be able to determine the intrinsic curvature without worrying how the surface is embedded, as was already explained in the preceding chapter. Knowing the distance between two infinitesimally separated points, we can find the distance between two points far apart by integrating ds along any path connecting the two points. We can then define a "straight line" between two points as that path that minimizes the distance between them. This allows us to do geometry as Euclid had shown us. Shades of Newton, Leibniz, and Lie!

Indeed, we defined the Poincaré half plane by specifying ds^2, without having to say, or even to care about, how it is embedded.

To calculate the intrinsic curvature, we need to know only the metric $g_{\mu\nu}$; we don't need to know the embedding functions $X^A(x^\mu)$. When the great Gauss discovered this fact for curved surfaces in 1828, he was so struck by it that he called it the Theorema Egregium

---

* Since $B_{\mu\nu,\lambda\sigma}$ are nothing but the second order Taylor coefficients in an expansion of $g_{\mu\nu}$ around the point P.
[†] This is also confirmed by the intuitive example in (1) and (2).

(the outstanding or extraordinary theorem; the meaning of the Latin word has been much distorted in the English word "egregious"). Here's hoping that you find your Theorema Egregium some day.

The whole point of the story in the prologue is that, like the mites, we cannot get out of the universe, yet we can measure its curvature.

Bernhard Riemann, who was two years old when the Theorema Egregium was born and who as a student attended Gauss's lectures, took the profound step of extending differential geometry to arbitrary dimensions and definitively taught us how to calculate curvature. Given a metric $g_{\mu\nu}$, Riemann wanted the curvature.

Now that I have sketched it out for you, you could set yourself a challenge and see how you stack up compared to Riemann. The problem is easily stated and well posed. Construct a tensor $R.... \sim \partial\partial g..$ out of two partial derivatives and the metric that would measure intrinsic curvature. See how far you can get before reading further.

You will soon see that it is not so simple. Indeed, even for $D = 2$, looking at (3), and knowing the answer, it is not obvious what the combination should be. In that simplest of all cases, the Riemann curvature tensor has only one component and thus degenerates into a scalar. So given 3 functions g_{11}, g_{22}, and g_{12}, call them E, F, G say, each a function of two variables x, y, find an expression involving E, F, G, allowing yourself only two derivatives, such that the expression does not change under the transformation in (8). You recognize that this amounts to Gauss's problem. Challenge yourself!

A historical curiosity. After Riemann worked out the general treatment of curved spaces, he had, remarkably enough, some vague thoughts about curved spaces having something to do with gravity. Unfortunately for him, he was way too early. Special relativity and Minkowski's unification of space and time into spacetime (as we will see in part III) were still in the future. We now know that it is curved spacetime, not curved space, that has something to do with gravity (as we will see in part IV).

Appendix 1: Coordinate singularity, a simple version of the Einstein-Rosen bridge, and a wormhole

The metric discussed in the text $ds^2 = \frac{dr^2}{1-r^2} + r^2 d\varphi^2$ illustrates an important point. Here are the answers to the questions I asked you. Set $r = \sin\theta$, so that $dr^2 = \cos^2\theta d\theta^2$ and ds^2 becomes $d\theta^2 + \sin^2\theta d\varphi^2$. The singularity at $r = 1$ is merely due to our choice of coordinates going bad at the equator. In fact, the sphere is perfectly smooth there.

When we study black holes in parts VI and VII, we will encounter this kind of singularity, known as a coordinate singularity, in contrast to an actual or physical singularity, when the geometry itself becomes singular. As another example, consider the surface described by

$$ds^2 = \frac{1}{1 - \frac{r_S}{r}} dr^2 + r^2 d\varphi^2 \tag{10}$$

with r_S a positive constant. Again, the singularity at $r = r_S$ is merely a coordinate singularity. Indeed, this surface could be embedded into E^3 using the familiar cylindrical coordinates $ds^2 = dr^2 + r^2 d\varphi^2 + dz^2$ and setting $z^2 = 4r_S(r - r_S)$. (Let's verify this: $zdz = 2r_S dr$, so that $dr^2 + dz^2 = \left(1 + \frac{4r_S^2}{z^2}\right)dr^2 = \frac{r}{r-r_S}dr^2$.) Thus, the surface is perfectly smooth at $r = r_S$ (see figure 2). It consists of two planes connected by a "throat," known as the Einstein-Rosen bridge. Note that a mite geometer could perfectly well travel from the upper to the lower plane without noticing any singularity at all. So this type of geometry is sometimes known picturesquely as a wormhole.

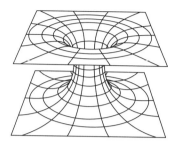

Figure 2 The Einstein-Rosen bridge, underlining the difference between a coordinate singularity and a physical singularity.

A line of fixed r is simply a circle. For $r \gg r_S$, the space becomes flat: $ds^2 \to dr^2 + r^2 d\varphi^2$. Incidentally, in the same cylindrical set-up, the S^2 metric we started this appendix with is given by the embedding $z^2 = 1 - r^2$. In fact, I am repeating myself.

Meanwhile, Confusio has flown off to the equator to investigate the $r = 1$ singularity there. "What singularity?" the locals ask him.

Appendix 2: Spheres

In the text, we studied the 2-dimensional sphere S^2. Clearly, it is not difficult to generalize our discussion to higher dimensional spheres. For instance, S^3 is embedded in E^4 by $X^2 + Y^2 + Z^2 + W^2 = 1$. Replace (X, Y, Z) by the usual spherical coordinates, so that $W^2 = 1 - r^2$ and $W dW = -r dr$, and eliminate W in $ds^2 = dX^2 + dY^2 + dZ^2 + dW^2 = dr^2 + r^2(d\theta^2 + \sin^2\theta d\varphi^2) + \frac{(-r dr)^2}{1-r^2}$. We thus obtain the metric for S^3:

$$ds^2 = \frac{dr^2}{1 - r^2} + r^2(d\theta^2 + \sin^2\theta d\varphi^2) = \frac{dr^2}{1 - r^2} + r^2 d\Omega_2^2 \tag{11}$$

The only difference from (7) is that here the angular element $d\Omega_2^2$ on S^2 appears rather than the angular element $d\Omega_1^2 = d\varphi^2$ on S^1 (namely the circle). Indeed, recalling what we just learned in the preceding appendix, we invite ourselves to write $r = \sin\psi$ in (11), thus obtaining $d\Omega_3^2 = d\psi^2 + \sin^2\psi d\Omega_2^2$, where, in line with the usual solid angle notation, we have renamed the line element ds^2 on S^3 as $d\Omega_3^2$.

Evidently, we can determine the line element on S^n iteratively,

$$d\Omega_n^2 = d\psi^2 + \sin^2\psi d\Omega_{n-1}^2 \tag{12}$$

thus recovering the result of exercise I.5.10. As noted in that exercise, this generalizes the elementary school observation that the curves of constant latitude on the globe form circles. Here, the subspaces of constant ψ form S^{n-1}.

I already mentioned in exercise I.5.9 that we may actually live in S^3. In (11), simply scale $r \to r/L$, and then multiply ds^2 by L^2 to obtain $ds^2 = \frac{dr^2}{1-(\frac{r}{L})^2} + r^2(d\theta^2 + \sin^2\theta d\varphi^2)$. All we have done is to restore the length dimension to r. One objective of observational cosmology is to determine L, or failing that, to set a lower bound on it.

Appendix 3: Hyperbolic spaces

The 2-dimensional hyperbolic space H^2 (or pseudosphere, as some people call it) is less intuitively accessible than S^3 and hence generally not mentioned in elementary schools. Define a 2-dimensional surface by $X^2 + Y^2 - W^2 = -1$, embedded (crucially) not in E^3 but in a "pseudo-Euclidean" space with the metric

$$ds^2 = dX^2 + dY^2 - dW^2 \tag{13}$$

Once again, replace (X, Y) by the usual polar coordinates, so that $W^2 = 1 + r^2$ and $W dW = r dr$. Writing $dW^2 = \frac{(r dr)^2}{W^2}$ in ds^2, we obtain the metric for H^2:

$$ds^2 = dr^2 + r^2 d\theta^2 - \frac{(r dr)^2}{1 + r^2} = \frac{dr^2}{1 + r^2} + r^2 d\theta^2 \tag{14}$$

Compare and contrast with (7) for S^2. That one sign makes all the difference, of course.

Now we are invited to write $r = \sinh \psi$, so that $ds^2 = d\psi^2 + \sinh^2 \psi d\theta^2$. Hyperbolic sine, hyperbolic space, got it? Curves of constant "latitude" (that is, ψ) also form circles, while $W^2 = 1 + r^2$ traces out a hyperbola in the $(W\text{-}r)$ plane.

Evidently, we can move up in dimension. The 3-dimensional hyperbolic space H^3 is described by the line element (not very imaginative notation) $dH_3^2 = d\psi^2 + \sinh \psi^2 (d\theta^2 + \sin^2 \theta d\varphi^2) = d\psi^2 + \sinh \psi^2 d\Omega_2^2$. More generally,

$$dH_n^2 = d\psi^2 + \sinh \psi^2 d\Omega_{n-1}^2 \tag{15}$$

Note that H^n is constructed from S^{n-1}, not H^{n-1}.

Again, one issue addressed by cosmology is whether we live in H^3 or S^3. Just as for spheres, we could restore the length dimension to r and write the metric for H^3 as $ds^2 = \frac{dr^2}{1 + (\frac{r}{L})^2} + r^2 d\Omega_2^2$. (Note that (14) looks like but is not to be confused with the stereographic metric of a sphere given in exercise I.5.15.) We will encounter hyperbolic spaces when we study de Sitter and anti de Sitter spacetimes in part IX.

Note that while the embedding space is pseudo-Euclidean, the hyperbolic space is clearly locally Euclidean. Indeed, around an arbitrary point on H^3, say $r = L$, $\theta = \pi/2$, and $\varphi = 0$, the mites living in the space would experience the perfectly Euclidean metric $ds^2 \simeq \frac{1}{2} dr^2 + L^2 (d\theta^2 + d\varphi^2)$. Indeed, if you want, you can define $x = r/\sqrt{2}$, $y = L\theta$, $z = L\varphi$, so that $ds^2 \simeq dx^2 + dy^2 + dz^2$. The mites don't know that the embedding space is not Euclidean and couldn't care less.

Appendix 4: A potential confusion over hyperbolic spaces

Consider another hyperbolic surface defined by $X^2 + Y^2 - W^2 = 1$ embedded in a space with the metric $ds^2 = dX^2 + dY^2 - dW^2$. You should draw the surface defined by $X^2 + Y^2 - W^2 = 1$ and compare it with the corresponding surface in appendix 3. Following the same steps as above, you find $ds^2 = dr^2 + r^2 d\theta^2 - \frac{(r dr)^2}{r^2 - 1} = \frac{dr^2}{1 - r^2} + r^2 d\theta^2$. This is just the sphere in (7). Surprise! (Or were you surprised?)

Well, I bet that you drew something like a cylinder but with a radius that grew toward infinity at both ends. If you didn't, hurray for you. In all likelihood, you drew the surface as if it were embedded in E^3, but it isn't. Indeed, you see that if you analytically continue $W \to iW$, the surface defines S^2.

Exercises

1 Find the transformation relating the coordinates used by the Eskimo mites in exercise I.5.2 to the coordinates in (2).

2 Calculate the curvature of the torus by the tangent plane method.

3 A civilization living in 2-dimensional space marks their world with the coordinates (κ, ζ) handed down eons ago by their ancestors. Careful measurements of distances between various points over time have shown that the metric of their world is given by $g_{\kappa\kappa} = 1 + \cdots$, $g_{\zeta\zeta} = 1 + 2\kappa + \cdots$, $g_{\kappa\zeta} = 0 + \cdots$, where the dots indicate terms quadratic in κ and ζ. The civilization is in fact planning to deploy a team of geometers to measure these quadratic terms. As they develop physics, they find the linear term in $g_{\zeta\zeta}$ terribly irksome.
   (a) One day, a bright young physics student points out that changing coordinates from (κ, ζ) to (ω, ϕ), defined by $\kappa = \omega + \frac{1}{2}\phi^2$, $\zeta = \phi - \omega\phi$, would cause the linear term in the metric to disappear. Show that this is in fact the case.

(b) Many years later, another bright young physics student realizes that the "crazy" coordinates (κ, ζ) are just remnants of polar coordinates left by advanced interstellar visitors, who had long since departed. The student writes down polar coordinates (r, θ) with $ds^2 = dr^2 + r^2 d\theta^2$ and shows that the civilization has been flourishing in a small neighborhood of the point P with coordinates $(r, \theta)_P = (r_*, 0)$. The mysterious coordinates (κ, ζ) turn out to be merely the deviation of r, θ from P, suitably scaled by r^*. Explain how this works. Of course, this is just a theory: the civilization would now have to measure the quadratic terms in their metric to be sure, but preliminary measurements indicate that this theory will very likely work. Measurements show that the origin of the polar coordinate system is incredibly far away; nevertheless, an expedition is planned to visit this mysterious place.

4 Find the locally flat coordinates on the Poincaré half plane.

5 Show that for $D = 2$, the combination $2B_{12,12} - B_{11,22} - B_{22,11}$ measures intrinsic curvature. In the simple example discussed in connection with the tangent plane, since the combination $dx^2 + dy^2$ is invariant under rotation, it is equal to $du^2 + dv^2$, and thus $ds^2 = dx^2 + dy^2 + dz^2 = du^2 + dv^2 + (\mu u du + v v dv)^2 = (1 + \mu^2 u^2)du^2 + (1 + v^2 v^2)dv^2 + 2\mu v\, uv\, dudv$. Work out $B_{\mu\nu,\lambda\sigma}$ and the combination specified here.

6 Calculate the combination $2B_{12,12} - B_{11,22} - B_{22,11}$ for the metric found in exercise 4.

7 Show that for $D = 1$, we can set g_{11} to 1 by a coordinate transformation and so curves have no intrinsic curvature.

8 It is easy to introduce a coordinate singularity by a poor choice of coordinates. Start with $ds^2 = dx^2 + dy^2$ and let $z = y^p$. Find the metric in terms of (x, z).

9 Note that the coordinates (x, y, z) introduced for H^3 in appendix 3 are not locally flat. Find the transformation to locally flat coordinates.

10 Two spaces described by the metric $\tilde{g}_{\mu\nu}$ and $g_{\mu\nu}$ are said to be conformally related if

$$\tilde{g}_{\mu\nu}(x) = \Omega^2(x)g_{\mu\nu}(x) \tag{16}$$

Show that, given two infinitesimal line segments originating from a point, the angle between them is preserved by this conformal transformation. In particular, that is why the Mercator map of exercise I.5.3 was popular with navigators. (Bad terminology alert: The term "conformal transformation" often suggests to students that the two metrics $\tilde{g}_{\mu\nu}$ and $g_{\mu\nu}$ are related by a coordinate transformation. In general, they are not. Thus, it is probably better to call "conformal transformation" a Weyl transformation instead.)

11 In the preceding exercise, if the metric $g_{\mu\nu}$ is flat, then the metric $\tilde{g}_{\mu\nu}$ is said to be conformally flat. In other words, a metric $g_{\mu\nu}$ (dropping the tilde) is said to be conformally flat if there exists an Ω such that $g_{\mu\nu}(x) = \Omega^2(x)\delta_{\mu\nu}$. (In fact, we have already encountered conformally flat spaces in exercise 14 in the preceding chapter.) In higher dimensions, a metric has to be very special (in particular, it must be characterized by a single function) to be conformally flat. But 2-dimensional surfaces are so "simple" that they are all (locally) conformally flat. Show this by a counting argument.

12 Show that the sphere and the Poincaré half plane are conformally flat. (Again, a bad terminology alert: The term "conformal flat space" misleads many students into thinking that the space is flat. In general, it is not. For example, consider the Mercator map of exercise I.5.3: the sphere S^2 is manifestly not flat.)

13 Find the curvature of the space described by $ds^2 = ydx^2 + xdy^2$.

14 Show that hyperbolic spaces are conformally flat.

Notes

1. Later in life, he also appears in *QFT Nut*, older but not wiser.
2. For pictures, *Toy/Universe*, p. 25.
3. Can any space be embedded in E^N? If so, what is the minimum value of N for a given space? These nontrivial questions were answered by John Nash, the mathematician portrayed in the film *A Beautiful Mind: Ann. Math.* 63 (1956), p. 20.
4. I purposely use the letter r to emphasize that we can use the same letter to describe different things in different situations. I trust you not to confuse this r with the r in the usual spherical coordinates and which we restricted to be 1 in the preceding chapter to obtain the metric for the unit sphere. The r here is the "polar" radial variable in polar coordinates.
5. For the purpose of this book, we call a Riemannian manifold a space whose metric is smooth enough to be differentiated an appropriate number of times. This may require finding an appropriate set of coordinates.
6. The rigor-minded reader realizes that we actually need to check this.

"Classical" differential geometry

I feel that it would be good for those readers seeing Riemannian geometry for the first time to work through some "classical" differential geometry[1] dealing with curves and surfaces, "real" stuff that you could actually see and "hold in your hands." Throughout this chapter, we will be living in good old 3-dimensional Euclidean space. I am going to tell you how the greats like Frenet and Gauss thought about curves and surfaces. None of the fancy tangent bundle talk for us; we will just do it. Action, not talk!

One advantage of this approach[2] is that you will gain a geometric feel for important concepts such as covariant differentiation, curvature, and the Christoffel symbol, which in some texts are introduced immediately in a more high powered and abstract fashion. We will, of course, also get to the more general and direct Riemannian approach[3] to curvature in due time. Hence, it is entirely possible for those who do not care as much as I do about "classical" mathematics to skip this chapter.

Curves

Consider a curve $\mathcal{C}$ given by $\vec{X}(l)$ parametrized by the length l along the curve. In other words, we start from an arbitrary point $\mathcal{O}$ on the curve, and pace off a distance equal to l along the curve. Our location is then specified by the vector $\vec{X}(l)$.

The unit tangent vector is $\vec{t} \equiv \dot{\vec{X}} \equiv \frac{d\vec{X}}{dl}$. Following Newton, we denote differentiation with respect to l by a dot, as was already done in chapter I.1. (All vectors in this chapter have 3 components and will be labeled with an arrow.) Since we said that l is the length, namely $dl^2 = d\vec{X} \cdot d\vec{X}$, we have $\vec{t} \cdot \vec{t} = 1$. Differentiating, we obtain $\vec{t} \cdot \dot{\vec{t}} = 0$. Thus, the unit vector $\vec{p}$ defined by

$$\dot{\vec{t}} = \kappa \vec{p} \tag{1}$$

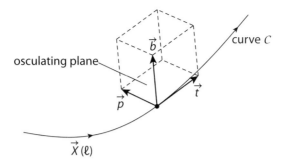

Figure 1 The moving trihedron and osculating plane of a smooth curve $\mathcal{C}$.

(known as the principal normal vector) is orthonormal to $\vec{t}$. You can see that κ, equal to $|\dot{\vec{t}}|$ by definition, has to do with the curvature of the curve $\mathcal{C}$ at the point $\vec{X}$.

Imagine driving a race car along the curve: $\vec{t}$ indicates the direction you are pointing in and $\dot{\vec{t}}$ how fast you have to turn the steering wheel. Anyway, we physicists recognize $\dot{\vec{t}}$ as the acceleration $\ddot{\vec{X}}$ and κ as a measure of the centrifugal force. In this analogy, l is time, and so speed is fixed to be 1. This is kind of a neat example of how physics and mathematics intertwine: the centrifugal force tells us about curvature.

Next define the unit vector $\vec{b}(l) \equiv \vec{t}(l) \times \vec{p}(l)$, known as the binormal vector. As we move along the curve, we have what 18th century mathematicians called a moving trihedron formed by the triplet $\vec{t}$, $\vec{p}$, and $\vec{b}$. The vectors $\vec{t}$ and $\vec{p}$ form a plane, named the osculating plane (from the Latin for "kissing") by D'Amondans Charles de Tinseau (1748–1822) in 1780. See figure 1.

By construction, $\vec{t} \cdot \vec{b} = \vec{t} \cdot (\vec{t} \times \vec{p}) = 0$. Differentiating this equation, we get $\dot{\vec{t}} \cdot \vec{b} + \vec{t} \cdot \dot{\vec{b}} = 0$. Using $\vec{p} \cdot \vec{b} = 0$, we find that $\dot{\vec{b}}$ is orthogonal to $\vec{t}$. Also, differentiating $\vec{b} \cdot \vec{b} = 1$, we obtain $\vec{b} \cdot \dot{\vec{b}} = 0$. Since $\dot{\vec{b}}$ is orthogonal to both $\vec{t}$ and $\vec{b}$, we conclude that

$$\dot{\vec{b}} = -\tau \vec{p} \tag{2}$$

Evidently, τ, known as the torsion of the curve $\mathcal{C}$ at the point $\vec{X}$, measures how the curve is twisting. If the discussion is unclear to you at any point, draw your own picture!

The next question is how the unit normal $\vec{p}$ changes as we move along. Noting that $\vec{p} = \vec{b} \times \vec{t}$, we differentiate and use (1) and (2) to obtain

$$\dot{\vec{p}} = \dot{\vec{b}} \times \vec{t} + \vec{b} \times \dot{\vec{t}} = \tau \vec{b} - \kappa \vec{t} \tag{3}$$

We can package the three equations, (1), (2), and (3), known as the Frenet-Serret equations (in memory of Jean Frenet (1816–1900) and Joseph Serret (1819–1885)), more compactly by introducing the 9-component object $\psi \equiv \begin{pmatrix} \vec{t} \\ \vec{p} \\ \vec{b} \end{pmatrix}$. Then

$$\dot{\psi} = A\psi, \quad \text{with } A = \begin{pmatrix} 0 & \kappa & 0 \\ -\kappa & 0 & \tau \\ 0 & -\tau & 0 \end{pmatrix} \tag{4}$$

The antisymmetry of A ensures that $\psi \cdot \dot{\psi} = 0$.

Surfaces

We now graduate from curves to surfaces in the 3-dimensional Euclidean space E^3 we were born into. A surface embedded in E^3 is defined by $\vec{X}(x^1, x^2)$. In contrast to a curve $\vec{X}(l)$ parametrized by l, the surface is parametrized by two coordinates x^μ, with the index μ taking on 2 values: this feature is of course what makes a surface a 2-dimensional object. Also, while the length along the curve l provides a natural parametrization, there is no comparable natural parametrization for a surface. If the discussion becomes too abstract for you at any point, you can always think of the familiar sphere (with $x^1 = \theta$, $x^2 = \varphi$) for which

$$\vec{X} = \begin{pmatrix} \sin\theta\cos\varphi \\ \sin\theta\sin\varphi \\ \cos\theta \end{pmatrix}$$

The two 3-component vectors $\vec{e}_\mu \equiv \partial_\mu \vec{X} = \frac{\partial \vec{X}}{\partial x^\mu}$, labeled by the index μ, form the basis vectors for the surface. For example, on the sphere,

$$\vec{e}_1 = \begin{pmatrix} \cos\theta\cos\varphi \\ \cos\theta\sin\varphi \\ -\sin\theta \end{pmatrix} \quad \text{and} \quad \vec{e}_2 = \begin{pmatrix} -\sin\theta\sin\varphi \\ \sin\theta\cos\varphi \\ 0 \end{pmatrix}$$

To make absolutely sure that there is no confusion, let me say again that $\vec{X}(x)$ lives in the ambient 3-dimensional Euclidean space E^3, and $\vec{e}_\mu(x)$ are two 3-vectors labeled by $\mu = 1, 2$. Note that x^1, x^2 are coordinates on the surface, not components of $\vec{X}$.

Tangent plane and metric

Linear combinations of the two basis vectors span the tangent plane at the point labeled by the coordinates x; in other words, the set of all points represented by $u^\mu \vec{e}_\mu(x) = u^1 \vec{e}_1(x) + u^2 \vec{e}_2(x)$, for u^1, u^2 any two real numbers, form the tangent plane. The tangent plane changes as we move around on the surface, of course. For example, on the sphere, at the point $(\theta = \pi/2, \varphi = 0)$, the tangent plane consists of $\begin{pmatrix} 0 \\ u^2 \\ -u^1 \end{pmatrix}$ as u^1 and u^2 range over the real numbers. For another example, again on the sphere, at the point $(\theta = \pi/2, \varphi = \pi/2)$, the tangent plane consists of $\begin{pmatrix} -u^2 \\ 0 \\ -u^1 \end{pmatrix}$ again as u^1 and u^2 range over the real numbers.

Another familiar example is the cylinder with radius a, for which (with the choice $x^1 = \varphi, x^2 = z$) $\vec{X} = \begin{pmatrix} a\cos\varphi \\ a\sin\varphi \\ z \end{pmatrix}$. It follows that $\vec{e}_1 = \begin{pmatrix} -a\sin\varphi \\ a\cos\varphi \\ 0 \end{pmatrix}$ and $\vec{e}_2 = \begin{pmatrix} 0 \\ 0 \\ 1 \end{pmatrix}$. Note that if we think of a as a length, $\vec{e}_1$ and $\vec{e}_2$ do not have the same dimension.

Since $d\vec{X} = \partial_\mu \vec{X} dx^\mu$, the distance squared between two neighboring points with coordinates x and $x + dx$ is given by $ds^2 = d\vec{X} \cdot d\vec{X} = (\partial_\mu \vec{X} \cdot \partial_\nu \vec{X})dx^\mu dx^\nu = \vec{e}_\mu \cdot \vec{e}_\nu \, dx^\mu dx^\nu$.

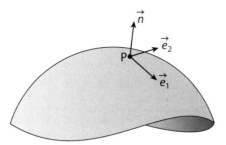

Figure 2 The tangent plane and the normal to a surface at some point P.

In other words, the metric on the surface is

$$g_{\mu\nu} = \vec{e}_\mu \cdot \vec{e}_\nu \tag{5}$$

We use the Einstein repeated summation convention: all repeated indices are summed over. For example, $d\vec{X} = \partial_\mu \vec{X} dx^\mu = \partial_1 \vec{X} dx^1 + \partial_2 \vec{X} dx^2$. You should check that (5) gives the familiar result $ds^2 = d\theta^2 + \sin\theta^2 d\varphi^2$ for the sphere. For the cylinder, $ds^2 = \vec{e}_1 \cdot \vec{e}_1 d\varphi^2 + \vec{e}_2 \cdot \vec{e}_2 dz^2 = a^2 d\varphi^2 + dz^2$.

As we move about on the surface, how does the tangent plane rock and roll?

We now ask how the two basis vectors change as we move about on the surface. Consider $\vec{e}_{\mu,\nu} \equiv \partial_\nu \vec{e}_\mu = \partial_\nu \partial_\mu \vec{X}$. Here for typographical convenience, we have introduced the standard notation $\mathcal{E}_{,\mu} = \partial_\mu \mathcal{E}$ for any expression $\mathcal{E}$. Thus, $\vec{e}_{\mu,\nu}$ denotes a 3-vector labeled by two indices μ and ν. In general, this vector will have a component pointing out of the surface.

Next, denote the unit normal to the surface by

$$\vec{n} = \frac{\vec{e}_1 \times \vec{e}_2}{|\vec{e}_1 \times \vec{e}_2|} \tag{6}$$

(not to be confused with $\vec{p}$ in our discussion of curves, of course). See figure 2. For example, for the cylinder, $\vec{n} = \begin{pmatrix} \cos\varphi \\ \sin\varphi \\ 0 \end{pmatrix}$.

As I just said, the vector $\vec{e}_{\mu,\nu}$, namely the change of $\vec{e}_\mu$ in the direction ν, sticks out of the surface and thus has a component along $\vec{n}$. So let us expand $\vec{e}_{\mu,\nu}$ in terms of the basis vectors $\vec{e}_\lambda$ and $\vec{n}$:

$$\vec{e}_{\mu,\nu} = \Gamma^\lambda_{\mu\nu} \vec{e}_\lambda + K_{\mu\nu} \vec{n} \tag{7}$$

Since $\partial_\mu \partial_\nu = \partial_\nu \partial_\mu$, the vectors $\vec{e}_{\mu,\nu}$ and the expansion coefficients $\Gamma^\lambda_{\mu\nu}$ and $K_{\mu\nu}$ are symmetric in their two lower indices. (For future use, we note that $\Gamma^\lambda_{\mu\nu}$ is known as the Christoffel symbol.) Contracting (7) by the normal vector, we obtain

$$K_{\mu\nu} = \vec{e}_{\mu,\nu} \cdot \vec{n} \tag{8}$$

known as Gauss's equation.

We hasten to give an example. For the sphere, with X, $\vec{e}_1$, and $\vec{e}_2$ as given before, we have $\vec{n} = \vec{X}$, $\vec{e}_{1,1} = \partial_1 \vec{e}_1 = -\vec{n}$, $\vec{e}_{2,1} = \partial_1 \vec{e}_2 = \vec{e}_{1,2} = \partial_2 \vec{e}_1 = \cot\theta \vec{e}_2$, and $\vec{e}_{2,2} = \partial_2 \vec{e}_2 = -\sin\theta\cos\theta \vec{e}_1 - \sin^2\theta \vec{n}$. From (7) we read off

$$K_{11} = -1, \qquad K_{22} = -\sin^2\theta \tag{9}$$

and

$$\Gamma^2_{12} = \cot\theta, \qquad \Gamma^1_{22} = -\sin\theta\cos\theta \tag{10}$$

with all other entries of K and Γ vanishing.

By drawing a picture, you can see that, as we move from one point to a neighboring point, the change of $\vec{e}_1$ and $\vec{e}_2$ projected into the tangent plane (the first term in (7)) tells us how the tangent plane is rotating around the normal vector $\vec{n}$, while the change projected in the direction of the normal (the second term in (7)) tells how the tangent plane is "rocking and rolling." Thus, the coefficients $K_{\mu\nu}$ tell us about how the surface is curving in the ambient 3-dimensional Euclidean space.

Covariant derivative

Let $W^\mu(x)$ be a vector field. In other words, at every point* x, two numbers, $W^1(x)$ and $W^2(x)$, are given, so that someone living in the ambient E^3 sees a vector field $\vec{W}(x) \equiv W^\mu(x)\vec{e}_\mu(x)$. Since $\vec{W}$ is a linear combination of the two basis vectors, it lives on the tangent plane. In other words, it does not stick out of the surface.

I now introduce one of the most basic concepts of differential geometry, that of covariant derivative. Unaccountably, some texts make covariant differentiation sound mysterious and complicated, when in fact it is intuitive and simple. Suppose we want to differentiate the vector field $W^\mu(x)$. It is not enough to ask how the components $W^1(x)$ and $W^2(x)$ change when we move from the point x to a neighboring point $x + dx$. The basis vectors $\vec{e}_\mu(x)$, against which the components are measured, are themselves changing. The covariant derivative simply takes into account this obvious geometric fact, namely the variation of the basis vectors. This effect does not occur in Euclidean space: once we set up the usual unit basis vectors pointing in the x, y, z directions, they do not change.

Mathematically, all we have to do is to differentiate using the product rule and follow our noses:

$$\partial_\nu \vec{W}(x) = \partial_\nu(W^\mu(x)\vec{e}_\mu(x)) = (\partial_\nu W^\mu(x))\vec{e}_\mu(x) + W^\mu(x)\partial_\nu \vec{e}_\mu(x)$$
$$= (\partial_\nu W^\mu)\vec{e}_\mu + W^\lambda \Gamma^\mu_{\lambda\nu}\vec{e}_\mu + W^\mu K_{\mu\nu}\vec{n} \tag{11}$$

In the first line, the second term expresses the effect we were just talking about: the basis

---

* The phrase "point x" is of course shorthand for "point described by x in our chosen coordinate system." We are a bit less precise, but remember, "brevity is the soul of wit."

vectors themselves vary as we move about. In the second line, we used (7) and renamed some dummy indices we sum over. The key point is that $\partial_\nu \vec{W}$ contains a component along $\vec{n}$, the normal to the surface.

Imagine ourselves members of the mite civilization in the prologue. We do not know about vectors sticking out of our universe: all we know and care about are vectors lying inside our universe. Thus, we invite ourselves to define a covariant derivative[4] by dropping the term proportional to $\vec{n}$ in (11):

$$D_\nu \vec{W} \equiv (\partial_\nu W^\mu + W^\lambda \Gamma^\mu_{\lambda\nu})\vec{e}_\mu \equiv (D_\nu W^\mu)\vec{e}_\mu \tag{12}$$

In the last step we defined $D_\nu W^\mu \equiv \partial_\nu W^\mu + \Gamma^\mu_{\nu\lambda} W^\lambda$. (Recall that Γ is symmetrical in its two lower indices.)

From my experience teaching, I know that some beginning students get confused here. But really, the concept of covariant derivative is at heart quite simple. Think of yourself as a mite, and you don't know about vectors sticking out of the surface that forms your world. So you just drop that component in the derivative, and you get the covariant derivative.

You the mighty mite do not know about $\partial_\nu \vec{W}$, only about $D_\nu \vec{W}$. The concept of covariant derivative is central to differential geometry and hence to Einstein gravity.

Parallel transport

Let me explain the covariant derivative in a slightly different way. If we are living in Euclidean space and we want to differentiate a vector field, we simply follow Newton and calculate the limit of $\vec{W}(x + \delta x) - \vec{W}(x)$. But this implies that we know how to subtract a vector defined at the point x from another vector defined at a different point $y = x + \delta x$. Recall how we learned as children to subtract one vector from another. We were taught to slide one vector over to the other, so that their feathered ends coincide; the gap between their sharp pointy ends is then the difference we want. See figure 3. But of course when we slide a vector over, we have to take care that we do not rotate it; we need to keep it pointing in the same direction.

Sliding a vector over taking care to keep it pointing in the same direction is known as "parallel transporting" the vector. Of course, in Euclidean space, parallel transport is trivial,

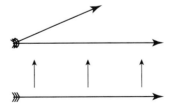

Figure 3 How we learned as children to subtract one vector from another.

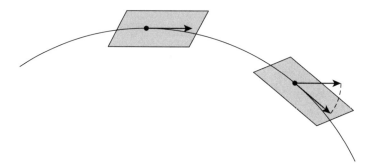

Figure 4 When a vector living in the tangent plane at one point on a curved surface is parallel transported to a nearby point, it will in general not live in the tangent plane at that point. The component sticking out of the tangent plane (dashed line) has to be projected away.

and we do it without giving it a second thought. But if we are living on a curved surface, it is not so simple to parallel transport. Given a vector $\vec{V}$ at the point x, how do we parallel transport it to some other point y?

To the more knowledgeable beings living in the ambient E^3, it is obvious. Just parallel transport $\vec{V}$ in the ambient Euclidean space. Any child could do it, they yell.

The trouble is that while the 3-vector $\vec{V}$ lives in the tangent plane at x, it doesn't necessarily live in the tangent plane at y: it will in general have a component sticking out of that tangent plane. Thus, we have to project $\vec{V}$ onto the tangent plane at y. See figure 4.

Well, we know how. Write $\vec{V}$ as a linear combination of $\vec{e}_1(y)$, $\vec{e}_2(y)$, and $\vec{n}(y)$, and then simply drop the piece proportional to $\vec{n}(y)$. The result is the vector $\vec{V}$ projected onto the tangent plane at y, which we denote by $(\vec{V} \to y)_\mathrm{P}$. You can work out $(\vec{V} \to y)_\mathrm{P}$ in an exercise, but we don't really need the explicit form here. Note that $(\vec{V} \to y)_\mathrm{P}$ is a vectorial function of the vector $\vec{V}$ and of the location y. The subscript P reminds us that we are projecting. I have to ask you to understand the notation before reading on. Again, I have encountered an occasional student who finds the notation confusing. In fact, the notation, which may appear cumbersome, is needed to make the discussion clear.

This discussion tells us not to blindly follow Newton and calculate the limit of $\vec{W}(x + \delta x) - \vec{W}(x)$. No, Sir Isaac, we don't want to compare $\vec{W}(x + \delta x)$ against $\vec{W}(x)$. No sir, we want to compare $\vec{W}(x + \delta x)$ against $(\vec{W}(x) \to x + \delta x)_\mathrm{P}$.

And, dear reader, what does this clunky but informative notation $(\vec{W}(x) \to x + \delta x)_\mathrm{P}$ mean? It means the result of the following procedure: we take the vector $\vec{W}(x)$ (corresponding to the $\vec{V}$ in the preceding paragraph), parallel transport it to $x + \delta x$, and then project by throwing away the component that sticks out of the tangent plane at $x + \delta x$. That is the beast we want to subtract from $\vec{W}(x + \delta x)$.

So, put the difference $\vec{W}(x + \delta x) - (\vec{W}(x) \to x + \delta x)_\mathrm{P}$ into Newton's limiting machine, that is, take the limit $\delta x \to 0$ of this difference.

But the result of this "geometrical" construction is effectively exactly the same as the previous construction, namely dropping the component proportional to $\vec{n}$ from the ordinary derivative $\partial_\nu \vec{W}$ to define the covariant derivative $D_\nu \vec{W}$, precisely as in (12).

Yet another way of saying this is that the covariant derivative $D_\nu \vec{W}$ does not express how $\vec{W}(x + \delta x)$ differs from $\vec{W}(x)$, but how $\vec{W}(x + \delta x)$ differs from $\vec{W}(x)$ parallel transported to $x + \delta x$, that is, how $\vec{W}(x + \delta x)$ differs from $(\vec{W}(x) \to x + \delta x)_P$. (Some students are perhaps confused because here, by parallel transport on a curved surface, we actually mean parallel transport in the ambient E^3 and then projection onto the tangent plane.)

A rough analogy may help some readers. When you think about how much your income has risen, you don't simply differentiate your income with respect to time. More meaningful is your income increase adjusted for inflation. It could be that your income is not increasing intrinsically, but the dollar (or whatever currency you get paid in) figure is rising because of inflation. Similarly, the covariant derivative $D_\nu \vec{W}$ is the ordinary derivative $\partial_\nu \vec{W}$ adjusted for the change in the reference frame.

Thus, if we have a vector field $\vec{V}(x)$ satisfying $D_\nu \vec{V} = 0$, then it is not changing intrinsically. It has simply been parallel transported all over space, a fact of considerable military significance.

The ancient art of war

Imagine that it is 300 BC and that you are the physicist-sorcerer attached to the army commanded personally by the Emperor the Son of Heaven. The other sorcerer, the astro guy whom you have always derided for staring at the sky, predicted a thick fog on the day of the battle, so that the soldiers would not be able to see which way was which. The Emperor ordered you to solve the problem. Having read this book, you immediately realized that your task was to parallel transport a vector so that it always pointed south. You quickly had an imperial south-pointing carriage constructed, on top of which was a statue of the Emperor pointing south. Indeed, the day was frighteningly foggy, a pea soup fog, as the Brits would say. The soldiers could barely see beyond an arm's length, and the enemy became totally confused. In contrast, as the south-pointing carriage moved around this way and that, the statue always pointed south, and so the Emperor scored a huge victory. The Emperor was so delighted that you (and that astro guy too) received tenure at the court and lived happily ever after.

Would I make something like this up? Obviously not. The south-pointing carriage was described in ancient Chinese chronicles. Unfortunately, the detailed construction plan was lost to posterity, but in the 20th century, a contest produced a modern reconstruction.* (See figure 5.) I won't bother to show the engineering drawing here but pose the design to you as a challenge.[5]

---

* Louis Grace of the physics department at the University of California, Santa Barbara, kindly built this war chariot for me, complete with a statue of Emperor Albert on top.

Figure 5 A modern version of the south-pointing carriage described in ancient Chinese chronicles.

Gauss's strategy

To determine the curvature of the surface at a given point P, Gauss's strategy was to study curves on the surface passing through P and calculate their curvatures. This sounds puzzling at first. How did Gauss expect to determine[6] the curvature of a surface by studying the curvature of curves lying on the surface?

To motivate his reasoning, consider the curvature at the saddle point P with coordinates ($x = 0$, $y = 0$) on the surface $z = \frac{1}{2}px^2 - \frac{1}{2}qy^2$, taking $p > 0$, $q > 0$ for definiteness (a special case of the surface considered in chapter I.5). Imagine yourself walking along the ridge defined by $y = 0$: on both sides of you, the land falls away (see figure 6). When you reach the lowest point $x = 0$ on the ridge, the land in front of you and behind you rises up. Consider the family of curves going through P. For a curve pointing in the y direction, the curvature is positive, while for a curve pointing in the x direction, the curvature is negative. Evidently, for a curve pointing in some other direction, the curvature is intermediate between two extremal values. Gauss proposed to find these two extremal values.

So for a given point P on a curved space, let's look at some curve going through that point and study its tangent vector $\vec{t} = t^\mu \vec{e}_\mu$ with $t^\mu = \frac{dX^\mu}{dl}$. Then

$$\frac{d\vec{t}}{dl} = \frac{dt^\mu}{dl}\vec{e}_\mu + t^\mu \frac{d\vec{e}_\mu}{dl} \tag{13}$$

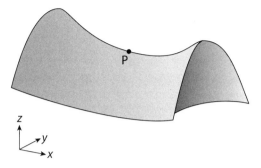

Figure 6 Determining the curvature of a surface with a saddle point P.

Write the right hand side as some linear combination of $\vec{e}_1$, $\vec{e}_2$, and $\vec{n}$, so that

$$\frac{d\vec{t}}{dl} = \kappa_g \vec{e} + \kappa_n \vec{n} \tag{14}$$

with $\vec{e}$ some unit vector given by some linear combination of $\vec{e}_1$ and $\vec{e}_2$. Note in passing that since $\vec{t} \cdot \vec{t} = 1$, we have $\vec{t} \cdot \frac{d\vec{t}}{dl} = 0$ and so $\vec{t} \cdot \vec{e} = 0$. The three unit vectors $\vec{n}, \vec{t}$, and $\vec{e}$ form an orthonormal triad. Note also that $\vec{e}_\mu$ and $\vec{n}$ pertain to the surface, while $\vec{t}$ and $\vec{e}$ pertain to the curve.

To get a feel for the two quantities κ_g and κ_n in (14), consider a couple of surfaces. Draw a couple of pictures as you read the next two paragraphs. That will make it easy to follow what I am talking about, which is not much more than common spatial sense.

First, a plane flat as Kansas. Through P draw a curve as curvy as you like, and you can make κ_g as large as you like. But try as you may, $\frac{d\vec{t}}{dl}$ will stay in the plane by definition, and κ_n will remain stubbornly equal to 0. Evidently, κ_g does not tell us about the curvature of the surface, merely the curvature of the curve.

Next, picture a sphere, and let P be Copenhagen, say. Following Gauss, consider a variety of curves going through Copenhagen. For example, think of the circle of constant latitude. Then $\frac{d\vec{t}}{dl}$ points toward the axis of the earth joining the north and south poles and can be written as in (14) with $\vec{e}$ pointing north along some street in Copenhagen and $\vec{n}$ pointing upward* at the sky (as always, independent of where we are). As we try different curves, κ_g and κ_n vary. If we choose the curve to be the circle of constant longitude (rather than latitude) going through Copenhagen, κ_g vanishes, while $|\kappa_n|$ is maximized, given by the inverse of the earth's radius. Indeed, $|\kappa_n|$ attains its maximum value for the two great circles going through Copenhagen. If you have trouble with this, try drawing a picture.

This example shows clearly that it is κ_n, not κ_g, that tells us about the curvature of the surface. You can further convince yourself by picturing other examples, such as the ridge you were walking on earlier. While there is an infinite number of surfaces with different

---

* With this convention (rather than pointing downward toward the center of the earth, as would be mathematically more natural), $\kappa_g > 0$ while $\kappa_n < 0$.

global properties, if we focus on a local patch of the surface around a point P, there are only a couple prototypical surfaces, sphere-like or ridge-like, so to speak.

From curvature of curves to curvature of surface

Using (13) and (7), we have (suppressing terms not of interest to us)

$$\frac{d\vec{t}}{dl} = \cdots + t^\mu \frac{d\vec{e}_\mu}{dl} = \cdots + t^\mu \frac{\partial \vec{e}_\mu}{\partial x^\nu} t^\nu = \cdots + t^\mu (K_{\mu\nu} \vec{n}) t^\nu \tag{15}$$

Comparing with (14), we can extract

$$\kappa_n = K_{\mu\nu} t^\mu t^\nu \tag{16}$$

A digressive word about what appears to physicists as rather quaint terminology. Strip the dl off $t^\mu = \frac{dx^\mu}{dl}$ in κ_n and we encounter the combination $K_{\mu\nu} dx^\mu dx^\nu$, known to classical differential geometers as the second fundamental form.[7] Sounds pretty important. Then what do these guys call the first fundamental form? None other than our beloved infinitesimal distance squared: $g_{\mu\nu} dx^\mu dx^\nu$.

Next, differentiate $\vec{t} \cdot \vec{n} = 0$ to obtain (using (14))

$$\vec{t} \cdot \frac{d\vec{n}}{dl} = -\vec{n} \cdot \frac{d\vec{t}}{dl} = -K_{\mu\nu} t^\mu t^\nu = -\kappa_n \tag{17}$$

known as Weingarten's equation after Julius Weingarten (1836–1910).

Gauss's idea was to look at all curves passing through the point P. For each curve, calculate κ_n. Following the great man, we are supposed to find the extremal values of κ_n. So, let us extremize $K_{\mu\nu} t^\mu t^\nu$. But t^μ is not entirely free to vary: it has to satisfy $g_{\mu\nu} t^\mu t^\nu = 1$. Thus, we have a constrained extremization problem.

But we know how to deal with this: introduce a Lagrange multiplier[8] k. In other words, vary $K_{\mu\nu} t^\mu t^\nu - k(g_{\mu\nu} t^\mu t^\nu - 1)$ with respect to t^μ and set the result to 0, thus obtaining $(K_{\mu\nu} - k g_{\mu\nu}) t^\nu = 0$. Multiplying by $g^{\lambda\mu}$, we obtain an equation

$$(g^{\lambda\mu} K_{\mu\nu} - k\delta^\lambda_\nu) t^\nu = 0 \tag{18}$$

for the eigenvalues of the 2-by-2 matrix $M^\lambda_\mu \equiv g^{\lambda\mu} K_{\mu\nu}$. Contracting (18) with $t_\lambda \equiv g_{\lambda\rho} t^\rho$, we find $k = K_{\mu\nu} t^\mu t^\nu = \kappa_n$. In other words, the two eigenvalues k_1 and k_2 give the extremal values of κ_n.

As usual, the eigenvalues are determined by

$$\det(g^{-1} K - kI) = 0 \tag{19}$$

which (since $g^{-1} K$ is a 2-by-2 matrix) amounts to the quadratic equation $k^2 - \mathrm{tr}(g^{-1} K)k + \det(g^{-1} K) = 0$. Thus, the product of the eigenvalues is given by

$$\mathcal{G} \equiv \det(g^{-1} K) = \frac{\det K}{\det g} \tag{20}$$

and the sum by

$$\mathcal{S} \equiv \mathrm{tr}(g^{-1} K) \tag{21}$$

Intrinsic versus extrinsic again

In story 2 in the prologue, and also in the preceding chapter, you learned to distinguish between extrinsic and intrinsic curvatures. A cylinder has extrinsic curvature by virtue of how it is embedded in the ambient Euclidean space, but it has no intrinsic curvature. We can unroll a cylinder into a flat piece of paper. Like the mite geometers of the prologue, we cannot get out of the universe we live in, and so we are interested in the intrinsic, not the extrinsic, curvature, as was mentioned in the preceding chapter.

The quantities $\mathcal{G}$ and $\mathcal{S}$ present us with two measures of curvature. How do we know which one is intrinsic? The easy way is to look at the cylinder, just as we did in the preceding chapter. It also gives us a chance to try out the machinery we just worked out. Recall from earlier in this chapter that $\vec{e}_1 = \begin{pmatrix} -a\sin\varphi \\ a\cos\varphi \\ 0 \end{pmatrix}$, $\vec{e}_2 = \begin{pmatrix} 0 \\ 0 \\ 1 \end{pmatrix}$, and $\vec{n} = \begin{pmatrix} \cos\varphi \\ \sin\varphi \\ 0 \end{pmatrix}$. First, using (5), we already worked out the metric $g_{11} = a^2$, $g_{12} = g_{21} = 0$, $g_{22} = 1$. Next, we take derivatives: $\partial_1\vec{e}_1 = -a\vec{n}$, $\partial_1\vec{e}_2 = \partial_2\vec{e}_1 = \partial_2\vec{e}_2 = 0$. Plugging these values into Gauss's equation (8) $K_{\mu\nu} = \vec{e}_{\mu,\nu} \cdot \vec{n}$, we obtain $K_{11} = -a$, with all other entries vanishing. Thus, $\det K = 0$, so that $\mathcal{G} = 0$. This immediately tells us (and told Gauss) that it is the product $\mathcal{G}$ that measures the intrinsic curvature of the surface at the point P. The sum $\mathcal{S}$ measures the extrinsic curvature. To be consistent with the previous chapter, we define the intrinsic curvature to be $\mathcal{G}$ and the extrinsic curvature to be $\mathcal{E} \equiv (\frac{1}{2}\mathcal{S})^2$.

For the surface we started the last section with (and on which you have been hiking), we have $\vec{X} = \begin{pmatrix} x \\ y \\ \frac{1}{2}px^2 - \frac{1}{2}qy^2 \end{pmatrix}$ with $x^1 = x$ and $x^2 = y$. It follows that $\vec{e}_1 = \begin{pmatrix} 1 \\ 0 \\ px \end{pmatrix}$ and $\vec{e}_2 = \begin{pmatrix} 0 \\ 1 \\ -qy \end{pmatrix}$. Note that to calculate the curvature at the point P, we need various quantities only at P. At our chosen point P, where $x = y = 0$, the arithmetic simplifies. We have $\vec{n} = \vec{e}_1 \times \vec{e}_2 = \begin{pmatrix} 0 \\ 0 \\ 1 \end{pmatrix}$, $ds^2 = \vec{e}_1 \cdot \vec{e}_1 dx^2 + \vec{e}_2 \cdot \vec{e}_2 dy^2 = dx^2 + dy^2$, so that $g_{11} = 1$, $g_{22} = 1$, and $g_{12} = g_{21} = 0$. Furthermore, $\partial_1\vec{e}_1 = p\vec{n}$, $\partial_2\vec{e}_2 = -q\vec{n}$, and $\partial_1\vec{e}_2 = \partial_2\vec{e}_1 = 0$, so that from $K_{\mu\nu} = \vec{e}_{\mu,\nu} \cdot \vec{n}$, we obtain $K_{11} = p$, $K_{22} = -q$, and $K_{12} = K_{21} = 0$. Thus, $\mathcal{G} = -pq$ and $\mathcal{S} = p - q$.

For a sphere of radius a, $p = -q = a$, and so $\mathcal{G} = a^2 = \mathcal{E}$. We have calculated these quantities only at $(x, y) = (0, 0)$, but of course for the sphere, this amounts to knowing the intrinsic and extrinsic curvatures everywhere.

If $\mathcal{G} > 0$, then the two eigenvalues k_1 and k_2 have the same[9] sign, and the surface at P is spherical. If $\mathcal{G} < 0$, the two eigenvalues have opposite signs. The surface is shaped like a saddle (or potato chip) and is hyperbolic at P. If $\mathcal{G} = 0$, the surface is cylindrical at P.

The whole point of Riemann's formalism (as will be developed in detail in part VI) is that we do not need to embed the space we are studying in some ambient Euclidean space. But we have lots of intuition about 3-dimensional Euclidean space (and, as I said before, who can blame us?), and so this 18th and 19th century differential geometry is a lot easier to visualize and to grasp.

With the benefit of hindsight, Gauss's strategy is also perfectly reasonable. To find the curvature at a point P on the surface, you build race tracks going through P. Find the race tracks with the largest and smallest possible curvature. The product and sum of these two curvatures measure the curvature, intrinsic and extrinsic, respectively.

Appendix: Spherical coordinates

Throughout this text, we will be often changing to spherical coordinates. You must have done this about a hundred times in courses on mechanics and on electromagnetism. So it might be good to collect some useful information here and to define $\omega^1 \equiv \sin\theta \cos\varphi$, $\omega^2 \equiv \sin\theta \sin\varphi$, and $\omega^3 \equiv \cos\theta$. Denote* the coordinates in E^3 by (v^1, v^2, v^3). Transform to spherical coordinates (r, θ, φ) by $v^i = f(r)\omega^i$, $i = 1, 2, 3$. Differentiating $\sum_i (\omega^i)^2 = 1$, we have $\sum_i \omega^i d\omega^i = 0$, where $d\omega^i = \partial_\theta \omega^i d\theta + \partial_\varphi \omega^i d\varphi$. Furthermore, $\sum_i \partial_\theta \omega^i \partial_\varphi \omega^i = 0$, $\sum_i (\partial_\theta \omega^i)^2 = 1$, and $\sum_i (\partial_\varphi \omega^i)^2 = \sin^2\theta$, so that $(d\omega^i)^2 = d\theta^2 + \sin^2\theta d\varphi^2$.

Thus, the metric is given by $ds^2 = \sum_i (dv^i)^2 = \sum_i (f'(r)dr\omega^i + f(r)d\omega^i)^2 = f'(r)^2 dr^2 + f(r)^2 \sum_i (d\omega^i)^2 = f'(r)^2 dr^2 + f(r)^2 (d\theta^2 + \sin^2\theta d\varphi^2)$. For example, to derive (X.1.20) and (X.1.21), we will need to use a slightly generalized version of this result.

Exercises

1   Calculate the curvature κ and the torsion τ of the exponential spiral $\vec{X} = \begin{pmatrix} f(\varphi)\cos\varphi \\ f(\varphi)\sin\varphi \\ g(\varphi) \end{pmatrix}$. Take $f(\varphi) = e^{a\varphi}$ and $g(\varphi) = 0$ for simplicity.

2   Find an expression for $\vec{V}_P(y)$.

3   Show that if $k_1 \neq k_2$, the two corresponding eigenvectors t_1 and t_2 are orthogonal in accordance with our geometrical intuition.

4   On a cylinder, draw the curve defined by $X^1 = a \cos l$, $X^2 = a \sin l$, and $X^3 = bl$, with l the length. Show that $a^2 + b^2 = 1$. Calculate the curvature and the torsion.

5   Calculate $\mathcal{G}$ for a unit sphere. (Anticipating a bit, we will see that $\mathcal{G}$ is equal to the Riemann curvature R_{1212}.)

6   Show that if the mite professors of geometry are given the metric, they can determine the Christoffel symbol, even though they don't know about $\vec{n}$.

Notes

1. E. Kreyszig, *Differential Geometry*, University of Toronto Press, 1959.
2. Several years ago, I gave some chapters from part I to a few University of California, Santa Barbara, undergrads to read. One of them emailed me a few days later. I cannot resist quoting, with insignificant edits, from his email.

   I'm very excited about your approach to introducing differential geometry first in E^3 before doing GR for multiple reasons. Conversing with another undergraduate physics major who took the undergraduate GR course here, we agreed that general relativity books and courses ought to spend more time describing intuitive and easily visualizable examples, and laying down as rigorously as possible (given

---

* This is to avoid confusion with the $\vec{X} = (X^1, X^2, X^3)$ in the text.

the constraints of GR courses needing to be about physics, not math) the meaning and definitions of curvature, geodesics, etc, at least partially through examples in E^2 and E^3. I found it spooky that a day after having this conversation, I read these chapters and found they did basically what I had hoped an introductory book to general relativity would do. I found this chapter personally more useful than the other chapters because the way I learned differential geometry was the more abstract approach.

Spooky indeed!

3. Indeed, you already had a first taste in the preceding chapter.

4. In an alternative view, which I do not like as much, we can think of the covariant derivative as acting only on objects carrying the vectorial arrow. Insisting that D_ν is distributive just like ∂_ν, some authors write $D_\nu \vec{W} = D_\nu(W^\mu \vec{e}_\mu) = (\partial_\nu W^\mu)\vec{e}_\mu + W^\mu(D_\nu \vec{e}_\mu)$. Note that in this formulation, D_ν acting on the "numbers" W^μ is given by the ordinary derivative by definition. Comparing with (12) then gives $D_\nu \vec{e}_\mu = \Gamma^\lambda_{\nu\mu}\vec{e}_\lambda$. Referring back to (7), we see that $D_\nu \vec{e}_\mu$ differs from $\partial_\nu \vec{e}_\mu$ by a term proportional to $\vec{n}$, as we might expect.

5. Here is a hint. Gears convert the rotation of the left wheel and the right wheel separately into rotations around the vertical axis. Another differential gear responds to the difference in the output from the two wheels and rotates the statue around the vertical axis. If the chariot is moving in a straight line, the statue is not rotated. But when the chariot moves along a curve, the right wheel (say) rotates more than the left, and this difference gets converted into a rotation of the statue around the vertical axis, compensating for the turning of the body of the chariot. In other words, as the chariot turns this way and that, the statue points in a fixed direction. To the extent that Riemannian surfaces are locally flat, the war chariot also more or less works on a surface that is not flat, provided that the chariot is much smaller than the radius of curvature, and after subtracting out the (negligible) vertical component of the vector represented by the Emperor's arm.

6. Need I remind you to distinguish between these two uses of the word "curvature"?

7. The word "form" here does not carry the same meaning as the word "form" in chapter IX.7.

8. For those readers who have forgotten the notion of a Lagrange multiplier from their course on calculus, here is a quick review. The problem is to extremize a function $f(x, y)$ with the constraint $g(x, y) = 0$. The brute force approach would be to solve $g(x, y) = 0$ to obtain $y(x)$, eliminate y in $f(x, y)$, and extremize $f(x, y(x))$. The same Lagrange you were introduced to in the text invented the following more symmetrical, and often better, method. Form the function $h(x, y) = f(x, y) + \lambda g(x, y)$, where λ is known as the Lagrange multiplier. Extremize $h(x, y)$ to determine x and y in terms of λ, and then impose the constraint $g(x, y) = 0$. An example: $f(x, y) = ax + by$ and $g(x, y) = \frac{1}{2}(x^2 + y^2 - 1)$. In other words, we are to extremize $f(x, y) = ax + by$, with x and y constrained to the unit circle. Following Lagrange, we obtain after the first step $x = -a/\lambda$, $y = -b/\lambda$. Imposing $x^2 + y^2 = 1$, we have $\lambda = \pm\sqrt{a^2 + b^2}$. For $a > 0$, $b > 0$, the plus root gives the maximum of $f(x, y)$ at $x = a/\sqrt{a^2 + b^2}$, $y = b/\sqrt{a^2 + b^2}$.

9. We are dealing with Euclidean surfaces here, so that $\det g > 0$.

Recap to Part I

An essential feature of Newton's force law is that it involves two derivatives. The presence of two derivatives will permeate almost everything we do.

For a given situation, a judicious choice of coordinates makes our lives a lot easier. Coordinates can of course be freely chosen, but the square of the separation between two neighboring points, which according to Pythagoras has the form $ds^2 = g_{\mu\nu}(x)dx^\mu dx^\nu$, must not depend on the coordinate choice. Once we master the changing of coordinates, it is but a short hop over to curved spaces.

Choosing coordinates wisely, we can always make our immediate neighborhood look flat. The extent to which we notice deviation from local flatness as we move away from our neighborhood is the measure of curvature. A simple counting argument tells us how many numbers we need to characterize the intrinsic curvature at a given point. To get a feel for curvature, it is good to spend some time back in the good old days with the likes of Gauss and play with some surfaces we can literally hold in our hands.

Part II | Action, Symmetry, and Conservation

II.1   The Hanging String and Variational Calculus

The hanging string

To understand the action principle, which to a large extent will permeate this book, we have to master a slight generalization of ordinary calculus known as variational calculus. In ordinary calculus, we take derivatives with respect to some variable, typically a real number. In variational calculus, we take derivatives with respect to a function. To learn what that means and to see how variational calculus arises in physics, let us start with a simple problem.

Throw a marble into a bowl. When you come back later, you expect it to be sitting at rest at the bottom of the bowl. This is formalized by saying that, if we denote the cross section of the bowl by $v(x, y)$ so that the potential energy of the marble is proportional to $v(x, y)$, the position of the marble is found by solving $\frac{\partial v}{\partial x} = 0$ and $\frac{\partial v}{\partial y} = 0$.

To explain variational calculus, let us tackle a problem in baby string theory. Consider an ideal elastic string tied down at two ends and hanging under the force of gravity. See figure 1 for how we set up our coordinates. Denote by $\phi(x)$ the amount by which the bit of string at x hangs below the horizontal line. That the string is tied down at the two ends gives us the boundary conditions $\phi(L/2) = \phi(-L/2) = 0$. We want to solve for the shape of the hanging string, which is of course determined by the tug of war between the downward pull of gravity and the elastic force trying to minimize the amount of stretch in the string.

The elastic energy is given by a constant T intrinsic to the string times the stretch, defined as the length of the string minus the original length, which, as you could verify later, does not come into our calculation, so that we might as well take it to be L. To find the length of the string, think of how Newton and Leibniz discovered calculus and imagine dividing the string up into little segments labeled by j. Pythagoras tells us that each segment has length given by (see figure 1) $\sqrt{\Delta x_j^2 + \Delta \phi_j^2} = \Delta x_j \sqrt{1 + \left(\frac{\Delta \phi_j}{\Delta x_j}\right)^2}$. Taking the Newton-Leibniz limit, we see that the length of the string is equal to $\int_0^L \sqrt{dx^2 + d\phi^2} = \int_0^L dx \sqrt{1 + \left(\frac{d\phi}{dx}\right)^2}$. (Henceforth, we will carry out this sort of manipulation without further

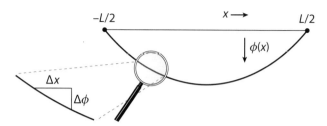

Figure 1 An elastic string tied down at two ends hangs under the force of gravity.

ado. We will also tend to suppress the integration limits.) Thus, the elastic energy is equal to $T(\int dx \sqrt{1 + \left(\frac{d\phi}{dx}\right)^2} - L) = T \int dx (\sqrt{1 + \left(\frac{d\phi}{dx}\right)^2} - 1)$.

For pedagogical clarity, we consider the case where $\frac{d\phi}{dx} \ll 1$, that is, when the string is stretched only by a little bit, so that $\sqrt{1 + \left(\frac{d\phi}{dx}\right)^2} \simeq 1 + \frac{1}{2}\left(\frac{d\phi}{dx}\right)^2$. (It turns out that this simplification is not necessary, and you can work things out without it. See exercise 1.)

The gravitational energy is given by $\int dx (-\sigma g \phi(x))$, where σ denotes the mass per unit length of our string. Note the minus sign: we have chosen $\phi(x)$ to point downward, as indicated in the figure. Again, we have assumed that the stretch is small.

Thus the string has energy

$$E(\phi) = \int_0^L dx \left(\frac{T}{2}\left(\frac{d\phi}{dx}\right)^2 - \sigma g \phi(x)\right) \tag{1}$$

Extremizing a functional

Note that E is not a function of a real variable named ϕ. Rather, E is known as a functional of a real valued function $\phi(x)$. When you plug a number x into a function f, you get out a number $f(x)$. Analogously, when you plug a function $\phi(x)$ into the functional E, you get out a number, the string energy $E(\phi)$. For total clarity, we could write $E(\phi(\cdot))$, with the dot emphasizing that ϕ is itself a function. Note that we should not write $E(\phi(x))$ in (1): x is a dummy integration variable on the right hand side.

Our task is to find the specific function $\phi(x)$ that minimizes the energy E of the string. In ordinary calculus, we differentiate a function with respect to its argument and then set the result to zero to find the extremum of the function. Analogously, in variational calculus, we differentiate a functional with respect to its argument, which is a function, and then set the result to zero to find the extremum of the functional.

But how do we differentiate with respect to a function?

For pedagogical clarity, let us go back to the marble in the bowl and write $\delta v = \frac{\partial v}{\partial x} \delta x + \frac{\partial v}{\partial y} \delta y$ to first order in δx and δy, some arbitrary and small variations in the position of the marble. We notice that δv vanishes if and only if $\frac{\partial v}{\partial x} = 0$ and $\frac{\partial v}{\partial y} = 0$. To first order in δx and δy, the variation of v vanishes if we happen to be sitting at an extremum. To second order, the variation of v is positive if we are at a minimum. In other words, we nudge the marble and if it costs us energy, we know that it is sitting at the bottom of the bowl. All of this is elementary and well understood by you. So now we do the same: we push the string slightly and ask if it costs us energy.

We vary the function $\phi(x)$ by changing it to $\phi(x) + \eta(x)$. We then compare $E(\phi + \eta)$ and $E(\phi)$. Take $\eta(x)$ small, so that it suffices to calculate the change in energy $\delta E = E(\phi + \eta) - E(\phi)$, expanding in η. To first order in η, δE should vanish. If the shape $\phi(x)$ minimizes the energy, δE would furthermore be positive to second order.

Let us go slow and first deal with the second term in (1): taking out the overall constant $-\sigma g$, we have $\int dx\{(\phi(x) + \eta(x)) - \phi(x)\} = \int dx\, \eta(x)$. That was easy!

The first term in (1) is only slightly more difficult to deal with. Since $\frac{d}{dx}(\phi + \eta) = \frac{d\phi}{dx} + \frac{d\eta}{dx}$, we find

$$\int dx \frac{1}{2} \left(\left(\frac{d\phi}{dx} + \frac{d\eta}{dx} \right)^2 - \left(\frac{d\phi}{dx} \right)^2 \right) \simeq \int dx \left(\frac{d\phi}{dx} \frac{d\eta}{dx} \right) = \left(\frac{d\phi}{dx} \eta \right) \Big|_{x=-L/2}^{x=L/2} + \int dx \left(-\frac{d^2\phi}{dx^2} \eta \right) \quad (2)$$

In the last step, we integrated by parts. Note that our boundary conditions tying down the string at the two ends, $\eta(x = L/2) = 0 = \eta(x = -L/2)$, imply that the boundary terms in (2) vanish.

Putting it together, we have

$$\delta E = \int dx \left(-T \frac{d^2\phi}{dx^2} - \sigma g \right) \eta(x) \quad (3)$$

Since $\eta(x)$ is arbitrary, δE can vanish only if the integrand in (3) vanishes. Thus, the shape of the hanging string is determined by the differential equation

$$T \frac{d^2\phi}{dx^2} = -\sigma g \quad (4)$$

namely, a graceful parabola described by $\phi(x) = \frac{\sigma g}{2T}\{(\frac{L}{2})^2 - x^2\}$. At the two ends, $\phi(\frac{L}{2}) = \phi(-\frac{L}{2}) = 0$, which just expresses the boundary conditions. In the middle, $\phi(0) = \frac{\sigma g}{2T}(\frac{L}{2})^2$. (Remember our convention that $\phi > 0$ means hanging down.)

The physics is simple, and the math merely describes the physics. In (1), the first term wants to make $(\frac{d\phi}{dx})^2$ small, that is to make $\phi(x)$ constant, and hence $\phi(x) = 0$ with the given boundary conditions. In contrast, the second term wants to make $\phi(x)$ as large and as positive as possible. The actual shape is a compromise between these two terms. This theme of compromise will pervade this book.

General lessons

We are less interested in the hanging string than in what general lessons we can learn from this simple example. Here are some remarks.

1. Evidently $\delta \int dx \phi^n = \int dx\{(\phi + \eta)^n - \phi^n)\} \sim \int dx\, n\phi^{n-1}\eta$, where for the sake of notational simplicity, we have suppressed the x dependence of $\phi + \eta$. Since any functional of ϕ can be expanded as a power series, we have, for instance, $\delta \int dx \cos \phi = \int dx(\cos(\phi + \eta) -$

$\cos \phi) \sim - \int dx (\sin \phi) \eta$. In general, $\delta \int dx V(\phi) = \int dx (V(\phi + \eta) - V(\phi)) \sim \int dx V'(\phi)\eta$, where $V'(\phi) = \frac{dV}{d\phi}$ is computed by pretending that $V(\phi)$ is a function of a real variable ϕ, ignoring the fact that ϕ is itself a function.

Now that the utility of η has come to an end, we might as well, in analogy with ordinary calculus, define the functional derivative by

$$\frac{\delta}{\delta\phi(y)} \int dx V(\phi(x)) = V'(\phi(y)) \tag{5}$$

Note that to be careful, we have restored the argument of the function ϕ. It is important to realize that here x is a dummy variable to be integrated over, but y is a "free" variable: we are free to vary the function ϕ at a point y of our choice. To our satisfaction, we see that the variation depends only on quantities evaluated at the point y. This states that the energy density is a local quantity and does not depend on what is happening far away from y.

2. Suppose we now have to vary $\int dx F(\frac{d\phi}{dx})$. By the same reasoning, we have

$$\delta \int dx F \left(\frac{d\phi}{dx} \right) = \int dx \left(F \left(\frac{d\phi}{dx} + \frac{d\eta}{dx} \right) - F \left(\frac{d\phi}{dx} \right) \right) \sim \int dx F' \left(\frac{d\phi}{dx} \right) \frac{d\eta}{dx}$$

$$= \int dx \left(\frac{d}{dx} F' \left(\frac{d\phi}{dx} \right) \right) \eta$$

where, as before, we have integrated by parts and dropped the boundary terms since $\eta(x)$ vanishes at the boundaries. Once again, F' is defined as if the argument of F were an ordinary real number. In our simple example, $F(u) = \frac{1}{2}u^2$ and so $F'(u) = u$. See exercise 1 for another example. Again, we can drop η and write $\frac{\delta}{\delta\phi(y)} \int dx F(\frac{d\phi}{dx}) = -\frac{d}{dy} F'(\frac{d\phi}{dy})$.

Thus, in general, if the energy functional is given by

$$E(\phi) = \int dx \left\{ F \left(\frac{d\phi}{dx} \right) + V(\phi) \right\} \tag{6}$$

with the boundary condition that $\phi(x)$ vanishes at the integration endpoints, we obtain the equation

$$\frac{d}{dx} F' \left(\frac{d\phi}{dx} \right) - V'(\phi(x)) = 0 \tag{7}$$

Even more generally, if the energy functional is given by

$$E(\phi) = \int dx \mathcal{E} \left(\frac{d\phi}{dx}, \phi \right) \tag{8}$$

again with the standard boundary condition that $\phi(x)$ vanishes at the integration endpoints, we obtain

$$\frac{d}{dx} \left(\frac{\delta\mathcal{E}}{\delta \frac{d\phi}{dx}} \right) - \frac{\delta\mathcal{E}}{\delta\phi} = 0 \tag{9}$$

Verify this: you will grasp the notation better. As in (5), we now pretend that $\mathcal{E}(a, b)$ is an ordinary function of two variables a and b. By $\frac{\delta\mathcal{E}}{\delta \frac{d\phi}{dx}}$ we mean $\frac{\partial\mathcal{E}(a,b)}{\partial a}$ with a subsequently set equal to $\frac{d\phi}{dx}$ and b to $\phi(x)$. I will leave it to the reader to figure out what $\frac{\delta\mathcal{E}}{\delta\phi}$ means.

The equation (9), known as the Euler-Lagrange equation, is of fundamental importance in theoretical physics.

3. Bad notation alert! The present discussion highlights the notational confusion I mentioned in chapter I.1 that bedevils some students. For the marble in the bowl problem, it would have been best to reserve x and y for spatial coordinates and to denote the position of the marble by q_1 and q_2, say. In the mechanics of point particles, it is standard to abuse notation and use $x, y, \ldots$ for both spatial coordinates and for the positions of particles. Generally there is no confusion.

But when we go to continuum mechanics, such as our hanging string problem, we must distinguish between dynamical variables (in our example, just the function $\phi(x)$) and spatial coordinates (in our example, the single coordinate x). Here x serves as a label to tell us which infinitesimal segment of the string we are talking about. The displacement of that particular string segment from where it would have been were gravity turned off is given by $\phi(x)$. (This is a mouthful for saying that $\phi(x)$ denotes the amount by which the string is hanging down at the point x, but I want to be precise and academic here.) Note that the 2-dimensional position of that particular string segment is given by $(x, \phi(x))$. Thus, in some sense, the letter x, seen this way, is doing double duty, both as a label and as a position. When we introduce time in a later chapter, it will become clear that x has no dynamics but $\phi(x)$ does. Some books use the horrendous notation $y(x)$ instead of $\phi(x)$, which has caused endless confusion. I belabor these rather obvious points because I have seen too many students getting confused, particularly when they encounter field theory, classical or quantum.

4. Another bad notation alert! In most textbooks, the variation of $\phi(x)$ is written as $\delta\phi(x)$. This confuses some students, because they think of δ as some operation acting on $\phi(x)$ and quite legitimately worry whether the two operations δ and $\frac{d}{dx}$ commute. Instead, I write $\eta(x)$ for $\delta\phi(x)$. In particular, a manipulation analogous to the first step in (2), $\frac{d}{dx}(\phi + \eta) - \frac{d\phi}{dx} = \frac{d\phi}{dx} + \frac{d\eta}{dx} - \frac{d\phi}{dx} = \frac{d\eta}{dx}$, proves that $\delta\frac{d\phi}{dx}$ really is equal to $\frac{d\delta\phi}{dx}$. Now that I have clarified this point, I will mostly lapse into the more explanatory notation $\delta\phi(x)$, which also avoids introducing yet another symbol.

5. The careful reader might worry that we have found only the extremum, rather than the minimum, of $E(\phi)$. In most problems, that the solution is a minimum of the energy should be physically clear, as is the case here. To show that our solution is indeed a minimum of $E(\phi)$, we would have to expand to second order in $\delta\phi(x) = \eta(x)$. This is especially easy in our example here, because $V(\phi)$ is linear in ϕ, so that we can read off from (2) that the second order variation of $E(\phi)$ is given by the manifestly positive quantity $\int dx (\frac{d\eta}{dx})^2$.

6. With some practice, you will be able to do variational calculus without having to go through the steps we went through here. When you go on to study quantum field theory, you will encounter these so-called functional derivatives all over the place.

7. Instead of gravity pulling down on the string uniformly, we could load the string unevenly. Indeed, for convenience, let's define $E(\phi)$ with an overall factor of T taken out in (1) and replace the constant $\sigma g / T$ by a specified function $\rho(x)$, so that

$$E(\phi) = \int dx \left(\frac{1}{2} \left(\frac{d\phi}{dx} \right)^2 - \rho(x)\phi(x) \right)$$

From string to membrane

Several possible generalizations immediately suggest themselves: we could increase the number of spatial coordinates, or we could increase the number of functions, or both. We consider the first possibility here and defer the second possibility to the next chapter.

Go from a hanging string to a hanging membrane (or "brane" for short). See figure 2. We generalize the energy functional (1) as amended above to

$$E(\phi) = \int dx dy \left(\frac{1}{2}\left[\left(\frac{\partial \phi}{\partial x} \right)^2 + \left(\frac{\partial \phi}{\partial y} \right)^2 \right] - \rho(x,y)\phi(x,y) \right)$$

with the boundary condition that $\phi(x,y)$ vanishes along some nice closed curve (such as a circle) in the $(x$-$y)$ plane. Note that $E(\phi)$ involves an integral over the two spatial coordinates (x,y).

While you can easily derive this expression for the energy by working out how much the membrane is stretched, we can simply use rotational invariance to fix the form of the integrand: the energy should not change when we rotate (x,y). Recall from exercise I.4.1 that $\left(\frac{\partial \phi}{\partial x} \right)^2 + \left(\frac{\partial \phi}{\partial y} \right)^2$ transforms like a scalar and the accompanying discussion in chapter I.4.

Again the physics behind the various terms is clear. To minimize the first two terms, we want $\left(\frac{\partial \phi}{\partial x} \right)^2$ and $\left(\frac{\partial \phi}{\partial y} \right)^2$ to be as small as possible, that is, to stretch the membrane as little as possible. In contrast, to minimize the third term in $E(\phi)$, we want ϕ to be positive (we are taking the load $\rho(x,y)$ to be positive) and as large as possible to lower the energy. Just as in the hanging string, it is the struggle between these two terms that determines the shape of the membrane. Note that once again we choose to have ϕ point downward; hence the minus sign in the potential term.

Varying E and going through the same steps as before, we generalize (2) trivially to

$$\int dx dy \frac{1}{2}\left(\left(\frac{\partial \phi}{\partial x} + \frac{\partial \eta}{\partial x} \right)^2 + \left(\frac{\partial \phi}{\partial y} + \frac{\partial \eta}{\partial y} \right)^2 - \left(\frac{\partial \phi}{\partial x} \right)^2 - \left(\frac{\partial \phi}{\partial y} \right)^2 \right) = -\int dx dy \left(\frac{\partial^2 \phi}{\partial x^2} + \frac{\partial^2 \phi}{\partial y^2} \right)\eta$$

Thus, we obtain Poisson's equation[1] $\nabla^2 \phi(x,y) = -\rho(x,y)$, with the Laplacian $\nabla^2 \equiv \frac{\partial^2}{\partial x^2} + \frac{\partial^2}{\partial y^2}$.

Figure 2 A hanging membrane.

Newton's gravitational potential as a field

A deep strand . . . was his total love of the idea of a field . . . which made him know that there had to be a field theory of gravitation, long before the clues to that theory were securely in his hand.

—Freeman Dyson speaking of Einstein

I went through this membrane example for a reason. Newton's gravitational potential Φ satisfies

$$\nabla^2 \Phi(x, y, z) = 4\pi G \rho(x, y, z) \tag{10}$$

with $\nabla^2 = \frac{\partial^2}{\partial x^2} + \frac{\partial^2}{\partial y^2} + \frac{\partial^2}{\partial z^2}$ and $\rho(x, y, z)$ the mass distribution. (If you are not aware of this, we will show explicitly, for ρ describing a point mass, that this equation gives for Φ the Newtonian gravitational potential.) What we just did for the membrane tells us that (10) for Newtonian gravity emerges from minimizing the energy functional

$$E(\Phi) = \int d^3x \left(\frac{1}{8\pi G} (\vec{\nabla}\Phi)^2 + \rho(\vec{x})\Phi(\vec{x}) \right) \tag{11}$$

Note the plus sign in (11), which reflects how the gravitational potential is defined so that a mass m located at $\vec{x}$ in the potential has energy $+m\Phi(\vec{x})$.

By the way, $E(\Phi)$ defines a classical field theory, and $\Phi(\vec{x})$ is known as a field,* as it pervades space, just like the familiar electromagnetic field. The hanging string and membrane allow us not only to introduce the variational principle, but importantly, also the notion of a field.

To verify that (11) leads to (10), we could simply invoke the membrane example or use the Euler-Lagrange equation (9). Alternatively, it is easy enough to vary (11) directly, going through the same steps as earlier in this chapter:

$$\delta E(\Phi) = \int d^3x \left(\frac{1}{4\pi G} (\vec{\nabla}\Phi \cdot \vec{\nabla}\delta\Phi) + \rho(\vec{x})\delta\Phi(\vec{x}) \right) = \int d^3x \left\{ \frac{1}{4\pi G} (-\nabla^2\Phi) + \rho(\vec{x}) \right\} \delta\Phi(\vec{x})$$

where we have integrated by parts. Setting the coefficient of $\delta\Phi$ to zero yields (10).

It suffices to solve (10) for a point mass at the origin,[†] that is, for $\rho = M\delta^3(\vec{x}) \equiv M\delta(x)\delta(y)\delta(z)$. Here we generalize the Dirac delta function defined in chapter I.1 to a 3-dimensional delta function, as indicated and discussed in exercise I.3.2. Recall that you can think of the delta function $\delta^3(x)$ as essentially a function sharply peaked at $x = 0$. Thus, $\rho(\vec{x})$ is sharply spiked at the origin $\vec{x} = 0$ and vanishes everywhere else. The total mass $\int d^3x \rho(x, y, z) = M(\int dx\delta(x))(\int dy\delta(y))(\int dz\delta(z))$ is equal to M.

---

* But, at this point, merely a static field without any dynamics, that is, without any dependence on time.
† For an arbitrary mass distribution, we can imagine $\rho(x, y, z)$ as being composed of point masses and add the contribution from each mass to Φ.

Dimensional analysis* determines the solution of (10) up to an overall constant: since ∇^2 goes like $1/L^2$ and $\delta^3(\vec{x})$ like $1/L^3$, the potential can only go like $1/r$. As you know, it is in fact $\Phi = -GM/r$. To verify the overall constant, we integrate the two sides of (10) over a ball of radius R centered at the origin, obtaining for the left hand side $\int d^3x \nabla^2 \Phi = \int d\vec{S} \cdot \vec{\nabla}\Phi = 4\pi R^2(GM/R^2)$, and for the right hand side $\int d^3x \, 4\pi GM\delta^3(\vec{x}) = 4\pi GM$. In other words, we have derived the useful identity

$$\nabla^2 \left(-\frac{1}{4\pi r} \right) = \delta^3(\vec{x}) \tag{12}$$

When you first learned about the inverse square law, did you not wonder where the inverse square comes from? This discussion shows that it is essentially a consequence of the form of the $(\vec{\nabla}\Phi)^2$ term in $E(\Phi)$, required by rotational invariance, which in turn leads to the Laplacian in (10). The inverse square then follows by dimensional analysis. You realize of course that the electrostatic potential satisfies the same equation (10) if we interpret the right hand side $G\rho$ as the charge density. This is not an accident, but is due to the same deep consequence of rotational invariance.

Anticipating, you will see that powerful invariance requirements also fix the form of Einstein gravity.

Appendix 1: The lion by his paw prints

The brachistochrone problem makes for a great physics story. One winter day in 1697, when Newton (1642–1727) was $97 - 42 = 55$ (not old by modern standards but old in the age he lived in and in any case, long past the creative brilliance of his youth), he received a letter from Johann Bernoulli (1677–1748) posing the following problem. Fashion a stiff wire into a curve connecting two points A and B, as shown in figure 3. Thread the wire through a bead of mass m, as shown. Gravity is acting downward as usual. Release the bead at rest at the higher point, say A, and let it slide down the wire. What should the shape of the wire be if the bead is to reach B in the least amount of time? (In Greek, brachistos means "shortest" and chronos means "time.")

Galileo had erroneously thought that the curve was a circular arc, but he can be excused because, unlike you, he did not know variational calculus.

Newton recognized this as a brazen attempt by one of the best minds on the continent to embarrass him. In the bitter and contentious controversy between Newton and Gottfried Wilhelm Leibniz (1646–1716) over who invented calculus, Bernoulli had sided with Leibniz. By this time, Newton was a high-level government official in charge of the Royal Mint. England was in the middle of issuing new coins in an effort to combat widespread counterfeiting. Newton's job was demanding and included catching and executing counterfeiters. The old man had had a long day at his day job, but he took up the gauntlet, working feverishly into the night, surprising his dedicated servants. Newton had the solution before dawn. By gosh,[2] he still had all his marbles together.

Now what did Newton do? He published his solution anonymously in the next issue of the *Philosophical Transactions of the Royal Society*. When Bernoulli saw the elegant solution, he exclaimed, "Tanquam ex ungue leonem!" ("One recognizes the lion by his paw prints!")

In fact, only three other physicists in Europe (besides Bernoulli) were able to solve what was at the time a fiendishly difficult problem: Bernoulli's older brother Jacob Bernoulli, the Marquis de l'Hospital[3] (of the rule you learned in calculus), and the great Gottfried Leibniz. The variational calculus was not invented until 1766 (almost 70 years after the lion chose to show his paw prints rather than to roar) by Leonhard Euler (1707–1783) and then subsequently refined by Joseph-Louis Lagrange (1736–1813).

It is a bit of a shame that anonymous scientific publication is no longer done these days. Lots of fun stories would result, no doubt.

---

* From $\int dx\delta(x) = 1$, it follows that $\delta(x)$ has dimension of an inverse length, like $1/L$, and thus $\delta^3(\vec{x})$ has dimension $1/L^3$.

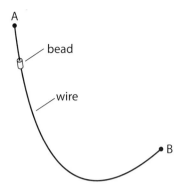

Figure 3 The brachistochrone problem. A bead is released at rest from A and slides down a stiff wire fashioned into a curve connecting A to B. What should the shape of the wire be if the bead is to reach B in the least amount of time?

Before you read on, see if you can go toe to toe and mano à mano with Isaac and solve the problem before dawn! The solution is given below.

Perhaps you can see intuitively that the correct shape looks like that given in figure 3. The dumb guess would be the straight line joining A and B. What you want to do is to start out dropping as vertically as possible to pick up speed. Depending on the position of B, it may actually pay to drop below B to attain more speed and then "coast" back up. With beads and wires, you could experiment by holding races.

Appendix 2: Another approach to functional variation

I mentioned that there are two approaches to variational calculus. The other approach is to discretize: replace the continuous variable x by the discrete variables $x_j = ja$, $j = -N, \ldots, N-1$, N with $Na = L/2$. We denote $\phi(x_j)$ by ϕ_j. In some sense, we go in the opposite direction from that taken by Newton and Leibniz when they invented calculus. As already alluded to above, we mentally imagine dividing the string up into tiny segments. In the limit $a \to 0$ and $N \to \infty$, we should recover our continuous string. Thus, we write $\int dx\, V(\phi(x)) \sim a \sum_j V(\phi_j)$, so that

$$\frac{\delta}{\delta\phi(y)} \int dx\, V(\phi(x)) \sim a \frac{d}{d\phi_k} \sum_j V(\phi_j) = a V'(\phi_k)$$

with k determined by $y \sim x_k$. Discretization reduces functional differentiation to ordinary differentiation. The derivative term requires a bit more work:

$$\int dx \left(\frac{d\phi}{dx}\right)^2 \sim a \sum_j ((\phi_{j+1} - \phi_j)/a)^2 \tag{13}$$

so that

$$\frac{\delta}{\delta\phi(y)} \int dx \left(\frac{d\phi}{dx}\right)^2 \sim \frac{1}{a} \frac{d}{d\phi_k} \sum_j (\phi_{j+1} - \phi_j)^2$$

$$= \frac{2}{a}(-(\phi_{k+1} - \phi_k) + (\phi_k - \phi_{k-1})) \sim 2a\left(-\frac{d^2\phi}{dy^2}\right) \tag{14}$$

I leave the reader to check that this reproduces all the results we derived before.

I mention one rather trivial technicality, which, however, might bother some fastidious readers. For discrete variables, clearly $\frac{\delta\phi_j}{\delta\phi_k} = \delta_{jk}$, where the Kronecker delta δ_{jk} is defined, as in part I, to be 1 if $j = k$ and 0 otherwise. For continuous variables, we would like to write

$$\frac{\delta\phi(x)}{\delta\phi(y)} = \delta(x - y) \tag{15}$$

with the Dirac delta function introduced in chapter I.1. In particular, this will reproduce (5). Since $\delta(x - y)$ has dimension of an inverse length, we shouldn't think of $\frac{\delta\phi(x)}{\delta\phi(y)}$ as something like an ill-defined $\frac{\Delta\phi(x)}{\Delta\phi(y)}$. To have the correct dimensions, we can either multiply the right hand side of (15) by the "short distance regulator" a, or more simply and preferably, absorb a into the definition of $\frac{\delta}{\delta\phi(y)}$.

Exercises

1 In the hanging string problem, if we did not make the approximation that the string is stretched only a little bit, we would have $F(u) = \sqrt{1 + u^2}$ (see remark 2 after (5)). Find the equation determining the shape of the string.

2 What is the analog of the inverse square law in D spatial dimensions? We will need this result in chapter X.2.

3 Consider the functional $S(a, b) = \int_0^\infty dr \, r(1 - b)a'$ of two functions $a(r)$ and $b(r)$ (with $a' \equiv da/dr$). Find the $a(r)$ and $b(r)$ that extremize S, with the boundary condition $a(\infty) = 1$ and $b(\infty) = 1$.

4 Denote the downward displacement of a hanging membrane by $\phi(x, y)$. Show that the amount of area by which the membrane is stretched is given by $\int dxdy \, \frac{1}{2}[(\frac{\partial\phi}{\partial x})^2 + (\frac{\partial\phi}{\partial y})^2]$. Hint: Use what you learned in chapters I.5 and I.6.

5 Solve the brachistochrone problem. By the way, the solution contains a "moral to the story."

Notes

1. There is a movie about a fish called Wanda, but not one about a fish called Poisson.
2. A mild anachronism: the euphemism was introduced circa 1750. Incidentally, I originally wrote "by Jehovah," but one reader thought that very few readers would know who Jehovah was. Too bad.
3. Incidentally, although the French often snickered at the ignorance of American physicists who talked about the hospital rule, the marquis himself spelled his name with an s, as in "the hospital."

II.2 | The Shortest Distance between Two Points

Variational calculus with several unknown functions

In the previous chapter we considered a functional of a single function ϕ, which could depend on more than one variable $x^1, x^2, \ldots, x^D$. We could also easily consider a functional $E(\phi_1, \phi_2, \ldots, \phi_n) = \int dx \mathcal{E}(\frac{d\phi_1}{dx}, \phi_1, \ldots, \frac{d\phi_n}{dx}, \phi_n)$ that depends on several functions $\phi_j(x)$, $j = 1, \ldots, n$, each of which is a function of a single variable x. Simply vary each function $\phi_j(x)$ to obtain n Euler-Lagrange equations (you should verify this):

$$\frac{d}{dx} \left(\frac{\delta \mathcal{E}}{\delta \frac{d\phi_j}{dx}} \right) - \frac{\delta \mathcal{E}}{\delta \phi_j} = 0 \tag{1}$$

to be solved for the n unknown functions $\phi_1, \ldots, \phi_n$. This is no different from the conceptual jump from the calculus of a single variable to that of many variables. Note that the formalism is completely general; in particular, E is just a functional, not necessarily having anything to do with energy. More generally, we could of course also consider a functional of several functions, each of which depends on several variables.

Armed with this understanding, you can now solve various classic problems. One of the most celebrated is that of finding the path of shortest distance between two given points, the so-called geodesic problem.

Reparametrization invariance

To start out easy, let us look at paths in flat 2-dimensional Cartesian space. A curve is described by a set of points labeled by two real functions $x(\lambda)$, $y(\lambda)$, which vary as the parameter λ varies. (See figure 1.) The choice of λ is arbitrary: the only requirement is that it increases monotonically from some initial value λ_i to some final value λ_f as we move along the curve from the starting point $A = (x(\lambda_i), y(\lambda_i))$ to the endpoint $B = (x(\lambda_f), y(\lambda_f))$. Imagine the curve as a highway on a map connecting city A with city B. We could

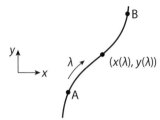

Figure 1 To describe a curve between two points, we have considerable freedom in choosing the parameter λ.

parametrize where we are on the highway by specifying the amount of gas consumed for example, or the number of songs sung, but it is clearly more sensible to use for λ the actual road distance l we have covered. As we will see, while the choice of parametrization is up to us, mathematics and common sense favors the particular parametrization $\lambda = l$.

The length of the curve is given by

$$\int \sqrt{dx^2 + dy^2} = \int_{\lambda_i}^{\lambda_f} d\lambda \sqrt{\left(\frac{dx}{d\lambda}\right)^2 + \left(\frac{dy}{d\lambda}\right)^2} \tag{2}$$

The length, being geometric, is manifestly reparametrization invariant, that is, independent of our choice of λ as long as it is reasonable. This is one of those "more obvious than obvious" facts, since $\int \sqrt{dx^2 + dy^2}$ on the left hand side of (2) is manifestly independent of λ.

Your calculus teacher probably told you not to write something like the left hand side of (2): a properly formulated integral should look like the right hand side of (2). But as I already said in the preceding chapter, it is perfectly kosher: just think of $\int \sqrt{dx^2 + dy^2}$ as the sum of infinitesimals $\sum_i \sqrt{(\Delta x)_i^2 + (\Delta y)_i^2}$.

If we insist, we can check the reparametrization invariance of (2). First, obviously the powers of $d\lambda$ match, so that if we scale $\lambda \to a\lambda$, the length is unchanged, as it should be. More generally, if we use another parameter η related to λ by $\lambda = \lambda(\eta)$, then $\frac{dx}{d\lambda} = \frac{d\eta}{d\lambda}\frac{dx}{d\eta}$ and $d\lambda = \frac{d\lambda}{d\eta}d\eta$. Plugging in, we have

$$\int_{\lambda_i}^{\lambda_f} d\lambda \sqrt{\left(\frac{dx}{d\lambda}\right)^2 + \left(\frac{dy}{d\lambda}\right)^2} = \int_{\eta_i}^{\eta_f} d\eta \sqrt{\left(\frac{dx}{d\eta}\right)^2 + \left(\frac{dy}{d\eta}\right)^2}$$

Finding the straight line

Of course, this particular example is trivial to solve, but that's the point, using a trivial example to illustrate general concepts. You have probably already realized that it would be sensible to put the starting point $A = (0, 0)$ at the origin and to rotate axes so that the endpoint B sits on the x-axis, in which case we might use the parametrization $\lambda = x$, so that $E = \int dx \sqrt{1 + (\frac{dy}{dx})^2}$. Obviously, the length is minimized by setting $\frac{dy}{dx} = 0$. Imposing the boundary conditions, we obtain the solution $y(x) = 0$.

We, however, do not want to solve the problem in the simplest possible way, but more generally, so as to learn about the calculus of variation. Well, all we have to do is to use (1), with the notational shift $\phi_1 \to x$, $\phi_2 \to y$, $x \to \lambda$. In this almost laughably simple example $E(x, y) = \int d\lambda \mathcal{E}$, with $\mathcal{E}(\frac{dx}{d\lambda}, \frac{dy}{d\lambda}) = \sqrt{(\frac{dx}{d\lambda})^2 + (\frac{dy}{d\lambda})^2}$, so that (1) gives two equations for two unknowns:

$$\frac{d}{d\lambda}\left(\frac{1}{\sqrt{(\frac{dx}{d\lambda})^2 + (\frac{dy}{d\lambda})^2}}\frac{dx}{d\lambda}\right) = 0 \quad \text{and} \quad \frac{d}{d\lambda}\left(\frac{1}{\sqrt{(\frac{dx}{d\lambda})^2 + (\frac{dy}{d\lambda})^2}}\frac{dy}{d\lambda}\right) = 0 \tag{3}$$

These are easy enough to solve by inspection: $\frac{dx}{d\lambda}$ and $\frac{dy}{d\lambda}$ are both constant, independent of λ. Hence we have shown, as expected, that the straight line gives the shortest distance between two points.

The more important lesson here, however, is to observe how (3) simplifies with the "preferred" or natural parametrization, namely to use for λ the length l along the curve, defined by $dl = \sqrt{dx^2 + dy^2}$, so that

$$\sqrt{\left(\frac{dx}{dl}\right)^2 + \left(\frac{dy}{dl}\right)^2} = \frac{\sqrt{dx^2 + dy^2}}{dl} = 1$$

Replace λ by l in (3). Watch the square roots disappear, so that (3) simplifies to $\frac{d^2x}{dl^2} = 0$ and $\frac{d^2y}{dl^2} = 0$. We will exploit this simplification ruthlessly when we get to Einstein's theory of gravity.

The astute reader might realize that there is a third equation, the definition of l: $(\frac{dx}{dl})^2 + (\frac{dy}{dl})^2 = 1$. Differentiating this equation with respect to l, we obtain $\frac{dx}{dl}\frac{d^2x}{dl^2} + \frac{dy}{dl}\frac{d^2y}{dl^2} = 0$, and thus this equation is not independent of the other two equations already given.

The world's most complicated description of a straight line

I prefer to go slow here, and so, instead of jumping into curved spaces immediately, let's stay on the flat plane but now use polar coordinates. The length of a curve connecting the two points is then given by

$$\int \sqrt{dr^2 + r^2 d\theta^2} = \int_{\lambda_i}^{\lambda_f} d\lambda \sqrt{\left(\frac{dr}{d\lambda}\right)^2 + r^2 \left(\frac{d\theta}{d\lambda}\right)^2} \equiv \int_{\lambda_i}^{\lambda_f} d\lambda L \tag{4}$$

with L now playing the role of $\mathcal{E}$ in (1).

A thorough understanding of this example will be important in mastering Einstein gravity later, so to make sure that you follow, let me spell out the steps here. First, we vary L with respect to $\frac{dr}{d\lambda}$ and obtain $\frac{1}{L}\frac{dr}{d\lambda}$ with L the square root defined in (4). Then we vary with respect to r and obtain $\frac{1}{L}r(\frac{d\theta}{d\lambda})^2$. Thus, we obtain $\frac{d}{d\lambda}(\frac{1}{L}\frac{dr}{d\lambda}) - \frac{1}{L}r(\frac{d\theta}{d\lambda})^2 = 0$ as in (1). Once again it is clear that we could make life easier by using the length parametrization.

Thus, setting λ to l, we have $L = 1$, so that this equation simplifies to

$$\frac{d^2r}{dl^2} - r\left(\frac{d\theta}{dl}\right)^2 = 0 \tag{5}$$

Similarly, repeating for θ what we did with r, we find

$$\frac{d}{dl}\left(r^2\frac{d\theta}{dl}\right) = 0 \Rightarrow \frac{d^2\theta}{dl^2} + \frac{2}{r}\frac{dr}{dl}\frac{d\theta}{dl} = 0 \tag{6}$$

In addition, we also have of course

$$\left(\frac{dr}{dl}\right)^2 + r^2\left(\frac{d\theta}{dl}\right)^2 = 1 \tag{7}$$

Confusio: "But if we have (7), isn't the integrand we are varying in (4) just 1? How do you vary 1?"

Dear reader, if you are confused also, just use some other parameter λ, obtain the variational equations with the square root in it (as in (3), for example), and then replace λ by l, as we did above.

Confusio: "But we have three equations (5), (6), and (7) to determine two unknown functions?"

You could choose any two of the three. The third then provides a consistency check, if you want.

Confusio: "Which two should I choose?"

Well, if I were you, Confusio, I would choose the two that make my life the easiest.

Since (7) is a first order differential equation, it is our clear-cut favorite. Of the two second order differential equations (5) and (6), the latter is clearly simpler to solve and hence is the more sensible* choice. Indeed, we now see there is no point in even differentiating, as we rather stupidly did in (6). The original form $\frac{d}{dl}(r^2\frac{d\theta}{dl}) = 0$ yields immediately $r^2\frac{d\theta}{dl} = a$ an unknown constant. Inserting this into (7) gives

$$\left(\frac{dr}{dl}\right)^2 + \frac{a^2}{r^2} = 1 \tag{8}$$

Indeed, you might have realized that, in terms of a mechanical analog (interpreting l as time), we have just used angular momentum conservation to eliminate $\frac{d\theta}{dl}$ in (7), so that (8) simply describes a particle with mass $= 2$ and energy $= 1$ moving in a repulsive "centrifugal" potential $V(r) = \frac{a^2}{r^2}$.

Integrating (8), we find $r^2 = l^2 + a^2$, where we absorbed an integration constant into l by setting $l = 0$ when $r = a$. Integrating $\frac{d\theta}{dl} = \frac{a}{l^2+a^2}$, we obtain $a\tan(\theta - \theta_0) = l$. A nice exercise in elementary geometry shows that this indeed describes a straight line.

The point of this is not to show that the author is capable of solving coupled differential equations and to obtain a rather complicated description of a straight line, but to

---

* As an undergraduate, I once took a math course for which the final exam consisted of ten problems, of which we were required to do only one. The ability to recognize which problem is doable is an extremely valuable skill in theoretical physics, and presumably in math also. The professor later told us that one of the ten problems had not yet been solved, and he was hoping that some bright undergrad would solve it "by chance."

understand that a straight line can apparently take on quite different forms in different coordinate systems. This will be a recurring theme in Einstein gravity.

Just to whet the reader's appetite, I might mention that when we study the motion of particles around a black hole, we will encounter the same type of equations as (5) and (7). Here we made our lives sweeter by favoring (7) over (5). When we get to Einstein gravity, we will use this "trick" again and again, namely tackling the analog of (6) first.

Great circles

We continue with a slightly more difficult problem: find the geodesics on a sphere. With the usual spherical coordinates, the length of a curve on a sphere is given by

$$\int \sqrt{d\theta^2 + \sin^2\theta \, d\varphi^2} = \int_{\lambda_i}^{\lambda_f} d\lambda \sqrt{\left(\frac{d\theta}{d\lambda}\right)^2 + \sin^2\theta \left(\frac{d\varphi}{d\lambda}\right)^2} \equiv \int_{\lambda_i}^{\lambda_f} d\lambda L \tag{9}$$

Here we set λ to l without further ado and obtain

$$\frac{d^2\theta}{dl^2} - \sin\theta \cos\theta \left(\frac{d\varphi}{dl}\right)^2 = 0 \tag{10}$$

and

$$\frac{d}{dl}\left(\sin^2\theta \frac{d\varphi}{dl}\right) = 0 \Rightarrow \frac{d^2\varphi}{dl^2} + \frac{2\cos\theta}{\sin\theta}\frac{d\theta}{dl}\frac{d\varphi}{dl} = 0 \tag{11}$$

Since L does not depend on φ, the equation for $\frac{d\varphi}{dl}$ is particularly easy* to obtain.

We all know that on a sphere great circles are geodesics. Indeed, we see that φ constant with θ a linear function of l is a solution. Lines of fixed longitude are great circles. In contrast, θ constant is not a solution unless that constant is $\pi/2$. Flying along a fixed latitude is not a fuel-economizing move[†] unless you are at the equator.

As in the simpler flat space problem above, we should remember that we have a third equation

$$\left(\frac{d\theta}{dl}\right)^2 + \sin^2\theta \left(\frac{d\varphi}{dl}\right)^2 = 1 \tag{12}$$

and just as in that case the reader could differentiate this equation and verify that it is not independent of (10) and (11). (The reader could also see that if l is interpreted as time, (12) expresses the conservation of energy in the motion described by (10) and (11).)

With what you have learned here, you could now immediately determine the geodesics on the Poincaré half plane discussed in chapter I.5. Try doing it, and then look at appendix 4 if you need help.

---

[*] The analog of this will also hold in Einstein gravity, associated with the notion of Killing vectors (see chapter V.4).

[†] As you will see, this foreshadows the discussion in chapter V.1.

The geodesic equation

After these three examples, we are now ready to formulate the general case. The strategy is now familiar: we do what the ants mentioned in the prologue do. Without the benefit of advanced math, they wander off from some "tried and true" path, hoping to find a better one.

Consider a space with the metric $g_{\mu\nu}(x)$ and a curve $X^\mu(\lambda)$, where λ is, once again, any parameter that varies monotonically along the curve. (As in chapter I.5, we use Greek indices to label the coordinates, and repeated indices are summed over.) Once again, we deftly avoid the notational confusion inflicted on the reader by some books. Beware of the distinction between x and X! We are to minimize the length of the curve fixed at some initial and final positions:

$$\int dl = \int \sqrt{g_{\mu\nu} dX^\mu dX^\nu} = \int d\lambda \sqrt{g_{\mu\nu}(X(\lambda)) \frac{dX^\mu}{d\lambda} \frac{dX^\nu}{d\lambda}} \equiv \int d\lambda \, L \tag{13}$$

where $g_{\mu\nu}$ is evaluated at X of course. (At any point, if you are confused, you could always refer to the example of the sphere in (9), for which $g_{\theta\theta} = 1$, $g_{\theta\varphi} = 0$, and $g_{\varphi\varphi} = \sin^2 \theta$.)

Here we could have simply plugged this functional of $X^\mu(\lambda)$ into the general equation (1), but in the interest of pedagogy, it is worthwhile (and easy enough) to go through the Euler-Lagrange steps again. Setting the variation of $\int d\lambda \, L$ to zero, we obtain

$$\delta \int d\lambda \, L = \int d\lambda \, \delta L$$

$$= \int d\lambda \left(\sqrt{g_{\mu\nu}(X(\lambda) + \delta X(\lambda)) \frac{d(X^\mu + \delta X^\mu)}{d\lambda} \frac{d(X^\nu + \delta X^\nu)}{d\lambda}} - \sqrt{g_{\mu\nu}(X(\lambda)) \frac{dX^\mu}{d\lambda} \frac{dX^\nu}{d\lambda}} \right)$$

$$= \int d\lambda \frac{1}{L} \left(2 g_{\mu\nu} \frac{dX^\mu}{d\lambda} \frac{d\delta X^\nu}{d\lambda} + \partial_\sigma g_{\mu\nu} \frac{dX^\mu}{d\lambda} \frac{dX^\nu}{d\lambda} \delta X^\sigma \right) = 0 \tag{14}$$

(If you are confused, read the paragraph following (4) again.)

We emphasize that $g_{\mu\nu}$ and $\partial_\sigma g_{\mu\nu}$ are evaluated at $X(\lambda)$. Indeed, the second term arises from the dependence of $g_{\mu\nu}$ on X. Integrating the first term by parts (with $\delta X^\sigma = 0$ at the endpoints as usual), we obtain

$$L \frac{d}{d\lambda} \left(\frac{1}{L} 2 g_{\mu\sigma} \frac{dX^\mu}{d\lambda} \right) - \partial_\sigma g_{\mu\nu} \frac{dX^\mu}{d\lambda} \frac{dX^\nu}{d\lambda} = 0 \tag{15}$$

We have learned only too well that to simplify (15) we should exploit the freedom in choosing λ and use length parametrization. Set $d\lambda$ to dl so that $L = 1$. Then we have

$$\frac{d}{dl} \left(g_{\mu\sigma} \frac{dX^\mu}{dl} \right) - \frac{1}{2} \partial_\sigma g_{\mu\nu} \frac{dX^\mu}{dl} \frac{dX^\nu}{dl} = 0 \tag{16}$$

where, to repeat, $g_{\mu\sigma}$ and $\partial_\sigma g_{\mu\nu}$ are to be evaluated at $X(\lambda)$. (For example, for the sphere, for $\sigma = \theta$ we have, noting that the metric is diagonal, $\frac{d}{dl}(g_{\theta\theta} \frac{d\theta}{dl}) - \frac{1}{2} \partial_\theta g_{\varphi\varphi}(\frac{d\varphi}{dl})^2 = 0$, which is just (10).)

In specific cases, rather than pushing through the differentiation in the first term in (16), we are better off restraining ourselves. The geodesic equation in this form corresponds to the first halves of (6) and (11). In particular, if $g_{\mu\nu}$ does not depend on x^σ, the second term in (16) vanishes, and we learn immediately that $g_{\mu\sigma}\frac{dX^\mu}{dl}$ is constant along the geodesic curve.

For further analysis, however, we should, and could, push ahead and carry out the differentiation in the first term of (16), just as in the second halves of (6) and (11). We obtain $g_{\mu\sigma}\frac{d^2X^\mu}{dl^2} + \partial_\nu g_{\mu\sigma}\frac{dX^\nu}{dl}\frac{dX^\mu}{dl} - \frac{1}{2}\partial_\sigma g_{\mu\nu}\frac{dX^\mu}{dl}\frac{dX^\nu}{dl} = 0$, which on multiplication by $g^{\rho\sigma}$ becomes*

$$\frac{d^2X^\rho}{dl^2} + \frac{1}{2}g^{\rho\sigma}(2\partial_\nu g_{\mu\sigma} - \partial_\sigma g_{\mu\nu})\frac{dX^\mu}{dl}\frac{dX^\nu}{dl} = 0 \tag{17}$$

Defining the Christoffel symbol by

$$\Gamma^\rho_{\mu\nu}(x) \equiv \frac{1}{2}g^{\rho\sigma}(x)\left(\partial_\mu g_{\nu\sigma}(x) + \partial_\nu g_{\mu\sigma}(x) - \partial_\sigma g_{\mu\nu}(x)\right) \tag{18}$$

we can write (17) as

$$\frac{d^2X^\rho}{dl^2} + \Gamma^\rho_{\mu\nu}(X(l))\frac{dX^\mu}{dl}\frac{dX^\nu}{dl} = 0 \tag{19}$$

Note that $\Gamma^\rho_{\mu\nu}$ is defined to be symmetric in $\mu\nu$, since in (19) it multiplies the symmetric combination $\frac{dX^\mu}{dl}\frac{dX^\nu}{dl}$.

It is often useful to deal with the auxiliary quantity

$$\Gamma_{\mu\nu\cdot\sigma} \equiv \frac{1}{2}(\partial_\mu g_{\nu\sigma} + \partial_\nu g_{\mu\sigma} - \partial_\sigma g_{\mu\nu}) \tag{20}$$

We put a dot in the group of three indices to remind us that this object is symmetric under the exchange of the first two indices, and that the last index σ is the one who came downstairs, and when needed, could be sent upstairs again.

Comparing (19) with (5) and (6), you could read off the nonvanishing Christoffel symbols for polar coordinates:

$$\Gamma^r_{\theta\theta} = -r \quad \text{and} \quad \Gamma^\theta_{r\theta} = \frac{1}{r} \tag{21}$$

(Note the factor of 2 in (5) disappears because of the symmetry of the Christoffel symbol in its two lower indices.) Similarly, you could read off from (10) and (11) the nonvanishing Christoffel symbols for the sphere:

$$\Gamma^\theta_{\varphi\varphi} = -\sin\theta\cos\theta \quad \text{and} \quad \Gamma^\varphi_{\theta\varphi} = \frac{\cos\theta}{\sin\theta} \tag{22}$$

You should verify that you obtain the same results plugging the metric into (18).

Decide for yourself which method of calculating $\Gamma^\rho_{\mu\nu}$ is computationally simpler for you. I prefer to vary (13) directly to using (18). With practice the variation could be done mentally. For really simple examples, such as the ones here, it is pretty much a toss up.

Note that the geodesic equation (19), when written in terms of the tangent vector $V^\mu \equiv \frac{dX^\mu}{dl}$ to the curve, actually looks quite simple:

$$\frac{dV^\rho}{dl} + \Gamma^\rho_{\mu\nu}V^\mu V^\nu = 0 \tag{23}$$

---

* Recall from chapter I.5 that $g^{\rho\sigma}g_{\mu\sigma} = \delta^\rho_\mu$.

Keep in mind also that often it is easier to solve (16), which can now be written as

$$\frac{d}{dl}(g_{\mu\sigma}V^{\mu}) - \frac{1}{2}(\partial_{\sigma}g_{\mu\nu})V^{\mu}V^{\nu} = 0 \tag{24}$$

I have intentionally derived the geodesic equation in a slightly cumbersome way, dragging L along, to emphasize that the length (13) is a geometric quantity independent of the parametrization λ used. Only at the end do I exploit our parametrization freedom and set λ to l to obtain (16). I will do this throughout. In contrast, the authors of some texts vary an alternative quantity $\int dl\, g_{\mu\nu}\frac{dX^{\mu}}{dl}\frac{dX^{\nu}}{dl}$ (which is emphatically not parametrization independent) and use the Euler-Lagrange equation (1) to arrive at (16) in one step. The derivation looks cleaner but comes with the cost of potential conceptual misunderstanding. Once you understand all this, you can, however, safely use this method. At the least, it serves as a useful mnemonic.

Connection to classical differential geometry

At this point Confusio suddenly speaks up. "Hey, didn't we encounter the Christoffel symbol already back when we discussed classical differential geometry in chapter I.7?"

Excellent! Confusio has been paying attention and is not as confused as he looks! But what is the connection? In chapter I.7 the Christoffel symbol appeared in the variation of the basis vectors $\partial_{\mu}\vec{e}_{\nu}$. Here it has to do with the first derivative of the metric.

In fact, you already encountered the key in exercise I.7.6, namely that

$$\partial_{\mu}g_{\nu\sigma} = \Gamma_{\mu\nu\cdot\sigma} + \Gamma_{\mu\sigma\cdot\nu} \tag{25}$$

Applying a lemma given in chapter I.4, we can invert (20) to obtain precisely this.

A straight line does not have to look straight

Professor Flat ambles by again, saying, "It would be enlightening to give an alternative derivation of the geodesic equation (19)."

"Let's guess," you and I reply in unison, "We go to locally flat coordinates, right?" By now, we are catching on.

PF: "Excellent! We all know that in a locally flat region the geodesics are just straight lines, from stretched linen, you know."

It sounds vaguely plausible; Professor Flat is also an amateur etymologist.

So, consider locally flat coordinates $y^{\rho}(x)$, related to our curved coordinates as indicated. The metric is $ds^2 = \delta_{\rho\sigma}dy^{\rho}dy^{\sigma} = g_{\mu\nu}(x)dx^{\mu}dx^{\nu}$ so that

$$g_{\mu\nu}(x) = \delta_{\rho\sigma}\frac{\partial y^{\rho}}{\partial x^{\mu}}\frac{\partial y^{\sigma}}{\partial x^{\nu}} \tag{26}$$

A straight line is a curve $y^{\rho}(s)$ for which the tangent vector $\frac{dy^{\rho}}{ds}$ does not change, that is, $\frac{d^2 y^{\rho}}{ds^2} = 0$.

Professor Flat interjects: "That's what 'straightforward' means: you keep moving forward. This derivation will be straightforward also, just plug and chug. We replace y by x in the simple equation $\frac{d^2 y^\rho}{ds^2} = 0$ to obtain a complicated looking equation!"

So, $\frac{dy^\rho}{ds} = \frac{\partial y^\rho}{\partial x^\lambda} \frac{dx^\lambda}{ds}$, and then

$$0 = \frac{d^2 y^\rho}{ds^2} = \frac{\partial y^\rho}{\partial x^\lambda} \frac{d^2 x^\lambda}{ds^2} + \frac{\partial^2 y^\rho}{\partial x^\mu \partial x^\nu} \frac{dx^\mu}{ds} \frac{dx^\nu}{ds} \tag{27}$$

Multiplying by $\frac{\partial x^\sigma}{\partial y^\rho}$ (and using $\frac{\partial x^\sigma}{\partial y^\rho} \frac{\partial y^\rho}{\partial x^\lambda} = \delta^\sigma_\lambda$), we obtain (after renaming the index σ)

$$\frac{d^2 x^\lambda}{ds^2} + \frac{\partial x^\lambda}{\partial y^\rho} \frac{\partial^2 y^\rho}{\partial x^\mu \partial x^\nu} \frac{dx^\mu}{ds} \frac{dx^\nu}{ds} = 0 \tag{28}$$

As promised, we manage to replace the simple equation $\frac{d^2 y^\rho}{d^2 s} = 0$ by the complicated-looking* (28).

Upon identifying

$$\Gamma^\lambda_{\mu\nu} = \frac{\partial x^\lambda}{\partial y^\rho} \frac{\partial^2 y^\rho}{\partial x^\mu \partial x^\nu} \tag{29}$$

we recognize (28) as just the geodesic equation (19).

PF: "See how easy it is? We just have to show off our mastery of the chain rule!"

A straight line does not have to look straight in curved coordinates. Rather, it is described by a curve determined by (28). Nothing wrong with the straight line, but you have chosen funny coordinates. A bit like the fun house mirrors in amusement parks.

In this formalism, how do we relate the Christoffel symbol in (29) to the first derivative of the metric, as given in (18) and (20)? The way to do this is as follows. Differentiating[1] the metric in (26) gives

$$\partial_\lambda g_{\mu\nu}(x) = \delta_{\rho\sigma} \left(\frac{\partial^2 y^\rho}{\partial x^\lambda \partial x^\mu} \frac{\partial y^\sigma}{\partial x^\nu} + \frac{\partial y^\rho}{\partial x^\mu} \frac{\partial^2 y^\sigma}{\partial x^\lambda \partial x^\nu} \right)$$

Using the identity (I.4.14) you derived in chapter I.4, you could solve for $\delta_{\rho\sigma} \frac{\partial^2 y^\rho}{\partial x^\lambda \partial x^\mu} \frac{\partial y^\sigma}{\partial x^\nu}$. Lowering an index in (29) by using the metric in (26), you obtain $\Gamma_{\mu\nu,\sigma}$ as given in (20).

Appendix 1: Drowning in a sea of indices

The appearance of the 3-indexed Christoffel symbol $\Gamma^\rho_{\mu\nu}$ has detonated the explosion of indices that gives Einstein gravity its (undeserved) reputation of being difficult, and the reader is hereby warned that it will get much worse in subsequent chapters when the 4-indexed Riemann curvature tensor[†] appears. On first exposure, some students could easily feel that they are "drowning in a sea of indices." Some sophisticated types favor a fancy-schmancy index-free notation.[‡] This is analogous to the vector notation $\vec{v}$ that you are fluent with, instead of the index notation v^i. But it takes considerable effort to learn the index-free notation, and when push comes to shove, in an

---

* This peculiar replacement of a simple equation by a complicated one foreshadows Einstein's deep insight about gravity. See the discussion of the equivalence principle in part V.

† The astute reader might sense that the appearance of 3- and 4-indexed objects was foreshadowed by the expansion $g_{\mu\nu}(x) = g_{\mu\nu}(0) + A_{\mu\nu,\lambda} x^\lambda + B_{\mu\nu,\lambda\sigma} x^\lambda x^\sigma + \cdots$ we used in chapter I.6 to go to locally flat coordinates.

‡ Which we will eventually get to in chapter IX.7.

actual calculation,* even a sophisticate might have to descend to indices. Besides, you have to learn to walk before you can fly, and I think that for a first introduction to Einstein gravity, grappling with indices is an essential and ennobling experience.

Different people have different ways of remembering how the indices go. My advice is to remember the symmetry properties of various objects we encounter, for example, that $\Gamma^{\rho}_{\mu\nu}$ is defined to be symmetric in $\mu\nu$. I remember the schematic form $\Gamma^{\cdot}_{\cdot\cdot} \sim g^{\cdot\cdot}\partial_{\cdot}g_{\cdot\cdot}$, which enables me to reconstruct the precise expression (18).

Appendix 2: How the Christoffel symbol transforms

The beginning student should skip this appendix. Although it contains a result we will need later, it also contains "a sea of indices"!

Knowing how $g_{\mu\nu}$ transforms under a coordinate transformation (I.5.14), you could immediately plug that transformation law into (18) to determine how $\Gamma^{\lambda}_{\mu\nu}$ transforms. It is easier, however, to use (29) and grind ahead.

Following Professor Flat's suggestion, we went from the locally flat coordinates y to the coordinates x, obtaining the $\Gamma^{\lambda}_{\mu\nu}$ in (29). What if somebody comes along and changes from y to x'? She would obtain some other Christoffel symbols $\Gamma'^{\lambda}_{\mu\nu}$. How are they related to our symbols?

To relate $\Gamma'^{\lambda}_{\mu\nu} = \frac{\partial x'^{\lambda}}{\partial y^{\rho}}\frac{\partial^2 y^{\rho}}{\partial x'^{\mu}\partial x'^{\nu}}$, we simply replace the derivatives with respect to x' by those with respect to x. The calculation looks messy, but hey, it's only the chain rule once again.

Start with $\frac{\partial y^{\rho}}{\partial x'^{\nu}} = \frac{\partial x^{\sigma}}{\partial x'^{\nu}}\frac{\partial y^{\rho}}{\partial x^{\sigma}}$, and so

$$\frac{\partial^2 y^{\rho}}{\partial x'^{\mu}\partial x'^{\nu}} = \frac{\partial x^{\omega}}{\partial x'^{\mu}}\frac{\partial}{\partial x^{\omega}}\left(\frac{\partial y^{\rho}}{\partial x'^{\nu}}\right) = \frac{\partial x^{\omega}}{\partial x'^{\mu}}\frac{\partial}{\partial x^{\omega}}\left(\frac{\partial x^{\sigma}}{\partial x'^{\nu}}\frac{\partial y^{\rho}}{\partial x^{\sigma}}\right)$$

$$= \frac{\partial x^{\omega}}{\partial x'^{\mu}}\frac{\partial x^{\sigma}}{\partial x'^{\nu}}\left(\frac{\partial^2 y^{\rho}}{\partial x^{\omega}\partial x^{\sigma}}\right) + \frac{\partial x^{\omega}}{\partial x'^{\mu}}\frac{\partial^2 x^{\sigma}}{\partial x^{\omega}\partial x'^{\nu}}\left(\frac{\partial y^{\rho}}{\partial x^{\sigma}}\right)$$

The other factor in $\Gamma'^{\lambda}_{\mu\nu}$ is much easier to deal with: $\frac{\partial x'^{\lambda}}{\partial y^{\rho}} = \frac{\partial x'^{\lambda}}{\partial x^{\eta}}\frac{\partial x^{\eta}}{\partial y^{\rho}}$. Putting it together, we obtain

$$\Gamma'^{\lambda}_{\mu\nu} = \frac{\partial x'^{\lambda}}{\partial y^{\rho}}\frac{\partial^2 y^{\rho}}{\partial x'^{\mu}\partial x'^{\nu}} = \frac{\partial x'^{\lambda}}{\partial x^{\eta}}\frac{\partial x^{\omega}}{\partial x'^{\mu}}\left\{\frac{\partial x^{\sigma}}{\partial x'^{\nu}}\left(\frac{\partial x^{\eta}}{\partial y^{\rho}}\frac{\partial^2 y^{\rho}}{\partial x^{\omega}\partial x^{\sigma}}\right) + \frac{\partial^2 x^{\sigma}}{\partial x^{\omega}\partial x'^{\nu}}\left(\frac{\partial x^{\eta}}{\partial y^{\rho}}\frac{\partial y^{\rho}}{\partial x^{\sigma}}\right)\right\}$$

$$= \frac{\partial x'^{\lambda}}{\partial x^{\eta}}\frac{\partial x^{\omega}}{\partial x'^{\mu}}\frac{\partial x^{\sigma}}{\partial x'^{\nu}}\Gamma^{\eta}_{\omega\sigma} + \frac{\partial x'^{\lambda}}{\partial x^{\eta}}\frac{\partial^2 x^{\eta}}{\partial x'^{\mu}\partial x'^{\nu}} \tag{30}$$

where we used $\left(\frac{\partial x^{\eta}}{\partial y^{\rho}}\frac{\partial y^{\rho}}{\partial x^{\sigma}}\right) = \delta^{\eta}_{\sigma}$ once again.

The Christoffel symbol does not transform as a tensor, and hence, is not a tensor!

Referring to chapters I.4 and I.5, we see that the first term in (30) is precisely what is needed for the Christoffel symbol to transform as a tensor under the coordinate change $x \to x'$, but that is spoiled by the presence of the inhomogeneous term $\frac{\partial x'^{\lambda}}{\partial x^{\eta}}\frac{\partial^2 x^{\eta}}{\partial x'^{\mu}\partial x'^{\nu}}$. To repeat, the important point here is that the Christoffel symbol is not a tensor.

That the Christoffel symbol carries three indices and that it fails to transform nicely like a tensor are among the two root causes of some technical complications of general relativity. We can handle something carrying two indices: a square array of numbers may naturally be treated as a matrix, but not a cubical array. Note that we did not go looking for this 3-indexed beast; it came looking for us.

Using the notation $S^{\mu}_{\nu} = \frac{\partial x'^{\mu}}{\partial x^{\nu}}$, $(S^{-1})^{\mu}_{\rho} = \frac{\partial x^{\mu}}{\partial x'^{\rho}}$, and $\partial'_{\mu} = (S^{-1})^{\nu}_{\mu}\partial_{\nu}$ introduced in chapters I.4 and I.5, we could write (30) in a slightly nicer form:

$$\Gamma'^{\lambda}_{\mu\nu} = S^{\lambda}_{\eta}(S^{-1})^{\omega}_{\mu}(S^{-1})^{\sigma}_{\nu}\Gamma^{\eta}_{\omega\sigma} + S^{\lambda}_{\eta}(S^{-1})^{\rho}_{\mu}\partial_{\rho}(S^{-1})^{\eta}_{\nu} \tag{31}$$

Appendix 3: Finding locally flat coordinates

Professor Flat is getting visibly more and more excited. We gently inquire why.

---

* See the story in endnote 23 to the preface.

PF: "Don't you see? Given how the flat coordinates y depend on the curved coordinates x, you can calculate Γ. Now just reverse the math. Suppose you are in a curved space at some point P with coordinates x_*, and you know Γ at that point, then you can reverse the steps above and find the desired flat coordinates y."

In chapter I.6, I showed you by simple counting that it should be possible by changing coordinates to make the region around any point locally flat. But I didn't show you how to do it explicitly. Now we see that all we have to do is to solve (29) for y. The solution is

$$y^\rho(x) = K^\rho_{\ \lambda}\{(x^\lambda - x^\lambda_*) + \tfrac{1}{2}\Gamma^\lambda_{\ \mu\nu}(x_*)(x^\mu - x^\mu_*)(x^\nu - x^\nu_*) + \cdots\} \tag{32}$$

with the shorthand $K^\rho_{\ \lambda} = \frac{\partial y^\rho}{\partial x^\lambda}|_{x_*}$ and with the dots indicating corrections of order $(x - x_*)^3$. We simply insert this into (29) and verify that we recover $\Gamma^\lambda_{\ \mu\nu}(x_*)$. (You might want to compare with the discussion in chapter I.6.) Of course, after obtaining y as given by (32), we still have the freedom of applying an arbitrary rotation followed by a translation without affecting (26) and (29). We didn't include these to avoid cluttering up (32).

To see how all this works, it is best to go to an example. Imagine you are living at latitude* θ_* and longitude φ_*. Since the metric does not depend on φ, we could set[†] $\varphi_* = 0$, but you might want to drag it along nevertheless. Simply plug into (22) and (32).

Appendix 4: Geodesics on the Poincaré half plane

Let us now find the geodesics on the Poincaré half plane defined by $ds^2 = (dx^2 + dy^2)/y^2$ in chapter I.5. First, you may wish to recall the discussion there. We concluded that, to go from one point to another, we would want to curve away from the edge at $y = 0$, but we were not able to determine the actual curve. Now yes, we can. We are to minimize

$$\mathcal{D} = \int_A^B \sqrt{\frac{dx^2 + dy^2}{y^2}} \tag{33}$$

Rudolf Peierls famously said[2] to the young Hans Bethe, "Erst kommt das Denken,[3] dann das Integral." (Roughly, "First think, then calculate.") Now that we have done the thinking in chapter I.5, we could simply plug into the geodesic equation derived here. But wait, how about a tiny bit more Denken? Let's exploit parametrization invariance and choose the parameter that would minimize not only the integral but also our labor.

So, what did you choose? Inspired by your solution of the brachistochrone problem in exercise II.1.5 when you went mano à mano with Isaac Newton, you chose y as the parameter! Then[‡] $\mathcal{D} = \int_A^B dy \sqrt{1 + (\frac{dx}{dy})^2/y^2} \equiv \int_A^B dy\, L$. The key observation is that with this choice the integrand L is independent of the "dynamical" variable x. In other words, the Euler-Lagrange equation simplifies:

$$\frac{d}{dy}\left(\frac{\delta L}{\delta \frac{dx}{dy}}\right) = \frac{\delta L}{\delta x} = 0 \implies \frac{d}{dy}\left(\frac{\frac{dx}{dy}}{y\sqrt{1 + (\frac{dx}{dy})^2}}\right) = 0 \tag{34}$$

We obtain immediately $\frac{dx}{dy} = \frac{y}{b}\sqrt{1 + (\frac{dx}{dy})^2}$, with b an integration constant. This elementary differential equation is solved by $x - x_* = \pm\sqrt{b^2 - y^2}$. The second integration constant x_* reflects the translation invariance in x.

The geodesics are semi-circles of radius b centered at any point $(x_*, 0)$ on the x-axis. (The $\pm$ sign we encountered in the solution corresponds to the two halves of the semi-circle.) Note also the "vertical" lines $x = $ constant also solve the geodesic equation. See figure 2.

Thus, the geodesic going from point A to point B could be determined by a geometric construction. Draw a circle centered on the x-axis going through the two points. The circular arc between A and B is the desired geodesic, in full accordance with our intuition in chapter I.5.

---

* We mean physicists' latitude, of course, with $\theta = 0$ at the north pole.

[†] The story of France's losing battle to set $\varphi = 0$ for Paris makes for an interesting bit of history. Fortunately, the British were not so arrogant as to set the latitude of Greenwich to 0 also. More recently, GMT, meaning Greenwich Mean Time, was replaced by the compromise term UTC, an acronym neither in English nor in French.

[‡] In fact, you can see that this problem bears a striking resemblance to the brachistochrone problem but is not at all the same.

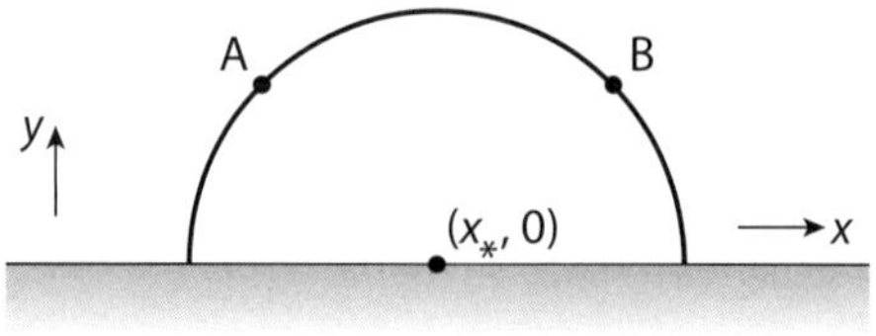

Figure 2 Geodesics on the Poincaré half plane.

Of course, if you chose the length l along the curve instead of y as the parameter, you could still solve the problem readily. See exercise 4. After all, Euler and Lagrange already did the Denken for you.

Appendix 5: Coordinates from a family of geodesics

Given a space with a coordinate system and a metric, we can determine the geodesics. Conversely, a family of geodesics can also lead us to a "natural" coordinate system.

Consider 2-dimensional space and a family of geodesics. We restrict ourselves to a region in which the geodesics do not intersect. (The coordinate system fails where the geodesics intersect. For example, the usual spherical coordinates on the sphere are not well defined at the north pole, where lines of constant longitudes intersect.) Label each geodesic by the parameter θ continuously. In other words, each value of θ uniquely specifies a geodesic, and values of θ infinitesimally close to each other specify geodesics that neighbor each other. On each geodesic, mark off the distance l and call this the coordinate r; in other words $r = l$. This is equivalent to saying $1 = g_{\mu\nu}\frac{dx^\mu}{dl}\frac{dx^\nu}{dl} = g_{rr}(\frac{dr}{dl})^2 = g_{rr}$: the first equality is the definition of l, the second follows from our construction that θ is constant along each geodesic, while the third is due to the choice $r = l$. Hence we have $g_{rr} = 1$.

Next, let's see if we can get rid of the cross-term $drd\theta$. Change coordinate by setting $r = \tilde{r} + h(\theta)$ so that $dr = d\tilde{r} + h'(\theta)d\theta$. In the new coordinates $(\tilde{r}, \theta)$, we have the cross-term $2(g_{r\theta}(r, \theta) + h'(\theta))d\tilde{r}d\theta$. Of course, $g_{r\theta}(r, \theta)$ can be equivalently written as some function $k(\tilde{r}, \theta)$. Thus, in general, we cannot get rid of the cross-term by suitably choosing $h(\theta)$.

But we are able to get rid of the cross-term at a specific value of $\tilde{r}$, call it $\tilde{r}_0$. In other words, we choose $h(\theta)$ so that $h'(\theta) = -k(\tilde{r}_0, \theta)$.

To proceed further, we will be kind to ourselves and drop the tilde sign, that is, we rename $\tilde{r}$ and call it r. To summarize, at this point we have $ds^2 = dr^2 + 2g_{r\theta}drd\theta + g_{\theta\theta}d\theta^2$, with $g_{r\theta}(r_0, \theta) = 0$.

We now plug in the geodesic equation for the coordinate r and obtain $0 = -\frac{d^2r}{dl^2} = \Gamma^r_{\mu\nu}\frac{dx^\mu}{dl}\frac{dx^\nu}{dl} = \Gamma^r_{rr}$. The second equality is the geodesic equation, while the first equality follows from $r = l$, the third from the fact that along a geodesic θ is constant and $r = l$. Thus, we have learned that $\Gamma^r_{rr} = \frac{1}{2}g^{r\theta}(2\partial_r g_{r\theta}) = 0$. Here the first equality follows from $g_{rr} = 1$. So, either $g^{r\theta} = 0$ or $\partial_r g_{r\theta} = 0$. In the first case, the 2-by-2 matrix $g^{\mu\nu}$ is diagonal and so its inverse is also diagonal, meaning that $g_{r\theta} = 0$. In the second case, $\partial_r g_{r\theta} = 0$. In other words, $g_{r\theta}$ does not vary as r varies. But we already know that $g_{r\theta}(r_0, \theta) = 0$ for some r_0, and hence we can conclude that $g_{r\theta}(r, \theta) = 0$ for all r.

To conclude, for 2-dimensional spaces, we have found, at least in some region, a class of coordinate systems with the metric

$$ds^2 = dr^2 + F(r, \theta)d\theta^2 \tag{35}$$

Note that in general we cannot simplify further.

You know some examples from this class. In particular, consider the subclass with metric $ds^2 = dr^2 + f(r)d\theta^2$. Two examples are $ds^2 = dr^2 + r^2d\theta^2$, polar coordinates for the plane, and $ds^2 = d\theta^2 + \sin^2\theta d\varphi^2$, spherical coordinates for the sphere.

Exercises

1 As explained in the text, to solve for the geodesics on the sphere, we could choose two of the three equations (10), (11), and (12). The wise person chooses (11) and (12) (of course!). Solve (11) immediately and plug into (12) to obtain

$$\left(\frac{d\theta}{dl}\right)^2 + \frac{K^2}{\sin^2\theta} = 1 \tag{36}$$

where K is an integration constant. Show that this equation could be interpreted as that of a particle moving in the potential $V(\theta) = \frac{K^2}{\sin^2\theta}$. Discuss.

2   Show that for diagonal metrics, at most one of the three terms on the right hand side of (18) defining the Christoffel symbol actually survives.

3   Show that (18) and (20) follow from (29).

4   Determine the geodesics on the Poincaré half plane.

5   Use the transformation law (30) of the Christoffel symbol to determine $\Gamma^r_{\theta\theta}$ and $\Gamma^\theta_{r\theta}$ in polar coordinates. Hint: Consider the coordinate transformation from $(x^1, x^2) = (x, y)$ to $(x'^1, x'^2) = (r, \theta)$.

6   Instead of varying the integral $\int d\lambda \sqrt{g_{\mu\nu}(X(\lambda))\frac{dX^\mu}{d\lambda}\frac{dX^\nu}{d\lambda}}$ in (13), we could, as an alternative, vary the integral $\int d\lambda g_{\mu\nu}(X(\lambda))\frac{dX^\mu}{d\lambda}\frac{dX^\nu}{d\lambda}$. Show that this leads to the same geodesic equations as in the text. Although this approach is arithmetically simpler, it is, as explained in the text, conceptually opaque, since this integral is not parametrization invariant and does not represent a geometric quantity, such as the length of the curve.

7   In chapter I.6, in proceeding to locally flat coordinates, after the first step, with the metric already in the form $g_{\mu\nu}(x) = \delta_{\mu\nu} + A_{\mu\nu,\lambda}x^\lambda + \cdots$, we claimed that by using the transformation $x^\mu = x'^\mu + L^\mu_{\nu\lambda}x'^\nu x'^\lambda + \cdots$, we could get rid of the linear terms in the metric. Using the transformation property of the Christoffel symbol, determine $L^\mu_{\nu\lambda}$. (Just from the index structure, you could probably guess the answer.)

8   Find the equation for geodesics in conformally flat spaces (which, as you might recall, were defined in an exercise in chapter I.6).

Notes

1. A tiny bit of subtlety gets glossed over here. See S. Weinberg, *Gravitation and Cosmology*.
2. *QFT Nut*, p. 365.
3. As in gedanken experiment.

II.3 Physics Is Where the Action Is

The action principle

You realize that in the last couple of chapters, I have been setting you up for a return to the prologue—to Fermat with his least time principle and to Feynman choosing the right path.

Next time you are invited to a dinner party at the home of a philosophy professor, say the word "teleological" in the middle of the main course. After these guys have stopped clawing at each other, utter, with nonchalant total self-assurance, "The ontological is distinct from the epistemological, while the tautological is antithetical to the logical," and watch the fun start again. That statement is of course what is known in polite circles as "utter nonsense" and in less polite circles as total BS, but it gives you an idea of how some academics talk.

The philosophies-R-us version, which I could give you for no charge, is that things are teleological if they have a purpose, or at least act as if they have a purpose. That's a big no-no in Western science. You see, Fermat's least time principle[1] has a strongly teleological flavor: that light, and particularly daylight, somehow knows how to save time—a flavor totally distasteful to the post-rational palate. In contrast, at the time of Pierre Fermat (1601 or* 1607/08?–1665), there was lots of quasi-theological talk about Divine Providence and Harmonious Nature, so there was no question that light would be guided to follow the most prudent path.

Thus, after the success of the least time principle for light, physicists naturally wanted to find a similar principle for material particles. Something is minimized, but what? Matter appears to behave quite differently from light, and this puzzled physicists for centuries.†

---

* The heavy academic controversy over Fermat's birth year stems from his father marrying twice and naming two sons from two different wives Pierre.[2]

† See the discussion in chapter III.5. The ultimate resolution had to wait until the advent of quantum field theory, but that's another story I've told elsewhere.

From the hanging string to the falling apple: Time makes its grand entrance

Now that you have mastered variational calculus, you can easily figure out whether New-ton's $F = ma$ follows from a variational principle. Consider the falling apple. Newton taught us to determine the path $q(t)$ of the apple by solving

$$m\frac{d^2q}{dt^2} = -mg \tag{1}$$

The question is: what variational principle can give us this differential equation?

I certainly dropped a trail of hints when I discussed the variational calculus. Recall (II.1.1) and (II.1.4): we saw that the Euler-Lagrange equation for a hanging string

$$T\frac{d^2\phi}{dx^2} = -\sigma g, \tag{2}$$

follows from extremizing the energy functional

$$E(\phi) = \int dx \left(\frac{T}{2}\left(\frac{d\phi}{dx}\right)^2 - \sigma g\phi(x)\right) \tag{3}$$

At this point, you don't have to be a genius to see that (1) and (2) have the same form. Simply replace $x \to t$, $\phi \to q$. Pretty nifty, eh? Remarkably, we simply flip figure II.1.1 over and relabel the horizontal axis by t instead of x, to obtain figure 1.

Voilà, from the hanging string to the falling apple by replacing space with time! In other words, extremizing the integral $S(q) = \int dt\{\frac{1}{2}m(\frac{dq}{dt})^2 - mgq(t)\}$ gives us (1). Thus far, in part II, we have had no time. Now time makes its grand entrance.

More generally, we have discovered, with no further ado, that extremizing

$$S(q) = \int dt \left\{\frac{1}{2}m\left(\frac{dq}{dt}\right)^2 - V(q)\right\} \tag{4}$$

yields the equation of motion

$$m\frac{d^2q}{dt^2} = -V'(q) \tag{5}$$

for a particle moving in a potential V. The integral $S(q)$ is known as the action.

The action $S(q)$ is a functional of the path $q(t)$. With each path, we assign a real number, namely the action $S(q)$ evaluated for that particular path $q(t)$. We are then instructed to

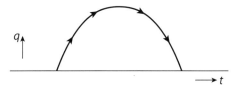

Figure 1 To go from the hanging string to the falling apple, we simply flip figure II.1.1 over and relabel the horizontal axis by t instead of x.

vary $S(q)$ subject to the boundary conditions (which, if we feel pedantic, we should perhaps refer to as the initial and final conditions, since we are dealing with time) that $q(t_i) = q_i$ and $q(t_f) = q_f$. The particular path $q(t)$ that extremizes the action $S(q)$ satisfies Newton's law (5).

The Lagrangian

The action principle states that the movement of particles is determined by extremizing an action. In general, the action functional is given by

$$S(q) = \int dt\, L\left(\frac{dq}{dt}, q\right) \tag{6}$$

The quantity $L(\frac{dq}{dt}, q)$ is known as the Lagrangian. The action is to be varied with the initial and final positions $q(t_i)$ and $q(t_f)$ held fixed to some q_i and q_f, respectively.

The variation of the action vanishes if

$$\frac{d}{dt}\left(\frac{\delta L}{\delta \frac{dq}{dt}}\right) - \frac{\delta L}{\delta q} = 0 \tag{7}$$

To be precise and pedantic, I stress, as in chapter II.1, that the notation means the following. We pretend that $L(a, b)$ is an ordinary function of two variables a and b. By $\frac{\delta L}{\delta \frac{dq}{dt}}$, we mean $\frac{\partial L(a,b)}{\partial a}$ with a subsequently set equal to $\frac{dq}{dt}$ and b to $q(t)$. Similarly, by $\frac{\delta L}{\delta q}$, we mean $\frac{\partial L(a,b)}{\partial b}$ with a subsequently set equal to $\frac{dq}{dt}$ and b to $q(t)$.

Switching from Leibniz's notation to Newton's notation, we could write the Euler-Lagrange equation (7) in the elegantly compact form

$$\frac{\dot{\delta L}}{\delta \dot{q}} = \frac{\delta L}{\delta q} \tag{8}$$

At Princeton University, there is a church-like gothic building with stained glass windows, each of which is inlaid with a fundamental equation of physics. One of the equations is (8). This equation, suitably generalized to quantum fields, underlies all known dynamics (we will come back to this shortly) in the universe.

A minus sign

In nonrelativistic mechanics, the Lagrangian $L(\frac{dq}{dt}, q)$ has the additive form

$$L\left(\frac{dq}{dt}, q\right) = \frac{1}{2}m\left(\frac{dq}{dt}\right)^2 - V(q) \tag{9}$$

(but that is not necessarily the case in general). Notice that the Lagrangian is equal to the kinetic energy minus the potential energy. In a way, there is a "what else could it be?" quality to what the Lagrangian turns out to be.

The minus sign is needed to generate the correct equation of motion (5), as you saw. In the energy functional (3) for the string, the variation of ϕ in space adds to the potential energy in the energy functional. In contrast, in the action functional (4), the variation of q in time goes against the potential energy: the crucial minus sign. Consider the simple harmonic oscillator, for which the action would be $S = \int dt \left[\frac{1}{2} m \left(\frac{dq}{dt} \right)^2 - \frac{1}{2} k q^2 \right]$. This is a first hint that time and space work against each other. Time differs from space by a sign, so to speak.

Choosing a path as a metaphor for life

Let us see how the action principle works in a specific case. Consider an object falling from a height of h to the ground in time T. The Lagrangian is then $L(\frac{dq}{dt}, q) = \frac{1}{2} m (\frac{dq}{dt})^2 - mgq(t)$. To keep irrelevant symbols from cluttering the page, let us choose units so that $g = 1$. Then Newton's $ma = F$ reads $\frac{d^2 q}{dt^2} = -1$ to be solved with the initial and final conditions $q(0) = h$ and $q(T) = 0$. Dear reader, I know that you can solve this equation practically in your sleep, thus finding the familiar parabola

$$q(t) = h + \left(\frac{T}{2} - \frac{h}{T} \right) t - \frac{1}{2} t^2 \tag{10}$$

Note one important difference between Fermat's least time principle governing light and the action principle governing material particles. In the least time principle, we are to minimize, duh, the transit time. In the action principle the transit time T is specified in advance. The particle is required to get to its destination in time T, the ultimate "on-time company." Notice that in the particular solution $q(t)$, the coefficient of the term linear in t switches sign when T becomes larger than $\sqrt{2h}$: Given too much time, the particle has to "waste" some time moving up before coming down.

Just as in the discussion of the hanging string following (II.1.9), the potential in this particular case is linear in q, and so, under $q(t) \to q(t) + \eta(t)$, the second order variation of S is given by $\frac{1}{2} m (\frac{d\eta}{dt})^2$ and thus is manifestly positive. The path in (10) in fact minimizes the action S. This is not true in general. (See exercise 1.) A simple calculation shows that, for the path actually followed, namely the path in (10), the action is equal to (henceforth we will factor out m and drop it) $S_{\min}(q) = \frac{h^2}{2T} - \frac{hT}{2} - \frac{T^3}{24}$.

It is intriguing that in the least action formulation of mechanics, the particle gives the impression of choosing a path to follow, as if it could sample all possible paths* before finding the one with the minimum action. "Mmm, this path is no good. Let me try another." Let's have fun seeing how this actually works.

What does the particle decide to do? Should the particle sit around and suddenly make a mad dash? Or should it get going promptly and then coast to the destination? What sort of "personality" does it have?

Perhaps the particle does something in between. Suppose that it decides to just forget about Newton and to fall down with constant speed, that is, following the path $q_1(t) =$

---

* This is a profound statement. To some extent, one could argue that the existence of the action principle in classical physics foreshadows the quantum world. See "Local versus global" later in this chapter.

$h - \frac{h}{T}t$. A steady kind of guy going at a steady pace. You could check that the action is then $S(q_1) = \frac{h^2}{2T} - \frac{hT}{2}$, indeed more than $S_{\min}(q)$ given above.

In everyday life, a falling object, especially if it is fragile and valuable, appears to hesitate for a moment or so, almost as if it is saying "Catch me if you can!" before gathering speed and crashing to the floor. That's Galileo's law of acceleration in action of course. From the action point of view, you can understand what is going on. The object, by staying at high altitude for "as long as possible," maximizes its potential energy and thus lowers the action. But then it has to rush at the end to get to the floor in the allotted time T, and hence pays the price of a larger kinetic energy. You could easily compute that the particle, by choosing the actual path $q(t)$ rather than the alternate path $q_1(t)$, pays an extra time-integrated kinetic energy equal to $\frac{T^3}{24}$, but it also raises its time-integrated potential energy by $2\left(\frac{T^3}{24}\right)$, thus managing to lower its action. It has figured out how to get the best action deal.

Some would see in the action principle a metaphor for life. You want to live life maximizing something, perhaps the total time-integrated happiness. You could either party now, dude, or you could study the action principle and party later in life. Of course, physics is so much simpler than real life, for which the quantity corresponding to the Lagrangian consists of a multitude of terms, each with zillions of parameters that vary from individual to individual. For example, for some geeks, studying physics has got to be way more fun than partying. There is also the minor detail that T is not known in advance.

Newton versus Aristotle

I have restricted the discussion to a single particle moving in 1 dimension. As in chapter I.1, through the magic of indices, we can immediately jump to the case of many particles moving in any dimension you desire. For instance, for a particle moving in 3-dimensional space, the position of the particle is specified by a 3-vector $\vec{q}(t)$. By rotational invariance (that is, "no direction is special"), the Lagrangian $L(\frac{d\vec{q}}{dt}, \vec{q})$ must be a scalar. This pretty much restricts it to have the form

$$L\left(\frac{d\vec{q}}{dt}, \vec{q}\right) = \frac{1}{2}m\left(\frac{d\vec{q}}{dt}\right)^2 - V(|\vec{q}|) \tag{11}$$

where $(\frac{d\vec{q}}{dt})^2$ represents the dot product $\frac{d\vec{q}}{dt} \cdot \frac{d\vec{q}}{dt}$ of course. You might also think of adding the term $\vec{q} \cdot \frac{d\vec{q}}{dt}$, but this is equal to $\frac{1}{2}\frac{d}{dt}\vec{q}^2$ and so contributes to the action $S = \int dt\, L$ merely a boundary term $\frac{1}{2}(\vec{q}^2(T) - \vec{q}^2(0))$. The equation of motion is not affected.

Aristotle thought that force is equal to velocity. It took the full genius of Newton to point out that no, force is in fact equal to acceleration. That, $F = ma$, is certainly one of the least obvious ideas in the history of theoretical physics: just try explaining it to a medieval peasant struggling to keep his cart moving.

When you first studied physics, didn't you wonder why Nature went against Aristotle and chose acceleration, and not velocity? Here is the answer. In some sense, rotational invariance and the action principle dictate the form of the kinetic term in (11). The Lagrangian contains two powers of $\frac{d}{dt}$. No way one of these is to disappear on our way to the equation of motion. It has to be a second order differential equation. This fundamental truth in fact continues to hold as we move onward to more advanced physics, all the way to the action of fundamental forces governed by quantum field theory.

I don't claim that this is a proof that would satisfy mathematicians. For example, we could have a term like $(\frac{d\vec{q}}{dt} \cdot \frac{d\vec{q}}{dt})^{\frac{1}{4}}$ in the Lagrangian, but this sure would not lead to mechanics as we know it.

Local versus global

Newton's equation of motion is described as local in time: it tells us what is going to happen in the next instant. In contrast, the action principle is global: one integrates over various possible trajectories and chooses the best one. While the two formulations are mathematically entirely equivalent, the action principle offers numerous advantages over the equation of motion approach. We mention some interesting points here.

1. The action leads directly to an understanding of quantum mechanics via the so-called Dirac-Feynman path integral[3] formulation. Indeed, the discussion here gives a premonition of the emergence of probability in the quantum world. Which path would the particle choose? Betting odds, anybody?

2. Intriguingly, while the energy functional is unequivocally asking to be minimized, the action principle merely tells us to extremize, rather than to maximize or minimize, a functional. In exercise 1, you will show that for the harmonic oscillator, the actual path can correspond to either a maximum or a minimum of the action. Within classical mechanics, this would appear somewhat puzzling, at least to me when I was a student. In the Dirac-Feynman path integral, this fact emerges naturally: classical paths correspond to the stationary phase in the sum over amplitudes. I consider this insight to be one of the great triumphs of quantum physics. Unfortunately, this point is obscured in the more familiar Schrödinger or Heisenberg formalisms.

3. The action principle gives a deeply satisfying and unifying understanding of conservation laws, as we will discuss in the next chapter.

4. The fundamental interactions we know about—the strong, weak, electromagnetic, and gravitational—can all be described by the action principle.* As you will see in this book, the action principle provides a natural route to special relativity, electromagnetism, and Einstein gravity. The action, rather than equations of motion, furnishes the language of quantum field theory. For instance, in perturbative field theory, we can go directly from the action to Feynman diagrams without ever mentioning equations of motion.

5. A practical, but a relatively minor, advantage is that since the action involves only first derivatives with respect to time, rather than second derivatives as in the equation of motion, it saves us computational labor in changing variables. An example is provided by Newton's solution

---

* Why this should be so represents a profound mystery.

for planetary orbits. In changing coordinates $x, y \to r, \theta$, we do not have to differentiate twice to get to (I.1.11); we could stop at (I.1.10), square, and add to obtain $\dot{x}^2 + \dot{y}^2 = \dot{r}^2 + r^2\dot{\theta}^2$. The action then becomes

$$S(r, \theta) = m \int dt \left\{ \frac{1}{2}(\dot{x}^2 + \dot{y}^2) + \frac{\kappa}{r} \right\} = m \int dt \left\{ \frac{1}{2}(\dot{r}^2 + r^2\dot{\theta}^2) + \frac{\kappa}{r} \right\} \tag{12}$$

Indeed, what we are effectively doing is replacing $dx^2 + dy^2$ by $dr^2 + r^2d\theta^2$, or more generally, $g_{\mu\nu}dx^\mu dx^\nu$. (Note that we are using the notation of chapters I.5 and I.6, in particular, Greek indices.) Here S is a functional of two functions, $r(t)$ and $\theta(t)$, so that varying, we obtain two Euler-Lagrange equations: $\ddot{r} = r\dot{\theta}^2 - \frac{\kappa}{r^2}$ and $\frac{d}{dt}(r^2\dot{\theta}) = 0$. Note that the conservation of angular momentum pops out, without our having to derive (I.1.13) and stare at it to recognize its more compact form.

6. Relating our discussion here to that in chapters I.5 and I.6, we see that the action allows us to immediately formulate mechanics in any coordinate system and in curved space. Simply replace (11) by

$$L\left(\frac{dq}{dt}, q\right) = \frac{1}{2}m\left(g_{\mu\nu}(q)\frac{dq^\mu}{dt}\frac{dq^\nu}{dt}\right) - V(q) \tag{13}$$

as in (12).

7. In the treatment in chapter I.1 involving equations of motion, the mass of the planet m drops out. In the action formalism here, this corresponds to m appearing merely as an overall factor for the action in (12). As I mentioned in chapter I.1, this fact will play a central role in Einstein gravity.

A particle at rest will remain at rest

For a pedagogical exercise, imagine that you and I were around in the early 18th century. How could we, doused with a liberal dose of hindsight, have developed the principle that governed the motion of material particles? What I will describe is not how it actually happened, but what we can readily imagine as how it could have happened, a sort of alternative physics history, perhaps in another civilization far far away in another galaxy.

We have heard of Fermat's marvelous principle, but material particles obviously can't also follow a least time principle, since their speeds can vary according to their energies.

Consider the simplest case of a particle moving in 1-dimensional space. To get started, suppose that there is no force and the particle just sits there: the couch potato problem. What is the particle minimizing? Obviously, it is keeping its kinetic energy as small as possible.

Define the quantity $S(q) = \int dt \frac{m}{2}(\frac{dq}{dt})^2$, which we will call the action. The integral over t is evaluated from some initial time, call it 0, to some final time T. The particle starts from some initial position $q(0)$ and ends up at some final position $q(T)$. To ease writing, choose the coordinate such that $q(0) = 0$. We also impose $q(T) = q(0) = 0$. Evidently, S is minimized for $\frac{dq}{dt} = 0$, which when solved with the stated boundary conditions gives us

$q(t) = 0$. Indeed, we have successfully described, by design, a particle sitting at rest. We have solved the couch potato problem, and perhaps we can even ask for tenure.

The law of inertia

So far so good. Next, we are emboldened to ask what would happen if at time T, we require the particle to be in some other position* $q(T) = Q$. Is our action principle smart enough to tell us that the particle will get there with constant speed?

Consider the path $q(t) = \frac{Q}{T}t + \sigma(t)$. Evidently, $\sigma(t)$, subject to the boundary conditions $\sigma(0) = \sigma(T) = 0$, describes deviation from "the straight and narrow path." Plug $\frac{dq}{dt} = \frac{Q}{T} + \frac{d\sigma}{dt}$ into our action:

$$S = \frac{m}{2}\int_0^T dt \left(\frac{Q}{T} + \frac{d\sigma}{dt}\right)^2 = \frac{mQ^2}{2T} + \frac{mQ}{T}\int_0^T dt \frac{d\sigma}{dt} + \frac{m}{2}\int_0^T dt \left(\frac{d\sigma}{dt}\right)^2$$

$$= \frac{mQ^2}{2T} + \frac{m}{2}\int_0^T dt \left(\frac{d\sigma}{dt}\right)^2$$

where in the last step we used the boundary conditions. This is obviously minimized by $\frac{d\sigma}{dt} = 0$, which together with the boundary conditions implies that $\sigma(t) = 0$, and thus the actual path $q(t) = \frac{Q}{T}t$.

A small triumph: we have recovered the law of inertia! In the absence of an external force, the particle continues to move† at a constant velocity $\frac{Q}{T}$. No force: no speeding up, no slowing down.

When we first learned to solve Newton's law, we were typically given the initial position $q(t = 0)$ and the initial velocity $\frac{dq}{dt}(t = 0)$, rather than the initial and final positions, as in the action principle. Of course we can also solve Newton's law with specified initial and final positions. Mathematically, we have a second order differential equation, so in any case, we need two conditions to fix two integration constants.

Getting the potato off the couch

Let us now go back to the couch potato and see how we could nudge him into motion. Turn on the gravitational potential $V(q) = mgq$. Note that q is a vertical coordinate pointing up: larger q corresponds to higher potential energy. Our boundary conditions $q(0) = q(T) = 0$ imply that the initial velocity $\frac{dq}{dt}(0)$ must be positive: the particle shoots upward and eventually falls back down. With a negative initial velocity, the boundary conditions could not be satisfied.

The question is how to modify the free particle action $S(q) = \int dt \frac{m}{2}(\frac{dq}{dt})^2$. Indeed, by dimensional analysis, our desired action principle is pretty much fixed: in the integrand,

---

* Compare *3:10 to Yuma*. The problem there is that you have to get to Yuma, Arizona, at 3:10 PM.
† Which is of course a highly nontrivial statement that essentially got physics started.

we can either add or subtract the potential energy. As mentioned earlier, adding would just mean the integrand becomes the total energy. So we try subtracting and write $S(q) = m \int dt \{ \frac{1}{2}(\frac{dq}{dt})^2 - gq \}$.

What the minus sign does is that now the particle can get a better deal on the action by moving to positive q, and thus lowers the action via the second term, compensating for the gain in the first term. We can easily estimate how high h the particle has to rise to get the best deal. Dropping the overall m, we have $S \sim T(\frac{h}{T})^2 - Tgh$: as anticipated, the kinetic term wants the particle to stay on the couch ($h = 0$), while gravity urges it to go up ($h > 0$). Extremizing in h, we obtain the familiar $h \sim gT^2$.

I am somewhat surprised that Newton didn't discover the action principle. Indeed, I see the brachistochrone problem, with its flavor of least time, as a bridge between the least time and the action principles. Leibniz, Newton's constant rival, apparently almost had it.

The Hamiltonian

After Lagrange invented the Lagrangian, Hamilton invented the Hamiltonian.*

Given a Lagrangian $L(\dot{q}, q)$, define the momentum by $p = \frac{\delta L}{\delta \dot{q}}$ and the Hamiltonian by

$$H(p, q) = p\dot{q} - L(\dot{q}, q) \tag{14}$$

where it is understood that $\dot{q}$ on the right hand side is to be eliminated in favor of p.

Let us illustrate this procedure by a simple example. Given the Lagrangian $L(\dot{q}, q) = \frac{1}{2}m\dot{q}^2 - V(q)$ in (9), we have $p = m\dot{q}$, which is precisely what we normally mean by momentum. The Hamiltonian is then given by

$$H(p, q) = p\dot{q} - L(\dot{q}, q) = p\dot{q} - \frac{1}{2}m\dot{q}^2 + V(q) = \frac{p^2}{2m} + V(q) \tag{15}$$

where in the last step we wrote $\dot{q} = p/m$. You should recognize the final expression as the total energy, namely the sum of the kinetic energy $\frac{p^2}{2m}$ and the potential energy $V(q)$. The Hamiltonian represents the total energy of the system.

Lagrange and Feynman

We close with two small stories about two towering figures.

Starting when he was 18, Joseph-Louis, the Comte de Lagrange (who, by the way, was born Giuseppe Lodovico Lagrangia before the term "Italian" existed), worked on the problem of the tautochrone, which nowadays we would describe as the problem of finding the extremum of functionals. A year or so later, he sent a letter to Leonhard Euler, the leading mathematician of the time, to say that he had solved the isoperimetrical problem: for curves of a given perimeter, find the one that would maximize the area enclosed. Euler had been struggling with the same problem, but he generously gave the teenager full credit.

---

* We mention this here for future use in part III. There is of course a lot more to the Hamiltonian than given here, but this is all we need.

Later, he recommended that Lagrange should succeed him as the director of mathematics at the Prussian Academy of Sciences.

Richard Feynman (1918–1988) recalled that when he first learned of the action principle, he was blown away. Indeed, the action principle underlies some of Feynman's deepest contributions to theoretical physics. In particular, his formulation of quantum mechanics depends very much on the action. I have left a clue in one sentence in the text for the curious student on how quantum physics can be understood without writing down the Schrödinger equation. Here you have learned how to do classical mechanics without Newton's equation.

Appendix 1: Particles and fields: Each telling the other how to behave

The action principle is now practically begging us to add the action (11) telling us how masses move in a gravitational potential to the energy functional (II.1.11) in chapter II.1 telling us how masses generate the gravitational field.

First of all, let us generalize (11) trivially to incorporate a whole bunch of masses through the magic trick of indices:

$$S_{\text{matter}} = \int dt \sum_a m_a \left\{ \frac{1}{2} \left(\frac{d\vec{q}_a}{dt} \right)^2 - \Phi(\vec{q}_a(t)) \right\} \tag{16}$$

We have used the fact that the potential V felt by the ath particle is given by $m_a \Phi$, with the gravitational potential $\Phi(x)$ evaluated at $x = q_a(t)$. (We now suppress the arrow on various vector quantities.)

In chapter II.1, we learned that the gravitational potential $\Phi(x)$ is determined by minimizing (II.1.11) $E(\Phi) = \int d^3x \left(\frac{1}{8\pi G}(\nabla\Phi)^2 + \rho(x)\Phi(x) \right)$, where the mass density $\rho(x, t) = \sum_a m_a \delta^3(x - q_a(t))$ is given by a sum of spikes centered on each of the particles. But wait, this is an energy functional, not an action. We now turn it into an action by the simple expedient of integrating its negative over time (the overall minus sign is because $E(\Phi)$ is a potential energy):

$$S_{\text{gravity}} = - \int dt \int d^3x \left(\frac{1}{8\pi G}(\nabla\Phi)^2 + \rho(x)\Phi(x) \right) \tag{17}$$

This is of course the infamous instantaneous action at a distance of Newtonian gravity: at time t, the gravitational potential $\Phi(x, t)$ is determined by $\rho(x, t)$ at the same time. (Needless to say, you should distinguish between the two different uses of the word "action"!) You may have also noticed that I have already stealthily snuck a time dependence into the mass density $\rho(x, t)$: after all, the location $q_a(t)$ of the particles changes with time. It is a free country: the particles are allowed to move around.

Confusio: "Do we now add the two actions, S_{matter} and S_{gravity}, together?"

No!

We see that, after we substitute the expression above for $\rho(x, t)$ as a sum of delta functions into the second term in S_{gravity} in (17) and then integrate over space, we obtain for this term

$$- \int dt \int d^3x \sum_a m_a \delta^3(x - q_a(t))\Phi(x) = - \int dt \sum_a m_a \Phi(q_a(t))$$

But this is precisely the second term in S_{matter} given in (16). So Confusio, if you add the two actions together, you would be double counting.

Thus, to obtain the total action for this Newtonian world, we should merge, rather than add, S_{matter} and S_{gravity}. The correct action is

$$S = \int dt \left\{ \sum_a \frac{1}{2} m_a \left(\frac{dq_a}{dt} \right)^2 - \int d^3x \frac{1}{8\pi G}(\nabla\Phi)^2 \right.$$
$$\left. - \int d^3x \left(\sum_a m_a \delta^3(x - q_a(t)) \right) \Phi(x, t) \right\} \tag{18}$$

The first term describes the dynamics of the particles, the second term describes the field $\Phi(x, t)$, and the third term couples the particles and the field together.

In Newtonian gravity, the field dictates how the particles move, and the particles in turn generate the field.

If you think that the action (18) is an ugly and awkward mess, you are completely right. The gravitational field pervades space but has no dynamics in time, merely reacting instantaneously to the location of the particles, while the particles are treated as points. How these two issues are resolved comprises two magnificent achievements in physics.

Dear reader, perhaps you have already seen a way of endowing the field with dynamics. By analogy with the particle's kinetic term $(\frac{dq_a}{dt})^2$, which is quadratic in time derivative, you might add to the second term in (18) a term quadratic in time derivative also, thus changing $(\nabla \Phi)^2$ to $-\frac{1}{C^2}(\frac{\partial \Phi}{\partial t})^2 + (\nabla \Phi)^2$, with C some unknown constant with the dimension of length over time (that is, the dimension of speed) to make the dimensions come out right. Varying S and repeating the same steps that led to II.1.10, we see that the equation determining Φ is changed to

$$\left(-\frac{1}{C^2}\frac{\partial^2}{\partial t^2} + \nabla^2 \right)\Phi = G\rho \tag{19}$$

Now Φ no longer reacts instantaneously to ρ at the same time t. No more Newton's spooky action at a distance! The gravitational potential now has dynamics and takes time to propagate from one point in space to another. Indeed, in empty space, with $\rho = 0$, the solution to (19) has the familiar wave form $\Phi(\vec{x}, t) = A \sin(\omega t - \vec{k} \cdot \vec{x}) + B \cos(\omega t - \vec{k} \cdot \vec{x})$, with $\omega^2 = C^2\vec{k}^2$. The unknown constant C measures the speed of propagation.

If you indeed saw all this, congratulations, you are some kind of a genius. You had foreseen Lorentz invariance and predicted the existence of gravitational waves. We will return to this in chapter IX.4.

But still, classical physics cannot remove the dichotomy between the field pervading space and the particles localized as points. Only by going into the quantum realm, where particles can be realized as excitations in fields, do we have a pleasingly unified description of Nature. This dichotomy between field and particle in fact provides one of the driving motivations[4] for quantum field theory. As you will eventually learn (but not in this text), in quantum field theory, the form $\frac{1}{C^2}(\frac{\partial \Phi}{\partial t})^2 - (\nabla \Phi)^2$ in the modified action is equivalent to the statement that the graviton, the particle associated with the gravitational field, is massless.

Appendix 2: The string in action and light cone coordinates

In chapter II.1, the string is just hanging there limply, far from being an action hero. With the action principle, it is a cinch to make it spring into action. The displacement $\phi(x)$ of the string segment labeled by x now depends on time and so has to be promoted to $\phi(t, x)$. Denote by ζ the mass per unit length, so that our little string segment has mass ζdx. Add a kinetic energy term $\int dx \frac{1}{2}\zeta(\frac{\partial \phi}{\partial t})^2$ to the energy functional of the string given in chapter II.1. The action becomes

$$\begin{aligned}
S &= \int dt \int dx \left[\frac{1}{2}\zeta \left(\frac{\partial \phi}{\partial t} \right)^2 - \left(\frac{T}{2} \left(\frac{\partial \phi}{\partial x} \right)^2 - \sigma g \phi(t, x) \right) \right] \\
&= \frac{1}{2}\zeta \int dt \int dx \left[\left(\frac{\partial \phi}{\partial t} \right)^2 - c_s^2 \left(\frac{\partial \phi}{\partial x} \right)^2 + \kappa \phi(t, x) \right]
\end{aligned} \tag{20}$$

Note the relative signs between the terms: as we now realize, the energy functional defined in chapter II.1 should really be called the potential energy functional and as we learned in this chapter, the kinetic energy and potential energy work against each other in the action. In the second step, since the Euler-Lagrange equation of motion does not depend on the overall normalization of the action, we took out a factor of $\frac{1}{2}\zeta$ and defined $c_s^2 = \frac{T}{\zeta}$ and $\kappa = \frac{2\sigma g}{\zeta}$.

Indeed, you recognize that we give life to the string in the same way we gave life to the gravitational field in the preceding appendix.

Varying ϕ and integrating by parts (once in time and once in space), we obtain the equation of motion

$$\left(\frac{\partial^2}{\partial t^2} - c_s^2 \frac{\partial^2}{\partial x^2} \right)\phi = \frac{1}{2}\kappa \tag{21}$$

One important point here is that c_s has dimensions of speed and thus, as you might expect, controls the speed with which vibrations on the string propagate.

The physics we want to emphasize here could be made more clear by turning off gravity, that is, by setting κ to 0. Then the equation of motion for the string has the general solution $\phi(t, x) = f_L(c_s t + x) + f_R(c_s t - x)$, where f_L and f_R are two arbitrary smooth functions describing a wave propagating to the left and to the right, respectively (a fact you can see by sketching $\phi(t, x)$ as a function of x at different values of t). Our subsequent discussion is a bit cleaner if we use units so that the constant c_s is effectively equal to 1. (This is completely analogous to the astronomical practice of using light years as a measure of distance.)

The general solution $\phi(t, x) = f_L(t + x) + f_R(t - x)$ is practically begging us to use the coordinates

$$x^\pm \equiv t \pm x \tag{22}$$

instead of (t, x). We then have $\partial_\pm \equiv \frac{\partial}{\partial x^\pm} = \frac{1}{2}(\partial_t \pm \partial_x)$. To see the last equality, simply act with both sides on $x^\pm$. (To save writing, we use the notation $\partial_t \equiv \frac{\partial}{\partial t}$ and so forth.) We thus have the string action (up to an irrelevant overall constant and with gravity turned off)

$$S = \int dx dt [(\partial_t \phi)^2 - (\partial_x \phi)^2] = 2 \int dx^+ dx^- \frac{\partial \phi}{\partial x^+} \frac{\partial \phi}{\partial x^-}$$

and the equation of motion $(\frac{\partial^2}{\partial t^2} - \frac{\partial^2}{\partial x^2})\phi = 0$, or even simpler with the $x^\pm$ coordinates, $\frac{\partial^2 \phi}{\partial x^+ \partial x^-} = 0$. By the way, for reasons that will become clear later, the coordinates $x^\pm$ are known as light cone coordinates.

As I already mentioned, a dynamical variable such as $\phi(t, x)$ that depends on both time and space is known as a field, just as an electric field or a magnetic field depends on time and space. Much more on fields later. Here let's recall the bad notation alert in chapter I.1. In this example, in particular, you see clearly that x is not a dynamical variable but a label telling which segment of the string we are talking about.

Appendix 3: Baby string theory and a sneak peek at the Lorentz transformation

Recall that when we discussed the energy functional of the membrane in chapter II.1, we used an observation from chapter I.4 that the expression $(\frac{\partial \phi}{\partial x})^2 + (\frac{\partial \phi}{\partial y})^2$ transforms like a scalar (see exercise I.4.1) under the rotation $x' = \cos\theta\, x + \sin\theta\, y$, $y' = -\sin\theta\, x + \cos\theta\, y$.

The expression $(\partial_t \phi)^2 - (\partial_x \phi)^2$ we encountered here in the action for the string looks similar except for a crucial minus sign. We naturally wonder if there might be a transformation similar to rotation that would leave this expression invariant.

A clue is provided by the fact that to show $x'^2 + y'^2 = x^2 + y^2$, we need the trigonometric identity $\cos^2\theta + \sin^2\theta = 1$. Let me remind you that the hyperbolic functions, $\cosh\varphi$, $\sinh\varphi$, and $\tanh\varphi$, are defined in analogy to the trigonometric functions (and are in some respects even simpler):

$$\cosh\varphi = \frac{1}{2}(e^\varphi + e^{-\varphi}), \quad \sinh\varphi = \frac{1}{2}(e^\varphi - e^{-\varphi}), \quad \tanh\varphi = \frac{\sinh\varphi}{\cosh\varphi} \tag{23}$$

from which the hyperbolic identity $\cosh^2\varphi - \sinh^2\varphi = 1$ follows.

Let's try transforming t and x using hyperbolic cosine and sine instead of trigonometric cosine and sine. We see immediately that with

$$t' = \cosh\varphi\, t + \sinh\varphi\, x \quad \text{and} \quad x' = \sinh\varphi\, t + \cosh\varphi\, x \tag{24}$$

we have $t'^2 - x'^2 = t^2 - x^2$. Furthermore, $\partial_t \phi = \frac{\partial t'}{\partial t} \partial_{t'} \phi + \frac{\partial x'}{\partial t} \partial_{x'} \phi = \cosh\varphi\, \partial_{t'} \phi + \sinh\varphi\, \partial_{x'} \phi$. Similarly, $\partial_x \phi = \sinh\varphi\, \partial_{t'} \phi + \cosh\varphi\, \partial_{x'} \phi$. We then verify that, indeed, $(\partial_{t'} \phi)^2 - (\partial_{x'} \phi)^2 = (\partial_t \phi)^2 - (\partial_x \phi)^2$. The string action is left invariant by the transformation (24). Everything parallels the corresponding discussion for rotation of the action for the membrane.

We will see in chapter III.3 that although the string discussed here obeys Newtonian physics and has nothing to do with special relativity, the same transformation, known as the Lorentz transformation, appears in Einstein's theory of relativity. It is amusing that the transformation is foreshadowed by baby string theory.

To show the power of the $x^\pm$ coordinates, we note that the $\frac{\partial \phi}{\partial x^+} \frac{\partial \phi}{\partial x^-}$ is obviously left unchanged if we multiply x^+ and divide x^- by the same factor, call it e^φ. But the transformation $x'^+ = e^\varphi x^+$, $x'^- = e^{-\varphi} x^-$ when written in terms of t and x immediately translates into the transformation above. (For example, $t' = \frac{1}{2}(x'^+ + x'^-) = \frac{1}{2}(e^\varphi x^+ + e^{-\varphi} x^-) = \cosh \varphi\, t + \sinh \varphi\, x$.) With the right choice of coordinates, we don't even need an inspired guess.[*]

Appendix 4: Particle on a sphere

Here is a fun problem. Determine the motion of a particle on a sphere with unit radius. Some features of the calculation we do here will come in handy when we get to anti de Sitter spacetime in part IX. Denote the position of the particle by $\vec{X} = (X^1, X^2, X^3)$, satisfying the constraint $\vec{X}^2 = 1$. (Henceforth, we suppress the arrow to minimize clutter.)

The problem is best solved with a double dose of Lagrange, using the Lagrangian and the Lagrange multiplier[5] to implement the constraint. So, write $L = \frac{1}{2}\dot{X}^2 + \frac{1}{2}\lambda(X^2 - 1)$, with λ the Lagrange multiplier. (We recognize this as a special case of (11). We have also chosen units in which the mass of the particle is unity.) Here L is to be regarded as dependent on X and λ.

Plug this into the Euler-Lagrange equation (7). For q taken to be λ, we recover the constraint $X^2 = 1$, while for q taken to be X, we obtain $\ddot{X}^i = \lambda X^i$ (with $i = 1, 2, 3$). Differentiating the constraint, we have $X \cdot \dot{X} = 0$ (where we indicate the dot in the dot product to remind us that we are dealing with vectors).

Now notice by direct differentiation that the quantity $J^{ij} \equiv X^i \dot{X}^j - X^j \dot{X}^i$ does not depend on time: $\dot{J}^{ij} = 0$. In other words, it is conserved. Do you recognize it? Angular momentum of course! (We will have a lot more to say about it in the next chapter.) Define $2J^2 = J^{ij} J^{ij} = 2(X^i \dot{X}^j - X^j \dot{X}^i) X^i \dot{X}^j = 2\dot{X}^2$. Hence $\dot{X}^2 = J^2$ is also constant. Recognize it? Total energy.

This last equation has the solution $X^i = a^i e^{iJt} + b^i e^{-iJt}$, with $\vec{a}$ and $\vec{b}$ two constant complex vectors satisfying $a^2 = b^2 = 0$, $2a \cdot b = 1$. Note that for X^i to be real, $\vec{b} = \vec{a}^*$. (Of course, if you prefer, you can express the solution in terms of sine and cosine.) The constraint $1 = X^2$ is also satisfied. Hence, the motion of the particle is completely solved.

Geometrically, it is clear that the particle travels along great circles. Let's verify this. What is the distance $\mathcal{D}$ traversed by the particle between times t_1 and t_2? We have

$$\mathcal{D} = \int \sqrt{(d\vec{X})^2} = \int dt \sqrt{\dot{X}^2} = J \int dt = J(t_2 - t_1)$$

In contrast, at time t_1, $X_1 = ae^{iJt_1} + be^{-iJt_1}$. The position X_2 at time t_2 is given by a similar expression with t_2 replacing t_1. The angle θ_{12} between the two vectors X_1 and X_2 is then given by $\cos\theta_{12} = X_1 \cdot X_2 = 2a \cdot b \cos J(t_1 - t_2) = \cos \mathcal{D}$, since $2a \cdot b = 1$. We obtain, as expected, $\mathcal{D} = \theta_{12}$.

The attentive reader might have noticed that we need not determine the Lagrange multiplier, but it is easy to do. Using $X^2 = 1$ and $X \cdot \dot{X} = 0$, we find $J^{ij} \dot{X}^j = J^2 X^i$ and $J^{ij} X^j = -\dot{X}^i$. But if we differentiate the last equation, we obtain $-\ddot{X}^i = J^{ij} \dot{X}^j = J^2 X^i$. Comparing with the equation of motion obtained earlier, we find $\lambda = -J^2$.

Exercises

1   Show that for a harmonic oscillator, the actual path can be either the maximum or the minimum of the action.

2   Suppose the falling particle in (1) follows an inverted parabola $q_2(t) = h(t - T)^2 / T^2$. By the argument given in the text, the corresponding action must be higher than the actual action. Calculate $S(q_2)$ and show that it is indeed higher.

3   Show that for the action $S(q) = \int dt\{\frac{1}{2}(\frac{dq}{dt})^2 - gq\}$ discussed in the text, with initial and final conditions $q(0) = q(T) = 0$, the action evaluates to $-\frac{1}{24}g^2 T^3$ for the actual path $q(t) = -\frac{1}{2}gt(t - T)$. If the particle stayed on the couch, its action would have been 0.

[*] If $g = 0$, the physics is invariant under a much larger class of transformations, as you will see in exercise 5 and later in chapter IX.9.

4   Vary $S = 2 \int dx^+ dx^- \frac{\partial \phi}{\partial x^+} \frac{\partial \phi}{\partial x^-}$ to obtain the equation of motion $\frac{\partial^2 \phi}{\partial x^+ \partial x^-} = 0$.

5   Show that for $g = 0$, the dynamics of the string is invariant under $x'^+ = f_+(x^+)$, $x'^- = f_-(x^-)$, with f_+ and f_- two arbitrary smooth functions.

6   Solve the isoperimetrical problem. You know the solution is a circle.

Notes

1. If it ever comes to a priority dispute, Fermat would have to cede to Heron of Alexandria (circa 65 AD).
2. K. Barner, "How Old Did Fermat Become?" *NTM* 9 (2001), p. 209.
3. R. P. Feynman and A. R. Hibbs, *Quantum Mechanics and Path Integrals*; also, *QFT Nut*, chapter I.2.
4. *QFT Nut*.
5. Let me remind those readers who are a bit shaky about the Lagrange multiplier that I gave a brief review in endnote 8 in chapter I.7.

II.4 | Symmetry and Conservation

"Spiritual formulas"

Pure mathematics is, in its way, the poetry of logical ideas. One seeks the most general ideas of operation which will bring together in simple, logical and unified form the largest possible circle of formal relationships. In this effort toward logical beauty spiritual formulas are discovered necessary for the deeper penetration into the laws of nature.

— Albert Einstein, writing about Amalie Emmy Noether
(1882–1935)

Symmetry[1] and conservation played, and continue to play, intertwined and central roles in physics. A set of transformations that leaves physics unchanged is said to be a symmetry. The example of angular momentum conservation, as discussed in chapter I.2, strongly indicates that symmetry and conservation are intimately related. In this chapter, we will have a general discussion showing that this is indeed the case.

I give immediately two concrete examples.

Example A

A particle moves in 2-dimensional space under a rotationally invariant potential, that is, a potential without a preferred direction:

$$L = \frac{1}{2}m\left[\left(\frac{dx}{dt}\right)^2 + \left(\frac{dy}{dt}\right)^2\right] - V(r) \tag{1}$$

where, as usual, $r = \sqrt{x^2 + y^2}$.

Example B

Two particles move in 1-dimensional space interacting with a translation invariant potential, that is, a potential that depends only on the distance between them:

$$L\left(\frac{dq_a}{dt}, q_a\right) = \frac{1}{2}m\left[\left(\frac{dq_1}{dt}\right)^2 + \left(\frac{dq_2}{dt}\right)^2\right] - V(|q_1 - q_2|) \tag{2}$$

Let us consider a Lagrangian $L\left(\frac{dq_a}{dt}, q_a\right)$ depending on a certain number of qs, which we denote as q_a. You see that our notation is flexible enough so that the index a can have entirely different meanings in different examples. In example A, it labels the x and y coordinates of a single particle. In example B, it labels two different particles. In general, we can have n particles moving in D-dimensional space, so that $a = (\alpha, i)$, with $\alpha = 1, \cdots, n$ and $i = 1, \cdots, D$. In other words, the index a labels the different degrees of freedom, and $L\left(\frac{dq_a}{dt}, q_a\right)$ denotes $L\left(\frac{dq_1}{dt}, \frac{dq_N}{dt}, q_1, \cdots, q_N\right)$ with $N = nD$ degrees of freedom.

With this rather general setup, we proceed. We say that our Lagrangian exhibits a symmetry if it remains invariant under an infinitesimal transformation $q_a \to q_a + \delta q_a$. It suffices to specify an infinitesimal transformation, since a noninfinitesimal transformation can be built up by repeating infinitesimal transformations, as explained in chapter I.3.

Again, let us hasten to our concrete examples. The Lagrangian in example A does not change under the transformation $x \to x + \epsilon y$ and $y \to y - \epsilon x$. Recall from chapter I.3 that this is just a rotation through an infinitesimal angle ϵ. The Lagrangian in example B does not change under the transformation $q_1 \to q_1 + \epsilon$ and $q_2 \to q_2 + \epsilon$.

Profundity and simplicity

We are now ready to prove Noether's theorem. As is often the case with the most profound theorems in theoretical physics, the proof is astonishingly simple. The statement that a Lagrangian does not change under an infinitesimal transformation $q_a(t) \to q_a(t) + \delta q_a(t)$ can be written as

$$0 = \delta L = \sum_a\left[\left(\frac{\delta L}{\delta\frac{dq_a}{dt}}\right)\delta\left(\frac{dq_a}{dt}\right) + \frac{\delta L}{\delta q_a}\delta q_a\right] \tag{3}$$

Under the transformation $q_a \to q_a + \delta q_a$, we have, by differentiating, $\frac{dq_a}{dt} \to \frac{d}{dt}(q_a + \delta q_a) = \frac{dq_a}{dt} + \frac{d}{dt}\delta q_a$, so that $\delta(\frac{dq_a}{dt}) = \frac{d}{dt}\delta q_a$. Thus, (3) becomes

$$0 = \sum_a\left[\left(\frac{\delta L}{\delta\frac{dq_a}{dt}}\right)\frac{d}{dt}\delta q_a + \frac{\delta L}{\delta q_a}\delta q_a\right] \tag{4}$$

To keep things uncluttered so that you can see more clearly what is going on, I ask you to hold the summation $\sum_a$ in your head. I now suppress the index a and write (4) as

$$0 = \left(\frac{\delta L}{\delta\frac{dq}{dt}}\right)\frac{d}{dt}\delta q + \frac{\delta L}{\delta q}\delta q \tag{5}$$

What do we do now? The only weapon at our disposal is the Euler-Lagrange equation

$$\frac{\delta L}{\delta q} = \frac{d}{dt}\left(\frac{\delta L}{\delta \frac{dq}{dt}}\right) \tag{6}$$

Putting this into (3), we have

$$0 = \left(\frac{\delta L}{\delta \frac{dq}{dt}}\right)\frac{d}{dt}\delta q + \frac{d}{dt}\left(\frac{\delta L}{\delta \frac{dq}{dt}}\right)\delta q \tag{7}$$

Lo and behold! The two terms on the right hand side combine, and we obtain

$$0 = \frac{d}{dt}\left(\frac{\delta L}{\delta \frac{dq}{dt}}\delta q\right) \tag{8}$$

In other words, the motion is such that

$$Q = \frac{\delta L}{\delta \frac{dq}{dt}}\delta q \tag{9}$$

does not change in time. Q is conserved! (Restoring the sum in your head, we have $Q = \sum_a \frac{\delta L}{\delta \frac{dq_a}{dt}}\delta q_a$.)

This is Noether's theorem, proved in that momentous year for physics 1915: for every transformation that leaves the Lagrangian unchanged, there is a conserved quantity.

Applying Noether's theorem

I am sure that you are eager to see how this actually works in practice. So let us jump to our two examples on the double. We just have to plug in (9).

Example A

$$Q = \epsilon\left(\frac{\delta L}{\delta \frac{dx}{dt}}y + \frac{\delta L}{\delta \frac{dy}{dt}}(-x)\right) = \epsilon m\left(y\frac{dx}{dt} - x\frac{dy}{dt}\right) \tag{10}$$

Don't forget the sum $\sum_a$ that you were holding in your head! That is why there are two terms: the sum runs over x and y.

A trivial remark is that we can drop the overall constant ϵ now that its job is done. The more important remark is that the conserved quantity $m(y\frac{dx}{dt} - x\frac{dy}{dt})$ is just the angular momentum. Thus, the conservation of angular momentum follows from rotational symmetry, as we suspected in chapter I.2.

Example B

$$Q = \epsilon m\left(\frac{\delta L}{\delta \frac{dq_1}{dt}} + \frac{\delta L}{\delta \frac{dq_2}{dt}}\right) = \epsilon m\left(\frac{dq_1}{dt} + \frac{dq_2}{dt}\right) \tag{11}$$

We recognize this as just the conservation of total momentum! The conservation of momentum follows from translational symmetry.

Utter generality

With practice, you will soon get the hang of it. You are given some Lagrangian, and after staring at it, you realize that it is invariant under some transformation. Then you simply plug the change of q_a, namely δq_a, into Noether's formula. There you are! The conserved quantity is $\sum_a \frac{\delta L}{\delta \frac{dq_a}{dt}} \delta q_a$. Here I have restored the sum over a.

Note that the derivation does not refer to the specific form and content of the Lagrangian, except for the fact that the Euler-Lagrange equation holds.

Noether's formula is utterly general and holds, all the way up, for quantum field theory, for string theory, and in fact, for any theory that respects the action principle. For the simple case of Newtonian mechanics, the Lagrangian has the generic form $L = \sum_a \frac{1}{2} m_a \left(\frac{dq_a}{dt}\right)^2 - V(q_1, q_2, \ldots)$, in which case the conserved quantity corresponding to translation invariance is given simply by $\sum_a m_a \left(\frac{dq_a}{dt}\right) \delta q_a$. Note that I have allowed for the possibility of the particles having different masses. Thus, for example, if the two particles in example B have different masses, the conserved quantity is $m_1 \frac{dq_1}{dt} + m_2 \frac{dq_2}{dt}$.

Energy conservation

Staring at (3), the astute reader might have noticed that even if δL does not vanish, but as long as it is a time derivative (namely that $\delta L = \frac{dK}{dt}$ for some K), we can still proceed. When we get to (8), we now have $\frac{dK}{dt}$ on the left hand side instead of 0:

$$\frac{dK}{dt} = \frac{d}{dt}\left(\frac{\delta L}{\delta \frac{dq}{dt}} \delta q\right) \tag{12}$$

We still have a conserved quantity:

$$Q = \frac{\delta L}{\delta \frac{dq}{dt}} \delta q - K \tag{13}$$

You might think that this is an obscure clause, a technicality fit for some nattering mathematical nitpickers. What kind of transformation would change L into this special form? But you would be wrong. Energy conservation provides an example of this phenomenon.

Consider a Lagrangian without any explicit time dependence, such as the Lagrangian in our examples A and B. Let the transformation be an infinitesimal translation in time: $q(t) \to q(t + \epsilon) \simeq q(t) + \epsilon \frac{dq}{dt}$. We are just shifting the argument of $q(t)$ and $\frac{dq}{dt}(t)$ inside L by ϵ and thus, obviously, $\delta L = \epsilon \frac{dL}{dt}$; in other words, we have $K = \epsilon L$. Plugging $\delta q = \epsilon \frac{dq}{dt}$ into (13) for the simplest case of a particle moving in a 1-dimensional potential, we find

$$Q = \epsilon \left(m \frac{dq}{dt}\frac{dq}{dt} - L\right) = \epsilon \left(m\left(\frac{dq}{dt}\right)^2 - \left(\frac{1}{2}m\left(\frac{dq}{dt}\right)^2 - V(q)\right)\right) = \epsilon \left(\frac{1}{2}m\left(\frac{dq}{dt}\right)^2 + V(q)\right) \tag{14}$$

which is, up to the no-longer-relevant overall constant ϵ, just the total energy. Note that in contrast to chapter I.2, here we simply turn the crank.

To understand better what is going on, it is instructive to look at this derivation from a somewhat different point of view. Start with the action

$$S = \int_0^T dt \left\{ \frac{1}{2} m \dot{q}(t)^2 - V(q(t)) \right\} \tag{15}$$

Now consider

$$S' \equiv \int_0^T dt \left\{ \frac{1}{2} m \dot{q}(t+\epsilon)^2 - V(q(t+\epsilon)) \right\} \tag{16}$$

It is crucial to realize that we are holding the limits of integration fixed, so this is not just a trivial shift of the dummy integration variable by ϵ. Thus, $S' \neq S$. Expanding the difference to linear order in ϵ, we have

$$\delta S = S' - S = \epsilon \int_0^T dt \, (m \dot{q} \ddot{q} - V'(q) \dot{q}) \tag{17}$$

We now evaluate δS in two different ways. First, using the equation of motion, we can write (17) as

$$\delta S = \epsilon \int_0^T dt \, (m\ddot{q} - V'(q)) \dot{q} = \epsilon \int_0^T dt \, (-2V'(q)) \dot{q} = -2\epsilon \int_0^T dt \, \frac{dV}{dt}$$

Second, without using the equation of motion, we write (17) as

$$\delta S = S' - S = \epsilon \int_0^T dt \, \frac{d}{dt} \left(\frac{1}{2} m \dot{q}^2 - V(q) \right)$$

Equating these two expressions for δS, we obtain $\int_0^T dt \, \frac{d}{dt} (\frac{1}{2} m \dot{q}^2 + V(q)) = 0$. (Note how the relative sign between the kinetic and potential energy terms flips at this point.) Doing the trivial integral and defining $E(t) \equiv \frac{1}{2} m \dot{q}^2 + V(q)$, we obtain $E(T) = E(0)$.

Perhaps Noether's infinitesimal transformations here reminded you of Lie's infinitesimal transformations in chapter I.3. Clearly, there is a fruitful connection to be exploited. Just a remark to whet your appetite for more.

Exercise

1   Derive the conservation of angular momentum in 3 dimensions using Noether's theorem.

Note

1. *Fearful.*

Recap to Part II

We find the shortest path by the incredibly clever trick of comparing the length of different paths, basically the same method used by a colony of ants. The ants send out zillions to try out different paths to the honey. We adopt the same idea. They exploit pheromone evaporation; we use the variational calculus. Euler and Lagrange proposed to change things a little and see what happens.

Mysteriously, all of fundamental physics is governed by the action, from which the equations of motion follow.

A profound truth is that the conservation laws are due to symmetries.

Part III | Space and Time Unified

I am convinced that the philosophers have had a harmful effect upon the progress of scientific thinking in removing certain fundamental concepts from the domain of empiricism, where they are under our control, to the intangible heights of the a priori. . . . This is particularly true of our concepts of time and space, which physicists have been obliged by the facts to bring down from the Olympus of the a priori in order to adjust them and put them in a serviceable condition.

—A. Einstein[1]

Galilean transformation

Go back to the prelude, in which Galileo's ship was updated to Einstein's train. The observer on the train, Ms. Unprime, ascribed to some event the spatial coordinates (x, y, z) and temporal coordinate t. To the same event, the observer on the ground, Mr. Prime, assigns the coordinates (x', y', z') and t'. Denote the speed of the train by u, and choose the axis so that the train moves along the x-axis. Then the two sets of coordinates are related by

$$t' = t$$
$$x' = x + ut$$
$$y' = y$$
$$z' = z \tag{1}$$

a set of relations known as the Galilean transformation. Consider a point on the train with $x = 0$. Plugging this into (1), we see that, for Mr. P on the ground, this point moves along according to $x' = ut = ut'$.

The innocuous looking equalities $y' = y$ and $z' = z$ actually represent an important consequence of Galileo's relativity principle. Call the y direction the vertical direction. We can supply sticks of a standard length L to Ms. U and Mr. P to build a fence.

To make sure that the sticks supplied to the two observers are identical, we can arrange for the woodcutter to ride in a train going by at speed $\frac{1}{2}u$ relative to Mr. P and $-\frac{1}{2}u$ relative to Ms. U. In other words, the coordinates of the woodcutter are given by $t' = t_w$

and $x' = x_w + \frac{1}{2}ut_w$. As far as the woodcutter is concerned, he is at rest, and Mr. P and Ms. U are going by him at the same speed but in opposite directions.[2] The woodcutter can toss the pre-cut sticks in identical ways to the two observers and their helpers. This long-winded digression is to answer any objection that the tossing of sticks from Mr. P to Ms. U, say, could have done something to the lengths of the sticks.

The top of the two fences is then given by $y = L$ and $y' = L$, respectively. The two lengths must agree, because as the two fences sweep past each other, the two observers could see whether one fence is taller than the other. In either case, Galileo's relativity principle, stating that two observers in relative uniform motion could not decide who is moving relative to the other, would be violated. Thus, we must have $y' = y$. Similarly, $z' = z$. The coordinates perpendicular to the direction of motion are unaffected by the motion.

The relation $x' = x + ut$ certainly does not violate Galileo's principle, since $x = x' + (-u)t'$. To Ms. U, she is at rest, but relative to her, Mr. P, sitting at $x' = 0$, is moving with speed $-u$ in the x direction.

We have set up the coordinates so that when $t' = t = 0$, we have $x' = x = 0$. Just as in chapter I.3, we can avoid having to line up the origins of the two coordinate systems by considering the separation between two events E_1 and E_2 in spacetime located at (t_1, x_1, y_1, z_1), (t'_1, x'_1, y'_1, z'_1), (t_2, x_2, y_2, z_2), and (t'_2, x'_2, y'_2, z'_2). Writing $\Delta t = t_2 - t_1$, $\Delta x = x_2 - x_1$, and so on, and $\Delta t' = t'_2 - t'_1$, $\Delta x' = x'_2 - x'_1$, and so forth, we have

$$\Delta t' = \Delta t$$
$$\Delta x' = \Delta x + u\Delta t$$
$$\Delta y' = \Delta y$$
$$\Delta z' = \Delta z \tag{2}$$

Since the y and z coordinates are just going along for the ride, we omit writing the transformation equations for them henceforth. Again, just as for rotations in chapter I.3, we can replace the finite differences Δt, Δx, and so on by infinitesimals dt, dx, and so forth:

$$dt' = dt$$
$$dx' = dx + udt \tag{3}$$

Adding velocities

The addition of velocities is so physically intuitive that almost everybody grasps it in everyday life. You are in a car speeding down the highway at 70 miles an hour. A fly trapped in the car flies forward at 3 miles an hour. To a hitchhiker standing by the roadside, the fly evidently moves forward at $70 + 3 = 73$ miles an hour, even though flies normally can't fly that fast. Indeed, if the hitchhiker also sees the fly moving forward at 3 miles an hour, it would have smashed into the rear window in an instant.

To formalize this intuitively obvious understanding, let us go back to the train. Ms. U tosses an object forward with velocity v; in other words, the object's trajectory is described by $x = vt$. (See figure 1, showing Ms. U as a stoker on Einstein's train.)

Figure 1 A lump of coal is tossed forward on a moving train. (Illustration adapted from *Fearful*.)

Simply plug this into (1) and we obtain, very slowly and carefully, the velocity seen by Mr. P:

$$v' \equiv \frac{dx'}{dt'} = \frac{d}{dt}(x + ut) = \frac{dx}{dt} + u = v + u \tag{4}$$

We just add the velocity of the object to the velocity of the train, as everybody would have felt intuitively. We can obtain the same result, perhaps a tad quicker, by going to (3) and dividing dx' by dt' to obtain

$$v' = \frac{dx'}{dt'} = \frac{dx + udt}{dt} = \frac{dx}{dt} + u = v + u \tag{5}$$

The calculus book I read in high school warns the reader sternly that $\frac{dx}{dt}$ is a holistic (but of course that word did not become fashionable until much later) symbol of a single mathematical entity and is not to be thought of as dx divided by dt. I am telling you that at the level of rigor of theoretical physics it is okay. Just think of the differential dx as the difference Δx, divide by Δt, and then take the Newton-Leibniz limit. When we get to general relativity, we will be constantly manipulating differentials.

We now see that the invariance of Newtonian mechanics under the Galilean transformation follows merely because Newton's law involves the second derivative, so that

$$m\frac{d^2x'}{dt'^2} = m\frac{d}{dt'}\left(\frac{dx}{dt} + u\right) = m\frac{d^2x}{dt^2} \tag{6}$$

An important point here is that this derivation even tells us when Galilean invariance of Newtonian mechanics fails. If u changes in magnitude or in direction (we had chosen $\vec{u}$ to point in the x direction, but $\vec{u}$ is really a vector!), then (6) is changed to

$$\vec{F}' = m\vec{a}' = m\vec{a} + m\frac{d\vec{u}}{dt} \tag{7}$$

An ancient part of our brains interprets this extra term as an apparent additional force: our body feels it when the driver of the car (remember, the one with the fly trapped in it) speeding down the highway suddenly slams on the brake or zips around a sharp curve.

Even someone as dumb as a fly would feel the additional force $m\frac{d\vec{u}}{dt}$ as it smashes into the windshield. Unfortunate as well as dumb.

But for now, what the fly knows is advanced stuff for us; we will get to it when we discuss gravity. Let us check that the action for Newtonian mechanics is Galilean invariant. First, for simplicity, look at the action for a single free particle in one dimension:

$$S = \int dt' \tfrac{1}{2} m \left(\frac{dx'}{dt'}\right)^2 = \int dt \tfrac{1}{2} m \left(\frac{dx}{dt} + u\right)^2$$

$$= \int dt \tfrac{1}{2} m \left(\frac{dx}{dt}\right)^2 + u \int dt\, m \frac{dx}{dt} + u^2 \int dt \tfrac{1}{2} m \tag{8}$$

The extra term linear in u in the Lagrangian is proportional to the integral of the derivative $\frac{dx}{dt}$. With fixed initial and final conditions, it is just an irrelevant additive constant. The term quadratic in u is also an additive constant. In other words, the change in the action S is just some additive constant whose variation vanishes.

This simple demonstration can be immediately generalized to the many-particle case with

$$S = \int dt \left\{ \sum_a \tfrac{1}{2} m_a \left(\frac{dx_a}{dt}\right)^2 - \sum_{a \neq b} V(x_a - x_b) \right\} \tag{9}$$

Note that it is necessary for the interaction potential to depend on the difference $x_a - x_b = x_a' - x_b'$. The generalization to higher dimensional space is trivial.

Incidentally, you might have noticed that implicit in the argument is the assumption that the two observers in relative motion agree on the same mass. I have underlined this by writing m explicitly in (6) and (8). There is no m'. Galilean relativity requires that different observers measure the same mass.

Contrary to what the guy in the street might think, the principle of relativity did not start with Einstein, but, in a sense, was reestablished by Einstein's special relativity.

Showdown between Galileo and Maxwell

While the addition of velocities (4) is so intuitively obvious, even to a layperson not versed in physics (as in my everyday example of a speeding car), it came to play a central role in the looming crisis that confronted physics toward the end of the 19th century. In his monumental work, Maxwell finally gave a precise elucidation of the mystery of light, revealing it to be an undulating electromagnetic field. An electric field varying in space and time generates a magnetic field varying in space and time, which in turn generates an electric field varying in space and time, and thus the wave propagates through space and time. The speed of propagation c depends only on how oscillating electric and magnetic fields generate each other, and that, as the reader may recall or have heard, does not depend on the observer.

On that occasion with the fly in the car, I was riding in the back seat, and I had a camera[3] with me. I took a picture of a friend riding in the front seat next to the driver and the flash went off. Telling my friend that the speed of light is* $186,000 \times 3,600 = 669,600,000$ miles per hour, I asked my friend how fast a hitchhiker standing by the roadside would have seen the flash of light go by. Her answer, indeed the only intuitively reasonable and incontrovertible answer, was $70 + 669,600,000 = 669,600,070$ miles per hour.

But this contradicts Maxwell's equations.

To read this book, for the most part you do not need to have completely mastered Maxwell's theory of electromagnetism (although it would help). I will even derive it later. At this point in our development, the single most important point is that light does not obey the law of addition of velocities (4) that everyone took to be totally obvious. For light, both observers measure the same speed:

$$c = c \tag{10}$$

As I mentioned in chapter I.1, in the showdown between these two equations, (4) and (10), the law of addition of velocities blinked and had to be modified.

This great antinomy made him stuck

Various eminent physicists in the late 19th century realized that they could reconcile the contradiction between Maxwell's theory and the law of addition of velocities if they postulated that light, just like sound, had to propagate in a medium, an ether pervading the universe. The speed of light c determined by Maxwell's theory is the speed of light as seen by an observer at rest with respect to the ether. As the earth moves through the ether, the speed of light measured on earth would vary.

Notice that the existence of the ether would have profound implications for the foundation of physics, namely, that absolute rest could be defined as rest with respect to the ether. Ms. U and Mr. P could determine who is at rest and who is moving.

As you may have heard, the experimental evidence was against the infamous ether. In 1887 (when Einstein was 8 years old), Michelson and Morley performed a famous experiment to detect the ether and failed. By the way, Einstein claimed that he was guided solely by Maxwell's equations and had never heard of the experiment.

Indeed, Einstein even contemplated his own experimental setup to look for the ether. In an impromptu speech given in December 1922 in Kyoto, Japan, describing how he had discovered special relativity, he said that he had not doubted the existence of the ether and that he had even thought of an experiment using two thermocouples to measure the difference in the heat generated by two light rays, one moving in the same direction as the earth, the other in the opposite direction.

A pair of thermocouples to measure the difference in the heat generated by the two light rays, yeah right! You might have smiled: good old Albert was a far better theorist than experimentalist. Michelson and Morley had a far better idea, to interfere the two light rays. Putting that aside, you could sense Einstein's frustration. In 1922, he said something

---

* Since this true story took place in southern California, we use "royal" rather than "revolutionary" units here.

to the effect that this "great antinomy," between Maxwell's equations and the addition of velocities, had really made him stuck.

Theoretical physicists love nothing better than a major contradiction between two well-established results, each seemingly beyond reproach. In this case, it was a shoot-out between electrodynamics and the addition of velocities. Einstein and his contemporaries were inclined to blame electromagnetic theory. The law of addition of velocities seemed rock solid. It took the cumulative genii of Lorentz, Fitzgerald, Poincaré, Einstein, and others to suspect that something was wrong with (1).

Appendix: Galilean invariance and fluid dynamics

Most texts pass over the Galilean transformation in a headlong rush toward special relativity. I like to mention that in fact, Galilean invariance offers us a powerful and often useful constraint[4] on Newtonian physics.

You may or may not know that much of fluid dynamics is governed by the Navier-Stokes equation (with ν denoting the viscosity, ρ the density, and P the pressure):

$$\frac{\partial \vec{v}}{\partial t} + (\vec{v} \cdot \vec{\nabla})\vec{v} = \nu\nabla^2\vec{v} - \frac{1}{\rho}\vec{\nabla}P \tag{11}$$

The pressure gradient ∇P provides the driving force, and the appearance of the mass density ρ comes from the m in $F = ma$. I will now give you a quick derivation using Galilean invariance.

Suppose that Ms. Unprime wants to study fluid dynamics but has never heard of the Navier-Stokes equation. She proceeds to write an equation for $\frac{\partial \vec{v}}{\partial t}$, where $\vec{v}(t, \vec{x})$ is the fluid velocity at the point $\vec{x}$ at time t. What are the possible terms in this equation? By rotation invariance, we are to construct vectors out of what is available, namely $\vec{v}(t, \vec{x})$ and $\vec{\nabla} = (\frac{\partial}{\partial x}, \frac{\partial}{\partial y}, \frac{\partial}{\partial z})$.

The key to the symmetry approach presented here is to require that whatever equation Ms. Unprime writes down has to be the same as what Mr. Prime writes down. Mr. Prime sees the fluid moving with the velocity $\vec{v}'(t', \vec{x}') = \vec{v}(t, \vec{x}) + \vec{u}$. In what follows, it is sometimes convenient to take $\vec{u}$ to point in the x direction, but it is also easy to write (1) a bit more generally: $t' = t$, $\vec{x}' = \vec{x} + \vec{u}t$. First, $\frac{\partial}{\partial x'} = \frac{\partial t}{\partial x'}\frac{\partial}{\partial t} + \frac{\partial x}{\partial x'}\frac{\partial}{\partial x} = \frac{\partial}{\partial x}$, that is, $\vec{\nabla}' = \vec{\nabla}$. Next, we have $\frac{\partial}{\partial t'} = \frac{\partial t}{\partial t'}\frac{\partial}{\partial t} + \frac{\partial \vec{x}}{\partial t'} \cdot \vec{\nabla} = \frac{\partial}{\partial t} - \vec{u} \cdot \vec{\nabla}$ (as usual, the symbol $\frac{\partial}{\partial t'}$ indicates that the partial derivative is to be taken with $\vec{x}'$ held fixed, and so in the last step, since $\vec{x} = \vec{x}' - \vec{u}t'$, we have $\frac{\partial \vec{x}}{\partial t'} = -\vec{u}$).

Let us now express what Mr. Prime writes down in terms of what Ms. Unprime would write down. First, $\frac{\partial \vec{v}'}{\partial t'} = \frac{\partial(\vec{v}+\vec{u})}{\partial t} - (\vec{u} \cdot \vec{\nabla})(\vec{v} + \vec{u}) = \frac{\partial \vec{v}}{\partial t} - (\vec{u} \cdot \vec{\nabla})\vec{v}$, since $\vec{u}$ is constant by assumption. Next, observe that

$$(\vec{v}' \cdot \vec{\nabla}')\vec{v}' = ((\vec{v}+\vec{u}) \cdot \vec{\nabla})(\vec{v}+\vec{u}) = (\vec{v} \cdot \vec{\nabla})\vec{v} + (\vec{u} \cdot \vec{\nabla})\vec{v}$$

Thus, we learn that the differential operator

$$\frac{\partial \vec{v}'}{\partial t'} + (\vec{v}' \cdot \vec{\nabla}')\vec{v}' = \frac{\partial \vec{v}}{\partial t} - (\vec{u} \cdot \vec{\nabla})\vec{v} + (\vec{v} \cdot \vec{\nabla})\vec{v} + (\vec{u} \cdot \vec{\nabla})\vec{v} = \frac{\partial \vec{v}}{\partial t} + (\vec{v} \cdot \vec{\nabla})\vec{v} \tag{12}$$

is invariant under Galilean transformation. Thus, Galilean invariance mandates that the combination $\frac{D\vec{v}}{Dt} \equiv \frac{\partial \vec{v}}{\partial t} + (\vec{v} \cdot \vec{\nabla})\vec{v}$ appears in the equation for fluid flow. We also note, more trivially, $\nabla'^2\vec{v}' = \nabla^2\vec{v}$.

Therefore, requiring that Mr. Prime and Ms. Unprime observe the same physics, we arrive at the Navier-Stokes equation[5] (11).

One final comment based on symmetry: under time reversal, $\frac{\partial}{\partial t}$ and $\vec{v}$ change sign, but not $\vec{\nabla}$. Hence, the term $\nabla^2\vec{v}$ in (11) violates time reversal. Since in Newtonian physics, time reversal is violated by friction, we can identify the coefficient ν as a measure of viscosity. If $\nu = 0$, then (11) is known as the Euler equation.

Exercise

1   In solving problems in mechanics, when we go from the lab frame to the center of mass frame, or vice versa, we are invoking Galilean invariance of Newton's laws without saying so explicitly. Here is a classic example. Let a billiard ball hit another billiard ball at rest elastically head-on. Show that the two balls move off at right angles to each other, as every pool shark knows.

Notes

1. A. Einstein, *The Meaning of Relativity*, p. 2.
2. We are implicitly assuming that even if the tossing of sticks might have done something to their lengths, this effect does not depend on whether a stick is tossed to the right or to the left. Alternatively, Mr. P and Ms. U could toss pre-cut sticks to each other.
3. By the time this book was finished, the camera had morphed into a cell phone.
4. For application to a problem on surface growth, see *QFT Nut*, chapter VI.6.
5. It is instructive to compare this symmetry-driven derivation with the standard textbook derivation, for example, J. S. Trefil, *Introduction to the Physics of Fluids and Solids,* Pergamon Press, 1975, pp. 5 and 127.

Alice said, "In *our* country, there's only one day at a time." To which the Red Queen responds, "That's a poor thin way of doing things. Now *here*, we mostly have days and nights two or three at a time, and sometimes in the winter we take as many as five nights together—for warmth, you know."

—Lewis Carroll

The patent clerk invents a clock

So, how are we to modify the Galilean transformation laws so that the speed of light c does not obey the everyday understanding of how velocities add? Somehow, particles of matter (my friends and I, and the fly, in the story from the previous chapter) and particles of light (the camera's flash) do not tally the passage of time and space in the same way.

In the prologue, we saw how Einstein, through a thought experiment, showed that simultaneity must fail. In another elegant thought experiment, Einstein proposed a clock consisting of a pulse of light bouncing between two mirrors separated by distance L (figure 1a). He was, after all, a patent examiner living in a time of technological innovations[1] of all sorts, including ever-better chronometers.[2] Ms. Unprime has one of these high-tech clocks with her. For each tick-tock, three events occur: A = light leaves the lower mirror, B = light bounces off the top mirror, and C = light arrives back at the lower mirror.

Let us write down the separation between events A and C in space and time. Evidently, $\Delta x = 0$, $\Delta y = 0$, $\Delta z = 0$, since the pulse of light gets back to where it started. By construction, $\Delta t = 2L/c$.

Mr. Prime, the observer on the ground, watches the train carrying Ms. Unprime move by with speed u in the x direction and sees a pulse of light bouncing up and down in the y direction. What is the separation between A and C as seen by Mr. Prime? Since he sees the clock moving along the x-axis, he notes that $\Delta y' = 0$, $\Delta z' = 0$ (that's what "moving along the x-axis" means). But $\Delta x'$, unlike $\Delta x = 0$, is nonzero and is given by $\Delta x' = u \Delta t'$ (that's what "moving with speed u" means).

But how do we determine $\Delta x'$ and $\Delta t'$ separately?

Use the fabulously astonishing equation $c = c$!

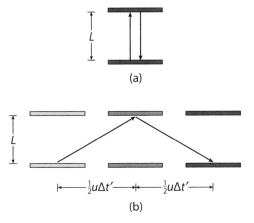

Figure 1 Einstein's clock in its rest frame (a) and in a moving frame (b).

It follows then that $\Delta t'$ is the distance traveled by the light pulse divided by c. But what is the distance traveled? Ask Mr. Pythagoras for help! We have two right-angled triangles back to back, each with right sides (figure 1b) with lengths $\frac{1}{2}u\Delta t'$ and L, and hypotenuse $\sqrt{\left(\frac{1}{2}u\Delta t'\right)^2 + L^2}$. So, between tick and tock, light travels a distance of $2\sqrt{\left(\frac{1}{2}u\Delta t'\right)^2 + L^2}$, and hence

$$c\Delta t' = 2\sqrt{\left(\frac{1}{2}u\Delta t'\right)^2 + L^2} \tag{1}$$

Anybody who got a passing grade in high school algebra could solve this equation to determine $\Delta t'$ and hence obtain $\Delta x'$. (See exercise 1.) But a much more clever strategy is to recall that $\Delta x' = u\Delta t'$ and to substitute this into (1), obtaining $c\Delta t' = 2\sqrt{(\frac{1}{2}\Delta x')^2 + L^2}$. Now square this equation to obtain $(c\Delta t')^2 = 4[(\frac{1}{2}\Delta x')^2 + L^2]$. Lo and behold, we have

$$(c\Delta t')^2 - (\Delta x')^2 = 4\left[\left(\frac{1}{2}\Delta x'\right)^2 + L^2\right] - (\Delta x')^2 = 4L^2 = (c\Delta t)^2 = (c\Delta t)^2 - (\Delta x)^2 \tag{2}$$

since $\Delta x = 0$. A fortiori, since $\Delta y' = \Delta y$ and $\Delta z' = \Delta z$ this also implies $(c\Delta t')^2 - (\Delta x')^2 - (\Delta y')^2 - (\Delta z')^2 = (c\Delta t)^2 - (\Delta x)^2 - (\Delta y)^2 - (\Delta z)^2$.

We can now consider an observer named Double Prime, with respect to whom the mirrors are moving at some other speed along the x-axis. By the same reasoning, $(c\Delta t'')^2 - (\Delta x'')^2 - (\Delta y'')^2 - (\Delta z'')^2 = (c\Delta t)^2 - (\Delta x)^2 - (\Delta y)^2 - (\Delta z)^2$. Thus, we conclude that the quadratic form $(c\Delta t)^2 - (\Delta x)^2 - (\Delta y)^2 - (\Delta z)^2$ must be the same[3] for all observers in uniform motion relative to one another.

By this clever thought experiment, Einstein used the Pythagoras theorem for space to obtain a sort of generalized Pythagoras theorem for space and time.

Distinction between a very good physicist and a great physicist! A very good physicist knows math (high school algebra in our case) and can solve equations (solve for $\Delta t'$ in our example), but a great physicist listens to what the equations are telling him or her (that Nature likes Pythagoras theorem so much that she wants to generalize it!).

Lorentz's transformation

> He meant more than all the others I have met on life's journey.
> —Einstein speaking of Lorentz

What a guy! Did he mean to include family and friends in "all the others"?

Let us now see how we can modify the Galilean transformation (III.1.1), so that $(c\Delta t)^2 - (\Delta x)^2 - (\Delta y)^2 - (\Delta z)^2$ does not depend on the observer.

First, the relation between (t', x', y', z') and (t, x, y, z) must be linear, since nothing prevents us from scaling $\{t', x', y', z', t, x, y, z\}$ by a common multiplicative factor (in other words, $\{t', x', y', z', t, x, y, z\} \rightarrow \{\lambda t', \lambda x', \lambda y', \lambda z', \lambda t, \lambda x, \lambda y, \lambda z\}$). Thus, we can't have something like t' equal to $t + ax^2$ with some constant a. The relation has to be linear.*

Second, we have the seemingly innocuous requirement that as $u \rightarrow 0$, the transformation must reduce to the Galilean transformation. But, importantly, notice that before the realization that c is a universal quantity of the universe, dimensional analysis alone would have stopped our effort cold at this point. Without c, we have only x and u to play with, and so the only quantity with dimension of time is x/u. The linearity requirement plus dimensional analysis dictates the form $t' = t + ax/u$ with some numerical constant a, but this makes no sense as $u \rightarrow 0$. We are forced to $t' = t$.

If c is not a universal constant, we are stuck with the Galilean transformation. But with c now off the bench and on the field, suddenly we have a new ball game: the combination ux/c^2 has dimension of time.

Now we can write $t' = t + \zeta ux/c^2$, with ζ some function of $\frac{u}{c}$ to be determined. But this is not yet the most general relation. We could write $t' = w(t + \zeta ux/c^2)$, with w also some function of $\frac{u}{c}$ to be determined, provided that $w(\frac{u}{c} = 0) = 1$ so that we recover the Galilean transformation.

Similarly, we can modify the Galilean relation $x' = x + ut$ to $x' = \tilde{w}(x + ut)$, where $\tilde{w}$ is also some unknown function of $\frac{u}{c}$ such that $\tilde{w}(\frac{u}{c} = 0) = 1$. Notice that we do not write $x' = \tilde{w}(x + \tilde{\zeta}ut)$, with $\tilde{\zeta}$ yet another function of $\frac{u}{c}$, because we could simply give the name u to the combination $\tilde{\zeta}u$. The relative velocity u between the two observers is defined by the statement that $x' = 0$ implies $x = -ut$. And of course we still have $y' = y$ and $z' = z$.

Let us impose the requirement that for observers in uniform relative motion, the combination $(c\Delta t)^2 - (\Delta x)^2 - (\Delta y)^2 - (\Delta z)^2$ does not depend on the observer. As already mentioned in the prologue, clearly it would be a good idea not to use some dumb English king's foot to measure distance, but instead to use something such as the lightsecond, so that[†] the speed of light $c = 1$. The algebra becomes cleaner.

---

* Another argument is that if the relation were not linear, free particle motion would look different to different observers.

[†] In other words, we want to use the same units along the t-axis and x-axis. Similarly, in studying rotations, it is a good idea (and obvious common sense) to use the same units along the x-axis and along the y-axis to measure length. Think about what rotations would look like using centimeters along the x-axis and kilometers along the y-axis.

I might add that some beginning students are nervous about c suddenly disappearing. Please be reassured that you can always restore c easily using dimensional analysis.

Since we have written our transformation such that the origins of the primed and unprimed coordinates coincide, we can simply demand $t'^2 - x'^2 = t^2 - x^2$. Expanding $w^2(t + \zeta ux)^2 - \tilde{w}^2(x + ut)^2 = t^2 - x^2$, we obtain 3 equations,* which determine the 3 unknowns to be $w = \tilde{w} = \frac{1}{\sqrt{1-u^2}}$ and $\zeta = 1$.

We have thus derived the Lorentz transformation[†] for a boost in the x direction:

$$ct' = \frac{ct + \frac{u}{c}x}{\sqrt{1 - \frac{u^2}{c^2}}}$$

$$x' = \frac{x + \frac{u}{c}ct}{\sqrt{1 - \frac{u^2}{c^2}}}$$

$$y' = y$$

$$z' = z \tag{3}$$

with c restored for the reader's convenience. You are invited to write down the Lorentz transformation for a boost in an arbitrary direction $\vec{u}$.

By construction, the Lorentz transformation reduces to the Galilean transformation in the domain of everyday experience, namely in the limit $u \ll c$. Simply take the $c \to \infty$ limit.

Since $\sqrt{1 - \frac{u^2}{c^2}}$ becomes imaginary for $u > c$, we learned that a universal speed limit $u \le c$ exists. The train cannot go faster than the speed of light without all of our equations breaking down.

Note that $cdt' = \frac{cdt + \frac{u}{c}dx}{\sqrt{1 - \frac{u^2}{c^2}}} \ne cdt$! Our fallacy was that we thought for sure that when 1 second passed for my friends and I, and the fly, 1 second also passed for the hobo hitchhiker. This assumption went into the derivation of Galileo's common sense addition law of velocities, which is so common sensical that we invoke it in everyday life without ever feeling the need to prove it.

There is no universal clock in the universe ticking off the same universal time for everyone.

---

* Namely $w^2 - \tilde{w}^2 u^2 = 1$, $w^2 \zeta^2 u^2 - \tilde{w}^2 = -1$, and $w^2 \zeta = \tilde{w}^2$, upon equating the coefficients of t^2, x^2, and tx.

[†] Interestingly, in 1887, the German physicist W. Voigt came close to having this transformation. In Voigt's transformation, the right hand side of (3) was divided by $\sqrt{1 - \frac{u^2}{c^2}}$. Not knowing Voigt's work, in 1895, Lorentz derived the transformation in a better form than Voigt's. J. Larmor found the exact form in 1900. Not knowing Larmor's work, Lorentz discovered the exact form in 1904. In 1905, H. Poincaré, knowing only of Lorentz's work, developed the transformation further and named it the Lorentz transformation. As for Einstein, he only knew the 1895 version of the Lorentz transformation. The term "Lorentz transformation" is an example of the Matthew principle: Whoever has will be given more. . . . Whoever does not have, even what he has will be taken from him (Matthew 13:12).

Light cone coordinates

So there we have it, the Lorentz transformation that replaces the Galilean transformation:

$$t' = \frac{t + ux}{\sqrt{1 - u^2}}$$
$$x' = \frac{x + ut}{\sqrt{1 - u^2}}$$
$$y' = y$$
$$z' = z \tag{4}$$

(We have set $c = 1$ once again.)

Remarkably, once we require that $t'^2 - x'^2 = t^2 - x^2$, the derivation takes only a few lines of high school algebra. In fact, the derivation becomes even simpler if we use the so-called* light cone coordinates $x^{\pm} \equiv t \pm x$ I introduced back in chapter II.3. Indeed, there I already gave you a sneak preview of the Lorentz transformation, in the context of the purely Newtonian problem of a vibrating string, not anywhere near an electromagnetic wave and its universal speed c! One of the most appealing features of theoretical physics is its unified perspective.

The key observation is the identity $a^2 - b^2 = (a + b)(a - b)$, so that $t^2 - x^2 = (t + x)(t - x) = x^+ x^-$. (As usual, we consider two observers in relative uniform motion along the x direction, with y and z merely going along for the ride and hence asking to be suppressed.) Evidently, $x^+ x^-$ is left invariant if we multiply x^+ by some factor and divide x^- by the same factor. The Lorentz transformation is strikingly simple in these coordinates:

$$x'^+ = e^{\varphi} x^+ \quad \text{and} \quad x'^- = e^{-\varphi} x^- \tag{5}$$

with φ some real parameter.

From (5) you can immediately recover (4): $t' = \frac{1}{2}(x'^+ + x'^-) = \frac{1}{2}(e^{\varphi} x^+ + e^{-\varphi} x^-) = (\cosh \varphi)t + (\sinh \varphi)x$, and similarly $x' = (\sinh \varphi)t + (\cosh \varphi)x$. It is easy to relate the boost parameter or angle φ to the relative velocity u. From the condition that $x' = 0$ implies $x = -ut$, we discover that

$$u = \frac{\sinh \varphi}{\cosh \varphi} = \tanh \varphi \tag{6}$$

(Purists might frown at physicists calling φ an angle, since it ranges from $-\infty$ to $+\infty$ as u ranges from -1 to $+1$, but the terminology has the virtue of emphasizing the connection[†] with the rotation angle.) Using the identity $\cosh^2 \varphi - \sinh^2 \varphi = 1$ mentioned back in chapter II.3, we then obtain

$$\cosh \varphi = \frac{1}{\sqrt{1 - u^2}} \quad \text{and} \quad \sinh \varphi = \frac{u}{\sqrt{1 - u^2}} \tag{7}$$

---

* The terminology will become clear in the next chapter.
[†] See appendix 1 of chapter III.3 for further discussion.

The light cone coordinates thus provide a "10-second derivation" of the Lorentz transformation. In many situations (for example, the development[4] of string theory), the coordinates $x^\pm$ are much more convenient than t and x. In the same way, an earlier generation found that rotations are more easily handled by going to complex and polar coordinates $z = x + iy = re^{i\theta}$ (and $z^* = x - iy = re^{-i\theta}$).

How velocities actually add

We can now easily derive the correct law of addition of velocities. I will let you work out the more general case (exercise 3); here we consider the simple case of an object moving in the x direction. The relevant part of (4) reads

$$t' = \frac{t + ux}{\sqrt{1 - u^2}}$$

$$x' = \frac{x + ut}{\sqrt{1 - u^2}} \tag{8}$$

Let the velocity of an object as seen by Ms. U be $v = \frac{dx}{dt}$ and as seen by Mr. P on the ground be $v' = \frac{dx'}{dt'}$. Then, dividing dx' by dt' as given by (8), we obtain

$$v' = \frac{dx'}{dt'} = \frac{u\,dt + dx}{dt + u\,dx} = \frac{\frac{dx}{dt} + u}{1 + u\frac{dx}{dt}} = \frac{v + u}{1 + uv} \tag{9}$$

Instead of $v' = v + u$, the correct law of addition of velocities contains a crucial denominator:

$$v' = \frac{v + u}{1 + uv} \tag{10}$$

The function $v' \equiv f_u(v)$ has some remarkable properties. It is symmetric under $u \leftrightarrow v$, as it should be. If the object is slowly moving with $v \ll 1$, then $v' \simeq v + u$, in accordance with everyday intuition. But if the object happens to be a particle of light so that $v = 1$, then $v' = \frac{1+u}{1+u} = 1$ independent of u, contrary to everyday intuition. If we solve (10) for $v = f_u^{-1}(v')$, we obtain $v = \frac{v'-u}{1-uv'} = f_{-u}(v')$, consistent with $f_{-u}(f_u(v)) = v$, of course. If Mr. P sees Ms. U going by with velocity u, then of course Ms. U would see Mr. P going by with velocity $-u$.

To "feel" how counterintuitive (10) is, imagine yourself carrying the ball in a game of American football, running toward the goal line at 9 meters per second. Behind you is the safety, chasing after you at 10 meters per second. You feel the safety gaining on you at a relatively benign 1 meter per second. But suppose the safety had dropped way back toward the goal line and is now coming at you at -10 meters per second. You see him fast approaching at a bone-crunching 19 meters per second, a factor of 19 faster! (See figure 2.) Now suppose the safety has strapped on a rocket moving at near light speed. Then, regardless of whether he is chasing after you or coming toward you, you see him closing in at $v = \frac{v'-u}{1-\frac{uv'}{c^2}}\big|_{v'\simeq\pm c} \simeq \pm c$. The rate of approach is almost the same in the two cases, and becomes the same as v' reaches light speed.

(a)

(b)

Figure 2 Relativity in a game of American football.

"As the ancients dreamed"

> In a certain sense, therefore, I hold it true that pure thought can grasp reality, as the ancients dreamed.
> —A. Einstein, 1954

I do not want to go into the historical controversy of whether Einstein knew about the Michelson-Morley experiment when he worked out special relativity. I am inclined to believe his statement that he didn't. The independence of c on the observer follows from Maxwell's theory, while begging the question of what medium electromagnetic waves propagate in. Instead, I discuss the existence of a speed limit.

Let us imagine how a philosopher (or perhaps an "ancient" in a civilization far far away) could have argued. Suppose there is no speed limit. Then something could have gotten from here to anywhere else in the universe in an instant. This is clearly absurd. So suppose there is a speed limit c. An observer sees this thing moving at the speed c. But another observer moving at a speed u in a direction opposite to this thing would, according to the Galilean transformation, see it moving at a speed $c + u$, but this would contradict c being a speed limit. Ergo, the common sense Galilean transformation has to be modified.

This is not how the Lorentz transformation was discovered in our civilization, but it could have happened this way elsewhere. To me, the logical fallacy in this argument by pure thought is that it may take an infinite amount of energy to make something go at infinite speed. Indeed, that is what happens in a Newtonian universe, which is logically consistent if you don't ask disallowed questions such as how the universal clock was set up. (Or, who set it up?)

This brief digression shows why it may be wise to focus on physics.

Exercises

1 Derive the Lorentz transformation directly by solving (1).

2 Show that the inverse transformation giving t, x, y, z in terms of t', x', y', z' is given by (3) with $u \to -u$.

3 Derive the law of addition of velocities in general, when $\vec{u}$ and $\vec{v}$ point in different directions. The law must obey rotational invariance.

4 Consider a series of observers, with each observer seeing the preceding observer moving away along the x-axis with speed u. Let the 0th observer see a particle moving with speed v_0 in her frame. Then the $(k+1)$st observer sees the particle moving with speed $v_{k+1} = \frac{v_k + u}{1 + u v_k}$. Find the limiting value of v_k as k tends to infinity.

Notes

1. According to the literary scholar Dame Gillian Beer, around 1865, when Lewis Carroll, an early practitioner of photography, wrote *Alice in Wonderland*, photography "froze or made portable a moment and a place." To me, that could have easily led to the concept of events in spacetime, as we will discuss in the next chapter. Carroll was notoriously concerned with the notion of time, with for example the white rabbit constantly consulting his pocket watch, an affectation and necessity when railways, with timetables and Einstein's trains, came into common use. To a physicist like myself, the two Alice books are full of allusions to concepts from physics: gravity, scale transformation, and mirror reflection, to name a few.
2. P. Galison, *Einstein's Clocks, Poincaré's Maps*.
3. Some authors state that the invariance of this quadratic form follows immediately from $(c\Delta t)^2 = (\Delta x)^2 + (\Delta y)^2 + (\Delta z)^2$ and $(c\Delta t')^2 = (\Delta x')^2 + (\Delta y')^2 + (\Delta z')^2$. At best, this argument is highly misleading and incomplete: if you know only that $(c\Delta t')^2 - (\Delta x')^2 - (\Delta y')^2 - (\Delta z')^2 = (c\Delta t)^2 - (\Delta x)^2 - (\Delta y)^2 - (\Delta z)^2$ when both sides vanish, you cannot conclude that they are equal in general. You need Einstein and his clock.
4. See, for example, B. Zwiebach, *A First Course in String Theory*, 2009.

Unifying time and space

> Henceforth space by itself, and time by itself, are doomed to fade away into mere shadows, and only a kind of union of the two will preserve an independent reality.
>
> —Hermann Minkowski

In a far reaching move, Hermann Minkowski (1864–1909) introduced geometry into special relativity. Some of the notions commonly attributed to Einstein, such as a 4-dimensional* spacetime, are actually due to Minkowski.

In Euclidean space, the invariance of the combination $dl^2 = dx^2 + dy^2 + dz^2$ under rotation allows us to define dl as the distance between two points. Two observers whose coordinate systems are related by a rotation measure the same distance. Indeed, as I emphasized in chapter I.3, the invariance of dl^2 defines rotation.

With profound insight, Minkowski realized in 1907 (a mere[1] 2 years after Einstein introduced special relativity) that the invariance of the combination $dt^2 - dx^2 - dy^2 - dz^2$ allows us to talk about the "distance" or "separation" between two points in spacetime. We saw in chapter III.2 that this invariance determines the Lorentz transformation. Similar to the case of rotation, two observers in uniform motion relative to each other can now agree on the spacetime distance[†] between two points.

---

* In fact, psychologists tell us that some of our difficulties in life stem from a natural tendency to view time as if it were like a spatial dimension, as reflected in many languages. In English, one says that, for example, the past is behind us, we are rushing toward the future, and so on.[2]

[†] As an example of a lyrical confounding of space and time, consider Rudyard Kipling's line "Damned from here to eternity," which subsequently lent itself to the title of a famous novel and film, not to mention a Yale drinking song. Sounds so much better than "from now to eternity," something that a Galilean physicist might have said.

Euclidean geometry is specified by the distance between two nearby points in space, given by $dl^2 = dx^2 + dy^2 + dz^2$, while Minkowskian geometry is specified by the distance between two nearby points in spacetime, given by $ds^2 = -dt^2 + (dx^2 + dy^2 + dz^2)$. The quantity ds^2 naturally generalizes Pythagoras's dl^2 and allows us to do geometry in space-time.[*]

The "modern" way of looking at special relativity is to emphasize the geometry of spacetime, an approach which will lead us naturally to general relativity and Einstein gravity.

Distance in Minkowskian geometry

With this singular minus[†] sign in ds^2, the geometry of Minkowski spacetime is definitely and defiantly not Euclidean. In particular, the infinitesimal quantity $ds^2 = -dt^2 + (dx^2 + dy^2 + dz^2)$ between two nearby points may not even be positive. Conceptually, it may be preferable to think of ds^2 as a peculiar symbol in its own right, not necessarily as the square of a real quantity. We say that the separation between two nearby points in spacetime is timelike if $ds^2 < 0$, spacelike if $ds^2 > 0$, and lightlike or null if $ds^2 = 0$. The term "lightlike," to be used interchangeably with "null" in this text, is evidently due to the fact that light traces out a straight line path in spacetime given by $dt^2 = (dx^2 + dy^2 + dz^2)$.

$$dt^2 > (dx^2 + dy^2 + dz^2) \quad \text{timelike}$$
$$dt^2 = (dx^2 + dy^2 + dz^2) \quad \text{lightlike or null}$$
$$dt^2 < (dx^2 + dy^2 + dz^2) \quad \text{spacelike} \tag{1}$$

The classification timelike, lightlike or null, and spacelike, can obviously be applied to curves in Minkowski spacetime, not just straight lines. A curve is timelike if the separation between any two infinitesimally nearby points on the curve is timelike. In other words, the tangent vector on a timelike curve is a timelike vector. The worldline of a massive particle is timelike. Similarly, we can define spacelike curves. The worldline of a massless particle (like a photon) is lightlike or null.

It is hardly surprising, then, that many geometrical facts we take for granted no longer hold true in Minkowskian spacetime. In particular, a straight line between two points in spacetime is not necessarily the path of shortest distance.

Define the straight line "distance" between two points separated by Δt, Δx, Δy, and Δz to be $(\Delta t)^2 - (\Delta x)^2 - (\Delta y)^2 - (\Delta z)^2$. Consider the triangle in the $(t\text{-}x)$ plane formed by the three points A $= (0, 0)$, C $= (2, 0)$, and B $= (1, x)$ for $x < 1$. See figure 1. The three sides have "lengths" $d_{\text{AC}} = \sqrt{2^2 - 0^2} = 2$ and $d_{\text{AB}} = d_{\text{BC}} = \sqrt{1 - x^2}$. Notice that

---

[*] Perhaps a more compact word is "zaum," made up from the German words "Raum" and "Zeit," which form the title of the classic book by Hermann Weyl on space and time.

[†] Imagine telling Pythagoras that time has something to do with flipping a sign in his magical formula. You would have been certified as a total nut. We now know that time differs from space by a sign, but that hardly means we understand time. Physicists' time is not the same as psychological time, whatever that is.

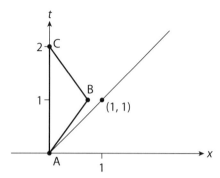

Figure 1 A straight line between two points in spacetime is not necessarily the path of shortest distance. In the triangle shown, the sum of the length of the two sides, AB and BC, is less than the length of the side AC.

$d_{AB} + d_{BC} = 2\sqrt{1 - x^2}$ is always less than $d_{AC} = 2$. Indeed, as $x \to 1$, the distances $d_{AB} = d_{BC}$ approach 0, becoming null or lightlike. That's what a minus sign can do for you!

This little example captures quite a bit of the geometry of Minkowski spacetime, as we will see. In exercise 10, you will generalize this example.

"A stubbornly persistent illusion" (?)

> With this most valiant piece of chalk I might project upon the blackboard four world-axes. Since merely one chalky axis, as it is, consists of molecules all a-thrill, and moreover is taking part in the earth's travels in the universe, it already affords us ample scope for abstraction; the somewhat greater abstraction associated with the number four is for the mathematician no infliction. . . . Then we obtain, as an image, so to speak, of the everlasting career of the substantial point, a curve in the world a *world-line*. . . . The whole universe is seen to resolve itself into similar world-lines, and I would fain anticipate myself by saying that in my opinion physical laws might find their most perfect expression as reciprocal relations between these world-lines.
>
> —Hermann Minkowski[3]

That piece of chalk was certainly valiant. In Minkowski spacetime, time and space are distinguished only by a sign, but what a sign! No doubt one of the most significant* in all of physics.

---

* Like objects repel in electromagnetism and attract in gravity. This amazing fact, which to a large extent is responsible for the physical world as we know it, can be explained in terms of this sign. See *QFT Nut*, p. 37.

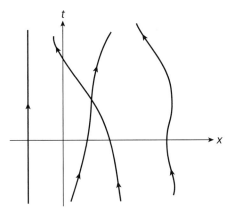

Figure 2 The worldlines of several particles.

Since an event is specified by where and when it happened, we call a point in spacetime an event. As a particle moves about in spacetime, it traces out a curve known as the worldline. (See figure 2, in which we see the worldlines of several particles, including one at rest.)

The distinguished physicist George Gamow wrote a charming autobiography attractively titled *My World Line*. You can imagine the world as consisting of a tangle of worldlines, with some coming together and intertwined with one another for a while and then moving apart. Somehow, we can experience this tangle only one time slice at a time.

This picture of a tangle of worldlines, perhaps with a reality that goes beyond time, has prompted many pseudo-philosophical ramblings. Einstein himself gave in to this temptation. After the sudden death of his school friend Michele Besso, who had helped him understand time in special relativity, Einstein wrote, only weeks before his own death (as it would turn out) to Besso's son: "Now he has departed from this strange world a little ahead of me. That signifies nothing. For us, physicists in the soul, the distinction between past, present, and future is only a stubbornly persistent illusion."[4]

Light cone

Light propagates at the speed of light, that is, with $dt^2 = dx^2 + dy^2 + dz^2$, as already mentioned above. Thus, light rays emitted from the origin of spacetime span a cone, known as the future light cone, in Minkowski space, defined by $t^2 = x^2 + y^2 + z^2$ and $t \geq 0$. Similarly, light rays that reach the origin span the past light cone defined by $t^2 = x^2 + y^2 + z^2$ and $t \leq 0$. (See figure 3, in which we have to suppress the z-axis.) Note that every point in spacetime has its own future and past light cones. Light cones everywhere!

Since a material object can't move faster than c, its worldline is subject to the constraint $dt^2 \geq dx^2 + dy^2 + dz^2$, with the equality allowed only if the object is massless. In other words, at all points along the worldline of a massive particle, its slope has to be greater

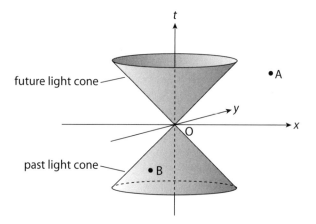

Figure 3 The future and past light cones.

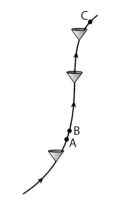

Figure 4 A particle must move inside its future
light cone at all points along its worldline.

than 45°. The particle must move inside its future light cone at all points, as indicated in figure 4.

Causality thus states that what happens at a point O in spacetime (see figure 3) can only influence what happens inside its future light cone but not what happens outside its future light cone (such as the event A in figure 3). Similarly, only events that occur inside its past light cone (such as the event B in figure 3) can influence what happens at O.

Note that if we restore c, the light cone flattens out as $c \to \infty$, so that the future light cone encompasses all of future $t \geq 0$, and we are back to the Galilean view of space and time. (In figure 5, call where we are sitting the origin of spacetime; then the entire shaded region is in our past "light cone.")

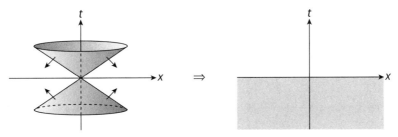

Figure 5 The Galilean limit of the past light cone.

Proper time

Consider two events A and B, infinitesimally separated, on the worldline shown in figure 4. Our two observers Ms. Unprime and Mr. Prime agree on the Minkowskian distance between A and B, namely $dt^2 - dx^2 - dy^2 - dz^2 = dt'^2 - dx'^2 - dy'^2 - dz'^2$. That's the whole point of special relativity! I will go almost painfully slow here for reasons that will become clear. First, let the figure shown be the one Ms. Unprime would actually draw, using her clocks and rulers. Mr. Prime would draw a different, but analogous, figure (which we are not showing) using his clocks and rulers.

Suppose the worldline in figure 4 is actually that traced out by a Dr. D, using coordinates t'', x'', y'', and z''. Since the worldline is curved, Dr. D is actually accelerating this way and that, definitely not an inertial kind of guy. To Dr. D, the spatial separation between A and B is given by $dx'' = 0$, $dy'' = 0$, $dz'' = 0$, of course. You are not going anywhere in your rest frame, by definition.

Now, special relativity informs us that the quantity $dt''^2 - dx''^2 - dy''^2 - dz''^2 = dt^2 - dx^2 - dy^2 - dz^2 = dt'^2 - dx'^2 - dy'^2 - dz'^2$ is the same for all observers. Thus, it makes sense for Ms. Unprime to define

$$d\tau^2 \equiv dt^2 - dx^2 - dy^2 - dz^2 \qquad (2)$$

Since $dx'' = 0$, $dy'' = 0$, $dz'' = 0$, we have $d\tau^2 = dt''^2 - dx''^2 - dy''^2 - dz''^2 = dt''^2$, so that we can interpret $d\tau$ as the actual biological time elapsed between A and B as experienced by Dr. D, were he or she a biological organism. We call $d\tau$ the proper time interval* for Dr. D.

Notice that $d\tau$ is not the proper time lapse as experienced by Ms. Unprime. Nor is it the proper time as experienced by Mr. Prime. The point of special relativity is that Ms. Unprime and Mr. Prime agree that $d\tau$ is the proper time experienced by Dr. D.

---

* The term "proper" time is meant to refer to the time felt by Dr. D him- or herself, but to me hints of a lesson in etiquette. A better term would have been "eigenzeit" or "eigentime," as in "eigenvalue" or "eigenvector."

Distance measure in spacetime

The proper time naturally provides a "distance" measure in spacetime along an observer's worldline. What is the proper time experienced by Dr. D between two events A and C far apart on his or her worldline? In physics, we assume that time is additive, and so we simply sum up the infinitesimal proper time lapse to obtain $\int_A^C d\tau$. The statement, inherent in Riemannian geometry and Einstein gravity, is that space and time can be experienced in infinitesimal bite-sized segments. I don't see how we could even do physics without this, but of course it is still an assumption.

We have (compare this with the length of a curve given in (II.2.2))

$$\int_A^C d\tau = \int_A^C \sqrt{dt^2 - d\vec{x}^2} = \int_A^C d\tau \sqrt{\left(\frac{dt}{d\tau}\right)^2 - \left(\frac{d\vec{x}}{d\tau}\right)^2} = \int_A^C dt \sqrt{1 - \left(\frac{d\vec{x}}{dt}\right)^2} \tag{3}$$

I have purposely given four different expressions for the spacetime distance between A and C. The second expression emphasizes that it is completely parametrization invariant. The third expression uses the proper time itself as the integration parameter. The fourth expression uses coordinate time as the parameter. Whose coordinate time? Ms. Unprime's.

The fourth expression in (3) also explains our observation about the triangle in figure 1. The line AC is in fact the path of longest distance between A and C, since $\frac{d\vec{x}}{dt} = 0$ maximizes the square root in that fourth expression in (3). Any curve (as long as it doesn't have any spacelike segment, for which the proper time interval in the integrand would be imaginary) joining A and C will be shorter in length, thanks to the minus sign in (3).

One must exercise considerable care in looking at spacetime diagrams such as those in the next chapter. It is easy to fall into the trap of thinking Euclidean.

Motion of a free particle

Students sometimes feel that the equation of motion of a particle can be derived somehow. To be contrary, it has to be abstracted from empirical observations and then enunciated by some great physicist, made great by said enunciation. Newton enunciated that, in the absence of external forces, a particle will maintain a constant velocity. In particular, if the particle is at rest, it will remain at rest. Hence the equation of motion $\frac{d^2\vec{X}}{dt^2} = 0$, or in other words, $\frac{d\vec{X}}{dt}$ stays unchanged.

How does a free particle move in special relativity? Again, the answer had to be enunciated by Einstein and then verified by experiments. But since Einstein came after Newton, the postulated equation must reduce to Newton's equation for a slowly moving particle. In addition, there is the very stringent requirement that Ms. Unprime and Mr. Prime have to agree that the particle is free. Both of these requirements are satisfied by

$$\frac{d^2 X^\mu}{d\tau^2} = 0 \tag{4}$$

Here $X^\mu(\tau) = (X^0(\tau), X^i(\tau))$, with $i = 1, 2, 3$, denotes the location of the particle in spacetime as measured by Ms. Unprime, and τ is the proper time determined by $d\tau = \sqrt{(dX^0)^2 - (d\vec{X})^2}$. To see what equation Mr. Prime subscribes to, we simply note that X' is related to X by a linear transformation, $X'^\mu = \Lambda^\mu{}_\nu X^\nu$, and so

$$\frac{d^2 X'^\mu}{d\tau^2} = \Lambda^\mu{}_\nu \frac{d^2 X^\nu}{d\tau^2} = 0 \tag{5}$$

and that $d\tau$ is the same for Ms. Unprime and Mr. Prime. Note that all we need here is that the relation between X' and X must be linear and that $\Lambda^\mu{}_\nu$ must depend only on the relative velocity between the two observers, so that it can pass through $\frac{d^2}{d\tau^2}$ untouched. We will come back to the issue of how physical quantities observed by Ms. Unprime and Mr. Prime are related in chapter III.6.

The Minkowskian metric for spacetime

Clearly, a more compact notation would be desirable. Let us write $d\tau^2 = dt^2 - dx^2 - dy^2 - dz^2 = -\eta_{\mu\nu} dx^\mu dx^\nu$, with $x^0 \equiv t$ and $\eta_{\mu\nu}$ defined by

$$\eta_{00} = -1, \quad \eta_{11} = \eta_{22} = \eta_{33} = +1, \quad \text{and} \quad \eta_{\mu\nu} = 0 \quad \text{if } \mu \neq \nu \tag{6}$$

As always, the Einstein summation convention holds unless stated otherwise.

You should be reminded of the distance squared between two nearby points in generically curved spaces $ds^2 = g_{\mu\nu} dx^\mu dx^\nu$, with the metric $g_{\mu\nu}$ that we studied in chapters I.5 and I.6. We may regard $\eta_{\mu\nu}$ as the flat Minkowskian metric of spacetime, just as we regard δ^{ij} as the Euclidean metric of ordinary flat space. Geometry was originally the science of measuring the earth; here we are measuring spacetime.

From here on, the discussion parallels completely the discussion of rotation in chapter I.3 and of curved spaces in chapters I.5 and I.6, except that $\eta_{\mu\nu}$ replaces δ^{ij} and $g_{\mu\nu}$, respectively. Inevitably, I will repeat some of the earlier discussions, as I will be talking about vectors and tensors, upper and lower indices, all that good stuff. Given the confusion that some beginning students have, I feel quite strongly that some repetition is worthwhile. Indeed, my pedagogical strategy in this text is to proceed as follows:

rotation → coordinate transformation → curved space → Minkowskian spacetime
→ curved spacetime

These five topics can, and should, be treated as an organic whole.

As in our earlier discussions, $dx^\mu = (dt, dx, dy, dz)$ defines the basic or ur-vector. A vector p^μ in spacetime is defined as a set of four numbers $p^\mu = (p^0, p^1, p^2, p^3)$ that transform in the same way as dx^μ transforms under the Lorentz transformation. It is sometimes called a 4-vector to distinguish it from ordinary 3-vectors in space. Evidently, the 4-vector p^μ contains the 3-vector $p^i = (p^1, p^2, p^3)$.

The square of the length of the 4-vector p is defined as $p^2 \equiv p \cdot p \equiv \eta_{\mu\nu} p^\mu p^\nu$. (The dot will be often omitted henceforth.) Indeed, just as rotations can be defined as linear transformations that leave the lengths of 3-vectors unchanged, Lorentz transformations are defined as linear transformations that leave the (Minkowskian) lengths of 4-vectors unchanged. In particular, $d\tau^2 = \eta_{\mu\nu}dx^\mu dx^\nu$ is left invariant.

For two arbitrary 4-vectors p and q, consider the vector $p + \alpha q$ (for α an arbitrary real number) and its length squared $p^2 + 2\alpha p \cdot q + \alpha^2 q^2$. Since α is arbitrary and since Lorentz transformations leave lengths unchanged, they also leave the scalar dot product between two 4-vectors

$$p \cdot q \equiv \eta_{\mu\nu}p^\mu q^\nu = -p^0 q^0 + p^1 q^1 + p^2 q^2 + p^3 q^3 \tag{7}$$

unchanged. (Recall that we used a similar argument in chapter I.3.)

Indices upstairs and downstairs

Earlier, in our discussion of curved surfaces in chapter I.6, I snuck in lower indices by writing the indices on $g_{\mu\nu}$ as subscripts. Here, I have done the same, writing $\eta_{\mu\nu}$ as an object carrying lower indices.

Thus far, $\eta_{\mu\nu}$ is the only object with lower indices. When we want to sum over two indices μ and ν, the rule is that we multiply by $\eta_{\mu\nu}$ and invoke Einstein's repeated summation convention. We say that we have contracted the two indices. For example, given two vectors p^μ and q^μ, we might want to contract the indices μ and ν in $p^\mu q^\nu$ and obtain $p \cdot q \equiv \eta_{\mu\nu}p^\mu q^\nu$. Another example: given $p^\mu q^\nu r^\rho s^\sigma$, suppose we want to contract μ with ρ. Easy, just write $\eta_{\mu\sigma}\eta_{\nu\rho}p^\mu q^\nu r^\rho s^\sigma = (p \cdot s)(q \cdot r)$. Savvy readers will recognize that I am going painfully slowly here for the sake of those who have never seen this material before.

So far so good. All vectors carry upper indices, and the only object that carries lower indices is η.

The next step is purely for the sake of notational brevity. To save ourselves from constantly writing the Minkowski metric $\eta_{\mu\nu}$, we define, when we are given a vector p^μ, a vector with a lower index

$$p_\nu \equiv \eta_{\mu\nu}p^\mu \tag{8}$$

In other words, if $p^\mu = (p^0, \vec{p})$ then $p_\mu = (-p^0, \vec{p})$. Thus, $p \cdot q = p_\mu q^\mu = -p^0 q^0 + \vec{p} \cdot \vec{q}$. (Notice that the same dot in this last equation carries two different meanings: the scalar product between two 4-vectors on the left hand side and between two 3-vectors on the right hand side, but there should be no confusion.) With this notation, we can write $p \cdot q = p_\nu q^\nu = p^\nu q_\nu$. Similarly, an expression $\eta_{\mu\nu}p^\mu q^\nu \eta_{\rho\sigma}r^\rho s^\sigma$ can be written more simply as $p_\nu q^\nu r_\sigma s^\sigma$. The Minkowski metric has been folded into the indices, so to speak.

Just a convenient notation

Unaccountably, some students are twisted out of shape by this trivial act of notational sloth. "What?" they say, "There are two kinds of vectors?" Yes, fancy people speak of contravariant vectors (p^μ for example) and covariant vectors (p_μ for example), but let me assure the beginners that there is nothing terribly profound* going on here. Just a convenient notation.

Let us immediately clear up some potential questions about the notation. Some students have asked why there isn't a distinction between upper and lower indices for ordinary vectors. The answer is that we could have, if we wanted to, written the Euclidean metric δ_{ij} with lower indices back in chapter I.3 and risked confusing the reader at that early stage. But there is no strong incentive for doing that: the Euclidean metric does not contain any minus signs, while the Minkowskian metric necessarily has one negative sign and three positive signs to distinguish time from space. The upper and lower index notation serves to keep track of the minus signs. In the Euclidean case, if we define $p_i = \delta_{ij} p^j$, the vector p_i would be numerically the same as the vector p^i. In Minkowski space, $p_1 = p^1$, $p_2 = p^2$, and $p_3 = p^3$, but $p_0 = -p^0$.

The next question might be: given p_μ, how do we get back to p^μ?

Here is where I think beginners can get a bit confused. If you have any math sense at all, you would expect that we use the inverse of η, and you would be absolutely right. Surely, if you use η to move indices downstairs, you would use the inverse of η to move them upstairs. But the inverse of the matrix

$$\begin{pmatrix} -1 & 0 & 0 & 0 \\ 0 & +1 & 0 & 0 \\ 0 & 0 & +1 & 0 \\ 0 & 0 & 0 & +1 \end{pmatrix}$$

is itself. So traditionally, the inverse of η is denoted by the same symbol, but with two upper indices, like this: $\eta^{\mu\nu}$. We define $\eta^{\mu\nu}$ by $\eta^{00} = -1$, $\eta^{11} = \eta^{22} = \eta^{33} = +1$, and $\eta^{\mu\nu} = 0$ if $\mu \neq \nu$.

Indeed, $\eta^{\mu\nu}$ is the inverse of $\eta_{\mu\nu}$ regarded as a matrix: $\eta^{\mu\nu}\eta_{\nu\lambda} = \delta^\mu_\lambda$, where the Kronecker delta δ^μ_λ is defined, as usual, to be 1 if $\mu = \lambda$ and 0 otherwise. It is worth emphasizing that while $\eta^{\mu\nu}$ and $\eta_{\mu\nu}$ are numerically the same matrix, they should be distinguished conceptually. Let us check the obvious, that the inverse metric $\eta^{\mu\nu}$ raises lower indices: $\eta^{\mu\nu} p_\nu = \eta^{\mu\nu}\eta_{\nu\lambda} p^\lambda = \delta^\mu_\lambda p^\lambda = p^\mu$. Yes, indeed.

Confusio: "Ah, I get it. The same symbol η is used to denote a matrix and its inverse, distinguished by whether η carries lower or upper indices."

From this we see that the Kronecker delta δ^μ_λ has to be written with one upper and one lower index. In contrast, η^μ_ν does not exist. And there is no such thing as $\delta^{\mu\nu}$ or $\delta_{\mu\nu}$. Also note that the Kronecker delta δ does not contain any minus signs, unlike the Minkowski metric η.

It follows that the shorthand ∂_μ for $\frac{\partial}{\partial x^\mu}$ has to carry a lower index, because $\partial_\mu x^\nu = \frac{\partial x^\nu}{\partial x^\mu} = \delta^\nu_\mu$. In other words, for the indices to match, ∂_μ must be written with a lower index. This

---

* Of course, if you woke up one day and discovered that you were a mathematician or a mathematician-want-to-be, you should and could read more profound books.

makes sense, since the coordinates x^μ carry an upper index but in $\frac{\partial}{\partial x^\mu}$, it appears in the denominator, so to speak. We will use this fact repeatedly. Once again, we already know this from chapters I.5 and I.6.

Let me emphasize again an extremely useful feature of this notational device. In the Einstein convention, a lower index is always contracted with an upper index (that is, summed over), and vice versa. Never never sum two lower indices together, or two upper indices together! If you ever encounter an expression in which two lower (or two upper) indices are summed over, you know that there is a mistake* somewhere.

At the risk of repeating the obvious, remember, there is nothing profound[†] going on here. The whole business of introducing upper and lower indices is just for notational convenience. You are to read (8) as merely a definition introduced to save writing.

Spacelike and null surfaces

Now that space is married to time, what do we mean by the term "space"? Evidently, a $t =$ constant slice of Minkowski spacetime can be regarded as space. But the time and space axes of another observer would in general be tilted with respect to yours, so a tilted slice should also count. Such considerations lead us to generalize the concept of space to spacelike surfaces to be defined below. Well, we learned how to characterize surfaces embedded in Euclidean space in chapters I.6 and I.7, defining tangent vectors lying in the surface and normal vectors (usually only one of them) perpendicular to the surface. Here we do the same for surfaces embedded in Minkowski spacetime.

So, a surface in Minkowski spacetime consists of the set of points satisfying some (reasonable) equation of the form $F(x^0, x^1, x^2, x^3) = 0$ (a special case of which is $x^0 = f(x^1, x^2, x^3)$). Here the word "surface" is used in a generalized sense, not necessarily a 2-dimensional surface of the kind we encounter in everyday life. The Jargon Guy yells, "Call it a hypersurface!" but we ignore him. A surface is called spacelike if the separation between any two infinitesimally nearby points on the surface is spacelike. The three tangent vectors are everywhere spacelike. The normal to the surface is then a timelike vector. Notice that this definition allows for the possibility that the surface is curved. Thus, when we get to curved spacetime in part V, the same definition is still serviceable and provides what we mean by "space."

---

* In practice, this evident truth is used as follows: people can afford to be sloppy in intermediate stages of a calculation, but then they move indices up and down at the end to satisfy this rule.

[†] Once, when I taught special relativity, I surveyed the students to find out what, if anything, they found confusing or deficient in the textbook I used. One student, a kind of super-Confusio, told me that the textbook never explicitly said that $p_\mu q^\mu$ and $p_\lambda q^\lambda$ are actually the same and it took him, poor fellow, a long time to figure it out. Let it be recorded that this textbook explicitly states that $p_0 q^0 + p_1 q^1 + p_2 q^2 + p_3 q^3 = \sum_\kappa p_\kappa q^\kappa = p_\sigma q^\sigma = \eta_{\sigma\rho} p^\rho q^\sigma = p^\rho \eta_{\sigma\rho} q^\sigma = p^\rho \eta_{\rho\sigma} q^\sigma = p^\rho q_\rho = p_\xi q^\xi = p_\pi q^\pi$ (with the last choice, although perfectly okay, not generally advisable, given the cultural baggage[5] associated with π). Got that? The so-called dummy summation variables are just dummies to keep track of things.

Surfaces generated by light rays, not surprisingly, form another important class of surfaces known as null surfaces. Write Minkowski spacetime in spherical coordinates, with $ds^2 = -dt^2 + dr^2 + r^2 d\theta^2 + r^2 \sin^2 \theta d\varphi^2$. In these coordinates, the metric is given by $g_{tt} = -1$, $g_{rr} = 1$, $g_{\theta\theta} = r^2$, $g_{\varphi\varphi} = r^2 \sin^2 \theta$, with all other components equal to 0 (in other words, exactly the same as in chapter I.5 except for the addition of the time coordinate). Now consider the light cone defined by the equation $t = r$ in these coordinates (t, r, θ, φ) and formed physically by light rays going radially outward from the origin, along the trajectory $(dt, dr, d\theta, d\varphi) \propto l^\mu = (1, 1, 0, 0)$. This tangent vector along the light cone is clearly null, since $g_{\mu\nu}l^\mu l^\nu = g_{tt}l^t l^t + g_{rr}l^r l^r = -1 + 1 = 0$.

The other two tangent vectors are given by

$$h^\mu = (0, 0, r^{-1}, 0) \quad \text{and} \quad k^\mu = (0, 0, 0, (r \sin \theta)^{-1})$$

(We have normalized them to $g_{\mu\nu}h^\mu h^\nu = 1$ and $g_{\mu\nu}k^\mu k^\nu = 1$.) Evidently, they are spacelike and orthogonal to l^μ, since $g_{\mu\nu}l^\mu h^\nu = 0$ and $g_{\mu\nu}l^\mu k^\nu = 0$. They are also mutually orthogonal, since $g_{\mu\nu}h^\mu k^\nu = 0$. These three 4-vectors l, h, k furnish the three tangent vectors to the surface.

Now comes the fun part. What is a 4-vector normal to the null surface? The answer is evidently l itself! Since $g_{\mu\nu}l^\mu l^\nu = 0$, l is "Minkowski perpendicular" to itself, and furthermore, to h and k. In other words, l is Minkowski perpendicular to all three tangent vectors lying in the null surface.

We could utter the following peculiar-sounding statement: the normal to a null surface is a null vector that lies in the surface. Minus signs could do "funny tricks" for us that Euclid never dreamed of!

The null surface of the light cone has another peculiar property: it's a "one way" surface. Think of a massive particle moving along a timelike worldline. Once it enters into a given light cone, it can't get out again. If we think of the null surface as a membrane, it is permeable only in one direction just like a certain hotel (in California): you can check in, but you can never leave.

The relativistic Doppler shift

As a simple application of the 4-vector formalism and as a break from a rather formal discussion, let's derive the relativistic Doppler effect. Consider an electromagnetic wave observed by Ms. Unprime and described schematically by $\cos(\omega t - \vec{k} \cdot \vec{x}) = \cos kx = \cos \eta_{\mu\nu}k^\mu x^\nu$, where we have defined the 4-vector $k = (\omega, \vec{k})$ with the circular frequency ω and the wave vector $\vec{k}$. As usual, $\omega^2 = \vec{k}^2$. The physical requirement $k'x' = kx$ is satisfied if k transforms like x, that is, like a 4-vector.

What is the frequency and wave vector observed by Mr. Prime? The answer, namely the relativistic Doppler formula, follows almost instantly from the Lorentz transformation: $\omega' = (\omega + uk_x)/\sqrt{1 - u^2}$, $k'_x = (k_x + u\omega)/\sqrt{1 - u^2}$, $k'_y = k_y$, $k'_z = k_z$. We obtain thus

$$\omega' = \omega(1 + u \cos \theta)/\sqrt{1 - u^2} \tag{9}$$

where θ is the angle between $\vec{k}$ and $\vec{u}$. As the train is approaching Mr. Prime, $\theta \simeq 0$, $\omega' \simeq \omega\sqrt{\frac{1+u}{1-u}}$, and so the frequency observed by Mr. Prime increases; in other words, it is blue shifted. As the train recedes, $\theta \simeq \pi$, $\omega' \simeq \omega\sqrt{\frac{1-u}{1+u}}$, and the frequency is redshifted. Compare this derivation with the elementary nonrelativistic discussion involving, if the source is receding along the line of sight, the extra distance the next crest of the wave has to travel and so on and so forth. Note that we have an extra relativistic factor of $1/\sqrt{1-u^2}$, which we will identify in the next chapter as due to time dilation.

One unified language

We derived the Lorentz transformation in the preceding chapter, but it is instructive to derive it again. Let p and q be two arbitrary 4-vectors. Consider the linear transformation

$$p'^{\mu} = \Lambda^{\mu}_{\ \sigma} p^{\sigma} \quad \text{and} \quad q'^{\mu} = \Lambda^{\mu}_{\ \sigma} q^{\sigma} \tag{10}$$

Notice the upper-lower summation convention. For Λ to be a Lorentz transformation, we require $p' \cdot q' = p \cdot q$, that is,

$$p' \cdot q' = \eta_{\mu\nu} p'^{\mu} q'^{\nu} = \eta_{\mu\nu} \Lambda^{\mu}_{\ \sigma} p^{\sigma} \Lambda^{\nu}_{\ \rho} q^{\rho} = p \cdot q = \eta_{\sigma\rho} p^{\sigma} q^{\rho} \tag{11}$$

Since p^{σ} and q^{ρ} are arbitrary, Λ must satisfy

$$\eta_{\mu\nu} \Lambda^{\mu}_{\ \sigma} \Lambda^{\nu}_{\ \rho} = \eta_{\sigma\rho} \tag{12}$$

Just as we determined rotations as transformations that left $\vec{p} \cdot \vec{q}$ invariant (in chapter I.3), here we determine Lorentz transformations as those transformations that leave $p \cdot q$ invariant. You also may recognize that $\Lambda^{\mu}_{\ \sigma}$ is playing the same role as $S^{\mu}_{\ \sigma}$ in chapter I.5. Indeed, in parallel with the discussion there, let us define the transpose by $(\Lambda^{T})_{\sigma}^{\ \mu} = \Lambda^{\mu}_{\ \sigma}$ (note the position of the indices!), so that we may write (12) as $(\Lambda^{T})_{\sigma}^{\ \mu} \eta_{\mu\nu} \Lambda^{\nu}_{\ \rho} = \eta_{\sigma\rho}$, or more succinctly, $\Lambda^{T} \eta \Lambda = \eta$.

I find it rather pleasing to have one unified language to describe four apparently different subjects: rotation, change of coordinates, flat space, and Lorentz transformation. As I have alluded to and as you will soon see, the same language is used in studying curved spacetimes.

For the sake of the super-Confusio mentioned in the preceding footnote and first alluded to in chapter I.6, let me stress once again that repeated indices are summed over and so can be represented by any letter we wish, as long as it's a letter in whatever alphabet you are using that we haven't yet used in the same expression. For instance, we could write (11) just as well as $p' \cdot q' = \eta_{\varphi\psi} p'^{\varphi} q'^{\psi} = \eta_{\varphi\nu} \Lambda^{\varphi}_{\ \kappa} p^{\kappa} \Lambda^{\nu}_{\ \mu} q^{\mu} = p \cdot q = \eta_{\kappa\mu} p^{\kappa} q^{\mu}$. Since there are only so many letters in the alphabets commonly used in physics, you will often see the same expression written (as in the example here) with completely different letters, particularly when we get to general relativity.

Lorentz algebra

Following Sophus Lie once again, we consider an infinitesimal transformation $\Lambda^\mu_{\ \sigma} \simeq \delta^\mu_{\ \sigma} + \varphi \mathcal{K}^\mu_{\ \sigma}$ (with the Kronecker delta defined earlier), just as in chapter I.3 we considered an infinitesimal rotation $R \simeq I + \theta \mathcal{J}$. As indicated above for $(\Lambda^T)_{\ \sigma}^\mu$, we should also define $(\mathcal{K}^T)_{\ \sigma}^\mu = \mathcal{K}^\mu_{\ \sigma}$. Inserting this infinitesimal transformation into (12), we obtain, to leading order in φ,

$$\mathcal{K}^\mu_{\ \sigma} \eta_{\mu\rho} + \eta_{\sigma\nu} \mathcal{K}^\nu_{\ \rho} = 0 \tag{13}$$

which we can write as

$$\mathcal{K}^T \eta = -\eta \mathcal{K} \tag{14}$$

We are to solve (14) for the unknown 4-by-4 matrix $\mathcal{K}$, but actually this problem involves only 2-by-2 matrices. Consider a boost in the x direction. Since $y' = y$ and $z' = z$, there is no point in dragging them around, and we can focus on the 2-dimensional space spanned by t and x, so that effectively $\eta = \begin{pmatrix} -1 & 0 \\ 0 & +1 \end{pmatrix}$ and we are to solve (14) for the effectively 2-by-2 matrix $\mathcal{K}$. The solution is

$$\mathcal{K} = \begin{pmatrix} 0 & 1 \\ 1 & 0 \end{pmatrix} \tag{15}$$

(If you have taken a course on quantum mechanics and know what Pauli matrices σ_i are, you can see that $\mathcal{K}$ and $-\eta$ are just σ_1 and σ_3, the first and the third Pauli matrix, respectively. Finding the solution is a snap. Noting that σ_1 is symmetric and that it anticommutes with σ_3 gives us (15) immediately.)

You should appreciate, exactly as in chapter I.3 for rotations, how easy it is to solve the Lorentz condition for infinitesimal boosts. To leading order in φ, with $t' \simeq t + \varphi x$ and $x' \simeq x + \varphi t$, we have $t'^2 - x'^2 \simeq t^2 - x^2 + 2\varphi(tx - xt) = t^2 - x^2$.

Note a crucial difference between infinitesimal boosts and rotations: $\mathcal{K}$ is symmetric, while in contrast, $\mathcal{J}$ is antisymmetric (see I.3.7). Indeed, if we replace $\eta \to I$, $\mathcal{K} \to \mathcal{J}$ in (14), we obtain $\mathcal{J}^T = -\mathcal{J}$.

As Lie had assured us, just as in our discussion of rotation, once we have determined the infinitesimal boost, we can generate finite boosts by repeatedly boosting by an infinitesimal amount. For φ finite and N large, write $\Lambda(\frac{\varphi}{N}) \simeq I + \frac{\varphi}{N}\mathcal{K} \simeq e^{\frac{\varphi}{N}\mathcal{K}}$. For a finite boost, then $\Lambda(\varphi) = (\Lambda(\frac{\varphi}{N}))^N = e^{\varphi \mathcal{K}}$. Expanding the series, we obtain

$$\Lambda(\varphi) = e^{\varphi \mathcal{K}} = \sum_{n=0}^{\infty} (\varphi)^n \mathcal{K}^n / n! = \left(\sum_{k=0}^{\infty} \varphi^{2k}/(2k)! \right) I + \left(\sum_{k=0}^{\infty} \varphi^{2k+1}/(2k+1)! \right) \mathcal{K}$$

$$= \cosh \varphi \, I + \sinh \varphi \, \mathcal{K}$$

$$= \begin{pmatrix} \cosh \varphi & \sinh \varphi \\ \sinh \varphi & \cosh \varphi \end{pmatrix} \tag{16}$$

Note that we used $\mathcal{K}^2 = 1$ in crucial contrast to $\mathcal{J}^2 = -1$ for rotation. Thus, we obtain

$$t' = \cosh \varphi \, t + \sinh \varphi \, x$$
$$x' = \sinh \varphi \, t + \cosh \varphi \, x$$
$$y' = y$$
$$z' = z \qquad (17)$$

with y and z brought back in.

And thus, once again we have derived the Lorentz transformation. I hope that you appreciate how elegant Lie's method is compared to the brute force method we used in the preceding chapter.

For two successive boosts in the x direction, $\Lambda(\varphi_1)\Lambda(\varphi_2) = \Lambda(\varphi_1 + \varphi_2)$. The parameter φ, sometimes called rapidity, combines additively. This also implies that $\Lambda^{-1}(\varphi) = \Lambda(-\varphi)$.

Also, (13) can be rearranged as $\eta_{\rho\mu}\mathcal{K}^\mu_{\ \sigma}\eta^{\sigma\nu} = -\mathcal{K}^\nu_{\ \rho}$, which implies that $\eta e^{\varphi\mathcal{K}}\eta = e^{-\varphi\mathcal{K}}$ and hence (since $\Lambda = e^{\varphi\mathcal{K}}$) $\eta_{\rho\mu}\Lambda^\mu_{\ \sigma}\eta^{\sigma\nu} = (\Lambda^{-1})^\nu_{\ \rho}$. We could of course verify this explicitly using (16):

$$\begin{pmatrix} -1 & 0 \\ 0 & +1 \end{pmatrix} \begin{pmatrix} \cosh \varphi & \sinh \varphi \\ \sinh \varphi & \cosh \varphi \end{pmatrix} \begin{pmatrix} -1 & 0 \\ 0 & +1 \end{pmatrix} = \begin{pmatrix} \cosh \varphi & -\sinh \varphi \\ -\sinh \varphi & \cosh \varphi \end{pmatrix} \qquad (18)$$

We note for future use that since $\cosh^2 \varphi - \sinh^2 \varphi = 1$, the Jacobian det Λ for Lorentz transformations, just as for rotations, is equal to 1. Thus, $d^4x' = d^4x$: spacetime volume is unchanged.

It is instructive at this point to work out how p_ν transforms:

$$p'_\nu \equiv \eta_{\nu\mu}p'^\mu = \eta_{\nu\mu}\Lambda^\mu_{\ \sigma}p^\sigma = \eta_{\nu\mu}\Lambda^\mu_{\ \sigma}\eta^{\sigma\rho}p_\rho = (\Lambda^{-1})^\rho_{\ \nu}p_\rho \qquad (19)$$

Thus, while a vector with an upper index transforms with Λ, a vector with a lower index transforms with Λ^{-1}. They transform oppositely.

Note carefully the locations of various indices in the above discussion. Note also from (16) that Λ and Λ^{-1} are symmetric as matrices. As an exercise, we could verify explicitly that $\partial_\mu \equiv \frac{\partial}{\partial x^\mu}$ transform like a vector with a lower index, as we argued earlier. Since $x'^\mu = \Lambda^\mu_{\ \nu}x^\nu$, we have $x^\nu = (\Lambda^{-1})^\nu_{\ \rho}x'^\rho$ and hence $\partial'_\mu = \frac{\partial}{\partial x'^\mu} = \frac{\partial x^\nu}{\partial x'^\mu}\frac{\partial}{\partial x^\nu} = (\Lambda^{-1})^\nu_{\ \mu}\partial_\nu$.

Lorentz tensors

Henceforth, physics is required to be invariant not only under rotations but also under boosts. (Clearly, in addition to the boost along the x direction displayed in (17), we can also boost along the y and z directions.) The set of all rotations and boosts is known as the Lorentz group, which we will discuss in more detail in appendix 1. The rotation group is evidently a subgroup of the Lorentz group.

The concept of tensors as discussed in chapter I.4 can be immediately generalized. Mimicking the discussion for the rotation group, we immediately define a tensor with 2 upper indices $T^{\mu\nu}$ to be something that transforms as $T^{\mu\nu} \to T'^{\mu\nu} = \Lambda^\mu_{\ \sigma}\Lambda^\nu_{\ \omega}T^{\sigma\omega}$. The earlier discussion goes through; in particular, the symmetric and antisymmetric parts

($S^{\mu\nu} \equiv T^{\mu\nu} + T^{\nu\mu}$ and $A^{\mu\nu} \equiv T^{\mu\nu} - T^{\nu\mu}$) of $T^{\mu\nu}$ transform separately, furnishing a $4 \cdot 5/2 = 10$-dimensional and a $4 \cdot 3/2 = 6$-dimensional representation, respectively. Just as vectors can carry lower as well as upper indices, so can tensors. We may lower and raise indices at will, using $\eta_{\mu\nu}$ and $\eta^{\mu\nu}$, respectively. For example, $T^{\sigma\lambda}_{\nu\rho} \equiv \eta_{\nu\kappa} T^{\kappa\sigma\lambda}_{\rho}$.

Euclid did not forbid you to study curves

It is worthwhile to underline a deep-seated but common misunderstanding that exists among some students of special relativity. The subject is concerned with the physics seen by two observers in uniform motion relative to each other. Absolutely nothing says that the objects* they are studying have to move at constant speed. The confusion appears to stem from thinking that special relativity is only capable of dealing with objects that do not accelerate and that you need general relativity.[†]

Put differently, special relativity teaches us how dx'^{μ} and dx^{μ} are related, but nothing in the Lorentz transformation requires that $\frac{d^2x'^{\mu}}{d\tau^2}$ and $\frac{d^2x^{\mu}}{d\tau^2}$ vanish.

There are actually misguided people walking around talking about the twin "paradox." A guy takes off in a rocketship while his twin stays home. When he comes back, he finds that the stay-at-home twin has aged a lot more.[6]

So? The twin "paradox" is resolved by pointing out that it is not a paradox at all.[7] In ordinary space, nobody claims that the lengths of different paths connecting two points are necessarily the same. A guy drove from Los Angeles to San Francisco via Las Vegas. His twin drove directly from Los Angeles to San Francisco. When they met, the guy who went to Las Vegas found that he had burned up more gas than his twin did. That is no more a paradox than the twin paradox is a paradox. The lengths of different paths connecting two points in spacetime can of course be different. Indeed, it would be quite amazing if the lengths of entirely different paths connecting two points in spacetime turn out to be the same. If the twins were the same age when they met, now that would be quite a paradox indeed!

At big accelerators, unstable particles zip around the ring at speeds close to c. A particle of the same type sitting in the lab has long decayed, while its "twin" is still speeding around the ring. That the twin paradox is not a paradox is a solid experimental fact that has been verified countless times.

In the twin paradox, two observers (Ms. Unprime and Mr. Prime) in relative uniform motion observe the two twins. Notice that the stay-at-home twin is not required to sit still. What special relativity requires is that the two observers agree on the proper time that has elapsed for the stay-at-home twin when his wandering sibling returns. The two observers

---

* The theory of special relativity does not care whether these objects are animate or not: they could be charmed mesons or other observers.

[†] This is manifestly untrue: accelerators accelerate particles, but to master particle physics, students do not necessarily have to become proficient in general relativity.

also must agree on the proper time that has elapsed for the wandering twin. What is not required at all is that these two proper times agree.

Perhaps I should say something even more obvious. Euclidean geometry is left invariant by rotations, but this does not mean we are allowed to study only straight lines in Euclidean space. We are of course also free to study curves. What Euclid requires is that two observers whose coordinates are related by a rotation must agree on the length of a given curve, but it would be absurd to insist that they proclaim all curves to have the same length.

So, let me say it again: just as Euclid did not forbid you to study curves in his space, Einstein did not forbid you to study curves in his spacetime. Indeed, the equation of motion of a free particle (4) may be immediately generalized to that of a particle acted upon by an external force:

$$m\frac{d^2X^\mu}{d\tau^2} = F^\mu \tag{20}$$

This is just Newton's law $m\vec{a} = \vec{F}$ promoted to Einstein's world. (We will talk a lot more about promotion in chapter III.6.) The requirement that Ms. Unprime and Mr. Prime subscribe to the same equation means that the force has to be promoted from a 3-vector $\vec{F}$ to a 4-vector F^μ, so that Mr. Prime would see the force $F'^\mu = \Lambda^\mu_{\ \nu}F^\nu$. (We will see an explicit example of this when we discuss electromagnetism in chapter IV.1.)

To underline the fact that special relativity can be applied to observers undergoing arbitrary accelerations, I will let you prove a basic fact about acceleration in Minkowski spacetime, calculate how a constantly accelerating particle would move, and develop the concept of Fermi-Walker transport in the exercises.

Lorentz, Poincaré, and Einstein: "to not trouble . . . old habits"

The intellectual history of special relativity is exceptionally fascinating because, in contrast to general relativity, which was born largely through the labor of a single man, so many great minds participated in developing special relativity, with several coming to within a whisker of the final theory. Henri Poincaré in particular developed the Lorentz transformation into the form now known to us, but without taking that final leap of forcing the mathematics on physics.* Many felt that[8] perhaps Einstein got too much credit and Poincaré too little. The French physicist Thibault Damour had examined this point in depth and concluded that the cartoon history, giving Einstein most of the credit, is largely correct.[9] I think that Lorentz and Poincaré quite simply did not enjoy the boldness of youth: in 1905, both were in their early 50s, while[10] Einstein was 26. Indeed, months before his death in 1912, Poincaré wrote: "Today some physicists want to adopt a new convention. It is not that they are forced to; they judge this new convention to be more useful, that is all; and those who are not of the same opinion may legitimately keep the former convention

---

* Some theoretical physicists think that the pendulum has perhaps swung to the other extreme: these days, the leap may be leapt before doing anything else.

in order to not trouble their old habits. I think, between us, that this is what they shall do for a long time."[11]

I trust that you, the astute reader, after getting to this point, could explain to Poincaré that the work of Einstein and Minkowski (whom he did not refer to by name) was not merely a convention one could choose not to adopt. The telling phrase here, and a warning to you, is "to not trouble their old habits." So reader, whatever your age may be, remember "the boldness of youth."

I close with a "pregnant" quote from Minkowski: "The essence of this postulate may be clothed mathematically in a very pregnant manner in the mystic formula $3 \cdot 10^5$ km $= \sqrt{-1}$ secs."[12] I like his choice of words.

Appendix 1: Generalized "rotation" groups

Looking at the boost in the x direction (17), you might have been struck by its uncanny resemblance to a rotation in the $(t$-$x)$ plane, except that cosine and sine have been replaced by their hyperbolic counterparts and that a minus sign has disappeared. Surely, this is not an accident. Indeed, recall from chapter I.3 that the rotation group in $SO(D)$ is defined as the set of all linear transformations $d\vec{x}' = Rd\vec{x}$ on a collection of D real variables $(dx^1, dx^2, \cdots, dx^D)$, such that the quadratic form $ds^2 = \sum_{i=1}^{D}(dx^i)^2$ is left unchanged and $\det R = 1$ (this specifies the S in $SO(D)$).

Our experience with Minkowski geometry warmly invites us to generalize. Define the group $SO(m, n)$ as the set of all linear transformations $d\vec{x}' = Rd\vec{x}$ on a collection of D real variables $(dx^1, dx^2, \cdots, dx^{m+n})$, such that the quadratic form $ds^2 = \sum_{i=1}^{m}(dx^i)^2 - \sum_{i=m+1}^{m+n}(dx^i)^2 = \eta_{\mu\nu}dx^\mu dx^\nu$ is left unchanged and $\det R = 1$. (Here $\eta_{\mu\nu}$ denotes a generalized Minkowski metric, namely a diagonal $(m+n)$-by-$(m+n)$ matrix with m $(+1)$s and n (-1)s along the diagonal. The indices μ and ν range over $1, 2, \cdots, m+n$.) These transformations form a group, since if R_1 and R_2 leave the quadratic form invariant, then R_1R_2 also leaves the quadratic form invariant. The other defining requirements of a group are even more obviously satisfied. The two integers (m, n) are known as the signature of the group. (Incidentally, as mentioned in the text, quantities such as ds^2 and $d\tau^2$ are not necessarily positive.)

Clearly, the groups $SO(m, n)$ and $SO(n, m)$ are the same: if ds^2 is left invariant, then so is $-ds^2$. The Lorentz group is then simply $SO(3, 1)$. The rotation groups and the Lorentz group can thus be studied in a unified fashion as special cases of $SO(m, n)$.

Again, the Lie algebra of $SO(m, n)$ is obtained by studying the infinitesimal transformation $R \simeq I + i \sum_{\mu\nu} \theta^{\mu\nu} J_{\mu\nu}$, with real parameters $\theta^{\mu\nu}$, the analogs of the angles φ and θ in our earlier discussions. You can verify that the entire discussion in appendix 2 to chapter I.3 and in this chapter can be repeated, with the Kronecker delta δ_{mn} replaced by the generalized Minkowski metric $\eta_{\mu\nu}$. In particular, the commutation relations between the generators $J_{\mu\nu}$ can be carried over from (I.3.23):

$$[J_{\mu\nu}, J_{\rho\sigma}] = i(\eta_{\mu\rho}J_{\nu\sigma} + \eta_{\nu\sigma}J_{\mu\rho} - \eta_{\nu\rho}J_{\mu\sigma} - \eta_{\mu\sigma}J_{\nu\rho}) \tag{21}$$

We specialize to the Lorentz group $SO(1, 3)$ or $SO(3, 1)$. Reverting to standard physics notation, we have 3 boosts generated by $K_i \equiv J_{0i}$ and 3 rotations by $J_i = \frac{1}{2}\varepsilon_{ijk}J_{jk}$. We can read off from (21) the commutation relation between boosts and rotations:

$$[J_x, J_y] = [J_{23}, J_{31}] = -i\eta_{33}J_{21} = iJ_{12} = iJ_z \tag{22}$$

$$[J_z, K_x] = [J_{12}, J_{01}] = -i\eta_{11}J_{20} = iJ_{02} = iK_y \tag{23}$$

$$[K_x, K_y] = [J_{01}, J_{02}] = +i\eta_{00}J_{12} = -iJ_z \tag{24}$$

All other commutation relations can be gotten from cyclically permuting the ones displayed here.

The relation (22) generates the subalgebra corresponding to the rotation subgroup $SO(3)$ of $SO(3, 1)$, familiar from chapter I.3. The relation (23) tells us that the 3 boost generators (K_x, K_y, K_z) transform as a 3-vector under the rotation group, exactly as you would expect. (To see this, consider $K_x(\theta) \equiv e^{-i\theta J_z} K_x e^{i\theta J_z}$. Differentiating, we obtain

$$\frac{dK_x(\theta)}{d\theta} = e^{-i\theta J_z}(-i)[J_z, K_x]e^{i\theta J_z} = e^{-i\theta J_z}K_y e^{i\theta J_z} \equiv K_y(\theta)$$

Similarly, $\frac{dK_y(\theta)}{d\theta} = -K_x(\theta)$. Solving, we obtain $K_x(\theta) = \cos\theta\, K_x + \sin\theta\, K_y$ and $K_y(\theta) = -\sin\theta\, K_x + \cos\theta\, K_y$.)
The third relation (24) is the most interesting: it shows that successively boosting in the x and y directions can produce a rotation[13] about the z-axis.

This set of commutation relations underlies many interesting developments[14] in quantum field theory, such as the Dirac equation, the Weyl equation, parity violation in the weak interaction, helicity spinors, and twistors, to name a few.

We also note that the differential representation of the rotation generators mentioned in chapter I.3 may be immediately promoted to the differential representation of the generators of $SO(m, n)$

$$J_{\mu\nu} = i(x_\mu \partial_\nu - x_\nu \partial_\mu) \tag{25}$$

Note that since $J_{\mu\nu}$ is defined with two lower indices, our index convention requires $x_\mu = \eta_{\mu\nu}x^\nu$, rather than x^μ, to appear on the right hand side. Thus, in calculating the commutator in (21) when we push a $\partial_\rho = \frac{\partial}{\partial x^\rho}$, for example, past $x_\mu = \eta_{\mu\nu}x^\nu$, we will get $\eta_{\mu\rho}$. This is another way of seeing why the generalized Minkowski metric appears in (21).

Finally, we check explicitly that (17) analytically continues to a rotation. Write $t = x^0 = ix^4$ and $\varphi = i\theta$ and continue to x^4 and θ real. Then $\cosh\varphi = \frac{1}{2}(e^\varphi + e^{-\varphi}) \to \frac{1}{2}(e^{i\theta} + e^{-i\theta}) = \cos\theta$ and $\sinh\varphi = \frac{1}{2}(e^\varphi - e^{-\varphi}) \to \frac{1}{2}(e^{i\theta} - e^{-i\theta}) = i\sin\theta$. Then the relevant part of (17) becomes (we write x^1 for x)

$$x'^4 = \cos\theta\, x^4 + \sin\theta\, x^1$$
$$x'^1 = -\sin\theta\, x^4 + \cos\theta\, x^1 \tag{26}$$

precisely a rotation in the (1-4) plane. The Lorentz group $SO(3, 1)$ is intimately connected to the 4-dimensional rotation group $SO(4)$. Clearly, linear transformations that leave $(x^0)^2 - \vec{x}^2$ invariant upon analytic continuation $x^0 \to x^4$ leave $(x^4)^2 + \vec{x}^2 = (x^1)^2 + (x^2)^2 + (x^3)^2 + (x^4)^2$ invariant. Going from x^0 to x^4, known as Wick rotation, is a standard procedure in quantum field theory.[15]

Appendix 2: From the Lorentz algebra to the Poincaré algebra

The set of generators $J_{\mu\nu}$ can be supplemented by the generators of translation $P_\mu = i\partial_\mu$. We see that P_μ generates translation by acting with it: $(I - ia^\mu P_\mu)x^\lambda = (I + a^\mu \partial_\mu)x^\lambda = x^\lambda + a^\lambda$. This is of course the same way we see that $J_{\mu\nu}$ generates rotations and boosts. (Even farther back, in chapter I.3, we saw that $J_z = i(y\frac{\partial}{\partial x} - x\frac{\partial}{\partial y})$ generates rotation about the z-axis: act with $I + i\theta J_z$ on (x, y, z).)

Thus, the Lorentz algebra defined by (21) can be extended to the so-called Poincaré algebra, generated by $(J_{\mu\nu}, P_\mu)$. In addition to the commutation relation (21), we now have

$$[P_\mu, P_\nu] = 0, \quad [J_{\mu\nu}, P_\rho] = i(\eta_{\mu\rho}P_\nu - \eta_{\nu\rho}P_\mu) \tag{27}$$

Exercises

1   Write $d\tau^2$ in light cone coordinates.

2   Just as we are allowed to change coordinates in Euclidean space and in curved spaces, of course we are also free to change coordinates in Minkowski spacetime. Consider, for example (after going to the usual spherical coordinates $x = r\sin\theta\cos\varphi$, $y = r\sin\theta\sin\varphi$, $z = r\cos\theta$), the transformation $t = \rho\sinh T$, $r = \rho\cosh T$, introduced by Rindler, with T ranging from $-\infty$ to ∞ and ρ ranging from 0 to ∞. Show that $d\tau^2 = dt^2 - dx^2 - dy^2 - dz^2 = \rho^2 dT^2 - d\rho^2 - \rho^2\cosh^2 T(d\theta^2 + \sin^2\theta d\varphi^2)$. For fixed θ and φ, the lines of constant ρ trace out hyperbolas in the $(t$-$r)$ plane as T ranges from $-\infty$ to ∞. Note that, since $r > |t|$, the coordinates $(T, \rho, \theta, \varphi)$ cover only one quadrant or wedge of Minkowski spacetime (figure 6).

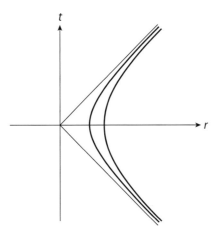

Figure 6 Rindler coordinates cover only one quadrant or wedge of Minkowski spacetime.

3 As emphasized in the text, we can certainly describe accelerating particles in special relativity. Consider the 4-velocity $V^\mu = \frac{dx^\mu}{d\tau}$ and the 4-acceleration $a^\mu = \frac{dV^\mu}{d\tau} = \frac{d^2x^\mu}{d\tau^2}$. Show that $a_\mu V^\mu = 0$, so that in the rest frame of the particle, $a^\mu = (0, \vec{a})$.

4 Show that the worldline of a particle with acceleration given by $a_\mu a^\mu = g^2$, with g a constant, is a hyperbola.

5 Fermi-Walker transport: Consider an observer undergoing arbitrary acceleration, carrying with her a vector W^μ. The vector W^μ is said to be Fermi-Walker transported if it varies along the observer's worldline according to

$$\frac{dW^\mu}{d\tau} = (V^\mu a^\nu - a^\mu V^\nu)W_\nu \tag{28}$$

(a) Show that the velocity V^μ is Fermi-Walker transported.
(b) Show that if U^μ and W^μ are both Fermi-Walker transported, the scalar product $U_\mu W^\mu$ is left unchanged.
   These results imply the physically sensible conclusion that an observer in an accelerating rocketship can perfectly well enjoy the benefits of having an orthonormal coordinate frame. Our observer can set up, at some proper time, an orthonormal coordinate frame consisting of her 4-velocity, namely the timelike vector V^μ and three spacelike unit 4-vectors e_a^μ (with $a = 1, 2, 3$), satisfying the orthonormal conditions $e_a \cdot e_b = \delta_{ab}$, $V \cdot e_a = 0$, and of course $V \cdot V = -1$. In her rest frame, $V^\mu = (1, 0, 0, 0)$, and she can choose $e_1^\mu = (0, 1, 0, 0)$, $e_2^\mu = (0, 0, 1, 0)$, $e_3^\mu = (0, 0, 0, 1)$. If she Fermi-Walker transports e_a^μ, then the result of this exercise guarantees that the orthonormal coordinate frame will remain orthonormal: the orthonormal conditions $e_a \cdot e_b = \delta_{ab}$, $V \cdot e_a = 0$, and $V \cdot V = -1$ are all preserved.

6 Work out explicitly how the components F^{0i} and F^{ij} of the antisymmetric tensor $F^{\mu\nu}$ introduced in chapter I.6 transform under a Lorentz transformation.

7 Show that the signature is an invariant. (This was known in the 19th century as Sylvester's law of inertia. Sylvester will appear again in chapter III.5.)

8 Follow a boost in the x direction with a boost in the y direction. Take the infinitesimal limit and compare with (24).

9 We observe an experimentalist moving by with 4-velocity u^μ and a particle zipping by with 4-momentum p^μ. Show that magnitude of the particle's 3-momentum as seen by the experimentalist is given by

$$|\vec{p}| = \left[(p \cdot u)^2 + (p \cdot p)^2\right]^{1/2}$$

10 Generalize the triangle shown in figure 1 by letting B $= (t, x)$ for $0 < x < t$, while keeping A $= (0, 0)$ and C $= (2, 0)$ fixed, so that all three sides are timelike. By symmetry, we can take $0 < t < 1$. Show that $d_{AB} + d_{BC}$ is always less than $d_{AC} = 2$. This exercise can be interpreted as the twin paradox. One twin stays at home, while the other goes from A to B and then from B to C. By the way, although both in the text and here, the side AC is aligned with the time axis, the situation analyzed is actually more general, since by a Lorentz transformation, we can always bring AC to align with the time axis.

Notes

1. It puzzles me somewhat that Lorentz and Einstein did not realize this. Poincaré apparently did. See T. Damour, *Once Upon Einstein*.
2. L. Boroditsky, *Cognition* 75 (2000), p. 1.
3. Translation of an address delivered at Cologne, 1908. Reprinted in *The Principle of Relativity: A Collection of Papers by A. Einstein, H. Lorentz, H. Weyl and H. Minkowski*, with Notes by A. Sommerfeld, Dover, 1952.
4. F. Dyson, *Disturbing the Universe*, Harper and Row, 1979, chapter 17.
5. See, for example, *Fearful*, p. 169.
6. Einstein had introduced in his 1905 paper what later became known as the clock paradox. The twins were introduced by P. Langevin in 1911.
7. This "resolution" of the twin "paradox" has already been emphasized in several well-known textbooks. See, for example, W. Rindler, *Relativity*, p. 77; J. B. Hartle, *Gravity*, p. 65.
8. Indeed, within days of writing this, I got into a heated argument on a social occasion with a Caltech physicist about this very point. I was arguing in Einstein's favor.
9. T. Damour, *Once Upon Einstein*, p. 49.
10. Note in this connection that Newton was 24 in 1666, his miraculous year.
11. H. Poincaré, "Space and Time," paper presented at a conference at the University of London, May 4, 1912 (*Scientia* 12 (1912), p. 159 [in French]).
12. H. Minkowski, "Space and Time," in A. Einstein et al., *The Principle of Relativity*.
13. This mathematical fact leads to the physical phenomena of Thomas precession and spin-orbit coupling in atomic physics.
14. See, for example, S. Weinberg, *The Quantum Theory of Fields; QFT Nut*.
15. For example, *QFT Nut*, pp. 12 and 23.

III.4 | Special Relativity Applied

Scarcely anyone who truly understands relativity theory can escape this magic.

—A. Einstein

Events and worldlines

Now that we have the Lorentz transformation*

$$t' = \frac{t + vx}{\sqrt{1 - v^2}}$$

$$x' = \frac{x + vt}{\sqrt{1 - v^2}} \tag{1}$$

and its inverse (obtained instantly by flipping v to $-v$)

$$t = \frac{t' - vx'}{\sqrt{1 - v^2}}$$

$$x = \frac{x' - vt'}{\sqrt{1 - v^2}} \tag{2}$$

we are ready to work out all kinds of problems involving special relativity. Hopefully, a few examples will suffice to give you the idea of how to proceed in tackling this kind of problem.

Many students, and not a few professionals, get easily confused by problems in special relativity. I recommend the following plodding, but almost foolproof, method. Make a list of all the relevant events and their given locations in spacetime. Recall from the preceding chapter that an event is specified by where and when it occurred, that is, by a point in spacetime. Even better, if necessary, work out the relevant worldlines. The locations of the relevant events are given sometimes in the primed frame, sometimes in the unprimed frame, and sometimes "in a mixture" with some events located in one frame and others located in the other frame. After all the locations are written down, then it is just a matter

---

* For the relative velocity between the two frames, we have switched from u, used in the preceding chapter, to v here.

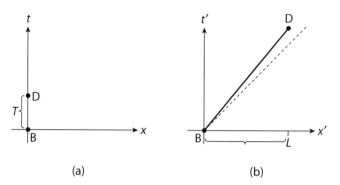

Figure 1 The tick (B) and the tock (D) of a clock as seen by two different observers.

of elementary algebra to work out the desired information using (1) and (2). You are free to call either frame the primed frame, and the other unprimed, whatever seems more natural or would make the arithmetic "look better." Often the solution can be found more quickly by using a clever observation, typically based on invariance. This will be illustrated below in our examples.

Time goes by

Perhaps the most astonishing prediction of special relativity is that time flows at different rates for different observers. The discussion in the preceding chapters already implies this, but let us now work out the effect carefully.

Provide two observers, Ms. Unprime and Mr. Prime, with identically manufactured clocks, going tick tock tick tock. Consider two spacetime events: Ms. Unprime's clock ticks, an event we call B, then her clock tocks, an event we call D. Denote by T the time between tick and tock. We now write down the spacetime location of these two events with pedantic care:

$$\text{B: } (t, x)_B = (0, 0)$$
$$\text{D: } (t, x)_D = (T, 0) \tag{3}$$

Note that $x_B = 0 = x_D$: in the unprimed frame, the clock did not move (figure 1a).

Mr. Prime sees Ms. Unprime go by with her clock. We simply plug in (1) to find the locations of these two events in the primed frame (figure 1b):

$$\text{B: } (t', x')_B = (0, 0)$$
$$\text{D: } (t', x')_D = \left(\frac{T}{\sqrt{1-v^2}}, \frac{vT}{\sqrt{1-v^2}} \right) \tag{4}$$

For example, $x'_D = \frac{x_D + vt_D}{\sqrt{1-v^2}} = \frac{vT}{\sqrt{1-v^2}}$, since $x_D = 0$ and $t_D = T$. Mr. Prime sees the time interval

$$t'_D - t'_B = \frac{T}{\sqrt{1-v^2}} > T \tag{5}$$

between a tick and a tock on Ms. Unprime's clock. This interval is larger than the interval T between a tick and a tock on his clock. Thus, we have shown that an observer sees a moving clock as running slow, an effect known as time dilation. The clock at rest in the observers frame has tocked before the moving clock tocks.

Each accuses the other of having a clock that runs slow

Now Confusio looks agitated. He cries out, "Wait a second here! But to Ms. Unprime, Mr. Prime is the one zipping by with his clock. The very principle of relativity states that either observer could claim to be at rest. How can two observers, each accusing the other of having a clock that runs slow, both be right? This flies in the face not only of common sense, but of logic itself!"

Ah, generations of students have run afoul of this point, a common confusion that has generated a seemingly endless stream of crackpot claims that Einstein must be wrong. "You mean to say that Ms. Unprime says that Mr. Prime's clock runs slow and Mr. Prime says that Ms. Unprime's clock runs slow, and yet both of them are absolutely correct?!"

To see that there is no contradiction, first note that a quicker way of deriving time dilation is to simply difference the first equation in (1) and set $\Delta x = 0$, thus concluding that $\Delta t' = \Delta t / \sqrt{1 - v^2}$ in agreement with (5). But now suppose we difference the first equation in (2) and set $\Delta x' = 0$, thus obtaining $\Delta t = \Delta t' / \sqrt{1 - v^2}$. Compare this with what we had a moment ago. The two conclusions, $\Delta t' = \Delta t / \sqrt{1 - v^2}$ in one case, and $\Delta t = \Delta t' / \sqrt{1 - v^2}$ in the other, do not contradict each other, because one is derived with $\Delta x = 0$ and the other with $\Delta x' = 0$. Logic still stands.

The lesson is simply that when we difference or differentiate we have to specify what we hold fixed.

When does Mr. Prime's clock tock?

Confusio looked convinced but still puzzled. We gently advised him to go read this chapter (up to this point) again. He came back the next day saying, "In the first section of this chapter, you said to make a list of all the relevant events. Then in the second section, you listed two events: event B when Ms. Unprime's clock ticks, and event D when her clock tocks. What about Mr. Prime's clock? When does it tick and tock?"

"Aha!" We all cried in unison, including Mr. Prime, Ms. Unprime, and Confusio.

Confusio grumbled, "There ought to be two other events: event B' when Mr. Prime's clock ticks, and event D' when his clock tocks. You didn't talk about those."

Mr. Prime said, "I can always set my clock to tick when Ms. Unprime's ticks, so that B = B'."

In other words, B and B' are one and the same event. There are actually 3 events to reckon with: B, D, and D'. The key question is then "When does Mr. Prime's clock tock?"

So in addition to (3) and (4), we should also write

$$B'(= B): (t', x')_{B'=B} = (0, 0)$$
$$D': (t', x')_{D'} = (T, 0) \tag{6}$$

Watch the primes and absence of primes like a hawk! The fact that the same T appears in (3) and (6) is part of the clock manufacturer's warranty.

Now, all we have to do is to plug (6) into (2) to find the spacetime locations of events $B'(= B)$ and D' in the unprimed frame:

$$B'(= B): (t, x)_{B'=B} = (0, 0)$$
$$D': (t, x)_D = \left(\frac{T}{\sqrt{1 - v^2}}, \frac{-vT}{\sqrt{1 - v^2}} \right) \tag{7}$$

Thus, Ms. Unprime sees the time interval

$$t_{D'} - t_{B'} = \frac{T}{\sqrt{1 - v^2}} > T \tag{8}$$

between a tick and a tock on Mr. Prime's clock.

Compare and contrast (5) and (8).

Confusio exclaims, "I see! The confusion that befuddled, and befuddles, generations of students is really a case of bad notation alert! The notations Δt, $\Delta t'$, and so forth correspond to time differences for different pairs of events."

Birth and death of particles

In everyday life, $v \ll 1$, so the effect of time dilation is minimal, but in high energy physics, particles typically move almost at light speed, and time dilation is of central importance to experimentalists. For instance, a cosmic ray particle (a proton, for example) may crash into a nucleus in a photographic emulsion, thus producing an unstable particle moving at speed v and disintegrating at a distance L downstream. What is the particle's lifetime T in its rest frame, that is, its intrinsic lifetime?

The two relevant events are B, the birth of the particle, and D, its death. Let the lab frame the experimentalist is sitting in be primed, and the rest frame of the particle be unprimed. The preceding analysis immediately applies. (Now you see why I used the letters B and D earlier in (3) and (4).)

In its own rest frame, the particle did not go anywhere: it died where it was born. That's what "at rest" means! Therefore, $\Delta x = 0$, and so from (1), we obtain the time elapsed between birth and death of a fast moving particle as seen by the experimentalist $T_{\text{lab}} \equiv t'_D - t'_B = \frac{T}{\sqrt{1-v^2}}$. As $v \to 1$, T_{lab} can become much larger than the particle's intrinsic lifetime T. The distance traversed (in the lab frame, of course) is given by $L = \Delta x' = \frac{vT}{\sqrt{1-v^2}}$. Knowing L and v, the experimentalist can solve to obtain the intrinsic lifetime $T = \sqrt{1 - v^2} L / v$.

Time dilation facilitated the measurement of particle lifetimes in the early days of particle physics: fast moving particles can live for quite a while in the lab frame, so that the track of a decaying high energy particle can be recorded in some medium. Experimentalists could literally measure L with a ruler. That there is indeed time dilation has been verified experimentally countless numbers of times.

A faster way of solving this problem is to use the invariance of the proper time interval as discussed in the preceding chapter: $(t_D - t_B)^2 - (x_D - x_B)^2 = (t'_D - t'_B)^2 - (x'_D - x'_B)^2$, namely $[(T, 0) - (0, 0)]^2 = T^2 = [(T_{lab}, vT_{lab}) - (0, 0)]^2 = T_{lab}^2 (1 - v^2)$, and so we obtain, once again, $T_{lab} = \frac{T}{\sqrt{1-v^2}}$. As indicated in figure 1, the line joining B and D looks longer in (1b) than in (1a), but because of the minus sign in the Minkowski metric, the two lines actually have the same "length."

Lorentz-Fitzgerald length contraction

After the discussion of time dilation, it is easy to understand that with two rulers moving relative to each other, each could see the other as having become shorter. Again, let's go slowly. Consider an observer watching a ruler going by. Let us call the rest frame of the ruler the unprimed frame.

The back end of the ruler traces out the worldline $(t, x)_B = (t, 0)$, in other words, a line parametrized by $t_B(t) = t$, $x_B = 0$. The statement that the length of the ruler is L means that the front end traces out $(t, x)_F = (t, L)$. (See figure 2a.) Here, t is to be thought of as a parameter that runs from $-\infty$ to ∞. In the primed frame, according to (1), these two worldlines are described by $(t', x')_B = (\frac{t}{\sqrt{1-v^2}}, \frac{vt}{\sqrt{1-v^2}})$ and $(t', x')_F = (\frac{t+vL}{\sqrt{1-v^2}}, \frac{L+vt}{\sqrt{1-v^2}})$. Again, t is to be thought of as a parameter that runs from $-\infty$ to ∞. To be utterly pedantic, let me state that we are here dealing with four functions of t: $t'_B(t) = \frac{t}{\sqrt{1-v^2}}$, $x'_B(t) = \frac{vt}{\sqrt{1-v^2}}$, $t'_F(t) = \frac{t+vL}{\sqrt{1-v^2}}$, and $x'_F(t) = \frac{L+vt}{\sqrt{1-v^2}}$. Let's plot (figure 2b) these two lines in the primed coordinates, just two parallel straight lines both with slope $\frac{\Delta x'}{\Delta t'} = v$.

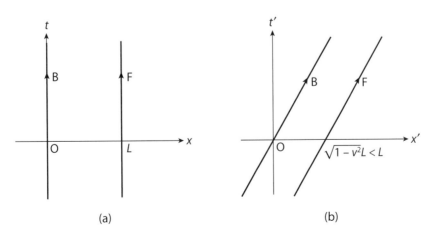

Figure 2 The back end (B) and the front end (F) of a ruler as seen by two different observers: (a) an observer at rest with the ruler and (b) an observer watching the ruler move by.

The key point is that as far as Mr. Prime is concerned, the length of the ruler is given by $x'_F - x'_B$ evaluated at the same time, that is, for $t'_F = t'_B$. Graphically, we see that we can choose any value of t'_F and t'_B, as long as they are equal. So choose $t'_F = \frac{t + vL}{\sqrt{1-v^2}} = 0$, which gives $t = -vL$ and thus $x'_F = \frac{L + v(-vL)}{\sqrt{1-v^2}} = \sqrt{1-v^2}L < L$. According to Mr. Prime, the ruler has length less than L: it has contracted!

Again, you can see that, pace the crackpots, there is no logical contradiction with the two observers each claiming the other's ruler has contracted. According to Ms. Unprime, the length of the ruler is defined by $x_F - x_B$ evaluated for $t_F = t_B$. For Mr. Prime, the length of the ruler is given by $x'_F - x'_B$ evaluated for $t'_F = t'_B$, most certainly not for $t_F = t_B$!

Just as in the time dilation discussion, the result for length contraction can be obtained almost instantly by differencing (1) or (2), evaluating Δx with $\Delta t = 0$, or $\Delta x'$ with $\Delta t' = 0$.

Historically, Fitzgerald had the clever idea that if moving rulers are contracted, then we could understand the puzzling result of the Michelson-Morley experiment.

Dueling theorists and the fall of simultaneity

For our next example, let us go back to the dueling theorists described in the prologue, on pages 7–9. We have three events, V = Professor Vicious at the rear of the carriage pushing the button to signal that she has finished her calculation, N = Dr. Nasty at the front end of the carriage pushing the button, and G = the gong bonging indicating that it has detected the arrival of two pulses of photons at the same instant. In the unprimed frame on the train, we write down

$$V : (t, x)_V = (0, -L)$$
$$N : (t, x)_N = (0, +L)$$
$$G : (t, x)_G = (L, 0) \tag{9}$$

We have set the length of the carriage to be $2L$, with the detector located in the middle. Since photons went from N to G, the invariant interval between G and N must vanish (and similarly for the invariant interval between G and V). This requirement fixes $t_G = L$ (indeed, the invariant interval between G and N is equal to $(L - 0)^2 - (0 - L)^2 = 0$).

Now that we have listed the three events, it is, as I said, more or less foolproof to find the primed coordinates for these events by simply plugging the unprimed coordinates given in (9) into the Lorentz transformation in (1): $t' = \frac{t + vx}{\sqrt{1-v^2}}$, $x' = \frac{x + vt}{\sqrt{1-v^2}}$. Thus, we have

$$V : (t', x')_V = \left(\frac{-vL}{\sqrt{1-v^2}}, \frac{-L}{\sqrt{1-v^2}} \right)$$

$$N : (t', x')_N = \left(\frac{vL}{\sqrt{1-v^2}}, \frac{L}{\sqrt{1-v^2}} \right)$$

$$G : (t', x')_G = \left(\frac{L}{\sqrt{1-v^2}}, \frac{vL}{\sqrt{1-v^2}} \right) \tag{10}$$

We see immediately, with no further ifs and buts, that in this frame, event N occurs after V : $t'_N - t'_V = 2vL/\sqrt{1-v^2} > 0$. The Swede standing on the platform has no doubt that Dr. Nasty does not get to go to Stockholm.

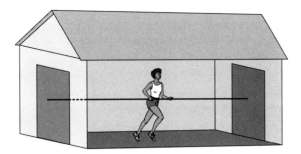

Figure 3 The pole in the barn.

As a check on our calculation, we can compute in the primed frame the velocity of light getting from the point V to the point G in spacetime: $\Delta t' = t'_G - t'_V = \frac{(1+v)L}{\sqrt{1-v^2}}$ and $\Delta x' = x'_G - x'_V = \frac{(v+1)L}{\sqrt{1-v^2}}$ giving $\frac{\Delta x'}{\Delta t'} = 1$. The reader can easily check that in getting from point N to point G, the velocity of light equals $\frac{\Delta x'}{\Delta t'} = -1$. These results of course just reflect the invariance of the interval $(\Delta t')^2 - (\Delta x')^2 = 0 = (\Delta t)^2 - (\Delta x)^2$.

Pole in the barn

Another apparent paradox that has confounded generations of students is the traditional pole in the barn problem, often given on exams.

One fine day, Ms. Unprime is possessed by a maniacal desire to run at almost light speed carrying a pole of length L toward a barn of length L. (See figure 3.) The farmer who owns the barn, Mr. Prime, calmly watches. To him, the pole has contracted and thus should fit easily inside the barn. But to Ms. Unprime, the barn, rushing toward her, has contracted alarmingly.

To dramatize the story further, suppose that Mr. Prime, also afflicted by some mental disorder, closes the front door of his barn while leaving the back door open. He has rigged things up with fancy electronics so that the front door won't fling open until the instant the back end of the pole gets inside the barn—that is, as soon as it passes the back door. For good measure, although not actually necessary for the narrative, at that instant, the back door will slam shut. Will there be a crash, or will Ms. Unprime sail right through?

Most importantly, stay calm. (That is, you the exam taker.) Write down the 4 relevant worldlines, that of the back end of the pole (Pb), the front end of the pole (Pf), the back door of the barn (Bb), and the front door of the barn (Bf):

$$(t, x)_{Pb} = (t, 0), \qquad (t, x)_{Pf} = (t, L) \tag{11}$$

and

$$(t', x')_{Bb} = (t', 0), \qquad (t', x')_{Bf} = (t', L) \tag{12}$$

Notice that we have written down these two pairs of parallel lines in their respective rest frames (figure 4a,b). Plugging (11) into (1), we obtain

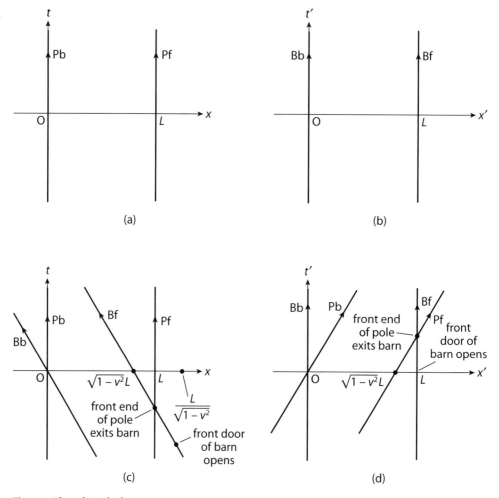

Figure 4 The pole in the barn as seen in spacetime.

$$(t', x')_{\text{Pb}} = \left(\frac{t}{\sqrt{1-v^2}}, \frac{vt}{\sqrt{1-v^2}} \right), \qquad (t', x')_{\text{Pf}} = \left(\frac{t+vL}{\sqrt{1-v^2}}, \frac{L+vt}{\sqrt{1-v^2}} \right) \qquad (13)$$

Plugging (12) into (2), we obtain

$$(t, x)_{\text{Bb}} = \left(\frac{t'}{\sqrt{1-v^2}}, \frac{-vt'}{\sqrt{1-v^2}} \right), \qquad (t, x)_{\text{Bf}} = \left(\frac{t'-vL}{\sqrt{1-v^2}}, \frac{L-vt'}{\sqrt{1-v^2}} \right) \qquad (14)$$

(Is this plodding enough for you? Yes, plodding but foolproof.)

Note that in (14), as t' varies, $(t, x)_{\text{Bb}}$ and $(t, x)_{\text{Bf}}$ trace out two parallel lines with slope $\Delta x / \Delta t = -v$ as indicated in figure 4c. (This provides a slight check against copying errors.) To construct the figure, we have to figure out where the line Bf intersects the x-axis. Simply set t_{Bf} in (14) to 0, thus giving $t' = vL$. Plugging that into (14), we obtain $x_{\text{Bf}}(\text{at } t_{\text{Bf}} = 0) = \frac{L(1-v^2)}{\sqrt{1-v^2}} = \sqrt{1-v^2}L < L$. We are of course just rediscovering the Lorentz-Fitzgerald length contraction.

Similarly, in (13), as t varies, $(t', x')_{\text{Pb}}$ and $(t', x')_{\text{Pf}}$ trace out two parallel lines with slope $\Delta x'/\Delta t' = v$ as indicated in figure 4d. You can verify that the line Pf intersects the x'-axis at $x'_{\text{Pf}}(\text{at } t'_{\text{Pf}} = 0) = \sqrt{1 - v^2}L < L$.

Now we go from worldlines to events. An important event in the story is when the back end of the pole (Pb) reaches the back door of the barn (Bb). Recall that at that instant, Mr. Prime flings open the front door of the barn.

You should be able to see right off when that occurs, but let us be plodding. Set $(t, x)_{\text{Pb}} = (t, 0)$ from (11) to $(t, x)_{\text{Bb}} = \left(\frac{t'}{\sqrt{1-v^2}}, \frac{-vt'}{\sqrt{1-v^2}}\right)$ from (14). It is important to note that we set coordinates equal in the same frame of course, namely $(t, x)_{\text{Pb}} = (t, x)_{\text{Bb}}$. The resulting two equations, namely $t = \frac{t'}{\sqrt{1-v^2}}$ and $0 = \frac{-vt'}{\sqrt{1-v^2}}$, give $t' = 0$ and $t = 0$. Thus, in fact, this event occurs at the origin in both the primed and unprimed frame, which reflects our wisdom in setting our coordinate systems. (Did you figure that out without solving the equations like a plodder? You can see it from figure* 4c,d.)

At that instant, the front door of the barn opens. For Ms. Unprime, this occurs at $(t, x)_{\text{Bf}}(\text{when front door of barn opens}) = \left(\frac{-vL}{\sqrt{1-v^2}}, \frac{L}{\sqrt{1-v^2}}\right)$. Note that $t_{\text{Bf}}(\text{when front door of barn opens}) < 0$. More importantly, $x_{\text{Bf}}(\text{when front door of barn opens}) = \frac{L}{\sqrt{1-v^2}} > L$, and she sails right through! (See figure 4c.)

It is interesting to check the spacetime location of another important event, the beginning of the pole's exit from the barn, namely when the front end of the pole (Pf) passes through the front door of the barn (Bf). In the unprimed coordinates, this occurs when $(t, x)_{\text{Pf}} = (t, L)$ is equal to $(t, x)_{\text{Bf}} = \left(\frac{t'-vL}{\sqrt{1-v^2}}, \frac{L-vt'}{\sqrt{1-v^2}}\right)$. Solving the two equations $t' - vL = \sqrt{1 - v^2}t$ and $L - vt' = \sqrt{1 - v^2}L$, we obtain $t = (\sqrt{1 - v^2} - 1)L/v < 0$ and $x = L$. Similarly, in the primed frame, the exit occurs at $t' = (1 - \sqrt{1 - v^2})L/v > 0$ and $x' = L$. See figure 4c,d.

Uncommon sense in, uncommon sense out

As I was finishing this book, I happened to have dinner with a distinguished condensed matter physicist. He mentioned that he was teaching a course on special relativity and that he didn't like the textbook, because it gave the students the impression that special relativity consisted of a series of paradoxes. I couldn't agree with him more. I told him that I was writing a textbook on special relativity and I had limited the number of paradoxes. In my opinion, the pedagogically correct way of presenting special relativity is to emphasize, as I tried to do in the preceding chapter, the geometry of spacetime, a point of view that generalizes naturally to the curved spacetime of general relativity.

---

* By the way, for the sake of clarity, I did not superpose the images in figure 4a–d. This is why this figure has four parts, a, b, c, and d. Also, notice that figure 2a,b (describing length contraction) is contained in figure 4c,d.

204 | III. Space and Time Unified

The paradoxes* are of course helpful in making students understand the subtle issues behind the Lorentz transformation, and they played an important clarifying role
historically. But once we accept that the speed of light is a universal constant, a statement that blatantly contradicts the velocity addition law of Newtonian physics, we can
expect to encounter situations that defy the common sense built on our everyday experiences. As already mentioned in the prologue, we could say, paraphrasing computer
scientists, uncommon sense in, uncommon sense out. All these apparent paradoxes contradict our Newtonian intuition, but they could not possibly contradict logic, as I have
emphasized here.

These puzzles are best regarded as refreshing reminders of how counterintuitive special
relativity actually is. In fact, they play almost no role in actual research in high energy
physics. Lorentz invariance is actually built into the grammar of the language used in
high energy theory, namely quantum field theory, from the start.

Causality and temporal ordering

Although simultaneity fails, causality still holds, as I emphasized in the preceding chapter.
Particles have to propagate inside their future light cones. In particular, temporal ordering
cannot depend on the observer.

More explicitly, consider a particle, either a material particle or a particle of light,
propagating from event A to event B, that is, event B is in the causal future of event A.
In other words, $\Delta t \equiv t_B - t_A > 0$ and $|\Delta x| \equiv |x_B - x_A| \leq \Delta t$. Is it possible to reverse the
temporal ordering by going to another frame?

We simply difference (1): $\Delta t' = (\Delta t + v\Delta x)/\sqrt{1 - v^2}$. Since $v^2 \leq 1$, $\Delta t'$ cannot go negative. Temporal ordering is maintained.

But perhaps this discussion suggests to you how temporal ordering might be reversed
under certain circumstances. What are these circumstances? See if you can figure it out!

Well, just by eyeball, we can see from $\Delta t' = (\Delta t + v\Delta x)/\sqrt{1 - v^2}$ that we can make $\Delta t'$
vanish by choosing $v = -\Delta t/\Delta x$. But $v^2 = (\Delta t/\Delta x)^2$ has to be less than 1, which is only
possible if $(\Delta t)^2 < (\Delta x)^2 \leq (\Delta x)^2 + (\Delta y)^2 + (\Delta z)^2$, that is, if the two events A and B are
spacelike with respect to each other. By continuity, if we can choose a v that makes $\Delta t'$
vanish, we can have a v that makes $\Delta t'$ go negative.

So, it is possible to reverse the temporal ordering of two events if they are spacelike with
respect to each other, which means precisely that they could not affect each other causally.
There is still sanity within the craziness, so to speak.

Next, what does a reversed time ordering imply for physics? The answer is in the
appendix.

---

* In high school math, you shouldn't spend all your time doing trick problems in a puzzle book. It's more
important to grasp the general principles.

Special relativity and starting your car

No doubt, time dilation and the creation of antiparticles (see the appendix) at high energy accelerators are plenty stunning, but still, it would appear that special relativity is totally remote from everyday life. Not so! In 2011, a mere 106 years after Einstein's paper, it was discovered[1] that relativistic effects account for about 1.8 volts out of the 2.1 volts produced by the common lead-acid battery. This is because the lead nucleus is so massive that the motion of the electrons around it is highly relativistic. So, the next time you hear a car start up somewhere, you can mutter to yourself, "Ah, the Lorentz transformation again!"

"From the very first line I am stopped by 'signs'"

On March 31, 1922, the new global celebrity Albert Einstein made a triumphant visit to Paris, greeted by headlines like "Time Does Not Exist, Says Einstein." A huge crowd tried to get in to hear his lecture. Among those caught up in the excitement was Marcel Proust, the famous author of the masterwork *À la recherche du temps perdu [In Search of Lost Time]*, with its wistful message that the passage of time was merely an illusion. Indeed, Proust even thought of time as a dimension analogous to space. He wrote[2] to a physicist friend:

> How I would love to speak to you about Einstein! Although it has indeed been written to me that I derive from him, or he from me, I do not understand a single word of his theories, not knowing algebra. And I doubt for my part that he has read my novels. It seems we have analogous ways of deforming Time. But I cannot figure it out for myself, because it is me, and we don't know each other, nor can I do so for him because he is a great mind in sciences that I am ignorant of, and from the very first line I am stopped by "signs" that I don't recognize.

Appendix: Appearance of antimatter

It is possible to skip this appendix upon a first reading, as subsequent material will not depend on it. The discussion in this chapter, indeed in most of this textbook, is entirely classical. But as was mentioned in the introductory chapters, our real world is fundamentally quantum. The only knowledge of the quantum world I require of you for the following discussion is Heisenberg's uncertainty principle. According to Heisenberg, due to fundamental limitations on what we can know about the microscopic world, we cannot measure all observables to arbitrary accuracy. As a result, it is possible to have spacelike propagation $|\Delta x| \gtrsim \Delta t$, but only for a very short time limited[3] by the mass of the particle. In figure 5a, we show an actual physical process in which (left half of the diagram) a proton turns into a neutron by emitting a pion (known as the π^+), which by charge conservation necessarily carries positive charge. This is event A. In event B, the π^+ is absorbed (right half of the diagram) by a neighboring neutron, which as a result turns into a proton. This process generates an attraction between the proton and the neutron, thus accounting for the strong or nuclear force.

Spacelike propagation is allowed, as Heisenberg said, only for a short time and over a small distance given by the inverse of the pion mass (in natural units). Indeed, since the range of the nuclear force was known, Hideki Yukawa was able to predict, in 1935, the mass of the hitherto unknown pion. We cannot go into further details here. Instead, let us ask what an observer zipping by would see.

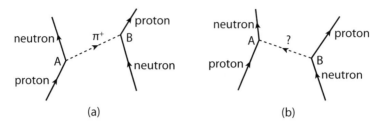

Figure 5 The need for antimatter. (a) A proton turns into a neutron by emitting a positively charged pion, which is subsequently absorbed by a neighboring neutron, which as a result turns into a proton. (b) The same process as seen by a different observer: a neutron turns into a proton by emitting a negatively charged pion, which is subsequently absorbed by a neighboring proton, which as a result turns into a neutron. (The time axis is along the vertical direction.) This figure, which describes a physical process in spacetime, is known as a Feynman diagram.

Since the pion propagates over a spacelike interval, it is possible for this observer to see a temporal ordering in which event A occurs after event B, as was explained in the text. Thus, she would see (figure 5b) the neutron turning into a proton (right half of the diagram) by emitting something (event B). But by charge conservation, this something necessarily has to carry a negative charge!

In other words, given the π^+, theoretical physics has predicted a negatively charged pion with exactly the same mass, known as the π^-. To finish the story, the π^- is then absorbed (event A) by the neighboring proton, which as a result turns into a neutron.

The π^- is the antiparticle of the π^+. In essence, this is the kind of physical reasoning that led Dirac to predict the existence of antimatter in 1928. I hope to give you, by this brief heuristic argument, some flavor of what might happen when you marry quantum mechanics to special relativity. In our story, the key is electric charge conservation. In the microscopic world, several other charges are also conserved, and thus a particle and its antiparticle necessarily carry opposite charges.

Incidentally, you can see that the discussion here is a generalization of the Vicious versus Nasty story in the prologue.

Notes

1. R. Ahuja, A. Blomqvist, P. Larsson, P. Pyykk, and P. Zaleski-Ejgierd, "Relativity and the Lead-Acid Battery," *Phys. Rev. Lett.* 106 (2011) 018301.

2. T. Damour, *Once Upon Einstein*, p. 34.

3. The uncertainty principle states that $\Delta E \Delta t \sim \hbar$. Here $\Delta E > m$, which implies $\Delta t < \hbar/m$. For details, consult a textbook on quantum field theory, such as *QFT Nut*.

The Worldline Action and the Unification of Material Particles with Light

A child's way of calculating square roots

Imagine yourself a bright young theoretical physicist toward the end of the 19th century. You felt annoyed about how light and material particles were treated differently. You admired Fermat's least time principle for light beams, so elegantly stated.

Such simplicity, light in a hurry! In contrast, look at the Euler-Lagrange action for material particles

$$S = \int dt \left[\frac{1}{2}m \left(\frac{d\vec{x}}{dt} \right)^2 - V(x) \right] \tag{1}$$

Clunky in comparison.

Then you put this thought aside and went on with your day-to-day research—you did need to get tenure. In spite of what people like deans say, day-to-day research was what led to all the good stuff, not the very best stuff, but the good stuff, in academic life. One day, while sitting in your office daydreaming, you remembered the first time you learned about the concept of a square root. You learned that the square root of 25 is 5, of 36 is 6, and so on. But soon, since you were one of those smart kids who grew up to be theoretical physicists, you wondered about the square root of a number that was not obviously the square of an integer. What is the square root of 24, for example? You used the time honored method of trial and error. So you multiplied 4.9 by itself, 4.8 by itself, and so forth. Pretty soon you could guesstimate square roots quite well. Some time later, you learned about the brilliant idea* of representing numbers by letters. The formula you wanted turned out to be

$$\sqrt{a^2 - \varepsilon^2} \simeq a - \frac{\varepsilon^2}{2a} + \cdots \tag{2}$$

---

* In 820, while working in the House of Wisdom in Bagdad, the Persian Al-Khwārizmī, whose name gave us the words "algorithm"[1] and "logarithm," proposed a method of calculation he called al-jabr.

which you verified by squaring the right hand side $\left(a - \frac{\varepsilon^2}{2a}\right)^2 = a^2 - \varepsilon^2 + \frac{\varepsilon^4}{4a^2}$: the error in (2) is of higher order.

A few days after your daydream, you look at the action for material particles again. Even without the external potential $V(x)$, the action still does not look like it could be simplified to look like anything as elegant as Fermat's least time principle. You do what A. Zee in chapter III.1 of his gravity book said the calculus textbook he had in high school told him never to do: think of derivatives as fractions. You do precisely that and cancel off one power of dt:

$$S = \int dt \tfrac{1}{2}m \left(\frac{d\vec{x}}{dt}\right)^2 = \tfrac{1}{2}m \int \frac{(d\vec{x})^2}{dt} \tag{3}$$

Now you are offended by the totally different ways in which space and time are treated. The undemocratic treatment of $d\vec{x}$ and dt irritates your liberal ideal deeply. Yes, the dean keeps reminding the physics faculty that every subject on campus is equally worthy of respect. That $d\vec{x}$ appears squared in (3) more or less follows from rotational invariance, but why does dt deserve only one power? What a strange combination, $\frac{(\Delta \vec{x})^2}{\Delta t}$, the square of something divided by something else!

Then your subconscious nudges you: you have seen this combination before, the square of something divided by something else! Oh dear reader, where but where have you seen this before?

Speak, memory

Aha! You rewrite (3) as $\frac{\varepsilon^2}{2a} \simeq -\sqrt{a^2 - \varepsilon^2} + a + \cdots$. In other words,

$$\frac{(\Delta \vec{x})^2}{2\Delta t} = c\frac{(\Delta \vec{x})^2}{2c\Delta t} = -c\sqrt{(c\Delta t)^2 - (\Delta \vec{x})^2} + c^2 \Delta t \tag{4}$$

To get the dimensions to come out right, you are forced to introduce a constant c with the dimension of a speed. What could it be?

You realize that the only speed around with any intrinsic significance is the speed of light. Notice that this c literally muscles its way in for dimensional reasons. We didn't go looking for the speed of light, the speed of light came looking for us. So you write the action for a point particle as

$$S = -mc \int \sqrt{(cdt)^2 - (d\vec{x})^2} + mc^2 \int dt$$

The relativistic point particle action

The term $mc^2 \int dt = mc^2(t_{\text{final}} - t_{\text{initial}})$ in the action S treats dt and $d\vec{x}$ differently, and thus would negate your entire philosophy. But fortunately, its variation vanishes, since the initial and final times are fixed. Hence, this term in the action does not contribute to the equation of motion, and so the action principle allows you to drop this offending term.

All looks great then, and so you quickly publish the action $S = -mc \int \sqrt{(cdt)^2 - (d\vec{x})^2}$, which in units with $c = 1$ reads

$$S = -m \int \sqrt{dt^2 - (d\vec{x})^2} \tag{5}$$

This action, compared to (3), treats space and time much more democratically.

It didn't happen this way in our civilization, but it could have happened this way in some other civilization far far away. I like writing alternative physics history.

Of course, what actually happened was that your paper was rejected by a succession of referees. One of them told you that you should brush up on your calculus. Didn't you know that integrals are supposed to have the form $\int dt$ of some function of t? This guy is easy to take care of: you write

$$S = -m \int dt \sqrt{1 - \left(\frac{d\vec{x}}{dt}\right)^2} \tag{6}$$

To show another referee that you can reproduce Newtonian mechanics, you expand (6) and rearrange slightly to obtain

$$S = \int dt \left\{ \frac{m}{2} \left(\frac{d\vec{x}}{dt}\right)^2 - m + \cdots \right\} \tag{7}$$

Heavens to Betsy, you even get that most famous $\frac{1}{2}$ in physics history, as in $\frac{1}{2}mv^2$, which in hindsight has been whispering "Square root square root" for centuries.

The one formula even the person in the street knows

You don't know how to incorporate a potential; just adding $-\int dt\, V(x)$ to the action (5), as in (1), would again favor dt. But if you had a potential as in (1), then in the nonrelativistic limit of (7), m would just be added to the potential $V(x)$. You submit another paper interpreting the extra term m as a rather peculiar kind of potential energy.

Even a particle just sitting there has energy, in fact an enormous amount of energy compared with the kinetic energy $\frac{1}{2}mv^2$ it could acquire in everyday life, with $v \ll c$. Let's restore c and repeat the derivation: $S = -mc \int \sqrt{(cdt)^2 - d\vec{x}^2} = -mc \int dt \sqrt{c^2 - (\frac{d\vec{x}}{dt})^2} = -m \int dt \{c^2 - \frac{1}{2}(\frac{d\vec{x}}{dt})^2 + \cdots\} = \int dt \{\frac{1}{2}mv^2 - mc^2 + \cdots\}$. The referee snarled that this author didn't even know that an additive constant in $V(x)$ mattered not a whit. Paper rejected.

Phew! It's a good thing that you got tenure first before pursuing this stuff. Meanwhile, in a civilization in another galaxy far far away, a man named Albert discovered what is probably the most famous equation of all time[2]

$$E = mc^2 \tag{8}$$

More likely that the proverbial guy in the street has heard of this equation than of $F = ma$!

At first sight, it may seem strange that the two terms in S, with opposite signs, both correspond to energy. Hamilton was clever enough to resolve this apparent problem. Recall from chapter II.3 that for the Lagrangian $L(\dot{q}, q)$, the Hamiltonian is given by

$H(p, q) = p\dot{q} - L(\dot{q}, q)$ with $p = \frac{\delta L}{\delta \dot{q}}$. Hence, if we were given $L = \frac{1}{2}m\dot{q}^2 - mc^2$, we would have $p = m\dot{q}$, thus giving $H = p\dot{q} - (\frac{1}{2}m\dot{q}^2 - mc^2) = \frac{p^2}{2m} + mc^2$, which is exactly right for the energy of the system: kinetic energy plus mc^2. Well, all we're saying is that mc^2 should be counted as part of the potential energy.

A symmetry pops out

Meanwhile, a mathematician friend of yours—could have been James Joseph Sylvester, a rather astute fellow, since he demanded that his salary from Johns Hopkins University be paid in gold before accepting its invitation to move from England to a scientifically impoverished but economically upstart country called the United States—told you about some fancy-pants math called matrix theory. He pointed out that if you defined a 4-by-4 diagonal matrix $\eta_{\mu\nu}$ with $(-1, +1, +1, +1)$ along the diagonal, you could write the combination $dt^2 - d\vec{x}^2$ in your action (5) more compactly as $-\eta_{\mu\nu}dx^\mu dx^\nu$, defining $x^0 = t$.

At first you dismissed this as mere notational dressing, but after studying this matrix theory, you realized that your action did not change under the linear transformation

$$dx^\mu \to \Lambda^\mu_{\ \sigma} dx^\sigma \tag{9}$$

provided that $\eta_{\mu\nu}\Lambda^\mu_{\ \sigma}\Lambda^\nu_{\ \rho} = \eta_{\sigma\rho}$, a matrix equation to be solved for Λ. Nature has a "hidden" symmetry that extends and generalizes rotation! Of course, by this point, you did not even dream of publishing your discovery any more. You just showed it to Sylvester, who thought it might be somehow related to invariant theory, later developed into group theory.

Ah, the joy of hindsight![3] If you were around earlier and felt the ugliness[4] of the Newtonian action (1), you too could have discovered special relativity, and perhaps even the group theory of linear transformations!

Action, geometry, and mass as a conversion factor

The action for a relativistic point particle has the elegant form

$$S = -m \int \sqrt{-\eta_{\mu\nu}dx^\mu dx^\nu} \tag{10}$$

Recognizing $\eta_{\mu\nu}dx^\mu dx^\nu$ as the Minkowskian distance squared between neighboring points, we see that the particle's action is (up to an overall constant) the distance it has travelled in spacetime. An appealingly geometric* picture!

Indeed, if we were told to construct the action for a point particle, the only coordinate invariant quantity we have available is the "length" or proper time duration of the worldline

---

* Invite yourself at this point to generalize this action to that for a relativistic string. See, for example, *QFT Nut*, chapter IV.4 and a later section in this chapter.

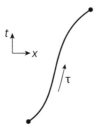

Figure 1 The action for a point particle. The only coordinate invariant quantity we have available is the "length" or proper time duration of the worldline traced out by the particle.

traced out by the particle (figure 1), namely $\int d\tau = \int \sqrt{-\eta_{\mu\nu}dx^\mu dx^\nu}$. We call the proportionality factor the mass of the particle. If you like, that provides one cool definition, a rather profound one at that, of mass: mass is the conversion factor between geometry (the length of the worldline) and physics (the action).

Bad notation alert! In fact, it is the same alert as in chapter II.1. The symbol x^μ refers to the spacetime coordinates of the particle traversing spacetime, not spacetime itself. Again, this is laid bare by considering the case of many particles labeled by an index a with $S = -\sum_a m_a \int \sqrt{-\eta_{\mu\nu}dx_a^\mu dx_a^\nu}$. The bad notation is best avoided by denoting the spacetime coordinates of the point particle by X^μ (as we already did in passing in chapter III.3): here, unlike in chapter II.1, q^μ would be nonstandard and a bit pedantic.

Let us now go back to a single particle merely for ease of writing; you could add the summation sign if you want. With ζ any parameter that varies monotonically along the worldline so that we can write $X^\mu(\zeta)$ as a function of ζ, we can write the action as

$$S = -m \int \sqrt{-\eta_{\mu\nu}dX^\mu dX^\nu} = -m \int d\zeta \sqrt{-\eta_{\mu\nu}\frac{dX^\mu}{d\zeta}\frac{dX^\nu}{d\zeta}} = \int d\zeta\, L \tag{11}$$

The length of the worldline, being a geometric quantity, is manifestly reparametrization invariant, that is, independent of our choice of ζ as long as it is reasonable. As already remarked in chapter II.2, this is one of those "more obvious than obvious" facts, since $\int \sqrt{-\eta_{\mu\nu}dX^\mu dX^\nu}$ is manifestly independent of ζ. Indeed, everything is the same as in chapter II.2, except that here the metric signature is spacetime rather than space, that is, Minkowskian rather than Euclidean.

The Lagrangian has a rather odd-looking square root form (just as in chapter II.2)

$$L = -m \left(-\eta_{\mu\nu}\frac{dX^\mu}{d\zeta}\frac{dX^\nu}{d\zeta}\right)^{\frac{1}{2}} \tag{12}$$

but that does not stop us from sticking the Euler-Lagrange variation to it. Since L is independent of X, we obtain the equation of motion

$$\frac{d}{d\zeta}\left(\frac{\delta L}{\delta\frac{dX^\lambda}{d\zeta}}\right) = 0 = m\frac{d}{d\zeta}\left(\frac{1}{L}\eta_{\mu\lambda}\frac{dX^\mu}{d\zeta}\right) \tag{13}$$

To simplify (13), we exploit the freedom in choosing ζ and set $d\zeta$ to $d\tau$, so that $L = -m$. Then (13) becomes

$$\frac{d^2 X^\mu}{d\tau^2} = 0 \tag{14}$$

as we already learned in chapter III.3 (and in some sense even in chapter II.2).

Indeed, the geometrical action (11) was already foreshadowed by the triangle shown in figure III.3.1 and discussed in chapter III.3.

Unification of material particles with light

At the beginning of this chapter, you felt annoyed about how light and material particles were treated differently. Now you can actually do something about it. The nonrelativistic action (1) is manifestly incapable of describing a photon, the ultrarelativistic particle of light. The relativistic action (11), in contrast, might yet have a fighting chance.

At first glance, things do not look good, since the action (11) $S = -m \int d\zeta \sqrt{-\eta_{\mu\nu} \frac{dX^\mu}{d\zeta} \frac{dX^\nu}{d\zeta}}$ does not make sense for a massless particle. To remedy this, consider another action:

$$\tilde{S} = -\frac{1}{2} \int d\zeta \left(\sigma(\zeta) \left(\frac{dX}{d\zeta} \right)^2 + \frac{m^2}{\sigma(\zeta)} \right) \tag{15}$$

where (as always) $(\frac{dX}{d\zeta})^2 = -\eta_{\mu\nu} \frac{dX^\mu}{d\zeta} \frac{dX^\nu}{d\zeta}$.

This action looks strange at first sight: not only does it depend on the spacetime trajectory $X^\mu(\zeta)$ of our point particle, it also contains another dynamical variable $\sigma(\zeta)$. Notice, however, that $\frac{d\sigma}{d\zeta}$ does not appear in the action. Thus, the Euler-Lagrange equation for $\sigma(\zeta)$, namely $\frac{d}{d\zeta} \left(\frac{\delta \tilde{S}}{\delta \frac{d\sigma}{d\zeta}} \right) - \frac{\delta \tilde{S}}{\delta \sigma} = 0$, collapses to $\frac{\delta \tilde{S}}{\delta \sigma} = 0$:

$$\frac{m^2}{\sigma(\zeta)^2} = \left(\frac{dX}{d\zeta} \right)^2 \tag{16}$$

This is an algebraic equation, not a differential equation, for $\sigma(\zeta)$. In other words, the dynamical variable $\sigma(\zeta)$ does not have dynamics of its own but rather is totally yoked[5] to $X^\mu(\zeta)$.

In fact, if we use (16) to eliminate σ in $\tilde{S}$, we recover S. Thus, the two actions, S and $\tilde{S}$, are equivalent in the sense that they yield the same equation of motion for the particle.

We can easily verify this equivalence explicitly. Varying $\tilde{S}$ with respect to X^λ, we obtain $\frac{d}{d\zeta} \left(\frac{\delta \tilde{S}}{\delta \frac{dX^\lambda}{d\zeta}} \right) = \frac{d}{d\zeta} (\sigma \eta_{\mu\lambda} \frac{dX^\mu}{d\zeta}) = 0$, since $\tilde{S}$ does not depend on X explicitly. (Compare this with (13).) Using the equation of motion for σ to eliminate it from this equation of motion for X, we obtain

$$\frac{d}{d\zeta} \left(\frac{m}{\sqrt{(\frac{dX}{d\zeta})^2}} \eta_{\mu\lambda} \frac{dX^\mu}{d\zeta} \right) = 0$$

As we have done many times by now, we use our freedom in choosing ζ and set $d\zeta$ to $d\tau$, which is defined by $(\frac{dX}{d\tau})^2 = 1$. Then this simplifies to $\frac{d^2 X^\mu}{d\tau^2} = 0$, and we recover (14).

You might say that the introduction of σ is a cheap[6] trick to get rid of the square root in the action S in (11). Indeed, (16) is trying to tell us that σ^{-1} is a proxy for the square root.

Massless particles

After this slightly long, although straightforward, manipulation, you might have lost sight of what we want to achieve. Let me remind you: we would like to have a unified treatment of massive and massless particles. Newtonian physics cannot deal with massless particles at all. It hardly makes sense to set $m = 0$ in $F = ma$. Nor is the action S in (11) up to the task.

Now, ta dah, setting $m = 0$ in $\tilde{S}$, we obtain an action

$$S_{\text{massless}} = \frac{1}{2} \int d\zeta \left(\sigma \eta_{\mu\nu} \frac{dX^\mu}{d\zeta} \frac{dX^\mu}{d\zeta} \right) \tag{17}$$

which makes perfect sense. Varying with respect to $\sigma(\zeta)$ now gives

$$\eta_{\mu\nu} dX^\mu dX^\nu = 0 \tag{18}$$

for a massless particle, or in other words, $(d\vec{X})^2 = (dX^0)^2$. We recover what we have always known, that massless particles travel at the speed of light.

To be massless in the contemporary world

To me, one truly profound intellectual triumph of special relativity, with far-reaching impact on contemporary particle physics, is the notion of a massless particle. To appreciate how mysterious and alien this concept is, try explaining it to an intelligent person who happens not to be a physicist. One problem is the definition of mass in elementary physics texts: typically, mass is said to be the amount of substance contained in the object. You mean something without substance can still have energy and momentum?

But we could perfectly well set $m = 0$ in the Einsteinian $E = \sqrt{\vec{p}^2 + m^2}$, in contrast to the Newtonian $E_K = \frac{\vec{p}^2}{2m}$. In contemporary particle physics,[7] all the known particles—the various quarks, the electron, the electron neutrino, and their various cousins; the photon and its various cousins responsible for the strong and weak interactions; and the graviton, responsible for gravity—all of them start out in life* massless. (They acquire masses only later, through a phenomenon known as spontaneous symmetry breaking that does not concern us here.)

Before 1905 and special relativity, there was no need for a massless particle. Light was known to be a wave. But 1905, Einstein's annus mirabilis, was also the year Einstein came up with the Nobel Prize–winning idea of light as consisting of photons.

---

* More precisely, the quantum fields corresponding to all known particles appear in the action as massless fields.

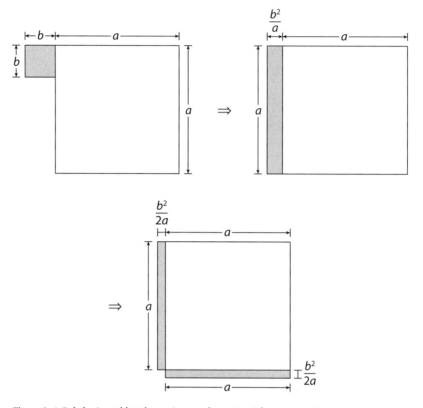

Figure 2 A Babylonian tablet (drawn in a modern pictorial representation).

Babylonian tablet

There is a Babylonian tablet from about 6,000 years ago on which figure 2 was inscribed.* Can you figure out what it says?

Take the small square of side b and replace it by a rectangle of equal area, of sides a and b^2/a. Cut the rectangle into two equal smaller rectangles, and paste them onto the sides, kitty-corner, of the large square of side a. The author of the tablet was trying to tell you that the result is almost a square of side $a + \frac{b^2}{2a}$.

Here is the algebraic translation of the Babylonian tablet

$$\sqrt{a^2 + b^2} \simeq a + \frac{b^2}{2a} + \cdots \tag{19}$$

Clever, no? That guy would have surely gotten the Fields Medal had it existed. That tablet blew me away when I saw it. I wondered whether that Babylonian could have thought, in his wildest imagination, that his discovery also held the secret of space and time. If so, he deserved the Nobel Prize in addition to the Fields Medal.

---

* What figure 2 shows is of course my attempt at copying the tablet, not the original tablet.

Appendix 1: The preferred parameter choice for massless particles

As before, if we use the parameter ζ, somebody else could use ζ' as long as the two parameters are related by a smooth monotonic function. We couldn't have lost the reparametrization invariance enjoyed by the action S by going to an equivalent form. To verify this, change variable $\zeta \to \zeta'$ in $\tilde{S}$ and write $\tilde{S} = -\frac{1}{2} \int d\zeta' \frac{d\zeta}{d\zeta'} \left(\sigma(\zeta)(\frac{d\zeta'}{d\zeta})^2 (\frac{dX}{d\zeta'})^2 + \frac{m^2}{\sigma(\zeta)} \right)$. Upon defining $\sigma(\zeta) \equiv \sigma'(\zeta') \frac{d\zeta}{d\zeta'}$, we recover the same form (15) of $\tilde{S}$ as expected.

So then, is there a "best" choice for ζ? For massive particles, we know the answer: proper time. But what is the best choice for a massless particle, for which proper time has no meaning?

When we vary $\tilde{S}$ with respect to X^μ, we obtain $\frac{d}{d\zeta}(\sigma \eta_{\mu\nu} \frac{dX^\nu}{d\zeta}) = 0$. Multiply this by σ and go to a parameter ζ' defined by $\sigma(\zeta) = \frac{d\zeta}{d\zeta'}$. Then this equation of motion becomes simply $\frac{d^2 X^\nu}{d\zeta'^2} = 0$. A parameter that makes the equation of motion take on this simple form is known as an affine parameter. (I must confess that I have always disliked this wishy-washy word "affine"—none of the strength of character of a word such as "entropy," for example.)

Let's see what this somewhat formal discussion is all about in the case of a photon propagating along, say the x direction. We don't have to solve any equation of motion to know that X^μ is proportional to $(1, 1, 0, 0)$. End of story. No need to parametrize this worldline going across Minkowski spacetime at $45°$. The equation (18) stands on its own merits: it does not need a parameter.

But if you insist, you could write $X^\mu = f(\zeta)(1, 1, 0, 0)$ with some monotonic function f. Then $\frac{d^2 X^\nu}{d\zeta^2} = (f''/f) X^\mu$. To make life easier, we should obviously choose f to be a linear function of ζ. I explain all this seemingly useless stuff here, because we will encounter similar considerations in curved spacetime. To me, the pages some texts spend on the affine parametrization of massless particles literally amounts to much ado about nothing.

Appendix 2: Baby string theory

The take-home message is that the action principle together with symmetry makes for a powerful combination. Indeed, so powerful that it enables us to generalize the action for a relativistic particle almost immediately to the action for a relativistic string.[8] The reader seeing all this for the first time might wish to skip this appendix.

The location of a point particle in d-dimensional spacetime is given by $X^\mu(\tau)$, with $\mu = 0, 1, \cdots, d-1$, and that is that. In contrast, the location of a string in d-dimensional spacetime is given by $X^\mu(\tau, \sigma)$, where σ is a parameter telling us where we are along the length of the string. For an open string, σ starts at 0 on one end of the string and ends at some σ_* at the other end. Thus, $X^\mu(\tau, 0)$ and $X^\mu(\tau, \sigma_*)$ give the locations of the two ends. For a closed string, σ is conventionally taken to range between 0 and 2π, with the two ends identified: $X^\mu(\tau, 0) = X^\mu(\tau, 2\pi)$.

As τ varies, $X^\mu(\tau)$ sweeps out a worldline. In precisely the same way, the location $X^\mu(\tau, \sigma)$ of a string sweeps out a world sheet, an open sheet (figure 3a) with a boundary for the open string, and a cylindrical tube (figure 3b) for the closed string.

You learned way back in chapter I.6 how to embed a curved surface in a higher dimensional space. Now is the time to put your knowledge to good use! Denote* $\sigma^0 = \tau, \sigma^1 = \sigma$. Then σ^α ($\alpha = 0, 1$) provides a coordinate system on the surface swept out by the string. Recall from chapter I.6 that a metric is induced on the surface, in this context by the ambient Minkowski metric $\eta_{\mu\nu}$:[†] $G_{\alpha\beta} = \eta_{\mu\nu} \partial_\alpha X^\mu \partial_\beta X^\nu$, where $\partial_\alpha X^\mu = \frac{\partial X^\mu}{\partial \sigma^\alpha}$.

In the case of the point particle, the only quantity with intrinsic geometric significance is the length of the worldline. For the string, the corresponding quantity is the area of the world sheet. But you know from chapter I.5 how to write down the area of a surface, namely $\int d^2\sigma \sqrt{\det G}$, where $\det G$ denotes the determinant of the 2-by-2 matrix $G_{\alpha\beta}$ and $d^2\sigma \equiv d\tau d\sigma$. The basic action for string theory

$$S_{\text{Nambu-Goto}} = T \int d\tau d\sigma \sqrt{\det G} \tag{20}$$

---

* Sometimes σ^α is called x^α, but we want to avoid confusion with the x^μ appearing earlier in this chapter.
[†] To avoid confusion, I call the induced metric G, not g as in chapter I.6.

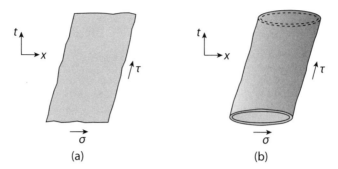

Figure 3 A string sweeps out a world sheet: (a) an open sheet with a boundary for an open string and (b) a cylindrical tube for a closed string.

was first proposed by Nambu and Goto. The overall constant T, with dimensions of $1/(\text{length})^2$ or mass/length, may be interpreted as the string tension (as in chapter II.1).

We can readily verify that if we change the coordinates σ^α on the world sheet, we leave the string action invariant. Under a "world sheet" reparametrization $\sigma^\alpha \to \sigma'^\alpha(\sigma)$, the integration measure changes by $d^2\sigma = d^2\sigma' J$, where the Jacobian J is the determinant of the 2-by-2 matrix $s^\alpha_\beta \equiv \frac{\partial \sigma^\alpha}{\partial \sigma'^\beta}$. Meanwhile, $G_{\alpha\beta} = \eta_{\mu\nu} \frac{\partial X^\mu}{\partial \sigma'^\gamma} \frac{\partial X^\nu}{\partial \sigma'^\varepsilon} \frac{\partial \sigma'^\gamma}{\partial \sigma^\alpha} \frac{\partial \sigma'^\varepsilon}{\partial \upsilon^\beta} = G'_{\gamma\varepsilon}(s^{-1})^\gamma_\alpha (s^{-1})^\varepsilon_\beta$, so that $\det G = J^{-2} \det G'$. Thus, finally $\int d^2\sigma \sqrt{\det G} = \int d^2\sigma' \sqrt{\det G'}$, and the action is indeed geometrical.

Exercises

1 In a precise parallel with the discussion for the point particle, we can avoid the square root in the Nambu-Goto action and instead use the action

$$S = \tfrac{1}{2} T \int d\tau d\sigma \gamma^{\tfrac{1}{2}} \gamma^{\alpha\beta} (\partial_\alpha X^\mu \partial_\beta X_\mu) \tag{21}$$

with $\gamma_{\alpha\beta}$ ($= \det \gamma_{\alpha\beta}$) an auxiliary variable playing the same role as the auxiliary variable in (15). Show that S is equivalent classically to $S_{\text{Nambu-Goto}}$.

2 Show that the string action is invariant under an arbitrary local rescaling

$$\gamma_{\alpha\beta}(\tau, \sigma) \to e^{2\omega(\tau, \sigma)} \gamma_{\alpha\beta}(\tau, \sigma)$$

known as a Weyl transformation. As a result, the equations of motion determine $\gamma_{\alpha\beta}$ only up to this rescaling.

Notes

1. No, Al Gore did not invent "algorithm."
2. In 1908, Einstein wrote to Johannes Stark complaining that the latter did not properly acknowledge his priority in deriving $E = mc^2$. Stark wrote back, and Einstein was appropriately apologetic, writing "People who have been privileged to contribute something to the advancement of science should not let such things becloud their joy over the fruits of common endeavor." By the way, Stark later called Werner Heisenberg "a white Jew" for defending Einstein's theory of relativity.
3. Psychologists have quantified this phenomenon of feeling that in hindsight things always seem so easy (B. Fischoff, *J. Exp. Psych.* 3 (1977), p. 349).

4. The ugliness can be quantified by comparing the symmetry groups of nonrelativistic and relativistic physics, the Galilean group versus the Lorentz group. See F. Dyson, "Missed Opportunities," *Bull. Am. Math. Soc.* 78, (1972), p. 635.

5. In quantum field theory, something like $\sigma(\zeta)$ is known as an auxiliary field.

6. Albeit one used quite often in quantum field theory and string theory!

7. I remember vividly arguing with my professor S. B. Treiman when I was an undergraduate. He had told me about the soft photon theorems in particle physics, which describe the interaction of the photon with charged particles in the limit that the photon's energy and momentum go to zero. I thought that the photon ceased to exist, but he reminded me that the photon's spin was still there. I simply could not understand how something with no mass, no energy, no momentum, and no quantum number could still be spinning.

8. J. Polchinski, *String Theory,* vol. 1.

Natural and unnatural quantities

We now know that nonrelativistic physics is but an approximation to a deeper truth. Since physics is Lorentz invariant, all physical quantities have to transform in a well-defined fashion under the Lorentz group, namely according to some definite representation of the group. The 3-vector $d\vec{x}$ has to be unified with the 3-scalar dt to form a 4-vector $dx^\mu = (dt, d\vec{x})$. Neither $d\vec{x}$ nor dt is relativistically complete. As Minkowski foresaw, space by itself, and time by itself, have now "faded away into mere shadows," at least in fundamental physics.

In this chapter, we will study how various quantities in nonrelativistic physics are to be "completed and promoted." Doing this, we will also discover the nature of the gravitational field and be one step closer to the main subject of this book.

Completion and promotion

Consider the 3-velocity $\vec{v}_N \equiv \frac{d\vec{x}}{dt}$. What an awkward quantity, a 3-vector $d\vec{x}$ divided by a 3-scalar dt that happens to be the "time component" of a 4-vector. Ugh! A 3-vector divided by a 3-scalar, with neither transforming nicely under the Lorentz group. Gimme a break, no way the resulting object $\vec{v}_N$ is going to transform nicely!

In contrast, consider a 3-vector $d\vec{x}$ divided by a Lorentz scalar, namely $d\tau$ (defined as always by $d\tau^2 = -\eta_{\mu\nu}dx^\mu dx^\nu = dt^2 - d\vec{x}^2$). Now we are talking: the resulting object $\frac{d\vec{x}}{d\tau}$ can be contained in a 4-vector. Indeed, let us define $\vec{v} \equiv \frac{d\vec{x}}{d\tau}$, a 3-vector whose relativistic completion is naturally the 4-vector $v^\mu \equiv \frac{dx^\mu}{d\tau}$. (In other words, $\vec{v}$ consists of the v^i components of the 4-vector v^μ.) The velocity v^μ is the right thing to study.

If you insist, as some authors do, on writing relativistic quantities in terms of "nonrelativistic quantities," you can do it (it's a free country), but it's best not to do so. For example, since $\frac{dt}{d\tau} = \frac{dt}{\sqrt{dt^2 - d\vec{x}^2}} = \frac{1}{\sqrt{1 - (\frac{d\vec{x}}{dt})^2}}$, we can write

$$v^\mu = \frac{dx^\mu}{d\tau} = \left(\frac{dt}{d\tau}, \frac{d\vec{x}}{dt}\frac{dt}{d\tau} \right) = \left(\frac{1}{\sqrt{1 - v_N^2}}, \frac{\vec{v}_N}{\sqrt{1 - v_N^2}} \right)$$

(with $v_N^2 = \vec{v}_N \cdot \vec{v}_N$) and other awkward looking relations such as $\vec{v} = v^0\vec{v}_N$. It may be convenient on occasions, but it sure ain't natural.

In theoretical physics, you are free to define any quantity you want, but two criteria are of prime concern: (1) whether the quantity you define is useful in the particular situation you are considering, and (2) whether the quantity you define is conceptually natural and thus serves to deepen, rather than confound, your understanding.

Our discussion makes clear that $\vec{v}_N \equiv \frac{d\vec{x}}{dt}$ and $\vec{v} \equiv \frac{d\vec{x}}{d\tau}$ represent different quantities. Some of the confusion over special relativity stems from confounding the two different velocities. Note that $\vec{v}_N$, not $\vec{v}$, is the velocity that appears in the factor $\sqrt{1 - v_N^2}$. To compound the confusion, people often omit the subscript N, a standard practice that we will also indulge in. Note also that while $\vec{v}_N^2 \leq 1$, the quantity $\vec{v}^2$ ranges from 0 to ∞. Never mind what the subscript N stands for (but if you insist, you could try "Newtonian"). (In this connection, some authors also adopt for $\vec{v}_N$ the unfortunate notation $\vec{v}_{NR}$ with NR standing for "nonrelativistic," even though $\vec{v}_{NR}$ can get up to light speed.)

Laws to be promoted

Concepts (such as v_N^i) appropriate for space and time should be promoted to concepts (such as v^μ) appropriate for spacetime. Correspondingly, the laws of physics have to be promoted also. Toward the end of chapter I.3 on rotation, I explained why physicists insist that physical quantities must transform "nicely." (The reader new to this might want to reread what I said there.) The niceness is not so much an aesthetic nicety but rather the fundamental requirement that physical laws should not depend on the observer. Physics must be independent of physicists!

Consider the law of momentum conservation. Multiplying v^μ by the mass, we have the 4-momentum $p^\mu = mv^\mu = m\frac{dx^\mu}{d\tau}$ of a particle of mass m moving with 4-velocity v^μ. The conservation of nonrelativistic 3-momentum $\vec{p}_N = m\vec{v}_N$ in nonrelativistic physics strongly suggests that the 4-momentum $p^\mu = m\frac{dx^\mu}{d\tau}$ is also conserved. (Ultimately, this statement has to be verified by empirical measurement rather than by philosophical pronouncements, of course.)

In particle physics, two particles of momentum p_1 and p_2 collide to produce a bunch of particles (could be two or could be two hundred, all with different masses). Momentum conservation states that $\sum_{(a\in I)} p_a^\mu = \sum_{(a\in F)} p_a^\mu$. Here the index a labels the different particles, and the subscripts on the sums, $(a \in I)$ and $(a \in F)$, instruct us to sum over all the particles in the initial and final states, respectively. This is just the direct generalization of the familiar $\sum_{(a\in I)} \vec{p}_{Na} = \sum_{(a\in F)} \vec{p}_{Na}$. Now let us define $K^\mu \equiv \sum_{(a\in I)} p_a^\mu - \sum_{(a\in F)} p_a^\mu$, so that momentum conservation states $K^\mu = 0$. The whole point is that since $K'^\mu = \Lambda^\mu_{\ \nu} K^\nu$, if $K^\mu = 0$, then $K'^\mu = 0$.

If Ms. Unprime has momentum conservation, then Mr. Prime better have momentum conservation also. This is the reason momentum must transform like a 4-vector in a Lorentz invariant world. We are not saying anything conceptually different from what we said at the end of chapter I.3. (Note also that this discussion does not depend in any way on the details of particle physics except that particles can be produced in collisions.) Indeed, we had already used this argument in chapter III.3 when we "guessed" what the equation of motion of a free particle must look like: one requirement was that Ms. Unprime and Mr. Prime subscribe to the same equation, which amounts to saying that $\frac{d^2 x^\mu}{d\tau^2}$ transforms like a 4-vector. (Notice that in this chapter, we are slipping back into the bad notation of using x^μ for the location of the particle.)

But now comes the important point, indicating that we have gone past rotational invariance. The power of Lorentz invariance is such that you cannot conserve p^i without also conserving p^0, since they transform into linear combinations of each other.

Let us expand $p^0 = m\frac{dt}{d\tau} = \frac{m}{\sqrt{1-v_N^2}}$ to find out what it is. Restoring c, we have $p^0 = mc^2 + \frac{1}{2}mv_N^2 + O(v_N^4)$. The big surprise (as we have already seen in the preceding chapter) is of course that the Newtonian kinetic energy $\frac{1}{2}mv_N^2$ is not the leading term in the expansion of p^0 but the second term. Even a particle at rest has energy!

The most famous formula in physics

What we have discovered is that p^0 is the energy of a particle of mass m moving with 4-velocity v^μ, not exactly the energy we knew, but the energy we knew plus mc^2 (and an infinite series besides)! So we officially give p^0 another name, namely $E \equiv p^0$.

A "real" relativistic physicist should of course not write the nonrelativistic formula* $E = mc^2$. You might tell your lay friends that the pros write $E = mc^2$ as

$$-p^2 = m^2 \tag{1}$$

since we have $p^2 = \eta_{\mu\nu}p^\mu p^\nu = -(p^0)^2 + (\vec{p})^2 = m^2\left(\eta_{\mu\nu}\frac{dx^\mu}{d\tau}\frac{dx^\nu}{d\tau}\right) = -m^2$, using the definition of $d\tau^2$ in the last step. We should think of (1) as a constraint on p, a constraint known as the mass shell or on shell condition, because in the 4-dimensional space spanned by $p^\mu = (p^0, \vec{p})$, (1) restricts the momentum to a hyperbolic shell defined by $(p^0)^2 - (\vec{p})^2 = m^2$.

---

* Einstein's famous paper[1] contained the result

$$K_0 - K_1 = \frac{L}{V^2}\frac{v^2}{2}$$

What! It doesn't look like $E = mc^2$ to you? Einstein is telling us that when an object moving at velocity v radiates, its kinetic energy K changes by (in modern notation) $\delta K = \frac{\delta E}{c^2}\frac{v^2}{2}$. (In his paper, L denotes the energy emitted in radiation and V the speed of light.) See appendix 1. He then goes on to say, a couple of paragraphs later, "It is not excluded that it will prove possible to test this theory using bodies whose energy content is variable to a high degree (e.g., radium salts)."

It is worth emphasizing that the relation (1) holds in any frame for any particle, including those for which $m = 0$. The same is not true of the more famous $E = mc^2$.

Strictly speaking, the conservation of p^μ (and hence of p^0) must be regarded as a theoretical suggestion, albeit a very strong suggestion, to be verified by careful experimental measurements. Indeed, so far we have only noninteracting free particles. At this stage, we do not even know how to include interactions between particles. The nonrelativistic expedient of introducing a potential $V(\vec{x}_1 - \vec{x}_2)$ does not work, since it is not Lorentz invariant.* Here the power of Noether's theorem shines through. Without having to specify the interaction, we know that p^μ conservation will follow as long as the interaction is invariant under spacetime translation; that is, it does not depend on any specific point in spacetime, which certainly seems reasonable. Recall from chapter II.4 that in nonrelativistic physics, conservation of 3-momentum and of energy follow from invariance under translation in space and in time, respectively. It is pleasing that these two translation invariances are now unified into a single invariance, just as space and time are unified into spacetime.

Logically, nothing in our discussion says that mass can change and the enormous amount of energy locked up in the rest energy mc^2 can be released. As you know, that is indeed possible. But input from atomic, nuclear, and particle physics is needed to tell us how and when mass can change.

Relativistic kinematics

To illustrate what 4-momentum conservation (henceforth, just momentum conservation) can do, let us go through Compton scattering, in which a photon[†] of momentum k hits an electron at rest and goes off with momentum k'. See figure 1 depicting this process in the lab frame. Let us find the frequency ω' of the outgoing photon as a function of the scattering angle θ in the lab frame as measured from the direction of the incoming photon.

Momentum conservation gives $k + p = k' + p'$, with p the initial and p' the final momentum of the electron. Note that we are always talking about 4-momentum unless otherwise stated. The desired quantity can be extracted from the relativistic invariant $k' \cdot p$, which is equal to $-m\omega'$ when evaluated in the lab frame in which $p = (m, \vec{0})$. (Notice that we do not simply say that $k' \cdot p$ is equal to $-m\omega'$. We must specify the frame, since ω' is not a Lorentz invariant quantity.) We have $k' \cdot p = k' \cdot (k' + p' - k) = k' \cdot p' - k' \cdot k$. In the first equality, we used momentum conservation; in the second, we used the fact that the photon is massless to set $k'^2 = 0$ invoking (1). From $k + p = k' + p'$, it follows that $(k + p)^2 = (k' + p')^2$. Since $(k + p)^2 = 2k \cdot p - m^2$ and $(k' + p')^2 = 2k' \cdot p' - m^2$, we obtain $k \cdot p = k' \cdot p'$. Combining this with the previous relation, we find $k' \cdot p = k \cdot p - k' \cdot k$. Evaluating this in

---

* To introduce interactions correctly, we have to use fields, as we will see. Classical field theory suffices; no need for quantum field theory yet.

† In Einstein's other great 1905 paper, he introduced the concept of the photon and showed that its energy and 3-momentum is given by $E = \hbar\omega$ and $\vec{p} = \hbar\vec{k}$, respectively. We are using natural units with $\hbar = 1$ and denoting the photon 4-momentum by $k = (\omega, \vec{k})$.

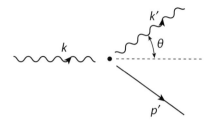

Figure 1 Compton scattering. A photon of momentum k hits an electron at rest and goes off with momentum k' at an angle θ.

the lab frame, we obtain $m\omega' = m\omega - (\omega'\omega - \vec{k}' \cdot \vec{k}) = m\omega - \omega\omega'(1 - \cos\theta)$, and thus[2]

$$\omega' = \frac{\omega}{1 + \frac{\omega}{m}(1 - \cos\theta)} \tag{2}$$

Note the downward shift of the photon frequency as a function of scattering angle. (Relativistic kinematics alone cannot tell you the range of angles the photon prefers to come out in, namely the differential cross section.)[3]

This example provides a prototype for how such problems, typical of particle physics and astrophysics, should be approached. First, write the quantity you want to determine in terms of some Lorentz invariant. Then use the equations you have, namely the mass shell condition for each particle and momentum conservation. It is best to keep the calculation Lorentz invariant until the last stage, at which point you can evaluate the result in any frame you like.

The relativistic Doppler shift again

The preceding discussion gives an alternative derivation of the relativistic Doppler shift. Suppose a particle of momentum p emits a photon of momentum k, which is then absorbed by a particle of momentum p'.

To obtain the frequency shift, our strategy is to evaluate the Lorentz scalar $k \cdot p'$ in two frames and demand that the results agree. In the rest frame of the emitting particle, $p = (m, \vec{0})$, $k = (\omega, \vec{k})$, and $p' = m'\left(\frac{1}{\sqrt{1-\vec{v}^2}}, \frac{\vec{v}}{\sqrt{1-\vec{v}^2}}\right)$, where $\vec{v}$ is the velocity of the absorbing particle in the rest frame of the emitting particle, so that* $k \cdot p' = -m'(\omega - \vec{v} \cdot \vec{k})/\sqrt{1 - \vec{v}^2}$. (Note that we do not assume that the emitting particle and the absorbing particle necessarily have the same mass.) However, in the rest frame of the absorbing particle, $p' = (m', \vec{0})$ and $k = (\omega', \vec{k}')$, where ω' is the frequency of the photon seen by the absorbing particle, we have $k \cdot p' = -m'\omega'$. Since the scalar $k \cdot p'$ has the same value in all frames, we equate these two expressions for $k \cdot p'$ to obtain $\omega' = (\omega - \vec{v} \cdot \vec{k})/\sqrt{1 - \vec{v}^2}$. Writing $\vec{v} \cdot \vec{k} = \omega|\vec{v}|\cos\theta$, we

---

* Note that here $\vec{v}$ is the concept previously known as $\vec{v}_N$. I got tired of writing the clarifying subscript. A simple rule (as has already been mentioned): the $\vec{v}$ inside the famous square root $\sqrt{1 - \vec{v}^2}$ is $\vec{v}_N$.

see that this agrees with our result in (III.3.9). With the convention here, $\theta = 0$ means that the receiver is receding and hence sees a redshift.

The particles in this "particle physics" derivation could be replaced by observers of course. Let each observer "carry" a 4-vector U, which has the form $U^\mu = (1, \vec{0})$ in his or her rest frame. Then simply replace p and p' in the derivation by U and U', respectively, since the masses m and m' are not relevant to this discussion. We will encounter U^μ again later in this chapter.

Currents

Our next example of relativistic completion is a bit more involved. Consider the number density (of atoms, molecules, particles, or any objects which cannot disappear into thin air) $n(t, \vec{x})$. This quantity, the number of particles per unit volume, is a rotational scalar. Think of a bunch of particles sitting inside a box. Rotations change neither the volume of the box nor the number of particles inside.

Now the question: Does $n(t, \vec{x})$ relativistically complete into a Lorentz scalar or something else? In general, there is no algorithm such that you can simply turn the crank and answer this question. To obtain the answer, you need to exercise some physical insight or mathematical savvy. For this simple example, I will give a physical argument, to be followed by a more formal mathematical analysis.

To an observer moving by, the box is Lorentz contracted (figure 2) in the direction of motion by $\sqrt{1 - \vec{v}^2}$ and thus its volume is diminished by this factor. (Incidentally, as already mentioned earlier, the $\vec{v}^2$ that appears in this square root factor stands for $\vec{v}_N^2$; we will henceforth suppress the subscript N.) Since the number of particles inside is unchanged, we conclude that the number density as seen by this observer is larger than the number density seen by an observer at rest with the box by $1/\sqrt{1 - \vec{v}^2}$. In other words, $n(t, \vec{x})$ transforms like the time component of a 4-vector.

Physically, it is obvious what the other 3 components are. This observer sees the particles moving and thus observes a current density. The most naive guess that the rotational

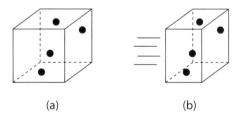

(a)   (b)

Figure 2 (a) A box containing a certain number of particles inside. (b) To an observer moving by, the box is Lorentz contracted, and thus the number density as seen by this observer is larger than the number density seen by an observer at rest with the box.

scalar $n(x)$ is promoted to a Lorentz scalar is wrong. Rather, it is promoted to be a component of a Lorentz vector $n^\mu(x) = (n^0(x), n^i(x))$, a 4-current. Nothing strange or unusual here: after all, energy, a rotational scalar, is promoted to be a component of the momentum p^μ. As a check, apply the transformation law for a Lorentz vector. To the observer at rest with the box, $n^\mu(x) = (n, \vec{0})$. Thus, to the observer watching the box go by (figure 2) along the 1-axis say, Lorentz transformation gives $n'^0 = \frac{n^0 + |\vec{v}|n^1}{\sqrt{1-\vec{v}^2}} = \frac{n}{\sqrt{1-\vec{v}^2}} \geq n$ and $n'^1 = \frac{|\vec{v}|n^0 + n^1}{\sqrt{1-\vec{v}^2}} = \frac{n|\vec{v}|}{\sqrt{1-\vec{v}^2}} \geq n|\vec{v}|$, as we argued physically. The inequalities indicate Lorentz contraction.

We next give a more mathematical treatment, reaching the same conclusion in the end. For simplicity, suppose we have only one particle sitting at the origin. Then $n(t, \vec{x}) = \delta^{(3)}(\vec{x})$, with the 3-dimensional delta function introduced in chapter II.1. In other words, $n(t, \vec{x})$ is concentrated at the origin, vanishing everywhere else and unchanging in time. Check to make sure that we have one particle: $\int d^3x\, n(t, \vec{x}) = \int d^3x\, \delta^{(3)}(\vec{x}) = \int dx \int dy \int dz\, \delta(x)\delta(y)\delta(z) = 1^3 = 1$ as expected. More importantly, this shows that $\delta^{(3)}(\vec{x})$ is rotationally invariant,* since d^3x is rotationally invariant. Our challenge is now to write $n(t, \vec{x}) = \delta^{(3)}(\vec{x})$ in a relativistic form.

Introduce the worldline of the particle traced out by $q^\mu(\tau)$. For a particle just sitting there, $q^0(\tau) = \tau$, $\vec{q}(\tau) = 0$, and $d\tau^2 = -\eta_{\mu\nu}dq^\mu dq^\nu$. (Now is a good time to recall the bad notation alerts I sounded repeatedly in part I concerning the distinction between x and q, between spacetime and some particle's location. In other words, x is where you are and q is where the particle is. Here the word "when" is subsumed into "where" (in spacetime) as per Minkowski.) Write $\delta^{(3)}(\vec{x}) = \int d\tau\, \delta(x^0 - q^0(\tau))\delta^{(3)}(\vec{x}) = \int d\tau\, \frac{dq^0}{d\tau}\delta(x^0 - q^0(\tau))\delta^{(3)}(\vec{x} - \vec{q}(\tau)) = \int d\tau\, \frac{dq^0}{d\tau}\delta^{(4)}(x - q(\tau))$. Here we have introduced the 4-dimensional delta function $\delta^{(4)}(x) \equiv \delta(x^0)\delta^{(3)}(\vec{x})$, which (in analogy with the discussion for $\delta^{(3)}(\vec{x})$ being a rotational scalar) we argue is a Lorentz scalar, since $\int d^4x\, \delta^4(x) = 1$ and d^4x is Lorentz invariant (recall chapter III.3).

For the abecedarian struggling to follow this, it may seem a totally pointless academic exercise in which we make things progressively more complicated. For example, in the second equality, we rewrote 1 as $\frac{dq^0}{d\tau}$ and introduced $0 = \vec{q}(\tau)$. But of course there is a point; otherwise this wouldn't appear in a textbook. The point is that the expression we ended up with is manifestly the time component of the Lorentz 4-vector $n^\mu(x) = \int d\tau\, \frac{dq^\mu}{d\tau}\delta^{(4)}(x - q(\tau))$. Everything on the right hand side is a Lorentz scalar except for the 4-vector $\frac{dq^\mu}{d\tau}$. Indeed, by Lorentz, this expression, now that we have it in this form, holds for any worldline $q^\mu(\tau)$, not just the simple form we had above. Furthermore, we can sum over an arbitrary number of particles. See figure 3.

To summarize, we define the number current (4-current, to be pedantic) as

$$n^\mu(x) = \sum_a \int d\tau_a \frac{dq_a^\mu}{d\tau_a}\delta^{(4)}(x - q_a(\tau_a)) \tag{3}$$

---

* You have essentially shown this already in exercise I.3.2.

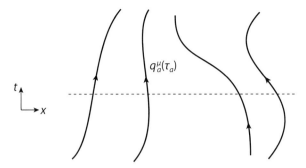

Figure 3 Defining the number current as a 4-vector in spacetime; the dashed line indicates a moment in time.

with $d\tau_a^2 = -\eta_{\mu\nu}dq_a^\mu dq_a^\nu$. It transforms correctly and reduces in the appropriate limit to something with vanishing space components and a time component equal to what we would call a number density. What more could you want?

In chapter I.4, I explained that an irreducible representation of a group, upon restriction to a subgroup, will become reducible and break up in general into a direct sum of a number of smaller representations of the subgroup. Thus, the 4-dimensional representation of the Lorentz group $SO(3, 1)$, upon restriction to the rotation group $SO(3)$, breaks up as $4 \to 3 + 1$, a 3-vector and a 3-scalar. Here we are asking the reverse: given a rotational scalar (what we also call a 3-scalar), which representation of $SO(3, 1)$ does it come from? In general, there is no unique answer. We have to appeal to physical considerations as is done here.

Current conservation

Another nice feature is that the conservation of the number of particles we started out with, $\frac{d}{dt}n(t, \vec{x}) = \frac{d}{dt}\delta^{(3)}(\vec{x}) = 0$, is instantly generalized to

$$\partial_\mu n^\mu \equiv \frac{\partial n^\mu}{\partial x^\mu} = \frac{\partial n^0}{\partial t} + \frac{\partial n^i}{\partial x^i} = \frac{\partial n^0}{\partial t} + \vec{\nabla} \cdot \vec{n} = 0 \tag{4}$$

known as the continuity equation. You no doubt encountered this when you first learned what a divergence $\vec{\nabla} \cdot \vec{n}$ was. You were probably taught to draw a little cube and to add up the number current flowing in and out of the six faces of the cube. Under rotations, $\partial_0 n^0$ and $\partial_i n^i$ both transform as 3-scalars, but under Lorentz transformations, they are revealed as parts of the same package $\partial_\mu n^\mu$.

Admire the power of Lorentz invariance: it mandates that $\frac{dn}{dt}$ must be promoted to $\partial_\mu n^\mu$.

Nevertheless, it may also be instructive to verify (4) laboriously. Acting with ∂_μ on (3), we encounter $\partial_\mu \delta^{(4)}(x - q_a(\tau_a)) = -\frac{\partial}{\partial q_a^\mu}\delta^{(4)}(x - q_a(\tau_a))$ and hence the integral

$$-\int d\tau_a \frac{dq_a^\mu}{d\tau_a}\frac{\partial}{\partial q_a^\mu}\delta^{(4)}(x - q_a(\tau_a)) = -\int d\tau_a \frac{d}{d\tau_a}\delta^{(4)}(x - q_a(\tau_a)) = -\delta^{(4)}(x - q_a(\tau_a))|_{\tau_a=-\infty}^{\tau_a=\infty} = 0$$

since[4] $q_a^0(\pm\infty) = \pm\infty$, while $t \equiv x^0$ is some fixed instant in time.

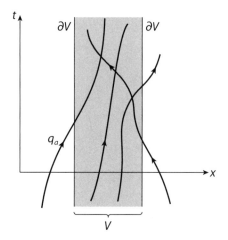

Figure 4 Counting the number of particles inside some 3-volume V at some time t. Here ∂V indicates the boundary of V.

Another instructive exercise is to integrate $n^0(t, \vec{x})$ over some 3-volume V at some time t (figure 4). We encounter $\delta(t - q_a^0(\tau_a)) \int_V d^3x \delta^{(3)}(\vec{x} - \vec{q}_a(\tau_a))$, with the integral giving 1 if the particle is inside V at that time and 0 if not. Note the time delta function "slices" the worldline and fixes τ_a to be that value at which $q_a^0(\tau_a) = t$. Thus,

$$\int_V d^3x \, n^0(x) = \sum_{(a\epsilon V)} \int d\tau_a \frac{dq_a^0}{d\tau_a} \delta(t - q_a^0(\tau_a)) = \sum_{(a\epsilon V)} \int dq_a^0 \delta(t - q_a^0) = \sum_{(a\epsilon V)} 1 = N_V(t)$$

giving precisely the number of particles inside V. You may find this all quite involved, but it is actually just telling you the obvious. Furthermore, applying (4), we have $\frac{d}{dt} N_V(t) = \int_V d^3x \frac{d}{dt} n^0(x) = -\int_V d^3x \frac{\partial n^i}{\partial x^i} = -\int_{\partial V} dS_i \, n^i$ (using the divergence theorem): the change in $N_V(t)$ with time is of course given by integrating the number current n^i flowing through the surface element dS_i forming the surface ∂V enclosing V.

We went through the number current in detail to save work later. Indeed, we can now write down the relativistic form of the electromagnetic current without further ado by simply including the charge e_a carried by the ath particle:

$$J^\mu(x) = \sum_a e_a \int d\tau_a \frac{dq_a^\mu}{d\tau_a} \delta^{(4)}(x - q_a(\tau_a)) \tag{5}$$

By the same considerations as above, we obtain the conservation of the electromagnetic current

$$\partial_\mu J^\mu = 0 \tag{6}$$

Energy momentum tensor

Moving on from the number density, we can now easily study energy density. As before, consider a bunch of particles sitting in a box. As is obvious by now, in nonrelativistic physics, energy density, which we denote by $\rho(t, \vec{x})$, is a rotational scalar, since energy and the volume of the box are both rotationally invariant. How does it transform under a Lorentz transformation?

We simply invoke our physical argument again. Our moving observer not only sees the box contracted, but he also sees the particles moving; thus $\rho(t, \vec{x})$ is enhanced by not one, but two factors of $1/\sqrt{1 - v^2}$. From what we learned in chapter III.3, it transforms like the time-time component $T^{00}(x)$ of a 4-tensor $T^{\mu\nu}(x)$, known as the energy momentum tensor (also called the stress energy tensor by some). In other words, this observer sees $T'^{00} = \Lambda^0{}_\mu \Lambda^0{}_\nu T^{\mu\nu} = \Lambda^0{}_0 \Lambda^0{}_0 T^{00} = \left(\frac{1}{\sqrt{1-v^2}}\right)^2 T^{00}$ (since for the unprimed observer the only nonvanishing component of $T^{\mu\nu}$ is T^{00}).

Just as for the number current (3), for which we count the particles, and for the electromagnetic current (5), for which we add up the charges, here we tally the 4-momentum p_a^ν carried by each particle. We write down instantly the energy momentum tensor

$$T^{\mu\nu}(x) = \sum_a \int d\tau_a \frac{dq_a^\mu}{d\tau_a} p_a^\nu(\tau_a) \delta^{(4)}(x - q_a(\tau_a)) = \sum_a \int d\tau_a \left(m_a \frac{dq_a^\mu}{d\tau_a} \frac{dq_a^\nu}{d\tau_a} \right) \delta^{(4)}(x - q_a(\tau_a)) \quad (7)$$

We already know that we need a tensor. Gratifyingly, $\frac{dq_a^\mu}{d\tau_a} \frac{dq_a^\nu}{d\tau_a}$, the product of two vectors, is manifestly a tensor. The second form in (7) emphasizes that $T^{\mu\nu}(x)$ is a symmetric tensor and thus possesses $\frac{4 \cdot 5}{2} = 10$ independent components.

We now have the energy momentum conservation law (exercise 3)

$$\partial_\mu T^{\mu\nu}(x) = 0 \quad (8)$$

in parallel with the current conservation law (6). Just as before when we counted the number of particles in a volume V, the amount of energy momentum contained in the volume V is given by

$$\int_V d^3x \, T^{0\nu}(x) = \sum_{(a \in V)} \int d\tau_a \frac{dq_a^0}{d\tau_a} p_a^\nu \delta(x^0 - q_a^0(\tau_a)) = \sum_{(a \in V)} p_a^\nu \equiv P_V^\nu(t) \quad (9)$$

In the second equality, we again use $d\tau_a \frac{dq_a^0}{d\tau_a} = dq_a^0$ to convert the integral over τ_a into an integral over q_a^0 to knock off the remaining δ function. (As they say, the first time a philosopher, the second time a connoisseur. In physics, you are a world expert the second time you use a trick.) We obtain the total 4-momentum $P_V^\nu(t)$ contained inside the 3-volume V at time t.

Stress

Again, to the beginner, it may seem at first sight a bit odd that we would take something like $T^{0\nu}(x)$, which carries an explicit time index, and integrate it over the Lorentz noninvariant 3-volume. But d^3x is precisely looking for something that transforms like dt to "complete itself."

Confusio nods: "Yes, I can see from (9) that $T^{0v}(x) = (T^{00}(x), T^{0i}(x))$ describes the spatial densities of energy and momentum, but I have a harder time picturing $T^{ij}(x)$."

Well, Confusio, you are not alone. Most people have trouble. It might help to note that (as before for the number current) we have $\frac{d}{dt}P_V^v(t) = \int_V d^3x \frac{\partial T^{0v}(x)}{\partial t} = -\int_V d^3x \frac{\partial T^{iv}}{\partial x^i} = -\int_{\partial V} dS_i\, T^{iv}$. Set $v = j$. Then

$$\frac{d}{dt}P_V^j(t) = -\int_{\partial V} dS_i\, T^{ij} \tag{10}$$

tells us that the time rate of change of the 3-momentum $P_V^j(t)$ (hence a force) has to do with T^{ij} acting on the surface element dS_i. In other words, T^{ij} is a force per unit area, and hence must be a pressure pushing in the jth direction, exerted on an area element pointing in the ith direction. Note that since $T^{ij} = T^{ji}$, this pressure is the same as the pressure pushing in the ith direction, exerted on an area element pointing in the jth direction. This physical picture also underlines the tensorial character of T^{ij}: there are two directions involved when you press against a surface. Does that make sense, Confusio?

"Yes indeed, one for the direction of the force, the other for the orientation of the surface."

Thus, some authors call the energy momentum tensor the stress energy tensor.

Incidentally, this discussion shows explicitly that upon restriction (recall chapter I.4) to the rotation subgroup, the 10-dimensional representation of the Lorentz group decomposes as $10 \to 1 + 3 + 6$, namely $T^{\mu v} = \{T^{00}, T^{0i}, T^{ij}\}$.

At a specific time

We could do the integrals in (3), (5), and (7), if our little hearts desire. To do this, I need to teach you an identity. Since the delta function is a big spike with total area under the peak $\int_{-\infty}^{\infty} dx\, \delta(x) = 1$, we have $\int_{-\infty}^{\infty} dx\, \delta(x)s(x) = s(0)$ for a sufficiently smooth function $s(x)$. Also,

$$\int_{-\infty}^{\infty} dx\, \delta(bx)s(x) = \int_{-\infty}^{\infty} dx\, \frac{\delta(x)}{|b|}s(x) = \frac{s(0)}{|b|}$$

where the factor of $1/b$ follows from dimensional analysis. (To see the need for the absolute value, simply note that $\delta(bx)$ is a positive function. Alternatively, change the integration variable from x to $y = bx$: for b negative, we have to flip the integration limits.) A trivial generalization is $\int_{-\infty}^{\infty} dx\, \delta(b(x - a))s(x) = \frac{s(a)}{|b|}$. Once we know how to deal with a linear function inside the delta function, we can handle any smooth function $f(x)$ inside the delta function. Denote the zeroes of $f(x)$ by x_A (in other words, $f(x_A) = 0$) and write $f'(x_A) = \frac{df}{dx}(x_A)$. Expand the function $f(x)$ around its zeroes. Break the integral $\int_{-\infty}^{\infty} dx\, \delta(f(x))s(x)$ into pieces, each containing a zero x_A. Hence the identity

$$\int_{-\infty}^{\infty} dx\, \delta(f(x))s(x) = \sum_A \frac{s(x_A)}{|f'(x_A)|} \tag{11}$$

(I trust you to deal with nongeneric cases, such as what happens if $f'(x_A)$ vanishes for some A.)

We now focus on an integral $\int d\tau_a \delta(t - q_a^0(\tau_a)) s(\tau_a)$ in (3), singling out the time delta function as the one we are knocking off and calling the rest of the integrand $s(\tau_a)$. Applying the identity (11), we find that the integral equals $s(\tau_a)/|(dq_a^0/d\tau_a)|$ evaluated at the value of τ_a that solves the equation $q_a^0(\tau_a) = t$. Once again, this may seem rather sophisticated to the reader seeing it for the first time, but it is simply determining the proper time τ_a of particle a as it crosses the specified time slice t. (See figure 3.) Since worldlines cannot go backward, the sum in (11) reduces to just one term. Thus, we encounter $(dq_a^\mu/d\tau_a)/(dq_a^0/d\tau_a) = (1, v_{N,a}^i)$ as defined earlier and (3) breaks up into

$$n^0(x) = \sum_a \delta^{(3)}(\vec{x} - \vec{q}_a(\tau_a))$$

$$n^i(x) = \sum_a v_{N,a}^i \delta^{(3)}(\vec{x} - \vec{q}_a(\tau_a)) \tag{12}$$

(with τ_a defined by $q_a^0(\tau_a) = t$) precisely as we would expect. (Reneging on our earlier promise, we put back the subscript N here for emphasis.)

Similarly, when we do the integral in (7), we encounter $p_a^\nu(\tau_a)(dq_a^\mu/d\tau_a)/(dq_a^0/d\tau_a)$, which we write as $p_a^\mu(\tau_a)p_a^\nu(\tau_a)/E_a(\tau_a)$ (since $E_a = m\frac{dq_a^0}{d\tau_a} = p_a^0$ is just another name for p_a^0). Thus,

$$T^{\mu\nu}(x) = \sum_a \frac{p_a^\mu(\tau_a)p_a^\nu(\tau_a)}{E_a(\tau_a)} \delta^{(3)}(\vec{x} - \vec{q}_a(\tau_a)) \tag{13}$$

The denominator E_a may seem a bit off to you, but you will soon see that it plays a necessary role.

Perfect fluids and the comoving observer

Consider a system of many particles. If the spatial separation between particles and the mean time between collisions are much less than the length and time scales we are interested in, we have a fluid (a term used loosely here to include gases). The various currents we have been discussing all become smooth functions of x. At a given point in spacetime, the fluid moves with a 4-velocity $U^\mu(x)$ normalized to

$$\eta_{\mu\nu}U^\mu U^\nu = U^\nu U_\nu = -1 \tag{14}$$

For instance, the number current would then be given by $n^\mu(x) = n(x)U^\mu(x)$.

For an observer going with the flow so to speak, known as a comoving observer, $U^0 = 1$, $U^i = 0$, and thus $n(x)$ is just the number density seen by the comoving observer.

If the fluid is isotropic as seen by the comoving observer (that is, the fluid does not have a special direction in its local rest frame), it is said to be perfect.

Since $U^i = 0$ and there is no other 3-vector available to construct T^{0i} out of, T^{0i} vanishes, by rotational invariance. Furthermore, the only thing we have available to construct the symmetric rotational tensor T^{ij} is the Kronecker delta δ^{ij}. Hence the stress energy tensor of a perfect fluid at that point has the form $T^{00}(x) = \rho(x)$, $T^{0i}(x) = 0$, and $T^{ij}(x) = P(x)\delta^{ij}$, with ρ and P some function of x. Written out as a matrix, we have

$$T^{\mu\nu} = \begin{pmatrix} \rho & 0 & 0 & 0 \\ 0 & P & 0 & 0 \\ 0 & 0 & P & 0 \\ 0 & 0 & 0 & P \end{pmatrix} \tag{15}$$

Let us invite ourselves to express this in terms of U^μ and $\eta^{\mu\nu}$. As you can quickly verify component by component,

$$T^{\mu\nu}(x) = (\rho(x) + P(x))U^\mu(x)U^\nu(x) + P(x)\eta^{\mu\nu} \tag{16}$$

Note that U^μ may vary from point to point, of course. As another easy check, note that in the comoving frame, with the form (16), the conservation law $\partial_\mu T^{\mu\nu} = 0$ reduces to $\frac{\partial\rho}{\partial t} = 0$ for $\nu = 0$ and $\frac{\partial P}{\partial x^i} = 0$ for $\nu = i$.

Moving fluids

What if the fluid is moving, as fluids are wont to do?

Behold the power of Lorentz invariance. Simply go to the frame of an observer moving relative to our comoving observer. To this observer, the perfect fluid is moving. All we have to do is Lorentz boost the 4-vector $U^\mu = (1, \vec{0})$ to $U^\mu = \left(\frac{1}{\sqrt{1-\vec{v}^2}}, \frac{v^i}{\sqrt{1-\vec{v}^2}}\right)$. As we have been saying until we are practically hoarse, if a tensor equation like (16) holds in one frame, it holds in all frames. Thus, for example, the energy density of a moving fluid is

$$T^{00}(x) = (\rho + P)(U^0)^2 - P = \frac{\rho + P}{1 - \vec{v}^2} - P = \frac{\rho + \vec{v}^2 P}{1 - \vec{v}^2} \tag{17}$$

Without the power of a symmetry argument, it would be somewhat challenging to work out the relativistic corrections embodied in (17). Note that relativistically, the pressure P contributes to the energy density T^{00}. I won't deprive you of the fun of working out the other components. (See exercise 4, from the result of which you can also see when T^{ij} for $i \neq j$ might be nonzero.) As should be clear from our discussion, here ρ, P, and $\vec{v}$ can all depend on x.

Two gases for cosmology

We now go back to (13) and calculate ρ and P for two important cases. To say that we have a fluid means that we are to average the factor $\frac{p_a^\mu(\tau_a)p_a^\nu(\tau_a)}{E_a(\tau_a)}$ in (13) over the multitude of particles. By definition, a perfect fluid means that there is no preferred direction in the comoving frame: the 3-vector $\vec{p}_a(\tau_a)$ points in all possible directions, so that $p_a^0(\tau_a)p_a^i(\tau_a)$ averages to 0, giving $T^{0i} = 0$, thus confirming our earlier conclusion using rotational invariance. Next, recall exercise I.3.5, which showed that $p_a^i(\tau_a) p_a^j(\tau_a)$ becomes $\frac{1}{3}\vec{p}_a^2(\tau_a)\delta^{ij}$ when averaged over all directions. Thus, for a perfect fluid,

$$\rho = \left\langle \sum_a E_a(\tau_a)\delta^{(3)}(\vec{x} - \vec{q}_a(\tau_a)) \right\rangle$$

$$P = \left\langle \sum_a \frac{\vec{p}_a^2(\tau_a)}{3E_a(\tau_a)}\delta^{(3)}(\vec{x} - \vec{q}_a(\tau_a)) \right\rangle \tag{18}$$

where the angled brackets indicate averaging over particles (an averaging implied by the sum already for a macroscopically large number of particles). Since for each particle, $E^2 = \vec{p}^2 + m^2 \geq \vec{p}^2$, we have $\rho \geq 3P \geq 0$. We can evaluate this result in two extreme cases.

For a nonrelativistic gas, $E = m + \frac{\vec{p}^2}{2m} + \cdots$. Statistical mechanics teaches us that the quantity called temperature T is twice the energy* possessed by each degree of freedom. Each (monoatomic) gas particle has three kinetic degrees of freedom, so that its average kinetic energy is just $\frac{3}{2}T$. Thus, using (18), we obtain the ideal gas law $\rho = n(m + \frac{3}{2}T)$ and, since $\langle \frac{\vec{p}_a^2(\tau_a)}{3E_a(\tau_a)} \rangle \simeq \frac{\vec{p}^2}{3m} = \frac{2}{3}\frac{\vec{p}^2}{2m} = \frac{2}{3}(\frac{3}{2}T) = T$, we have $P = nT$, or $PV = NT$ in more elementary notation. A nice derivation of the equation of state of an ideal gas, no? As we have suspected all along, P is in fact the pressure.

For a highly relativistic gas, $E = |\vec{p}| + \cdots$ and so $\rho \simeq 3P$. In particular, the pressure of a photon gas is given by

$$P = \frac{1}{3}\rho \tag{19}$$

Note from (15) that in this case, the energy momentum tensor is traceless: $\eta_{\mu\nu}T^{\mu\nu} = -\rho + 3P = 0$.

In modern cosmology, the content of the universe is typically treated as a perfect fluid, as we will see in chapter VIII.1.

Nature of the gravitational field

Our simple intuitive argument involving moving particles in a contracting box (which shows that the energy density T^{00}, a rotational scalar, is promoted to a component of a 2-indexed tensor) gives us a tantalizing hint of the nature of the gravitational field. Already in chapters II.1 and II.3, I reminded you that in Newtonian gravity, the gravitational potential Φ satisfies Poisson's equation $\vec{\nabla}^2\Phi(\vec{x}) = 4\pi G\rho(\vec{x})$.

But if ρ is promoted, then it appears that Φ should also be promoted to the time-time component of a 2-indexed tensor, so that the two sides of whatever equation we have in Einstein gravity to determine Φ in terms of ρ would transform in the same way. This strongly suggests that the gravitational field is a tensor field. We will see in chapter VI.1 that this guess is correct. Furthermore, $\vec{\nabla}^2$ is clearly not Lorentz invariant and should be

---

* We omit the historical conversion factor k between degree and the unit of energy introduced by Boltzmann.

promoted to $-\partial^2 = \eta^{\mu\nu}\partial_\mu\partial_\nu = -\partial_t^2 + \vec{\nabla}^2$. We will develop this line of thought further in chapter IX.5.

Appendix 1: Einstein's derivation of the amusing and seductive $E = mc^2$

Did you know that Einstein didn't have $E = mc^2$ in his first paper on special relativity? This famous relation[5] appeared a few months later in a very brief note. Einstein wrote to a friend excitedly: "One more consequence of the paper on electrodynamics has also occurred to me. . . . The argument is amusing and seductive; but for all I know the Lord might be laughing over it and leading me around by the nose."

As we all know, the Lord did not lead Einstein around by the nose.

The derivation I gave in this chapter of this relation is "modern" and more or less standard. Here I will describe an elegant derivation* given by Einstein in 1946, which surprisingly, is omitted from most textbooks[6] and so is in danger of being forgotten.

Suppose Ms. Unprime observes an atom of mass M at rest emitting two photons with equal and opposite momenta, thus leaving the "daughter atom" at rest. See figure 5a,b. Let Mr. Prime move with velocity $-v$ in a direction (call it the x-axis) perpendicular to the direction defined by the motion of the photons. We will take $v \ll c$ so that we can use Newtonian mechanics to describe the motion of the atom before and after the emission. Thus, Mr. Prime sees, before the emission, an atom moving along the x-axis with velocity v, and, after the emission, the daughter atom moving along the x-axis with velocity v, together with two photons with the x components of their velocities equal to v. See figure 5c,d. Given that the speed of light is c, it follows that the two photons move away from the x-axis at an angle θ given by $\cos\theta = v/c$, as indicated in the figure. (We are taking $v \ll c$ so the figure is not to scale.)

The key ingredient in the argument is that a photon carrying energy E_γ has momentum $p_\gamma = E_\gamma/c$, a result that goes back in some sense to Einstein's own Nobel Prize–winning work on the photoelectric effect. (Here is a nifty derivation. By dimensional analysis, p_γ must be a constant times E_γ/c. But Mr. Maxwell already calculated the momentum carried by an electromagnetic wave (using the Poynting vector, remember?) to be equal to the energy of the wave divided by c. Thus, if we think of the wave as a stampede of photons, we could argue that the constant must be 1.)

Momentum conservation holds trivially for Ms. Unprime. Now watch Mr. Prime impose momentum conservation in the x direction. The x component of the photon momentum is $p_\gamma^x = \cos\theta \, E_\gamma/c = v E_\gamma/c^2$. So momentum conservation requires

$$Mv = mv + 2\frac{vE_\gamma}{c^2} \tag{20}$$

Here, just to keep an open mind, we write the mass of the daughter atom as m, which may well be equal to M, the mass of the atom before emission. As you know, before Einstein, most physicists would have thought that $m = M$. But now we see immediately that momentum conservation cannot be satisfied unless $m < M$. Cool, eh? From this argument, it already follows that during emission the atom must lose mass.

Next, Mr. Prime applies energy conservation, as any decent physics student would. Before emission, the atom has energy $A + \frac{1}{2}Mv^2$. Again, to be open minded, we add a constant A, which Mr. Newton did not know about, to his kinetic energy. The atom loses mass, and so might end up possessing less of everything. Similarly, we suppose that after emission, the atom has energy $a + \frac{1}{2}mv^2$. Energy conservation now requires

$$A + \frac{1}{2}Mv^2 = a + \frac{1}{2}mv^2 + 2E_\gamma = a + \frac{1}{2}mv^2 + (M - m)c^2 \tag{21}$$

where in the last step we used (20). By assumption $v \ll c$, so we can drop the v^2 terms (the energy conservation equation is consistent as written, since in the derivation we already have dropped terms suppressed by v/c). We thus obtain $A - a = (M - m)c^2$, or in other words, $\delta E = (\delta m)c^2$. Following Einstein, we integrate this to obtain $E = mc^2$.

Nowadays, I could have used the decay of the neutral pion into two photons $\pi^0 \to \gamma + \gamma$ instead of the radiative emission of an atom and simplified the derivation slightly (since the π^0 meson has no daughter, so to speak).

---

* I like Einstein's 1946 derivation much better than his original 1905 derivation, which invoked the Lorentz transformation and was unnecessarily complicated.

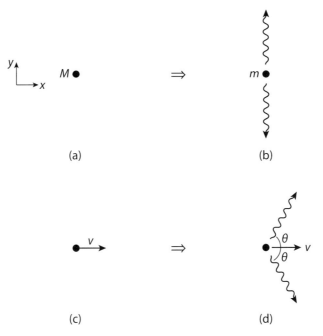

Figure 5 Einstein's nearly forgotten 1946 derivation of $E = mc^2$. (a) Ms. Unprime observes an atom of mass M at rest. (b) The atom subsequently emits two photons with equal and opposite momenta, thus leaving the daughter atom of mass m at rest. (c) Mr. Prime observes an atom moving along the x-axis with velocity v. (d) The atom subsequently emits two photons. To Mr. Prime, the daughter atom continues to move along the x axis with velocity v, and the two photons have velocities with x-components equal to v.

But it would be somewhat unfair, since the possibility of a massive particle disappearing into two poufs of energy was hardly conceivable back then. Of course, if we are allowed to use the entire formalism of Lorentz vectors, we could simply write down the conservation of 4-momentum $p_\pi = k_\gamma + q_\gamma$ and evaluate it in the rest frame of the pion.

Appendix 2: Conservation and relativistic fluid dynamics

This appendix may be omitted upon first reading. Fluid dynamics describes how stuff, energy, and momentum flow from place to place as a function of time and is thus governed by the two conservation laws (4) and (8).

Written out more explicitly, $\partial_\mu n^\mu = 0$ becomes

$$\partial_0 \left(\frac{n}{\sqrt{1 - \vec{v}^2}} \right) + \partial_i \left(\frac{n v^i}{\sqrt{1 - \vec{v}^2}} \right) = 0 \tag{22}$$

Plugging the perfect fluid form (16) into $\partial_\mu T^{\mu\nu} = 0$, we have

$$\{\partial_\mu[(\rho + P)U^\mu]\}U^\nu + (\rho + P)U^\mu\partial_\mu U^\nu = -\eta^{\mu\nu}\partial_\mu P \tag{23}$$

It takes some work to massage this into shape. Divide the four equations in (23) into one "time equation," obtained by setting $\nu = 0$, and three "space equations," obtained by setting $\nu = i$. Solve the time equation for the quantity in the curly brackets, $\partial_\mu[(\rho + P)U^\mu] = \frac{1}{U^0}[\partial_0 P - (\rho + P)U^\mu\partial_\mu U^0]$, and plug this back into the space equations. We encounter* $v^i \equiv \frac{U^i}{U^0}$ and the differential operator $\frac{U^\mu}{U^0}\partial_\mu = \frac{\partial}{\partial t} + \vec{v} \cdot \vec{\nabla}$, which the reader familiar with elementary fluid dynamics will recognize as the convective derivative. (Indeed, we derived it in the appendix to chapter III.1 using Galilean invariance.)

After the dust settles, we obtain the relativistic Euler equation

$$\left(\frac{\partial}{\partial t} + \vec{v} \cdot \vec{\nabla} \right) \vec{v} = - \left(\frac{1 - \vec{v}^2}{\rho + P} \right) \left(\vec{v} \frac{\partial P}{\partial t} + \vec{\nabla} P \right) \tag{24}$$

The reader just alluded to would notice that the usual nonrelativistic Euler equation (also known as the Navier-Stokes equation and referred to as such in chapter III.1) ($\frac{\partial}{\partial t} + \vec{v} \cdot \vec{\nabla})\vec{v} = -\frac{1}{\rho}\vec{\nabla}P$ emerges in the limit $v \ll 1$ and $P \ll \rho$.

We have thus far extracted 3 equations, namely (24), out of the 4 equations contained in (23). To extract the remaining equation, contract (23) with U_ν and use $0 = \partial_\mu(U^\nu U_\nu) = 2(\partial_\mu U^\nu)U_\nu$ obtained by differentiating (14). We find

$$\partial_\mu[(\rho + P)U^\mu] = U^\mu \partial_\mu P \quad \text{"mystery equation"} \tag{25}$$

which you may or may not recognize.

What do we do with this? The clever trick is to go back to (4) $\partial_\mu n^\mu = \partial_\mu(nU^\mu) = 0$ and write the left side of this "mystery equation" as $\partial_\mu[(\rho + P)U^\mu] = \partial_\mu[(\frac{\rho+P}{n})nU^\mu] = nU^\mu\partial_\mu(\frac{\rho+P}{n}) = nU^\mu\partial_\mu(\frac{\rho}{n}) + nPU^\mu\partial_\mu(\frac{1}{n}) + U^\mu\partial_\mu P$. Thus, the mystery equation becomes $U^\mu\partial_\mu(\frac{\rho}{n}) + PU^\mu\partial_\mu(\frac{1}{n}) = 0$.

Still don't recognize this? We have already exploited our knowledge of statistical mechanics; now we invoke thermodynamics. First, notice that $\frac{\rho}{n}$ and $\frac{1}{n}$ are energy and volume per particle, respectively. Second, recall the first law of thermodynamics $dE + PdV = TdS$, with S the entropy. Define s as the entropy per particle. Since $U^\mu\partial_\mu$ is proportional to the convective derivative, we see that the mystery equation tells us that, as the fluid flows along, the changes in energy and in volume per particle are related by[7]

$$d\left(\frac{\rho}{n}\right) + Pd\left(\frac{1}{n}\right) = Tds \tag{26}$$

In other words, the mystery equation says $TU^\mu\partial_\mu s = 0$, that is,

$$\left(\frac{\partial}{\partial t} + \vec{v} \cdot \vec{\nabla} \right) s = 0 \tag{27}$$

We obtain the convective conservation of specific entropy and have shown that the flow is adiabatic. Very satisfying! No dissipation in a perfect fluid.

The set of equations, continuity (22), Euler (24), entropy conservation (27), together with an equation of state relating P and ρ and thus specifying the fluid, allows us to solve for the motion of the fluid.

When I see elegant relativistic equations, (4) $\partial_\mu n^\mu = 0$ and (8) $\partial_\mu T^{\mu\nu} = 0$ in this case, split up brutally[†] into their space and time components (22), (24), and (27), I must say that I am reminded of the biblical injunction "What therefore God hath joined together, let not man put asunder."

Appendix 3: The speed of sound

We now use the formalism of the preceding appendix to calculate the speed of sound in a static relativistic fluid. We will need the result when we discuss cosmology in chapter VIII.3.

Consider a density wave described by $n = \bar{n} + \delta n$, $\rho = \bar{\rho} + \delta\rho$, $P = \bar{P} + \delta P$, $s = \bar{s} + \delta s$, and $\vec{v} = \vec{0} + \delta\vec{v}$. The equilibrium quantities, indicated by an overbar, do not depend on space and time. We have also indicated that the fluid velocity vanishes before the density wave comes along. You probably know what to do: simply expand the relevant equations in the preceding appendix to first order in the small quantities δn, $\delta\rho$, δP, δs, and $\delta\vec{v}$.

---

* A question for you: is this $\vec{v}$ or $\vec{v}_N$ from the second section of this chapter?
† Yes, I know. In the real world, plenty are put asunder.

First, continuity (22) gives $\frac{\partial}{\partial t} \delta n + \bar{n} \vec{\nabla} \cdot \delta \vec{v} = 0$. Second, Euler (24) tells us that $\frac{\partial}{\partial t} \delta \vec{v} = -\frac{1}{\bar{\rho}+\bar{P}} \vec{\nabla} \delta P$. Third, (27) says that to leading order $\frac{\partial}{\partial t} \delta s = 0$: the specific entropy does not change, which by the first law of thermodynamics (26) implies $\bar{n} \delta \rho = (\bar{\rho} + \bar{P}) \delta n$. Define*

$$c_s^2 = \left(\frac{\partial P}{\partial \rho} \right)_s = \frac{\delta P}{\delta \rho} \tag{28}$$

Note that the derivation given here specifies that $\frac{\partial P}{\partial \rho}$ is to be evaluated at fixed specific entropy. This enables us to eliminate $\delta P = c_s^2 \delta \rho = c_s^2 (\bar{\rho} + \bar{P}) \delta n / \bar{n}$ in the equation for $\delta \vec{v}$, so that $\frac{\partial}{\partial t} \delta \vec{v} = -(c_s^2 / \bar{n}) \vec{\nabla} \delta n$. Putting this into the equation for δn, we finally obtain the wave equation

$$\frac{\partial^2}{\partial^2 t} \delta n - c_s^2 \vec{\nabla}^2 \delta n = 0 \tag{29}$$

(Recall that you have encountered the 1-dimensional version of this equation in appendix 2 to chapter II.3.) For example, for a plane wave propagating along the x direction, a particular solution is $\delta n \propto \sin\{k(x - c_s t)\}$, thus showing that c_s as defined in (28) is in fact the speed of sound.

Plugging in (19), we find that for a highly relativistic gas

$$c_s = \frac{1}{\sqrt{3}} \tag{30}$$

It is instructive to compare the differential operator $\frac{\partial^2}{\partial^2 t} - c_s^2 \vec{\nabla}^2$ that appears in (29) with the Lorentz invariant operator $\partial^2 = \eta^{\mu\nu} \partial_\mu \partial_\nu = \partial_t^2 - c^2 \vec{\nabla}^2$ mentioned earlier. We have restored the speed of light c to emphasize that the two operators have the same form, with c_s playing the role of c, as was already foreshadowed by the discussion of the baby string in appendix 3 in chapter II.3.

Appendix 4: The current in string theory

This appendix may also be omitted upon first reading. Here we work out the current associated with a point particle as it traces out a worldline $q^\mu(\tau)$. It is more or less straightforward to generalize this to an extended object like a string. Quite aside from string theory, there is also the possibility that our universe may contain what are known as cosmic strings.

In the preceding chapter, we worked out the string action. Recall that at a given value of τ, we also have to specify where we are along the string by another parameter σ. In other words, the spacetime location of a point on the string is specified by[†] $q^\mu(\tau, \sigma)$. In contrast to the case of a particle, q^μ now depends on two parameters τ and σ.

Think about how to generalize the current $J^\mu(x) = \int d\tau \frac{dq^\mu}{d\tau} \delta^{(4)}(x - q(\tau))$ associated with a point particle to the current associated with a string. Try not to read ahead immediately. See if you can write it down.

I now give you hints galore. The current should of course also contain $\delta^{(4)}(x - q(\tau, \sigma))$: no current if you are not where the string is. We need to integrate over both τ and σ. In other words, the current should have the form $\sim \int d\tau d\sigma \mathcal{M} \delta^{(4)}(x - q(\tau, \sigma))$. It remains to identify the mystery factor $\mathcal{M}$, the generalization of $\frac{dq^\mu}{d\tau}$ in the particle case. We now have at our disposal $\partial_\tau q^\mu \equiv \frac{\partial q^\mu}{\partial \tau}$ and $\partial_\sigma q^\mu \equiv \frac{\partial q^\mu}{\partial \sigma}$.

Geometry and symmetry provide the guiding lights once again. As in the discussion of the string action in the preceding chapter, we insist that the current must be unchanged if we choose a different parametrization $\tau \to \tau'(\tau, \sigma), \sigma \to \sigma'(\tau, \sigma)$. As I said earlier, the second time around you are already an expert. Since the integration measure $d\tau d\sigma = d\tau' d\sigma' J$ changes by a Jacobian (given as usual by a determinant), as discussed

---

* Note that the first subscript s refers to "sound" and the second to "specific entropy."

[†] To conform to the notation used in this chapter, we use q instead of X. For a certain type of reader, I can only offer Ralph Waldo Emerson's famous dictum: "A foolish consistency is the hobgoblin of little minds, adored by little statesmen and philosophers and divines." I will assume henceforth that the reader is neither a little statesman nor a divine.

in the preceding chapter, we need $\mathcal{M}$ to be a determinant to counteract J. Evidently, the factor $\frac{dq^\mu}{d\tau}$ in the particle current is generalized to the determinant of the 2-by-2 matrix $\begin{pmatrix} \partial_\tau q^\mu & \partial_\tau q^\nu \\ \partial_\sigma q^\mu & \partial_\sigma q^\nu \end{pmatrix}$.

Thus, the current associated with a string is given by

$$J^{\mu\nu}(x) = \int d\tau d\sigma \; \det \begin{pmatrix} \partial_\tau q^\mu & \partial_\tau q^\nu \\ \partial_\sigma q^\mu & \partial_\sigma q^\nu \end{pmatrix} \delta^{(d)}(x - q(\tau, \sigma)) \tag{31}$$

for $\mu, \nu = 0, 1, \cdots, d - 1$.

The determinant is antisymmetric under the exchange $\mu \leftrightarrow \nu$: we are led to an antisymmetric tensor current $J^{\mu\nu}$. Thus, the analog of the electromagnetic potential $A_\mu(x)$ coupling to the particle current J^μ has to be an antisymmetric tensor field $B_{\mu\nu}(x)$ coupling to the string current $J^{\mu\nu}$.

Exercises

1   As you probably know, the universe is suffused with a cosmic microwave background. A high-energy charged particle (such as an electron or a proton) traversing this background will occasionally hit one of these microwave photons, transferring its energy to the photon in what is known as inverse Compton scattering. Indeed, observationally, it is often by detecting these high-energy photons that we deduce the presence of sources of high-energy electrons. Show that, upon impact by a highly energetic electron (say) of energy E, the maximum energy the photon can have is given by $\omega' \simeq \frac{E}{1 + \frac{m^2}{4\omega E}}$, with m the mass of the electron and ω the energy of the microwave photon.

2   Consider a process in which two particles go into two particles $p_1 + p_2 \rightarrow p_3 + p_4$. All 4 particles may have different masses $p_a^2 = -m_a^2$ for $a = 1, 2, 3, 4$. Apparently, we can form 3 Lorentz invariants, namely $s \equiv (p_1 + p_2)^2 = (p_3 + p_4)^2$, $t \equiv (p_1 - p_3)^2 = (p_2 - p_4)^2$, and $u \equiv (p_1 - p_4)^2 = (p_2 - p_3)^2$. But we have known since childhood that there are only two kinematic variables in 2-to-2 scattering: the total energy and the scattering angle. Show that a Lorentz invariant identity connects $s, t,$ and u.

3   Verify (8) directly by plugging in (7).

4   Work out T^{0i} and T^{ij} for a moving perfect fluid.
   (a) Apply the result to a fluid moving in the x direction. Show that $T^{xx} = \frac{\rho v^2 + P}{1 - v^2}$, which deviates from the nonrelativistic result $T^{xx} = P$. Also, show that $T^{xy} = 0$ and $T^{yy} = P$. Are you surprised?
   (b) More generally, under what circumstances would $T^{ij} \neq 0$ for $i \neq j$?

5   Upon restriction of the Lorentz group to its rotation subgroup, the symmetric tensor $T^{\mu\nu}$ decomposes as $10 \rightarrow 1 + 3 + 6$, as was shown in the text. Consider the antisymmetric tensor $F^{\mu\nu} = -F^{\nu\mu}$. Show that it has 6 components and that it decomposes as $6 \rightarrow 3 + 3$.

6   Verify that the string current $J^{\mu\nu}$ does not depend on the coordinate choice on the world sheet or world tube.

7   Show that $\mathcal{M}^{\mu\nu} = \int d^3x(x^\mu T^{0\nu} - x^\nu T^{0\mu})$ describes the angular momentum of the system.

Notes

1. A. Einstein, "Does the Inertia of a Body Depend on Its Energy Content?" *Ann. Phys.* 18 (1905), p. 639.
2. For the historical importance of this result, see R. Baierlein, *Newton to Einstein*.
3. For that you need quantum field theory (for example, *QFT Nut*, chapter II.8), but way back when, the result (2) empirically verified sufficed for a Nobel Prize. Of course, the prize was actually for discovering the effect.

4. By the same token, if a worldline terminates, then the number of particles is not conserved at that instant. In particle physics, one particle can decay into other particles, for example, in the decay of a negatively charged pion into an electron and an antineutrino $\pi^- \to e^- + \bar{\nu}$. The worldline of the pion would then terminate at some point P in spacetime, while the worldlines for the electron and the antineutrino would commence there. For the level of discussion in this chapter, this is hardly something the reader needs to be concerned about. Note, however, that this notion of worldlines ending and beginning contains the seeds of Feynman diagrams. See, for example, *QFT Nut*.

5. Interestingly, there was a history of speculations in the 19th century concerning the energy contained in mass. See J. Stachel's commentary in A. Einstein, *Einstein's Miraculous Year*.

6. A notable exception is R. Baierlein, *Newton to Einstein*. I am grateful to R. Baierlein for providing me the original reference: A. Einstein, *Technion Yearbook* 5 (1946) p. 16.

7. Some technical assumptions, the discussion of which would take us too far afield, have been implicitly made here. What we have shown here is that the flow is adiabatic; hence the fluid does not exchange heat as it flows. To say that the right hand side of (26) is $T ds$ requires an additional assumption: the flow is quasistatic between consecutive states of thermodynamic equilibria, for which entropy is defined. This assumption can be justified under some circumstances.

Recap to Part III

In the showdown between $t' = t$ and $c' = c$, the former blinked, and lost.

With t' no longer chained to t, Lorentz found a nicer transformation than the one Galileo found, a transformation that was more symmetrical and associated with a better group, and hence is more pleasing to the eye. The group consists of "rotations" in a $(3 + 1)$-dimensional spacetime.

Instead of approaching special relativity as a series of would-be paradoxes, we should learn to appreciate the geometry of Minkowskian spacetime in which a straight line between two points may be the longest, rather than the shortest, path. The key to the action governing the motion of particles in this spacetime was hidden in how Babylonians, and smart school children, figure out the square root of a number that is almost, but not quite, a perfect square.

Various physical quantities had to be promoted and completed. Not only did this reveal a terrifying secret about the energy locked up in mass, it also gave us a clue about the true nature of the gravitational field.

Incidentally, we might as well drop the prime and simply write $c = c$.

Part IV | Electromagnetism and Gravity

IV.1 You Discover Electromagnetism and Gravity!

A bright young theoretical physicist

Once again, imagine yourself a bright young theoretical physicist in some civilization far far away, perhaps the same guy in chapter III.5 or perhaps a successor who discovered that guy's obscure work. Who knows what the environment is like, perhaps the civilization is floating in some molecular cloud, and who knows what the order of physics discoveries might be in an environment radically different from ours? Perhaps the speed of light was established to be independent of observers in relative uniform motion before electromagnetism was understood.

Every morning (assuming such a phenomenon exists over there), you admire the elegance of the action

$$S = -m \int \sqrt{-\eta_{\mu\nu} dx^\mu dx^\nu} = -m \int \sqrt{dt^2 - d\vec{x}^2} \tag{1}$$

Elegant indeed! But how do you get the particle to interact with the rest of the world?

You look at the nonrelativistic action

$$S_{\text{NR}} = \int dt \left(\frac{1}{2} m \left(\frac{d\vec{x}}{dt} \right)^2 - V(x) \right) \tag{2}$$

for inspiration. The point particle interacts with the world through the external potential $V(x)$. How do you include $V(x)$ in the relativistic point particle action (1)?

Two options: Outside or inside

One day, you realize that you could put $V(x)$ either outside or inside the square root in (1). So, you excitedly write a paper proposing not one, but two, possible actions:

Option E: $S = -\displaystyle\int \{m\sqrt{-\eta_{\mu\nu}dx^{\mu}dx^{\nu}} + V(x)dt\}$ $\qquad(3)$

or

Option G: $S = -m\displaystyle\int \sqrt{\left(1 + \frac{2V}{m}\right)dt^2 - d\vec{x}^2}$ $\qquad(4)$

In option G, you have to expand twice to get back to the nonrelativistic action. First, take $|d\vec{x}| \ll dt$ (that is, the distance the particle traverses, $|d\vec{x}|$, is tiny compared to cdt) so that

$$S \simeq -m\int \left\{\sqrt{1 + \frac{2V}{m}}\,dt - \frac{d\vec{x}^2}{2\sqrt{1 + \frac{2V}{m}}\,dt}\right\} \qquad(5)$$

Second, take the potential energy V to be much smaller than m (that is, the rest energy mc^2), so that $\sqrt{1 + \frac{2V}{m}} \simeq 1 + \frac{V}{m}$. Note that in (5), the second term is already much smaller than the first term, so we do not have to keep the $\frac{V}{m}$ correction to the second term. Thus,

$$S \simeq -m\int \left\{\left(1 + \frac{V}{m}\right)dt - \frac{d\vec{x}^2}{2dt}\right\} = \int dt \left\{\tfrac{1}{2}m\left(\frac{d\vec{x}}{dt}\right)^2 - V - m\right\} \qquad(6)$$

The action in option G leads to the Newtonian action (2) in the appropriate limits, but with a mysterious additive constant $-m$ that does not figure in the equation of motion. Well, the real you knows what that is.

You excitedly submit to a journal, and this time, remarkably, you actually get a perspicacious referee, who rejects the paper saying that the added term, in both option E and option G, is manifestly not Lorentz invariant: dt plays a more privileged role than $d\vec{x}$. Boy, didn't think of that! Dumb!

Symmetry, completion, and promotion

A clarifying comment about symmetry and invariance: consider the harmonic oscillator, that is, (2) with $V(x) = -\tfrac{1}{2}kx^2$. It is not translation invariant, but in a trivial way. We have simply excluded the much heavier mass that the spring is anchored to. Including that heavy mass, we replace (2) by $S_{\mathrm{NR}} = \int dt(\tfrac{1}{2}m(\frac{d\vec{x}}{dt})^2 - V(x - X) + \tfrac{1}{2}M(\frac{d\vec{X}}{dt})^2)$, or, if we insist on focusing on the small mass tied to the spring, by $S'_{\mathrm{NR}} = \int dt(\tfrac{1}{2}m(\frac{d\vec{x}}{dt})^2 - V_X(x))$ with an external potential $V_X(x) \equiv V(x - X)$ that depends on some parameter X. Translation invariance holds if we transform both $x \to x + a$ and $X \to X + a$.

Any beginning student of physics understands all this. Similarly here, if we think of the external potential $V(x)$ as imposed from the outside and fixed, then of course the action can never be made invariant. It is understood that we also have to transform $V(x)$.

You think hard, but confound it! Option E in (3) and option G in (4) are the only two possibilities you can think of. Either the added interaction term is inside the square root or it is outside. Inside or outside. Where else could it be? Puzzled, you file the manuscript away in a drawer. Years later, you decide to come back to it and see if you can make it Lorentz invariant.

Dear reader, who may or may not be the same person as the you with the two marvelous actions, can you see how? The key is the relativistic completion we learned in chapter III.6.

Electromagnetism and gravity

You stare at option E in (3) for the longest time, and suddenly it becomes obvious! You have to relativistically complete the action. Promote $V(x)$ to be the time component $A_0(x)$ of a Lorentz vector field $A_\mu(x)$, and $V(x)dt$ could be just the first term in $A_\mu(x)dx^\mu = A_0(x)dt + A_i(x)dx^i$. You merely have to introduce a vector field $A_\mu(x)$ into the universe! You propose the action

$$\text{Option E improved:} \quad S = \int \{-m\sqrt{-\eta_{\mu\nu}dx^\mu dx^\nu} + A_\mu(x)dx^\mu\} \tag{7}$$

Comparing with (2), we see that we should identify $A_0 = -V$. When we Lorentz transform x^μ, we must Lorentz transform $A_\mu(x)$ as well, just as in the example with the spring tied to an external massive object. (It is also implicitly understood that the argument x in $A_\mu(x)$ now includes time as well as space.) The expression $A_\mu(x)dx^\mu$, the contraction of two Lorentz vectors, is manifestly a Lorentz scalar. With this understanding, the action S in (7) is Lorentz invariant.

After this great triumph, you immediately try to relativistically complete option G also. Staring at the expression $(1 + \frac{2V}{m})dt^2 - d\vec{x}^2$ inside the square root in (4), you understand that the key is democracy between dt and $d\vec{x}$. If dt^2 is multiplied by some function, then $d\vec{x}^2$ should be too. Denoting $(1 + \frac{2V}{m})$ by g, you write something like $g(x)dt^2 - \tilde{g}(x)d\vec{x}^2$, with g and $\tilde{g}$ depending on spacetime. But Lorentz transformations "mix up" g and $\tilde{g}$. Even worse, dt^2 gets transformed into a linear combination of dt^2, dx^i, and $dtdx^i$. It would seem that you can't get away without including $dtdx^i$ inside the square root.

Eventually, you realize that (up to a sign) $dt^2 - d\vec{x}^2$ started out in life as the Lorentz tensor $dx^\mu dx^\nu$ contracted with the Minkowski metric $\eta_{\mu\nu}$, and so the answer has been staring you in the face: you must relativistically complete by promoting $(1 + \frac{2V}{m})$ to be the time-time component of a Lorentz tensor $g_{\mu\nu}(x)$. In other words, you should promote $\eta_{\mu\nu}dx^\mu dx^\nu$ to $g_{\mu\nu}(x)dx^\mu dx^\nu$, that is, promote the fixed numerical matrix $\eta_{\mu\nu}$ to a matrix field $g_{\mu\nu}(x)$ varying in spacetime.

So you introduce the tensor field $g_{\mu\nu}(x)$ into the universe, and quickly publish another action:

$$\text{Option G improved:} \quad S = -m \int \sqrt{-g_{\mu\nu}(x)dx^\mu dx^\nu} \tag{8}$$

You now understand (4) as a mere special case of (8), upon restricting $g_{\mu\nu}(x)$ to the special form $g_{00} = -(1 + \frac{2V}{m})$, $g_{0i} = g_{i0} = 0$, and $g_{ij} = \delta_{ij}$.

With these three papers, you are bound for the extragalactic version of Stockholm for sure!

Dear reader, you might have realized that the extragalactic version of you has just discovered electromagnetism and gravity. A double whammy! If not, read on.

Electromagnetism pops out of special relativity

Let us look at electromagnetism (option E) first and postpone gravity (option G) until chapter IV.3.

Excitedly, you vary the action in (7) to see what you get. As always, we parametrize with the proper time:

$$S = -m \int d\tau \sqrt{-\eta_{\mu\nu} \frac{dx^\mu}{d\tau} \frac{dx^\nu}{d\tau}} + \int d\tau A_\mu(x(\tau)) \frac{dx^\mu}{d\tau} \tag{9}$$

Note that in the second term, A_μ is evaluated at the spacetime position $x^\mu(\tau)$ of the particle. In other words, the field $A_\mu(x)$ pervades spacetime, but the particle only samples the field at its particular location.

The variation of the first term in (9) is easy and was done in chapter III.5, which we repeat here for convenience:

$$\delta \left(-m \int d\tau \sqrt{-\eta_{\mu\nu} \frac{dx^\mu}{d\tau} \frac{dx^\nu}{d\tau}} \right) = m \int d\tau \, \eta_{\mu\nu} \frac{dx^\mu}{d\tau} \frac{d\delta x^\nu}{d\tau} = -m \int d\tau \, \eta_{\mu\rho} \frac{d^2 x^\mu}{d\tau^2} \delta x^\rho \tag{10}$$

Notice that in the last step, for later convenience, we have renamed a dummy.

The variation of the second term in (9) is a bit more involved:

$$\delta \int d\tau A_\mu(x) \frac{dx^\mu}{d\tau} = \int d\tau \left\{ A_\mu(x) \frac{d\delta x^\mu}{d\tau} + [\partial_\nu A_\mu(x) \delta x^\nu] \frac{dx^\mu}{d\tau} \right\} \tag{11}$$

We have to remember that $A_\mu(x)$ depends on x in order not to miss the $\partial_\nu A_\mu(x)$ term in (11). Integrate the first term in (11) by parts:

$$\int d\tau A_\mu(x) \frac{d\delta x^\mu}{d\tau} = -\int d\tau \frac{dA_\mu(x)}{d\tau} \delta x^\mu = -\int d\tau \partial_\nu A_\mu(x) \frac{dx^\nu}{d\tau} \delta x^\mu \tag{12}$$

Putting the two terms together and renaming indices, we find

$$\delta \int d\tau A_\mu(x) \frac{dx^\mu}{d\tau} = \int d\tau (\partial_\mu A_\nu - \partial_\nu A_\mu) \frac{dx^\nu}{d\tau} \delta x^\mu \tag{13}$$

The antisymmetric tensor field

$$F_{\mu\nu}(x) \equiv \partial_\mu A_\nu(x) - \partial_\nu A_\mu(x) \tag{14}$$

just popped out! Some readers may recognize this as the electromagnetic field; if you don't, once again read on.

Putting (10) and (13) together, we have

$$\delta S = \int d\tau \left(-m\eta_{\mu\rho} \frac{d^2 x^\mu}{d\tau^2} \delta x^\rho + F_{\mu\nu} \frac{dx^\nu}{d\tau} \delta x^\mu \right) \tag{15}$$

Now define $F^\mu_{\ \nu} \equiv \eta^{\mu\lambda} F_{\lambda\nu}$ so that the last term in (15) can be written as $F^\mu_{\ \nu} \frac{dx^\nu}{d\tau} \eta_{\mu\rho} \delta x^\rho$. Setting the coefficient of $\eta_{\mu\rho} \delta x^\rho$ to zero, we obtain the equation of motion

$$m \frac{d^2 x^\mu}{d\tau^2} = +F^\mu_{\ \nu}(x) \frac{dx^\nu}{d\tau} \tag{16}$$

which has precisely the form given in (III.3.20).

Lo and behold, we have discovered the Lorentz force law (more below) for a charged particle moving in an electromagnetic field! We have discovered electromagnetism!

Electromagnetism came looking for us

We did not go looking for electromagnetism; electromagnetism came looking for us. Some readers may need to be reminded that $F_{\mu\nu}$ represents the electromagnetic field and that (16) describes a charged particle in an electromagnetic field. Before showing that, let's do something simpler: we check that we recover Newton's equation in the nonrelativistic limit. Since $dt = dx^0 \gg dx^j$ and $(dx^0)^2 - d\vec{x}^2 = d\tau^2$, we have $\frac{dt}{d\tau} \simeq 1 \gg \frac{dx^j}{d\tau}$. Setting μ to i in (16) gives $m\frac{d^2x^i}{dt^2} \simeq F^i_{\ \nu}(x)\frac{dx^\nu}{d\tau} \simeq F^i_{\ 0}(x)$. Comparing (2) and (7), we are reminded that the only nonzero component of $A_\mu(x)$ is $A_0(x) = -V(\vec{x})$, so that $F^i_{\ 0} = F_{i0} = \partial_i A_0 = -\partial_i V$. We recover Newton's second law $m\frac{d^2\vec{x}}{dt^2} = -\vec{\nabla}V$, hardly surprising given how we went from (2) to (7).

Next, we identify the usual electric $\vec{E}$ and magnetic $\vec{B}$ field as follows:

$$B^3 \equiv F^{12} = F_{12}, \qquad B^1 \equiv F^{23} = F_{23}, \qquad B^2 \equiv F^{31} = F_{31},$$

$$E^1 \equiv F^{01} = -F_{01}, \qquad E^2 \equiv F^{02} = -F_{02}, \qquad E^3 \equiv F^{03} = -F_{03} \tag{17}$$

We also write $\vec{B} = (B^1, B^2, B^3)$ and $\vec{E} = (E^1, E^2, E^3)$. (At this stage in the book, I hardly need to remind the reader that while $\vec{B}$ and $\vec{E}$ transform like 3-vectors under rotation, they are not the spatial components of two 4-vectors. In particular, the index i on B^i (and ditto on E^i) is not to be lowered and raised by the standard rules for Lorentz indices. They are just convenient labels.)

To show that (16) indeed reproduces the Lorentz force law, let us write it in terms of the 4-momentum $p^\mu = m\frac{dx^\mu}{d\tau}$:

$$\frac{dp^\mu}{d\tau} = F^\mu_{\ \nu}(x)\frac{dx^\nu}{d\tau} \tag{18}$$

Look first at the spatial components. Set μ to 1: then $\frac{dp^1}{d\tau} = -F^{10}\frac{dx^0}{d\tau} + F^{12}\frac{dx^2}{d\tau} + F^{13}\frac{dx^3}{d\tau}$. (Here we raised the lower index on $F^\mu_{\ \nu}$ using the Minkowski metric $\eta^{\lambda\nu}$ and used the antisymmetry of $F^{\mu\nu}$.) With the identification in (17), we obtain $\frac{dp^1}{d\tau} = E^1\frac{dx^0}{d\tau} + \frac{dx^2}{d\tau}B^3 + \frac{dx^3}{d\tau}B^2$. Multiply throughout by $\frac{d\tau}{dt}$ to convert the derivatives with respect to τ to derivatives with respect to t. (This step is optional and in fact best omitted.) By rotational symmetry, we can write this as

$$\frac{d\vec{p}}{dt} = \vec{E} + \vec{v} \times \vec{B} \tag{19}$$

with $\vec{v} = \frac{d\vec{x}}{dt}$. The Lorentz force law just popped out!

Good. We just found out what the spatial components of (18) say. What about the time component? Three guesses.

The time component $\mu = 0$ of (18) tells us that $\frac{dp^0}{d\tau} = F^0_{\ i}(x)\frac{dx^i}{d\tau}$. But because p^0 is just the energy E of the particle, we learn, upon multiplying by $\frac{d\tau}{dt}$ and defining $\vec{v} = \frac{d\vec{x}}{dt}$, that

$$\frac{dE}{dt} = \vec{E} \cdot \vec{v} \tag{20}$$

describing how a charged particle in an electric field $\vec{E}$ gains energy E. It is worth emphasizing that (19) and (20) are fully, but not manifestly, relativistic.

What I have done here is write a bit of an alternative history of physics in a galaxy far far away. Imagine that in this civilization, we knew nothing about electromagnetism, but somehow, by doing an experiment, our savants discovered to everybody's astonishment that the speed of light is a universal constant independent of the observer. Then by studying the addition of velocities, we discovered special relativity and then electromagnetism. This was not how it happened in our civilization, but it could have.

The notion of charge

Once again, it is time for a bad notation alert! As before, it would be best to denote the position of the particle not by the generic x, but by q or X. Let us choose X and write

$$S = -m \int d\tau \sqrt{-\eta_{\mu\nu} \frac{dX^\mu}{d\tau} \frac{dX^\nu}{d\tau}} + \int d\tau A_\mu(X(\tau)) \frac{dX^\mu}{d\tau} \tag{21}$$

This makes clear that in (16), the electromagnetic field $F^\mu_{\ \nu}(x)$ is to be evaluated at the position of the particle, as was already emphasized in the discussion following (9).

As always, the point becomes glaringly clear if we think about the case of many particles labeled by $a = 1, 2, \cdots, N$, possibly with different masses m_a. Indeed, denote the space-time position of particle a by X_a. In the generalization of (21), $A_\mu(x)$ is to be evaluated at $x = X_a(\tau)$. In other words, while $A_\mu(x)$ exists throughout spacetime, in the action, it "knows" about particle a only through $X_a(\tau)$. This is what we mean by saying that the action is local. Of course, the electromagnetic field could acquire dynamics of its own (as we will see in the next chapter), in which case the physical effects of the electromagnetic field would be propagated throughout spacetime.*

So, write the action

$$S = -\sum_a m_a \int d\tau_a \sqrt{-\eta_{\mu\nu} \frac{dX_a^\mu}{d\tau_a} \frac{dX_a^\nu}{d\tau_a}} + \sum_a e_a \int d\tau_a A_\mu(X_a(\tau_a)) \frac{dX_a^\mu}{d\tau_a} \tag{22}$$

When we do that, we bump up against another important point. We see that if we have more than one particle, then when we add the interaction term between particles and field to the action, we can allow, as in (22), each particle to "couple" to the new field A_μ with a different strength[1] e_a, which we will call charge. In contrast, in (21), there is no point in introducing e, since it could be absorbed into the definition of A_μ. The action produces the equations of motion:

$$\frac{dp_a^\mu}{d\tau_a} = e_a F^\mu_{\ \nu}(X_a(\tau_a)) \frac{dX_a^\nu}{d\tau_a} \tag{23}$$

---

* Verily, this is why we need fields: we like to have a local action, but at the same time, also physical effects that can propagate from one point to another.

I prefer to write (22) in a parametrization independent form:

$$S = \int \sum_a \left\{ -m_a \sqrt{-\eta_{\mu\nu} dX_a^\mu dX_a^\nu} + e_a A_\mu(X_a) dX_a^\mu \right\} \tag{24}$$

Relativistic unification

In our study of physics, we typically encounter the two vector fields $\vec{E}$ and $\vec{B}$ in stages and are then told that they are components of an electromagnetic field tensor $F^{\mu\nu}$. In fact, if we are allowed an antihistorical perspective, we see that, given special relativity, we could have anticipated the packaging of two 3-vectors into an antisymmetric tensor with a combination of physical and mathematical considerations.

Einstein's special relativity forcefully unifies $\vec{E}$ and $\vec{B}$. In chapter III.6, we spoke of relativistic completion. Suppose at the dawn of special relativity, you were given $\vec{E}$ and $\vec{B}$ and asked to complete them. Your first thought, that the 3-vector $\vec{E}$ gets promoted to be part of a 4-vector, cannot be correct: this would mean that under a boost, the change of $\vec{E}$ is a rotational scalar, but since Øersted's great discovery in 1820, it has been known that $\vec{E}$ transforms into a linear combination of itself and $\vec{B}$. Thus, the simplest guess is that the two 3-vectors $\vec{E}$ and $\vec{B}$ get unified into an antisymmetric Lorentz tensor. (Also recall exercise III.6.5.)

The electric field and the magnetic field are unified when time and space get unified. To many theoretical physicists, the identification in (17), important though it is to make contact with experimental physics, is akin to taking an exquisite object of art and pulling it apart.

Exercise

1 Show that the Lorentz force law (18) is consistent with p^2 being a constant.

Note

1. Note that the e_a can be arbitrary real numbers of either sign. To understand why the electron charge is precisely equal and opposite to the proton charge, you would have to learn quantum field theory. See *QFT Nut*, chapter VII.5.

The electromagnetic field in action

After your triumph in the preceding chapter, people naturally ask you how the field $A_\mu(x)$ is generated. Electrodynamics should be a mutual dance between particles and field. The field causes the charged particles to move, and the charged particles should in turn generate the field.

We had the first half of this dynamics in the preceding chapter. Now we have to describe the second half; in other words, we are going to look for the action governing the dynamics of $A_\mu(x)$.

To construct the action, as we have seen many times by now, it is imperative that we understand all the relevant symmetries first. Lorentz invariance has to be imposed of course. But then a brilliant young physicist, not you this time for a change, notices a peculiar invariance of the term you added to the action

$$\int A_\mu(x)dx^\mu \tag{1}$$

Recall that $x^\mu(\tau)$ here represents the trajectory of a charged particle.

Gauge invariance

Consider the transformation

$$A_\mu(x) \to \tilde{A}_\mu(x) = A_\mu(x) + \partial_\mu \Lambda(x) \tag{2}$$

for some function $\Lambda(x)$. Then (1) changes by

$$\int \partial_\mu \Lambda(x) dx^\mu = \int_{\tau_i}^{\tau_f} d\tau \frac{dx^\mu}{d\tau} \partial_\mu \Lambda(x) = \int_{\tau_i}^{\tau_f} d\tau \frac{d\Lambda(x)}{d\tau} = \Lambda(x(\tau))\Big|_{\tau_i}^{\tau_f} \tag{3}$$

Formally, we take the beginning and end of the particle's trajectory in the far past and future and assume that $\Lambda(x)$ vanishes at infinity. Then the action does not change under the transformation in (2), known as a gauge transformation. We have discovered a hidden symmetry of the action, called gauge invariance.

Strictly speaking, a gauge symmetry of the type discussed here is not a symmetry, but rather a redundancy in the description. The statement here is that $A_\mu(x)$ and $\tilde{A}_\mu(x)$ describe the same physics; in other words, $A_\mu(x)$ contains degrees of freedom that are not physical, which could be removed by appropriate choices of $\Lambda(x)$. There is a great deal more I could say about this rather subtle subject, but for now I am content to refer you to various field theory texts.[1]

In the preceding chapter, the electromagnetic field strength tensor

$$F_{\mu\nu}(x) \equiv \partial_\mu A_\nu(x) - \partial_\nu A_\mu(x) \tag{4}$$

which contains the familiar electric and magnetic fields, naturally emerged when we varied (1). Now we understand on symmetry grounds why this particular combination must appear. Whatever emerges from varying a gauge invariant action has to be gauge invariant. The gauge potential A is not gauge invariant, but the field strength F is. Under a gauge transformation, we have

$$F_{\mu\nu} \to \partial_\mu[A_\nu(x) + \partial_\nu \Lambda(x)] - \partial_\nu[A_\mu(x) + \partial_\mu \Lambda(x)] = F_{\mu\nu} + \partial_\mu \partial_\nu \Lambda - \partial_\nu \partial_\mu \Lambda = F_{\mu\nu} \tag{5}$$

so that $F_{\mu\nu}$ does not change.

Thus, the action for electrodynamics we are searching for should be constructed out of $F_{\mu\nu}(x)$. To obtain a Lorentz invariant object, the simplest possibility would be to "square" F and contract the indices, obtaining $F^{\mu\nu}F_{\mu\nu}$. Note that this term also contains two powers of the time derivative $\frac{\partial}{\partial t}$, in line with all the actions we have encountered thus far. We integrate this Lorentz scalar over spacetime $\int d^4x(-\frac{1}{4}F^{\mu\nu}F_{\mu\nu})$ and add it to the action we had in (IV.1.21). (The factor $-\frac{1}{4}$ is conventional.)*

Discovering Maxwell

We have discovered Maxwell's Lagrangian!

The Maxwellian Lagrangian[†] $\mathcal{L} = -\frac{1}{4}F^{\mu\nu}F_{\mu\nu}$ when expanded out contains a number of terms, in particular $(\partial_0 A_i)^2$. The two powers of ∂_0 are the same as the two powers of time or proper time derivative in the Newtonian Lagrangian $(\frac{d\vec{x}}{dt})^2$ or the Lorentzian Lagrangian $\eta_{\mu\nu}\frac{dx^\mu}{d\tau}\frac{dx^\nu}{d\tau}$.

The bad notation that I keep harping on is particularly bothersome here. It is important to carefully distinguish between dynamical variables and mere labels. In Newtonian mechanics, the position of the particle $\vec{x}(t)$, or better, $\vec{q}_a(t)$ or $\vec{X}_a(t)$, is the dynamical variable. We have also indicated the possibility of having several particles, labeled by a. Here

---

* Once we have discovered the possibility of including charge, as in (IV.1.23), we are free to scale $A_\mu \to \lambda A_\mu$ and the charges accordingly, and thus set the coefficient of $F^{\mu\nu}F_{\mu\nu}$ in the action to be any real number we like.

† Some purists insist on calling $\mathcal{L}$ a Lagrangian density, since it is $\int d^3x\mathcal{L}$ that has the same status as the point particle Lagrangian $L = \frac{m}{2}(\frac{d\vec{q}}{dt})^2$. I would rather abuse terminology than clutter the page.

the dynamical variable is a field $A_\mu(x)$, perhaps better written more explicitly as $A_\mu(\vec{x}, t)$, whose time dependence we want to study. We see clearly that the $\vec{x}$, which the dynamical variable A_μ depends on, is a label, not a dynamical variable: it tells us which A_μ we are talking about, namely the A_μ at the spatial location $\vec{x}$. Thus, the translation table[2] from the mechanics of several particles to a field theory (such as Maxwell's Lagrangian) contains $a \to \vec{x}$ and $\vec{X}_a(t) \to A_\mu(\vec{x}, t)$.

Because of gauge invariance (2), the vector field A_μ contains fewer degrees of freedom than meet the eye, reflecting the redundancy I mentioned earlier. As you already know from classical electromagnetism, you have to practice the arcane art of fixing the gauge,[3] after which an electromagnetic wave has only two polarizations. In quantum field theory, you learn to quantize the electromagnetic field. When you do this, the electromagnetic field is described in terms of photons. Even though the photon carries spin 1, it has only[4] two spin states, corresponding to the two polarizations of the classical wave.

Coupling of field and particles

So now we have the complete action for charged particles and the electromagnetic field:

$$S = \left\{ \int \sum_a \left(-m_a \sqrt{-\eta_{\mu\nu} dX_a^\mu dX_a^\nu} + e_a A_\mu(X_a) dX_a^\mu \right) \right\} - \int d^4x \, \tfrac{1}{4} F^{\mu\nu} F_{\mu\nu} \tag{6}$$

I have purposely written it in this peculiar form to emphasize that the nature of the integral differs significantly for the three terms. The first term describes free massive particles. The third term describes a field. The world of particles and the world of the field would be forever estranged were it not for the second term, which couples particles and field.

This may remind you of something you have seen. Indeed, back in chapter II.3, we wrote down the action for Newtonian gravity consisting also of three terms:

$$S = \int dt \left\{ \sum_a \tfrac{1}{2} m_a \left(\frac{dq_a}{dt} \right)^2 - \int d^3x \left(\sum_a m_a \delta^{(3)}(x - q_a(t)) \Phi(x, t) \right) - \int d^3x \frac{1}{8\pi G} (\nabla \Phi)^2 \right\} \tag{7}$$

The first term describes the dynamics of the particles, the third term describes the gravitational field $\Phi(x, t)$, and the second term couples the particles and the field together. In Newtonian gravity, the field dictates how the particles move, and the particles in turn generate the field.

In exactly the same way, in electromagnetism, the field dictates how the particles move, and the particles in turn generate the field.

How the field moves

To obtain the equation of motion for the electromagnetic field, we vary S with respect to the field $A_\mu(x)$. But we already learned how to vary with respect to a field, also in chapter II.3, when we discussed the dynamical string described by the field $\phi(t, x)$. The only complication for the electromagnetic field is the presence of Lorentz indices, but they mostly just go along for the ride, so to speak.

Let's start by varying the integrand of the third term in (6): $\delta(F^{\mu\nu}F_{\mu\nu}) = 2F^{\mu\nu}\delta F_{\mu\nu} = 4F^{\mu\nu}\partial_\mu\delta A_\nu$. (The reader should understand the two factors of 2 and hence the conventional choice of $\frac{1}{4}$ in (6).) Thus, the variation of the third term is

$$\delta \int d^4x (-\tfrac{1}{4}F^{\mu\nu}F_{\mu\nu}) = \int d^4x \, \partial_\mu F^{\mu\nu}(x)\delta A_\nu(x) \tag{8}$$

where we have integrated by parts. The first term in (6) does not depend on the field and so may be ignored. Thus, were the second term not present, we would have obtained, by setting the variation to zero, the free Maxwell's equations

$$\partial_\mu F^{\mu\nu} = 0 \tag{9}$$

Return of the current

But the second term in (6), indicating the presence of charged particles, is usually there. It is written as an integral over the particle trajectories. In contrast, the third term is written as an integral over spacetime. To compare the variation of the second term with the variation of the third term, we have to write the second term also as an integral over d^4x, so that the two terms have the same form. But we know how to do this. Indeed, (7) provides a strong hint. Use the Dirac delta function introduced in chapter I.1!

Recall from chapter III.6 that the delta function is defined by $\int dx \, \delta(x - y)f(y) = f(x)$ for any reasonably smooth function $f(x)$. There we also generalized to the 4-dimensional delta function, defined in an obvious way by $\delta^{(4)}(x - X_a) \equiv \delta(x^0 - X_a^0)\delta(x^1 - X_a^1)\delta(x^2 - X_a^2)\delta(x^3 - X_a^3)$. In other words, $\delta^{(4)}(x - X_a)$ vanishes unless x^μ and X_a^μ are equal component by component.

With this quick review, we now write $A_\mu(X_a) = \int d^4x \, \delta^{(4)}(x - X_a)A_\mu(x)$, and thus the second term in (6) as

$$\sum_a e_a \int d\tau_a A_\mu(X_a)\frac{dX_a^\mu}{d\tau_a} = \int d^4x \sum_a e_a \int d\tau_a \, \delta^{(4)}(x - X_a)A_\mu(x)\frac{dX_a^\mu}{d\tau_a} \tag{10}$$

You might recognize that this form is analogous to the form of the second term in (7), except that the physics is relativistic here and nonrelativistic there.

Now vary $A_\mu(x)$ to obtain

$$\int d^4x \sum_a e_a \int d\tau_a \, \delta^{(4)}(x - X_a)\frac{dX_a^\mu}{d\tau_a}\delta A_\mu(x) \tag{11}$$

Very nicely, we see the return of the 4-current defined in chapter III.6

$$J^\mu(x) = \sum_a e_a \int d\tau_a \, \delta^{(4)}(x - X_a)\frac{dX_a^\mu}{d\tau_a} \tag{12}$$

We can then write (11) more compactly as $\int d^4x \, J^\mu(x)\delta A_\mu(x)$.

Maxwell's equations

Combining (8) and (11), we find

$$\partial_\mu F^{\mu\nu}(x) = -J^\nu(x) \tag{13}$$

These are of course Maxwell's equations governing the dynamics of the field, telling us how the electromagnetic field is generated by the movement of charged particles in spacetime.

To write these equations in terms of the familiar electric $\vec{E}$ and magnetic $\vec{B}$ fields, we simply use the identification given in IV.1.17.

First, look at the time component ($\nu = 0$) of (13). We have $\partial_i F^{i0} = -\partial_i E^i = -\vec{\nabla} \cdot \vec{E} = -J^0$. Calling J^0 the charge density ρ, we recover the familiar*

$$\vec{\nabla} \cdot \vec{E} = \rho \tag{14}$$

Next, look at a spatial component by setting ν to 3, for example. Note that $\partial_\mu F^{\mu 3} = \partial_0 F^{03} + \partial_1 F^{13} + \partial_2 F^{23} = \partial_0 E^3 - \partial_1 B^2 + \partial_2 B^1$. Thus, we obtain

$$\vec{\nabla} \times \vec{B} = \frac{\partial \vec{E}}{\partial t} + \vec{J} \tag{15}$$

Seeing (14) and (15) emerge so naturally, you naturally wonder where "the other half of Maxwell's equations" are, namely the ones that do not involve the current. The answer is that they are actually identities.[5]

The point is that the electromagnetic field $F_{\lambda\sigma}$ is not any garden variety antisymmetric tensor, but the relativistic curl of a 4-vector: $F_{\lambda\sigma} = \partial_\lambda A_\sigma - \partial_\sigma A_\lambda$. To exploit this bit of information, let us define, in analogy with the 3-dimensional antisymmetric symbol ϵ^{ijk}, the 4-dimensional antisymmetric symbol[†] $\epsilon^{\mu\nu\lambda\sigma}$ by $\epsilon^{0123} = 1$ and the rest determined by antisymmetry. (For example, $\epsilon^{2031} = -\epsilon^{2013} = +\epsilon^{0213} = -\epsilon^{0123} = -1$.) Newton and Leibniz tell us that derivatives commute (as in (5)). Therefore,

$$\epsilon^{\mu\nu\lambda\sigma}\partial_\nu F_{\lambda\sigma} = \epsilon^{\mu\nu\lambda\sigma}(\partial_\nu\partial_\lambda A_\sigma - \partial_\nu\partial_\sigma A_\lambda) = 2\epsilon^{\mu\nu\lambda\sigma}\partial_\nu\partial_\lambda A_\sigma = 0 \tag{16}$$

If we write out (16) explicitly in terms of $\vec{E}$ and $\vec{B}$, we obtain the "missing Maxwell's equations." Set μ to 3 to obtain $-\partial_0 F_{12} + \partial_0 F_{21} + 2(\partial_1 F_{02} - \partial_2 F_{01}) = 0$. Translating, we find

$$\vec{\nabla} \times \vec{E} = -\frac{\partial \vec{B}}{\partial t} \tag{17}$$

Similarly, setting μ to 0, we find[6]

$$\vec{\nabla} \cdot \vec{B} = 0 \tag{18}$$

---

* We could also run the argument the other way. After reading chapter III.6 on relativistic completion, you could have contemplated the equation of electrostatics (14). Since $\vec{\nabla}$ and ρ are both being promoted to 4-vectors, $\vec{E}$ has to be part of a 4-tensor. We could then discover $F_{\mu\nu}$.

† Also known as the Levi-Civita symbol. Note that in d-dimensional spacetime, $\epsilon^{\mu\nu\lambda\cdots\sigma}$ carries d indices with $\epsilon^{012\cdots d-1} = 1$.

One appealing feature of this approach is that the correct form of the electromagnetic current (12) emerges naturally. Furthermore, applying ∂_ν to (13), we recover current conservation

$$\partial_\nu J^\nu = 0 \tag{19}$$

since $F^{\mu\nu} = -F^{\nu\mu}$ is antisymmetric.

The reader may be bothered, yet again, by the asymmetry between particles and field as manifested here in (6). The integration runs over the worldlines of the particles and over all of spacetime for the field. We need quantum field theory to tell us how to remove this asymmetry regarding how particles and field are treated.

If you have taken a course on electromagnetism, you would recognize that this elegant treatment has captured the essence of the subject.

Einstein's legacy to physics

When I was in high school, I came across a popular account of Einstein's theories. Like the typical layperson, I was captivated by the outlandish and bizarre aspects of Einstein's universe. Later, in college, after I had mastered enough physics and mathematics to understand Einstein's work, I marveled at the mathematical subtleties involved, and I saw Einstein's strange conclusions as perfectly logical consequences of his theory. But as I learned more physics and started doing research, I finally realized the true intellectual legacy[7] Einstein bequeathed to later generations of physicists amounted to nothing less than a new way of doing physics.

To appreciate Einstein's insight, let us review the schema followed in developing that quintessential 19th century theory, the theory of electromagnetism. By fooling around with frog's legs and wires, physicists saw that Nature behaves in a certain pattern, summarized by Maxwell's equations. The equations, once written down, sing out a song, waiting patiently for someone with ears to hear. Finally, a bright young fellow comes along and hears the equations saying that they are Lorentz invariant. This fellow then realizes that the symmetry demands a revision of all of physics.

After Einstein worked out special relativity, it dawned on him and some of his contemporaries, Minkowski in particular, that the logical arrows in this schema may be reversible. Suppose that it was secretly revealed to us, in the dark of night, that the world is Lorentz invariant. Knowing this, can we deduce Maxwell's theory and hence, the facts of electromagnetism, without ever stepping inside a laboratory?

To a large extent, we can! The requirement of Lorentz invariance is a powerful constraint on Nature. Maxwell's equations are so intricately interrelated by this invariance that, given one of the equations, we can deduce the others. Start with, say, Coulomb's law describing how the electric field produced by a charge decreases as one moves away from the charge.

We are given a symmetry that relates space to time, the electric to the magnetic. So, not surprisingly, we also would know how a magnetic field would vary in space.

Einstein taught us to deduce physics from symmetry, instead of symmetry from physics. In Philip Roth's *The Ghostwriter,* one of the characters, a famous writer, tells another character that he always writes one sentence before lunch. After lunch, he turns the sentence around, and he spends his life turning sentences around and around in his head. In much the same way, theoretical physicists turn logical structures around and around in their heads. Einstein and Minkowski realized that one can turn the logical arrows of the 19th century around.

Having grasped the power of symmetry, Einstein put it to use in developing his theory of gravity. Instead of laboriously distilling this theory from a motley collection of experimental facts and then extracting a symmetry, he formulated a symmetry empowering him to write down his theory of gravity in one fell swoop. To appreciate this, let us imagine what would happen if physicists followed the 19th century schema in studying gravity, as some physicists tried to do. After years of carefully studying planetary orbits, astronomers would have noticed absolutely minute deviations of the orbits from the Newtonian prediction. To account for this, physicists would add a tiny correction to Newton's law of gravity. More careful study would reveal that this is still inadequate, and physicists then would be compelled to correct Newton's law by an even tinier amount. In practice, this program would quickly grind to a halt. But even if we imagine that physicists were able to determine as many correction terms as they like, it would take a stroke of mathematical genius to see that the corrections would all combine to produce a rather different theory. The theory in the intermediate stage would be a complicated mess.

I regard Einstein's understanding of how symmetry dictates design as one of the truly profound insights in the history of physics. Fundamental physics is now conducted largely according to Einstein's schema rather than that of 19th century physics. Physicists in search of the fundamental design begin with a symmetry, then check to see whether its consequences accord with observation. But how is a physicist to get to square one in playing Einstein's game? the reader might ask. Presumably, no one is going to come in the dark of the night and whisper to us the symmetries Nature has woven into her tapestry. If an architect's client wants to have symmetrical designs but won't tell the architect what symmetry he has in mind, how is the architect to find out?

Well, physicists can extract the symmetry from known experimental facts. That is what Einstein did. The difficult part is to decide on the one most relevant fact that allows formulation of a symmetry. Out of the many facts known about gravity, Einstein fastened onto, as we will see, the fact that objects fall at the same rate, regardless of mass. He did not use, for example, the fact that the gravitational attraction between two objects varies inversely as the square of the distance between them. This and all other known facts emerge, as we will see in detail in Book 2, as consequences of the symmetry imposed on gravity.

An interesting historical fact is that some of Einstein's contemporaries, such as Lorentz, who had been struggling to produce a dynamical theory of the electron and of the ether, thought Einstein had cheated. Einstein simply imposed the principle of relativity and deduced the consequences. These other physicists felt that the principle of relativity should emerge from the dynamics.

Perhaps some of the biggest puzzles of contemporary physics are waiting for a principle—a principle to be imposed, not derived.

Exercises

1 Show that $\mathcal{L} = \frac{1}{2}(\vec{E}^2 - \vec{B}^2)$.

2 (a) Show that the symmetric tensor defined by

$$T^{\mu\nu} = F^\mu_\lambda F^{\nu\lambda} - \frac{1}{4}\eta^{\mu\nu}F_{\sigma\rho}F^{\sigma\rho} \tag{20}$$

satisfies the conservation law $\partial_\mu T^{\mu\nu} = 0$, in the absence of charged particles of course.

(b) Since charged particles and the electromagnetic field interact, that is, the particles and the field can exchange energy and momentum, we would not expect $\partial_\mu T^{\mu\nu} = 0$ to hold in the presence of charged particles. Add the energy momentum tensor $T^{\mu\nu}_{\text{particles}}$ of point particles defined in (III.6.7) to the energy momentum tensor in (20), which we will now call $T^{\mu\nu}_{\text{electromagnetic}}$. Show that the resulting energy momentum tensor $T^{\mu\nu} = T^{\mu\nu}_{\text{particles}} + T^{\mu\nu}_{\text{electromagnetic}}$ satisfies $\partial_\mu T^{\mu\nu} = 0$.

Later, in chapter VI.4, when we get to energy momentum in curved spacetime, this will become much clearer with a more powerful formalism.

3 Show that, in the absence of charged particles, $T^{00} = \frac{1}{2}(\vec{E}^2 + \vec{B}^2)$, the standard expression for the energy density of an electromagnetic field. This suggests that $T^{\mu\nu}$ is the energy momentum tensor of the electromagnetic field. We will show that this is in fact the case in chapter VI.4.

4 Calculate $\partial_0 T^{00} = \frac{1}{2}\frac{\partial}{\partial t}(\vec{E}^2 + \vec{B}^2)$ using the standard Maxwell's equations.

5 Evaluate the dual electromagnetic tensor $\tilde{F}_{\mu\nu} = -\frac{1}{2}\epsilon_{\mu\nu\lambda\sigma}F^{\lambda\sigma}$. Explain why it is called dual.

6 Derive the identity $\eta_{\lambda\sigma}(F^{\mu\lambda}F^{\nu\sigma} - \tilde{F}^{\mu\lambda}\tilde{F}^{\nu\sigma}) = \frac{1}{2}\eta^{\mu\nu}F^{\rho\tau}F_{\rho\tau}$.

7 Use the identity in the preceding exercise to show that the energy momentum tensor of the electromagnetic field can be written in the symmetric form

$$T^{\mu\nu} = \frac{1}{2}\eta_{\lambda\sigma}(F^{\mu\lambda}F^{\nu\sigma} + \tilde{F}^{\mu\lambda}\tilde{F}^{\nu\sigma})$$

8 Show that we can construct only two scalars that are quadratic in the electromagnetic field. Identify them in terms of $\vec{E}$ and $\vec{B}$.

9 Show that the $T^{\mu\nu}$ for the electromagnetic field as given in exercise 2 has zero trace.

10 If you remember what the virial theorem in classical mechanics[8] is, you may be wondering about its relativistic generalization. Consider a system consisting of charged particles interacting with the electromagnetic field. Assume that the motion of the particles is confined to a finite region and that the electromagnetic field vanishes at spatial infinity. Physically, this means that we do not allow electromagnetic radiation to escape to infinity. Mathematically, we can then freely integrate by parts. Calculate the trace of the energy momentum and from this deduce the relativistic virial theorem. Hint: This is not an easy problem; you need to use some results from chapter III.6.

Notes

1. For example, S. Weinberg, *The Quantum Theory of Fields*; *QFT Nut*; and so forth.
2. See *QFT Nut*, p. 19.
3. We will do this for gravitational waves in Part IX.

4. Normally in quantum mechanics, a spin 1 particle has 3 spin states. For a discussion of why a massless spin 1 particle has only 2 spin states, see, for example, *QFT Nut,* chapter III.4.

5. Whether they are equations or identities depends on whether you regard A_μ or $F_{\mu\nu}$ as fundamental. In the quantum world, you are forced to treat A_μ as fundamental.

6. See *QFT Nut,* chapter IV.4.

7. This section is adapted from pp. 95–100 of my popular book *Fearful,* written for the educated public.

8. See, for example, H. Goldstein, *Classical Mechanics.*

IV.3 | Gravity Emerges!

Forced to a tensor field

Now that we have dealt with electromagnetism, we turn to gravity, or rather option G in chapter IV.1. Recall that you obtained

$$\text{Option G improved:} \quad S = -m \int \sqrt{-g_{\mu\nu}(x)dx^\mu dx^\nu} \tag{1}$$

with $g_{00} = -(1 + \frac{2V}{m})$, $g_{0i} = g_{i0} = 0$, and $g_{ij} = \delta_{ij}$ as a special case.

Remarkably, if you put the Newtonian potential term $V dt$ outside the square root, you are led to a vector field A_μ, but if you put it inside, you are forced to a tensor field: two indices are needed to match $dx^\mu dx^\nu$. In a sense, we have to thank Pythagoras for this tensor field.

Time and gravity

For the moment, let's treat the special case

$$S = -m \int \sqrt{\left(1 + \frac{2V}{m}\right) dt^2 - d\vec{x}^2} \tag{2}$$

The fraction V/m looks a little strange, but then you suddenly realize that if the particle, of mass m, is living in a gravitational potential $V = -GMm/r$, m would cancel out, so that

$$S = -m \int \sqrt{\left(1 - \frac{2GM}{r}\right) dt^2 - d\vec{x}^2} \tag{3}$$

That the mass m cancels out depends of course on the profound observational fact that the inertial mass and the gravitational mass are equal, as we have alluded to already several times, in chapters I.1, II.3, and so forth.

Suppose this particle is actually a clock, sitting still in the potential (so that $d\vec{x} = 0$). Then $S = -m \int \sqrt{\left(1 - \frac{2GM}{r}\right) dt^2} \simeq -m \int (1 - \frac{GM}{r}) dt$. But for a particle at rest, this is just $-m \int d\tau$ with τ the proper time. Hence, $d\tau = (1 - \frac{GM}{r}) dt$ or $\Delta t = \Delta\tau/(1 - \frac{GM}{r}) > \Delta\tau$. You have discovered that in a gravitational field, a clock runs slow.

Gravity affects the flow of time! An astounding statement.

Universality of gravity

In sounding a bad notation alert, in several instances, we used the trick of considering several particles instead of a single particle to render the notational defect glaringly obvious. For example, using this trick in the discussion leading up to (IV.1.21,23), we were led to the notion of electric charge. So, consider a bunch of particles with different masses. Instead of (2), we have

$$S = -\sum_a m_a \int \sqrt{\left(1 + \frac{2V(x_a)}{m_a}\right) dt_a^2 - d\vec{x}_a^2} \tag{4}$$

A serious conceptual problem becomes apparent: unless the potential evaluated at x_a is proportional to m_a, particles with different masses would experience the passage of time differently. Ultimately, we have to ask our experimental colleagues, of course, if they know of such an effect. They don't, and so we can say with some confidence that (4) does not describe the physical world as we know it unless $V(x_a)$, the potential experienced by the particle, is proportional to m_a. Remarkably, the gravitational potential has precisely this property.

Curved spacetime came looking for us

Of course, this assertion that the gravitational mass and inertial mass are equal has to be tested by performing experiments to ever increasing accuracy. Our experimental colleagues have assured us (and continue to assure[1] us) that, yes, this is indeed the case; therefore, we can describe a bunch of relativistic particles in a gravitational field by the action

$$S = -\sum_a m_a \int \sqrt{-g_{\mu\nu}(x_a) dx_a^\mu dx_a^\nu} \tag{5}$$

with $g_{\mu\nu}(x_a)$ independent of the properties of the particle a, such as its mass.

Now comes your (actually, Einstein's) profound insight. Aha, you say, this looks just like the length of different curves in curved spaces that we discussed in part I, except that here we have both space and time. So particles in a gravitational field move as if they were in curved spacetime!

We did not go looking for curved spacetime. Curved spacetime came looking for us!

This represents one of the quickest ways I know of introducing curved spacetime. Curved spacetime follows from your desire to stick the Newtonian potential V inside the square root in (2).

To summarize, the interpretation of gravity as the effect of curved spacetime is possible only because gravity is universal. Thanks to the equality of gravitational mass and inertial mass, the effect of a gravitational field on the motion of particles does not depend on the particle, whether it is an apple or a rock.

This is not how gravity was discovered on the planet Terra. But as I mentioned earlier, I can imagine a civilization (in a molecular cloud?) without a Newton, without apples and rocks, which somehow discovers that light travels at a universal speed. Along comes some bright young theorist who tries to stick the potential term inside the square root. He or she (or whatever) would then discover universal gravity.

Gravitational redshift

Let's go back to the prediction that gravity slows down the flow of time. The universality of gravity means that all clocks, regardless of manufacturer, slow down by the same amount.

The Smart Experimentalist* pipes up: "But how can you observe this effect if all physical processes at a given point slow down by the same factor?"

Hmm, well yes, we are stumped. But she is just thinking out loud, and continues, "We compare the flow of time at different points! We could send a signal, say a photon, from here to there in a gravitational field. If clocks run at different rates at different places in a gravitational potential well, then the frequency of a photon climbing out of a gravitational well would be shifted toward the red."

Excellent suggestion! More on this prediction of gravitational redshift by Einstein later!

A recurring theme of modern physics

Let's summarize how you discovered electromagnetism in the preceding chapters. Generalizing the potential term $V(\vec{x})dt$ to $A_\mu(x)dx^\mu$, you uncovered a hidden gauge invariance, which then completely fixes the form of the electromagnetic field $F_{\mu\nu}(x) \equiv \partial_\mu A_\nu(x) - \partial_\nu A_\mu(x)$ and subsequently the action governing its dynamics. The long trudge (not to mention the drudge) through the electromagnetic courses you took "merely" amounts to studying this dynamics for ever more involved situations involving wires, conducting plates, frog's legs, and so forth.

In discovering gravity, you are led by Lorentz invariance to the action (5). Being a graduate of part I of this book, you immediately recognize that this action enjoys an even richer "hidden" invariance: we can transform $x^\mu \to x^\mu(x')$ so that $g_{\mu\nu}(x)dx^\mu dx^\nu = g'_{\lambda\rho}(x')dx'^\rho dx'^\sigma$ and leave the action invariant. Much of general relativity is concerned

---

* Like Confusio, also a character from *QFT Nut*.

with this freedom to make coordinate transformations. Your next task, in analogy with the electromagnetic story, is to exploit this invariance to find the action governing the dynamics of the gravitational field, the analog of Maxwell's action for the electromagnetic field.

A recurrent theme of modern theoretical physics has been unification and the resulting discovery of ever deeper invariances in the laws of physics.

Note

1. Read about the work of E. Adelberger and the Eöt-Wash group at the University of Washington: http://www.npl.washington.edu/eotwash/. See also the discussion and the endnotes in chapter I.1.

Recap to Part IV

Living in a galaxy far far away, you admire the beauty of the relativistic free particle action and contemplate how to deprive the particle of its freedom.

As far you can see, there are only two options: put the potential either outside, or inside, the square root. Two, and (apparently) only two, options.

With the first option, you discover how charged particles hear the electromagnetic field, and even better, you also discover the gauge principle. Understanding the gauge principle, you can understand how the electromagnetic field responds to the movement of charged particles.

Intoxicated by your success, you go on to discover "half of gravity" by sneaking the potential into the square root. You then come to the astonishing insight that gravity slows down the flow of time. You sure are a smart guy, no doubt about it.

Hindsight is oh so easy.

BOOK TWO

From the Happiest Thought to the Universe

Prologue to Book Two

The Happiest Thought

The happiest thought of his life

> I was sitting in a chair in the patent office in Bern when all of a sudden a thought occurred to me: "If a person falls freely he will not feel his own weight." I was startled. This simple thought made a deep impression on me. It impelled me toward a theory of gravitation.
>
> —A. Einstein

One November day in 1907, Einstein had what he later called the happiest thought[1] of his life. In 1905, he had his annus mirabilis, producing five papers[2] that shook physics to its foundation, including not only the papers founding the theory of special relativity but also the paper on the photoelectric effect that helped establish quantum physics and introduced the concept of the photon. You would have thought that Einstein would have been made a full professor on the spot if the physics community had any sense at all. Well, he was indeed promoted, a year later, to technical expert second class in the patent office.

An April Fools' prank

To understand what the daydreaming second class expert was happily thinking about, let's play an elaborate April Fools' Day prank on one of our friends. While the guy is asleep, put him in a spacious box elaborately furnished inside to look exactly like his living room. We then drop the box from a high-flying airplane (see figure 1). When our friend

Figure 1 A living room falling.

wakes up, he thinks that he is in his living room. Curiously, though, he feels that he is floating.[3] To an observer on the ground, our friend and his living room are hurtling toward a crunching rendezvous with the ground. Our friend, however, is blissfully unaware of the impending disaster. Since he is accelerating downward at the same rate as the box and all the objects contained inside, he feels that he is not moving downward at all relative to his surroundings. A slight spring in his step and he finds himself drifting toward the ceiling. He feels that he is floating. But this action is interpreted by the ground observer quite differently: our friend, by stepping on the floor, has at the same time decreased slightly his downward velocity and increased slightly the box's downward velocity. He thinks he is floating upward but in reality his downward plunge is accelerating at the same rate as before.

Indeed, this awfully unethical April Fools' joke has already been tried: we put astronauts inside a box called a spaceship and drop it out of the sky. To be humane, we give the box a forward motion so that as soon as the box drops, the ground would have the good sense of curving away by just the right amount, so the box stays up at the same altitude. When you see on TV an astronaut floating in space, with the announcer commenting in the background that the astronaut is in the zero-g environment of space 100 miles above our heads, you of course know better, since you are capable of reading this book. The astronaut is in a 0.95-g environment, subject to only 5% less gravity than we are. He is floating because he is falling, and because he is falling, he does not feel that he is falling, just as the young technical expert second class thought.

A birthday toy

On Einstein's 76th and last birthday in 1955, his neighbor Eric Rogers presented him with a toy[4] constructed for the occasion. (In figure 2a, I show the engineering drawing for the toy, and in figure 2b, a photo of the toy constructed for me by Louis Grace.) Basically, a spring tries to pull a ball hanging limply outside into a cup but is too weak to do so. The challenge is to get the ball into the cup. The historian of science I. Bernard Cohen visited Einstein not long after, and he wrote:

> At last I was taking my leave. Suddenly Einstein turned and called "Wait. Wait. I must show you my birthday present." Back in the study I saw Einstein take from the corner of the room what looked like a curtain rod five feet tall, at the top of which was a plastic sphere about four inches in diameter. "You see," said Einstein, "this is designed as a model to illustrate the equivalence principle. . . . " A big grin spread across his face and his eyes twinkled with delight as he said, "And now the equivalence principle." Grasping the gadget in the middle of the long brass curtain rod, he thrust it upwards until the sphere touched the ceiling. "Now I will let it drop," he said, "and according to the equivalence principle there will be no gravitational force. So the spring will now be strong enough to bring the little ball into the plastic tube." With that he suddenly let the gadget fall freely and vertically, guiding it with his hand, until the bottom reached the floor. The plastic sphere at the top was now at eye level. Sure enough, the ball rested in the tube.[5]

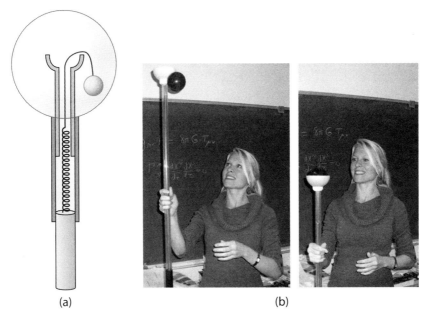

(a) (b)

Figure 2 (a) An engineering drawing of the old man's toy. (b) The toy constructed for the author, shown in its two states.

Much more about the equivalence principle later, but for now, note that the ball is just as easily fooled as an astronaut. When Einstein let his toy fall, the ball, precisely because it was falling, did not feel any gravity; the ball was the stand-in for the falling person in the patent clerk's daydream.[6] The spring, normally too weak to pull the ball up against gravity, now seized the chance to yank the ball into the cup.

The falling candle

Einstein loved to pop playful little puzzles on his visitors. He was equally delighted whether or not they knew the answers. If they didn't, he would get a big kick out of explaining it. Let's see if you figure this one out. Suppose you have just lighted a candle in an elevator when, unfortunately, the cable breaks. The elevator falls freely. What happens to the candle flame? Try to answer the grinning old man looking at you with a twinkle in his eyes.

"In proportion to its quantity"

> After dinner, the weather being warm, we went into the garden and drank thea, under the shade of some apple trees,* only he and myself. Amidst other discourse, he told me he was just in the same situation, as when formerly, the notion of gravitation came into his mind. It was occasion'd by the fall of an apple, as he sat in a contemplative mood. Why should that apple always descend perpendicularly to the ground, thought he to himself. . . . Assuredly, the reason is, that the earth draws it. There must be a drawing power in the matter: and the sum of the drawing power in the matter of the earth must be in the earths center, not in any side of the earth. Therefore dos this apple fall perpendicularly, or towards the center of the earth. If matter thus draws matter, it must be in proportion of its quantity. Therefore the apple draws the earth, as well as the earth draws the apple. That there is a power, like that we here call gravity, which extends its self thro' the universe. [W. Stukeley,[7] in his memoir of Sir Isaac Newton]

To understand gravity in more detail, let us consider our April Fools' prank again. For the prank to work, it is crucial that all objects fall at exactly the same rate. Suppose to the contrary that the box falls faster than our friend. Then our friend would find himself pinned to the ceiling, which he would interpret as being due to the presence of a force pushing him up. Conversely, if the box were to fall slower, our friend would feel a force pulling him to the floor. The extreme case in which the box is not falling at all is of course the normal situation, with the box resting on the house foundation.

That objects all fall at the same rate regardless of their composition is contrary to everyday intuition, but as Galileo suspected, our everyday experiences are distorted by air resistance. As you know, and as I explained in chapter I.1, inertial mass is equal to

---

* Supposedly, a descendant of Newton's apple tree[8] now stands outside Trinity College in Cambridge, England.

gravitational mass, and so the motion of an object in a gravitational field does not depend on its mass: in a vacuum, a feather and a cannonball fall at the same rate.[9]

Physics students generally identify Einstein as the person who brought fame to various gedanken experiments. But in fact thought experiments go way back to Galileo, at least. The following is taken straight from his "Discorsi e dimostrazioni matematiche" (1628):

Salviati: If then we take two bodies whose natural speeds are different, it is clear that on uniting the two, the more rapid one will be partly retarded by the slower, and the slower will be somewhat hastened by the swifter. Do you not agree with me in this opinion?

Simplicio: You are unquestionably right.

Salviati: But if this is true, and if a large stone moves with a speed of, say, eight while a smaller moves with a speed of four, then when they are united, the system will move with a speed less than eight; but the two stones when tied together make a stone larger than that which before moved with a speed of eight. Hence the heavier body moves with less speed than the lighter; an effect which is contrary to your supposition. Thus you see how, from your assumption that the heavier body moves more rapidly than the lighter one, I infer that the heavier body moves more slowly.[10]

Universality of gravity and a ball of whiskey

Let's try to imagine the patent clerk's train of thought. A falling person does not know he is falling, because everything around him is falling at the same rate, in other words, because of the universality of gravity. Can I turn this around? Gravity must be universal because a falling person does not know he is falling. In a way, falling cancels out gravity. Hmmm, suppose I somehow reverse falling by thrusting upward. Can I then produce gravity? Aha!

To understand what Einstein had in mind, let us inflict an even more elaborate April Fools' joke on our friend. This time, while he is asleep, we put him inside a box and fly him deep into intergalactic space, far away from any gravitational field of force. Now rev up the engine and accelerate the whole contraption at a constant rate. When he wakes up, he notices nothing unusual at all. No floating sensation this time. He drops an apple, and it promptly falls to the floor (figure 3). But to an outside observer, floating in space and watching the spaceship go zinging by, the dropped apple is actually floating in space in happy ignorance of the fact that the floor is rushing at it with ever-increasing speed. If we accelerate the rocket at precisely the right rate, our friend would see the apple falling to the floor exactly as if he were back on earth. Keep in mind the key phrase "as if" for future reference.

By accelerating the rocket—in effect, reversing free fall—we can produce gravity. Clearly, if our friend had dropped a stone as well as the apple from the same height from the floor, the stone and the apple will "fall" and hit the floor at exactly the same instant.

But what is to him a mysterious universality is laughably obvious to the observer floating about outside: the floor is moving up to meet the floating apple and stone and so obviously arrives at the two objects at the same time.

Figure 3 The floor rushing up to meet the apple.

In one of the Tintin stories, Captain Haddock has smuggled on board a spaceship a bottle of whiskey hidden inside a hollowed-out book on cosmology.[11] Just as he was about to set lips to glass, a bumbling character named Thomson accidentally turns off the spaceship's engine. The spaceship stops accelerating. The whiskey, suddenly feeling no gravity, has no further compunction to stay inside the glass: it exploits surface tension to curl itself up into a ball and floats out of the glass. Tintin then manages to turn the engine back on. The spaceship accelerates. Gravity comes back on, and Captain Haddock and the ball of whiskey crash to the floor.

Confusio: "I get it! So, an apple and a stone dropped from the Leaning Tower of Pisa did not fall, but were suspended motionless in space. It was actually the ground which rushed up to meet them! That would explain why the apple and the stone hit the ground at exactly the same instant. The relativity of motion!"

Indeed, a mite on the ground looked up at the enormous apple coming down and saw his entire life flash by him in an instant, but a mite on the apple, equally terrorized, watched the ground rushing up to crush her.

"As if" is good enough

But yet, this sounds like total nonsense. The earth carrying the Leaning Tower and the entire town of Pisa rushing up toward the apple and the stone? How could you explain gravity with that peculiar hallucination? Besides, all around the world, people are dropping things, ripe fruits are falling down from trees, and nerdy physicists are tripping all over themselves. The earth would have to be rushing this way and that.

Nevertheless, the notion of the ground rushing up to meet the apple and the stone is such an amazingly simple explanation of why the apple and the stone hit the ground at exactly the same instant—there must be some element of truth to it.

To make sense out of nonsense, the key insight is that "as if" is good enough. The ground does not literally have to rush up. It is enough to say that gravity behaves as if the ground were rushing up. We can formulate this more academically by calling the "as if" an equivalence.

Einstein's equivalence principle

Dear reader, together we have arrived at Einstein's equivalence principle. This profound and fundamental principle states that, in a small enough region of spacetime, no experiment can tell us whether we are in a gravitational field or in an accelerating frame.

Note the caveat of a "small enough region of spacetime," where by small enough we mean a region small compared to the characteristic size of the gravitational field. This is easy to comprehend. Suppose our friend dropped the stone and the apple on earth rather than in a box accelerating in empty space. We eternal students of physics, not to mention Newton and his friend Stukeley, know that the stone and the apple do not fall down, but that they fall toward the center of the earth (recall chapter I.2). Indeed, as the stone and the apple fall, they approach each other slightly, the effect being suppressed by the ratio of the separation between the stone and the apple to the radius of the earth. By careful measurement of this so-called tidal effect (recall chapter I.4), we can in fact determine whether we are in the earth's gravitational field or in an accelerating box.

The equivalence principle is a statement about physics in a small region of spacetime.

Exercise

1   What happens to the flame of a falling candle?

Notes

1. This prologue is based in part on A. Zee, *An Old Man's Toy*.
2. Einstein, *Einstein's Miraculous Year*.
3. You can now experience this for yourself if you are willing to pay. Einstein's happy thought is being exploited commercially. See www.gozerog.com.
4. A. Zee, *An Old Man's Toy*. See also A. Zee, in *E = Einstein*, ed. D. Goldsmith and M. Bartusiak, Sterling, 2007, p. 223.
5. I. Bernard Cohen, "An Interview with Einstein," *Scientific American*, July 1955, p. 73.
6. For a brief discussion of acrophobia, elevator phobia, and daydreams, see *Toy/Universe*, pp. 17, 257.
7. See *Toy/Universe*.
8. For a sketch of Newton's life and his encounter with that famous apple, see *Toy/Universe*, pp. xv–xvii.
9. Now you can see this amazing fact on the web.
10. S. Drake, *Galileo Galilei, Two New Sciences*. Copyright © 1974 by the Regents of the University of Wisconsin System. Reprinted by permission of The University of Wisconsin Press.
11. By Lemaître? Could have been a book on gravity, I'm not sure. See *Toy/Universe*, p. 15.

Part V | Equivalence Principle and Curved Spacetime

V.1 | Spacetime Becomes Curved

A mysterious force emanating from the Bering Strait

Imagine flying from Los Angeles to Taipei. Flipping idly through the back of an in-flight magazine (or more likely the flight map on the video by the time I finish this book), you might notice that the plane follows a curved path arcing toward the Bering Strait. Is the Bering Strait exerting a mysterious attractive force on the plane?

On your next trip you try another airline. This pilot follows exactly the same curved path. Don't these pilots have any sense of personality or originality? Why don't they sometimes, just for the heck of it, swing south and fly over Hawaii, say? They seem to prefer to fly over[1] grim and unsuspecting Inuit hunters rather than cheerful Polynesian maidens.

Not only is the mysterious force attractive, it is universal, independent of the make of the airplane. Should you seek enlightenment from the guy sitting next to you? Dear reader, surely you are chuckling. You know perfectly well that the Mercator projection distorts the earth, and pilots follow scrupulously the shortest possible path between Los Angeles and Taipei. The answer to the universality of the mystery force is to be sought, not in the physics, but in the economics, department.

But is it so laughably obvious? Consider the leading theoretical physicists before Einstein came along. They knew the well-verified experimental fact that all things fall at the same rate, be it an apple or a stone. Perhaps the fact that an apple and a stone would fall in exactly the same way in a gravitational field is no more amazing than different airlines, regardless of national or political affiliation, would choose exactly the same path getting from Los Angeles to Taipei. An apple or a stone traverses the same path in spacetime, just as a commercial flight follows the same path on the curved earth regardless of the airline.[2] In hindsight, we might see an "obvious" connection, but hindsight[3] is of course way too easy. For three hundred years, the universality of gravity has been whispering "curved spacetime" to us.

As I said in chapter IV.3, we did not go looking for curved spacetime, curved spacetime came looking for us!

No gravity, merely the curvature of spacetime

Just as there is no mysterious force emanating from the Bering Strait, one could say that there is no gravity, merely the curvature of spacetime. The gravity we observe is due to the curvature of spacetime. More accurately, gravity is equivalent to the curvature of spacetime, or gravity and the curvature of spacetime are really the same thing.

To summarize and to underline the point, Einstein says that spacetime is curved and that objects take the path of least distance in getting from one point to another in spacetime. Environment dictates motion. The curvature of spacetime tells the apple and the stone to follow the same path from the top of the tower to the ground. The curvature of the earth tells the pilots to follow the same path from Los Angeles to Taipei.

This amazing revelation about the role of spacetime offers an elegantly simple explanation of the universality of gravity. Gravity curves spacetime. That's it. Spacetime is curved and gravity's job is done. It's now up to every particle in the universe to follow the best path in this curved environment. This explains why gravity acts indiscriminately on every particle in exactly the same way. Next time you take a nasty fall, whether on the ski slope or in the bathtub, just think, every particle in your body is merely trying to get the best deal for itself.

Acceleration

In the prologue to book two, you read about Einstein's happy thought that being in an accelerated frame is equivalent to being in a gravitational field. The parable here suggests that gravity is a manifestation of curved spacetime. Let us now substantiate these analogies.

Warm up with the simplest example of a freely moving Newtonian particle in one spatial dimension obeying $\frac{d^2y}{dt^2} = 0$. (To avoid writing primes in the subsequent discussion, we call the spatial coordinate y instead of x'.) Let us now transform to an accelerated frame. Instead of the linear Galilean transformation, we now have $y = x - \frac{1}{2}at^2$. Differentiating twice, we obtain

$$\frac{d^2y}{dt^2} = \frac{d^2x}{dt^2} - a \tag{1}$$

Thus, the observer in the accelerated frame insists that there is a force, given by the mass m of the particle times

$$\frac{d^2x}{dt^2} = a \tag{2}$$

Note that the force is proportional to the inertial mass. Simple yet profound!

This is all familiar stuff, experienced often in daily life. Riding in a speeding car, we are thrown forward when the driver suddenly slams on the brake. Beginning physics students learn this as the "effect of inertia." A wet dog shaking itself dry knows how to exploit this effect.

Repeat this little discussion for a relativistic particle. Suppose that an observer, living in Minkowskian spacetime with $d\tau^2 = -\eta_{\rho\sigma}dy^\rho dy^\sigma$, sees no force acting on a particle, that is,

$$\frac{d^2y^\rho}{d\tau^2} = 0 \tag{3}$$

What does the other observer see?

Instead of the simple relation between y, x, and t in our warm-up Newtonian example, we now let the coordinates y^ρ be related to the other observer's coordinates x^μ by a general coordinate transformation specified by 4 functions $y^\rho(x)$. Now it's just a matter of arithmetic, albeit highbrow arithmetic, to work out what (3) implies for $\frac{d^2x^\rho}{d\tau^2}$. Just plug in, then chug away. Differentiating y^ρ once, we get

$$\frac{dy^\rho}{d\tau} = \frac{\partial y^\rho}{\partial x^\mu}\frac{dx^\mu}{d\tau} \tag{4}$$

Differentiating y^ρ twice, we get

$$\frac{d^2y^\rho}{d\tau^2} = \frac{\partial y^\rho}{\partial x^\mu}\frac{d^2x^\mu}{d\tau^2} + \frac{\partial^2 y^\rho}{\partial x^\mu \partial x^\nu}\frac{dx^\mu}{d\tau}\frac{dx^\nu}{d\tau} \tag{5}$$

Thus, if one observer sees a freely moving particle $\frac{d^2y^\rho}{d\tau^2} = 0$, the other sees

$$\frac{d^2x^\lambda}{d\tau^2} + \left(\frac{\partial x^\lambda}{\partial y^\rho}\frac{\partial^2 y^\rho}{\partial x^\mu \partial x^\nu}\right)\frac{dx^\mu}{d\tau}\frac{dx^\nu}{d\tau} = 0 \tag{6}$$

We have multiplied by $\frac{\partial x^\lambda}{\partial y^\rho}$ and used $\frac{\partial x^\lambda}{\partial y^\rho}\frac{\partial y^\rho}{\partial x^\mu} = \frac{\partial x^\lambda}{\partial x^\mu} = \delta^\lambda_\mu$. The "$x$ observer" sees a force acting on the particle: $\frac{d^2x^\lambda}{d\tau^2}$ does not vanish. Compare (2) and (6). The latter is just a more complicated version of the former; the physics involved is essentially the same.

Curved space and curved spacetime

"But wait a minute!" you exclaim. "It all looks familiar. Didn't we see this somewhere already?"

Yes indeed, way back in chapter II.2, Professor Flat explained that, in a locally flat region, the geodesic equation for a curved space reduces to the equation for a straight line, as anybody would expect. There we didn't include time, and the geodesic is parametrized by the length defined by $ds^2 = \delta_{\rho\sigma}dy^\rho dy^\sigma$; here the path followed by the particle is parametrized by the proper time defined by $d\tau^2 = \eta_{\rho\sigma}dy^\rho dy^\sigma$. The role of Euclid's δ is played by Minkowski's η.

The metric for curved spacetime can hardly wait to pop out. Given that the "y observer" sees the Minkowski metric, what metric does the "x observer" see? The invariance of the proper time interval gives instantly

$$d\tau^2 = \eta_{\rho\sigma}dy^\rho dy^\sigma = \eta_{\rho\sigma}\frac{\partial y^\rho}{\partial x^\mu}\frac{\partial y^\sigma}{\partial x^\nu}dx^\mu dx^\nu \tag{7}$$

Thus, the "x observer" sees the metric

$$g_{\mu\nu} = \eta_{\rho\sigma} \frac{\partial y^\rho}{\partial x^\mu} \frac{\partial y^\sigma}{\partial x^\nu} \tag{8}$$

Define the Christoffel symbol as

$$\Gamma^\lambda_{\mu\nu} \equiv \frac{\partial x^\lambda}{\partial y^\rho} \frac{\partial^2 y^\rho}{\partial x^\mu \partial x^\nu} \tag{9}$$

and you literally see (6) morph into the geodesic equation (II.2.19) found by minimizing the length of a curve (II.2.13) before your very eyes.

I cordially invite you to go back to chapter II.2 and compare our discussion of curved spacetime here with the discussion of curved space there. Everything looks the same. For example, the connection between $\Gamma^\lambda_{\mu\nu}$ and the metric goes through as before. Observe that we are on the right track: $\Gamma^\lambda_{\mu\nu}$ depends on the second derivative of y with respect to x. Thus, if the relationship between y and x is linear, as given by a Lorentz transformation or a simple rotation, then indeed $\frac{d^2 x^\lambda}{d\tau^2} = 0$, and there is no gravitational force. Comparing (8) and (9), you see that $\Gamma^\lambda_{\mu\nu}$ can be constructed out of the metric and its first derivatives $\partial_\omega g_{\mu\nu}$, just as in chapter II.2.

The astute reader would recognize that the discussion of the geodesic equation in chapter II.2 is just the mathematician's version of the physicist's equivalence principle that we can always go to an inertial frame in which there is no gravity. Translation: locally flat coordinates = inertial frame, and Christoffel symbol = force attributed to gravity.

We will explore motion in curved spacetime in detail in the next two chapters.

Since the laws of arithmetic are reversible, we can reverse the logic. Here we start with a particle happily cruising in flat spacetime (3), free from the demand of any force. We then make an arbitrary coordinate transformation, writing $y(x)$ for y. Apply the chain rule of elementary calculus, and we discover the geodesic equation (6) and curved spacetime (8). Now reverse the logic: start with a particle in a gravitational field, go to a locally flat region of spacetime (sometimes called a locally inertial frame) in which $\Gamma^\lambda_{\mu\nu}$ vanishes, and watch (6) simplify dramatically to the motion of a freely moving particle (3). This is of course Einstein's happy thought again: a freely falling person does not feel any gravity.

Fictitious forces

Some older books call the force in (2) an "inertial force," but wait, particle theorists assure us that there are only the strong force, the weak force, the electromagnetic force, and the gravitational force. So, what is an inertial force?

In high school, I was also terribly confused[4] by the centrifugal force. The book said the force was "fictitious," but yet I remembered that as a little kid riding the Merry-Go-Round, I was told in no uncertain terms that I must hold on tight. I certainly felt all these fictitious forces. Even more puzzling: the book went on to mention the centripetal force. Why couldn't the book make up its mind? Is it fugal* or is it petal[†]?

You know the resolution of all this confusing talk: the book was just moving bits and pieces of $\frac{d^2\vec{x}}{dt^2}$ back and forth between the right hand side and the left hand side of $ma = F$. What you call force I could call a piece of the acceleration, and what you call centripetal I could call centrifugal. Einstein's insight was that the most commonly experienced force of all, the gravitational force, may be an example of a "fictitious force."

Exercise

1   A helium balloon is attached to a child's seat in the back of a car. When the speeding car suddenly brakes to a stop, how does the balloon move?

Notes

1. I am abusing geography slightly in the same way I occasionally abuse notation.
2. Referring back to (IV.3.4), we see that if the gravitational mass is not equal to the inertial mass, this would correspond to, in our analogy, different airplanes seeing a different curvature of the earth.
3. Staircase wit, l'esprit d'escalier, Treppenwitz; firing the cannon after the cavalry had already charged by you.
4. I can't blame the teacher since I did not get to take a physics course.
5. The next line is "Sic transit gloria mundi".
6. The next line is "Gather ye rosebuds while ye may."
7. The root "petere," for "to desire" or "to seek," appears in words like "appetite" and "petition."

———

  * As in[5] "Tempus fugit."
  [†] As in[6] "All flowers wilt"—just kidding.[7]

V.2 | The Power of the Equivalence Principle

The equivalence principle predicts

We learned in the prologue to book two (and in the preceding chapter) that, in a small enough region of spacetime, no experiment can tell us whether we are in a gravitational field or in an accelerating frame.

Einstein's theory of gravity is built on this equivalence principle. As I suggested heuristically, through thought experiments and parables, the equivalence principle leads us directly to an understanding of the gravitational field as a manifestation of curved spacetime.[1] At this point, many textbook authors start to wring their hands, fretting about the long road ahead, and warning their readers about the considerable mathematical machinery involved in mastering Riemannian geometry. Of course, all this is true; to say the contrary would be like saying that you could master Newtonian mechanics without learning calculus. But by arranging the material so that you started by learning to hop from coordinate transformations to curved surfaces, I hope that by now you have already absorbed enough of the relevant mathematics so that the rest of the road we have to travel will not look so formidable.

Before we start developing Einstein gravity, I want to mention that two of its most striking predictions, namely the deflection of light and the gravitational redshift, follow directly from the equivalence principle.

Recall from the prologue to book two that there are two distinct April Fools' thought experiments we can contemplate doing. In one, we put our friend in a box in empty space, far from any gravitational field, and accelerate the box. In the other, we drop the box in some gravitational field, such as that of the earth. It would be instructive to derive both predictions using the two different thought experiments, which I will refer to as "Accelerated" and "Dropped," respectively, sort of how a funding agency would file away these two kinds of experiments.

Bending of light

Thought experiment "Accelerated"

Our friend drops an apple, and it promptly falls to the floor. The apple does what it has always done. But to the observer floating outside, the floor is rushing, "rushing up" as we are almost tempted to say, with ever increasing haste, to meet the apple. Then our friend, quite a quirky person, fires a laser gun at a wall. He notices that the red spot is located below the mark he aimed at. No question about it, light falls in a gravitational field! The laser beam bends in a graceful parabola, like any material object thrown at that wall.

But the outside observer would describe what happened quite differently. He sees the laser beam moving in a straight line, since there is no gravitational field around. But the wall had moved "upward" in the time it took the beam to cross the room. See figure 1a. Amazingly simple and beautiful argument! The equivalence principle settles, once and for all, the question whether light falls.

It is worth remarking that, while this argument, often given in popular physics books,[2] establishes that light falls, it does not determine the actual amount precisely. The reason is that it does not take the intrinsically relativistic nature of light into account; the argument would apply even if the laser beam consists of a stream of tiny particles obeying Newtonian mechanics and moving at speed c, as in Newton's corpuscle theory of light. In chapter VI.3 we will do the calculation in Einstein's theory and show that the amount of bending is twice the Newtonian value.

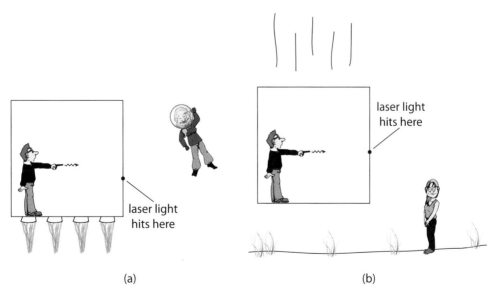

(a)	(b)

Figure 1 Firing a laser gun at a wall in (a) an accelerating box deep in space and (b) a box dropped from a great height above the earth.

Thought experiment "Dropped"

Now consider the other thought experiment of dropping the box with our friend in it from a great height above the earth (figure 1b). A dropped apple floats in space, and our friend is blissfully unaware of any gravitational field. He fires the laser gun, and sure enough, it hits the exact spot he aimed for. Why shouldn't it? There is no gravitational field around. A freely falling person does not know gravity!

But the outside observer sees the entire box falling. The spot marked on the wall for target practice has dropped in the time it took the laser beam to get there. For the laser beam to hit the spot, it must have fallen exactly as much as the wall. To her, standing on earth, there is a gravitational field and light falls.

Very nicely, both thought experiments reach the same conclusion.

Gravitational redshift

Thought experiment "Accelerated"

Back in the box accelerating in deep space, our friend now fires his laser gun at the ceiling. (More than being quirky, he might have some personality disorder.) First, what does our all-seeing outside observer see? By the time the light gets up to the ceiling, since the box is accelerating, the detector attached to the ceiling is moving faster than when the shot was fired. Plain old Doppler effect tells us that the detector will see light of a lower frequency. Our friend, since he is convinced he is in some gravitational field, concludes that light redshifts as it "climbs" from a point of lower gravitational potential to a point of higher gravitational potential. In other words, the outside observer sees a Doppler redshift, while our friend the inside observer sees a gravitational redshift (figure 2a).

Not only is the equivalence principle argument direct and convincing, it allows us to calculate the effect using literally freshman physics! Let the height of the ceiling from the floor be h. So light took time $\Delta t = h/c$ to get from the floor to the ceiling, by which time the ceiling is moving with speed $v = g\Delta t = gh/c$ in a frame in which it was at rest at time $t - \Delta t$. For $gh/c \ll c$, we don't even need the fancy relativistic Doppler result derived in chapter I.3, merely the elementary Doppler shift result $\Delta\omega/\omega = -v/c = -(gh/c)/c$. The situation $gh/c \ll c$ corresponds to a weak gravitational potential $\Phi = gh$ (recall that Φ has been defined as the potential energy per unit mass ever since part I). We thus conclude that $\Delta\omega/\omega = -(gh/c)/c = -(\Phi_{\text{ceiling}} - \Phi_{\text{floor}})/c^2$.

The equivalence principle tells us that this holds in any weak gravitational field:

$$\left(\frac{\omega_{\text{receiver}} - \omega_{\text{emitter}}}{\omega_{\text{emitter}}}\right) = -\left(\frac{\Phi_{\text{receiver}} - \Phi_{\text{emitter}}}{c^2}\right) \tag{1}$$

To check the sign in (1), remember that $\Phi_{\text{higher}} > \Phi_{\text{lower}}$ where Φ_{higher} is the potential at the tree branch where the apple was hanging and Φ_{lower} is the potential at Newton's head. Note that the terms "higher" and "lower" are in accordance with everyday usage. Thus, if

(a)   (b)

Figure 2 Firing a laser gun at the ceiling in (a) an accelerating box deep in space and (b) a box dropped from a great height above the earth.

the emitter is on the floor and the receiver is on the ceiling, the right hand side in (1) is negative, and the frequency received is lowered toward the red.

Thought experiment "Dropped"

What about the other thought experiment, in which the box with our friend inside is dropped from a great height? (See figure 2b.) He invokes the daydreaming clerk's happy thought, that a falling person does not feel gravity. Indeed, he is happily floating, in the idealized freely moving inertial frame that elementary physics is described in. So, of course, the light detector in the ceiling registers the same frequency: why would the light change its frequency propagating in an inertial frame in the absence of the gravitational field? In fact, being an accomplished experimentalist, he rigs up the detector to flash a signal indicating "Yes, same frequency!"

The observer standing on earth sees the signal and is puzzled. She observes the detector rushing down toward the light and so should see a Doppler blue shift and register a higher frequency. Being an insightful theorist, she eventually suspects that there must be another effect that cancels the Doppler blue shift: the earth's gravitational field must shift the light's frequency toward the red by precisely the same amount as the Doppler shift.

Instructively, in these thought experiments, although the two observers disagree on what is going on, they come to the same conclusion: when a photon "climbs up" a gravitational potential, its frequency redshifts.

Gravity affects the flow of time

Actually, we already had a hint of gravitational redshift back in chapter IV.3 when the bright young physicist tried to incorporate the physical notion of a potential into the relativistic action for a point particle. In option G, the action is modified to

$$S = \int \sqrt{(1 + 2\Phi(\vec{x}))dt^2 - d\vec{x}^2}$$

Let me remind you that, in particular, for a particle at a fixed position, its proper time interval is given by $d\tau = \sqrt{(1 + 2\Phi(\vec{x}))}dt \simeq (1 + \Phi(\vec{x}))dt$. Gravity affects the flow of time. The change in this flow of time translates into a change in frequency: frequency effectively depends on where you are according to $\omega(\vec{x}) \propto 1/(1 + \Phi(\vec{x})) \simeq 1 - \Phi(\vec{x})$. In other words, $\frac{\omega_{\text{receiver}}}{\omega_{\text{emitter}}} \simeq \frac{1 - \Phi_{\text{receiver}}}{1 - \Phi_{\text{emitter}}} \simeq 1 - \Phi_{\text{receiver}} + \Phi_{\text{emitter}}$. Restoring a factor of c^2 by dimensional analysis, we recover precisely (1).

Incidentally, this also resolves an apparent puzzle. When you first heard about the gravitational redshift, you might have wondered how counting the number of waves that pass by per unit time could be affected by gravity. The answer is that gravity affects the running of the clock used to define "unit time."

The Smart Experimentalist

As we will discuss in chapter VI.3, the deflection of light was observed soon after Einstein proposed his theory of gravity in 1915. In contrast, almost 50 years had to pass before gravitational redshift was verified. While the two effects are conceptually almost equally easy to understand, one effect challenges the experimentalist much more severely.

For the sun, $\Phi(\text{surface of sun})/c^2 = GM/(Rc^2) \sim 10^{-6}$. How do you disentangle this tiny frequency shift of one part in a million from the standard Doppler shift due to the thermal motion of the emitting atom on the solar surface? With $GM/(Rc^2) \sim 10^{-9}$ for the earth, terrestrial experiments seemed even more out of reach until the discovery in 1958 of the Mössbauer effect. Normally, emission lines from an atom in a crystal are broadened by recoil and by the interaction of that particular atom with its neighbors, all in thermal agitation. Mössbauer discovered (while a graduate student) that under certain circumstances, the atoms are all locked together so that the recoil is transferred to the crystal as a whole and the emission lines are much sharpened. In a famous experiment performed in a tower* of the Harvard physics building, Pound and Rebka in 1960 exploited the Mössbauer effect to verify the gravitational redshift.

Here comes our friend the Smart Experimentalist. "The tower is only about $h = 20$ meters high, and so the effect is only $(GM/(Rc^2))(h/R) \sim 10^{-15}$! How would one of you

---

* The tower was built as part of a deal to recruit Edwin Hall (1855–1938) from Johns Hopkins University, where he had discovered the effect bearing his name. Hall believed that a similar effect might exist for gravity, hence the tower.

theorists do the experiment?" Think for a moment, particularly if you are a theorist. The answer is in the appendix.

The power of the equivalence principle

Let us appreciate the far-reaching power of the equivalence principle. Suppose we have mastered the physics of a certain class of phenomena in the absence of gravity. In other words, we understand the physics in flat Minkowskian spacetime. It doesn't matter what kind of physics; it could be the physics of quarks interacting with gluons, for example. Thanks to the equivalence principle, we know immediately the corresponding physics in the presence of gravity. All we have to do is to go to an accelerating frame. But this amounts to a change of coordinates, and we know how to do that in general. As we saw in the preceding chapter, to write down the physics in the presence of gravity, all we have to do is replace flat spacetime with curved spacetime.

In the simplest case of a point particle, we merely have to replace the Minkowski metric $\eta_{\mu\nu}$ in the action $S = -m \int \sqrt{-\eta_{\mu\nu}dx^\mu dx^\nu}$ by the general metric $g_{\mu\nu}(x)$. There, we have it without doing any work! The action for a particle moving in a gravitational field is

$$S = -m \int \sqrt{-g_{\mu\nu}(x)dx^\mu dx^\nu} \tag{2}$$

Well, well, it is precisely option G in chapter IV.3.

Later in chapters V.4 and V.6, you will see more examples of the equivalence principle in action.

A matter of words: Gravitational field "versus" curved spacetime

We now understand that the gravitational field is a manifestation of curved spacetime, or perhaps more accurately, that the gravitational field and curved spacetime are effectively the same thing. In the parable given in chapter V.1, the Bering Strait does not exert a mysterious force on a plane flying from Los Angeles to Taipei. Rather, the plane is following the curvature of the earth. We could say that there is no such thing as gravity, only curved spacetime.

But you could say with equal justification that spacetime does not exist; there is only the gravitational field. To me, it is just a matter of words, and the only relevant issue is which language you find more useful to think in. Some authors[3] like to make dramatic pronouncements, something along the following: space has disappeared, time has disappeared, spacetime has disappeared! Yes, indeed, with a gravitational field and hence the geodesics of test particles, you can determine where and when without "referring" to an underlying spacetime. Einstein himself, in his more philosophical moments, adopted this point of view, writing in 1916 that "the requirement of general covariance takes away from space and time the last remnant of physical objectivity." In other words, there is no spacetime, only a bunch of fields interacting with one another,* with the gravitational first among equals. As a quantum field theorist, this picture appeals to me also, with the gravitational field providing an arena for the other fields to play in.

However, I think that most physicists, myself included, find it more natural to think of particles and fields moving in a curved spacetime seeking the best action deal for themselves. But as I said, it is merely a matter of words, and in the end, it is the equations that matter.

A misconception

I conclude by mentioning a common misconception. Some textbooks state that Einstein gravity is based on the principle of general covariance. If we interpret this principle as stating that we are free to choose whatever coordinate system we like, then this statement by itself is empty of content, or misleading at best. We've always had that freedom, even in Newtonian physics; perhaps we have forgotten, because we usually choose coordinates that make the equations look the simplest.

The correct statement is that Einstein gravity is based on the equivalence principle, which relates two physical situations, one with a gravitational field and one without. This is quite different from a symmetry, such as rotation or special relativity, which tells us that, under rotation or Lorentz transformation, respectively, physical laws are left invariant. I cannot emphasize this point enough.

More precisely, we note that for each of the effects described in this chapter, the funding agency was generous enough to support two different experiments, one filed under "Accelerated," the other called "Dropped." For each of these experiments, two principal investigators are involved: an "inside" observer and an "outside" observer. Consider again the gravitational redshift. In the Accelerated experiment, we could rig up the detector on the ceiling to email the frequency reading to the two observers. No doubt about it, both observers write down that the frequency has shifted toward the red.

Confusio: "I see that the confusion some students have may have stemmed from an inherent sloppiness in the English language. You wrote that the outside observer sees a Doppler redshift, while the inside observer sees a gravitational redshift. What do you mean by the word 'sees'?"

Confusio is absolutely right. At the cost of being more wordy, we should have said that the outside observer thinks that he sees a Doppler redshift, while the inside observer thinks that he sees a gravitational redshift, even though they agree on the actual amount of frequency shift.

The equivalence principle is not a statement that "physics" does not depend on the observer (as in the corresponding statements for rotation invariance or special relativity). It does! One observer sees a gravitational field, the other does not. But now the Talmudic quibble could be over what we mean by the word "physics." The two observers see the same

---

* Quantum field theory teaches us that particles are ultimately manifestations of fields.

frequency shift: they receive the same email from the detector. But they offer different theoretical interpretations for the same redshift, as due to acceleration for the outside observer, and to gravity for the inside observer.

Similarly, in the Dropped experiment, both observers are told by the detector that there isn't any frequency shift. But while the inside observer thinks that he is freely moving, the outside observer publishes an explanation that the gravitational redshift has canceled a Doppler blue shift.

Appendix: The "meaning" of gravity?

Okay, did you figure how those smart experimentalists did it way back when?* The clever idea is to move the source (or the receiver) up and down at precisely calibrated speeds, thus exploiting the Doppler redshift to alternately add to or subtract from the gravitational redshift.

The gravitational redshift was later verified to a much higher degree of accuracy by launching a rocket carrying a maser to a height of 10^4 km, comparable to the radius of the earth. Experimental physicists like to say that yesterday's discovery is today's calibration and tomorrow's background. Nowadays, the gravitational redshift appears as a necessary correction term in the global positioning system (GPS) that many use routinely without giving it a second thought.[†] General relativity has entered into everyday life.

Incidentally, the 19th century tower was not equipped with an elevator. Pound and his assistants had to carry the heavy equipment up the narrow tower in mountaineering backpacks. Years later, I heard Pound joke that by performing this experiment, he had truly learned the meaning of gravity.

Exercise

1   A high precision atomic clock is carried on a plane flying eastward around the world. Calculate the fractional time shift between the time elapsed as measured by this clock versus a clock kept on the ground (and versus a clock on a plane flying westward around the world). (The experiment has been carried out.) Hint: Both special and general relativity come in. Do the calculation for the idealized case (of course): plane flying along equator at a constant altitude, earth's spin axis perpendicular to the equatorial plane, and so on.

Notes

1. My wife and I bought our son Max a few balls when he was about 1 year old. I expected that he would learn about Newtonian, or at least Aristotelian, mechanics, but no, he graduated right off to Einsteinian mechanics. The newly installed wood floor in our house, while level and flat to the eye, is in fact not: the balls follow noticeably curved paths. See also *Toy/Universe*, figure 2.4 on p. 26.
2. For example, *Toy/Universe*.
3. For example, C. Rovelli, in *Quantum Gravity*.
4. C. W. Chou et al., *Science* 329 (2010), p. 1630.

---

* The technology of clocks has improved so much in 50 years that now one merely has to raise one clock by 33 centimeters relative to another to detect[4] a difference. Incidentally, with the same technology, time dilation in special relativity has been measured down to speeds of about 35 kilometers per hour.

[†] And let's not forget that the laser and the sensor used in the system are based, respectively, on the concept of stimulated emission set forth in Einstein's 1917 paper on radiation and on the photoelectric effect explained in his 1905 Nobel Prize–winning paper.

V.3 The Universe as a Curved Spacetime

From curved space to curved spacetime

I set up this book so that it is now a cinch for you to jump from curved space to curved spacetime. Given that you have played around with curved spaces, understood Minkowski's geometry of spacetime, and heard what Professor Flat said about local flatness, you are more than ready to play with curved spacetimes.

You understood Professor Flat's explanation that in a small enough region around any given point, it is always possible to go to locally flat coordinates, a definition of Riemannian manifold if you like. Easy: a simple matter of diagonalizing a matrix and counting the degrees of freedom we have in changing coordinates to cancel off the linear deviations from flatness.

Well, we can jump from curved spaces to curved spacetimes immediately, if by flat we now mean Minkowskian flat, rather than Euclidean flat. The entire discussion in chapter I.5 can be taken over; you merely have to mentally replace $\delta_{\mu\nu}$ by $\eta_{\mu\nu}$.

A $d = (D + 1)$-dimensional curved spacetime, with D spatial dimensions and one temporal dimension, is defined by

$$ds^2 = g_{\mu\nu}(x)dx^\mu dx^\nu \tag{1}$$

with $\mu, \nu = 0, 1, \cdots, D$ taking on $d = D + 1$ values, such that at any given point x_* we can transform the coordinates so that $g_{\mu\nu}(x_*)$ becomes $\eta_{\mu\nu}$ (with, as usual, $\eta_{00} = -1$, and $\eta_{ii} = +1$, $i = 1, \cdots, D$, and $\eta_{ij} = 0$ otherwise). We normally take $D = 3$ for the spacetime we are living in.

My pedagogical strategy is to let you play around with a couple of examples of curved spacetime instead of giving you a long formal exposition.

The simple spacetime we quite likely live in

Our first example is the spacetime described by

$$ds^2 = -dt^2 + a(t)^2 d\vec{x}^2 = -dt^2 + a(t)^2(dr^2 + r^2 d\Omega^2) \tag{2}$$

where $d\Omega^2 = d\theta^2 + \sin^2\theta \, d\varphi^2$ represents the usual spherical coordinates. At any given time t, the spatial geometry $dS^2 \equiv a(t)^2 d\vec{x}^2$ is just the familiar flat Euclidean space E^3 we deal with every day, homogeneous (that is, every point is just as good as every other point) and isotropic (every direction is as good as any other direction).

The proper distance between two points $(t, \vec{x})$ and $(t, \vec{x} + d\vec{x})$ at the same time but separated infinitesimally by $d\vec{x}$ is then given by $a(t)|d\vec{x}|$. Thus, with $a(t)$ (known as the scale factor of the universe) some function of t, this spacetime could describe an expanding or contracting universe. Eventually, we will learn about the dynamics driving the time evolution of $a(t)$, but for the moment, we are just going to describe and explore this particular spacetime with a given $a(t)$. Not only is (2), which I will refer to as the expanding universe* for short, the simplest curved spacetime I know of, but it is also quite likely the spacetime we live in.

For phenomena with a characteristic time scale small compared to the time scale on which $a(t)$ varies, the spacetime is effectively locally Minkowskian. Almost trivially, at any point in spacetime, we simply define y^i by $dy^i = a(t_*)dx^i$, and we have $g_{\mu\nu}(y_*) = \eta_{\mu\nu}$ after changing coordinates from $(t, \vec{x})$ to $(t, \vec{y})$.

A historical note. Metrics akin to (2), now almost universally and erroneously attributed to Willem de Sitter (1872–1934), were first written down in 1925 by Georges Lemaître (1894–1966). In 1917, de Sitter wrote down a metric that was not homogeneous in space, an error corrected later by Kornel Lanczos (1893–1974) and Hermann Weyl (1885–1955) independently. It has been said[1] that "Lanczos had the key to an expanding universe in his hands, but he did not unlock the door." Let that be a lesson to the reader! Lemaître arrived at the metric[2] (2) independently, but in contrast to Lanczos and Weyl, understood the physics of the expanding universe. (Hence the quote about Lanczos.)

Motion in curved spacetime

Life is easy. Since you already know how to determine geodesics in curved space, you know how to determine geodesics in curved spacetime and hence the paths followed by free particles. Indeed, I even used the notation $g_{\mu\nu}$ in chapter II.2. Hence, as was already discussed in the preceding chapter, you could lift the geodesic equation $\frac{d^2 X^\lambda}{d\tau^2} + \Gamma^\lambda_{\mu\nu}\frac{dX^\mu}{d\tau}\frac{dX^\nu}{d\tau} = 0$ in its entirety from chapter II.2 and use it in curved spacetime. As was

---

* A common misconception[3] is that cosmic expansion causes all distances to scale up, which is manifestly false. For example, locally, the earth and the moon are bound to each other and the distance between them is not expanding.

explained in chapter II.2, it is actually simpler to extremize the action

$$S = -m \int d\tau \left\{ \left(\frac{dt}{d\tau} \right)^2 - a(t)^2 \left(\frac{d\vec{x}}{d\tau} \right)^2 \right\}^{\frac{1}{2}} \tag{3}$$

directly. We obtain

$$\frac{d^2 t}{d\tau^2} + a(t)\dot{a}(t) \left(\frac{d\vec{x}}{d\tau} \right)^2 = 0 \tag{4}$$

and

$$\frac{d}{d\tau} \left(a(t)^2 \frac{d\vec{x}}{d\tau} \right) = 0 \Rightarrow \frac{d^2 \vec{x}}{d\tau^2} + \frac{2\dot{a}(t)}{a(t)} \frac{dt}{d\tau} \frac{d\vec{x}}{d\tau} = 0, \tag{5}$$

plus of course

$$\left(\frac{dt}{d\tau} \right)^2 - a(t)^2 \left(\frac{d\vec{x}}{d\tau} \right)^2 = 1 \tag{6}$$

Except for a sign here and there, everything is the same as in chapter II.2.

As before, we are to solve two out of three equations. Let us choose (5) and (6). When I was an undergrad, I had a professor who liked to say that some equations are so simple that you only need your eyeballs, not your brain, to solve them. Speaking more academically, we can solve (5) and (6) immediately by inspection.

Lines of constant $\vec{x}$ with $\tau = t$ are geodesics, and thus, in this universe, lazily going along with the flow is your best bet for extremizing the action. More seriously, on cosmological scales with galaxies treated as points, we could label individual galaxies appropriately and use the labels as the $\vec{x}$ coordinates. A coordinate system based on a network of geodesics is known as comoving coordinates. (Recall that we already met comoving observers in our discussion of perfect fluids in chapter III.6.) See the appendix.

For future use, we read off the Christoffel symbols from (4) and (5):

$$\Gamma^0_{ij} = a\dot{a}\delta_{ij}, \qquad \Gamma^i_{0j} = \Gamma^i_{j0} = \frac{\dot{a}}{a}\delta^i_j \tag{7}$$

with all other components vanishing.

Definition of spatial distance in a general curved spacetime

Many things are said by many physics texts to be obvious. My self-proclaimed goal is to try not to say something is obvious unless it really is obvious by some community standard. In a previous section, I said that the proper distance between two points separated infinitesimally by $d\vec{x}$ is given by $a(t)|d\vec{x}|$. That was pretty obvious, but perhaps we ought to be more careful. After all, with spacetime warped this way or that, our intuition might fail.

Instead of the simple spacetime in (2), we now consider the general spacetime described by $ds^2 = g_{\mu\nu}(x)dx^\mu dx^\nu = g_{00}(x)dt^2 + 2g_{0i}(x)dtdx^i + g_{ij}(x)dx^i dx^j$. To keep things specific and focused, imagine that we are sitting at a point with spatial coordinates $x^i + dx^i$

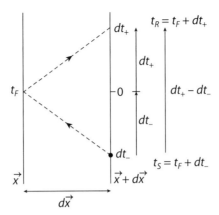

Figure 1 The operational definition of the distance to a nearby point involves sending a light beam to that point and waiting for it to bounce back to us.

and our friend is sitting at a nearby point with coordinates x^i. Intuitively, you might feel that the distance dl between us and our friend is given by $dl^2 \sim g_{ij}dx^i dx^j$. But with $g_{0i} \neq 0$, our intuition may be a bit shaky. Besides, what role does $g_{00} \neq -1$ play?

The Smart Experimentalist pops up again and says, "Distance is not some fancy theoretical construct but the result of an actual measurement." She instructs us that the operational definition of distance involves sending a light beam to our friend and waiting for it to bounce back to us. What else could we do?

Let's set up the situation with a bit of care (figure 1). Denote by t_F the coordinate time when our friend received the signal, by $t_S = t_F + dt_-$ the coordinate time when we sent the signal, and by $t_R = t_F + dt_+$ the coordinate time when we received the return signal. (The peculiar notation $dt_\pm$ will become clear soon.) The event of our friend receiving the signal has spacetime coordinates (t_F, x^i). The two events, sending the signal and receiving the return signal, have spacetime coordinates $(t_F + dt_-, x^i + dx^i)$ and $(t_F + dt_+, x^i + dx^i)$, respectively. Thus, these two spacetime events have coordinates differing by $(t_F + dt_+, x^i + dx^i) - (t_F + dt_-, x^i + dx^i) = (dt_+ - dt_-, 0)$, and hence the elapsed proper time (defined as usual by $d\tau^2 = -ds^2$) between the two events is given by $d\tau = \sqrt{-g_{00}}(dt_+ - dt_-)$. Note that our, and our friend's, spatial coordinates have not changed: that's what the word "sitting" means. The distance dl between us and our friend is defined to be $d\tau$ divided by $2c$, in other words, the elapsed proper time we experienced (not some unphysical coordinate time!) divided by light speed and a factor of 2 (to account for the round trip our light signal took).

Indeed, we measure the distance between the earth and the moon, or for that matter, between an aircraft and a control tower, in precisely this way using radar ranging. And of course these days anybody using the global positioning system (GPS) is, knowingly or unknowingly, exploiting this method to pinpoint locations, as already mentioned in chapter V.2.

We now also understand what "nearby" means: the spacetime interval during which the bouncing of light (or radar, laser, whatever) takes place has to be sufficiently small so that we can neglect any variation in the metric. Treating $g_{\mu\nu}(x)$ as constants will really simplify the calculation. Since $ds = 0$ for light, the path traced by our light ray satisfies

$$0 = g_{00}(x)dt^2 + 2g_{0i}(x)dt\,dx^i + g_{ij}(x)dx^i dx^j \tag{8}$$

This quadratic equation for dt determines how much coordinate time has elapsed when light traverses dx^i. Solving, we obtain the two roots

$$dt_\pm = \frac{1}{g_{00}}\left(-g_{0i}dx^i \pm \sqrt{(g_{0i}dx^i)^2 - g_{00}g_{ij}dx^i dx^j}\right)$$

Hence the square of the operationally defined distance dl is given by[4]

$$dl^2 = -g_{00}\left(\tfrac{1}{2}(dt_+ - dt_-)\right)^2 = \left(g_{ij} - \frac{g_{0i}g_{0j}}{g_{00}}\right)dx^i dx^j \tag{9}$$

We see the role played by g_{00} and the off-diagonal components of the metric g_{0j}. As indicated, they correct the naive answer $dl^2 \sim g_{ij}dx^i dx^j$. Fortunately, g_{0j} vanishes in most of the spacetimes we will look at (see, however, exercise 2), including various universes (17) to be explored later in this chapter and the Schwarzschild spacetime to be studied in the next chapter. In particular, for the expanding universe (2), our naive supposition $dl^2 = (a(t)d\vec{x})^2$ is indeed correct. Later, in chapter VII.5, when we study rotating black holes, we will encounter a spacetime with $g_{0j} \neq 0$ and $g_{00} \neq -1$.

Distances in an expanding universe

The result (9) holds only for nearby observers. To calculate the distance between two distant observers, we will in general have to integrate along lightlike geodesics. For (2), the metric is so simple that we can do this radar ranging calculation explicitly.

Since space is completely homogeneous and isotropic, we can, with no loss of generality, suppose that we are sitting at the origin $r = 0$ and our friend is living in a distant galaxy at $r = R$. We saw earlier that an inertial observer can stay at rest at a fixed value of $\vec{x}$. Indeed, this follows from a symmetry argument, since there is no special direction for the inertial observer to move in. Since space is homogeneous and isotropic, why should this observer move to $\vec{x} + \Delta\vec{x}$ any more than to $\vec{x} - \Delta\vec{x}$?

Let us determine the distance to our friend. Send a signal at time t_S and wait for the return signal at t_R. Denote by $t(R; t_S)$ the time when our friend receives the signal. Note that, in an expanding universe, this time depends not only on R but also on when we send the signal. Setting $ds = 0$ in (2) for light travel and taking the root $dt = a(t)dr$, we have for the outbound trip from $(t_S, 0)$ to $(t(R; t_S), R)$ the relation

$$R = \int_0^R dr = \int_{t_S}^{t(R;t_S)} \frac{dt}{a(t)} \tag{10}$$

The second equality determines $t(R; t_S)$. For the return trip from $(t(R; t_S), R)$ back to $(t_R, 0)$, we take the other root and obtain $-\int_R^0 dr = \int_{t(R;t_S)}^{t_R} \frac{dt}{a(t)}$. Adding, we obtain $2R = \int_{t_S}^{t_R} \frac{dt}{a(t)}$. We define the distance between us and our friend by

$$D(R; t_S) = \tfrac{1}{2}(t_R - t_S) \tag{11}$$

Communication in an expanding universe

The scale factor $a(t)$ of the universe* will have to be determined by observation and by theory (which we will get to in chapter VI.3). Until a decade or so ago, it was thought that the universe was expanding like a power law $a(t) \propto t^\alpha$. But as you may have heard, observational evidence now suggests that our universe is well described by (10) with $a(t) = e^{Ht}$ for some constant H, known as the Hubble parameter. (Note that we have set the origin of time.)

If somebody gives us $a(t)$, then we can evaluate $t(R; t_S)$ and $D(R; t_S)$ using (10) and (11), respectively. To illustrate what is going on, I will do the exponential case and let you do the power law case as an exercise. Integrating (10), we obtain $R = \frac{1}{H}(e^{-Ht_S} - e^{-Ht(R;t_S)})$. Evidently, we should measure distance and time in units of the Hubble distance or Hubble time $R_H \equiv H^{-1}$; that is, we could set $H = 1$ to lessen clutter. Without pausing to solve $R = e^{-t_S} - e^{-t(R;t_S)}$, we can see that as R increases, $t(R; t_S)$ increases. This makes sense: the farther away our friend is, the later she will receive our signal.

But now comes an interesting observation: there exists a $R_{max} = e^{-t_S}$ at which $t(R; t_S)$ reaches infinity. Thus, if our friend is located farther away from us than that, she will never receive our signal. In other words, if $R > R_{max}$, our signal will not be able to catch up with her. The universe, expanding too fast for light to keep up, is said to have an event horizon, known as the de Sitter horizon. The terminology is in analogy with the fact that we cannot see beyond the (everyday) horizon. Note that the later we send the signal, the smaller is R_{max}. With the passage of time, the horizon closes in on us.

Next, solving $2R = e^{-t_S} - e^{-t_R}$, we obtain the distance between us and our friend

$$D(R; t_S) = \tfrac{1}{2}(t_R - t_S) = -\tfrac{1}{2}\log(1 - 2Re^{t_S}) \tag{12}$$

Note that the distance depends on t_S, as it should since the universe is changing. For $R \ll 1$ (that is, for nearby objects), we recover $D(R; t_S) \simeq Re^{t_S}$, as we would expect from (9). In contrast, for far away objects, as $Re^{t_S} \to \frac{1}{2}$ from below, $D(R; t_S) \to \infty$. In other words, in the regime $R_{max} > R > \frac{1}{2}R_{max}$, we never hear back from our friend, even though she receives our message. This makes sense: between the time we sent our message and the time our friend sends her response, the universe has expanded some more. Light can get from us to her, but won't make it back from her to us.

---

* We will discuss cosmology in more detail later. Clearly, the metric (2) is applicable to the universe only on cosmic distance scales, over which the universe appears to be homogeneous and isotropic.

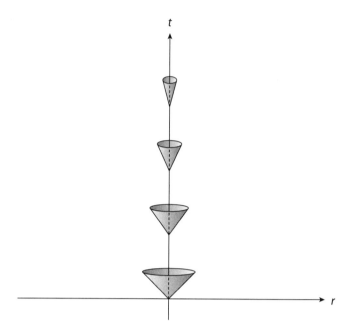

Figure 2 In an expanding universe, the light cones get ever narrower and sharper as time goes on.

Referring back to (10), we see that the existence of an event horizon hinges on the integral $\int^{\infty} \frac{dt}{a(t)}$ being finite: an exponentially growing $a(t)$ certainly renders the integral convergent.

Observers in an exponentially expanding universe do not move faster than the speed of light

Confusio: "Doesn't that mean that for $R > R_{\max}$, our friend is moving away faster than the speed of light, like* Ms. Bright? Doesn't that violate special relativity?"

No, all that special relativity requires is that the worldline of physical particles lies inside the local light cone. In other words, physical particles only have to compare themselves with the light rays around them. At any given point, the trajectories of outgoing light rays in the radial direction are defined by $dt = a(t)dr = e^{Ht}dr$: the light cones are getting ever narrower and sharper as time goes on (figure 2). Nevertheless, the worldlines of physical particles, and of our friend in particular, stay within the light cone. General relativity cannot possibly violate special relativity; after all, one is built on the other.

In a universe with $a(t) = e^{Ht}$, galaxies will pass out of our horizon one after another. Eventually, we will be left all alone, like unfortunates marooned on a desert island watching the pirate ship about to pass over the horizon.

---

* You've probably read about her. No? Here is her story:[5] There was a young lady named Bright / Whose speed was far faster than light / She went out one day / In a relative way / And returned on the previous night.

Cosmological redshift

Suppose that a proper time interval Δt_S after we sent our initial pulse to our friend, we send another pulse. To lessen clutter, write T for $t(R; t_S)$, the time she receives the initial pulse. For her, the proper time interval Δt_R between the two pulses follows immediately from (10): $R = \int_{t_S}^{T} \frac{dt}{a(t)} = \int_{t_S+\Delta t_S}^{T+\Delta t_R} \frac{dt}{a(t)}$, which for a time interval much smaller than the characteristic time scale of $a(t)$ is given by $\frac{\Delta t_R}{a(T)} = \frac{\Delta t_S}{a(t_S)}$. We can translate this immediately into a frequency shift for an electromagnetic wave emitted with frequency ω_e:

$$1 + z \equiv \frac{\omega_e}{\omega_r} = \frac{a(T)}{a(t_S)} \tag{13}$$

where we have defined the redshift factor z used by astronomers. In an expanding universe, $a(T) > a(t_S)$ and so $z > 0$, corresponding to a redshift. The frequency ω_r at receipt is less than the frequency at emission ω_e.

I must emphasize that, in this chapter, for pedagogical clarity I have often specialized to the case $a(t) = e^{Ht}$ to illustrate the point being made, but evidently our discussion often holds for other forms of $a(t)$. For example, the derivation of the redshift formula (13) does not depend at all on the assumed form of $a(t)$. As I said, you will eventually learn how to determine $a(t)$ given the content of the universe.

Light rays at 45°

It is rather inconvenient to have the shape of the light cone depend on where we are. Now that we have built up some intuition about Minkowski spacetime, we would prefer radial light rays to always travel along the 45° lines. You can readily accomplish this by a coordinate change. I will take you through the rather simple steps involved. Change t to η and require $ds^2 = dt^2 - a(t)^2 d\vec{x}^2 = b(\eta)^2(d\eta^2 - d\vec{x}^2)$ with some unknown function $b(\eta)$. The coefficient of $d\vec{x}^2$ tells us that $b(\eta) = a(t)$, which allows us to determine η in terms of t by the relation $d\eta = \frac{dt}{b(\eta)} = \frac{dt}{a(t)}$.

The variable η is sometimes known as cosmic time. At a given cosmic time η, spacetime is Minkowskian, and light propagates along 45° lines $d\eta = \pm|d\vec{x}|$ in the $(\eta, \vec{x})$ plane.

For any given $a(t)$, we can work everything out explicitly. For example, suppose $a(t) = (t/T)^{\frac{1}{2}}$, with T some characteristic time scale, which as we will see in chapter VIII.1, may have been the case in the early universe. Then $\eta = 2(Tt)^{\frac{1}{2}}$ (an irrelevant integration constant has been absorbed into η) and

$$ds^2 = \left(\frac{\eta}{2T}\right)^2 (-d\eta^2 + d\vec{x}^2) \tag{14}$$

For $a(t) = e^{Ht}$, the variable $\eta = -e^{-Ht}/H$ increases from $-\infty$ toward 0 as t goes from $-\infty$ to $+\infty$. Note that our sign choices are such that dt and $d\eta$ have the same sign and that $\eta < 0$. We obtain

$$ds^2 = \frac{1}{(H\eta)^2}(-d\eta^2 + d\vec{x}^2) \tag{15}$$

Closed and open universes

"Sandage, can you really envisage curved space and the beauties of Riemannian geometry, so necessary for relativity?" I replied, "No, Father, I have tried and tried, using all the tricks known to visualize curved space, but my visualizations have so far failed." Lemaître then sighed and said, "I understand, but it is a pity because the visualization is so beautiful. Perhaps it might be best for you to change fields." He said it gently, like a father to a son.[6]

—the distinguished cosmologist Allan Sandage,
recalling his encounter with Georges Lemaître
in 1961 in Santa Barbara, California

Whew! Perhaps we should change fields. Well, the only way to gain some intuitive feel for curved spacetimes is to be exposed to several examples, and that's what we start to do here. Incidentally, Lemaître was a Catholic priest, hence the form of address used by Sandage.

After Lemaître proposed (2) in 1925, he was unhappy about the Euclidean character of space. Spacetime is curved, but space is flat. Then in 1927, he managed to write down the modern form with space compactified into a 3-sphere S^3. It should not take you 2 years, however, given that you were exposed to S^3 back in chapter I.6. Simply replace the metric $(dr^2 + r^2 d\Omega^2)$ for Euclidean 3-space E^3 in (2) by the metric

$$\left(\frac{1}{1 - \frac{r^2}{L^2}} dr^2 + r^2 d\Omega^2 \right)$$

for S^3. Indeed, from chapter I.6, you even know how to write down the metric for the hyperbolic space H^3 (I.6.15). Thus, we can combine these three cases into

$$ds^2 = -dt^2 + a(t)^2 \left(\frac{1}{1 - k\frac{r^2}{L^2}} dr^2 + r^2 d\Omega^2 \right) \tag{16}$$

with the integer $k = 1$, 0, or -1 corresponding to a (spatially) closed, flat, or open universe, respectively. Much of the discussion in this chapter can then be repeated for the $k = \pm 1$ cases, which I will leave to you in the exercises.

As we will see in part VIII, the universes described by (16), commonly called Friedmann-Robertson-Walker universes, form the basis of modern cosmology. More correctly, they should be called Friedmann-Lemaître-Robertson-Walker, or perhaps simply Friedmann-Lemaître universes, since the work of Robertson and of Walker[7] was considerably later, in the 1930s.

You might have noticed that we can absorb* the length scale L into r. Actually, some authors do precisely that and write $ds^2 = dt^2 - a(t)^2\left(\frac{1}{1-kr^2}dr^2 + r^2 d\Omega^2\right)$, so that r is dimensionless. But I think this confuses some people, because it now seems that $k = \pm 1, 0$ are three discrete possibilities. One often hears the assertion that "the universe is now known to be flat" as if it were an absolute statement.[†] In fact, physical observation can only give us a (possibly very large) lower bound on L. For this reason, the reader will see me often laboriously dragging L around when I could have chucked it.

It is often convenient to transform coordinates by setting $L\psi = L \int d\psi = \int \frac{dr}{\sqrt{1-k\frac{r^2}{L^2}}}$ in (16). Integrating, we have $r = \sin \psi$ (closed), $r = \psi$ (flat), $r = \sinh \psi$ (open) for the three cases. Thus, the closed, flat, and open universe are described by

$$ds^2 = -dt^2 + L^2 a(t)^2 (d\psi^2 + \sin^2 \psi\, d\Omega^2) \quad \text{(closed)} \tag{17}$$

$$ds^2 = -dt^2 + L^2 a(t)^2 (d\psi^2 + \psi^2 d\Omega^2) \quad \text{(flat)} \tag{18}$$

and

$$ds^2 = -dt^2 + L^2 a(t)^2 (d\psi^2 + \sinh^2 \psi\, d\Omega^2) \quad \text{(open)} \tag{19}$$

The reader may recall from chapter I.6 that the spatial section for the closed and open universe describes the 3-dimensional sphere S^3 and hyperbolic space H^3, respectively.

Proper distances in cosmology and a "cosmic conspiracy"

In the literature, people often invite themselves to define, perhaps a bit sloppily, a "proper distance" $d(t, R)$ between two distant points $(t, 0, \theta, \varphi)$ and (t, R, θ, φ) by integrating the length segment $dl = \sqrt{g_{rr}(t, r)}dr$ derived in (9):

$$d(t, R) \equiv a(t) \int_0^R \frac{dr}{\sqrt{1 - k\frac{r^2}{L^2}}} \tag{20}$$

But we just did this integral. Thus, we have

$d(t, R) = a(t)L \sin^{-1}(R/L)$ (closed)

$d(t, R) = a(t)R$ (flat)

$d(t, R) = a(t)L \sinh^{-1}(R/L)$ (open) $\tag{21}$

As expected, for small R, $d(t, R) \simeq a(t)R$. For the closed universe, $d(t, R)$ is only defined for $R \leq L$.

You might wonder why $d(t, R)$ is so proper: as we have seen in the derivation of (9), $a(t)|\Delta \vec{x}|$ is the actual operational distance between two neighboring points separated

---

* For a more careful treatment, see chapter X.1.

† Next time you hear this statement, you will know to ask how a physical measurement could possibly give an exact result without any error bars. Typically, the speaker will mumble something about k being a discrete variable that can only take on three integer values $k = 1, 0$, and -1.

by $\Delta\vec{x}$ only if the separation is infinitesimal. Nevertheless, $d(t, R)$ is useful because for $R \ll L$, it agrees with the variety of distance measures* used by observational cosmologists.

As emphasized by Weinberg,[8] for example, to interpret $d(t, R)$ as a physical distance requires what he called a cosmic conspiracy. We need to line up, between $r = 0$ and $r = R$, many comoving observers, each separated from the next by some infinitesimal dr. At some agreed-upon cosmic time t, they would bounce light off a nearby observer to measure the proper distance $\sqrt{g_{rr}(t, r)}dr$ to that proximate neighbor and report the result to some central authority, who sums up $\sqrt{g_{rr}(t, r)}dr$ and then multiplies by $a(t)$ to form $d(t, R)$ as per (20).

Appendix: Comoving coordinates

The reader can skip this appendix on comoving coordinates upon a first reading.

Before we can describe what they are, our friend the Jargon Guy pops up and says that comoving coordinates are also known as Gaussian normal. Thank you, but let's give a physical rather than mathematical description. As mentioned in the text, a collection of freely falling particles (such as galaxies in the context of cosmology, or the dust in a collapsing cloud on its way to form a star or a black hole) naturally provides a set of coordinates that makes particularly good sense to physicists. We use some suitable labels on the particles as the $\vec{x}$ coordinates. For t, we use the proper time experienced by the particle. In other words, imagine each particle carrying a clock, the reading on which we take to be t.

This last statement implies that $d\tau^2 = -g_{\mu\nu}dx^\mu dx^\nu$ evaluated for $d\vec{x} = 0$ is equal to dt^2, that is, $-g_{00}dt^2 = dt^2$. Thus, in comoving coordinates, $g_{00} = -1$.

That the particles are freely moving means that constant $\vec{x}$ corresponds to geodesics, with $t = \tau$. Setting $\vec{x}$ constant in the geodesic equation $\frac{d^2x^i}{d\tau^2} + \Gamma^i_{\mu\nu}\frac{dx^\mu}{d\tau}\frac{dx^\nu}{d\tau} = 0$, we learn immediately that $\Gamma^i_{00} = 0$, which by the definition of $\Gamma^\rho_{\mu\nu}$ implies $g^{ij}\partial_t g_{0j} = 0$. For the sort of spacetimes we will deign to consider, g_{ij} is nonsingular, so we can conclude that

$$\partial_t g_{0j} = 0 \tag{22}$$

What we would like to get is $g_{0j}(t, \vec{x}) = 0$. Notice that (22) tells us that if we could fulfill our heart's desire at one particular instant in time t, we have it for all time.

By now you know the trick we have at our disposal: find a coordinate transformation to get rid of g_{0j}. We are free to go around resetting the clocks on each particle, namely $t = t' - f(\vec{x}')$ and $\vec{x} = \vec{x}'$, so that $g'_{0j} = g_{\mu\nu}\frac{\partial x^\mu}{\partial x'^0}\frac{\partial x^\nu}{\partial x'^j} = g_{0j} - \frac{\partial f}{\partial x^j}$.

The trouble is that to get $g'_{0j} = 0$, we have three (in general, $d - 1$) equations for one unknown function f. So we can't do it in general. But we can do it in two cases.

1. *Spherical symmetry.* Since $d\theta$ and $d\varphi$ must appear in the combination $d\theta^2 + \sin^2\theta d\varphi^2$, the components $g_{0\theta}$ and $g_{0\varphi}$ vanish, and the three equations collapse to one: $\frac{\partial f}{\partial r} = g_{0r}$, with the solution $f(r) = \int^r dr' g_{0r}(r')$, possible because g_{0j} does not depend on t by (22). Dropping primes, we arrive at the comoving coordinates

$$ds^2 = -dt^2 + B(t, r)dr^2 + C(t, r)(d\theta^2 + \sin^2\theta d\varphi^2) \tag{23}$$

This metric, depending on two unknown functions of t and r, would be particularly suitable for studying the gravitational collapse of a spherical cloud of dust.

2. *Polar-like coordinates.* At $t = 0$, around a specified particle, we can always go to locally flat coordinates y^μ. We have $g_{0j}(0, \vec{x}) = \eta_{\mu\nu}(\frac{\partial y^\mu}{\partial x^0}\frac{\partial y^\nu}{\partial x^j})_{t=0}$. Suppose these coordinates have the following two properties:

---

* For example, they define a luminosity distance by measuring the apparent brightness of standard candles.

(a) $(\frac{\partial y^0}{\partial x^i})_{t=0} = 0$ and (b) $(\frac{\partial y^i}{\partial t})_{t=0} = 0$. Condition (a) says that the separation $y^\mu(0, \vec{x} + \delta\vec{x}) - y^\mu(0, \vec{x}) \simeq \delta\vec{x}^i(\frac{\partial y^\mu}{\partial x^i})_{t=0}$ between the specified particle and its neighbors is purely spatial, that is, it vanishes when $\mu = 0$. Condition (b) says that the movement of the particle in spacetime is purely temporal. It then follows that $g_{0j}(0, \vec{x}) = 0$ and hence by (22), $g_{0j}(t, \vec{x}) = 0$. We arrive at

$$ds^2 = -dt^2 + g_{ij}(t, \vec{x})dx^i dx^j \tag{24}$$

Note that in the text, for the universe, we have the additional requirements of homogeneity and isotropy, which restrict the spatial metric g_{ij} further and license us to write (16).

To help you understand conditions (a) and (b) better, let me reveal that the familiar polar coordinates and spherical coordinates are Gaussian normal coordinates for E^2 and E^3, respectively. To see this, simply let $t \to r$, $x^i \to \theta^i$, and flip a sign to rewrite (24) as

$$ds^2 = dr^2 + g_{ij}(r, \theta^1, \cdots, \theta^{D-1})d\theta^i d\theta^j \tag{25}$$

In going from spacetime to space, we see that "a collection of freely falling particles" gets translated into "a collection of straight lines," the locally flat coordinates y^μ into "Cartesian coordinates" centered at the point we are focusing on, with y^r pointing in the radial direction and y^i in the angular direction. Condition (a) says that $\left(\frac{\partial y^r}{\partial \theta^i}\right)_{r=a} = 0$, and condition (b) that $\left(\frac{\partial \theta^i}{\partial r}\right)_{r=a} = 0$. Draw a picture for polar coordinates to see for yourself that it all makes sense. We have replaced the arbitrary setting $t = 0$ on the clock in the spacetime discussion by $r = a$ in the polar discussion to avoid the inconvenient fact that even at physicists' level of rigor, polar coordinates are ill defined at the origin.

Physicists have generally borrowed the principle of presumed innocence from the Anglo-American legal system. As implied in the introduction to this chapter, we will happily presume, unless proven otherwise, that what we know about space works for spacetime also (except for the obvious stuff due to the crucial flip of sign). Here is an interesting example of going the other way, using the physics of freely falling dust to illuminate something about space, something that presumably even Gauss had to work a little to figure out.

Exercises

1 Show that (5) and (6) imply (4) by differentiating $(\frac{dt}{d\tau})^2 - a(t)^2(\frac{d\vec{x}}{d\tau})^2$ with respect to τ.

2 Consider $ds^2 = -dt^2 - 2\sin x\, dt\, dx + \cos^2 x\, dx^2 + dy^2 + dz^2$. Using (9), calculate dl between two points separated by dx. Can you explain the result you obtain?

3 Explore the behavior of $D(R; t_s)$ for two cases of power law $a(t) \propto t^\alpha$: $\alpha = \frac{2}{3}$ for a universe dominated by cold matter, and $\alpha = \frac{1}{2}$, by radiation (as we will see in chapter VIII.1).

4 Evaluate the redshift formula for a universe with an exponentially growing $a(t)$ and for a universe with a power law $a(t)$.

5 Derive (17).

6 Derive (21).

7 Show that the relation between redshift and scale factor derived in the text for the flat universe holds just as well for curved universes.

8 Extend the discussion in the text for $k = 0$ to the cases $k = \pm 1$.

Notes

1. H. Nüssbaumer and L. Bieri, *Discovering the Expanding Universe*, p. 196.
2. For all historical remarks in this chapter, see H. Nüssbaumer and L. Bieri, *Discovering the Expanding Universe*, particularly pp. 195–199.
3. See http://www.youtube.com/watch?v=5U1-OmAICpU.
4. We encounter $g_{00}/(g_{00})^2 = -1/g_{00}$.
5. See *Fearful*, pp. xx and 68.
6. H. Nüssbaumer and L. Bieri, *Discovering the Expanding Universe*, p. xvii.
7. I talked to a few cosmologists while writing this book. They were unanimous that the terminology "Robertson-Walker universe" should be dropped.
8. S. Weinberg, *Gravitation and Cosmology*, p. 415.

V.4 | Motion in Curved Spacetime

Presence of external forces

I explained in chapter V.1 that we can, with no further work, simply lift the geodesic equation

$$\frac{d^2 X^\lambda}{d\tau^2} + \Gamma^\lambda_{\mu\nu} \frac{dX^\mu}{d\tau} \frac{dX^\nu}{d\tau} = 0 \tag{1}$$

from part II of this book and study motion in curved spacetime. Indeed, we did precisely that in the preceding chapter. Unfortunately, the frequent appearance of (1) leads some students to a misconception that material particles and observers are obliged to follow geodesics. To the contrary, as a young observer, you are certainly free to strap a rocket pack to your back and blast off in this direction or that.

The geodesic equation (1) describes the motion of a particle in the absence of any other force besides gravity, as is clear from the derivation in chapter V.1. With another force present, the particle follows

$$\frac{d^2 X^\lambda}{d\tau^2} + \Gamma^\lambda_{\mu\nu} \frac{dX^\mu}{d\tau} \frac{dX^\nu}{d\tau} = f^\lambda(X) \tag{2}$$

In particular, in an electromagnetic field, the force is given by

$$f^\lambda(X(\tau)) = -\frac{e}{m} F^\lambda{}_\nu(X(\tau)) \frac{dX^\nu}{d\tau}$$

for a particle of charge e and mass m. Here $F^\lambda{}_\nu = g^{\lambda\mu} F_{\mu\nu}$. (See chapter V.6 for further discussion.) Note the glaring contrast between the gravitational force and the electromagnetic force: one is universal, the other not. In other words, the left hand side of (2) makes no reference to any properties of the particle, while the right hand side depends on the ratio e/m, which varies enormously from particle to particle.

That Einstein could write down the equation of motion in the combined presence of a gravitational and electromagnetic field, rather than spend years looking for a more general

general relativity, testifies to the power of the equivalence principle. We simply promote (IV.1.16) to (2). Conversely, setting $g_{\mu\nu}$ to $\eta_{\mu\nu}$ in (2), we revert back to (IV.1.16).

What is a free particle?

Physicists are fond of speaking of free particles: a particle moving in the absence of any external forces is said to be free. But in Einstein gravity or general relativity, the concept is subtly different. In Einstein gravity, a particle following a geodesic in spacetime described by (1) is said to be free. In other words, a free particle* is not acted upon by any external forces except for gravity. In Einstein's theory, the gravitational field is equivalent to curved spacetime.

Part of the confusion can be regarded as semantic. We could say that there is no gravity in Einstein gravity, only curved spacetime. Perhaps the best policy is to dispense with words and look at only the equations. If the motion of a particle satisfies (1), it is free. If the motion of a particle satisfies (2), it is not free.

Let's put it more dramatically to underline the point. As you fall, at ever increasing speed, into the unrelenting clutch of a black hole, perhaps never again to be released, you are free. But if you turn on your rocket pack as you approach the horizon, and blast your way back out to infinity to live out the rest of your days, you are not free. While a layperson might find these statements paradoxical, you and I happily know that they originate in the patent clerk's happy thought, that a freely falling particle does not feel gravity.

Recovery of Newtonian motion in a gravitational field

Einstein's description of a particle moving in a gravitational field—that the particle is seeking the "best" possible path in spacetime—is at first sight strikingly different from Newton's—that the particle is acted on by a force. One advantage of the action principle treatment given in part IV of this book is that it renders going from Newton to Einstein more natural and makes clear that Einstein's description must reduce to Newton's.

Let's recover Newton's equation from Einstein's equation. This is not entirely obvious, since the geodesic equation (1) appears to give the 4-acceleration $\frac{d^2 X^\lambda}{d\tau^2}$ in terms of two powers of the 4-velocity $\frac{dX^\mu}{d\tau}$.

To recover the Newtonian limit, three conditions must be met:

1. The particle moves slowly: $\frac{dX^i}{d\tau} \ll \frac{dX^0}{d\tau}$.

---

* As you know, the term "test particle" is often used to emphasize that not only is the particle small enough for its internal structure not to be relevant but it is also insignificant enough not to affect whatever is producing the external forces. In the case of Einstein gravity, when we say particle, we assume that the particle does not affect and modify the curved spacetime it is in. As you also know, this idealizing and simplifying assumption often does not hold in physically interesting situations, such as two black holes circling each other. See chapter X.4 for a first step away from (1).

2. The gravitational field is weak, so that the metric is almost Minkowskian: $g_{\mu\nu} \simeq \eta_{\mu\nu} + h_{\mu\nu}$, with h small in the sense that we can neglect terms quadratic in h.

3. The gravitational field $h_{\mu\nu}$ does not depend on time.

Condition 1 means $\frac{d^2X^\lambda}{d\tau^2} + \Gamma^\lambda_{00}\left(\frac{dX^0}{d\tau}\right)^2 \simeq 0$, while 2 and 3 imply $\Gamma^\lambda_{00} \simeq -\frac{1}{2}\eta^{\lambda\rho}\partial_\rho h_{00}$, so that $\Gamma^0_{00} \simeq -\frac{1}{2}\partial_0 h_{00} = 0$ and $\Gamma^i_{00} \simeq \frac{1}{2}\partial_i h_{00}$. The geodesic equation (1) then reduces to $\frac{d^2X^0}{d\tau^2} \simeq 0$ (which implies that $\frac{dX^0}{d\tau}$ is a constant) and $\frac{d^2X^i}{d\tau^2} + \frac{1}{2}\partial_i h_{00}\left(\frac{dX^0}{d\tau}\right)^2 \simeq 0$, which since $X^0 = t \simeq \tau$ (because of (1)) becomes $\frac{d^2X^i}{dt^2} \simeq -\frac{1}{2}\partial_i h_{00}$. Thus, if we identify the gravitational potential Φ by $h_{00} = 2\Phi = -\frac{2GM}{r}$, we obtain Newton's equation $\frac{d^2\vec{X}}{dt^2} \simeq -\vec{\nabla}\Phi$. As you see, the secret to Newton's equation emerging is that in the "force term" $\frac{dX^\mu}{d\tau}\frac{dX^\nu}{d\tau}$, the time component dominates the space components.

This derivation shows that far from a spherically symmetric mass distribution, the spacetime metric must be such that $g_{00} \to 1 - \frac{2GM}{r}$. Notice also that our derivation does not depend on h_{ij}, nor on h_{0j}, as long as they are time independent.

The result we just obtained, that $g_{00} \simeq 1 - \frac{2GM}{r}$ in the Newtonian limit, is entirely consistent with option G in chapter IV.1.

Gravitational redshift

In chapter V.2, we derived gravitational redshift using the equivalence principle. Let us derive this result again using curved spacetime. For pedagogical clarity, we assume that the spacetime is static, in other words, that the metric does not depend on the time coordinate t (for example, the metric to be presented in the next section).

Suppose an observer (call him the emitter) located at some fixed $\vec{x}_E$ sends light signals at regular intervals, separated by $\Delta\tau_E$, as measured by his proper time, of course, to another observer (call her the receiver) at some fixed $\vec{x}_R$. Consider a particular light signal traveling from $\vec{x}_E$ to $\vec{x}_R$ following some trajectory (which, using the metric, we could determine, but which, as we will see presently, we don't need to know). In a static spacetime, physics is invariant under translation in time. Thus, the next signal would travel (see figure 1) by the same trajectory simply displaced by coordinate time $\Delta t_E = \Delta\tau_E/\sqrt{-g_{00}(\vec{x}_E)}$. (Since this observer is fixed at $\vec{x}_E$, for him $d\tau^2 = -g_{\mu\nu}dx^\mu dx^\nu = -g_{00}(\vec{x}_E)dt^2$.)

Here is the question for you to work out before reading on. When will the receiver receive the next signal?

In coordinate time, she receives this signal Δt_E after the preceding signal (according to the time translation invariant argument just given). But we have to express this in terms of the receiver's proper time (namely the time she experiences). For the receiver, the proper time interval between the two signals is given by

$$\Delta\tau_R = \sqrt{-g_{00}(\vec{x}_R)}\,\Delta t_E = \Delta\tau_E\left(\sqrt{g_{00}(\vec{x}_R)}/\sqrt{g_{00}(\vec{x}_E)}\right) \tag{3}$$

We can translate this result into the frequency shift for an electromagnetic wave. Just think of the successive crests of the wave as the signals. Thus, the frequency ω_R seen by

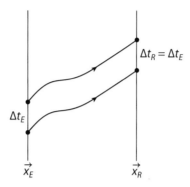

Figure 1 In a static spacetime, physics is invariant under translation in time. A light signal sent after another light signal would travel by the same trajectory simply displaced by some coordinate time.

the receiver is related to the frequency ω_E of the emitter by

$$\omega_R = \omega_E \sqrt{\frac{g_{00}(\vec{x}_E)}{g_{00}(\vec{x}_R)}} \tag{4}$$

We assumed that the emitter and receiver are fixed at $\vec{x}_E$ and $\vec{x}_R$, respectively, but for (4) to hold, all that is required is that the emitter and receiver do not move appreciably during a time $\sim 1/\omega$.

Note that all we used is that the metric is time translation invariant. In particular, in the weak field limit, we have the fractional frequency shift $(\omega_R - \omega_E)/\omega_E \simeq \sqrt{\frac{1+2\Phi(\vec{x}_E)}{1+2\Phi(\vec{x}_R)}} - 1 \simeq \Phi(\vec{x}_E) - \Phi(\vec{x}_R)$. A receiver located in a region with a higher gravitational potential sees a lower frequency. We have recovered the gravitational redshift, which we derived using the equivalence principle in chapter V.2. A useful mnemonic is that a photon of energy $\hbar\omega$ loses energy when climbing out of a gravitational potential well, but this mnemonic does not amount to a correct argument,[1] since it confounds a quantum relation with a Newtonian concept (namely the gravitational potential, which cannot be applied to a massless particle anyway).

As this derivation makes clear, and as was already mentioned in chapter V.2, what is commonly called gravitational redshift should more accurately be called gravitational time dilation. The phenomenon involved does not have to be characterized by a frequency at all.

Spacetime around a spherically symmetric mass distribution

We are not yet in a position to determine the metric in any given physical situation, but from symmetry alone, we can learn a lot about the spacetime metric $g_{\mu\nu}$. The expanding universe in the preceding chapter is a good example. Here we examine the spacetime around a spherically symmetric mass distribution. The mass distribution may depend on time, such as that of a pulsating spherical star, for example.

Because of general coordinate invariance, we have considerable freedom in choosing the coordinates. This corresponds to picking a gauge in electromagnetism. At this point, the rich man with his or her wealth of fancy terms starts talking about Killing vectors, and possibly even foliation. We will get to all that later. But for the moment, it is pedagogically more transparent to follow the poor man's way,* using explicit nuts and bolts, wearing no fancy pants.

By assumption, space is isotropic: there is no privileged direction. Thus, the differential elements we use must be rotation invariant, and there are three such elements: dt, $d\vec{x}^2 = d\vec{x} \cdot d\vec{x}$, and $\vec{x} \cdot d\vec{x}$. Go to spherical coordinates so that as usual, $d\vec{x}^2 = dr^2 + r^2(d\theta^2 + \sin^2\theta d\varphi^2) = dr^2 + r^2 d\Omega^2$. Differentiating $r^2 = \vec{x} \cdot \vec{x}$, we have $rdr = \vec{x} \cdot d\vec{x}$ and so $(\vec{x} \cdot d\vec{x})^2 = r^2 dr^2$.

Pythagoras (as generalized to spacetime) requires the line element ds^2 to be quadratic in dt and $d\vec{x}$. The inventory just given shows that we have four quadratic differentials to construct ds^2 with, namely dt^2, $dtdr$, dr^2, and $d\Omega^2$. Isotropy means that the coefficients of these quadratic differentials in ds^2 cannot depend on θ and φ, and they are thus functions of only t and r. Putting it all together, we obtain $ds^2 = -U(t, r)dt^2 - 2V(t, r)dtdr + W(t, r)dr^2 + (X(t, r))^2 d\Omega^2$ with four arbitrary functions of t and r.

Spherical symmetry has gotten us quite far, but we still have the freedom of changing coordinates. First, define a new radial coordinate $\tilde{r} = X(t, r)$, so that $d\tilde{r} = \partial_t X(t, r)dt + \partial_r X(t, r)dr$. Eliminate r and dr in terms of $\tilde{r}$, t, $d\tilde{r}$, and dt in ds^2 to obtain a mess of the form $ds^2 = -\tilde{U}(t, \tilde{r})dt^2 - 2\tilde{V}(t, \tilde{r})dtd\tilde{r} + \tilde{W}(t, \tilde{r})d\tilde{r}^2 + \tilde{r}^2 d\Omega^2$. We have effectively gotten rid of $X(t, r)$.

There is no need to work out the mess; we merely note that it has the indicated form. Now we simply rename functions and variables by dropping the twiddles to obtain $ds^2 = -U(t, r)dt^2 - 2V(t, r)dtdr + W(t, r)dr^2 + r^2 d\Omega^2$. (These are of course not the same functions we started out with, but there is no point in using up more letters.)

Suppose somebody gives us this ds^2 with these three functions U, V, and W. We now show that we still have enough freedom to get rid of the nasty $dtdr$ term. Define a new time coordinate $\tilde{t}$ by $d\tilde{t} = \zeta(t, r)(Udt + Vdr) = \partial_t \Phi(t, r)dt + \partial_r \Phi(t, r)dr$, where the unknown function $\zeta(t, r)$ is determined by the condition that the second equality holds for some Φ. In other words, we require that $\zeta(Udt + Vdr)$ be a total differential, so that the first equality makes sense. (The reader familiar with partial differential equations will recognize this as an often used trick.) Evaluating $\partial_t \partial_r \Phi = \partial_r \partial_t \Phi$, we find $\partial_t(\zeta V) = \partial_r(\zeta U)$. Given $U(t, r)$, $V(t, r)$, and the initial value $\zeta(0, r)$, we can determine $\zeta(t, r)$ by integrating this equation in t. In any case, we don't care about all the details, merely that in principle there exists a $\tilde{t}$ with the stated property $d\tilde{t} = \zeta(t, r)(Udt + Vdr)$. Eliminating dt in $U(t, r)dt^2 + 2V(t, r)dtdr$, we find that this expression becomes $(\zeta^2 U)^{-1}d\tilde{t}^2 - Y(\tilde{t}, r)dr^2$, where $Y(\tilde{t}, r)$ is some function we don't need to determine. All we care about is good riddance to the nasty cross term.

---

* In theoretical physics, we also have the smart man and the dumb man. It may be swell to be a smart rich man, but I would venture that being a dumb rich man may be worse than being a dumb poor man.

Putting it altogether, dropping twiddles, and renaming functions, we finally obtain

$$ds^2 = -A(t,r)dt^2 + B(t,r)dr^2 + r^2 d\Omega^2 \tag{5}$$

To summarize, we have used the spherical symmetry and exploited our freedom to change coordinates to reduce $g_{\mu\nu}(x)$, potentially 10 functions each of 4 variables, to $A(t,r)$ and $B(t,r)$, 2 functions each of 2 variables. This enormous simplification is typical of many problems in general relativity.

The geometrical meaning of this metric is easy to grasp without any fancy math. Space is "foliated" by spheres S^2, each with area $4\pi r^2$, and with the gap between "successive" spheres, that is, the distance between (r, θ, φ) and $(r+dr, \theta, \varphi)$, given by $B(t,r)dr$. When coordinate time changes by dt, the elapsed proper time felt by different observers fixed at different values of r is given by $A(t,r)dt$.

Motion in a static isotropic spacetime

Let us restrict the mass distribution further to be static, so that $A(r)$ and $B(r)$ do not depend on time. Thus, the most general static and isotropic metric

$$ds^2 = -A(r)dt^2 + B(r)dr^2 + r^2 d\Omega^2 \tag{6}$$

can be written in terms of two functions[2] $A(r)$ and $B(r)$ of a single variable r. At this point, we do not know anything about these two functions, except that, far away from the mass, $A(r) \simeq (1 - \frac{2GM}{r})$ as $r \to \infty$, so that we recover the Newtonian gravitational potential $\Phi(r) = -\frac{GM}{r}$. To determine $A(r)$ and $B(r)$, we would need to master the dynamics of the gravitational field, and we won't get to that until part VI. But meanwhile, we can start studying the motion of a particle, such as a planet, in this curved spacetime by varying the Lagrangian

$$L = \left[A(r) \left(\frac{dt}{d\tau} \right)^2 - B(r) \left(\frac{dr}{d\tau} \right)^2 - r^2 \left(\frac{d\theta}{d\tau} \right)^2 - r^2 \sin^2\theta \left(\frac{d\varphi}{d\tau} \right)^2 \right]^{\frac{1}{2}} \tag{7}$$

By now, we can practically vary in our heads and immediately write down the 4 equations:

$$\frac{d}{d\tau} \left(A(r) \frac{dt}{d\tau} \right) = 0 \tag{8}$$

$$\frac{d}{d\tau} \left(B(r) \frac{dr}{d\tau} \right) + \frac{1}{2} A'(r) \left(\frac{dt}{d\tau} \right)^2 - \frac{1}{2} B'(r) \left(\frac{dr}{d\tau} \right)^2 - r \left(\frac{d\theta}{d\tau} \right)^2 - r \sin^2\theta \left(\frac{d\varphi}{d\tau} \right)^2 = 0 \tag{9}$$

$$\frac{d}{d\tau} \left(r^2 \frac{d\theta}{d\tau} \right) - r^2 \sin\theta \cos\theta \left(\frac{d\varphi}{d\tau} \right)^2 = 0 \tag{10}$$

$$\frac{d}{d\tau} \left(r^2 \sin^2\theta \frac{d\varphi}{d\tau} \right) = 0 \tag{11}$$

These are of course precisely the equations contained in (1). For example, (10) is just $\frac{d}{d\tau}(g_{\theta\theta} \frac{d\theta}{d\tau}) = \frac{1}{2}(\frac{\partial}{\partial\theta} g_{\varphi\varphi})(\frac{d\varphi}{d\tau})^2$.

It is important to remember, from our discussion in chapter II.2, that we are entitled to trade any one of the 4 equations (8–11), which all involve second derivatives, for the defining equation for $d\tau$,

$$A(r)\left(\frac{dt}{d\tau}\right)^2 - B(r)\left(\frac{dr}{d\tau}\right)^2 - r^2\left(\frac{d\theta}{d\tau}\right)^2 - r^2\sin^2\theta\left(\frac{d\varphi}{d\tau}\right)^2 = 1 \tag{12}$$

which only involves first derivatives. Of course, if we had any sense, we would trade (9) for (12). With this trade, these equations are not that difficult to solve.

That math professor I referred to in the preceding chapter would dismiss most theorems as being so obvious that they are "self-proving." In the same sense, (8) and (11) are self-solving, yielding

$$\frac{dt}{d\tau} = \frac{\epsilon}{A(r)} \tag{13}$$

and

$$\frac{d\varphi}{d\tau} = \frac{l}{r^2\sin^2\theta} \tag{14}$$

with ϵ and l two integration constants. (Do these two equations say anything to you?) Furthermore, we can solve (10) by setting $\theta(\tau) = \frac{\pi}{2}$. This is of course another consequence of the rotational symmetry of the problem: the planet stays in the equatorial plane.

Plugging all this into (12), we obtain

$$\frac{\epsilon^2}{A(r)} - B(r)\left(\frac{dr}{d\tau}\right)^2 - \frac{l^2}{r^2} = 1 \tag{15}$$

Cleaning up and rearranging a bit, we find that, remarkably enough, we can cast this equation for an Einsteinian particle moving in curved spacetime in the form of an equation for a Newtonian particle (of unit mass) moving in a potential $v(r)$ with zero total energy (with τ playing the role of time):

$$\frac{1}{2}\left(\frac{dr}{d\tau}\right)^2 + v(r) = 0 \tag{16}$$

with

$$v(r) = \frac{1}{2B(r)}\left(1 + \frac{l^2}{r^2}\right) - \frac{\epsilon^2}{2A(r)B(r)} \tag{17}$$

Once we are given $A(r)$ and $B(r)$, we merely have to solve a Newtonian problem in an unfamiliar potential!

How light moves

Earlier in this chapter, we recovered Newtonian motion for a slowly moving particle. Now let us treat the opposite limit and ask how light, or an ultrarelativistic particle, would move in curved spacetime.

That's easy to do. Back in chapter III.5, we unified the action principle for material particles and the action principle for light, which I regard as one of the great achievements of special relativity. No more Lagrange on the one hand, and Fermat on the other!

The equivalence principle now makes our life sweet: we merely have to take what we did in chapter III.5 and promote the Minkowski metric $\eta_{\mu\nu}$ to $g_{\mu\nu}$.

So write the action $S = -m \int \sqrt{-g_{\mu\nu}(X)dX^\mu dX^\nu}$ (look at the manifest parametrization independence since there is no parameter!) for a particle of mass m in the form $\tilde{S} = -\frac{1}{2} \int d\zeta \left(\sigma(\frac{dX}{d\zeta})^2 + \frac{m^2}{\sigma}\right)$, where $(\frac{dX}{d\zeta})^2 = -g_{\mu\nu}\frac{dX^\mu}{d\zeta}\frac{dX^\nu}{d\zeta}$, so that we can take the massless limit to obtain

$$S_{\text{massless}} = \frac{1}{2}\int d\zeta \, \sigma(\zeta)g_{\mu\nu}(X(\zeta))\frac{dX^\mu}{d\zeta}\frac{dX^\nu}{d\zeta} \tag{18}$$

Varying with respect to $\sigma(\zeta)$ tells us that for a massless particle,

$$g_{\mu\nu}dX^\mu dX^\nu = 0 \tag{19}$$

Varying with respect to X, we obtain (just as back in (II.2.16))

$$\frac{d}{d\zeta}\left(\sigma g_{\mu\rho}\frac{dX^\mu}{d\zeta}\right) - \frac{1}{2}(\partial_\rho g_{\mu\nu})\sigma\frac{dX^\mu}{d\zeta}\frac{dX^\nu}{d\zeta} = 0$$

As in appendix 1 in chapter III.5, it is clearly advantageous to define an affine parameter by $d\zeta = \sigma(\zeta)d\zeta'$. Dropping the prime, we obtain

$$\frac{d^2X^\lambda}{d\zeta^2} + \Gamma^\lambda_{\mu\nu}\frac{dX^\mu}{d\zeta}\frac{dX^\nu}{d\zeta} = 0, \tag{20}$$

which looks superficially the same as (1). The difference is in the parameter choice. As usual, we trade the most complicated equation among (20) for (19), as has been explained ad nauseum starting in chapter II.2.

It is perhaps good to summarize this business about natural parametrization. For curves in space, the length along the curve provides the natural parameter. For the worldline of a particle in spacetime, the proper time, namely the elapsed time in the rest frame of the particle, is the obvious candidate. For a massless particle such as the photon, no natural candidate presents itself, and we choose whichever parameter will make life easier.

So, for light moving in the spacetime described by the metric (6), we have the same equations as (8–11) but with the proper time τ replaced by the "affine parameter" ζ. For a photon moving in the equatorial plane (so that (10) is solved by $\theta = \frac{\pi}{2}$), we have, once again, $\frac{dt}{d\zeta} = \frac{\epsilon}{A(r)}$ and $\frac{d\varphi}{d\zeta} = \frac{l}{r^2}$, with ϵ and l two integration constants. Inserting this into (18), we obtain $\frac{\epsilon^2}{A(r)} - B(r)(\frac{dr}{d\zeta})^2 - \frac{l^2}{r^2} = 0$ (which, of course, is just (15) with the right hand side set to 0 and with $\tau \to \zeta$). After rearranging, we obtain

$$\frac{1}{l^2}\left(\frac{dr}{d\zeta}\right)^2 + \frac{1}{B(r)}\left(\frac{1}{r^2} - \frac{\epsilon^2}{l^2A(r)}\right) = 0 \tag{21}$$

Once again, this looks like a Newtonian problem in an unfamiliar potential. But we still have the freedom of scaling the affine parameter $\zeta \to \zeta/l$, and thus we learn that the physics does not depend on ε and l separately, but only on $b^2 \equiv \frac{l^2}{\epsilon^2}$.

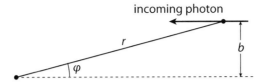

Figure 2 The impact parameter b in a scattering process.

A better way of saying this is to eliminate the affine parameter ζ by dividing $\frac{dr}{d\zeta} / \frac{d\varphi}{d\zeta} = \frac{dr}{d\varphi}$, so that

$$\left(\frac{dr}{d\varphi}\right)^2 = \frac{r^4}{B(r)}\left(\frac{1}{b^2 A(r)} - \frac{1}{r^2}\right) \tag{22}$$

To identify this mysterious quantity b, we let $r \to \infty$, where space is nice and flat so high school geometry applies. Then (22) becomes $(\frac{dr}{d\varphi})^2 \simeq \frac{r^4}{b^2}$, which has the solution $r\varphi \simeq b$. Thus, we see that b is what in a scattering process is called the impact parameter. See figure 2.

The fact that in a static isotropic spacetime, the motion of material particles and of light both reduce to a Newtonian problem can be traced back to the metric $ds^2 = g_{\mu\nu}dx^\mu dx^\nu$ being a quadratic form, and so in a sense, ultimately to Pythagoras.

Parametrized post-Newtonian approximation

We still have some distance to go before we learn how to determine $A(r)$ and $B(r)$ (in chapter VI.3), but meanwhile, dimensional analysis can take us quite far. You have probably heard of the celebrated solar system tests of Einstein gravity (which we will come to in chapter VI.3), such as the precession of the planet Mercury and the bending of starlight as it passes by the sun. Since G and the mass of the sun M always occur in the combination GM, as was already mentioned in part 0, $A(r)$ and $B(r)$ can only be functions of $\frac{GM}{r}$. The gravitational field in the solar system is so weak ($\frac{GM}{r} \ll 1$) that it is entirely adequate for these classic tests to expand and keep only the leading terms in the so-called parametrized post-Newtonian (PPN to those who love acronyms) approximation:

$$A(r) = 1 - \frac{2GM}{c^2 r} + 2(\beta - \gamma)\left(\frac{GM}{c^2 r}\right)^2 + \cdots$$

$$B(r) = 1 + 2\gamma\left(\frac{GM}{c^2 r}\right) + \cdots \tag{23}$$

I have purposely restored c, so that you can see that the expansion parameter is the ratio of the Newtonian gravitational potential energy of a unit mass test particle to its rest mass. It is also illuminating to take the large c limit:

$$ds^2 = -A(r)c^2 dt^2 + B(r)dr^2 + r^2 d\theta^2 + r^2\sin^2\theta d\varphi^2$$

$$\to -c^2 dt^2 + d\vec{x}^2 + \frac{2GM}{r}dt^2 + O\left(\frac{1}{c^2}\right) \tag{24}$$

Recall chapter IV.1.

As you will see in chapter VI.3, Einstein gravity gives $\beta = 1$ and $\gamma = 1$. Over the years since 1915 there have been competing theories of gravity giving other values, but they have generally fallen by the wayside. Current observations bound the deviation of β and γ from 1 by two parts in 10^4 and in 10^5, respectively. Einstein gravity is in excellent agreement[3] with observation, at least in this post-Newtonian approximation.

Conservation laws and Killing vectors

You may have recognized (13) and (14) as the general relativistic versions of energy conservation and angular momentum conservation, respectively. Going back, you see that these two conservation laws follow from the fact that the metric $g_{\mu\nu}$ in (6) does not depend on t and on φ, respectively.

In general, if the metric $g_{\mu\nu}$ does not depend on a particular coordinate x^λ, then the geodesic equation for that coordinate, obtained as always (of course) by varying $\int (g_{\mu\nu} dx^\mu dx^\nu)^{\frac{1}{2}}$ with respect to x^λ, simplifies immediately to

$$\frac{d}{d\tau}\left(g_{\lambda\mu}\frac{dx^\mu}{d\tau}\right) = 0 \tag{25}$$

(In our example, (25) corresponds to (8) and (11).) In other words, $g_{\lambda\mu}\frac{dx^\mu}{d\tau}$ does not change as the particle moves along the geodesic. (For a massless particle, we merely replace the proper time by an affine parameter.)

Since we are just applying the action principle, Noether's theorem (as discussed in chapter II.4) directly implies these conservation laws. The action here $\int (g_{\mu\nu} dx^\mu dx^\nu)^{\frac{1}{2}}$ does not change upon shifting x^λ by a constant.

This discussion can be rendered more formal as follows. Let the metric be invariant upon $x^\mu \to x^\mu + \epsilon\xi^\mu$, with ϵ some infinitesimal. Then $\xi \cdot \frac{dx}{d\tau} = g_{\mu\nu}\xi^\mu\frac{dx^\nu}{d\tau}$ is conserved. In general, there may be several such ξs, known as Killing vectors. In our example, $\xi_e^\mu = (1, 0, 0, 0)$ and $\xi_l^\mu = (0, 0, 0, 1)$.

I rather dislike such apparently useless formal manipulations, but later in chapter IX.6, we will see that Killing vectors describe isometries of spacetime, an important and useful concept. For now, however, simply think of writing $\xi \cdot \frac{dx}{d\tau}$ as a shorthand for the more descriptive $g_{\lambda\mu}\frac{dx^\mu}{d\tau}$. Another way of saying this is that the momentum of a particle as usually defined $\left(p^\mu = m\frac{dx^\mu}{d\tau}\right)$ is not conserved, but its component $\xi \cdot p$ along a Killing vector is. With our sign convention, the energy of the particle is given by $E = -\xi_e \cdot p$ and the angular momentum by $L = \xi_l \cdot p = mr^2 \sin^2\theta\frac{d\varphi}{d\tau}$.

Appendix: Christoffel symbols around a time independent spherically symmetric mass distribution

We can read off the Christoffel symbols from the geodesic equations. For instance, (8) works out to be $\frac{d^2t}{d\tau^2} + \frac{A'(r)}{A(r)}\frac{dr}{d\tau}\frac{dt}{d\tau} = 0$, from which we read off $\Gamma^t_{tr} = \Gamma^t_{rt} = \frac{A'}{2A}$. We will now list all of them:

$$\Gamma^t_{tr} = \frac{A'}{2A}, \qquad \Gamma^r_{tt} = \frac{A'}{2B}, \qquad \Gamma^r_{rr} = -\frac{B'}{2B}, \qquad \Gamma^r_{\theta\theta} = -\frac{r}{B}, \qquad \Gamma^r_{\varphi\varphi} = -\frac{r \sin^2\theta}{B},$$

$$\Gamma^\theta_{r\theta} = \frac{1}{r}, \qquad \Gamma^\varphi_{r\varphi} = \frac{1}{r},$$

$$\Gamma^\theta_{\varphi\varphi} = -\sin\theta\cos\theta, \qquad \Gamma^\varphi_{\theta\varphi} = \cot\theta \tag{26}$$

with all other components not related to these by symmetry vanishing.

We can understand a lot by symmetry considerations. For example, because the metric is invariant under $t \to -t$, Christoffel symbols with an odd number of t indices must vanish. Also, $\Gamma^\theta_{\varphi\varphi} = -\sin\theta\cos\theta$, $\Gamma^\varphi_{\theta\varphi} = \cot\theta$ are the same as those for S^2 found way back in chapter II.2 and merely reflect the spherical symmetry of the metric.

See also exercise 3.

Exercises

1 Work out the potential $v(r)$ in the parametrized post-Newtonian approximation, assuming that β and γ are of order unity, and sketch its general form.

2 Calculate $\Gamma^\nu_{\mu\nu}$ for the Christoffel symbols in (26) and verify an identity derived in chapter II.2.

3 Find the Christoffel symbols for the time dependent spherically symmetric spacetime in (5). Show that we simply have to add[4]

$$\Gamma^t_{tt} = \frac{\dot{A}}{2A}, \qquad \Gamma^t_{rr} = \frac{\dot{B}}{2A}, \qquad \Gamma^r_{tr} = \frac{\dot{B}}{2B} \tag{27}$$

(with the dot indicating $\frac{\partial}{\partial t}$) to the list in (26). Note that this is consistent with the transformation $t \to t$.

4 Show that in the spacetime $ds^2 \equiv dt^2 + \frac{1}{1 - \frac{2GM}{r}} dr^2 + r^2 d\Omega^2$, particles with constant r, θ, φ are actually freely falling.

5 Show that in the spacetime $ds^2 \equiv \frac{1}{1 - \frac{2GM}{r}} dt^2 + \frac{1}{1 - \frac{2GM}{r}} dr^2 + r^2 d\Omega^2$, freely falling particles with constant θ, φ starting at $r > 2GM$ actually fall toward larger r.

6 Show that the geodesic followed by a massless particle is also the geodesic followed by a massless particle in a conformally equivalent spacetime.

Notes

1. Beware of some textbooks on this point!
2. A historical note: until the 1950s, the notation $g_{00} = -e^{\mu(r)}$ and $g_{rr} = e^{v(r)}$ (or this form with some other "suitable" letters) was used. The "nonexponential" notation used here appeared later.
3. Periodically, there are reports of deviation from our understanding of gravity, both Newtonian and Einsteinian. Most of these eventually either go away with better measurements or are found to be due to "mundane" causes. (No doubt some 19th century physicists could say the same about Mercury's perihelion precession (see chapter VI.3) before Einstein came along.) One interesting anomaly is the so-called Pioneer anomaly regarding the observed accelerations of the Pioneer 10 and Pioneer 11 spacecrafts after they passed out of the solar system. As this book was being completed, this anomaly had been determined to be due to mundane causes.
4. In S. Weinberg, *Gravitation and Cosmology* (1972 edition), Γ^t_{rr} is missing a time derivative on p. 336.

The mother of all vectors

I have insisted again and again that the laws of physics should be expressed in terms of vectors and tensors, so that what different observers see can be simply related. I will now talk about vectors and tensors for the third time.

My pedagogical philosophy in this book can be expressed as "one baby step at a time." We started with a discussion of vectors and tensors under rotations in chapter I.3. Then we discussed vectors and tensors in special relativity, and encountered the novelty of having to keep track of upper and lower indices. The coordinates transform as $x'^\mu = \Lambda^\mu_{\ \nu} x^\nu$, or better, the coordinate differentials transform as $dx'^\mu = \Lambda^\mu_{\ \nu} dx^\nu$. In this case, the insistence on talking about differentials hardly matters, since $\Lambda^\mu_{\ \nu}$ is a constant matrix and so the differential version of the transformation follows trivially upon differentiating: dx^μ and x^μ transform in the same way.

In both these cases, rotation and Lorentz transformation, vectors are defined as objects that transform like dx^μ: we could say that dx^μ is the "ur-vector" or "the mother of all vectors." Tensors are then defined as objects that transform as if they were built out of vectors.

Now that we have mastered these two baby steps, we are ready to leap to vectors and tensors in general relativity!

Under a general coordinate transformation $x \to x'(x)$, the coordinate differentials transform as

$$dx'^\mu = \left(\frac{\partial x'^\mu}{\partial x^\nu}\right)dx^\nu \equiv S^\mu_{\ \nu}(x)dx^\nu \tag{1}$$

To save writing, we have defined the transformation matrix $S^\mu_{\ \nu}(x)$, which plays the same role as $\Lambda^\mu_{\ \nu}$ in a Lorentz transformation, but with one huge difference: unlike Λ, S depends on x. The transformation law changes from place to place.

Another important point is that our insistence on coordinate differentials, which seemed so academic and nit-picking, has finally become de rigueur: by definition dx^μ, but not x^μ, transforms like a vector. In general, the transformation $x^\mu \to x'^\mu$ is arbitrary and nonlinear.

You must master this point: dx^μ transform linearly, but not x^μ. There is all the difference in the world between these two different mathematical creatures.

Indeed, I have prepared you for these two steps also in chapters I.5 and I.6, discussing coordinate transformation and curved spaces. The formalism here is exactly the same, except the signature is that of spacetime rather than that of space. You have already encountered essentially everything we will discuss presently. As I anticipated in chapter I.5, the concepts needed for curved spacetimes have their seeds in the conceptually simple transformation from Cartesian coordinates to spherical coordinates that every physicist has dealt with since childhood. The point we just emphasized can already be seen in making a coordinate transformation: while x, y, z and r, θ, φ are related to each other nonlinearly, dx, dy, dz and dr, $d\theta$, $d\varphi$ are related to each other linearly. Granted, the matrix that relates dx, dy, dz and dr, $d\theta$, $d\varphi$ may have elements that involve highly nontrivial functions (such as trigonometric and inverse trigonometric functions), but the important property is the linearity. To repeat, although the functions $S^\mu_{\ \nu}(x)$ may be algebraically quite complicated, the relation $dx'^\mu = S^\mu_{\ \nu}(x)dx^\nu$ is nevertheless linear and hence easy to manipulate.

Vectors and the construction of tensors

A vector $W^\mu(x)$ (or more precisely a vector field: a vector that depends on x) is defined as something that transforms under a general coordinate transformation like the ur-vector dx^μ does in (1):

$$W'^\mu(x') = S^\mu_{\ \nu}(x)W^\nu(x) \tag{2}$$

Notice that W'^μ is evaluated at x', while W^ν is evaluated at x, but these are just different coordinate values describing the same point P.

Remember the student in chapter I.4 who was puzzled by the statement that a tensor is something that transforms like a tensor. In fact, he should already have been puzzled by the statement that a vector is something that transforms like a vector, an example of which is the ur-vector dx^μ.

Just as before, we define a 2-indexed tensor $T^{\mu\nu}(x)$ as something that transforms like

$$T'^{\mu\nu}(x') = S^\mu_{\ \rho}(x)S^\nu_{\ \sigma}(x)T^{\rho\sigma}(x) \tag{3}$$

We can now go on to define tensors with as many indices as we want, transforming like

$$T'^{\mu\nu\cdots\omega}(x') = S^\mu_{\ \rho}(x)S^\nu_{\ \sigma}(x)\cdots S^\omega_{\ \tau}(x)T^{\rho\sigma\cdots\tau}(x) \tag{4}$$

As I warned you, I am literally saying some of these things for the third time.

Upper and lower indices

The transformation of the metric is fixed by the invariance of the proper length, exactly as in chapter I.5, when we rather innocently first changed coordinates,

$$ds^2 = g'_{\rho\sigma}(x')dx'^{\rho}dx'^{\sigma} = g_{\mu\nu}(x)dx^{\mu}dx^{\nu} = g_{\mu\nu}(x)\frac{\partial x^{\mu}}{\partial x'^{\rho}}\frac{\partial x^{\nu}}{\partial x'^{\sigma}}dx'^{\rho}dx'^{\sigma} \tag{5}$$

so that

$$g'_{\rho\sigma}(x') = g_{\mu\nu}(x)(S^{-1})^{\mu}_{\ \rho}(x)(S^{-1})^{\nu}_{\ \sigma}(x) \tag{6}$$

with the definition $(S^{-1})^{\mu}_{\ \rho} \equiv \frac{\partial x^{\mu}}{\partial x'^{\rho}}$ once again. For your enlightenment, I even used the same notation in chapter I.5. And, yes, yet once again, using the chain rule discovered by our forefathers who invented calculus, we verify $(S^{-1})^{\mu}_{\ \rho}S^{\rho}_{\ \nu} = \frac{\partial x^{\mu}}{\partial x'^{\rho}}\frac{\partial x'^{\rho}}{\partial x^{\nu}} = \frac{\partial x^{\mu}}{\partial x^{\nu}} = \delta^{\mu}_{\nu}$, so that S^{-1} is indeed the inverse of S.

A few points are worth emphasizing here. The need to maintain the summation convention of always summing an upper index with a lower index forces $g_{\mu\nu}$ to carry lower indices, since it is by definition yoked to $dx^{\mu}dx^{\nu}$. The transformation is then thrust upon us.

Tensors with upper indices transform as in (4) with S, while here a tensor with lower indices, namely the metric itself, transforms "oppositely" with S^{-1}, namely like $g'_{\rho\sigma} = g_{\mu\nu}(S^{-1})^{\mu}_{\ \rho}(S^{-1})^{\nu}_{\ \sigma}$. No surprise here at all: the upper and lower indices are contracted in the invariant $ds^2 = g_{\mu\nu}(x)dx^{\mu}dx^{\nu}$. Exactly as in chapter III.3, we can define the transpose of S^{-1} by $((S^{-1})^T)^{\ \mu}_{\rho} \equiv (S^{-1})^{\mu}_{\ \rho}$. (Notice that, just as in chapter III.3, when we transpose we do not move anybody up and down stairs.) We can then write $g' = (S^{-1})^T g S^{-1}$, regarding the metric as a matrix.

Thus far, I have carefully indicated that unprimed tensors depend on x and primed tensors depend on x'. (Of course, since x and x' are related, we can always regard a function of x' as a function of x and vice versa, but this way, in which things are usually written, is more natural.) We adopt the convention of thinking of S and S^{-1} as functions of x. To avoid clutter, we will henceforth often suppress the x and x' dependence of various objects if there is no risk of confusion.

Once again, as in our earlier discussion of coordinate changes and curved spaces, we can use the metric $g_{\mu\nu}(x)$ to lower indices. Given a vector W^{ν}, we invite ourselves to construct a vector with lower index

$$W_{\mu} \equiv g_{\mu\nu}W^{\nu} \tag{7}$$

No prize for your correct guess on how W_{μ} transforms:

$$W'_{\rho} \equiv g'_{\rho\sigma}W'^{\sigma} = g_{\mu\nu}(S^{-1})^{\mu}_{\ \rho}(S^{-1})^{\nu}_{\ \sigma}S^{\sigma}_{\ \omega}W^{\omega} = W_{\mu}(S^{-1})^{\mu}_{\ \rho} \tag{8}$$

Speaking colloquially, we can say that when a lower index and an upper index (σ in this example) are summed over as in the Einstein convention, the associated transformation matrices S^{-1} and S knock each other off. Compare (8) with (2).

Given two vectors W and U, we can form, in analogy with ordinary Euclid 3-vectors and Lorentz 4-vectors, the dot product $W_\mu U^\mu = g_{\mu\nu} W^\nu U^\mu$. In light of the preceding remark, we expect $W_\mu U^\mu$ to transform like a scalar, that is, it does not transform at all: indeed,

$$W'_\rho U'^\rho = W_\mu (S^{-1})^\mu_{\ \rho} S^\rho_{\ \nu} U^\nu = W_\mu U^\mu \tag{9}$$

As far as the transformation properties are concerned, the summed-over pair of indices effectively disappear. Indeed, we can define $\phi(x) \equiv W_\mu(x) U^\mu(x)$, and it transforms like a scalar $\phi'(x') = \phi(x)$.

Now that we know how a vector with a lower index transforms, we can define tensors with lower indices. For example, the tensor $T_{\mu\nu\rho}$ transforms as if it were (even though it is not) built up of three vectors $W_\mu V_\nu U_\rho$. Indeed, we can clearly define tensors with arbitrary numbers of upper and lower indices, transforming like

$$T'^{\mu\nu\cdots\omega}_{\eta\cdots\kappa} = S^\mu_{\ \rho} S^\nu_{\ \sigma} \cdots S^\omega_{\ \tau} T^{\rho\sigma\cdots\tau}_{\zeta\cdots\psi} (S^{-1})^\xi_{\ \eta} \cdots (S^{-1})^\psi_{\ \kappa} \tag{10}$$

You get the idea before I run out of Greek letters! As before, fancy people who like big words call the upper indices contravariant and the lower indices covariant. Upper and lower indices transform oppositely.

Certainly, there are deep mathematical reasons underlying the appearances of upper and lower indices, but at a pedestrian level, just as in our discussion of Lorentz vectors and tensors, you can simply regard lower indices as a notational device to avoid writing $g_{\mu\nu}$ all the time. Also not surprisingly, since we use $g_{\mu\nu}$ to lower indices, we might expect to use its inverse to raise them.

We define the inverse metric $g^{\sigma\mu}$ by $g^{\sigma\mu} g_{\mu\nu} = \delta^\sigma_\nu$. In other words, define the inverse metric as the inverse of the metric regarded as a matrix. (As I remarked in connection with the Minkowski metric $\eta_{\mu\nu}$ and its inverse $\eta^{\sigma\mu}$, the spacetime metric $g_{\mu\nu}$ and its inverse $g^{\sigma\mu}$ are also denoted by the same letter g but distinguished by the position of their indices, a potential source of confusion for some seeing this for the first time. Things are just as in chapter III.3, where we used the Minkowski metric $\eta_{\mu\nu}$ and its inverse to lower and raise indices, respectively.)

We can check that the inverse metric raises indices by contracting $g^{\sigma\mu}$ with W_μ: $g^{\sigma\mu} W_\mu = g^{\sigma\mu} g_{\mu\nu} W^\nu = \delta^\sigma_\nu W^\nu = W^\sigma$; indeed, we get W^σ back. Once again, as in our discussion of Lorentz tensors, we can lower and raise indices at will using $g_{\mu\nu}$ and $g^{\mu\nu}$, respectively. For example, $T^{\sigma\lambda}_{\ \nu\rho} \equiv g_{\nu\kappa} T^{\kappa\sigma\lambda}_{\ \ \ \rho}$. Incidentally, it is common practice to use the same letter T to denote entirely different tensors, distinguished by the number and kind of indices they carry.

Taking the inverse of $g' = (S^{-1})^T g S^{-1}$, we see that $g'^{-1} = S g^{-1} S^T$, or written out more explicitly,

$$g'^{\lambda\rho}(x') = S^\lambda_{\ \kappa} g^{\kappa\omega}(x) (S^T)^\rho_{\ \omega} = S^\lambda_{\ \kappa} S^\rho_{\ \omega} g^{\kappa\omega}(x) \tag{11}$$

We can, if you wish, check the obvious, that the transformed $g'^{\lambda\rho}$ is indeed the inverse of the transformed $g'_{\rho\sigma}$ in (6):

$$g'^{\lambda\rho} g'_{\rho\sigma} = S^\lambda_{\ \kappa} S^\rho_{\ \omega} g^{\kappa\omega} g_{\mu\nu} (S^{-1})^\mu_{\ \rho} (S^{-1})^\nu_{\ \sigma} = S^\lambda_{\ \kappa} g^{\kappa\mu} g_{\mu\nu} (S^{-1})^\nu_{\ \sigma} = S^\lambda_{\ \kappa} \delta^\kappa_\nu (S^{-1})^\nu_{\ \sigma} = \delta^\lambda_\sigma$$

By now, you should be familiar with how this works; for example, in the second equality, $S^\rho_{\ \omega}$ and $(S^{-1})^\mu_{\ \rho}$ knock each other off.

To summarize, we can define tensors with however many upper and lower indices we like. Each upper index transforms with S, and each lower index with S^{-1}. For example, a tensor with one upper and three lower indices transforms like

$$W'^\lambda_{\ \rho\mu\nu} = S^\lambda_{\ \sigma} W^\sigma_{\ \omega\eta\kappa} (S^{-1})^\omega_{\ \rho} (S^{-1})^\eta_{\ \mu} (S^{-1})^\kappa_{\ \nu} \tag{12}$$

In the repeated index summation convention, we are allowed to contract an upper index with a lower index, namely to set them equal and sum over them. For example, setting λ equal to μ in (12) and summing over them, we obtain

$$W'^\mu_{\ \rho\mu\nu} = S^\mu_{\ \sigma} W^\sigma_{\ \omega\eta\kappa} (S^{-1})^\omega_{\ \rho} (S^{-1})^\eta_{\ \mu} (S^{-1})^\kappa_{\ \nu} = \delta^\eta_{\ \sigma} W^\sigma_{\ \omega\eta\kappa} (S^{-1})^\omega_{\ \rho} (S^{-1})^\kappa_{\ \nu} = W^\eta_{\ \omega\eta\kappa} (S^{-1})^\omega_{\ \rho} (S^{-1})^\kappa_{\ \nu}$$

In other words, as you might expect, $W_{\omega\kappa} \equiv W^\eta_{\ \omega\eta\kappa}$ transforms like a tensor with two lower indices: in (12), the S knocks off an S^{-1} as explained earlier.

You are never allowed to contract an upper index with an upper index, or a lower index with a lower index. The reason is obvious. Suppose you set μ and ν equal in (12) and sum. You would encounter, instead of an S and an S^{-1} knocking each other off, $\sum_\mu (S^{-1})^\eta_{\ \mu} (S^{-1})^\kappa_{\ \mu}$, which is not anything in particular. If you want to set two upper indices (or two lower indices) equal and sum over them, the correct procedure is to multiply by the metric (or the inverse metric). For example, $W^\lambda_{\ \rho\mu\nu} g^{\mu\nu}$ is a legitimate tensor. We can regard the contraction with the metric (or the inverse metric) as a two-step process: we use the metric to lower (or the inverse metric to raise) one of the two indices, and then contract an upper index with a lower index. Thus, in our example, first define $W^{\lambda\omega}_{\ \ \rho\mu} \equiv W^\lambda_{\ \rho\mu\nu} g^{\omega\nu}$ and then evaluate $W^{\lambda\mu}_{\ \ \rho\mu}$.

I keep belaboring the obvious, but as I said earlier in chapter I.5, I want to make sure that the rest of the book will not pose any difficulty for you. We can of course also contract an upper index from one tensor (for example $A^{\mu\nu}_{\ \ \sigma}$) with a lower index from another tensor (for example $B^\tau_{\ \lambda\omega\rho\eta}$). This is a trivial statement, as we can always define the tensor $T^{\mu\nu\tau}_{\ \ \ \sigma\lambda\omega\rho\eta} \equiv A^{\mu\nu}_{\ \ \sigma} B^\tau_{\ \lambda\omega\rho\eta}$ and then contract any upper index with any lower index on T (for example, $T^{\mu\nu\tau}_{\ \ \ \sigma\lambda\omega\nu\eta}$ producing the tensor $T^{\mu\tau}_{\ \ \sigma\lambda\omega\eta}$).

It might be worth emphasizing that I always write the coordinates x^μ with an upper index. Formally, if we run into the combination $g_{\mu\nu} dx^\nu$, we could call it dx_μ if we insist, but the symbol x_μ by itself is meaningless, or at least not useful.

Quotient theorem

I close by mentioning an obvious truth that is sometimes elevated to the status of a theorem. As I just said, if you multiply two tensors together and contract some upper indices with lower indices, the result is evidently a tensor. For example, if Q and W are tensors, then $P^\sigma_{\ \eta} = Q^{\omega\kappa} W^\sigma_{\ \omega\eta\kappa}$ is a tensor. The proof is straightforward: when we go from unprimed to primed coordinates, for each pair of contracted indices, S and S^{-1} knock each other off, and the various Ss and S^{-1}s left dangling are precisely what are needed to make the left hand side a tensor. The quotient theorem states that given two tensors P and W and $P^{\sigma}_{\ \eta} = Q^{\omega\kappa}W^{\sigma}_{\ \omega\eta\kappa}$, we can conclude that Q is a tensor. I will leave it to you to prove this more than plausible assertion by simply writing down the transformed versions of this equation and "peeling off" Ss and S^{-1}s. Think of tensors as delicate contraptions "turned" by various factors of Ss and S^{-1}s upon a coordinate transformation. Unless Q also gets turned by the appropriate factors of Ss and S^{-1}s, there is no way that P will transform correctly. I will often use this theorem implicitly.

Lorentz transformation, change of coordinates, curved space, and curved spacetime: Not quite the same deal

I keep emphasizing that one unified formalism can be used to discuss rotation, Lorentz transformation, change of coordinates, curved space, and curved spacetime. But it is also important to understand the crucial differences among them. Let us summarize and contrast.

A Lorentz transformation Λ transforms the Minkowski metric into itself (as stated in (III.3.8):

$$\eta_{\rho\sigma} = (\Lambda^T)_{\ \rho}^{\ \mu}\eta_{\mu\nu}\Lambda^{\nu}_{\ \sigma} \tag{13}$$

This requirement of invariance imposes a restriction on Λ and defines the Lorentz group. Rotations may be treated as a special case of this. We simply mentally replace η by the unit matrix and Λ by R. The restriction on R defines the rotation group.

When we transform coordinates, or study curved space, we have

$$g'_{\rho\sigma}(x') = \left((S^{-1})^T\right)_{\ \rho}^{\ \mu}g_{\mu\nu}(x)(S^{-1})^{\nu}_{\ \sigma} \tag{14}$$

While (14) looks deceptively similar to (13) with $S^{-1}(x)$ playing the role of Λ, it conveys quite a different message. This equation is not an invariance requirement like (13), but rather informs us about how the metric $g_{\mu\nu}$ transforms under a coordinate transformation described by $S(x)$: it tells us what $g'_{\rho\sigma}(x')$ is. However, there is also a subtext about invariance, namely a statement of how two metrics g' and g should be related in order to describe the same geometric entity. A key difference is that in the left hand side of (13), primed quantities are nowhere to be found. In contrast to (14), (13) informs us that a very special metric, written down by a certain Mr. Minkowski, and before him a certain Mr. Pythagoras, is left unchanged by a group of linear transformations, Lorentz transformations in one case, and rotations in the other.

The equivalence principle is not a statement about symmetry

Next, as we proceed from curved space to curved spacetime, we need Mr. Einstein's additional insight of linking the metric to the gravitational field. Formally, we have the same equation (14), but now it informs us that the physics of one observer under the influence of a gravitational field is related to the physics of another observer under the influence of some other gravitational field. In particular, according to the equivalence principle, one observer could even feel no gravitational field; that is, $g_{\mu\nu}$ may happen to be equal to $\eta_{\mu\nu}$. In this sense, the equivalence principle is reminiscent of the statement that we can always transform to locally flat coordinates.

In short, a coordinate transformation in general relativity relates the physics seen by two different observers, even if one of them reports seeing a gravitational field and the other not. If you wish, you could say that there is no such thing as a gravitational field, only curved spacetime, or vice versa, as discussed in chapter V.2.

Differentiating scalars, vectors, and tensors

Let $\phi(x)$ be a scalar under general coordinate transformation, so that $\phi'(x') = \phi(x)$. How does $\partial_\mu \phi(x)$ transform? Again, by now, you would have guessed that it transforms like a vector with a lower index, and indeed

$$\partial'_\mu \phi'(x') \equiv \frac{\partial \phi'(x')}{\partial x'^\mu} = \frac{\partial x^\nu}{\partial x'^\mu} \frac{\partial \phi(x)}{\partial x^\nu} = \partial_\nu \phi(x)(S^{-1})^\nu{}_\mu \tag{15}$$

Speaking loosely, we might say that in the definition $\partial_\mu \equiv \frac{\partial}{\partial x^\mu}$, since the upper index μ appears in the denominator, it acts effectively as a lower index. Another way of thinking about this is to regard ∂_μ, the ur-vector with a lower index, as the "dual" of dx^μ, the ur-vector with an upper index.* It is useful to remember that

$$\partial'_\mu = (S^{-1})^\nu{}_\mu \partial_\nu \tag{16}$$

We write $(S^{-1})^\nu{}_\mu$ to the left of ∂_ν to make clear that the derivative does not act on S^{-1}.

An important question: given a vector $W^\mu(x)$, do you expect $\partial_\lambda W^\mu(x)$ to transform as a tensor? Think about this for a moment before reading on.

Some key things to know about tensors

Let us collect together some key things you have learned about tensors in general relativity:

1. The indices on a tensor transform independently, that is, as if the other indices were not there.

2. Tensors in general relativity work pretty much the same way as the tensors you are familiar with in connection with the rotation group except for two important differences:

    a. The transformation matrix $S^\mu{}_\nu$ changes from place to place.
    b. There are two floors, and you have to use $g_{\mu\nu}$ and $g^{\mu\nu}$ to move indices upstairs and downstairs.

---

*Some readers may recognize that I am sneaking in some notions of a more modern approach to vector and tensor analysis by speaking of dx^μ and ∂_μ as dual ur-vectors.

3. Always contract an upper index with a lower index. It doesn't make sense to contract an upper index with an upper index. You can multiply two upper indices by $g_{\mu\nu}$, but that amounts to moving one of them downstairs first, and then contracting.

4. The coordinate x^μ is not a vector, but dx^μ is.

5. The two ur-vectors dx^μ and ∂_μ transform oppositely.

6. The ordinary derivative of a tensor is not a tensor. You must use the covariant derivative.

Statement 6 answers the question I asked you in the previous section. The next chapter is devoted to explaining and elaborating this statement, in particular, to finding out what the covariant derivative is.

Appendix: Index-free representation of vector fields

Statement 5 suggests that dx^μ and ∂_μ could be used as basis vectors. We could exploit this observation to develop an index-free formalism along the following line. Given a vector field $A^\mu(x)$, we can define an index-free object $A(x) \equiv A^\mu(x)\partial_\mu$. An abecedarian might take some time to get used to this notion of regarding vector fields as differential operators, but it leads us naturally to consider the commutator $C = [A, B]$. More explicitly, $C^\nu\partial_\nu = (A^\mu\partial_\mu)(B^\nu\partial_\nu) - (A \leftrightarrow B) = A^\mu(\partial_\mu B^\nu)\partial_\nu - (A \leftrightarrow B)$, since $\partial_\mu\partial_\nu = \partial_\nu\partial_\mu$. Thus, $C^\nu = A^\mu(\partial_\mu B^\nu) - B^\mu(\partial_\mu A^\nu)$. In other words, we differentiate the vector field B in the direction of the vector field A, interchange A and B, and then compare the two results.

If you are reminded of the commutators in the Lie algebra introduced way back in appendix 2 to chapter I.3, you might have also suspected that a deep connection exists. Indeed, the notion of representing generators by differential operators instead of matrices also appeared there. For example, we had $-iJ_z = (y\frac{\partial}{\partial x} - x\frac{\partial}{\partial y})$ for the generator of rotations about the z-axis. In the notation used in the preceding paragraph, what we did in chapter I.3 amounts to writing $-iJ_z = A^\mu\partial_\mu$ with the vector field A^μ defined by $(y, -x, 0)$. This provides another way of calculating the Lie algebra for the rotation group, for example, $[J_x, J_y] = iJ_z$. You can readily verify this relation, regarding J_x, J_y, and J_z as differential operators and using elementary calculus.

Similarly, given a vector field $A_\mu(x)$ with a lower index, we can define an index-free object $A(x) \equiv A_\mu(x)dx^\mu$. We will come back to this in chapter IX.3.

V.6 | Covariant Differentiation

"How do you transform?"

In our continued effort to honor the fundamental principle that physics does not depend on the physicist, we have to ask, sort of as in daily life, every new object or expression we encounter, "How do you transform?"[1] Physical laws are to be formulated in terms of objects that transform properly.

In the preceding chapter, we verified that, given a scalar $\phi(x)$, its derivative or "gradient" $\partial_\mu \phi(x)$ transforms like a vector with a lower index. This also justifies the shorthand notation $\partial_\mu \equiv \frac{\partial}{\partial x^\mu}$. Taking the partial derivative of an object, we add a lower index to the object. Quite naturally, we would like to go from scalar to vector and beyond.

At the end of the preceding chapter, I asked you a crucial question: given a vector $W^\mu(x)$, how does $\partial_\lambda W^\mu(x)$ transform? Naively (actually, very naively), you might guess, just by looking at the indices the object $\partial_\lambda W^\mu$ carries, that it transforms like a tensor T^μ_λ with one upper and one lower index. But you can see that can't be true just by looking at the transformation law for $W^\mu(x)$

$$W'^\mu(x') = S^\mu_{\ \nu}(x) W^\nu(x) \tag{1}$$

The object $\partial_\lambda W^\mu(x)$ transforms to $\partial'_\lambda W'^\mu(x')$. We have $\partial'_\lambda = (S^{-1})^\rho_{\ \lambda} \partial_\rho$ by the chain rule. So act with ∂_ρ on (1) using the product rule. We watch, in horror, ∂_ρ hitting $S^\mu_{\ \nu}$, thus wrecking the nice tensor transformation law. Instead, we obtain

$$\partial'_\lambda W'^\mu(x') = \frac{\partial W'^\mu(x')}{\partial x'^\lambda} = \frac{\partial x^\rho}{\partial x'^\lambda} \frac{\partial}{\partial x^\rho} (S^\mu_{\ \nu}(x) W^\nu(x))$$
$$= (S^{-1})^\rho_{\ \lambda} S^\mu_{\ \nu} \partial_\rho W^\nu + ((S^{-1})^\rho_{\ \lambda} \partial_\rho S^\mu_{\ \nu}) W^\nu \tag{2}$$

The fact that the transformation S varies from place to place has negated the naive guess. If the second term, which comes from differentiating S, were absent, $\partial_\lambda W^\mu$ would transform like a tensor!* The naive guess would be valid.

Wanted: A derivative that transforms properly

What is happening is quite clear: as the vector W varies from a given point to a neighboring point, the coordinate axes that define the components of W also change. This suggests that we can define a more suitable derivative, known as the covariant derivative and written as $D_\lambda W^\mu$, to take this effect into account, so that $D_\lambda W^\mu$ would transform like a tensor with one upper and one lower index.

It is also instructive here to compare (2) with something from way back, (I.4.3). There the rotation matrix R, the analog of S here, sails right past the derivative, so that, under rotations, the derivative of a vector field is a tensor. The other difference is that $R^{-1} = R^T$, which is not true for S.

We already encountered a similar problem in chapter I.5. The simple minded divergence $\partial_\mu W^\mu(x)$ does not transform properly: $\partial_\mu W^\mu(x) \neq \partial'_\mu W'^\mu(x')$. It has to be corrected by an additive term to

$$D_\mu W^\mu = \partial_\mu W^\mu + \left(\frac{1}{\sqrt{-g}} \partial_\mu \sqrt{-g} \right) W^\mu \tag{3}$$

This offers a strong hint about what to do. Construct $D_\lambda W^\mu$ by adding something to $\partial_\lambda W^\mu$ to cancel the second term in (2). We want the covariant derivative $D_\lambda W^\mu$ to have many of the properties enjoyed by the ordinary derivative $\partial_\lambda W^\mu$, for example, linearity in W (so that multiplying W by 2, say, doubles $D_\lambda W^\mu$), which requires that the added term must be linear in W just as in (3). The most general expression with the correct index structure is then $D_\lambda W^\mu \equiv \partial_\lambda W^\mu + \tilde{\Gamma}^\mu_{\lambda\nu} W^\nu$. We need an object $\tilde{\Gamma}$ with one upper index and two lower indices, and we specifically want it not to be a tensor.

But we are already acquainted with such an object, the Christoffel symbol Γ from way back in chapters I.7, II.1, II.2, and V.3. Our notation $\tilde{\Gamma}$ is intentionally suggestive! To see that Γ might work, we recall its definition:

$$\Gamma^\mu_{\lambda\nu} \equiv \tfrac{1}{2} g^{\mu\sigma} (\partial_\lambda g_{\nu\sigma} + \partial_\nu g_{\sigma\lambda} - \partial_\sigma g_{\lambda\nu}) \tag{4}$$

It involves ordinary partial derivatives of the metric tensor $g_{\mu\nu}$ and thus, for the same reason as in (2), Γ can't possibly transform like a tensor. We can thus hope that the nontensorial piece in the transformation law of Γ will cancel precisely the unwanted second term in (2). This "smells right": the derivatives of $g_{\mu\nu}$ in (4) describe the very effect we are worried about, namely the variation from point to point of the coordinate axes relative to which the components of our vector W^μ are defined.

---

* One of my professors, the distinguished theoretical physicist Murph Goldberger, was fond of shouting, "If my aunt had balls, she would be my uncle!" Believe me, this made a deep impression on a Chinese kid fresh from Brazil.

Overheard at a party: "What, you're also not a tensor? Me neither!" "Maybe we could hook up and become a tensor?" There is a Chinese saying that those who are similarly afflicted empathize with one another. And so $\partial_\lambda W^\mu$ and $\Gamma^\mu_{\lambda\nu}W^\nu$ could join hands to form one of those ideal couples in which one person's character defects cancel the other person's.

Canceling the nontensorial pieces

So let's define the covariant derivative

$$D_\lambda W^\mu \equiv \partial_\lambda W^\mu + \Gamma^\mu_{\lambda\nu}W^\nu \tag{5}$$

and show that it indeed transforms like a tensor. (At this point, we actually do not know that the two terms in (5) should be added with relative coefficient 1. But instead of cluttering things up with an arbitrary constant in front of the second term, we will show that the expression as written works.)

In fact, we already know how the Christoffel symbol transforms! Go back to what Professor Flat taught us in chapter II.2. For convenience, I copy (II.2.31) here (after relabeling some indices):

$$\Gamma'^\mu_{\lambda\kappa} = S^\mu_\eta (S^{-1})^\omega_\lambda (S^{-1})^\sigma_\kappa \Gamma^\eta_{\omega\sigma} + S^\mu_\eta (S^{-1})^\rho_\lambda \partial_\rho (S^{-1})^\eta_\kappa \tag{6}$$

We now plug this into (5) and show that $D_\lambda W^\mu$ transforms nicely like a tensor.

Deep down in our hearts we already know* that it must work. Still, it is fun to see how the different pieces come together and knock each other out.

From (6), we see that we need to determine ∂S^{-1}. For any (invertible) matrix M, differentiate $MM^{-1} = I$ to obtain $(\partial M)M^{-1} + M\partial M^{-1} = 0$, which allows us to relate the derivative of M^{-1} to the derivative of M:

$$\partial M^{-1} = -M^{-1}(\partial M)M^{-1} \tag{7}$$

Notice that this generalizes what you learned in a calculus course on how to differentiate the inverse of a function $d(\frac{1}{f}) = -\frac{df}{f^2}$.

Using the identity (7), we write the second term in (6) as

$$S^\mu_\eta (S^{-1})^\rho_\lambda \partial_\rho (S^{-1})^\eta_\kappa = -(S^{-1})^\rho_\lambda (\partial_\rho S^\mu_\sigma)(S^{-1})^\sigma_\kappa$$

Plugging this into (6) and multiplying $\Gamma'^\mu_{\lambda\kappa}$ by $W'^\kappa = S^\kappa_\tau W^\tau$, we finally obtain (after quickly renaming indices)

$$\Gamma'^\mu_{\lambda\kappa}W'^\kappa = S^\mu_\nu (S^{-1})^\rho_\lambda \Gamma^\nu_{\rho\tau}W^\tau - ((S^{-1})^\rho_\lambda \partial_\rho S^\mu_\tau)W^\tau \tag{8}$$

Again, for convenience, I copy the nasty (2) we started this chapter with here:

$$\partial'_\lambda W'^\mu(x') = (S^{-1})^\rho_\lambda S^\mu_\nu \partial_\rho W^\nu + ((S^{-1})^\rho_\lambda \partial_\rho S^\mu_\nu)W^\nu \tag{9}$$

---

* From a review of my book *QFT Nut* for the American Mathematical Society: "It is often deeper to know why something is true rather than to have a proof that it is true."

Adding (8) and (9), we see that the character defect in each of them cancel each other, so that indeed $D'_\lambda W'^\mu(x') = (S^{-1})^\rho_{\ \lambda} S^\mu_{\ \nu} D_\rho W^\nu$ transforms like a tensor with one upper and one lower index.

The covariant derivative from geometry

Another motif from an earlier part of this book resonates with the present discussion. Recall that, in chapter I.7 on classical differential geometry, the mite professors came to the concept of covariant derivative by simply dropping from the vector $\partial_\nu \vec{W}(x)$ the component sticking out of the surface. Also recall that in exercise I.7.6, you showed that $\vec{e}_\rho \cdot \vec{e}_{\mu,\nu} = \Gamma_{\rho \cdot \mu\nu}$. From the definition of the basis vectors $\vec{e}_\mu = \partial_\mu \vec{X} = \frac{\partial \vec{X}}{\partial x^\mu}$, we see that under the coordinate transformation $x \to x'$, we have $\vec{e}\,'_\mu = \partial'_\mu \vec{X} = (S^{-1})^\nu_{\ \mu} \vec{e}_\nu$ and so $\Gamma'_{\rho \cdot \mu\nu} = \vec{e}\,'_\rho \cdot \vec{e}\,'_{\mu,\nu} = (S^{-1})^\sigma_{\ \rho} (S^{-1})^\eta_{\ \nu} \vec{e}_\sigma \cdot \partial_\eta ((S^{-1})^\omega_{\ \mu} \vec{e}_\omega)$. We can see the transformation law of Christoffel symbols emerging.

Although the discussion of classical differential geometry in chapter I.7 is nominally only for surfaces, it clearly generalizes to curved spaces and spacetimes. Some readers might prefer this derivation, which is more geometrical and intuitive.

In contrast, the derivation given in the preceding section, based on requiring that $D_\lambda W^\mu$ transform properly, is more abstract and high powered. As I mentioned, this requirement of proper transformation pervades theoretical high energy physics in recent decades. In this sense, this derivation might be considered more modern and general.

A wildly varying vector field?

At this point, our friend the rich man could start spouting fancy talk about the covariant derivative, presumably without writing down a single index and disdaining such "quaint old-fashioned notions" as transformation, and thus cause our other friend the Jargon Guy to become flush with joy. Instead, let's be more modest and, together with our friend the poor man, try to understand what the covariant derivative really means by working out a simple example. Again, a tale best told through a fable.

A civilization of mites used the coordinates r, θ with the metric $g_{rr} = 1$, $g_{r\theta} = 0$, $g_{\theta\theta} = r^2$. One day they discovered a wild and woolly vector field $W^\mu(x)$, which they eventually determined to be given by $W^r(r, \theta) = \cos\theta$, $W^\theta(r, \theta) = -\frac{1}{r}\sin\theta$ (measured with error bars of course but well described phenomenologically by these expressions). For example, a scientific expedition sent to the point $(r, \theta) = (3, 30°)$ measured $W^r(3, 30°) = \frac{\sqrt{3}}{2}$, $W^\theta(3, 30°) = -\frac{1}{6}$. Another expedition sent elsewhere reported vastly different values for W^μ, and so on. Eventually, a table of the 4 quantities $\partial_\lambda W^\mu$ was published to guide travelers.

One day, a bright young guy pointed out that the mite savants should have calculated the covariant derivatives $D_\lambda W^\mu = \partial_\lambda W^\mu + \Gamma^\mu_{\lambda\nu} W^\nu$, instead of the ordinary derivatives $\partial_\lambda W^\mu$. The symbol $\Gamma^\mu_{\lambda\nu}$ had already been determined (way back in chapter II.2 by studying geodesics) to be $\Gamma^r_{\theta\theta} = -r$ and $\Gamma^\theta_{r\theta} = \frac{1}{r}$. For example, $D_r W^\theta = \partial_r W^\theta + \Gamma^\theta_{r\theta} W^\theta = \frac{1}{r^2} \sin\theta + \frac{1}{r}(-\frac{1}{r}\sin\theta) = 0$. You should now go on and verify that all 4 quantities $D_\lambda W^\mu$ vanish.

Quickly, another young guy showed the elderly savants that if they were to transform coordinates to $x = r\cos\theta$, $y = r\sin\theta$, then in these newfangled coordinates, $W^x = \frac{\partial x}{\partial r}W^r + \frac{\partial x}{\partial \theta}W^\theta = \cos\theta\, W^r - r\sin\theta\, W^\theta = 1$ and (as you should show) $W^y = 0$. After the fact, the savants plotted $W^r(r,\ \theta) = \cos\theta$, $W^\theta(r,\ \theta) = -\frac{1}{r}\sin\theta$ in polar coordinates and saw that, indeed, the vector field W^μ was constant. That $\partial_\lambda W^\mu$ did not vanish was merely due to the coordinate basis vectors $\vec{e}_r$ and $\vec{e}_\theta$ varying.

The young guys explained: "Saying that a vector field is constant must mean that the covariant derivatives, instead of the ordinary derivatives, vanish." Since $D_\lambda W^\mu$ is a tensor, if it vanishes in one coordinate system, it vanishes in all coordinate systems. The same statement cannot be made of a nontensor like $\partial_\lambda W^\mu$.

Covariant derivative of tensors

We have determined the covariant derivative of a vector with an upper index. What about the covariant derivative of a vector with a lower index?

Here I will switch to the more compact notation $W^\mu_{,\lambda} \equiv \partial_\lambda W^\mu$ and $W^\mu_{;\lambda} \equiv D_\lambda W^\mu$ (the first of which you already encountered in chapter I.7), also commonly used.

As we have seen, the covariant derivative of a scalar is simply the ordinary derivative: suffices to fix the covariant derivative of a vector with a lower index. Insisting that the covariant derivative, just like the ordinary derivative, satisfies the product rule, we have $(U_\mu W^\mu)_{;\lambda} = U_{\mu;\lambda} W^\mu + U_\mu W^\mu_{;\lambda}$. In contrast, since $U_\mu W^\mu$ is a scalar, we have

$$(U_\mu W^\mu)_{;\lambda} = (U_\mu W^\mu)_{,\lambda} \tag{10}$$

For convenience, let us rewrite (5) in this semicolon notation as

$$W^\mu_{;\lambda} = W^\mu_{,\lambda} + \Gamma^\mu_{\lambda\nu} W^\nu \tag{11}$$

We see that the condition (10) is satisfied if

$$U_{\mu;\lambda} = U_{\mu,\lambda} - \Gamma^\sigma_{\mu\lambda} U_\sigma \tag{12}$$

Contrast the minus sign in (12) with the plus sign in (11). We can readily check that the opposite signs ensure that the Christoffel symbols cancel out, thus giving us (10):

$$(U_\mu W^\mu)_{;\lambda} = (U_{\mu,\lambda} - \Gamma^\sigma_{\mu\lambda} U_\sigma)W^\mu + U_\mu(W^\mu_{,\lambda} + \Gamma^\mu_{\lambda\nu}W^\nu) = U_{\mu,\lambda}W^\mu + U_\mu W^\mu_{,\lambda} = (U_\mu W^\mu)_{,\lambda}$$

The covariant derivative of tensors with more indices, such as $T^{\mu\nu}_\rho$, can be worked out by pretending that $T^{\mu\omega}_\rho = W^\mu Y^\omega U_\rho$ is made up of three vectors, so that we can use (11) and (12) repeatedly. The pretense works because all we care about in this context is the transformation property of T rather than its true nature. Thus,

$$T^{\mu\omega}_{\rho;\lambda} = T^{\mu\omega}_{\rho,\lambda} + \Gamma^\mu_{\lambda\nu}T^{\nu\omega}_\rho + \Gamma^\omega_{\lambda\nu}T^{\mu\nu}_\rho - \Gamma^\sigma_{\rho\lambda}T^{\mu\omega}_\sigma \tag{13}$$

Once you know how to differentiate vectors, you know how to differentiate tensors.

The general rule should be obvious: colloquially, for each upper index, attach a $+\Gamma$, and for each lower index, attach a $-\Gamma$. We now know how to repeatedly take the covariant derivative. For example, since $U_{\mu;\lambda}$ is a tensor, we can apply the rule and write

$$U_{\mu;\lambda;\rho} = U_{\mu;\lambda,\rho} - \Gamma^{\sigma}_{\mu\rho} U_{\sigma;\lambda} - \Gamma^{\sigma}_{\lambda\rho} U_{\mu;\sigma} \tag{14}$$

Starting from (11), we defined the covariant derivative in such a way that the product rule is clearly satisfied. For example, $(S^{\nu}_{\tau} T^{\mu\omega}_{\rho})_{;\lambda} = S^{\nu}_{\tau;\lambda} T^{\mu\omega}_{\rho} + S^{\nu}_{\tau} T^{\mu\omega}_{\rho;\lambda}$.

Mirror mirror on the wall, what is the most special tensor of them all? I trust that the magic mirror would say the metric tensor $g_{\mu\nu}$. Would you be surprised to learn that the covariant derivative of the metric tensor vanishes? You should check that indeed

$$g_{\mu\nu;\lambda} = 0 \tag{15}$$

Of course, the ordinary derivative $g_{\mu\nu,\lambda}$ is assuredly not zero. In other words, the metric tensor is not (necessarily) constant, but it is always covariantly constant. That sure makes sense.*

In the preceding chapter, I defined the commutator $C = [A, B]$ of two vector fields A^{μ} and B^{ν}. Since the Christoffel symbol is symmetric in its two lower indices, the ordinary derivative in the definition can in fact be replaced by covariant derivatives:

$$C^{\nu} = [A, B]^{\nu} = A^{\mu}(\partial_{\mu} B^{\nu}) - B^{\mu}(\partial_{\mu} A^{\nu}) = A^{\mu}(D_{\mu} B^{\nu}) - B^{\mu}(D_{\mu} A^{\nu}) \tag{16}$$

To get a feel for the covariant derivative, you should practice writing down a few more examples.

Electromagnetism in curved spacetime

Notice that the covariant curl is equal to the ordinary curl

$$U_{\mu;\lambda} - U_{\lambda;\mu} = U_{\mu,\lambda} - U_{\lambda,\mu} \tag{17}$$

The Christoffel terms cancel. In particular, in curved spacetime, the electromagnetic field strength is still[†] given by $F_{\mu\nu} = \partial_{\mu} A_{\nu} - \partial_{\nu} A_{\mu}$, a fact we can now exploit.

In chapter V.2, I extolled the power of the equivalence principle. As another example, the equivalence principle tells us that we can immediately obtain the action of an electromagnetic field in the presence of a gravitational field by promoting the Maxwell action $-\frac{1}{4} \int d^4 x \, F^{\mu\nu} F_{\mu\nu} = -\frac{1}{4} \int d^4 x \, \eta^{\lambda\mu} \eta^{\rho\nu} F_{\lambda\rho} F_{\mu\nu}$ in (IV.2.6) to

$$S_{\text{Maxwell}} = -\frac{1}{4} \int d^4 x \, \sqrt{-g} g^{\lambda\mu} g^{\rho\nu} F_{\lambda\rho} F_{\mu\nu} \tag{18}$$

---

* Some people prefer the following slightly more mathematical approach to the covariant derivative. After defining $D_{\lambda} W^{\mu} \equiv \partial_{\lambda} W^{\mu} + \tilde{\Gamma}^{\mu}_{\lambda\nu} W^{\nu}$ with an unknown object $\tilde{\Gamma}$, as was done earlier in this chapter, extend the definition to the covariant derivative of a tensor. Then impose the condition $D_{\lambda} g_{\mu\nu} = 0$ to determine $\tilde{\Gamma}$.

† Indeed, in chapter V.4, you might have wondered how $F_{\mu\nu}$ is defined in curved spacetime.

In light of the preceding remark, in (18) the effect of the gravitational field on electromagnetism has been explicitly displayed.

A matrix identity, the patented "1-2" test, and the covariant divergence

We have one loose end to tie up. As noted in (3), we have already encountered the covariant divergence $D_\mu W^\mu = \frac{1}{\sqrt{-g}}\partial_\mu(\sqrt{-g}W^\mu)$ way back in chapter I.5. However, we can also obtain the covariant divergence by contracting the indices in (5): $D_\mu W^\mu = \partial_\mu W^\mu + \Gamma^\mu_{\mu\nu} W^\nu$. From (4), we have $\Gamma^\mu_{\mu\nu} = \frac{1}{2}g^{\mu\sigma}\partial_\nu g_{\mu\sigma}$.

For these two forms to agree with each other, so that the laws of arithmetic are upheld, we must have $g^{\mu\sigma}\partial_\nu g_{\mu\sigma} = \frac{1}{g}\partial_\nu g$. There must be a cool matrix identity[2] involving the determinant, and indeed there is!

For any diagonalizable matrix M

$$\log \det M = \operatorname{tr} \log M \tag{19}$$

(The logarithm of a matrix can be formally defined by a power series $\log M = \log(I - (I - M)) = \sum_{k=1}^\infty (I - M)^k/k$.) To prove (19), simply diagonalize $M = A^{-1}DA$ with D a diagonal matrix with entries $d_1, d_2, \cdots, d_{\text{dimension of } M}$. Then $\log \det M = \log \det D = \log \prod_j d_j = \sum_j \log d_j = \operatorname{tr} \log D = \operatorname{tr} A^{-1}(\log D)A = \operatorname{tr} \log M$, and we have proved (19).

Differentiate (19) to get* $(\det M)^{-1}\partial \det M = \partial(\operatorname{tr} \log M) = \operatorname{tr}(\partial \log M) = \operatorname{tr}(M^{-1}\partial M)$. In particular, substituting the metric $g_{\mu\sigma}$ for $M_{\mu\sigma}$, we obtain the desired equality

$$\frac{1}{\sqrt{-g}}\partial_\nu\sqrt{-g} = \frac{1}{2g}\partial_\nu g = \frac{1}{2}\partial_\nu \log g = \frac{1}{2}g^{\sigma\mu}\partial_\nu g_{\mu\sigma} = \Gamma^\mu_{\mu\nu} \tag{20}$$

Using (20), we readily verify that the usual identities involving the partial derivative also work for the covariant derivative, provided that the correct integration measure $d^4x\sqrt{-g}$ is used instead of d^4x. For example, from (20), it follows that

$$\int d^4x\sqrt{-g}D_\mu W^\mu = \int d^4x\sqrt{-g}(\partial_\mu W^\mu + \Gamma^\mu_{\mu\nu} W^\nu)$$

$$= \int d^4x\sqrt{-g}\left(-\left(\frac{1}{\sqrt{-g}}\partial_\mu\sqrt{-g}\right)W^\mu + \Gamma^\mu_{\mu\nu} W^\nu\right) = 0$$

(Of course, this also follows directly from the covariant divergence

$$D_\mu W^\mu = \frac{1}{\sqrt{-g}}\partial_\mu(\sqrt{-g}W^\mu)$$

which we started this discussion with.)

Integration by parts also works in the same way. For example,

$$\int d^4x\sqrt{-g}K^{\lambda\mu}D_\lambda W_\mu = -\int d^4x\sqrt{-g}(D_\lambda K^{\lambda\mu})W_\mu \tag{21}$$

This follows from what we just learned ($\int d^4x\sqrt{-g}D_\lambda(K^{\lambda\mu}W_\mu) = 0$) and the product rule.

---

* By the way, to verify matrix identities of this type, you can always apply my patented "1-2" test: check it for 1-by-1 and 2-by-2 matrices.

Appendix 1: Differentiation along a curve

Given a curve $\mathcal{C}$ parametrized by ζ and described by $X(\zeta)$, we might be interested in how various quantities defined on the curve change as we move along the curve. An everyday example might be the temperature along a highway we are driving on. More relevant to physics, an example might be the direction in which a macroscopic object (like a gyroscope) or a microscopic object (like an elementary particle) is spinning, with $\mathcal{C}$ its worldline. In fact, let us focus on a vector $W^\mu(\zeta)$, with the word "vector" implying that under a coordinate transformation,

$$W'^\mu(\zeta) = S^\mu_{\ \nu}(X(\zeta))W^\nu(\zeta) \tag{22}$$

It is important to realize that we are not talking about a vector field $W^\mu(x)$, as in the text. Our vector $W^\mu(\zeta)$ is meaningful only on $\mathcal{C}$. For example, the spin of an electron is defined only on its worldline.

Note also that I did not say that the curve $\mathcal{C}$ is necessarily a geodesic. Our gyroscope could be inside some rocketship in full throttle blasting by some black hole, for example, and not in free fall.

The question is how to differentiate $W^\mu(\zeta)$ along $\mathcal{C}$. By now you have caught on that the naive proposal $\frac{dW^\mu(\zeta)}{d\zeta}$ is not going to cut it: it does not transform like a vector. From (22) we have

$$\frac{dW'^\mu(\zeta)}{d\zeta} = S^\mu_{\ \nu}(X(\zeta))\frac{dW^\nu(\zeta)}{d\zeta} + (\partial_\lambda S^\mu_{\ \nu}(X(\zeta)))V^\lambda(\zeta)W^\nu(\zeta) \tag{23}$$

where $V^\lambda(\zeta) \equiv \frac{dX^\lambda(\zeta)}{d\zeta}$ is the tangent or velocity vector to the curve $\mathcal{C}$ at the point $X(\zeta)$. If only the first term on the right hand side of (23) were present, then $\frac{dW^\nu(\zeta)}{d\zeta}$ would transform like a vector.

But by now you also know how to fix this problem. Define the covariant derivative $\frac{DW^\mu(\zeta)}{D\zeta}$ along the curve by

$$\frac{DW^\mu(\zeta)}{D\zeta} = \frac{dW^\mu(\zeta)}{d\zeta} + \Gamma^\mu_{\lambda\nu}(X(\zeta))V^\lambda(\zeta)W^\nu(\zeta) \tag{24}$$

Using (6), you can check that $\frac{DW^\mu(\zeta)}{D\zeta}$ indeed transforms like a vector: $\frac{DW'^\mu(\zeta)}{D\zeta} = S^\mu_{\ \nu}(X(\zeta))\frac{DW^\nu(\zeta)}{D\zeta}$.

In part IX, we will come back to this covariant derivative along a curve.

Appendix 2: Lie derivative

Given a vector field $V^\mu(x)$, we physicists can readily picture it as the local velocity field of a fluid, albeit in spacetime rather than in space. Speaking loosely, we can mentally "fill in" the flow by connecting the "feathered ends" of $V^\mu(x)$ and construct the trajectories of an infinitesimal fluid element. See figure 1. More formally, integrate the first order equation $\frac{dX^\mu}{d\tau} = V^\mu(X(\tau))$ for $X^\mu(\tau)$.

Now suppose we are given a tensor field $W^{\cdots}_{\cdots}(x)$ in addition to $V^\mu(x)$. We could differentiate $W^{\cdots}_{\cdots}(x)$ by comparing its value at two nearby points P and Q. More precisely, let the coordinates of P and Q be x and $\tilde{x}$,

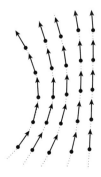

Figure 1 A vector field visualized as a fluid.

respectively; then $W^{...}_{...}(\tilde{x}) - W^{...}_{...}(x) \to (\tilde{x} - x)^\nu \partial_\nu W^{...}_{...}(x)$, as Newton and Leibniz taught us. No mystery there, we have known this as the ordinary derivative since childhood.

Sophus Lie now invites us to do something different. Let $\tilde{x}$ be the location that the fluid element at x flows to: $x \to \tilde{x}^\mu = x^\mu + d\tau V^\mu(x(\tau))$. For pedagogical clarity, let us specialize to the case of a vector field $W^\mu(x)$ instead of $W^{...}_{...}(x)$; you can always fill in the dots on $W^{...}_{...}(x)$ to your heart's content later. Lie says, "Going with the flow, we 'drag' $W^\mu(x)$ to the point $\tilde{x}$ by regarding $x \to \tilde{x}^\mu = x^\mu + d\tau V^\mu(x(\tau))$ as a coordinate transformation." In other words, we define $\tilde{W}^\mu(\tilde{x}) = W^\nu(x)\frac{\partial \tilde{x}^\mu}{\partial x^\nu} = W^\mu(x) + d\tau W^\nu(x)\partial_\nu V^\mu(x(\tau))$. We are now supposed to compare this with the vector field W^μ evaluated at Q, namely $W^\mu(\tilde{x})$. Got that?

Lie tells us to do something different from what Newton and Leibniz told us to do: instead of comparing $W^\mu(\tilde{x})$ with $W^\mu(x)$, compare $W^\mu(\tilde{x})$ and $\tilde{W}^\mu(\tilde{x})$. So, take the limit $W^\mu(\tilde{x}) - \tilde{W}^\mu(\tilde{x}) \simeq d\tau V^\nu(x(\tau))\partial_\nu W^\mu(x) - d\tau W^\nu(x)\partial_\nu V^\mu(x(\tau))$.

To underline what we are doing, speak colloquially for a moment. "Let's be cool and go with the flow, but hmm, somehow this vector we are carrying is not as relaxed as we are and is not pointing in the direction we expect; something must be acting on this vector." This difference, between actual (namely $W^\mu(\tilde{x})$) and expected (namely $\tilde{W}^\mu(\tilde{x})$), is what we want to measure.[3]

Given two vector fields $V^\mu(x)$ and $W^\mu(x)$, define the Lie derivative of $W^\mu(x)$ in the direction of $V^\mu(x)$ by

$$\mathcal{L}_V W^\mu(x) \equiv V^\nu(x)\partial_\nu W^\mu(x) - W^\nu(x)\partial_\nu V^\mu(x) = V^\nu(x)D_\nu W^\mu(x) - W^\nu(x)D_\nu V^\mu(x) \qquad (25)$$

In the last step, we replace ∂_ν by D_ν, which you can verify is allowed, since the Christoffel symbol is symmetric in its two lower indices. This should remind you of a similar step in the text when we discussed the commutator $C = [A, B]$ of two vector fields. Indeed, the connection between the two discussions should leap out at you: the Lie derivative $\mathcal{L}_V W^\mu$ is just the μ component of the commutator $[V, W]$, namely $\mathcal{L}_V W^\mu = [V, W]^\mu$. Those readers who know that Lie algebras are constructed out of commutators (as we saw in chapter I.3) would not be surprised that the same person was responsible for the Lie derivative and Lie algebra. Mathematically, the Lie derivative is regarded as being a more "primitive" concept than the covariant derivative, since, as shown in (25), it can be defined without referring to the Christoffel symbol.

The last step in (25) suggests defining the covariant derivative in the direction of a given vector V by $D_V \equiv V^\nu(x)D_\nu$. (The notation ∇_V is also often used.) In fact, we now see that, in (24), if W^μ is a vector field, then the derivative along a curve is just $D_V W^\mu$, with V the tangent vector of the curve.

Misconception alert! We can replace the ordinary derivative by the covariant derivative in $[V, W]^\mu$ (as shown in (25)), but not in $[V, W]$. In other words, $[V, W] \neq [D_V, D_W]$.

Going through the same steps for a vector field with a lower index $U_\mu(x)$, we obtain

$$\mathcal{L}_V U_\mu(x) \equiv V^\nu(x)\partial_\nu U_\mu(x) + U_\nu(x)\partial_\mu V^\nu(x) = V^\nu(x)D_\nu U_\mu(x) + U_\nu(x)D_\mu V^\nu(x) \qquad (26)$$

Various properties of the Lie derivative follow. For example, it satisfies the product rule: $\mathcal{L}_V(U_\mu Y_\lambda) = (\mathcal{L}_V U_\mu)Y_\lambda + U_\mu(\mathcal{L}_V Y_\lambda)$. This allows us to define the Lie derivative of tensors immediately: as in the discussion surrounding (13), simply pretend that the tensor is a product of vectors with upper and lower indices. For example,

$$\mathcal{L}_V W_{\mu\lambda} = V^\nu \partial_\nu W_{\mu\lambda} + W_{\nu\lambda}\partial_\mu V^\nu + W_{\mu\nu}\partial_\lambda V^\nu \qquad (27)$$

A trivial example is that the Lie derivative of a scalar is just $\mathcal{L}_V \phi = V^\nu \partial_\nu \phi$.

A tensor $W^{...}_{...}$ is said to be Lie transported along a curve if its Lie derivative along the curve vanishes, namely $\mathcal{L}_V W^{...}_{...} = 0$, with V the tangent vector to the curve. To understand physically what the mathematician is talking about, think of the curve as your geodesic as you move through spacetime. Set up coordinates so that x^0 is just your proper time and $x^1, \cdots, x^{d-1}$ are constant along your geodesic. (Indeed, set them all to 0; you are at the center of your universe.) Then $V^\mu = \frac{dX^\mu(\tau)}{d\tau} = (1, 0, \cdots, 0)$. Look at (27), for example: with this coordinate choice, since $\partial_\mu V^\nu = 0$, we have $0 = \mathcal{L}_V W_{\mu\lambda} = V^\nu \partial_\nu W_{\mu\lambda} = \partial_0 W_{\mu\lambda}$. The math types make it sound mysterious, but a tensor Lie transported along your geodesic is simply a tensor that does not change in time (your proper time, that is).

The curve in the definition of Lie transportation does not necessarily have to be a geodesic: I just pick it as an example. It could be a curve in a flow field $V(x)$. People in the New Age talk about going with the flow, but what should they do with the vectors and tensors they want to carry with them? Lie transport them, that's what.

Appendix 3: Transforming the Christoffel symbol by brute force

You can skip this appendix and the next one upon a first reading of the text. They are for those readers who thirst for explicit computation, not believing in anything until they have ground it out.

To work out how $\Gamma^{\mu}_{\lambda\nu}$ transforms, you can just plug the transformation law $g'_{\nu\sigma}(x') = g_{\omega\tau}(x)(S^{-1})^{\omega}_{\nu}(S^{-1})^{\tau}_{\sigma}$ into (4) and proceed by brute force. It will be convenient in this calculation to keep referring to the collection of formulas in the back of the book.

As I explained, if we could sail the various factors of S^{-1} past the derivatives in (4), Γ would transform like a tensor with one upper index and two lower indices. In fact, we can't, and there are extra terms involving the derivatives acting on S^{-1}, so that we expect

$$\Gamma'^{\mu}_{\lambda\nu} = S^{\mu}_{\sigma}(S^{-1})^{\rho}_{\lambda}(S^{-1})^{\tau}_{\nu}\Gamma^{\sigma}_{\rho\tau} + M^{\mu}_{\lambda\nu} \tag{28}$$

where $M^{\mu}_{\lambda\nu}$ involves ∂S^{-1}. The claim is that $M^{\mu}_{\lambda\nu}W^{\nu}$ will cancel the unwanted term in (2).

Let us now take a deep breath and churn through this straightforward though somewhat laborious calculation, starting from the definition $\Gamma^{\mu}_{\lambda\nu} \equiv \frac{1}{2}g^{\mu\sigma}(\partial_{\lambda}g_{\nu\sigma} + \partial_{\nu}g_{\sigma\lambda} - \partial_{\sigma}g_{\lambda\nu})$. Perhaps you should try to do it first. I should warn you though, you have to slog through a swamp of indices. Don't give up too soon! It is merely an exercise in multivariable calculus after all, keeping track of various partial derivatives.

Focus on a piece of $\Gamma^{\mu}_{\lambda\nu}$ (say, $g^{\mu\sigma}\partial_{\lambda}g_{\nu\sigma}$), and ask how it transforms. Let's first suppress indices to get oriented: $g^{\cdot\cdot}\partial g_{\cdot\cdot}$ transforms into $\sim SSg^{\cdot\cdot}S^{-1}\partial(g_{\cdot\cdot}S^{-1}S^{-1})$. Looking only at the terms generated when ∂ hits S^{-1}, we have $\sim SSS^{-1}S^{-1}\partial S^{-1} \sim S^{-1}(\partial S)S^{-1}$.

I am now going to keep careful track of the indices, which of course makes the calculation seem clunky (and confusing when it in fact isn't). First,

$$\partial_{\lambda}g_{\nu\sigma} \rightarrow \partial'_{\lambda}g'_{\nu\sigma} = \partial'_{\lambda}[g_{\omega\tau}(S^{-1})^{\omega}_{\nu}(S^{-1})^{\tau}_{\sigma}] \tag{29}$$

There are two kinds of terms: those in which the derivative hits $g_{\omega\tau}(x)$ and those in which it hits the S^{-1}s. Clearly, the first kind of terms $(S^{-1})^{\rho}_{\lambda}\partial_{\rho}g_{\omega\tau}(S^{-1})^{\omega}_{\nu}(S^{-1})^{\tau}_{\sigma}$, those that are present even if S does not vary from place to place, will take care of themselves. As the discussion after (2) made clear, we should focus on the troublemakers, namely the second kind of terms contained in $\partial'_{\lambda}g'_{\nu\sigma}$ due to the variation of S, for which we invent on the spot a double bracket notation $\{\{\partial'_{\lambda}g'_{\nu\sigma}\}\}$. To save writing, define $K^{\omega}_{\lambda\nu} \equiv \partial'_{\lambda}(S^{-1})^{\omega}_{\nu} = \partial'_{\lambda}\frac{\partial x^{\omega}}{\partial x^{\nu}} = \frac{\partial^2 x^{\omega}}{\partial x'^{\lambda}\partial x'^{\nu}}$. Then, corresponding to ∂'_{λ} in (29) hitting one or the other of the two S^{-1}s in the square bracket, we obtain two terms $\{\{\partial'_{\lambda}g'_{\nu\sigma}\}\} = g_{\omega\tau}\{K^{\omega}_{\lambda\nu}(S^{-1})^{\tau}_{\sigma} + K^{\tau}_{\lambda\sigma}(S^{-1})^{\omega}_{\nu}\}$.

You may not recall instantly, but this has precisely the form of the little lemma you proved way way back in exercise I.4.8, namely $H_{\lambda\cdot\nu\sigma} = G_{\lambda\nu\cdot\sigma} + G_{\lambda\sigma\cdot\nu}$, where $H_{\lambda\cdot\nu\sigma} \equiv \{\{\partial'_{\lambda}g'_{\nu\sigma}\}\}$ is manifestly symmetric under $\nu \leftrightarrow \sigma$, and $G_{\lambda\nu\cdot\sigma} \equiv g_{\omega\tau}K^{\omega}_{\lambda\nu}(S^{-1})^{\tau}_{\sigma}$ is manifestly symmetric under $\lambda \leftrightarrow \nu$. To determine how $\Gamma^{\mu}_{\lambda\nu}$ transforms, we need precisely the combination found in that exercise: $H_{\lambda\cdot\nu\sigma} + H_{\nu\cdot\sigma\lambda} - H_{\sigma\cdot\lambda\nu} = 2G_{\lambda\nu\cdot\sigma} = 2g_{\omega\tau}K^{\omega}_{\lambda\nu}(S^{-1})^{\tau}_{\sigma}$.

Putting it altogether, we obtain $\{\{\Gamma'^{\mu}_{\lambda\nu}\}\} = \frac{1}{2}S^{\mu}_{\rho}S^{\sigma}_{\kappa}g^{\rho\kappa}(2g_{\omega\tau}K^{\omega}_{\lambda\nu}(S^{-1})^{\tau}_{\sigma}) = S^{\mu}_{\rho}K^{\rho}_{\lambda\nu} = S^{\mu}_{\rho}(S^{-1})^{\rho}_{\lambda}\partial_{\kappa}(S^{-1})^{\rho}_{\nu} = (S^{-1})^{\kappa}_{\lambda}(S\partial_{\kappa}S^{-1})^{\mu}_{\nu} = -(S^{-1})^{\kappa}_{\lambda}((\partial_{\kappa}S)S^{-1})^{\mu}_{\nu}$. In the last step, we used the matrix identity (7). This is precisely what we heuristically guessed for $M'^{\mu}_{\lambda\nu} \equiv \{\{\Gamma'^{\mu}_{\lambda\nu}\}\}$. Note that the factor $\frac{1}{2}$ in the definition of the Christoffel symbol is needed here.

Thus, $\{\{\Gamma'^{\mu}_{\lambda\nu}W'^{\nu}\}\} = -((S^{-1})^{\kappa}_{\lambda}\partial_{\kappa}S^{\mu}_{\eta})W^{\eta}$, precisely the negative of the unwanted second term in (2). So indeed we have $D'_{\lambda}W'^{\mu} = (S^{-1})^{\rho}_{\lambda}S^{\mu}_{\nu}D_{\rho}W^{\nu}$, and $D_{\lambda}W^{\mu}$ transforms as a tensor as desired. It all seems a bit complicated, but in fact it is simple: as I explained in the text, the two terms in $D_{\lambda}W^{\mu}$ each produce unwanted terms, but they cancel each other.

Incidentally, if we write out S, S^{-1}, and K explicitly, we have found here that

$$\Gamma'^{\mu}_{\lambda\nu} = \frac{\partial x'^{\mu}}{\partial x^{\sigma}}\frac{\partial x^{\rho}}{\partial x'^{\lambda}}\frac{\partial x^{\tau}}{\partial x'^{\nu}}\Gamma^{\sigma}_{\rho\tau} + \frac{\partial x'^{\mu}}{\partial x^{\eta}}\frac{\partial^2 x^{\eta}}{\partial x'^{\lambda}\partial x'^{\nu}} \tag{30}$$

which agrees with (6), namely what we had back in chapter II.2, of course.

Note that the first term is the "uninteresting" part, gathered up from those terms that we said would take care of themselves. It tells us how the Christoffel symbol would have transformed had it been a tensor. We worked hard to obtain the second term, which, as anticipated, depends on the second derivative $\frac{\partial^2 x^{\eta}}{\partial x'^{\lambda}\partial x'^{\nu}}$, in other words, on the variation of S^{-1} from place to place. The presence of the second term indicates, as emphasized again and again, that the Christoffel symbol is not a tensor.

Appendix 4: Arguing from the geodesic equation

The computation in the preceding appendix involves a fair amount of work. In this appendix, we will try to avoid work by winging it as much as possible.

Our starting point is the geodesic $X^\mu(\tau)$, which we first met way back in chapter II.2, and then again in chapter V.1, determined by

$$\frac{dV^\rho}{d\tau} + \Gamma^\rho_{\mu\nu} V^\mu V^\nu = 0 \tag{31}$$

with $V^\mu \equiv \frac{dX^\mu}{d\tau}$.

First a notational clarification and a bad notation alert! The geodesic is a curve defined by $X^\mu(\tau)$, with $d\tau^2 = -g_{\mu\nu}(X(\tau))dX^\mu dX^\nu$. The velocity vector is defined by $V^\mu(\tau) = \frac{dX^\mu(\tau)}{d\tau}$. Strictly speaking, it is not correct to write $V^\mu(X(\tau))$, as some authors sometimes somewhat sloppily do. This notation suggests that there is a vector field $V^\mu(x)$ (and there isn't) defined all over spacetime, and $V^\mu(X(\tau))$ is equal to $V^\mu(x)$ evaluated at $x = X(\tau)$. In fact, $V^\mu(\tau)$ is only defined on the particle trajectory (be it a geodesic or not).

Under a coordinate transformation $x \to x'(x)$, the ur-vector $dx^\mu \to dx'^\mu = (\frac{\partial x'^\mu}{\partial x^\nu})dx^\nu \equiv S^\mu_\nu(x)dx^\nu$. The velocity V^μ is most certainly a vector, since $dX^\mu \to dX'^\mu = S^\mu_\nu(X)dX^\nu$ is a vector and $d\tau$ is a scalar. So $V'^\mu(\tau) = S^\mu_\nu(X(\tau))V^\nu(\tau)$. Note that the transformation matrix $S^\mu_\nu(x)$ is evaluated at $x = X(\tau)$.

There are levels of the game[4] in theoretical physics as in any other substantive endeavor. After we learn the geodesic equation (31), we could go on happily solving for geodesics in any given spacetime until we are blue in the face. On a deeper level, however, we can ask what (31) teaches us about curved spacetime.

The left hand side of (31) carries a single free index ρ and thus looks like a vector. It better be a vector, since otherwise what one observer sees as a geodesic would not be a geodesic to another observer. Suppose Ms. Unprime writes (31). If the whole package on the left hand side of (31) transforms as claimed, Mr. Prime would have $\frac{dV'^\rho}{d\tau} + \Gamma'^\rho_{\mu\nu} V'^\mu V'^\nu = S^\rho_\sigma(\frac{dV^\sigma}{d\tau} + \Gamma^\sigma_{\eta\omega} V^\eta V^\omega) = 0$. The two observers agree that particles move along geodesics in spacetime. We have insisted again and again that the laws of physics must transform appropriately, and here is yet another example.

Instead of spacetime, we can talk about curved space (in which case τ in (31) would denote length rather than proper time). A geodesic curve as the path of shortest distance between two points has intrinsic geometric meaning and so cannot possibly depend on the coordinate system we use to describe it. For example, a great circle on the globe does not depend on the particular system of latitude and longitude we happen to use because of British naval power. The geodesic has to satisfy (31) in all coordinates.

But, for the same kind of reason as in (2), we see that while V^ρ is a vector, $\frac{dV^\rho}{d\tau}$ assuredly is not. Indeed, $\frac{dV'^\rho(\tau)}{d\tau} = \frac{d}{d\tau}(S^\rho_\nu(X(\tau))V^\nu(\tau)) = S^\rho_\nu(X(\tau))\frac{d}{d\tau}V^\nu(\tau) + (\frac{d}{d\tau}S^\rho_\nu(X(\tau)))V^\nu(\tau)$. The derivative $\frac{d}{d\tau}$ also acts on the transformation matrix S as it varies along the geodesic. Sound familiar? Thus, $\frac{dV^\rho}{d\tau}$ would have been a vector, had it not been for the second term $(\frac{d}{d\tau}S^\rho_\nu)V^\nu = \frac{dX^\lambda}{d\tau}(\partial_\lambda S^\rho_\nu)V^\nu = (\partial_\lambda S^\rho_\nu)V^\lambda V^\nu$.

As in appendix 3, use the notation $\{\{\frac{dV^\rho}{d\tau}\}\} \equiv \partial_\lambda S^\rho_\sigma V^\lambda V^\sigma$ to indicate the extra term that prevents $\frac{dV^\rho}{d\tau}$ from being a vector. Using the notation introduced in (28), we also have for the second term in (31) $\{\{\Gamma^\rho_{\mu\nu} V^\mu V^\nu\}\} = M^\rho_{\mu\nu} S^\mu_\lambda S^\nu_\sigma V^\lambda V^\sigma$. Thus, the requirement $\{\{\frac{dV^\rho}{d\tau} + \Gamma^\rho_{\mu\nu} V^\mu V^\nu\}\} = 0$ gives us $(\partial_\lambda S^\rho_\sigma + M^\rho_{\mu\nu} S^\mu_\lambda S^\nu_\sigma)V^\lambda V^\sigma = 0$. Here comes the nonrigorous part, which renders the argument heuristic. We argue that $\partial_\lambda S^\rho_\sigma + M^\rho_{\mu\nu} S^\mu_\lambda S^\nu_\sigma = 0$, even though we have only shown that this quantity vanishes when multiplied by $V^\lambda V^\sigma$ and evaluated on a geodesic. We could of course show by explicit and laborious calculation, as was done in appendix 3, that this is in fact true. But it is nevertheless highly plausible, since it holds for any geodesic.

So we feel that it must be true. Once again, I can appeal to what the American Mathematical Society said, as recounted in a footnote in this chapter. Multiplying by S^{-1} (and relabeling indices), we have $(S^{-1})^\rho_\lambda \partial_\rho S^\mu_\sigma + M^\mu_{\lambda\nu} S^\nu_\sigma = 0$. In contrast, looking at (5), (28), and (2), we have $\{\{D_\lambda W^\mu\}\} = ((S^{-1})^\rho_\lambda \partial_\rho S^\mu_\sigma + M^\mu_{\lambda\nu} S^\nu_\sigma)W^\sigma = 0$, so that as claimed, the covariant derivative $D_\lambda W^\mu$ transforms like a tensor.

You perhaps see that this heuristic argument is actually quite simple, even though writing it all out involves way too many words.

At the risk of repeating myself, I close this appendix with two important clarifying remarks:

1. The geodesic equation is guaranteed to transform like a vector, because the action it was derived from is manifestly a scalar; in other words, the left hand side of (31) comes from varying an action invariant under general coordinate transformation.

2. Beginning students are puzzled by the statement "a tensor is something that transforms like a tensor" (as mentioned in chapter I.4), because they rarely encounter something that does not transform like a tensor. To understand something, it is often easier to understand what would negate that something. Here we encounter two objects $\partial_\lambda W^\mu$ and $\Gamma^\mu_{\lambda\nu} W^\nu$ that are not tensors, even though they have indices and everything. When they are added together, their separate nontensorial characters cancel out, kind of like in the ideal couple alluded to earlier.

Appendix 5: A constant vector field

It is often illuminating to look at important concepts from different angles. In this spirit, let me introduce you to another friend of mine, the naive guy.

He tells us excitedly, "I have discovered a constant vector field. I spent years measuring its 16 derivatives $\partial_\mu W^\lambda(x)$ in my lab and found that they all vanish within experimental error!"

We explain, "Well, first, your statement is local to your lab, but more importantly, your concept of constant vector field is special to your coordinate system. In our coordinates, $\partial'_\mu W'^\lambda(x') = (S^{-1})^\nu_{\ \mu} \partial_\nu [S^\lambda_{\ \rho} W^\rho(x)] = (S^{-1})^\nu_{\ \mu} (\partial_\nu S^\lambda_{\ \rho})(S^{-1})^\rho_{\ \sigma} W'^\sigma(x') \equiv -\Gamma'^\lambda_{\mu\sigma}(x') W'^\sigma(x')$ are not zero."

The naive guy lapses into stunned silence. But we soothe his disappointment by pointing out that his concept of a constant vector field is still useful but has to be defined as a vector field whose covariant derivatives $D'_\mu W'^\lambda(x') \equiv \partial'_\mu W'^\lambda(x') + \Gamma'^\lambda_{\mu\sigma}(x') W'^\sigma(x')$ vanish.

Together with the naive guy, we work out (the third equality follows from (7))

$$\Gamma'^\lambda_{\mu\sigma}(x') = -(S^{-1})^\nu_{\ \mu}(\partial_\nu S^\lambda_{\ \rho})(S^{-1})^\rho_{\ \sigma} = -(S^{-1})^\nu_{\ \mu}[(\partial_\nu S)(S^{-1})]^\lambda_{\ \sigma}$$

$$= (S^{-1})^\nu_{\ \mu}(S\partial_\nu S^{-1})^\lambda_{\ \sigma} = \frac{\partial x'^\lambda}{\partial x^\rho} \frac{\partial^2 x^\rho}{\partial x'^\mu \partial x'^\sigma} \qquad (32)$$

in agreement with (30). This shows that the naive guy's concept of constancy holds only in those coordinates related linearly to his, namely $x^\mu = ax'^\mu + b^\mu$. Then we have $\Gamma'^\lambda_{\mu\sigma}(x') = 0$. He happens to have chosen locally flat coordinates!

Appendix 6: Lie derivative once more

It may be worthwhile to approach the Lie derivative from a slightly different direction. In elementary physics, you learned about the gradient of a function $\vec{\nabla} f$ and the rate of change of f in the direction of a given vector $\vec{v}$, namely $\vec{v} \cdot \vec{\nabla} f$.

Given a vector field $V^\mu(x)$ and a scalar field $\phi(x)$, we use this notion to define the Lie derivative $\mathcal{L}_V \phi \equiv V^\nu \partial_\nu \phi$. We then attempt to generalize this to the Lie derivative of a vector field $W^\mu(x)$.

For our first try, we write down $V^\nu(x) \partial_\nu W^\mu(x)$, but this does not transform properly. In the text, by promoting ∂_ν to the covariant derivative D_ν we could make $V^\nu(x) D_\nu W^\mu(x)$ transform like a vector. But another possibility suggests itself: $W^\nu(x) \partial_\nu V^\mu(x)$ also transforms badly, and as described somewhat picturesquely in the text, two objects both with character defects could join together to form a couple in which the defects cancel out. Thus, we are led to define

$$\mathcal{L}_V W^\mu \equiv V^\nu \partial_\nu W^\mu - W^\nu \partial_\nu V^\mu = -\mathcal{L}_W V^\mu \qquad (33)$$

and then to show that it transforms properly.

Insisting on $\mathcal{L}_V(W^\mu U_\mu) = V^\nu \partial_\nu(W^\mu U_\mu)$ (since $W^\mu U_\mu$ is a scalar) and the product rule $\mathcal{L}_V(W^\mu U_\mu) = (\mathcal{L}_V W^\mu) U_\mu + W^\mu(\mathcal{L}_V U_\mu)$, we are forced to

$$\mathcal{L}_V U_\mu = V^\nu \partial_\nu U_\mu + U_\nu \partial_\mu V^\nu \qquad (34)$$

Compare the opposite signs in (33) and (34) with the opposite signs in (11) and (12).

Now that we have (33) and (34), we can define $\mathcal{L}_V$ acting on any tensor by pretending that it is composed by multiplying vectors together (for example, pretending that $T^\mu_{\nu\rho}$ is equal to $W^\mu U_\nu Y_\rho$) and invoking the product rule. Thus, $\mathcal{L}_V T_{\mu\nu} = V^\lambda \partial_\lambda T_{\mu\nu} + T_{\lambda\nu} \partial_\mu V^\lambda + T_{\mu\lambda} \partial_\nu V^\lambda$.

In chapter IX.6, we will pose the following question. Given a metric $g_{\mu\nu}$, does there exist a vector field $\xi(x)$ such that its Lie derivative acting on the metric vanishes? If so, then $\xi(x)$ is known as a Killing field. We see that the condition for a Killing field to exist is

$$\mathcal{L}_\xi g_{\mu\nu} = \xi^\lambda \partial_\lambda g_{\mu\nu} + g_{\lambda\nu}\partial_\mu \xi^\lambda + g_{\mu\lambda}\partial_\nu \xi^\lambda = 0 \tag{35}$$

We will study this equation in detail in chapter IX.6.

Appendix 7: Curved spacetime in the lab?

The subject of this appendix is somewhat off the subject of this chapter, but I wish to introduce you to an interesting and amusing area of physics, namely that of constructing analog curved spacetime in the lab. Before you say "what?", note the word "analog."

To set up the discussion, consider a scalar field $\varphi(x)$ in Minkowski spacetime governed by the action $S = -\int d^d x (\frac{1}{2}\eta^{\mu\nu}\partial_\mu\varphi\partial_\nu\varphi + V(\varphi))$, with $V(\varphi)$ some function of φ. (You might recall that we first encountered an action of this form way back in appendix 2 of chapter II.3, before we even got to special relativity.) To keep the discussion simple, we will ignore $V(\varphi)$ or simply set it to 0.

In the discussion leading up to (18), we showed the power of the equivalence principle. We obtained the action of an electromagnetic field in the presence of a gravitational field by simply replacing the Minkowski metric $\eta_{\mu\nu}$ in the Maxwell action by $g_{\mu\nu}$. We now repeat this amazing feat: we obtain the action governing a scalar field in curved spacetime, namely $S = -\frac{1}{2}\int d^d x \sqrt{-g}\, g^{\mu\nu}\partial_\mu\varphi\partial_\nu\varphi$.

Phew, that was easy! Exactly: the equivalence principle is powerful stuff.

Now that we have followed Einstein to the scalar field action in curved spacetime, we follow Euler and Lagrange to the corresponding equation of motion, namely $\partial_\mu\left(\frac{\delta\mathcal{L}}{\delta\partial_\mu\varphi}\right) = 0$. This works out to be

$$\frac{1}{\sqrt{-g}}\partial_\mu(\sqrt{-g}\,g^{\mu\nu}\partial_\nu\varphi) = 0 \tag{36}$$

Applying what we learned in this chapter, we recognize this as just $D_\mu D^\mu \varphi = 0$, the curved spacetime version of the flat spacetime equation of motion $\partial_\mu\partial^\mu\varphi = 0$.

Fine—now what? The point is that (36), when written out in terms of $\frac{\partial}{\partial t}$, $\frac{\partial}{\partial x}$, $\frac{\partial}{\partial y}$, and so on (with a slight abuse of notation here), is just a second order partial differential equation. But second order partial differential equations appear in many areas of physics, and some of these could be written as (36) for some $g_{\mu\nu}$. For example, consider the Bose-Einstein condensate as a fluid.[5] Its phase angle $\varphi(t, \vec{x})$ satisfies a second order partial differential equation; when written in the form given in (36), the equation can be interpreted as a scalar field moving in curved spacetime. For instance, terms that involve both $\frac{\partial}{\partial t}$ and $\frac{\partial}{\partial x}$ correspond to the entry g^{tx} in $g^{\mu\nu}$ in (36). The name of the game is to set up some flow in the lab (such as that of a fluid rushing down a drain) that corresponds to an interesting analog curved spacetime, such as that of a black hole!

Incidentally, we will come back to the scalar field action in curved spacetime in chapter VIII.4, when we discuss inflationary cosmology.

Exercises

1 Show that the divergence of a tensor is given by $D_\mu T^{\mu\nu} = \partial_\mu T^{\mu\nu} + \Gamma^\mu_{\mu\lambda}T^{\lambda\nu} + \Gamma^\nu_{\mu\lambda}T^{\mu\lambda}$.

2 Given the covariant derivative $D_\nu W^\mu$ in (5), integrate $\int d^D x\sqrt{-g}\,T^\nu_\mu D_\nu W^\mu$ to obtain the covariant divergence of the tensor T^ν_μ. Verify that it agrees with the result of (1).

3 Using the explicit expression for the Christoffel symbol in terms of the metric, show that $D_\lambda g_{\mu\nu} = 0$. Note that in contrast $\partial_\lambda g_{\mu\nu}$ definitely does not vanish. Thus, the metric is a very special tensor: it is the tensor with two lower indices that is covariantly constant. Also, note that the condition $D_\lambda g_{\mu\nu} = 0$ can be used to determine the Christoffel symbol. Check this for the sphere. Show also that $D_\lambda g^{\mu\nu} = 0$.

4   Add to the electromagnetic action a term coupling A_μ to a current and vary to find Maxwell's equations in curved spacetime.

5   Evaluate the electromagnetic action in the expanding universe discussed in chapter V.3 in terms of the electric and magnetic fields.

6   Define the covariant derivative of a scalar along a curve.

7   Define the covariant derivative of a tensor along a curve.

Notes

1. The question "How do you do?"—perhaps short for "How do you do it day after day?"—could be stated as "How do you transform under time translation?"
2. Incidentally, you will also need this identity repeatedly when doing quantum field theory.
3. To use a stock market analogy, it is not the change in earnings, but the difference between actual earnings and expected earnings, that counts.
4. J. McPhee, *Levels of the Game,* Farrar Straus Giroux, 1979.
5. See the interesting work of Luis Garay, in particular a talk he gave in Leiden in 2007.

Recap to Part V

If you hold a ketchup bottle upside down and wait for the ketchup to come out, you are applying gravity, but if you shake the bottle or hit it, you are applying the inertial principle, though in both cases merely in the Newtonian limit. The truly amazing thing is that the second strategy, involving accelerating frames, contains the seed of the first strategy!

Remarkably, any layperson familiar with airline maps can grasp Einstein's equivalence principle, one of the deepest and most powerful principles in physics.

The message is that if you know how to change coordinates, you almost know curved space and curved spacetime, and once you know how to find "straight" lines in curved space, you know how to track the motion of particles in curved spacetime!

Honoring the fundamental principle that physics should not depend on the physicist, we have to understand how things transform. Understanding that, we know how to differentiate.

Interestingly, analog curved spacetimes may appear in an actual lab.

Part VI Einstein's Field Equation Derived and Put to Work

VI.1 To Einstein's Field Equation as Quickly as Possible

Years of intense longing

[Now] the happy achievement seems almost a matter of course. . . . But the years of anxious searching in the dark, with their intense longing, their alternations of confidence and exhaustion, and the final emergence into the light;—only those who have experienced it can understand that.

—A. Einstein[1]

Traditionally, students of general relativity often feel like foot soldiers in the Napoleonic army on an interminable march[2] toward Moscow. After conquering tensors, there is the battle of differential geometry, and on, and on. I certainly felt that way. For many, even learning Einstein gravity could be characterized as "intense longing, alternating with confidence and exhaustion." In this chapter, I will attempt the pedagogical equivalent of airlifting you, given that you now know how to differentiate a vector, directly to Einstein's action for gravity.

Let me take you to Einstein's field equation as quickly as possible, starting with what you already know. My pedagogical philosophy is to keep things as simple as possible. I will necessarily have to take shortcuts, but when I do, I will alert you. I could elaborate and expand on each point, but it is better to come back and do that later. Remember, it took Einstein 10 years to get there. Rather than deriving everything at once, we will wing it at times; you will see what I mean.

Before we start, let's take stock of our situation. Thanks to the equivalence principle (whose essence I claim that Galileo could have understood), you and I know that gravity amounts to curved spacetime. Thanks to how I set up this book, starting with coordinate changes leading immediately to curved space, and with the action principle capable of incorporating immediately a curved background, you and I have come a very long way indeed. If somebody gives us a curved spacetime, we could jump to it and figure out how particles, massive or massless, move in this curved spacetime.

But you and I do not know how to generate this spacetime. Yes, we know how to use symmetry to restrict the form of the metric, as in chapters V.2 and V.3, so that it depends on merely one or two functions in the more symmetric cases ($a(t)$ for the expanding universe, and $A(r)$, $B(r)$ for the Schwarzschild metric). But how to determine these functions?

Let's understand where we are by analogy with what we know already. In Newtonian gravity, we first learned in chapter I.1 how particles move in a gravitational field $\Phi(\vec{x})$. But it was not until chapter II.1 that we, suitably inspired by the hanging membrane, figured out how to determine $\Phi(\vec{x})$: simply add $\int dt d^3x \, (\vec{\nabla}\Phi)^2$ to the action, that is, $(\vec{\nabla}\Phi)^2$ to the Lagrangian. In electromagnetism, our present situation with regard to gravity is analogous to our knowing, in chapter IV.1, the Lorentz force law telling us how particles move in an electromagnetic field, but not Maxwell's equations. This was remedied in chapter IV.2 by adding $F_{\mu\nu}F^{\mu\nu} = (\partial_\mu A_\nu - \partial_\nu A_\mu)(\partial_\lambda A_\rho - \partial_\rho A_\lambda)\eta^{\mu\lambda}\eta^{\nu\rho}$ to the Lagrangian.

Searching for something containing two derivatives

In all these cases, we add to the Lagrangian a term quadratic in the field (Φ in one case, A_μ in the other) and quadratic in derivatives (spatial or temporal). In fact, all this parallels the point particle Lagrangian $\frac{1}{2}m(\frac{d\vec{q}}{dt})^2$. As Einstein said, it now "seems almost a matter of course," at least in hindsight. To describe the dynamics of the gravitational field, we are evidently invited to add a term involving two powers of derivative acting on the metric, a term that reduces to $(\vec{\nabla}\Phi)^2$ in the nonrelativistic weak field limit.

The search for actions containing two powers of $\frac{\partial}{\partial t}$ has served as a "golden" guiding principle in theoretical physics: golden because it has worked[3] from Newtonian mechanics to grand unified theory, and because theoretical physicists do not know how to handle the inherent instability[4] in dynamics with higher powers of time derivative, a little known instability discovered in 1850 by the Russian M. V. Ostrogradsky. We have already encountered this principle as far back as chapter II.3 (see appendix 1), and we will encounter it again on a number of occasions. In this chapter we will see that it works for Einstein gravity.

We also immediately notice some crucial differences. In Newtonian gravity and in Maxwellian electrodynamics, the relevant fields Φ and A_μ vanish* in the absence of the gravitational field and the electromagnetic field, respectively. In contrast, in the absence of a gravitational field, we know that $g_{\mu\nu}$ reduces to $\eta_{\mu\nu}$. Indeed, we know from our discussions in chapters IV.1 and V.2 that in the weak field limit, $g_{00} \simeq -(1 + 2\Phi)$. In general, let's write $g_{\mu\nu} = \eta_{\mu\nu} + h_{\mu\nu}$; it is $h_{\mu\nu}$ that measures the deviation of the spacetime from flat spacetime and that may play a role analogous to Φ and A_μ. Furthermore, we expect that the inverse metric $g^{\mu\nu}$ will also appear in the Lagrangian (since in Maxwell's Lagrangian, $\eta^{\mu\nu}$ already enters, and certainly there is no reason for $g^{\mu\nu}$ not to appear). If

---

* Up to a trivial additive constant in the case of Φ and a gauge transform in the case of A_μ.

so, since $g^{\mu\nu}$ is the inverse of $g_{\mu\nu} = \eta_{\mu\nu} + h_{\mu\nu}$ and hence an infinite series in the field $h_{\mu\nu}$, we might anticipate that the gravitational Lagrangian we are searching for may be considerably more complicated than quadratic in $h_{\mu\nu}$. (In other words, it may only be in the weak field limit that the Lagrangian has the schematic form $\sim (\partial h)^2$. For further discussion along this line, see chapter IX.5.)

An important consideration is the restriction imposed by symmetry. In Newtonian gravity, rotational symmetry forbids us to write, instead of the familiar and nice

$$\left(\vec{\nabla} \Phi \right)^2 = \left(\frac{\partial \Phi}{\partial x} \right)^2 + \left(\frac{\partial \Phi}{\partial y} \right)^2 + \left(\frac{\partial \Phi}{\partial z} \right)^2$$

something dreadful like

$$a_x \left(\frac{\partial \Phi}{\partial x} \right)^2 + a_y \left(\frac{\partial \Phi}{\partial y} \right)^2 + a_z \left(\frac{\partial \Phi}{\partial z} \right)^2 + b_{xy} \left(\frac{\partial \Phi}{\partial x} \frac{\partial \Phi}{\partial y} \right) + \cdots$$

with arbitrary coefficients $a_x, a_y, \cdots$. (In other words, we know from experiments that space does not pick out a special direction, and so $a_x = a_y$, $b_{xy} = 0$, and so forth to a high degree of accuracy.) In the case of electromagnetism, we have Lorentz symmetry and gauge invariance, which together fix the form $F_{\mu\nu} F^{\mu\nu}$. (I have already emphasized in chapter IV.2 that gauge invariance is not a symmetry as such, but a redundancy in the description: A_μ and $A_\mu + \partial_\mu \Lambda$ describe the same physics.) Thus, for example, we can't have something like $\partial_\mu A_\nu \partial_\lambda A_\rho \eta^{\mu\lambda} \eta^{\nu\rho}$ for the electromagnetic Lagrangian. In the case of gravity, the requirement is even more stringent: the action must be invariant under coordinate transformations $x \to x'(x)$. Again, this indicates a redundancy in the description: different coordinate systems can describe the same spacetime.

Einstein's search for action and Riemann's quest for curvature

At this point, you might suddenly realize that Einstein's search for an action for gravity is more than intimately linked to Riemann's quest for an invariant or a covariant description of curvature. We kept saying in part I that the Riemann curvature tensor must involve two powers of derivatives acting on the metric and must transform properly as a tensor. Here we want to find an action. Indeed, in this chapter, we will solve both problems at once: find the Riemann curvature tensor and the action for gravity. Two birds with one stone!

In the history of the intersection between physics and mathematics, this realization, that gravity and curvature are one and the same, represents one of the most profound insights ever. Perhaps we are reminded of the search for a mechanics of motion and the quest for a calculus to describe infinitesimal change. There the solution was essentially provided by one single person.

Incidentally, even with all this background, the correct form for what we seek is far from obvious. You could challenge yourself by finding it without reading on. The invariance of the action under coordinate transformations suggests an expression with two powers of the covariant derivative D_λ and the metric, but we also know that $D_\lambda g_{\mu\nu}$ vanishes identically. So the correct expression must involve terms with the ordinary derivative acting on the metric but arranged in such a way that the entire package is invariant under coordinate transformations.

Breaking the Newton-Leibniz rule

After all this preamble, setting the stage as it were, we now roll up our sleeves and get to work. Start with what you know: the covariant derivative of a vector with a lower index:

$$D_\nu S_\rho = \partial_\nu S_\rho - \Gamma^\sigma_{\nu\rho} S_\sigma \tag{1}$$

Let us first go back to elementary calculus. Given the value of a function $f(x)$ and its derivative at some point x, then the value of the function at the point $x + \delta x$ a short distance away is of course given by $(1 + \delta x \frac{d}{dx}) f(x) \simeq f(x + \delta x)$. Think of the operation $(1 + \delta x \frac{d}{dx})$ as a translation operator: it moves or translates the function from the point x to the point $x + \delta x$. Generalize to 2-dimensional flat space. Consider a small rectangle with opposite corners at (x, y) and $(x + \delta x, y + \delta y)$. To find the value of a function at $(x + \delta x, y + \delta y)$, we can translate first in the x direction and then in the y direction:

$$\left(1 + \delta y \frac{\partial}{\partial y}\right)\left(1 + \delta x \frac{\partial}{\partial x}\right) f(x, y) = \left(1 + \delta x \frac{\partial}{\partial x} + \delta y \frac{\partial}{\partial y} + \delta x \delta y \frac{\partial}{\partial y}\frac{\partial}{\partial x}\right) f(x, y) \tag{2}$$

We apply the translation operator in the x direction, followed by the translation operator in the y direction.

Now let us ask a seemingly pointless question: suppose we translate first in the y direction and then in the x direction. We could travel from the corner (x, y) to the diagonally opposite corner $(x + \delta x, y + \delta y)$ along the edges of the rectangle in two different ways. You say, we would get the same answer, of course. Indeed, the difference between the two ways of getting $f(x + \delta x, y + \delta y)$ is equal to

$$\left(\left\{1 + \delta x \frac{\partial}{\partial x} + \delta y \frac{\partial}{\partial y} + \delta x \delta y \frac{\partial}{\partial y}\frac{\partial}{\partial x}\right\} - (x \leftrightarrow y)\right) f(x, y) = \delta x \delta y \left[\frac{\partial}{\partial x}, \frac{\partial}{\partial y}\right] f(x, y) = 0 \tag{3}$$

where I have introduced the commutator[5] $[\frac{\partial}{\partial x}, \frac{\partial}{\partial y}] \equiv \frac{\partial}{\partial x}\frac{\partial}{\partial y} - \frac{\partial}{\partial y}\frac{\partial}{\partial x}$. In the last step, I used the Newton-Leibniz rule that the order of taking derivatives does not matter.

Now that you have mastered this laughably easy stuff, we are ready to move on to curved space. Instead of a simple function $f(x)$, we will translate an arbitrary vector $S_\rho(x)$. We apply the translation operator to $(1 + (\delta_1 x)^\nu D_\nu)$, since we have just learned that in curved space, the ordinary derivative ∂_ν is to be replaced by the covariant derivative D_ν. Here $(\delta_1 x)$ denotes an infinitesimal displacement with components $(\delta_1 x)^\nu$ (with $\nu = 0, 1, \cdots, d$).

Let us now play the same game as before and consider a curved "rectangle" with $(\delta_1 x)$ along one edge, and some other infinitesimal displacement $(\delta_2 x) = (\delta_2 x)^\nu$ along the other edge. See figure 1. We translate $S_\rho(x)$ first along $(\delta_1 x)$ and then along $(\delta_2 x)$. The result is then $(1 + (\delta_2 x)^\mu D_\mu)(1 + (\delta_1 x)^\nu D_\nu)S_\rho(x)$. We then go the other way around, with the result $(1 + (\delta_1 x)^\nu D_\nu)(1 + (\delta_2 x)^\mu D_\mu)S_\rho(x)$. The difference in the two results is then given by $(\delta_2 x)^\mu (\delta_1 x)^\nu [D_\mu, D_\nu]S_\rho(x)$. Were we in flat space, this quantity would have been zero.

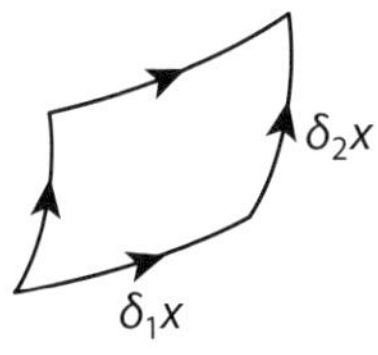

Figure 1 Displacing a vector in two different ways to the opposite corner of a curved rectangle.

Thus, the nonvanishing of this quantity measures the curvature at the point described by x.

So, let's compute this commutator of two covariant derivatives acting on our vector S: $[D_\mu, D_\nu]S_\rho = D_\mu D_\nu S_\rho - D_\nu D_\mu S_\rho$.

Before we drown in a sea of indices, let us anticipate the structure of what we will get by schematically doing the calculation, suppressing all but the two most essential indices: $[D_\mu, D_\nu]S_\rho \sim (\partial_\mu + \Gamma^{\cdot}_{\mu \cdot})(\partial_\nu + \Gamma^{\cdot}_{\nu \cdot})S. - (\mu \leftrightarrow \nu)$. Remarkably, we will not end up with any partial derivative ∂ acting on S. First, obviously $\partial_\mu \partial_\nu - (\mu \leftrightarrow \nu)$ vanishes, once again as Newton and Leibniz assured us. But what about one power of ∂ acting on S? The relevant terms are $\Gamma^{\cdot}_{\nu \cdot}\partial_\nu S. + \Gamma^{\cdot}_{\nu \cdot}\partial_\mu S. - (\mu \leftrightarrow \nu)$, which vanishes. (We will shortly put in all the indices and do it more carefully.) Thus, impressionistically, we obtain

$$[D., D.]S. \sim [\partial. + \Gamma^{\cdot}_{\cdot \cdot}, \partial. + \Gamma^{\cdot}_{\cdot \cdot}]S. \sim \left((\partial.\Gamma^{\cdot}_{\cdot \cdot} + \Gamma^{\cdot}_{\cdot \cdot}\Gamma^{\cdot}_{\cdot \cdot}) - (\cdot \leftrightarrow \cdot)\right) S. \tag{4}$$

The result turns out to be S multiplied by a tensor:

$$[D_\mu, D_\nu]S_\rho = -R^\sigma{}_{\rho\mu\nu} S_\sigma \tag{5}$$

(The minus sign is to conform to convention.) As the use of the capital letter R may suggest, $R^\sigma{}_{\rho\mu\nu}$ is the celebrated Riemann curvature tensor. It is manifestly a tensor, since the left hand side of (5) is a tensor: $D_\mu D_\nu S_\rho$ and $D_\nu D_\mu S_\rho$ are both tensors, and the difference between two tensors is a tensor.

Riemann curvature tensor

What are the differential laws which determine the Riemannian metric (i.e. $g_{\mu\nu}$) itself? . . . [The] solution obviously needed invariant differential systems of the second order taken from $g_{\mu\nu}$. We[6] soon saw that these had already been established by Riemann (the tensor of curvature). We had already considered the right field equation for gravitation for two years before the publication of the general theory of relativity, but we were unable to see how they could be used in physics. On the contrary I felt sure that they could not do justice to experience. Moreover I believed that I could show on general considerations that a law of gravitation invariant in relation to any transformation of coordinates whatever was inconsistent with the principle of

causation. These were errors of thought which cost me two years
of excessively hard work, until I finally recognized them as such
at the end of 1915 and succeeded in linking the question up
with the facts of astronomical experience, after which I ruefully
returned to the Riemannian curvature.

—A. Einstein[7]

Now let's compute for real, keeping track of all indices. We learned in (V.6.14) that the
second covariant derivative is given by

$$D_\mu D_\nu S_\rho = \partial_\mu (D_\nu S_\rho) - \Gamma^\sigma_{\mu\nu} D_\sigma S_\rho - \Gamma^\sigma_{\mu\rho} D_\nu S_\sigma \tag{6}$$

To calculate $[D_\mu, D_\nu]S_\rho$, we are to subtract from (6) the expression we obtain by interchanging μ and ν in (6). We notice that the second term is symmetric in μ and ν and so can be
dropped. Next, for $D_\nu S_\rho$ in the first and third terms, we insert the expression in (1). Thus,

$$[D_\mu, D_\nu]S_\rho = \partial_\mu(\partial_\nu S_\rho - \Gamma^\sigma_{\nu\rho} S_\sigma) - \Gamma^\sigma_{\mu\rho}(\partial_\nu S_\sigma - \Gamma^\kappa_{\nu\sigma} S_\kappa) - (\mu \leftrightarrow \nu)$$

$$= \partial_\mu \partial_\nu S_\rho - (\partial_\mu \Gamma^\sigma_{\nu\rho}) S_\sigma - \Gamma^\sigma_{\nu\rho} \partial_\mu S_\sigma - \Gamma^\sigma_{\mu\rho} \partial_\nu S_\sigma + \Gamma^\sigma_{\mu\rho} \Gamma^\kappa_{\nu\sigma} S_\kappa - (\mu \leftrightarrow \nu) \tag{7}$$

Indeed, the calculation is even simpler than shown here if we trust the preceding argument
and do not even bother to write down any term involving $\partial.S.$ and $\partial.\partial.S.$. But let's be careful.

Happily, lots of terms knock each other off when we antisymmetrize in μ and ν. The first
term goes away, and the third and fourth terms go away together, in accordance with our
earlier "sloppier" argument. Thus we are left with (for convenience, we have interchanged
the dummy indices κ and σ in the fifth term in (7))

$$[D_\mu, D_\nu]S_\rho = -(\partial_\mu \Gamma^\sigma_{\nu\rho}) S_\sigma + \Gamma^\kappa_{\mu\rho} \Gamma^\sigma_{\nu\kappa} S_\sigma - (\mu \leftrightarrow \nu)$$

$$= - \left(\partial_\mu \Gamma^\sigma_{\nu\rho} - \Gamma^\kappa_{\mu\rho} \Gamma^\sigma_{\nu\kappa} - (\mu \leftrightarrow \nu) \right) S_\sigma \tag{8}$$

in agreement with (4). Comparing with (5), we obtain the defining expression for the
Riemann curvature tensor:[*]

$$R^\sigma_{\rho\mu\nu} = (\partial_\mu \Gamma^\sigma_{\nu\rho} + \Gamma^\sigma_{\mu\kappa} \Gamma^\kappa_{\nu\rho}) - (\partial_\nu \Gamma^\sigma_{\mu\rho} + \Gamma^\sigma_{\nu\kappa} \Gamma^\kappa_{\mu\rho}) \tag{9}$$

Note that once we arrive here, we could care less about S_ρ: it is just a convenient crutch.

In summary, as anticipated, curvature expresses the failure of the Newton-Leibniz rule
for covariant derivatives.[†] Since the Christoffel symbol has the schematic form $\Gamma^{\cdot}_{\cdot\cdot} \sim g^{\cdot\cdot}\partial.g..$, the curvature tensor $R^{\cdot}_{\cdot\cdot\cdot}$ involves[‡] two derivatives acting on the metric, as we
anticipated, here and as far back as in chapter I.6.

---

[*] In part IX, we will give two alternative derivations of $R^\sigma_{\rho\mu\nu}$, one based on parallel transport (and closely
related to the derivation given here) and one based on geodesic deviation.

[†] This derivation has the added advantage that when and if you study Yang-Mills theory, you will see essentially
the same argument. The Yang-Mills field strength is also given by the commutator of two covariant derivatives.
Indeed, if you are familiar[8] with the gauge invariant derivative in electromagnetism $D_\mu \equiv \partial_\mu - iA_\mu$, we also have
$[D_\mu, D_\nu] = -iF_{\mu\nu}$, the electromagnetic field strength.

[‡] The curvature tensor clearly involves $\sim \partial\partial g$ and $\sim \partial g \partial g$. When evaluating $\partial.g^{\cdot\cdot}$, note that $g^{\cdot\cdot}$ is the inverse
of $g..$ and recall the identity (V.6.7).

In a locally flat coordinate system, the Riemann curvature tensor expresses the second order deviation of the metric from being flat, as we discussed in part I. In fact, $R^{\sigma}_{\rho\mu\nu}$ is the only tensor we can form out of two derivatives acting on the metric.

Professor Flat pops up, declaring, "That's easy to prove! Suppose there is another tensor with the same properties as the Riemann curvature tensor. Form the difference $\Delta^{\sigma}_{\rho\mu\nu}$. Since Δ vanishes in a locally flat system and since it is a tensor by assumption, it vanishes in all frames." This uniqueness is of crucial importance when we come to construct the action for gravity.

Symmetry properties of the Riemann tensor

The Riemann curvature tensor $R^{\lambda}_{\rho\mu\nu}$ is a formidable object, but fortunately it enjoys various symmetry properties upon interchange of indices that will make our lives a lot easier. Already, from the derivation in (8) and (9), we know that it is antisymmetric under the interchange of μ and ν:

$$R^{\lambda}_{\rho\mu\nu} = -R^{\lambda}_{\rho\nu\mu} \tag{10}$$

To discover further symmetries, we first let the four indices have, at least nominally, equal status. Thus, let us lower the index λ: $R_{\tau\rho\mu\nu} = g_{\tau\lambda}R^{\lambda}_{\rho\mu\nu}$.

What makes us think that there are additional symmetries? Our fingers. We count. Way back when in chapter I.6, we counted that 20 numbers were required to specify curvature in 4-dimensional spacetime. Here we count $4 \times 4 \times \frac{1}{2}(4 \times 3) = 96$ components in $R_{\tau\rho\mu\nu}$ thus far, which have to be reduced to 20 by symmetries.

Professor Flat ambles by again, just in time to save us a lot of work. He says: "Since you are looking for symmetry properties of a tensor, you could simply go to a locally flat coordinate system around some generic point P, just as in chapter I.6."

So, translate our coordinate system so that P is at the origin $x = 0$. Expanding, we write

$$g_{\tau\mu}(x) = \eta_{\tau\mu} + B_{\tau\mu,\lambda\sigma}x^{\lambda}x^{\sigma} + \cdots \tag{11}$$

where the Taylor coefficient $B_{\tau\mu,\lambda\sigma}$ is, by construction, symmetric under the interchange of τ and μ or of λ and σ. (Recall from chapter I.6 that the comma on the quantity B is purely for typographical convenience: it helps us separate mentally two sets of indices $\tau\mu$ and $\lambda\sigma$ that appear for different reasons.)

Good! Plug this into $\Gamma^{\lambda}_{\rho\nu} \equiv \frac{1}{2}g^{\lambda\tau}(\partial_{\rho}g_{\tau\nu} + \partial_{\nu}g_{\tau\rho} - \partial_{\tau}g_{\rho\nu})$ to find

$$\Gamma^{\lambda}_{\rho\nu} = \eta^{\lambda\tau}(B_{\tau\nu,\kappa\rho} + B_{\tau\rho,\kappa\nu} - B_{\rho\nu,\kappa\tau})x^{\kappa} + \cdots \tag{12}$$

As expected, $\Gamma^{\lambda}_{\rho\nu}$ vanishes at the point P, and thus the expression (9) for the Riemann curvature tensor simplifies enormously to $R^{\lambda}_{\rho\mu\nu} = \partial_{\mu}\Gamma^{\lambda}_{\rho\nu} - (\mu \leftrightarrow \nu)$. Furthermore, the ∂_{μ} acting on $\Gamma^{\lambda}_{\rho\nu}$ merely removes x^{κ} from the right hand side of (12) and sets the index κ to μ in what is left. We thus find easily

$$R^{\lambda}_{\rho\mu\nu} = \eta^{\lambda\tau}(B_{\tau\nu,\mu\rho} + B_{\tau\rho,\mu\nu} - B_{\rho\nu,\mu\tau}) - (\mu \leftrightarrow \nu) \tag{13}$$

Note that the middle term in the parentheses is symmetric in $\mu\nu$ and hence goes away. Lowering the λ index, we obtain

$$R_{\tau\rho\mu\nu} = B_{\tau\nu,\mu\rho} - B_{\rho\nu,\mu\tau} - (\mu \leftrightarrow \nu) = B_{\tau\nu,\mu\rho} - B_{\rho\nu,\mu\tau} - B_{\tau\mu,\nu\rho} + B_{\rho\mu,\nu\tau}$$

$$= (B_{\tau\nu,\rho\mu} + B_{\rho\mu,\tau\nu}) - (B_{\rho\nu,\tau\mu} + B_{\tau\mu,\rho\nu}) \tag{14}$$

To our surprise, we have

$$R_{\tau\rho\mu\nu} = -R_{\rho\tau\mu\nu} \tag{15}$$

as we could already have seen from the second term in (14). The antisymmetry of the Riemann curvature tensor $R_{\tau\rho\mu\nu}$ in the last pair of indices is, as remarked already, obvious from its construction, but this antisymmetry in the first pair of indices $\tau\rho$ catches us by surprise. It was completely obscured in (8) and in (9). (In chapter IX.7, we will give a better understanding of this unexpected symmetry.)

Staring at (14) and remembering the symmetry properties of $B_{..,..}$, we next discover that the Riemann tensor is symmetric upon interchange of the first and second pair of indices:

$$R_{\tau\rho\mu\nu} = R_{\mu\nu\tau\rho} \tag{16}$$

Note that (10) and (16) imply (15). In the same way, looking at (14), you can prove cyclicity in the last three indices: $R_{\tau\rho\mu\nu} + R_{\tau\mu\nu\rho} + R_{\tau\nu\rho\mu} = 0$.

Professor Flat: "Let me stress it again. Since these symmetry relations are tensor equations, they hold in any coordinate system, even though they are derived in a locally flat coordinate system."

As an exercise, you can show by explicit counting that, in 4-dimensional spacetime, the 96 components in $R_{\tau\rho\mu\nu}$ indeed reduce to 20 independent components, thanks to the symmetry relations just proved.

Onward to the Einstein-Hilbert action

> The last month I have lived through the most exciting and the most exacting period of my life. . . . I saw clearly that a satisfactory solution could only be reached by linking [the theory of gravity] with Riemann variations.
>
> —A. Einstein, writing to Arnold Sommerfeld, late 1915

After all this math, let us not lose sight of what we are after: we want to construct an action for gravity, an action to describe how spacetime curves under the grip of gravity. The action is required to be invariant under general coordinate transformations. It certainly must not depend on the observer! Remarkably, this requirement determines the action uniquely.

As explained earlier in chapter I.5, the coordinate invariant volume element is not d^4x, but $d^4x\sqrt{-g}$ where $g = \det(g_{..})$ denotes the determinant of the metric tensor regarded as a matrix. (For spacetime, g is negative, hence the minus sign.) Thus, we demand an action of the form $S_{\text{gravity}} = \int d^4x\sqrt{-g(x)}A(x)$, where A is a scalar (in other words,

$A'(x') = A(x)$ under a general coordinate transformation). For the action to govern the dynamics of spacetime curvature, the unknown scalar A should contain two derivatives acting on the metric, as already mentioned in chapter IV.2 and as explained earlier in this chapter.

You say, now that we have the Riemann curvature tensor $R_{\tau\rho\mu\nu}$, we could simply contract all the indices to obtain a scalar, namely a tensor with no indices. This scalar would then allow us to construct the action.

More explicitly, the first step in the process would be to multiply the curvature tensor by the metric tensor and sum over indices, obtaining a tensor with two indices. For example, we could construct $g^{\tau\mu} R_{\tau\rho\mu\nu}$, which is evidently a tensor with two lower indices ρ and ν. We say that we have contracted the indices τ and μ. We then repeat the process, multiplying by $g^{\rho\nu}$ to obtain $g^{\rho\nu} g^{\tau\mu} R_{\tau\rho\mu\nu}$.

Now we see why we have to study the symmetry properties of the curvature tensor. We need to know how many distinct scalars we can construct.

We want to get to Einstein gravity as quickly as possible, but not any quicker! Pausing to study the symmetry properties of the curvature tensor was unavoidable.

Indeed, some of the possible contractions turn out to give zero. For example, if we start by contracting $\mu\nu$, the result $g^{\mu\nu} R_{\tau\rho\mu\nu}$ would evidently vanish, since $R_{\tau\rho\mu\nu}$ is anti-symmetric in $\mu\nu$ by construction, while $g^{\mu\nu}$ is symmetric. (To see this, note that $g^{\mu\nu} R_{\tau\rho\mu\nu} = g^{\nu\mu} R_{\tau\rho\mu\nu} = -g^{\nu\mu} R_{\tau\rho\nu\mu}$ but an object equal to its negative can only be zero. Remember exercise I.4.5?)

But from what we just learned in (15), we can't contract $\tau\rho$ either. We can contract $\tau\mu$ and define a 2-indexed tensor known as the Ricci* tensor by

$$R_{\rho\nu}(x) \equiv g^{\tau\mu}(x) R_{\tau\rho\mu\nu}(x) \tag{17}$$

But this is the only possibility! All the other contractions you can think of either vanish (as mentioned above) or give the Ricci tensor again up to an overall sign. For instance, $g^{\rho\mu} R_{\tau\rho\mu\nu} = -g^{\rho\mu} R_{\rho\tau\mu\nu} = -R_{\tau\nu}$. You can get only one single 2-indexed tensor by contracting two of the four indices of the Riemann tensor!

Notice that (16) implies that the Ricci tensor is symmetric.

Now there is only one way, duh, to contract the two indices on the Ricci tensor:

$$R(x) \equiv g^{\rho\nu}(x) R_{\rho\nu}(x) \tag{18}$$

This second contraction produces a scalar very imaginatively named the scalar curvature and also denoted by the letter R. I trust you not to be confused by this standard notation: there are three different tensors, all carrying the name R in honor of Riemann (or perhaps also Ricci!). The Riemann, Ricci, and scalar curvatures are distinguished by how many indices they carry: 4, 2, or 0, respectively.

In summary, out of the Riemann curvature tensor, we can form one and only one scalar, namely $R(x)$. Remarkable! Under a general coordinate transformation, the 20 components

---

* Gregorio Ricci-Curbastro shortened his last name when he published the most important paper of his career. Perhaps there is a lesson in there somewhere for the reader.

(for 4-dimensional spacetime) of the curvature tensor transform into linear combinations of each other with coefficients that depend on the transformation, but out of these, only one combination remains unchanged. The scalar curvature is the unique scalar we can form out of the metric and two powers of derivatives.

Thus, general coordinate invariance fixes the action for gravity uniquely to be $\mathcal{I} = \int d^4x \sqrt{-g(x)} R(x)$, the spacetime integral of $R(x)$, times some overall constant.

Let us do some simple dimensional analysis. The action has dimensions of mass times length, which we will write as ML. (To see this, simply recall the action for a point particle $S_{\text{point particle}} = -m \int d\tau$. We are of course using units in which $c = 1$.) In contrast, the metric $g_{..}$ is dimensionless and R, constructed out of the metric and two derivatives, has dimension $\frac{1}{L^2}$. Hence the integral $\mathcal{I}$ has dimension $L^4 \frac{1}{L^2} = L^2$. To obtain the action, we have to divide the integral by a constant with the dimensions of $\frac{L}{M}$ to get the dimension right. But recall from the introduction that Newton's constant G has precisely dimension $\frac{L}{M}$.

A highly satisfying fact! Einstein did not have to introduce any[9] new fundamental constants into physics to construct his theory of gravity. The action can only be $S = aG^{-1} \int d^4x \sqrt{-g} R$, with a some pure number fixed by the requirement that S reduces to Newtonian gravity in the appropriate limit. As we will see, $a = \frac{1}{16\pi}$, but we will not need this historical number for quite a while.

The action for gravity, known as the Einstein-Hilbert action after its two discoverers, is thus (trumpets please)

$$S_{\text{EH}} \equiv \frac{1}{16\pi G} \int d^4x \sqrt{-g} R \tag{19}$$

The Einstein-Hilbert action possesses a wonderfully unique quality. As Ludwig Beethoven declared about his composition, "Muß es sein? Es muß sein." [Must it be? It must be.] Art in its perfection must be a necessity.[10] (See figure 2.)

We conclude that the action[11] of the universe is given by

$$S = S_{\text{EH}} + S_{\text{matter}} \tag{20}$$

Figure 2 "Must it be? It must be." (Illustration adapted from *Fearful*.)

where S_{matter} is defined as the action for everything else (for example, including the electromagnetic field).

Einstein's field equation

After we finish celebrating our success in deriving the action for gravity, we realize that, as Euler and Lagrange remind us, we still have to vary the action S with respect to the gravitational field $g_{\mu\nu}$ to obtain Einstein's field equation. This variation turns out to involve a bit of work. For instance, to determine how R varies with respect to $g_{\mu\nu}$, we have to determine how the Riemann curvature tensor varies, and to get that, we have to determine how the Christoffel symbol varies. Not that difficult to do, but it involves some set up. We will postpone this until chapter VI.5.

Meanwhile, I will show you that we can get remarkably far doing practically no work. First, without sweating through the actual varying, we will give a name to the result. Define a tensor with two upper indices $K^{\mu\nu}(x)$ by

$$\delta\mathcal{I} = \delta \int d^4x \sqrt{-g(x)} R(x) \equiv \int d^4x \sqrt{-g(x)} K^{\mu\nu}(x)\delta g_{\mu\nu}(x) \tag{21}$$

In varying S in (20), we have to vary S_{matter} also, of course. Here, in the spirit of getting to Einstein's field equation as quickly as possible, I will take a shortcut. We content ourselves, for the moment, with Einstein's field equation for empty spacetime; that is, we drop S_{matter} from the action, so that we avoid having to vary S_{matter}. With that simplification, the field equation is given simply by $K^{\mu\nu}(x) = 0$.

As we will see, this suffices to derive the Schwarzschild metric and hence the physics of the three classic tests and of black holes.

So, let us try to get away with doing as little work as possible. Even though we don't know what $K^{\mu\nu}$ is, we know a lot about him. Since in (21), the two derivatives contained in $R(x)$ cannot disappear into thin air, $K^{\mu\nu}(x)$ must contain two derivatives, no more, no less. Also, $K^{\mu\nu}$ is manifestly a tensor, since in (21) $\delta\mathcal{I}$ is a scalar and $\delta g_{\mu\nu}$ is a tensor.

So list all the tensors with two upper indices we know of. The metric tensor $g^{\mu\nu}$ comes to mind (of course), but it does not contain derivatives. However, we could multiply $g^{\mu\nu}$ by the scalar curvature, which does contain two derivatives, just what we need. So $g^{\mu\nu}R$ is one candidate. Next up is the Ricci tensor $R^{\mu\nu}$, which also contains two derivatives. That's it. (You might suggest, for example, the tensor $R^{\rho\mu\sigma\nu}R_{\rho\sigma}$, but it contains four derivatives.) In other words, $K^{\mu\nu}$ must be a linear combination of $g^{\mu\nu}R$ and $R^{\mu\nu}$: $K^{\mu\nu} = A(R^{\mu\nu} + \alpha g^{\mu\nu}R)$, with A and α two unknown numerical constants we will have to work to determine.

Hence, with almost no work, we obtain Einstein's field equation in empty spacetime:

$$K^{\mu\nu} = A(R^{\mu\nu} + \alpha g^{\mu\nu}R) = 0 \tag{22}$$

The constant A cannot vanish, since that would imply that the proposed action is independent[12] of the metric. Next, multiplying (22) by $g_{\mu\nu}$, we obtain $(1 + 4\alpha)R = 0$. Unless we

Figure 3 Einstein writing his field equation.

are extremely unlucky and α turns out to be $-\frac{1}{4}$ (we will verify soon enough that it is not), we can conclude that $R = 0$. Hence (22) says that

$$R^{\mu\nu} = 0 \tag{23}$$

This is the Einstein field equation in empty spacetime. (In figure 3, we see Einstein writing[13] this famous equation.) It says that the Ricci tensor vanishes. We speak of the field equation in the singular, but in fact it consists of a set of equations according to the values taken by $\mu\nu$. It is important to realize that the vanishing of the Ricci tensor does not imply the vanishing of the Riemann curvature tensor. Evidently, $R_{\mu\nu}(x) = g^{\tau\sigma}(x)R_{\tau\mu\sigma\nu}(x)$, given by a sum over the components of the Riemann curvature tensor, can vanish without $R_{\tau\mu\sigma\nu}(x)$ having to vanish. Were the Riemann curvature tensor zero, spacetime would be flat. Einstein tells us that, in empty spacetime, a particular sum of the various components of the Riemann curvature tensor vanishes.

We are now in a position to derive the celebrated Schwarzschild metric around a mass distribution, study black holes, and even play with the universe under some circumstances. See the following chapters.

I will show by a simple calculation in appendix 1 that $\alpha \neq -\frac{1}{4}$. I could have done it here, but I did not want to slow us down on our way to the famous field equation for empty spacetime.

What we have done

Since this has been a lightning fast way—the fastest I know of—of deriving the Einstein field equation (albeit in empty spacetime), it is worthwhile summarizing what we have done.[*]

Ever since chapter IV.3, we have suspected that the action for the gravitational field has to contain two powers of derivatives acting on the dynamical variable, namely the metric. And even earlier, ever since chapter I.6, we have anticipated (more than once in fact) that the curvature is given by two powers of derivatives acting on the metric.

To determine the curvature tensor, we acted with the commutator $[D_\mu, D_\nu]$ on some arbitrary vector field $S_\rho(x)$. This involved two powers of derivatives and the metric tensor all over the place, and thus had the structure we were looking for. The beginner might have felt a bit overwhelmed by the plethora of indices, but in fact it only took two lines to get from (6) to (8). Once we obtained the curvature tensor $R^\sigma{}_{\rho\mu\nu}$, we followed Professor Flat, as always, to locally flat coordinates and worked out the symmetry properties of the tensor, which showed us that there was one, and only one, scalar we can form to put into the action.

Thus was the Einstein-Hilbert action S_{EH} uniquely determined.

If we dispense with matter for the time being, we don't even have to do any work varying to obtain the equation of motion $\frac{\delta S_{\text{EH}}}{\delta g_{\mu\nu}} = 0$. If a certain constant (α in (22)) does not have a particular value, we can wing it and argue by symmetry considerations that the field equation amounts to simply $R_{\mu\nu} = 0$. In the next chapter, we will solve this equation.

Riemann curvature tensor

During our headlong sprint to the field equation, we barely noticed that we derived the long-sought expression for the Riemann curvature tensor. From way back early in this book, already in part I, we have talked about curvature and have sought to calculate it. We discussed some intuitive methods. You might recall that one method, if the space is a surface, involves marking out a circle and measuring its radius and circumference. In another, we have to determine the tangent plane. More generally, we can go to locally

---

[*] After we had worked out the symmetry properties of the Riemann tensor, we could, of course, have argued directly from (22) without mentioning the action. See appendix 6 in chapter VI.5.

flat coordinates and calculate the quadratic deviation of the metric. All these methods, while conceptually clear, are rather unwieldy to implement in practice, adequate mostly for curved surfaces only. With this perspective, we can appreciate the power of the Riemann curvature tensor. Given a metric, determine the Christoffel symbols (most efficiently by using the action principle to obtain the geodesic equations). Then simply plug in (9) to obtain the curvature.

For fun, we can try it on the sphere with $ds^2 = d\theta^2 + \sin^2\theta \, d\varphi^2$. Note from chapter II.2 that the only nonvanishing components, $\Gamma^\theta_{\varphi\varphi} = -\sin\theta\cos\theta$ and $\Gamma^\varphi_{\theta\varphi} = \frac{\cos\theta}{\sin\theta}$, are independent of φ, and so $R^\varphi_{\theta\varphi\theta} = (\partial_\varphi\Gamma^\varphi_{\theta\theta} + \Gamma^\varphi_{\varphi\kappa}\Gamma^\kappa_{\theta\theta}) - (\partial_\theta\Gamma^\varphi_{\varphi\theta} + \Gamma^\varphi_{\theta\kappa}\Gamma^\kappa_{\varphi\theta}) = -(\partial_\theta\Gamma^\varphi_{\varphi\theta} + \Gamma^\varphi_{\theta\varphi}\Gamma^\varphi_{\varphi\theta}) = -(\partial_\theta\frac{\cos\theta}{\sin\theta} + (\frac{\cos\theta}{\sin\theta})^2) = +1$. We know from chapter I.6 that in 2 dimensions, the Riemann curvature tensor has only one component. So all the other components must either vanish or be related to the one we just calculated: for example, $R^\theta_{\varphi\theta\varphi} = g^{\theta\theta}R_{\theta\varphi\theta\varphi} = R_{\theta\varphi\theta\varphi} = R_{\varphi\theta\varphi\theta} = g_{\varphi\varphi}R^\varphi_{\theta\varphi\theta} = \sin^2\theta$. Thus, $R_{\theta\theta} = 1$ and $R_{\varphi\varphi} = \sin^2\theta$, giving $R = g^{\theta\theta}R_{\theta\theta} + g^{\varphi\varphi}R_{\varphi\varphi} = 2$, a constant, as might be expected.

Best of all, the procedure is straightforwardly algorithmic, and you can easily instruct a computer to produce the curvature tensor once you input the metric. Even with my rather rudimentary computer skills, I was able to do it.

Appendix 1: A scaling argument on the way to Einstein's field equation

In the text, instead of sweating through actually varying the action with respect to $g_{\mu\nu}$, we winged it, arguing from symmetry and other general considerations that the variation must have the form (21): $\delta\mathcal{I} = \delta\int d^4x\sqrt{-g}R = \int d^4x\sqrt{-g}A(R^{\mu\nu} + \alpha g^{\mu\nu}R)\delta g_{\mu\nu}$. In this expression, we must take for $\delta g_{\mu\nu}$ an arbitrary variation, as Euler and Lagrange had instructed us.

We now give a simple proof that α cannot be $-\frac{1}{4}$. The trick is to consider a specific, rather than an arbitrary, variation, and to pick an especially simple variation, namely $g_{\mu\nu}(x) \to \tilde{g}_{\mu\nu}(x) = \Omega^2 g_{\mu\nu}(x)$ with Ω a number. In other words, we scale* the metric. For an infinitesimal transformation, we write $\Omega^2 \simeq 1 + \varepsilon$. Then $\delta g_{\mu\nu}(x) \equiv \tilde{g}_{\mu\nu}(x) - g_{\mu\nu}(x) \simeq \varepsilon g_{\mu\nu}(x)$.

Under $g_{\mu\nu} \to \Omega^2 g_{\mu\nu}$, we have $\Gamma \sim g^{..}\partial.g.. \to \Omega^{-2}\Omega^2 g^{..}\partial.g.. \sim \Gamma$, that is, $\Gamma \to \Gamma$ and so $R^{\cdot}_{\cdots} \sim \partial\Gamma + \Gamma\Gamma \to R^{\cdot}_{\cdots}$, thus leading to $R_{..} \to R_{..}$ and $R = g^{\mu\nu}R_{\mu\nu} \to \Omega^{-2}R$. Since $g \to \Omega^8 g$ (being the determinant of a 4-by-4 matrix), we obtain $\mathcal{I} \to \Omega^4\Omega^{-2}\mathcal{I} = \Omega^2\mathcal{I}$ and hence $\delta\mathcal{I} = \Omega^2\mathcal{I} - \mathcal{I} \simeq \varepsilon\mathcal{I}$, where in the last step we have gone to the infinitesimal limit $\Omega^2 \simeq 1 + \varepsilon$. However, plugging $\delta g_{\mu\nu} = \varepsilon g_{\mu\nu}$ into (21), we have

$$\delta\mathcal{I} = \varepsilon\int d^4x\sqrt{-g}A(R^{\mu\nu} + \alpha g^{\mu\nu}R)g_{\mu\nu} = \varepsilon A(1 + 4\alpha)\int d^4x\sqrt{-g}R = \varepsilon A(1 + 4\alpha)\mathcal{I}$$

Equating our two results for $\delta\mathcal{I}$, we find $A(1 + 4\alpha) = 1$. Ta dah, $\alpha \neq -\frac{1}{4}$. In chapter VI.5, we will show that $\alpha = -\frac{1}{2}$.

Conceptually, it is important to realize that the transformation $g_{\mu\nu}(x) \to \tilde{g}_{\mu\nu}(x) = \Omega^2 g_{\mu\nu}(x)$ is not a general coordinate transformation. Of course not, since the general coordinate invariant object $\mathcal{I}$ actually varies under this transformation. But students are often confused, because it looks like a general coordinate transformation. Be careful! If we plug the coordinate transformation $x'^\mu = x^\mu/\Omega$ into the general formula $g'_{\rho\sigma}(x') = g_{\mu\nu}(x)\frac{\partial x^\mu}{\partial x'^\rho}\frac{\partial x^\nu}{\partial x'^\sigma}$, we obtain $g'_{\mu\nu}(x/\Omega) = \Omega^2 g_{\mu\nu}(x)$. The crucial difference is that the argument on the left hand side is not x, but x/Ω.

---

* See chapter IX.9 for a more extensive discussion of this and related transformations.

Appendix 2: A mnemonic

The expression for the Riemann curvature tensor in (9) is perhaps the most involved you have encountered thus far in your study of physics, and so I am moved to say a few words about how it is not that hard to remember. Of course, in general, there is not much point to memorizing physics formulas: it is much better to reconstruct them when needed, or failing that, to look them up.

For convenience, let me repeat (9) here:

$$R^{\sigma}{}_{\rho\mu\nu} = (\partial_{\mu}\Gamma^{\sigma}{}_{\nu\rho} + \Gamma^{\sigma}{}_{\mu\kappa}\Gamma^{\kappa}{}_{\nu\rho}) - (\partial_{\nu}\Gamma^{\sigma}{}_{\mu\rho} + \Gamma^{\sigma}{}_{\nu\kappa}\Gamma^{\kappa}{}_{\mu\rho}) \tag{24}$$

First, the general structure is clear from our schematic derivation $R^{\sigma}{}_{\rho\mu\nu} \sim [\partial + \Gamma, \partial + \Gamma] \sim \partial.\Gamma^{\cdot}{}_{\cdot\cdot} + \Gamma^{\cdot}{}_{\cdot\cdot}\Gamma^{\cdot}{}_{\cdot\cdot}$. No doubt, everybody could come up with a different mnemonic for where the indices go. We need only construct the first half of the expression in (24), since we can obtain the second half by interchanging $\mu \leftrightarrow \nu$ in the first half.

We have three lower indices but only one upper index on $R^{\sigma}{}_{\rho\mu\nu}$ so σ is "king." In the first term $\partial.\Gamma^{\sigma}{}_{\cdot\cdot}$, the only question is which of the three lower indices on $R^{\sigma}{}_{\rho\mu\nu}$ goes on the partial derivative? It can't be ρ, because $\Gamma^{\sigma}{}_{\mu\nu}$ is symmetric in $\mu\nu$, but $R^{\sigma}{}_{\rho\mu\nu}$ is antisymmetric by construction. It could be μ or ν, but μ "comes before" ν in $R^{\sigma}{}_{\rho\mu\nu}$, and so we pick* him, marking μ as "special." So of the 4 indices, we have separated two guys, "king" and "special." The first term in (24) is thus uniquely fixed as $\partial_{\mu}\Gamma^{\sigma}{}_{\nu\rho}$. In the second term $\Gamma^{\cdot}{}_{\cdot\cdot}\Gamma^{\cdot}{}_{\cdot\cdot}$, we have two upper slots and four lower slots for us to lodge one upper index σ and three lower indices $\rho\mu\nu$ into. Once we put σ in, since we don't have another upper index, the remaining upper slot in $\Gamma^{\sigma}{}_{\cdot}\Gamma^{\cdot}{}_{\cdot\cdot}$ has to be occupied by a dummy index κ to be summed with its lower counterpart: but this lower κ cannot be on the same Γ as the upper κ, since we know from chapter V.6 that $\Gamma^{\kappa}{}_{\nu\kappa}$ simplifies, and we don't remember that happening in the derivation. So for the second term, we have $\Gamma^{\sigma}{}_{\cdot\kappa}\Gamma^{\kappa}{}_{\cdot\cdot}$ thus far. We also know that $\Gamma^{\kappa}{}_{\cdot\cdot}$ cannot be $\Gamma^{\kappa}{}_{\mu\nu}$ by symmetry; thus, the indices μ and ν can't be on the same Γ. Since μ is "special," he clamors to be with the "king" on the same Γ. So we arrived at $\Gamma^{\sigma}{}_{\mu\kappa}\Gamma^{\kappa}{}_{\nu\rho}$. Finally, remember to antisymmetrize in the last two indices.

Anyway, that is how I do it, but the reader might come up with a better way. These days, of course, as I said, what I do is simply write an algebraic manipulation program once and for all. If you are marooned on a deserted island, probably your best tack would be to go through the simple derivation commuting two covariant derivatives.

In chapters IX.7 and IX.8, I will describe a more powerful method for calculating curvature using differential forms.

Exercises

1   Given two vector fields W_{ρ} and U^{ρ} on the sphere (with $\rho = \theta$, φ of course), calculate $D_{\nu}W_{\rho}$ and $D_{\nu}U^{\rho}$ explicitly. As a small check, show that $(D_{\nu}W_{\rho})U^{\rho} + W_{\rho}(D_{\nu}U^{\rho})$ is equal to $\partial_{\nu}(W_{\rho}U^{\rho})$.

2   Evaluate (5) explicitly on the sphere and thus obtain the Riemann curvature tensor for the sphere.

3   Derive from (14) another important property, namely that the curvature tensor has the cyclic symmetry

$$R_{\tau\rho\mu\nu} + R_{\tau\mu\nu\rho} + R_{\tau\nu\rho\mu} = 0$$

We hold the index τ fixed and cylically permute the triplet $\rho\mu\nu$. Again, this is hardly evident from (9). Show that this imposes $d(d-1)(d-2)(d-3)/24$ constraints.

4   Combine the definition of the Riemann curvature tensor in (5) $V_{\rho;\mu;\nu} - V_{\rho;\nu;\mu} = -R^{\sigma}{}_{\rho\mu\nu}V_{\sigma}$ with exercise 3 to show that, for any vector V,

---

* Some authors pick ν as special because he is last. A word of caution: the curvature tensors used in the literature can thus differ by an overall sign.

$$V_{\rho;\mu;\nu} - V_{\rho;\nu;\mu} + V_{\nu;\rho;\mu} - V_{\nu;\mu;\rho} + V_{\mu;\nu;\rho} - V_{\mu;\rho;\nu} = 0 \tag{25}$$

We will need this when we discuss isometries in chapter IX.6.

5 Show that (5), which says that $V_{\rho;\mu;\nu} - V_{\rho;\nu;\mu} = -R^{\sigma}{}_{\rho\mu\nu}V_{\sigma}$ for any vector, can be generalized to any tensor, for example, $T_{\mu\rho;\nu;\omega} - T_{\mu\rho;\omega;\nu} = -R^{\sigma}{}_{\mu\nu\omega}T_{\sigma\rho} - R^{\sigma}{}_{\rho\nu\omega}T_{\mu\sigma}$.

6 Given its various symmetry properties, the Riemann curvature tensor in 2 dimensions has only one independent component R_{1212}, with all other vanishing components related to it. Various components, such as R_{1122}, all vanish. Show that all these facts can be summarized compactly by the expression $R_{\tau\rho\mu\nu} = R_{1212}(g_{\tau\mu}g_{\rho\nu} - g_{\tau\nu}g_{\rho\mu})/g$, with $g = \det g = g_{11}g_{22} - g_{12}g_{21}$. Contract to find the Ricci tensor and the scalar curvature. Show that $R_{\tau\rho\mu\nu} = \frac{1}{2}R(g_{\tau\mu}g_{\rho\nu} - g_{\tau\nu}g_{\rho\mu})$.

7 In cases where the metric has the form (11), (14) gives an alternative to calculate the Riemann curvature tensor. Do this for the example in chapter I.6 with the metric $ds^2 = dx^2 + dy^2 + dz^2 = dx^2 + dy^2 + ((ax + cy)dx + (by + cx)dy)^2$. Show that it is given by

$$R_{1212} = 2B_{12,12} - B_{11,22} - B_{22,11}$$

Hint: The first term on the right hand side of (11) is $B_{11,22}$, which is the coefficient of y^2 in g_{11}, which we can read off as c^2.

8 Back in chapter I.5, I asked which of the two spaces described by $ds^2 = (1 + u^2)du^2 + (1 + 4v^2)dv^2 + 2(2v - u)dudv$ and $ds^2 = (1 + u^2)du^2 + (1 + 2v^2)dv^2 + 2(2v - u)dudv$ is curved. Now you can answer this question readily.

9 Consider a $(4 + 1)$-dimensional spacetime with $ds^2 = \eta_{\mu\nu}dx^\mu dx^\nu + \phi(x)^2 dy^2$. Note that the 4th spatial coordinate is called y and that the function $\phi(x)$ does not depend on y. Show that the scalar curvature is given by $R = -2\frac{\Box\phi}{\phi}$, where $\Box\phi \equiv \frac{1}{\sqrt{-g}}\partial_\mu(\sqrt{-g}g^{\mu\nu}\partial_\nu\phi)$ simplifies in the present instance to $\eta^{\mu\nu}\partial_\mu\partial_\nu\phi$. We will need the result of this exercise in chapter IX.1.

10 Petrov notation: group the four indices carried by $R_{\tau\rho\mu\nu}$ into two sets of two, in other words, write R_{AB} with the indices $A \equiv (\tau\rho)$ and $B \equiv (\mu\nu)$ and regard R_{AB} as a matrix. The various symmetry properties can then be interpreted as properties of this matrix.
(a) Show that the indices A and B each take on $\frac{1}{2}d(d - 1)$ values.
(b) Show that R_{AB} is a symmetric matrix, and count the number of independent components it contains due to this fact.
(c) Count the number of independent components after imposing the constraints mandated by exercise 3. Show that the resulting number of independent components contained in the Riemann curvature tensor agrees with that given in chapter I.6. Hint: Check your computation after each step for some small values of d.

11 Is there a dimension d in which the Riemann and the Ricci tensors have the same number of independent components? Hint: The answer is contained in the next exercise.

12 The result of exercise 11 suggests that, for $d = 3$, there exists a relation between the Riemann and the Ricci tensors. Using various properties of these tensors, you can practically write down this relation:

$$R_{\tau\rho\mu\nu} = g_{\tau\mu}R_{\rho\nu} - g_{\tau\nu}R_{\rho\mu} - g_{\rho\mu}R_{\tau\nu} + g_{\rho\nu}R_{\tau\mu} - \frac{1}{2}(g_{\tau\mu}g_{\rho\nu} - g_{\tau\nu}g_{\rho\mu})R$$

Prove this. Hint: Ask Professor Flat for help.

13 Given a metric $g_{\mu\nu}$, let's construct another metric $\tilde{g}_{\mu\nu}(x) = \Omega^2(x)g_{\mu\nu}(x)$. The two metrics are said to be conformally related. (Recall that way back in chapter I.5, you worked with conformally flat metrics in an exercise. In the present terminology, a metric is said to be conformally flat if it is conformally related to the flat metric.) Show that various quantities calculated for $\tilde{g}_{\mu\nu}$ can be expressed in terms of the corresponding quantities calculated for $g_{\mu\nu}$ as follows:

$$\tilde{\Gamma}^{\mu}_{\nu\lambda} = \Gamma^{\mu}_{\nu\lambda} + \frac{1}{\Omega}(\delta^{\mu}_{\nu}\partial_{\lambda}\Omega + \delta^{\mu}_{\lambda}\partial_{\nu}\Omega - g_{\nu\lambda}g^{\mu\rho}\partial_{\rho}\Omega) \tag{26}$$

$$\tilde{R}^{\mu}_{\ \nu\lambda\sigma} = R^{\mu}_{\ \nu\lambda\sigma} - \left(\delta^{\mu}_{\lambda}\delta^{\rho}_{\sigma}\delta^{\omega}_{\nu} - \delta^{\mu}_{\sigma}\delta^{\rho}_{\lambda}\delta^{\omega}_{\nu} + g_{\nu\sigma}g^{\mu\omega}\delta^{\rho}_{\lambda} - g_{\nu\lambda}g^{\mu\omega}\delta^{\rho}_{\sigma}\right)\frac{D_{\rho}\partial_{\omega}\Omega}{\Omega}$$

$$+ \left(2\delta^{\mu}_{\lambda}\delta^{\rho}_{\sigma}\delta^{\omega}_{\nu} - 2g_{\nu\lambda}g^{\mu\omega}\delta^{\rho}_{\sigma} + 2g_{\nu\sigma}g^{\mu\omega}\delta^{\rho}_{\lambda} - 2\delta^{\mu}_{\sigma}\delta^{\omega}_{\nu}\delta^{\rho}_{\lambda} + g_{\nu\lambda}g^{\rho\omega}\delta^{\mu}_{\sigma} - g_{\nu\sigma}g^{\rho\omega}\delta^{\mu}_{\lambda}\right)\frac{(\partial_{\rho}\Omega)(\partial_{\omega}\Omega)}{\Omega^2}$$

$$\tilde{R}_{\nu\lambda} = R_{\nu\lambda} - \left[(d-2)\delta^{\rho}_{\nu}\delta^{\omega}_{\lambda} + g_{\nu\lambda}g^{\rho\omega}\right]\frac{D_{\rho}\partial_{\omega}\Omega}{\Omega} + \left[2(d-2)\delta^{\rho}_{\nu}\delta^{\omega}_{\lambda} - (d-3)g_{\nu\lambda}g^{\rho\omega}\right]\frac{(\partial_{\rho}\Omega)(\partial_{\omega}\Omega)}{\Omega^2}$$

$$\tilde{R} = \frac{R}{\Omega^2} - 2(d-1)g^{\rho\omega}\frac{D_{\rho}\partial_{\omega}\Omega}{\Omega^3} - (d-1)(d-4)g^{\rho\omega}\frac{(\partial_{\rho}\Omega)(\partial_{\omega}\Omega)}{\Omega^4} \tag{27}$$

Here we can of course always write $D_{\rho}\partial_{\sigma}\Omega$ more symmetrically as $D_{\rho}D_{\sigma}\Omega$. Some of these expressions suggest that we write $\Omega = e^{\omega}$, as some authors prefer.

14 For a conformally flat metric $\tilde{g}_{\mu\nu}$, the results of the preceding exercise are particularly useful, since $R^{\mu}_{\ \nu\lambda\sigma}$ and all the curvature invariants derived from $g_{\mu\nu}$ vanish. Recall that the sphere is conformally flat, with $ds^2 = \left(1 + \frac{\rho^2}{4L^2}\right)^{-2}(d\rho^2 + \rho^2 d\varphi^2)$. Verify that the curvature of the sphere is in fact constant using the results of the preceding exercise.

15 Weyl tensor: the Weyl tensor in d-dimensional spacetime (or space) is defined by

$$C_{\mu\nu\rho\sigma} \equiv R_{\mu\nu\rho\sigma} + (d-2)^{-1}(g_{\mu\sigma}R_{\rho\nu} + g_{\nu\rho}R_{\sigma\mu} - g_{\mu\rho}R_{\sigma\nu} - g_{\nu\sigma}R_{\rho\mu})$$

$$+ ((d-1)(d-2))^{-1}(g_{\mu\rho}g_{\sigma\nu} - g_{\mu\sigma}g_{\rho\nu})R$$

(a) Show that the Weyl tensor has all the same symmetries as the Riemann tensor but that in addition it is traceless: if we contract any pair of indices carried by the Weyl tensor with the metric, we get nothing.

(b) Using the result of exercise 12, show that if two metrics $\tilde{g}_{\mu\nu}$ and $g_{\mu\nu}$ are conformally related, $\tilde{C}^{\mu}_{\ \nu\rho\sigma} = C^{\mu}_{\ \nu\rho\sigma}$ (note the one raised index). It follows that the Weyl tensor vanishes if the metric is conformally flat. Hence, the Weyl tensor is also known as the conformal tensor and can be used to test for conformal flatness (just as the Riemann tensor is used to test for plain old flatness).

16 Recall from chapter V.6 the definition of D_V associated with a vector field V. Show that for three vector fields U, V, W, we have

$$D_U D_V W^{\lambda} - D_V D_U W^{\lambda} = D_{[U,V]}W^{\lambda} + R^{\lambda}_{\ \sigma\mu\nu}U^{\mu}V^{\nu}W^{\sigma}$$

17 Show that the space described by $ds^2 = y^2 dx^2 + x^2 dy^2$ is actually flat (a) by direct calculation of the Riemann curvature and (b) by showing that the metric is conformally flat and then using the result of exercise 13.

Notes

1. A. Einstein, *Essays in Science*, p. 84.
2. Not to mention the Long March!
3. As explained in *QFT Nut*, and as we will briefly describe in an appendix to chapter IX.7, the Dirac action for spin $\frac{1}{2}$ fields provides an interesting "exception," but not truly an exception, as we could formulate the principle more cumbersomely by saying "two or fewer powers" rather than "two powers."
4. For a pedagogical discussion, see R. Woodard, arXiv:0907.4238, p. 31.
5. Recall chapter I.3 on rotations.
6. In the previous paragraph, Einstein wrote, "I worked on these problems from 1912 to 1914 together with my friend Grossmann."
7. A. Einstein, *Essays in Science*, p. 83.

8. The appearance of the imaginary unit i indicates that in electromagnetism, the covariant derivative is to be understood in the context of quantum mechanics. In contrast, in gravity, the covariant derivative is a purely classical construct. You may or may not know that the Schrödinger equation for a (nonrelativistic) charged particle in a magnetic field reads

$$i\frac{\partial \psi}{\partial t} = H\psi = -\frac{1}{2m}(\vec{\nabla} - i\vec{A})^2 \psi \tag{28}$$

which (as you can see) is obtained from the Schrödinger equation in the absence of the magnetic field by turning the ordinary derivative $\vec{\nabla}$ into the covariant derivative $\vec{\nabla} - i\vec{A}$. The relativistic completion of this is evidently $\partial_\mu - iA_\mu$. For those readers unfamiliar with (28), here is a quick derivation. Start with the classical Lagrangian for a charged particle in a magnetic field $L = \frac{m}{2}(\frac{d\vec{q}}{dt})^2 - \vec{A} \cdot \frac{d\vec{q}}{dt}$. (This is simply the nonrelativistic version of the Lagrangian, which can be read off from the action studied in chapter IV.1, with the proper time τ replaced by time t. We also denote the position of the particle by $\vec{q}$ to conform to standard usage in this context.) The conjugate momentum is then given by $\vec{p} = \frac{\delta L}{\delta \frac{d\vec{q}}{dt}} = m\frac{d\vec{q}}{dt} - \vec{A}$. Eliminating $\frac{d\vec{q}}{dt} = (\vec{p} + \vec{A})/m$ in the Hamiltonian (recall chapter III.5) $H(p, q) = \vec{p} \cdot \frac{d\vec{q}}{dt} - L(\vec{q}, \frac{d\vec{q}}{dt})$, we obtain $H = \frac{1}{2m}(\vec{p} + \vec{A})^2$. Finally, we go from classical mechanics to quantum mechanics by setting $\vec{p} \to i\vec{\nabla}$, thus obtaining (28).

9. In contrast, in contemporary particle theory, every time we turn around to construct a new action (following Einstein's lead in fact) to explain something or another, we run into what seem like 29 new and hitherto unmeasured constants. Imagine what the history of Einstein gravity would have been like if the action contained 7 constants and experimentalists had to go out and measure 6 of them.

10. *Fearful*, pp. 93–94.

11. To me, the Einstein-Hilbert action is just about the simplest, and hence the most beautiful, action in all of physics. Of course, as they say, simplicity is in the eyes of the beholder, and you the beholder have to know what the letter R stands for.

12. In that case, $\mathcal{I}$ would be a topological invariant; we are implicitly assuming that it is not.

13. From the back cover of the Japanese translation of my popular book *An Old Man's Toy*. As far as I know, Einstein did not write backward as a matter of habit. The photo was just printed this way, showing that my Japanese publisher did not know Einstein gravity. One of my distinguished colleagues quipped that Einstein would naturally write in Hebrew.

VI.2 To Cosmology as Quickly as Possible

The universe made as simple as possible, but not any simpler

[I had] again committed, in regards to gravity, something which puts me in danger of being shut up in an insane asylum.
—Einstein, writing to Paul Ehrenfest on February 4, 1917

Newton was bold enough to apply his mechanics to the celestial sphere where the planets reside. But the audacity of Einstein, thinking that the universe could[1] be described by an equation! Turned out that he was right and in no danger of being dragged away by men in white coats.

In the preceding chapter, we derived, with scandalously little work, Einstein's field equation

$$R_{\mu\nu}(x) = 0 \tag{1}$$

in empty spacetime. We first deduced on general grounds, invoking symmetry considerations with gleeful abandon, what the action for Einstein gravity must be. Then, instead of carefully varying the action as dutiful scholars would, we winged it like a lazy southern Californian, and arrived at (1). We could now derive the warped spacetime around a mass, a star or a black hole, but I will postpone that until the next chapter.

In the spirit of the preceding chapter, I would like to get you to cosmology as quickly as possible.

Filling the universe with a constant energy density

We obtained (1) for empty spacetime. Here, let's get the universe expanding by filling it with the "simplest stuff"[*] we know of, namely a positive[†] constant energy density Λ.

It is instructive to go all the way back to the most elementary example of an action, namely the action of a nonrelativistic particle in a potential

$$S_{\text{nonrelativistic particle}} = \int dt \left(\frac{1}{2} m \left(\frac{dq}{dt} \right)^2 - V \right)$$

and ask how we would include a constant energy density Λ pervading all of space. Well, the resulting total energy amounts to, duh, $\int d^3x \Lambda$, thus shifting the potential $V \to V + \int d^3x \Lambda$ and hence adding to the action the term $-\int dt \int d^3x \Lambda$. Before Einstein, nobody would care about an additive constant in the potential: the equation of motion is not sensitive to it. Indeed, the particle does not even know that we added something: the dynamical variable q does not appear anywhere in the added term.

But in curved spacetime, we know that the 4-volume element $dt d^3x = d^4x$ has to be modified to $d^4x \sqrt{-g}$.

So yes, gravity knows. No way to sneak around gravity and stealthily add a constant to the Lagrangian.

Thus, we add to the Einstein-Hilbert action S_{EH} the outrageously simple term

$$S_{\text{cosmological}} = -\int d^4x \sqrt{-g} \, \Lambda$$

as was first done by Einstein, who referred to Λ as the cosmological constant.[‡]

In the late 1990s, observational cosmologists discovered that the universe is suffused by a mysterious dark energy. The origin of this dark energy remains shrouded in mystery. But over the years, observational cosmologists have established, with ever diminishing uncertainty, that the density of dark energy is constant in spacetime. Consequently, many theoretical physicists believe that the dark energy may well be the fabled cosmological constant Λ introduced by Einstein. As I said, the spirit here is to get you to cosmology and our first application of Einstein's field equation as quickly as possible. Thus, we will postpone a more detailed discussion of observational cosmology until later in this chapter.

Instead, we rush headlong to the action

$$S = S_{\text{EH}} + S_{\text{cosmological}} = \int d^4x \sqrt{-g} \left(\frac{1}{16\pi G} R - \Lambda \right) \tag{2}$$

We now hold the action of a universe in our hands. ("Amazing!" I say.)

---

[*] To get to cosmology, albeit a purely hypothetical cosmology, even more quickly than as quickly as possible, fill the universe with only the gravitational field. See exercise 1.

[†] In chapter IX.11, you will learn that a negative constant energy density Λ would lead to a completely different kind of spacetime.

[‡] While the cosmological constant may be mathematically simple to describe, it may be one of the most mysterious concepts in theoretical physics. See chapter X.7.

An equation of motion for the universe

Instead of Einstein's field equation in empty spacetime (1), we now have Einstein's field equation in the presence of the cosmological constant, namely

$$\delta S = \delta(S_{\text{EH}} + S_{\text{cosmological}}) = 0$$

Dutiful scholars would now vary $S_{\text{cosmological}}$. Actually, that's very easy to do, and if you remember (V.6.20), you already know how to do it.

But in keeping with the spirit of the preceding chapter, we will keep things easier than easy and try to get away with doing as little work as possible. We argue that when we vary $S_{\text{cosmological}}$ with respect to $g_{\mu\nu}$, the result must be proportional to $\Lambda g^{\mu\nu}$ since there is no other tensor around: only the determinant of the metric appears in $S_{\text{cosmological}}$. Also, the spacetime derivative ∂ does not appear. Thus, we obtain

$$A(R^{\mu\nu} + \alpha g^{\mu\nu} R) = -16\pi G\beta\Lambda g^{\mu\nu} \tag{3}$$

where β is some numerical constant, which we will put off computing until chapters VI.4 and VI.5.

In the preceding chapter, the right hand side is equal to 0. Here we have sort of the next best thing: the right hand side, while not 0, is proportional to $g^{\mu\nu}$. Just as in the preceding chapter, we can clean up (3) by contracting it with $g_{\mu\nu}$. We get $A(1+4\alpha)R = R = -16\pi G(4\beta\Lambda)$, where in the second equality we used a result from the preceding chapter. The scalar curvature R is some constant times Λ. Eliminating R in (3), we arrive at

$$R_{\mu\nu} = +\tilde{\Lambda} g_{\mu\nu} \tag{4}$$

with $\tilde{\Lambda}$ equal to some numerical constant times Λ. This equation describes the dynamics of a universe filled with a constant energy density proportional to Λ.

An exponentially expanding universe

Let us now plug the Lemaître–de Sitter metric

$$ds^2 = -dt^2 + a(t)^2 d\vec{x}^2 \tag{5}$$

from chapter V.3 into (4). Note that there we studied the properties of the spacetime described by (5) for some assumed $a(t)$, but now we are in the much more powerful position of being able to determine this function.

In chapter V.3, we computed the nonvanishing Christoffel symbols to be

$$\Gamma^0_{ij} = a\dot{a}\delta_{ij} \quad \text{and} \quad \Gamma^i_{0j} = \Gamma^i_{j0} = \frac{\dot{a}}{a}\delta^i_j \tag{6}$$

Given this, we could now plow ahead and compute the Ricci tensor $R_{\mu\nu}$, plug it into (4), and solve for $a(t)$. As easy as pie, actually. But in keeping with our lifestyle, let's wing it first.

First, as we have been saying ever since chapter I.5, curvature involves two powers of derivatives, so $R_{\mu\nu}$ should involve $\ddot{a}$ and $\dot{a}^2$. Second, consider the coordinate transformation $\vec{x} = \lambda\hat{\vec{x}}$ for some arbitrary constant λ, keeping t unchanged, then $a(t) = \lambda^{-1}\hat{a}(t)$. But by the transformation law of tensors, $\hat{R}_{00} = R_{00}$, $\hat{R}_{ij} = \lambda^2 R_{ij}$. So R_{00} must have the form $\ddot{a}/a$ and $(\dot{a}/a)^2$, and hence the time-time component of (4) gives an equation like $\ddot{a}/a + (\dot{a}/a)^2 \sim \tilde{\Lambda}$ with various unknown numerical coefficients.

We see immediately that the solution is $a = e^{Ht}$ with $H^2 \sim \tilde{\Lambda}$. An exponentially expanding universe, as discussed in chapter V.3, pops out!

Solving Einstein's field equation

Actually, it is not hard at all to calculate properly like a decent hard-working physicist. Insert (6) into the formula for the Ricci tensor obtained by contracting the Riemann curvature tensor (VI.1.9):

$$R_{\mu\nu} = R^o_{\mu\sigma\nu} = (\partial_\sigma \Gamma^\sigma_{\mu\nu} + \Gamma^\sigma_{\kappa\sigma}\Gamma^\kappa_{\mu\nu}) - (\partial_\nu \Gamma^\sigma_{\mu\sigma} + \Gamma^\sigma_{\kappa\nu}\Gamma^\kappa_{\mu\sigma}) \tag{7}$$

Since the spatial slice of our metric is rotational invariant, we already know that R_{0i} vanish and $R_{ij} \propto \delta_{ij}$. We obtain

$$R_{00} = -(\partial_0 \Gamma^\sigma_{0\sigma} + \Gamma^\sigma_{\kappa 0}\Gamma^\kappa_{0\sigma}) = -\left[3\partial_0(\frac{\dot{a}}{a}) + 3(\frac{\dot{a}}{a})^2\right] = -3\frac{\ddot{a}}{a} \tag{8}$$

and

$$R_{ij} = (\partial_0 \Gamma^0_{ij} + \Gamma^l_{0l}\Gamma^0_{ij}) - (\Gamma^k_{0j}\Gamma^0_{ik} + \Gamma^0_{kj}\Gamma^k_{i0}) = \left[\partial_0(a\dot{a}) + (3-2)\frac{\dot{a}}{a}(a\dot{a})\right]\delta_{ij} = (2\dot{a}^2 + a\ddot{a})\delta_{ij} \tag{9}$$

Note that the scaling conditions $\hat{R}_{00} = R_{00}$, $\hat{R}_{ij} = \lambda^2 R_{ij}$ we derived are indeed satisfied by the computed $R_{\mu\nu}$.

We could now solve (4), consisting of the two equations

$$R_{00} = -3\frac{\ddot{a}}{a} = -\tilde{\Lambda} \tag{10}$$

and

$$R_{ij} = (2\dot{a}^2 + a\ddot{a})\delta_{ij} = \tilde{\Lambda}a^2\delta_{ij} \tag{11}$$

By eyeball, we find the solution $a = e^{Ht}$ with $H^2 = \tilde{\Lambda}/3$. Isn't it easy? (Actually, I cheated you a teeny bit here. You can either figure out my sleight of hand, or wait until chapter VI.5, where it will be revealed to you.)

Proceeding more carefully, we could use (10) to eliminate $\ddot{a}$ in (11), thus obtaining $\dot{a}^2 = H^2 a^2$ with the two roots $\dot{a} = \pm Ha$ related by the time reversal transformation $t \to -t$. These two equations are solved, respectively, by $a = e^{Ht}$ (describing an expanding universe) and $a = e^{-Ht}$ (describing a contracting universe). This agrees with the invariance under time reversal of Einstein gravity and more generally of the fundamental laws of physics.

You may have noticed an apparent "miracle" here: we have two differential equations for one unknown function and they "happen" to be compatible. In chapter VI.4, you will acquire a deeper understanding of this rather mysterious fact.

Dark energy

As you have surely heard, and as I mentioned earlier, observational cosmologists made the astonishing discovery that the dominant content of our universe consists of a previously unknown dark energy. They counted the number of a certain type of supernova (called type Ia) that had been established as "standard candles," namely objects whose intrinsic brightness was known. Thus, from its observed brightness, the distance to the supernova could be fixed. The observational data indicate, to everybody's surprise, that the expansion rate of the universe is accelerating, an amazing discovery made even more dramatic[2] by the fact that it was made almost simultaneously by two competing teams.

The expansion history of the universe is determined by its content, as we have seen in this chapter and as we will see in more detail in part VIII, and an accelerating expansion rate is precisely what is indicated by (10). Thus, the data and Einstein gravity suggest a constant energy density. As I mentioned, the cosmological constant provides the simplest (and most compelling) explanation. (The alternative is to throw in any number[3] of scalar fields that do not vary in space.)

Since the original discovery, more observations have been made. Phenomenologically, the data are fitted by assuming that the universe is filled with a component described by the equation of state of the form $P = w\rho$ relating pressure to energy density. The parameter $w = w(z)$ is taken to be a function of the redshift z. The data indicate that $w(z)$ is nearly constant and close to -1. As we will see in the next section, for the cosmological constant, $w = -1$.

In Planck's natural units, the observed* Λ is of order $(10^{-3}\,\text{eV})^4$. The dark energy accounts for something like $\sim74\%$ of the total mass content of the universe. The other 26% consists of mostly dark matter with a few percent of intergalactic gas and stars thrown in. Thus, the universe we have studied in this chapter, without having to break a sweat, could in fact provide a first approximation to our universe.

A major goal of observational cosmology over the next decade or so is to either establish or rule out the supposition that the density of dark energy is in fact constant in spacetime. In this golden age of cosmology, we should not be surprised, of course, if observations reveal more unexpected facts about the universe, but in this text, to streamline the theoretical presentation, we will assume that dark energy could be represented by Λ.

---

* One aspect of the mystery is that $\sim10^{-3}$ eV is also roughly the not-yet-understood scale characteristic of neutrino masses. Pure coincidence?

Negative pressure

The dark energy has some rather peculiar properties. The popular media often give the impression that the strange properties of dark energy somehow have something to do with Einstein, whom nobody could understand anyway. In fact, the strange properties follow from elementary physics and the statement that the energy density is constant.

Consider a container (think of a balloon) of volume V filled with some stuff, be it a gas or something else. Now squeeze on the container and change the volume by dV. Then the energy in the container increases by $dE = -PdV$, which in fact defines the pressure P exerted by the stuff. Note that $dV < 0$ (squeezing), $dE > 0$, and $P > 0$. This of course just states energy conservation*: the work you have done by squeezing is $-PdV$.

Once we say that the energy density is a positive constant Λ, then we are immediately led to something bizarre. Since $E = \Lambda V$, then $dE = \Lambda dV < 0$ since $dV < 0$. But with $dE < 0$ and $dV < 0$, $dE = -PdV$ tells us that the pressure P must be negative! (Indeed, you could see that P is just $-\Lambda$.) So instead of resisting your squeeze, the balloon or container would suck your hands in.

The "rich man" has a fancier but still elementary way of saying this. Recall (III.6.16) from way back, that the energy momentum tensor of a perfect fluid is given by $T^{\mu\nu} = (\rho + P)U^\mu U^\nu + P\eta^{\mu\nu}$ in flat spacetime, promoted to $T^{\mu\nu} = (\rho + P)U^\mu U^\nu + Pg^{\mu\nu}$ in curved spacetime. Since the dark energy is allegedly constant in spacetime, there is no U^μ available and so $T^{\mu\nu} = Pg^{\mu\nu}$. The energy density is thus given by $T^{00} = Pg^{00} = -P$. We conclude that, since Λ is the energy density, a positive Λ gives a negative pressure $P = -\rho = -\Lambda$, reaching the same conclusion[†] as the "poor man." The rich man might criticize the poor man's approach by asking him or her to find a container to contain the dark energy. The material scientists have yet to come up with such a container.

What is not forbidden is mandatory

Historically, Einstein added the cosmological constant Λ to his theory and then later removed it. Although Einstein, with his groundbreaking work on the photoelectric effect, was indisputably one of the founders of quantum mechanics, he was first and foremost a classical physicist. (Such was his greatness that he could be both at the same time.) In classical physics, one could include or exclude possible terms in the action as one pleases; the goal is to include enough terms to account for observations.

But quantum physics, with its probabilistic character and constant fluctuation, differs profoundly from classical physics. If you neglected a term not specifically forbidden by

---

* More generally, the first law of thermodynamics states $dE = TdS - PdV$, but here we are not increasing the entropy S and there is no temperature.

[†] In chapter VI.5, we will derive $T^{\mu\nu} = -\Lambda g^{\mu\nu}$ directly from $S_{\text{cosmological}}$, using the approach of an "even richer man."

a fundamental principle, quantum fluctuations would force that term on you. You must have a reason for why a given term should not be there. Roughly, that is because quantum physics is probabilistic. Physicists can only determine the probabilities of various processes occurring. Any process not explicitly forbidden* will occur, even though the probability of the process actually occurring may be very small.

Thus, in the quantum world, Einstein is no longer allowed to remove the cosmological constant Λ from his theory.

In part II, we touched upon the notion of a classical field theory, and in part IV we studied electromagnetic field theory. Fields exhibit waves schematically of the form $\sin(\omega(\vec{k})t - \vec{k} \cdot \vec{x})$. In classical physics, the waves could be quiescent. Just as a harmonic oscillator in quantum mechanics could never be at rest and thus has a minimum energy of $\frac{1}{2}\hbar\omega$, a quantum field could never be quiescent and thus contributes a minimum energy density $\sim \int d^3k \frac{1}{2}\hbar\omega(\vec{k})$ to spacetime. The puzzle for quantum field theorists is not whether Λ is present in the action, but why observationally it has the value it does. We will come back to this in chapter X.7.

Exercises

1   An even simpler, but less physical, universe than the one described here was discovered by E. Kasner in 1921 (*Am. J. Math.*, vol. 43, p. 217). Show that the metric $ds^2 = -dt^2 + (t^{2p}dx^2 + t^{2q}dy^2 + t^{2r}dz^2)$ with three constants p, q, r solves Einstein's equation $R_{\mu\nu} = 0$ provided that $p + q + r = p^2 + q^2 + r^2 = 1$. The Kasner universe expands or contracts at different rates along the three different spatial directions.

2   The Kasner universe generalizes nicely to higher dimensions. Let's go to a 5-dimensional metric by adding $t^{2s}dw^2$ to the ds^2 in the preceding exercise. Solve.

3   You might have noticed that the numbers 3 and 2 in (8) and (9), respectively, depend on the dimension of space. Determine these numbers for arbitrary dimensions.

Notes

1. I must confess that occasionally I am also beset by nameless doubts. See the Closing Words to this book.
2. For an entertaining and exciting account, see Y. Bhattacharjee's article "A week in Stockholm" (*Science* 2012, vol. 336, p. 28). The 2011 Nobel Prize in Physics was awarded for the discovery of dark energy.
3. That's because (at least in part) scalar fields are free, in the sense that they cost nothing.

———

   * In T. H. White's *The Once and Future King*, the boy Arthur dreams of visiting a kingdom governed on the principle that whatever is not forbidden is mandatory. The story inspired the physicist Murray Gell-Mann to quip that in quantum physics what is not taboo is a commandment.

Gravity in empty spacetime

In empty spacetime (around a star or a black hole for example), we learned in chapter VI.1 that the Ricci tensor vanishes:

$$R_{\mu\nu} = 0 \tag{1}$$

As remarked earlier, since the Ricci tensor is constructed by summing various components of the Riemann tensor, this does not necessarily mean that $R^{\sigma}_{\rho\mu\nu} = 0$, which would imply that spacetime is Minkowski flat.

Consider a spherical mass distribution, such as a star, of mass M and radius R. Outside the massive lump, we already know what the spacetime metric looks like from chapter V.4:

$$g_{tt} = -A(r), \quad g_{rr} = B(r), \quad g_{\theta\theta} = r^2, \quad g_{\varphi\varphi} = r^2 \sin^2 \theta \tag{2}$$

with all off-diagonal components vanishing. Indeed, we even listed all the Christoffel symbols in the appendix there. But back then, we had no idea what A and B were. Now we have (1).

Our job is conceptually straightforward. Compute the Riemann curvature tensor and thence the Ricci tensor. Then determine the two unknown functions A and B by solving Einstein's equation (1) with the boundary condition $A(r) \to (1 - \frac{2GM}{r})$ and $B(r) \to 1$ as $r \to \infty$.

Newton had his plague and Schwarzschild his heavy gunfire

As you see, the war treated me kindly enough, in spite of the heavy gunfire, to allow me to get away from it all and take this walk in the land of your ideas.

—Karl Schwarzschild, writing to A. Einstein

I hope that you have been enjoying your "walk in the land of ideas," hopefully in pleasant surroundings, without being bothered by heavy gunfire.

In 1915, the very same year that Einstein published his theory of general relativity, Karl Schwarzschild (1873–1916), an officer serving in the German army on the Russian front during World War I, wrote to Einstein saying that he had found[1] the solution for the spacetime metric around a spherical mass distribution. Interestingly, in Einstein's celebrated 1915 article, he only found an approximate solution valid for large r (using Cartesian coordinates!?), which was in fact adequate for his purpose of working out observational tests of his theory. By the way, the family name Schwarzschild means "black sign" or "black shield,"[2] not "black child," contrary to what many (non-German-speaking) students of general relativity believe. Tragically, Schwarzschild died a year later of a painful autoimmune disease contracted on the battlefield.

So, you should be able to work out the solution in the tranquility and privacy of your own home. Here is what you do. Since you already have the Christoffel symbols, you simply plug in the appropriate formulas and calculate the Riemann curvature tensor. Then sum over a pair of indices to find the Ricci tensor. Set the Ricci tensor to 0, and solve for A and B. In fact, you could even bypass the Riemann curvature tensor and compute the Ricci tensor directly. Do this before reading on and be glad you are not on the Russian front.

The Schwarzschild solution

You look up the formula for the Ricci tensor

$$R_{\nu\rho} = (\partial_\sigma \Gamma^\sigma_{\nu\rho} + \Gamma^\sigma_{\sigma\kappa} \Gamma^\kappa_{\nu\rho}) - (\partial_\nu \Gamma^\sigma_{\sigma\rho} + \Gamma^\sigma_{\nu\kappa} \Gamma^\kappa_{\sigma\rho}) \tag{3}$$

and plug in the Christoffel symbols from chapter V.4. It is a bit tedious but totally straightforward. For example,

$$R_{tt} = (\partial_\sigma \Gamma^\sigma_{tt} + \Gamma^\sigma_{\sigma\kappa} \Gamma^\kappa_{tt}) - (\partial_t \Gamma^\sigma_{\sigma t} + \Gamma^\sigma_{t\kappa} \Gamma^\kappa_{\sigma t}) = \partial_r \Gamma^r_{tt} + \Gamma^\sigma_{\sigma r} \Gamma^r_{tt} - 2\Gamma^t_{tr} \Gamma^r_{tt}$$

$$= \left(\frac{A'}{2B}\right)' + \left(\frac{A'}{2A} + \frac{B'}{2B} + \frac{2}{r}\right)\left(\frac{A'}{2B}\right) - 2\left(\frac{A'}{2A}\right)\left(\frac{A'}{2B}\right)$$

$$= \left(\frac{A''}{2B}\right) - \frac{A'}{4B}\left(\frac{A'}{A} + \frac{B'}{B}\right) + \frac{A'}{rB} \tag{4}$$

Proceeding in this way, you obtain

$$R_{tt} = \frac{A''}{2B} + \frac{A'}{rB} - \frac{A'}{4B}\left(\frac{A'}{A} + \frac{B'}{B}\right) \tag{5}$$

$$R_{rr} = -\frac{A''}{2A} + \frac{B'}{rB} + \frac{A'}{4A}\left(\frac{A'}{A} + \frac{B'}{B}\right) \tag{6}$$

$$R_{\theta\theta} = 1 - \frac{1}{B} - \frac{r}{2B}\left(\frac{A'}{A} - \frac{B'}{B}\right) \tag{7}$$

and $R_{\varphi\varphi} = \sin^2 \theta R_{\theta\theta}$. All other components vanish.

Again, some features are easy to understand by symmetry considerations. For example, consider the coordinate transformation $t = a\tilde{t}$. Then $\tilde{A} = a^2 A$, with B unchanged. That $R_{\mu\nu}$ should transform appropriately gives a check on (5–7). The vanishing of the off-diagonal components follows from rotational invariance and the $t \to -t$ symmetry of the metric.

At this point, you might be worried that you have 3 second order coupled ordinary differential equations, $R_{tt} = 0$, $R_{rr} = 0$, and $R_{\theta\theta} = 0$ for 2 unknown functions A and B. Einstein's beautiful theory might yet be inconsistent and fall flat on its face. Now you recall that a similar worry presented itself in the preceding chapter: there were 2 equations for 1 unknown function $a(t)$, but then an apparent "miracle" happened and things worked out okay. So let's proceed and see what happens.

Staring at (5) and (6), you realize that a good strategy is to get rid of the second derivative. So form $\frac{R_{tt}}{A} + \frac{R_{rr}}{B} = \frac{1}{rB}\left(\frac{A'}{A} + \frac{B'}{B}\right) = 0$. This instantly solves itself as $AB = 1$, where we fixed the integration constant by the boundary condition at $r \to \infty$. Eliminating $\frac{A'}{A} = -\frac{B'}{B}$ in (7), we obtain $r(\frac{1}{B})' + \frac{1}{B} = 1$, with the solution $\frac{1}{B} = 1 + \frac{b}{r}$, and so $A = 1 + \frac{b}{r}$. The boundary condition at infinity fixes the integration constant b to be $-2GM$.

An amazing identity?

It remains for us to hold our breath and plug the solution into, say, $R_{rr} = 0$ to see whether it is solved. It works! An apparent miracle happens again. Since we physicists do not believe in miracles, there must be an amazing identity we don't yet know about. Be patient. I will get to this identity in the next chapter. (Alternatively, you could try to discover this identity!) We are in good company, since for many years Einstein did not know about this identity either, and this ignorance was one of the reasons that it took him 10 years to get to his field equations.

How does the radius come in?

Meanwhile, let's try to imagine the excitement Schwarzschild must have felt in the trenches, discovering that the curved spacetime around a spherical mass distribution of mass M and radius R is described by this remarkable metric

$$ds^2 = -\left(1 - \frac{2GM}{r}\right) dt^2 + \frac{1}{(1 - \frac{2GM}{r})} dr^2 + r^2(d\theta^2 + \sin^2\theta d\varphi^2) \tag{8}$$

soon to be named after him.* Einstein was elated when he learned that his highly nonlinear field equations had such a simple solution. A priori, A and B could have been two totally complicated functions.

Confusio suddenly speaks up. "Where is the dependence on the radius R?"

---

* See appendix 5.

Good question! The answer is of course that the solution (8) only holds for $r > R$. For $r < R$, the empty spacetime field equation $R_{\mu\nu} = 0$ that we obtained in chapter VI.1 by winging it is obviously not valid. If we are talking about the sun, for example, we would have to take into account the hot gas in the solar interior.

The Schwarzschild coordinate singularity

After admiring this metric for a while, you might start worrying again even if you were not born a worrywart. What about g_{tt} vanishing and g_{rr} blowing up at the Schwarzschild radius $r_S \equiv 2GM$? And vice versa for $g^{tt} = 1/g_{tt}$ and $g^{rr} = 1/g_{rr}$? Indeed, both Schwarzschild and Einstein were alarmed by this and thought somewhat confusedly[3] about this problem.*

First of all, these metric coefficients, being components of a tensor in a particular coordinate system, depend on the coordinate system. Just as the usual spherical coordinate is no good at the north and south poles, we could merely have made a bad coordinate choice at $r = r_S$. Indeed, for the sphere, we also have chosen (in chapter I.6) coordinates in which the metric blows up on the equator. In chapter VII.2, we will show that this is indeed the case by exhibiting a set of coordinates, namely the Kruskal[4] coordinates, in which the Schwarzschild solution is not singular at r_S. In fact, allow me to remind you that, way back in chapter I.6, we already discussed the distinction between a coordinate singularity and an actual or physical singularity, where the geometry itself goes out of control. Remember the Einstein-Rosen bridge, a kind of tunnel or "wormhole" between two flat spaces?

There is a relatively quick way to allay your worry. Let us look at a scalar quantity such as[5] $R^{\mu\nu\rho\sigma} R_{\mu\nu\rho\sigma}$,[†] which turns out to be $\frac{12 r_S^2}{r^6}$, with a perfectly innocuous behavior around $r = r_S$.

Why look at a scalar? Because scalars transform like $S'(x') = S(x)$. Thus, if a scalar blows up in one coordinate system, it blows up in all coordinate systems. It follows that if a scalar blows up, then we are in trouble. Tensors, in contrast, can "catch" a singularity going from one coordinate system to another: they might get infected by various factors of $\frac{\partial x'^\mu}{\partial x^\nu}$ in their transformation law. Hence, if a component of a tensor, such as g_{rr} and g^{tt}, blows up in one coordinate, it does not mean that it will blow up in all coordinate systems. The Mercator map is no good at the poles, but it does not mean that the poles have anything singular about them.

You see that it was totally worth it to learn about scalars and tensors! Of course, calculating one scalar does not prove anything: we have to check out all the scalars. I won't digress by discussing how many "fundamental" scalars there are in a Riemannian spacetime. But verifying that at least one scalar, namely $R^{\mu\nu\rho\sigma} R_{\mu\nu\rho\sigma}$, is perfectly well behaved at r_S does allay our anxiety a bit. In fact, as we will see in chapter VII.2, the singularity at r_S exhibited in the metric is merely due to a poor coordinate system, of the type of singularity we encounter at the poles in a Mercator map.

---

* At one point, Einstein thought that a particle falling in would bounce back at $r = r_S$.
[†] The Jargon Guy tells us that this is known as the Kretschmann scalar.

Incidentally, that $R^{\mu\nu\rho\sigma} R_{\mu\nu\rho\sigma}$ blows up at $r = 0$ shows that a real physical singularity lurks there: Einstein theory as we know it must break down at infinite spacetime curvature.

Here is a simple but striking example of a coordinate singularity. Transform coordinates by plugging $x = 2\rho^{\frac{1}{2}}$ into $dl^2 = dx^2 + dy^2$ to obtain $dl^2 = (d\rho^2/\rho) + dy^2$. The metric appears to blow up at $\rho = 0$, but that's merely due to a bad coordinate choice, as you can see as clear as day. For ρ negative we appear to have pulled a spacetime out of a space hat! Let $w = -\rho > 0$ for ρ negative, and define $t = 2w^{\frac{1}{2}}$, so that[6] $ds^2 = dl^2 = -dt^2 + dy^2$! Of course, ρ negative is precisely where the original coordinate x makes no sense.

A whiff of the black hole

Although the singularity at $r = r_S$ is merely due to a bad choice of coordinates, it leads nevertheless to important physics, as we will explore in chapter VII.2. Our answer to Confusio's question provides a first hint: if $R > r_S = 2GM$, then we do not have to worry about this singularity. But this condition, as we discussed way way back in part 0, is precisely that given by Michell and Laplace for the mass M not to be a black hole!

Meanwhile, we will work out Einstein's two celebrated solar system tests.* For the sun, $r_S \sim 3$ km, which is tiny compared to its radius $R_\odot$. This again reflects the weakness of the gravitational force. For your information, for various astrophysical objects, the typical values of r_S/R are given by 10^{-9} (earth), 10^{-6} (sun), 10^{-4} (white dwarf), and 10^{-1} (neutron star).

The deflection of light and a factor of 2

> For the deviation of light by the sun I obtained twice the former amount.
>
> —A. Einstein, writing to Arnold Sommerfeld, late 1915

Newton himself wondered, "Do not Bodies act upon Light at a distance, and by their action bend its Rays?" In 1801, Johann Soldner used Newton's corpuscular theory supposing light to consist of a stream of miniscule particles and calculated the deflection of light by astronomical objects, thus obtaining the Newtonian value against which we now compare Einstein's value. Recall that you did this very same calculation as an exercise way back in chapter I.1.

History often takes curious turns. In 1911, Einstein, unaware of Soldner's calculation, predicted that light would bend in a gravitational field in his still-evolving theory of gravity.

---

* By now, there are a number of other tests. In appendix 2, we discuss radar echo delay. Some tests actually test the equivalence principle. For example, Nordtvedt noted that accurate measurements of the earth-moon distance would reveal whether the earth and the moon fall toward the sun at slightly different rates. Thus far, Einstein has triumphed. Otherwise, you would have heard about it, duh.

He followed a naive approach, reasoning that since energy was equivalent to mass, the photon could be thought of as having a tiny mass. In hindsight, we see that Einstein would simply recover the Newtonian value. In fact, the correct answer is a factor of 2 larger, as we will now derive.

Back in chapter V.4, we had already worked out the motion of light and material particles in a static isotropic Einstein spacetime described by two unknown functions $A(r)$ and $B(r)$. Now that we know what they are, all we have to do is plug these into the appropriate equations from that chapter, namely $AB = 1$ and $B^{-1} = 1 - \frac{2GM}{r}$.

Often, it is convenient to use units in which G is set to 1. Then, (V.4.22) in particular gives us $(\frac{dr}{d\varphi})^2 + r^2(1 - \frac{2M}{r}) = \frac{r^4}{b^2}$. As in chapter I.1, change variable from r to $u = 1/r$, so that $r' \equiv \frac{dr}{d\varphi} = -\frac{1}{u^2}u'$ with $u' = \frac{du}{d\varphi}$. Then,

$$u'^2 + u^2 - 2Mu^3 = \frac{1}{b^2} \tag{9}$$

Differentiating, we obtain the "analog Newtonian" equation

$$u'' + u = 3Mu^2 \tag{10}$$

with φ playing the role of time.

Without the M term, this is the harmonic oscillator equation, with the solution $bu = \sin \varphi$, which you recognize as just saying that light moves in a straight line. This suggests that we treat the M term as a perturbation. Plugging in $bu = \sin \varphi + bu_1$, we have, keeping only first order terms, $u_1'' + u_1 \simeq 3(M/b^2) \sin^2 \varphi$. The solution to this order is then $bu \simeq \sin \varphi + (M/b)(2 - \sin^2 \varphi)$. For a light ray grazing the sun, the expansion parameter $M/b \sim r_S/R_\odot$ is just the ratio of the sun's Schwarzschild radius to its actual radius, which, as we anticipated, is tiny.

To work out the deflection, refer to figure 1 and warm up with the no-deflection case (that is, with M set to 0) $bu = \sin \varphi$. As $r \to \infty$, $u \to 0$, which corresponds to $\varphi = 0$ and $\varphi = \pi$. No deflection, as expected. With $M \neq 0$, for $r \to \infty$, $u \to 0$, the resulting quadratic equation for $\sin \varphi$ yields two roots. We reject one of these roots as unphysical, with the other root being $\sin \varphi(r = \infty) = -2M/b$. As the light is coming in from infinity, $\varphi = -2M/b$, and as it is going out to infinity, $\varphi = \pi + (2M/b)$. Thus, Einstein, realizing his error in 1911, obtained in 1915 a deflection of $\Delta\varphi = 4M/b = 4GM_\odot/R_\odot \simeq 1.75''$, twice the Newtonian value.

Shortly after the end of World War I, the Royal Society financed two expeditions, one to Brazil led by Andrew Crommelin and one to Africa led by Arthur Eddington, to observe the total solar eclipse* of May 29, 1919, with the express purpose of testing Einstein's theory. If the light from a distant star bends as it glazes the edge of the sun (so that the deflection of light predicted by Einstein would be as large as possible), then the position of the star would appear to be shifted from its known position.

---

* As a naive theorist, Einstein wrote to George Hale, the director of the Mount Wilson Observatory, wanting to know "how close to the Sun fixed stars could be seen *in daylight*" (italics Einstein's).[7] Hale explained that exploiting a solar eclipse would be more promising.

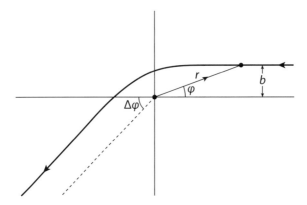

Figure 1 The deflection of light.

Mercury and heart palpitations

> Imagine my joy at . . . the result that the equations give the perihelion motion of Mercury correctly. For a few days I was beside myself with joyous excitement.
> —A. Einstein, writing to Paul Ehrenfest, 1916

For two centuries after Newton, due to the efforts of greats like Lagrange, Laplace, Bessel,[8] and Le Verrier, planetary orbits were calculated to astonishing accuracy. The perihelion of Mercury was observed to advance (as depicted, vastly exaggerated, in figure 2) by something like 5,600″ (seconds* of arc) per century. After all the known effects (for example, the pull of Jupiter accounted for 153″) were taken out, a troublesome discrepancy of 43″ per century remained. On the basis of a similar discrepancy in the orbit of Uranus, Urbain Le Verrier (1811–1877) had triumphantly predicted the existence of the previously unknown Neptune. A planet named Vulcan was similarly predicted to orbit between the sun and Mercury, but it was never found.

Then Einstein proposed his curved spacetime, and out pops the 43″ per century. Amazing! I still find it incredible that this clunk of rock would know, every time it completes a revolution around the sun, to move ahead by a teeny bit precisely as dictated by the curvature of spacetime. It's a tribute not only to Einstein, but also to all those celestial mechanicians from Tycho Brahe on, whose massive efforts allow us inhabitants of the third planet from the sun to understand the movement of the celestial sphere at this level of minute detail.

---

* Have you ever wondered what the term "second" used in measuring angles and time has to do with the notion of the ordinal number "second"? Well, the first is a corruption of a phrase containing the second. Ptolemy proposed subdividing the degree in mapping heaven and earth, and his subdivisions became known in Latin as "partes minutae primae" and "partes minutae secundae."

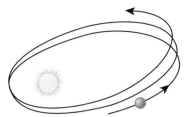

Figure 2 Mercury's perihelion advances (vastly exaggerated).

The calculation of the perihelion shift,[9] while a monument to theoretical physics, amounts "merely" to a beautiful exercise in Newtonian mechanics as I have already said, and so reluctantly I will relegate it to appendix 1.

Later, Einstein told his friend Adriaan Fokker[10] (1887–1972) that he had heart palpitations when he got the 43″ per century. He also wrote to his friend Sommerfeld saying "How helpful* to us here is astronomy's pedantic accuracy, which I often used to ridicule secretly!"[11]

Einstein's luck

> It is legitimate to speak of a pound of light as we speak of a pound of any other substance. . . . I have calculated that . . . an Electric Light Company would have to sell[†] light at the rate of £140,000,000 a pound.
>
> —Arthur Eddington

It's Eddington's deflection of light that made Einstein a worldwide celebrity—the general public could hardly be expected to care about the perihelion of Mercury. But space warp? Now that's another story! J. J. Thomson, the discoverer of the electron, presiding over a special meeting at the Royal Society convened to announce the result of the solar eclipse expeditions, hailed the result as the most important since Newton's work and Einstein's theory as "one of the highest achievements of human thought," which regrettably, he added, was incomprehensible. "No one can understand the new law of gravitation without a thorough knowledge of the theory of invariants and of the calculus of variations." Well, dear reader, I gave both of them to you already back in part I and part II, respectively.

I can now also tell you that it was not a reporter who asked Eddington the question in the famous story I recounted in the preface. It was Ludwik Silberstein, a Polish-American physicist who had studied Einstein's theory. According to one account, he was expecting

---

* Newton similarly benefited from the labor of Brahe and Kepler. By the way, the Danes say that they've had a "Tycho Brahe day," meaning that they've had a really bad day.

† At 1920 prices. Remember my ranting and raving in the introduction about using sensible units, not something like pound per pound.

Eddington to name him, Silberstein, as the third. Thus, Eddington's response could be seen as not only arrogant, but insultingly arrogant. Later, Silberstein claimed to have discovered a fatal flaw in Einstein's theory, thus provoking the 1935 Einstein-Silberstein debate, which evidently Einstein won.

Einstein was quite capable of pulling the leg of his friend Max Planck, once remarking:

> [Max Planck] was one of the finest people I have ever known . . . but he didn't really understand physics, [because] during the eclipse of 1919 he stayed up all night to see if it would confirm the bending of light by the gravitational field. If he had really understood [the general theory of relativity], he would have gone to bed the way I did.[12]

But that was staircase wit on Einstein's part. In fact, he was almost preternaturally lucky, as documented by Waller.[13] After Einstein's mistaken calculation in 1911, reproducing Soldner's 1801 Newtonian result, there was in 1912 an Argentinian eclipse expedition that encountered bad weather. Next, with Einstein still blissfully unaware of his error, he convinced his friend the astronomer Erwin Freundlich to organize an expedition, financed by the munitions manufacturer Krupp, to observe the deflection of light during a solar eclipse in the Crimea on August 21, 1914.[14] Not surprisingly, but fortunately for Einstein, the German astronomers, with all their telescopes and financing by Krupp, were promptly arrested by the Russians as spies.

Meanwhile, during the war, Einstein discovered his factor-of-2 error. Without these twists and turns of history, his celebrity-making triumph might have been a wet fizzle. It has also been suspected that Eddington, an enthusiast for Einstein's theory, might have fudged[15] the data in Einstein's favor. He was also an ardent pacifist, like Einstein, and might have been eager to show British support for the work of a German citizen.

Gravitational lensing

In less than a century, the deflection of light has come a long way, from a minute effect to a major tool in our exploration of the universe. As you have no doubt heard, and as was mentioned in chapter VI.2, matter in the universe appears to be dominated by an unseen dark matter, rather than the luminous matter we know and love, consisting of nucleons and electrons. Dark matter, while it does not emit or absorb light, interacts gravitationally and thus clumps. Indeed, it is now believed that the galaxies consist of enormous lumps of dark matter, each with an island of luminous matter sort of floating inside. Consider a distant light source (a quasar, a supernova, a galaxy—it does not matter what). Suppose that a large distribution of unseen dark matter is located between us and the light source. One way we could detect the presence of this distribution is by how it deflects the light from the distant source, known as gravitational lensing.

For several reasons, I will not go into a detailed discussion of this rapidly developing and important subject. Once we work out how a light ray deflects, the rest of the lensing calculation involves rather intricate though straightforward trigonometric and algebraic equations that have nothing to do with general relativity as such. For comparison with observational data, there are of course numerous effects to be taken into account, such as the possibility that both the light source and the dark matter distribution could be moving at close to light speed relative to us. I will however make the important and interesting remark that the deflection angle $\Delta\varphi$ increases with decreasing impact parameter b. With the kind of lens you are used to, such as the pair inside your head you are using to read this sentence, the deflection angle $\Delta\varphi$ decreases with decreasing impact parameter b and thus leads to focusing. Gravitational lensing works oppositely, thus producing in some circumstances ring-like images known as Einstein rings.

I could perhaps close this section by paraphrasing my colleague Tommaso Treu, a distinguished practitioner of gravitational lensing, that this exciting subject, including Einstein rings, is best appreciated by raising a wine glass filled with a fine white wine to the candles illuminating an elegant dinner.

Appendix 1: Planetary orbit in the Schwarzschild metric

We start by plugging A and B, as given in (8), into (V.4.15) to obtain (with a tiny notational change)

$$\frac{1}{2}\left(\frac{dr}{d\tau}\right)^2 + v(r) = \tfrac{1}{2}(\epsilon^2 - 1) \tag{11}$$

with the potential

$$v(r) = \frac{1}{2}\left[\left(1 + \frac{l^2}{r^2}\right)\left(1 - \frac{2M}{r}\right) - 1\right] = \frac{1}{2}\left[\frac{l^2}{r^2} - \frac{2M}{r} - \frac{2Ml^2}{r^3}\right] \tag{12}$$

Recall from (V.4.13) that ϵ is defined by $\frac{dt}{d\tau} = \frac{\epsilon}{1-\frac{2M}{r}}$; in other words, ϵ is the value of $\frac{dt}{d\tau}$ at $r = \infty$, namely the energy of the particle divided by its mass.

It is instructive to compare with the Newtonian potential in chapter I.1. The first term in $v(r)$ is the familiar centrifugal term, the second is the universal gravitational attraction. Remarkably, going from Newton to Einstein, we merely have to add to the potential an extra $\frac{1}{r^3}$ term. Also, we have $\frac{d}{d\tau}$ instead of $\frac{d}{dt}$. We now bring what we learned in classical mechanics to bear on (11).

To determine the shape of the orbit $r(\varphi)$ and hence the perihelion shift, we repeat what we did for the deflection of light and define $r'(\varphi) = \frac{dr}{d\varphi} = \frac{\frac{dr}{d\tau}}{\frac{d\varphi}{d\tau}} = r^2\dot{r}/l$, since, as you recall from (V.4.14), $\frac{d\varphi}{d\tau} = \frac{l}{r^2}$, with l the conserved angular momentum (per unit mass) of the particle. As in the deflection of light calculation, change variable from r to $u = 1/r$. Plugging all this into (11) and (12), we obtain[16] (with $u' = \frac{du}{d\varphi}$)

$$u'^2 + u^2 - 2\sigma u - \lambda u^3 = 2E \quad \text{(Einstein)} \tag{13}$$

where we have defined $\sigma \equiv M/l^2$, $\lambda \equiv 2M$, and $2E \equiv \epsilon^2 - 1$. What we should do is of course compare this Newtonian problem with the Newtonian problem you, yes you, solved back in chapter I.1, namely

$$u_0'^2 + u_0^2 - 2\sigma u_0 = 2E \quad \text{(Newton)} \tag{14}$$

with the obvious solution $u_0 = \sigma(1 + e\cos\varphi)$. As we discussed in chapters I.1 and I.4, not to precess is the exceptional case, valid only for Newton's inverse square force.

We now treat the λ term in (13) as a perturbation. (Incidentally, even though you already know that λ is tiny, you are invited to plug in the numbers and show that $\lambda \sim 10^{-8}$ for Mercury.) Thus, write $u = u_0 + u_1$, with u_1 of order λ, plug into (13), and collect terms of order λ. We obtain

$$-\sin\varphi\, u_1' + \cos\varphi\, u_1 = \frac{\lambda\sigma^2}{2e}(1 + e\cos\varphi)^3 \tag{15}$$

At this point, the typical student would fire up the computer and push a few buttons to obtain the solution $u_1(\varphi)$, and indeed, we could all do exactly that. But it is also rather neat to think through the problem.

The left hand side of (15) is linear in u_1, with the driving term on the right hand side given by a sum of 1, $\cos\varphi$, $\cos^2\varphi$, and $\cos^3\varphi$. We might expect u_1 to be given by an analogous sum of terms. But we are not interested in most of these terms. The constant term in u_1, for example, when added to u_0, would just shift σ by one part in 10^8. Periodic terms, such as $\cos^2\varphi$, also do not interest us; after $\varphi \to \varphi + 2\pi$, they would just return $u = u_0 + u_1$ and hence r to the same place. We want aperiodic terms, such as φ, $\varphi\sin\varphi$, and $\varphi\cos\varphi$. You can see by inspection that the first possibility doesn't work, but the second does (since $-\sin\varphi(\varphi\sin\varphi)' + \cos\varphi(\varphi\sin\varphi) = -\sin^2\varphi$ and the right hand side contains $\cos^2\varphi$) and the third doesn't. (If you must know, the complete solution to (15) has the form $u_1 = \alpha + \beta\cos\varphi + \gamma\cos^2\varphi + \frac{3}{2}\lambda e\sigma^2\varphi\sin\varphi$, where we do not give a flying nickel about α, β, γ, which, however, you can determine easily enough.)

We thus obtain, following Einstein, that $u \simeq \sigma\left(1 + e\cos\varphi + \frac{3}{2}\lambda e\sigma\varphi\sin\varphi\right) \simeq \sigma\left(1 + e\cos\left\{\left(1 - \frac{3}{2}\lambda\sigma\right)\varphi\right\}\right)$. For r to reach the same value it had at $\varphi = 0$, we need to have $\varphi = 2\pi/\left(1 - \frac{3}{2}\lambda\sigma\right)$. In other words, the perihelion advances by* $\Delta\varphi = 3\pi\lambda\sigma = 6\pi(M/l)^2$.

Appendix 2: Radar echo delay

To these two classic tests, deflection of light and perihelion shift, we can now add radar echo delay, proposed and pushed through by Shapiro in the 1960s. A radar beam is bounced off the planet Venus, and the time it takes for the echo to get back to the earth is carefully measured. The terminology "delay" is unfortunate. Like everything else, the photons in the beam get the best possible deal in the curved spacetime around the sun: they follow a geodesic of course. So what is the "delay"? The delay is in comparison with what would be expected in a Newtonian world.

By now you should be able to work out this problem by yourself without reading on. Just plug in the appropriate formulas in chapter V.4 and in this chapter. A hint: for the two classic tests, we need the expression for $d\varphi/dr$, but for the radar echo delay, we need dt/dr instead.

So, look up the expression (V.4.21) for $dr/d\zeta$ and the conservation law $dt/d\zeta = \epsilon/A(r)$. Eliminate the affine parameter ζ by dividing. We obtain

$$\left(\frac{dr}{dt}\right)^2 = \frac{A(r)}{B(r)}\left(1 - \frac{b^2 A(r)}{r^2}\right) = \left(1 - \frac{r_S}{r}\right)^2\left(1 - \frac{b^2}{r^2}\left(1 - \frac{r_S}{r}\right)\right) \tag{16}$$

(with the Schwarzschild radius $r_S = 2GM$, as you may recall). The first expression is general, the second is specific to the Schwarzschild solution. As explained in chapter V.4, physics does not depend on ϵ and l separately, but only on $b^2 \equiv \frac{l^2}{\epsilon^2}$. The radius r_0 at closest approach to the sun (see figure 3) is determined by $\frac{dr}{dt}|_{r=r_0} = 0$. Thus, from (16), we find $b^2 = r_0^2/\left(1 - \frac{r_S}{r_0}\right) \simeq r_0^2\left(1 + \frac{r_S}{r_0}\right)$. In the context of this problem, the notion of impact parameter is not relevant, and thus we trade b for r_0. Evidently, the effect is maximized if the beam gets as close to the sun as possible, which occurs when Venus and the earth are at opposite sides of the sun. Thus, in this problem we have $r_E, r_V \gg r_0 \gg r_S$, and so we expand $\frac{dt}{dr}$ to first order in r_S: $\frac{dt}{dr} \simeq \left(1 - \frac{r_0^2}{r^2}\right)^{-\frac{1}{2}}\left(1 + \frac{2r+3r_0}{2(r+r_0)}\frac{r_S}{r}\right)$.

The time $t(r_1, r_2)$ for the radar beam to get from r_1 to r_2 is given by $t(r_1, r_2) = \int_{r_1}^{r_2} dr\frac{dt}{dr}$ (by convention, for $r_2 > r_1$, since in the expression for $\frac{dt}{dr}$ given above we have taken the positive root in (16)). The time for a round trip from the earth to Venus is thus given by $T(r_E, r_V, r_0) = 2(t(r_0, r_V) + t(r_0, r_E))$. As a very small check, let us compute the time with r_S set to 0: $t_N(r_1, r_2) = \int_{r_1}^{r_2} dr\left(1 - \frac{r_0^2}{r^2}\right)^{-\frac{1}{2}} = \sqrt{r_2^2 - r_0^2} - \sqrt{r_1^2 - r_0^2}$ (for $r_1, r_2 > r_0$), in agreement with expectation. Talk about misleading terminology to call this the Newtonian time against which we define the "delay"; we might call it more accurately the Minkowskian time, or perhaps even the Pythagorean time.

In any case, the time delay is given by $\Delta T(r_E, r_V, r_0) = 2(\Delta t(r_0, r_V) + \Delta t(r_0, r_E))$ with

$$\Delta t(r_0, r_V) = r_S\int_{r_0}^{r_1} dr\left(1 - \frac{r_0^2}{r^2}\right)^{-\frac{1}{2}}\frac{2r+3r_0}{2r(r+r_0)} = r_S\left(\frac{1}{2}\sqrt{\frac{r-r_0}{r+r_0}} + \log\frac{r_1 + \sqrt{r_1^2 - r_0^2}}{r_0}\right) \tag{17}$$

---

* Recently a friend of mine remarked to me over dinner that anybody who has read a book on general relativity knows how to calculate the 43″ per century, but how many physicists can calculate the 5,600″ per century? Ouch, point well taken!

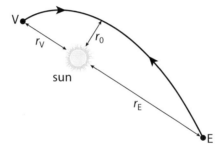

Figure 3 Radar echo delay. V stands for Venus, E for the earth.

(While we could evaluate the integral[17] exactly, to extract the leading logarithmic term $\Delta t(r_0, r_V) \sim r_S(\log(r_1/r_0) + \cdots)$ for $r_1 \gg r_0$, we can simply set r_0 to 0 in the integral.) I won't bother to put together a final expression for $\Delta T(r_E, r_V, r_0)$ for you, but I do wish to remark that, as usual, it takes herculean effort to realize the actual experiment with such a tiny effect. We have not even mentioned various necessary corrections, such as the propagation of the radar beam through the solar corona. Shapiro was able to verify Einstein's theory to a couple of percentage points, but over the decades, the accuracy has now been improved to about a tenth of a percent, using satellites carrying frequency dependent transponders rather than using Venus.

By the way, just about nothing sets off a stampede of crackpots saying that Einstein was wrong than a newspaper report about the latest radar echo delay measurement. The unfortunate word "delay" suggests to the uninformed that light does not actually move at the speed of light. You of course know better: (V.4.21) for $dr/d\zeta$ comes directly from $g_{\mu\nu}dx^\mu dx^\nu = 0$.

Appendix 3: Time dependent spherically symmetric mass distribution and the Jebsen-Birkhoff theorem

Jørg Tofte Jebsen in 1921 and George Birkhoff in 1923 showed that, remarkably, the Schwarzschild solution continues to hold outside a time dependent spherically symmetric mass distribution. This result, evidently of great relevance in studying the gravitational collapse of a spherically symmetric dust cloud to form a black hole, is known[18] as Birkhoff's theorem in most textbooks. I will walk you through a proof by direct computation.

In exercise V.4.3, you showed that time dependence leads to three more nonvanishing Christoffel symbols, namely $\Gamma^t_{tt} = \frac{\dot{A}}{2A}$, $\Gamma^t_{rr} = \frac{\dot{B}}{2A}$, $\Gamma^r_{tr} = \frac{\dot{B}}{2B}$, with odd numbers of the t index. This introduces one more nonvanishing component in the Ricci tensor:

$$R_{tr} = \frac{\dot{B}}{rB} \tag{18}$$

The three components we already had in (5–7) acquire additional terms as follows:

$$R_{tt} = \frac{A''}{2B} + \frac{A'}{rB} - \frac{A'}{4B}\left(\frac{A'}{A} + \frac{B'}{B}\right) - \frac{\ddot{B}}{2B} + \frac{\dot{B}}{4B}\left(\frac{\dot{A}}{A} + \frac{\dot{B}}{B}\right) \tag{19}$$

$$R_{rr} = -\frac{A''}{2A} + \frac{B'}{rB} + \frac{A'}{4A}\left(\frac{A'}{A} + \frac{B'}{B}\right) + \frac{\ddot{B}}{2A} - \frac{\dot{B}}{4A}\left(\frac{\dot{A}}{A} + \frac{\dot{B}}{B}\right) \tag{20}$$

$$R_{\theta\theta} = 1 - \frac{1}{B} - \frac{r}{2B}\left(\frac{A'}{A} - \frac{B'}{B}\right) \tag{21}$$

and of course we still have $R_{\varphi\varphi} = \sin^2\theta R_{\theta\theta}$.

This looks like a scary mess,* but fortunately here we only want to solve Einstein's equation in empty spacetime, outside the mass distribution. The equation $R_{tr} = 0$ tells us immediately that $\dot{B} = 0$. Happily for us, we see that, in R_{tt} and R_{rr}, $\dot{A}$ appears multiplied by $\dot{B}$ and so drops out.

Alternatively, the equation $R_{\theta\theta} = 0$ tells us that, since B does not depend on t, $(\log A)' = \frac{A'}{A}$ depends only on r. Thus, $\log A$ is the sum of a function of r and a function of t, and hence A is the product of a function of r and a function of t, namely $A = f(t)(1 - \frac{2GM}{r})$, with $f(t)$ some unknown function. But we can then simply define $d\tilde{t} = \sqrt{f(t)}dt$ to get rid of $f(t)$.

Thus, the Schwarzschild solution holds outside a spherically symmetric mass distribution, even if it varies in time. We will come back to the Jebsen-Birkhoff theorem in chapter IX.4. For now, note that the theorem is the general relativistic analog of Newton's two superb theorems mentioned way back in chapter I.1. To solve Einstein's field equation for the empty spacetime inside a spherical shell, simply go through all the same steps as in solving the Einstein field equation outside a spherically symmetric mass distribution, except that when you obtain $\frac{1}{B} = 1 + \frac{b}{r}$, you have to set the integration constant b to 0, since the spacetime must not be singular at $r = 0$.

Appendix 4: Weyl's shortcut to Schwarzschild

It is amusing to mention a quick, but not totally kosher, way[19] to the Schwarzschild solution given by Weyl, which Einstein professed in his writings to like.[20]

Weyl simply plugs the Ansatz for the metric (1) into the Einstein-Hilbert action $S_{\text{EH-Weyl}} = \int d^4x \sqrt{-g} R$ to obtain an "effective" action $S_{\text{effective}}(A, B)$. Any student who understands the variational principle could tell him this is not[†] quite legitimate. The correct procedure is of course to plug the Ansatz into the equations of motion obtained by varying the action. The rigorous mathematical justification[21] of what Weyl did took almost a century.[22]

The determinant of the metric $-g = ABr^4 \sin^2 \theta$ is easy. The scalar curvature

$$R = g^{\mu\nu} R_{\mu\nu} = -\frac{R_{tt}}{A} + \left(\frac{R_{rr}}{B} + \frac{1}{r^2} \left(R_{\theta\theta} + \frac{1}{\sin^2 \theta} R_{\varphi\varphi} \right) \right) = -\frac{R_{tt}}{A} + \left(\frac{R_{rr}}{B} + \frac{2}{r^2} R_{\theta\theta} \right) \tag{22}$$

is evaluated using (5–7). Weyl found that the substitution $A = a^2 b$ and $B(r) = 1/b(r)$ simplifies the resulting mess considerably.[‡] After integrating by parts, Weyl found that the action $S_{\text{effective}}(A, B)$ becomes

$$S_{\text{EH-Weyl}} = 8\pi \int dt \left[\int dr\, ra \left(b' + \frac{b}{r} - \frac{1}{r} \right) \right] = 8\pi \int dt \left\{ \int dr\, r(1-b)a' \right\} \tag{23}$$

where in the last step, we integrated $\int dr\, rab' = -\int dr\, (ra' + a)b$ by parts. In other words, Weyl dropped surface terms left and right. In fact, we can drop the integration over t just as we had integrated over θ and φ. Weyl was left with the amazingly simple effective action

$$S_{\text{effective}}(a, b) = -\int_0^\infty dr\, r(1-b)a' \tag{24}$$

Varying $S_{\text{effective}}$ with respect to b gives $a' = 0$, and with respect to a gives $(r(1-b))' = 0$. Fitting to the boundary conditions at spatial infinity gives $a = 1$ and $b = 1 - \frac{2M}{r}$, in other words, the Schwarzschild solution. (Recall an exercise you did back in chapter II.1.)

Actually, in my humble opinion, even by committing an illegitimacy, Weyl did not save all that much in arithmetic. In chapter VI.4, I will show that, if we are allowed to integrate by parts and throw away boundary terms with no questions asked, then we can write the Einstein-Hilbert action as $S_{\text{EH}} = \int d^4x \sqrt{-g} [\Gamma^\rho_{\sigma\lambda} \Gamma^\sigma_{\rho\nu} g^{\lambda\nu} - (\Gamma^\rho_{\nu\lambda} g^{\lambda\nu}) \Gamma^\eta_{\rho\eta}]$, after considerable formal manipulations. If we are allowed to start with this action and use Weyl's trick, then we can avoid calculating any of the curvature tensors and do save some arithmetical drudgery (which in any case we could foist on a computer, not to mention a competent student).

---

* As a check, note that R_{tt} and R_{rr} transform correctly under the scaling $t \to \lambda t$, $A \to \lambda^{-2} A$, and $B \to B$.

[†] To find the minimum of $f(x_1, \cdots, x_n)$ we should of course solve $\frac{\partial f}{\partial x_i} = 0$, $i = 1, \cdots, n$ with the appropriate Ansatz, instead of plugging some Ansatz for $x_1, \cdots, x_n$ into f first and then differentiating. But if both the Ansatz and the actual solution possess the same high degree of symmetry, it might perhaps be okay.

[‡] You might have noticed that this substitution is designed, with hindsight, to "deal with" the two combinations $(\frac{A'}{A} + \frac{B'}{B})$ and $(\frac{A'}{A} - \frac{B'}{B})$ that appear in the Ricci tensor.

Appendix 5: Droste's solution of Einstein's field equation

Amazing though Schwarzschild's story is, what happened to the obscure Dutchman Johannes Droste (1886–1963) is perhaps no less remarkable.[23] Droste, who received his doctorate in 1916 with Lorentz in Leiden, solved Einstein's 1915 field equations around a spherically symmetric mass, starting with the preliminary version of the field equation published by Einstein in 1913. His work[24] was communicated by Lorentz to the Royal Dutch Academy of Sciences on May 27, 1916, a few months after Einstein had communicated Schwarzschild's solution to the Prussian Academy of Sciences on January 13, 1916. In my opinion, Droste's paper is cleaner and less confused than Schwarzschild's, and furthermore contains an analysis of the motion of a particle in the spacetime. Interestingly, Droste also used Weyl's approach, which we explained in appendix 4 and which Einstein liked so much, long before Weyl. For some reason, the physics community totally ignored this work. Droste became a high school teacher, and later a professor of mathematics at Leiden University. (While writing this appendix, I asked a professor of physics at Leiden University, who said he had never heard of Johannes Droste. He did tell me, however, that "Droste" was well known in the Netherlands as a brand of cocoa powder, after which the Droste effect [an apparently infinite regression of pictures within pictures[25]] was named. He told me of his fond childhood memory[26] of drinking hot chocolate while being fascinated with infinity.)

I was quite astonished by this story. Somehow, Lorentz never mentioned his student's work to Einstein. Or perhaps he did, but Einstein chose to promote Schwarzschild, who after all died rather tragically. But why didn't Droste protest every time the Schwarzschild solution was mentioned? Perhaps we simply live in a noisier and more assertive era. Here is an interesting tale for a budding historian of physics to look into.

In the appendices to the chapter, I tell you about, not one, but two young guys getting shafted by the establishment.

Exercises

1   Calculate the Ricci tensor in terms of A and B.

2   Calculate the Riemann tensor in terms of A and B.

3   Show that the Schwarzschild metric can be written in the isotropic form

$$ds^2 = \left(\frac{1-\frac{GM}{2\rho}}{1+\frac{GM}{2\rho}}\right)^2 dt^2 - \left(1+\frac{GM}{2\rho}\right)^4 \left(d\rho^2 + \rho^2(d\theta^2 + \sin^2\theta\, d\varphi^2)\right) \tag{25}$$

Where is the horizon?

4   Show that the Schwarzschild metric can be written in the harmonic form

$$ds^2 = \left(\frac{1-\frac{GM}{R}}{1+\frac{GM}{R}}\right) dt^2 - \left(1+\frac{GM}{R}\right)^2 d\vec{x}^2 - \left(\frac{GM}{R^2}\right)^2 \left(\frac{1+\frac{GM}{R}}{1-\frac{GM}{R}}\right)(\vec{x}\cdot d\vec{x})^2 \tag{26}$$

with $R = \vec{x}^2$.

5   Show that in the parametrized post-Newtonian approximation described in chapter V.4, the deflection of light is given by $\Delta\varphi = \left(\frac{1+\gamma}{2}\right)(\Delta\varphi)_{\text{Einstein}}$, and the perihelion shift by $\Delta\varphi = (\frac{2-\beta+2\gamma}{3})(\Delta\varphi)_{\text{Einstein}}$.

6   Show that

$$ds^2 = \left(1 - \frac{2M}{r} - r^2\right)dt^2 - \left(\frac{dr^2}{1-\frac{2M}{r}-r^2} + r^2 d\Omega_2^2\right)$$

satisfies the Einstein field equation $R_{\mu\nu} = -3\Lambda g_{\mu\nu}$ with a cosmological constant. This is known as the Schwarzschild–de Sitter spacetime. We will come back to this in chapter IX.10.

7   Show that in $(4 + 1)$-dimensional spacetime, the analog of the Schwarzschild solution is given by

$$-ds^2 = -\left(1 - \frac{r_S^2}{r^2}\right) dt^2 + \left(1 - \frac{r_S^2}{r^2}\right)^{-1} dr^2 + r^2 d\Omega_3^2$$

where $d\Omega_3^2$ is the metric on the 3-sphere S^3.

Notes

1. It is less well known that while on the front, he also wrote a paper on the Stark effect. Perhaps it is only a slight exaggeration to say that these days there are professors of general relativity walking around proudly ignorant of atomic physics (and professors of atomic physics proudly ignorant of general relativity).
2. Some would say that this is a highly appropriate name for someone who discovered black holes.
3. Historically, the horizon was a source of great confusion, and Kruskal's contribution cannot be overestimated. For example, on p. 203 of Bergmann's textbook *Introduction to the Theory of Relativity (with a foreword by A. Einstein)* (1976 Dover edition, originally published in 1942), he quoted Robertson as concluding that "at least part of the singular character" of the metric at $r = 2GM$ must be attributed to the choice of coordinates. Curiously, people at the time did not follow the modern expedient of simply noting the smoothness of the Riemann curvature tensor, which Schwarzschild himself, at the very least, must have calculated. Bergmann then went on and cited a paper by Einstein (*Ann. Math.* 40 (1939), p. 922) purportedly showing that in a toy model of a spherical cluster of noninteracting particles, the Schwarzschild singularity could not form. The general feeling was that the Schwarzschild singularity could not occur in nature.
4. The second and third sentence in Kruskal's paper read: "Kasner, Lemaître, Einstein and Rosen, Robertson, Synge, Ehlers, Finkelstein, and Fronsdal have shown that the singularities at $r = 0$ and $r = 2GM$ are very different in character. Their conclusion—that there is no real singularity at $r = 2GM$—can be demonstrated by a choice of coordinates seemingly simpler and more explicit than any introduced so far to this end." The papers cited range from Kasner's in 1921 to Fronsdal's in 1959. I am assuredly not a historian, but this certainly indicates that after 44 years, the issue of the "spherical singularity" was about to be settled in 1960.
5. Notice also that you calculated only the Ricci tensor (which, being zero, manifestly does not blow up at r_S) but not the full Riemann tensor, which you can now do as a tedious exercise.
6. This is, of course, Minkowski's "mystical" substitution $x = it$.
7. A photograph of this letter (in German) is in the Huntington Digital Library (http://hdl.huntington.org).
8. Most physics students associate Friedrich Bessel with cylindrical coordinates, but in fact his work with Bessel functions (actually first discovered by Daniel Bernoulli) was largely in connection with perturbations of planetary orbits.
9. In 2012, astronomers discovered a star that has an orbital period of only 11.5 years around the Milky Way's central black hole. This will, in due time, give another test of Einstein's prediction of the perihelion shift.
10. Not to be confused with his cousin the aircraft maker.
11. Einstein to Sommerfeld, December 9, 1915: "Wie kommt uns da die pedantische Genauigkeit der Astronomie zu Hilfe, über die ich mich im Stillen früher oft lustig machte!" *The Collected Papers of Albert Einstein*, vol. 8, *The Berlin Years: Correspondence, 1914–1918*, ed. Robert Schulmann et al., Princeton University Press, 1998, p. 217; English translation by Ann Hentschel, cited from the companion translation volume, p. 159.
12. Quoted in A. Calaprice, ed. *The Expanded Quotable Einstein*, Princeton University Press, 2000.
13. J. Waller, *Einstein's Luck*.
14. If you've heard the phrase "the guns of August," which inspired a book with that title, you would know that the timing was optimal for Einstein, as it would turn out.
15. For a contemporary study exonerating Eddington, see D. Kennefick, "Testing Relativity from the 1919 Eclipse—A Question of Bias," *Physics Today*, March 2009, p. 37.
16. Note that by differentiating (13) for the perihelion motion, we obtain basically the same differential equation as in (10) for the deflection of light, perhaps not surprisingly.
17. Note Newton's greatness. In calculating the deviation from Newtonian physics, the mathematics we use is all Newtonian.
18. Perhaps this is because Birkhoff was a famous professor at Harvard, while Jebsen (1888–1922) died young (of tuberculosis) and obscure. This is another example of the Matthew principle cited in chapter III.2. For

the discovery of Jebsen's paper, see S. Deser, *Gen. Relativ. Gravit.* 37 (2005), p. 2251, and N. V. Johansen and F. Ravndal, arXiv 0508163.

19. H. Weyl, *Space-Time-Matter,* Dover, 1952; S. Deser and B. Tekin, *Class. Quantum Grav.* 20 (2003), pp. 4877–4883; S. Deser and J. Franklin, *Am. J. Phys.* March 2005, pp. 261–264.

20. "The derivation given by Weyl in his book "Raum-Zeit-Materie" is particularly elegant." A. Einstein, *The Meaning of Relativity,* Princeton University Press, 2004, p. 94.

21. R. S. Palais, *Comm. Math. Physics,* 69 (1979), p. 19.

22. Perhaps ironically one of the leading mathematicians of his time. Those who say that my textbooks are not rigorous enough for them, take note! Winging it often turns out to lead to the right answer.

23. I am grateful to Gary Gibbons for telling me about Droste during a visit to Trinity College.

24. J. Droste, reprinted in *Gen. Rel. Grav.* 34 (2002), p. 1545. See the historical notes by T. Rothman and by C. Beenakker, pp. 1541 and 1543.

25. If you search the web for the Droste effect, you will see why it is named after cocoa powder. I was tempted to make this an endnote inside an endnote.

26. Beats Proust any day. Recall also chapter III.4.

VI.4 | Energy Momentum Distribution Tells Spacetime How to Curve

The action of the world

In chapter VI.1, we arrived at the action of the world (whoa!) $S = S_{\text{EH}} + S_{\text{matter}}$. Here S_{EH} denotes the Einstein-Hilbert action and S_{matter} a sum of various matter actions, such as the action for point particles and the Maxwell action for the electromagnetic field. (As already mentioned in chapter VI.1, in Einstein gravity, the term "matter" is often used in an extended sense to include everything else besides gravity, such as the electromagnetic field, which we normally do not think of as matter.)

We have to vary $S = S_{\text{EH}} + S_{\text{matter}}$ with respect to the gravitational field $g_{\mu\nu}$ to obtain the full field equation for Einstein gravity. Thus far, we have avoided* varying S_{matter}. In this chapter, we will learn how to vary several different forms of S_{matter}.

First, how do we obtain S_{matter}, as for example the Maxwell action for the electromagnetic field in curved spacetime? As explained in chapters V.2 and V.6, when we discussed the power of the equivalence principle, we simply[†] take the flat spacetime actions we have known and loved, such as the Maxwell action, and promote them to curved spacetime by replacing the Minkowski metric $\eta_{\mu\nu}$ by $g_{\mu\nu}$.

The energy momentum tensor once again

Second, what do we get when we vary S_{matter}? Write the variation of S_{matter} as $\delta S_{\text{matter}} = \frac{1}{2} \int d^4x \sqrt{-g}\, T^{\mu\nu}(x)\delta g_{\mu\nu}(x)$. In other words, define

$$T^{\mu\nu}(x) \equiv \frac{2}{\sqrt{-g}} \frac{\delta S_{\text{matter}}}{\delta g_{\mu\nu}(x)} \tag{1}$$

---

* In chapter VI.2, we guessed what varying the exceptionally simple $S_{\text{cosmological}} = -\int d^4x \sqrt{-g}\,\Lambda$ would give us, rather than actually varying it.

[†] See appendix 3 in chapter IX.7 for an exception to this statement.

so that the field equation now has the form (recall the notation of chapter VI.1) $A(R^{\mu\nu} + \alpha g^{\mu\nu}R) = T^{\mu\nu}$.

By this point, you understand that $T^{\mu\nu}(x)$ defines a symmetric 2-indexed tensor at every point in spacetime. Furthermore, it appears in the field equation for Einstein gravity and determines the curvature of spacetime. So what could it be?

Remember the energy momentum tensor $T^{\mu\nu}$ we derived in chapter III.6? Indeed, you might accuse me of trying to sneak one past you by using the same notation. Yes, as you might have guessed, the $T^{\mu\nu}$ here is the curved spacetime generalization of the $T^{\mu\nu}$ there. Physically, it makes sense that the distribution of energy momentum determines the curvature. The energy momentum tensor $T^{\mu\nu}$ appears in the Einstein's equation as a source for the gravitational field, just as the electromagnetic current J^{μ} appears in Maxwell's equation as a source for the electromagnetic field. In fact, you can see that what we are doing is analogous to what we did in chapter IV.2; one difference is that we have to carry one more index around.

We have arrested the suspect, but how do we convict him? Already, we have mentioned a load of circumstantial evidence.* We will now identify $T^{\mu\nu}$ for a number of cases, show that the integral $\int_V d^3x \sqrt{-g}\,T^{0\nu}$ gives the energy and the momentum contained in the volume V, and verify that $T^{\mu\nu}$ reduces correctly to the energy momentum tensor we knew and loved in flat spacetime back in chapter III.6. Most importantly, we will show the court that the $T^{\mu\nu}$ defined in (1) is covariantly conserved

$$D_\mu T^{\mu\nu} = 0 \tag{2}$$

which reduces correctly to the familiar $\partial_\mu T^{\mu\nu} = 0$ in the flat spacetime limit. If it waddles and quacks like a duck, then it is a duck.

We did not go looking for the energy momentum tensor, but the energy momentum tensor came looking for us!

Energy and momentum of point particles

Let's see how (1) works for the simple case of a gas of point particles that do not interact with one another, known as dust in general relativity and cosmology. The action reads

$$S_{\text{particles}} = -\sum_a m_a \int d\tau_a \sqrt{-g_{\mu\nu}(X_a)\frac{dX_a^\mu}{d\tau_a}\frac{dX_a^\nu}{d\tau_a}} \tag{3}$$

You have encountered this action in flat spacetime several times already, in chapters III.5 and IV.2. The equivalence principle again roars with its awesome power: simply replace

---

* Another clue comes from the equation for the Newtonian gravitational potential: $\nabla^2\Phi = 4\pi G\rho$. In a relativistic theory, the mass density ρ is replaced by the energy density, and we can't speak of energy density without talking about momentum density as well. This suggests that the right hand side of Einstein's field equation should involve energy and momentum density. See chapters III.6 and IX.5.

$\eta_{\mu\nu}$ by $g_{\mu\nu}$ and behold the action in its full glory in curved spacetime. Varying $S_{particles}$ with respect to $g_{\mu\nu}(x)$ and using (1), we obtain

$$T^{\mu\nu}(x) = \frac{2}{\sqrt{-g(x)}} \frac{\delta}{\delta g_{\mu\nu}(x)} S_{particles}$$

$$= \frac{1}{\sqrt{-g(x)}} \sum_a m_a \int d\tau_a \frac{1}{\sqrt{-g_{\sigma\rho}(X_a)\frac{dX_a^\sigma}{d\tau_a}\frac{dX_a^\rho}{d\tau_a}}} \frac{dX_a^\mu}{d\tau_a} \frac{dX_a^\nu}{d\tau_a} \delta^4(x - X_a(\tau_a))$$

$$= \frac{1}{\sqrt{-g(x)}} \sum_a m_a \int d\tau_a \frac{dX_a^\mu}{d\tau_a} \frac{dX_a^\nu}{d\tau_a} \delta^4(x - X_a(\tau_a)) \tag{4}$$

where as usual the third equality follows from the sensible parametrization of the particle worldlines, namely defining τ_a by setting $g_{\mu\nu}(X_a)\frac{dX_a^\mu}{d\tau_a}\frac{dX_a^\nu}{d\tau_a} = -1$. As we suspected, $T^{\mu\nu}(x)$ is precisely the curved spacetime generalization of the energy momentum tensor we first encountered in Minkowskian spacetime. Setting $g_{\mu\nu}$ to $\eta_{\mu\nu}$, we recover (III.6.7): the "only" difference is the appearance of the density factor $1/\sqrt{-g}$, which is precisely what is needed to counteract the $\sqrt{-g}$ in the volume factor when we integrate over $T^{0\nu}$ to obtain the energy and momentum of the particles. Indeed, this provides a nice formal check of our more laborious, but more physical, derivation in chapter III.6. It also shows that the plus sign in (1) comes from the minus sign in (3), which was needed (as was first explained back in chapter III.5) to reproduce the Newtonian action for a point particle.

A common sign error

I must emphasize, once again, that the gravitational field is $g_{\mu\nu}$, not $g^{\mu\nu}$. The reason is that from the very beginning, we defined coordinates to carry upper indices, and so for point particles the dynamical variables X_a^μ carry an upper Lorentz index. Thus, particles couple to $g_{\mu\nu}$, not $g^{\mu\nu}$. In varying the action here, we are required to hold the dynamical variables X_a^μ fixed as we did in (4). Of course, once we obtain $T^{\mu\nu}$, we can lower indices at will using the metric and define $T_{\mu\nu} \equiv g_{\mu\lambda} T^{\lambda\rho} g_{\rho\nu}$.

People sometimes commit a sign error here. For an invertible matrix M, recall from (V.6.7) that $\delta M^{-1} = -M^{-1}\delta M M^{-1}$. Applying this to the metric, we have

$$\delta g^{\rho\sigma} = -g^{\rho\mu}\delta g_{\mu\nu}g^{\nu\sigma} \tag{5}$$

(Note the sign, which we can verify in the 1-by-1 case: $\delta x^{-1} = -x^{-2}\delta x$.)

Were we to define a $T_{\mu\nu}$ by varying with respect to $g^{\mu\nu}$ in (1), we would have produced an erroneous sign. I will show you this schematically, omitting inessential factors. Given a matter action $S(X^{\mu\nu} g_{\mu\nu}) = S(X_{\rho\sigma} g^{\rho\sigma})$, we have $\delta S = S' X^{\mu\nu}\delta g_{\mu\nu}$, but $\delta S = S'X_{\rho\sigma}\delta g^{\rho\sigma} = S'X_{\rho\sigma}(-1)g^{\rho\mu}\delta g_{\mu\nu}g^{\nu\sigma} = -S'X^{\mu\nu}\delta g_{\mu\nu}$. This slightly subtle point is sometimes not made sufficiently clear. The point is of course that in varying, you have to specify* what is being held fixed.

---

* As in thermodynamics.

Energy momentum in the electromagnetic field

Let me now show you the power of this definition of $T^{\mu\nu}$ by obtaining results that may be familiar to you about the electromagnetic field in flat spacetime. Promote the Maxwell action $S_{\text{Maxwell}} = \int d^4x \mathcal{L}_{\text{Maxwell}} = -\frac{1}{4} \int d^4x F_{\mu\nu} F^{\mu\nu} = -\frac{1}{4} \int d^4x \eta^{\sigma\zeta} \eta^{\lambda\rho} F_{\sigma\lambda} F_{\zeta\rho}$ for the electromagnetic field in Minkowskian spacetime introduced in chapter IV.2 to

$$S_{\text{Maxwell}} = -\frac{1}{4} \int d^4x \sqrt{-g} g^{\sigma\zeta} g^{\lambda\rho} F_{\sigma\lambda} F_{\zeta\rho} \tag{6}$$

in curved spacetime. Next, vary as in (1) and thus calculate $T^{\mu\nu}$ of the electromagnetic field.

The important point here is that A_μ, not $A^\mu \equiv g^{\mu\nu} A_\nu$, is the dynamical variable and is to be held fixed. One way to see this is to recall that in our discussion (chapter IV.2) of electromagnetic gauge invariance, A_μ goes with $\partial_\mu \equiv \frac{\partial}{\partial x^\mu}$. So the mnemonic is that X_a^μ and A_μ are the dynamical variables. Another important bit of information you should recall is that, as explained in chapter V.6, the covariant curl is equal to the ordinary curl $D_\mu A_\nu - D_\nu A_\mu = \partial_\mu A_\nu - \partial_\nu A_\mu$, so that $F_{\mu\nu}$ does not depend on the metric.

Here, we have to vary the determinant $g = \det g_{\mu\nu}$, but we did that already back in chapter V.6. For convenience, I will repeat it here. Using the identity $\log \det M = \text{tr} \log M$, we obtained $\delta \det M = \det M \delta(\text{tr} \log M) = \det M \, \text{tr}(M^{-1} \delta M)$ and thus

$$\delta \sqrt{-g} = \frac{1}{2} \sqrt{-g} g^{\mu\nu} \delta g_{\mu\nu} \tag{7}$$

(To check, we again go to the 1-by-1 case: $\delta \sqrt{-x} = \frac{1}{2} \sqrt{-x} x^{-1} \delta x$.)

Now we're ready to vary:

$$T^{\mu\nu}(x) = \frac{2}{\sqrt{-g(x)}} \frac{\delta}{\delta g_{\mu\nu}(x)} S_{\text{Maxwell}}$$

$$= -\frac{2}{4\sqrt{-g(x)}} \frac{\delta}{\delta g_{\mu\nu}(x)} \int d^4y \sqrt{-g(y)} g^{\sigma\zeta}(y) g^{\lambda\rho}(y) F_{\sigma\lambda}(y) F_{\zeta\rho}(y)$$

$$= F_\lambda^\mu(x) F^{\nu\lambda}(x) - \frac{1}{4} g^{\mu\nu}(x) F_{\sigma\rho}(x) F^{\sigma\rho}(x) \tag{8}$$

To obtain the last expression, we used (5) and (7). Note the local character of the energy momentum tensor.

The energy momentum tensor (8) of the electromagnetic field contains two pieces. Take the trace and watch the results from the two pieces cancel each other:

$$T \equiv g_{\mu\nu} T^{\mu\nu} = g_{\mu\nu} (F_\lambda^\mu F^{\nu\lambda} - \frac{1}{4} g^{\mu\nu} F_{\sigma\rho} F^{\sigma\rho}) = 0 \tag{9}$$

Note how various signs play a crucial role here.

The energy momentum tensor of the electromagnetic field is traceless.* Interestingly, in chapter III.6, we derived the tracelessness of the energy momentum tensor of a photon gas

---

* Looking ahead, we remark here that this is related to the invariance of the Maxwell action under scale transformation. We will discuss scale invariance in chapter IX.9.

coming from a rather different direction. This is an example of the "particle-field duality" that is at the core of quantum field theory.

Electromagnetism in flat spacetime

It is instructive to make contact with what you know about electromagnetism in flat spacetime, either from chapter IV.1 or IV.2. Recall that the field strength $F_{\mu\nu}$ is related to the electric and magnetic fields by $F_{0i} = E_i$ and $F_{ij} = \frac{1}{2}\varepsilon_{ijk}B_k$.

Well, simply demote $g^{\mu\nu}$ in (6) and (8) to $\eta^{\mu\nu}$. The Maxwell Lagrangian (6) becomes

$$\mathcal{L} = -\tfrac{1}{4}F_{\mu\nu}F^{\mu\nu} = -\tfrac{1}{4}\left(-2\sum_i F_{0i}^2 + \sum_i\sum_j F_{ij}^2\right) = \tfrac{1}{2}(\vec{E}^2 - \vec{B}^2) \tag{10}$$

where for clarity, I have reinstated the summation sign. Next, work out the different components of $T^{\mu\nu}$ and compare with what you know, either from a course on electromagnetism or with exercise IV.2.2. The energy density is given by

$$T^{00} = T_{00} - +F^0_{\lambda}F^{0\lambda} + \eta^{00}\mathcal{L} = +\vec{E}^2 - \tfrac{1}{2}(\vec{E}^2 - \vec{B}^2) = \tfrac{1}{2}(\vec{E}^2 + \vec{B}^2) \tag{11}$$

(Note that in Minkowski spacetime T^{00} and T_{00} are numerically the same.) That's comforting to see an energy density we've known from "childhood": the energy density is a rotational scalar to which the electric and the magnetic fields contribute equally.

Contrast the signs in (10) and (11). The signs work just as in Newtonian mechanics, where the energy is the sum of the kinetic and potential energies, while the Lagrangian is the difference. In electromagnetism, the electric field plays the kinetic role, while the magnetic field plays the potential role. Indeed, the electromagnetic field may be regarded as a collection of an infinite number of harmonic oscillators. This view provides one possible starting point for quantum field theory.

Next, calculate the momentum density

$$T_{0i} = F_{0\lambda}F_i{}^{\lambda} = F_{0j}F_{ij} = \varepsilon_{ijk}E_jB_k = (\vec{E} \times \vec{B})_i \tag{12}$$

The Poynting vector you learned in electromagnetism has just emerged! It is the simplest rotational vector you can form out of the electric and the magnetic fields with the correct transformation properties under reflections in space and in time.

Interaction among different matter sectors

Thus far, we have treated $S_{\text{particles}}$ and S_{Maxwell} in turn. But now consider the action $S_{\text{particles}} + S_{\text{Maxwell}} + S_{\text{interaction}}$, where, as we first saw in chapter IV.2,

$$S_{\text{interaction}} = \sum_a e_a \int d\tau_a \frac{dX_a^{\mu}}{d\tau_a} A_{\mu}(X_a) \tag{13}$$

with e_a the charge of particle a. All we have to do is to plug this action into (1) and turn the crank.

"But I don't see $g_{\mu\nu}$ anywhere in (13)," Confusio pipes up.

Indeed, for once Confusio is right: perhaps surprisingly, $S_{\text{interaction}}$ does not contribute to $T^{\mu\nu}$ according to (1). Thus, for charged particles interacting with one another, the energy momentum tensor that the gravitational field responds to is just the sum of the energy momentum tensors in (4) and (8):

$$T^{\mu\nu} = T^{\mu\nu}_{\text{particles}} + T^{\mu\nu}_{\text{Maxwell}}$$

$$\equiv \frac{1}{\sqrt{-g}} \sum_a m_a \int d\tau_a \frac{dX_a^\mu}{d\tau_a} \frac{dX_a^\nu}{d\tau_a} \delta^4(x - X_a) - F_\lambda^\mu F^{\nu\lambda} + \frac{1}{4} g^{\mu\nu} F_{\sigma\rho} F^{\sigma\rho} \tag{14}$$

However, as we would expect physically, since the charged particles and the electromagnetic field can exchange energy and momentum, the tensors $T^{\mu\nu}_{\text{particles}}$ and $T^{\mu\nu}_{\text{Maxwell}}$ are no longer separately conserved. Only $T^{\mu\nu}$ is conserved.[1] We will reach a deeper understanding of this point in appendix 1.

What exactly are energy and momentum, anyway?

You started studying physics by learning about mass, energy, momentum, all sorts of great stuff like that. But what exactly is energy and momentum anyway? I want to emphasize here that (1) gives us a fundamental definition of energy (and hence mass) and momentum. Energy momentum is what the graviton listens to, just as electric charge is what the photon listens to. (The graviton is of course the particle associated with the field $g_{\mu\nu}$.) In other words, energy momentum as embodied in $T^{\mu\nu}$ is what appears in the right hand side of Einstein's field equation: it is the stuff that tells spacetime how to curve.

As is evident from the electromagnetic example, this definition is useful even if we are not interested in curved spacetime per se. Given an action in flat spacetime, we can always temporarily promote $\eta_{\mu\nu}$ to $g_{\mu\nu}$, multiply d^4x by $\sqrt{-g}$, use (1) to find $T^{\mu\nu}(x)$, and then set $g_{\mu\nu}$ back to the Minkowski metric $\eta_{\mu\nu}$. We are guaranteed, as we will show shortly, to obtain an energy momentum tensor satisfying $D_\mu T^{\mu\nu}(x) = 0$, and hence $\partial_\mu T^{\mu\nu}(x) = 0$ in flat spacetime. In contrast to a formula like $E_K = \frac{1}{2}mv^2$, the definition (1) of energy momentum can be applied to any theory based on the action principle, such as quantum field theory.[2] You will explore this further in exercise 4.

More importantly, this fundamental definition of the energy momentum tensor leads us to a deep understanding of why energy and momentum are conserved. As the derivation is a bit long, I place it in appendix 1, where I will show that the conservation law $D_\mu T^{\mu\nu} = 0$ follows elegantly from the principle of general invariance. We would expect $T^{\mu\nu}$ to have "nice" properties; it is the variation not of any old piece of junk, but of an exquisite object that controls how things move and that does not change under coordinate transformations.

Appendix 1: Conservation of energy momentum

As I emphasized way back in chapter II.4, conservation laws and symmetries are intimately connected. Here we will exploit the general invariance of the matter action S_{matter} to prove energy momentum conservation. We could discuss this in the abstract, but just to be concrete, let us specialize to $S_{\text{Maxwell}} = -\frac{1}{4} \int d^4x \sqrt{-g} g^{\sigma\zeta} g^{\lambda\rho} F_{\sigma\lambda} F_{\zeta\rho}$.

In other words, we are taking the action for the world to be $S = S_{\text{EH}} + S_{\text{Maxwell}}$, namely a world with only gravity and electromagnetism, and nothing else.

Be forewarned. The following derivation may appear rather long to the novice, but it is completely general and hence quite profound. Every step may seem trivially true, but that might well be the way of the Zen master. Don't be lulled to sleep, and watch the primes, and even more importantly, the absence of primes, like a hawk.

General invariance means that S_{Maxwell} remains unchanged if we make these replacements:

$$x \to x', \ g_{\rho\sigma}(x) \to g'_{\rho\sigma}(x'), \ g^{\mu\sigma}(x) \to g'^{\mu\sigma}(x'), \ A_\rho(x) \to A'_\rho(x') \tag{15}$$

This may seem like a long list, but you understand that all we are doing is making a coordinate transformation. As always, $g'_{\rho\sigma}(x') = g_{\mu\nu}(x)(S^{-1})^\mu{}_\rho (S^{-1})^\nu{}_\sigma$, $g'^{\mu\sigma}(x') = S^\mu{}_\rho S^\sigma{}_\tau g^{\rho\tau}(x)$, $A'_\rho(x') = A_\mu(x)(S^{-1})^\mu{}_\rho$, with $S^\mu{}_\nu \equiv \frac{\partial x'^\mu}{\partial x^\nu}$ and $(S^{-1})^\mu{}_\rho \equiv \frac{\partial x^\mu}{\partial x'^\rho}$.

We presently specialize to an infinitesimal transformation $x'^\mu = x^\mu + \varepsilon^\mu(x)$ so that, to leading order, $S^\mu{}_\nu = \delta^\mu_\nu + \partial_\nu \varepsilon^\mu(x)$ and $(S^{-1})^\mu{}_\rho = \delta^\mu_\rho - \partial_\rho \varepsilon^\mu(x)$. Then $A'_\rho(x') = A_\mu(x)(S^{-1})^\mu{}_\rho = A_\rho(x) - A_\mu(x)\partial_\rho \varepsilon^\mu(x)$ and

$$g'_{\rho\sigma}(x') = g_{\mu\nu}(x)(S^{-1})^\mu{}_\rho (S^{-1})^\nu{}_\sigma = g_{\rho\sigma}(x) - g_{\mu\sigma}(x)\partial_\rho \varepsilon^\mu(x) - g_{\rho\nu}(x)\partial_\sigma \varepsilon^\nu(x) \tag{16}$$

Keep that in the back of your mind.

After we make those replacements in (15), we end up with

$$S_{\text{Maxwell}} = -\frac{1}{4} \int d^4x' \sqrt{-g'(x')} g'^{\sigma\zeta}(x') g'^{\lambda\rho}(x') F'_{\sigma\lambda}(x') F'_{\zeta\rho}(x')$$

(with $F'_{\mu\nu}(x') = \partial'_\mu A'_\nu(x') - \partial'_\nu A'_\mu(x')$). We have exactly the same S_{Maxwell} we started with, and so $\delta S_{\text{Maxwell}} = 0$.

Looks like we did nothing! We have merely verified that S_{Maxwell} is invariant under general coordinate transformation. But the magic trick is about to begin.

Since x' is a dummy integration variable, we can erase all the primes on x' in that integral for S_{Maxwell} displayed in the preceding paragraph. So go ahead and do it. I will wait for you.

You didn't erase the prime on g' and A', did you?

If you did, you need to review your calculus. The dummy x' can be replaced by anything you want, in particular x. But of course you and I have no right to erase the primes on the dynamical fields $g'_{\rho\sigma}$ and A'_ρ. (Here and henceforth, all statements made about $g'_{\rho\sigma}$ also apply mutatis mutandis to $g'^{\mu\sigma}$, which after all just denotes the inverse.) In other words, $S_{\text{Maxwell}} = -\frac{1}{4} \int d^4x \sqrt{-g'(x)} g'^{\sigma\zeta}(x) g'^{\lambda\rho}(x) F'_{\sigma\lambda}(x) F'_{\zeta\rho}(x)$. The net effect is that we have replaced $A_\rho(x) \to A'_\rho(x) = A'_\rho(x') - (A'_\rho(x') - A'_\rho(x))$ and $g_{\rho\sigma}(x) \to g'_{\rho\sigma}(x) = g'_{\rho\sigma}(x') - (g'_{\rho\sigma}(x') - g'_{\rho\sigma}(x))$. To leading order in ε, we made the change

$$\delta g_{\rho\sigma}(x)^{\text{specific}} \equiv g'_{\rho\sigma}(x) - g_{\rho\sigma}(x) = \{g'_{\rho\sigma}(x') - g_{\rho\sigma}(x)\} - \{g'_{\rho\sigma}(x') - g'_{\rho\sigma}(x)\}$$

$$= -\left(g_{\mu\sigma}(x)\partial_\rho \varepsilon^\mu(x) + g_{\rho\nu}(x)\partial_\sigma \varepsilon^\nu(x) + \varepsilon^\lambda \partial_\lambda g_{\rho\sigma}(x)\right) + O(\varepsilon^2) \tag{17}$$

In the last step, we simplified the first bracket using (16), which I told you to keep in the back of your mind, and the second bracket using calculus to the order indicated. We put a superscript on $\delta g_{\rho\sigma}(x)$ to remind us that this is a specific variation of the metric, given by the specific form in (18), not a general variation.

Recalling chapter V.6, you might have recognized that we can write this in terms of covariant derivatives of ε^μ:

$$\delta g_{\rho\sigma}^{\text{specific}} = -(\varepsilon_{\rho;\sigma} + \varepsilon_{\sigma;\rho}) + O(\varepsilon^2) \tag{18}$$

(Incidentally, the expression inside the parentheses could also be written in terms of the Lie derivative $\mathcal{L}_\varepsilon g_{\rho\sigma}$.)

Similarly, we can work out the change $\delta A_\rho(x)^{\text{specific}} \equiv A'_\rho(x) - A_\rho(x)$ to be some specific expression involving ε and its derivatives, the same kind of expression as in (18) but, as you will see, we don't need to know the specific form.

As I said, everything seems trivial step by step. But now let us put the pieces together: the variation $\delta S_{\text{Maxwell}}$ consists of two terms, one due to the variation $\delta A_\rho(x)^{\text{specific}}$, the other to $\delta g_{\rho\sigma}(x)^{\text{specific}}$. Since we also know that $\delta S_{\text{Maxwell}}$ vanishes by general invariance, we obtain

$$0 = \delta S_{\text{Maxwell}} = \int d^4x (\cdots)^\rho \delta A_\rho(x)^{\text{specific}} + \frac{1}{2} \int d^4x \sqrt{-g} \, T^{\rho\sigma}(x) \delta g_{\rho\sigma}(x)^{\text{specific}} \tag{19}$$

by virtue of (1).

In the first term, the expression denoted by $(\cdots)^\rho$ vanishes. Indeed, according to Euler and Lagrange, that is how we derive the equation of motion for the electromagnetic field $A_\rho(x)$. We vary the action S_{Maxwell} with respect to a general variation of $A_\rho(x)$ and set that variation of the action to 0 to obtain the equation of motion. If this variation of S_{Maxwell} vanishes for a general variation $\delta A_\rho(x)$, then a fortiori it vanishes for the specific variation $\delta A_\rho(x)^{\text{specific}}$. This point is worth emphasizing: the rest of this derivation can proceed only if the electromagnetic field satisfies Maxwell's equations. But of course: the energy momentum tensor (8) of the electromagnetic field can hardly be expected to be conserved if the electromagnetic field is not going to vary in time according to the rules!

To obtain the second term in (19), we use the definition (1), substituting for $\delta g_{\rho\sigma}$ the specific variation $\delta g_{\rho\sigma}(x)^{\text{specific}}$ in (18). We thus conclude that $0 = \int d^4x \sqrt{-g}\, T^{\rho\sigma}(x)\delta g_{\rho\sigma}(x)^{\text{specific}}$.

Confusio looks puzzled. We knew he would!

"You mean (1) is just zero?"

"No! The statement in (1) says that if you vary S_{matter} by varying $g_{\mu\nu}$ arbitrarily, the coefficient gives you $T^{\mu\nu}$. It is the definition of $T^{\mu\nu}$," you and I say in unison, "but the statement we just derived says that the variation of S_{matter} vanishes for a specific $\delta g_{\rho\sigma}(x)^{\text{specific}}$ as given in (18)."

So, we just derived

$$\int d^4x \sqrt{-g}\, T^{\rho\sigma}(\varepsilon_{\rho;\sigma} + \varepsilon_{\sigma;\rho}) = 0 \tag{20}$$

The expression in the parentheses in (20), since it is multiplied by a symmetric tensor, can be replaced by $2\varepsilon_{\sigma;\rho}$. A (covariant) integration by parts immediately gives (since $\varepsilon^\mu(x)$ is arbitrary) what we set out to prove:

$$D_\rho T^{\rho\sigma} = 0 \tag{21}$$

For the discussion in the next appendix, it is also illuminating to write (21) as

$$D_\rho T^{\rho\sigma} = \frac{1}{\sqrt{-g}}\partial_\rho(\sqrt{-g}T^{\rho\sigma}) + \Gamma^\sigma_{\rho\lambda}T^{\rho\lambda} \tag{22}$$

Some remarks follow.

1. It is clear from the derivation that we get covariant conservation of the energy momentum tensor only if the equations of motion for the matter degrees of freedom (here the electromagnetic field) are satisfied. This makes physical sense of course.

2. Here we set S_{matter} to S_{Maxwell} to be concrete, but clearly all the action—pardon, all the juice—is in the variation of $g_{\mu\nu}$. The electromagnetic field $A_\rho(x)$ just went along for the ride. We didn't even need to know in detail the expression for $\delta A_\rho(x)^{\text{specific}}$ and $(\cdots)^\rho$ in (19). Indeed, S_{matter}, the action for everything else in the universe besides the gravitational field, may very well contain 47 fields all madly interacting with one another. Then the first term in (19) would be replaced by the sum of 47 analogous terms, with Euler and Lagrange assuring us that every one of the 47 analogs of $(\cdots)^\rho$ would vanish.

3. I am assuming that the only field you the reader know about is the electromagnetic field, and that you are reading this book to learn about the gravitational field. At this stage, you might think of what we normally call matter as a collection of particles described by S_{gas}. But when you go on to quantum field theory, you will learn that everything in the universe is described by fields, hence the preceding remark. The ugly asymmetry in treating particles and fields at this level of physics in fact provides a strong motivation for the development of quantum field theory, as already alluded to in chapters II.3 and IV.2.

4. I leave it to you as an exercise to derive energy momentum conservation for S_{gas}.

5. Refer back to the point made in remark 2. The action for everything else in the universe besides the gravitational field, S_{matter}, contains many terms. Some terms describe interactions among different dynamical variables; for example, the term (13) contains both X^μ_a and $A_\mu(x)$. These terms contribute to the equations of motion of course and hence to the derivation of (20). Thus, the conservation law $D_\mu T^{\mu\nu} = 0$ indeed takes into account the interactions among different types of matter, as it must on physical grounds.

6. Notice that in determining the energy momentum tensor and in proving that it is conserved, we vary not the entire action of the world $S = S_{\text{EH}} + S_{\text{matter}}$, but only S_{matter}. This is crucial.

7. As I have said, in theoretical physics, often the more profound results have simple, almost trivial, derivations, at least in hindsight. Now that you understand energy momentum conservation in curved spacetime, you can see that much of the long discussion leading up to (19) could in fact have been

dispensed with. All we need is $\delta g_{\rho\sigma}^{\text{specific}} = -(\varepsilon_{\rho;\sigma} + \varepsilon_{\sigma;\rho})$ (see (18)) under an infinitesimal coordinate transformation.

Suppose we have a generic matter action S_{matter} with the dynamical variable Φ (whose indices, if any, we suppress). Then general coordinate invariance says, simply and clearly,

$$0 = \delta S_{\text{matter}} = \int d^4x (\cdots)\delta\Phi(x)^{\text{specific}} + \frac{1}{2}\int d^4x \sqrt{-g}\, T^{\rho\sigma}(x)\delta g_{\rho\sigma}(x)^{\text{specific}}$$

$$= \frac{1}{2}\int d^4x \sqrt{-g}\, T^{\rho\sigma}(x)\delta g_{\rho\sigma}(x)^{\text{specific}} \tag{23}$$

where the third equality follows from the matter equation of motion. (As noted in remark 3 above, the matter variation may involve a sum of terms.) Plugging in $\delta g_{\rho\sigma}^{\text{specific}}$, we obtain energy momentum conservation immediately.* See also exercise 5.

Appendix 2: Energy momentum of the gravitational field

The presence in (22) of the second term, mandated by the construction of the covariant derivative and the fact that the energy momentum tensor carries two indices, indicates that the conservation of energy momentum in general relativity is more subtle than you might have expected. Contrast this with the covariant conservation of a current (the electromagnetic current, for example) $D_\rho J^\rho = 0$, which if written out, reads $\frac{1}{\sqrt{-g}}\partial_\rho(\sqrt{-g}\, J^\rho) = 0$. Integrate this over a 4-dimensional spacetime region $\mathcal{V}$:

$$\int_{\mathcal{V}} d^4x \sqrt{-g}\, D_\rho J^\rho = 0 = \int_{\mathcal{V}} d^4x \partial_\rho(\sqrt{-g}\, J^\rho) = \int_{\partial\mathcal{V}} dS_\rho \sqrt{-g}\, J^\rho$$

where $\partial\mathcal{V}$ denotes the boundary of $\mathcal{V}$ and dS_ρ a 3-dimensional "surface" element. We see that the factors of $\sqrt{-g}$ are arranged in precisely such a way as to allow us to use the divergence theorem suitably generalized to 4-dimensional spacetime. This is in accord with our physical intuition that $D_\rho J^\rho = 0$ implies that current does not flow out of the 4-dimensional spacetime region.

The second term in (22) tells us that (21) no longer implies that $\int_{\partial\mathcal{V}} dS_\rho \sqrt{-g}\, T^{\rho\lambda}$ vanishes. But this apparently puzzling conclusion is in fact physically correct. Since the gravitational field itself carries energy momentum, we cannot possibly demand that $T^{\rho\lambda}$ does not flow out of a 4-dimensional spacetime region.

Indeed, the equivalence principle asserts forcefully that any definition of the energy momentum carried by the gravitational field cannot possibly be valid locally. We know that locally, we can always transform away the gravitational fields.

I might mention in passing, merely for the sake of completeness, that it is possible[3] to find an object with two indices $t^{\rho\lambda}$, known as the energy momentum pseudotensor, such that $\partial_\rho(T^{\rho\lambda} + t^{\rho\lambda}) = 0$. I strongly prefer to stay away from objects that are manifestly not tensors, and equations (such as the one just mentioned) that hold only in a specific coordinate system seem to be contrary to the very spirit of relativity. Suffice it to note that the ensuing discussion can become extremely involved.

Exercises

1   Show that $T_{ij} = -(E_i E_j + B_i B_j) + \frac{1}{2}\delta_{ij}(\vec{E}^2 + \vec{B}^2)$ and hence $T = 0$ in flat spacetime.

2   Show that the stress energy tensor obtained from $S_{\text{particles}}$ is covariantly conserved.

3   Verify explicitly that $T^{\mu\nu}$ in (14) for a collection of charged particles is covariantly conserved.

---

* Incidentally, this is very similar to the derivation of current conservation in electromagnetism using gauge invariance.

4   Show that for the action (introduced in appendix 7 in chapter V.6)

$$S_{\text{scalar}} = -\int d^4x \sqrt{-g} \left(\tfrac{1}{2}(\partial\varphi)^2 + V(\varphi) \right) \tag{24}$$

where $(\partial\varphi)^2 \equiv g^{\lambda\rho}\partial_\lambda\varphi\partial_\rho\varphi$, the energy momentum tensor is given by

$$T^{\mu\nu} = \partial^\mu\varphi\partial^\nu\varphi - g^{\mu\nu}\left(\left(\tfrac{1}{2}\partial\varphi\right)^2 + V(\varphi) \right) \tag{25}$$

As you will learn in a course on quantum field theory, S_{scalar} describes a self-interacting scalar field. Evaluate T^{00} in flat spacetime.

5   Pedagogically, the derivation of energy momentum conservation can be made even more transparent by using the cosmological action $S_c = \Lambda \int d^4x \sqrt{-g} g^{\sigma\zeta} g^{\lambda\rho}$ rather than S_{Maxwell}, since there is no electromagnetic gauge potential to keep track of. Work this out.

6   Note that in deriving (18), we did not use any specific property of $g_{\mu\nu}$. In other words, show that for any two-indexed tensor $s_{\mu\nu}(x)$, we have

$$\delta s_{\rho\sigma}(x)^{\text{specific}} \equiv s'_{\rho\sigma}(x) - s_{\rho\sigma}(x) = -\left(s_{\mu\sigma}(x)\partial_\rho\varepsilon^\mu(x) + s_{\rho\nu}(x)\partial_\sigma\varepsilon^\nu(x) + \varepsilon^\lambda\partial_\lambda s_{\rho\sigma}(x) \right) + O(\varepsilon^2) \tag{26}$$

You can readily generalize this expression to any tensor.

7   Suppose you are given the energy momentum tensor of a point particle $T^{\mu\nu}(x) = \frac{m}{\sqrt{-g(x)}} \int d\tau \frac{dX^\mu}{d\tau} \frac{dX^\nu}{d\tau} \delta^4(x - X(\tau))$. Show that energy momentum conservation $D_\mu T^{\mu\nu} = 0$ requires that the particle follows a geodesic, precisely as you would expect.

8   In cosmology, a ideal fluid that exerts no pressure is known as dust. Show that with $T^{\mu\nu} = \rho U^\mu U^\nu$, energy momentum conservation implies $U^\mu D_\mu U^\nu = 0$.

Notes

1. Historically, the realization that the sum of "different forms" of energy is conserved represents a tremendous conceptual advance for physics. It was enunciated by, among others, Count Rumford of the Holy Roman Empire. While supervising the boring of cannons in Bavaria, he noticed how hot the cannons became and theorized that the heat was put in there by the team of horses doing the boring work. Incidentally, Count Rumford was born Benjamin Thomson in Massachusetts: he fled to England on the eve of the American revolution, what we would call a traitor then and now. His nobility was bestowed by the ruler of Bavaria, whom he served. While professionally I benefited from his conservation principle, personally I benefited, when I lived in Munich, from the English garden he established there. Count Rumford later endowed the Rumford professorship at Harvard University, where he allegedly, as a poor boy growing up, audited physics courses without permission. Somehow they don't make physicists like that any more.

2. *QFT Nut*, p. 78.

3. See L. D. Landau and E. M. Lifschitz, *The Classical Theory of Fields*, Addison-Wesley, 1971.

VI.5 | Gravity Goes Live

Einstein was . . . one of the friendliest of men. I had the impression that he was also, in an important sense, alone. Many very great men are lonely.

—Freeman Dyson[1]

The Einstein tensor

In a mockery of a standard proverb, we kept putting off until tomorrow what we did not need today. Ever since chapter VI.1, we have been avoiding the labor of varying the Einstein-Hilbert action

$$S_{\text{EH}} \equiv \frac{1}{16\pi G} \int d^4x \sqrt{-g} R \tag{1}$$

with respect to $g_{\mu\nu}$ for as long as we could. Amusingly, we managed to get pretty far; without ever varying S_{EH}, we worked out an expanding universe in chapter VI.2 and the Schwarzschild solution around a star or a black hole in chapter VI.3. Finally, finally, we now do the heavy lifting and vary S_{EH}. But as we will see presently, thanks to an identity, the task is not as onerous as we feared. In fact, with our setup, it is downright easy.

So let us vary $\mathcal{I} \equiv \int d^4x \sqrt{-g}\, g^{\sigma\rho} R_{\sigma\rho}$. We need to vary $g^{\sigma\rho}$, $\sqrt{-g}$, and $R_{\sigma\rho}$ with respect to $g_{\mu\nu}$. There are thus three pieces that we will attend to in turn.

But wait, didn't we already learn how to vary $g^{\sigma\rho}$ and $\sqrt{-g}$ back in chapter V.6? Indeed, we did it again in the preceding chapter. Thus, two of the three pieces are really easy.

First, the easiest piece: $\delta_1 \mathcal{I} = \int d^4x \sqrt{-g} R_{\sigma\rho} \delta g^{\sigma\rho}$. Use (VI.4.5), $\delta g^{\sigma\rho} = -g^{\sigma\mu} \delta g_{\mu\nu} g^{\nu\rho}$, to obtain $\delta_1 \mathcal{I} = \int d^4x \sqrt{-g} R_{\sigma\rho}(-g^{\sigma\mu} \delta g_{\mu\nu} g^{\nu\rho}) = -\int d^4x \sqrt{-g} R^{\mu\nu} \delta g_{\mu\nu}$.

Next, vary the determinant $g = \det g_{\mu\nu}$. Use (VI.4.7), $\delta \sqrt{-g} = \frac{1}{2}\sqrt{-g} g^{\mu\nu} \delta g_{\mu\nu}$, to obtain $\delta_2 \mathcal{I} = \int d^4x \sqrt{-g} (\frac{1}{2} g^{\mu\nu} R) \delta g_{\mu\nu}$.

To keep count, we have thus far $(\delta_1 + \delta_2)\mathcal{I} = -\int d^4x \sqrt{-g} (R^{\mu\nu} - \frac{1}{2} g^{\mu\nu} R) \delta g_{\mu\nu}$. The particular combination of Ricci and scalar tensors that appears here is so important that it is known as the Einstein tensor[2]

$$E_{\mu\nu} \equiv R_{\mu\nu} - \frac{1}{2} g_{\mu\nu} R \tag{2}$$

Finally, we tackle the most frightening piece of all: $\delta_3 \mathcal{I} = \int d^4x \sqrt{-g}\, g^{\mu\nu} \delta R_{\mu\nu}$. We are to subtract the Ricci tensor $R_{\mu\nu}$ calculated from the metric $g_{\mu\nu}$ from $\tilde{R}_{\mu\nu}$ calculated from $\tilde{g}_{\mu\nu} = g_{\mu\nu} + \delta g_{\mu\nu}$ to obtain $\delta R_{\mu\nu}$. Given the rather complicated expression for the Riemann and Ricci tensors, this would seem to involve a lot of work.

Palatini identity

Fortunately, our lives are made easy by some key observations. To get oriented, let's not worry about indices and vary the schematic expression (with antisymmetrization understood) for the Riemann tensor $R^{\cdot}_{\ \dots} \sim \partial.\Gamma^{\cdot}_{\ \cdot} + \Gamma^{\cdot}_{\ \cdot}\Gamma^{\cdot}_{\ \cdot}$ to obtain $\delta R^{\cdot}_{\ \dots} \sim \partial.\delta\Gamma^{\cdot}_{\ \cdot} + \delta\Gamma^{\cdot}_{\ \cdot}\Gamma^{\cdot}_{\ \cdot} + \Gamma^{\cdot}_{\ \cdot}\delta\Gamma^{\cdot}_{\ \cdot}$. So the calculation depends on first determining $\delta\Gamma^{\rho}_{\mu\lambda}$.

Look at how the Christoffel symbols transform from (II.2.29) and (V.6.6):

$$\Gamma'^{\mu}_{\lambda\kappa} = S^{\mu}_{\eta}(S^{-1})^{\omega}_{\ \lambda}(S^{-1})^{\sigma}_{\ \kappa}\Gamma^{\eta}_{\omega\sigma} + S^{\mu}_{\eta}(S^{-1})^{\rho}_{\ \lambda}\partial_{\rho}(S^{-1})^{\eta}_{\ \kappa} \tag{3}$$

At the risk of repeating ourselves, we emphasize again what we mean by varying $\Gamma^{\rho}_{\mu\lambda}$. We are to calculate $\Gamma^{\rho}_{\mu\lambda}$ using the metric $g_{\mu\nu}$ and $\tilde{\Gamma}^{\rho}_{\mu\lambda}$ using a metric $\tilde{g}_{\mu\nu}$ slightly different from $g_{\mu\nu}$, and then calculate the difference $\delta\Gamma^{\rho}_{\mu\lambda} = \tilde{\Gamma}^{\rho}_{\mu\lambda} - \Gamma^{\rho}_{\mu\lambda}$.

Confusio: "So it is not about comparing $\Gamma'^{\rho}_{\mu\lambda}$ and $\Gamma^{\rho}_{\mu\lambda}$?"

No no no! We are varying $g_{\mu\nu}$, not transforming $g_{\mu\nu}$.

But that's a common confusion, because in fact we are going to use (3) right now. Under the same coordinate transformation as in (3), $\tilde{\Gamma}'^{\rho}_{\mu\lambda}$ and $\tilde{\Gamma}^{\rho}_{\mu\lambda}$ are going to be related by an equation obtained from (3) simply by putting tildes on the Γs, namely

$$\tilde{\Gamma}'^{\mu}_{\lambda\kappa} = S^{\mu}_{\eta}(S^{-1})^{\omega}_{\ \lambda}(S^{-1})^{\sigma}_{\ \kappa}\tilde{\Gamma}^{\eta}_{\omega\sigma} + S^{\mu}_{\eta}(S^{-1})^{\rho}_{\ \lambda}\partial_{\rho}(S^{-1})^{\eta}_{\ \kappa} \tag{4}$$

Subtracting (3) from (4), we obtain

$$\delta\Gamma'^{\mu}_{\lambda\kappa} \equiv \tilde{\Gamma}'^{\mu}_{\lambda\kappa} - \Gamma'^{\mu}_{\lambda\kappa} = S^{\mu}_{\eta}(S^{-1})^{\omega}_{\ \lambda}(S^{-1})^{\sigma}_{\ \kappa}(\tilde{\Gamma}^{\eta}_{\omega\sigma} - \Gamma^{\eta}_{\omega\sigma}) = S^{\mu}_{\eta}(S^{-1})^{\omega}_{\ \lambda}(S^{-1})^{\sigma}_{\ \kappa}\delta\Gamma^{\eta}_{\omega\sigma} \tag{5}$$

The inhomogeneous term $\sim SS^{-1}\partial S^{-1}$ in (3) that makes the Christoffel symbol not a tensor gets subtracted away. Remarkably, in contrast to $\Gamma^{\rho}_{\mu\lambda}$, which is assuredly not a tensor, the variation $\delta\Gamma^{\rho}_{\mu\lambda} = \tilde{\Gamma}^{\rho}_{\mu\lambda} - \Gamma^{\rho}_{\mu\lambda}$ is a tensor.

This exemplifies precisely what we talked about in chapter V.6: the "character defects" in $\Gamma^{\rho}_{\mu\lambda}$ and $\tilde{\Gamma}^{\rho}_{\mu\lambda}$ cancel out. Exploiting this fact, we can derive a nice identity for the variation of the Riemann curvature tensor.

Professor Flat pops up just in time, up to his usual trick. "Go to locally flat coordinates," he urges us. Look at $\delta R^{\cdot}_{\ \dots} \sim \partial.\delta\Gamma^{\cdot}_{\ \cdot} + \delta\Gamma^{\cdot}_{\ \cdot}\Gamma^{\cdot}_{\ \cdot} + \Gamma^{\cdot}_{\ \cdot}\delta\Gamma^{\cdot}_{\ \cdot}$. At a point where the coordinates are flat, $\Gamma^{\cdot}_{\ \cdot}(x) = 0$, and our expression simplifies to $\delta R^{\cdot}_{\ \dots} \sim \partial.\delta\Gamma^{\cdot}_{\ \cdot}$, so that

$$\delta R^{\rho}_{\mu\sigma\lambda}(x) = \partial_{\sigma}\delta\Gamma^{\rho}_{\mu\lambda}(x) - \partial_{\lambda}\delta\Gamma^{\rho}_{\mu\sigma}(x) \tag{6}$$

But at that point, since $\Gamma^{\cdot}_{\ \cdot} = 0$, the ordinary derivative ∂ is the same as the covariant derivative D, thus enabling us to write

$$\delta R^{\rho}_{\mu\sigma\lambda} = D_{\sigma}\delta\Gamma^{\rho}_{\mu\lambda} - D_{\lambda}\delta\Gamma^{\rho}_{\mu\sigma} \tag{7}$$

an equality known as the Palatini identity.

Professor Flat reminds us once again that since the Palatini identity is an equality between tensors, it holds not only for locally flat coordinates, but in general!

In particular, the variation of the Ricci tensor is given by

$$\delta R_{\mu\nu} = \delta R^{\rho}_{\mu\rho\nu} = D_{\rho}\delta\Gamma^{\rho}_{\mu\nu} - D_{\nu}\delta\Gamma^{\rho}_{\mu\rho} \tag{8}$$

and so $\delta_3\mathcal{I} = \int d^4x\sqrt{g}g^{\mu\nu}(D_{\rho}\delta\Gamma^{\rho}_{\mu\nu} - D_{\nu}\delta\Gamma^{\rho}_{\mu\rho})$. Now integrate by parts (as explained in chapter V.6) and use the identity (V.6.15) $D_{\rho}g^{\mu\nu} = 0$. Happily, we conclude that $\delta_3\mathcal{I}$ is a surface term and does not affect the equation of motion!

The Einstein field equation

Putting it altogether, we have $\delta\mathcal{I} = (\delta_1 + \delta_2 + \delta_3)\mathcal{I} = -\int d^4x\sqrt{-g}(R^{\mu\nu} - \frac{1}{2}g^{\mu\nu}R)\delta g_{\mu\nu} +$ a surface term.[3] Remember, the $\frac{1}{2}$ comes from varying the square root, and the relative minus sign from varying the inverse.

In chapter VI.1, we tried to avoid work and simply denote the relative coefficient between $R^{\mu\nu}$ and $g^{\mu\nu}R$ by α. This unknown constant α turns out to be $-\frac{1}{2} \neq -\frac{1}{4}$, as we had hoped all along, and so everything we did in chapters VI.2 and VI.3 is okay. (Also note that $A = -1$, and so the result $A(1 + 4\alpha) = 1$ we obtained in appendix 1 to chapter VI.1 is indeed satisfied.)

Noting that $S_{\text{EH}} = \frac{1}{16\pi G}\mathcal{I}$, we obtain

$$\delta S = \delta S_{\text{EH}} + \delta S_{\text{matter}} = \int d^4x\sqrt{-g}\left\{-\frac{1}{16\pi G}(R^{\mu\nu} - \frac{1}{2}g^{\mu\nu}R) + \frac{1}{2}T^{\mu\nu}\right\}\delta g_{\mu\nu} \tag{9}$$

We have derived the wondrous glorious stupendous tremendous Einstein's field equation

$$R^{\mu\nu} - \frac{1}{2}g^{\mu\nu}R = +8\pi G T^{\mu\nu} \tag{10}$$

A parade of greats, Euler, Lagrange, Riemann, Christoffel, Ricci, Einstein, Palatini, and many many others, have brought us to this, one of the most profound statements in physics: The distribution of energy in spacetime governs the curvature of spacetime.

Often $T^{\mu\nu}$ is given explicitly in a relative simple form, for example, $T^{\mu\nu} = (\rho + P)U^{\mu}U^{\nu} + Pg^{\mu\nu}$ for a perfect fluid. A trivial rewrite of (10) is thus more convenient: contract (10) with $g_{\mu\nu}$ to obtain $R = -8\pi G T$ so that

$$R^{\mu\nu} = 8\pi G\left(T^{\mu\nu} - \frac{1}{2}g^{\mu\nu}T\right) \tag{11}$$

Einstein remarked that the left hand side of his field equation (10) was born elegantly of geometry, while the right hand side seemed to have an ugly ad hoc quality, with one term after another thrown in* according to what kind of matter we chose to fill spacetime with.

---

* This was particularly true in Einstein's time, when matter was poorly understood. Hence Einstein's unrealized dream of a unified field theory with matter also described in geometrical terms.

Theoretical physicists have joked ever since that even Nature appears to prefer the left over the right.

The Newtonian limit

We have one remaining task: show that G is what Newton said it was. Recover Newtonian gravity in the appropriate limit. In other words, we will verify the overall coefficient in the Einstein-Hilbert action (1).

Consider the weak gravitational field limit in which spacetime deviates little from Minkowskian. Write $g_{\mu\nu} = \eta_{\mu\nu} + h_{\mu\nu}$ and expand $R_{\mu\nu}$ to linear order in h. First, $\Gamma^{\rho}_{\mu\nu} = \frac{1}{2}\eta^{\rho\lambda}(\partial_\mu h_{\nu\lambda} + \partial_\nu h_{\mu\lambda} - \partial_\lambda h_{\mu\nu}) + O(h^2)$, so that $R^{\sigma}_{\rho\mu\nu} = (\partial_\mu \Gamma^{\sigma}_{\nu\rho} + \Gamma^{\sigma}_{\mu\kappa}\Gamma^{\kappa}_{\nu\rho}) - (\partial_\nu \Gamma^{\sigma}_{\mu\rho} + \Gamma^{\sigma}_{\nu\kappa}\Gamma^{\kappa}_{\mu\rho}) = \partial_\mu \Gamma^{\sigma}_{\nu\rho} - \partial_\nu \Gamma^{\sigma}_{\mu\rho} + O(h^2)$. We obtain

$$R_{\mu\nu} = -\frac{1}{2}(\partial^2 h_{\mu\nu} - \partial_\mu \partial_\lambda h^{\lambda}_{\nu} - \partial_\nu \partial_\lambda h^{\lambda}_{\mu} + \partial_\mu \partial_\nu h^{\lambda}_{\lambda}) + O(h^2) \tag{12}$$

where evidently, to leading order, indices are raised by the Minkowski metric: $h^{\lambda}_{\nu} = \eta^{\lambda\mu} h_{\mu\nu}$.

Recall, from chapter V.4, that to recover Newtonian gravity we only need $h_{\mu\nu}$ not to depend on time. Then $R_{00} \simeq -\frac{1}{2}\nabla^2 h_{00} = \nabla^2 \Phi$, upon noting (from chapter V.4, and earlier, chapter IV.3) that the Newtonian potential is given by $\Phi = -\frac{1}{2}h_{00}$. (Recall also (V.4.24) in which we restored c and took the nonrelativistic limit.) For Newtonian matter, $T_{00} \simeq \rho \gg T_{ij}$, so that $T_{00} - \frac{1}{2}\eta_{00}T \simeq \frac{1}{2}\rho$. The field equation (11) thus reduces to $\nabla^2 \Phi = 4\pi G\rho$, showing that we have the right coefficient in (1).

Dark energy again

Back in chapter VI.2, I did not vary the action $S_{\text{cosmological}} = -\int d^4x \sqrt{-g}\,\Lambda$ but merely argued on symmetry grounds what the energy momentum tensor of an expanding universe must be, up to an overall constant. Using (VI.4.8), we can now vary instantaneously: $\delta S_{\text{cosmological}} = -\int d^4x \,(\frac{1}{2}\sqrt{-g}\,g^{\mu\nu})\delta g_{\mu\nu}\Lambda$ and thus by (VI.4.2),

$$T^{\mu\nu} = -\Lambda g^{\mu\nu} \tag{13}$$

In the flat spacetime limit, the energy density becomes $T^{00} = -\Lambda \eta^{00} = \Lambda$ as expected.

With $T = 4\Lambda$, the field equation (11) then reads

$$R^{\mu\nu} = 8\pi G\Lambda g^{\mu\nu} \tag{14}$$

The temporary symbol $\tilde{\Lambda}$ used in chapter VI.2 is in fact just $8\pi G\Lambda$, so that the Hubble parameter was determined to be

$$H^2 = \frac{8\pi G}{3}\Lambda \tag{15}$$

(The slight cheat I committed back in chapter VI.2 was assuming implicitly that $\tilde{\Lambda}$ and Λ have the same sign.) Note that the expanding universe (with its flat space*) we discussed in chapter VI.2 only makes sense with a positive cosmological constant.

There is a silly debate regarding whether in (14) the Λ term should be placed on the left or right hand side, silly because those of you who have mastered algebra certainly feel free (free country, remember?) to move it to the left hand side. But some people who write the Λ term on the left hand side then think of it as part of gravity, and go on to say that anti-gravity, or a new repulsive force, has been found, a language I strongly disfavor. The justification for this ill-advised language is that the only dynamical variable appearing in $S_{\text{cosmological}}$ is the metric. But as we saw in chapter VI.2, $S_{\text{cosmological}}$ is inevitably produced by fluctuating matter fields.

Bianchi identity

One reason that it took Einstein 10 arduous years, from 1905 to 1915, to derive the field equation was that he didn't know an identity due to Luigi Bianchi (1856–1928), which we will now derive. We've had an inkling of the existence of this identity ever since chapter VI.2.

Let us covariantly differentiate the Riemann curvature tensor $D_\nu R^\rho_{\mu\sigma\lambda} = D_\nu\{(\partial_\sigma\Gamma^\rho_{\mu\lambda} + \Gamma^\rho_{\kappa\sigma}\Gamma^\kappa_{\mu\lambda}) - (\partial_\lambda\Gamma^\rho_{\mu\sigma} + \Gamma^\rho_{\kappa\lambda}\Gamma^\kappa_{\mu\sigma})\}$. Naturally, our favorite person pops up. Go to locally flat coordinates, he mimes. Then at the chosen point, the Christoffel symbol vanishes, and this whole mess collapses dramatically to $\partial_\nu R_{\rho\mu\sigma\lambda} = \partial_\nu\partial_\sigma\Gamma_{\rho\cdot\mu\lambda} - \partial_\nu\partial_\lambda\Gamma_{\rho\cdot\mu\sigma}$. (Note that we have lowered the upper index.) Observe that the index structure on the right hand side has the form $(\nu\sigma\,\lambda) - (\nu\lambda\,\sigma)$. So, cyclically permute the three indices $(\nu\sigma\,\lambda)$ and add the results. Out pops the Bianchi identity

$$D_\nu R_{\rho\mu\sigma\lambda} + D_\sigma R_{\rho\mu\lambda\nu} + D_\lambda R_{\rho\mu\nu\sigma} = 0 \tag{16}$$

As always, because this is a tensor identity, it holds in general, even though it is derived in locally flat coordinates. You shouldn't even need Professor Flat to tell you that any more.

Contract this with $g^{\rho\sigma}$. Remembering that the covariant derivative of the metric tensor vanishes so that we can slip the metric tensor past the covariant derivative D, we obtain

$$D_\nu R_{\mu\lambda} + D^\rho R_{\rho\mu\lambda\nu} - D_\lambda R_{\mu\nu} = 0 \tag{17}$$

where we used the antisymmetry of the Riemann tensor in the last two indices. Contracting again with $g^{\mu\lambda}$, we find $D_\nu R - D^\rho R_{\rho\nu} - D^\mu R_{\mu\nu} = 0$, or written more compactly,

$$D^\mu(R_{\mu\nu} - \tfrac{1}{2}g_{\mu\nu}R) = D^\mu E_{\mu\nu} = 0 \tag{18}$$

---

* We will discuss the expanding universe with curved space when we explore cosmology in part VIII.

This identity satisfied by the Einstein tensor (and a direct consequence of (16)) is some-times also referred to as the Bianchi identity, although it should, more properly, be known as the contracted Bianchi identity.

Einstein's real blunder

Applying the contracted Bianchi identity (18) to the field equation (10), we obtain energy momentum conservation

$$D_\mu T^{\mu\nu} = 0 \tag{19}$$

In most applications, we deal with an energy momentum tensor known to be conserved, so that the Bianchi identity actually is telling us that the field equations are not independent: one linear combination of the derivatives of the different equations vanishes identically.

Let us go back to the apparent miracle in chapters VI.2 and VI.3. In both, when we solved for the metric, we had one more equation than unknowns but nevertheless we obtained a consistent solution. Now we understand what is going on, thanks to Bianchi. We over counted the number of equations by one: one linear combination is satisfied automatically. In practice, this provides us with a useful check on our arithmetic.

In his epoch-making 1915 paper on gravity, Einstein actually did not use the action principle, but wrote down the field equation directly by arguing what the left hand side must be. (We will come back to this in chapter IX.5.) For a number of years, he struggled with this approach, and at one point wrote down $R^{\mu\nu} = 8\pi G T^{\mu\nu}$, which did not work, since Bianchi identity applied to this would give $D^\mu T_{\mu\nu} \neq 0$.

There is a lot of quasi-nonsense written about Einstein's greatest blunder in introducing the cosmological constant* which I find rather annoying.[4] In my opinion, if the great man had blundered at all, it was in not using the action principle (see, however, appendix 5).

Appendix 1: Bianchi identity

The Bianchi identity can also be derived as a special case of the Jacobi identity $[A, [B, C]] + [B, [C, A]] + [C, [A, B]] = 0$, which you can prove by writing out all the terms on the left hand side. Here A, B, C are three operators. (Or, you could argue that there are $2 \times 2 \times 3 = 12$ terms on the left hand side, so that each of the 6 possible terms, for example ABC, must appear twice, once with a positive sign (in $[A, [B, C]]$ in the example) and once with a negative sign (in $[C, [A, B]]$ in the example).) In particular,

$$[D_\mu, [D_\nu, D_\lambda]] + [D_\nu, [D_\lambda, D_\mu]] + [D_\lambda, [D_\mu, D_\nu]] = 0 \tag{20}$$

Using (VI.1.5) $[D_\mu, D_\nu]S_\rho = -R^\sigma_{\rho\mu\nu}S_\sigma$ and the fact that the covariant derivative is distributive, we obtain the Bianchi identity (16).

We now see that the "other half of Maxwell's equations" $\epsilon^{\sigma\mu\nu\lambda}\partial_\mu F_{\nu\lambda} = 0$ discussed in chapter IV.2 are in fact Bianchi identities as we could see by setting the covariant derivatives in (20) to the quantum mechanical covariant

---

* Contrary to what is constantly reported, Einstein never said in print that the cosmological constant was his greatest blunder. It was George Gamow, a jokester of record, who wrote that Einstein told him as much.

derivative of electromagnetism in flat spacetime $D_\rho = (\partial_\rho - i A_\rho)$. (See an endnote in chapter VI.1 for a derivation of the i.)

Appendix 2: Another derivation of the contracted Bianchi identity

It is illuminating to give yet another derivation of the contracted Bianchi identity. We simply modify the discussion given in appendix 1 of chapter VI.4. There we derived energy momentum conservation using for the prototypical matter action the Maxwell action S_{Maxwell}. We are going to do unto the Einstein-Hilbert action S_{EH} here what we did unto S_{Maxwell} there. It would be helpful for you to review the appendix in question now, as I am not going to write out the steps in detail again.

We know that S_{EH} is invariant under the replacements $x \to x'$, $g_{\rho\sigma}(x) \to g'_{\rho\sigma}(x')$, and $g^{\mu\sigma}(x) \to g'^{\mu\sigma}(x')$. It is instructive to compare the consequence of general invariance for S_{EH} and S_{Maxwell}. There, in chapter VI.4, after a few steps we find that the general invariance of S_{Maxwell} implies (VI.4.19)

$$0 = \delta S_{\text{Maxwell}} = \int d^4x (\cdots)^\rho \delta A_\rho(x)^{\text{specific}} - \frac{1}{2} \int d^4x \sqrt{-g}\, T^{\rho\sigma}(x) \delta g_{\rho\sigma}(x)^{\text{specific}} \tag{21}$$

Here, applying the same reasoning, we find that the general invariance of S_{EH} implies

$$0 = \delta S_{\text{EH}} = \frac{1}{16\pi G} \int d^4x \sqrt{-g}\, (R^{\rho\sigma} - \tfrac{1}{2} g^{\rho\sigma} R)(x) \delta g_{\rho\sigma}(x)^{\text{specific}}$$

$$= \frac{1}{16\pi G} \int d^4x \sqrt{-g}\, E^{\rho\sigma} \delta g_{\rho\sigma}(x)^{\text{specific}} \tag{22}$$

Note two differences between (22) and (21). First, there is not a δA term in (22), of course, since here we don't even have an A field to vary. Second, while the variation of S_{Maxwell} with respect to the metric gives the energy momentum tensor of the electromagnetic field, the variation of S_{EH} with respect to the metric gives the Einstein tensor.

As explained in appendix 1 of chapter VI.4, in (21) we then invoke the equation of motion for A, use $\delta g_{\rho\sigma}^{\text{specific}} = -(\varepsilon_{\rho;\sigma} + \varepsilon_{\sigma;\rho})$, and integrate by parts, thus obtaining energy momentum conservation of the matter fields $D_\rho T^{\rho\sigma} = 0$.

In (22), using the form of $\delta g_{\rho\sigma}^{\text{specific}}$ and integrating by parts, we obtain the contracted Bianchi identity $D_\rho E^{\rho\sigma} = 0$.

While the derivations of the two equations $D_\rho T^{\rho\sigma} = 0$ and $D_\rho E^{\rho\sigma} = 0$ appear superficially similar, we must keep in mind an important conceptual difference. Energy momentum conservation of the matter fields follow only if the matter fields satisfy their respective equations of motion, as makes sense physically. In contrast, the contracted Bianchi identity is, well, an identity. This is rendered particularly clear in the derivation given here: in one case, we have to invoke the relevant equation of motion, in the other case, not.

The contracted Bianchi identity $D_\rho E^{\rho\sigma} = 0$ and Einstein's field equation $E^{\rho\sigma} = 16\pi G T^{\rho\sigma}$ imply energy momentum conservation $D_\rho T^{\rho\sigma} = 0$. Conversely, the field equation and energy momentum conservation demand the existence of an identity. As was mentioned in the text, Einstein's ignorance of the contracted Bianchi identity contributed to his difficulties in arriving at his field equation.

Appendix 3: The "total energy momentum tensor" vanishes

If we wish, we can extend the definition (VI.4.2) for the energy momentum tensor to the Einstein-Hilbert action (1) and define $T_{\text{gravity}}^{\mu\nu} \equiv \frac{2}{\sqrt{-g}} \frac{\delta S_{\text{EH}}}{\delta g_{\mu\nu}} = -\frac{1}{8\pi G}(R^{\mu\nu} - \frac{1}{2} g^{\mu\nu} R)$, so that the field equation can be written as

$$T_{\text{gravity}}^{\mu\nu}(x) + T_{\text{matter}}^{\mu\nu}(x) = 0 \tag{23}$$

Some authors like to say that Einstein's field equation tells us that the total energy momentum tensor is equal to zero. I do not find this formulation particularly useful.*

---

* For instance, we could move ma in $F = ma$ to the left hand side, define $-ma$ as the "inertial force," and say that the total force vanishes. Not a terribly useful way to think about the second law.

Appendix 4: More on the variation of the Ricci tensor

We give an alternative, and perhaps simpler, proof that $\delta R_{\mu\nu}$ is a total derivative. From general considerations, this tensor has to be constructed out of $\delta g_{..}$ and two powers of the covariant derivative. Let's classify all possible terms according to where the indices $\mu\nu$ go. There are three possibilities. Both indices are carried by δg: we have $D^2\delta g_{\mu\nu}$. One of them is carried by δg: we have $D_\mu D_\lambda \delta g_{\nu\lambda}$ (plus of course the term obtained by exchanging μ and ν). Neither index is carried by δg: we have $D_\mu D_\nu g^{\rho\lambda}\delta g_{\rho\lambda}$ (recall that $D_\nu g^{\rho\lambda} = 0$, and so it does not matter whether the inverse metric is inside or outside the covariant derivatives). Each of these terms is a total derivative.

Appendix 5: Palatini (actually Einstein) formalism

Here we write the action for Einstein gravity in the Palatini formalism

$$S = \frac{1}{16\pi G} \int d^4x \sqrt{-g}\, g^{\mu\nu} R_{\mu\nu}(\partial\Gamma, \Gamma) + S_{\text{matter}} \tag{24}$$

When shown this for the first time, your immediate reaction might be, "What? Isn't this the same thing as what we had in the text?" Indeed, your indignation is justified, but I have not yet specified for you the dynamical variables, which you must insist on knowing immediately when you are shown an action.

Attilio Palatini chose to regard the metric $g_{\mu\nu}$ and the Christoffel symbol $\Gamma^\sigma_{\mu\nu}$ as independent dynamical variables. You say, "How could that be? Isn't the Christoffel symbol defined in terms of the metric?" Yes, you are totally right, but only in the standard formalism. You are now invited to contemplate the action in (24), in which the symbol $R_{\mu\nu}(\partial\Gamma, \Gamma)$ is to be regarded as shorthand for

$$R_{\mu\nu}(\partial\Gamma, \Gamma) = (\partial_\sigma \Gamma^\sigma_{\mu\nu} + \Gamma^\sigma_{\kappa\sigma}\Gamma^\kappa_{\mu\nu}) - (\partial_\nu \Gamma^\sigma_{\mu\sigma} + \Gamma^\sigma_{\kappa\nu}\Gamma^\kappa_{\mu\sigma}) \tag{25}$$

with Γ some unknown object carrying 3 indices. Note that the first term in (24) now involves only one power of derivative; thus, the Palatini formalism is also known as the first order formalism for gravity.

Ask not what you can do for the Palatini formalism; ask what the Palatini formalism can do for you. What it can do is to render the variation of S with respect to $g_{\mu\nu}$ "trivially" easy, since we don't have to vary $R_{\mu\nu}(\partial\Gamma, \Gamma)$: it doesn't contain $g_{\mu\nu}$. We almost "instantly" obtain Einstein's field equation $R^{\mu\nu} - \frac{1}{2}g^{\mu\nu}R = +8\pi G T^{\mu\nu}$.

But not so fast! We haven't quite gotten Einstein's field equation yet, since at this point, $R_{\mu\nu}$ was just a symbol for the mess in (25) involving the unknown object Γ. We get Einstein's field equation only after we determine Γ in terms of the metric.

How do we do that? Remember that we are treating Γ as an independent dynamical variable. Euler and Lagrange tell us that we are obliged to also vary S with respect to Γ. Since S_{matter} does not depend on Γ, we only have to vary $g^{\mu\nu} R_{\mu\nu}(\partial\Gamma, \Gamma)$ in the first term in S.

To avoid drowning in a sea of indices, let's do it schematically, living what I call the unindexed life:

$$\delta \left(\sqrt{-g}\, g^{..}(\partial_. \Gamma^{..}_. + \Gamma^{..}_. \Gamma^{..}_.)\right) \sim \sqrt{-g}\, g^{..}(\partial_. \delta\Gamma^{..}_. + \delta\Gamma^{..}_. \Gamma^{..}_. + \Gamma^{..}_. \delta\Gamma^{..}_.) \sim \sqrt{-g}(\partial_. g^{..} + g^{..}\Gamma^{..}_.)\delta\Gamma^{..}_. \tag{26}$$

where in the last step, we integrated by parts, which we are effectively allowed to do since this variation is to be performed inside an integral. Thus, upon setting the coefficient of $\delta\Gamma^{..}_.$ to zero, we end up with an equation of the schematic* form $\Gamma \sim g\partial g$.

Three guesses on what the equation turns out to be when you keep track of the indices carefully. If you made it this far in this book, surely you guessed $\Gamma^\rho_{\mu\nu} = \frac{1}{2}g^{\rho\lambda}(\partial_\mu g_{\nu\lambda} + \partial_\nu g_{\mu\lambda} - \partial_\lambda g_{\mu\nu})$. What else could it be? Consider

---

* Ha! We are now so schematic that we even drop the dots.

this proved by the "what else could it be" method. You are of course urged to put all the indices back and check this statement by direct computation.

Appendix 6: "The wretchedness of humanity"

In this book, I emphasize the action. In contrast, most textbooks I know approach the subject by trying to find an equation of motion that reduces to Newton's $\nabla^2 \Phi = 4\pi G\rho$. I prefer the action approach, because contemporary theoretical physics at the fundamental level deals mostly with actions and Lagrangians, not equations of motion.* The equation of motion approach goes something like the following. We know that Φ measures the deviation of g_{00} from its Minkowski value, so let's look for something with two derivatives acting on $g_{\mu\nu}$ that would reduce to $\nabla^2\Phi$. We then argue that this something must be $R^{\mu\nu} + \beta g^{\mu\nu} R$, with some unknown coefficient β. But for this to be equal to $T^{\mu\nu}$, we must have energy momentum conservation and thus $D_\mu(R^{\mu\nu} + \beta g^{\mu\nu} R) = 0$, which then fixes β after some calculation. Arriving at Einstein's field equation (10) this way is of course entirely equivalent to the action approach. Each to his or her own taste, but if you are to move on to field theory and string theory, you better get used to where the action is.

Ever since I was a student, I wondered why Einstein, who was surely familiar with the action principle, did not follow the action principle, which would not have demanded that he knew the Bianchi identity and thus would have significantly lessened his struggle. The answer is that, in fact, he did!

So a bit of history[5] I learned while writing this text. Einstein was smarter than the textbooks that follow the equation of motion approach make him out to be. He and his friend Marcel Grossmann published a paper in 1914 about a variational principle for gravity and then wrote to Lorentz about it. Stimulated by this letter, Lorentz published a paper in 1915 varying a Lagrangian $\mathcal{L}(g, \partial g)$ without specifing what $\mathcal{L}$ was. Then, in a paper presented to the Royal Prussian Academy of Sciences on November 4, 1915, Einstein obtained a set of field equations using the action principle, but with an $\mathcal{L}$ that is not a scalar! Furthermore, he imposed the condition $\det(g_{\mu\nu}) = -1$. Three weeks later, on November 25, 1915, Einstein presented to the same academy his field equations, but without using the action principle.

But Einstein was scooped! On November 20, David Hilbert presented to the Göttingen Academy the gravitational field equations he derived by varying an action. This action, as you know, is now called the Einstein-Hilbert action. Quite rightly in the opinion of all physicists, Einstein is credited with this action, even though strictly speaking, he found the equations of motion that emerge from the action rather than the action itself. The theoretical physics community is not a court of law: it regards Hilbert, although he did find the action first, as playing second fiddle to Einstein.

Incidentally, historians of physics have come to a "belated decision in the Hilbert-Einstein priority dispute."[6]

But at that time, Einstein didn't know that history would be kind to him in this one respect. He was justifiably worried and, perhaps less justifiably, angry. In fact, he was sufficiently incensed as to dash off a letter on November 26 to a friend. In the letter, the great man also bitterly denounced his estranged wife for her influence on their children,[7] but before launching into a diatribe about his personal life, he first accused Hilbert of stealing his theory.

Einstein wrote, "The theory is of incomparable beauty. But only one colleague has really understood it, and he is trying, rather skillfully, to 'nostrify' it. That's Max Abraham's coinage. In my personal experience, I've hardly come to know the wretchedness of humanity better than in connection with this theory."[8] Well, dear reader, nostrification is not only still practiced in theoretical physics, but ever more skillfully.

One could fantasize that had Einstein mastered the action approach and Riemann's work, his travails over the 10 years from special to general relativity could have been replaced by an inspired guess. Hilbert had a tremendous advantage over the befuddled Einstein struggling to learn differential geometry: he was a leading mathematician who obviously already knew the subject forward and backward. Moreover, he had worked on the theory of invariants, a branch of mathematics concerned with the question of what is left unchanged by a given set of transformations. He knew that the scalar curvature does not change under general coordinate transformations. Thus, once the question was posed properly, Hilbert knew instantly that the sought-for action governing spacetime must be the scalar curvature.

---

* Just flip through any textbook on quantum field theory.

Here is a coda to the story. Now that Lorentz knew what $\mathcal{L}$ was, he studied* the action principle for gravity in a series of papers, using a more general S_{matter} than Hilbert did. As for Einstein, a year later, on November 26, 1916, he presented a paper titled "Hamilton's Principle and the General Theory of Relativity" in which he wrote pointedly, "We shall make as few specializing assumptions as possible, in marked contrast to Hilbert's treatment."

Now I come back to the Palatini formalism. It is nowhere to be found in the paper Palatini presented to the Circolo Matematico di Palermo on August 10, 1919! He improved Hilbert's calculation and in fact had the Palatini identity introduced in this chapter. What the textbooks now call the Palatini formalism was actually invented in 1925 by Einstein! As the years passed, apparently people mixed up the Palatini identity and the Palatini formalism, and various people, including Einstein,[†] referred to (24) as the Palatini formalism. After all, both the Palatini identity and the Palatini formalism absolve us of having to vary the Ricci tensor, so it is easy to mix up the two. This confusion has been perpetuated unwittingly in many textbooks (and wittingly in this one).

Appendix 7: An alternative form of the Einstein-Hilbert action

We can rewrite the Einstein-Hilbert action in an alternative form that I do not like for reasons that will become clear but that I mention here for the sake of completeness. Looking up the expression for the Ricci tensor given in (25), we write the action in (1) (suppressing the irrelevant overall constant) as

$$S = \int d^4x \sqrt{-g}\, g^{\rho\nu} R_{\rho\nu} = \int d^4x \sqrt{-g}\, g^{\rho\nu}(\partial_\sigma \Gamma^\sigma_{\nu\rho} - \partial_\nu \Gamma^\sigma_{\sigma\rho} + \Gamma^\sigma_{\sigma\kappa}\Gamma^\kappa_{\nu\rho} - \Gamma^\sigma_{\nu\kappa}\Gamma^\kappa_{\sigma\rho})$$
$$= S_{\text{I}} + S_{\text{II}} + S_{\text{III}} + S_{\text{IV}} \tag{27}$$

We next integrate by parts to get rid of the derivatives on the Christoffel symbols, assuming that all surface terms can be thrown away. First we have

$$S_{\text{I}} = \int d^4x \sqrt{-g}\, g^{\rho\nu} \partial_\sigma \Gamma^\sigma_{\nu\rho} = -\int d^4x\, \partial_\sigma(\sqrt{-g}\, g^{\rho\nu})\Gamma^\sigma_{\nu\rho}$$

Differentiating, we write $\partial_\sigma(\sqrt{-g}\, g^{\rho\nu}) = \sqrt{-g}(\Gamma^\eta_{\sigma\eta} g^{\rho\nu} + \partial_\sigma g^{\rho\nu})$. Invoking the identity $D_\sigma g^{\rho\nu} = 0 = \partial_\sigma g^{\rho\nu} + \Gamma^\rho_{\sigma\lambda} g^{\lambda\nu} + \Gamma^\nu_{\sigma\lambda} g^{\rho\lambda}$, we can then write $S_{\text{I}} = -\int d^4x \sqrt{-g}(\Gamma^\eta_{\sigma\eta} g^{\rho\nu} - \Gamma^\rho_{\sigma\lambda} g^{\lambda\nu} - \Gamma^\nu_{\sigma\lambda} g^{\rho\lambda})\Gamma^\sigma_{\nu\rho}$. Similarly, we obtain

$$S_{\text{II}} = -\int d^4x \sqrt{-g}\, g^{\rho\nu} \partial_\nu \Gamma^\sigma_{\sigma\rho} = \int d^4x\, \partial_\nu(\sqrt{-g}\, g^{\rho\nu})\Gamma^\sigma_{\sigma\rho}$$
$$= \int d^4x \sqrt{-g}(\Gamma^\eta_{\nu\eta} g^{\rho\nu} - \Gamma^\rho_{\nu\lambda} g^{\lambda\nu} - \Gamma^\nu_{\nu\lambda} g^{\rho\lambda})\Gamma^\sigma_{\sigma\rho}$$

Adding, we find $S_{\text{I}} + S_{\text{II}} = 2\int d^4x \sqrt{-g}[\Gamma^\rho_{\sigma\lambda} \Gamma^\sigma_{\nu\rho} g^{\lambda\nu} - (\Gamma^\rho_{\nu\lambda} g^{\lambda\nu})\Gamma^\sigma_{\sigma\rho}]$. We next observe that $S_{\text{III}} + S_{\text{IV}}$, as written in (27) with no further massaging needed, is just, interestingly, $-\frac{1}{2}$ times this expression. Thus, we end up with an alternative form for the Einstein-Hilbert action

$$S = \int d^4x \sqrt{-g}[\Gamma^\rho_{\sigma\lambda} \Gamma^\sigma_{\rho\nu} g^{\lambda\nu} - (\Gamma^\rho_{\nu\lambda} g^{\lambda\nu})\Gamma^\eta_{\rho\eta}] \tag{28}$$

The reason that I don't like this form of the action is now clear: the integrand is not a scalar under coordinate transformation, as is the case for the Einstein-Hilbert action. Throwing boundary terms away at will has done violence to the underlying invariance properties of the action. Nevertheless, this form of the action is useful in some situations.

---

* Lorentz mentioned that the Belgian Théophile Ernest de Donder (1872–1957) also contributed. History has not been kind to de Donder. (See, however, the footnote on p. 21 of *QFT Nut*.) The harmonic gauge to be introduced in chapter IX.4 is also known as the de Donder gauge. Perhaps you have seen the famous photograph of the post–quantum mechanics 1927 Solvay Conference? De Donder stood behind Dirac, who in turn was seated behind Lorentz and Einstein.

† By that time, Einstein, assured of his place in history, could afford to be generous to a minor figure like Palatini.

Appendix 8: Diffeomorphism

Our good friend the Jargon Guy has been lobbying us to use the word diffeomorphism. "How can you write a book on general relativity without saying diffeomorphism?" Fine. We will say it,[9] here being as good a place as anywhere. A map f of a manifold M onto itself that is smooth, differentiable, and invertible is called a diffeomorphism. (Think of a sphere S^2 being mapped onto itself.) Let a point P on the manifold be mapped to a point $Q = f(P)$. Suppose that associated with each point P is a number $T(P)$. In other words, the function T is defined on the manifold. Suppose also that the diffeomorphism moves the number $T(P)$ to the point Q. We define a new function $\tilde{T}$ by

$$\tilde{T}(Q) = T(P) = T(f^{-1}(Q)) \tag{29}$$

What thrills our friend the Jargon Guy is that all this could be done without "dirtying our hands" with coordinates. Physically, we can imagine an incompressible fluid flowing on the sphere S^2. A fluid element that is at the point P now will arrive at the point Q after some time Δt. The number $T(P)$ could then represent a physical property associated with the fluid element at P on the manifold. (For example, we could think of T as temperature. Assuming that there is no heat transfer between neighboring fluid elements, $\tilde{T}(Q)$ would then be the temperature at Q at a time Δt from now.) The diffeomorphism discussed here is also known as an active diffeomorphism.

Descending to physics, we are now crass enough to introduce coordinates to cover local patches on the manifold, as we have done since early on in this book. To please our friend, we put on our mathematical hats and regard the coordinates x^μ as an invertible smooth map $x : M \to R^d$, associating d real numbers $x^\mu(P)$ with each point P on the d-dimensional manifold M. We can now define $\hat{T}(x) = T(P(x))$, with $P(x)$ denoting the inverse map $x^{(-1)} : R^d \to M$; in other words, $\hat{T}(x)$ is the composition of this map with the map $T : M \to R$. Most physicists would probably drop the hat on $\hat{T}$ and simply write $T(x)$, but conceptually, $\hat{T}$ and T are entirely different creatures, and we will maintain the distinction for now to keep our friend happy.

Let us change coordinates $x \to x' = F(x)$ so that as usual

$$\hat{T}'(x') = \hat{T}(x) = \hat{T}(F^{-1}(x')) \tag{30}$$

The rigor minded refer to coordinate transformation as a passive diffeomorphism to distinguish it from an active diffeomorphism.

Now comes the key point: (29) and (30) are structurally the same with f "corresponding" to F. Hence, an action that does not distinguish between different coordinate choices is invariant under active diffeomorphisms. But that has been precisely what we've been doing all along, but not stating explicitly. The whole point of this appendix[*] is to formalize the obvious (and to show off our knowledge of the word "diffeomorphism" to our friend).

Exercises

1   Check that varying the Palatini action with respect to $\Gamma^\rho_{\mu\nu}$ gives the usual relation between the Christoffel symbol and the metric.

2   Show that the Einstein tensor $E_{\mu\nu}$ vanishes identically in 2-dimensional spacetime.

---

[*] After I wrote this appendix during final revision of this book, I sent it to a colleague, one of those distinguished physicists listed in the preface. He sent back a terse email, complaining that the inclusion of diffeomorphism is "like a scratch on a record playing the sublime music of gravity." Then he told me, if I must include this kind of stuff, to find a better place to hide it.

Notes

1. F. Dyson, in *New York Review of Books*.
2. The notation $G_{\mu\nu}$ is commonly used. In later chapters, we will use $G_{\mu\nu}$ to denote the metric in higher dimensional spacetime.
3. We are assuming that spacetime does not have a boundary, so that we can ignore this surface term. If spacetime has a boundary, then we can add to the action an additional term defined on the boundary, known as the Gibbons-Hawking-York boundary term, whose variation with respect to $g_{\mu\nu}$ is designed to cancel the surface term we encounter here.
4. I would like to see discussions of whether the cosmological constant represents Einstein's greatest blunder banished forever from any sensible discourse on gravity. There is little hope of that.
5. M. Ferraris, M. Francaviglia, and C. Reina, "Variational Formulation of General Relativity from 1915 to 1925 'Palatini's Method' Discovered by Einstein in 1925," *Gen. Rel. Grav.* 14 (1982), pp. 243–254. Since the authors are all Italians, I would surmise that their curt dismissal of Palatini was not driven by nationalistic pride.
6. From the abstract of the paper "Belated Decision in the Hilbert-Einstein Priority Dispute" by L. Corry, J. Renn, and J. Stachel, *Science* 278 (1997), p. 1270: "A close analysis of archival material reveals that Hilbert did not anticipate Einstein." Also, Hilbert apparently did not know how to get the $\frac{1}{2}$ in (10).
7. Einstein wrote:

   My son [the 11-year-old Hans Albert] still hasn't answered my inquiry about meeting. . . . That is surely the influence of the woman. . . . You'll see more and more, on which side goodwill and honesty are to be found. There are reasons that I couldn't abide staying with that woman, despite the tender love that binds me to my children. When we first separated, the thought of my children stabbed me like a dagger every morning when I woke up. Nonetheless, I never regret having taken the step.

   Quoted in *Physics Today*, October 2005, p. 18.
8. Quoted in A. Fölsing, *Albert Einstein: A Biography*, Viking, 1997.
9. I follow the discussion in *Quantum Gravity* by C. Rovelli, Cambridge University Press, 2004.

The initial value or Cauchy problem

The nonlinearity of Einstein's field equation renders exact solutions rather unlikely, except in situations endowed with a high degree of symmetry. The advent of computers has thus resulted in the booming field of numerical relativity.[1] It is definitely beyond the scope of this textbook to go into details of numerical relativity, with its highly sophisticated methods and approaches. Rather, the purpose here is to acquaint the reader with the formulation of the initial value problem (also known as the Cauchy problem) in Einstein gravity.[2]

We all know how the initial value problem works in Newtonian mechanics. If we know the position $\vec{q}$ of a particle and its velocity $\frac{d\vec{q}}{dt}$ at some time t_0, Newton's law allows us to determine $\frac{d^2\vec{q}}{dt^2}$ at that time. We thus know the position $\vec{q}$ of the particle and its velocity $\frac{d\vec{q}}{dt}$ at time $t_0 + \delta t$ for δt an infinitesimal. We then repeat this procedure, using a process known as integration.

The initial value problem is particularly well suited to numerical work. Given the initial data, namely the position $\vec{q}$ of the particle and its velocity $\frac{d\vec{q}}{dt}$ at an initial time t_0, the computer has to be instructed on how to generate the corresponding data at a later time $t_0 + \delta t$, with δt small but finite. Once instructed, the computer can then blast away and calculate the particle's position and velocity at some later time $t_0 + T$. It is of course a science (and an art) to determine the optimal choice of δt for a given T and to make sure that the roundoff errors at each step do not accumulate out of control.

This basic scheme of evolving the initial data in time can be immediately generalized, first to the case of many particles (with the initial data now consisting of the position $\vec{q}_a$ and the velocity $\frac{d\vec{q}_a}{dt}$ of the N particles, $a = 1, \cdots, N$) and then to fields. Consider, for example, the scalar wave equation (described in chapters II.3 and III.6) $(\partial_\mu \partial^\mu - m^2)\phi(x) = 0$, which we now write as $\partial_0^2 \phi(t, \vec{x}) = (\nabla^2 - m^2)\phi(t, \vec{x})$. Note that the conceptual jump from many particles to field involves promoting[3] the discrete index a to the continuous spatial variable $\vec{x}$ (and trivially changing notation $q \to \phi$).

The initial data now consist of the two functions $\phi(0, \vec{x})$ and $\partial_0 \phi(0, \vec{x})$. For ease of writing, we set $t_0 = 0$ now and henceforth. The scalar wave equation then allows us to evolve these data in time. Knowing $\partial_0 \phi(0, \vec{x})$, we can determine $\phi(\delta t, \vec{x})$. The wave equation then gives $\partial_0^2 \phi(0, \vec{x})$, which then allows us to determine $\partial_0 \phi(\delta t, \vec{x})$. The key here, as for Newtonian mechanics, is that the equation of motion is second order, that is, it contains $\partial_0^2 = \frac{\partial^2}{\partial t^2}$. Indeed, Newton's deep insight was that dynamics involve the second derivative, not the first.

Gauge freedom and initial value in Maxwell electromagnetism

All this is straightforward and elementary; the subtlety first arises in gauge theory. Consider Maxwell electromagnetism. We are to solve (IV.2.13)

$$\partial_\mu F^{\mu\nu} = -J^\nu \tag{1}$$

for the vector potential $A_\mu(x)$. (Since $F_{\mu\nu} \equiv \partial_\mu A_\nu - \partial_\nu A_\mu$, the other "half" of Maxwell's equations $\varepsilon^{\rho\sigma\mu\nu} \partial_\sigma F_{\mu\nu} = 0$ is identically satisfied.) Everything would appear to be the same as before. Given the initial data $A_\mu(0, \vec{x})$ and $\partial_0 A_\mu(0, \vec{x})$, plus $J^\nu(0, \vec{x})$, we can then use (1), which are again (apparently) second order differential equations in time, that is, equations containing ∂_0^2, to evolve the initial data. (Of course, we also have to include the equations that tell the charges how to move, that is, how $J^\nu(0, \vec{x})$ changes with time, but that half of the story is not the focus of our discussion here.) It would seem that the four equations in (1) determine the four functions $A_\mu(x)$.

Not so fast! This is a gauge theory, and hence $A_\mu(x)$ and $\tilde{A}_\mu(x) \equiv A_\mu(x) + \partial_\mu \Lambda(x)$ correspond to the same physics. In other words, the subsequent value of $A_\mu(x)$ should be determined only up to an arbitrary function $\Lambda(x)$ (taken to vanish smoothly together with its first time derivative $\partial_0 \Lambda(x)$ as $x^0 \to 0$, so that A_μ and $\tilde{A}_\mu$ have the same initial data).

The resolution of this apparent paradox is simply that the $\nu = 0$ equation in (1) does not involve ∂_0^2 and is thus not a time evolution equation. We can see this explicitly: $-J^0 = \partial_\mu F^{\mu 0} = \partial_i F^{i0} = \partial_i (\partial^i A^0 - \partial^0 A^i)$, and hence this equation contains only 1 power of ∂_0.

It would be good to express this more physically. If we write out this equation in more elementary notation, we recognize it as simply Gauss's law $\vec{\nabla} \cdot \vec{E} = \rho$. In the initial data, once we write down some initial charge distribution $\rho(0, \vec{x})$, we are not allowed to write down any old $A_\mu(0, \vec{x})$ and $\partial_0 A_\mu(0, \vec{x})$; these eight functions of $\vec{x}$ must lead to an electric field $\vec{E}(0, \vec{x})$ satisfying Gauss's law. This makes physical sense.

We conclude that of the 4 equations in (1), one merely imposes a constraint on the initial data $A_\mu(0, \vec{x})$ and $\partial_0 A_\mu(0, \vec{x})$. There are only three time evolution equations, which do not determine the 4 unknown functions $A_\mu(t, \vec{x})$ completely. But that is exactly right: $A_\mu(t, \vec{x})$ should be determined only up to the gauge function $\Lambda(t, \vec{x})$.

Let us give another demonstration that $\partial_\mu F^{\mu 0}$ contains only one power of ∂_0. The proof will be by contradiction. Consider the trivial identity $\partial_\nu \partial_\mu F^{\mu\nu} = 0$, which follows from the antisymmetry of $F^{\mu\nu}$ and the fact that ∂_ν and ∂_μ commute. Write out this identity* as $\partial_0(\partial_\mu F^{\mu 0}) = -\partial_i \partial_\mu F^{\mu i}$. Suppose $\partial_\mu F^{\mu 0}$ contains 2 powers of ∂_0. Then the left hand side of this identity would contain three powers of ∂_0, but the right hand side manifestly has no room for three powers of ∂_0. Thus, we have proved what we set out to prove.

This proof using a trivial identity might seem to be overkill, given that the property we want to prove, that $\partial_\mu F^{\mu 0}$ does not contain 2 powers of ∂_0, is something we can see by eyeball, as indicated above. However, this line of attack will turn out to be useful in the context of Einstein gravity.

If Gauss's law holds at the initial time, we expect it to continue to hold as the charges rush madly about with the electric field changing accordingly. To verify this, simply differentiate the quantity $\vec{\nabla} \cdot \vec{E} - \rho$ with respect to time: $\partial_0(\vec{\nabla} \cdot \vec{E} - \rho) = \vec{\nabla} \cdot \partial_0 \vec{E} - \partial_0 \rho = \vec{\nabla} \cdot (\vec{\nabla} \times \vec{B} - \vec{J}) + \vec{\nabla} \cdot \vec{J} = 0$, where we used a Maxwell equation and current conservation. Thus, if the quantity $\vec{\nabla} \cdot \vec{E} - \rho$ vanishes at the initial time, it will continue to vanish at later times.

This physical check also leads to the trivial identity we used earlier. Write the quantity we just calculated as $\partial_0(\partial_i F^{i0} + J^0)$, and you see that its relativistic completion is $\partial_\nu(\partial_\mu F^{\mu\nu} + J^\nu)$, which with $\partial_\nu J^\nu = 0$ is just $\partial_\nu \partial_\mu F^{\mu\nu}$.

Gauge freedom and initial value in Einstein gravity

We are now warmed up sufficiently to tackle Einstein's field equations

$$E^{\mu\nu} \equiv R^{\mu\nu} - \frac{1}{2} g^{\mu\nu} R = T^{\mu\nu} \tag{2}$$

(where we have dropped the $16\pi G$ for convenience). We expect that the initial data on the $t = 0$ slice of spacetime consist of $g_{\mu\nu}(0, \vec{x})$ and $\partial_0 g_{\mu\nu}(0, \vec{x})$. (In general, the initial data will be specified on some spacelike hypersurface, known as the Cauchy surface; we simply assume that we can choose the t coordinate so that the hypersurface is described by $t = 0$, at least locally.) Since the field equations contain two powers of spacetime derivative, as we have long known, we would think that, given the metric and how it is changing with time at $t = 0$, the field equations will give us $\partial_0^2 g_{\mu\nu}(0, \vec{x})$. If so, then we could time evolve the metric as in examples of Newtonian mechanics and the scalar wave equation. The 10 equations in (2) would then determine the 10 unknown functions $g_{\mu\nu}(x)$.

But we are forewarned by the example of Maxwell's equations that things may not be so simple. Indeed, the invariance of physics under general coordinate transformation $x^\mu \to x'^\mu(x)$ allows us to eliminate 4 of the 10 unknown functions $g_{\mu\nu}(x)$. In other words, Einstein's field equations should not determine $g_{\mu\nu}(x)$ completely. Our experience with the Maxwell case suggests that 4 of the 10 equations in (2) are not time evolution equations but merely constraints on the initial data.

---

* Note that this identity together with Maxwell's equation implies that the current must be conserved: $\partial_\nu(\partial_\mu F^{\mu\nu} - J^\nu) = 0$ gives $\partial_\nu J^\nu = 0$.

Now that I have explained to you how things work in the Maxwell case, it should be fairly clear to you how things work in the Einstein case. But historically, the great Einstein was confused at this point. He concluded that he was faced with a logical choice: either (a) the field equations are not deterministic, or (b) invariance of physics under general coordinate transformation is too strong a requirement. Unfortunately, in 1914, he chose to abandon invariance of physics under general coordinate transformation. Ouch, Albert, the wrong choice! Fortunately, in 1915, with Hilbert breathing down his neck, Einstein's brainpower kicked into high gear, and he realized that the correct choice was (a). The field equations do not, and should not, determine $g_{\mu\nu}(x)$ completely.

So, give me your best guess before reading on: which four of the equations in (2) are not time evolution equations, that is, do not contain ∂_0^2 acting on the metric?

Right, the claim is that $E^{0\nu} = T^{0\nu}$ amounts to constraints on $g_{\mu\nu}(0, \vec{x})$ and $\partial_0 g_{\mu\nu}(0, \vec{x})$. Of course, you could verify this claim by working out $E^{0\nu}$ directly, but things here are not as simple as seeing by eyeball that $\partial_\mu F^{\mu 0}$ does not contain ∂_0^2. As suggested by the Maxwell example, we need the analog of the identity $\partial_\nu \partial_\mu F^{\mu\nu} = 0$, namely the contracted Bianchi identity $D_\mu E^{\mu\nu} = 0$. Let us write this out in longhand: $\partial_0 E^{0\nu} = -\partial_i E^{i\nu} +$ terms involving the Christoffel symbols Γ times various Es.

Again, the proof is by contradiction. Suppose that $E^{0\nu}$ contains ∂_0^2. Then the left hand side of the preceding equation contains ∂_0^3. But nowhere on the right hand side can we find ∂_0^3. (For example, the Christoffel symbols contain ∂_0 and the Es contain two powers of ∂_0.) Contradiction and QED. In other words, the two powers of ∂_0 contained in $E^{0\nu}$ must appear in the form $g \partial_0 g .. \partial_0 g ...$ The four equations $E^{0\nu} = T^{0\nu}$ merely constrain the initial data. (We are implicitly assuming that $T^{0\nu}$ does not contain ∂_0^2, as is true of the standard[4] energy momentum tensor we normally encounter.)

Confusio appears, well, confused. "Why must the right hand side also contain ∂_0^3 if the left hand side contains ∂_0^3? Newton's equation $F = ma$ has ∂_0^2 on one side but not the other." You explain to Confusio patiently, "We must distinguish between equations and identities.[5] The equations of motion are satisfied only by the actual solutions, while the contracted Bianchi identity must be satisfied for any $g_{\mu\nu}$."

Again, this makes physical sense: after writing down some initial $T^{0\nu}(0, \vec{x})$, you can't write down any old $g_{\mu\nu}(0, \vec{x})$. You have to figure out how $g_{\mu\nu}(0, \vec{x})$ and $\partial_0 g_{\mu\nu}(0, \vec{x})$ are constrained by $T^{0\nu}(0, \vec{x})$. For Maxwell electromagnetism, you have a similar problem, but writing down the charge distribution, you simply sum up the electric field due to each charge. Here the constraint equations $E^{0\nu} = T^{0\nu}$ are highly nonlinear, and in general would require numerical methods to solve. Thus, before you solve the initial value problem numerically, you may already have a nontrivial numerical problem to solve just to set up the numerical problem! Before we start our lives as adults, we already have to solve problems almost as hard as the problems we have to solve as adults.

By the way, as you recall from chapter VI.5, the contracted Bianchi identity plays the same role as the identity $\partial_\mu \partial_\nu F^{\mu\nu}$ mentioned above: it implies energy momentum conservation $D_\mu T^{\mu\nu} = 0$.

Sometimes the initial data are specified on only a patch on the Cauchy surface. Then the Cauchy problem is valid only in a bounded spacetime domain, namely the region that is in causal contact with the initial data. The boundary of that domain is called the Cauchy horizon.

Appendix: Einstein's confusion

The astute reader might notice that if we were to regard $\vec{E}$ and $\vec{B}$ as the dynamical variables rather than A_μ, there would have been no hand-wringing over the initial value problem in electrodynamics. Given the initial values of $\vec{E}$ and $\vec{B}$, and the initial positions and velocities of the charged particles, the two Maxwell's equations

$$\frac{\partial \vec{E}}{\partial t} = (\vec{\nabla} \times \vec{B}) - \vec{J} \tag{3}$$

$$\frac{\partial \vec{B}}{\partial t} = -(\vec{\nabla} \times \vec{E}) \tag{4}$$

specify how $\vec{E}$ and $\vec{B}$ are to evolve and the Lorentz force law tells the charges how to move. The other Maxwell's equations $\vec{\nabla} \cdot \vec{E} = \rho$ and $\vec{\nabla} \cdot \vec{B} = 0$ constrain the initial data.

Historically, up until 1915, Einstein was confused, because he thought that the metric $g_{\mu\nu}$ was analogous to $\vec{E}$ and $\vec{B}$, rather than to A_μ. He presented an infamous "hole argument" as follows. Given some initial data, allow the field equations (which of course he was still searching for at the time) to evolve the metric $g_{\mu\nu}$ to later time. Imagine a spacetime region in the future, which Einstein referred to as a hole. Now perform, inside the hole, a general coordinate transformation that goes over smoothly to the identity transformation outside the hole. Reasoning from the analogous problem in electromagnetism with $\vec{E}$ and $\vec{B}$ specified at some initial data, Einstein was perplexed that the same initial data would lead to two different metrics inside the hole. As mentioned in the text, he concluded that either the field equation cannot be covariant under coordinate transformation, or the metric is not physical. Eventually it dawned on him that, in the context of the initial value problem, the metric is actually analogous to A_μ, rather than $\vec{E}$ and $\vec{B}$.

Given our present understanding of physics today, it is difficult to imagine that such a great mind could be confused on "so elementary" a point. One possible explanation is that A_μ did not come to the forefront of physics until the advent of quantum mechanics.[6] Indeed, luminaries such as Oliver Heaviside[7] had thundered that A_μ should be consigned to the dustbins of history and that physics needed only $\vec{E}$ and $\vec{B}$.

By the way, now you can understand Einstein's rueful confession that I quoted in chapter VI.1: "I believed that I could show on general considerations that a law of gravitation invariant in relation to any transformation of coordinates whatever was inconsistent with the principle of causation . . . errors of thought which cost me two years of excessively hard work."[8]

Exercises

1   Verify by brute force computation that $E^{0\mu}$ does not contain ∂_0^2. Hint: To save labor, ask Professor Flat for help.

2   As in the electromagnetic case, once the constraint $E^{0\nu} = T^{0\nu}$ is imposed at the initial time, the time evolution equations guarantee that it will continue to hold, which makes physical sense. Verify this. In other words, show that $\partial_0(E^{0\nu} - T^{0\nu})$ vanishes.

Notes

1. Currently, a major effort is under way to understand binary black hole merger by a combination of numerical and analytical methods. We will touch upon one aspect of this in chapter X.4.
2. I have benefited from a discussion with T. Jacobson.
3. See *QFT Nut*, p. 19.

4. If you want to mess with nonstandard theories of matter and of gravity, then you would be obliged to examine the initial value problem anew. See T. Jacobson, arXiv:1108.1496.

5. Not to belabor a point, but in my experience, there might still be a student who is confused. Tell him or her to compare and contrast the equation $(x - 1)(x + 1) = 2$ and the identity $(x - 1)(x + 1) = x^2 - 1$.

6. See *QFT Nut*, p. 245.

7. Oliver Heaviside, described by his best friend as a "first rate oddity," was actually responsible for Maxwell's equations in the form we know them today. Self-educated, he never held an academic position. See B. J. Hunt, *Physics Today*, November 2012, p. 48.

8. A. Einstein, *Essays in Science*, p. 83.

Recap to Part VI

Here we finally get to the heart of the matter, or rather, of the field: the gravitational field is striving to extremize the scalar curvature. Remarkably, once we decide to regard the metric as the dynamical variable, the action governing the gravitational field is uniquely determined.

Arguing by symmetry, and essentially without doing any work, we can write down Einstein's field equation, and with that, start to unlock the expanding universe and the secrets of the black hole.

Surprisingly, or perhaps not so surprisingly, the motion of particles of matter and of light around a massive body can be determined by analog problems in Newtonian mechanics. The second derivative, not the first nor the third, rules.

As lifelong students of physics, we have talked about energy and momentum for a long time, but finally we know what they are. Energy and momentum are what the gravitational field listens to.

Part VII | Black Holes

Ambling around a black hole

You have surely read popular accounts of black holes, without question one of the most fascinating features of Einstein gravity. Indeed, already in the introduction, I gave the heuristic Michell-Laplace 18th-century argument that a (spherical) object of mass M and radius R is a black hole if

$$R < 2GM \tag{1}$$

You learned in chapter VI.3 that the empty spacetime around a spherically symmetric mass distribution is described by

$$ds^2 = -\left(1 - \frac{r_S}{r}\right) dt^2 + \frac{1}{\left(1 - \frac{r_S}{r}\right)} dr^2 + r^2 d\Omega \quad \text{(for } r > R) \tag{2}$$

with the Schwarzschild radius $r_S \equiv 2GM$. As we noted, the radius R of this spherical object does not appear explicitly in ds^2 but only implicitly, as we have indicated here. Since we only solved Einstein's field equation $R_{\mu\nu} = 0$ in empty spacetime, the Schwarzschild solution (2) holds only for $r > R$.

The Schwarzschild radius of an ordinary massive object, the sun for example, is much less than its characteristic size R and so would be located inside the object, where the Schwarzschild solution is not relevant, as was already mentioned in chapter VI.3. Thus, we don't have to worry about the apparent singularity in ds^2 at $r = r_S$ for stars and planets and almost everything else.

By the same token, we now recognize that the prescient Michell-Laplace criterion (1) amounts to saying that if a massive object is so compact that its actual size R is smaller than its Schwarzschild radius, it is a black hole. In other words, an object small for its mass, or equivalently, massive for its size, is a black hole. For a black hole, the surface defined by $r = r_S$, known as the horizon, is situated outside the black hole in empty spacetime, where the Schwarzschild solution certainly holds. As we will see in detail in this chapter and the next chapter, if an intrepid explorer reaches the Schwarzschild radius of such an object and crosses into the region where $r < r_S$, he or she can never get back out to spatial infinity. In short, for our purposes here, a black hole is defined as a massive object with an accessible horizon.

The mystery of the black hole also deepens with the realization that spacetime is perfectly smooth there (as indicated by the behavior of $R_{\mu\nu\lambda\rho}R^{\mu\nu\lambda\rho}$, for example). In the next chapter, we will confirm our suspicion that the singularity in the metric at the horizon is merely due to a poor choice of coordinates.

In this chapter, we focus on the motion of massive and massless particles around a black hole, leaving various other issues for the next chapter.

A common misconception about black holes

A common misconception is that, around a black hole, an irresistible mysterious force sucks everything in.* But in fact, physicists do not know of any additional force besides the ones they usually enumerate. Gravity is gravity, and whether outside a regular star or a black hole, we have the very same Schwarzschild metric. Thus, even if some evil power, in an implausible sci fi movie, somehow manages to turn the sun suddenly into a black hole, our earth, though deprived of its main energy source, would still calmly cruise along the same orbit. What is true, as we will see in this and the following chapter, is that near a black hole, spacetime can be warped so much that once trapped, even light cannot escape.

We will start by studying the motion of massive and massless particles, such as planets and photons, around a black hole. Indeed, in chapter V.4, we already worked out the equations of motion for both particles and light in a general spherically symmetric static spacetime described by two unknown functions $A(r)$ and $B(r)$. After we found the Schwarzschild metric, all we had to do was plug in $A(r) = 1 - \frac{r_S}{r}$ and $A(r)B(r) = 1$, which was precisely what we did in chapter VI.3 to study the deflection of light and the perihelion shift of Mercury. So we are all set and ready to go. Here, for convenience, I will list again the relevant equations, which, I emphasize again, are precisely the same around a black hole as around any other massive object. The only difference is whether or not the equations are relevant all the way down to r_S.

An unfamiliar Newtonian potential

We first study the case of a massive particle moving in the equatorial plane, with the two conservation laws $\frac{dt}{d\tau} = \frac{\epsilon}{1 - \frac{r_S}{r}}$ and $\frac{d\varphi}{d\tau} = \frac{l}{r^2}$, where ϵ and l denote the energy and angular momentum per unit mass of the particle, respectively. The motion of the radial coordinate is governed by

$$\left(\frac{dr}{d\tau}\right)^2 - \frac{r_S}{r} + \frac{l^2}{r^2} - \frac{l^2 r_S}{r^3} = \epsilon^2 - 1 \tag{3}$$

---

* Mark Twain allegedly said that the trouble with people is not that they know so little, but that what they know is largely not true.

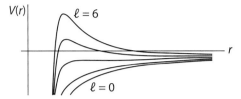

Figure 1 The potential $V(r)$ as a function of $\frac{r}{r_S}$ for $\left(\frac{l}{r_S}\right)^2 = 0, 1, 3, 5, 6$. Note that on the scale of this plot, the minimum of $V(r)$ is hardly visible.

As already mentioned back in chapter V.4, we merely have to solve a Newtonian (except that coordinate time has been replaced by proper time) problem in the potential*

$$V(r) = -\frac{r_S}{r} + \frac{l^2}{r^2} - \frac{l^2 r_S}{r^3} = -\left(\frac{r_S}{r}\right) + \left(\frac{l}{r_S}\right)^2\left\{\left(\frac{r_S}{r}\right)^2 - \left(\frac{r_S}{r}\right)^3\right\} \tag{4}$$

The second form shows that if we measure r and l in sensible units, the potential is surprisingly simple, controlled by the single parameter $\left(\frac{l}{r_S}\right)^2$.

After all the Riemannian geometry, with space and time unified and curved this way and that, the bottom line boils down, remarkably[†] enough, to a "pretend" Newtonian problem.

Life is sweet then: as long as you have mastered Newtonian mechanics, you can blast (3) any which way you like, and many texts fill page after page with exhaustive and exhausting studies of the resulting equations.

The first two terms in the potential V are our old friends, representing gravitational attraction and centrifugal repulsion, respectively. The third term, call it the Einstein term, represents a novel and unfamiliar effect. In the real Newtonian problem, for $l = 0$, we fall into the singularity at $r = 0$, but for $l \neq 0$, the centrifugal term keeps us from falling in (as we discover every morning that the earth is still going around the sun). In contrast, in the pretend Newtonian problem, for small r, the Einstein term $\left(-\frac{l^2 r_S}{r^3}\right)$ kicks in and totally dominates the centrifugal term. Under the right circumstances, we could fall into the singularity even with $l \neq 0$.

Let us plot $V(r)$ as in figure 1, $\left(\frac{l}{r_S}\right)^2 = 0, 1, 3, 5, 6$. For $l = 0$, we have simply $V(r) = -\frac{r_S}{r}$, but as we increase l, the potential soon develops a minimum and a maximum. Our revered Newton taught us how to find them: set $V'(r) = 0$ to obtain

$$r_S r^2 - 2l^2 r + 3l^2 r_S = 0 \tag{5}$$

with the solutions

$$r_{\min} = \frac{l^2}{r_S}\left(1 + \sqrt{1 - \frac{3r_S^2}{l^2}}\right) \quad \text{and} \quad r_{\max} = \frac{l^2}{r_S}\left(1 - \sqrt{1 - \frac{3r_S^2}{l^2}}\right) \tag{6}$$

---

* Note that in this "pretend" Newtonian problem, I define the "kinetic energy" $\left(\frac{dr}{d\tau}\right)^2$ without the usual factor of $\frac{1}{2}$ to remove various factors of 2 in (VI.3.14) and (VI.3.15), rendering later expressions somewhat cleaner.

† Again, we have Pythagoras to thank.

The critical value for these extrema to appear is $\left(\frac{l_c}{r_S}\right)^2 = 3$. The shape of $V(r)$ differs according to whether $l < l_c$ or $l > l_c$. I have intentionally omitted some of the labels in figure 1. So, which curve in the figure corresponds to $l_c = \sqrt{3}r_S$? Note also that $r_{\min}$ approaches $\simeq 2l^2/r_S$ for large l and that the minimum is very shallow.

Radial plunge

For $l < l_c$, the particle has no option but to fall in.

Consider the simplest case of an intrepid observer plunging into the black hole along the radial direction, so that $d\varphi = 0$, which by $\frac{d\varphi}{d\tau} = \frac{l}{r^2}$ corresponds to $l = 0$. To simplify further, let the observer start with vanishing energy at $r = \infty$ (so that $\frac{dt}{d\tau}|_{r=\infty} = 1$, which by $\frac{dt}{d\tau} = \frac{\epsilon}{1-\frac{r_S}{r}}$ implies $\epsilon = 1$). Then (3) implies $\frac{dr}{d\tau} = -\left(\frac{r_S}{r}\right)^{\frac{1}{2}}$, which literally anybody could integrate to give $r^{\frac{3}{2}} = r_0^{\frac{3}{2}} - \left(\frac{3}{2}\right)r_S^{\frac{1}{2}}\tau$, with r_0 the observer's position at $\tau = 0$.

The point is not our ability to solve a differential equation but that the observer reaches the horizon r_S at some finite proper time starting at some $r_0 > r_S$ (a fact we can see directly from the differential equation without integrating). Not only does the observer suffer no harm as he crosses the horizon, he also gets there soon enough according to his clock. After he passes the horizon, he eventually reaches the origin $r = 0$, also in finite proper time, at which point he is crushed by infinite tidal forces as measured by (recall chapter VI.3) $R^{\mu\nu\rho\sigma} R_{\mu\nu\rho\sigma} = \frac{12r_S^2}{r^6}$.

Nevertheless, even though nothing appears to be singular at the horizon, the equation $\frac{dt}{d\tau} = \frac{\epsilon}{1-\frac{r_S}{r}}$ indicates that something strange does occur there: as $r \to r_S^+$, the coordinate time $t \to \infty$. Our observer crosses the horizon at infinite coordinate time. To his friend stationed at $r = \infty$ ("stationed" means that his friend has to fire small rockets occasionally to avoid slowly falling toward the black hole) the observer appears to approach but never quite cross the horizon. (The time experienced by the friend, namely her proper time, coincides with coordinate time, since for her, $\frac{dt}{d\tau} = 1$.)

A small puzzle here for you: what happens to t as the observer crosses the horizon?

Recall our analysis of the gravitational redshift in chapter V.4, showing that the proper time interval between the two signals as seen by the receiver is related to the proper time interval as seen by the emitter $\Delta\tau_R = \Delta\tau_E(g_{00}(r_R)/g_{00}(r_E))^{\frac{1}{2}}$. So indeed, for the observer stationed at $r = \infty$, the interval between signals sent by the infalling emitter gets infinitely time dilated by the factor $1/\sqrt{1 - \frac{r_S}{r_E}}$ as he approaches the horizon $r_E \to r_S$.

Orbits with substantial angular momentum

By substantial angular momentum, I mean $l > l_c = \sqrt{3}r_S$, so that $V(r)$ has a maximum and a minimum. As shown in figure 2, we now have three possible cases, depending on the effective energy ($\epsilon^2 - 1$) as per (3):

1. For[1] $(\epsilon^2 - 1) > V(r_{\max})$, the particle sails over the top of the potential and thus spirals into the black hole, as shown in figure 2a.

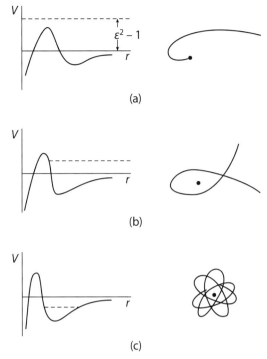

Figure 2 For $l > l_c = \sqrt{3}r_S$, there are three possible types of orbit (shown at right), depending on the value of the effective energy $\epsilon^2 - 1$. The potential $V(r)$ is plotted schematically at left (compare figure 1) to emphasize its two extrema. (a) Effective energy $\epsilon^2 - 1 > V(r_{\max})$; (b) $V(r_{\max}) > \epsilon^2 - 1 > 0$; (c) $0 > \epsilon^2 - 1 > V(r_{\min})$.

2. For $V(r_{\max}) > (\epsilon^2 - 1) > 0$, the particle bumps into the potential barrier and retreats back to infinity. The shape of the orbit is shown in figure 2b.

3. For $^2 0 > (\epsilon^2 - 1) > V(r_{\min})$, the particle is trapped in the potential, and follows an "elliptical" orbit with a shifting perihelion (figure 2c).

If you want, you can work out the shape of these orbits by solving $r(\tau)$ and $\varphi(\tau)$: just a matter of showing off your ability to solve differential equations. You could always integrate them numerically on a computer.

Circular orbits and Kepler

Looking at $V(r)$, we see that there are two circular orbits with radius given by (6). The orbit at $r_{\max} = \frac{l^2}{r_S}\left(1 - \sqrt{1 - \frac{3r_S^2}{l^2}}\right)$, perched at the maximum of $V(r)$, is obviously unstable. Any perturbation will either cause the orbiting particle to fall into the black hole or to move toward the stable orbit at $r_{\min} = \frac{l^2}{r_S}\left(1 + \sqrt{1 - \frac{3r_S^2}{l^2}}\right)$. In contrast, the stable circular orbit lives comfortably at the minimum of $V(r)$.

Interestingly, Kepler's third law continues to hold, as we now show. The 4-velocity $V^\mu = (V^t, 0, 0, V^\varphi)$ of the planet, or whatever, is given by the conservation laws (V.4.13–14): $V^t \equiv \frac{dt}{d\tau} = \epsilon/\left(1 - \frac{r_S}{r}\right)$ and $V^\varphi \equiv \frac{d\varphi}{d\tau} = l/r^2$, with the two conserved quantities l determined by (5)

$$l^2 = \frac{1}{2}r_S r\left(1 - \frac{3r_S}{2r}\right)^{-1}$$

and ϵ determined by (3)

$$\epsilon^2 = V(r) + 1 = \left(1 - \frac{r_S}{r}\right)^2 \left(1 - \frac{3r_S}{2r}\right)^{-1}$$

(Here various quantities are to be evaluated at the radius of the orbit.)

Define the angular velocity $\Omega \equiv \frac{d\varphi}{dt} = \frac{d\varphi}{d\tau}/\frac{dt}{d\tau} = V^\varphi/V^t$. Note that Ω is defined in terms of coordinate time, not proper time. Plugging in what we had, we obtain*

$$\Omega^2 = \frac{r_S}{2r^3} = \frac{GM}{r^3} \tag{7}$$

Kepler's third law survives Einstein.

Accretion disks: Mightier than nuclear fusion

As l decreases toward $l_c = \sqrt{3}r_S$, the stable orbit with radius $r_{\min} = \frac{l^2}{r_S}\left(1 + \sqrt{1 - \frac{3r_S^2}{l^2}}\right)$ keeps shrinking until $r_{\min}$ reaches its minimum value of

$$r_{\mathrm{ISCO}} = r_{\min}\Big|_{l=l_c} = 3r_S \tag{8}$$

The orbit with radius r_{ISCO} is known as the innermost stable circular orbit in relativistic astrophysics. Note that it sits well outside the Schwarzschild radius.

These simple remarks underlie the essential physics of accretion disks around black holes. Around a black hole, infalling debris, consisting of matter from a companion star, for example, forms a disk. Through dissipative processes, such as collisions between particles and electromagnetic radiation, particles in the disk gradually lose energy and angular momentum and move inward until they reach r_{ISCO}. From there, further loss of angular momentum will cause them to fall in. By angular momentum conservation, this process will cause the black hole to rotate, so that eventually the Schwarzschild solution is no longer adequate. In chapter VII.5 we will discuss rotating black holes.

Black holes power some of the most spectacular processes known to astrophysics. What fraction of the rest energy mc^2 does a particle lose as it crashes inward to the innermost stable circular orbit? As always, see if you can figure it out before reading on.

What is the energy of a particle in the innermost stable circular orbit? Using (4), we evaluate $V(r = r_{\mathrm{ISCO}}, l = l_c) = -\frac{1}{3} + \frac{3}{9} - \frac{3}{27} = -\frac{1}{9}$. Setting $\frac{dr}{d\tau} = 0$ in (3), we find $\epsilon =$

---

* A quick reminder of Kepler's third law in Newtonian physics: $v^2/r = GM/r^2$, $\Omega^2 = ((2\pi)/(2\pi r/v))^2 = GM/r^3$.

$\sqrt{1 - \frac{1}{9}} = \frac{2\sqrt{2}}{3}$. But from $\frac{dt}{d\tau} = \frac{\epsilon}{1 - \frac{r_S}{r}}$, we see that the particle started out at $r = \infty$ with $\epsilon = 1$. Thus, the fraction of energy ultimately lost to electromagnetic radiation amounts to $1 - \frac{2\sqrt{2}}{3} \approx 0.06$.

Is this a lot? What do we compare the 6% to? Consider nuclear fusion, which powers the stars and the hydrogen bomb. Four protons are converted to a helium nucleus accompanied by the emission of two positrons and two neutrinos. The fraction of energy release is calculated easily by comparing the mass of the helium nucleus to 4 times the mass of the proton, according to what we learned back in part III. The fabled equation $E = mc^2$, yes! It turns out that in nuclear fusion, the energy released amounts to 0.7%. The black hole is almost ten times more efficient. Later, we will see that a rotating black hole is even more efficient!

Massless particle

For a massless particle, we have to replace the proper time τ by the affine parameter ζ. Then, for a photon moving in the equatorial plane, we have (in parallel with the case of a massive particle) $\frac{dt}{d\zeta} = \frac{\epsilon}{1 - \frac{r_S}{r}}$ and $\frac{d\varphi}{d\zeta} = \frac{l}{r^2}$, with ϵ and l two integration constants. The radial coordinate of the worldline satisfies

$$\frac{1}{l^2}\left(\frac{dr}{d\zeta}\right)^2 + \frac{1}{r^2}\left(1 - \frac{r_S}{r}\right) = \frac{\epsilon^2}{l^2} \equiv \frac{1}{b^2} \tag{9}$$

From the displayed equations, we can see explicitly our freedom to scale the affine parameter by $\zeta \to \zeta/l$. As explained earlier, physics does not depend on ϵ and l separately, but only on the impact parameter squared: $b^2 \equiv \frac{l^2}{\epsilon^2}$.

Alternatively, eliminate the affine parameter ζ by dividing $\frac{dr}{d\zeta} / \frac{d\varphi}{d\zeta} = \frac{dr}{d\varphi}$, so that

$$\left(\frac{dr}{d\varphi}\right)^2 = \frac{r^4}{b^2} - r\left(r - r_S\right) \tag{10}$$

(which we used in discussing light deflection) or by dividing $\frac{dr}{d\zeta} / \frac{dt}{d\zeta} = \frac{dr}{dt}$, so that

$$\left(\frac{dr}{dt}\right)^2 = \frac{(r - r_S)^2}{r^2}\left(1 - \frac{b^2}{r^3}(r - r_S)\right) \tag{11}$$

(which we used in discussing radar echo delay, in chapter VI.3).

The qualitative features of light moving around a black hole can be immediately read off from (9), which with a particular choice of the affine parameter becomes $(\frac{dr}{d\zeta})^2 + U(r) = \frac{1}{b^2}$. Plot the effective potential

$$U(r) = \frac{1}{r^2} - \frac{r_S}{r^3} \tag{12}$$

For large r, $U(r) \simeq \frac{1}{r^2}$ and goes up as r decreases, reaching a maximum value of $U_{\max} = 4/(27r_S^2)$ at $r = 3r_S/2$, then plunging downward to $-\infty$. Thus, the motion of light can be divided into three cases (figure 3):

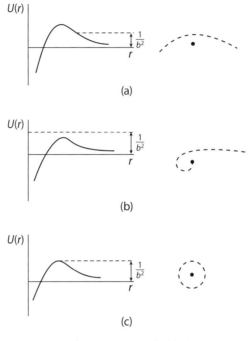

Figure 3 For light moving around a black hole, there are three possible types of orbit (shown at right), depending on the impact parameter. The potential $U(r)$ is plotted schematically (left). (a) Large impact parameter $1/b^2 < U_{\max}$; (b) small impact parameter $1/b^2 > U_{\max}$; (c) impact parameter just right $1/b^2 = U_{\max}$.

1. Large impact parameter $1/b^2 < U_{\max}$: light comes in and then goes back out, corresponding to the deflection of light we studied in chapter VI.3.

2. Small impact parameter $1/b^2 > U_{\max}$: light comes in and plunges into the black hole.

3. The impact parameter is just right ($1/b^2 = U_{\max}$) to trap light going around in an (unstable) circular orbit.

The precise shape of the orbit can be obtained by integrating (10).

A common confusion about plunging into a black hole

Confusio speaks up: "I have learned that the fundamental laws in classical physics (and also quantum physics) are time reversal invariant,[3] that is, they are unchanged upon $t \to -t$. I read that if we take a movie depicting a microscopic* process and run it backward, the reversed process must also be allowed by the laws of physics. So why can't I run the film of the observer radially plunging into a black hole and watch him come flying out?"

---

* Thus excluding processes involving a macroscopic number of particles, with the attendant discussion about entropy, second law, and so on and so forth. Indeed, you can't make an egg out of an omelette.

Well, well, that Confusio is more astute than we think. Indeed, the Lagrangian

$$L = \left[\left(1 - \frac{r_S}{r}\right)\left(\frac{dt}{d\tau}\right)^2 - \left(1 - \frac{r_S}{r}\right)^{-1}\left(\frac{dr}{d\tau}\right)^2 - r^2\left(\frac{d\theta}{d\tau}\right)^2 - r^2\sin^2\theta\left(\frac{d\varphi}{d\tau}\right)^2 \right]^{\frac{1}{2}}$$

governing the motion of a particle in Schwarzschild spacetime is manifestly invariant under $t \to -t$. So where is the catch in the standard arguments[4] about time reversal invariance?

The catch, as I have already mentioned, is that the coordinate time t increases to $+\infty$ as $r \to r_S^+$ and then decreases from $+\infty$ after the observer crosses the horizon. Indeed, as is evident from the Lagrangian just displayed, t and r exchange roles for $r < r_S$. The letter "t" no longer denotes time! Much more on this in the next chapter.

The standard arguments about time reversal invariance work perfectly well as long as $r > r_S$. Thus, if we could somehow install a trampoline at r_S^+ just outside the black hole, the observer in radial plunge could bounce back[5] out to $r = \infty$, retracing his trajectory.

Appendix: Painlevé-Gullstrand coordinates

An interesting set of coordinates was introduced by Paul Painlevé* and Allvar Gullstrand† in 1921. In the Schwarzschild metric $ds^2 = -(1 - v^2)dt^2 + \frac{1}{(1-v^2)}dr^2 + r^2d\Omega$ with $v^2 = \frac{r_S}{r}$, let $dt = dT - h(r)dr$. We could take $h(r)$ to be any reasonable function of r, but a particularly nice choice is to make the coefficient of dr^2 equal to 1, which fixes $h = \frac{v}{(1-v^2)}$. We obtain

$$ds^2 = -dT^2 + \left(dr + \sqrt{\frac{r_S}{r}}dT\right)^2 + r^2d\Omega \tag{13}$$

This shows conclusively that spacetime is not singular at $r = r_S$, a point that was not widely appreciated until the early 1960s. (More on this in the next chapter.) Recall from the text that an observer in radial plunge starting at rest at infinity follows $\frac{dr}{d\tau} = -\sqrt{\frac{r_S}{r}}$. (See exercises 3 and 4.) Note also that in Painlevé-Gullstrand coordinates, a slice of spacetime at fixed T corresponds to flat space.

Exercises

1   Show that in the derivation of Kepler's third law, the results we obtained for V^t and V^φ satisfy the consistency check $g_{\mu\nu}V^\mu V^\nu = -1$.

2   Determine the shape of the orbit of light in the presence of a black hole.

3   Verify that in Painlevé-Gullstrand coordinates, an observer in radial plunge starting at rest at infinity follows $\frac{dr}{d\tau} = -\sqrt{\frac{r_S}{r}}$, as must be the case, since in transforming from the Schwarzschild coordinates, we did not change r. Show also that $\frac{dr}{dT} = -\sqrt{\frac{r_S}{r}}$.

---

* Twice the prime minister of France.
† Nobel Laureate in Physiology or Medicine 1911, he opposed giving Einstein the Nobel Prize for his theory of special relativity.

4   The preceding result shows that the velocity $\frac{dr}{dT}$ of the radially plunging observer reaches 1 at the horizon and then goes to ∞ at the physical singularity at $r = 0$. You are of course sophisticated enough by now not to dash off some crackpot claim that Einstein was wrong about the speed of light setting the ultimate speed limit. Verify that $\frac{dr}{dT}$ is still less than the speed of light.

Notes

1. For the record, $V(r_{\max}) = \dfrac{h\left(h - \sqrt{(h-3)h} - 2\right)}{\left(\sqrt{(h-3)h} - h\right)^3}$, with $h \equiv \left(\dfrac{l}{r_S}\right)^2$.

2. Again, for the record, $V(r_{\min}) = -\dfrac{h\left(h + \sqrt{(h-3)h} - 2\right)}{\left(h + \sqrt{(h-3)h}\right)^3}$.

3. The only known exceptions occur in the decay of certain elementary particles.

4. I recommend J. J. Sakurai, *Invariance Principles and Elementary Particles*, chapter 4.

5. As I mentioned in chapter VI.3, Einstein did think erroneously that a particle falling into a black hole would bounce back at the Schwarzschild radius.

Poor choice of coordinates

In the Schwarzschild solution, which I reproduce here for convenience,

$$ds^2 = -\left(1 - \frac{r_S}{r}\right) dt^2 + \frac{1}{\left(1 - \frac{r_S}{r}\right)} dr^2 + r^2 d\Omega^2 \tag{1}$$

the components of the metric $g_{00} = -\left(1 - \frac{r_S}{r}\right)$ and $g_{rr} = \left(1 - \frac{r_S}{r}\right)^{-1}$ change sign at $r = r_S \equiv 2GM$, a place known as the horizon, as was mentioned in the preceding chapter. (The reason for the term "horizon" will become clear in this chapter.) What we call time t and what we regard as the radial coordinate r exchange roles* inside the horizon, leading us to expect that something extraordinary happens at the horizon. However, we learned in chapter VI.3 (by computing $R_{\mu\nu\lambda\rho} R^{\mu\nu\lambda\rho}$, for example) that spacetime is perfectly smooth there. We suspected that the singularity at the horizon was merely due to a nasty coordinate choice (recall appendix 1 of chapter I.6). In this chapter, we will confirm this suspicion by exhibiting a better behaved coordinate system near and inside the horizon.

Let's look at radial light rays, for which $d\theta = 0$ and $d\varphi = 0$, so that we can effectively suppress the angular coordinates. Their paths are determined, as usual, by $ds^2 = 0 = -\left(1 - \frac{r_S}{r}\right) dt^2 + \frac{1}{\left(1 - \frac{r_S}{r}\right)} dr^2$, that is, by $dt = \pm \frac{1}{1 - \frac{r_S}{r}} dr = \pm \frac{r}{r - r_S} dr$. For $r > r_S$, the plus sign describes outgoing light rays ($dr > 0$ for $dt > 0$), and the minus sign incoming light rays ($dr < 0$ for $dt > 0$). Infinitely far away from the black hole, $dt = \pm dr$ and light rays move at 45°. But as we move in close to the black hole, the angle the light rays make with the r-axis increases with decreasing r until it reaches 90° at the horizon, as illustrated in figure 1. The light cone closes up like a clam.

---

* Because of this role exchange, it has been suggested that the region $r < r_S$ be called not "inside the horizon" as it is usually called, but "after the horizon." By the same token, the Schwarzschild singularity at $r = 0$ is a moment in time, not a place you visit.

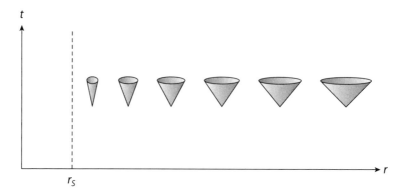

Figure 1 The light cone closes up like a clam as r decreases toward the horizon.

Tilting light cones spill particles into the black hole

Let's look for a better set of coordinates than (t, r) by massaging the Schwarzschild solution as follows:

$$ds^2 = -\left(1 - \frac{r_S}{r}\right)\left(dt^2 - \frac{1}{(1 - \frac{r_S}{r})^2}dr^2\right) + r^2 d\Omega^2$$

$$= -\left(\frac{r - r_S}{r}\right)\left(dt + \frac{r}{r - r_S}dr\right)\left(dt - \frac{r}{r - r_S}dr\right) + r^2 d\Omega^2$$

$$= -\left(\frac{r - r_S}{r}\right)(d\bar{t} + dr)\left(d\bar{t} - \frac{r + r_S}{r - r_S}dr\right) + r^2 d\Omega^2 \tag{2}$$

where we have defined $d\bar{t} \equiv dt + \frac{r_S}{r - r_S}dr$. (We could easily integrate this to determine $\bar{t}(t, r)$ up to an irrelevant additive constant, but it is not needed for our purposes here.) Note that the angular part of the metric is just going along for the ride.

We now show that the coordinates $(\bar{t}, r)$ are more suitable than (t, r) for describing outgoing and incoming radial light rays. In the $(\bar{t}, r)$ plane, the radial light rays follow $d\bar{t} + dr = 0$ (incoming, since $dr < 0$ for $d\bar{t} > 0$) or $d\bar{t} = \frac{r + r_S}{r - r_S}dr$ (outgoing for $r > r_S$, since then $dr > 0$ for $d\bar{t} > 0$). Thus, the incoming light rays always move at 45°. In contrast, the angle the outgoing light ray makes with the r-axis varies with r, starting out at 45° for $r \gg r_S$ far from the black hole, slowly increasing with decreasing r until it reaches 90° at the horizon, as shown in figure 2.

The light cone gradually tilts over. Once $r < r_S$, the outgoing light ray no longer deserves the name "outgoing": the relation $d\bar{t} = \frac{r + r_S}{r - r_S}dr = -\frac{r + r_S}{r_S - r}dr$ now implies that dr and $d\bar{t}$ have opposite signs, so that as $\bar{t}$ increases, r decreases. See figure 2.

The extraordinary feature is that for $r < r_S$, we no longer have any outgoing light rays!

All light rays are ingoing. A fortiori, material particles cannot escape, since their worldlines have to lie inside the light cone. Inside the horizon, the tilting light cone appears to "spill" the material particles contained inside toward $r = 0$.

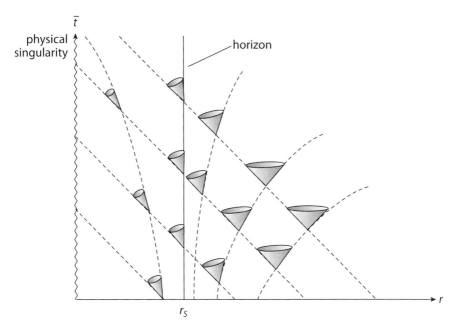

Figure 2 In the $(\bar{t}, r)$ plane, the light cone gradually tilts over. Inside the horizon, the tilting light cone appears to "spill" the material particles contained inside toward $r = 0$.

As already remarked, invariant measures of spacetime curvature behave smoothly at the horizon, and spacetime appears to be perfectly normal. What changes at the horizon is the causal structure of spacetime, as we will see in more detail shortly. The $(\bar{t}, r)$ coordinates make clear that inside the horizon, light rays and particles can perfectly well reach $r = 0$; the closing up of the light cone in the (t, r) plane as $r \to r_S^+$ merely shows the inadequacy of t as a coordinate.

Eternal versus actual black holes

Our description here is as if somebody manufactured a black hole a really long time ago, somehow, and placed it at $r = 0$ at the beginning of time $t = -\infty$. But the universe started in a Big Bang. Thus, as a description of a black hole, the Schwarzschild solution in its entirety, including the region inside the horizon, represents a mathematical curiosity, not an actual physical situation. In contrast to this so-called eternal black hole, an actual physical black hole represents a possible final state of stellar evolution. (If you don't know this, I will touch upon this fascinating story[1] briefly in chapter VII.4.)

A realistic description of black hole formation can be enormously complicated, but in theoretical physics, the process is often idealized as a spherically symmetric cloud of dust collapsing. The technical term "dust" refers to a collection of particles, each following a geodesic, that do not interact with one another directly. The worldline of a particle on the surface of the dust ball is indicated schematically in figure 3a, which shows the interior

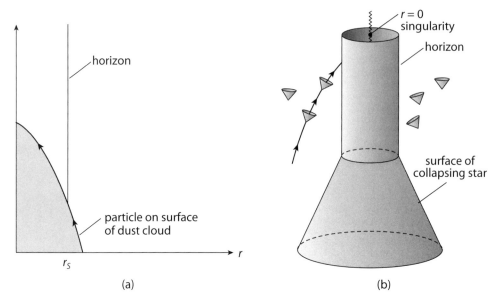

Figure 3 A dust ball collapsing into a black hole. (a) The worldline of a particle on the surface of the dust ball; the interior of the dust ball as a shaded region as shown in the $(\bar{t}, r)$ plane. (b) The formation of a black hole, with the angle θ suppressed.

of the dust ball as a shaded region. (Strictly speaking, inside the dust ball, the $\bar{t}$ coordinate may be inappropriate and so should be used only after the formation of the black hole.) As usual, this depiction in the $(\bar{t}, r)$ plane is $(1 + 1)$-dimensional, with θ and φ suppressed. In figure 3b, we show the same process in a $(2 + 1)$-dimensional depiction, with θ suppressed.

An escape attempt that barely failed

Imagine that after the formation of the black hole, a spherical shell of matter centered at the origin comes crashing into the black hole. In other words, the dust ball was actually enclosed by a spherical shell of dust. The collapse of the shell increases the mass of the black hole from M to $M + \Delta M$, with a corresponding increase of the Schwarzschild radius from r_S to $r_S + \Delta r_S$. See figure 4. The original horizon and the new horizon are indicated by the dashed and solid lines, respectively.

Now consider a light ray emitted from inside the dust cloud, thinking to itself, "Phew, I'm going to escape from this black hole!" but then just barely getting trapped by the more massive black hole. This story of a barely failed escape attempt makes clear that the horizon should be thought of as a surface formed by light rays in the $(\bar{t}, r)$ plane moving "vertically," that is, at an angle of $90°$ with the r-axis, in other words, moving along the line $r = r_S$. It is a null surface (as first defined in chapter III.3).

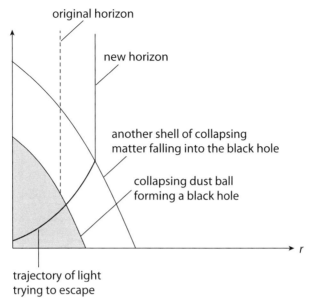

original horizon

new horizon

another shell of collapsing
matter falling into the black hole

collapsing dust ball
forming a black hole

r

trajectory of light
trying to escape

Figure 4 After the formation of a black hole, a spherical shell of
matter centered at the origin comes crashing in, thus forming a
more massive black hole. A light ray emitted from inside the dust
cloud that would have escaped from the less massive black hole is
now trapped by the more massive one.

Light rays moving at 45°

In the $(\bar{t}, r)$ coordinates, ingoing radial light rays always move at 45° from the vertical,
suggesting to us that it might be nice to have outgoing radial light rays also move at 45° from
the vertical. Instead of this light cone that tilts as we approach the horizon, we would have a
fixed light cone, just as in Minkowskian spacetime. Kruskal[2] and Szekeres independently
found[3] the desired coordinates. In my experience, students are often confused, and so I
will go at a perhaps excruciatingly slow pace.

Once again, look at the second line in (2): $ds^2 = -\left(\frac{r-r_S}{r}\right)\left(dt + \frac{r}{r-r_S}dr\right)\left(dt - \frac{r}{r-r_S}dr\right) +$
$r^2d\Omega^2$, practically begging us to define

$$dp = dt + \frac{r}{r - r_S}dr \quad \text{and} \quad dq = dt - \frac{r}{r - r_S}dr \tag{3}$$

Then

$$ds^2 = -\left(\frac{r - r_S}{r}\right)dpdq + r^2d\Omega^2 \tag{4}$$

with r to be regarded as a function of p and q. Indeed, $d(p + q) = 2dt$ and
$d(p - q) = \frac{2r}{r-r_S}dr = 2\left(1 + \frac{r_S}{r-r_S}\right)dr$, so that $p + q = 2t$ and $p - q = 2r + 2r_S \log \frac{|r-r_S|}{r_S}$,

with a convenient choice of integration constants. Note the need for the absolute value: the integral of $1/x$ is $\log |x|$, not $\log x$!

For $r \gg r_S$, we recover Minkowski spacetime, of course, with $dp, \, dq \to dt \pm dr$.

This suggests yet another change of coordinates: $P = e^{p/2r_S}$ and $Q = -e^{-q/2r_S}$, so that (4) becomes

$$ds^2 = -\frac{4r_S^3}{r} e^{-r/r_S} \operatorname{sign} (r - r_S) \, dP dQ + r^2 d\Omega^2 \tag{5}$$

The appearance of the sign function stems from the appearance of the absolute value. The only singularity is now at $r = 0$, which we know to be physical. (Note that while p and q have the same dimension as r, the coordinates (P, Q) are dimensionless.)

Had we sloppily neglected the absolute value in integrating $\frac{dr}{r-r_S}$ and thus omitted the sign function in (5), we would have been tempted to write $V = \frac{1}{2}(P + Q)$, $U = \frac{1}{2}(P - Q)$, so that $dV^2 - dU^2 = dP dQ$. But being careful, we see that if we want to have the nice form

$$ds^2 = -\frac{4r_S^3}{r} e^{-r/r_S} \left(dV^2 - dU^2 \right) + r^2 d\Omega^2 \tag{6}$$

we have to require $dV^2 - dU^2 = \operatorname{sign}(r - r_S) dP dQ$ with the sign function. We will determine V, U in a minute, but for now, let's admire (6).

Radial light rays are determined by $dU = \pm dV$. We have accomplished our goal of having light rays move always at $45°$ from the vertical, so that material particles always move at less than $45°$ from the vertical. Note that V is always the timelike coordinate; none of this funny business that t sometimes denotes a timelike coordinate and sometimes a spacelike coordinate.

Most of all, we see that, ta dah, the metric is not singular at all[4] at the horizon $r = r_S$, but still singular at the origin $r = 0$, as it should be.

But as some of you know, and may even know very well, there is no free lunch. The coordinate singularity at the horizon cannot simply vanish into thin air. Where is it? The answer will be revealed in appendix 4 if you can't figure it out in the mean time.

Kruskal-Szekeres coordinates

The requirement $dV^2 - dU^2 = \operatorname{sign}(r - r_S) dP dQ$ indicates that we should define, for $r > r_S$, $V = \frac{1}{2}(P + Q)$, $U = \frac{1}{2}(P - Q)$, and for $r < r_S$, $V = \frac{1}{2}(P - Q)$, $U = \frac{1}{2}(P + Q)$. Note the interchange between V and U outside and inside the horizon.

It is now simple and straightforward to determine V and U in terms of t and r in the two regions. First,

$$V^2 - U^2 = \operatorname{sign}(r - r_S) PQ = -\operatorname{sign}(r - r_S) e^{(p-q)/2r_S}$$
$$= -\operatorname{sign}(r - r_S)(|r - r_S|/r_S) e^{r/r_S} = \left(1 - \frac{r}{r_S} \right) e^{r/r_S} \tag{7}$$

The sign function disappears. (Also, note the factor $(1 - \frac{r}{r_S}) = (r_S - r)/r_S$, not $(1 - \frac{r_S}{r}) = (r - r_S)/r$ as in the Schwarzschild metric!)

Second, for $r > r_S$, $V/U = (P + Q)/(P - Q) = (e^{(p+q)/2r_S} - 1)/(e^{(p+q)/2r_S} + 1) = \tanh \frac{t}{2r_S}$, while for $r < r_S$, $V/U = (P - Q)/(P + Q) = \coth \frac{t}{2r_S}$. Evidently, the relation of the Kruskal-Szekeres coordinates (V, U) and the Schwarzschild coordinates (t, r) depends on the sign of $(r - r_S)$.

Outside the horizon, that is, for $r > r_S$, since $V/U = \tanh \frac{t}{2r_S}$, we have

$$V = \left(\frac{r}{r_S} - 1\right)^{1/2} e^{r/2r_S} \sinh \left(\frac{t}{2r_S}\right), \qquad U = \left(\frac{r}{r_S} - 1\right)^{1/2} e^{r/2r_S} \cosh \left(\frac{t}{2r_S}\right) \tag{8}$$

Inside the horizon, that is, for $r < r_S$, since $V/U = \coth \frac{t}{2r_S}$, we have

$$V = \left(1 - \frac{r}{r_S}\right)^{1/2} e^{r/2r_S} \cosh \left(\frac{t}{2r_S}\right), \qquad U = \left(1 - \frac{r}{r_S}\right)^{1/2} e^{r/2r_S} \sinh \left(\frac{t}{2r_S}\right) \tag{9}$$

Note that the factor $(\frac{r}{r_S} - 1)^{1/2}$ in one region and $(1 - \frac{r}{r_S})^{1/2}$ in the other are both real, otherwise (8) and (9) would not make sense.

Kruskal-Szekeres diagram of the Schwarzschild black hole

We can now describe spacetime around a black hole using (V, U) coordinates. See figure 5, known as a Kruskal-Szekeres diagram.

Lines of constant t correspond to straight lines with some fixed slope as given by

$$V/U = \Theta \left(r - r_S\right) \tanh \frac{t}{2r_S} + \Theta \left(r_S - r\right) \coth \frac{t}{2r_S}$$

as plotted in figure 5a. (The step function is defined as usual by $\Theta(x) = 1$ for $x > 0$ and $\Theta(x) = 0$ for $x < 0$.)

From (7), we see that the lines of constant r correspond to hyperbolas in the (V, U) plane, "vertically oriented" hyperbolas for $r > r_S$ and "horizontally oriented" hyperbolas for $r < r_S$. In particular, the physical singularity at $r = 0$ corresponds to the horizontally oriented hyperbola $V = +\sqrt{U^2 + 1}$, with the plus sign mandated by (9). The horizon at $r = r_S = 2M$ degenerates into the two straight lines $V = \pm U$, as we could have deduced from the fact that at the horizon, vertically oriented hyperbolas transition into horizontally oriented hyperbolas.

From the Kruskal-Szekeres diagram, a small puzzle that might have bothered you in the radial plunge discussion in the preceding chapter, namely what happens to t after the observer passes the horizon, resolves itself. As he approaches the horizon, t increases, reaching ∞ when he reaches the horizon, and after he passes the horizon, t decreases from ∞. By now you know better than to claim, as some crackpots do, that he gets to live his life backward: it is proper time that counts.

The moral of the story is that no single coordinate system is perfect. You would not want to calculate the perihelion shift of Mercury using the (V, U) coordinates; the (t, r) coordinates are clearly superior.

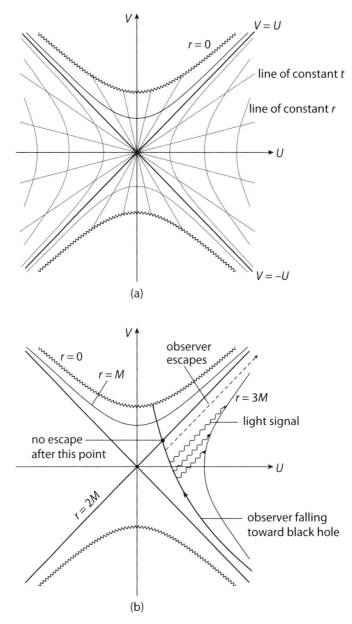

Figure 5 Kruskal-Szekeres diagram of the Schwarzschild black hole. (a) Note the lines of constant t and r. (b) As indicated by the dashed line, an observer falling into a black hole can escape before reaching the horizon $V = U$.

A common misconception alert! The lines of constant r are not geodesics. To hover at constant r outside a black hole requires constant and careful firing of a rocket pack strapped to your back. (Keep in mind also that the angular coordinates θ, φ are suppressed, and so each point in figure 5 corresponds to a unit sphere.)

However, the observer indulging in the extreme sport "radial plunge" is following a geodesic, as shown in figure 5b. Note that the angle the curve makes with the vertical has to be less than 45° at all points along the curve. As you can see, his worldline will eventually end at the hyperbola $V = +\sqrt{U^2 + 1}$, where infinite tidal forces await him.

Observers in theoretical physics, however, are endowed with free will. Dear reader, as a "young observer," you could always strap a rocket pack on your back and fire it whenever you wish. As indicated by the dashed line in figure 5b, you can always escape from the black hole by firing your rocket pack before you reach the horizon $V = U$. But once you pass the horizon, then no amount of firing would allow you to come back out. Indeed, even light traveling along the 45° lines ($V = U$ + positive constant) will eventually also end up (as indicated by the dotted line) at the physical singularity (as indicated by the jagged line), so that you can no longer send signals to your friends outside the black hole.

The gravitational time dilation discussed earlier also becomes clear pictorially. The infalling observer sends off signals at regular proper time intervals, as indicated by the wavy lines in the figure. As you can see, for the observer hovering at some constant r outside the black hole, these signals arrive with ever increasing intervals between them. It is worth emphasizing again that, once fallen through the horizon, the observer is not in any way obliged to follow a geodesic. He could certainly fire his rocket pack and zip off this way or that, frolicking inside a black hole, as long as his worldline makes an angle of less than 45° with the vertical.

The Kruskal-Szekeres diagram is drawn for an eternal black hole. For an actual physical black hole, the Kruskal-Szekeres diagram is physically relevant only to the right of the solid line, which could also be taken to depict the geodesic of a particle on the surface of the collapsing star.

Penrose diagrams

As this discussion makes clear, it is really advantageous to have radial light rays always move along 45° lines. To this, Roger Penrose added another attractive feature of having the range for the coordinates be finite. The resulting spacetime diagram is known[5] as a Penrose diagram and is extraordinarily useful for seeing the causal structure of the spacetime.

To see how this works, consider the easiest case of Minkowski spacetime $ds^2 = -dt^2 + dr^2 + r^2 d\Omega^2$. Write $p = t + r$, $q = t - r$ (namely, go to the light cone coordinates mentioned in chapter III.3). Then $ds^2 = -dt^2 + dr^2 + r^2 d\Omega^2 = -dpdq + r^2 d\Omega^2$. (Note that our convention is consistent with the $r_S \to 0$ limit of what we had in (3) and (4).) Since t ranges over $(-\infty, \infty)$ and r over $(0, \infty)$, Minkowski spacetime covers the half plane $r > 0$, each point of which corresponds to a sphere described by the suppressed angular variables θ and φ. Lines of constant p correspond to $t = -r + p$, and of constant q to $t = r + q$; the (p, q) coordinates are just the (t, r) coordinates rotated by 45°. Note that, since $p - q = 2r > 0$, the region ($p < 0, q > 0$) is not allowed. The half plane is divided into three regions, with ($p > 0, q > 0$), ($p > 0, q < 0$), and ($p < 0, q < 0$), separated by the two lines defined by $q = 0$ and $p = 0$, respectively.

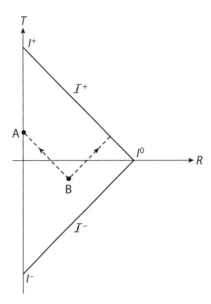

Figure 6 Penrose diagram of Minkowskian spacetime: the future and past null infinities are denoted by $\mathcal{I}^+$ and $\mathcal{I}^-$, respectively, the future and past timelike infinities by I^+ and I^-, respectively, and the spacelike infinity by I^0. Causal relationships can now be determined at a glance. For example, if the observer at B wants to send a message to an observer at rest at $r = 0$, the earliest the message could reach her would be at point A.

Since (p, q) range over $(-\infty, \infty)$, we could compactify them by a simple change of variable $p = \tan P$, $q = \tan Q$, so that (P, Q) range over the finite range $(-\frac{\pi}{2}, \frac{\pi}{2})$. Again, spacetime consists of a triangle bounded by the three straight lines $P = \pi/2$, $Q = -\pi/2$, and $P = Q$, divided into three regions, with $(P > 0, Q > 0)$, $(P > 0, Q < 0)$, and $(P < 0, Q < 0)$. Finally, we can "rotate back" by writing $T = P + Q$, $R = P - Q$. The resulting diagram is shown in figure 6. Minkowskian spacetime is represented by a triangle bounded by the three straight lines $T = \pi - R$, $T = -\pi + R$, and $R = 0$.

Light rays (or null lines) propagate at 45°, and so the future and past light cones are easily drawn. (For example, the future light cone of the observer at B is indicated by the dashed lines.)

As indicated in the figure, it is customary to denote the future and past null infinities by $\mathcal{I}^+$ and $\mathcal{I}^-$, respectively,* where null lines originate and end up; the future and past timelike infinities by I^+ and I^-, respectively; and the spacelike infinity by I^0. Keep in mind that the angular coordinates have been suppressed, so that I^0 actually represents "the sphere at infinity" often mentioned in physics.

---

* In case some people get bent out of shape, I might mention that for null infinity, I use the "calligraphic I" $= \mathcal{I}$ (following, for example, S. Hawking and R. Penrose,[6] while many authors use some version of the "calligraphic J" $= \mathcal{J}$ (as defined by the T$_E$X typesetting program). It is a trivial distinction, dependent merely on who your teacher was in penmanship class.

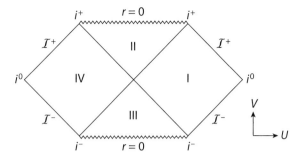

Figure 7 Penrose diagram of the Schwarzschild black hole.

Note that the specific "compacting" function tan used to relate a variable (P or Q, for example) with a finite range to a variable (p or q, for example) with an infinite range hardly matters in the present context. It is the causal structure of spacetime that we are after. For Minkowski spacetime, evidently, every timelike worldline will end up at $\mathcal{I}^+$ (excepting one line that ends up at I^+). Causal relationships can now be determined at a glance. For example, if the observer at B wants to send a message to an observer at rest at $r = 0$, the earliest the message could reach her would be at point A. By the same token, if the observer at B wants to get to $r = 0$, there is no way he could get there before point A.

We will have a bit more to say about Minkowski spacetime in appendix 5.

Now it is easy to draw the Penrose diagram for Schwarzschild spacetime: we simply bring the various infinities in, just as we brought the various infinities in Minkowski spacetime in. The result is shown in figure 7. If you want, you could work through the arithmetic following the same procedure as for Minkowski spacetime, rotate by 45°, compactify variables, and then rotate back by 45°, but there is no point in doing this.

Sewing spacetimes together

I now describe the formation of a black hole under the simplest circumstances that theoretical physicists have come up with. The black hole is formed by the collapse of a thin spherical shell of photons in Minkowski spacetime, all moving radially inward toward a point. The description of the formation process involves "sewing" two distinct spacetimes together. Roughly speaking, before the collapse of the shell of photons, spacetime is Minkowski, while after the collapse, spacetime becomes Schwarzschild. Thus, we have to join two distinct spacetimes together.

Let us start by envisaging the thin spherical shell of photons (a) in a Minkowski space-time, and (b) in an eternal Schwarzschild spacetime. These scenarios are depicted in figure 8a and figure 8b, respectively, with the shell represented by a double line. These two figure panels represent situations in which the shell contains very little energy and has a negligible effect on the existing spacetimes, Minkowski in one case and Schwarzschild in the other.

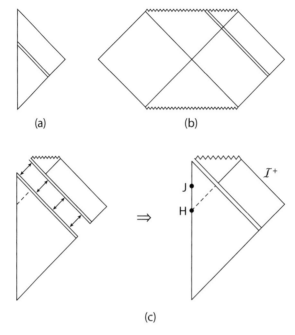

(a)   (b)

⇒

(c)

Figure 8 A thin spherical shell of photons, represented by a double line, (a) in a Minkowski spacetime, and (b) in an eternal Schwarzschild spacetime. (c) After excising the physically irrelevant regions from the spacetime in panels a and b, we sew together the two physically relevant regions to form one single spacetime. Before reaching point H, you could still reach the future timelike infinity. But after point H, your fate eventually is to meet the singularity represented by the jagged line.

But now suppose that the shell of light contains sufficient energy to form a black hole. Then the Minkowski spacetime above and to the right of the double line in figure 8a is no longer relevant: a black hole has formed! That region of Minkowski spacetime should be excised.

The situation is quite different in figure 8b. We are trying to describe the formation of a black hole due to an incoming shell of light. Before the shell arrives, spacetime is supposed to be Minkowskian. Thus, the region below and to the left of the double line in figure 8b is physically irrelevant and should be excised. Also, we don't think eternal black holes exist physically, so in any case, the spacetime depicted in figure 8b should not be taken in its entirety.

So, we cut off the parts of figure 8a and figure 8b that are irrelevant. Let us then "sew together" the two physically relevant regions left in these two figures to form one single spacetime, as shown in figure 8c. Before the shell arrives, we have flat spacetime, and afterward, a Schwarzschild spacetime containing a black hole. Thus, this cutting and sewing construction provides us with a spacetime description of an idealized black hole formation process. The resulting Penrose diagram shows a black hole forming in an initially flat spacetime, causing a horizon (indicated by the solid line in the Schwarzschild portion of the composite spacetime, continued as a dotted line into the Minkowski spacetime before the shell of light arrives) and a physical singularity (indicated by the jagged line) to form.[7] Note that all angles in this figure are at either 45° or 90°.

This spacetime diagram tells an intriguing story. Suppose you are living a contented life at the origin (indicated by the vertical line in the figure, which serves to represent your worldline) of a Minkowski spacetime. You have no idea that a monster shell of light is coming at you at light speed. Before point H, you could still blast off and reach the future timelike infinity denoted by $\mathcal{I}^+$, if your rocket were fast enough. But after point H, you are totally doomed: your fate is to meet the jagged line sooner or later.

This description indicates that the horizon is a global, not a local, concept characterizing the causal structure. At point J, after you pass H, no signal from the monster shell of light can have reached you yet, and you could well be minding your own business. But already it is too late for you! No matter what you do, you would still be headed toward the physical singularity.

Let me emphasize again that the coordinate independent measure of curvature $R^{\mu\nu\rho\sigma}R_{\mu\nu\rho\sigma} = 12r_S^2/r^6$ evaluated at the Schwarzschild radius r_S goes like $1/r_S^4$. Thus, by taking $r_S = 2M$ large, we can make this measure of curvature as small as we like, so that spacetime around a black hole could appear to be arbitrarily close to everyday flat spacetime. Nevertheless, the global causal structure of spacetime has been changed essentially and irrevocably.

Appendix 1: Eddington-Finkelstein coordinates

I describe here a coordinate system first used by Eddington in 1924, and then rediscovered by Finkelstein in 1958. Speaking loosely, we can think of the Eddington-Finkelstein coordinates as something halfway between the Schwarzschild and the Kruskal-Szekeres coordinates. Go back to the second line in (2), which I reproduce here for convenience:

$$ds^2 = -\left(\frac{r-r_S}{r}\right)\left(dt + \frac{r}{r-r_S}dr\right)\left(dt - \frac{r}{r-r_S}dr\right) + r^2 d\Omega$$

Define $dp = dt + \frac{r}{r-r_S}dr$ as before but not dq. In terms of this new coordinate, we have $dt - \frac{r}{r-r_S}dr = dp - \frac{2r}{r-r_S}dr$. Thus,

$$ds^2 = -\left(\frac{r-r_S}{r}\right)dp^2 + 2dpdr + r^2 d\Omega \tag{10}$$

Radial light rays follow paths determined by solving $\left(\frac{r-r_S}{r}\right)dp^2 = 2dpdr$. Light rays along $dp = 0$ are always ingoing. Light rays along $\left(\frac{r-r_S}{r}\right)dp = 2dr$ are outgoing for $r > r_S$ and ingoing for $r < r_S$.

Recall that $d\bar{t} \equiv dt + \frac{r_S}{r-r_S}dr$. Be careful to distinguish $d\bar{t}$ from dp! In fact, $dp = d\bar{t} + dr$, so that the $dp = 0$ light rays are just the $d\bar{t} + dr = 0$ light rays discussed earlier. We can also integrate $dp = d\bar{t} + dr = dt + \left(1 + \frac{r_S}{r-r_S}\right)dr$ to obtain $p = \bar{t} + r = t + r + r_S \log\frac{|r-r_S|}{r_S}$.

Appendix 2: Area of the horizon

We ask Confusio, "What is the dimension of the horizon of a black hole?" He responds, "Let's see. Set $r = r_S + \varepsilon$ in the Schwarzschild solution (1) to get $ds^2 \simeq \frac{\varepsilon}{r_S}dt^2 - r_S^2 d\Omega^2$ (since $dr = 0$). Sure looks like it's $2 + 1$ dimensional."

Good. As we've noticed, Confusio is getting less confused by the day. Indeed, most people think of the horizon as a mathematical 2-sphere surrounding the black hole. You add time and it's $2+1$ dimensional for sure.

But now set $\varepsilon = 0$. Time goes. Not just time goes by, but time literally goes. We are left with $-ds^2 = r_S^2 d\Omega^2$; the actual horizon, contrary to what the naive might think, is actually 2-dimensional and has an area

$$A = 4\pi r_S^2 = 16\pi (GM)^2 \tag{11}$$

That this 2-step discussion is even necessary points to the deficiency of the Schwarzschild coordinates. If we set $r = r_S$ in the (p, q) coordinates (4), or $V = U$ in the Kruskal-Szekeres coordinates (6), or $r = r_S$ in the Eddington-Finkelstein coordinates (10), we obtain immediately $-ds^2 = r_S^2 d\Omega^2$ and the area law (11).

Appendix 3: Misleading to show the black hole as a funnel or as a rubber sheet

You've probably seen a picture of a black hole depicted as a kind of funnel, or alternatively as a rubber sheet depressed by a heavy round mass. Far away from the funnel, or the depression in the rubber sheet, the surface is supposed to be flat. I will pointedly not show this picture (you can draw it yourself based on the mathematical description given below), but it and its variants have appeared in countless magazines, newspapers, popular books, and even on the cover of a textbook. In many science museums, visitors are invited to toss a small ball onto the surface of an actual funnel shaped construction. If you toss the ball with sufficient speed in an angular direction, it will orbit around the central funnel, slowly spiraling into the dark "bottomless" pit in the center. And of course, if you toss the ball in the radial direction, it will fall right in, "sucked in by the irresistible force" of the black hole, often thought of as a "source of evil" in the visitor's mind. You know of course that this display depicts the sun equally well.

This museum display entertains the visitors and educates them to some extent, but D. Marolf has pointed out that it is misleading at best. For sure, it has seriously confused some students.

This popular picture and the display that goes with it are obtained by setting t equal to some constant and $\theta = \pi/2$ in the Schwarzschild metric (1) to obtain

$$ds^2 = \frac{1}{\left(1 - \frac{r_S}{r}\right)} dr^2 + r^2 d\varphi^2 \tag{12}$$

which is then embedded in 3-dimensional Euclidean space E^3. (The museum staff could hardly do otherwise.) Using the usual cylindrical coordinates (z, ρ, φ) for E^3, we specify the embedding by writing $z = f(r)$, $\rho = r$, $\varphi = \varphi$. You can work out $f(r)$ if you want, but it is not necessary here. In science museums, they don't use the actual $f(r)$, but instead, use an $f(r)$ such that $f(r) \to$ constant for large r and $f'(a) = -\infty$ for some small value of a. So you see why I don't need to draw a picture for you!

Marolf's point is that this picture represents a slice in time and is not directly connected to the gravitational attraction of the black hole. (The actual force "sucking" the ball into the funnel is of course supplied externally, by the earth.) In fact, there are spacetimes with the same $t - \varphi$ slice as (12) but with totally different gravitational fields as in the Schwarzschild case (as you saw in exercises V.4.4 and V.4.5).

To obtain a more appropriate representation of the black hole, we should take a slice of (1) with θ and φ both constant (in contrast to the funnel picture based on a slice with t and θ constant) and then embed the slice in $(2+1)$-dimensional Minkowski spacetime $M^{2,1}$. The resulting picture contains two flanges.[8]

Appendix 4: Wormholes and such

Confusio grumbles, "Did we not cheat? I can see that the metric in the Kruskal-Szekeres coordinates

$$ds^2 = -\frac{4r_S^3}{r} e^{-r/r_S} \left(dV^2 - dU^2\right) + r^2 d\Omega^2 \tag{13}$$

[reproduced here for convenience] is free of the coordinate singularity at $r = r_S$, but the transformation (8) and (9) from (t, r) to (V, U) is not smooth as we cross over $r = r_S$ (that is, dV/dr, dU/dr are singular at $r = r_S$)."

Indeed, Confusio is right, but that's just the law of calculus: if we transform from singular to nonsingular coordinates, the transformation necessarily must be singular. Think about it this way. Imagine a civilization far far away in which the metric (13) just fell on the head of a physicist who had never heard of the usual Schwarzschild metric written in (t, r) coordinates. Or perhaps more likely, a mathematician simply presented it to a physicist, saying, "Lo, behold this metric: it solves Einstein's field equation in empty spacetime." (The symbol r, defined earlier in (7) by $\left(1 - \frac{r}{r_S}\right)e^{r/r_S} = V^2 - U^2$, is a perfectly well-defined function of $V^2 - U^2$ for $0 \leq V^2 - U^2 \leq 1$.)

What we should ask is how the original spacetime maps into the Kruskal spacetime. Indeed, (8) shows that every point (t, r), for $\infty > r > r_S$ and $\infty > t > -\infty$, maps to a unique point (V, U) in quadrant I in figure 7, and (9) shows that every point (t, r), for $r_S > r > 0$ and $\infty > t > -\infty$, maps to a unique point (V, U) in quadrant II. So the original spacetime maps only into half of the Kruskal spacetime, namely the half defined by $V \geq -U$. The crucial question is, what do we make of quadrants III and IV? They do not correspond to anything in the original spacetime. For example, take the negative U-axis defined by $V = 0$, $U < 0$. Examining (9), we see that this does not exist in the original spacetime. Mathematicians say that the Kruskal coordinates define an extension of the Schwarzschild solution.

Physicists, including astrophysicists, typically take the attitude that for an actual black hole formed from a collapsing star, a solution to Einstein's field equation in empty spacetime is relevant only to the right of the worldline of a massive particle in figure 5b. In other words, we now think of what we previously called the worldline of an observer falling into an eternal black hole as the worldline of a particle on the surface of a collapsing star. Quadrants III and IV in figure 7 are physically irrelevant (at least until further discoveries in physics*).

In contrast, mathematicians and speculators can certainly invite themselves (free country, remember?) to study the spacetime described mathematically by (13). Looking at (13), we see that, since quadrant III is related to I by $U \to -U$, the two quadrants are the same: III also describes the outside of a black hole, approaching an asymptotically flat spacetime. Quadrant IV is more peculiar, with a physical singularity at $V = -\sqrt{U^2 + 1}$. Classical general relativity cannot tell us anything about what actually happens at a physical singularity except that the theory breaks down. We need a theory of quantum gravity. Naively, we can draw lines coming out of this physical singularity with light at 45° and so on, capable of propagating into I and III, so that it looks like what we might call a "white hole," whatever that means—a place where particles could come streaming out. Clearly, our present understanding of physics does not allow us to say anything meaningful, which of course does not deter people from publishing any number of speculative papers.

It is somewhat interesting to look at the $V = 0$ slice connecting the two asymptotically flat spacetimes I and III and described by the 3-dimensional line element $ds^2 = \frac{4r_S^3}{r}e^{-r/r_S}dU^2 + r^2 d\Omega^2$. By (8), $V = 0$ implies that $U = \left(\frac{r}{r_S} - 1\right)^{1/2}e^{r/2r_S}$ and $dU = \frac{r}{2r_S^2(\frac{r}{r_S}-1)^{1/2}}e^{r/2r_S}dr$. Inserting into (13), we see that this spatial slice is described by

$$ds^2 = \frac{1}{1 - \frac{r_S}{r}}dr^2 + r^2 d\Omega^2 \tag{14}$$

The reader with a long memory would recognize this as the Einstein-Rosen bridge discussed way back in chapter I.6, which John Wheeler[9] picturesquely described as a wormhole.

The question naturally arises, if an eternal black hole exists somewhere, whether one could get through the wormhole to another asymptotically flat universe. Inspection of the Penrose diagram in figure 8 shows that it is not possible to get from I to III even if you were to travel at the speed of light. However, an observer starting in I, after falling through the horizon into II, could receive signals originating from within III. In other words, our intrepid observer, while unable to get to III, can look at part of III.

An important point is that while the Einstein-Rosen bridge can be studied as a static 3-dimensional space, the wormhole is a dynamic entity evolving in time. Indeed, let's take the $V = V_0 > 0$ slice of the Kruskal spacetime. (Recall that V is the timelike variable.) From (7), we have $U = \pm\sqrt{V_0^2 - \left(1 - \frac{r}{r_S}\right)e^{r/r_S}}$. Substituting this and $V = V_0$ into (13), we find

$$ds^2 = \frac{1}{1 - \frac{r_S}{r}\left(1 - V_0^2 e^{-r/r_S}\right)}dr^2 + r^2 d\Omega^2 \tag{15}$$

We see that the throat of the wormhole, determined by the value of r where $g_{rr} \to \infty$, decreases from r_S at $V_0 = 0$, approaching 0 as $V_0 \to 1$, and reaches 0 when we hit the physical singularity at $V_0 = 1$. The wormhole closes up at the physical singularity. By dimensional analysis, since r_S is the only dimensionful parameter around, we see that the wormhole closes up[†] on a time scale of the order of r_S.

---

* I subscribe to this attitude. What attitude you take is of course up to you.

† This classical analysis, like the rest of this chapter, completely ignores the increasingly large fluctuations due to quantum gravity as we approach the physical singularity.

Appendix 5: A bit more on Minkowski spacetime

In the text, we constructed the Penrose diagram for Minkowski spacetime as follows: rotate (by 45°), compactify, rotate back, that is, by the sequential changes of variables* $t = \frac{1}{2}(p+q)$, $r = \frac{1}{2}(p-q)$, $p = \tan P$, $q = \tan Q$, $T = P + Q$, $R = P - Q$. The metric is transformed as

$$
\begin{aligned}
ds^2 &= -dt^2 + dr^2 + r^2 d\Omega^2_{d-2} \\
&= -dp\,dq + \tfrac{1}{4}(p-q)^2 d\Omega^2_{d-2} \\
&= \frac{1}{4\cos^2 P \cos^2 Q} \left(-dT^2 + dR^2 + R^2 d\Omega^2_{d-2} \right)
\end{aligned}
\tag{16}
$$

As explained in the text and as shown in figure 6, spacetime consists of a triangle. (Indeed, the factor $(\cos^2 P \cos^2 Q)^{-1}$ indicates that the coordinates end at $P = \pi/2$, $Q = -\pi/2$.) We see that ds^2 is conformally related (see exercise I.5.14) to $d\tilde{s}^2 = -dT^2 + dR^2 + r^2 d\Omega^2_{d-2}$. Recall that R runs between 0 and π, and thus in spite of appearances, this spacetime, while conformally related to the flat Minkowski spacetime, is not flat. Space consists of the sphere S^{d-1}, with $R = 0$ and $R = \pi$ corresponding to the north and south poles, respectively. Indeed, we might want to rename R as θ, the familiar latitude. (Note that we have generalized slightly from the text to consider $M^{d-1,1}$ with no cost to us; as usual, the angular coordinates just go along for the ride.)

But if somebody handed us $d\tilde{s}^2$ with T and R restricted to the triangular region, we could invite ourselves to extend this spacetime outside the triangle: simply let T run from $-\infty$ to $+\infty$. Without the factor $(\cos^2 P \cos^2 Q)^{-1}$, we can wander outside the triangle in figure 6 with impunity. The resulting spacetime has the topology of $R \times S^{d-1}$ and is known as the maximal extension of the Minkowski spacetime we started out with, which now corresponds to a patch in this spacetime.

We now note that $(1+1)$-dimensional Minkowski spacetime is a special case: as we can see in (16), $d\Omega^2_{d-2}$ degenerates for $d = 2$. Instead of starting out with (t, r) with $0 \le r < \infty$, we have (t, x) with $-\infty < x < \infty$ and thus

$$
ds^2 = -dt^2 + dx^2 = -dp\,dq = \frac{1}{4\cos^2 P \cos^2 Q} \left(-dT^2 + dX^2 \right)
$$

Now the coordinate X ranges between $-\pi$ and π, in contrast to the coordinate R, which ranges between 0 and π in the higher dimensional cases. Instead of a triangle, spacetime now consists of a diamond shaped region, that is, a square rotated by 45°. (Another way of saying this is that spatial boundary S^0 is not connected and contains 2 points. In contrast, S^{d-2} is connected for $d > 2$.)

For the statement that the maximal extension of Minkowski spacetime $M^{d-1,1}$ has the topology of $R \times S^{d-1}$ to be applicable also for $d = 2$, we could identify the two points $(T = 0, X = \pi)$ and $(T = 0, X = -\pi)$. Then $M^{1,1}$ is maximally extended to $R \times S^1$, familiarly known as the cylinder. We will come back to this in chapter IX.11.

Exercise

1   Show that for the actual Schwarzschild solution, the science museums should use $f(r) = 2\sqrt{r_S(r - r_S)}$.

Notes

1. For readers utterly unfamiliar with the story, I recommend chapters 20–22 in R. Freedman and W. Kauffman, *The Universe*.
2. Kruskal was a distinguished plasma physicist. This episode in the history of physics reminds us that there is a huge gulf between "nonexperts" and crackpots.
3. In hindsight, it might seem somewhat surprising that the Kruskal coordinates were found only in 1960. Allegedly, John Wheeler had to compel a reluctant Martin Kruskal to publish his work by writing it up for

---

* Note the factors of 2.

him. M. Kruskal's paper (*Physical Review* 119 (1960), p. 1743) carries a note stating "This work was reported in abbreviated form by J. A. Wheeler on behalf of the author," suggesting that Wheeler did indeed force Kruskal to write it up.

4. In hindsight, it really is puzzling how the confusion over the nonexistent Schwarzschild singularity persisted for so long. Apparently, G. Lemaître (recall chapter V.3; see also chapter VIII.1) had already shown in 1933 that the "singularity" could be removed by a coordinate transformation. But his paper, published in a little-read Belgian journal, was roundly ignored. Later, in 1950, J. L. Synge also clarified the nature of this nonsingularity. See A. Gsponer, arXiv:physics/0408100. I understand that in the former Soviet Union, I. Novikov had long understood that there is only a coordinate singularity at the horizon.

5. Perhaps more accurately, as a Carter-Penrose diagram.

6. S. Hawking and R. Penrose, *The Nature of Space and Time*, Princeton University Press, 1996, pp. 42 and 43.

7. For a nice pedagogical treatment of how the horizon appears as a black hole forms, see the not terribly well-known work by R. Adler, J. Bjorken, P. S. Chen, and J. S. Liu, "Simple Analytical Models of Gravitational Collapse," *American Journal of Physics* 73 (2005), p. 1148.

8. See D. Marolf, arXiv:gr-qc/9806123 for more details.

9. As an undergraduate, I was a devotee of John Wheeler. In an article titled "John Wheeler's mentorship: An enduring legacy," *Physics Today* 62 (2009), p. 55, T. M. Christensen wrote, "Among the eminent physicists who were influenced as undergraduates by personal contact with Wheeler are James Hartle, David Sharp, Bruce Partridge, Anthony Zee, and Gary Horowitz." See also my letter in the same volume.

Quantum fluctuations can set you free

Nothing can get out of black holes. Spacetime is warped in such a way that, once inside the horizon, even light can never emerge, as the Kruskal diagram shows for a Schwarzschild black hole. We worked this picture out in detail in the preceding chapters.

But as you have no doubt heard, that picture, painted exclusively with classical physics, no longer holds true when quantum effects are turned on, as we have already discussed in the introduction to this text. Black holes radiate as black bodies, each with a temperature characteristic of the specific black hole. Indeed, we were even able to determine, purely by dimensional analysis supplemented by a bit of basic knowledge about gravity, that the Hawking temperature for a Schwarzschild black hole is given by

$$T_{\mathrm{H}} \sim \frac{1}{GM} \sim \frac{\hbar c^3}{GM} \tag{1}$$

with M the mass of the black hole. You may wish to go back to the introduction to review how we did that. There we already noted that this simply derived result indicates that black hole radiation ends explosively. As M decreases, T goes up.[1]

From (1), using the thermodynamic definition of entropy $dM = T dS$, we immediately determined the entropy to be

$$S \sim GM^2 \sim \frac{GM^2}{\hbar c^3} \tag{2}$$

I have used dimensional analysis to restore $\hbar$ in (1) and (2), thus showing clearly that for $\hbar = 0$, we have $T = 0$ and $S = \infty$, so that classically, black holes do not radiate.

We now try to understand how quantum effects could change the picture so drastically. At the most handwaving and heuristic level, with quantum fluctuations, a photon can no longer be sure which side of the horizon it is on, and thus there is some chance it could get out. Let's put substance on this basically correct explanation by learning about the phenomenon of the restless vacuum in quantum field theory.

The essence of quantum field theory in five minutes

What is quantum field theory? Why is quantum field theory necessary?

We need quantum field theory[2] when we confront simultaneously the two great physics innovations of the last century of the previous millennium: special relativity and quantum mechanics. Consider a rocket ship moving close to light speed. You need special relativity but not quantum mechanics to study its motion. In contrast, to study a slow electron scattering off of a proton, you must invoke quantum mechanics, but you don't have to know a thing about special relativity.

It is in the peculiar confluence of special relativity and quantum mechanics that a new set of phenomena arises: particles can be born and particles can die. It is this matter of birth, life, and death that requires the development of a new subject in physics, that of quantum field theory.* Let me explain presently how special relativity and quantum mechanics together can lead to dramatically novel physics.

Consider empty spacetime. The vacuum, which we normally think of as vacuous, is (according to quantum field theory) rather astonishingly actually a boiling sea of fluctuating pairs of particles and antiparticles, containing, for example, pairs consisting of an electron and a positron (the electron's antiparticle).

Before special relativity came along, you couldn't simply conjure up the mass of the electron and of the positron out of the vacuum. But with Einstein's gold-plated equation $E \sim mc^2$, you could if you have enough energy.

However, without quantum mechanics, the process is still forbidden by energy conservation. Where would the necessary energy come from?

The gold-plated equation of quantum mechanics, Heisenberg's uncertainty principle, $\Delta t \sim \hbar / \Delta E$, comes to the rescue. When Nature balances her accounts, she can tolerate briefly a certain amount of fuzziness.

Thus, in a world with both special relativity and quantum mechanics, an electron and a positron can pop out of the vacuum, but only for a characteristic time of at most[†] order $\Delta t \sim \hbar / (m_e c^2)$, a very small time interval by everyday standards, after which the electron and the positron must annihilate each other and disappear back into the vacuum. Quantum electrodynamics was invented partly to deal with this sort of vacuum fluctuation.

This heuristic discussion indicates that the fluctuations are universal and involve all particle species, including the photon and the graviton (which happen to be their own antiparticles). For massless particles such as the photon and the graviton, the denominator in the estimate for Δt should be replaced by their characteristic energies.[3]

---

* I remark in passing that you have already seen repeatedly, in chapters II.3, IV.1–IV.3, VI.4, and VI.5 for example, another need for quantum field theory when we write down actions describing the interaction between particles and the electromagnetic and gravitational fields. There was always an unbearable and unsightly dichotomy between point particles on one hand, and spacetime-pervading fields on the other hand. It would be intellectually more satisfying to treat all the elementary particles, the electron and all the rest, on the same footing as fields.

[†] Since the minimum energy the electron and the positron can have is of order m_e.

This boiling vacuum with all its agitation is irrelevant for a large chunk of physics. The reason is that the time scale Δt over which a fluctuation occurs is much shorter than the typical time scales explored in many areas of physics. In a collision between particles, however, the fluctuating pair can borrow the required energy from the colliding particles and thus evade the Heisenberg bound on Δt. The electron and positron pair does not have to annihilate each other but can escape to infinity. Thus, we can scatter an electron at high energy off a proton and produce an electron-positron pair, a process known as pair production in quantum field theory.

In contrast, write down the Schrödinger equation for an electron scattering off a proton. The equation describes the wave function of one electron, and no matter how you shake and bake the mathematics of the partial differential equation, you will always have one and only one electron. The Schrödinger equation is simply incapable of describing pair production. Nonrelativistic quantum mechanics must break down under these circumstances.

Not quantum field theory in five minutes, of course, but the essence of quantum field theory in five minutes!

Vacuum fluctuations near a black hole

Pairs of particles and antiparticles pop out of the vacuum for an instant and then vanish. These incessant but ephemeral fluctuations were largely of interest only to particle physicists until Hawking came along.[4]

But what if the fluctuations occur near the horizon of a black hole?

As we have learned, Riemann curvature near the horizon scales like $\sim 1/r_S^2 \propto 1/(GM)^2$, and spacetime can be almost flat for M large. Unlike particle physics, Einstein gravity is not normally concerned with high energies. But it's not the curvature, rather the causal structure, that matters!

At the horizon $r = 2GM$, the coefficients of dt^2 and dr^2 change sign, indicating that time and space, and hence energy and momentum, are interchanged. A pair pops out near the horizon. During the short time the pair can exist, one of them, say the antiparticle to be definite, could fall through the horizon, at which point its energy becomes a momentum component! The particle, liberated from the constraints of energy conservation and Heisenberg's principle, can now exist forever and escape to infinity, where it tries to live happily ever after without its partner.

In particle physics, colliding particles supply the energy needed to balance the books. In Einstein gravity, while Nature compulsively balances her energy budget, we fool her by dumping one of the particles of the pair down a black hole. The Heisenberg restriction on Δt is evaded by changing what we mean by energy as the particle crosses the horizon.

In a Kruskal diagram, you can easily depict this process, showing the antiparticle falling to its doom at $r = 0$ and the particle escaping to $\mathcal{I}^+$. To balance the energy momentum budget, the black hole would have to lose a bit of mass and recoil a little. For a black hole with mass M much greater than the typical energy of the escaping particle, these effects are negligible. These fluctuations occur ceaselessly around the horizon, and thus we conclude that the black hole radiates universally: all particles are involved.

A detailed quantum field theoretic calculation* should reveal to us the energy distribution of the radiated particles, which is precisely what Hawking did.

But even without doing the calculation, we can anticipate, if we are willing to play fast and loose,[†] what the distribution has to be: the only universal energy distribution known to physics is the Boltzmann distribution, and thus we expect that the probability for the radiated particle to have energy E is proportional to e^{-E/T_H} for some temperature T_H characteristic of the black hole.

The kind of discussion given here is clearly meant to be heuristic. One caveat: a photon in the Hawking radiation has characteristic energy $\omega \sim T_H \sim (GM)^{-1}$ and thus a wavelength $\lambda \sim GM$ comparable to the size of the black hole. The very concept of a particle may be a bit dicey.

Although many people believe that Hawking radiation provides a crucial clue to the eventual understanding of quantum gravity, it is worth emphasizing that the calculation leading to Hawking radiation does not involve quantum gravity as such. The role of the gravitational field is to provide a spacetime with a peculiar causal structure for the other fields to do their quantum fluctuating in. The Schwarzschild solution is still treated classically.[‡]

An important clue is provided by the black hole information paradox, first articulated by Hawking. Put at the most elementary level, the question is: what happened to the information contained in the material that fell in to form a black hole? Eventually, we end up with thermal radiation, which, according to standard considerations, does not contain any information at all. The paradox may be sharpened as follows. Consider an initial distribution of matter described by a pure state in quantum mechanics, which collapses to form a black hole. After we wait long enough, this evolves into a thermal state described by a density matrix in quantum mechanics. But quantum mechanical evolution is governed by a unitary operator, which cannot possibly turn a pure state into a thermal state. Thus, there appears to be a basic contradiction with quantum mechanics, hence a paradox. This

---

* At least with the benefit of hindsight, the calculation is not as difficult as you might think. Consider a black hole radiating electrons and positrons, to be specific. We are not interested in the interaction of the electron and positron with the electromagnetic field and with each other. In other words, we don't need a full blown mastery of quantum electrodynamics, but instead, we can treat the electron field as a free field propagating in the Schwarzschild metric, that is, free except for the influence of gravity.[5] Nevertheless, I choose not to do the calculation here, as it involves a number of concepts from quantum field theory. Instead, I give in appendix 1 a slick derivation of T_H, which may well turn out to be more profound than the actual nitty-gritty calculation. For those who want to go through an actual calculation, a good place to start is the paper by W. G. Unruh.[6]

[†] Objections to this kind of handwaving argument come readily to mind. Why are things in thermal equilibrium? The Boltzmann distribution presupposes some kind of heat bath. Where is it?

[‡] In the introduction in part 0, in discussing the cube of physics, I associated the corner with $G \neq 0$, $\hbar \neq 0$, and $c \neq 0$ with quantum gravity. I mentioned, in a footnote, a slight caveat to this statement. Here it is. While G, $\hbar$, and c all appear in (1) and (2), the calculation leading to them was done without quantizing gravity. Even if we were to include the radiation of gravitons, the gravitons could be treated as small fluctuations superposed on the classical gravitational field.

subject has a long and controversial history that cannot possibly be covered here; you are invited to trace back this history starting with the recent literature.[7] One possibility is that the naive view that, in a region in which the Riemann curvature tensor could be made arbitrarily small, physics would be indistinguishable from flat spacetime, may be wrong. The limit may turn out not to be smooth.

A semi-quantitative argument for Hawking radiation

Our friend the Smart Experimentalist suddenly speaks up, "Without knowing quantum field theory, we should still be able to make the heuristic argument about the infalling particle semi-quantitative. Think of an experimentalist at rest close to the horizon at $r = r_S + a$. She observes in her lab a particle-antiparticle pair popping out. The entire lab falls freely and crosses the horizon. The horizon is not marked by a line or anything; inside the lab is merely almost-flat, almost-empty spacetime."

Confusio catches on enthusiastically. "The smaller a, the sooner we cross the horizon, the shorter is Δt, and, so according to Heisenberg, the escaping particle could have a higher energy. Oops, it seems that the characteristic energy of the particle detected at infinity increases as a decreases."

SE smiles, "Confusio, you forgot the gravitational redshift! Recall that energy is redshifted down by a factor given by the square root of g_{00} evaluated at $r = (r_S + a)$, which almost by definition vanishes as we approach the horizon, as $a \to 0$."

Confusio is delighted. "Let's hope that the two effects cancel out."*

I say to both of them, "We will let the attentive reader find out if the a dependence does indeed cancel out." Challenge yourself. See exercise 1.

"The consequences of my crime echo down to the end of time"

One afternoon in 1970, . . . I told [Bekenstein] of the concern I always feel when a hot cup of tea exchanges heat energy with a cold cup of tea. By allowing that transfer of heat . . . I increase [the universe's] microscopic disorder, its information loss, its entropy. "The consequences of my crime, Jacob, echo down to the end of time," I noted. "But if a black hole swims by, and I drop the teacups into it, I conceal from all the world the evidence of my crime. How remarkable!" Bekenstein, a man of deep integrity, takes the lawfulness of creation as a matter of the utmost seriousness. Several months later he came back with a remarkable idea. "You don't destroy entropy when you drop those teacups into the black hole. The black hole already has entropy and you only increase it!"

—John Archibald Wheeler[8]

---

* This indicates that an observed photon in the Hawking radiation may have originated near the horizon with trans-Planckian energy—a fact that you may or may not find disturbing.

As this story told by Wheeler indicates, his student Jacob Bekenstein was the first to recognize that black holes have entropy. In fact, as (2) shows, in classical general relativity, not only does the Schwarzschild black hole have entropy, it also actually has an infinite amount of entropy. This makes sense, since entropy is the logarithm of the number[9] of microstates[10] that correspond to a single equilibrium macrostate, and we can make a Schwarzschild black hole of a given mass M in an infinite number of ways, by throwing any amount and any variation of stuff into it, provided that the total mass adds up to M.

For something as fundamental as the entropy of black holes, we politely decline to use ludicrous units, such as joules per degree centigrade, and so once again in this chapter, we go back to the introduction to this text and recall the other profound concept mentioned there, namely Planck's insight into measurement. Recall that we have three fundamental units to do physics with: the Planck mass $M_P = \sqrt{\frac{\hbar c}{G}}$, the Planck length $l_P = \sqrt{\frac{\hbar G}{c^3}}$, and the Planck time $t_P = \sqrt{\frac{\hbar G}{c^5}}$. By now we understand well how length and time can be measured with the same unit, so set $c = 1$ and write

$$G = \frac{l_P^2}{\hbar} = \frac{t_P^2}{\hbar} = \frac{\hbar}{M_P^2} \tag{3}$$

Note also that $M_P l_P = \hbar$.

In the introduction, we already derived in these natural units the entropy of a Schwarzschild black hole: $S \sim GM^2/\hbar \sim R^2/\hbar G \sim A/l_P^2$, with the surface area the black hole $A \sim R^2 \sim r_S^2 \sim (GM)^2$. I also told you there that you should be shocked, shocked, shocked. The entropy of a physical system is normally extensive* and proportional to its volume. It is as if the entropy of a black hole were to reside completely on its surface. Indeed, imagine laying down a grid on the surface of a black hole. Somehow, each Planck-sized cell contains one unit of entropy. This mysterious property of black holes, which represents one of the deepest puzzles in theoretical physics, led 't Hooft and Susskind separately to formulate the so-called holographic principle (see chapter IX.11).

In appendix 1, we derive the precise expression $T_H = \frac{\hbar c^3}{8\pi GM}$ for the Hawking temperature. Given this, we can use elementary physics to write down a precise expression for the entropy: $d(Mc^2) = T_H dS = \hbar c^3 dS/8\pi GM$, which implies that $S = 4\pi GM^2/\hbar c$. Using $A = 4\pi r_S^2$ and $r_S = 2GM/c^2$, we obtain

$$S = \frac{A}{4l_P^2} \tag{4}$$

(You could of course absorb the factor of 4 into the definition of l_P if you want.)

When we take the classical limit by letting $\hbar \to 0$, we hold G, not M_P, fixed. Indeed, M_P is not a concept in classical physics. In the classical treatment of black holes, the entropy $S \propto \hbar^{-1}$ is formally infinite (as I have mentioned twice already), since the black hole can

---

* This is proved for systems with short-ranged interactions.

be made in an unlimited number of ways. For $dM = TdS$ to be satisfied, it is consistent to set $T = 0$, which means the black hole does not radiate.

This suggests another handwaving (be warned!) argument for Hawking radiation, due to Gibbons and Hawking. For the entropy S of a black hole to be finite, quantum physics must somehow limit the number of ways a black hole can be made. Let us focus on the difference in entropy between two black holes of mass M and $M + dM$. Consider the relation $dM = TdS$: if dS is infinite, T would be 0, and we would have no radiation. As you know, an elementary fact of quantum physics is that the size of a particle is characterized by its de Broglie wavelength. A particle whose wavelength is much smaller than the Schwarzschild radius r_S can be regarded as a point particle and would fall in (depending on its velocity and impact parameter, and so forth), but a particle whose wavelength is larger than r_S could simply pass the black hole by. Thus, a particle whose wavelength is larger than GM but smaller than $G(M + dM)$ is less likely to fall into the smaller black hole. We thus argue that dS is actually finite when quantum mechanics is turned on. Once you admit that dS is not infinite, then the relation $dM = TdS$ no longer forces T to vanish, and once you admit that $T \neq 0$, we can then run our dimensional analysis argument.

The entropy of a black hole is finite, and so Wheeler was not able to violate the second law of thermodynamics by throwing cups of tea into a passing black hole. As Bekenstein explained to him and to the rest of us, he had merely increased the entropy of the black hole. If Wheeler were right, we could all help to decrease the disorder in the universe by simply dumping our mess into passing black holes.

't Hooft's bound

The mass and surface area of a Schwarzschild black hole are closely related. A rotating black hole, however, has another dimensionful parameter, the angular momentum J, so that its surface area $A = A(M, J)$ is a function of its mass M and angular momentum J, as we will see in chapter VII.5. Classically, the mass of a Schwarzschild black hole always increases, and so by free association, one might be tempted to think, as people did around the time of Bekenstein's insight, that it is related to another quantity in physics that always increases, namely the entropy. For a rotating black hole, however, as Penrose discovered in 1969 and as we will explain in chapter VII.5, we can decrease M by a physical process. But remarkably, the decrease in M is always accompanied by a decrease in J in precisely such a way so that A always increases. This indicates that we should associate the entropy of a black hole with its surface area.

Let's go back to the Schwarzschild black hole. Imagine letting two black holes with masses M_1 and M_2 slowly coalesce into a single black hole with mass $M_1 + M_2$, neglecting[11] the energy carried away by gravitational waves. Indeed, the surface area always increases: $(M_1 + M_2)^2 > M_1^2 + M_2^2$.

Since the Planck area l_P^2 is so ludicrously small (numerically, $\sim 2.6 \times 10^{-66}$ cm^2), any macroscopic black hole has an enormous entropy, which as you might expect, greatly exceeds the entropy of other physical systems.

For comparison, consider a box of volume V filled with relativistic matter, for example photons, characterized by a temperature T. By relativistic matter, we mean matter consisting of particles whose masses are negligible compared to their energies. The entropy and energy of a photon gas are worked out in textbooks on statistical mechanics, but for our purposes, we can simply use dimensional analysis. In natural units, temperature T has dimensions of energy or inverse length. In contrast, entropy S is dimensionless and proportional to the volume of the box $V \sim L^3$, with L the characteristic size of the system. So, the entropy can only be $S \sim L^3 T^3$. Similarly, the energy density ε has dimensions of mass over length cubed, or mass to the fourth power, and thus by dimensional analysis $\varepsilon \sim T^4$, leading to a total energy $E \sim L^3 T^4$. As you will see presently, the overall numerical factors here do not matter at all.

An almost universally accepted (but not yet mathematically proven) folk belief is that if a physical system has a Schwarzschild radius $r_S \sim M$ larger than its size L, it will collapse into a black hole. (The obesity index in the introduction!) Now consider a box of photons so hot that, if we throw in just a bit more energy, the box will collapse and become a black hole. The condition of being on the verge translates into $E \sim L^3 T^4 \lesssim L$, that is $T \lesssim 1/L^{\frac{1}{2}}$. The entropy of the box is thus

$$S \sim L^3 T^3 \lesssim L^{\frac{3}{2}} \sim A^{\frac{3}{4}} \tag{5}$$

A box of electromagnetic radiation hot enough to be almost a black hole has an entropy that can grow at most like the $\frac{3}{4}$ power of area, rather than like the area, as is the case for a black hole. Remember that we are using the Planck area to measure area with. Thus, for $A \gg l_P^2$, this entropy is tiny compared to that of a black hole. This bound was obtained by 't Hooft in 1993.

When do we need quantum gravity?

It is important to emphasize that in deriving Hawking radiation, we don't have to quantize the gravitational field. What we have to quantize is the particle being emitted: it and its antipartner are the ones that are quantum fluctuating out of the vacuum. Gravity's task is to change the causal structure of spacetime, and Einstein's classical theory is entirely up to the job. No quantum gravity is needed.[12]

This may be an appropriate occasion to give a handwaving argument[13] regarding when we have to worry about the quantum nature of a field. Consider an object of mass M, for example, you. As you walk around, you are surrounded by a gravitational field that in reality consists of a swarm of gravitons. Let's estimate N, the number of quanta in the swarm. If the number of quanta in the field is of order 1, then we would certainly have to deal with the quantum nature of the field. But if $N \gg 1$, then the field can be treated classically. To estimate N, let the object be spherical,[14] and imagine the swarm of gravitons spread out in a spherical distribution with a characteristic size L. By the uncertainty principle, the characteristic energy of a graviton is then of order $\varepsilon \sim \hbar/L$. The total energy contained in the gravitational potential $\phi = -GM/r$ is, according to the Newton-Einstein-Hilbert action, given by

$$E \sim G^{-1} \int d^3x \, (\nabla \phi)^2 \sim G^{-1} \int dr \, r^2 \, (GM/r^2)^2 \sim GM^2/L$$

Thus, the number of quanta equals $N \sim (GM^2/L)/(\hbar/L) \sim GM^2/\hbar \sim (M/M_P)^2$, where I used (3) in the last step.

This is a pleasing result and presumably accords with your intuition: unless the mass M is comparable to the Planck mass M_P, you don't need to lose any sleep over quantum gravity at all. You certainly did not expect that the field surrounding you could not be treated classically, did you? This heuristic argument applies to all masses, including black holes. Thus, you only have to worry when the mass of the black hole drops to $\sim M_P$ as it approaches its explosive end.[15]

The origin of the Bekenstein-Hawking entropy (4) poses a deep mystery. As already mentioned, and as you know from a course on statistical physics, entropy measures the number of microstates that correspond to a given macrostate. But no amount of staring at the Schwarzschild metric, which is just a solution of some coupled differential equations, is going to let you count the microstates. To address this mystery, a theory of quantum gravity is no doubt needed. Indeed, one triumph of string theory as a candidate theory for quantum gravity is to provide this counting. This was accomplished by Strominger and Vafa in 1996 for a class of 5-dimensional extremal* black holes in string theory. The reasoning[16] is highly technical and involves, for example, concepts such as supersymmetry. Roughly speaking, the strategy involves adiabatically lowering the gravitational constant in a thought experiment to the point when the black hole dissociates into a bunch of objects specific to string theory known as D-branes, whose degrees of freedom one can count using highly nontrivial techniques. Remarkably, the counting yields precisely the area-entropy relation (4). Since then, much progress had been made, and now people understand what is going on in $(3 + 1)$-dimensional spacetime, including some cases without supersymmetry.[17] At present, a straightforward accounting of the entropy of a plain Schwarzschild black hole has not yet been accomplished.

I should warn you that the three appendices to this chapter are exceptionally demanding. Some minimal knowledge of quantum and statistical mechanics is required to read these appendices. Those readers who have never heard of quantum mechanics may wish to skip the first two appendices, or at least read them with the appropriate attitude to get merely a flavor of these more advanced topics.

Appendix 1: Determining the Hawking temperature

As I've long promised, ever since the introduction, we will now calculate the Hawking temperature T_H, including all the factors of 2 and π. These factors are not essential for our understanding of Einstein gravity, but we

---

* The term "extremal" will be explained in chapter VII.6.

physicists, in contrast to talkers, do have to subscribe to the Feynman "shut up and calculate" dictum, at least occasionally.

First, I have to tell you about a mysterious correspondence between quantum statistical mechanics and quantum field theory. You have probably learned that in Heisenberg's formulation of quantum mechanics, the evolution of a quantum state after time T is governed by the evolution operator e^{-iHT}, with H the Hamiltonian. The probability amplitude for an initial state $|I\rangle$ to end up in the final state $|F\rangle$ is then given by

$$\mathcal{Z} = \langle F | e^{-iHT} | I \rangle \tag{6}$$

Much of the work in quantum mechanics and in quantum field theory involves massaging (or beating) this quantity into a form we can work with. For example, in the Dirac-Feynman path integral formulation,[18] one follows Newton and Leibniz by breaking up $\langle F | e^{-iHT} | I \rangle$ into infinitesimal factors and then expressing the resulting product as an integral over all possible paths the classical system could have followed going from the initial to the final state.

However, Boltzmann taught us that, at temperature T, the relative probability of a state $|n\rangle$ of energy E_n occurring is given by $e^{-E_n/T} = e^{-\beta E_n}$, where $\beta \equiv 1/T$, as usual. (You should not confuse the temperature T with the time T in the preceding paragraph of course: the same letter for two different concepts in two different areas of physics! The introduction of the inverse temperature β helps in this context; we won't return to temperature until later.) We define the partition function of a quantum mechanical system with the Hamiltonian H by

$$Z = \sum_n \langle n | e^{-\beta H} | n \rangle = \sum_n e^{-\beta E_n} = \mathrm{Tr}\, e^{-\beta H} \tag{7}$$

The sum over states is represented by a trace, with $e^{-\beta H}$ regarded as a matrix. As is probably well known to you, various physical qualities, such as the expected value of energy $E \equiv \sum_n E_n e^{-\beta E_n}/Z$, can be extracted from the partition function Z.

Evidently, there is a potentially profound correspondence between the two fundamental equations (6) and (7). To go from (6) to (7), we simply replace the time T by $-i\beta$, set $|I\rangle = |F\rangle = |n\rangle$, and sum over $|n\rangle$. What a mysterious procedure! First, we make time imaginary. Then we force every state $|n\rangle$ to go back to itself. But how can we make sure that every quantum state does this? We can if time is somehow cyclic, so that what is past is the future. I have no idea what that means. The inverse temperature β is equal to the recurrence period in this strange world with imaginary time.

Well, we don't have to understand what any of this means, but we can certainly regard this as a devilishly nifty computational trick. Consider a quantum field, be it the field of a photon, an electron, or whatever, propagating in spacetime. Suppose it discovers that time is actually imaginary and cyclic. The field is fooled into thinking that it is living in a temperature bath, to use a term from statistical mechanics, with the temperature determined by the inverse of the recurrence period β of this bizarre imaginary time.

Amazingly, we can now use this strange observation to determine the temperature of the Hawking radiation from a Schwarzschild black hole. Consider the electromagnetic field, for instance, governed by the action $S = \int d^4x \sqrt{-g}\left(-\frac{1}{4} g^{\mu\rho} g^{\nu\sigma} F_{\mu\nu} F_{\rho\sigma}\right)$, propagating in the Schwarzschild spacetime described by

$$ds^2 = -\left(1 - \frac{r_S}{r}\right) dt^2 + \left(1 - \frac{r_S}{r}\right)^{-1} dr^2 + r^2 d\theta^2 + r^2 \sin^2\theta\, d\phi^2 \tag{8}$$

with $r_S = 2GM$. Near the horizon, $ds^2 \simeq -\frac{r - r_S}{r_S} dt^2 + \frac{r_S}{r - r_S} dr^2 + r_S^2 d\Omega^2$. Change variables from r to ρ given by $\rho^2 = 4r_S(r - r_S)$. Then $\rho d\rho = 2r_S dr$, so that $\rho^2 d\rho^2 = 4r_S^2 dr^2$ or $(r - r_S)d\rho^2 = r_S dr^2$. Plugging this into ds^2, we find that spacetime near the horizon is described by

$$ds^2 \simeq -\frac{\rho^2}{4r_S^2} dt^2 + d\rho^2 + r_S^2 d\Omega^2 \to \frac{\rho^2}{4r_S^2} dt_E^2 + d\rho^2 + r_S^2 d\Omega^2$$

where in the last step, we set* $t = -it_E$ as per the mysterious procedure outlined above. If we now change variable, setting $t_E = 2r_S \psi$, we obtain

$$ds^2 \simeq d\rho^2 + \rho^2 d\psi^2 + r_S^2 d\Omega^2 \tag{9}$$

We recognize that the first two terms describe a plane with polar radius ρ and polar angle ψ. The $(3 + 1)$-dimensional spacetime has been analytically continued into a 4-dimensional Euclidean space consisting of a

---

* The subscript E stems from the terminology used in quantum field theory; upon time becoming imaginary, $(3 + 1)$-dimensional Minkowskian spacetime morphs into 4-dimensional Euclidean space.

plane, at every point of which is attached a sphere of radius r_S. More importantly, since ψ is an angular variable, we see that the "imaginary time" $t_E = 2r_S\psi$ has a recurrence period of $2r_S(2\pi) = 4\pi r_S$. Thus, according to the preceding discussion, the electromagnetic field propagating near the horizon of the Schwarzschild black hole thinks that it is living in a heat bath with temperature

$$T_H = \frac{1}{4\pi r_S} = \frac{1}{8\pi GM} = \frac{\hbar c^3}{8\pi GM} \tag{10}$$

This is the Hawking temperature* of the black hole!

In Wigner's influential essay "The Unreasonable Effectiveness of Mathematics in Physics," he told a story about two men sitting next to each other on a plane.[19] One asked the other, "What do you do for a living?" The man answered, "I work for the insurance company and I use math to predict how long people will live." The first man said, "You are pulling my leg; I don't believe that you can do that." So the second man pulled out a report on which was written the Gaussian distribution. The first man pointed to the letter π, saying "But isn't that the ratio of the circumference of a circle to its diameter?" "Exactly." The first man then exclaimed with a touch of displeasure, "Now I know you were fooling around with me. What does the circle has to do with how long a man will live?" In an updated version of this story, I imagine you answering, "I am a theoretical physicist and I figure out how hot black holes are." As your flight companion expresses a mixture of admiration and disbelief, you show him (10). After he points to the π in the equation, you can tell him that it comes in because time moves in a circle!

Appendix 2: The Unruh effect

A dutiful reader with a good memory might have recognized that the form $ds^2 \simeq -\frac{\rho^2}{4r_S^2}dt^2 + d\rho^2 + r_S^2 d\Omega^2$ of the near-horizon Schwarzschild metric described in the preceding appendix looks like the Rindler metric $ds^2 = -\rho^2 dT^2 + d\rho^2 + \rho^2\cosh^2 T\,d\Omega^2$ worked out as an exercise back in chapter III.3. After appropriate rescaling, the near-horizon Schwarzschild metric and the Rindler metric are in fact the same for fixed θ, φ (that is, for $d\Omega^2 = 0$). Recall that we obtained the Rindler metric by changing the standard coordinates (t, r, θ, φ) for Minkowski spacetime to the Rindler coordinates $(T, \rho, \theta, \varphi)$ by $t = \rho\sinh T$, $r = \rho\cosh T$ and then plugging these transformations into the Minkowski metric $ds^2 = -dt^2 + dr^2 + r^2 d\Omega^2$. It is important to note that the coordinate transformations just given have the ranges $-\infty < T < \infty$ and $0 < \rho < \infty$. Thus, the new coordinates only cover the quadrant defined by $r > |t|$, as shown in figure III.3.6.

For fixed θ and φ, the lines of constant ρ trace out hyperbolas in the (t, r) plane as T varies from $-\infty$ to ∞. It can now be revealed that these hyperbolas, as you may have already seen, are in fact the worldlines of accelerating observers in Minkowski spacetime. Suppressing θ and φ and writing $q^0 = \rho\sinh T$, $q^1 = \rho\cosh T$ for the spacetime location of an observer labeled by the parameter ρ, we have for the proper time of the observer $d\tau = \sqrt{(dq^0)^2 - (dq^1)^2} = \rho dT$. Thus, $v^\mu \equiv \frac{dq^\mu}{d\tau} = \frac{dq^\mu}{dT}/\rho = (\cosh T, \sinh T)$. (As expected, $\eta_{\mu\nu}v^\mu v^\nu = -1$, as per the definition of proper time.) The acceleration is then $a^\mu \equiv \frac{dv^\mu}{d\tau} = \rho^{-1}(\sinh T, \cosh T)$, where I display only the 2 nonzero components of the vector v^μ. Hence $\eta_{\mu\nu}a^\mu a^\nu = \rho^{-2}$, the Lorentz invariant measure of the acceleration squared, is a constant independent of T. Note that, for a given ρ, the minimum value of q^1 is ρ, attained when $T = 0$. It makes sense that the most highly accelerated observers, namely those with the smallest ρ, manage to get the closest to the surface $r = |t|$, which defines the light cone centered at the origin.

With these preliminaries, we are now ready for the point of this appendix. Bill Unruh[20] discovered that, as a result of quantum fluctuations, an accelerated observer in Minkowski spacetime would perceive a bath of thermal radiation. As we will see, this so-called Unruh effect is closely related to the Hawking effect. Just like the Hawking effect, a proper derivation of the Unruh effect requires some knowledge of quantum field theory, which as I said, I do not presume the typical reader of this book to have. Instead, let me give a handwaving argument.[21]

Let our accelerated observer carry a detector designed to detect quantum fluctuations in, say, the electromagnetic field. The detector might consist of a quantum mechanical system with energy levels E_i, $i = 0, 1, 2, \cdots$. Every time the electromagnetic field causes a transition from some level i to level j, the detector will beep. Now

---

* Surely you would hit it big with mystical types if you tell them that temperature is equivalent to cyclic imaginary time. At the arithmetic level, this connection merely comes from the fact that the central objects in quantum physics e^{-iHT} and in thermal physics $e^{-\beta H}$ are formally related by analytic continuation. Some physicists, including me, feel that there may be something profound here that we have not quite understood.

if the detector is being carried by a uniformly moving observer, and by Lorentz invariance it might as well be sitting at rest, we know that nothing will happen. The reason is that a fluctuation that causes a transition from i to j would be quickly followed, in a time of order $\hbar/|E_i - E_j|$ (which we assume to be much shorter than the reaction time of the detector), by a counterfluctuation that would cause a transition from j back to i. (Some readers might know that, in quantum field theory, fluctuations at different points in spacetime are correlated, as quantified by the 2-point Green's function of that field.) But if the detector is being accelerated, then by the time the counterfluctuation comes along, it would be moving at a different velocity from before, that is, its rest frame would differ from what it was before. The electromagnetic field $\vec{E}(t, \vec{x})$ and $\vec{B}(t, \vec{x})$, when Lorentz transformed to the new frame, would not be quite right to cause the transition from j back to i. As a result of this mismatch, the detector would indicate the presence of a bath of radiation. (What? You're not convinced? Well, I did tell you that the argument was going to involve hand waving.)

To me, a far more convincing heuristic argument is the essential equality between the near-horizon Schwarz-schild metric and the Rindler metric, as indicated above. A quantum field only knows about the environment it finds itself in through its knowledge of the metric. How does the detector "know" that it is being accelerated rather than cruising near the horizon of a black hole? Thus, the Hawking effect and the Unruh effect are both likely to be true (or, far less likely, to both be false).

A crucial feature common to both effects is the presence of a horizon. As you can see from figure III.3.6, nothing from the region $r < t$ for t positive could reach, even if it were traveling at the speed of light, the accelerated observer. Thus, the surface defined by $r = t$, which we previously identified as the forward light cone centered at the origin, effectively acts as a horizon. Indeed, figure III.3.6 resembles the Kruskal-Szekeres diagram for the spacetime around a Schwarzschild black hole. Hence, we can invoke the argument given in this chapter in support of Hawking radiation: quantum fluctuation produces a particle and an antiparticle, and before they can come together, one of them goes beyond the horizon, leaving the other free to escape. These escaping particles would constitute the Unruh radiation.

The temperature of the Unruh radiation can be estimated to be

$$T_U \sim a \sim \frac{\hbar a}{c} \tag{11}$$

by dimensional analysis, since the only quantity with the dimension of energy, or equivalently an inverse length, is the magnitude of the acceleration $a \sim (\eta_{\mu\nu}a^\mu a^\nu)^{\frac{1}{2}}$.

Next, I will sketch, using broad brushstrokes, the serious derivation first presented by Unruh. Readers are forewarned that this will require some knowledge of the quantum world, and those without this knowledge are urged to skip the rest of this appendix.

In quantum mechanics, the position operator $q(t)$ of a harmonic oscillator is expressed in terms of annihilation and creation operators a and $a^\dagger$ as follows: $q(t) \sim ae^{-iEt} + a^\dagger e^{iEt}$ with $E > 0$ the characteristic energy of the oscillator. The ground state of the harmonic oscillator, denoted by $|0\rangle$ (in a notation already used in this chapter), is annihilated by the annihilation operator a in the sense that $a|0\rangle = 0$. We generate the excited states $|n\rangle \sim (a^\dagger)^n|0\rangle$ by repeatedly acting with the creation operator $a^\dagger$ on the ground state. (Hence it is actually more accurate, in the context of quantum mechanics, to speak of a and $a^\dagger$ as lowering and raising operators.) How do we know, of a and $a^\dagger$, which one is the annihilation operator and which the creation operator? The answer goes back to the fundamental requirement that the creation operator is to create a state with positive energy. Thus, a is always associated with the positive energy wave function e^{-iEt} and $a^\dagger$ with e^{iEt}.

In quantum field theory, these notions are generalized in a straightforward fashion. The generic field $\phi(t, \vec{x})$, namely the analog of $q(t)$, depends on space as well as time, and so the positive energy wave function is generalized to $e^{-i(Et-\vec{k}\cdot\vec{x})}$, that is, a wave in space and time, characterized by energy E and momentum $\vec{k}$. Correspondingly, $a(\vec{k})$ and $a^\dagger(\vec{k})$ now depend on $\vec{k}$, and an integral over $\vec{k}$ is required. Thus, we end up writing

$$\phi(t, \vec{x}) \sim \int d^3k \left(a(\vec{k})e^{-i(Et-\vec{k}\cdot\vec{x})} + a^\dagger(\vec{k})e^{i(Et-\vec{k}\cdot\vec{x})} \right)$$

All this is baby quantum field theory and is explained in any book[22] on the subject. Bottom line: the field $\phi(t, \vec{x})$ is a linear combination of $a(\vec{k})$ and $a^\dagger(\vec{k})$, associated with e^{-iEt} and e^{iEt}, respectively. In fact, a quantum field can be thought of as a collection of harmonic oscillators. An important conceptual difference between quantum field theory and quantum mechanics is that the ground state $|0\rangle$ is now more properly called the vacuum state, a state in which no particle is present and the field is quiescent. Acting with $a^\dagger(\vec{k})$ on $|0\rangle$ produces a state with a particle carrying momentum $\vec{k}$: the operator $a^\dagger(\vec{k})$ is said to create a particle out of the vacuum.

For our purpose here, let us write, more schematically, $\phi \sim \sum_\alpha (a_\alpha f_\alpha + a_\alpha^\dagger f_\alpha^*)$. (Here, $*$ denotes complex conjugation.) The important point to take away is merely that the quantum field ϕ can be written as a linear sum of annihilation and creation operators a_α and $a_\alpha^\dagger$ capable of annihilating and creating particles. The subscript α

labels the properties (such as momentum) the particles may carry. The corresponding wave functions are written as f_α and f_α^*, and the integral over $\vec{k}$ is replaced by a sum over α. The vacuum state $|0_A\rangle$ is defined as the state annihilated by a_α: $a_\alpha |0_A\rangle = 0$.

Relativity brings a new twist to this framework for studying quantum fields: different observers can disagree politely over what they regard as time. For the case at hand, while an observer sitting at rest in Minkowski spacetime uses t for his time coordinate, the accelerated observer would insist on using T for her time coordinate. More generally, another observer could use as wave functions g_β and g_β^*, instead of f_α and f_α^*, and decompose the quantum field as $\phi \sim \sum_\beta (b_\beta g_\beta + b_\beta^\dagger g_\beta^*)$. She would use b_β and $b_\beta^\dagger$ as her annihilation and creation operators, and define her vacuum state $|0_B\rangle$ as the state annihilated by b_β: $b_\beta |0_B\rangle = 0$.

This discussion shows that the concept of particles, and even that of the vacuum, depends on the observer.

In general, the wave functions g_β and g_β^* can be written as linear combinations of f_α and f_α^*. Since ϕ is the same old ϕ regardless of observer, comparison of the two expressions for ϕ implies that a_α and $a_\alpha^\dagger$ can be written as linear combinations of b_β and $b_\beta^\dagger$, and vice versa. This relationship between the two sets of annihilation and creation operators is known as a Bogoliubov transformation.

We are getting close to the punchline! Suppose observer A says, "We are in the state $|0_A\rangle$, and there aren't any particles around." Observer B would disagree. To her, since a_α is given as a linear combination of b_β and $b_\beta^\dagger$, schematically, $a_\alpha \sim \sum_\beta (U_{\alpha\beta} b_\beta + V_{\alpha\beta} b_\beta^\dagger)$, the condition $a_\alpha |0_A\rangle = 0$ amounts to, schematically, $\sum_\beta U_{\alpha\beta} b_\beta |0_A\rangle \sim \sum_\beta V_{\alpha\beta} b_\beta^\dagger |0_A\rangle$. In other words, $b_\beta ||0_A\rangle$, far from being 0, is actually related to a linear combination of $b_\beta^\dagger |0_A\rangle$. Observer B would say that the state $|0_A\rangle$ contains particles as defined by her. This may appear as a long winded way of saying that $|0_A\rangle$ is not equal to $|0_B\rangle$, but it goes beyond that by indicating that the number of b-type particles contained in $|0_A\rangle$ can be calculated in terms of the coefficients in the Bogoliubov transformation.

Now we apply this to the situation at hand. An observer sitting at rest in Minkowski spacetime can insist that no particle is present, and yet the accelerated observer will see a bath of particles. In other words, the Unruh effect!

I hope that I have given you a flavor of the derivation and prepared you to read Unruh's paper. For those readers who know that the wave functions of a nonrelativistic single particle in a 1-dimensional box of length L are given by, for $n = 1, 2, \cdots$, $\psi_n(x) = \sin(n\pi x/L)$ for $0 \le x \le L$, and $\psi_n(x) = 0$ otherwise, I can offer a toy example that may or may not help.

Suppose our nonrelativistic particle is sitting in the ground state $\psi_1(x)$. The probability of finding the particle in an excited state $\psi_{n>1}(x)$ is strictly zero. Now suppose the box is suddenly expanded to twice its former size. The wave functions are now given by, for $n = 1, 2, \cdots$, $\Psi_n(x) = \sin(n\pi x/(2L))$ for $0 \le x \le 2L$, and $\Psi_n(x) = 0$ otherwise. Note that $\Psi_n(x)$ is not the same as $\psi_n(x)$. The initial wave function $\psi_1(x)$ can be expressed, according to Fourier, as a linear combination of the new wave functions $\Psi_n(x)$, namely $\psi_1(x) = \sum_n c_n \Psi_n(x)$. The probability of finding the particle in an excited state with $n > 1$ is now nonvanishing, $\propto |c_n|^2$. The sudden expansion of the box has excited the particle. Note that in nonrelativistic quantum mechanics, if we start with a single particle, we end with a single particle, albeit in an excited state.

When we go to quantum field theory, the role of the particle is played by a quantum field, and the particle jumping into an excited state gets translated into the quantum field becoming excited and hence capable of creating particles. (Do not confuse the notion of particles in nonrelativistic quantum mechanics with the notion of particles in quantum field theory, which correspond to "excitations" in the quantum field. When excited, a nonrelativistic particle jumps to a higher energy level; when excited, a quantum field creates particles.) I hope that this toy example of an expanding box does not confuse you too much and that it conveys to you the possibility that an expanding universe is able to create particles and antiparticles.

Appendix 3: Thermodynamics of spacetime and Einstein's field equation

One intriguing, and possibly fruitful, approach, proposed by Jacobson,[23] is to regard the entropy formula (4) as fundamental and to derive Einstein's field equation from it. To see how this is possible, consider an ideal gas in a container of volume V and total energy E. Given $S(E, V)$, thermodynamics teaches us how to find the equation of state. In general, $dE = TdS - PdV$, or $dS = T^{-1}(dE + PdV)$. In other words, $\frac{1}{T} = \left(\frac{\partial S}{\partial E}\right)_V$ and $\frac{P}{T} = \left(\frac{\partial S}{\partial V}\right)_E$. For an ideal gas, the entropy is given by the logarithm of the number of possible states. Since each molecule can roam over the volume V, the number of accessible states is proportional to V^N, and so we argue that the entropy goes like $S = N \log V + f(E)$, with some function $f(E)$. The second of the thermodynamic relations just given then yields $\frac{P}{T} = \frac{N}{V}$, that is, the well-known equation of state $PV = NT$.

At the level of this book, I can only sketch Jacobson's argument in the crudest possible terms, just to show how it might be possible for Einstein's field equation to come out of the entropy formula. By necessity, I will gloss over a great many technicalities and am intentionally vague at times. The following should be read more as an enticement to look into the original literature than as an explanation.

Consider an infinitesimal amount of heat δQ going through a causal horizon: δQ is given by integrating the energy flux, which depends on $T^{\mu\nu}$, over the area A of the horizon. As a result, the area A changes, but the change in area δA is determined geometrically by light rays near the horizon converging toward or diverging away from one another. In chapter IX.3, starting from the geodesic equation, we will derive an equation (known as the Raychaudhuri equation) governing the amount by which light rays converge or diverge. You would expect this to be determined by the curvature of spacetime. Indeed, the Ricci tensor $R^{\mu\nu}$ appears in the Raychaudhuri equation. The entropy formula $S = A$ (suppressing the irrelevant overall constant or choosing sensible units) tells us how δS (which is related to δQ) is related to δA, and thus how $T^{\mu\nu}$ is related to $R^{\mu\nu}$. The relation, perhaps not surprisingly, turns out to be Einstein's field equation.

I have brushed over a host of technicalities, but should have persuaded you that it is at least conceivable that Einstein's field equation could come out of the entropy formula and thermodynamics. Let me say it more colloquially. The formula $S = A$ (rather than $S \propto V$) is incredibly special and weird; how could the entropy possibly be proportional to the area!!? Well, the physics of gravity has to be arranged in precisely such a way so that it holds. (Jacobson intended his argument to hold for any spacetime, but for pedagogical clarity, I have focused here on a black hole.)

Oh dear, if this view is correct,[24] then Einstein's field equation is demoted to the status of $PV = NT$, a mere equation of state. If so, it may have important consequences. To quote Jacobson, "This perspective suggests that it may be no more appropriate to quantize the Einstein equation than it would be to quantize the wave equation for sound in air."[25]

Exercise

1   Work out the heuristic calculation outlined in the text by SE and Confusio, thus obtaining another estimate of T_H.

Notes

1. This fact is sometimes somewhat misleadingly presented as something amazing about black holes. Actually, it follows essentially from the virial theorem (see exercise 10 in chapter IV.2) and is generic to gravitating systems, including stars. As a star loses its energy through radiation, it generically heats up.

2. This discussion is taken from p. 3 of *QFT Nut*, to which you are referred for more details.

3. To learn how quantum electrodynamics deals with the fluctuating photon, see *QFT Nut*, or any other reputable quantum field theory text.

4. A definitive and detailed history of the discovery of Hawking radiation has yet to be written, as far as I know. At the time, a Russian group consisting of Y. Zel'dovich, A. Starobinsky, and others was actively working on radiation from rotating black holes (which we will discuss in chapter VII.5.) Unfortunately for them, they had convinced themselves that Schwarzschild black holes do not radiate, an entirely plausible supposition, since the Schwarzschild solution is static. Furthermore, even classically, rotating black holes emit particles through the Penrose process (also to be discussed in chapter VII.5). Meanwhile, Don Page, a graduate student at the California Institute of Technology, was also working on radiation from black holes and discussing his calculations with R. Feynman. Page independently discovered that rotating black holes radiate, and Feynman agreed with the conclusion, but then they discovered that Zel'dovich et al. had beaten them to it. Several others, including Bill Unruh and Larry Ford, were also working on similar ideas. I was told that had Hawking not found the radiation from the Schwarzschild black hole, Unruh probably would have. Incidentally, Hawking's original motivation was actually to prove that Bekenstein's proposal that black holes had entropy was wrong. I am grateful to Gary Gibbons and Don Page for personal accounts of the events surrounding the discovery of black hole radiation.

5. This suggests another way of understanding Hawking radiation, in terms of quantum tunneling. Work out the wave equation of a quantum particle in the Schwarzschild metric. Classically, a particle inside the horizon trying to get out is faced with a potential barrier, but a quantum particle could tunnel through the barrier.

6. W. G. Unruh, *Phys. Rev.* D14 (1976), p. 870.

7. See A. Almheiri, D. Marolf, J. Polchinski, and J. Sully, "Black Holes: Complementarity or Firewalls?" arXiv:1207.3123v2. Note also the papers generated in response to this paper.

8. J. A. Wheeler, *A Journey into Gravity and Spacetime,* Scientific American Library, W. H. Freeman, 1990, p. 221.

9. Strictly speaking, since counting is involved, entropy is a concept of quantum statistical mechanics and does not make complete sense in classical physics.

10. See, for example, R. Feynman, *Statistical Mechanics.*

11. Or surround a black hole of mass M with a spherical shell consisting of a large number of black holes with masses M_j, which we allow to fall into the central black hole, giving us a black hole of mass $M + \sum_j M_j$. Then $\left(M + \sum_j M_j \right)^2 > M^2 + \sum_j M_j^2$.

12. Even in the Hawking radiation of gravitons from a black hole, we can imagine cutting the metric into two pieces, a classical piece plus a small quantum piece, small in the sense that we can ignore the interaction between the gravitons. This type of procedure will be used in discussing gravitational waves in chapter IX.4.

13. I heard this argument from G. Dvali.

14. That is, in the spirit of the famous book *Consider a Spherical Cow* by J. Harte, consider a spherical you.

15. Indeed, at one point, a contentious subject revolved around what you would expect to see: peculiar remnants or nothing.

16. A. Strominger and C. Vafa, *Phys. Lett.* B 379 (1996), pp. 99–104, arXiv:hep-th/9601029.

17. A. Strominger, private communication.

18. This sentence is not intended to make sense in the context of this book. For a detailed explanation, see, for example, chapter I.2 of *QFT Nut.*

19. Wigner had them sitting at a bar.

20. W. Unruh, *Phys. Rev.* D 14 (1976), p. 870.

21. I heard this from Bill Unruh (private communication).

22. For example, *QFT Nut*, p. 63.

23. T. Jacobson, *Phys. Rev. Lett.* 75 (1995), p. 1260; arXiv: 9504004v2, 1112.6215v2. See also T. Padmanabhan, arXiv: 0911.5004. For more recent work, see E. P. Verlinde, *JHEP* 1104:029 (2011).

24. There are skeptics. One relativist I talked to scoffed that this merely proved that some people were able to run the relevant equations backward. You judge for yourself.

25. T. Jacobson, *Phys. Rev. Lett.* 75 (1995), p. 1260.

Interior of stars

In this chapter we study what general relativity has to say[1] about the interiors of stars. We are going to deal with only the most idealized situation. The magnificent complications of stellar interior dynamics are far beyond the scope of this book.

In the simplest model, the star is assumed to be perfectly spherical (and hence nonrotating), with its interior consisting of a perfect fluid, a notion we defined way back in chapter III.6. There we derived the energy momentum tensor of a perfect fluid in flat spacetime $T^{\mu\nu} = (\rho + P)U^\mu U^\nu + P\eta^{\mu\nu}$, with U^μ the local 4-velocity of the fluid. Once again, behold the power of the equivalence principle! We merely have to promote $\eta^{\mu\nu}$ to $g^{\mu\nu}$ to obtain

$$T^{\mu\nu} = (\rho + P)U^\mu U^\nu + P g^{\mu\nu} \tag{1}$$

for curved spacetime.

Let's plug this into Einstein's field equation

$$R_{\mu\nu} = +\kappa \left(T_{\mu\nu} - \tfrac{1}{2} g_{\mu\nu} T \right) \tag{2}$$

where we have introduced the shorthand $\kappa = 8\pi G$. With $T = -\rho + 3P$, we have

$$R_{\mu\nu} = +\kappa \left[(\rho + P)U_\mu U_\nu + \tfrac{1}{2}(\rho - P)g_{\mu\nu} \right] \tag{3}$$

Assume a static spherically symmetric interior described by the metric

$$ds^2 = -A(r)dt^2 + B(r)dr^2 + r^2 d\Omega^2 \tag{4}$$

The resulting Ricci tensor was calculated back in chapter VI.3. Recall that $R_{\mu\nu}$ is diagonal and that $R_{\varphi\varphi} = \sin^2\theta\, R_{\theta\theta}$.

Solving for the spacetime inside the star

The static spherical symmetry implies physically that the fluid can't flow, so that $U^i = 0$. This is also forced upon us by (3) and the vanishing of R_{0i}. The normalization condition $g_{\mu\nu}U^\mu U^\nu = -1 = -A(r)(U^0)^2$ gives $U^0 = A^{-\frac{1}{2}}$ and $U_0 = g_{00}U^0 = -AA^{-\frac{1}{2}} = -A^{\frac{1}{2}}$. Thus, the field equation (3) implies

$$R_{tt} = \frac{A''}{2B} + \frac{A'}{rB} - \frac{A'}{4B}\left(\frac{A'}{A} + \frac{B'}{B}\right) = \tfrac{1}{2}\kappa(\rho + 3P)A \tag{5}$$

$$R_{rr} = -\frac{A''}{2A} + \frac{B'}{rB} + \frac{A'}{4A}\left(\frac{A'}{A} + \frac{B'}{B}\right) = \tfrac{1}{2}\kappa(\rho - P)B \tag{6}$$

$$R_{\theta\theta} = 1 - \frac{1}{B} - \frac{r}{2B}\left(\frac{A'}{A} - \frac{B'}{B}\right) = \tfrac{1}{2}\kappa(\rho - P)r^2 \tag{7}$$

We are to solve these three coupled ordinary differential equations.[2]

Back in chapter VI.3, we had the easier problem of solving (5–7) with their right hand sides set to zero. Let us form the same combination $\frac{R_{tt}}{A} + \frac{R_{rr}}{B} + \frac{2R_{\theta\theta}}{r^2}$ that served us well there. We find the equation $\left(1 - \frac{1}{B} + \frac{rB'}{B^2}\right) = \kappa r^2 \rho$, with a right hand side, though no longer zero, depending only on ρ. Inspired by the Schwarzschild solution, we define a mass function $\mathcal{M}(r)$ by

$$\frac{1}{B} \equiv 1 - \frac{2G\mathcal{M}(r)}{r} \tag{8}$$

Inserting this into the equation for B, we find

$$\frac{d\mathcal{M}(r)}{dr} = 4\pi r^2 \rho(r) \tag{9}$$

which we can integrate immediately to obtain

$$\mathcal{M}(r) \equiv 4\pi \int_0^r dr' r'^2 \rho(r') \tag{10}$$

According to the Bianchi identity (as explained in chapter VI.5), we can trade one of the field equations (5–7) for $D_\mu T^{\mu\nu} = 0$. Plugging in the perfect fluid energy momentum tensor, noting that $D_\mu g^{\lambda\nu} = 0$, and using the expression for the covariant divergence of a tensor $D_\mu T^{\mu\nu} = \frac{1}{\sqrt{-g}}\partial_\mu(\sqrt{-g}T^{\mu\nu}) + \Gamma^\nu_{\mu\lambda}T^{\mu\lambda}$, we have

$$0 = D_\mu \left\{(\rho + P)U^\mu U^\nu + Pg^{\mu\nu}\right\}$$
$$= \frac{1}{\sqrt{-g}}\partial_\mu\left\{\sqrt{-g}(\rho + P)U^\mu U^\nu\right\} + \Gamma^\nu_{\mu\lambda}(\rho + P)U^\mu U^\lambda + g^{\mu\nu}\partial_\mu P$$
$$= \frac{\rho + P}{A}\Gamma^\nu_{00} + g^{rv}\frac{dP}{dr} \tag{11}$$

The last equality follows since the only nonzero component of U^μ is U^0, and the assumed spherical symmetry implies that various quantities, such as the pressure P, depend on r only.

Looking up the list of Christoffel symbols, we see that the only nonvanishing Γ^{ν}_{00} is $\Gamma^{r}_{00} = \frac{A'}{2B}$, so that we obtain the condition of hydrostatic equilibrium on the pressure gradient $P' = \frac{dP}{dr}$:

$$\frac{P'}{\rho + P} = -\frac{A'}{2A} \tag{12}$$

Stellar equilibrium

Let's keep count. Of the three field equations (5–7), we have effectively used two, massaging them into (10) and (12). We choose (7) as our final equation, from which we eliminate B and A using (8) and (10), and (12), respectively. After some algebra, we arrive at the Tolman-Oppenheimer-Volkoff[3] equation for relativistic stellar structure

$$\frac{dP}{dr} = -\frac{G\mathcal{M}(r)\rho(r)}{r^2} \left(1 + \frac{P(r)}{\rho(r)}\right) \left(1 + \frac{4\pi r^3 P(r)}{\mathcal{M}(r)}\right) \left(1 - \frac{2G\mathcal{M}(r)}{r}\right)^{-1} \tag{13}$$

We also have to specify what the star is made of by giving the equation of state $P = P(\rho)$, relating pressure to density.

We can now work out the stellar structure in this idealized model. Given some equation of state, eliminate ρ in terms of P, and then integrate the two coupled first order differential equations (13) and (9) for $P(r)$ and $\mathcal{M}(r)$. For some simple $P = P(\rho)$, analytic solutions may be found, but in general, it is necessary to numerically integrate outward from $r = 0$ with the boundary condition $\mathcal{M}(r = 0) = 0$ (obviously) and some chosen value of the central pressure $P(r = 0)$ or equivalently, some central density $\rho(r = 0)$. From (13), we see that $P(r)$ steadily decreases until it vanishes at some radius R, which defines the radius of the star. (The pressure vanishes in empty space outside the star, and so, if $P(R) \neq 0$, there would be an infinite pressure gradient at the surface, which is not physically acceptable.) In other words, the radius R of the star is determined by $P(R) = 0$. The mass of the star is then given by $M = \mathcal{M}(R)$. Thus, there is a one-parameter family of solutions, with the mass M and radius R of the star dependent on $P(r = 0)$.

At this point, we can also determine the spacetime inside the star. Already, $B(r)$ is given by (8). We insert (13) into (12) to obtain

$$\frac{A'}{A} = \frac{2G\mathcal{M}(r)}{r^2} \left(1 + \frac{4\pi r^3 P(r)}{\mathcal{M}(r)}\right) \left(1 - \frac{2G\mathcal{M}(r)}{r}\right)^{-1} \tag{14}$$

which typically we would have to integrate numerically inside the star.

Outside the star, however, $\mathcal{M}(r) = M$ and $P(r) = 0$, so we can integrate (14) almost instantly to give $A = 1 - \frac{2GM}{r}$. Nicely, the interior solution joins on with the Schwarzschild solution with $r_S = 2GM = 2G\mathcal{M}(R)$. This verifies the Newton-Jebsen-Birkhoff theorem yet again.

The Tolman-Oppenheimer-Volkoff equation is written in a particularly attractive form in (13) to exhibit the Newtonian limit explicitly. To see this, restore c by high school dimensional reasoning. For example, denoting the dimension of P by $[P]$, we have $[P] = [\text{force/area}] = [(ML/T^2)/L^2] = [M/(LT^2)]$. Similarly, $[\rho] = [M/L^3]$. Thus, $[P/\rho] =$

$[(L/T)^2]$, so that with nonrelativistic units, the expression in the first parentheses in (13) should be written as $\left(1 + \frac{P}{\rho c^2}\right)$, which tends to 1 as $c \to \infty$. (Actually, you know this already if you recall from chapter III.6 that for a nonrelativistic gas, the pressure is negligible compared to the mass density.) You can convince yourself that the other two expressions in parentheses in (13) also tend to 1 as $c \to \infty$. Therefore, in the nonrelativistic limit the Tolman-Oppenheimer-Volkoff equation reduces to Newton's equation for stellar structure

$$\frac{dP}{dr} = -\frac{GM\rho}{r^2} \tag{15}$$

which just says that the outward force due to pressure on an infinitesimal volume of stellar material balances the inward force due to gravity. To see this, visualize a thin slab of stellar material of cross-sectional area dA and bounded between r and $r + dr$. The net force due to pressure is given by $P(r)dA - P(r + dr)dA = -\frac{dP}{dr}dr\,dA$, and the force due to gravity by $GM(\rho\,dr\,dA)/r^2$. Note that here we have to invoke Newton's two "superb theorems" explained way back in our very first chapter, chapter I.1.

Quite remarkably, there is none of this talk about forces in the field equation (3), just a statement about how the energy momentum tensor curves spacetime. A lot of physics lurks secretly inside (3). To me, that's part of the magic of theoretical physics.

Buchdahl's theorem

A particularly simple (but somewhat unphysical) equation of state is that of an incompressible fluid, namely ρ equal to a constant independent of the pressure. Then (10) may be trivially evaluated, giving $\mathcal{M}(r) = (4\pi/3)r^3\rho = (r/R)^3M$, where the radius of the star R is determined by $P(R) = 0$, as explained earlier. The mass of the star M is equal to $\mathcal{M}(R)$. Evidently, we will encounter the combination GM, and so it is convenient to introduce the symbol $r_S \equiv 2GM$, even though we are talking about a star, rather than a black hole, here.

Things are now sufficiently simple for us to integrate the Tolman-Oppenheimer-Volkoff equation (13) analytically. We obtain

$$P(r) = \rho\,\frac{\left(1 - \frac{r_S}{R}\left(\frac{r}{R}\right)^2\right)^{\frac{1}{2}} - \left(1 - \frac{r_S}{R}\right)^{\frac{1}{2}}}{3\left(1 - \frac{r_S}{R}\right)^{\frac{1}{2}} - \left(1 - \frac{r_S}{R}\left(\frac{r}{R}\right)^2\right)^{\frac{1}{2}}} \tag{16}$$

The scale of $P(r)$ is set by the constant density ρ.

For the central pressure $P(0) = \rho\,\frac{1 - \left(1 - \frac{r_S}{R}\right)^{\frac{1}{2}}}{3\left(1 - \frac{r_S}{R}\right)^{\frac{1}{2}} - 1}$ to be positive, we must have $3\left(1 - \frac{r_S}{R}\right)^{\frac{1}{2}} > 1$, with $P(0)$ blowing up when the inequality becomes an equality. Thus, we require

$$R > \tfrac{9}{8}r_S = \tfrac{9}{4}GM \tag{17}$$

Buchdahl's theorem states that for any "reasonable" equation of state, the inequality (17) holds. Recall the criterion $r_S > R$ for the star to become a black hole. Thus, Buchdahl's star can never become a black hole.

Stellar collapse into black holes

> Chandrasekhar . . . shows that a star of mass greater than a
> certain limit M . . . has to go on radiating and radiating and
> contracting and contracting until, I suppose, it gets to a few
> km. radius, when gravity becomes strong enough to hold in the
> radiation, and the star can at last find peace. . . . I think, there
> should be a law of nature to prevent a star from behaving in this
> absurd way.
> —Arthur S. Eddington, comments at the Royal Astronomical
> Society Meeting, on January 11, 1935

As I have already said, a detailed analysis of stellar equilibrium using various equations
of state $P(\rho)$ is beyond the scope of this text, even though, as the reader probably knows,
the results from such an analysis constitute some of the most spectacular highlights of
stellar astrophysics. For example, if the pressure is supplied by the quantum motion of the
electrons in the star, namely the Fermi pressure of degenerate electrons (see any text on
statistical mechanics), the stellar mass M cannot exceed an upper limit of about $1.4 M_\odot$,
known as the Chandrasekhar limit.

To study the collapse of a spherical cloud of matter into a black hole requires generalizing
the metric (4) used here to the time dependent metric mentioned in appendix 3 to chapter
VI.3. The analysis of the resulting Einstein field equation becomes considerably more
complicated.

I am content to point out a key physical feature of the relativistic equation for stellar
equilibrium. We see that the expression in the first parentheses in (13) effectively changes
the mass density ρ in Newtonian gravity (15) to $\rho + P$. This important piece of physics can
be traced all the way back to special relativity: pressure, being an energy density, counts
also as a mass density. Thus, as infalling matter piles onto a superdense star and squeezes
it gravitationally, the star resists by increasing its internal pressure P, which only adds to
the mass density bearing in. In essence, this vicious cycle is at the root of the physics of
black hole formation.

Gravitational binding energy and Einstein getting almost run over

Now that you are a budding relativist familiar with the Schwarzschild solution, weren't you
pleased to see (8) and (9) appearing? But at the same time, did you not find the expression
for the total mass

$$M = \mathcal{M}(R) = 4\pi \int_0^R dr r^2 \rho(r) \tag{18}$$

a bit odd? At first sight, it looks like the sum of the infinitesimal mass elements that make
up the star, but then you realize that space is curved!

At a fixed instant in time t, space as a slice of spacetime is described (from (4)) by the line element $B(r)dr^2 + r^2 d\Omega^2$. Thus, some authors choose to define the integrated mass

$$\tilde{M} = 4\pi \int_0^R dr\, r^2 \frac{\rho(r)}{\sqrt{1 - \frac{2G\mathcal{M}(r)}{r}}} \tag{19}$$

and regard the difference $\tilde{M} - M > 0$ as the gravitational binding energy of the star. I must emphasize, though, that I know of no simple experiment that would measure $\tilde{M}$. In contrast, since the exterior Schwarzschild geometry is determined by M, a distant astronomer knowing Kepler's laws and measuring the period of a planet orbiting the star would end up calculating M.

I end this chapter by recounting a story told by George Gamow[4] in his autobiography. While crossing a street, Gamow mentioned to Einstein that Pascual Jordan had realized that a star could be made of nothing if its negative gravitational energy balances its positive rest mass energy. According to Gamow, "Einstein stopped in his tracks and . . . several cars had to stop to avoid running us down."

Appendix: The expanding universe again

I'd like to mention something amusing here. An astute reader might notice that the setup in this chapter, with the metric $ds^2 = -A(r)dt^2 + B(r)dr^2 + r^2 d\Omega^2$ in (4), also allows us to solve the equation $R_{\mu\nu} = -8\pi G\Lambda g_{\mu\nu}$ that we solved in chapters VI.2 and VI.5. We can solve for a universe filled with the cosmological constant (which may well be the dark energy, as mentioned in chapter VI.2). For $\rho = \Lambda$, we have immediately $\mathcal{M} = 4\pi \Lambda r^3/3$. Furthermore, with $P = -\Lambda$, the field equation (7) gives $A = 1/B$, where we have absorbed an integration constant. After some algebra, we find that the metric in (4) works out to be

$$ds^2 = -\left(1 - H^2 r^2\right) dt^2 + \left(1 - H^2 r^2\right)^{-1} dr^2 + r^2 d\Omega^2 \tag{20}$$

with $H^2 = 8\pi G\Lambda/3$ (as in chapter VI.5).

"What is going on?" you exclaim. Back in chapters VI.2 and VI.5, filling the universe with a cosmological constant gave an exponentially expanding universe described by

$$ds^2 = -dt^2 + e^{2Ht}\left(dx^2 + dy^2 + dz^2\right) \tag{21}$$

But what is expanding in (20)? That metric does not even depend on time! I will let you think about that one for a moment.

It is the magic of coordinate transformation, of course! You can transform (20) into (21). Try it. The precise relationship between these two apparently entirely different metrics will be revealed in chapter IX.10: they both describe what is called de Sitter spacetime.

Another astute reader might notice that the solution here is not the most general. The equation (9) $\frac{d\mathcal{M}(r)}{dr} = 4\pi r^2 \Lambda$ also allows the solution $\mathcal{M} = M + 4\pi \Lambda r^3/3$, with an arbitrary additive constant* M. Using (7), you can check that $A = 1/B$ continues to hold. Thus, another solution is given by

$$A = \frac{1}{B} = 1 - \frac{2GM}{r} - H^2 r^2 \tag{22}$$

Interestingly, we can put[†] a black hole in de Sitter spacetime.

---

*Not to be confused with the mass of the star, of course. We are now talking about an entirely different physical situation. This additive constant is not allowed in the context of the stellar problem, since spacetime is required to be nonsingular at the center $r = 0$ of the star.

[†] This fact turns out to be of great relevance to recent developments in theoretical physics.

Exercises

1 Derive (16).

2 Find some analytic solutions of the Tolman-Oppenheimer-Volkoff equation. For help, read section 3 of Tolman's 1939 paper cited in the endnotes.

Notes

1. Actually, most stars never evolve into structures dense enough for general relativity to play a role.
2. It's perhaps worth remarking somewhere, so it might as well be here, that Einstein's quip regarding the left hand side of his field equation versus the right hand side (mentioned in chapter VI.5) is not as clear-cut as it sounds upon first hearing: in general, $T_{\mu\nu}$ depends on the metric, so that geometry does not appear exclusively on the left side. Here A and B appear explicitly on the right side of (5) and (6).
3. R. C. Tolman, *Phys. Rev.* 55 (1939), p. 364; J. R. Oppenheimer and G. M. Volkoff, *Phys. Rev.* 55 (1939), p. 374.
4. G. Gamow, *My World Line*.

VII.5 | Rotating Black Holes

Rotating bodies: General considerations

Rotating black holes are important for practical and theoretical reasons.

Astrophysical objects invariably rotate due to the chaotic way they are formed. For normal stars, such as the sun, the amount of rotation, as measured by the rotational speed v at the surface divided by c, is negligible. That's why we are able to use the Schwarzschild metric to describe the spacetime outside the sun. Around a black hole, however, infalling debris invariably causes the black hole to spin, as we saw in chapter VII.1. Several important astrophysical processes appear to be powered by rotating black holes. One of our goals in this chapter is to understand how rotating black holes can be such powerful sources of radiation.

Historically, the Schwarzschild horizon bothered the founding fathers of general relativity so much that some of them suggested that its presence is an artifact of the spherical symmetry and that a rotating black hole would be free of such bizarre features. The discovery of a rotating black hole solution of Einstein's field equation by Roy Kerr in 1963 finally put this supposition to rest.

This chapter may be skipped over upon first reading; a first understanding of Einstein gravity does not require mastering the Kerr solution. Like the Schwarzschild spacetime, the Kerr spacetime is a solution of Einstein's field equation $R_{\mu\nu} = 0$ in empty spacetime. We anticipate that there could be a horizon. A rotating object small enough to fit inside its own horizon is known as a Kerr black hole.

Before we look at Kerr's specific solution, let us see how far we can get with general considerations. We assume stationary and cylindrical symmetry. Stationary means that the object rotates with a constant angular velocity, so that the spacetime does not change with time. With the usual coordinates (t, r, θ, φ), the metric components $g_{\mu\nu}(r, \theta)$ are functions only of r and θ, but not of t and φ. Furthermore, the solution must be unchanged under the discrete transformation $t \to -t$ together with $\varphi \to -\varphi$. This rules out in ds^2 cross terms such as $dt\,dr$, $dr\,d\varphi$, and so forth, allowing only $dt\,d\varphi$. (Convince yourself that by a coordinate transformation (see exercise 1), you can also eliminate the cross term $drd\theta$.) Hence in general, the spacetime outside is described by

$$ds^2 = g_{tt}dt^2 + g_{rr}dr^2 + g_{\theta\theta}d\theta^2 + g_{\varphi\varphi}d\varphi^2 + 2g_{t\varphi}dtd\varphi \tag{1}$$

with $g_{tt}, g_{rr}, g_{\theta\theta}, g_{\varphi\varphi}, g_{t\varphi} = g_{\varphi t}$ five functions of r and θ. The appearance of the off-diagonal component $g_{t\varphi} = g_{\varphi t}$ in the metric will lead to fascinating new physics.

The term $g_{t\varphi}dtd\varphi$ in ds^2 means that ds^2 is no longer invariant under $t \to -t$, as was the case with the Schwarzschild solution. That the spacetime is invariant only under the combined transformation $t \to -t$ and $\varphi \to -\varphi$ is a hallmark of rotation about the z-axis.

Note that we can also write

$$ds^2 = \left(g_{tt} - \frac{g_{t\varphi}^2}{g_{\varphi\varphi}} \right) dt^2 + g_{\varphi\varphi}\,(d\varphi - \omega dt)^2 + g_{rr}dr^2 + g_{\theta\theta}d\theta^2 \tag{2}$$

with $\omega(r, \theta) \equiv -g_{t\varphi}/g_{\varphi\varphi}$.

That the metric $g_{\mu\nu}(r, \theta)$ does not depend on t and φ immediately implies that two of the equations of motion amount to conservation laws. Varying the action for a point particle

$$S = \int \left(g_{tt}dt^2 + g_{rr}dr^2 + g_{\theta\theta}d\theta^2 + g_{\varphi\varphi}d\varphi^2 + 2g_{t\varphi}dtd\varphi \right)^{\frac{1}{2}} \tag{3}$$

with respect to t gives the geodesic equation $\frac{d}{d\tau}(g_{tt}\frac{dt}{d\tau} + g_{t\varphi}\frac{d\varphi}{d\tau}) = 0$, and with respect to φ gives another geodesic equation $\frac{d}{d\tau}(g_{\varphi t}\frac{dt}{d\tau} + g_{\varphi\varphi}\frac{d\varphi}{d\tau}) = 0$, which we recognize as energy and angular momentum conservation, respectively. In other words, the quantities

$$\epsilon = -\left(g_{tt}\frac{dt}{d\tau} + g_{t\varphi}\frac{d\varphi}{d\tau} \right) \tag{4}$$

and

$$l = g_{\varphi t}\frac{dt}{d\tau} + g_{\varphi\varphi}\frac{d\varphi}{d\tau} \tag{5}$$

do not change along geodesics. More explicitly, a particle of mass m has momentum $p^\mu = m\frac{dx^\mu}{d\tau}$, and its motion in spacetime conserves the components $p_t = g_{t\nu}p^\nu$ and $p_\varphi = g_{\varphi\nu}p^\nu$ (note the lowered indices).

More formally, our spacetime is isometric* under $t \to t + \text{constant}$ and $\varphi \to \varphi + \text{constant}$, and thus possesses two Killing vectors† $\xi_e = (1, 0, 0, 0)$ and $\xi_l = (0, 0, 0, 1)$. The two quantities $E = -\xi_e \cdot p = -\xi_e^\mu p_\mu = -p_t = -(g_{tt}p^t + g_{t\varphi}p^\varphi)$ and $L = \xi_l \cdot p = \xi_l^\mu p_\mu = p_\varphi = (g_{\varphi t}p^t + g_{\varphi\varphi}p^\varphi)$, corresponding to energy and angular momentum, respectively, are conserved for particles moving around in the spacetime. (The quantities E and L are simply ϵ and l, respectively, multiplied by m.)

It is worthwhile to comment on the signs in (4), (5), and the expressions just given for E and L. In the $(- + + +)$ convention used here, $g_{\varphi\varphi} > 0$. We define the angular momentum

---

* We will explore isometry in detail in chapter IX.6. For now the term "iso + metry" simply means that the geometry stays the same.

† The rich man would want to write $\xi_e = \xi_e^\mu \partial_\mu = \frac{\partial}{\partial t}$ and $\xi_l = \xi_l^\mu \partial_\mu = \frac{\partial}{\partial \varphi}$, as I explained back in chapter V.5.

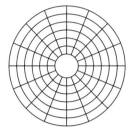 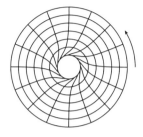

Figure 1 Frame dragging due to spacetime being deformed by a rotating body.

of a particle whose φ coordinate increases with increasing proper time τ as positive. Hence, angular momentum is $+p_\varphi$.

Frame dragging

From this general form, we can immediately deduce an interesting consequence. The novel feature is that the angular momentum $l \equiv g_{t\varphi}\frac{dt}{d\tau} + g_{\varphi\varphi}\frac{d\varphi}{d\tau}$ now consists of two terms, thanks to the presence of the nondiagonal term $g_{t\varphi}$.

Drop a particle, massive or massless,* from far away, with vanishing initial angular momentum[†] $l = 0$, toward the rotating body (notice that, thus far in the discussion, nothing requires that the metric be that of a black hole). Far away, we expect spacetime to approach Minkowskian, so that $g_{t\varphi} \to 0$, $g_{\varphi\varphi} \to 1$. Thus $l = 0$ means that $\frac{d\varphi}{d\tau} \to 0$ far away, as expected.

But as this particle approaches the rotating body, since the angular momentum l, being conserved, stays at 0, the particle picks up, according to (5), a position dependent angular velocity

$$\omega(r, \theta) \equiv \frac{d\varphi}{dt} = \frac{d\varphi}{d\tau}\bigg/\frac{dt}{d\tau} = -\frac{g_{t\varphi}}{g_{\varphi\varphi}} \tag{6}$$

Angular velocity without angular momentum!

Note that this angular velocity is defined by the rate of change of φ with respect to coordinate time, not proper time. Furthermore, recall that $\omega(r, \theta)$ has already been defined in (2): this discussion reveals its physical meaning.

We interpret this peculiar phenomenon, known as frame dragging, as due to spacetime being deformed by the rotating body (see figure 1). We fix the direction of rotation by taking $g_{t\varphi} < 0$ so that $\omega > 0$ (since $g_{\varphi\varphi} > 0$ in our $(-++)$ convention).

---

* For a massless particle, the parameter τ should be interpreted as an affine parameter, not proper time, as has already been explained a number of times.

[†] As always, we prefer to be less wordy at the cost of some loss of precision. Thus, we generally eschew saying things like "angular momentum per unit mass" if we can get away with it without confusing anybody.

It is worth emphasizing that frame dragging is not a mysterious effect associated somehow only with black holes, as some people confusedly think. It is a general relativistic effect, which does not occur in Newtonian gravity, generated by any rotating massive body. In particular, the earth drags the spacetime frame around it.

The Kerr solution turns out to be arithmetically complex, and so it would be advisable to do a back-of-the-envelope estimate of this effect at this point. Let the body rotate with angular velocity $\Omega \sim v/R$, where R denotes the object's characteristic size. Through gravity the object curves the spacetime around it, which causes the particle, in seeking the best deal (namely the geodesic), to be[1] "dragged along." The strength of gravity is characterized by the dimensionless ratio GM/Rc^2, as we have seen many times. Thus, we might expect

$$\omega \sim \frac{GM}{Rc^2}\Omega \sim \left(\frac{GM}{Rc^2}\right)\left(\frac{v}{R}\right) \tag{7}$$

Let us also anticipate how the Kerr solution would be parametrized. In units with $G = 1$ and $c = 1$, M has dimensions of length, and angular momentum $J \sim MvR$ has dimensions of length squared. Let us measure rotation by the length $a = J/M$. Indeed, it is convenient to continue using the length $r_S \equiv 2GM = 2M$, even though we are not dealing with the Schwarzschild solution here. We then have the dimensionless measure of rotation $2a/r_S = J/M^2$, which is of order $\sim v/c$ if we set $R \sim r_S \sim M$.

Stationary limit surface

Consider a light ray emitted, initially with $dr = 0$ and $d\theta = 0$, from some point. We have initially $0 = ds^2 = g_{tt}dt^2 + 2g_{t\varphi}dtd\varphi + g_{\varphi\varphi}d\varphi^2$. Solving this quadratic equation for $d\varphi$, we obtain

$$\frac{d\varphi}{dt} = \frac{-g_{t\varphi} \pm \sqrt{g_{t\varphi}^2 - g_{tt}g_{\varphi\varphi}}}{g_{\varphi\varphi}} = -\frac{g_{t\varphi}}{g_{\varphi\varphi}} \pm \sqrt{\left(\frac{g_{t\varphi}}{g_{\varphi\varphi}}\right)^2 - \frac{g_{tt}}{g_{\varphi\varphi}}} \tag{8}$$

To save writing, it is customary to define, as in (6), $\omega \equiv -g_{t\varphi}/g_{\varphi\varphi} > 0$ (which we note is a function of r and θ).

Far away from the rotating body or black hole, with $g_{tt} < 0$ and $g_{\varphi\varphi} > 0$, we have two roots

$$\Omega_+ = \left(\frac{d\varphi}{dt}\right)_+ = \omega + \sqrt{\omega^2 + \left|\frac{g_{tt}}{g_{\varphi\varphi}}\right|} > 0 \tag{9}$$

and

$$\Omega_- = \left(\frac{d\varphi}{dt}\right)_- = \omega - \sqrt{\omega^2 + \left|\frac{g_{tt}}{g_{\varphi\varphi}}\right|} < 0 \tag{10}$$

(Recall that $\omega > 0$.) These two quantities, Ω_+ and Ω_-, one positive and one negative, describe two light rays, known as corotating and counterrotating, respectively, emitted in the same and in the opposite direction as the direction of rotation. Note that, except

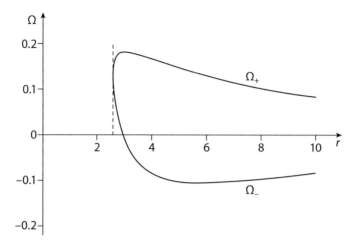

Figure 2 A schematic plot of Ω_+ and Ω_- in the equatorial plane as a function of r for a Kerr black hole with $r_S = 3$ and $a = 1$ (and hence $r_{S+} = r_S = 3$ and $r_+ \simeq 2.618$). The upper and lower curves correspond to Ω_+ and Ω_-, respectively. On the stationary limit surface $r = r_{S+}$, g_{tt} vanishes, so that $\Omega_+ = 2\omega$ and $\Omega_- = 0$: the lower curve crosses the horizontal axis. The rotating body has caused the counterrotating light on the stationary limit surface to stand still. On the outer horizon, $r = r_+$, $\Omega_+ = \Omega_-$, and the upper and lower curves meet.

in the equatorial plane $\theta = \pi/2$, light rays do not maintain $d\theta = 0$; hence in general, $\Omega_\pm$ denotes the angular velocities at emission only.

It is worthwhile to note that the discussion here, which involves light, is not be confused with the discussion of frame dragging in the preceding section, which applies to massive as well as massless particles. In particular, there we specialized to $l = 0$ for simplicity. In contrast, l is not specified here. Students sometimes confound these two distinct discussions, since some of the same metric components, $g_{t\varphi}$ and $g_{\varphi\varphi}$, are involved. Note, however, that g_{tt}, which appears in (9) and (10), did not enter into the discussion in the preceding section.

Indeed, we are now going to talk about g_{tt}. Far away, $g_{tt} \sim -1 < 0$. Suppose that, as we come in closer, there exists a surface, known as a stationary limit surface, on which $g_{tt} = 0$. Then on that surface, $\Omega_+ = 2\omega$ and $\Omega_- = 0$. The rotating body has caused the counterrotating light on the stationary limit surface to stand still!

As we will see, in the Kerr solution, g_{tt} does vanish as we come in. In figure 2, we plot Ω_+ and Ω_- for a particular Kerr black hole.

By the discussion back in chapter V.4, the $g_{tt} = 0$ surface is also the surface of infinite redshift. It is worth emphasizing that for a rotating black hole, there is no a priori reason why g_{rr} must blow up where g_{tt} vanishes, as in the Schwarzschild solution.

Even closer in, g_{tt} turns positive. Both

$$\Omega_+ = \omega + \sqrt{\omega^2 - \left|\frac{g_{tt}}{g_{\varphi\varphi}}\right|} \quad \text{and} \quad \Omega_- = \omega - \sqrt{\omega^2 - \left|\frac{g_{tt}}{g_{\varphi\varphi}}\right|}$$

are now positive. Even light can no longer move in the direction opposite to that of the rotation. All particles are swept along, hence the term "stationary limit surface." Inside this surface, you are swept along with the flow no matter how powerful your rocket pack might be. Everything is moving in the same direction as the central body rotates.

A clarifying remark here might be helpful. The angular velocities $\Omega_\pm = (\frac{d\varphi}{dt})_\pm$ refer to how the test particle moves through the coordinate φ fixed with respect to the stars. Relative to the frame being dragged along, $\Omega_+ - \omega = (\frac{d\varphi}{dt})_+ - \omega$ and $\Omega_- - \omega = (\frac{d\varphi}{dt})_- - \omega$ remain positive and negative, respectively.

Moving ever closer in, we may reach a point at which $g_{t\varphi}^2 - g_{tt}g_{\varphi\varphi} = 0$: on this surface, corotating and counterrotating light are emitted with the same angular velocity $\Omega_H \equiv \Omega_+ = \Omega_-$. You have no choice as an observer: your angular velocity $\Omega_{observer}$, squeezed between Ω_+ and Ω_-, must be equal to Ω_H. Guess what the subscript H signifies.

Three regimes

In summary, we have described three regimes: (I) with Ω_+ positive and Ω_- negative, you can move left or move right within limits; (II) with Ω_+ and Ω_- both positive, you are forced to move right; and finally, (III) with $\Omega_+ = \Omega_-$ both positive, you are forced to move right in lockstep with everybody else.

Regarding $g_{\mu\nu}$ as the matrix

$$\begin{pmatrix} g_{tt} & g_{t\varphi} & 0 & 0 \\ g_{t\varphi} & g_{\varphi\varphi} & 0 & 0 \\ 0 & 0 & g_{rr} & 0 \\ 0 & 0 & 0 & g_{\theta\theta} \end{pmatrix}$$

we readily recognize the combination

$$D \equiv g_{t\varphi}^2 - g_{tt}g_{\varphi\varphi} \tag{11}$$

that appears in (8) as minus the determinant of the 2-by-2 submatrix in the upper left corner. Note how D also appears in (2). The inverse matrix $g^{\mu\nu}$ is given by

$$g^{\mu\nu} = \begin{pmatrix} -D^{-1}g_{\varphi\varphi} & D^{-1}g_{t\varphi} & 0 & 0 \\ D^{-1}g_{t\varphi} & -D^{-1}g_{tt} & 0 & 0 \\ 0 & 0 & 1/g_{rr} & 0 \\ 0 & 0 & 0 & 1/g_{\theta\theta} \end{pmatrix} \tag{12}$$

Thus, at $D = 0$, the inverse metric $g^{\mu\nu}$ ceases to exist. Note also that $g^{rr} = 1/g_{rr}$.

Falling into a rotating black hole

We have tracked the angular velocity of a test particle of mass m falling into a rotating black hole. What about the other conserved quantity, its energy E?

One subtlety is that, because of the presence of the unfamiliar cross term $g_{t\varphi}$ in the metric, we have to clarify in our own heads whether it is p_t or p^t that is conserved. The conserved guy is in fact p_t (which we identified as $-E$), as we have mentioned in passing.

We can solve the mass-shell condition $p^2 = g_{\mu\nu}p^\mu p^\nu = g^{\mu\nu}p_\mu p_\nu = -m^2$ for E, as usual. The key is to use the form $g^{\mu\nu}p_\mu p_\nu$ for p^2, since as I just explained, p_t (and p_φ) are the conserved quantities we want to work with. Using the inverse metric (12), we write $g^{\mu\nu}p_\mu p_\nu = -m^2$ as

$$-g_{\varphi\varphi}p_t^2 + 2g_{t\varphi}p_t p_\varphi - g_{tt}p_\varphi^2 + D\left(\frac{p_r^2}{g_{rr}} + \frac{p_\theta^2}{g_{\theta\theta}} + m^2\right) = 0 \tag{13}$$

with $D = g_{t\varphi}^2 - g_{tt}g_{\varphi\varphi}$, as defined in (11). Note the oddly mismatched indices. Since neither $p_r = g_{rr}p^r$ nor p^r is conserved, we might as well write $\frac{p_r^2}{g_{rr}} = g_{rr}(p^r)^2 = m^2 g_{rr}\left(\frac{dr}{d\tau}\right)^2$. In this way, we express the last three terms in (13) as $K \equiv Dm^2\left(g_{rr}\left(\frac{dr}{d\tau}\right)^2 + g_{\theta\theta}\left(\frac{d\theta}{d\tau}\right)^2 + 1\right)$.

Solving the quadratic equation (13) for p_t, we obtain

$$E = -p_t = -\frac{g_{t\varphi}}{g_{\varphi\varphi}}p_\varphi + \sqrt{\frac{1}{g_{\varphi\varphi}^2}\left(\left(g_{t\varphi}^2 - g_{tt}g_{\varphi\varphi}\right)p_\varphi^2 + g_{\varphi\varphi}K\right)} \tag{14}$$

Note that, importantly, we have chosen the $+$ root, since far from the black hole (or for a weakly rotating body), where spacetime becomes Minkowskian and sanity is restored, this expression corresponds* to the correct $E = +\sqrt{\vec{p}^2 + m^2}$, rather than to[2] $E = -\sqrt{\vec{p}^2 + m^2}$. Once again, the weird-looking feature in (14), namely a term outside the square root, is due to the presence of the cross term $g_{t\varphi}$ in the metric. We will return to this expression in appendix 1.

The Kerr black hole

Thus far, we have been squeezing physics out of the general stationary cylindrically symmetric spacetime in (1). To go further and to see that what we say would happen actually happens, we need the specific metric of a rotating black hole.

In 1963, Kerr found a solution[3] of Einstein's field equation $R_{\mu\nu} = 0$ characterized by two parameters r_S and a with dimension of length:

$$ds^2 = -\left(1 - \frac{rr_S}{\rho^2}\right)dt^2 - \frac{2r_S ar\sin^2\theta}{\rho^2}dt d\varphi + \frac{\rho^2}{\Delta}dr^2 + \rho^2 d\theta^2$$
$$+ \left(r^2 + a^2 + \frac{r_S a^2 r\sin^2\theta}{\rho^2}\right)\sin^2\theta d\varphi^2 \tag{15}$$

with

$$\rho^2 = r^2 + a^2\cos^2\theta \quad \text{and} \quad \Delta = r^2 + a^2 - rr_S \tag{16}$$

---

* Take a particle at rest far away, so that $\frac{dr}{d\tau} = 0$, $\frac{d\theta}{d\tau} = 0$, $p_\varphi = 0$, $g_{\varphi\varphi} = 1$, $g_{t\varphi} = 0$, $g_{tt} = -1$, $D = 1$, $K = m^2$, and so E reduces to $+\sqrt{m^2}$.

Note that in this chapter, r_S is merely a convenient shorthand borrowed from our discussion of the Schwarzschild black hole: nothing much happens to (15) at $r \sim r_S$.

Far away from the black hole, as $r \to \infty$, spacetime approaches Minkowskian flat, as we would expect. In particular, $g_{00} \to -\left(1 - \frac{r_S}{r}\right)$. Using a result we learned as far back as chapter IV.3, we identify $M = \frac{1}{2}r_S$ as the mass of the rotating black hole.

Our friend the Smart Experimentalist explains, "The mass of an astrophysical object is not some symbol in a theoretical expression, but a quantity we should inquire of the astronomer peering through the telescope. The astronomer tells us that the mass is deduced from the orbit of a test particle, typically a planet, circling that object. Ultimately, that's related to the $O(1/r)$ term in g_{00}."

Excellent! That's the operational definition of mass. Similarly, the astronomer could deduce the angular momentum of the astrophysical object in principle (if not in practice) by watching how a gyroscope* circling the object precesses. That precession is governed by the asymptotic behavior of $g_{t\varphi}$; in particular, for the Kerr solution, we detect a deviation from Minkowski spacetime given by

$$-2g_{t\varphi}dtd\varphi = \frac{2r_S a r \sin^2 \theta}{\rho^2}dtd\varphi \to \frac{2r_S a \sin^2 \theta}{r}dtd\varphi = \frac{2r_S a}{r^3}dt(xdy - ydx) \tag{17}$$

upon reverting back, in the last step, to the usual Cartesian coordinates. The angular momentum of the object is defined to be $J = \frac{1}{2}r_S a = Ma$. Putting this into the form of the metric in (2), we have $ds^2 = \left(\cdots + r^2 \sin^2 \theta \left(d\varphi - \frac{2J}{r^3}dt\right)^2 + \cdots\right)$, since $\omega = -g_{t\varphi}/g_{\varphi\varphi} \to (r_S a \sin^2 \theta/r)/r^2 \sin^2 \theta = r_s a/r^3 = 2J/r^3$. (Note that as remarked earlier, J has dimensions of length squared.) We should show at some point that this definition of J reduces in the appropriate limit to what we commonly understand to be angular momentum; we will do this in chapter IX.4. Thus,

$$a = \frac{J}{M} = \frac{2J}{r_S} \tag{18}$$

In short, the Kerr solution is characterized by two lengths, r_S and a, corresponding to mass and angular momentum, respectively.

Considering that it took[4] almost 50 years for this solution to be found (while the Schwarzschild solution was found within a year of 1915), we realize that the Kerr metric represents a highly nontrivial accomplishment.[5] Unfortunately, there is not a simple[6] derivation[7] of the Kerr metric comparable to the straightforward derivation of the Schwarzschild metric. See appendix 2 for a possible approach. Of course, it is straightforward, particularly with the help of a computer, to verify that (15) is in fact a solution.

For this text, I am content to merely introduce you to some key features.

1. Let's check that our estimate of frame dragging in (7) is on the money: for a slowly rotating body, $\rho \sim r$ and $a = J/M \sim Mvr/M \sim vr$, and so indeed $\omega \sim g_{t\varphi}/g_{\varphi\varphi} \sim (r_S ar/\rho^2)/r^2 \sim Ma/r^3 \sim Mv/r^2 \sim (GM/Rc^2)(v/R)$, in agreement with (7). In the last step, we took R to be

---

* We will discuss the precession of gyroscopes in chapter IX.2.

the size of a black hole and restored G and c. If you like, you can regard this as a verification of (18) up to an overall factor, at least for a slowly rotating body.

2. After admiring (15), we check that it reduces appropriately in various limits.

  a. As $r \to \infty$, the metric becomes asymptotically flat, as we have already noted.

  b. As $a \to 0$, we recover the Schwarzschild solution:

$$ds^2_{\text{Kerr}} = ds^2_{\text{Schwarzschild}} - \left(2r_S a \sin^2 \theta / r \right) dt d\varphi + O\left(a^2 \right) \tag{19}$$

  c. As a particularly interesting limit, take $M \to 0$ and $J \to 0$, keeping the ratio $a = J/M$ fixed. We obtain

$$ds^2 = -dt^2 + \left(\frac{r^2 + a^2 \cos^2 \theta}{r^2 + a^2} dr^2 + \left(r^2 + a^2 \cos^2 \theta \right) d\theta^2 + \left(r^2 + a^2 \right) \sin^2 \theta d\varphi^2 \right) \tag{20}$$

  Are you surprised that you do not recover flat space? But perhaps after a moment you remember appendix 2 from way way back in chapter I.5: (20) in fact describes Minkowskian spacetime heavily disguised!

3. The general discussion in the preceding sections, in particular the result (6), holds for the Kerr solution of course, with the off-diagonal metric component $g_{t\varphi}(r, \theta) = -r_S a r \sin^2 \theta / \rho^2$. A particle dropped with $l = 0$ from far away attains the angular velocity

$$\omega(r, \theta) = -\frac{g_{t\varphi}}{g_{\varphi\varphi}} = \frac{r_S a r}{\rho^2 \left(r^2 + a^2 \right) + r_S a^2 r \sin^2 \theta} = \frac{r_S a r}{\left(r^2 + a^2 \right)^2 - \Delta a^2 \sin^2 \theta} \tag{21}$$

with Δ defined in (16).

4. The surface of infinite redshift $g_{tt} = 0$, on which counterrotating light stands still, is given by $\rho^2 = r r_S$, which has two solutions

$$r_{S\pm} = \tfrac{1}{2} \left(r_S \pm \sqrt{r_S^2 - 4a^2 \cos^2 \theta} \right) = M \pm \sqrt{M^2 - a^2 \cos^2 \theta} \quad \text{(stationary limit } g_{tt} = 0\text{)} \tag{22}$$

There are thus two surfaces of infinite redshift, an outer and an inner. (Note that the S in $r_{S\pm}$ is for stationary, while the S in r_S is for Schwarzschild.)

5. In accordance with our general discussion earlier, the Kerr metric is invariant under $t \to t + \text{constant}$ and $\varphi \to \varphi + \text{constant}$ and thus possesses two Killing vectors $\xi_e = (1, 0, 0, 0)$ and $\xi_l = (0, 0, 0, 1)$.

6. The Kerr metric can be written in a number of different forms (see exercises 4 and 5). Define $\Sigma^2 \equiv (r^2 + a^2)\rho^2 + r_S r a^2 \sin^2 \theta = (r^2 + a^2)^2 - \Delta a^2 \sin^2 \theta$, so that $g_{\varphi\varphi} = \frac{\Sigma^2}{\rho^2} \sin^2 \theta$. Then, for example, using (2), we can write

$$ds^2 = -\frac{\rho^2 \Delta}{\Sigma^2} dt^2 + \frac{\rho^2}{\Delta} dr^2 + \rho^2 d\theta^2 + \frac{\Sigma^2}{\rho^2} \sin^2 \theta (d\varphi - \omega dt)^2 \tag{23}$$

Another form is given by

$$ds^2 = -\frac{\Delta}{\rho^2} \left(dt - a^2 \sin^2 \theta d\varphi \right)^2 + \frac{\rho^2}{\Delta} dr^2 + \rho^2 d\theta^2 + \frac{1}{\rho^2} \sin^2 \theta \left((r^2 + a^2) d\varphi - a dt \right)^2 \tag{24}$$

Physical and coordinate singularities

We see that the Kerr metric (15) is singular at $\rho = 0$ and at $\Delta = 0$.

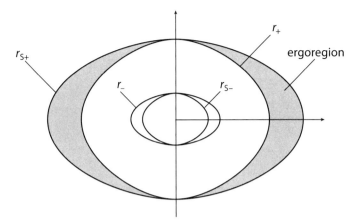

Figure 3 The surfaces $r = r_\pm$ (as explained in the text, surfaces of constant r are not spheres) and the surfaces of infinite redshift $r = r_{S\pm}(\theta)$ are shown schematically. As indicated in the text, the surfaces touch pairwise at the north and south poles. The region enclosed between the stationary limit surface r_{S+} and the horizon r_+ is known as the ergosphere or ergoregion. Inside the stationary limit surface of a Kerr black hole, the coordinate t becomes spacelike and energy morphs into momentum, but if you are outside the horizon, you can still get out if you want.

In the limit $a \to 0$, $\rho = 0$ reduces to $r = 0$, and $\Delta = 0$ to $r = r_S$. Thus, our experience with the Schwarzschild black hole suggests that $\rho = 0$ represents a physical singularity and $\Delta = 0$ a coordinate singularity, a highly plausible supposition that we can check by calculating $R_{\mu\nu\rho\sigma} R^{\mu\nu\rho\sigma}$, for example.

The physical singularity corresponds to, according to (16), $r^2 + a^2 \cos^2\theta = 0$, that is, $r = 0$ and $\theta = \frac{1}{2}\pi$. But according to appendix 2 of chapter I.5, this describes a ring of radius a.

In contrast, the coordinate singularities at $\Delta = r^2 + a^2 - rr_S = 0$, that is, at

$$r_\pm = \frac{1}{2}\left(r_S \pm \sqrt{r_S^2 - 4a^2}\right) = M \pm \sqrt{M^2 - a^2} \tag{25}$$

describe two ellipsoids. See figure 3.

Extremal black hole

Note that (25) suggests, but does not prove, that $|a| \le M = \frac{1}{2}r_S$, which according to (18), corresponds to the maximum angular momentum

$$|J| \le M^2 \tag{26}$$

An extremal Kerr black hole is one with angular momentum $|J| = M^2$. Most astrophysical black holes are observed to be nearly extremal, as would be expected if infalling debris tends to increase the angular momentum. Theoretically, extremal black holes also play an important role in string theory. Note that, heuristically, extremality is attained for $MvR \sim M^2v \sim M^2$, that is $v \sim 1$. So physically, it seems plausible that the angular momentum of a black hole is bounded. For a given mass, you can't keep on pumping angular momentum into the system.

The outer horizon

Let us locate ourselves at some point well outside the black hole (for $r > r_+$ say, so that $\Delta > 0$) and watch an outgoing photon. Setting $ds = 0$ in (15), we find that dr is given by the positive square root of the quantity

$$\frac{\Delta}{\rho^2}\left\{\left(1 - \frac{rr_S}{\rho^2}\right)dt^2 + \frac{2r_Sar\sin^2\theta}{\rho^2}dt d\varphi - \rho^2 d\theta^2 - \left(r^2 + a^2 + \frac{r_Sa^2r\sin^2\theta}{\rho^2}\right)\sin^2\theta d\varphi^2\right\}$$

We want to see whether the photon manages to get out. So, we want to maximize dr. Evidently, to do this, we should set $d\theta = 0$. But because of the cross term $g_{t\varphi}$, we should not set $d\varphi = 0$, that is, restrict our attention to radial light rays, as we did for the Schwarzschild black hole. Physically, this is clear.

We now move in closer and closer to the black hole. At some point, the photon will not be able to get out. For $r > r_S/2$, the quantity $\Delta = r^2 + a^2 - rr_S$ decreases as r decreases, eventually vanishing, at which point $dr = 0$. The photon can no longer get out; this defines the horizon. According to (25), the vanishing of Δ occurs at r_+. Our suspicion, based on applying the "correspondence principle" between the Kerr and Schwarzschild black holes, turns out to be valid.

For definiteness, let us focus on what happens at r_+ (which, as we can see from (25), is indeed $> r_S/2$). For further analysis, it is more convenient to use (23) rather than (15). With $ds = 0$ and $dr = 0$ in (23), set $\Delta = 0$ to obtain

$$0 = \rho_+^2 d\theta^2 + \frac{\Sigma_+^2}{\rho_+^2}\sin^2\theta\left(d\varphi - \omega_+ dt\right)^2 \tag{27}$$

where the subscript $+$ indicates that the various quantities are to be evaluated at $r = r_+$:

$$\rho_+^2 = r_+^2 + a^2\cos^2\theta, \quad \omega_+ = \frac{r_Sar_+}{\left(r_+^2 + a^2\right)^2} = \frac{r_Sar_+}{\left(r_+r_S\right)^2} = \frac{a}{r_+r_S}, \quad \Sigma_+ = r_+^2 + a^2 = r_+r_S \tag{28}$$

Now we easily solve (27) to find $d\theta = 0$ and

$$\frac{d\varphi}{dt} = \omega_+ = \frac{a}{r_+r_S} \tag{29}$$

In other words, on the horizon, light rays move along the trajectory

$$(dt, dr, d\theta, d\varphi) \propto l^\mu \equiv \left(1, 0, 0, \frac{a}{r_+r_S}\right) \tag{30}$$

The horizon is a null surface spanned by this null vector l^μ and two spacelike vectors, which we can take to be $h^\mu = (0, 0, 1, 0)$ and $k^\mu = (0, 0, 0, 1)$. Note that these vectors are orthogonal to l^μ: for instance, $g_{\mu\nu}l^\mu k^\nu = g_{t\varphi} + g_{\varphi\varphi}\omega_+ = 0$ by virtual of the definition of ω. Note that the normal to this null surface is l^μ itself. (Recall the discussion of the light cone in Minkowski spacetime back in chapter III.3.)

Take a constant t slice of this null surface. We obtain from (27) a 2-dimensional surface with the line element

$$dS^2 = \rho_+^2 d\theta^2 + \frac{\Sigma_+^2}{\rho_+^2} \sin^2\theta d\varphi^2 = \left(r_+^2 + a^2\cos^2\theta\right)d\theta^2 + \frac{\left(r_+^2 + a^2\right)^2}{r_+^2 + a^2\cos^2\theta}\sin^2\theta d\varphi^2 \quad (31)$$

Those readers with a good memory will recognize this as the squashed sphere you worked out in exercise I.5.12. For an extremal black hole, the distance around a circle of fixed longitude through the poles works out to be $\simeq 3.82 r_S$, considerably less than the distance (extremal or not) around the equator $2\pi r_S$, in accordance with our intuition about how spacetime might be squashed around a rotating black hole. Notice that all of this is happening to empty spacetime; what is being squashed is not a spinning material sphere. The area of this squashed sphere is given by

$$A = \int d\theta d\varphi \sqrt{g} = 4\pi\left(r_+^2 + a^2\right) = 8\pi\left(M^2 + \sqrt{M^4 - J^2}\right) \quad (32)$$

(with g the determinant of the metric in (31), as explained back in chapter I.5). As $J \to 0$, we recover the Schwarzschild result $A = 16\pi M^2$.

Also, note that the combination $D = g_{t\varphi}^2 - g_{tt}g_{\varphi\varphi}$ that appeared inside the square root in (8) works out nicely to be $\Delta\sin^2\theta$ in the Kerr solution. Thus, the horizon (where Δ vanishes) acquires another significance: as we have learned, it is where the corotating and counterrotating light beams have the same angular velocity:

$$\Omega_H \equiv \Omega_+ = \Omega_-$$

$$= \omega\left(r = r_+\right) = \frac{a}{r_+^2 + a^2} = \frac{a}{r_S r_+} = \frac{2a}{r_S\left(r_S + \sqrt{r_S^2 - 4a^2}\right)}$$

$$= \frac{J}{2M\left(M^2 + \sqrt{M^4 - J^2}\right)} \quad (33)$$

Ergoregion and the Penrose process

To summarize, there are two surfaces of infinite redshift, an outer and an inner,

$$r_{S\pm} = \frac{1}{2}\left(r_S \pm \sqrt{r_S^2 - 4a^2\cos^2\theta}\right) = M \pm \sqrt{M^2 - a^2\cos^2\theta} \quad \text{(stationary limit, } g_{tt} = 0) \quad (34)$$

There are also two horizons, an outer and an inner,

$$r_\pm = \frac{1}{2}\left(r_S \pm \sqrt{r_S^2 - 4a^2}\right) = M \pm \sqrt{M^2 - a^2} \quad \text{(horizon, } \Delta = 0) \quad (35)$$

I also remind the reader that in the Kerr solution $D = g_{t\varphi}^2 - g_{tt}g_{\varphi\varphi} = \Delta\sin^2\theta$ vanishes at the horizon. Hence, $g_{rr} = -\rho^2/\Delta = -\infty$ and $g^{rr} = 0$ at the horizon.

For the rest of our discussion, we will focus on the outer stationary limit surface and the outer horizon. For the Kerr black hole, these two surfaces, r_{S+} and r_+, no longer coincide, as is the case for the corresponding surfaces for the Schwarzschild black hole. Comparing (34) and (35), we see that $r_+ \le r_{S+}$, with equality attained at the two poles. Thus, the outer horizon lies inside the outer stationary limit surface, touching at the two poles, while the gap between the two surfaces is largest in the equatorial plane $\theta = \pi/2$. We drop the modifier "outer" henceforth.

The region enclosed between r_+ and r_{S+} is known as the ergosphere ("ergo" being Greek for work, as in ergonomics and erg). The somewhat more accurate but awkward term ergoregion is also used. While particles cannot stay at rest in the ergoregion, they can still escape to infinity since they are still outside the outer horizon. See figure 3.

The existence of a region between the stationary limit surface r_{S+} and the horizon r_+ allows exotic new physics not possible with the Schwarzschild black hole. Inside the stationary limit surface of a Kerr black hole, strange things start happening: in particular, the coordinate t becomes spacelike, and energy morphs into momentum. But if you are outside the horizon, you can still get out to tell the tale! Much cooler to fall into a Kerr black hole than into a Schwarzschild black hole.

Rather craftily, Penrose realized that these considerations allow us to extract energy from a rotating black hole. Consider a process in which a particle called "infalling" falls in freely from ∞. In the ergoregion, it goes into two particles called "outgoing" and "doomed," with their momenta arranged in such a way that while the doomed particle falls through the horizon, the outgoing particle escapes, moving freely along its geodesic to ∞. This could be a subnuclear process, such as $\pi^+ \to \mu^+ + \nu$ (a pion decaying into a muon plus a neutrino), or an entirely classical process in which* we throw a bad guy out of our rocketship. Conservation of 4-momentum $p_{in} = p_{out} + p_{doomed}$ holds, of course. Furthermore, along the geodesics of the various particles, the quantities $\xi \cdot p_{in}$, $\xi \cdot p_{out}$, and $\xi \cdot p_{doomed}$ are conserved, that is, they are constants of the motion. Here ξ can be either ξ_e or ξ_l. (I remind you that $\xi_e = (1, 0, 0, 0)$ and $\xi_l = (0, 0, 0, 1)$ denote the two Killing vectors that the Kerr spacetime possesses.) In other words, for each of the three particles, we have energy and angular momentum conservation.

Confusio looks a bit bewildered. "It seems like we are talking about conservation in two distinct ways."

Indeed! The discussion in some textbooks appears to be confused on this issue. That the sums of the momenta are the same before and after some local process is a direct consequence of the equivalence principle, which says that in a small enough region of spacetime (in a neighborhood around where the pion decays, for example), we can choose coordinates so that physics is exactly as it would be in flat spacetime. More mathematically, it follows from $D_\mu T^{\mu\nu} = 0$, which in turn follows from general covariance (as was discussed in chapter VI.5). In contrast, the constancy of $\xi \cdot p$ along various geodesics follows from specific properties of the spacetime we are in, namely its invariance under translation in t and φ.

Let us contract the Killing vector ξ_e with momentum conservation to obtain

$$\xi_e \cdot p_{in} = \xi_e \cdot p_{out} + \xi_e \cdot p_{doomed} \tag{36}$$

Write $E_{in}(\infty) \equiv -\xi_e \cdot p_{in}$ and $E_{out}(\infty) \equiv -\xi_e \cdot p_{out}$ to emphasize that these are the energies that the infalling and outgoing particles have at ∞, far away from the Kerr black

---

* This reminds me of action movies or kung fu stories in which, as you and a bad guy fall off a cliff, you, being a physics student, give the bad guy a downward shove, and exploiting Newton's law of action and reaction, bounce back onto the cliff.

hole. Of course, as always, being energies of physical particles, $E_{in}(\infty)$ and $E_{out}(\infty)$ are necessarily positive. In sharp contrast, we write simply $\varepsilon_{doomed} \equiv -\xi_e \cdot p_{doomed}$, without any (∞) symbol, to indicate that the doomed particle never gets out to flat spacetime, and so ε_{doomed} can be either positive or negative. Indeed, since $g_{\mu\nu}\xi_e^\mu \xi_e^\nu = g_{tt} = -\left(1 - \frac{rr_S}{\rho^2}\right) > 0$ for $r < r_{S+}$, in the ergoregion, the Killing vector ξ_e is spacelike, and ε_{doomed} is actually a momentum rather than an energy. From (36), we have

$$E_{out}(\infty) = E_{in}(\infty) - \varepsilon_{doomed} \tag{37}$$

and so we conclude that indeed it is possible, with $\varepsilon_{doomed} < 0$, to have $E_{out}(\infty) > E_{in}(\infty)$, that is, to get out more energy than we put in. The black hole would end up losing some of its mass M in the deal. Note that, in contrast[8] to the Hawking radiation discussed in the preceding chapter, the Penrose process is completely classical.

Thus, at least in principle,[9] we could solve[10] both the world's energy crisis and garbage problem, haha.

Angular momentum loss

Confusio lights up: "Wonderful! But what about the other Killing vector ξ_l?"

Excellent question! Confusio is getting smarter by the day. Consider an observer inside the stationary limit surface. Let his 4-velocity be given by $U^\mu = U^0(1, 0, 0, \Omega_{observer})$. In other words, for this observer, $\frac{dt}{d\tau} = U^0$ and $\frac{d\varphi}{d\tau} = U^0\Omega_{observer}$. The angular velocity $\Omega_{observer}$ must be positive, since like everybody else, the observer has to move in the direction of rotation, as explained earlier. We now use basic linear algebra to write $U^\mu = U^0(\xi_e + \Omega_{observer}\xi_l)$. It is worth emphasizing that the observer is not necessarily moving along a geodesic; he is certainly free to purchase a rocket pack and attach it to his back.

What is the energy of the doomed particle as seen by this observer? As usual, this is given by

$$-U \cdot p_{doomed} = -U^0(\xi_e \cdot p_{doomed} + \Omega_{observer}\xi_l \cdot p_{doomed}) = U^0(\varepsilon_{doomed} - \Omega_{observer}L_{doomed}) \tag{38}$$

where $L_{doomed} \equiv +\xi_l \cdot p_{doomed}$ is the angular momentum (with the plus sign, as explained earlier) of the doomed particle. But the energy of the doomed particle as measured by this observer must be positive, and hence $\varepsilon_{doomed} \geq \Omega_{observer}L_{doomed}$, which, since $\Omega_{observer}$ is positive, can be written as $\varepsilon_{doomed}/\Omega_{observer} \geq L_{doomed}$.

In the Penrose process, ε_{doomed} is negative, and so L_{doomed} must also be negative. By angular momentum conservation, as the doomed particle falls into the black hole, it reduces the black hole's angular momentum. In other words, the mass and angular momentum of the black hole change by $\delta M = \varepsilon_{doomed} < 0$ and $\delta J = L_{doomed} < 0$, respectively.

Not only does the black hole lose mass in the deal, but it also loses angular momentum by an amount δJ satisfying

$$\frac{\delta M}{\Omega_{observer}} \geq \delta J \tag{39}$$

By extracting energy from a Kerr black hole, we also decrease its angular momentum and hence reduce it eventually to a Schwarzschild black hole.

It is important to note that even if the process is not Penrose, namely if $\varepsilon_{\text{doomed}}$ is positive, we still have $\frac{\delta M}{\Omega_{\text{observer}}} \geq \delta J$, with δM positive in this case.

Area theorem

Recall the area $A = 8\pi \left(M^2 + \sqrt{M^4 - J^2} \right)$ of the black hole from (32). In a Penrose process, both M and J decrease. How does this area change? Varying, we find

$$\frac{1}{8\pi} \delta A = \frac{2M \left(M^2 + \sqrt{M^4 - J^2} \right) \delta M - J \delta J}{\sqrt{M^4 - J^2}} = \frac{J}{\sqrt{M^4 - J^2}} \left(\frac{\delta M}{\Omega_H} - \delta J \right) \tag{40}$$

where we used (33) in the last step.

But since the inequality (39) holds for any observer, it also holds for an observer hovering just outside the horizon, an observer whose angular velocity, as we saw in the discussion leading up to (33), is equal to Ω_H. Inserting $\frac{\delta M}{\Omega_H} \geq \delta J$ (which, as we had emphasized, holds regardless of whether the process is Penrose or not) into (40), we conclude that

$$\delta A \geq 0 \tag{41}$$

Remarkably, the area always increases! This result is often stated as the second law of black hole thermodynamics: no classical process can decrease the area of a black hole. As mentioned in the preceding chapter, results like this inspired Bekenstein to conjecture that the surface area of black holes should be associated with an entropy.

Appendix 1: First and second laws of black hole thermodynamics

Let's now follow a particle falling through the horizon of a Kerr black hole. Return to (14), which we rewrite here in a slightly more concise form for convenience:

$$E = -p_t = +\omega p_\varphi + \sqrt{\frac{D p_\varphi^2}{g_{\varphi\varphi}^2} + \frac{K}{g_{\varphi\varphi}}} \tag{42}$$

with $K = Dm^2 (g_{rr}(\frac{dr}{d\tau})^2 + g_{\theta\theta}(\frac{d\theta}{d\tau})^2 + 1)$. Since p_t is conserved along the particle's geodesic, we can calculate it at the instant the particle crosses the horizon, where $D = \Delta \sin^2\theta$ vanishes. Things simplify enormously! Do calculate before reading on.

Dear reader, if you drop the square root in (42), you would have made a hasty error. You should have checked if anything is blowing up as $D \to 0$. Indeed, as I noted after (35), $g_{rr} = \rho^2/\Delta \to \infty$, such that $Dg_{rr} \to \rho_+^2 \sin^2\theta$. In contrast, nothing much happens to (see (23), for example) $g_{\theta\theta} = \rho^2$ and $g_{\varphi\varphi} = \frac{\Sigma^2}{\rho^2} \sin^2\theta$. Hence, $K \to \rho_+^2 \sin^2\theta (m\frac{dr}{d\tau})_+^2$, so that $\frac{K}{g_{\varphi\varphi}} \to \frac{\rho_+^4}{\Sigma_+^2} (m\frac{dr}{d\tau})_+^2$. Thus, evaluating (42) on the horizon, we determine the energy of the infalling particle to be $E = \Omega_H L + \sqrt{\left(\frac{\rho_+^2}{\Sigma_+} m \left(\frac{dr}{d\tau} \right)_+ \right)^2}$ with $L = p_\varphi$. The mass and angular momentum of the black hole change according to (recall (9))

$$\delta M = \Omega_H \delta J + \frac{r_+^2 + a^2 \cos^2\theta}{r_+^2 + a^2} m \left| \frac{dr}{d\tau} \right|_+ \tag{43}$$

You of course recall that we proved the area theorem $\delta A \geq 0$ by inserting the key inequality $\frac{\delta M}{\Omega_H} \geq \delta J$ into (40). But now we learn more. We have identified what caused the area theorem to be an inequality rather than an equality: radial movement.

This result also serves to convince us of the physically motivated conclusion we reached in the text, that the angular momentum $|J|$ cannot exceed M^2. For an extremal black hole with $J = M^2$, we see from (33) that the angular velocity at the horizon $\Omega_H = \frac{J}{2M(M^2 + \sqrt{M^4 - J^2})}$ evaluates to $\Omega_H = \frac{1}{2M}$. Let us try to crank up J of an extremal black hole past M^2. But we just showed that $\delta M \geq \Omega_H \delta J = \frac{\delta J}{2M}$, that is, $\delta M^2 \geq \delta J$. Try as we may, we can't get J to exceed M^2.

We have already stated the second law of black hole thermodynamics. This conclusion allows us to formulate also the first law. Write (40) as

$$\delta M = \frac{\sqrt{M^4 - J^2}}{16\pi M(M^2 + \sqrt{M^4 - J^2})} \delta A + \Omega_H \delta J \tag{44}$$

The coefficient of $\frac{1}{8\pi}\delta A$ in δM is known as surface gravity and is denoted by κ. Thus, $\kappa = \frac{\sqrt{M^4 - J^2}}{2M(M^2 + \sqrt{M^4 - J^2})}$ for the Kerr black hole (and $\kappa = (4M)^{-1}$ for a Schwarzschild black hole).

If we make the correspondence $E \leftrightarrow M$, $T \leftrightarrow \kappa/2\pi$, and $S \leftrightarrow A/4$ (as in chapter VII.3), then (44) has the same form as the usual first law of thermodynamics $dE = TdS + dW = TdS - PdV$ relating the change in energy dE of a system to the work dW done. We could have used (33) to eliminate Ω_H in (44), but the form as written serves to show that the angular velocity Ω_H is dual to the angular momentum J in the same way that pressure P is dual to the volume V. The work done on the black hole by the infalling particle is $\Omega_H \delta J$.

In thermodynamics, a process is reversible if $dS = 0$. Here too, a process with $\delta A = 0$ is said to be reversible. Setting $\delta A = 0$ in (44) gives us a differential equation for $M(J)$, which we can integrate to obtain

$$2M_I^2 = M^2 + \sqrt{M^4 - J^2} \tag{45}$$

with M_I an integration constant. Physically, we can use the Penrose process to decrease both M and J, taking care to ensure that in (43), $\left|\frac{dr}{d\tau}\right|_+ = 0$. In particular, we can start with a Kerr black hole, let $J \to 0$, and end up with a Schwarzschild black hole with mass M_I. Conversely, we can start with a Schwarzschild black hole with mass M_I and crank up the angular momentum until we get an extremal black hole of mass $\sqrt{2}M_I$.

Appendix 2: The Weyl approach to the Kerr black hole

In chapter VI.3 I mentioned in an appendix Weyl's short-cut derivation of the Schwarzschild solution by plugging the Schwarzschild metric directly into the Einstein-Hilbert action and varying, a procedure justified mathematically only decades later. Following Weyl, Deser and Franklin[11] proposed plugging $ds^2 = g_{tt}dt^2 + g_{rr}dr^2 + g_{\theta\theta}d\theta^2 + g_{\varphi\varphi}d\varphi^2 + 2g_{t\varphi}dtd\varphi$ into the action and varying. As you may recognize, even with the short-cut, the situation is enormously more complicated than for the Schwarzschild case: we end up with five coupled partial differential equations for the five functions g_{tt}, g_{rr}, $g_{\theta\theta}$, $g_{\varphi\varphi}$, $g_{t\varphi}$ of r and θ. Deser and Franklin were able to make further progress only by using symmetry and gauge arguments to restrict these five functions.

Appendix 3: Rotating black holes are powerful sources of radiation

We finally come to our stated goal of understanding why rotating black holes are such powerful sources of radiation. In chapter VII.1, we calculated the amount of energy radiated by a particle in the accretion disk around a Schwarzschild black hole as it falls in. Here we will do the analogous calculation for a Kerr black hole.

In chapters V.4, VI.3, and VII.1, the motion of particles, massive or massless, was worked out in Schwarzschild spacetime. So by now you should be able to work things out for the Kerr spacetime,[12] but as you might suppose, the computations become considerably more involved. We now have only cylindrical, not spherical, symmetry, so that for the motion of a particle, only the component of its angular momentum along the direction of rotation is conserved. Thus, in general, the motion will not be confined to a plane, so that the orbits may be quite complicated. The exception is for motion entirely within the equatorial plane defined by $\theta = \pi/2$: then angular momentum conservation guarantees that the particle will stay in the plane.

Confusio looks puzzled for a moment, "But in chapter I.1, and in chapters V.4, VI.3, and VII.1, for both the Newtonian and the Schwarzschild problems, we also kept the particle in the equatorial plane."

Yes, but the difference is that in those cases, we could do that with no loss of generality, while here the equatorial plane is singled out as special.

So set $\theta = \pi/2$ and proceed as in chapter V.4. Compared to (V.4.11–12), the conservation laws are now necessarily more complicated due to the nondiagonal term $g_{t\varphi}$ in the metric. Write (4) and (5) as

$$\begin{pmatrix} g_{tt} & g_{t\varphi} \\ g_{\varphi t} & g_{\varphi\varphi} \end{pmatrix} \begin{pmatrix} \dot{t} \\ \dot{\varphi} \end{pmatrix} = \begin{pmatrix} -\epsilon \\ l \end{pmatrix} \tag{46}$$

where I have used the shorthand $\dot{t} = \frac{dt}{d\tau}$ and $\dot{\varphi} = \frac{d\varphi}{d\tau}$. As we have discussed almost ad nauseum since chapter II.2, together with two conservation laws, we also have what amounts to the definition of proper time, which after setting $\dot{\theta} = 0$ reads $g_{tt}\dot{t}^2 + 2g_{t\varphi}\dot{t}\dot{\varphi} + g_{\varphi\varphi}\dot{\varphi}^2 + g_{rr}\dot{r}^2 = -1$. Calling the matrix in (46) G, we can write the first three terms in this equation as

$$(\dot{t}\ \dot{\varphi})G \begin{pmatrix} \dot{t} \\ \dot{\varphi} \end{pmatrix} = (-\epsilon, l)G^{-1}GG^{-1}\begin{pmatrix} -\epsilon \\ l \end{pmatrix} = (-\epsilon, l)G^{-1}\begin{pmatrix} -\epsilon \\ l \end{pmatrix} = (\det G)^{-1}\left(g_{\varphi\varphi}\epsilon^2 + 2g_{t\varphi}\epsilon l + g_{tt}l^2\right)$$

But we have already evaluated $\det G = g_{tt}g_{\varphi\varphi} - g_{t\varphi}^2$. Proceeding thus, we arrive at

$$\dot{r}^2 - \frac{r_S}{r} + \frac{l^2 + a^2\left(1 - \epsilon^2\right)}{r^2} - \frac{r_S(l - a\epsilon)^2}{r^3} - \epsilon^2 + 1 = 0 \tag{47}$$

Of course, as an immediate check, we can verify that for $a = 0$, we recover (VII.1.1) for the Schwarzschild black hole. Remarkably, even though the Kerr metric is so much more complicated than the Schwarzschild metric, for this special equatorial plane case, the effective Newtonian potential still consists of a $1/r$, a $1/r^2$, and a $1/r^3$ term.

Confusio is quick to point out that, strictly speaking, this is no longer a standard Newtonian mechanics problem, since the potential also depends on the effective energy $\epsilon^2 - 1$.

But Confusio, this is no objection at all. We are merely using Newtonian mechanics as a pedagogical aid in solving an ordinary differential equation. In fact, let's write (47) as $\dot{r}^2 + V(r; l, a, r_S, \epsilon) = 0$ and think of a Newtonian particle with zero total energy moving in a potential $V(r)$ that depends on a bunch of parameters l, a, r_S, and ϵ.

From this point on, the physics is conceptually the same as in the Schwarzschild case in chapter VII.1, and I urge you to review the steps there. Physically, as before, a particle in the accretion disk, starting from far away, crashes through the other particles in the disk and eventually elbows its way into the innermost stable circular orbit, heating up the accretion disk and radiating away energy in the process. We thus have to find the radius r_{ISCO} of the innermost stable circular orbit and evaluate $V(r_{\text{ISCO}})$ to find out what fraction of the rest mass of the particle was lost to radiation.

For clarity, break the calculation into 3 steps.

1. As in chapter VII.1, we first solve $dV(r; l, a, r_S, \epsilon)/dr = 0$ to determine

$$r_{\text{min}}(l, a, r_S, \epsilon) \quad \text{and} \quad r_{\text{max}}(l, a, r_S, \epsilon)$$

the locations of the minimum and maximum of the potential, respectively. Since, as we already observed, V consists of a $1/r$, a $1/r^2$, and a $1/r^3$ term, this step only requires solving a quadratic equation.

2. The term "innermost" in the astrophysicist's acronym ISCO refers to $r_{\text{min}}(l, a, r_S, \epsilon)$ decreasing until the minimum of the potential disappears at a critical value $l = l_c = l_c(a, r_S, \epsilon)$ determined by solving $r_{\text{min}}(l, a, r_S, \epsilon) = r_{\text{max}}(l, a, r_S, \epsilon)$. Remarkably, after all the talk about stationary limit surface and frame dragging, the astrophysically relevant calculation involves only high school level algebra (but very messy algebra!).

3. The third and final step requires solving $V(r_{\text{ISCO}}; l_c, a, r_S, \epsilon) = 0$. Since r_{ISCO} has been determined in terms of (l_c, a, r_S, ϵ), and l_c in terms of (a, r_S, ϵ), this equation determines the dimensionless number $\epsilon(a, r_S)$ in terms of the two lengths (a, r_S); hence ϵ depends on the ratio a/r_S. As explained in chapter VII.1, the fraction of energy lost to radiation is then given by $1 - \epsilon(a/r_S)$.

You are invited to carry out this calculation (perhaps numerically), which I have so kindly outlined for you. You will discover that the particle, just as the light ray studied earlier in this chapter, can be corotating or counterrotating. Intuition probably tells you that by corotating, you can get in closer to the black hole and hence lose more energy. You could make up some nice plots of how the fraction of energy lost $1 - \epsilon(a/r_S)$ varies as a varies.

As for me, I am content to work out the extremal case $a = r_S/2 = M$, for which the expressions involved simplify quite a bit. Going through the steps outlined here, we find readily that* the fraction of energy lost is given by $1 - \frac{1}{\sqrt{3}} \simeq 0.42$, a whopping 42% compared to the 6% for a Schwarzschild black hole, and the pitiful 0.7% for the thermonuclear processes that power the stars!

Exercises

1   Consider the terms $a(r, \theta)dr^2 + 2c(r, \theta)drd\theta + b(r, \theta)d\theta^2$ in ds^2 in (1). Show that by redefining $r = f(\tilde{r}, \theta)$, you can get rid of the cross term.

2   Evaluate the Kerr metric in the limit $r \to \infty$.

3   Evaluate the Kerr metric in the limit $a \to 0$.

4   Write $g_{\varphi\varphi} = \Sigma^2 \sin^2 \theta / \rho^2$. Show that $\Sigma^2 = (r^2 + a^2)^2 - \Delta a^2 \sin^2 \theta$ and that the Kerr metric can be written in the form

$$ds^2 = -\frac{\Delta - a^2 \sin^2 \theta}{\rho^2} dt^2 - \frac{2r_S ar \sin^2 \theta}{\rho^2} dt d\varphi + \frac{\rho^2}{\Delta} dr^2 + \rho^2 d\theta^2 + \frac{\Sigma^2}{\rho^2} \sin^2 \theta d\varphi^2 \tag{48}$$

5   Show that the Kerr metric can be written in the form (23).

6   Plot $\omega(r, \theta)$ as a function of r for various values of θ and a.

7   Using r_S as the unit of length, show that

$$r_S \Omega_\pm(\theta = \pi/2) = \left(g \pm x\sqrt{g^2 + x(x-1)} \right) / \left(x^3 + g^2(x+1) \right)$$

with $x \equiv r/r_S$ and $g \equiv a/r_S$. Note that

$$x_{S+} = r_{S+}(\theta = \pi/2)/r_S = 1 \text{ and } x_+ = r_+(\theta = \pi/2)/r_S = \frac{1}{2}\left(1 + \sqrt{1 - 4g^2}\right)$$

Plot $r_S \Omega_\pm(\theta = \pi/2)$.

8   Show that in the equatorial plane, light rays follow

$$\frac{1}{l^2}\left(\frac{dr}{d\zeta}\right)^2 + \frac{1}{r^2}\left\{1 - \frac{a^2}{b^2} - \left(1 - \text{sign}(l)\frac{a}{b}\right)^2 \frac{r_S}{r}\right\} = \frac{1}{b^2} \tag{49}$$

with as usual the impact parameter $b = \frac{l}{\epsilon}$. Compare with the corresponding Schwarzschild potential $\frac{1}{r^2}\left(1 - \frac{r_S}{r}\right)$ in (VII.1.12), which we can recover from the potential here by setting $a = 0$. An interesting new feature here is the appearance of the sign function $\text{sign}(l)$, reminding us that corotating and counterrotating light behave differently.

9   For completeness, I write Kerr's two original forms[13] here: (I)

$$ds^2 = -\left(1 - \frac{rr_S}{\rho^2}\right)\left(du + a \sin^2 \theta d\varphi\right)^2$$

$$+ 2\left(du + a \sin^2 \theta d\varphi\right)\left(dr + a \sin^2 \theta d\varphi\right) + \rho^2 \left(d\theta^2 + \sin^2 \theta d\varphi^2\right) \tag{50}$$

---

* Also $l_c = \frac{r_S}{\sqrt{3}}$ and $r_{ISCO} = \frac{r_S}{2} = M$.

with $\rho^2 = r^2 + a^2 \cos^2 \theta$ as in the text, and (II)

$$ds^2 = -dt^2 + dx^2 + dy^2 + dz^2 + \frac{r_S r^3}{r^4 + a^2 z^2} \left(dt + \frac{r(xdx + ydy)}{r^2 + a^2} - \frac{a(ydx - xdy)}{r^2 + a^2} + \frac{z}{r}dz \right)^2 \tag{51}$$

with the function $r(x, y, z)$ defined by

$$\frac{x^2 + y^2}{r^2 + a^2} + \frac{z^2}{r^2} = 1 \tag{52}$$

Find the coordinate transformations that bring these into the Boyer-Lindquist form (15).

10 The Kerr-Schild form is obtained by noting that in (50), the dependence on the mass M of the black hole can be explicitly split off by writing the metric as $g_{\mu\nu} = g^0_{\mu\nu} + \frac{2Mr}{\rho^2} l_\mu l_\nu$, with $g^0_{\mu\nu}$ independent of M and with the vector $l_\mu = (1, 0, 0, a \sin^2 \theta)$ in the basis $x^\mu = (u, r, \theta, \varphi)$. Show that l_μ is null (that is, lightlike) with respect to both $g_{\mu\nu}$ and $g^0_{\mu\nu}$.

11 According to the preceding exercise, the Kerr metric in the limit $M \to 0$ should give the Minkowski metric heavily disguised as $g^0_{\mu\nu}$. Calculate the Riemann curvature tensor for $g^0_{\mu\nu}$.

Notes

1. Frame dragging: Consider a number of related slang expressions in various unrelated languages, for example "draguer" in French (originally, to fish with a drag net), and "to drag or pull a girl along" in Cantonese.
2. A deep and involved song and dance in quantum field theory shows that the negative root, when correctly interpreted, leads to the existence of antimatter. See *QFT Nut* or any other quantum field theory text.
3. As expressed in Boyer-Lindquist coordinates. Kerr originally wrote the solution in another form. See exercise 9.
4. For an interesting discussion of the history, see D. L. Wiltshire, M. Visser, and S. M. Scott, eds., *The Kerr Spacetime: Rotating Black Holes in General Relativity*, Cambridge University Press, 2009.
5. For a first-person account of the events leading up to the Kerr solution (and how the young Kerr was allowed only 10 minutes at the conference where he presented his solution), see G. Dautcourt, "Race for the Kerr Field," arXiv:0807.3473. According to this author, the construction of the Berlin Wall affected the race.
6. For a detailed and rather technical analysis, see N. Straumann, *General Relativity*, pp. 432ff.
7. In 1968, B. Carter showed how, with various assumptions (such as separability), one could obtain the Kerr metric.
8. There is, however, a formal similarity, with the role of the infalling particle in the Penrose process played by a vacuum fluctuation in the Hawking process.
9. The interested reader can find a drawing in C. W. Misner, K. S. Thorne, and J. A. Wheeler, *Gravitation*, showing an advanced civilization, having erected a spherical metal framework around a Kerr black hole, dumping its garbage through the horizon.
10. Of course, if we were able to get ourselves near a Kerr black hole, we might as well navigate close to any old sunlike star.
11. S. Deser and J. Franklin, arXiv: 1002.1066 (2010).
12. I should mention that for rotating spacetimes the analog of the Newton-Jebsen-Birkhoff theorem does not exist. Outside a rotating star or planet that is not a black hole, the spacetime does not have to be the Kerr spacetime; it merely has to approach the Kerr spacetime far away. (In chapter IX.4, we will see that the spacetime outside a mass distribution could be given as a multipole expansion in terms of the $T^{\mu\nu}$ of the mass distribution. Within some constraints, you have the freedom to arrange $T^{\mu\nu}$ and hence modify the spacetime outside. Far away, however, the higher multipoles fall away, and the spacetime must approach Kerr asymptotically.)
13. For a useful list of relevant results for the Kerr black hole, see the article by M. Visser in Wiltshire et al. ibid.

Black holes with electric charge

Reissner in 1916 and Nordström in 1918 discovered independently a spacetime with the same setup as in the Schwarzschild solution, except that the central mass carries an electric charge* Q. Since it is difficult[†] to imagine an astrophysical object with a large electric charge, the solution is of theoretical, rather than practical, interest.

As is the case for Schwarzschild spacetime, the metric has the form $ds^2 = -A(r)dt^2 + B(r)dr^2 + r^2 d\Omega^2$ with $A(r)$ and $B(r)$ to be determined. In addition to Einstein's equation, we now also have to solve Maxwell's equation $D_\mu F^{\mu\nu} = \frac{1}{\sqrt{-g}}\partial_\mu(\sqrt{-g}F^{\mu\nu}) = 0$. Spherical symmetry implies that the electric field has only a radial component, which we will call $E = F_{0r} = -F_{r0}$, with the identification in terms of the field strength given in chapter VI.4. Thus, $F^{0r} = g^{00}g^{rr}F_{0r} = E/(AB)$. Also, $g = -ABr^4\sin^2\theta$. Maxwell's equation thus reduces to $\partial_r(r^2 E/\sqrt{AB}) = 0$, with the solution

$$E = \frac{Q\sqrt{AB}}{r^2} \tag{1}$$

The electric charge Q is defined by the boundary condition $E(r) \to Q/r^2$ as $r \to \infty$. Since the energy density contained in the electric field dies off rapidly, we expect spacetime to be asymptotically flat, that is, $A(r) \to 1$, $B(r) \to 1$, as $r \to \infty$. With $F_{0r} = -F_{r0}$ the only nonzero components of the field strength $F_{\mu\nu}$, the other Maxwell's equation $\varepsilon^{\rho\lambda\mu\nu}D_\lambda F_{\mu\nu} = 0$ is trivially satisfied.

Next, we have to solve Einstein's equation, now with a nonvanishing $T_{\mu\nu}$ from the energy momentum contained in the electromagnetic field. From chapter VI.4, we have $T_{\mu\nu} = F_{\mu\lambda}F_\nu^\lambda - \frac{1}{4}g^{\mu\nu}F_{\sigma\rho}F^{\sigma\rho}$. Recall that the energy momentum tensor of the electromagnetic

---

* The Kerr solution also can be endowed with an electric charge, in which case it is known as the Kerr-Newman solution.

† Since any such object, if, say, positively charged, would attract electrons from its environment and repel protons and so quickly neutralize its charge.

field is traceless, so that Einstein's equation (VI.5.10) reduces to

$$R_{\mu\nu} = 8\pi G T_{\mu\nu} \tag{2}$$

where the Ricci tensor $R_{\mu\nu}$ was computed in chapter VI.3. As before, (2) contains three equations, for $\mu\nu = 00$, rr, and $\theta\theta$, but due to the Bianchi identity, only two of these equations are independent and serve to determine the two unknown functions A and B.

Let us now compute $T_{00} = g^{rr}F_{0r}F_{0r} - \frac{1}{4}g_{00}(2F_{0r}F_{0r})g^{00}g^{rr} = (E^2/B) + \frac{1}{4}(-2E^2)/B = E^2/(2B)$. Similarly, we find $T_{rr} = -E^2/(2A)$ and $T_{\theta\theta} = r^2E^2/(2AB)$. Thus, $BT_{00} + AT_{rr} = 0$, and so $BR_{00} + AR_{rr} = 0$. At this point, either we look up the expression for R_{00} and R_{rr} in chapter VI.3, or we remember, if we are endowed with a great memory, that $BR_{00} + AR_{rr}$ is proportional to the combination $\left(\frac{A'}{A} + \frac{B'}{B}\right)$. Either way, we find $\left(\frac{A'}{A} + \frac{B'}{B}\right) = 0$ with the instant solution $AB = 1$, just as in the Schwarzschild solution.

Things now simplify: in particular, $T_{\theta\theta} = Q^2/(2r^2)$. Putting this into the $\mu\nu = \theta\theta$ equation in (2), looking up $R_{\theta\theta}$ in chapter VI.3, and eliminating $B = 1/A$, we obtain $A + rA' = (rA)' = 1 - (4\pi G Q^2/r^2)$ with the solution

$$A(r) = 1 - \frac{2GM}{r} + \frac{4\pi GQ^2}{r^2} \tag{3}$$

Also, as in the Schwarzschild case, the total mass M appears as an integration constant fixed by the boundary condition as $r \to \infty$.

You might have thought that solving the coupled Einstein and Maxwell equations would be rather difficult, but thanks to spherical symmetry, the solution pops out easily and has a remarkably simple form.[1]

Spacetime structure of the charged black hole

We now regard the Reissner-Nordström spacetime as a charged black hole (rather than a charged star). I will merely sketch some salient features of the spacetime, referring you to more specialized treatments. Indeed, the rest of this chapter may be omitted upon a first reading and is not needed for the rest of this book. For the sake of clarity, let us use units in which $G = 1$ and absorb the 4π in (3) into the definition of Q, so that we write

$$ds^2 = -\left(1 - \frac{2M}{r} + \frac{Q^2}{r^2}\right)dt^2 + \left(\frac{1}{1 - \frac{2M}{r} + \frac{Q^2}{r^2}}\right)dr^2 + r^2 d\Omega^2 \tag{4}$$

Consider the function $A(r) = -g_{00}(r) = \left(1 - \frac{2M}{r} + \frac{Q^2}{r^2}\right) = (r - r_+)(r - r_-)/r^2$ with

$$r_\pm = M \pm \sqrt{M^2 - Q^2} \tag{5}$$

Evidently, charged black holes fall into three categories: (a) subextremal, with $Q < M$; (b) extremal, with $Q = M$; and (c) transextremal or "naked," with $Q > M$. The terminology will become clear shortly.

Subextremal black hole

For $Q = 0$, $r_+ = 2M$ and we recover the Schwarzschild black hole, of course. As r decreases from infinity, the function $A(r) = 1 - \frac{2GM}{r}$ decreases from 1 to $-\infty$, crossing zero at the Schwarzschild radius $2M$. But as soon as we crank up Q, the $+Q^2/r^2$ electric term in (3) takes over for small r, arresting the plunging $A(r)$ and pulling it back up to infinity. (Plot this!) The curious feature is that while t becomes a spacelike coordinate for $r_- < r < r_+$, it goes back to being a timelike coordinate again once we get below r_-. Indeed, in sharp contrast to the Schwarzschild black hole, the physical singularity at $r = 0$ is timelike. In other words, for small r, $ds^2 \rightarrow -\frac{Q^2}{r^2}dt^2 + \frac{r^2}{Q^2}dr^2 + r^2 d\Omega^2$.

As we crank up Q further, at $Q = M$, the function $A(r)$ just barely touches the r-axis and the two roots $r_\pm$ merge, with $r_+ = r_- = M$. This is known as an extremal black hole. We will come back to it later. For now, let's ask what happens if we crank up Q even further.

Naked singularity and cosmic censorship

For $Q > M$, the roots $r_\pm$ disappear. The function $A(r) = -g_{00}(r)$ does not vanish. It goes from 1 at $r = \infty$ to ∞ at $r = 0$, staying positive the whole time. Similarly, $1/A(r) = -g_{00}(r)$ stays positive. Thus, in contrast to the subextremal black hole (which includes the Schwarzschild black hole), t and r are perfectly respectable timelike and spacelike coordinates, respectively, with a spacetime described by the Penrose diagram in figure 1.

Recall that for the Schwarzschild black hole, because of the horizon, signals from the vicinity of the physical singularity at $r = 0$ cannot get out to an observer stationed at $r = \infty$. Observers outside the horizon cannot see the singularity. General relativists rather picturesquely say that the singularity is clothed by the horizon.

In contrast, here we have what is known as a naked singularity, visible to the outside world. Signals from the vicinity of the physical singularity at $r = 0$ can get out to $r = \infty$.

There was a long history of hand-wringing over the appearance of naked singularity in classical general relativity, culminating in the cosmic censorship conjecture. The conjecture states that for reasonable initial conditions, a naked singularity cannot form. This does not mean that Einstein's field equation does not allow naked singularities: verily, the Reissner-Nordström black hole for $Q > M$ offers an example. But it is an eternal black hole just sitting there; the conjecture addresses the issue of whether it could have formed. I direct the interested reader to the vast literature devoted to the conjecture.

You may catch some flavor of the conjecture by noting that, in the Reissner-Nordström example, M governs how the metric approaches Minkowskian as $r \rightarrow \infty$ and thus by definition is the total mass of the black hole. Because the electric force is repulsive, we expect the electric field to contribute positively to M. In contrast, the gravitational force is attractive, so we expect that it will contribute negatively. The transextremal condition $Q > M$ says that the black hole is not very massive for its charge. This implies a large

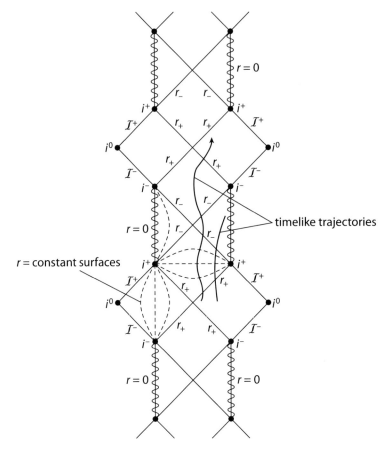

Figure 1 Penrose diagram for a charged black hole.

negative gravitational contribution needed to cancel the positive electric energy. At issue is whether the needed negative contribution is physically reasonable.

Although the conjecture sounds plausible, it has never been proved. The proof of the cosmic censorship conjecture constitutes a difficult mathematical challenge, but regardless, the presence of naked singularities indicating a breakdown in our conception of spacetime is surely a problem for classical general relativity. Opinions differ on whether naked singularities pose a problem for theoretical physics. Most people believe that quantum gravity would "smooth out" a naked singularity; indeed, even our classical conception of spacetime may disappear in whatever theory of quantum gravity we end up with.

One subject we do not go into in this book consists of the various rigorous singularity theorems[2] telling us in general under what circumstances various types of singularities can and cannot occur. These theorems[3] are typically proved by assuming that the metric satisfies Einstein's field equation with a physically reasonable* $T^{\mu\nu}$.

---

* For example, that it must satisfy the strong energy condition to be discussed in appendix 3 of chapter IX.3.

Extremal black holes neither attract nor repel

Let's return to extremal black holes.[4] You might have noticed that the extremal condition $Q^2 = M^2$ suggests that the gravitational attraction $-M^2/r^2$ and the electric repulsion Q^2/r^2 between two extremal Reissner-Nordström black holes balance each other. (Recall that we have set $G = 1$ and absorbed a factor of 4π.) That there is no net force between two extremal black holes indicates that we could put a bunch of these objects down at arbitrary locations and they would just sit there. We could not do this with Newtonian masses: they would fall toward one another. Nor could we do this with Coulomb charges. But with extremal black holes, yes! Newton balances Coulomb.

This physical intuition, if true, implies an amazing mathematical fact: there must exist an entire family of static solutions of the coupled Einstein and Maxwell equations describing a bunch of extremal black holes just sitting there. As you will see in appendix 1, it is not entirely trivial to find these solutions, but without the physical motivation, the solutions would appear to be miraculous.[5]

Extremal black holes are in some sense exceptional objects,[6] poised on the dividing line between subextremal black holes and naked abominations, perhaps reminiscent of a pencil balanced on its tip at the very edge of a table.

No-hair theorems

> In my entire scientific life . . . the most shattering experience has been the realization that an exact solution of Einstein's equations of general relativity, discovered by . . . Kerr, provides the absolutely exact representation of untold numbers of massive black holes that populate the universe.
>
> —S. Chandrasekhar

When you read the preceding chapter on the Kerr black hole, were you as shattered as Chandrasekhar was? What, you are still in one piece?

Consider two massive stars circling each other. To characterize the system completely, we would have to give the mass and size of each of the stars, the chemical composition, the temperature, the orbital parameters, and on and on—you get the idea. Eventually, they approach each other, and after radiating some electromagnetic and gravitational waves, form a rotating black hole. Now only two numbers, the mass M and angular momentum J, suffice to characterize the system.

Amazingly, spacetime, in swallowing up matter and curling over to form a black hole, manages to obliterate almost all the evidence, as it were. The Schwarzschild black hole is characterized completely and exactly by its mass M; the Reissner-Nordström black hole by M and Q; the Kerr black hole by M and J; and the Kerr-Newman black hole by M, J, and Q. (We omit mentioning magnetic charge here; see exercise 3.) In practical terms, if you were the commander of a spaceship approaching a planet, your second-in-command might have to give you a massive computer file listing the location and height of every mountain and so forth, but if you were approaching a black hole, only a tiny slip of paper with a couple of numbers written on it would suffice.

So, what are so special about M, J, and Q? The answer should be clear to you: they couple to the two infinite ranged fields we know about, namely the gravitational and the electromagnetic field. All the other physical quantities that went into the making of the black hole are subsequently hidden behind the horizon. Wheeler has summarized this state of affairs by quipping that "Black holes have no hair."

Appendix 1: Extremal black holes just sitting there

Let us now verify the physical argument that there must exist an entire family of static solutions of the coupled Einstein and Maxwell equations describing a bunch of extremal black holes "just sitting there." Start with a single extremal black hole. With $Q = M$, the metric in (4) becomes $ds^2 = -\left(1 - \frac{M}{r}\right)^2 dt^2 + \left(\frac{1}{1-\frac{M}{r}}\right)^2 dr^2 + r^2 d\Omega^2$. Spherical coordinates are dandy for one black hole, but not so good when we have a whole bunch of them. So, exploit our freedom to change coordinates and set $r = \rho + M$. A few lines of arithmetic lead us to $ds^2 = -f(\rho)^{-2}dt^2 + f(\rho)^2(d\rho^2 + \rho^2 d\Omega^2)$, with $f(\rho) = 1 + \frac{M}{\rho}$. We next introduce Cartesian coordinates by setting $d\rho^2 + \rho^2 d\Omega^2 = dx^2 + dy^2 + dz^2$.

Now consider the Ansatz

$$ds^2 = -f(x, y, z)^{-2}dt^2 + f(x, y, z)^2(dx^2 + dy^2 + dz^2) \tag{6}$$

with $f(x, y, z)$ some unknown function of x, y, z, not necessarily $f(\rho)$. We are supposed to plug this into Einstein's equation.

We also need an Ansatz for the electromagnetic field. For a single extremal black hole, the only nonvanishing component of A_μ (the factor of $\sqrt{4\pi}$ in the following expression comes from our scaling Q so that extremal means $Q = M$) is the time component $A_0 = Q/(\sqrt{4\pi}r) = M/(\sqrt{4\pi}(\rho + M)) = (1 - f(\rho)^{-1})/\sqrt{4\pi}$, with $f(\rho) = 1 + \frac{M}{\rho}$. Our inspired guess is to set the only nonvanishing component of A_μ to be

$$A_0(t, x, y, z) = \frac{1}{\sqrt{4\pi}}\left(1 - f(x, y, z)^{-1}\right) \tag{7}$$

with the same unknown function $f(x, y, z)$ as in (6).

A priori, it would seem hopeless that the Ansatz (6) and (7) with a single time independent function $f(x, y, z)$ could solve the numerous coupled Einstein and Maxwell equations, but as I said, we are buoyed by our faith in our physical picture. Start with $F_{0i} = -\partial_i A_0 \propto (\partial_i f)/f^2$ (where evidently, $\partial_i \equiv \frac{\partial}{\partial x^i}$ and $(x^1, x^2, x^3) = (x, y, z)$). Since $g = -f^4$, Maxwell's equation $\frac{1}{\sqrt{-g}}\partial_\mu(\sqrt{-g}F^{\mu\nu}) = \frac{1}{\sqrt{-g}}\partial_i(\sqrt{-g}F^{i\nu}) = 0$ reduces to $\partial_i(f^2 F_{0i}) = \nabla^2 f = 0$. The unknown function f satisfies Laplace's equation.

Next, to solve Einstein's equation, we first have to compute the energy momentum tensor

$$T_{\mu\nu} = F_{\mu\lambda}F_\nu^\lambda - \tfrac{1}{4}g_{\mu\nu}F_{\sigma\rho}F^{\sigma\rho} \tag{8}$$

which I copy here for convenience. First, note that $g_{00} = -1/f^2$, $g_{11} = f^2$, $g^{00} = -f^2$, and $g^{11} = 1/f^2$, so that $F^{01} = g^{00}g^{11}F_{01} = -(\partial_1 f)/f^2$ and $F_{\sigma\rho}F^{\sigma\rho} = -2(\partial_i f)^2/f^4$ (where $(\partial_i f)^2 = \sum_i \partial_i f \partial_i f$). Then $T_{00} = F_{0i}F_0^i - \tfrac{1}{4}g_{00}F_{\sigma\rho}F^{\sigma\rho} = (\partial_i f)^2/(2f^6)$. Similarly, we have $T_{11} = \{(\partial_i f)^2 - 2(\partial_1 f)^2\}/(2f^2)$ and of course also the corresponding expressions for T_{22} and T_{33}. Clearly, $T_{0i} = 0$, but $T_{12} = -(\partial_1 f)(\partial_2 f)/f^2$ does not vanish. A simple check on the arithmetic is that the trace $T = g^{\mu\nu}T_{\mu\nu}$ vanishes.

Onward to Einstein's equation $R_{\mu\nu} = 8\pi T_{\mu\nu}$. After some work, we find $R_{00} = \{(\partial_i f)^2 - f\nabla^2 f\}/f^6$, $R_{11} = \{(\partial_i f)^2 - 2(\partial_1 f)^2 - f\nabla^2 f\}/f^2$, and $R_{12} = -2(\partial_1 f)(\partial_2 f)/f^2$. Plugging in, we find, remarkably enough, that

Einstein's equation collapses to $\nabla^2 f(x, y, z) = 0$, in agreement with what Maxwell's equation demands. As I explained, it is amazing, but perhaps a bit less amazing, given our physical motivation.

Well, we all know how to solve Laplace's equation in empty 3-dimensional space. The general solution is

$$f(\vec{x}) = 1 + \sum_{a=1}^{N} \frac{M_a}{|\vec{x} - \vec{x}_a|} \qquad (9)$$

with the additive constant chosen so that $A_0 \to 0$ as $\vec{x} \to \infty$. As promised, we have found a time independent solution of the coupled Einstein-Maxwell equations describing N extremal black holes with arbitrary mass $M_a = Q_a$ sitting at arbitrary locations $\vec{x}_a$. They are not moving, because the force between any pair of black holes vanishes! Note that (9) implies a highly nontrivial electric field $\vec{E}$.

There are no doubt more elegant ways to arrive at the solution, but in an introductory text, I prefer an explicit calculation. The discussion provides an example of physical intuition leading to a mathematical result that would otherwise be unsuspected.

Appendix 2: The interior of a subextremal Reissner-Nordström black hole

I sketch here what happens to an observer falling into a subextremal Reissner-Nordström black hole with $Q^2 < M^2$. (Take the observer to be electrically neutral, so that there is no Lorentz force acting on him.) As he falls past the horizon at r_+, the physics is much the same as experienced by an observer falling past r_S in a Schwarzschild black hole. Indeed, for $r \lesssim r_+$, we have $-g_{00}(r) = (r - r_+)(r - r_-)/r^2 \simeq (r - r_+)(r_+ - r_-)/r_+^2$, which as expected, is close to the corresponding Schwarzschild expression $(r - r_+)/r_+$ for $r_+ \gg r_-$.

The key physics is that r has turned itself into a time coordinate, and the observer is obliged to keep moving in the direction of decreasing r toward the physical singularity $r = 0$. But unlike the Schwarzschild case, the observer falling into a subextremal Reissner-Nordström black hole is not doomed: once he gets past r_-, the r coordinate turns back into a space coordinate! In the region $r < r_-$, the observer can move however he wants. He could lazily fall in toward $r = 0$, but he could also move in the direction of increasing r by firing his rocket. Once he gets past r_-, the r coordinate turns into a time coordinate again, but this time, since he was moving in the direction of increasing r, he is obliged to keep on moving in the same direction. Eventually, he zooms past r_+ and escapes from the black hole.

Just as in the Schwarzschild case, where we saw that (t, r) are not good coordinates to use, to make sense of the story here, we also have to work and replace (t, r) by more sensible coordinates so as to eventually arrive at the analog of the Kruskal extension of the Schwarzschild metric. We won't go through that here, but have shown the result in the form of a diagram (see figure 1), the analog of figure VII.2.7. In the Schwarzschild case, the physical regions I and II are extended to regions III and IV. Here the physical regions we started out with are repeated indefinitely. As indicated in the figure, our intrepid observer, when he zooms past r_+, will actually enter a different asymptotically flat spacetime than the one he started out from.

If you feel that the eternal Schwarzschild black hole with its Kruskal extension is more of a mathematical construct than a physical entity, you would feel even more strongly about the subextremal Reissner-Nordström black hole with its extension.

Exercises

1   Show that a photon moving in a radial direction in a subextremal Reissner-Nordström black hole follows a path determined by $\frac{dt}{dr} = \pm \frac{r^2}{(r-r_+)(r-r_-)}$. Integrate this equation.

2   Show that by defining $d\bar{t} = dt + (A(r)^{-1} - 1)dr$ (in complete analogy to what we did in chapter VII.2), we can write the Reissner-Nordström metric in the form

$$ds^2 = -A d\bar{t}^2 + 2(1 - A)d\bar{t}dr + (2 - A)dr^2 + r^2 d\Omega \qquad (10)$$

3   Find a solution describing a black hole endowed with a magnetic charge. In fact, you can write a solution with both electric and magnetic charges. This merely reflects what is known as electromagnetic duality.

Notes

1. Here is a handwaving understanding of the solution. Since $A(r) \to 1 - \frac{2GM}{r}$ as $r \to \infty$, M represents the total mass, including the electromagnetic contribution. As r decreases, by Newton's second superb theorem (see chapter I.1), we should subtract off the electromagnetic contribution $\sim 4\pi \int_r^\infty dr\, r^2 \frac{1}{2} \frac{Q^2}{r^4} = \frac{2\pi Q^2}{r}$ and define, heuristically, an effective mass $\tilde{M}(r) \sim M - \frac{2\pi Q^2}{r}$. We then guess that $A(r) \sim 1 - \frac{2G\tilde{M}}{r} = 1 - \frac{2GM}{r} + \frac{4\pi G Q^2}{r^2}$.

2. Proved by B. Carter, G. F. R. Ellis, S. Hawking, R. Penrose, and many others. I refer you to more advanced monographs by some of these authors.

3. Just to give you a flavor of this kind of theorem, let me mention that one theorem states that, with the strong energy condition, if a metric has a trapped surface from which light rays cannot escape, then either a singularity or a closed timelike curve is present.

4. They have also played an important role in string theory.

5. This is why I give you the physics first, unlike some other authors.

6. The interested reader is referred to the literature, which can be traced from E. Poisson and W. Israel, *Phys. Rev.* D 41 (1990), p. 1796, and D. Marolf, arXiv:1005.2999. Poisson and Israel found an instability associated with the inner horizon of close-to-extremal but still subextremal black holes. Marolf, referring to this as the "dangers of extremes," suggests that the more remarkable features of the interior spacetime of extremal black holes would in fact not survive any quantum fluctuation.

Recap to Part VII

Surprisingly, that innocuous gravitational action that leapt out at us is capable of altering the causal structure of spacetime.

The 18th century fantasy of Michell and Laplace is realized by a global alteration of spacetime. Even more amazingly, when quantum fluctuations are turned on, a hapless member of a fluctuating pair could fall through the horizon, allowing its partner to escape to infinity.

A rotating black hole can drag spacetime around with it and can convert mass into energy at an efficiency almost 100 times higher than that of nuclear processes.

Charged and extremal black holes are fun objects to play around with, but perhaps more importantly, the relativistic equation for stellar interiors can also be written down without much fuss.

Part VIII | Introduction to Our Universe

VIII.1 | The Dynamic Universe

The universe comes to life

The Newtonian universe offers a rigid space for stuff to move in, but as we already saw in chapter VI.2, the Einsteinian universe enjoys a life of its own, bending and curving, reacting to the stuff that fills it. Stuff tells spacetime how to curve. The actors act back on the stage.

Back in chapter VI.2 we made a mad dash to cosmology, but at that point, we didn't know enough and were able to fill the universe with only dark energy, presumed to be a manifestation of Einstein's cosmological constant. Later, in chapter VI.4, we learned how to obtain, given a matter action, the corresponding $T^{\mu\nu}$. We can now plug our favorite $T^{\mu\nu}$ into the right hand side of Einstein's field equation (VI.5.10)

$$R^{\mu\nu} = 8\pi G S^{\mu\nu} \equiv 8\pi G(T^{\mu\nu} - \tfrac{1}{2}g^{\mu\nu}T) \tag{1}$$

thus filling up the universe with one ingredient or another, and watch the universe evolve as dictated by (1).

Also, in chapter VI.2, we took the universe to be described by

$$ds^2 = -dt^2 + a^2(t)((dx^1)^2 + (dx^2)^2 + (dx^3)^2)$$

While spacetime is curved, space itself is flat. Let us now generalize this description to

$$ds^2 = -dt^2 + a^2(t)\tilde{g}_{ij}(\vec{x})dx^i dx^j \tag{2}$$

where $\tilde{g}_{ij}(\vec{x})$ denotes a 3-dimensional metric associated with the space we live in. In other words, the universe is regarded as a curved space described by $dl^2 = \tilde{g}_{ij}dx^i dx^j$, stretched at any given instant by the function $a(t)$, the scale factor of the universe. For the moment, we leave $\tilde{g}_{ij}$ unspecified.

It is conventional to normalize the scale factor by setting $a(t_0) = 1$, with t_0 the time at present. Recall also from chapter V.3 that we can relate $a(t)$ to the redshift z commonly used by astronomers by

$$a(t) = \frac{1}{1+z} \tag{3}$$

(You can verify that the derivation given earlier goes through regardless of $\tilde{g}_{ij}$.) A large redshift $z > 0$ corresponds to a time when the universe was smaller by a factor $a(t) < 1$.

The curvature of the universe

One goal of cosmology is to find out what the universe is filled with, and as a result, how the universe expands. To study cosmic expansion, we have to calculate the Ricci tensor. Resort to our usual trick of extracting the Christoffel symbols we need from the geodesic equations obtained by varying $\int [dt^2 \quad a^2(t)\tilde{g}_{ij}(\vec{x})dx^i dx^j]^{\frac{1}{2}}$, with $d\tau^2 = dt^2 - a^2(t)\tilde{g}_{ij}(\vec{x})dx^i dx^j$, as has been explained ad nauseum in this text starting in chapter II.2. By now you can probably do this with your eyes half closed.

For example, varying with respect to x^l, we write the resulting Euler-Lagrange equation as

$$\tfrac{1}{2}a^2(t)\partial_l\tilde{g}_{ij}\frac{dx^i}{d\tau}\frac{dx^j}{d\tau} = \frac{d}{d\tau}\left(a^2(t)\tilde{g}_{li}(\vec{x})\frac{dx^i}{d\tau}\right)$$

$$= a^2(t)\tilde{g}_{li}(\vec{x})\frac{d^2x^i}{d\tau^2} + 2a\dot{a}(t)\tilde{g}_{li}(\vec{x})\frac{dt}{d\tau}\frac{dx^i}{d\tau} + a^2(t)\partial_j\tilde{g}_{li}(\vec{x})\frac{dx^i}{d\tau}\frac{dx^j}{d\tau} \tag{4}$$

Multiplying by $\tilde{g}^{nl}$ and cleaning up, we obtain

$$\frac{d^2x^l}{d\tau^2} + \frac{2\dot{a}}{a}\frac{dt}{d\tau}\frac{dx^l}{d\tau} + \tilde{\Gamma}^l_{ij}\frac{dx^i}{d\tau}\frac{dx^j}{d\tau} = 0 \tag{5}$$

The Christoffel symbol for the spatial metric $\tilde{g}_{ij}$

$$\tilde{\Gamma}^l_{ij} \equiv \tfrac{1}{2}\tilde{g}^{lk}\left(\partial_j\tilde{g}_{ik} + \partial_i\tilde{g}_{jk} - \partial_k\tilde{g}_{ij}\right) \tag{6}$$

emerged rather nicely, but of course from general considerations, we knew that the derivatives of $\tilde{g}_{ij}$ that appeared in (4) had to self-assemble appropriately. Note that the derivation of (5) makes clear that $\tilde{g}^{lk}$ in (6) denotes the inverse of $\tilde{g}_{ij}$, not the lk component of $g^{\mu\nu}$.

Varying with respect to t yields

$$\frac{d^2t}{d\tau^2} + a\dot{a}\tilde{g}_{ij}\frac{dx^i}{d\tau}\frac{dx^j}{d\tau} = 0 \tag{7}$$

From (5) and (7), we read off the following nonvanishing Christoffel symbols:

$$\Gamma^0_{ij} = a\dot{a}\tilde{g}_{ij}, \qquad \Gamma^l_{0i} = \frac{\dot{a}}{a}\delta^l_i, \qquad \Gamma^l_{ij} = \tilde{\Gamma}^l_{ij} \tag{8}$$

It is instructive to compare with what we had in chapter VI.2: $\Gamma^0_{ij} = a\dot{a}\delta_{ij}$, $\Gamma^i_{0j} = \Gamma^i_{j0} = \frac{\dot{a}}{a}\delta^i_j$, and $\Gamma^l_{ij} = 0$ for flat space (but curved spacetime!). Indeed, given these results, we could have almost guessed (8).

From (8), a straightforward and not so tedious calculation then gives the following nonvanishing components of the Ricci tensor:

$$R_{00} = -\frac{3\ddot{a}}{a}, \qquad R_{ij} = \tilde{R}_{ij} + \left(2\dot{a}^2 + a\ddot{a}\right)\tilde{g}_{ij} \tag{9}$$

where $\tilde{R}_{ij}$ is the Ricci tensor for the spatial metric $\tilde{g}_{ij}$. As in chapter VI.2, we can understand the general features of these results, notably that R_{00} cannot depend on $\tilde{g}$. If we set $\tilde{g}_{ij}$ to the flat metric, we should recover the result $R_{ij} = (2\dot{a}^2 + a\ddot{a})\delta_{ij}$ we had before. But if we set a to a constant, we must have $R_{ij} = \tilde{R}_{ij}$. Hence, the result for R_{ij} could also have been anticipated.

Cosmological principle

A working assumption of cosmology is that on scales much larger than galaxies, the universe is homogeneous and isotropic: it has neither a special location nor a special direction. This so-called cosmological principle, a direct intellectual descendant of the Copernican principle, has been verified to remarkable accuracy observationally, notably by measurements of the cosmic microwave background.* Of course, this "perfect" cosmological principle may have to be modified at any moment by new and unexpected observations,[†] but accepting it, we can then fix $\tilde{g}_{ij}(\vec{x})$.

Intuitively, we readily understand that 3-dimensional spaces without special location and direction more or less have to be the Euclidean 3-space E^3, the 3-sphere S^3, or its hyperbolic cousin H^3, discussed really way back in chapter I.6. (In chapter IX.6, we will make this expectation precise with a full-fledged discussion of isometry and maximally symmetric spaces.) Indeed, in chapter V.3, we already explained that our universe could be (spatially) closed, flat, or open, described by

$$ds^2 = -dt^2 + a(t)^2 \left(\frac{1}{1 - k\frac{r^2}{L^2}}dr^2 + r^2 d\Omega\right) \tag{10}$$

with the integer $k = 1, 0$, and -1, respectively, known as a Friedmann-Lemaître-Robertson-Walker universe. As was also explained there, we will be often tempted to absorb the length scale L into r, so that r is then dimensionless and $ds^2 = -dt^2 + R(t)^2(\frac{1}{1-kr^2}dr^2 + r^2 d\Omega)$, with $R(t) \equiv La(t)$. With the convention $a(t_0) = 1$, we have $L = R(t_0) \equiv R_0$. Do not confuse

---

* We assert here that a decade or two ago, it was possible to write a text on gravity and include a more or less complete discussion of observational cosmology. But by now, considering that we are living in a golden age of cosmology, such a discussion would be either annoyingly brief or soon hopelessly out of date. Thus, I cannot possibly do justice to observational cosmology and have to refer you to the standard texts on cosmology by Dodelson, by Mukhanov, and by Weinberg. This remark applies to all the chapters in part VIII.

† Periodically, there are disturbing hints[1] that the cosmological principle might fail.

$R(t)$, which is evidently a length, with the scalar curvature R. In any case, this is becoming standard usage.

Plugging the spatial metric $\tilde{g}_{ij}$ defined by $dl^2 = \tilde{g}_{ij}dx^i dx^j = \frac{1}{1-kr^2}dr^2 + r^2 d\Omega$ into the formula for the Ricci tensor, we obtain, after a straightforward calculation,

$$\tilde{R}_{ij} = \frac{2k}{L^2}\tilde{g}_{ij} \tag{11}$$

Indeed, we could have anticipated that the Ricci tensor $\tilde{R}_{ij}$ would turn out to be proportional to $\tilde{g}_{ij}$. What else could it be, for a space with no special direction and location? No other symmetric 2-indexed tensor is lurking around. (In chapter IX.6, we will prove that this property holds for any maximally symmetric space.) Furthermore, the Ricci tensor vanishes when $k = 0$. Dimensional analysis nails down the $1/L^2$. Thus, you could say that the calculation you just did (didn't you?) is merely to get the 2 in (11).

Thus, in the end, the message is that life is simple: the spatial components of the spacetime Ricci tensor (see (9) and (11)) are given by

$$R_{ij} = \left(2\dot{a}^2 + a\ddot{a} + \frac{2k}{L^2}\right)\tilde{g}_{ij} = \left(2\dot{R}^2 + R\ddot{R} + 2k\right)\tilde{g}_{ij}/L^2 \tag{12}$$

Note the distinction between the spatial components of the spacetime Ricci tensor R_{ij} and the spatial Ricci tensor $\tilde{R}_{ij}$. They are of course not the same.

Filling the universe with a perfect fluid

Let's fill the universe up with a perfect fluid: it deserves no less.

We first worked out the energy momentum tensor $T^{\mu\nu} = (\rho + P)U^\mu U^\nu + P\eta^{\mu\nu}$ of a perfect fluid in flat spacetime back in chapter III.4, and then, for our discussion of stellar interiors in chapter VII.4, promoted it, by invoking the equivalence principle, to the form

$$T^{\mu\nu} = (\rho + P)U^\mu U^\nu + Pg^{\mu\nu} \tag{13}$$

appropriate for curved spacetime. We now use (13) to source the geometry of the universe.

Let's write this out for a diagonal metric (such as the one we have here) in the comoving frame (in which $\vec{U}$ vanishes). First, the normalization condition $g_{\mu\nu}U^\mu U^\nu = -1$ reduces to $g_{00}(U^0)^2 = -1$, so that $U^\mu = (1, \vec{0})/\sqrt{-g_{00}}$. Thus, $T^{00} = (-g_{00})^{-1}(\rho + P) + Pg^{00} = -\rho/g_{00}$ and $T^{ij} = Pg^{ij}$.

In the present context, $g_{00} = -1$ and so $T^{00} = \rho$ and $T^{ij} = \frac{P}{a^2}\tilde{g}^{ij}$ in the comoving frame. Let's next see how energy momentum conservation

$$D_\mu T^{\mu\nu} = \partial_\mu T^{\mu\nu} + \Gamma^\mu_{\mu\lambda}T^{\lambda\nu} + \Gamma^\nu_{\mu\lambda}T^{\mu\lambda} = 0 \tag{14}$$

works in our expanding universe.

For $\nu = 0$, using the diagonal character of $T^{\mu\nu}$, the assumed spatial homogeneity of the universe, and the list (8) of Christoffel symbols, we obtain $D_\mu T^{\mu 0} = \partial_0 T^{00} + \Gamma^\mu_{\mu 0}T^{00} + \Gamma^0_{ij}T^{ij}$. Write this out more explicitly. First, look up in (8) that $\Gamma^\mu_{\mu 0} = \frac{3\dot{a}}{a}$ and $\Gamma^0_{ij} = a\dot{a}\tilde{g}_{ij}$. Plugging in our result $T^{00} = \rho$ and $T^{ij} = \frac{P}{a^2}\tilde{g}^{ij}$ in the comoving frame, we find $\Gamma^0_{ij}T^{ij} = a\dot{a}\frac{P}{a^2}\tilde{g}_{ij}\tilde{g}^{ij} = \frac{3P\dot{a}}{a}$, and thus

$$\dot{\rho} + \frac{3\dot{a}}{a}\rho + \frac{3P\dot{a}}{a} = 0 \Rightarrow \partial_0\left(\rho a^3\right) = -P\partial_0 a^3 \tag{15}$$

It is instructive to verify an identity we derived in chapter V.6: $\Gamma^\mu_{\mu\lambda} = \frac{1}{\sqrt{-g}}\partial_\lambda\sqrt{-g}$, where $g \equiv \det g_{\mu\nu}$. Recall from way back in chapter I.5 that g does not transform as a scalar, but that $d^4x\sqrt{-g}$ does and so measures volume. Physically, suppose the comoving observer marks out a spatial box measuring $\Delta x^1\Delta x^2\Delta x^3$; then the volume of the box is actually $\Delta x^1\Delta x^2\Delta x^3\sqrt{-g}$. Here $(-g) = a(t)^6\tilde{g}$, and thus the volume is proportional to $a(t)^3\sqrt{\det \tilde{g}_{ij}}$. The first factor $a(t)^3$ takes into account the expansion of the universe, the second the effect of spatial curvature. Now evaluate the $\lambda = 0$ component of the identity: $\Gamma^\mu_{\mu 0} = \frac{1}{\sqrt{-g}}\partial_0\sqrt{-g} = \frac{1}{a^3}\partial_0 a^3 = \frac{3\dot{a}}{a}$ (since $\tilde{g}$ does not depend on t), in agreement with what we had in (8). The identity works, of course; the pedagogical point, rather, is that this little exercise sheds light on the physical meaning of the term $\frac{3\dot{a}}{a}\rho$ in the conservation law above: the energy density changes partly because the volume seen by the comoving observer changes due to the expansion of the universe.

Now we also understand (15): the energy density also changes partly due to the pressure acting on the comoving volume. It is satisfying to see the adiabatic version of the first law of thermodynamics $dE = -PdV$ recovered in (15), as already explained in chapter VI.2.

Next, we look at the $\nu = j$ component of (14). What could it possibly tell us? Go ahead, take a guess before reading on.

Well, once again looking up in (8) the Christoffel symbols we need, we have $D_\mu T^{\mu j} = \partial_\mu T^{\mu j} + \Gamma^\mu_{\mu\lambda}T^{\lambda j} + \Gamma^j_{\mu\lambda}T^{\mu\lambda} = \partial_i T^{ij} + \tilde{\Gamma}^i_{il}T^{lj} + \tilde{\Gamma}^j_{il}T^{il}$. Plugging in $T^{ij} = \frac{P}{a^2}\tilde{g}^{ij}$, we see that $D_\mu T^{\mu j} = 0$ means that

$$\partial_i\left(\frac{P}{a^2}\tilde{g}^{ij}\right) + \frac{P}{a^2}\left(\tilde{\Gamma}^i_{il}\tilde{g}^{lj} + \tilde{\Gamma}^j_{il}\tilde{g}^{il}\right) = \frac{1}{a^2}\left[\tilde{g}^{ij}\partial_i P + P\left(\partial_i\tilde{g}^{ij} + \tilde{\Gamma}^i_{il}\tilde{g}^{lj} + \tilde{\Gamma}^j_{il}\tilde{g}^{il}\right)\right] = 0$$

Since we are assuming spatial homogeneity, P (and anything else, for that matter) cannot depend on x^i. Physically, the pressure gradient $\partial_i P$ must vanish in the comoving frame, since otherwise some sort of counterflow must occur to cancel out the pressure gradient. The three terms in the round parentheses multiplying P collect nicely into the (spatial) covariant derivative $\tilde{D}_i\tilde{g}^{ij}$ of the spatial metric, which vanishes identically, as we learned back in chapter V.6.

Thus, the correct guess is that $D_\mu T^{\mu j} = 0$ tells us nothing at all: it is identically satisfied. Did you pass the test? I know that all we are doing here is verifying the laws of arithmetic, but nevertheless, I find it quite satisfying to see all the pieces coming together to form 0 identically.

Closed, open, and flat universes

We are now ready to solve the field equation (1). Using (13), we find $T = \rho - 3P$ and hence $S_{\mu\nu} = (\rho + P)U_\mu U_\nu - \frac{1}{2}(\rho - P)g_{\mu\nu}$, with $S_{00} = \frac{1}{2}(\rho + 3P)$, and $S_{ij} = \frac{1}{2}(\rho - P)g_{ij} = \frac{1}{2}(\rho - P)a^2\tilde{g}_{ij}$. Recalling (9) and (12), we obtain

$$R_{00} = -\frac{3\ddot{R}}{R} = 4\pi G(\rho + 3P) \tag{16}$$

and

$$R_{ij} = \left(2\dot{R}^2 + R\ddot{R} + 2k\right)\tilde{g}_{ij}/L^2 = 4\pi G(\rho - P)\tilde{g}_{ij} \tag{17}$$

As might be expected, $\tilde{g}_{ij}$ cancels out, leaving behind

$$2\dot{R}^2 + R\ddot{R} + 2k = 4\pi G R_0^2(\rho - P) \tag{18}$$

You now understand perfectly well that, thanks to the Bianchi identity, only one linear combination of (16) and (18) is independent. Obviously, we would rather not deal with second derivatives if we can help it. Using (16) to eliminate $\ddot{R}$ in (18), we finally end up with a remarkably simple first order differential equation

$$\dot{R}^2 + k = \frac{8\pi G}{3}\rho R^2 \tag{19}$$

Note that P does not appear in (19).

To determine cosmic expansion, solve (19) together with the conservation of energy and momentum (15), which I reproduce here for convenience:

$$d(\rho R^3) = -P\,dR^3, \quad \text{or} \quad \frac{d\rho}{\rho + P} = -3\frac{dR}{R} \tag{20}$$

(Note that dt cancels out.) Of course, we also have to say what the universe is filled with by specifying an equation of state $P(\rho)$. Once given this, we can then solve (20) for ρ as a function of R, which we can then plug into (19) to solve for $R(t)$.

A Newtonian mnemonic

Remarkably, a pseudo-derivation of the central equation (19) of Einsteinian cosmology can be concocted using Newtonian mechanics. In a Newtonian universe filled with a constant mass density (never mind that such a universe does not really make sense), consider a large sphere of radius $R(t)$ and an infinitesimal unit mass on the surface of the sphere. The unit mass has kinetic energy $\frac{1}{2}\dot{R}^2$ and, by Newton's superb theorems, potential energy $-G(4\pi R^3/3)\rho/R$. By energy conservation, its total energy $\frac{1}{2}\dot{R}^2 - G(4\pi R^3/3)\rho/R$ should be conserved. Calling this constant $-\frac{1}{2}k$, we obtain (19) and even understand where the $8\pi/3$ comes from!

For $k = -1$, the total energy is positive, indicating that the Newtonian sphere could expand indefinitely, roughly corresponding to an open universe. For $k = +1$ and negative total energy, the sphere would ultimately have to yield to gravity and contract.

I do not take this Newtonian pseudo-derivation seriously but value it as a highly useful mnemonic that could also serve to motivate pedagogically the subtle physics contained in (19).

The universe expands according to what it is full of

To me, it is amazing that cosmic expansion is governed by two simple equations (19) and (20). (Of course, this is largely due to the perfect cosmological principle.) In the next chapter, we will solve these two equations in detail. To get oriented, let's solve them here in various simple situations. Keep in mind that the dimensionless a and the dimensional $R = R_0 a$ are related trivially, and we can easily pass from one to the other.

First of all, we will see a posteriori that in studying the early universe, we can neglect the curvature term k in (19), which thus becomes

$$\dot{a}^2 \propto \rho a^2 \tag{21}$$

Secondly, as we will see a posteriori, to a good approximation, we can study the universe filled with only one kind of stuff at a time.

Fill the universe with nonrelativistic matter, sometimes referred to as dust. As explained back in chapter III.6, the equation of state is simply $P = 0$. Plug this into (20) to obtain

$$\rho \propto \frac{1}{a^3} \quad \text{matter} \tag{22}$$

which, when inserted into (21), gives $a^{\frac{1}{2}} \dot{a} \propto 1$, which implies that

$$a \propto t^{\frac{2}{3}} \quad \text{matter} \tag{23}$$

As another example, fill the universe with radiation (perhaps more accurately referred to as relativistic matter), characterized by $P = \rho/3$, once again as explained back in chapter III.6. Plug this into (20) to obtain $d(\rho a^3) + \frac{1}{3} da^3 = 0$, giving

$$\rho \propto \frac{1}{a^4} \quad \text{radiation} \tag{24}$$

which, when inserted into (21), gives $a\dot{a} \propto 1$, which implies that

$$a \propto t^{\frac{1}{2}} \quad \text{radiation} \tag{25}$$

With ρ going like either $\frac{1}{a^4}$ or $\frac{1}{a^3}$, the right hand side of (19) blows up like either $\frac{1}{a^2}$ or $\frac{1}{a}$, respectively, as $a \to 0$. Thus, our neglect of the curvature contribution k in the early universe is entirely justified. You can solve (19) with the curvature term (see exercises 1–4) and verify this claim. In any case, the present observation evidence favors a flat universe, as was first described in (V.3.2).

You can understand both (22) and (24) easily using elementary physics.

For nonrelativistic matter, think of a bunch of nucleons or atoms sitting in a box of linear dimension of order a. The energy density ρ is entirely due to the mass density. (The kinetic energy is negligible by comparison, hence the pressure is negligible.) As the box expands, the energy density ρ decreases like $1/a^3$, merely because the volume of the box has increased like a^3. Hence (22).

For relativistic matter or radiation, think of a photon gas characterized by a temperature T. The properties of a photon gas are derived in detail in textbooks on statistical mechanics, but for our purposes, we can simply use dimensional analysis, as we have done already in chapter VII.3. There we showed that, in natural units, energy density $\rho \sim T^4$ and the entropy $S \sim V T^3$. Since S is conserved as the box expands adiabatically, $T \propto 1/V^{\frac{1}{3}} \sim 1/a$. Hence $\rho \sim 1/a^4$, in agreement with (24). This dimensional argument underlines the fact that the conclusion $T \propto 1/a$ holds only for a strictly massless particle.[2]

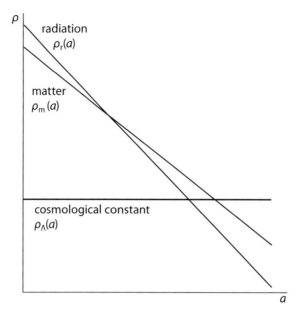

Figure 1 A schematic log-log plot of ρ versus the universe's scale factor a for radiation, matter, and the cosmological constant. As the universe evolves, matter eventually dominates over radiation. As the universe evolves further, the cosmological constant, which had been insignificant all along, eventually dominates over matter. The cosmic coincidence puzzle is: Why now, when we are around?

The universe dominated

As we go back into the early universe, $a \to 0$. Since radiation density goes like $\rho_r \propto \frac{1}{a^4}$ while matter density goes like $\rho_m \propto \frac{1}{a^3}$, radiation eventually dominates over matter. See figure 1. The universe started in a radiation dominated era and expanded into a matter dominated era.

Since the temperature of the radiation $T \sim 1/a$, as we go back into the early universe, T keeps on increasing. As we will see in more detail in chapter VIII.3, in the early universe, atoms and molecules were dissociated into nucleons and electrons, and even earlier, nucleons in turn were dissociated into quarks and gluons. As T increases, the masses of various particles become negligible compared to their kinetic energies of motion, so that everybody becomes "radiation," or more accurately, relativistic matter. Eventually, the temperature formally reaches infinity, our equations become singular, and we have reached the Big Bang.

Conversely, as the universe expanded, the radiation dominated era eventually gave way to the matter dominated era. Thus, as we anticipated, during its evolution, the universe is dominated by one kind of stuff or another.[3]

The preceding analysis indicates that we can cover both matter and radiation with the generic equation of state $P = w\rho$, where evidently $w = 0$ for matter and $w = \frac{1}{3}$ for radiation.

Indeed, this equation of state even covers the cosmological constant, for which $P = -\rho$, as was first mentioned in chapter VI.2. Again, plugging this equation of state into (20) to obtain $d(\rho a^3) + w\rho da^3 = 0$, we find $\rho \propto \frac{1}{a^{3(1+w)}}$, and hence $a \propto t^{\frac{2}{3(1+w)}}$ by (21), thus recovering our previous results as special cases.

In particular, for the cosmological constant, $w = -1$, and the universe expands according to $a \propto t^\infty$, which is just code for the exponential e^{Ht} behavior we found in chapter VI.2. Correspondingly, $\rho_\Lambda \propto a^0$, which is of course just a check, since we defined the cosmological constant to be, duh, a constant. An important remark: in the early universe, as $a \to 0$, in light of (22) and (24), ρ_Λ becomes negligible compared to ρ_r and ρ_m.

Critical density

In the previous section, to get oriented in cosmology and in the cosmos, we solved the cosmological equation $\dot{R}^2 + k = \frac{8\pi G}{3}\rho R^2$ for the flat $k = 0$ universe. But it is not difficult to see qualitatively what goes on in a closed universe

$$\dot{R}^2 + 1 = \frac{8\pi G}{3}\rho R^2 \quad \text{closed universe} \tag{26}$$

or an open universe

$$\dot{R}^2 - 1 = \frac{8\pi G}{3}\rho R^2 \quad \text{open universe} \tag{27}$$

Once you reach a qualitative understanding, a quantitative understanding is merely a matter of showing off your ability to solve a first order ordinary differential equation.

Define the critical density

$$\rho_c \equiv \frac{3\dot{R}^2}{8\pi G R^2} \tag{28}$$

Note that $\rho_c(t)$, which evidently is always nonnegative, depends on time in general. Beware: by the term "critical density," some people mean exclusively its present value $\rho_c(t_0)$.

Next, divide (26) and (27) by $\dot{R}^2$ to obtain

$$\frac{\rho}{\rho_c} = 1 + \dot{R}^{-2} \quad \text{closed universe} \tag{29}$$

for a closed universe and

$$\frac{\rho}{\rho_c} = 1 - \dot{R}^{-2} \quad \text{open universe} \tag{30}$$

for an open universe. Thus, to close the universe, ρ must be greater than the critical density ρ_c. That sure makes sense: you need lots of stuff to curl space around to make it close upon itself. In contrast, for $\rho < \rho_c$, the universe is open.

Consider a universe filled with only matter. Then from (22) $\rho = \rho_0 R_0^3/R^3$, where ρ_0 evidently denotes the present density. For a closed universe, $\dot{R}^2 = \frac{8\pi G\rho_0 R_0^3}{3R} - 1$, and we see that as the universe expands and R increases, the right hand side will eventually decrease to 0, so that $\dot{R} = 0$. The universe stops expanding and starts to contract, as described by the negative root in (26): $\dot{R} = -\sqrt{\dfrac{8\pi G\rho_0 R_0^3}{3R} - 1}$.

In contrast, a matter filled open universe, obeying $\dot{R} = \sqrt{\dfrac{8\pi G\rho_0 R_0^3}{3R} + 1}$, will expand forever, eventually reaching $\dot{R} \simeq 1$, with a curvature driven expansion, even as the matter density dilutes to practically nothing.

Next, consider a universe that is empty except for the cosmological constant. Of the three kinds of stuff to fill the universe considered here, the energy density $\rho_\Lambda = \Lambda$ can be negative, in contrast to matter density ρ_m and radiation density ρ_r.

From (29), we see that in a closed universe, $\rho_\Lambda > \rho_c$ and so a fortiori cannot be negative. But (30) tells us that in an open universe ρ_Λ can be either positive, in which case $\rho_\Lambda < \rho_c$, or negative with no restriction. From (27), we have $\dot{R}^2 = 1 + \frac{8\pi G}{3}\Lambda R^2$. We see that if $\Lambda > 0$, the universe will expand forever, while if $\Lambda < 0$, the universe expands until $\dot{R}$ reaches 0 and then starts to contract. As has already been mentioned, the evidence at present points to a positive Λ.

Big Bang: From no space to space

We have avoided dealing with the time-time component (16) of Einstein's field equation $-\frac{3\ddot{R}}{R} = 4\pi G(\rho + 3P)$, seeing that it involves a second derivative. However, it does convey an important message: as long as $\rho + 3P > 0$, the acceleration $\ddot{R} < 0$, so that $\dot{R}$ always decreases, regardless of whether the universe is closed, flat, or open. Hence the curve $R(t)$ is convex downward. At present, $\dot{R} > 0$, since we see redshifts. Extrapolating the curve backward, we have thus proved that $R(t) = R_0 a(t)$ must vanish at some point in the past. (See figure 2.) The metric in (2) $d\tau^2 \equiv -ds^2 = dt^2 - a^2(t)\tilde{g}_{ij}(\vec{x})dx^i dx^j$ degenerates to $d\tau^2 = dt^2$. No space! This spacetime singularity at which space disappears is known as the Big Bang.*

As long as there is a component of ρ that increases faster than $1/R^2$ as $R \to 0$, we can use (19) to reach the same conclusion we just proved. As we go back in time into the early universe, the right hand side blows up, the curvature term in (19) becomes irrelevant, $\dot{R} \to \infty$, and so $R(t)$ eventually must vanish. This argument also indicates that a universe containing only the cosmological constant could evade this argument and avoid having a Big Bang in its past (as we have seen in chapter VI.2, and as we will see in chapter IX.10).

As authors of popular physics books on the universe know, the most common misconception of the proverbial man in the street regarding the Big Bang is that it describes some kind of terrific primordial explosion, spewing matter every which way into space. You of course know better. The Big Bang is actually the creation of space: from no space to space, stretched by the factor $a(t)$ ever since.

---

* Originally a derogatory term used by Fred Hoyle to champion the steady state theory of the universe.

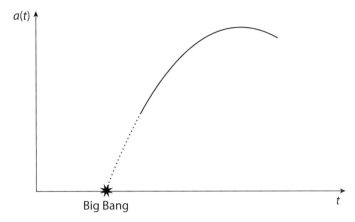

Figure 2 If $\rho + 3P > 0$, the scale factor of the universe is concave downward, and thus must vanish at some point in the past.

Before the discovery of dark energy, it was thought that $\rho + 3P > 0$ imposes a rather weak condition that surely the content of the universe must satisfy. Certainly, both relativistic matter $P = \rho/3$ and nonrelativistic matter $P = 0$ satisfy this condition with room to spare. Well, the dark energy, with $P = -\rho = -\Lambda$, is able to violate precisely this condition. Indeed, the simple universe discussed in chapter VI.2 with $a(t) = e^{Ht}$ does not have a Big Bang in its past.

In the next chapter we will study the cosmological equation in more detail.

Coincidence problem

With $\rho_r \propto 1/a^4$, $\rho_m \propto a^3$, and $\rho_\Lambda \propto a^0$, the universe passes through three epochs: a radiation dominated epoch early on, followed by a matter dominated epoch, which will eventually give way to a dark energy dominated era. This brief history of the cosmos immediately poses a coincidence puzzle. In the vast sweep of cosmic time, the period during which ρ_m is comparable to ρ_Λ, as is the situation now, represents but a blink. Is it a coincidence that we happen to inhabit the universe just as the two curves $\rho_m(a)$ and $\rho_\Lambda(a)$ are crossing each other? Or is there a deeper reason? Perhaps more likely, in my humble opinion, our understanding of cosmology is simply incomplete.

An apparent paradox: More stuff makes the universe expand faster

Readers of popular physics books, and quite a few beginning students as well, are often puzzled by Einstein's cosmological equation $\dot{R}^2 + k = \frac{8\pi G}{3} \rho R^2$: it says that more stuff (a larger ρ) would make the universe expand faster (a larger $\dot{R}$). You might have thought that the gravitational attraction exerted by a larger ρ would hold everybody back, thus slowing down the expansion rather than speeding it up. What gives?

I already alluded to the resolution of this apparent paradox. A particularly illuminating discussion invokes time reversal invariance. Einstein gravity is time reversal invariant, that is, the Einstein-Hilbert action is unchanged upon $t \to -t$. If we take a movie of the expanding universe and run it backward, the plot of the backward-running movie, namely the story of a contracting universe, must also be allowed by the laws of physics. Mathematically, this is implied by the appearance of $\dot{R}^2$ in (19) and hence the two solutions $\dot{R} = \pm\sqrt{\frac{8\pi G}{3}\rho R^2 - 1}$. Thus, if your intuition tells you that more stuff should speed up the contraction, then it also tells you that more stuff also speeds up the expansion.

The common confusion is basically between velocity and acceleration,[4] between $\dot{R}$ and $\ddot{R}$. A somewhat less illuminating resolution of the apparent paradox is to invoke the acceleration equation (16) we eliminated, namely $\frac{3\ddot{R}}{R} = -4\pi G(\rho + 3P)$. If matter is normal,[5] that is, $P > 0$ and increases with increasing ρ, then more stuff (large ρ) decelerates the universe more.

Einstein should do penance

In a recent historical study, Nüssbaumer and Bieri recounted the early history of the universe at considerable variance from the cartoon history given in many popular accounts. They made clear that Lemaître deserved much more credit than he had traditionally received, and others less. They concluded their book by imagining, amusingly, a dinner[6] gathering Einstein, de Sitter, Lemaître, Eddington, and Hubble. Lemaître emerged as a triple winner, for his* expanding universe,[7] for his seminal idea on what developed into the Big Bang, and for associating the cosmological constant with the vacuum energy. While the party toasted the tragically departed Friedmann,[8] Einstein should, according to Nüssbaumer and Bieri, "do penance."[9]

A small story from this laudably balanced history is illuminating. After his friends Eddington and de Sitter had both converted to the expanding universe, Einstein changed his opinion also. In 1932, Einstein and de Sitter were both visiting the California Institute of Technology, and they coauthored a paper that by all accounts would not have passed the refereeing system[10] had it not been authored by two big names. Nothing they said had not already been said earlier by Friedmann, Lemaître, and Robertson. Eddington later wrote:[11] "Einstein came to stay with me shortly afterwards, and I took him to task about it. He replied, 'I did not think the paper very important myself, but de Sitter was keen on it.' Just after Einstein had gone, de Sitter wrote to me announcing a visit. He added: 'You will have seen the paper by Einstein and myself. I do not myself consider the result of much importance, but Einstein seemed to think that it was.'"

---

* It is worth quoting from a letter from M. Way and H. Nüssbaumer to *Physics Today*, August 2011, p. 8. "It is widely held that in 1929 Edwin Hubble discovered the expanding universe and that his discovery was based on his extended observations of redshifts in spiral nebulae. Both statements are incorrect. . . . There is a great irony in these falsehoods still being promoted today. Hubble himself never came out in favor of an expanding universe; on the contrary, he doubted it to the end of his days." I am among those who oppose the continual promotion of falsehoods in physics.

Exercises

1. Solve the cosmological equation $\dot{R}^2 + 1 = \frac{8\pi G}{3} \rho R^2$ (19) for a radiation dominated closed universe. Plot your result. Verify in exercises 1–4 that the curvature is negligible in the early universe.

2. Solve the cosmological equation for a radiation dominated open universe. Plot your result.

3. Solve the cosmological equation for a matter dominated closed universe. Plot your result.

4. Solve the cosmological equation for a matter dominated open universe. Plot your result.

5. For a universe containing only a cosmological constant, show that $R(t) = H^{-1} \cosh Ht$ if the universe is closed, and $R(t) = H^{-1} \sinh Ht$ if the universe is open, with $H^2 = 8\pi G \Lambda / 3$. Note that in the closed case, the universe does not have a Big Bang; that is, $R(t)$ never does vanish. Recall also the result $R(t) = R_0 e^{Ht}$ for a flat universe.

Notes

1. See for example, E. D. Kovetz, A. Ben-David, and N. Itzhaki, "Giant Rings in the CMB Sky," *Astrophys. J.* 724 (2010), pp. 374–378.
2. Consider a box of neutrinos, which are known to have a very small mass m. As the box expands, elementary quantum mechanics shows that the momentum $p \propto 1/a$. When p drops below m, the neutrinos become nonrelativistic, with an average kinetic energy of $p^2/(2m) \propto 1/a^2$. The temperature of the neutrino gas, defined to be the average kinetic energy, would then drop like $T \propto 1/a^2$.
3. Another useful way of plotting the behavior of the universe in different eras, which I learned from J. Bjorken, is suggested by (21) and the various dependences of ρ on a. Plot $\log \dot{a}$ versus $\log a$, that is, do a log-log plot $\dot{a}$ versus a. For dark energy (the cosmological constant), $\log \dot{a} = \log a + $ constant; for radiation, $\log \dot{a} = -\log a + $ constant; for matter, $\log \dot{a} = -\frac{1}{2} \log a + $ constant.
4. When I was a freshman, it was announced that John Wheeler would give an experimental (in the sense of pedagogy rather than physics) course to a handpicked group of beginning students. Wheeler asked the assembled students a series of questions to separate the goats from the elect, so to speak. I still remember the question that eliminated the largest number of hopefuls. Does a tossed ball have zero acceleration at the top of its flight?
5. Or, somewhat less restrictively, assume the equation of state $P = w\rho$ and $(1 + 3w) > 0$.
6. H. Nüssbaumer and L. Bieri, *Discovering the Expanding Universe*, p. 187.
7. Indeed, there is some shady business for a budding historian of physics to look into and clarify. Since Lemaître's seminal 1927 paper was published in French in an obscure Belgian journal, Eddington arranged for it to be republished in English in 1931. But the two crucial pages containing Lemaître's estimate of the so-called Hubble constant were omitted in the English translation. Smells rather fishy to say the least. Some reader should track down the person responsible for this omission.

   By the time I was finishing this book, a couple of years after I wrote the words above, I learned that M. Livio (*Nature* 479 (2011), p. 171, http://www.nature.com/nature/journal/v479/n7372/full/479171a.html) had indeed tracked down the relevant documents and concluded that it was Lemaître himself who deleted the crucial pages. One of Livio's conclusions was disputed by S. van der Bergh in a letter to the editor (*Nature* 480 (2011), p. 321).

   Based on what I read while writing this book and also my earlier popular book (*An Old Man's Toy*), I feel that the kindest thing I can say about Hubble is that he went out of his way not to acknowledge the contributions of his contemporaries. I hope that Hubble's status in cosmology will be reevaluated in the future.
8. For the story of how Aleksandr Friedmann died at the young age of 37, see *Toy/Universe* p. 85.
9. H. Nüssbaumer and L. Bieri, *Discovering the Expanding Universe*, p. 187.
10. Ibid., p. 148.
11. Ibid., p. 128.

A cosmic diagram

The goal of this chapter is to derive the diagram shown in figure 1, which is sort of a phase diagram describing the overall history of the universe according to what it contains. The two axes are labeled by $\Omega_{m,0}$ and $\Omega_{\Lambda,0}$, two parameters that we will define in this chapter and that measure how much matter and how much dark energy, respectively, the universe contains at present. Here the term "matter" includes both dark and baryonic matter, with the bulk (more than 80%) of it in dark matter, as we will see. As in chapter VI.2 and the preceding chapter, dark energy is presumed to be the manifestation of Einstein's cosmological constant. Thus, to a first approximation, you can think of the diagram as describing the struggle between dark matter and dark energy.

In this highly simplified picture, the universe is specified by $\Omega_{m,0}$ and $\Omega_{\Lambda,0}$. Notice that this cosmic diagram contains two straight lines and two curved lines, dividing different types of cosmic behavior.

Above the straight line $\Omega_{\Lambda,0} = \frac{1}{2}\Omega_{m,0}$, cosmic expansion accelerates: the universe will expand faster and faster. Dark energy overwhelms dark matter. Below this line, cosmic expansion decelerates.

Next, look at the line defined by $\Omega_{m,0} + \Omega_{\Lambda,0} = 1$. The universe is spatially closed above this line and open below it. A universe sitting right on this line is spatially flat. As explained in the preceding chapters, in Einstein gravity, stuff curves space, so space can curl up on itself. Lots of stuff in the universe closes it, while not enough stuff leaves it open.

The Big Bang is defined as the singularity in spacetime when the scale factor of the universe a vanishes (as described in the preceding chapter). In figure 1, a curved line starts from the point $(\Omega_{m,0}, \Omega_{\Lambda,0}) = (0, 1)$. Below this line, the Big Bang banged. Above this line, no Bang.

Indeed, back in chapter VI.2, we studied the universe described by $(\Omega_{m,0}, \Omega_{\Lambda,0}) = (0, 1)$, a flat universe with no matter, only a cosmological constant. We found there and in

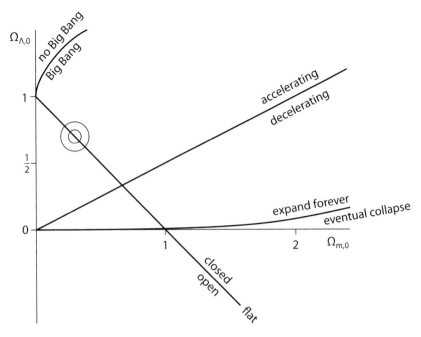

Figure 1 A cosmic diagram describing the overall history of the universe according to how much matter and how much dark energy the universe contains at present.

chapter VI.5 that $a(t) = e^{Ht}$ grows exponentially with the constant $H = 8\pi G \Lambda / 3$. Thus, a never does vanish, and there was no Big Bang. In other words, the point $(\Omega_{m,0}, \Omega_{\Lambda,0}) = (0, 1)$ belongs to the no Big Bang phase. The curved line tells us how much matter has to be put in to produce a Big Bang.

You may have noticed another curve starting from the point $(\Omega_{m,0}, \Omega_{\Lambda,0}) = (1, 0)$ and barely curving upward. Now consider a curve consisting of two pieces joined together, namely the curve just described and a straight line segment consisting of the portion of the $\Omega_{m,0}$ axis between the point $(\Omega_{m,0}, \Omega_{\Lambda,0}) = (1, 0)$ and the origin $(\Omega_{m,0}, \Omega_{\Lambda,0}) = (0, 0)$. This composite curve defines the boundary between two phases. Above this line, the universe will expand forever. Below it, the universe will eventually stop its expansion and contract.

Study this figure and decide if it makes sense to you. Observational evidence suggests that our universe lies inside the circle around $(\Omega_{m,0}, \Omega_{\Lambda,0}) = (0.3, 0.7)$. Thus, according to the cosmic diagram, our universe is flat and accelerating, with a Big Bang in its past and never-ending expansion in its future.

The cosmological equation

The derivation of the cosmic diagram starts with the cosmological equation

$$\dot{R}^2 + k = \frac{8\pi G}{3} \rho R^2 \tag{1}$$

given in the preceding chapter. Define the Hubble* parameter

$$H \equiv \frac{\dot{a}}{a} = \frac{\dot{R}}{R}$$ (2)

Notice that we said Hubble parameter, not Hubble constant. Unless $a(t)$ is a pure exponential (as in chapter VI.2), H will vary with time. In general, we will indicate the present value of a variable by the subscript 0. Thus, H_0 is the present value of the Hubble parameter. Using (2), we rewrite (1) as

$$H^2 = \frac{8\pi G}{3}\rho - \frac{k}{R^2}$$ (3)

What time is it over there?

For the cosmic clock, we have several choices.[1] We could use the time from the Big Bang t, or equivalently, the scale factor $a(t)$. Physically, a better choice is the ambient temperature T of the universe during the event under discussion (for example, radiation decoupling, which occurred at $T \simeq 0.3$ eV, as will be explained in the following chapter). But observational cosmologists quite naturally use the redshift z. In chapter V.3 we derived for light emitted at time t_e the relation

$$1 + z = \frac{1}{a(t_e)}$$ (4)

where we have set $a(t_0) = 1$ by convention.

Also, in chapter V.3, in defining proper distance, we encountered the integral $\mathcal{R} \equiv \int_{t_e}^{t_0} \frac{dt}{a(t)}$, which we can now convert to an integral over redshift as used by cosmologists:

$$\mathcal{R} = \int_{a(t_e)}^{1} \frac{da}{a\dot{a}} = \int_{a(t_e)}^{1} \frac{da}{a^2 H} = \int_0^z \frac{dz'}{H(z')}$$ (5)

Filling up the universe

The energy density $\rho = \sum_j \rho_j$ may consist of several components. For example, the index j can take on several values: $j = r, m$, and Λ, indicating radiation, matter, and cosmological constant, respectively. For the purpose of this chapter, baryonic matter, luminous or not, is lumped in with dark matter and collectively referred to as matter. As already mentioned, we assume that the dark energy is a manifestation of the cosmological constant. It is sometimes convenient to lump the curvature term in (3) into the energy density by defining $\frac{8\pi G}{3}\rho_k \equiv -\frac{k}{R^2}$. With these definitions, we may write (3) as

$$H^2 = \frac{8\pi G}{3}\sum_j \rho_j - \frac{k}{R^2} = \frac{8\pi G}{3}\sum_n \rho_n$$ (6)

---

* Regarding Hubble's discovery of the expanding universe, see the footnote on page 500 in the preceding chapter. For the story of an unschooled mule driver who contributed to Hubble's discovery, see *Toy/Universe*, p. 52.

where the index n runs over the set the index j runs over plus k. (Does the last phrase make sense? If not, read the preceding sentence.)

A key physical feature of our present understanding of cosmology is that these different ingredients do not interact with one another and are tied together only by gravity. You might think that stars produce light, thus converting matter into radiation, but on the cosmic scale, this effect is totally negligible, so that ρ_m and ρ_r evolve independently.

Let us divide (6) by H^2 and define

$$\Omega_j \equiv \frac{8\pi G}{3H^2}\rho_j = \frac{\rho_j}{\rho_c} \tag{7}$$

and

$$\Omega_k \equiv -\frac{k}{H^2 R^2} = -\frac{k}{\dot{R}^2} \tag{8}$$

(Note that the sign of Ω_k is opposite to that of k.) In (7), we have recalled that in the preceding chapter, we defined the critical density[2] $\rho_c \equiv \frac{3\dot{R}^2}{8\pi G R^2} = \frac{3H^2}{8\pi G}$. Thus, Ω_j has the pleasing interpretation as the ratio of the density of the "jth kind of stuff" to the critical density.

As we will see, there is some arithmetical advantage to regarding the curvature term as a kind of density, but physically, you should keep in mind that it originates in geometry. With the definition

$$\Omega \equiv \sum_j \Omega_j \tag{9}$$

(notice that Ω does not include the curvature contribution), we may rewrite (6) as

$$1 = \Omega + \Omega_k = \sum_j \Omega_j + \Omega_k = \sum_n \Omega_n \tag{10}$$

We can think of this as telling us that stuff plus curvature equals unity.

As discussed in the preceding chapter, the parameter Ω determines whether our universe is closed, flat, or open, according to whether $\Omega > 1$, $\Omega = 1$, or $\Omega < 1$, since correspondingly k could be equal to $+1, 0$, or -1, respectively. For many years, it was believed that the universe was closed, but recent observational evidence indicates that Ω is very close to 1, so that $\Omega_k \simeq 0$ and the universe seems quite flat.

Thus far, we have been merely defining and rewriting. Much of this defining and rewriting is to connect the terminology used by theoretical physicists to that used by observational cosmologists. Recall that in the preceding chapter, we found that as the universe expands, ρ_j varies according to $\rho_j \propto \frac{1}{a^{3(1+w_j)}} \equiv \frac{1}{a^{\gamma_j}}$. Let me remind you that

$$w_r = 3, \qquad w_m = 0, \qquad w_\Lambda = -1 \tag{11}$$

so that

$$\gamma_r = 4, \qquad \gamma_m = 3, \qquad \gamma_\Lambda = 0 \tag{12}$$

Hence, setting $a_0 = 1$, we have

$$\Omega_j = \left(\frac{H_0}{H}\right)^2 \frac{1}{a^{\gamma_j}} \Omega_{j,0} \tag{13}$$

(As usual, the subscript 0 indicates the present value of various quantities.) It is also convenient to define the analogous expression for Ω_k with $\gamma_k = 2$. We can then rewrite (10) as

$$1 = \left(\frac{H_0}{H}\right)^2 \sum_n \frac{\Omega_{n,0}}{a^{\gamma_n}} \tag{14}$$

or

$$H^2 = H_0^2 \sum_n \frac{\Omega_{n,0}}{a^{\gamma_n}} = H_0^2 \left(\frac{\Omega_{m,0}}{a^3} + \frac{\Omega_{r,0}}{a^4} + \Omega_{\Lambda,0} + \frac{\Omega_{k,0}}{a^2}\right) \tag{15}$$

The observed values for these cosmological parameters are

$$\Omega_{\Lambda,0} \sim 0.7, \qquad \Omega_{m,0} \sim 0.3, \qquad \Omega_{r,0} \sim 5 \times 10^{-5} \tag{16}$$

and $H_0 \sim 70$ km/sec/Mpc $\sim 2 \times 10^{-18}$ sec, where 1 Mpc $= 10^6$ parsec or $\sim 3 \times 10^{19}$ km.

The matter contribution $\Omega_{m,0}$ to Ω consists of dark matter $\Omega_{dm,0} \sim 0.25$ and baryonic matter $\Omega_{b,0} \sim 0.04$. The surprising discovery has been that the baryonic matter we know and love comprises only a teeny contribution to the energy budget of the universe. Another remarkable cosmological fact is that only a small fraction of the baryonic matter $\Omega_* \sim 0.008$ resides in stars; the rest appears to be in interstellar and intergalactic gases.

Keep in mind that our index j runs over r, m, and Λ, while the index n runs over the range of j plus the curvature term k.

Constructing a 2-dimensional map of universes

It may seem like the height of hubris, but within the context of this discussion, we can characterize our universe in terms of three present-day values $\{\Omega_{r,0}, \Omega_{m,0}, \Omega_{\Lambda,0}\}$. In fact, since $\Omega_{r,0} \ll \Omega_{m,0}, \Omega_{\Lambda,0}$ we can make do with a 2-dimensional parameter space with the two axes $\Omega_{m,0}$ and $\Omega_{\Lambda,0}$, which we could happily plot (see figure 1) on a piece of paper. Note that the universe considered in chapter VI.2 sits at the point $\{\Omega_{m,0} = 0, \Omega_{\Lambda,0} = 1\}$.

Draw the line

$$\Omega_{m,0} + \Omega_{\Lambda,0} = 1 \tag{17}$$

From (10), $\Omega_{m,0} + \Omega_{\Lambda,0} + \Omega_{k,0} = 1$, we see that the universe is closed above this line and open below it. A universe sitting right on this line is flat.

Acceleration or deceleration?

At this point, let us also make use of the other Einstein field equation

$$\frac{\ddot{R}}{R} = \frac{\ddot{a}}{a} = -\frac{4\pi G}{3}(\rho + 3P) = -\frac{4\pi G}{3}\sum_j (1 + 3w_j)\rho_j \tag{18}$$

which describes the acceleration of the cosmic expansion. Define the deceleration parameter

$$q \equiv -\frac{a\ddot{a}}{\dot{a}^2} = -\frac{\ddot{R}/R}{\dot{R}^2/R^2} = +\tfrac{1}{2}\sum_j (1+3w_j)\,\Omega_j$$

$$= \tfrac{1}{2}\left(2\Omega_r + \Omega_m - 2\Omega_\Lambda\right) \tag{19}$$

(Note that for many cosmological parameters, we could freely choose to use a or R.) That q is defined with a minus sign (and thus known as the deceleration parameter) is historical: before the discovery of the dark energy, it was thought that all possible values of w_j were positive and that the cosmic expansion was thus decelerating. You are of course free to define the acceleration parameter $Q \equiv \frac{a\ddot{a}}{\dot{a}^2}$ if you like.

As explained in the preceding chapter, in the absence of the cosmological constant, or if the cosmological constant Λ is negative, then q is manifestly positive, and the expansion of the universe will slow down.

Since from (19), we have

$$-q_0 = \Omega_{\Lambda,0} - \tfrac{1}{2}\Omega_{m,0} \tag{20}$$

cosmic expansion is accelerating above the line

$$\Omega_{\Lambda,0} = \tfrac{1}{2}\Omega_{m,0} \tag{21}$$

and decelerating below this line.

The fate of the universe

Will the universe expand forever? Did it have a Big Bang?

By now, you should have enough understanding to answer the following questions qualitatively. If dark energy overwhelms dark matter, did it bang? Yes or no?

Let us now quantify the word "overwhelm" by determining the two curved lines in figure 1. Write (15) as

$$\frac{1}{H_0^2}\dot{a}^2 - \left(\frac{\Omega_{m,0}}{a} + \frac{\Omega_{r,0}}{a^2} + \Omega_{\Lambda,0}a^2 + \Omega_{k,0}\right) = 0 \tag{22}$$

which we can interpret as a Newtonian problem of particle of mass $m \equiv \frac{2}{H_0^2}$ with zero total energy moving in a potential $V(a) \equiv -\left(\frac{\Omega_{m,0}}{a} + \frac{\Omega_{r,0}}{a^2} + \Omega_{\Lambda,0}a^2 + \Omega_{k,0}\right)$. We can eliminate $\Omega_{k,0} = 1 - (\Omega_{m,0} + \Omega_{r,0} + \Omega_{\Lambda,0})$ by evaluating (15) at the present time.

Perhaps astonishingly, the Newtonian mechanical analog keeps popping up in general relativity, whether we are studying the motion of a particle around a black hole or figuring out how the universe evolves. By now you surely understand why this is so. In the present context, Einstein's field equation involves, by construction, two powers of derivatives. Because of the perfect cosmological principle, the entire metric is described by one function of a single variable, so that we have ordinary, rather than partial, differential equations. Thanks to Bianchi's identity, we can eliminate the second derivative $\ddot{a}$. As a result, we happily end up with a problem in Newtonian mechanics, which we can readily solve

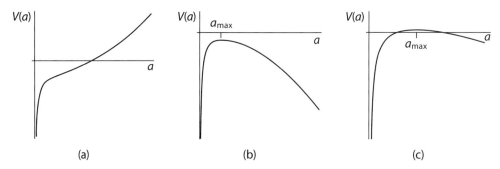

Figure 2 The cosmic potential for some representative values of $(\Omega_{m,0}, \Omega_{\Lambda,0})$ plotted against the scale factor of the universe. (a) For $(\Omega_{m,0}, \Omega_{\Lambda,0}) = (0.3, -0.8)$. The universe started in a Big Bang, with the present expansion headed toward an eventual collapse back into a Big Crunch. (b) For $(\Omega_{m,0}, \Omega_{\Lambda,0}) = (0.3, 0.8)$. The universe started out in a Big Bang and then proceeds to expand forever. (c) For $(\Omega_{m,0}, \Omega_{\Lambda,0}) = (2.2, 0.03)$. Note that this is for large $\Omega_{m,0}$ and small $\Omega_{\Lambda,0}$, and that the maximum of the potential barely sticks above the horizontal axis. Either the universe started in a Big Bang but will eventually collapse back, or the universe never did have a Big Bang and will expand forever.

numerically in the general case. In fact, setting $\Omega_{r,0} = 0$, we can analyze the problem completely, as we will now show.

To a good approximation, then, we have a Newtonian particle with zero total energy moving in the potential (see figure 2)

$$V(a) = -\left(\frac{\Omega_{m,0}}{a} + \Omega_{\Lambda,0}a^2\right) - \left(1 - \Omega_{m,0} - \Omega_{\Lambda,0}\right) \tag{23}$$

Note that this cosmic potential is the sum of three terms: a $-1/a$ term; an a^2 term, whose coefficient can take either sign; and a constant.

For small a, the potential $V \sim -\frac{\Omega_{m,0}}{a}$ is attractive like an inverse square law; for large a, $V \sim -\Omega_{\Lambda,0}a^2$ is repulsive or attractive, according to whether $\Omega_{\Lambda,0} > 0$ or < 0, like an inverted or a normal harmonic oscillator, respectively. The constant term in V moves the potential up or down. The boundary condition is that at present, $a = 1$ and $\dot{a} > 0$.

We can analyze the negative cosmological constant $\Omega_{\Lambda,0} < 0$ case instantly. The potential $V(a) = -\frac{\Omega_{m,0}}{a} + |\Omega_{\Lambda,0}|a^2 - (1 - \Omega_{m,0} + |\Omega_{\Lambda,0}|)$ is entirely attractive. See figure 2a. The particle is at present climbing the hill, but when it reaches the point where $V(a) = 0$, its velocity vanishes ($\dot{a} = 0$), since it has zero total energy, as you recall. It then turns around and slides back down the hill. In other words, the universe started in a Big Bang, with the present expansion headed toward an eventual collapse back into a Big Crunch, when $a(t)$ will once again vanish. This takes care of the entire lower half plane $\Omega_{\Lambda,0} < 0$ in the cosmic diagram.

We now take $\Omega_{\Lambda,0} > 0$. The potential now reaches a maximum at some a_{max}. There are two possibilities, as shown in figure 2b and figure 2c, according to whether the maximum of the potential sits below the horizontal axis (that is, $V(a_{max}) < 0$), or sticks up above the horizontal axis (that is, $V(a_{max}) > 0$).

Again, recall that the Newtonian analog particle has zero total energy. In the situation described in figure 2b, it has enough energy to reach the top of the potential and then cruise down. The universe started out in a Big Bang and then proceeds to expand forever.

We have to divide the situation described in figure 2c further into two cases, according to whether $a_{max} > 1$ or $a_{max} < 1$.

Suppose that $a_{max} > 1$. Since at present, $a = 1$, by definition, with $\dot{a} > 0$ by observation, then, as represented by the Newtonian particle, we have not yet gotten to the top of the hill, which we are determinedly climbing. But when we reach $V(a) = 0$, we will run out of steam; not having enough oomph to reach the top, we will fall back in toward $a = 0$. In other words, the universe will eventually collapse.

Now suppose that $a_{max} < 1$. Then we, presently at $a = 1$ with $\dot{a} > 0$, are already on the other side of the hill and are happily rolling downhill with ever increasing speed. We have never even been to $a = 0$: there was no Big Bang in our past. The universe never did have a Big Bang and will expand forever.

Note that in both of these cases shown in figure 2c, the Newtonian particle, with its zero total energy, can never have gotten above the horizontal axis. The entire history of the universe is described by the piece of $V(a)$ below the horizontal axis, either the piece on the left or that on the right.

We thus see that the dividing lines between these different scenarios are determined by first finding out if $V'(a_{max}) = 0$ has a solution, and if it does, then setting $V(a_{max}) = 0$. We can use one equation to eliminate a_{max} in the other, thus obtaining a relation between $\Omega_{\Lambda,0}$ and $\Omega_{m,0}$, leading to the curves shown in the cosmic diagram. To determine the behavior on the two sides of these curves, we have to further ascertain whether $a_{max} > 1$ or $a_{max} < 1$.

Quite remarkably, at this stage, to work out the cosmic diagram requires no more than high school algebra, not even solving a differential equation. You should challenge yourself before reading the solution, which I will relegate to the appendix. Observationally, as I have already mentioned, the favored region forms a small circle* centered at $(\Omega_{m,0}, \Omega_{\Lambda,0}) = (0.3, 0.7)$, far from the two curves we just discussed (see figure 1). Nevertheless, theoretically it is quite interesting to work out these two curves.

Einstein's static universe and his second greatest blunder

In the era when Einstein ever so boldly[†] ventured to apply his theory of gravity to the entire universe, physicists were philosophically prejudiced in favor of a static universe. Indeed, the expanding universe that we all, including the proverbial person in the street, take for granted was inconceivable once upon a time. When Einstein found, to his horror, that his field equation implied an expanding universe, he solved the perceived difficulty by introducing a cosmological constant and showing that he could have a static universe if

---

* The size of the circle changes as observation improves.
† Indeed, fearing for the asylum. See the opening of chapter VI.2.

$\Lambda = \frac{1}{2}\rho_m$ (as you can also show in exercise 3). He thus missed a tremendous opportunity to predict that the universe expands.

For those who delight in yakking about Einstein's greatest blunder, an idle preoccupation that strikes me as somewhat unseemly, I have already expressed my humble opinion that the greatest blunder was not the introduction of the cosmological constant, as the popular press would have it (which is in fact required by quantum field theory, as I explained in chapter VI.2 and will discuss in more detail in chapter X.7, and in any case appears observationally to be here to stay). Rather, it was his failure to use the action principle. With your indulgence, we will now talk about Einstein's second greatest blunder, which is not to check whether his solution was stable.

In light of the Newtonian analog potential in (23), this instability is starkly evident. Differentiating and setting $V'(a) = \frac{\Omega_{m,0}}{a^2} - 2\Omega_{\Lambda,0}a^2$ to 0 and a to 1, we recover Einstein's condition $2\Omega_{\Lambda,0} = \Omega_{m,0}$ expressed in the cosmologist's language. Einstein's static universe corresponds to sitting at the maximum of the potential in figure 2b.

Flow in the cosmic diagram

It is perhaps worth emphasizing that all relevant physics within the present context is contained in the cosmological equation

$$\dot{R}^2 + k = \frac{8\pi G}{3}\rho R^2 \tag{24}$$

and energy conservation

$$\frac{d}{dt}\rho a^3 = -P\frac{da^3}{dt} \tag{25}$$

The discussion here is merely expressing the same physics in a notation particularly convenient for observational cosmology.

Thanks to the Bianchi identity, we can recover the rest of Einstein's field equation from (24) and (25). It is instructive to verify this. First, rewrite (25) as

$$\dot{\rho} + 3(\rho + P)\frac{\dot{R}}{R} = 0 \tag{26}$$

Next differentiate (24) to obtain an equation for $\ddot{R}$, eliminating $\dot{R}$ by using (24) and $\dot{\rho}$ by using (26). Not surprisingly, we get the time-time component of Einstein's field equation:

$$\frac{\ddot{R}}{R} = -\frac{4\pi G}{3}(\rho + 3P) \tag{27}$$

You are of course free to keep on massaging these equations every which way. For example, we might ask how Ω_j varies with time. Recalling the definition $\Omega_j = \frac{8\pi G}{3H^2}\rho_j$, we have

$$\dot{\Omega}_j = \Omega_j\left(\frac{\dot{\rho}_j}{\rho_j} - 2\frac{\dot{H}}{H}\right) = -\Omega_j\left(3(1+w_j)H + 2\frac{\dot{H}}{H}\right) \tag{28}$$

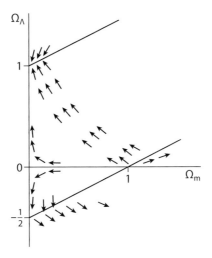

Figure 3 Neglecting radiation, we can picture the time evolution of the universe as a fluid flowing in the 2-dimensional space spanned by $(\Omega_m, \Omega_\Lambda)$, plotted here for a universe in an expanding phase. Of the three fixed points, $(\Omega_m, \Omega_\Lambda) = (0, 1)$, $(0, 0)$, and $(1, 0)$, only the one at $(0, 1)$ is a stable attractor.

where in the last step, we used (26). Now

$$\frac{\dot{H}}{H^2} = \frac{1}{H^2} \frac{d}{dt}\left(\frac{\dot{R}}{R}\right) = \frac{1}{H^2}\left(\frac{\ddot{R}}{R} - H^2\right) = -q - 1 = -1 - \frac{1}{2}\sum_i \left(1 + 3w_i\right)\Omega_i$$

using (19) in the last step. Inserting this into (28), we obtain the nifty equation

$$\dot{\Omega}_j = +H\Omega_j\left(-3w_j - 1 + \sum_i \left(1 + 3w_i\right)\Omega_i\right) \tag{29}$$

The rate of change of Ω_j depends on the other Ω_is. We can think of this as defining a flow in the space spanned by the Ω_js.

For example, if we neglect Ω_r as before, we have

$$\dot{\Omega}_m = H\Omega_m\left(\Omega_m - 2\Omega_\Lambda - 1\right) \quad \text{and} \quad \dot{\Omega}_\Lambda = H\Omega_\Lambda\left(\Omega_m - 2\Omega_\Lambda + 2\right) \tag{30}$$

We can think of this as defining a velocity field $\vec{v} = (\dot{\Omega}_m, \dot{\Omega}_\Lambda)$ for a fluid flowing in the 2-dimensional space spanned by $(\Omega_m, \Omega_\Lambda)$, which we plot in figure 3, assuming that the universe is in an expanding phase $H > 0$. To facilitate plotting the velocity field, notice that $\dot{\Omega}_m < 0$ above the line $\Omega_\Lambda = \frac{1}{2}(\Omega_m - 1)$ and > 0 below. Similarly, $\dot{\Omega}_\Lambda < 0$ above the line $\Omega_\Lambda = \frac{1}{2}\Omega_m + 1$ and > 0 below. There are three fixed points, $(\Omega_m, \Omega_\Lambda) = (0, 1)$, $(0, 0)$, and $(1, 0)$, defined as places where the velocity field vanishes. The fixed point at $(0, 1)$ is stable, known variously as an attractor or a sink in different areas* of physics, in the sense

---

* In quantum field theory and in condensed matter physics, this kind of flow is known as a renormalization group flow. The quantum and thermal fluctuations responsible for the flow are, however, completely absent in the present context.

that a particle flowing nearby would end up there. In contrast, we have two unstable fixed points at $(0, 0)$ and $(1, 0)$. In particular, a universe at $(0.3, 0.7)$ will eventually end up at $(0, 1)$. But we know all this already: dark energy in the form of a cosmological constant will eventually overwhelm the dark matter.

We see the essential role played by the cosmological constant. In its absence, we have a 1-dimensional flow along the Ω_m axis, as shown in figure 3: if $\Omega_m < 1$, the universe flows toward $\Omega_m = 0$, and if $\Omega_m > 1$, the universe flows to arbitrarily large values. Of the two unstable fixed points in the 2-dimensional flow, the one at $(0, 0)$ becomes stable if we restrict the flow to be along the Ω_m axis. When a positive Ω_Λ, no matter how small, is introduced, the fluid flows away from the Ω_m axis: Ω_Λ is known as a relevant perturbation. Again, this is just old knowledge repackaged: as the universe expands, matter density is diluted to nothing, while the cosmological constant remains constant. In a contracting phase, as indicated by (30), everything is reversed, of course.

Is flat stable?

It is interesting to apply this flow language to the curvature density $\Omega_k = -\frac{k}{H^2 R^2}$. Note that, in contrast to k, the quantity Ω_k is continuous rather than discrete, and so it makes sense to talk about its rate of change. Going through similar steps as above, we obtain

$$\dot{\Omega}_k = -2\Omega_k \left(\frac{\dot{H}}{H} + H \right) = 2\Omega_k H q = H \Omega_k \left(2\Omega_r + \Omega_m - 2\Omega_\Lambda \right) \tag{31}$$

Is a flat universe stable? If Ω_k is strictly 0, that is, if the integer $k = 0$, then $\dot{\Omega}_\Lambda = 0$ and Ω_k stays at 0. The issue is whether a universe with $\Omega_k \simeq 0$ flows toward or away from $\Omega_k = 0$. Again, assume that the universe is in an expanding phase, so that $H > 0$.

As you can see from (31), if $2\Omega_r + \Omega_m > 2\Omega_\Lambda$ (which is a fortiori satisfied if $\Omega_\Lambda < 0$ or if there is no cosmological constant $\Omega_\Lambda = 0$), then $\Omega_k = 0$ is an unstable fixed point.

In contrast, if $2\Omega_r + \Omega_m < 2\Omega_\Lambda$, then $\Omega_k = 0$ is a stable fixed point. But since Ω_r and Ω_m vanish rapidly, given enough time, this condition will eventually be satisfied. We will return to this point in the next chapter.

Age of the universe

Since H_0 has dimensions of inverse time, a rough estimate of the age of the universe is simply $t_{age} = 1/H_0$, but given the accuracy to which the cosmological parameters are now known, we can do better. Using (15), we obtain

$$t_{age} = \int_0^{t_{age}} dt = \int_0^1 \frac{da}{a(\frac{\dot{a}}{a})} = \int_0^1 \frac{da}{aH}$$

$$= \frac{1}{H_0} \int_0^1 \frac{da}{[\Omega_{m,0}a^{-1} + \Omega_{r,0}a^{-2} + \Omega_{\Lambda,0}a^2 + (1 - \Omega_{m,0} - \Omega_{r,0} - \Omega_{\Lambda,0})]^{1/2}} \tag{32}$$

In general, this will have to be integrated numerically. For a flat universe, if we neglect radiation, the integral can be done exactly:

$$t_{\text{age}} = \frac{2}{3H_0\sqrt{\Omega_{\Lambda,0}}} \log(\frac{\sqrt{\Omega_{\Lambda,0}} + \sqrt{\Omega_{m,0} + \Omega_{\Lambda,0}}}{\sqrt{\Omega_{m,0}}}) \simeq 1.02H_0^{-1} \tag{33}$$

Historically, the age of the universe presented a thorny issue for a long time, as it came out shorter than the age of certain stars and galaxies estimated reliably using other methods. Since a universe with only $\Omega_{\Lambda,0} > 1$ has no Big Bang and hence infinite t_{age}, the discovery of dark energy has evidently served to resolve this problem. Indeed, we can now turn the argument around, so that we can use the inferred age of a galaxy observed at such and such redshift to set a lower bound on $\Omega_{\Lambda,0}$.

Appendix: Phase boundaries in the cosmic diagram

To determine the curves in figure 1, it turns out that we have to solve a cubic equation. So let's first recall the (hyperbolic) cosine and sine method for solving the cubic. A cubic equation can always be cast in the form $4x^3 - 3x - s = 0$. For our problem, we want the real positive root (and in case there are more than one, the smaller of the two). For $s > 1$, there is one real positive root. Use the identity $4(\cosh \beta)^3 = 3 \cosh \beta + \cosh 3\beta$. Then the solution is evidently $x = \cosh \beta$, with β determined by $s = \cosh 3\beta$. For $0 < s < 1$, change the hyperbolic cosine in the preceding to a cosine. For $s < 0$, use the identity $4(\sin \beta)^3 = 3 \sin \beta - \sin 3\beta$. Then the solution is $x = \sin \beta$, with β determined by $s = -\sin 3\beta$. In particular, as $s \to 0$, the solution $x \to 0$.

The case $\Omega_{m,0} = 0$ was treated in chapter VI.5. The case $\Omega_{\Lambda,0} = 0$ is also easily analyzed by looking at a plot of the potential $V(a) = -\frac{\Omega_{m,0}}{a} - (1 - \Omega_{m,0})$. For $\Omega_{m,0} < 1$, the particle in the Newtonian analogy is unbound, and the universe expands forever. For $\Omega_{m,0} > 1$, the particle is bound, and the universe expands and then collapses.

Taking $\Omega_{m,0} \neq 0$, we divide $V(a)$ by $\Omega_{m,0}$ for convenience, so that we can write the potential effectively as $V(a) = -a^{-1} - 4x^3a^2 - (s - 4x^3)$ after defining $x \equiv \left(\frac{\Omega_{\Lambda,0}}{4\Omega_{m,0}}\right)^{\frac{1}{3}}$ and $s \equiv \frac{1-\Omega_{m,0}}{\Omega_{m,0}}$. The condition $V'(a_{\text{max}}) = 0$ then gives $a_{\text{max}} = \frac{1}{2x}$, which when substituted back, yields $V(a_{\text{max}}) = 4x^3 - 3x - s$. The condition $V(a_{\text{max}}) = 0$ then produces the cubic solved above.

For $\Omega_{m,0} \simeq 0$, $s \gg 1$, and we have the solution[3]

$$\Omega_{\Lambda,0} = 4\Omega_{m,0} \left(\cosh\left[\frac{1}{3} \text{arccosh}\left(\frac{1-\Omega_{m,0}}{\Omega_{m,0}}\right)\right]\right)^{\frac{1}{3}}$$

$$= 1 - 3(\Omega_{m,0}/2)^{\frac{2}{3}} - \Omega_{m,0} + 3(\Omega_{m,0}/2)^{\frac{4}{3}} + \cdots \tag{34}$$

This gives the curve starting at the point $(\Omega_{m,0}, \Omega_{\Lambda,0}) = (0, 1)$ separating universes that banged from those that did not. Note that since $s \gg 1$, implying $\beta \gg 1$, then $a_{\text{max}} \ll 1$ and the universe expands forever on both sides of this phase boundary.

For $\Omega_{m,0} \simeq 1$, we have the solution[4]

$$\Omega_{\Lambda,0} = 4\Omega_{m,0} \left(\sin\left[\frac{1}{3} \arcsin\left(\frac{\Omega_{m,0} - 1}{\Omega_{m,0}}\right)\right]\right)^{\frac{1}{3}}$$

$$= \frac{4}{27}(\Omega_{m,0} - 1)^3 - \frac{8}{27}(\Omega_{m,0} - 1)^4 + \cdots \tag{35}$$

This gives the curve starting at the point $(\Omega_{m,0}, \Omega_{\Lambda,0}) = (1, 0)$ separating universes that will expand forever from those that will collapse. As indicated in the cosmic diagram, all these universes had a Big Bang in their past.

I envisage a titanic struggle between dark energy and dark matter. Dark matter hardly has a chance against the explosively exponential expansion of a positive cosmological constant, but what we just learned is that for a tiny $\Omega_{\Lambda,0} > 0$, a sliver of opportunity still exists for dark matter. For example, (35) tells us that for $\Omega_{m,0} = 1.1$, if $\Omega_{\Lambda,0} < \frac{4}{27} \times 10^{-3}$, then dark matter could still reverse the expansion.

Note that before the discovery of dark energy, $\Omega_{\Lambda,0}$ was presumed to vanish, and so cosmologists were restricted to the $\Omega_{m,0}$-axis. We can use the cosmic diagram to verify the results in the previous chapter. Thus, along the $\Omega_{m,0}$-axis, for $\Omega_{m,0} < 1$, the universe is open and expands forever, while for $\Omega_{m,0} > 1$, it is closed and will eventually collapse. All these universes decelerate and had a Big Bang.

Exercises

1 For a flat universe filled with only nonrelativistic matter, show that $t_{\text{age}} = \frac{2}{3H_0}$ and $\rho = 1/(6\pi G t^2)$.

2 For a flat universe filled with only relativistic matter, show that $t_{\text{age}} = \frac{1}{2H_0}$ and $\rho = 3/(32\pi G t^2)$.

3 Obtain Einstein's static universe directly from his field equation and determine the radius of his static universe.

4 Derive the expression for the curvature density used by astronomers

$$\Omega_k(z) = \frac{\Omega_{k,0}}{\Omega_{m,0}(1+z) + \Omega_{r,0}(1+z)^2 + \Omega_{\Lambda,0}(1+z)^{-2} + \Omega_{k,0}} \tag{36}$$

5 Show that a flat universe containing any amount of $\Omega_{m,0}$, but with $\Omega_{\Lambda,0} < 0$, will reach what is known as a Big Crunch, a moment when a vanishes. Calculate the time to the end.

Notes

1. *What Time Is It over There?* is an interesting film by Ming-liang Tsai.
2. Note that if we define the Hubble length by $L \equiv H^{-1}$ and regard it as some kind of radius of the universe, we can massage the definition of the critical density $\rho_c = \frac{3H^2}{8\pi G}$ into the following: $2G(4\pi L^3/3)\rho_c = L$. If we think, rather loosely, of the universe as a Euclidean ball of radius L and mass density ρ_c, and hence mass $M = (4\pi L^3/3)\rho_c$, this tells us that the Schwarzschild radius of the universe is equal to its radius, that the universe is on the verge of being a black hole. This is of course just a heuristic way of interpreting what the critical density means.
3. For those who find this functional form indigestible, I offer (and prefer) the alternate form $\Omega_{\Lambda,0} = \frac{1}{2}\Omega_{m,0}(k + 3k^{\frac{1}{3}} + 3k^{-\frac{1}{3}} + k^{-1})$, with $k \equiv (1 - \Omega_{m,0} + \sqrt{1 - 2\Omega_{m,0}})/\Omega_{m,0}$.
4. Again, as in the preceding endnote, I offer the alternate form $\Omega_{\Lambda,0} = \frac{i}{2}\Omega_{m,0}(w - 3w^{\frac{1}{3}} + 3w^{-\frac{1}{3}} - w^{-1})$, with $w \equiv (\sqrt{2\Omega_{m,0} - 1} + i(\Omega_{m,0} - 1))/\Omega_{m,0}$.

VIII.3 | The Gamow Principle and a Concise History of the Early Universe

A physical history of the universe

At one time, textbooks on Einstein gravity devoted[1] a substantial fraction of their expositions to physical cosmology. In the intervening decades, cosmology has grown by leaps and bounds, and any serious discussion of the subject requires a textbook[2] of its own, as I've said in chapter VIII.1. Here I can only give you a sketch of the physical history of the universe, largely in qualitative[3] terms. Thus, this chapter will consist of mostly talk.

But as a subscriber to Feynman's aversion to all talk and no action, I feel compelled to do one small calculation toward the end of this chapter. We will determine the position of the first acoustic peak in the fluctuation of the cosmic microwave background.

The once-hot universe

Imagine that someone had filmed the universe's evolution. In what follows, we will play the movie backward and forward, rewind and fast forward. Let us start by playing the movie backward, starting from the present.

The crucial feature of our universe is that it expands and hence cools. As was derived in chapter VIII.1, the temperature of the radiation filling the universe varies with the cosmological scale factor $a(t)$ like $T \sim 1/a$, so that, as we go back into the early universe, T steadily increases. Yesterday was hotter than today by about $10^{-11}\%$, not by much, but a billion years ago, the universe was hotter by 7%. And so it goes. As Boltzmann and others taught us, temperature measures the average energy of the particles in a thermal distribution. Keep in mind that $1\,\text{eV} \sim 1.16 \times 10^4$ K, and so the photons in the cosmic microwave background, nowadays at 2.7 K, are almost negligibly feeble on the scale of atomic physics. But as we go back in time, these now frail photons once rampaged, ripping atoms apart.

Figure 1 In the primeval universe, photons rushed about vigorously trying to prevent the electrons from attaching themselves to the protons. A pictorial representation.

Next, play the movie forward, starting at a time when the temperature was way too hot for atoms to exist. You might recall that the binding energy of the hydrogen atom is 13.6 eV. So let us suppose that the typical energy of the photons far exceeds that. The universe was a hot soup of electrons, protons, and so forth. (As we will see later, there were a small number of deuterons, helium nuclei, and the like around, but to keep the story simple, we will ignore them.)

The photons are constantly scattering off the charged particles: the photons and the charged particles are said to be tightly coupled to each other. Occasionally, a proton and an electron would come together and form a hydrogen atom, but almost immediately, a photon would come along and knock the proton and the electron away from each other. See figure 1.

Recombination delayed and decoupling

But the universe keeps on cooling, and the average energy of the photons keeps on dropping. Eventually, the photons become too weak to ionize the hydrogen atom, at which point, protons and electrons rush to pair off with each other.

This milestone in the life of the universe, the combination of protons and electrons into hydrogen atoms, is known as recombination.* You might think that the recombination would occur as soon as the temperature drops below ∼13.6 eV, but two physical effects delay recombination. When we specify the temperature of a photon gas, we are talking about the average energy of the photons. The photons in the tail end of the thermal distribution are much more energetic (by definition of "tail end"). Furthermore, the universe contains vastly more photons than protons and electrons, by a factor of about 10^{10}. Thus, through the sheer numbers of photons, even if only a tiny fraction of them have energies in excess of 13.6 eV, the hydrogen atoms are still ripped apart as soon as they form. Consequently, recombination does not occur until the universe has cooled to about 0.3 eV.

---

* Recombination surely ranks as one of the least appropriate physics terms: the protons and the electrons had never been combined earlier.

Since the photon interacts oppositely with the positively charged protons and the negatively charged electrons in an atom, its interaction with an atom is vastly reduced compared to its interaction with the proton or with the electron. (The net interaction with the atom does not add up to exactly zero, because the electrons are spread out, while the protons are concentrated in the nucleus.)

After recombination, the mean free path of the photons increases dramatically. There are hardly any charged particles for the photons to scatter off, and the interaction of photons with the atoms, as explained above, is relatively weak. Soon, the mean free path of the photons exceeds the Hubble radius of the universe, and the photons are effectively free. They are said to decouple.*

Pale fire

Decoupling was a crucial event in the evolution of the universe, as was pointed out in 1948 by George Gamow, Ralph Alpher, and Robert Hermann. They realized that a pale shadow of the fire that filled the early universe should still be visible.

After decoupling, radiation (photons) and matter (atoms) more or less go their separate ways. These primeval photons interact so little with matter that through the eons, they merely drift through the universe, getting redshifted down to a cosmic microwave background permeating the universe. Gamow and his collaborators proposed the detection of these "relic photons" as a crucial test of the Big Bang. Finally, in 1965, Arno Penzias and Robert Wilson of Bell Telephone Company fortuitously detected these telltale photons.

That the actual detection of these primeval photons was not made until 17 years after the initial prediction, and then only by chance, poses something of a puzzle for historians of physics. The technology was available in 1948. Why then had experimenters not tripped over one another to look for the glow from the Big Bang?[4]

Primeval nucleosynthesis

As we go farther back in time, the universe gets even hotter. When the average energy gets to about one tenth of an MeV, nuclei are ripped apart. (Again, when the average energy of the particles is only one tenth of an MeV, quite a few photons already have energies in excess of a few MeV.) The universe was too hot for nuclei to exist and consisted of a hot cauldron of protons, neutrons, and electrons.

Now run the movie forward. The universe cools. Soon photons no longer have enough energy to break up a deuteron, so that the deuteron, once formed, could stick around. The number of deuterons increases rapidly. Violence fades as the universe ages. As the deuterons drift around, some of them are hit by protons and neutrons. When a proton sticks to a deuteron, a helium 3 nucleus results. When a neutron sticks to a deuteron, a

---

* Recombination and decoupling happen to occur at roughly the same time in our universe, but the events are conceptually distinct.

tritium nucleus results. And so on and so forth. The net result is that as the universe cools, protons and neutrons stick to one another. The primeval soup of protons and neutrons is converted into nuclei of various kinds, laying the foundation for the modern world. The construction of atomic nuclei in the early universe, known as primeval nucleosynthesis, is easy to understand qualitatively.

The Gamow principle

In 1948, the U.S. government declassified nuclear reaction rates—information about how readily a proton or a neutron would stick to a given nucleus to form a larger nucleus. George Gamow realized that with this information, he could calculate the relative abundance of the elements in the universe.

I refer to this insight as the Gamow principle: If you understand the physics at the energy scale E, then you can describe the evolution of the universe at temperature E.

As we travel back to the early universe in our mind, we go through the standard curriculum of physics. After atomic physics comes nuclear physics. After nuclear physics comes particle physics. After the known particle physics comes grand unified physics, applicable when the universe was a soup of quarks, leptons, and grand unified gauge bosons. After that comes string theory.

An obvious corollary follows: If you don't understand the physics at a given energy scale, then you can only speculate. For example, if you think that you understand physics above the Planck scale $M_P \sim 10^{19}$ GeV, then you could describe trans-Planckian cosmology. But if you don't, then all the talk amounts to mere speculation.

Perhaps paradoxically to the layperson (but not to you), the later the epoch, the more detailed and involved the physics you would have to master to work out the cosmology of the epoch. For example, around 100 million years after the Big Bang, hydrogen molecules began to form, and it is essential to understand the difference between the excitation spectrum of the hydrogen molecule versus that of hydrogen atoms. Much of what happened since then has to be worked out by massive numerical computation.[5]

Stellar nucleosynthesis

> The universe is a spiraling Big Band in a polka-dotted speakeasy,
> effectively generating new light every one-night stand.
>
> —Ishmael Reed

Gamow originally thought that all nuclei were formed in primeval nucleosynthesis. It later became clear, however, that nucleosynthesis essentially came to a halt shortly after helium was formed. By that time, the expansion of the universe had reduced drastically the numbers of protons, deuterons, and helium nuclei per unit volume. The collisions between them were so infrequent that nuclear processes by and large came to a halt.

But as the universe cools further, the electrons move ever more slowly. With the passing of time, the universe becomes cool enough for the products of nucleosynthesis—the protons, deuterons, and helium nuclei—to grab some passing electrons to form atoms. After atoms form, the nuclei can no longer get close to one another. When two atoms get close, the buzzing clouds of electrons keep the two nuclei far apart.

After primordial nucleosynthesis, the universe thus settles into a relatively dull existence, permeated by enormous clouds of gas, cool enough for atoms and molecules to exist. However, gravity has already been hard at work, pulling together neighboring globs of gas. Soon the first stars condense out of the primeval gas clouds. As the gas atoms rush together to form a star, they crash into one another with such abandon that they rip electrons off one another, thus allowing the nuclei to approach one another once again and restart nuclear reactions. The universe is suddenly lit with lights beyond measure.

Inside the stars, a helium nucleus bumps into another helium nucleus, which stick to each other to form a beryllium nucleus. Yet another helium nucleus wanders by, sticks to the beryllium nucleus, and produces a carbon nucleus. Out of starfires we humans become a possibility. Note the crucial difference between primeval nucleosynthesis and stellar nucleosynthesis. In the primeval setting, nuclei were drifting farther and farther apart in the expanding universe. But when they were confined inside stars, they were bound to bump into one another. Thus deuterium, helium, and a little bit of lithium were produced in the primeval universe, while the more massive nuclei were formed later in stars. When some of the first-generation stars exploded, they ejected into space these higher nuclei, among other things. Out of this ejected debris, a second generation of stars soon condensed. These stars started out containing heavier nuclei like carbon, out of which more and more complicated nuclei are manufactured. Eventually, these stars in turn exploded and splattered themselves over the cosmos.

You can't make much out of only hydrogen and helium, but with carbon, silicon, iron, and so forth, the possibilities become endless. You can make rocks, for instance. Bits of rocks come together to form rock piles, laughably minute specks of dust in the cosmic scheme of things. On one of these specks, carbon atoms started connecting up with hydrogen, oxygen, and so forth. Somehow, these bunches of atoms suddenly came alive. Eons and lots of self-improvement courses later, this moving, eating, reproducing ooze turned into what are known as human beings, who eventually end up writing and reading textbooks on Einstein gravity.

The rich get richer

Gravity plays an all-important role in the formation of structures in the universe.

Let me first tell the story without dark matter.

About 40,000 years after the Big Bang, bits of matter started to come together, forming enormous structures that eventually condensed into galaxies. Galaxy formation marked the first step in the emergence of structures in our universe: Within the galaxies, protostars soon formed.

Long ago, Newton had already identified the basic physics responsible for the emergence of structure: the inherent instability of gravity. In 1692, a certain Reverend Dr. Richard Bentley wrote to Newton, arguing that the universal presence of gravity proved the existence of God, a view with which Newton was much in sympathy. A lively correspondence followed. In one of his letters to Bentley, Newton suggested how structures could emerge in the universe, as follows.

Imagine space filled uniformly with matter. Newton made the point that any irregularity, no matter how minute, would grow larger. Consider a region with more matter per unit volume than the surrounding regions. Being denser, the matter in this region would pull matter in from the surrounding regions by the force of gravity. As a result, the matter distribution in this region becomes even denser. The process accelerates—it is the cosmic equivalent of the often-observed phenomenon that the rich get richer and the poor get poorer.* "And thus might the sun and fixed stars be formed," concluded Newton. (Galaxies were unknown in Newton's time.)

Newton's scenario[6] makes such obvious sense that it remains the basic explanation of how structures emerged in the early universe. Small fluctuations in the density of matter grew and became amplified.

In its intrinsic instability, gravity is dramatically different from the other forces. The electromagnetic force, the other long-ranged force in Nature, is intrinsically stable, because it acts oppositely on positive and negative charges. To see this stability, consider a gas consisting of equal numbers of protons and electrons. An excessive concentration of electrons in one region would be immediately smoothed out by the mutual repulsion among the electrons. Unlike gravity, the electromagnetic force tends to smooth matter out, counteracting gravity's tendency to dump matter together.

Dissipative collapse

In 1902, the English physicist Sir James Jeans tried to calculate the size of the actual lumps that would form. However, he did not know about the universe's expansion. Clearly, cosmic expansion, by thinning out the distribution to matter, works to slow down the formation of lumps. In our analogy, cosmic expansion acts like taxation: as the rich get richer, part of their wealth is continuously removed. But the tax rate is more or less flat: regions sparse with matter are stretched out at essentially the same rate as those regions dense with matter. A calculation[7] including the effect of cosmic expansion, first done by the Soviet physicist E. Lifschitz in 1946, shows that lumps will still form but at a far slower rate than would have been the case were the universe static. With taxation, the rich continue to get richer; they are merely slowed down.

---

* Introduced into sociology as the Matthew principle by R. K. Merton. We already cited this principle in chapters III.2 and VI.3.

As lumps of matter form, more atoms rush in toward the nearest lump. As they rush in, they collide with one another, emitting photons that, since they hardly interact with atoms, escape from the mad rush, thus carrying away energy. In this way, the atoms lose energy and move ever more slowly, less and less able to resist the inward pull of gravity. And thus matter collapses into increasingly compact lumps, in a process known as dissipative collapse.* That atoms can radiate photons and dissipate energy is essential to the story. Were there no mechanism for dissipation, the kinetic energy of the atoms would prevent them from collapsing into lumps. They would simply bounce off one another.

Without gravity, we would not be

Thanks, gravity. How wonderful gravity is! Without it, we would not be. The universe might have been a thin haze without much to recommend it. But gravity couldn't have done it alone. To me, it is awe inspiring how only the intricate intertwining of all four forces manages to bring it off. As gravity strives to bring structures out of the haze, the electromagnetic force is needed to carry the excess energy away. Once the particles quiet down and gravity brings the primeval hydrogen and helium nuclei face to face, the strong and the weak forces step in. The strong force causes nuclei to react with one another, thus igniting the nuclear fire that brings warmth to the vast void. The weak force is crucial lest stars become as uncivilized as nuclear bombs. Certain nuclear reactions can only proceed through the weak force. Because the weak force is, well, weak, these reactions proceed extremely slowly. As a result, the nuclear fires in stars burn at a stately pace. Meanwhile, gravity is busily collecting the ejecta left by the dying stars of the first stellar generation into planet-sized bits of interstellar dirt. The electromagnetic force is keeping busy, too. It is transporting energy from the stars to warm these bits of dirt, and it is running all kinds of chemical reactions, bonding one atom to another, so more and more interesting structures can be built. It's a team effort.

The problem of not enough time

The rich get richer, but there still is a problem that they often feel acutely: it takes time to become rich. Similarly, it takes time for the density fluctuations to grow to the point when structures can form. As explained above, we know when the fluctuations can start growing (not until recombination) and how fast the fluctuations grow once they get going. But in addition, we also know how large the initial density fluctuations were, thanks to the cosmic microwave background. The photons that compose the cosmic microwave background, having traveled unperturbed since the time of decoupling, tell us about the matter distribution at the time.

---

* For an illustration of dissipative collapse, see figure 8.3 on p. 129 of *Toy/Universe*.

Picture the primeval fluid sloshing back and forth in the universe. Photons leaving a denser, and hence hotter, region would be more energetic, and photons leaving a less dense region would be cooler. Thus, any density fluctuation in the universe at the time of decoupling would be imprinted as a temperature fluctuation in the observed microwave background. Indeed, at one time, this line of reasoning was used to predict the temperature fluctuation in the microwave background.

We are going to discuss one aspect of this temperature fluctuation in some detail later, but we might as well introduce the notation now. Let $T(\hat{n}) \equiv T(\theta, \varphi)$ be the observed temperature of the microwave background in the direction $\hat{n}$ defined by the angles θ and φ in spherical coordinates. As mentioned earlier, the average temperature $\langle T \rangle = \frac{1}{4\pi} \int d\theta d\varphi \, \sin \theta \, T(\theta, \varphi)$ was measured to be $\simeq 2.725$ K. Consider the fractional deviation from the mean $\frac{\delta T}{T}(\hat{n}) \equiv (T(\hat{n}) - \langle T \rangle)/\langle T \rangle$. One measure of the fluctuation is given by the root-mean-square of the fractional deviation $\langle (\frac{\delta T}{T})^2 \rangle$, which was observed to be $\simeq 10^{-5}$. This observed value turned out to be significantly less than the predicted value, thus indicating that the universe at decoupling was smoother than previously thought. But if the density fluctuation at recombination started small, there was not enough time for it to grow into the lumpy universe we know today. At one time, this posed a serious problem for cosmology, a problem now resolved, as we will see presently.

Dark matter

Thus far, I have told the story of the early universe without dark matter. As was mentioned earlier on various occasions, observational evidence indicates that the universe contains a lot of dark matter particles, sometimes known as wimps.* For example, stellar movements indicate that the dark matter in various galaxies outweighs the total collection of stars in those galaxies. This astonishing conclusion completely revises our picture of galaxies. The luminous matter we know and love, consisting of nucleons and electrons, is now seen as bits of flotsam bobbing about in a sea of dark matter.

Dark matter now comes to the rescue of cosmologists perplexed by the "not enough time" problem. By definition, dark matter does not interact electromagnetically. Since the wimps do not interact with photons, unlike the charged protons and electrons, they can start coming together gravitationally long before decoupling. Regions that happen to have a somewhat denser distribution of wimps could start getting even denser by pulling in wimps from the surrounding regions through gravity.

Meanwhile, the photons ignore the wimps as they struggle mightily to smooth out the distribution of ordinary matter. All that labor would prove to be in vain. The wimps are already condensing into lumps of dark matter, which tugged at the ambient ordinary matter through gravity, urging the protons and electrons to fall in. As soon as decoupling occurred, atoms formed, and ordinary matter, now neutral, fell into the ready-made dark matter

---

* An acronym for "weakly interacting massive particles." It should be mentioned that as of this writing, the actual particle (or particles) that dark matter consists of has not been directly observed and identified.

lumps. With dark matter present, the formation of structures in the universe could start earlier, without waiting for the formation of atoms.

Not only do the rich get richer, but the rich also can count on inheriting from the wimps.

First acoustic peak

After all this talk, let us do an actual calculation, as promised. The correlation function defined by[8]

$$C(\theta) \equiv \left\langle \frac{\delta T}{T}(\hat{n}_1) \frac{\delta T}{T}(\hat{n}_2) \right\rangle_{\cos\theta = \hat{n}_1 \cdot \hat{n}_2} \tag{1}$$

evidently measures how the fractional deviations in temperature in two regions of the sky separated by angle θ are correlated. We can expand this angular function in terms of Legendre polynomials: $C(\theta) = \frac{1}{4\pi} \sum_{l=0}^{\infty} C_l P_l(\cos\theta)$.

The observational data, with C_l plotted against l (which you can think of roughly as the conjugate of θ), is shown in figure 2. Earlier you were asked to picture the primeval fluid sloshing back and forth; hence the peaks in the plot are known as acoustic peaks. Note the position of the first peak at $l_1 \simeq 180$, corresponding to the angle $\theta_1 = \pi/l_1$.

Since the smaller values of l correspond to larger values of θ, the value of θ_1 tells us about the maximal angular size of the primeval density fluctuations. Let us use this observation to estimate the position of the first acoustic peak. We will do it for a flat universe for simplicity, and come back later to the question of how the curvature of the universe affects the position.

Back in school, we learned that the angular size of an object is given by $\theta = \lambda/d$, where λ denotes the linear size of the object, that is, the distance between its two ends, and d the distance from us to the object along the line of sight. Thus, to determine θ_1 and hence l_1,

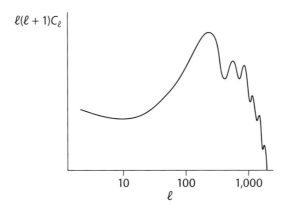

Figure 2 Schematic representation of the observational data on angular correlation in the fluctuations of the cosmic microwave background; see the text for details.

our task reduces to calculating two distances. Since we are calculating a ratio λ/d, we can afford to be sloppy and drop common overall factors.

As explained in detail in chapter V.3, distance should be defined operationally by bouncing light back and forth off the object of interest, as in radar ranging. For flat ($k = 0$) spacetime, the relevant integral for the coordinate distance between two events 1 and 2 is given by

$$d_{12} \equiv R_{12} = \int_{r_1}^{r_2} dr = c \int_{t_1}^{t_2} \frac{dt}{a(t)} = c \int_{a_1}^{a_2} \frac{da}{a\dot{a}} \equiv c\mathcal{J}_{12} \tag{2}$$

(see (V.3.10), with the notation used there). Note that we have restored the speed of light. In chapter VIII.1, we learned that $a(t) \propto t^{\frac{2}{3}}$ during the matter dominated era, and $a(t) \propto t^{\frac{1}{2}}$ during the radiation dominated era. Write $a \propto t^{\gamma}$, and hence $\dot{a} \propto t^{\gamma-1} \propto a^{\frac{\gamma-1}{\gamma}}$. The quantity $\mathcal{J}_{12}$ defined in (2) for convenience thus turns out to be

$$\mathcal{J}_{12} \propto \int_{a_1}^{a_2} \frac{da}{a} \, a^{\frac{1-\gamma}{\gamma}} \propto \left(a_2^{\frac{1-\gamma}{\gamma}} - a_1^{\frac{1-\gamma}{\gamma}} \right) \tag{3}$$

During the matter dominated era, $\mathcal{J}_{12} \propto \left(a_2^{\frac{1}{2}} - a_1^{\frac{1}{2}} \right)$, and during the radiation dominated era, $\mathcal{J}_{12} \propto (a_2 - a_1)$.

Recall from chapter VIII.2 that the scale factor a is related to the redshift z by $a = 1/(1 + z)$. The relevant numbers we need are $z_{\text{mdom}} \simeq 8{,}800$ when matter started dominating over radiation, $z_{\text{dec}} \simeq 1{,}100$ when radiation decoupled from matter, and of course $z_0 = 0$ at present. Or equivalently $a_0 = 1$, $a_{\text{dec}} \sim 10^{-3}$, and $a_{\text{mdom}} \sim 10^{-4}$. Since $a_0 \gg a_{\text{dec}} \gg a_{\text{mdom}}$, the evaluations of d and λ simplify, as we will see presently.

Now that I have set things up and told you the relevant numbers, we are ready to calculate. First, let us calculate the coordinate distance d between decoupling and the present. Since dark energy is only starting to take over now, it is an excellent approximation in calculating d to assume a matter dominated universe in the eons between decoupling and now. Thus, we have $d \propto c\left(a_0^{\frac{1}{2}} - a_{\text{dec}}^{\frac{1}{2}} \right) \simeq ca_0^{\frac{1}{2}} = c$.

The calculation of λ_1 is slightly more involved. The maximum size of the density fluctuation is limited by the distance that sound could have traveled since the Big Bang, and hence by the speed of sound c_s.

Normally, when we think of a sound wave, it can have any wavelength we like, but for matter to oscillate together as a density wave, information has to be conveyed from one end of the fluctuation to the other. In ordinary circumstances, compared to the characteristic time scale of oscillation, a sound wave can be taken to have existed forever; that is, it can be treated as a standing wave. In early cosmology, we have the extraordinary situation that time itself had started only a little while earlier. Sound, or more accurately a density wave, could not have gotten farther than a certain maximum distance* determined by the time elapsed since the Big Bang.

---

*What we are discussing here is known as the sound horizon. The concept of a light horizon will be discussed in the next chapter.

Hence, the maximum size of the fluctuations in the primeval fluid when the cosmic microwave photons decoupled is given by $\lambda_1 = c_s \int_0^{t_{dec}} \frac{dt}{a(t)} = c_s \int_0^{a_{dec}} \frac{da}{a\dot{a}}$. The crucial point is that the integral here is multiplied by c_s, rather than c, as in the calculation of the distance d.

This integral for λ_1 is naturally divided into two pieces according to the behavior of a: (a) from the Big Bang until matter dominance (call it BB to MD), that is, from $t = 0$ until t_{mdom}, and (b) from matter dominance to decoupling (call it MD to DEC). Using our results for $\mathcal{J}_{12}$ during the matter dominated era and during the radiation dominated era, and the numerical values of a given earlier, we see that the piece (MD to DEC) contributes much more to the integral than does the piece (BB to MD), since $a_{dec}^{\frac{1}{2}} - a_{mdom}^{\frac{1}{2}} \simeq a_{dec}^{\frac{1}{2}} \gg a_{mdom}^{\frac{1}{2}} \gg a_{mdom}$.

Putting it altogether, and using $c_s = c/\sqrt{3}$ from chapter III.6, we obtain

$$\theta_1 = \frac{c_s a_{dec}^{\frac{1}{2}}}{c a_0^{\frac{1}{2}}} \simeq \frac{1}{\sqrt{3(1 + z_{dec})}} \tag{4}$$

In other words, $l_1 = \pi/\theta_1 \simeq \pi \sqrt{3(1 + z_{dec})} \simeq 180$, in excellent agreement with the observational data shown in figure 2.

Effect of curvature on fluctuations in the cosmic microwave background

We did the calculation for a flat universe, but we could easily take into account the effect of spatial curvature. (Recall exercise 6 in chapter V.3.) Since curvature hardly plays a role in the very early universe, its effect is mainly in the calculation of d. Instead of $d = R = \int_0^R dr = c \int_{t_{dec}}^{t_0} \frac{dt}{a(t)} \equiv c\mathcal{J}$ as in (2), we have for the closed universe

$$\int_0^R \frac{dr}{\sqrt{1 - \frac{r^2}{L^2}}} = L \sin^{-1}(R/L) = c\mathcal{J}$$

so that $d = R = L \sin(c\mathcal{J}/L) < c\mathcal{J}$. Thus, d is smaller, so that θ_1 is larger. Hence in a closed universe, l_1 is smaller than what it would be in a flat universe.

This effect is also easy to understand pictorially. Think of yourself at the north pole looking at a stick aligned east-west at some latitude. Picture the two geodesics reaching you starting at the two ends of the stick. The angle between the two geodesics is larger than it would be were the earth flat. See figure 3.

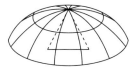

Figure 3 The effect of curvature on the position of the first acoustic peak.

For the open universe, replace the integral

$$\int_0^R \frac{dr}{\sqrt{1 - \frac{r^2}{L^2}}} \quad \text{by} \quad \int_0^R \frac{dr}{\sqrt{1 + \frac{r^2}{L^2}}}$$

or effectively sine in the preceding paragraph by hyperbolic sine. Hence in an open universe, l_1 is larger than what it would be in a flat universe.

We conclude that

$$l_1(\text{open}) > l_1(\text{flat}) > l_1(\text{closed}) \tag{5}$$

The first acoustic peak is shifted to a smaller value of l for a closed universe, and to a larger value of l for an open universe. The observational data on the position of the first acoustic peak favor a flat universe.

Appendix: Baryogenesis and leptogenesis

In this appendix, we wade into particle physics, and the discussion will get somewhat involved. It is fine if you choose to skip over this.

Imagine playing the cosmic movie backward from the time when the universe consists of a soup of protons, neutrons, and electrons. Soon, the protons and neutrons in their turn are dissociated into quarks and gluons (the "cousins" of the photon, which are responsible for the strong interaction), and a knowledge of quantum chromodynamics, the theory of the strong interaction, is needed to work out the physical cosmology of this epoch, in accordance with the Gamow principle.

At this point, you might well wonder where the quarks come from. While you are at it, how about the electrons?

To answer these questions, we will have to venture into more speculative areas of particle physics. Since the requisite knowledge[9] lies far outside the scope of this book, I will have to give an exceedingly brachylogous[10] account meant only to give you an overview rather than understanding.

Protons, neutrons, and their various cousins are called baryons and carry what is known as baryon number B. Similarly, electrons, electron neutrinos, and their various cousins are called leptons and carry what is known as lepton number L. After much travail, particle physicists now understand that each baryon is made out of three quarks. Since baryon number is additive, each quark carries baryon number $\frac{1}{3}$.

For a long time, it was believed that baryon number B and lepton number L are separately conserved. We now even understand why. When an electron emits or absorbs a photon, it remains an electron. Similarly, when a quark emits or absorbs a photon, it remains a quark, in fact, exactly the same kind of quark. In other words, the photon does not change the charged particle it interacts with. In contrast, the W boson, the cousin of the photon that is responsible for the weak interaction, changes the electron into a neutrino, and a down quark into an up quark, and so on. But in these weak interaction processes, B and L are still separately conserved. In the late 1960s, it was realized that the electromagnetic interaction and the weak interaction could be unified into a single electroweak interaction. Meanwhile, the gluon, which I already alluded to as being responsible for the strong interaction, changes an up quark of one color into an up quark of another color, and hence does not change the baryon number B. (I am assuming that you have heard somewhere that quarks carry a quantum number particle physicists call color, hence the name quantum chromodynamics for the theory of the strong interaction.) The gluon does not touch leptons at all.

What I have done here is outrageous, brushing over entire books on particle physics in a couple of short paragraphs.[11] Lest the reader lose sight of what we are doing, let me recap. Very roughly speaking, we would like to understand why the universe contains this many electrons, this many protons, this many neutrons, and so on. But electron number and proton number are not conserved. As a specific example, the neutron decays into a proton, an electron, and an antineutrino of the electron type (written as $n \rightarrow p + e^- + \bar{\nu}_e$). The antineutrino is assigned lepton number -1 (so that the neutrino has lepton number $+1$). In this process, neutron number changes from 1 to 0, while proton number changes from 0 to 1. In contrast, we start out with $B = 1$, $L = 0$, and end with $B = 1$, $L = 0$. Thus, it is not particularly useful to talk about electron number, proton number, neutron number, and the like, since they are liable to change, but it does make good sense to count with baryon number B and lepton number L.

At this point, the astute reader surely realizes that, unless the universe is finite in extent, it is more appropriate to talk about baryon number density n_B (namely the number of baryons per unit volume) and lepton number density n_L rather than the total number B and L. Furthermore, it is more convenient to talk about the dimensionless ratios n_B/n_γ and n_L/n_γ, since n_γ is easily determined by the well-measured temperature of the cosmic microwave background.

Incidentally, n_γ provides a good measure of the entropy of the universe. Recall that I have already quoted the observed value $n_B/n_\gamma \sim 10^{-10}$; in other words, the matter we know and love is truly a pitifully small contamination of an otherwise pristine universe. Equivalently, we could also say that the entropy per baryon is huge.

So, to understand the matter content of the universe, we would like to be able to calculate these two cosmological quantities n_B/n_γ and n_L/n_γ from scratch. But starting with what?

Good question. The most appealing supposition is to imagine that right after the Big Bang, the universe is pristine, with $B = 0$ and $L = 0$. The observed baryons and leptons are somehow generated later.

But, thus far in the story, B and L are separately conserved. Hence, if the universe started out with $B = 0$ and $L = 0$, it would always have $B = 0$ and $L = 0$. In other words, as long as B and L are conserved, the baryons and leptons of the universe have to be put in at the beginning: they are part of the initial conditions. While this possibility may be theologically appealing to some people, the typical physicist would much prefer to be able to calculate as many observed quantities as possible.

Thus, for an ab initio calculation of the baryon and lepton number of the universe at present, a necessary ingredient is baryon and lepton number nonconservation.

Now again a lightning summary of the relevant particle physics. Until the early 1970s, particle physicists thought of the four fundamental forces—the strong, the weak, the electromagnetic, and gravity—as unrelated. First, the electromagnetic and the weak interactions were merged into a single electroweak interaction, as mentioned above. Later, the strong and the electroweak interactions were further unified into a single interaction, known as the grand unified interaction, with a characteristic energy scale of about 10^{16} GeV. In other words, the grand unified theory predicts that in processes in which particles with energies of about 10^{16} GeV collide with one another, the strong, the weak, and the electromagnetic forces become the same force. While experimental confirmation remains lacking at present, there are various compelling theoretical reasons for believing in the grand unified theory.[12]

So once again, we invoke the Gamow principle: if we think that we understand grand unified theory, then we can discuss the universe in the grand unified era, when the temperature was of order 10^{16} GeV.

Reading the preceding description of the strong, the weak, and the electromagnetic interactions, you may have realized that, with these interactions, quarks are always transformed into quarks and leptons always into leptons. Before grand unification, you could take a piece of paper, draw a line down the middle, and write down the names of all the quarks on the left side of the line, and the names of all the leptons on the right side of the line. The fundamental forces act on these particles, the quarks and the leptons, transforming one particle into another. But in all these transformations, a particle on one side of the line is never changed into a particle on the other side of the line. A quark is never transformed into a lepton, and a lepton is never transformed into a quark.

Grand unification erases the line drawn on that piece of paper. The worlds of quarks and leptons can no longer be separated; the two worlds are unified. With grand unification come new transformations that change quarks into leptons and vice versa.*

With baryon and lepton number nonconservation, it is now possible to start the universe with $B = 0$ and $L = 0$, and then through various processes during the grand unified era to generate nonvanishing B and L. After the grand unified era, the universe is too cool for these baryon and lepton number nonconserving processes to proceed, and we end up with the number of baryons and leptons that are observed at present.

---

* In *Fearful*, I spoke of a magician whose art is limited to transforming one animal into another animal, one fruit into another. A rabbit and an apple are on the stage. The magician waves his cape, and, whoosh, the rabbit and the apple are transformed into a fox and some sour grapes. The audience bursts into applause. Whoosh! The fox and the grapes are gone, replaced by a mouse and a watermelon. But no matter how fantastic the transformations, there always will be one animal and one fruit on stage. So, too, the fundamental forces can only transform one quark into another quark, one lepton into another lepton. You may recognize that this implies baryon conservation: the three quarks that made up a baryon can be transformed only into three other quarks. There always will be three quarks, just as there always will be one animal on stage in the analogy. Onto the stage struts a new magician, the mysterious and amazing Mr. Grand Unification. Applause, and whoosh! The rabbit is transformed into an orange. No more animal on stage. So too in grand unified theory. Whoosh! No more baryon on stage. The three quarks inside a proton can be transformed into leptons. Baryon number is no longer conserved, and the proton can decay.

At this point, we realize that there is another problem.

Back in chapter III.4, I mentioned one triumph of special relativity: when combined with quantum physics, it necessarily leads to the existence of antimatter. For every particle, there is a corresponding antiparticle. Again, in chapter VII.3, in discussing Hawking radiation, we talked about an electron and a positron (the antielectron) popping out of the vacuum. Similarly, a proton and an antiproton can pop out of the vacuum. Particle physicists define the antiproton to have $B = -1$, and the positron to have $L = -1$, so that these processes in which pairs of particles and antiparticles pop out of the vacuum conserve B and L.

Furthermore, it was discovered that the fundamental laws of physics are invariant under an operation (known as CP: charge conjugation followed by parity) that interchanges particles and antiparticles. In other words, physics does not favor matter over antimatter, and vice versa.

So, here is the problem. Seen in this light, the problem of understanding the baryon and lepton content of this universe becomes a problem of understanding the matter-antimatter asymmetry of the universe. Why does matter dominate over antimatter in our universe? In other words, starting with $B = 0$ and $L = 0$, how would the universe know to generate a positive baryon number, rather than a negative baryon number? Similarly for the lepton number. If we truly understand what is going on, we should be able to calculate the sign as well as the magnitude of n_B/n_γ (and similarly for n_L/n_γ). Therefore, to cook up some baryons and leptons for the universe, we have to introduce yet another ingredient into the story: we need to invoke physical processes that would favor matter over antimatter.

Hence I need to mention one more thing about particle physics. In 1964, the belief that the laws of physics must be invariant under CP was shattered by J. Cronin, V. Fitch, and collaborators.[13] They found experimentally that, in the decay (mediated by the weak interaction) of certain mesons, particles and antiparticles behave differently by a tiny amount.

With CP violated experimentally and B and L violated theoretically, we finally have all the ingredients to generate the baryon and lepton content of the universe during the grand unified era.[14] The details do not concern us, but the relevant processes involve the decay and interaction of various cousins of the photons present only in the grand unified theory. Conceptually, our ability to calculate the matter content of the universe is not much different from our ability to calculate the hydrogen and helium content of the universe, but it is much shakier in accordance with the Gamow principle.

I end this rather long appendix by mentioning another twist to this story. Later, it was discovered theoretically that the electroweak interaction also violates B and L, through nonperturbative effects (that explains why it was not known earlier), but conserves $B - L$. Hence, people now claim that the baryons and leptons generated in the grand unified era would get washed out in the electroweak era (that is, the universe would relapse back to a state with $B = 0$ and $L = 0$).

Instead, according to one scenario known as leptogenesis, during the electroweak era, processes involving neutrinos are supposed to generate L, which, since $B - L$ is conserved, would also generate B as a kind of collateral damage. Some people swear by this leptogenesis scenario, but to some others, the original scenario of a grand unified birth seems much simpler and cleaner.

Notes

1. For example, S. Weinberg, *Gravitation and Cosmology: Principles and Applications of the General Theory of Relativity*, Wiley, 1972.

2. S. Weinberg, *Cosmology*, Oxford University Press, 2008, and ibid.; V. Mukhanov, *Physical Foundations of Cosmology*, Cambridge University Press, 2005.

3. Indeed, much of the exposition is adapted from *Toy/Universe*. The reader who is totally ignorant of cosmology might find this popular book an easy introduction to the subject.

4. Steve Weinberg has given a fascinating analysis of this question. More often than not, history does not develop in a straight line. For one thing, Gamow botched the details of primordial nucleosynthesis. He arbitrarily supposed that the early universe contained neutrons but not protons. For this and other reasons, the Big Bang cosmology of Gamow was not taken seriously and gradually faded from the general consciousness. Penzias and Wilson were totally unaware of Gamow's prediction that a faint glow from the Big Bang ought to be observable; they were trying to eliminate an annoying hum in an antenna they were working on. Quite remarkably, not 50 miles from them but unbeknownst to them, a group of physicists at Princeton University consisting of Robert Dicke, P. G. Roll, and David Wilkinson, were setting up an experiment to detect whether the universe had once been hotter. They had also forgotten Gamow's calculation. At Dicke's suggestion, a young theorist named James Peebles worked out primeval nucleosynthesis all over again. He was thus able to

predict the expected average energy of the microwave photons. Unfortunately for the Princeton group, they were beaten to the punch. When they heard about the persistent hum picked up at the telephone company lab, they were thunderstruck and immediately realized the magnitude of what Penzias and Wilson had discovered serendipitously.

5. For an easy account of what happened to the universe starting from about 400,000 years, see T. Abel, *Physics Today*, April 2011, p. 51.

6. Incidentally, Historians know about Newton's letters because in 1756, Bentley's heirs published them under the title "Four Letters from Sir Isaac Newton to Doctor Bentley Containing Some Arguments in Proof of a Deity."

7. For an easy pedagogical introduction to structure formation, see A. Zee, *Unity of Forces in the Universe*, volume II, chapter XII. A simple version of the calculation referred to here is given in the appendix.

8. The angle brackets denote averaging over an ensemble of realization of the fluctuation. There is a slight subtlety here, since we have only one universe. In practice, $\frac{\delta T}{T}(\hat{n})$ is expanded into multiple moments and the average is taken over different azimuthal moments. We won't go into such details of observational cosmology in this text.

9. See, for example, *QFT Nut*, part VII.

10. Recall the brachistochrone problem from chapter II.1.

11. For a more leisurely account at the level of popular physics books, see *Fearful*.

12. I discuss grand unified theory in considerable detail at the level of popular physics in *Fearful*.

13. Incidentally, both Jim Cronin and Val Fitch influenced my formation as a physicist.

14. The early work in the context of grand unified theory was done by M. Yoshimura, S. Dimopoulos, L. Susskind, D. Toussaint, S. Treiman, F. Wilczek, A. Zee, and others. Later, it became known that in the Soviet Union, A. D. Sakharov had discussed the general framework for generating the baryon content of the universe in 1967, long before the invention of grand unified theory. See A. Zee, *Unity of Forces*, chapter XI.

VIII.4 | Inflationary Cosmology

Traditional cosmology beset by problems

Before the discovery of dark energy in 1999, it was presumed that the cosmological constant was zero and cosmic expansion was driven entirely by matter and radiation. By the late 1970s, it gradually became clear that Big Bang cosmology as understood at that time was beset by several serious problems, known as the horizon problem, the homogeneity and isotropy problem, the flatness problem, and the relic problem. We will first discuss these problems in turn[1] before turning to inflationary cosmology.

Horizon problem

At any given time t after the Big Bang, light, even traveling at the universe's ultimate speed limit, could not have gotten arbitrarily far. There had not been enough time to have gotten farther than a certain horizon[2] distance $d_{horizon}(t)$ to be defined below. Thus, two points farther apart than $d_{horizon}(t)$ could not have been in causal contact with each other. Using the cosmic time η introduced in chapter V.3, we can make this starkly clear pictorially (figure 1). As in Minkowski spacetime, light moves along 45° lines. As indicated, events A and B are in the future light cone of O, and B and C are in the future light cone of P. Events A and B are causally correlated: events at η_0 could have influenced both of them. Similarly, events B and C are causally correlated, but not events A and C. I daresay that even the proverbial guy in the street could understand this point.[3]

Consequently, the early universe can be regarded as cut up into small patches called causal domains, with different causal domains uncorrelated with one another. True, the universe has expanded a great deal since then, and you might think that these primeval casual domains are now huge, but a power law expansion driven by matter and radiation proved to be insufficient. In this traditional Big Bang scenario, we would expect the present universe to look like patches of causal domains, rather than the homogeneous smooth universe that it actually is.

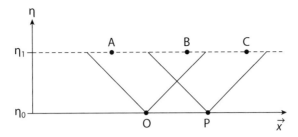

Figure 1 Causal domains in the early universe: Events A and B are causally correlated; similarly, events B and C are causally correlated, but not events A and C.

Homogeneity and isotropy problem

The universe is homogeneous and isotropic, and that's of course why cosmic expansion can be studied so simply with an ordinary differential equation, as in chapter VIII.1. But the universe is way too homogeneous and isotropic!

As mentioned in the preceding chapter, the temperature variation of the cosmic microwave background amounts to only $\langle (\frac{\delta T}{T})^2 \rangle \simeq 10^{-5}$ across the sky. How can causally uncorrelated domains end up having almost the same temperature? The size of these patches should be determined by $d_{\text{horizon}}(t)$ at photon decoupling, because ever since that time, photons have been streaming freely. How did the universe get to be so smooth?*

Before the late 1970s, some physicists would argue that this homogeneity problem is a matter of the initial conditions the universe started with, and hence outside the purview of theoretical physics.

Flatness problem

From (VIII.2.29), we learned that Ω_k, which measures the curvature of the universe, evolves according to $\dot{\Omega}_k = H\Omega_k(2\Omega_r + \Omega_m - 2\Omega_\Lambda)$. At one time, it was thought that Ω_Λ is strictly zero, and so

$$\dot{\Omega}_k = H\Omega_k(2\Omega_r + \Omega_m) \tag{1}$$

Since $(2\Omega_r + \Omega_m) > 0$, any deviation away from $\Omega_k = 0$ is unstable. As long as Ω_k is not 0, its magnitude would grow regardless of its sign. How did the universe get to be so flat?

---

* As explained also in the preceding chapter, the homogeneity problem implies there was not enough time for the irregularities at the time of decoupling to grow into the structures we see today. That problem was solved by dark matter getting a head start on growing structures.

Relic problem

One consequence of the grand unified theory mentioned in the preceding chapter is that it should contain a certain number of magnetic monopoles and antimonopoles per unit volume left over as a relic of the grand unified era, at a level totally in contradiction with observation. Indeed, as you no doubt know, the magnetic monopole has never been observed. If we claim that we understand the grand unified theory, then the Gamow principle tells us that the absence of relics from the grand unified era poses a serious problem. After all, the relic from the atomic era, namely the cosmic microwave background, is well measured. But a conservative would say that we understand atomic physics so much better than grand unified physics.

Distance to the horizon

Let us now calculate the distance to the horizon. In chapter V.3, we introduced the proper distance[4]

$$d(t, R) \equiv a(t) \int_0^R \frac{dr}{\sqrt{1 - k \frac{r^2}{L^2}}}$$

between two distant points with coordinates $(t, 0, \theta, \varphi)$ and (t, R, θ, φ). Evidently, it is a function of two variables t and R. The horizon distance $d_{\text{horizon}}(t)$, a function of one variable, is defined by choosing R to be the coordinate distance light could have traversed by time t, starting at the Big Bang. Since light rays follow paths determined by $ds = 0$, that is, by $dt/a(t) = dr/\sqrt{1 - k \frac{r^2}{L^2}}$, the horizon distance is given by

$$d_{\text{horizon}}(t) = a(t) \int_0^R \frac{dr}{\sqrt{1 - k \frac{r^2}{L^2}}} = a(t) \int_0^t \frac{dt'}{a(t')} \tag{2}$$

In other words, it is the coordinate distance $\int_0^t dt'/a(t')$ expanded by the scale factor $a(t)$ at time t.

Recall also from chapter VIII.1 that $a(t) \propto t^\gamma$, with $\gamma = \frac{1}{2}$ for a radiation dominated early universe and $\gamma = \frac{2}{3}$ for a matter dominated early universe. (The overall constant A in $a(t) = At^\gamma$ obviously cancels out in $d_{\text{horizon}}(t)$.) So, evaluate the integral in (2): $\mathcal{J}(t, t_0) \equiv \int_{t_0}^t dt'/a(t') = A^{-1}(1 - \gamma)^{-1}(t^{1-\gamma} - t_0^{1-\gamma})$.

For $(1 - \gamma) > 0$, as is the case for the traditional radiation or matter dominated early universe, $\mathcal{J}(t, t_0)$ receives most of its contribution from late times. In fact, for our purposes here, we can let $t_0 \to 0$, so that

$$d_{\text{horizon}}(t) = (1 - \gamma)^{-1} t \tag{3}$$

grows linearly with time.

For $(1 - \gamma) < 0$, that is, if $\gamma > 1$, the situation is obviously reversed: the regime around t_0 contributes the most to $\mathcal{J}(t, t_0)$. Evidently, the steep decrease of the integrand as time increases is responsible. In the extreme case, $a(t) \propto e^{Ht} \sim t^\infty$, $\mathcal{J}(t, t_0) = (e^{-Ht_0} - e^{-Ht})/H$ so that

$$d_{\text{horizon}}(t) = \frac{1}{H} \left(e^{H(t-t_0)} - 1 \right) \tag{4}$$

grows exponentially.

Inflationary epoch

With the blinding clarity of hindsight, we now see that all these problems can be solved if the early universe went through an inflationary epoch,[5] during which it expanded exponentially. As was just shown, $d_{\text{horizon}}(t)$ would then grow exponentially, thus solving[6] the horizon and homogeneity problems, and as we will see below, also the flatness problem. In addition, this exponential expansion would greatly dilute the density of any hitherto unobserved and hence undesirable relics.

Before I go further, let me warn the reader that as this book goes to press, there is considerable debate regarding whether inflationary cosmology is still viable.[7] You will have to form your own opinion over the coming years.

As you have known since chapter V.3, the desired exponential expansion can be produced readily by a constant energy density corresponding to some effective cosmological constant. With a cosmological constant, (1) is replaced by

$$\dot{\Omega}_k = H\Omega_k(2\Omega_r + \Omega_m - 2\Omega_\Lambda) \tag{5}$$

With $(2\Omega_r + \Omega_m - 2\Omega_\Lambda) < 0$, the flow around small Ω_k goes from being unstable to stable, and the flatness problem is solved. Any initial value Ω_k gets driven to 0.

One triumph of inflation is that it can account for the origin of the density fluctuations needed (as explained in the preceding chapter) for the growth of structures. Where did these density fluctuations come from? How did they get put in at the Big Bang? These are all questions you might have asked. You might even have thought of quantum fluctuations, which, since we live in a quantum universe, are inevitably present. Before the inflationary universe was proposed, people would have immediately dismissed the notion of quantum fluctuations being responsible for the density fluctuations in the primeval universe; the quantum length scale on which these fluctuations occur would seem to be irrelevantly minuscule compared to cosmological distance scales. However, in an inflationary scenario, the fluctuations could have stretched out enormously. Furthermore, after this inflationary stretching, the resulting spectrum of density fluctuations would end up having no characteristic length scale, giving rise to the scale-free spectrum proposed by E. R. Harrison and Y. B. Zeldovich long before inflation was invented. I will leave detailed calculations of the primeval density fluctuation to more specialized texts. Remarkably, fluctuations begotten by the quantum and stretched by inflation could have led to the structures we see around us.

An inflationary universe

That the universe went through some sort of inflationary epoch during which it expanded exponentially seems quite compelling, but the specifics of any of the proposed mechanisms should probably be taken with a grain of salt and a smile.

Numerically, to accord with observation, the scale factor $a(t)$ needs to have expanded by a factor of 10^{30}, starting from $\sim 10^{-36}$ second after the Big Bang to $\sim 10^{-35}$ second after the Big Bang. The required number of e-foldings is thus $30 \log 10 \sim 70$. Note that these numbers do not come out of the theory but are required to accord with experiment.

In Einstein gravity, it is easy to cause the universe to expand exponentially: all you have to do, as we saw in chapter VI.2, is to introduce a constant energy density in the vacuum, effectively a cosmological constant. In particle theory, it is easy to produce a constant energy density: a scalar field that does not vary in spacetime would do that, as you learned in chapter VI.4. Thus, with the benefit of hindsight, it is doubly easy to make the universe go through an inflationary epoch.

Amusingly, Einstein's introduction of the cosmological constant, far from a blunder as the uninformed called it, turns out to be essential for cosmology, both observationally and theoretically.

Indeed, you worked out in exercise VI.4.4 that, for a scalar field governed by the action $S_{\text{scalar}} = -\int d^4x \sqrt{-g}(\frac{1}{2}(\partial\phi)^2 + V(\phi))$ (with $(\partial\phi)^2 = g_{\mu\nu}\partial_\mu\phi\partial_\nu\phi$), the energy momentum tensor is given by $T^{\mu\nu} = \partial^\mu\phi\partial^\nu\phi - g^{\mu\nu}(\frac{1}{2}(\partial\phi)^2 + V(\phi))$. For $\phi(x)$ constant, $T^{\mu\nu} = -g^{\mu\nu}V(\phi)$. Actually, we don't even need to calculate the energy momentum tensor; we can see directly from the action that, as explained in chapter VI.2, a constant in $V(\phi)$ corresponds to a constant energy density.

One difficulty confronting inflationary theories is known as the graceful exit problem: how do you end inflation once you have had your fill of the 70 e-foldings? Another puzzle is why the effective cosmological constant was once large (so as to dominate the contribution of the relativistic matter, namely radiation in the generalized sense, to the energy density) but is now so incredibly small.*

Since we know almost nothing about the origin of this scalar field, called rather unimaginatively the inflaton, and the physics that goes with it, people feel free to draw whatever $V(\phi)$ would produce a desired outcome, for example, the "slow roll" potential shown in figure 2. In this pictorial analogy, the inflaton ϕ is supposed to start out on the nearly flat plateau and slowly roll downhill (see the appendix), ending up in a minimum that is many orders of magnitude smaller.

Not surprisingly, this has generated in the theory literature a multitude of scenarios that go under such names as old inflation, new inflation, chaotic inflation, eternal inflation, and stochastic inflation. Invent your own! In chaotic inflation, one does not need to fine-tune the potential $V(\phi)$; instead the universe is vast and inhomogeneous, with $V(\phi)$ taking on

---

* We will return to the cosmological constant in chapter X.7.

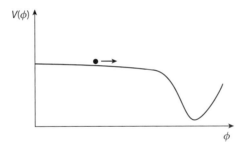

Figure 2 The inflaton potential.

random values in different regions, so that some regions inflate rapidly, while others may inflate less rapidly or even shrink. One then argues that the probability is overwhelming that we would find ourselves in a region that has undergone an inflationary phase, just because such regions occupy exponentially more volume than others.

As I said earlier, in recent years, inflationary cosmology has been faced with mounting difficulties and increasing criticism. I close by mentioning that there are a number of interesting alternatives.[8]

Appendix: Slow roll scenario

Here I sketch the popular slow roll scenario. Start with the energy momentum tensor for a scalar field ϕ, the inflaton, namely $T^{\mu\nu} = \partial^\mu\phi\partial^\nu\phi - g^{\mu\nu}(\frac{1}{2}(\partial\phi)^2 + V(\phi))$. It takes on the form of a perfect fluid $T^{\mu\nu} = (\rho + P)U^\mu U^\nu + P g^{\mu\nu}$, with $U_\mu = (\partial_\mu\phi/\sqrt{-(\partial\phi)^2}, \vec{0})$,

$$\rho = -\tfrac{1}{2}(\partial\phi)^2 + V(\phi) \tag{6}$$

and

$$P = -\tfrac{1}{2}(\partial\phi)^2 - V(\phi) \tag{7}$$

If we imagine that $\phi(t, \vec{x}) = \phi(t)$ could be independent of space, then $\rho = \frac{1}{2}\dot\phi^2 + V(\phi)$ and $P = \frac{1}{2}\dot\phi^2 - V(\phi)$, and the inflaton obeys the equation of motion

$$\ddot\phi + 3H\dot\phi + V'(\phi) = 0 \tag{8}$$

and the expansion of the universe is

$$H^2 = \tfrac{1}{3}\left(\tfrac{1}{2}\dot\phi^2 + V(\phi)\right) \tag{9}$$

in units with $8\pi G = 1$. Note that since we suppressed the spatial dependence by decree, the equation of motion (8) is that of a point particle rolling in the potential $V(\phi)$. Crucially, the expansion of the universe provides a friction term $\sim H\dot\phi$.

We can obtain an inflationary epoch if the inflaton varies sufficiently slowly in time so that $\dot\phi^2 \ll V(\phi)$. Then (9) becomes

$$H^2 \simeq \tfrac{1}{3}V(\phi) \tag{10}$$

with the Hubble parameter approximately constant in time, producing an effective cosmological constant. If we also[9] impose $\ddot\phi \ll V'(\phi)$, then we obtain from (8)

$$3H\dot\phi \simeq -V'(\phi) \tag{11}$$

Using (11) and (10), we have $\dot{\phi}^2 \sim (V'/H)^2 \sim V'^2/V$, and so we can write the condition $\dot{\phi}^2 \ll V(\phi)$ as $(V'/V)^2 \ll 1$. Furthermore, since $\dot{\phi} \sim -V'/V^{\frac{1}{2}}$, we have $\ddot{\phi} \sim (V'V''/V) - (V'^3/(2V^2))$, and thus the other condition $\ddot{\phi} \ll V'(\phi)$ gives $V''/V \ll 1$. Thus, it is customary to define two parameters $\varepsilon \equiv \frac{1}{2}(V'/V)^2$ and $\eta \equiv V''/V$. For the slow roll approximation to hold, we need both $\varepsilon \ll 1$ and $\eta \ll 1$. Note that the second condition also amounts to saying that the variation of $\varepsilon(\phi)$ with respect to ϕ is small.

Notes

1. As mentioned in the preceding chapter, the interested reader should also consult specialized textbooks on cosmology.
2. Some authors introduce two distinct terms, calling this horizon a "particle horizon" and the horizon around black holes an "event horizon." I think that it is easy enough to distinguish them by context.
3. And if not, replace light by messengers in ancient times.
4. We also mentioned there that the concept is slightly sloppy and involves a "cosmic conspiracy."
5. The early history of inflation is too involved to go into here. A detailed exposition may be found in A. Guth, M. Mukhanov, and S. Weinberg. We might mention here early work by A. Starobinsky, B. Chirikov, M. Mukhanov, R. Brandenburger, A. Guth, H. Tye, A. Zee, K. Sato, and M. Einhorn, among others. See in particular p. 180 of Guth.
6. It has been pointed out that there is a hidden assumption about the measure of the initial pre-evolutionary data. Mathematically, an arbitrary present configuration of the universe could be evolved backward in time to some initial configuration. The correct statement is that inflation can take a generic initial configuration and iron it out.
7. For instance, Max Tegmark, writing in *New Scientist* in 2012, states that the inflationary scenario should be abandoned. He puts it humorously as follows: "You know how sometimes you meet somebody and they're really nice, so you invite them over to your house and you keep talking with them and they keep telling you more and more cool stuff? But then at some point you're like, maybe we should call it a day, but they just won't leave and they keep talking and as more stuff comes up it becomes more and more disturbing and you're like, just stop already? That's kind of what happened with inflation."
8. For example, the bounce theory can solve all the problems that inflation can solve (R. H. Brandenberger, private communication). See R. H. Brandenberger, "Introduction to Early Universe Cosmology," PoS ICFI2010, 001 (2010), arXiv:1103.2271 [astro-ph.CO].
9. Some texts assert erroneously that the condition $\ddot{\phi} \ll V'(\phi)$ follows upon differentiating the condition $\dot{\phi}^2 \ll V(\phi)$. It is known to any student of calculus that $f(t) \ll g(t)$ does not imply that $f'(t) \ll g'(t)$. For example, $f(t) = \sin(100t)/10$, $g(t) = 1$.

Recap to Part VIII

The stuff contained in the universe causes the universe to expand, and the expansion of the universe affects the density of the stuff contained in the universe. Knowing Einstein's field equation and the properties of the stuff allows us to work out the expansion history of the universe.

To first approximation, the universe can be described as a struggle between dark energy and dark matter. This cosmic contest can be mapped out in a 2-dimensional diagram.

In speculating about the history of the universe, we should keep in mind Gamow's principle: if we understand the physics characteristic of a certain energy scale, then we can work out, in the grand tradition of physics, how the universe behaves when its temperature is of that scale.

BOOK THREE

Gravity at Work and at Play

Part IX | Aspects of Gravity

IX.1 | Parallel Transport

Keep on pointing in the same direction

Way back in chapter I.7 (titled "Differential Geometry Made Easy, but Not Any Easier" as you may recall), I introduced the notion of parallel transport in the context of curved surfaces. While some uninitiates might find the notion a bit difficult to grasp, parallel transport is in fact firmly rooted in common everyday intuition. Think of a patent clerk in Bern walking along a closed path, carrying a spear and pointing it in the same direction the whole time. For a vector on a surface, we simply parallel transport it in the ambient Euclidean space, keeping its "feathered end" in the surface, and then chop off the component sticking out of the surface. The result, as we saw back in chapter I.7, is closely related to the concept of covariant derivative.

In this chapter, we generalize the notion of parallel transport from the surfaces of chapter I.7 to Riemannian spacetimes. The key is the covariant derivative discussed in chapter V.6. Consider a curve $\mathcal{C}$ defined by $x^\mu(\tau)$ and parametrized by τ. The curve $\mathcal{C}$ is not necessarily a geodesic, just some curve. Let a vector S^μ be given at some point on the curve located by τ_I. You may recall that in appendix 1 of chapter V.6, the notion of covariant derivative along a curve was introduced. Here we simply set this derivative along the curve $\mathcal{C}$ to zero. We then determine the vector $S^\mu(\tau)$ at other points along the curve by integrating the first order differential equation

$$\frac{dS^\mu(\tau)}{d\tau} + \Gamma^\mu_{\rho\sigma}(x(\tau))V^\rho(\tau)S^\sigma(\tau) = 0 \tag{1}$$

with the given vector $S^\mu(\tau_I)$ providing the initial condition. Here $V^\sigma \equiv \frac{dx^\sigma}{d\tau}$ denotes the tangent vector along the curve. The vector S^μ is said to be parallel transported along the curve $\mathcal{C}$.

We should be careful not to write $S^\mu(\tau)$ as $S^\mu(x(\tau))$! The erroneous notation $S^\mu(x(\tau))$ would suggest that $S^\mu(x)$ exists, that is, as a vector field having a value at any point x in spacetime, and that $S^\mu(x(\tau))$ is equal to the vector field $S^\mu(x)$ evaluated on the curve. This may sound like nitpicking, but it's not: we must keep straight what we are doing. Similarly, $V^\rho(\tau)$ is only defined on the curve.

In contrast, the metric and the Christoffel symbol are defined (with possibly more than one coordinate patch needed) all over the manifold. Thus, in (1), $\Gamma^\mu_{\rho\sigma}(x(\tau))$ denotes $\Gamma^\mu_{\rho\sigma}(x)$ evaluated at the point $x(\tau)$ on the curve.

It is useful and natural to ask what differential equation $S_\mu = g_{\mu\nu}S^\nu$ would satisfy. Simply differentiate and use (1):

$$\frac{dS_\mu(\tau)}{d\tau} = \frac{dx^\rho}{d\tau}\left(\partial_\rho g_{\mu\sigma}\right)S^\sigma + g_{\mu\nu}\frac{dS^\nu(\tau)}{d\tau} = \left(\partial_\rho g_{\mu\nu} - g_{\mu\nu}\Gamma^\nu_{\rho\sigma}\right)V^\rho S^\sigma$$

Inserting the definition of $\Gamma^\nu_{\rho\sigma}$, we find that the expression in parentheses simplifies to $\Gamma_{\sigma\cdot\mu\rho}$, that is, the Christoffel symbol with its upper index lowered. Thus, we obtain

$$\frac{dS_\mu(\tau)}{d\tau} - \Gamma^\nu_{\mu\rho}(x(\tau))V^\rho(\tau)S_\nu(\tau) = 0 \tag{2}$$

Notice that (1) and (2) differ by a sign: we parallel transport S^μ and S_μ with a crucial difference in sign. This immediately implies that if we parallel transport two different vectors S^μ and T_μ, then

$$\frac{d(S^\mu T_\mu)}{d\tau} = \left(-\Gamma^\mu_{\rho\sigma}S^\sigma T_\mu + S^\mu\Gamma^\nu_{\mu\rho}T_\nu\right)V^\rho = \left(-\Gamma^\mu_{\rho\sigma}S^\sigma T_\mu + \Gamma^\mu_{\sigma\rho}S^\sigma T_\mu\right)V^\rho = 0 \tag{3}$$

since Γ is symmetric in its two lower indices. This result makes sense: parallel transporting two vectors S^μ and T_μ, carefully keeping them "pointing in the same direction," so to speak, we would have every right to expect that their scalar product $S^\mu T_\mu$ would not change.

Covariant derivative and parallel transport

At this point, bells should be ringing. You might recall that we went through an entirely similar discussion for covariant derivatives in chapter V.6. Given two vector fields $W^\mu(x)$ and $U_\mu(x)$, the covariant derivatives $D_\lambda W^\mu$ and $D_\lambda U_\mu$ are defined in (V.6.11) and (V.6.12), respectively, with opposite signs in precisely such a way that $D_\lambda(W^\mu U_\mu)$ simplifies to $\partial_\lambda(W^\mu U_\mu)$.

Imagine moving through a vector field $W^\mu(x)$ along a curve $\mathcal{C}$ (again, not necessarily a geodesic) defined by $x(\tau)$. Now it makes sense to define $W^\mu(\tau)$ to be equal to $W^\mu(x(\tau))$: it is the value of the vector field you experience at the point parametrized by τ as you move along $\mathcal{C}$. Then $\frac{dW^\mu(\tau)}{d\tau} = \frac{dx^\rho}{d\tau}\partial_\rho W^\mu(x(\tau)) = V^\rho(\tau)\partial_\rho W^\mu$. In other words, $\frac{dW^\mu}{d\tau} + \Gamma^\mu_{\rho\sigma}V^\rho W^\sigma = V^\rho D_\rho W^\mu$. We conclude that if vector field $W^\mu(x)$ is (covariantly) constant in the sense that its covariant derivative $D_\rho W^\mu$ vanishes, then the vector $W^\mu(x(\tau))$ you experience as you glide through the vector field is precisely parallel transported.

Everything is coming together. Already, back in chapter I.7, when we were discussing surfaces, Riemannian spaces that you can hold in your hands, or at least in your mind's eye, we went from the ordinary derivative (I.7.11) to the covariant derivative (I.7.12) by dropping the component sticking out of the surface, as already alluded to in the opening of this chapter.

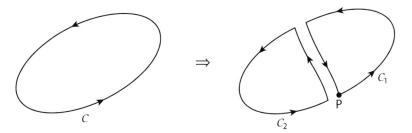

Figure 1 Parallel transport of a vector around a curve can be reduced to parallel transport of a vector around two smaller curves.

Shortest distance and straight lines

Indeed, when you saw the parallel transport equation (1), another bell might have rung. It should have reminded you of the geodesic equation

$$\frac{dV^{\mu}(\tau)}{d\tau} + \Gamma^{\mu}_{\rho\sigma}(x(\tau))V^{\rho}(\tau)V^{\sigma}(\tau) = 0 \tag{4}$$

A geodesic is a curve whose tangent vector $V^{\mu}(\tau)$ is parallel transported as we move along the curve. This is just the curved space generalization of the man-in-the-street statement that a straight line gives the shortest distance between two points. The word "straight" can only mean that the tangent vector keeps pointing in the same direction.

Riemann curvature and parallel transport

This discussion suggests yet another way for the mite geometers (namely us) to measure curvature. Consider a closed curve C (in general not a geodesic) starting and ending at the point P. Let's parallel transport a vector S_{μ} along C starting at P.

We ask whether the vector S_{μ} comes back to itself when we go around the closed curve. In flat space, it will. Thus, the extent to which it does not provides us with a measure of the curvature. In other words, we want to calculate ΔS_{μ} as we go around C.

The first remark is that we can take C to be infinitesimal. The argument is that given a curve C, we can always decompose it into smaller pieces.* As shown in figure 1, we can write $C = C_1 + C_2$ as the sum of two smaller curves C_1 and C_2. Pick a point P lying on both C_1 and C_2. Parallel transporting S_{μ} around C is equivalent to parallel transporting S_{μ} first around C_1, starting and ending at P, and then parallel transporting S_{μ} around C_2, again starting and ending at P. Clearly, the curved segment shared by C_1 and C_2 are traversed twice in opposite directions, producing canceling contributions to ΔS_{μ}. Repeating the argument, we can cut any given closed curve into smaller and smaller closed curves.

---

* You are probably familiar with this sort of argument from a variety of contexts in physics (most likely in a course on electromagnetism) and mathematics.

Alternatively, argue that any irregular shaped area enclosed by a closed curve can be approximated by many small rectangles. Shades of integral calculus. Thank you, Newton and Leibniz!

So, we take $\mathcal{C}$ to be infinitesimal, describing it by $x(\tau)$, so that the endpoint $x_F \equiv x(\tau_F)$ is equal to the starting point $x_I \equiv x(\tau_I)$, with some initial τ_I and final τ_F. Our discussion evidently goes through for both space and spacetime. A trivial notation alert: My use of τ here does not suggest a connection with time; it is just a parameter along the curve, which we are in fact taking to be closed.

Taking the limit in which the closed curve shrinks to zero, we determine the curvature of the manifold at the location of the closed curve.

The Riemann curvature tensor emerges

So, boys and girls, let us calculate away. Around a closed curve,

$$\Delta S_\mu \equiv S_\mu(\tau_F) - S_\mu(\tau_I) = \int_{\tau_I}^{\tau_F} d\tau \, \frac{dS_\mu}{d\tau} = \int_{\tau_I}^{\tau_F} d\tau \, \Gamma^\rho_{\mu\sigma}(x(\tau)) V^\sigma(\tau) S_\rho(\tau)$$

$$= \int_{x_I}^{x_F} dx^\sigma \, \Gamma^\rho_{\mu\sigma}(x(\tau)) S_\rho(\tau) \tag{5}$$

where we used $V^\sigma = \frac{dx^\sigma}{d\tau}$ in the last step.

In theoretical physics, when faced with a fairly involved calculation, it is always a good habit to anticipate the answer. Since parallel transport is linear in S, we expect ΔS to be proportional to S. Also, ΔS vanishes as the closed curve shrinks to zero. Do we expect it to be proportional to the perimeter of the closed curve or to the area enclosed by the closed curve? By following a curve for a bit and then backtracking along the same curve to get back to the starting point, we have traced a closed curve with a nonvanishing perimeter but enclosing no area. But $\Delta S = 0$ for such a closed curve, since the changes in S are reversed on the return trip. Thus, ΔS must be proportional to the area enclosed, not to the perimeter.

Since area is quadratic in length, we expect an infinitesimal area element to be given by something vaguely like $\delta x^\sigma \delta x^\lambda$. We don't quite know what that would be, but it must carry two indices like a 2-indexed tensor $a^{\sigma\lambda}$. Therefore, ΔS_μ has to be proportional to $S_\rho a^{\sigma\lambda}$. We need a 4-indexed tensor of the form $\mathcal{R}^\rho_{\mu\sigma\lambda}$ to convert the 3-indexed right hand side to the 1-indexed left hand side. Hence we anticipate the answer to have the form $\Delta S_\mu = \mathcal{R}^\rho_{\mu\sigma\lambda} S_\rho a^{\sigma\lambda}$.

Well, let's not be coy about it. You know, and I know, that if there is any justice in this world, $\mathcal{R}^\rho_{\mu\sigma\lambda}$ has to be none other than the Riemann curvature tensor $R^\rho_{\mu\sigma\lambda}$. We now verify our suspicion by doing the integral in (5).

Before arithmetic overwhelms us, let us pause and reflect that, for an infinitesimal closed curve, the integrand $\mathcal{I}$ in (5) can be Taylor expanded from the value it has at the starting point $\mathcal{I}(x(\tau)) = \mathcal{I}(x_I) + \partial_\lambda \mathcal{I}(x_I)(x(\tau)^\lambda - x_I^\lambda) + \cdots$. The constant term contributes nothing to the integral, since $\int_{x_I}^{x_F} dx^\sigma = x_F^\sigma - x_I^\sigma = 0$ for a closed curve. This also makes sense, since ΔS_μ obviously must vanish as the closed curve shrinks to nothing.

The contribution of the linear term is proportional to $\oint dx^\sigma (x - x_I)^\lambda = \oint dx^\sigma x^\lambda$, since $\oint dx^\sigma x_I^\lambda = x_I^\lambda \oint dx^\sigma = 0$. The object $a^{\sigma\lambda} \equiv \oint dx^\sigma x^\lambda$ carries information about the size $\sim(x - x_I)$ of the loop and has dimension of length squared. So you should not be surprised that it gives the infinitesimal area enclosed by the closed curve. To verify this, simply evaluate it for a small rectangle (see exercise 1). Notice that $a^{\sigma\lambda}$ may be positive or negative according to whether we go around the curve clockwise or counterclockwise, and it is antisymmetric in its indices. To see this, write $a^{\sigma\lambda}$ as an integral over τ and integrate by parts:

$$a^{\sigma\lambda} \equiv \int d\tau \frac{dx^\sigma}{d\tau} x^\lambda = -\int d\tau \frac{dx^\lambda}{d\tau} x^\sigma = -a^{\lambda\sigma} \tag{6}$$

Now that we know that only the linear term matters, we Taylor expand the two factors $\Gamma^\rho_{\mu\sigma}(x(\tau))$ and $S_\rho(\tau)$ that make up $\mathcal{I} = \Gamma^\rho_{\mu\sigma}(x(\tau))S_\rho(\tau)$:

$$\Gamma^\rho_{\mu\sigma}(x(\tau)) = \Gamma^\rho_{\mu\sigma}(x_I) + \partial_\lambda \Gamma^\rho_{\mu\sigma}(x_I)(x - x_I)^\lambda + \cdots \tag{7}$$

and

$$\begin{aligned} S_\rho(\tau) &= S_\rho(\tau_I) + \Gamma^\kappa_{\rho\omega}(x_I)V^\omega(\tau_I)S_\kappa(\tau_I)(\tau - \tau_I) + \cdots \\ &= S_\rho(\tau_I) + \Gamma^\kappa_{\rho\lambda}(x_I)S_\kappa(\tau_I)(x - x_I)^\lambda + \cdots \end{aligned} \tag{8}$$

where we have used $V^\omega = \frac{dx^\omega}{d\tau}$ (and changed the dummy index ω to λ). Thus, the linear term in $\Gamma^\rho_{\mu\sigma}(x(\tau))S_\rho(\tau)$ is $[\partial_\lambda \Gamma^\rho_{\mu\sigma}(x_I)S_\rho(\tau_I) + \Gamma^\rho_{\mu\sigma}(x_I)\Gamma^\kappa_{\rho\lambda}(x_I)S_\kappa(\tau_I)](x - x_I)^\lambda$.

Finally, putting things together, we see that the integral in (5) gives

$$\Delta S_\mu = [\partial_\lambda \Gamma^\rho_{\mu\sigma}(x_I) + \Gamma^\kappa_{\mu\sigma}(x_I)\Gamma^\rho_{\kappa\lambda}(x_I)]S_\rho(\tau_I)a^{\sigma\lambda} \tag{9}$$

(By now you might have caught on that one of the abilities you need to learn general relativity is to change dummy indices in your head.)

This important result can be written elegantly as

$$\Delta S_\mu \equiv \tfrac{1}{2}R^\rho_{\mu\sigma\lambda}S_\rho a^{\sigma\lambda} \tag{10}$$

as we warmly welcome the natural emergence of our beloved Riemann curvature tensor

$$R^\rho_{\mu\sigma\lambda} \equiv \left(\partial_\sigma \Gamma^\rho_{\mu\lambda} + \Gamma^\rho_{\kappa\sigma}\Gamma^\kappa_{\mu\lambda}\right) - \left(\partial_\lambda \Gamma^\rho_{\mu\sigma} + \Gamma^\rho_{\kappa\lambda}\Gamma^\kappa_{\mu\sigma}\right) \tag{11}$$

This amounts to an alternative derivation of the Riemann curvature tensor, a derivation showing clearly how the $\partial\Gamma$ and $\Gamma\Gamma$ terms come about: the former from the variation of Γ, the latter from the variation of the vector being transported, as we move around the infinitesimal loop. Notice that the Riemann curvature tensor is automatically defined to be antisymmetric in its last two indices, since the area element $a^{\sigma\lambda}$ is antisymmetric. You might have also realized that this derivation is intimately connected with the derivation given in chapter VI.1 based on the commutator of two covariant derivatives $[D_\mu, D_\nu]$. See also the appendix.

The amount by which a vector parallel transported around a closed loop does not come back to itself measures the local curvature.

Appendix: Transporting vectors via alternative routes

Here I give another derivation of the Riemann curvature tensor, a derivation closely related to the one given in the text (in fact, essentially the same derivation repackaged). Let us parallel transport a vector S_μ from x to $x + \delta a$, and then from there to $x + \delta a + \delta b$. Along the first leg, we have, according to (2), the change $\delta S_\mu(\tau) = \Gamma^\nu_{\mu\rho}(x(\tau))\delta a^\rho S_\nu(\tau)$. To find the change along the second leg from $x + \delta a$ to $x + \delta a + \delta b$, we use this nifty formula again but evaluated at the new starting point $x + \delta a$ instead of x (of course). We thus obtain the change

$$\Gamma^\nu_{\mu\rho}(x(\tau) + \delta a)\,\delta b^\rho S_\nu(\tau + \delta\tau) \simeq \left(\Gamma^\nu_{\mu\rho}(x(\tau)) + \delta a^\lambda \partial_\lambda \Gamma^\nu_{\mu\rho}\right)\delta b^\rho \left(S_\nu(\tau) + \Gamma^\sigma_{\nu\lambda}\delta a^\lambda S_\sigma\right)$$

$$\simeq \Gamma^\nu_{\mu\rho}(x(\tau))\delta b^\rho S_\nu(\tau) + \delta b^\rho \delta a^\lambda \left(\partial_\lambda \Gamma^\nu_{\mu\rho} S_\nu + \Gamma^\nu_{\mu\rho}\Gamma^\sigma_{\nu\lambda}S_\sigma\right) \qquad (12)$$

where in the last step we used the nifty formula yet once again, namely the result just obtained for the change along the first leg. The total change in S_μ, after being transported from x to $x + \delta a + \delta b$ via $x + \delta a$, is thus given by

$$\delta S_\mu(\text{from } x \text{ to } x + \delta a + \delta b \text{ via } x + \delta a) = \Gamma^\nu_{\mu\rho}\delta a^\rho S_\nu + \Gamma^\nu_{\mu\rho}\delta b^\rho S_\nu + \delta b^\rho \delta a^\lambda \left(\partial_\lambda \Gamma^\nu_{\mu\rho} + \Gamma^\sigma_{\mu\rho}\Gamma^\nu_{\sigma\lambda}\right)S_\nu \qquad (13)$$

Suppose we parallel transport S_μ from x to $x + \delta a + \delta b$ via $x + \delta b$ instead. The difference in the resulting S_μ going by the two different routes is given by subtracting from (13) the same expression with $\delta a \leftrightarrow \delta b$, namely

$$\Delta S_\mu = \delta b^\rho \delta a^\lambda S_\nu \left(\left(\partial_\lambda \Gamma^\nu_{\mu\rho} + \Gamma^\sigma_{\mu\rho}\Gamma^\nu_{\sigma\lambda}\right) - \left(\lambda \leftrightarrow \rho\right)\right) \qquad (14)$$

As expected, the terms linear in δa and δb in (13) drop out. We have derived the Riemann curvature tensor yet one more time.

Exercises

1   Evaluate $a^{\sigma\lambda}$ for a rectangle.

2   Evaluate $a^{\sigma\lambda}$ for a small circle.

3   In elementary discussions of curved surfaces, the reader is often invited to draw a triangle on a sphere and to observe that the three angles add up to more than 180°. (Indeed, we already mentioned this fact in the prologue to this book.) Show by parallel transporting a vector around such a triangle that the angular excess measures the curvature.

IX.2 | Precession of Gyroscopes

Parallel transport in action

We now apply the notion of parallel transport to study the precession of a gyroscope in curved spacetime. In 2004, a precision gyroscope[1] was launched in a satellite moving in an earth orbit, giving us yet another test of Einstein's theory. For a textbook treatment, we take the orbit to be circular and ignore the rotation of the earth, so that we can calculate the precession in Schwarzschild spacetime

$$ds^2 = -\left(1 - \frac{r_S}{r}\right) dt^2 + \left(1 - \frac{r_S}{r}\right)^{-1} dr^2 + r^2 \left(d\theta^2 + \sin^2\theta \, d\varphi^2\right)$$

This effect, known as de Sitter precession or geodetic precession, was first calculated in 1916 by Willem de Sitter.

Parallel transport of the spin vector gives

$$\frac{dS^\mu}{d\tau} + \Gamma^\mu_{\nu\lambda} V^\nu S^\lambda = 0 \tag{1}$$

As usual, we will work in the equatorial plane ($\theta = \pi/2$), so that the 4-velocity is given by $V^\mu = (V^t, 0, 0, V^\varphi)$. We take $S^\mu = (S^t, S^r, 0, S^\varphi)$. In the rest frame of the gyroscope, the spin vector is purely spatial, $S^\mu = (0, \vec{S})$, while V^μ is purely temporal, and hence $g_{\mu\nu} S^\mu V^\nu = 0$. Since this orthogonality condition $g_{\mu\nu} S^\mu V^\nu = 0$ equates two scalars, it holds in all frames.

Back in chapter VII.1, we determined the circular orbit around a massive object, obtaining $V^t \equiv \frac{dt}{d\tau} = \epsilon/(1 - \frac{r_S}{r})$ and $V^\varphi \equiv \frac{d\varphi}{d\tau} = l/r^2$, with r the (constant) radius of the circular orbit. Furthermore, we found that the two conserved quantities ϵ and l are given by $\epsilon^2 = \left(1 - \frac{r_S}{r}\right)^2 (1 - \frac{3r_S}{2r})^{-1}$ and $l^2 = \frac{1}{2} r_S r \left(1 - \frac{3r_S}{2r}\right)^{-1}$ (see the discussion around (VII.1.7)). We also learned, somewhat to our surprise, that the angular velocity $\Omega = \frac{V^\varphi}{V^t} = \frac{d\varphi}{dt}$ still obeys Kepler's third law

$$\Omega^2 = \frac{r_S}{2r^3} = \frac{GM}{r^3} \tag{2}$$

(I remind you that Ω, evidently a constant, is defined in terms of coordinate time t, not proper time.)

Having reviewed the properties of the orbit, we now impose the condition

$$g_{\mu\nu}S^\mu V^\nu = 0 = V^t \left(g_{tt}S^t + g_{\varphi\varphi}\Omega S^\varphi \right)$$

to obtain a relation between S^t and S^φ, namely $S^t = r^2 \left(1 - \frac{r_S}{r} \right)^{-1} \Omega S^\varphi$.

The gyroscope precesses

You should now be able to work out how the spin vector S^μ precesses as the particle goes around its orbit. Simply plug into the various components of (1) and chug. Try it before reading on.

Insert $V^\varphi = \Omega V^t$ into (1) and note that dividing through by V^t converts the τ derivative to a t derivative. Referring to the Christoffel symbols listed in the collection of formulas at the end of the book, you see that only a few terms in (1) do not vanish. We find

$$\frac{dS^\varphi}{dt} + \frac{1}{r}\Omega S^r = 0 \quad \text{and} \quad \frac{dS^r}{dt} - r\left(1 - \frac{3r_S}{2r} \right)\Omega S^\varphi = 0 \tag{3}$$

The $\mu = t$ component of (1) merely provides a consistency check that parallel transport maintains the condition $g_{\mu\nu}S^\mu V^\nu = 0$, while the $\mu = \theta$ component simply shows that it is consistent with the symmetry of the situation to set $S^\theta = 0$. Combining the two equations in (3), we obtain $\frac{d^2 S^r}{dt^2} + \Omega_s^2 S^r = 0$ with

$$\Omega_s = \Omega \left(1 - \frac{3r_S}{2r} \right)^{\frac{1}{2}} \tag{4}$$

From the elementary solution $S^r \propto \sin \Omega_s t$, we see that after one orbital revolution,

$$S^r \propto \sin \left(\frac{2\pi \Omega_s}{\Omega} \right) = \sin \left(2\pi \left(\frac{\Omega_s}{\Omega} - 1 \right) \right)$$

fails to return to 0. The precession angle is thus given by

$$\Delta\varphi = 2\pi \left(1 - \frac{\Omega_s}{\Omega} \right) = 2\pi \left(1 - \left(1 - \frac{3r_S}{2r} \right)^{\frac{1}{2}} \right) \simeq 2\pi \left(\frac{3GM}{2r} \right) \tag{5}$$

for $r_S \ll r$.

Appendix: Lense-Thirring precession

The de Sitter precession must be distinguished from the Lense-Thirring precession (calculated in 1918) of S^μ caused by the rotation of the massive body, the earth in this case. To calculate the latter, we invoke frame dragging from chapter VII.5. Far from the rotating body, the frame rotates, and hence the gyroscope precesses, at an angular velocity (for an orbit in an equatorial plane) of

$$\omega = -\frac{g_{t\varphi}}{g_{\varphi\varphi}} \to \frac{ar_S}{r^3} = \frac{2GMa}{r^3} = \frac{2GJ}{r^3} \tag{6}$$

where I used the result you obtained in exercise VII.5.2. (Plugging in the numbers for a satellite around the earth, the Lense-Thirring precession amounts to ∼0.05 arcsec/year, while the de Sitter precession comes out to be ∼7 arc sec/year.)

We can also determine the Lense-Thirring precession using (1), plugging in the Christoffel symbols appropriate for a rotating body. Note that we do not need the full Kerr metric, only the asymptotic behavior of $g_{t\varphi}$ for large r. We will see in chapter IX.4 that this leading term, which determines the Lense-Thirring precession, can be fixed by general considerations. The logic actually goes full circle: as we explained in chapter VII.5, an astronomer could determine the angular momentum of a rotating body, be it a star or a black hole, by measuring the Lense-Thirring precession of a gyroscope orbiting that body.

Note

1. The experiment, known as Gravity Probe B, was first conceived in 1959. The technological marvels involved in constructing a working gyroscope of the required precision are breathtaking. I urge you to search for "Gravity Probe B" on the web and read about the experiment, including the controversy it generated. The experiment ultimately took 50 years and cost $760 million. See *Physics Today,* July 2011, p. 14.

Separation between geodesics

Euclid asserted that parallel straight lines will never meet in his space. Indeed, as is well known, the failure of Euclid's famous axiom ushered in the development of modern geometry, and the extent of the failure measures the curvature of the space. A familiar example is of course the globe we live on. Suppose Ms. U and Mr. P (remember them from chapter I.3 and part III?) start out in two neighboring towns on the same latitude and fly due south along geodesics, namely lines of constant longitude. We all know that the separation between them will change as they move along, eventually vanishing at the South Pole. In contrast to Euclidean geometry, two parallel straight lines could eventually intersect.

Let $x^\mu(\tau)$ and $y^\mu(\tau)$ be two nearby geodesics on a Riemannian manifold. In the example just given, they could be two lines of constant longitudes, with τ given by the latitude times a constant. Write $y^\mu(\tau) = x^\mu(\tau) + \epsilon^\mu(\tau)$ and study how $\epsilon^\mu(\tau)$ varies with τ. Subtract one geodesic equation

$$\frac{d^2 x^\mu}{d\tau^2} + \Gamma^\mu_{\nu\lambda}(x(\tau)) \frac{dx^\nu}{d\tau} \frac{dx^\lambda}{d\tau} = 0 \tag{1}$$

from the other

$$\frac{d^2 y^\mu}{d\tau^2} + \Gamma^\mu_{\nu\lambda}(y(\tau)) \frac{dy^\nu}{d\tau} \frac{dy^\lambda}{d\tau} = 0 \tag{2}$$

and expand to first order in ϵ. See figure 1.

Just as we are about to plug and grind, we see Professor Flat sauntering toward us, mumbling "Tsk tsk," and we immediately realize why. We hurriedly say, "Yes, Professor, we could go to locally flat coordinates and save ourselves a lot of work!"

So, let the coordinates be locally flat at the point P described by $x^\mu(\tau)$ for some specified value of τ. Hence $\Gamma^\mu_{\nu\lambda}(x(\tau)) = 0$ and $\Gamma^\mu_{\nu\lambda}(y(\tau)) = \Gamma^\mu_{\nu\lambda}(x(\tau) + \epsilon(\tau)) = \epsilon^\rho \partial_\rho \Gamma^\mu_{\nu\lambda}(x(\tau))$. Then

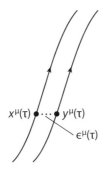

Figure 1 Two nearby geodesics can deviate from or approach each other.

the difference between (1) and (2) collapses in leading order in ϵ to

$$\frac{d^2\epsilon^\mu}{d\tau^2} + \epsilon^\rho \partial_\rho \Gamma^\mu_{\nu\lambda}(x(\tau)) \frac{dx^\nu}{d\tau} \frac{dx^\lambda}{d\tau} = 0 \tag{3}$$

But we want the covariant second derivative $\frac{D^2\epsilon^\mu}{D\tau^2}$, not the ordinary second derivative $\frac{d^2\epsilon^\mu}{d\tau^2}$. As mentioned in chapter V.6, the covariant derivative of ϵ along the geodesic is defined by

$$\frac{D\epsilon^\mu}{D\tau} = \frac{d\epsilon^\mu}{d\tau} + \Gamma^\mu_{\nu\lambda} \frac{dx^\nu}{d\tau} \epsilon^\lambda \tag{4}$$

which, unlike $\frac{d\epsilon^\mu}{d\tau}$, is assuredly a vector. The covariant second derivative of ϵ along the geodesic, $\frac{D^2\epsilon^\mu}{D\tau^2}$, is defined by the covariant derivative of $\frac{D\epsilon^\mu}{D\tau}$ along the geodesic, treating $\frac{D\epsilon^\mu}{D\tau}$ as a vector (of course). In other words,

$$\begin{aligned}
\frac{D^2\epsilon^\mu}{D\tau^2} &= \frac{d}{d\tau}\left(\frac{d\epsilon^\mu}{d\tau} + \Gamma^\mu_{\nu\lambda}\frac{dx^\nu}{d\tau}\epsilon^\lambda\right) + \Gamma^\mu_{\nu\lambda}\frac{dx^\nu}{d\tau}\left(\frac{d\epsilon^\lambda}{d\tau} + \Gamma^\lambda_{\omega\kappa}\frac{dx^\omega}{d\tau}\epsilon^\kappa\right) \\
&= \frac{d^2\epsilon^\mu}{d\tau^2} + \left(\partial_\rho\Gamma^\mu_{\nu\lambda}\right)\frac{dx^\rho}{d\tau}\frac{dx^\nu}{d\tau}\epsilon^\lambda \\
&= -\partial_\rho\Gamma^\mu_{\nu\lambda}\frac{dx^\nu}{d\tau}\frac{dx^\lambda}{d\tau}\epsilon^\rho + \partial_\rho\Gamma^\mu_{\nu\lambda}\frac{dx^\rho}{d\tau}\frac{dx^\nu}{d\tau}\epsilon^\lambda \\
&= \left(\partial_\rho\Gamma^\mu_{\sigma\lambda} - \partial_\lambda\Gamma^\mu_{\sigma\rho}\right)\frac{dx^\sigma}{d\tau}\frac{dx^\rho}{d\tau}\epsilon^\lambda
\end{aligned} \tag{5}$$

where we have exploited local flatness in the second equality and used (3) in the third equality. In the last equality, we merely rearranged the dummies.

Now we recognize the expression $(\partial_\rho\Gamma^\mu_{\sigma\lambda} - \partial_\lambda\Gamma^\mu_{\sigma\rho})$ in (5) as just what the Riemann curvature tensor $R^\mu_{\sigma\rho\lambda} = (\partial_\rho\Gamma^\mu_{\lambda\sigma} + \Gamma^\mu_{\rho\kappa}\Gamma^\kappa_{\lambda\sigma}) - (\partial_\lambda\Gamma^\mu_{\rho\sigma} + \Gamma^\mu_{\lambda\kappa}\Gamma^\kappa_{\rho\sigma})$ reduces to when evaluated in locally flat coordinates. Thus, we have derived, quick and fast, the equation[1] of geodesic deviation:

$$\frac{D^2\epsilon^\mu}{D\tau^2} = R^\mu_{\sigma\rho\lambda}\frac{dx^\sigma}{d\tau}\frac{dx^\rho}{d\tau}\epsilon^\lambda \tag{6}$$

Although we derived this result in locally flat coordinates, the by-now standard argument asserts that since both sides of (6) transform in the same way, it holds in any coordinate system.

The Riemann curvature tensor pops out, as if by magic. Of course, at least in hindsight, we know that it must: the vector $\frac{D^2\epsilon^\mu}{D\tau^2}$ has to be linear in ϵ^λ, and it must depend on the tangent or velocity vector $\frac{dx^\mu}{d\tau}$ at that point. The tangent vector must appear twice, not once, since we don't have a tensor with an odd number of indices that measures curvature. We also knew from the start that $\partial\Gamma$ must be involved. According to Euclid, the separation between two straight lines could only grow linearly; the second derivative $\frac{D^2\epsilon^\mu}{D\tau^2}$, sometimes called an "acceleration," reveals the presence of curvature.

For most purposes in physics, we naturally deal with timelike geodesics, but clearly (6) also applies to spacelike geodesics if we replace τ by the proper length along the geodesic. This hardly merits a remark since, after all, we obtained the formalism for determining geodesics with curves in space in the first place, not curves in spacetime.

Geodesic deviation, tidal force, and congruence of geodesics

We can now make contact with the Newtonian tidal force discussed way back in chapter I.4. Remember the ring of balls falling? The separation between two nearby balls evolves according to (I.4.9)

$$\frac{d^2 s^i}{dt^2} = -\mathcal{R}^{ij}s^j \tag{7}$$

which you now recognize as the Newtonian analog of (6).

We studied the separation between two geodesics, but more generally, we could consider a collection of geodesics $x^\mu(\tau, \sigma)$, distinguished from each other by a label σ. For instance, in cosmology, on distance scales such that galaxies could be treated as idealized mass points, with each galaxy tracing out a geodesic, the entire collection of geodesics could serve as a coordinate system for the universe. The label σ would then be 3-dimensional. Indeed, we have already mentioned this possibility when we discussed comoving coordinates back in chapter V.3.

In Italo Calvino's masterpiece *Cosmicomics,* the narrator, a man named Qfwfq, falls through spacetime, with his worldline tantalizingly close to that of a beautiful woman named Ursula H'x and that of a man named Fenimore. He is desperately in love with Ursula, but try as he may, he can't decrease the separation between his geodesic and her geodesic, seething in dismay and anger as he watches her geodesic and Fenimore's geodesic getting closer and closer to each other, or so it seems to him. The whole thing is told in the mind of Qfwfq.

This rather short chapter is now followed by four appendices. The first three are devoted to, respectively, a mathematically more sophisticated (not necessarily better in my opinion) derivation of geodesic deviation, the behavior of a bundle of timelike geodesics, and what can be proved about that behavior given various assumptions about the energy momentum tensor. The fourth appendix is somewhat tediously long and devoted to Fermi normal coordinates. The reader encountering Einstein gravity for the first time could certainly skip over this, or perhaps read only the first section, in which I explain intuitively why what is said to be true* is in fact true.

Appendix 1: Lie derivative and geodesic deviation

I will use the concept of Lie derivative introduced in chapter V.6 to give another derivation of (6). Consider a "dense" collection of timelike geodesics $x^\mu(\tau, \sigma)$, labeled by a real variable σ and defining a surface. Mathematicians say that the geodesics are knitted together to form a surface. Define the tangent vectors $V^\mu = \frac{dx^\mu}{d\tau}$ and the deviation vectors $\epsilon^\mu = \frac{dx^\mu}{d\sigma}$. Evaluate the Lie derivative

$$\mathcal{L}_V \epsilon^\mu = D_V \epsilon^\mu - D_\epsilon V^\mu = [V, \epsilon]^\mu$$

$$= V^\nu \partial_\nu \epsilon^\mu - \epsilon^\nu \partial_\nu V^\mu = \frac{d\epsilon^\mu}{d\tau} - \frac{dV^\mu}{d\sigma} = \frac{d^2 x^\mu}{d\sigma d\tau} - \frac{d^2 x^\mu}{d\tau d\sigma} = 0 \tag{8}$$

The first two equal signs merely state two alternative expressions for $\mathcal{L}_V$. Now act with D_V on the equation $D_V \epsilon^\mu = D_\epsilon V^\mu$ we just derived:

$$D_V D_V \epsilon^\mu = D_V D_\epsilon V^\mu = D_\epsilon D_V V^\mu + D_{[V,\epsilon]} V^\mu + R^\mu{}_{\nu\sigma\lambda} V^\nu V^\sigma \epsilon^\lambda = R^\mu{}_{\nu\sigma\lambda} V^\nu V^\sigma \epsilon^\lambda \tag{9}$$

You derived the second equality in exercise VI.1.16, namely that for three vector fields U, V, W, we have $D_U D_V W^\lambda - D_V D_U W^\lambda = D_{[U,V]} W^\lambda + R^\lambda{}_{\sigma\mu\nu} U^\mu V^\nu W^\sigma$. For the third equality, we used $D_V V^\mu = V^\nu D_\nu V^\mu = 0$ (which you recognize as the geodesic equation) and $[V, \epsilon] = 0$ (which we just derived in (8)). Now note that $D_V \epsilon^\mu = V^\mu D_\mu \epsilon^\mu = \frac{D\epsilon^\mu}{D\tau}$ and so $D_V D_V \epsilon^\mu = \frac{D^2 \epsilon^\mu}{D\tau^2}$. Thus, (9) is precisely (6). Even quicker and faster!

Appendix 2: The Raychaudhuri equation

Mathematically, a bundle of timelike geodesics $x^\mu(\tau, \sigma^1, \sigma^2, \sigma^3)$, labeled by three real variables σ, is known as a congruence in a certain region if, in that region, each point lies on one and only one geodesic. This implies that the geodesics in our bundle do not intersect. As soon as they intersect, the "congruence" is over. As mentioned in the text, a congruence of timelike geodesics could serve to coordinatize that region.

Pick a point P on one specific geodesic and denote the tangent vector by $V^\mu = \frac{dx^\mu}{d\tau}$. We have $V^\mu V_\mu = -1$, the definition of τ, and $V^\mu D_\mu V^\nu = 0$, the geodesic equation. Notation alert! In the definition of V^μ, the differentiation with respect to τ is clearly to be done holding the labels σ fixed; we want to follow a specific geodesic. But throughout this book, the tangent vector to a geodesic has always been denoted by $\frac{dx^\mu}{d\tau}$ (since we have always considered one single geodesic at a time, or at most two geodesics, as in this chapter), and it would be odd to suddenly start writing $\frac{\partial x^\mu}{\partial \tau}$ for the tangent vector. I think that it would be best, in this appendix and in appendix 4, to ask you to keep in mind what you are holding fixed by following the physics, rather than to have vertical bars all over the place indicating what is being held fixed, particularly in appendix 4 with its abundance of vertical bars (as you will see).

The 3 vectors $W^\mu = \frac{dx^\mu}{d\sigma}$ (for the 3 possible choices of σ; again, when differentiating with respect to one of the σs, we hold the other two σs and τ fixed) orthogonal to V^μ span a 3-dimensional subspace. The matrix $P^{\mu\nu} = g^{\mu\nu} + V^\mu V^\nu$ clearly projects into this subspace: $P^{\mu\nu} V_\nu = 0$, $P^{\mu\nu} P_\nu{}^\lambda = P^{\mu\lambda}$, and $P^{\mu\nu} P_{\nu\mu} = 3$. Then

$$\frac{DW^\mu}{D\tau} = V^\nu D_\nu W^\mu = W^\nu D_\nu V^\mu \equiv B^\mu{}_\nu W^\nu \tag{10}$$

---

* In this connection, I quote from a review of *QFT Nut* for the American Mathematical Society: "It is often deeper to know why something is true rather than to have a proof that it is true." See http://www.kitp.ucsb.edu/members/PM/zee/revMath.html.

where we used (8) with a trivial change of notation. Think of $B^\mu_{\ \nu}$ as a matrix acting on W to tell us the rate of change of W.

The idea is to derive an equation for $B_{\mu\nu}$. First, note that $V^\mu B_{\mu\nu} = V^\mu D_\nu V_\mu = \frac{1}{2} D_\nu (V^\mu V_\mu) = 0$ and $B_{\mu\nu} V^\nu = (D_\nu V_\mu) V^\nu = V^\nu D_\nu V_\mu = 0$. In other words, $B_{\mu\nu}$ lives in the 3-dimensional subspace. Recall how we decomposed 2-indexed tensors in chapter I.4. That piece of knowledge now comes in handy. Decompose $B_{\mu\nu}$ as

$$B_{\mu\nu} = \sigma_{\mu\nu} + \frac{1}{3}\theta P_{\mu\nu} + \omega_{\mu\nu} \tag{11}$$

Think of the geodesics as describing the motion of particles that form a cloud or fluid. You can see that each of the terms in (11) corresponds to a property of the flow. The trace $\theta = P^{\mu\nu} B_{\mu\nu} = D_\mu V^\mu$ describes expansion, the symmetric traceless part $\sigma_{\mu\nu} = \frac{1}{2}(B_{\mu\nu} + B_{\nu\mu}) - \frac{1}{3}\theta P_{\mu\nu}$ shear, and the antisymmetric part $\omega_{\mu\nu} = \frac{1}{2}(B_{\mu\nu} - B_{\nu\mu})$ rotation. (If you think about the spatial components of the corresponding quantities for a vector field in flat spacetime, for example $\omega_{ij} = \frac{1}{2}(\partial_i V_j - \partial_j V_i)$, you would understand the origin of the names[2] expansion, shear, and rotation.)

We are now ready to differentiate:

$$\frac{DB_{\mu\nu}}{D\tau} = V^\lambda D_\lambda B_{\mu\nu} = V^\lambda D_\lambda D_\nu V_\mu = V^\lambda D_\nu D_\lambda V_\mu + V^\lambda [D_\lambda, D_\nu] V_\mu$$

$$= D_\nu \left(V^\lambda D_\lambda V_\mu\right) - \left(D_\nu V^\lambda\right)\left(D_\lambda V_\mu\right) - V^\lambda R^\sigma_{\ \mu\lambda\nu} V_\sigma$$

$$= -B_{\mu\lambda} B^\lambda_{\ \nu} - R_{\sigma\mu\lambda\nu} V^\sigma V^\lambda \tag{12}$$

In the last step we used the geodesic equation and the definition of $B_{\mu\nu}$. This equation, known as the Raychaudhuri equation, governs how $B_{\mu\nu}$ varies as we move along a geodesic. Not surprisingly, as explained in the text, the Riemann curvature tensor appears. Spacetime curvature, aka the gravitational field, changes $B_{\mu\nu}$.

Often, we are mostly interested in knowing about whether a bundle of geodesics converges or diverges, a question determined by the expansion parameter θ. Since $\frac{D\theta}{D\tau} = g^{\mu\nu} \frac{DB_{\mu\nu}}{D\tau}$, we could extract an equation for $\frac{D\theta}{D\tau}$ by contracting (12) with $g^{\mu\nu}$.

Contracting the first term on the right hand side of (12) calls for a bit of work:

$$g^{\mu\nu} B_{\mu\lambda} B^\lambda_{\ \nu} = B_{\mu\nu} B^{\nu\mu} = \left(\sigma_{\mu\nu} + \frac{1}{3}\theta P_{\mu\nu} + \omega_{\mu\nu}\right)(\sigma^{\mu\nu} + \frac{1}{3}\theta P^{\mu\nu} - \omega^{\mu\nu})$$

$$= \sigma_{\mu\nu}\sigma^{\mu\nu} + \frac{1}{3}\theta^2 - \omega_{\mu\nu}\omega^{\mu\nu} \tag{13}$$

Notice, in the last step, the wisdom of decomposing tensors into pieces with different symmetry properties, as was advocated in chapter I.4.

Contracting the second term in (12) with $g^{\mu\nu}$, we watch the Ricci tensor pop out.

We thus obtain the desired result

$$\frac{D\theta}{D\tau} = -\frac{1}{3}\theta^2 - \sigma_{\mu\nu}\sigma^{\mu\nu} + \omega_{\mu\nu}\omega^{\mu\nu} - R_{\mu\nu} V^\mu V^\nu \tag{14}$$

This equation could be used to prove various theorems. To see how, first suppose that the geodesics are not rotational, namely that $\omega_{\mu\nu} = 0$. Since V^μ is timelike, the conditions $V^\mu B_{\mu\nu} = 0$ and $B_{\mu\nu} V^\nu = 0$ tell us that $B_{\mu\nu}$ is a purely spatial tensor, just like $P_{\mu\nu}$. Hence $\sigma_{\mu\nu}$ is also a purely spatial tensor. (This is also easily seen by going to a frame in which $V^\mu = (1, \vec{0})$ at the point in question: B_{ij} and P_{ij} are the only nonzero components of the tensor B and P, respectively, and hence σ_{ij} are the only nonzero components of the tensor σ.) Then we have $\sigma_{\mu\nu}\sigma^{\mu\nu} = \sigma_{ij}\sigma^{ij} \geq 0$.

Thus, with the assumption $\omega_{\mu\nu} = 0$, we obtain the inequality

$$\frac{D\theta}{D\tau} \leq -\frac{1}{3}\theta^2 - R_{\mu\nu} V^\mu V^\nu \tag{15}$$

To go further, we have to deal with the last term, which we can write, using Einstein's field equation, as

$$R_{\mu\nu} V^\mu V^\nu = 8\pi G \left(T_{\mu\nu} - \frac{1}{2} g_{\mu\nu} T\right) V^\mu V^\nu = 8\pi G \left(T_{\mu\nu} V^\mu V^\nu - \frac{1}{2} T V \cdot V\right)$$

Appendix 3: Energy conditions—weak, dominant, and strong

By making various debatable assumptions about $T_{\mu\nu}$, one can prove various debatable theorems. Depending on their inclinations, various authors regard one or more of these assumptions, known as energy conditions, as "self-evident," or at least "plausible." We can most easily understand the energy conditions listed in the literature by supposing that $T_{\mu\nu}$ has the perfect fluid form $T^{\mu\nu} = (\rho + P)U^\mu U^\nu + P g^{\mu\nu}$ (as has been discussed repeatedly, for example, in chapters III.6, VII.4, and VIII.1). We mention only a few here; you can make up your own.

1. The weak energy condition states that $T_{\mu\nu}V^\mu V^\nu \geq 0$ for all timelike Vs, which implies* that $\rho \geq 0$ and $(\rho + P) \geq 0$.

2. The dominant energy condition presupposes the weak energy condition and requires in addition that $T_{\mu\nu}V^\nu$ is not spacelike for all timelike Vs, namely that $g^{\mu\rho}(T_{\mu\nu}V^\mu)(T_{\rho\sigma}V^\sigma) \leq 0$, which amounts to $\rho^2 \geq P^2$.

3. The strong energy condition states that $T_{\mu\nu}V^\mu V^\nu \geq \frac{1}{2}T(V^\mu V_\mu)$ for all timelike Vs, which implies $(\rho + P) \geq 0$ and $(\rho + 3P) \geq 0$.

As an exercise, you could verify the stated implications for these three energy conditions.

Going back to the inequality (15)

$$\frac{D\theta}{D\tau} \leq -\tfrac{1}{3}\theta^2 - 8\pi G\left(T_{\mu\nu}V^\mu V^\nu - \tfrac{1}{2}TV \cdot V\right)$$

we see that if we have the strong energy condition, we can conclude that $\frac{D\theta}{D\tau} \leq 0$, so that the geodesics approach each other. The congruence of geodesics "focuses."[3]

The dark energy, aka the cosmological constant, is an interesting case. It violates the strong energy condition, since $(\rho + 3P) = (\rho - 3\rho) \not\geq 0$. Notice, however, that a positive cosmological constant, with $P = -\rho$ and $\rho \geq 0$, satisfies both the weak and dominant energy conditions (barely). At one time, most physicists would say that the strong energy condition evidently holds, since both matter and radiation (recall chapter VIII.1) satisfy it. But what is self-evident to one person may not be so obvious to another!

So what is a "reasonable" $T^{\mu\nu}$? Quantum field theorists would probably say that whatever $T^{\mu\nu}$ is produced by a field theory satisfying basic principles (such as unitarity and causality) is physically reasonable.

This is a good place to mention one apparently easy way to generate solutions of Einstein's equation. Write down a spacetime metric that you like. Calculate its Ricci tensor and plug into Einstein's equation, which specifies the $T_{\mu\nu}$ that would produce that particular spacetime. It's a cinch! The catch is that the $T_{\mu\nu}$ you obtain this way would most likely not satisfy the various energy conditions. You don't necessarily have the right stuff to produce the spacetime you like.

Appendix 4: Fermi normal coordinates

Intuitive motivation

Way way back in chapter I.6, we discussed the fairly obvious fact that at a given point P we could always choose locally flat coordinates. In fact, we could have coordinates, known as Fermi[†] normal coordinates, that are locally flat not only at a single point but also along an entire geodesic. The mathematical demonstration, that these coordinates are always available to us, is rather long[‡] and involved. So it would be best if I first describe a physical picture that makes the result more or less obvious.

Consider an observer moving along a timelike geodesic γ with the tangent vector V^μ.

Our friend the Smart Experimentalist interrupts, "It's obvious. I work in a lab attached to some planet going around some star, acted upon by gravity, so my lab is moving along a timelike geodesic. First, with gyroscopes I make sure that my lab is not spinning. Then I use my watch, namely my proper time, to mark the time coordinate,

---

* One way to show this is to go to the rest frame of U^μ and write $V^\mu = (\cosh \varphi, \sinh \varphi, 0, 0)$.

† Published in 1922.

‡ As I warned you, this appendix is longer than the main text!

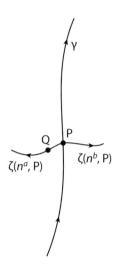

Figure 2 Construction of Fermi normal coordinates that are locally flat, not only at a single point but also along an entire geodesic.

and the walls of the lab to set up the space coordinates. Right there I have the locally flat coordinates of my choice. With due respect to my great experimental colleague Fermi, isn't that it?"

Yes, that's more or less it. But the long discussion we are about to embark on will yield more. Not only will we confirm SE's intuitive picture, we will also obtain precise expressions (see (30) below) for the second derivatives of the metric evaluated on γ.

So, to continue, we are given the coordinates x^μ, and we want to find the Fermi normal coordinates $y^\mu = (y^0, y^i) \equiv (T, y^i)$. For convenience, we gave y^0 the nickname T. Focus on a point P on γ and assign to point P the time $y^0 \equiv T = \tau$, the proper time τ elapsed since proper time started ticking at some point in the past. See figure 2.

"But that's exactly what I said!" exclaims SE. Yes indeed, we reassure her. Now we set up a set of four orthonormal vectors e_α^μ, with $\alpha = (0, a) = (T, a)$. We revel in notational "redundancy," writing the subscript T instead of 0 for emphasis. As SE suggested, we use $e_T^\mu = V^\mu$, with orthonormality

$$g_{\mu\nu} e_\alpha^\mu e_\beta^\nu = \eta_{\alpha\beta} \tag{16}$$

Once this is set up at P, we parallel transport (our gyroscopes are of high quality!) the 3 e_a^μs (that is, $e_{a;\nu}^\mu V^\nu = 0$) so that orthonormality always holds. The parallel transport of e_T^μ is automatic by virtue of γ being a geodesic.

Tentacles consisting of spacelike geodesics

To coordinatize the spacetime surrounding γ, we, sitting at P, now send out tentacles[4] consisting of spacelike geodesics $\zeta(n^a, P)$ with n^a a unit vector (that is, a vector satisfying $\delta_{ab} n^a n^b = 1$), determining the direction in which a particular geodesic $\zeta(n^a, P)$ emanates from P. See figure 2. In other words, the geodesic $\zeta(n^a, P)$ is characterized by the point P and the direction n^a in which it sallies forth from P.

More precisely, let W^μ denote the tangent vector along this geodesic $\zeta(n^a, P)$; then

$$W^\mu|_\gamma = e_a^\mu n^a \tag{17}$$

We shall henceforth use the notation $X|_\gamma$ to indicate that the expression X is evaluated on the geodesic γ. SE mumbles, "That's a lot of mathematical mumbling for what should be obvious to a child." I agree.

Onward to Fermi normal coordinates. Suppose that, after a proper distance σ, this particular geodesic $\zeta(n^a, P)$ reaches some point Q. Then we assign to the point Q the coordinates

$$y^0 = T, \qquad y^i = \sigma n^a \delta_a^i \tag{18}$$

Note that all these quantities depend on P as well as on Q: for example, σ is the proper distance along the geodesic $\zeta(n^a, P)$ from P to Q.

A note about notational clarity versus precision: in (18) I have chosen to suppress the dependence on P and assume that you know that y^μ denotes the coordinates of a particular point Q. I prefer to make it somewhat more mentally challenging to the reader (who has to quickly remind himself or herself of what depends on what) than to produce a page bristling with even more subscripts and superscripts than it already has. In any case, if you don't like it, grab a pen and mark the dependence on P and Q. As it is, I am already being excessively pedantic in writing in (18) $n^a \delta_a^i$ instead of just plain n^i: what is meant is clearly $y^1 = \sigma n^1$, $y^2 = \sigma n^2$, $y^3 = \sigma n^3$.

SE exclaims, "I can't stand all this math; it's just common sense. Something happens in my lab. I point to it, the direction of my finger is $\vec{n}$, the distance to that something is σ, and the time on my watch is T."

I said, "Yup, that's it. Enrico Fermi is all for common sense! I am also all for common sense! The Fermi normal coordinates of Q are given by y^μ."

To summarize, to determine the Fermi normal coordinates y^μ of a point Q, we have to ascertain the point P on γ from which a spacelike geodesic ζ would reach Q. We measure the proper length σ between P and Q along this geodesic ζ, and the direction n^a in which ζ emerges from P. The spacelike geodesics might eventually intersect, as discussed in appendix 2, but as long as they don't, every point Q in a finite region around the geodesic γ is uniquely characterized by (T, n^a, σ), the time on our friend's watch, the direction her finger is pointing in, and the geodesic distance from her to the point Q. Note that in the Fermi normal coordinates, the geodesic γ is described by $y^\mu = (T, 0)$, and the geodesic ζ by $y^\mu = (T, \sigma n^a \delta_a^i)$.

The metric in Fermi normal coordinates

Now the "hard" part: determine the metric in Fermi normal coordinates

$$g_{\lambda\rho}^{\rm F} = g_{\mu\nu} \frac{\partial x^\mu}{\partial y^\lambda} \frac{\partial x^\nu}{\partial y^\rho} \tag{19}$$

First, we have to relate y to x. For an arbitrary event Q in spacetime, its y coordinates are given by (18). What about its x coordinates $x^\mu(T, n^a; \sigma)$? (A friendly reminder: x^μ are the coordinates we started with, and y^μ are the Fermi normal coordinates.) It is the solution of the spacelike geodesic equation* $\frac{d^2 x^\mu}{d\sigma^2} + \Gamma_{\nu\lambda}^\mu \frac{dx^\nu}{d\sigma} \frac{dx^\lambda}{d\sigma} = 0$ with the initial position specified by T on γ and initial "velocity" in the direction n^a. (This is conceptually the same as the freshman physics problem of solving Newton's equation to determine the position of a particle after a certain time, starting with some initial position and velocity.) To get oriented, note that $x^\mu(T, n^a; 0)$ is just the point P and that $\frac{\partial x^\mu}{\partial y^0}|_\gamma = \frac{\partial x^\mu}{\partial T}|_\gamma = e_T^\mu$ is just the tangent vector V^μ to the geodesic γ at P. Also, the tangent vector of the spacelike geodesic on which the point Q sits is given by

$$W^\mu(T, n^a; \sigma) = \left(\frac{\partial x^\mu}{\partial \sigma} \right)_{T, n^a} \tag{20}$$

with $W^\mu(T, n^a; 0) = n^a e_\alpha^\mu$.

To relate[5] $x^\mu(T, n^a; \sigma)$ to y^μ, namely $y^\mu = (T, \sigma n^a)$, we note that the geodesic equation is invariant upon rescaling $\sigma \to f^{-1}\sigma$, with f an arbitrary real factor, under which $W^\mu \to f W^\mu$ and $n^a \to f n^a$. Thus,

$$x^\mu(T, n^a; \sigma) = x^\mu(T, fn^a; f^{-1}\sigma) = x^\mu(T, \sigma n^a; 1) = x^\mu(y) \tag{21}$$

Going back to (17) and (20), we have

$$e_a^\mu n^a = W^\mu|_\gamma = \left(\frac{\partial x^\mu}{\partial \sigma} \right)_{T, n^a}^{\sigma=0} = \left(\frac{\partial x^\mu}{\partial y^a} \frac{\partial y^a}{\partial \sigma} \right)_{T, n^a}^{\sigma=0} = \frac{\partial x^\mu}{\partial y^a}|_\gamma n^a \tag{22}$$

from which we conclude $\frac{\partial x^\mu}{\partial y^a}|_\gamma = e_a^\mu$. Plugging this and $\frac{\partial x^\mu}{\partial y^0}|_\gamma = e_T^\mu$ into (19), we find

$$g_{\lambda\rho}^{\rm F}|_\gamma = \eta_{\lambda\rho} \tag{23}$$

It's just (16)!

---

* Here I am faced with the notational dilemma of whether to use X or to stick with x, as per a bad notation alert way back in chapter I.1. On balance, I think sticking with x in this context is a bit clearer, as I have already done in the text and appendix 2 of this chapter.

To impress the Smart Experimentalist, we need to go further!

But our friend the Smart Experimentalist snickers, "You theorists, so much huffing and puffing for a result long obvious to me."

We agree. "Yes, but this is only zeroth order. The point is, now that the formalism is set up, we can calculate the metric to second order."

SE brags, "It is obvious; I already know what the metric to second order will depend on."

How about you? Did you guess that the Riemann curvature tensor evaluated on γ will have to come in?

As a warm up, tackle the metric to first order, namely the Christoffel symbols. Intuitively, both SE and we expect them to vanish. Normally, we use the geodesic equation to determine the geodesics. In this context, we already know the geodesics in the Fermi normal coordinates, namely ζ described by $y^\mu = (T, \sigma n^a \delta_a^i)$ and γ by $y^\mu = (T, \vec{0})$, which we will now plug into the appropriate geodesic equation to get information on the Christoffel symbols.

So, plug the geodesic ζ into $\frac{d^2 y^\mu}{d\sigma^2} + \Gamma^\mu_{\nu\lambda} \frac{dy^\nu}{d\sigma} \frac{dy^\lambda}{d\sigma} = 0$. (See the notation alert in appendix 2; we are following a single specific geodesic here. Later, when we want to be extra clear, we will use partial derivatives and specify what variables are being held fixed.) Since y^μ is at most linear in σ, we learn that $\Gamma^\mu_{\nu\lambda} \frac{dy^\nu}{d\sigma} \frac{dy^\lambda}{d\sigma} = 0$, which when evaluated on the timelike geodesic γ gives $\Gamma^\mu_{ij}|_\gamma n^i n^j = 0$ (with* $n^i \equiv n^a \delta_a^i$, clearly). Since the direction of $\vec{n}$ is arbitrary, we conclude (show this!) that $\Gamma^\mu_{ij}|_\gamma = 0$.

Similarly, plug the geodesic γ into $\frac{d^2 y^\mu}{d\tau^2} + \Gamma^\mu_{\nu\lambda} \frac{dy^\nu}{d\tau} \frac{dy^\lambda}{d\tau} = 0$ to obtain $\Gamma^\mu_{TT}|_\gamma = 0$. Also, by construction we parallel transport $e^\mu_\alpha = \left(e^\mu_T, e^\mu_a\right)$ along γ, so that $\left(\frac{de^\mu_\alpha}{dT} + \Gamma^\mu_{\nu\lambda} e^\nu_T e^\lambda_\alpha\right)|_\gamma = 0$. On γ, $e^\mu_\alpha = \delta^\mu_\alpha$ and so $\Gamma^\mu_{Tj}|_\gamma = 0$ and $\Gamma^\mu_{TT}|_\gamma = 0$. As expected, the Christoffel symbols vanish on γ.

The Riemann curvature tensor evaluated on γ

That was easy, but now let's determine the first derivatives of the Christoffel symbols on γ. The idea is to determine these derivatives in terms of the Riemann curvature tensor evaluated on γ.

First, the vanishing of the Christoffel symbols on γ allows us to conclude immediately that

$$\Gamma^\mu_{\nu\lambda, T}|_\gamma = 0 \tag{24}$$

Furthermore, the Riemann curvature tensor simplifies to $R^\mu_{\ \kappa\rho\lambda}|_\gamma = \left(\Gamma^\mu_{\kappa\lambda, \rho} - \Gamma^\mu_{\kappa\rho, \lambda}\right)|_\gamma$. In particular, using (24) we obtain

$$\Gamma^\mu_{\kappa T, \rho}|_\gamma = R^\mu_{\ \kappa\rho T}|_\gamma \tag{25}$$

To determine the other derivatives of the Christoffel symbols, we finally have to invoke the geodesic deviation equation (6) we derived in the text and which we rewrite in the form (which I now ask you to derive as an exercise)

$$\frac{d^2 \epsilon^\mu}{d\sigma^2} + 2\Gamma^\mu_{\nu\lambda} W^\nu \frac{d\epsilon^\lambda}{d\sigma} + (\Gamma^\mu_{\nu\lambda, \rho} - \Gamma^\mu_{\lambda\kappa} \Gamma^\kappa_{\rho\nu} + \Gamma^\mu_{\nu\kappa} \Gamma^\kappa_{\rho\lambda} - R^\mu_{\ \nu\rho\lambda}) W^\rho W^\nu \epsilon^\lambda = 0 \tag{26}$$

with the tangent vector $W^\nu = \frac{dy^\nu}{d\sigma}$ for spacelike geodesics.

Consider the collection of geodesics emanating from P and described by $x^\mu(T, n^a; \sigma)$. Let's compare two geodesics with the same $\vec{n}$ but emanating from P and P' slightly separated on γ and study $\epsilon^\mu \equiv \left(\frac{\partial y^\mu}{\partial T}\right)_{n^a, \sigma} = (1, 0) = \delta^\mu_T$ (since $y^\mu = (T, \sigma n^a \delta_a^i)$). Plugging this into (26), we see that the first two terms vanish. Also, $W^\nu = (0, n^a \delta_a^i)$. Evaluating what remains of (26) on γ, we obtain $(\Gamma^\mu_{\nu\lambda, \rho} - R^\mu_{\ \nu\rho\lambda})|_\gamma W^\rho W^\nu \epsilon^\lambda = 0$, and thus $\Gamma^\mu_{iT, j}|_\gamma = R^\mu_{\ ijT}|_\gamma$, something we already know from (25).

We can also compare two geodesics both emanating from P but in slightly different directions and study the separation between them, $\epsilon^\mu_{(a)} \equiv \left(\frac{\partial y^\mu}{\partial n^a}\right)_{T, \sigma} = (0, \sigma \delta_a^i) = \sigma \delta_a^\mu$. Plugging this into (26), we see that this time only the first term vanishes. Reminding ourselves that $W^\nu = (0, n^a \delta_a^i) = n^a \delta_a^\nu$, we obtain

$$2\Gamma^\mu_{\nu\lambda} n^b \delta_b^\nu \delta_c^\lambda + (\Gamma^\mu_{\nu\lambda, \rho} - \Gamma^\mu_{\lambda\kappa} \Gamma^\kappa_{\rho\nu} + \Gamma^\mu_{\nu\kappa} \Gamma^\kappa_{\rho\lambda} - R^\mu_{\ \nu\rho\lambda}) n^a \delta_a^\rho n^b \delta_b^\nu \delta_c^\lambda \sigma = 0 \tag{27}$$

---

* We finally succumb to sloppy notation.

Next, note that $\Gamma^\mu_{\nu\lambda} = \Gamma^\mu_{\nu\lambda}|_\gamma + \sigma\,(n^a\delta^i_a)\Gamma^\mu_{\nu\lambda,i}|_\gamma + O(\sigma^2) = \sigma\,(n^a\delta^\rho_a)\Gamma^\mu_{\nu\lambda,\rho}|_\gamma + O(\sigma^2)$. Setting σ to 0 in (27), we get $0 = 0$, but then extracting the terms linear in σ, we find $(3\Gamma^\mu_{\lambda\nu,\rho} - R^\mu_{\nu\rho\lambda})|_\gamma\, n^a\delta^\rho_a n^b\delta^\nu_b\delta^\lambda_c = 0$. Thus, we obtain

$$\left(\Gamma^\mu_{ki,j} + \Gamma^\mu_{kj,i}\right)|_\gamma = \frac{1}{3}\left(R^\mu_{ijk} + R^\mu_{jik}\right)|_\gamma \tag{28}$$

To solve this for the Christoffel symbols, generate two other versions of (28) by cycling the indices $(kij) \to (ijk)$, version A:

$$\left(\Gamma^\mu_{ij,k} + \Gamma^\mu_{ik,j}\right)|_\gamma = \frac{1}{3}\left(R^\mu_{jki} + R^\mu_{kji}\right)|_\gamma$$

and $(kij) \to (jki)$, version B:

$$\left(\Gamma^\mu_{jk,i} + \Gamma^\mu_{ji,k}\right)|_\gamma = \frac{1}{3}\left(R^\mu_{kij} + R^\mu_{ikj}\right)|_\gamma$$

Add version A to and subtract version B from (28) to obtain,[6] using various symmetry properties of the Christoffel symbol and the Riemann curvature tensor:

$$\Gamma^\mu_{ki,j}|_\gamma = \frac{1}{3}\left(R^\mu_{ijk} + R^\mu_{kji}\right)|_\gamma \tag{29}$$

Phew! We have finally nailed down, in (25) and (29), all the relevant quantities evaluated on γ.

Now recall from (II.2.25) that $g_{\mu\nu,\rho} = \Gamma_{\mu\cdot\nu\rho} + \Gamma_{\nu\cdot\mu\rho}$ where $\Gamma_{\mu\cdot\nu\rho} = g_{\mu\kappa}\Gamma^\kappa_{\nu\rho}$. Using (23), $g_{\mu\kappa}|_\gamma = \eta_{\mu\kappa}$, and $\Gamma^\mu_{\nu\lambda}|_\gamma = 0$, we have $g_{\mu\nu,\rho\omega}|_\gamma = (\Gamma_{\mu\cdot\nu\rho,\omega} + (\mu \leftrightarrow \nu))|_\gamma$.

Using (24), (25), and (29), we finally determine the second derivatives of the metric as follows:

$$g_{TT,ij}|_\gamma = 2\Gamma_{T\cdot Ti,j}|_\gamma = -2R_{TiTj}|_\gamma$$

$$g_{ij,kl}|_\gamma = \left(\Gamma_{i\cdot jk,l} + (i \leftrightarrow j)\right)|_\gamma = -\frac{1}{3}\left(R_{ikjl} + R_{iljk}\right)|_\gamma$$

$$g_{Ti,jk}|_\gamma = \left(\Gamma_{T\cdot ij,k} + \Gamma_{i\cdot Tj,k}\right)|_\gamma = \left[\frac{1}{3}\left(R_{Tjki} + R_{Tikj}\right) + R_{Tkji}\right]|_\gamma = \frac{2}{3}\left(R_{Tjki} + R_{Tkji}\right)|_\gamma \tag{30}$$

In the last step, we used the cyclic identity of the Riemann curvature tensor. Finally, using (25), we have easily $g_{\mu\nu,T\lambda} = (\Gamma_{\mu\cdot\nu T,\lambda} + \Gamma_{\nu\cdot\mu T,\lambda})|_\gamma = (R_{\mu\nu\lambda T} + R_{\nu\mu\lambda T})|_\gamma = 0$. So, in the second order expansion of $g_{\mu\nu}$, terms involving $x^0 x^0$ and $x^0 x^i$ do not appear.

The metric in Fermi normal coordinates

By now, we have long forgotten the coordinates that we started out with and traded for Fermi normal coordinates. So we might as well denote Fermi normal coordinates by x^μ, and also drop the nickname T for the more respectable 0. The results of this rather long analysis are then summarized by

$$g_{00} = -1 - R_{0i0j}|_\gamma x^i x^j$$

$$g_{ij} = \delta_{ij} - \frac{1}{3}R_{ikjl}|_\gamma x^k x^l$$

$$g_{0i} = \frac{2}{3}R_{0jki}|_\gamma x^j x^k \tag{31}$$

(For example, $g_{0i} = \frac{1}{2}g_{0i,jk}x^j x^k = \frac{1}{3}\left(R_{Tjki} + R_{Tkji}\right)|_\gamma x^j x^k = \frac{2}{3}R_{0jki}|_\gamma x^j x^k$.) By construction, the Riemann curvature tensor evaluated on γ depends only on x^0, not on x^i.

That sure was a load of work. So what did we get after all that? To summarize, for a given geodesic γ, we have now determined, following the great Fermi, the metric in a small tube around γ, good to second order.

You could now work out the metric $g_{\mu\nu}$ in Fermi normal coordinates to your heart's content for various freely falling observers, for example an observer falling radially into a Schwarzschild black hole. Note that γ has to be a geodesic. Along an arbitrary curve, it is not possible to arrange for all the Christoffel symbols to vanish.

Exercises

1   Verify (6) for the sphere.

2   The price of not heeding Professor Flat: derive the equation of geodesic deviation the hard way, without going to locally flat coordinates.

3   Derive

$$\frac{d^2\epsilon^\mu}{d\tau^2} + 2\Gamma^\mu_{\nu\lambda} V^\nu \frac{d\epsilon^\lambda}{d\tau} + \left(\Gamma^\mu_{\nu\lambda,\rho} - \Gamma^\mu_{\lambda\kappa}\Gamma^\kappa_{\rho\nu} + \Gamma^\mu_{\nu\kappa}\Gamma^\kappa_{\rho\lambda} - R^\mu_{\ \nu\rho\lambda}\right) V^\rho V^\nu \epsilon^\lambda = 0 \tag{32}$$

with the tangent vector $V^\nu = \frac{dx^\nu}{d\tau}$ as usual.

4   Verify the implications of the three stated energy conditions, strong, dominant, and weak (see appendix 3), for ρ and P.

Notes

1. I now go back to (1) and (2) to point out a slight subtlety. Each of the two geodesics has its own proper time, given by $d\tau_x = \sqrt{-g_{\mu\nu}(x)dx^\mu dx^\nu}$ and $d\tau_y = \sqrt{-g_{\mu\nu}(y)dy^\mu dy^\nu}$. We could of course trivially set $\tau_x = 0$ and $\tau_y = 0$ at the starting point, and define $\epsilon^\mu(\delta) \equiv y^\mu(\tau_y = \delta) - x^\mu(\tau_x = \delta)$. In other words, we study the separation between the two observers when each of their watches register the same time. The geodesic deviation equation describes how $\epsilon^\mu(\delta)$ changes with δ.
2. Recall how the concepts of divergence and curl were explained when you first encountered them.
3. It is sometimes said that, in Einstein gravity, the common person-in-the-street statement that gravity attracts requires the strong energy condition. This statement in itself is somewhat misleading: the geodesics here represent the movement of point particles in a background spacetime produced by an energy momentum tensor satisfying the strong energy condition.
4. Having been John Wheeler's student, I am now under the impression, these many years later, that this is the kind of phrase Wheeler would have used.
5. My discussion is based on the work of F. K. Manasse and C. W. Misner, *J. Math. Physics* (1963), vol. 4, p. 735.
6. Alternatively, regard $\Gamma^\mu_{ki,j}$ as a 3-by-3 matrix labeled by μ, k, with (28) telling us its symmetric part. Work out its antisymmetric part. This is essentially equivalent to the procedure given in the text.

Linearized Gravity, Gravitational Waves, and the Angular Momentum of Rotating Bodies

Einstein's unfinished symphony

Einstein gave life to spacetime. Previously rigid, spacetime could now curve and move. Certainly no surprise, then, that Einstein gravity predicts the existence of ripples crisscrossing the fabric of spacetime, what one writer refers to as Einstein's unfinished symphony.[1] Massive detectors have been built, with more to come, in an effort to tune in to the "song of the cosmos."

Theoretical physicists do not doubt* that gravitational waves exist. Indeed, there is already strong indirect evidence for gravitational waves with the discovery of a binary pulsar in 1974 by Hulse and Taylor. The change in the orbital period due to emission of gravitational waves could be accurately measured, and the data agreed extremely well with the prediction of Einstein gravity. The only question is when they will be detected: given our detectors, are there sufficiently powerful sources relatively nearby? An exciting new era will dawn with gravitational wave astronomy: hopefully, much will be revealed about the universe that we cannot see with electromagnetic wave astronomy.

Consider a small deviation from the Minkowski metric and write $g_{\mu\nu} = \eta_{\mu\nu} + h_{\mu\nu}$. In chapter VI.5, we already worked out that to leading order

$$R_{\mu\nu} = -\frac{1}{2}\left(\partial^2 h_{\mu\nu} - \partial_\mu \partial_\lambda h^\lambda_\nu - \partial_\nu \partial_\lambda h^\lambda_\mu + \partial_\mu \partial_\nu h^\lambda_\lambda\right) + O(h^2) \tag{1}$$

All we have to do is to plug this into Einstein's field equation and watch the ripples.

---

* Easy for me to say that now! Einstein famously clashed in 1936 with the editor of *Physical Review* and an anonymous referee over the existence of gravitational waves, even though he himself had introduced that notion back in 1916. Having moved to the United States, Einstein submitted a paper, together with his new American assistant Nathan Rosen, to *Physical Review,* claiming that gravitational waves did not exist. The grand old man, not used to having his papers rejected, wrote back angrily, vowing never to submit a paper to that journal again. The referee was later revealed to be H. P. Robertson.[2]

Weak field and harmonic gauge

By making a coordinate transformation $x^\mu \to x'^\mu = x^\mu + \varepsilon^\mu(x)$, we can simplify (1) considerably. Let $\partial_\mu \varepsilon_\nu$ be small, of the same order as $h_{\mu\nu}$, so that $\frac{\partial x^\mu}{\partial x'^\rho} = \delta^\mu_\rho - \partial_\rho \varepsilon^\mu + O(\varepsilon^2)$. The coordinate transformation $g'_{\rho\sigma}(x') = g_{\mu\nu}(x)\frac{\partial x^\mu}{\partial x'^\rho}\frac{\partial x^\nu}{\partial x'^\sigma}$ then reduces[3] to

$$h'_{\mu\nu} = h_{\mu\nu} - \partial_\mu \varepsilon_\nu - \partial_\nu \varepsilon_\mu \tag{2}$$

Notice the structural similarity to the electromagnetic gauge transformation $A'_\mu = A_\mu - \partial_\mu \Lambda$. Very nice! More on this in the next chapter. In electromagnetism, by choosing Λ, we can fix a gauge, typically the Lorenz gauge[4] $\partial_\mu A^\mu = 0$, to simplify Maxwell's equations. Conceptually, we have the same situation here.

Indeed, let's exploit the freedom in (2) to impose the so-called harmonic or de Donder gauge condition

$$\partial_\mu h^\mu_\nu = \tfrac{1}{2}\partial_\nu h \tag{3}$$

where we define the trace $h = \eta^{\mu\nu} h_{\mu\nu} = h^\lambda_\lambda$.

To see that this is always possible with a judicious choice of ε^μ, simply use (2) to compute $(\partial'_\mu h'^\mu_\nu - \tfrac{1}{2}\partial'_\nu h') \simeq (\partial_\mu h^\mu_\nu - \tfrac{1}{2}\partial_\nu h) - \partial^2 \varepsilon_\nu$ (where we dropped second order terms, while noting that $\partial'_\mu = \partial_\mu - (\partial_\mu \varepsilon^\lambda)\partial_\lambda$). Thus, if somebody gives you an h^μ_ν that does not satisfy (3), you can always choose an ε_ν, namely a solution of the equation $\partial^2 \varepsilon_\nu = (\partial_\mu h^\mu_\nu - \tfrac{1}{2}\partial_\nu h)$, so that h'^μ_ν satisfies (3) to the order considered. Then drop the prime. Note that to this order, we raise and lower indices with the Minkowski metric η.

Degrees of polarizations in a gravitational wave

We started out with a symmetric tensor $h_{\mu\nu}$ with $4 \cdot 5/2 = 10$ components. After imposing the 4 conditions in (3), we are left with $10 - 4 = 6$ components. The key point here is to realize that, even with $h_{\mu\nu}$ satisfying the harmonic gauge, we can still make a "residual" transformation (2). Since $\partial_\mu h'^\mu_\nu = \partial_\mu h^\mu_\nu - \partial^2 \varepsilon_\nu - \partial_\nu(\partial \cdot \varepsilon)$ and $\tfrac{1}{2}\partial_\nu h' = \tfrac{1}{2}\partial_\nu(h - 2\partial \cdot \varepsilon)$, we have $(\partial_\mu h'^\mu_\nu - \tfrac{1}{2}\partial_\nu h') = (\partial_\mu h^\mu_\nu - \tfrac{1}{2}\partial_\nu h) - \partial^2 \varepsilon_\nu$, and we see that, as long as $\partial^2 \varepsilon_\nu = 0$, the harmonic gauge (3) will continue to be satisfied. This brings the number of physical degrees of freedom in $h_{\mu\nu}$ down to $10 - 4 - 4 = 2$.

It is instructive to compare the familiar story in electromagnetism. By a gauge transformation, we can impose the Lorenz gauge so that Maxwell's equations $\partial^\mu F_{\mu\nu} = \partial^\mu(\partial_\mu A_\nu - \partial_\nu A_\mu) = 0$ simplify to $\partial^2 A_\nu = 0$. We can make a "residual" gauge transformation satisfying $\partial^2 \Lambda = 0$ and stay within the Lorenz gauge. Thus, the number of physical degrees of freedom in the electromagnetic field A_μ is $4 - 1 - 1 = 2$. You know very well that electromagnetic waves come in two polarizations.

Inspecting (1), we see that the harmonic gauge is designed to knock off the last 3 terms, so that $R_{\mu\nu} = -\tfrac{1}{2}\partial^2 h_{\mu\nu}$. In the weak field or linear approximation, Einstein's field equation

$R_{\mu\nu} - \frac{1}{2}g_{\mu\nu}R = +8\pi G T_{\mu\nu}$ reduces to*

$$\partial^2 h_{\mu\nu} - \frac{1}{2}\eta_{\mu\nu}\partial^2 h = -16\pi G T_{\mu\nu} \tag{4}$$

or equivalently

$$\partial^2 h_{\mu\nu} = -16\pi G \left(T_{\mu\nu} - \frac{1}{2}\eta_{\mu\nu}T\right) \tag{5}$$

with $T = \eta^{\mu\nu}T_{\mu\nu}$.

In vacuo, this simplifies[5] further to

$$\partial^2 h_{\mu\nu} = \left(-\frac{\partial^2}{\partial t^2} + \nabla^2\right) h_{\mu\nu} = 0 \tag{6}$$

which we recognize as the standard relativistic wave equation. (The reader with a long memory may recall that we already encountered this in chapter II.3.) Since the equation is linear, the general solution can be constructed as a linear superposition of plane waves[†] $h_{\mu\nu}(x) = \varepsilon_{\mu\nu}(k)\sin(k \cdot x)$, with $k \cdot x = \eta_{\mu\nu}k^{\mu}x^{\nu}$ and $k^2 = 0$ as required by (6). The polarization tensor $\varepsilon_{\mu\nu} = \varepsilon_{\nu\mu}$ (not to be confused with ε_{μ}!) satisfies $k_{\mu}\varepsilon_{\nu}^{\mu} = \frac{1}{2}k_{\nu}\varepsilon$ (with $\varepsilon \equiv \varepsilon_{\lambda}^{\lambda}$), the Fourier transform version of (3).

We now use the 4 degrees of freedom embodied in the residual transformation as explained above to impose 4 more conditions: $\varepsilon_{0i} = 0$, for $i = 1, 2, 3$, and $\varepsilon_{\lambda}^{\lambda} = 0$. (You should check that this is indeed possible!) With the latter condition, the harmonic condition collapses to $k^{\mu}\varepsilon_{\mu\nu} = 0$. The gauge defined by these $3 + 1 + 4 = 8$ conditions is known as the transverse-traceless or TT gauge.

We can now check that gravitational waves do indeed come in only 2 polarizations, just like electromagnetic waves. Let the wave propagate along the third axis so that $k^{\mu} = \omega(1, 0, 0, 1)$ (recall that $k^2 = 0$). The harmonic condition $k^{\mu}\varepsilon_{\mu\nu} = 0$ implies

$$\varepsilon_{0\nu} = \varepsilon_{3\nu} \Rightarrow \varepsilon_{00} = \varepsilon_{30} \quad \text{and} \quad \varepsilon_{3i} = \varepsilon_{0i} = 0, \quad \text{for } i = 1, 2, 3 \tag{7}$$

Thus, $\varepsilon_{33} = \varepsilon_{32} = \varepsilon_{31} = 0$. By the symmetry of $\varepsilon_{\mu\nu}$, we have $\varepsilon_{30} = \varepsilon_{03} = 0$, which implies $\varepsilon_{00} = \varepsilon_{30} = 0$. The traceless condition $\varepsilon_{\lambda}^{\lambda} = 0$ then collapses to $\varepsilon_{11} + \varepsilon_{22} = 0$.

You should check that this collection of simple but somewhat confusing statements we derived merely says that the zeroth and third rows and columns of the symmetric traceless matrix $\varepsilon_{\mu\nu}$ vanish. Thus, the polarization tensor

$$\varepsilon_{\mu\nu} = \begin{pmatrix} 0 & 0 & 0 & 0 \\ 0 & \varepsilon_+ & \varepsilon_\times & 0 \\ 0 & \varepsilon_\times & -\varepsilon_+ & 0 \\ 0 & 0 & 0 & 0 \end{pmatrix} \tag{8}$$

is characterized by 2 real numbers ε_+ and $\varepsilon_\times$, describing two independent degrees of polarizations. When we quantize electromagnetic waves, we obtain photons.[‡] Similarly,

---

* There is a hidden subtlety to this equation that will be made clear in exercise 3.

† The more sophisticated reader recognizes that more generally, we can also include a $\cos(k \cdot x)$ wave, or even better, write $h_{\mu\nu}(x) = \varepsilon_{\mu\nu}(k)e^{ik \cdot x}$.

‡ As shown in any textbook on quantum field theory.

when we quantize gravitational waves, we obtain gravitons. The 2 polarizations, found here for a classical gravitational wave, correspond to the 2 helicity states of the graviton in quantum field theory.*

Detection of gravitational waves

To understand what the "plus" and "cross" polarizations ε_+ and $\varepsilon_\times$ mean physically, we allow a gravitational wave to wash over a cloud of particles.

It is instructive to consider first a single particle initially at rest, that is, with 4-velocity $V^\mu = (1, 0, 0, 0)$. A gravitational field will cause it to move according to

$$\frac{dV^\rho}{d\tau} + \Gamma^\rho_{\mu\nu} V^\mu V^\nu = 0$$

Thus, at the initial instant, $\frac{dV^\rho}{d\tau} = -\Gamma^\rho_{00}$. Now compute

$$\Gamma^\rho_{00} \simeq \frac{1}{2}\eta^{\rho\lambda}\left(\partial_0 h_{0\lambda} + \partial_0 h_{0\lambda} - \partial_\lambda h_{00}\right)$$

but this vanishes in the TT gauge in which $\varepsilon_{0\lambda} = \varepsilon_{\lambda 0} = 0$ (see (8)). Thus, a moment of proper time later, V^μ is still $(1, 0, 0, 0)$. Repeating the argument, we see that a single particle initially at rest will remain at rest, completely ignoring the passing gravitational wave.

Confusio looks worried. "If a particle does not feel the gravitational wave passing by, is the wave real?"[6]

In fact, this apparently counterintuitive conclusion makes physical sense, since a Riemannian spacetime is by construction locally flat at any given point.

Our friend the Smart Experimentalist remarks, "Well, you simply need to go beyond a single point, to explore its neighborhood, to detect gravitational waves. So throw in a bunch of particles!"

Consider two particles separated by $\zeta^\mu = x_1^\mu - x_2^\mu = (0, \vec{\zeta})$. Let's see if the proper distance between them $\Delta l = \sqrt{\vec{\zeta}^2 + h_{ij}\zeta^i\zeta^j} = |\vec{\zeta}|(1 + \frac{1}{2}h_{ij}\zeta^i\zeta^j/\vec{\zeta}^2 + O(h^2))$ changes. Obviously it does if $h_{\mu\nu}$ varies. For a plane wave propagating along the third axis, rocking and rolling in the (1-2) plane, we see from (8) that in the TT gauge, we want to set $\vec{\zeta}$ in the (1-2) plane. We read off from (8) the fractional change in physical distance between the two particles (up to a factor of $\frac{1}{2}$ and with the unit 3-vector $\hat{\zeta}^i \equiv \zeta^i/|\vec{\zeta}|$):

$$h_{ij}\hat{\zeta}^i\hat{\zeta}^j = \left\{\varepsilon_+\left[\left(\hat{\zeta}^1\right)^2 - \left(\hat{\zeta}^2\right)^2\right] + 2\varepsilon_\times\hat{\zeta}^1\hat{\zeta}^2\right\}\sin(k \cdot x) \tag{9}$$

Thus, when a plus polarization wave comes along, a pair of particles separated along the 1-axis would move 180° out of phase compared to a pair of particles separated along the 2-axis, but would not respond at all to a cross-polarization wave. In contrast, a cross polarization wave would excite a pair of particles separated by $\vec{\zeta} \propto (1, \pm 1, 0)$, something that a plus polarization wave cannot do.

---

* The existence of 2 helicity states for massless particles of spin j, be it the photon or the graviton, follows in general from the Lorentz group and the CPT theorem. See, for example, S. Weinberg, *The Quantum Theory of Fields*, or *QFT Nut*, pp. 186 and 446.

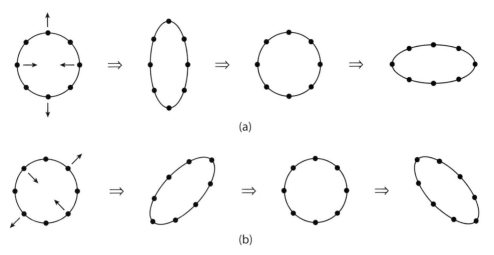

Figure 1 A ring of particles responding to a gravitational wave propagating perpendicular to the paper: (a) a plus wave; (b) a cross wave.

A few pictures are worth a thousand words, even with some Greek symbols thrown in. See figure 1. Panels a and b show how a ring of particles responds to a plus and cross polarization wave, respectively, propagating perpendicular to the paper. The pattern is characteristic of a tidal force, as was discussed way back in chapter I.4.

Gravitational wave detectors use laser interferometry to measure minute shifts in the distance between massive objects. Given how feeble a force gravity is, you can imagine the engineering feats involved. Several major projects[7] are either under way or are being planned. Hopefully, gravitational waves will be detected soon.

Emission of gravitational waves

Thus far, we have studied the propagation of gravitational waves in vacuo. To study their production, we include a source. We are invited by (3) and (4) to define $\tilde{h}_{\mu\nu} \equiv h_{\mu\nu} - \frac{1}{2}\eta_{\mu\nu}h$, so that

$$\partial^2 \tilde{h}_{\mu\nu} = -16\pi G T_{\mu\nu} \tag{10}$$

We see that the harmonic gauge condition $\partial^\mu \tilde{h}_{\mu\nu} = 0$ is consistent with $\partial^\mu T_{\mu\nu} = 0$, as it had better be.

Comparing this with the equation $\partial^2 A_\mu = -J_\mu$ governing the production of electromagnetic waves, we see that nothing much conceptually new is involved here, merely an extra index going along for the ride. The readers who have studied electromagnetism know what to do: define a Green's function (for those readers who don't know this, I give a brief explanation in appendix 2) by

$$\partial^2 \mathcal{G}(x) = \delta^{(4)}(x) \tag{11}$$

which simply says that $\mathcal{G}(x)$ is the (scalar) wave due to a unit point source at the origin of spacetime. The solution to (10) is then given by

$$\tilde{h}_{\mu\nu}(x) = -16\pi G \int d^4y\, \mathcal{G}(x-y) T_{\mu\nu}(y) \tag{12}$$

(plus, trivially, an arbitrary solution of the homogeneous equation, namely (10) with the right hand side set to zero). Physically, the Green's function approach merely says that since (10) is linear, we can solve it for a point source, and then, as Christiaan Huygens taught us long ago, add up the resulting waves from each point that makes up $T_{\mu\nu}$. That's precisely what (12) instructs us to do.

Note that $\mathcal{G}(x)$ is exactly the same Green's function you would use for electromagnetism: there are no indices in (11) after all. You may even know it off the top of your head as

$$\mathcal{G}(t, \vec{x}) = -\frac{\theta(t)\delta(t-r)}{4\pi r} \tag{13}$$

with $r = |\vec{x}|$ and the step function $\theta(t) = 1$ if $t > 0$ and 0 otherwise. This expression makes perfect sense: the first factor $\theta(t)$ tells us that the wave propagates only into the future (causality!), the second factor $\delta(t-r)$ says that the wave propagates at the speed of light, and the third factor $(-\frac{1}{4\pi r})$ satisfies* $\nabla^2(-\frac{1}{4\pi r}) = \delta^{(3)}(\vec{x})$.

Plugging (13) into (12), we obtain

$$\tilde{h}_{\mu\nu}(t, \vec{x}) = 4G \int d^4y\, \frac{\theta(t-y^0)\delta(t-y^0-|\vec{x}-\vec{y}|)T_{\mu\nu}(y)}{|\vec{x}-\vec{y}|}$$

$$= 4G \int d^3y\, \frac{T_{\mu\nu}(t-|\vec{x}-\vec{y}|, \vec{y})}{|\vec{x}-\vec{y}|} \tag{14}$$

Note that $T_{\mu\nu}$ is evaluated at the retarded time $t_R(t, \vec{x}, \vec{y}) \equiv t - |\vec{x}-\vec{y}|$. Just as for electromagnetic waves, a gravitational wave reaching $\vec{x}$ at time t had to be emitted at $\vec{y}$ at time t_R. You plug in whatever $T_{\mu\nu}$ you want into (14) and out pops $\tilde{h}_{\mu\nu}$, as simple as that.

Multipole expansion and compact source approximation

We can repeat many of the things we do in electromagnetism, as I said. For example, we can expand $\frac{1}{|\vec{x}-\vec{y}|}$ as a Taylor series in y^i and obtain the multipole expansion

$$\tilde{h}_{\mu\nu}(t, \vec{x}) = 4G \left\{ \frac{1}{r} \int d^3y\, T_{\mu\nu}\left(t_R, \vec{y}\right) + \frac{x^i}{r^3} \int d^3y\, y^i T_{\mu\nu}\left(t_R, \vec{y}\right) \right.$$

$$\left. + \frac{3x^i x^j - r^2\delta^{ij}}{2r^5} \int d^3y\, y^i y^j T_{\mu\nu}\left(t_R, \vec{y}\right) + \cdots \right\} \tag{15}$$

Note that we have not yet expanded the $|\vec{x}-\vec{y}| \simeq r\left(1 - \frac{x^i y^i}{r^2} + \cdots\right)$ lurking inside t_R. You see how all that stuff you learned about rotational tensors back in part I could be useful. For example, the traceless 3-tensor $(3x^i x^j - r^2\delta^{ij})$ pops up. Recall from chapter VI.3 that the analog of Newton's theorem, namely the Jebsen-Birkhoff theorem, continues to be valid

---

* Recall (II.1.12) from way way back.

in Einstein gravity. Around a spherically symmetric static (that is, time independent) mass distribution, only the first term in (15) survives.

If r is much larger than the size of the source, as is typically the case, we need to keep only the first term in this expansion, so that $\tilde{h}_{\mu\nu}(t, \vec{x}) = \frac{4G}{r} \int d^3y \, T_{\mu\nu}(t_R, \vec{y})$. For an astrophysical source, $\int d^3y \, T^{00}(t_R, \vec{y}) = M$ is its energy or mass, and $\int d^3y \, T^{0i}(t_R, \vec{y}) = P^i$ its momentum. If the source is not moving relative to the observer, so that $P^i = 0$, then we have simply $\tilde{h}_{00} = \frac{4GM}{r}$ and $\tilde{h}^{0j} = 0$, as expected.

Similarly, $\tilde{h}_{ij}$ is given in terms of $\int d^3y \, T_{ij}(t_R, \vec{y})$. Since our intuition about T_{ij} is rather weak, as was discussed in chapter III.6, it is preferable to use the conservation law $\partial_\nu T^{\mu\nu} = 0$ to eliminate T^{ij} in favor of T^{00}. As an exercise, you can derive

$$\tilde{h}^{ij}(t, \vec{x}) = \frac{2G}{r} \frac{d^2 Q^{ij}(t)}{dt^2}\bigg|_{t=t_R} \tag{16}$$

with $Q^{ij}(t) \equiv \int d^3y \, y^i y^j T^{00}(t, \vec{y})$ the quadrupole moment of the source.

Weak field around gravitating sources

We are now able to work out the spacetime around a gravitating source in the weak field limit. Consider stationary sources. Recall from chapter VII.5 on rotating black holes that a static source is not moving at all, while a stationary source could be moving, but without changing in time, for example a mass distribution rotating at constant angular velocity. With $T^{\mu\nu}$ in (14) not changing in time, we obtain immediately

$$\tilde{h}_{\mu\nu}(\vec{x}) = 4G \int d^3y \, \frac{T_{\mu\nu}(\vec{y})}{|\vec{x} - \vec{y}|} \tag{17}$$

No need to mess with retardation! The expansion in (15) is then a true multipole expansion.

Furthermore, if we have a nonrelativistic stationary source so that the typical velocity v of its components is much less than c, then recall from chapter III.6 that T^{0i} and T^{ij} are respectively a factor of v/c and v^2/c^2 down compared to T^{00}. So the leading term is $\tilde{h}_{00}$, followed by $\tilde{h}_{0i}$ and then $\tilde{h}_{ij}$.

Again, for $r = |\vec{x}|$ much larger than the size of the source, we obtain from (17) $\tilde{h}_{00} \simeq -4\Phi$, with

$$\Phi(\vec{x}) = -\frac{G}{r} \int d^3y \, T_{00}(\vec{y}) = -\frac{GM}{r} \tag{18}$$

the good old Newtonian potential. You might worry about the factor of 4 in $\tilde{h}_{00} = -4\Phi$, but notice that we still have to convert to $h_{\mu\nu}$. Since $\tilde{h} = \eta^{\mu\nu}\tilde{h}_{\mu\nu} = -\eta^{\mu\nu}h_{\mu\nu} = -h$, we have $h = -\tilde{h} \simeq \tilde{h}_{00} = -4\Phi$. Then $h_{\mu\nu} = \tilde{h}_{\mu\nu} - \frac{1}{2}\eta_{\mu\nu}\tilde{h}$, and so

$$h_{00} = \tilde{h}_{00} + \frac{1}{2}\tilde{h} = -4\Phi + 2\Phi = -2\Phi, \quad h_{11} = \tilde{h}_{11} - \frac{1}{2}\tilde{h} = 0 - 2\Phi = -2\Phi$$

and so on.* Thus, we finally obtain, within the approximations made, the perturbed spacetime metric

$$ds^2 \simeq (\eta_{\mu\nu} + h_{\mu\nu})dx^\mu dx^\nu \simeq -(1 + 2\Phi)dt^2 + (1 - 2\Phi)d\vec{x}^2 + 2h_{0i}dt dx^i \tag{19}$$

---

* Note that a common error is to suppose that $\tilde{h}_{00} \gg \tilde{h}_{0i} \gg \tilde{h}_{ij}$ implies that the same holds for $h_{\mu\nu}$.

For a static source, T_{0i} and hence h_{0i} vanish, and we obtain a form which the Schwarzschild metric must, and does, satisfy to this order. See also exercise 3.

It is worthwhile clarifying a point some texts appear to be confused about. When we look at the asymptotic behavior of $g_{00} \simeq -(1 - \frac{2GM}{r})$ to determine the mass of the Schwarzschild and of the Kerr black hole, we are not using the weak field approximation of this chapter. Far away from the black hole, where the observer sensing the asymptotic behavior is located, the gravitational field is indeed weak. But the field is definitely not weak near or inside the black hole, where $h_{\mu\nu}$ is in fact of $O(1)$. Schematically, we can write the full Einstein field equation as $\partial^2 h \sim GT + (h\partial^2 h + h\partial h\partial h + \cdots)$ with the expression in the parentheses, together with the $\partial^2 h$ on the left hand side, equal to the infinite series expansion of $R_{\mu\nu} - \frac{1}{2}g_{\mu\nu}R$. In other words, the mass M in $g_{00} \simeq -(1 - \frac{2GM}{r})$, as deduced by the hard-working astronomer measuring orbits (as explained by the Smart Experimentalist in chapter VII.5), is given by something like

$$\int d^3x \left\{ T + \left(h\partial^2 h + h\partial h\partial h + \cdots \right) / G \right\}$$

according to the theory. The gravitational self-interaction is automatically included. (See also the discussion in chapter VII.4.) We will explore this important conceptual point further in appendix 3.

Slowly rotating bodies

What about h_{0i}? As discussed in chapter VII.5 on the Kerr black hole, a term $g_{t\varphi}$ in the metric indicates that the source is rotating.

Let us consider a slowly rotating body whose energy momentum tensor is dominated by $T^{\mu\nu} \simeq \rho U^{\mu}U^{\nu}$, with ρ the mass density and with the pressure term negligible. Let the rotation be about the z-axis, and let the angular velocity be ω, so that $U^0 \simeq 1 \gg U^i$, $U^x = -\omega y$, $U^y = +\omega x$, and $U^z = 0$. (Confusing notation alert: we mix up two different notations $\vec{x} = (x^1, x^2, x^3) = (x, y, z)$ here for the sake of clarity!) The total angular momentum is then $J = \int d^3x \rho(\vec{x})(xU^y - yU^x) = \omega \int d^3x \rho(\vec{x})(x^2 + y^2)$.

Next, evaluate $\tilde{h}_{01}(\vec{x}) = -4G \int d^3x' \frac{T^{01}(\vec{x}')}{|\vec{x}-\vec{x}'|}$ by plugging in $T^{01}(\vec{x}') \simeq -\rho\omega y'$ (again, notation alert: $\vec{x}' = (x'^1, x'^2, x'^3) = (x', y', z')$) and expanding

$$\frac{1}{|\vec{x} - \vec{x}'|} \simeq (1/r) \left(1 + \vec{x} \cdot \vec{x}'/r^2 + \cdots \right)$$

We obtain

$$h_{01} = \tilde{h}_{01} \simeq -(4G/r) \int d^3x' \, \rho\omega y' \left(\vec{x} \cdot \vec{x}'/r^2 \right)$$

$$= -\left(4G/r^3 \right) y \int d^3x' \, \rho\omega \left(x'^2 + y'^2 \right)/2 = -\left(2GJ/r^3 \right) y \tag{20}$$

where we have invoked rotational invariance around the third axis (you've probably done similar calculations in electromagnetism). We obtain $h_{0i}dx^i = (2GJ/r^3)(x^1 dx^2 - x^2 dx^1) = -(2GJ/r)\sin^2\theta d\varphi$, in agreement with what we had in chapter VII.5.

Indeed, we are now able to evaluate the angular momentum of the Kerr black hole in the limit of slow rotation. Expand the Kerr metric (VII.5.15) for $r \gg r_S$, a:

$$ds^2 \simeq -\left(1 - \frac{r_S}{r}\right) dt^2 - \frac{2r_S a \sin^2 \theta}{r} dt d\varphi + \frac{1}{\left(1 - \frac{r_S}{r}\right)} dr^2 + r^2 \left(d\theta^2 + \sin^2 \theta d\varphi^2\right) \tag{21}$$

Comparing (20) and (21), we deduce that $r_S a = 2GJ$. Thus, we have verified, as promised, that the definition $J = Ma$ given in chapter VII.5 reduces in the small a limit to our usual understanding of angular momentum.

People often define $\tilde{h}_{0i} \equiv -4A^i$, with a notation intentionally suggestive of electromagnetism: if we think of GT^{00} and GT^{0i} as the charge and current density, respectively, then Φ and A^i correspond to the scalar and vector potentials of electromagnetism, respectively. Some physicists rightfully call the field $\vec{\nabla} \times \vec{A}$ the gravitomagnetic field, produced by a mass current, just like the magnetic field produced by a charge current.[8]

Quadrupole radiation

Back in chapter VI.3, I mentioned the Jebsen-Birkhoff theorem and its analogy to Newton's two superb theorems. You might have been slightly puzzled. Outside a pulsating spherically symmetric mass distribution, the spacetime remains stubbornly Schwarzschild, heedless of the pulsation. In Newtonian gravity, the absence of gravitational waves means that the pulsation does not result in any radiation, but in Einstein gravity, it would seem at first sight that the gravitational wave can communicate the pulsation to the spacetime outside. But now we understand why it can't. We learned from (16) that, far away from the source, the gravitational wave is generated by the quadrupole moment $Q^{ij}(t) \equiv \int d^3y \, y^i y^j T^{00}(t, \vec{y})$, and a spherical symmetric mass distribution simply doesn't have one. The classical statement that only quadrupole and higher moments can generate gravitational waves corresponds to the quantum statement that the graviton carries spin 2. You are probably familiar with the analogous statements in electromagnetism. A pulsating spherically symmetric charge distribution also cannot radiate. The classical statement that only dipole and higher moments can generate electromagnetic waves corresponds to the quantum statement that the photon carries spin 1.

Gravity is nonlinear

I have pointed out that much of what we learned about electromagnetic waves can be taken over for gravitational waves, but we must keep in mind that gravity is fundamentally different from electromagnetism. While Maxwell theory is linear, Einstein gravity is highly nonlinear, as we discussed after (19) and will emphasize in the next chapter. Within the linear approximation used in this chapter, things are simple since we are in a Minkowskian background. But beyond this approximation, the background will feel the energy and momentum of the gravitational wave and will deviate from Minkowskian. We will have to take into account the curvature of the background, and then the calculation of gravitational wave propagation becomes significantly more involved. A detailed treatment that does full justice to the subject is beyond the scope of this introductory text. For more, see review articles and more advanced texts.

Appendix 1: Determining the weak field action without Riemann

Time for another extragalactic fable. The smart young physicist that is you has been thinking about gravity along the same line as that Einstein in a civilization far far away. You understand that gravity is mediated by a tensor field $h_{\mu\nu}$ describing small deviations of the metric $g_{\mu\nu}$ from the Minkowski metric $\eta_{\mu\nu}$. You also realize the importance of coordinate transformation for a theory of gravity, and you get as far as understanding that the action must not change when $h_{\mu\nu}$ changes by $\delta h_{\mu\nu} = \partial_\mu \varepsilon_\nu + \partial_\nu \varepsilon_\mu$.

Unfortunately for you, your civilization has not yet produced a Riemann, and you have no idea how to construct an action out of $g_{\mu\nu}$.

What to do? You can still construct an action in the weak field regime. First, list all possible terms quadratic in h and quadratic in ∂. Lorentz invariance tells us that there are 4 possible terms:

$$S = \int d^4x \left(a\partial_\lambda h^{\mu\nu}\partial^\lambda h_{\mu\nu} + b\partial_\lambda h^\mu_\mu \partial^\lambda h^\nu_\nu + c\partial_\lambda h^{\lambda\nu}\partial^\mu h_{\mu\nu} + d h^\lambda_\lambda \partial^\mu \partial^\nu h_{\mu\nu} \right) \tag{22}$$

with 4 unknown constants a, b, c, and d. Note that we are talking about an action in Minkowski spacetime here. (To see that these are the only terms, first write down terms with the indices on the two ∂ matching, then the terms with the index on a ∂ matching an index on an h, and so on.)

Now impose your invariance requirement. Vary S with $\delta h_{\mu\nu} = -(\partial_\mu \varepsilon_\nu + \partial_\nu \varepsilon_\mu)$, integrating by parts freely. For example, $\delta(\partial_\lambda h^{\mu\nu}\partial^\lambda h_{\mu\nu}) = -2(\partial_\lambda(2\partial^\mu \varepsilon^\nu))(\partial^\lambda h_{\mu\nu})$ "=" $4\varepsilon^\nu \partial^2 \partial^\mu h_{\mu\nu}$. Since there are 3 objects linear in h, linear in ε, and cubic in ∂ (namely $\varepsilon^\nu \partial^2 \partial_\nu h$ and $\varepsilon^\nu \partial_\nu \partial^\lambda \partial^\mu h_{\lambda\mu}$ in addition to the one already displayed), the condition $\delta S = 0$ gives 3 equations, just enough to fix the action up to an overall constant, corresponding to Newton's constant. You work out, by high school algebra, that the combination

$$\mathcal{I} \equiv \tfrac{1}{2}\partial_\lambda h^{\mu\nu}\partial^\lambda h_{\mu\nu} - \tfrac{1}{2}\partial_\lambda h^\mu_\mu \partial^\lambda h^\nu_\nu - \partial_\lambda h^{\lambda\nu}\partial^\mu h_{\mu\nu} + \partial^\nu h^\lambda_\lambda \partial^\mu h_{\mu\nu} \tag{23}$$

is invariant.

You triumphantly publish a new theory of gravity* with the action

$$S_{\text{your name here}} = \int d^4x \left(-\frac{1}{32\pi G}\mathcal{I} + \frac{1}{2}h_{\mu\nu}T^{\mu\nu} \right)$$

with the coefficient of $\mathcal{I}$ fixed by comparing with Newtonian gravity. Indeed, the same story could have been told in our civilization. Without a little help from his friends, Einstein might have never heard of Riemannian geometry.

In other words, even if we had never heard of the Einstein-Hilbert action, we can still determine the action for gravity in the weak field limit by requiring the action to be invariant under the transformation (2), hardly surprising, since coordinate invariance determines the Einstein-Hilbert action uniquely.[9] Still, it is nice to construct gravity from scratch.

Note that invariance of $S_{\text{your name here}}$ under $\delta h_{\mu\nu} = \partial_\mu \varepsilon_\nu + \partial_\nu \varepsilon_\mu$ also tells us that the tensor $T^{\mu\nu}$ that $h_{\mu\nu}$ couples to must satisfy $\partial_\mu T^{\mu\nu} = 0$.

We have already noted that the electromagnetic gauge transformation $A'_\mu = A_\mu - \partial_\mu \Lambda$ is structurally similar. Using this observation in chapter IV.2, we immediately constructed the invariant field strength $F_{\mu\nu} = \partial_\mu A_\nu - \partial_\nu A_\mu$, which we squared to obtain the action. But suppose we didn't know about $F_{\mu\nu}$. Following the same procedure here, we list the two possible terms quadratic in A and quadratic in ∂ allowed by Lorentz and form the combination $\mathcal{L} = a\partial^\mu A^\nu \partial_\mu A_\nu + b\partial^\mu A^\nu \partial_\nu A_\mu$. The requirement that this is invariant under gauge transformation fixes $b = -a$, so that $\mathcal{L} = a\partial^\mu A^\nu (\partial_\mu A_\nu - \partial_\nu A_\mu) = a\partial^\mu A^\nu F_{\mu\nu} = \tfrac{1}{2}aF^{\mu\nu}F_{\mu\nu}$. Out pop $F_{\mu\nu}$ and the Maxwell action, not just in some weak field approximation, but in their full splendor.

---

* This discussion also connects with that in the next chapter.

Even though the following is slightly out of the scope of this text, I cannot resist remarking on how the harmonic gauge condition $\partial^\mu h_{\mu\nu} - \frac{1}{2}\partial^\nu h^\lambda_\lambda = 0$ is imposed in quantum field theory. We add the square of the left hand side $(\partial^\mu h_{\mu\nu} - \frac{1}{2}\partial^\nu h^\lambda_\lambda)^2$ to $\mathcal{I}$ and observe, in clever satisfaction, that this knocks off the last two terms in $\mathcal{I}$, so that the weak field action effectively becomes

$$S_{\text{weak field}} = \int d^4 x \frac{1}{2}\left[-\frac{1}{32\pi G}\left(\partial_\lambda h^{\mu\nu}\partial^\lambda h_{\mu\nu} - \frac{1}{2}\partial_\lambda h\partial^\lambda h\right) + h_{\mu\nu}T^{\mu\nu}\right] \tag{24}$$

This crucial step allows us to construct the Feynman propagator for the graviton.[10] If you knew just a tiny bit of field theory, you could now derive Einstein's famous calculation of the deflection of light directly[11] from $S_{\text{weak field}}$ without having to say a word about Riemann or curvature.

Appendix 2: Green's function

As promised, I briefly explain Green's function for those readers who have never heard of it. Others can skip this appendix.

First, the poor man's approach. Recall from chapter II.1 that the solution of

$$\nabla^2 G(\vec{x}) = \delta^{(3)}(\vec{x}) \tag{25}$$

is $G(\vec{x}) = -\frac{1}{4\pi r}$. We want to solve

$$\partial^2 \mathcal{G}(x) = \delta^{(4)}(x) \tag{26}$$

Away from the origin, this equation becomes, in spherical coordinates, $-\frac{\partial^2 \mathcal{G}}{\partial t^2} + \frac{1}{r^2}\frac{\partial}{\partial r}(r^2\frac{\partial \mathcal{G}}{\partial r}) = 0$. Writing $\mathcal{G} = g(t, r)/r$, we obtain $\left(-\frac{\partial^2}{\partial t^2} + \frac{\partial^2}{\partial r^2}\right)g = 0$, which, as we recall from chapter II.3, has the solution $g(t, r) = f_{\text{out}}(t - r) + f_{\text{in}}(t + r)$, with f_{out} and f_{in} two arbitrary functions corresponding to outgoing and incoming spherical waves. Physically, we keep only the outgoing piece, so that $\mathcal{G} = f(t - r)/r$, with some unknown function f.

But so far, we have only solved (26) without the delta function source, you protest—of course it's easy! The solution $\mathcal{G} = f(t - r)/r$ is only valid for $r > 0$. Now the poor man makes a clever observation. Evaluate $\partial^2 \mathcal{G} = \partial^2(f(t - r)/r)$ and take the limit $r \to 0$. In this limit, a spatial derivative hitting $1/r$ makes the result more singular but a time derivative hitting $f(t - r)$ does not. Hence, we can drop the time derivatives in ∂^2 and write, as $r \to 0$,

$$\partial^2 \mathcal{G}(x) = \left(-\partial_t^2 + \nabla^2\right)\mathcal{G}(x) \to f(t)\nabla^2 \frac{1}{r} \tag{27}$$

If we choose $f(t) = -\delta(t)/(4\pi)$ and use the known solution of (25), then this becomes $\delta(t)\delta^{(3)}(x)$ as desired. It follows that $\mathcal{G} = -\delta(t - r)/(4\pi r)$. We can multiply this expression by $\theta(t)$ for free, since it vanishes for $t < 0$ anyway, as $r > 0$ by definition. This gives the result cited in the text.

The rich man recognizes that (25) and (26) are members of a large class of linear equations easily solved by Fourier analysis. He or she also knows the Fourier representation of the (1-dimensional) delta function $\delta(x) = \int \frac{dk}{2\pi}e^{ikx}$. Thus, (25) and (26) are immediately solved by $G = -\int \frac{d^3k}{(2\pi)^3}\frac{e^{ikx}}{k^2}$ and $\mathcal{G} = -\int \frac{d^4k}{(2\pi)^4}\frac{e^{ikx}}{k^2}$, respectively, as you can see by formally plugging the appropriate expression into (25) and (26). But there is a huge difference between these two integrals, hidden by the highly compact notation. The expression k^2 in the denominator is evaluated with the Euclidean metric for (25) and thus is equal to $\delta_{ij}k^i k^j = \vec{k}^2$, but it is evaluated with the Minkowski metric for (26) and thus is equal to $\eta_{\mu\nu}k^\mu k^\nu = -(k^0)^2 + \vec{k}^2$. In the first case, the integral over d^3k can be done without too much difficulty and reproduces the result $G(\vec{x}) = -\frac{1}{4\pi r}$. In the second case, in integrating over k^0, we have to specify what to do with the poles at $k^0 = \pm|\vec{k}|$. For a detailed explanation, see any book on mathematical physics or field theory.[12] It turns out that the correct procedure is to interpret k^2 as $-(k^0)^2 + \vec{k}^2 + i\varepsilon\, \text{sign}(k^0)$, with the sign function defined as usual and with ε a positive infinitesimal to be set to 0 after the integral has been done. This procedure reproduces the result $\mathcal{G} = -\theta(t)\delta(t - r)/(4\pi r)$.

My advice for those not that familiar with the preceding is to simply go with the poor man's approach.[13]

Appendix 3: The gravitational field far from a possibly strong stationary source

This appendix is rather involved and can be skipped upon a first reading of this book.

As I emphasized in the text, the discussion in this chapter assumes that the gravitational field is weak everywhere. Here we want to study the gravitational field far from a strong source, say, a black hole. Near the black hole, the gravitational field is anything but weak and $h_{\mu\nu}$ is of order unity, so it would seem that linearized gravity would not have much to say. But the point is that, for an observer located sufficiently far away from the black hole, the local gravitational field she measures is weak enough for linearized gravity to hold. Far away, we can still exploit the field equation $\partial^2 \bar{h}_{\mu\nu} = 0$ and the harmonic gauge condition $\partial^\mu \bar{h}_{\mu\nu} = 0$. We assume in addition that the source is stationary, so that the field equation reduces to

$$\nabla^2 \bar{h}_{\mu\nu} = 0 \tag{28}$$

We now show that general considerations, based on rotational invariance, time reversal, parity, and so on, severely restrict the possible form of $h_{\mu\nu}$. The goal is now to determine $\bar{h}_{00}, \bar{h}_{0i}, \bar{h}_{ij}$ using general principles. Given that the metric does not depend on time, our manipulations will be strictly limited to everyday 3-dimensional quantities (as you will see). Hence we can suspend temporarily, for ease of writing, the distinction between upper and lower indices. Or, if you prefer, we raise and lower spatial indices with the Kronecker delta, without any funny signs. You will see what I mean.

A preliminary remark. Starting with the solution $1/r$ of Laplace's equation, namely $\nabla^2 (1/r) = 0$, we can generate more solutions simply by differentiating, since

$$\partial_i (1/r) = -x^i / r^3, \; \partial_i \partial_j (1/r) = \frac{3x^i x^j - \delta^{ij} r^2}{r^5}$$

and so on all solve Laplace's equation.

The subsequent discussion will be in two parts. The second part is more general than the first part.

First, we assume that our source defines an angular momentum vector $\vec{J}$. We now go way back to the discussion of rotational invariance in chapter I.3 to construct $\bar{h}_{00}, \bar{h}_{0i}$, and $\bar{h}_{ij}$ out of what we have available, namely the vectors $\vec{x}$ and $\vec{J}$.

A clarifying word or two about time reversal and parity. Under time reversal, $t \to -t, \vec{x} \to \vec{x}, \vec{J} \to -\vec{J}$. To leave ds^2 invariant, we must have $\bar{h}_{00} \to \bar{h}_{00}, \bar{h}_{0i} \to -\bar{h}_{0i}$, and $\bar{h}_{ij} \to \bar{h}_{ij}$. That's obvious enough. Parity is normally defined as the operation $t \to t, \vec{x} \to -\vec{x}$, but since $\vec{x} \to -\vec{x}$ is equal to a reflection in a mirror placed in the (y-z) plane $(x, y, z) \to (-x, y, z)$ followed by a rotation, we can equivalently think of mirror reflection. Think of an object rotating around the z-axis, and you can see that mirror reflection takes $\vec{J} \to -\vec{J}$.

So, write down the most general form consistent with these symmetries and impose the gauge condition. The result is

$$\bar{h}_{00} = \frac{A}{r} + O\left(\frac{1}{r^3}\right)$$

$$\bar{h}_{0i} = \frac{2\epsilon_{ijk} x^j J^k}{r^3} + O\left(\frac{1}{r^3}\right)$$

$$\bar{h}_{ij} = O\left(\frac{1}{r^3}\right) \tag{29}$$

To see how this goes, take $\bar{h}_{0i}$, for instance. Let $\vec{J}$ point in the z-direction. Time reversal flips h_{0i} and hence $\bar{h}_{0i}$. Thus, $\bar{h}_{0i}$ must be odd in $\vec{J}$. Similarly, for reflection as defined above to leave $ds^2 = (\cdots + 2(h_{01} dt dx + h_{02} dt dy + h_{03} dt dz) + \cdots)$ invariant, we must flip the sign of h_{01} and hence $\bar{h}_{01}$, but leave $\bar{h}_{02}$ and $\bar{h}_{03}$ alone. This mandates the appearance of the 3-dimensional antisymmetric symbol and fixes the form of $\bar{h}_{0i}$ up to an overall constant.

It is worth emphasizing that assuming rotational symmetry is not the same as assuming spherical symmetry, which is broken explicitly by the presence of the vector $\vec{J}$.

Second, we will be more general and not assume that the source provides a vector $\vec{J}$ around which we have cylindrical symmetry. The most general form that solves Laplace's equation we can write down is then

$$\tilde{h}_{00} = \frac{A}{r} + \frac{A^i x^i}{r^3} + O\left(\frac{1}{r^3}\right)$$

$$\tilde{h}_{0i} = \frac{B^i}{r} + \frac{B^{ij} x^j}{r^3} + O\left(\frac{1}{r^3}\right)$$

$$\tilde{h}_{ij} = \frac{C^{ij}}{r} + \frac{C^{ijk} x^k}{r^3} + O\left(\frac{1}{r^3}\right) \tag{30}$$

(the one assumption, I remind you, is that $\partial_0 g_{\mu\nu} = 0$). The various unknown rotational tensors characteristic of the source have obvious symmetry properties, for example $C^{ijk} = C^{jik}$.

Now that we have taken care of Einstein's field equation, we impose the harmonic gauge condition $\partial_\mu \tilde{h}_\nu{}^\mu = 0$. First, for $\nu = 0$, the condition $0 = -\partial_\mu \tilde{h}_0{}^\mu = -\partial_i \tilde{h}_{0i} = \frac{B^i x^i}{r^3} + \frac{B^{ij}(\delta^{ij} r^2 - 3x^i x^j)}{r^5} + O\left(\frac{1}{r^4}\right)$ implies that $B^i x^i = 0$ and $B^{ij}(\delta^{ij} r^2 - 3x^i x^j) = 0$. The first condition gives immediately $B^i = 0$. To solve the second condition, we triumphantly use what we learned in chapter I.3 about rotational tensors. Decompose B^{ij} into a traceless symmetric tensor, a trace, and an antisymmetric tensor. The stated condition knocks out the traceless symmetric tensor, so that $B^{ij} = B\delta^{ij} + 2\epsilon^{ijk} J^k$ with some unknown scalar B and vector $\vec{J}$ (which will turn out to be the angular momentum vector; let's not be coy about it).

Onward soldiers! Setting $\nu = i$ in the gauge condition, we find that $0 = \partial_\mu \tilde{h}_i{}^\mu = \partial_j \tilde{h}_{ij}$ gives $C^{ij} = 0$ and $C^{ijk}(\delta^{jk} r^2 - 3x^j x^k) = 0$. Now, in this round, solving the second equation really challenges us to show off our mastery of chapter I.3. We have a 3-indexed tensor C^{ijk} symmetric in the first two indices. We deal with these two indices in analogy with how we dealt with B^{ij} in the preceding paragraph. First, take out the trace: $C^{ijk} = \hat{C}^{ijk} + \delta^{ij} D^k$, where $\hat{C}^{ijk}$ is symmetric and traceless in its first two indices, so that $\delta^{ij} \hat{C}^{ijk} = 0$. Readers conversant with group theory would now recognize that the first two indices on $\hat{C}^{ijk}$ transform like a 5 of $SO(3)$, and the third index like a 3, so that our problem is solved by the decomposition $5 \times 3 = 3 + 5 + 7$. Those who know quantum mechanics would know furthermore that this is the problem of combining an angular momentum 2 object and an angular momentum 1 object. Those familiar with neither group theory nor quantum mechanics, fear not! You can simply contemplate the decomposition

$$\hat{C}^{ijk} = \left(E^i \delta^{jk} + F^{ih} \epsilon^{jkh}\right) + (i \leftrightarrow j) + G^{ijk} \tag{31}$$

where F^{ih} and G^{ijk} are both totally symmetric and traceless. In other words, if we contract any two indices on G^{ijk} with the Kronecker delta, we get zero.

Let us count to provide one check. The first two indices on $\hat{C}^{ijk}$ take on 5 different values, while the third index takes on 3 different values, giving us $5 \cdot 3 = 15$ independent components. On the other side of the ledger, E^i contains 3 components, and F^{ih} 5 components. How many components does G^{ijk} have? Readers with an elephant's memory would remember that we counted the number of symmetric triplets of indices back in chapter I.6: $\frac{1}{6} D(D+1)(D+2)$ if each index can take on D values; for $D = 3$, we have $\frac{1}{6}(3 \cdot 4 \cdot 5) = 10$. Remembering the three traceless conditions $G^{ijk} \delta^{ij} = 0$, we conclude that G^{ijk} has 7 components. Indeed, $15 = 3 + 5 + 7$; this is "exactly the same" equation as what we had written above, except for the conceptual fact that the earlier equation refers to representations of $SO(3)$.

We should not forget that all this work is needed to solve $C^{ijk}(\delta^{jk} r^2 - 3x^j x^k) = 0$, which now becomes $(\delta^{ij} D^k + E^i \delta^{jk} + E^j \delta^{ik} + F^{ih} \epsilon^{jkh} + F^{jh} \epsilon^{ikh} + G^{ijk})(\delta^{jk} r^2 - 3x^j x^k) = 0$, from which we obtain $D^i = -E^i$, $F^{ih} = 0$, and $G^{ijk} = 0$.

Let's take stock and summarize what we have wrought thus far:

$$\tilde{h}_{00} = \frac{A}{r} + \frac{A^i x^i}{r^3} + O\left(\frac{1}{r^3}\right)$$

$$\tilde{h}_{0i} = \frac{B x^i + 2\epsilon^{ijk} x^j J^k}{r^3} + O\left(\frac{1}{r^3}\right)$$

$$\tilde{h}_{ij} = \frac{\left(\delta^{ij} D^k x^k - D^i x^j - D^j x^i\right)}{r^3} + O\left(\frac{1}{r^3}\right)$$

You might be disappointed that after all this work we are still left with quite a mess. But in Einstein gravity, we have yet another trick up our sleeves: we can perform a coordinate transformation, otherwise known as a gauge transformation. From (2), we are free to change $\tilde{h}_{\mu\nu} \to \tilde{h}'_{\mu\nu} = \tilde{h}_{\mu\nu} - (\partial_\mu \varepsilon_\nu + \partial_\nu \varepsilon_\mu - \eta_{\mu\nu} \partial \cdot \varepsilon)$. Choose $\varepsilon_\mu = (B/r, 0, 0, 0)$ and we can knock off the B term in $\tilde{h}_{0i}$. Next, choose $\varepsilon_0 = 0$ and $\varepsilon_i = -D^i/r$ to knock all 3 D terms from $\tilde{h}_{ij}$. But oops, you might worry, because in this step $\tilde{h}_{00} \to \tilde{h}'_{00} = \tilde{h}_{00} + \partial_i \varepsilon_i = \tilde{h}_{00} + \frac{D^i x^i}{r^3}$, and

D^i pops its ugly head up somewhere else. Again, fear not, it can be absorbed into the A^i term. Finally, since $\frac{1}{|\vec{x}+\vec{a}|} = \frac{1}{r}\left(1 - \frac{a^i x^i}{r^2} + \cdots\right)$, we can knock off the A^i term in $\tilde{h}_{00}$ by suitably translating our coordinate system.

The bottom line is that, quite remarkably, the far field of any gravitating system, as long as it is stationary, whether a black hole or a collection of rocks, can be beaten down from (30) to

$$\tilde{h}_{00} = \frac{A}{r} + O\left(\frac{1}{r^3}\right)$$

$$\tilde{h}_{0i} = \frac{2\epsilon^{ijk} x^j J^k}{r^3} + O\left(\frac{1}{r^3}\right)$$

$$\tilde{h}_{ij} = O\left(\frac{1}{r^3}\right) \tag{32}$$

The far field depends on a number and a vector, just as in (29), even though we made fewer assumptions. Of course, it also agrees with the multipole expansion (15) and (17), discussed in the text; the quadrupole and higher moments, which depend on the shape of the mass distribution, are hidden precisely in the $O\left(\frac{1}{r^3}\right)$ terms in (32).

Our final task is to convert to the more physically relevant $h_{\mu\nu} = \tilde{h}_{\mu\nu} - \frac{1}{2}\eta_{\mu\nu}\tilde{h}$. In a calculation understood to be accurate up to but not including $O\left(\frac{1}{r^3}\right)$ terms, and similar to one performed in the text, we have $\tilde{h} = -A/r$; then $h_{00} = A/(2r)$ and $h_{ij} = \delta_{ij}A/(2r)$, leading to the metric*

$$ds^2 = -\left(1 - \frac{2M}{r}\right)dt^2 + \frac{2J}{r^3}(xdy - ydx)dt + \left(1 + \frac{2M}{r}\right)\left(dx^2 + dy^2 + dz^2\right) \tag{33}$$

(with the z-axis defined by the direction of $\vec{J}$). We have identified $A = 4M$. The vector $\vec{J}$ as it appears naturally here, in the tensor decomposition of B^{ij}, should be regarded as the definition of angular momentum. As I emphasized in chapter VII.5, it may be used to define the angular momentum of a rotating black hole. As shown by (20), it reduces to what we mean by angular momentum for a slowly rotating object.

Exercises

1 Use the equation of geodesic deviation derived in chapter IX.3 to describe the behavior of a pair of particles when a gravitational wave passes by. Recover the result derived in the text. Hint: Use the TT gauge. To leading order, everything simplifies.

2 Derive the quadrupole formula (16). Hint: Integrate by parts to obtain

$$\int d^3y \, y^j \partial_k T^{\mu k}(t, \vec{y}) = -\int d^3y \, T^{\mu j}(t, \vec{y}) + \int d^3y \, \partial_k\left[y^j T^{\mu k}(t, \vec{y})\right]$$

where the second integral on the right hand side can be converted by Gauss's theorem to a surface integral. If we integrate over a large enough region enclosing the source, then $T^{\mu k} = 0$ over the surface and the surface integral can be dropped. We thus obtain the identity

$$\int d^3y \, T^{\mu j}(t, \vec{y}) = -\int d^3y \, y^j \partial_k T^{\mu k}(t, \vec{y})$$

Use this identity and energy momentum conservation $\partial_\nu T^{\mu\nu} = 0$ (that is, $\partial_0 T^{\mu 0} + \partial_k T^{\mu k} = 0$).

3 Solve (10) far outside a static spherically symmetric mass distribution. You should be able to recover the asymptotic (that is, large r) form of the Schwarzschild metric. By the way, you could determine the answer by invoking some of the general results in the text, but that is not the point of the exercise. After all, we already know the Schwarzschild metric; the point is to see how its asymptotic behavior can be obtained directly from (10).

---

* Note that this metric for $J = 0$ does not reduce to the far field of the usual Schwarzschild solution, but of the solution written in the form given in exercise VI.3.3.

Notes

1. M. Bartusiak, "Einstein's Unfinished Symphony," 2000.
2. See D. Kennefick, "Einstein versus the Physical Review," *Physics Today,* September 2005, p. 43.
3. Some students are rightfully concerned about the meaning of the expansion. The fastidious may want to write $g_{\mu\nu} = \eta_{\mu\nu} + \lambda h_{\mu\nu}$ and $x^\mu \to x'^\mu = x^\mu + \lambda \varepsilon^\mu(x)$, expand equations such as $g'_{\rho\sigma}(x') = g_{\mu\nu}(x) \frac{\partial x^\mu}{\partial x'^\rho} \frac{\partial x^\nu}{\partial x'^\sigma}$ as a series in λ, and equate powers of λ.
4. See *QFT Nut,* chapters II.7 and III.4, particularly the footnote on p. 144 about Lorenz versus Lorentz.
5. Hence the name harmonic.
6. Historically, a confusing debate on whether gravitational waves could be removed by a coordinate transformation went on for some time. See the first footnote in this chapter. One subtlety is that if we consider a plane wave, as in standard treatments of electromagnetism, the infinite amount of energy contained in the wave would curl up spacetime, so localized wave packets must be used. We ignore all such subtleties in this introductory text.
7. At the moment, we have LIGO (short for the Laser Interferometer Gravitational Wave Observatory) in the United States, VIRGO in Italy, and GEO in Germany. Since the reader can easily find a list of projects on the web, I refrain from giving a more complete list that may be outdated soon. For example, in an earlier draft of this chapter, I had mentioned LISA (the Laser Interferometer Space Antenna) consisting of three spacecraft in orbits around the sun, but currently it is not funded, and even the proof-of-concept mission, LISA Pathfinder, is not scheduled until 2014.
8. The mathematical correspondence can be pursued further. With the correspondence $h_{00} \sim \Phi$, $h_{0i} \sim A_i$, that is, $h_{0\mu} \sim A_\mu$, you can work out Einstein's field equation in the post-Newtonian approximation and show that it has exactly the same form as Maxwell's equation. (Indeed, you can see that the leading Newtonian approximation $\nabla^2 \Phi \sim \rho$ is just Coulomb's law; generations of students have probably noticed that the partial differential equation for the gravitational potential and the electrostatic potential are identical in form.) You could then go on and indulge in some rather far-out speculations. For example, since we can add an as-yet-unobserved magnetic monopole to Maxwell's equation, we could ask if there is a gravitational analog of the magnetic monopole in Einstein gravity. See A. Zee, *Phys. Rev. Lett.* 55 (1985), p. 2379.
9. We see explicitly that the action contains two powers of ∂, and so the cosmological term is excluded. For the same reason, the higher derivative terms that we will discuss in chapter X.3 are also excluded.
10. For more, see chapter VIII.1 in *QFT Nut.*
11. See *QFT Nut,* p. 439. Indeed, this is essentially how Feynman does it.
12. For example, see *QFT Nut,* pp. 23–24.
13. The "poor man" I followed here is Landau. While he may not be rigorous enough for the jungle patrol on the Amazon, he is plenty rigorous for me.

IX.5 | A Road Less Traveled

"So great an absurdity"

Many roads lead to Einstein gravity. Back in chapter VI.1, I "air lifted" you over one of the shortest ways I know of to Einstein's celebrated field equation. Here I show you a road less traveled.

In chapters II.1 and II.3, I reminded you that in Newtonian gravity, the gravitational potential Φ is determined in terms of the mass distribution ρ by

$$\nabla^2\Phi(\vec{x}, t) = 4\pi G\rho(\vec{x}, t) \tag{1}$$

Look at this equation: any change in the mass distribution will be instantaneously communicated to the Newtonian potential. The gravitational potential Φ is slavishly yoked to the matter distribution.

Newton himself worried about this action at a distance. How could a planet know instantly any change in the position of its star? In the *Principia*, he left* this conundrum "to the consideration of the reader." But he did fret, and in a 1693 letter to his friend Richard Bentley, he opined:

> That gravity should be innate, inherent and essential to matter so that one body may act upon another at a distance through a vacuum without the mediation of anything else by and through which their action or force may be conveyed from one to another is to me so great an absurdity that I believe no man who has in philosophical matters any competent faculty of thinking can ever fall into it.[1]

Tell me, when you first learned about the inverse square law, did you not find it bizarre? Would Newton have described you as lacking in "faculty of thinking"?

---

* There is perhaps a lesson here somewhere for the young theoretical physicists reading this book. Newton was content to postulate the inverse square law and then explore its consequences. He left its dynamical origin[2] to others like Descartes, whose theory of vortices sweeping the planets along was swept into the dustbin of history.

Bringing time to gravity

By now, with your vast knowledge of Lorentz invariance and of relativistic completion, you know that we should bring time into the picture and thus promote ∇^2 to $\partial^2 = -\partial_t^2 + \nabla^2$, so that (1) is promoted to $\partial^2 \Phi = 4\pi G\rho$.

One important remark is that this modification immediately implies the existence of gravitational waves: something has to propagate. In empty spacetime, far from any matter distribution, the equation $\partial^2 \Phi(x) = 0$ has the wave solution $\Phi(x) = a \cos(\omega t - \vec{k} \cdot \vec{x}) + b \sin(\omega t - \vec{k} \cdot \vec{x})$, with $\omega^2 = \vec{k}^2$. Indeed, we already encountered gravitational waves in the preceding chapter (and in chapter II.3).

Historically, Laplace did have the foresight and insight to speculate about the speed of propagation c_G of the effect of gravity. Unfortunately, he concluded erroneously that $c_G \gg c$. These days, particle theorists subscribe to something known as the naturalness dogma[3] (or doctrine if you prefer), saying that fundamental constants with the same dimension should have roughly the same order of magnitude.* Otherwise, we would be confronted by a "hierarchy problem." So perhaps nowadays the default view would be that $c_G \sim c$. Of course, we now understand that the speed of propagation is a universal constant, a property of spacetime rather than the individual interaction. But before this understanding, it would seem strange, perhaps even bizarre, that gravitational and electromagnetic waves would propagate at precisely the same speed c. Conceivably, some bright young guy in another civilization far far away could have proposed the existence of gravitational waves with $c_G = c$ long before a complete understanding of curved spacetime was established.

But once you promote the Laplacian to the d'Alembertian ∂^2, you are obliged to also promote the mass density ρ. Here, as Robert Frost[4] said, we are at a fork faced with two roads. As we saw in chapter III.6, by studying how ρ transforms under a Lorentz boost, we would naturally promote ρ to an energy density T^{00}, the time-time component of an energy momentum tensor $T^{\mu\nu}$. As we will see, traveling down this road, we will arrive at Einstein gravity.

Traveling down the wrong road

The Finnish physicist Gunnar Nordström (1881–1923) pointed out another possibility, namely that ρ could be promoted to the Lorentz scalar $T \equiv T^\mu_\mu = \eta_{\mu\nu} T^{\mu\nu}$. In the non-relativistic limit, since $T^{ij} \ll T^{00}$, $-T$ reduces to T^{00}. So, $-T$ and T^{00} are both suitable role models for ρ to aspire to grow up into. In either case, we will recover Newtonian gravity. In Nordström's theory, the field equation for gravity reads

$$\partial^2 \Phi = -4\pi G T \qquad (2)$$

and the gravitational field $\Phi(x)$ is a Lorentz scalar.

* More in chapter X.3.

Alas for Nordström, he chose the wrong road.* Nature does not† make use of this possibility. Incidentally, he did this work[5] before Einstein formulated his theory of general relativity.

After special relativity was established, which after all was to make mechanics compatible with electromagnetism and its Lorentz invariance, Einstein was not the only one to realize that Newtonian gravity (1) also had to be made compatible with Lorentz invariance. Others in the race included Max Abraham (1875–1922) and Gustav Mie (1869–1957).

A less traveled road to Einstein gravity

> Compared with understanding gravity, the special theory of relativity was mere child's play.
> —A. Einstein writing to Arnold Sommerfeld, 1912

While ρ becomes a component of the tensor $T^{\mu\nu}$, the left hand side of $\partial^2\Phi = 4\pi G\rho$, for the equation to make sense, is also compelled to be a component of a tensor. We are thus forced to promote $\Phi(x)$ to a symmetric tensor field $h^{\mu\nu}(x)$ and write

$$\partial^2 h_{\mu\nu} = 8\pi G T_{\mu\nu} \tag{3}$$

where we have defined $h_{00} = -2\Phi$ to agree with our earlier discussion.

After the preceding chapter, it does not take much to guess that the field $h_{\mu\nu}$ will rather naturally turn out to be the deviation of the metric $g_{\mu\nu} = \eta_{\mu\nu} + h_{\mu\nu}$ from the Minkowski metric. Thus, once we decide not to wander off with Nordström, we are practically committed to curved spacetime. Indeed, as soon as we write down (3), we are led inexorably to Einstein gravity, as was shown by Stanley Deser, collaborating with David Boulware and echoing various earlier and later works of Suraj Gupta, Robert Kraichnan,[6] Richard Feynman,[7] Steve Weinberg,[8] and others.

I merely sketch how we will end up with Einstein gravity, suppressing indices for clarity. The action that would lead to (3) has the form $S_1 \sim \int d^4x(\frac{1}{G}\partial h \partial h + hT)$, with the (invisible) indices contracted by $\eta_{\mu\nu}$. Indeed, $\delta S_1 \sim \int d^4x(\frac{1}{G}\partial h \partial \delta h + \delta h T) \sim \int d^4x(-\frac{1}{G}\partial^2 h + T)\delta h = 0$, giving us $\frac{1}{G}\partial^2 h \sim T$. I must confess that living the unindexed life has its charms.

Keep on iterating

But this action can't be the end of the story. The very fact that the field $h_{\mu\nu}(x)$ is endowed with dynamics—that it can wriggle in spacetime—means that it carries energy and mo-

---

* Had Nature chosen this road, you wouldn't have to learn Riemannian geometry to master gravity.

† This is one of my favorite examples of the need to read with sophistication Einstein's dictum about making physics as simple as possible. Simple does not necessarily mean less math. Nature couldn't care less about how much or how little math you learned in school.

mentum. In S_1, we include in $T^{\mu\nu}$ only the contribution of the matter fields. Here the word "matter" is used in the same sense as in chapter VI.4 and includes anything but the gravitational field $h_{\mu\nu}$. We are now forced to include the contribution of $h_{\mu\nu}$ to $T^{\mu\nu}$ as well.

In chapter VI.4, we learned, given an action, how to determine its contribution to $T^{\mu\nu}$, even if the action is given merely in flat spacetime. We pretend momentarily that $\eta_{\mu\nu}$ is actually $g_{\mu\nu}$, we vary $g_{\mu\nu}$ to obtain $T^{\mu\nu}$, and then quickly set $g_{\mu\nu}$ back to $\eta_{\mu\nu}$, doing all this in our heads.

Apply this procedure to the term $\frac{1}{G}\partial h \partial h$ in S_1. Since this term carries 6 indices (which we have suppressed), there are actually 3 "invisible" ηs lurking in this term, in the schematic form $\sim \frac{1}{G}\eta\eta\eta\partial h\partial h$. Promoting $\eta_{\mu\nu}$ to $g_{\mu\nu}$, varying $g_{\mu\nu}$, and then setting $g_{\mu\nu}$ back to $\eta_{\mu\nu}$, we obtain a contribution to $T^{\mu\nu}$ of the schematic form $\frac{1}{G}\eta\eta\partial h\partial h$, where two indices are left dangling to match the $\mu\nu$ indices on $T^{\mu\nu}$. In other words, we have to shift $hT \to h(T + \frac{1}{G}\eta\eta\partial h\partial h)$. Including this contribution, we are forced to the action $S_2 \sim \int d^4x(\frac{1}{G}\partial h\partial h + hT + \frac{1}{G}h\partial h\partial h)$. (Our notation is evidently such that after every step, the letter T in the schematic form of the action once again includes only the contribution of the matter fields. We also suppress the many ηs lurking in the action.)

You see how the game goes: we keep on iterating. The term $\frac{1}{G}h\partial h\partial h$ in S_2 will now force us to include a term of the form $\frac{1}{G}h\partial h\partial h$ in T. Hence we are led by the nose to the action $S_3 \sim \int d^4x(\frac{1}{G}\partial h\partial h + hT + \frac{1}{G}h\partial h\partial h + \frac{1}{G}h^2\partial h\partial h)$.

Before you can shout "Here comes S_4!", you see that the action will iterate to an infinite series $S \sim \int d^4x\{\frac{1}{G}(\partial h\partial h + h\partial h\partial h + h^2\partial h\partial h + \cdots) + hT\}$. This much is easy to understand and makes sense physically, as I will explain presently. The hard part is to show, as Deser and company did, that the series $\frac{1}{G}(\partial h\partial h + h\partial h\partial h + h^2\partial h\partial h + \cdots)$ sums to the Lagrangian $\frac{1}{G}\sqrt{-g}R$ in Einstein gravity.*

In other words, the claim is that, upon substituting $g_{\mu\nu} = \eta_{\mu\nu} + h_{\mu\nu}$ into the Einstein-Hilbert Lagrangian and expanding in h, we will obtain the series we found by iterating. But in a sense this is hardly surprising, since general covariance forces us to the Einstein-Hilbert Lagrangian uniquely (as we saw in chapter VI.1).

In chapter V.2, we extolled the far-reaching power of the equivalence principle. If we know the action governing any interaction in flat Minkowskian spacetime, we immediately know the action governing that interaction in curved spacetime, in the presence of gravity: all we have to do, we learned, is to promote $\eta_{\mu\nu}$ to $g_{\mu\nu}$. But there is one interaction we cannot apply this wondrous stupendous trick to, so take back the word "any" in the preceding sentence. That very special interaction is the gravitational interaction itself! We knew gravity only in Newtonian, not Minkowskian, spacetime. What we have learned in this chapter is that if we try to construct the gravitational action in Minkowskian spacetime iteratively, we end up with the Einstein-Hilbert action in curved spacetime.

Instead of working with the action, we could also have worked with the equation of motion (3). Our task would then be to find a combination, involving two derivatives and the metric, with which to replace $\partial^2 h_{\mu\nu}$ on the left hand side, as we had already anticipated way way back in part II. Gauss, Riemann, and Ricci solved this highly nontrivial problem for us.

---

* To keep things simple, we did not mention that starting with hT, we generate an infinite series that will also sum up appropriately.

Geometry emerges

This discussion underlines the importance of the geometrical view of special relativity that we have emphasized in part III, in contrast to the view of special relativity as a series of apparent paradoxes. We could certainly work through the theory of special relativity applying the Lorentz transformation to various apparent paradoxes, without once mentioning the word "geometry." But that would be impoverishing physics.

Imagine yourself in a galaxy far far away, where people have never heard of Einstein gravity. But people know about the Newtonian equation (1) $\nabla^2 \Phi(\vec{x}, t) = 4\pi G \rho(\vec{x}, t)$, and Lorentz invariance has just been discovered. Suppose you then try to make this equation Lorentz invariant, following the road less traveled outlined here, and thus promoting Φ to $h_{\mu\nu}$. If you had understood special relativity in terms of the geometry of spacetime (that is, understood $\eta_{\mu\nu} dx^\mu dx^\nu$ as the generalized distance between two nearby points), then you would naturally interpret $(\eta_{\mu\nu} + h_{\mu\nu}) dx^\mu dx^\nu$ as an even more generalized distance between two nearby points. Geometry of curved spacetime naturally emerges. But the poor sap who knows special relativity as a series of paradoxes may have a hard time seeing curved spacetime.

The graviton interacts with itself

The physics behind the infinite series $\frac{1}{G}(\partial h \partial h + h \partial h \partial h + h^2 \partial h \partial h + \cdots)$ is easy to understand with a minimal knowledge[9] of quantum field theory. You have no doubt heard, and I have already mentioned, that when we quantize the electromagnetic field, we obtain the photon, and when we quantize the gravitational field, we obtain the graviton. A huge difference between the photon and the graviton is that the photon couples to charged fields, such as the electron field, but is not charged itself. In contrast, the graviton couples to anything carrying energy and momentum, and since it certainly carries energy and momentum, it couples to itself. In the electromagnetic action given in chapter IV.2, the photon couples to the charged fields that comprise the electromagnetic current J^μ via the term $A_\mu J^\mu$ in the action. Similarly, in the Einstein-Hilbert action, the graviton couples to matter fields that comprise the energy momentum tensor $T^{\mu\nu}$ via the term $h_{\mu\nu} T^{\mu\nu}$. But in addition, there are an infinite[10] number of terms of the form $h \cdots h \partial h \partial h$ describing the complicated interactions of many gravitons with one another. This partly accounts for the intractability of quantum gravity at present.

To say all this in a slightly different way, we recall from chapter VI.4 that $T^{\mu\nu}$ is not locally conserved, in contrast to the electromagnetic current J^μ. The physics behind this fact is the ability of the gravitational field to exchange energy momentum with $T^{\mu\nu}$.

Appendix 1: From electrostatics to Maxwell

It is instructive to repeat for electromagnetism what we did in this chapter. Suppose we start with electrostatics, with Poisson's equation $\nabla^2 \Phi(\vec{x}, t) = \rho(\vec{x}, t)$ determining the electrostatic potential Φ in terms of the charge density ρ. Indeed, it is essentially the same equation as (1) that we started this chapter with.

Again, we complete relativistically, except that now ρ is promoted to be the time component of a Lorentz vector J^μ instead of the time-time component of a Lorentz tensor $T^{\mu\nu}$. This forces us to promote Φ to be the time component of a Lorentz vector A^μ, leading us to the equation $\partial^2 A^\mu = -J^\mu$. The next step is motivated by our desire to have current conservation fall out of the equation of motion. This would happen if we change[11] our equation to $\partial_\nu(\partial^\nu A^\mu - \partial^\mu A^\nu) = -J^\mu$, from which $\partial_\mu J^\mu = 0$ follows as an identity. You might realize that this last remark in fact foreshadows energy momentum conservation falling out of the Bianchi identity, as we saw in chapter VI.5.

Appendix 2: Gravity is feeble, and the Planck mass is huge

As we have noted since the very beginning of this text, the immensity of the Planck mass $M_P \equiv \sqrt{\frac{1}{G}} \sim 10^{19}$ GeV directly reflects the extreme feebleness of gravity: G is teeny, so M_P is humongous. This is merely a simple matter of high school algebra, but it is rendered particularly clear by the weak field discussion of this chapter. Write the action in the text as $S \sim \int d^4x \{M_P^2(\partial h \partial h + h \partial h \partial h + h^2 \partial h \partial h + \cdots) + hT\}$. Now scale $h \equiv \hat{h}/M_P$, so that

$$S \sim \int d^4x \left\{ \left(\partial\hat{h}\partial\hat{h} + M_P^{-1}\hat{h}\partial\hat{h}\partial\hat{h} + M_P^{-2}\hat{h}^2\partial\hat{h}\partial\hat{h} + \cdots \right) + M_P^{-1}\hat{h}T \right\} \tag{4}$$

The gravitational field $\hat{h}$ is conventionally normalized in the sense that the "kinetic energy" term $\partial\hat{h}\partial\hat{h}$ has coefficient unity. But now we see that the interaction of $\hat{h}$ with the rest of the world (as represented by T) and with itself scales like inverse powers of M_P, as expected.

Appendix 3: A shorter road less traveled

Using the Palatini formalism introduced in chapter VI.5 (which you may wish to review now), we can shorten the road less traveled described in this chapter. I am content to do it schematically, omitting the indices. A flat spacetime version of Palatini would have started with two fields $h_{..}$ and $\Gamma^{.}_{..}$ governed by the action $S \sim \int (h\partial\Gamma + hT)$, where the appearance of the Minkowski metric when needed is understood. Upon varying h, we get $\partial\Gamma \sim T$. But we don't know what Γ is. To remedy this, add a quadratic term $\Gamma\Gamma$ to the action, so that $S \sim \int (h\partial\Gamma + hT + \Gamma\Gamma)$. Varying Γ, we obtain $\Gamma \sim \partial h$, which when plugged into the previous equation, leads us to $\partial^2 h \sim T$, namely (3). By the argument in the text, we now have to introduce the cubic term $h\Gamma\Gamma$, which together with $\Gamma\Gamma$, we then recognize as the first 2 terms in the expansion of $\sqrt{-g}\,g^{..}\Gamma^{.}_{..}\Gamma^{.}_{..}$. We thus recover the Palatini version of the Einstein-Hilbert action $S_{EH} \sim \int d^4x \sqrt{-g}\,g(\partial\Gamma + \Gamma\Gamma)$, namely the Palatini action given in chapter VI.5.

Notes

1. R. Bentley, *Works of Richard Bentley*, vol. 3, Francis Macpherson, 1838.
2. I might call the Descartes approach the "all or nothing approach," which some theoretical physicists still indulge in. At any stage in the development of physics, certain questions are not appropriate; for instance, somebody could always demand of Newton, "Hey Isaac, so why inverse square?"
3. We will come back to this idea in chapter X.7.
4. Written in that famous year 1915, by the way.
5. For the controversial relationship between Nordström and Einstein, see the letters by P. Freund and E. L. Schucking in *Physics Today*, August 2009, p. 8.
6. Incidentally, Kraichnan did his work as an 18-year-old undergraduate at the Massachusetts Institute of Technology. As a postdoc at the Institute for Advanced Study, he showed his work to Einstein, who was appalled by this so-called "particle physics" approach, in contrast with the geometrical approach. He delayed publication for 8 years and ended up publishing after Gupta. Perhaps partly as a result of this encounter with Einstein, Kraichnan left the field and later became an eminent authority on turbulence.
7. Feynman's work came out of his effort to quantize gravity. In one story, when Feynman told his colleague Murray Gell-Mann about his research, the latter told him to try quantizing Yang-Mills theory as a warm-up

exercise. Feynman wrote his wife Gweneth a famous letter from the Warsaw conference on gravity in 1962, in which he said, "Remind me not to come to any more gravity conferences." Also, "Now I will show that I too can write equations that nobody can understand." These two remarks more or less summed up the attitude of particle theorists to Einstein gravity until the mid-1970s.

8. In this particle physics approach, as championed by Feynman and Weinberg, curved spacetime and Riemannian geometry are not put in but rather fall out. See the preface and introduction to S. Weinberg, *Gravitation and Cosmology*, and R. P. Feynman et al., *Feynman Lectures on Gravitation*.

9. See *QFT Nut*, chapter VIII.1.

10. The Yang-Mills field is intermediate between the electromagnetic and the gravitational fields in complexity. The analog of the infinite series we encounter here terminates in Yang-Mills theory. See, for example, *QFT Nut*, chapter IV.5.

11. For more details, see *QFT Nut*, pp. 38 and 188.

Why do we love the sphere so much?

Think about the vast amount of theoretical physics you have learned and you realize the enormous role spheres and other symmetrical situations have played in enhancing your understanding. Even with the ubiquity of numerical computation these days, analytically soluble examples still provide valuable, perhaps indispensable, railings for us to hold on to. So too in Einstein gravity: the most intensely studied spacetimes, as you might expect, are the most symmetrical. In this chapter, we explore symmetry in the context of space and spacetime.

We love the sphere, obviously because its high degree of symmetry makes it easy to work with. Indeed, every point on the sphere is equivalent to every other point. But somebody could have given you the metric on the sphere in some awful and unfamiliar coordinates, and you may not recognize that it describes the sphere. In fact, were the metric $ds^2 = d\theta^2 + \sin^2\theta d\varphi^2$ unfamiliar to you, how would you go about discovering that it possesses the maximal amount of symmetry? The two coordinates θ and φ are treated so differently. Thus, we definitely need to develop some machinery to uncover any symmetry that might be masked by a poor choice of coordinates.

The isometry condition

In Riemannian geometry, because of the freedom in choosing coordinates, symmetry is not always glaringly advertised.

If the geometry at point P and the geometry at Q are the same, the two points P and Q are said to be isometric. The metric for the sphere in the standard coordinates is independent of φ, so obviously, two points related by $\varphi' = \varphi + c$ for an arbitrary constant c are isometric. But the isometry in the θ direction is not so evident, and the isometry in an arbitrary direction is even less so.

As always, $g'_{\rho\sigma}(x') = g_{\mu\nu}(x)\frac{\partial x^\mu}{\partial x'^\rho}\frac{\partial x^\nu}{\partial x'^\sigma}$. Now suppose we require that the metrics g' and g, as functions of their respective arguments, are the same, namely that $g'_{\rho\sigma}(x') = g_{\rho\sigma}(x')$. Thus, the question of whether the space enjoys any isometry amounts to asking whether the set of equations

$$g_{\rho\sigma}(x') = g_{\mu\nu}(x)\frac{\partial x^\mu}{\partial x'^\rho}\frac{\partial x^\nu}{\partial x'^\sigma} \tag{1}$$

has any solutions.

Watch the primes like a hawk here! Note carefully that this condition of isometry differs by a single prime from the far more commonly seen equation $g'_{\rho\sigma}(x') = g_{\mu\nu}(x)\frac{\partial x^\mu}{\partial x'^\rho}\frac{\partial x^\nu}{\partial x'^\sigma}$, which we just cited and which you first encountered in chapter I.5 telling us how the metric g' is determined by the metric g. (Indeed, the right hand side in (1) is none other than $g'_{\rho\sigma}(x')$.) In contrast, the isometry condition (1) compares a given metric $g_{\mu\nu}$ evaluated at x and x': it imposes a requirement on $g_{\mu\nu}$ that most metrics in fact fail to satisfy.

In general, the condition (1) consists of a set of formidable equations that are difficult to solve, but mathematicians have studied them in depth. We are content, however, to follow the German minister and mathematician Wilhelm Killing (1847–1923) and analyze (1) when the two points are related infinitesimally $x'^\mu = x^\mu + \varepsilon\xi^\mu(x)$, in the same spirit we adopted when studying Lie algebra. Indeed, Killing also discovered Lie algebras independently of Sophus Lie and even anticipated some later developments by Élie Cartan. Lie, however, bitterly disputed Killing's claim to Lie algebras.

So, let's expand (1) out to linear order in the small parameter ε. Setting $\frac{\partial x^\mu}{\partial x'^\rho} = \delta^\mu_\rho - \varepsilon\partial_\rho\xi^\mu(x) + O(\varepsilon^2)$ and collecting terms of order ε, we find

$$g_{\mu\sigma}\partial_\rho\xi^\mu + g_{\rho\nu}\partial_\sigma\xi^\nu + \xi^\lambda\partial_\lambda g_{\rho\sigma} = 0 \tag{2}$$

Indeed, if we write the isometry condition (1) as $0 = g_{\rho\sigma}(x') - g'_{\rho\sigma}(x') = (g_{\rho\sigma}(x') - g_{\rho\sigma}(x)) + (g_{\rho\sigma}(x) - g'_{\rho\sigma}(x'))$, the expression in the first parentheses provides the $\xi\partial g$ term in (2), while the expression in the second parentheses gives the $g\partial\xi$ terms. Note also that we have already encountered this combination in chapter VI.5.

Using the definition of the covariant derivative, we could write (2) more compactly as

$$\xi_{\sigma;\rho} + \xi_{\rho;\sigma} = 0 \tag{3}$$

Here we use the semicolon notation introduced in chapter V.6.

One potential source of confusion for the beginner is that the isometry condition, (2) or (3), can be looked at in two different ways. Given a metric, we could solve the isometry condition for the vector ξ. Alternatively, we could be given a bunch of ξs and ask how these isometries restrict the metric.

A vector field $\xi(x)$ satisfying (2) or (3)—the two conditions are equivalent—is called a Killing vector field. I will often be sloppy and omit the word "field" when it is clearly understood from the context that ξ depends on x.

Indeed, you might recall that you already encountered this term in chapter V.4. There we learned that the metric for a general static isotropic spacetime has the two Killing vectors $\xi = (1, 0, 0, 0)$ and $\xi = (0, 0, 0, 1)$. Referring to (2), which in this case reduces to $\xi^\lambda \partial_\lambda g_{\rho\sigma} = 0$, we see that speaking of Killing vectors is just an extra fancy way of saying that the static isotropic metric in chapter V.4 does not depend on the coordinates t and φ.

But already back in chapter V.4, you might have seen that there is a theoretical issue. The Killing vectors were obvious, because we chose a nice symmetric form for the metric, but in principle, as already mentioned, the metric could have been presented to us in some poorly chosen coordinates. The condition (2) or (3) tells you how to find the Killing vectors. However, by now, you realize that usually we start out with the isometries we want, which then guide us to the form of the metric, rather than the other way around of having to find the isometries for a given metric.

For further theoretical analysis, (3) is more suitable, but in an actual search for Killing vectors, (2) is simpler. By the way, we know that the ordinary derivatives in (2) must metamorphose into covariant derivatives as in (3), since the existence of Killing vectors is a coordinate independent statement, and so the existence condition must transform properly.

Those readers with a long memory might make a connection with the concept of Lie derivative introduced in chapter V.6. Indeed, the condition (2) had already appeared in appendix 2 of chapter V.6, and, rather nicely, asserts that the Lie derivative of the metric in the direction of the Killing vector ξ vanishes:

$$\mathcal{L}_\xi g_{\mu\nu} = 0 \tag{4}$$

Killing vectors

In general, (2) may have a number of solutions; we then label the various Killing vectors $\xi^\mu_{(a)}$ by an index a. Obviously, any linear combination of Killing vectors $\sum_a c_a \xi^\mu_{(a)}$ is also a Killing vector. That (3) is a tensor equation ensures that the number of linearly independent Killing vector fields does not depend on the coordinates used, evidently, since isometry reflects an intrinsic property of the space.

As explained in chapter V.4, with any vector V^μ, we can associate the differential operators $V^\mu \partial_\mu$. Indeed, in more advanced treatments, Killing vectors are thought of as differential operators, namely as $\xi_{(a)} = \xi^\mu_{(a)} \partial_\mu$. Thus, given a set of Killing vectors, we can study the isometry algebra generated by commuting the $\xi_{(a)}$s with one another. You can see how close this is to Lie's idea. See appendix 5. No wonder there was controversy over priority.

Let us try out our formalism on a laughably simple case, Euclidean 3-space E^3. Then (2) simplifies to the 6 equations $\partial_x \xi^x = 0$, $\partial_x \xi^y + \partial_y \xi^x = 0$, and so forth. These are easily solved. For example, acting with ∂_x on the second of the preceding equations and using the first equation, we obtain $\partial_x^2 \xi^y = 0$, showing that the y-component of the Killing vector ξ^y can depend on x and z linearly but not on y. We find 6 Killing vectors: $\xi_{(1)} = (1, 0, 0)$, $\xi_{(2)} = (0, 1, 0)$, $\xi_{(3)} = (0, 0, 1)$, $\xi_{(4)} = (y, -x, 0)$, $\xi_{(5)} = (0, z, -y)$, and $\xi_{(6)} = (-z, 0, x)$. As we would expect, the isometry algebra consists of the usual Euclidean algebra generating translations and rotations. For example, $\xi^\mu_{(4)} \partial_\mu = y \frac{\partial}{\partial x} - x \frac{\partial}{\partial y}$ generates rotations about the z-axis, as discussed in detail in chapter I.3.

A slightly more involved case is the familiar S^2. Write out (2) with $g_{\theta\theta} = 1$, $g_{\varphi\varphi} = \sin^2\theta$:

$$\partial_\theta \xi^\theta = 0, \quad 2\sin^2\theta\,\partial_\varphi \xi^\varphi + \xi^\theta \partial_\theta \sin^2\theta = 0, \quad \partial_\varphi \xi^\theta + \sin^2\theta\,\partial_\theta \xi^\varphi = 0 \tag{5}$$

These are easily solved to give (writing the Killing vectors conveniently as differential operators $\xi^\mu_{(a)}\partial_\mu$)

$$\xi_{(1)} = \sin\varphi\,\frac{\partial}{\partial\theta} + \cot\theta\cos\varphi\,\frac{\partial}{\partial\varphi}, \quad \xi_{(2)} = \cos\varphi\frac{\partial}{\partial\theta} - \cot\theta\sin\varphi\frac{\partial}{\partial\varphi}, \quad \xi_{(3)} = \frac{\partial}{\partial\varphi} \tag{6}$$

As I said at the start of this chapter, you could have guessed $\xi_{(3)}$ easily, but $\xi_{(1)}$ and $\xi_{(2)}$ are less obvious. These generators should look familiar to the reader who has taken courses on electromagnetism and quantum mechanics. At any given point on the sphere, we can translate in two directions and rotate around one axis, hence the 3 Killing vectors.

How many Killing vectors can we have?

From these two elementary examples, it is clear that a D-dimensional Riemannian manifold can have at most $\frac{1}{2}D(D+1)$ Killing vectors, corresponding locally to D translations and $\frac{1}{2}D(D-1)$ rotations. In both examples, this maximum number is attained, 6 in the case of E^3, and 3 in the case of S^2.

The Smart Experimentalist exclaims: "Right! Forget about fancy proofs: D-dimensional Euclidean space has this many Killing vectors; how could a space possibly be more symmetric?" Go ahead, be my guest, prove it if you must.

A D-dimensional Riemannian manifold with $\frac{1}{2}D(D+1)$ Killing vectors is said to be maximally symmetric.

Some definitions. A space is homogeneous if there exist translational Killing vectors to take any point to any other point in its vicinity. A space is isotropic at a given point X if there exist rotational Killing vectors that leave the point X fixed, that is, $\xi^\mu(X) = 0$, whose derivatives $\xi_{\sigma;\rho}(X)$ span the basis of D-by-D antisymmetric matrices. This merely means that every possible rotation about X is an isometry, as befits our intuitive definition of isotropy. (To see this, regard the derivatives $\xi_{\sigma;\rho}(X)$ as matrices with indices $\sigma\rho$. Note that (3), that is, $\xi_{\sigma;\rho} = -\xi_{\rho;\sigma}$, implies that these matrices are antisymmetric. The statement that the space is isotropic means that the generator of any rotation about X can be written as a linear combination of $\xi_{\sigma;\rho}$ evaluated at X.) A homogeneous space isotropic about some point is obviously isotropic about all points and so is maximally symmetric. All these statements are fairly straightforward to prove.

For definiteness, we talk about D-dimensional spaces, but clearly, everything we say here can be applied to spacetimes.

Maximally symmetric spaces

Indeed, if we are told that the space is maximally symmetric, we have $\frac{1}{2}D(D+1)$ constraints on the Riemann curvature tensor, enough to fix it uniquely. We will prove this in appendix 4, but for now, let us take the lazy man's way and try to wing it.

A maximally symmetric space is utterly featureless, so to speak. Every point, every direction, looks the same. So, what could the Riemann curvature tensor be? There is only the metric tensor kicking around.

Confusio asks: "What about the derivative of the metric tensor?"

This is to be an equality between tensors, so, Confusio, we must have the covariant derivative, not the ordinary derivative. But $g_{\mu\nu;\lambda} = 0$. So, only the metric tensor is available.

To construct the curvature tensor, which carries 4 indices, we need something like $g_{\tau\mu}g_{\rho\nu}$. Taking into account the many symmetry properties of the curvature tensor, we see that it must be given by

$$R_{\tau\rho\mu\nu} = K(g_{\tau\mu}g_{\rho\nu} - g_{\tau\nu}g_{\rho\mu}) \tag{7}$$

with K some constant. We will prove in appendix 4 that this is correct. Later, in chapters IX.7 and IX.8, we will see that, in the language of differential forms, the Riemann curvature looks even simpler.

Summing over pairs of indices, we have

$$R_{\rho\nu} = (D - 1)Kg_{\rho\nu}$$
$$R = D(D - 1)K \tag{8}$$

Maximally symmetric spaces have constant scalar curvature. This holds for the two almost trivial examples we know, Euclidean spaces and spheres.

Killing vectors and conservation laws

Physicists love isometries for another reason: Killing vectors are associated with conservation laws, a fact we have exploited and remarked upon several times, notably in chapter V.4. We can now easily prove this connection between symmetry and conservation, which goes way back to Noether's theorem in chapter II.4. Consider a geodesic described by $X^{\mu}(\tau)$, with the tangent or velocity vector $V^{\mu}(\tau) = \frac{dX^{\mu}}{d\tau}$. Let $\xi(x)$ be a Killing vector field of the spacetime. Then $\xi_{\mu}(X^{\mu})V^{\mu}$ is conserved along the geodesic.

To prove this, act with the covariant derivative along the geodesic (recall appendix 1 of chapter V.6) on the alleged conserved quantity:

$$V^{\nu}D_{\nu}\left(\xi_{\mu}V^{\mu}\right) = V^{\nu}V^{\mu}\xi_{\mu;\nu} + \xi_{\mu}\left(V^{\nu}D_{\nu}V^{\mu}\right) = V^{\nu}V^{\mu}\xi_{\mu;\nu} = 0 \tag{9}$$

We used the product rule for the first equality and the geodesic equation for the second equality. The isometry condition (3) tells us that $\xi_{\mu;\nu}$ is antisymmetric in $\mu\nu$, hence the last equality.

What isometry is all about: Would the view stay the same?

I conclude by giving you an intuitive account of what isometry is all about. If you find yourself in an unfamiliar landscape, you might want to ask yourself, "If I move a bit in that direction, would the view stay the same? If I turn around a bit, would the view stay the same?" To find out about the landscape, you look around. If the view stays the same as you turn around, the space is isotropic. If the view stays the same as you move without turning, the space is homogeneous.

Hiking in a maximally symmetric space would be kind of boring. But in fact, the spatial part of our universe appears to be maximally symmetric, as we saw in chapters V.3 and VIII.1. Indeed, the future of our universe might approximate a maximally symmetric spacetime. Or perhaps maximally symmetric spaces, like the sphere (and the de Sitter and anti de Sitter spacetimes we will study in chapters IX.10 and IX.11, respectively), represent what we can handle analytically.

A quick summary of this chapter. You are given a metric $g_{\mu\nu}(x)$. The prickly issue here is the freedom to change coordinates, so that the intrinsic qualities of the space may be totally obscured in the metric, which may look like a mess merely because of a poor choice of coordinates. The isometry condition (2) or (3) tells you the different ways you can move or turn without changing the view. Each of these ways is associated with a Killing vector, the number of which is limited by the dimension of the space. If we have the maximum number of possible Killing vectors, then, as you might expect, the space is pretty much featureless, so that the Riemann curvature tensor is completely fixed up to an overall constant K, and all curvature invariants, such as the scalar curvature R or the square of the curvature tensor $R_{\tau\rho\mu\nu}R^{\tau\rho\mu\nu}$, are all constants given in terms of K.

Appendix 1: Coset manifolds

This might be a good place to mention the concept of coset manifolds. Start with a Lie group G and a subgroup H of G. Let us say that two group elements g_1 and g_2 are equivalent if there exists an element h of H such that $g_1 = g_2 h$. Equivalence classes are then defined in the usual way: g_1 and g_2 belong to the same equivalence class if they are equivalent. The next step is to define a space or manifold by associating each equivalence class with a point. The resulting manifold is known as the coset manifold G/H.

As an example, the coset manifold $SO(3)/SO(2)$ is the familiar sphere S^2. Why? Let us go slow here. Every point P on the sphere S^2 is uniquely associated with a unit vector $\hat{u}$ pointing from the center of the sphere to P. Denote by $\hat{z}$ the unit vector pointing to the north pole, that is, the unit vector pointing in the z direction. Our first thought might be to associate the point P with the rotation g (that is, an element of $G = SO(3)$) that rotates $\hat{z}$ into $\hat{u}$. The problem is that the rotation g is not uniquely determined. Denote by $H = SO(2)$ the subgroup of G consisting of all rotations about the z-axis. Then two rotations g_1 and g_2 related by $g_1 = g_2 h$, with h an arbitrary element of H, would both rotate $\hat{z}$ into $\hat{u}$. In other words, $\hat{u} = g_1\hat{z} = g_2 h\hat{z} = g_2\hat{z}$. Thus, the point P is not to be associated with the element g_1, but with the entire equivalence class g_1 belongs to. In other words, P does not specify uniquely the rotation that would take $\hat{z}$ into the direction vector $\hat{u}$ associated with P.

Incidentally, the notation G/H is apt; we consider the elements g of G but with h factored out, so to speak.

In our example, we need 3 parameters to specify an $SO(3)$ rotation, and 1 parameter to specify an $SO(2)$ rotation. Hence, we need $3 - 1 = 2$ parameters to specify the equivalence classes and hence the points P in G/H. Indeed, S^2 is 2-dimensional. In general, the dimension[1] of G/H is equal to the $n(G) - n(H)$, with $n(G)$ and $n(H)$ the number of generators for the groups G and H, respectively.

This discussion can be immediately extended: for example, $SO(4)/SO(3) = S^3$. More generally, the sphere S^d can be identified as the coset manifold $SO(d+1)/SO(d)$ with dimension $\frac{1}{2}(d+1)d - \frac{1}{2}d(d-1) = d$. We will come across this coset construction again when we discuss de Sitter and anti de Sitter spacetimes in chapters IX.10 and IX.11.

Appendix 2: Hyperbolic spaces as coset manifolds

We first encountered the hyperbolic spaces H^n in chapter I.6. Given what you just learned about the spheres S^n, perhaps it is not a huge surprise that the hyperbolic spaces are also coset manifolds: $H^n = SO(n, 1)/SO(n)$.

Let us verify this explicitly for $n = 2$. The group $G = SO(2, 1)$ is the Lorentz group for $(2 + 1)$-dimensional Minkowski spacetime. The role of the north pole is played by $\hat{t} = (1, 0, 0)$ (the column vector written as a row vector for typographical reasons), left invariant by the subgroup $H = SO(2)$ consisting of rotations in the $(2 + 1)$-dimensional Minkowski spacetime. Thus, the coset manifold is generated by acting with boosts on $\hat{t}$; in other words, using the notation of appendix 3 of chapter I.6, we have $(W, X, Y) = (\cosh \psi, \sinh \psi \cos \theta, \sinh \psi \sin \theta)$. Then $ds^2 = dX^2 + dY^2 - dW^2 = d\psi^2 + \sinh^2 \psi d\theta^2$ in agreement with what we had in chapter I.6.

Appendix 3: From Killing vectors to Lie algebra

In chapter V.6, we introduced the commutator between two vector fields and also the Lie derivative. Let us apply what we learned there to the Killing vectors. Take the Killing vectors for the sphere given in (6) and calculate their commutators. For example, $[\xi_{(3)}, \xi_{(1)}] = [\frac{\partial}{\partial \varphi}, \sin \varphi \frac{\partial}{\partial \theta} + \cot \theta \cos \varphi \frac{\partial}{\partial \varphi}] = \xi_{(2)}$. You can verify that in fact, $[\xi_{(a)}, \xi_{(b)}] = \epsilon_{abc} \xi_{(c)}$, with ϵ_{abc} the antisymmetric symbol defined by $\epsilon_{123} = 1$. No wonder Lie was upset about Killing.

On a coset manifold, the isometry group is evidently just G. Then the discussion above goes through with ϵ_{abc} replaced by the structure constants f_{abc} of the Lie algebra of the group G, not surprisingly. In other words, on the coset manifold G/H, the Killing vectors satisfy

$$[\xi_{(a)}, \xi_{(b)}] = f_{abc} \xi_{(c)} \tag{10}$$

The reader who did not quite follow this need not worry; we will only use this fact in passing in chapter X.1.

An important special case occurs when H is the trivial subgroup consisting of only the identity element. Then G/H is just the group manifold of G: g_1 and g_2 are equivalent only if they are the same element. Each point on the group manifold corresponds to a distinct element of G.

Appendix 4: Constraints on the Riemann curvature tensor

We are now going to do what we postponed doing in the text, namely analyze the conditions (2) and (3) in detail. If you are seeing this for the first time, the analysis might seem a bit involved, bristling with indices. It is okay to skip this appendix.

For any vector V, you derived in exercise VI.1.4 that $V_{\rho;\mu;\nu} - V_{\rho;\nu;\mu} + V_{\nu;\rho;\mu} - V_{\nu;\mu;\rho} + V_{\mu;\nu;\rho} - V_{\mu;\rho;\nu} = 0$. For V, we now take a Killing vector. Since the Killing vector obeys (3), $\xi_{\rho;\mu} = -\xi_{\mu;\rho}$, the 6 terms in the identity reduce to 3 terms:

$$\xi_{\rho;\mu;\nu} - \xi_{\rho;\nu;\mu} - \xi_{\nu;\mu;\rho} = 0 \tag{11}$$

Using the defining expression for the curvature tensor $V_{\rho;\mu;\nu} - V_{\rho;\nu;\mu} = -R^\sigma{}_{\rho\mu\nu} V_\sigma$, we then obtain

$$\xi_{\mu;\rho;\nu} = -R^\sigma{}_{\nu\rho\mu} \xi_\sigma \tag{12}$$

Suppose we know the curvature tensor. Then, given a Killing vector $\xi_\rho(X)$ and its first derivative $\xi_{\rho;\mu}(X) = -\xi_{\mu;\rho}(X)$ at some point X, we know its second derivative, thanks to (12), and, by repeatedly differentiating (12), all of its higher derivatives. (We are of course talking about covariant derivatives.) Hence, we can construct $\xi_\rho(x)$ as a Taylor series in $(x^\mu - X^\mu)$. The result, $\xi_{(a)\rho}(x)$, for some specific a, evidently depends linearly on the $D + \frac{1}{2}D(D - 1) = \frac{1}{2}D(D + 1)$ initial values $\xi_{(a)\rho}(X)$ and $\xi_{(a)\rho;\mu}(X)$. It follows that the number of linearly independent $\xi_{(a)\rho}(x)$ cannot exceed $\frac{1}{2}D(D + 1)$. The maximum number of Killing vectors is equal to 3, 6, 10, for $D = 2, 3, 4$, respectively, as is clear intuitively.

In exercise VI.1.5, you derived $T_{\mu\rho;\nu;\omega} - T_{\mu\rho;\omega;\nu} = -(R^\sigma{}_{\mu\nu\omega} T_{\sigma\rho} + R^\sigma{}_{\rho\nu\omega} T_{\mu\sigma})$ for any tensor $T_{\mu\rho}$. Now apply this to $\xi_{\mu;\rho}$:

$$\xi_{\mu;\rho;\nu;\omega} - \xi_{\mu;\rho;\omega;\nu} = -\left(R^\sigma{}_{\mu\nu\omega} \xi_{\sigma;\rho} + R^\sigma{}_{\rho\nu\omega} \xi_{\mu;\sigma}\right) \tag{13}$$

But we can also compute the left hand side of (13) by differentiating (12): $\xi_{\mu;\rho;\nu;\omega} = -\left(R^\sigma{}_{\nu\rho\mu;\omega} \xi_\sigma + R^\sigma{}_{\nu\rho\mu} \xi_{\sigma;\omega}\right)$. Plug this into (13), and we get a longish equation

$$R^\sigma{}_{\mu\nu\omega} \xi_{\sigma;\rho} + R^\sigma{}_{\rho\nu\omega} \xi_{\mu;\sigma} = \left(R^\sigma{}_{\nu\rho\mu;\omega} \xi_\sigma + R^\sigma{}_{\nu\rho\mu} \xi_{\sigma;\omega}\right) - \left(R^\sigma{}_{\omega\rho\mu;\nu} \xi_\sigma + R^\sigma{}_{\omega\rho\mu} \xi_{\sigma;\nu}\right) \tag{14}$$

involving the curvature tensor and its derivatives, and the Killing vector and its derivatives.

Let's step back for a moment and think how best to organize the mess. We will collect all the terms involving ξ_σ on one side and all those involving $\xi_{\sigma;\kappa}$ on the other:

$$\left(R^\sigma_{\;\omega\rho\mu;\nu} - R^\sigma_{\;\nu\rho\mu;\omega}\right)\xi_\sigma = R^\sigma_{\;\mu\nu\omega}\xi_{\sigma;\rho} + R^\sigma_{\;\rho\nu\omega}\xi_{\mu;\sigma} - R^\sigma_{\;\nu\rho\mu}\xi_{\sigma;\omega} + R^\sigma_{\;\omega\rho\mu}\xi_{\sigma;\nu}$$

$$= \left(R^\sigma_{\;\mu\nu\omega}\delta^\kappa_\rho - R^\sigma_{\;\rho\nu\omega}\delta^\kappa_\mu - R^\sigma_{\;\nu\rho\mu}\delta^\kappa_\omega + R^\sigma_{\;\omega\rho\mu}\delta^\kappa_\nu\right)\xi_{\sigma;\kappa} \tag{15}$$

This way of collecting terms is possible, because ξ is not just any garden variety vector, but the very special Killing vector. In particular, due to (3), we can convert $\xi_{\mu;\sigma}$ in (13) to $-\xi_{\sigma;\mu}$. Incidentally, equations of this type in differential geometry are in fact much less forbidding than they look if we remain cognizant of various symmetry properties of the expressions involved; for example, that (13) is antisymmetric in ($\nu \leftrightarrow \omega$).

We study (15) in the following appendix.

Appendix 5: Maximal symmetry fixes the curvature tensor

We can exploit (15) by using our knowledge of the Killing vectors to place a powerful constraint on the curvature tensor. Note that for each Killing vector we know about, we have one constraint on the curvature tensor.

At a given point, we can take ξ to be a translation-type Killing vector, for which $\xi_{\sigma;\kappa}$ vanishes at that point. Since we have D linearly independent ξs, the coefficient of ξ_σ in (15) must vanish for each value of σ:

$$R^\sigma_{\;\omega\rho\mu;\nu} = R^\sigma_{\;\nu\rho\mu;\omega} \tag{16}$$

If we take ξ to be a rotation-type Killing vector, then ξ_σ vanishes at that point, and thus the left hand side of (15) vanishes. Since by definition, $\xi_{\sigma;\kappa}$ regarded as a matrix spans the basis of D-by-D antisymmetric matrices, the vanishing of the right hand side of (15) forces the coefficient of $\xi_{\sigma;\kappa}$ to be symmetric under ($\sigma \leftrightarrow \kappa$):

$$R^\sigma_{\;\mu\nu\omega}\delta^\kappa_\rho - R^\sigma_{\;\rho\nu\omega}\delta^\kappa_\mu - R^\sigma_{\;\nu\rho\mu}\delta^\kappa_\omega + R^\sigma_{\;\omega\rho\mu}\delta^\kappa_\nu = R^\kappa_{\;\mu\nu\omega}\delta^\sigma_\rho - R^\kappa_{\;\rho\nu\omega}\delta^\sigma_\mu - R^\kappa_{\;\nu\rho\mu}\delta^\sigma_\omega + R^\kappa_{\;\omega\rho\mu}\delta^\sigma_\nu \tag{17}$$

We now have all the information we need from (17). Contract κ with ρ and obtain

$$R^\sigma_{\;\mu\nu\omega}D - R^\sigma_{\;\mu\nu\omega} - R^\sigma_{\;\nu\omega\mu} + R^\sigma_{\;\omega\nu\mu} = R^\sigma_{\;\mu\nu\omega} - R_{\nu\mu}\delta^\sigma_\omega + R_{\omega\mu}\delta^\sigma_\nu$$

where we used $R^\rho_{\;\rho\nu\omega} = 0$. Invoking the cyclic identity in exercise VI.1.3, we obtain $(D-1)R^\sigma_{\;\mu\nu\omega} = R_{\omega\mu}\delta^\sigma_\nu - R_{\nu\mu}\delta^\sigma_\omega$, or, after lowering indices, $(D-1)R_{\sigma\mu\nu\omega} = R_{\omega\mu}g_{\sigma\nu} - R_{\nu\mu}g_{\sigma\omega}$. Contracting with $g^{\sigma\nu}$, we find that $R_{\sigma\nu} = \frac{1}{D}Rg_{\sigma\nu}$. Inserting back, we learn that the Riemann curvature tensor is given by $R_{\sigma\mu\nu\omega} = \frac{R}{D(D-1)}(g_{\omega\mu}g_{\sigma\nu} - g_{\nu\mu}g_{\sigma\omega})$. We are almost there: we will have (7) if we can show that R is necessarily a constant. Intuitively, that is more or less obvious, because a maximally symmetric space is featureless (every other point is as good as every other point), and R is a scalar independent of coordinate choice.

We have one more card up our sleeves, the Bianchi identity $(R^{\sigma\nu} - \frac{1}{2}Rg^{\sigma\nu})_{;\nu} = 0$. Inserting $R_{\sigma\nu} = \frac{1}{D}Rg_{\sigma\nu}$, we obtain $(\frac{1}{D} - \frac{1}{2})\partial_\nu R = 0$. Thus, for $D \neq 2$, the scalar curvature R is a constant, which we write conventionally as $R = D(D-1)K$, and we are done.

The special case $D = 2$ is easily dispatched, since the curvature tensor then has only one component, R_{1212}. Indeed, according to exercise VI.1.6, regardless of whether the space is maximally symmetric, the Riemann curvature tensor always has the form $R_{\sigma\mu\nu\omega} = \frac{R}{2}(g_{\omega\mu}g_{\sigma\nu} - g_{\nu\mu}g_{\sigma\omega})$. But for a maximally symmetric space, we have another equation up our sleeves we haven't used, namely (16). Plugging the form of the curvature tensor into (16), we find that R is also constant for a $D = 2$ maximally symmetric space.

So, for a maximally symmetric space of any dimension, we have, as we had more or less guessed, the highly appealing result $R_{\tau\rho\mu\nu} = K(g_{\tau\mu}g_{\rho\nu} - g_{\tau\nu}g_{\rho\mu})$.

Appendix 6: Form invariant tensors

We can apply a condition analogous to (1) to tensors other than the metric tensor. We say that a tensor $T_{\mu\nu\cdots\omega}$ is form invariant if

$$T_{\rho\sigma\cdots\tau}(x') = T_{\mu\nu\cdots\omega}(x)\frac{\partial x^\mu}{\partial x'^\rho}\frac{\partial x^\nu}{\partial x'^\sigma}\cdots\frac{\partial x^\omega}{\partial x'^\tau} \tag{18}$$

The infinitesimal form reads

$$T_{\mu\sigma\ldots\tau}\partial_\rho\xi^\mu + T_{\rho\nu\ldots\tau}\partial_\sigma\xi^\nu + \cdots + \xi^\lambda\partial_\lambda T_{\rho\sigma\ldots\tau} = 0 \tag{19}$$

This has the same form as (2): a sum of terms involving the derivative of $\xi^{\cdot}$ is equal to a term involving the derivative of $T_{\ldots}$. With this definition, the existence of isometry amounts to requiring that the metric be form invariant.

The concept of a form invariant tensor makes sense regardless of whether the space itself is maximally symmetric. If the space is maximally symmetric, a tensor that satisfies (19) for all $\frac{1}{2}D(D+1)$ Killing vectors is said to be maximally form invariant. This is clearly quite restrictive; see exercise 3.

Exercises

1 Solve for the Killing vectors on the sphere.

2 Evaluate the 3 Killing vectors ξ in (6) at various points on the sphere.

3 Establish the following facts for a maximally symmetric space. (a) A maximally form invariant scalar is necessarily constant. (b) There is no maximally form invariant vector for $D \neq 1$. (c) For $D \neq 2$, a maximally form invariant 2-indexed tensor $T_{\mu\nu}$ must be equal to a constant times the metric tensor $g_{\mu\nu}$, as you might expect. Intuitively, a maximally symmetric space is completely featureless (think of the familiar sphere). So what can the maximally form invariant vector possibly depend on? Answer the same question for the maximally form invariant 2-indexed tensor.

Note

1. Coset manifolds also enter into the concept of spontaneous symmetry breaking in quantum field theory, and the dimension of G/H has to do with the number of Nambu-Goldstone bosons. See *QFT Nut*, p. 229.

Many legs

I now teach you a fancier, but in fact easier, method of calculating curvature than the traditional method given in chapter VI.1. Namely, I will tell you about vielbein (German[1] for "many legs": dreibein = three legs, vierbein[2] = four legs, and so on) and differential forms, which definitely do not belong to the "fancy but useless" category in which I file away many things at the more mathematical end of theoretical physics. Indeed, both the vielbein and differential forms have their origins in humble physical considerations.

At this point, Professor Flat ambles by again, mumbling, "Even though the world is round, locally we can still erect orthonormal frames* of reference, Descartes's good old x-, y-, and z-axes, now called legs."

You and I respond, "Yes, professor, we have learned that Riemannian manifolds are locally flat. Everyday life is flat."

The idea is to write the metric as

$$g_{\mu\nu}(x) = \eta_{\alpha\beta} e^{\alpha}_{\mu}(x) e^{\beta}_{\nu}(x) \tag{1}$$

As Professor Flat explained in chapter I.6 (in our demonstration that we can always choose locally flat coordinates at a given point x), in the first step we turn $g_{\mu\nu}(x)$ into $\eta_{\alpha\beta}$ by a similarity transformation. In (1), we have merely denoted the matrix that appears in the similarity transformation by $e^{\alpha}_{\mu}(x)$.

The metrics we commonly encounter are so simple that we can even write down $e^{\alpha}_{\mu}(x)$ by inspection. I hasten to give an example: the familiar 2-sphere with $ds^2 = d\theta^2 + \sin^2\theta \, d\varphi^2$ (that is, with $g_{\theta\theta} = 1$, $g_{\varphi\varphi} = \sin^2\theta$), from which we read off $e^1_\theta = 1$ and $e^2_\varphi = \sin\theta$ (all other components are zero).[3] In (1), the indices α and β, called Lorentz indices, take on d values

---

* Indeed, we have already encountered orthonormal frames when we discussed Fermi normal coordinates in chapter IX.3.

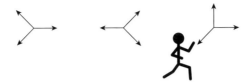

Figure 1 Running around erecting orthonormal frames.

in d-dimensional spacetime. The usual indices μ, ν are called world indices to distinguish them from the Lorentz indices. Even though the two kinds of indices are conceptually quite different (see below), we can still think of the square array $e(x)$ numerically as a square matrix. Then we can write (1) as a matrix equation $g = e^T \eta e$, and, as was just mentioned, think of e as a similarity transformation that diagonalizes $g_{\mu\nu}$ and scales it to the unit matrix.

Since we are physicists, most of the manifolds we deal with will be locally Minkowskian, and hence the Minkowski metric $\eta_{\alpha\beta}$ appears here. If the manifold is locally Euclidean, I should have written $g_{ij} = \delta_{ab} e_i^a e_j^b$, but I am not that fussy. For both cases, I will use Greek rather than Latin letters for the indices, and trust you, when appropriate, to replace the Minkowski metric by the Euclidean metric in your head. In either case, we picture a little man running around erecting orthonormal frames (dreibein, as shown in figure 1). If the metric $g_{\mu\nu}$ describes the universe, all we are doing is setting up local Lorentz frames at each point in spacetime.

Lorentz indices versus world indices

Lorentz indices α, β, $\cdots$ are contracted with the Minkowski metric $\eta_{\alpha\beta}$ (or the Euclidean metric, as the case may be), which, consisting of ±1s and 0s, is much easier to deal with than $g_{\mu\nu}$. (That is the point of the formalism!) In contrast, world indices μ, ν, $\cdots$ are contracted with the metric $g_{\mu\nu}$, as usual. As $g_{\mu\nu}(x)$ varies from point to point, so does the vielbein $e_\mu^\alpha(x)$, as indicated in the figure. (Note that we use the Greek letters early in the alphabet for Lorentz indices and the Greek letters later in the alphabet for world indices.)

The transformation of the metric

$$g'_{\lambda\sigma}(x') = g_{\mu\nu}(x)\frac{\partial x^\mu}{\partial x'^\lambda}\frac{\partial x^\nu}{\partial x'^\sigma} \tag{2}$$

under a coordinate transformation translates into

$$e'^\alpha_\lambda(x') = e^\alpha_\mu(x)\frac{\partial x^\mu}{\partial x'^\lambda} \tag{3}$$

Verify this by calculating $g'_{\lambda\sigma}(x') = \eta_{\alpha\beta} e'^\alpha_\lambda(x') e'^\beta_\sigma(x')$.

In a sense, the vielbein represents the square root* of the metric. Taking the determinant of (1), we have $-g = e^2$, where e denotes the determinant of e^α_μ regarded as a matrix, and thus the pesky square root in the volume factor $\sqrt{-g} = e$ goes away.

In taking an ordinary square root, we are free to take either the plus or minus sign. Similarly, we are free to Lorentz transform (rotate, in the Euclidean case) the vielbein: if you use the vielbein e^α_μ, I am free to use some other vielbein $\tilde{e}^\alpha_\mu$ instead, as long as mine is related to yours by a Lorentz transformation

$$e^\alpha_\mu(x) = \Lambda^\alpha_{\ \beta}(x)\tilde{e}^\beta_\mu(x) \tag{4}$$

Let us check that indeed $g_{\mu\nu}(x) = e^\alpha_\mu(x)\eta_{\alpha\beta}e^\beta_\nu(x) = \tilde{e}^\alpha_\mu(x)\eta_{\alpha\beta}\tilde{e}^\beta_\nu(x) = g_{\mu\nu}(x)$ if $\Lambda^T\eta\Lambda = I$. Note that the transformation in (4), in contrast to that in (3), leaves x untouched. What we are discussing here is not a coordinate transformation, but the freedom to orient our vielbein whichever way we like. To emphasize this, I have used twiddle instead of prime. (Also, note that Λ here can depend on x, unlike the discussion in special relativity; in field theory, this is known as a local or gauge transformation.)

Again thinking of $e(x)$ as a square matrix, we can consider its inverse, which we write as $e^\nu_\alpha(x)$. The standard result from linear algebra states that the left and right inverses of a nonsingular matrix are the same, and hence we have $e^\nu_\alpha(x)e^\beta_\nu(x) = \delta^\beta_\alpha$ and $e^\nu_\alpha(x)e^\alpha_\mu(x) = \delta^\nu_\mu$. For diagonal $g_{\mu\nu}$, $e^\alpha_\mu(x)$ is also diagonal, and the inverse vielbein $e^\nu_\alpha(x)$ may be written down by inspection. For example, for the 2-sphere, $e^\theta_1 = 1$ and $e^\varphi_2 = \frac{1}{\sin\theta}$ (all other components are zero).

The transformation properties (3) and (4) of the vielbein show that it straddles the domain of the Lorentz indices and the domain of the world indices. The vielbein can be used to convert one type of index to the other type. For example, given a world vector $J^\mu(x)$, we can construct $J^\alpha(x) = e^\alpha_\mu(x)J^\mu(x)$, which is in fact a Lorentz vector, as you are invited to verify. Under a local Lorentz transformation, $J^\alpha(x)$ transforms as a Lorentz vector, but under a coordinate transformation, it transforms as a world scalar. Similarly, given the Riemann curvature tensor $R^\sigma_{\ \rho\mu\nu}$, we can form $R^\gamma_{\ \delta\alpha\beta} = R^\sigma_{\ \rho\mu\nu}e^\gamma_\sigma e^\rho_\delta e^\mu_\alpha e^\nu_\beta$. We can of course also consider mixed objects, such as $R^\gamma_{\ \delta\mu\nu} = R^\sigma_{\ \rho\mu\nu}e^\gamma_\sigma e^\rho_\delta$. Again, keep in mind the distinction between early indices, such as γ and δ, and late indices, such as μ and ν.

Differential forms

I now introduce the language of differential forms. Fear not, we will need only a few elementary concepts. Let x^μ be d real variables (thus, the index μ takes on d values) and $A_\mu(x)$ be d functions of the xs. (In this purely mathematical section, we do not for the moment specify what A_μ is.) We could discuss everything in the abstract like mathematicians, but in fact, in our applications, x^μ represent coordinates, and as we will see, forms have natural geometric interpretations.

---

* Readers of my field theory book will recognize that this is one of three ways of the warrior theorist. See "New Closing Words" in *QFT Nut*, p. 522.

We call the object $A \equiv A_\mu dx^\mu$ a 1-form. The differentials dx^μ are treated following Newton and Leibniz. If we change coordinates $x \to x'$, then as usual, $dx^\mu = \frac{\partial x^\mu}{\partial x'^\nu} dx'^\nu$. As we said in chapter I.3, dx^μ is the ur-vector. (If you confuse the d here with the d in the preceding paragraph, you are in trouble!)

The form A does not carry any indices and so does not transform. Indeed, we insert the transformation of dx^μ just mentioned to obtain $A \equiv A_\mu dx^\mu = A_\mu \frac{\partial x^\mu}{\partial x'^\nu} dx'^\nu \equiv A'_\nu dx'^\nu$. The last step defines A'_ν and thus reproduces the standard transformation law of vectors with a lower index under coordinate transformation: $A'_\nu = A_\mu \frac{\partial x^\mu}{\partial x'^\nu}$. Hence $A_\mu(x)$ is a vector field. The important point here is that the form A does not depend on any specific coordinate system: it is coordinate free. But you are welcome to express it as a linear combination of dx^μ in a coordinate system of your choice.

It was already explained in part I that A_μ transforms like the dual ur-vector ∂_μ. As an example, consider the 1-form $A = \cos\theta \, d\varphi$. Regarding θ and φ as angular coordinates on the 2-sphere, we have $A_\theta = 0$ and $A_\varphi = \cos\theta$. Note that the natural union of A_μ and dx^μ, a marriage made in heaven so to speak, was already foreordained in chapter I.5.

Similarly, we define a p-form as $H = \frac{1}{p!} H_{\mu_1\mu_2\cdots\mu_p} dx^{\mu_1} dx^{\mu_2} \cdots dx^{\mu_p}$. (Repeated indices are summed, as always.) The degenerate example is that of a 0-form, call it Λ, which is just a scalar function of the coordinates x^μ. An example of a 2-form is $F = \frac{1}{2!} F_{\mu\nu} dx^\mu dx^\nu$. We can add two p-forms together in the obvious way:

$$H + K = \frac{1}{p!} \left(H_{\mu_1\mu_2\cdots\mu_p} + K_{\mu_1\mu_2\cdots\mu_p} \right) dx^{\mu_1} dx^{\mu_2} \cdots dx^{\mu_p}$$

In contrast, we cannot add a p-form to a q-form unless $p = q$. But we can naturally multiply any two forms together: for example,

$$AF = \left(A_\lambda dx^\lambda \right) \left(\frac{1}{2!} F_{\mu\nu} dx^\mu dx^\nu \right) = \frac{1}{2!} A_\lambda F_{\mu\nu} dx^\lambda dx^\mu dx^\nu$$

The product of a p-form and a q-form is evidently a $(p + q)$-form. In the example just given, the product of a 1-form and a 2-form is a 3-form.

The reason for anticommuting

We now face the question of how to think about the products of differentials $dx^\mu dx^\nu$. In an elementary course on calculus, we learned that $dxdy$ represents the area of an infinitesimal rectangle with length dx and width dy. At that level, we more or less automatically regard $dydx$ as the same as $dxdy$. The order of writing the differentials does not matter.

Think, however, about making a coordinate transformation $(x, y) \to (x', y')$, with $x' = x'(x, y)$ and $y' = y'(x, y)$ two functions of the coordinates x and y. Now look at

$$dx'dy' = \left(\frac{\partial x'}{\partial x} dx + \frac{\partial x'}{\partial y} dy \right) \left(\frac{\partial y'}{\partial x} dx + \frac{\partial y'}{\partial y} dy \right) \tag{5}$$

Notice that the coefficient of $dxdy$ is $\frac{\partial x'}{\partial x} \frac{\partial y'}{\partial y}$ and that the coefficient of $dydx$ is $\frac{\partial x'}{\partial y} \frac{\partial y'}{\partial x}$. We see that things work out neatly if we treat the product between differentials as anticommuting, so that $dy dx = -dx dy$. Then the $dxdy$ and $dydx$ terms in (5) combine into

$$\left(\frac{\partial x'}{\partial x} \frac{\partial y'}{\partial y} - \frac{\partial x'}{\partial y} \frac{\partial y'}{\partial x} \right) dx dy$$

and we recognize the expression in parentheses as the determinant of the matrix

$$\begin{pmatrix} \frac{\partial x'}{\partial x} & \frac{\partial x'}{\partial y} \\ \frac{\partial y'}{\partial x} & \frac{\partial y'}{\partial y} \end{pmatrix}$$

namely, the Jacobian $J(x', y'; x, y)$ for the transformation $(x, y) \rightarrow (x', y')$. Furthermore, if the product of differentials anticommutes, we would have $dx dx = -dx dx = 0$, and similarly $dy dy = 0$. Consequently, (5) simplifies nicely to

$$dx' dy' = \left(\frac{\partial x'}{\partial x} \frac{\partial y'}{\partial y} - \frac{\partial x'}{\partial y} \frac{\partial y'}{\partial x} \right) dx dy = J(x', y'; x, y) dx dy \tag{6}$$

We obtain the correct Jacobian for transforming the area element $dx dy$ to the area element $dx' dy'$.

The product between differentials $dx^\mu dx^\nu$ is known as the wedge product and is written as $dx^\mu \wedge dx^\nu$ in many texts. We will omit the wedge—no need to clutter up the page, at least for our purposes. Alternatively, we can regard the differentials dx^μ formally as anticommuting objects,[4] so that by definition, $dx^\mu dx^\nu = -dx^\nu dx^\mu$.

The little exercise given above motivates the natural emergence of anticommutation in this context. Otherwise, it would appear to be totally arbitrary. Our little exercise also makes clear the geometrical origin of the "extra" sign. The area element $dx dy$ is directional: $dx dy$ and $dy dx$ span the same area but point in opposite directions. You can see that this makes sense by recalling, for example, your first encounter with the notion of the divergence of some vector field, for example, the divergence of a current $\vec{J}(\vec{x})$. You were taught to think of an infinitesimal cube and multiply the current by the area element on each of the six faces of the cube. To obtain $\vec{\nabla} \cdot \vec{J}$, you clearly have to treat the area elements on opposite faces as pointing in opposite directions.

The anticommuting property $dx^\mu dx^\nu = -dx^\nu dx^\mu$ indicates that in d-dimensions, we can have p-forms only for $p \leq d$.

The exterior derivative

We now define a differential operation d (known as the exterior derivative) to act on any form. Acting on a p-form H, it gives by definition a $(p + 1)$-form

$$dH = \frac{1}{p!} \partial_\nu H_{\mu_1 \mu_2 \cdots \mu_p} dx^\nu dx^{\mu_1} dx^{\mu_2} \cdots dx^{\mu_p}$$

Thus, $d\Lambda = \partial_\nu \Lambda dx^\nu$ and $dA = \partial_\nu A_\mu dx^\nu dx^\mu = \frac{1}{2}(\partial_\nu A_\mu - \partial_\mu A_\nu) dx^\nu dx^\mu$.

We see that this mathematical formalism is almost tailor made to describe* electromagnetism. If we call $A \equiv A_\mu dx^\mu$ the potential 1-form and think of A_μ as the

---

* For a simple formulation of Yang-Mills theory using differential forms, see *QFT Nut*.

electromagnetic potential, then $F = dA$ is in fact the field 2-form. If we were to write F out in terms of its components $F = \frac{1}{2!}F_{\mu\nu}dx^\mu dx^\nu$, then $F_{\mu\nu} = \partial_\nu A_\mu - \partial_\mu A_\nu$ is indeed the electromagnetic field. We see that the exterior derivative is just the sophisticate's name for what the common people would call the gradient or the curl.

Note that x^μ is not a form, and dx^μ is not d acting on a form.

If you like, you can think of differential forms as "merely" an elegantly compact notation. The point is to think of physical objects like A and F as entities, without having to commit to any particular coordinate system. This is particularly convenient when you have to deal with objects more complicated than A and F, for example in string theory (see appendix 2). By using differential forms, we avoid drowning in a sea of indices.

Consider the product of two 1-forms A and B. Now act with d on the 2-form $AB = A_\mu B_\nu dx^\mu dx^\nu$. We have

$$d(AB) = \partial_\lambda(A_\mu B_\nu)dx^\lambda dx^\mu dx^\nu = \left((\partial_\lambda A_\mu)B_\nu + A_\mu(\partial_\lambda B_\nu)\right)dx^\lambda dx^\mu dx^\nu$$
$$= (\partial_\lambda A_\mu)dx^\lambda dx^\mu B_\nu dx^\nu - A_\mu dx^\mu(\partial_\lambda B_\nu)dx^\lambda dx^\nu = (dA)B - A(dB)$$

The first equality comes from the definition of d; the second from the product rule in ordinary calculus; the third from writing out the previous expression and moving dx^λ past dx^μ in the second term; and finally, the fourth from grouping everything back into the forms. The important point here is the minus sign that appeared due to our moving dx^λ past dx^μ.

This result generalizes readily. Let A be a p-form and B a q-form. Then we have

$$d(AB) = (dA)B + (-1)^p A(dB) \tag{7}$$

Evidently, the sign $(-1)^p$ appears because we moved dx^λ past $dx^{\mu_1}dx^{\mu_2}\cdots dx^{\mu_p}$. (Note that q does not appear explicitly in (7).)

An important identity is

$$dd = 0 \tag{8}$$

This says that acting with d on any form twice gives zero. This fundamental identity is easy to prove by direct evaluation: $ddH = \frac{1}{p!}\partial_\lambda(\partial_\nu H_{\mu_1\mu_2\cdots\mu_p})dx^\lambda dx^\nu dx^{\mu_1}dx^{\mu_2}\cdots dx^{\mu_p} = 0$, since $dx^\lambda dx^\nu = -dx^\nu dx^\lambda$, while Newton and Leibniz told us that $\partial_\lambda\partial_\nu = \partial_\nu\partial_\lambda$. In particular, $dF = ddA = 0$. Writing this out in components, you will recognize this as a standard identity (the Bianchi identity) in electromagnetism.

Note also that the square of a p-form A satisfies $A^2 = (-1)^{p^2}A^2$. (Write $A^2 = AA = \frac{1}{p!}A_{\mu_1\mu_2\cdots\mu_p}dx^{\mu_1}dx^{\mu_2}\cdots dx^{\mu_p}\frac{1}{p!}A_{\nu_1\nu_2\cdots\nu_p}dx^{\nu_1}dx^{\nu_2}\cdots dx^{\nu_p}$ and mentally move the dx^μs past all the dx^νs. For each of them, we get a factor of $(-1)^p$. Thus, the stated result follows.) Therefore, for p odd, $A^2 = 0$.

Relating connection 1-forms

After this excursion into differential forms, I can at long last tell you Élie Cartan's formulation of the differential geometry of Riemannian manifolds. The transformation law (3) immediately suggests packaging the vielbein into a 1-form $e^\alpha = e^\alpha_\mu dx^\mu$ called, naturally, the vielbein 1-form. In our simple example, $e^1 = d\theta$ and $e^2 = \sin\theta d\varphi$. Carrying no world index, e^α should transform like a scalar under coordinate transformation. We can readily check this: $e'^\alpha(x') = e'^\alpha_\lambda(x')dx'^\lambda = e^\alpha_\mu(x)\frac{\partial x^\mu}{\partial x'^\lambda}\frac{\partial x'^\lambda}{\partial x^\nu}dx^\nu = e^\alpha_\mu(x)\delta^\mu_\nu dx^\nu = e^\alpha(x)$.

We studied infinitesimal rotations in chapter I.3 and found that the generators are antisymmetric matrices. Similarly, we studied infinitesimal Lorentz transformations in chapter III.3 and found that the generators are also antisymmetric, except that we have to watch out for signs when raising and lowering indices. Recall that we considered (with some suitable changes in notation) an infinitesimal transformation $\Lambda^\alpha_{\ \beta} \simeq \delta^\alpha_{\ \beta} + \varphi\omega^\alpha_{\ \beta}$. To leading order in φ, the condition that Λ is in fact a Lorentz transformation reduces to (see (III.3.13)) $\omega^\alpha_{\ \beta}\eta_{\alpha\gamma} + \eta_{\beta\alpha}\omega^\alpha_{\ \gamma} = 0$, that is, $\omega_{\gamma\beta} + \omega_{\beta\gamma} = 0$.

On a curved manifold, as we move from point x to a nearby point $x + dx$, we expect that the local frame will rotate or Lorentz transform, depending on whether the manifold is locally Euclidean or Minkowskian. In other words, an infinitesimal translation has the effect of rotating the form $e^\alpha(x)$ infinitesimally. Thus, the result of differentiating (that is, applying the exterior derivative d to) $e^\alpha(x)$ should be given by

$$de^\alpha = -\omega^\alpha_{\ \beta}e^\beta \tag{9}$$

for some antisymmetric $\omega_{\alpha\beta}$. (Here the minus sign is conventional and is part of the definition of ω.) Since e is a 1-form, it follows that de is a 2-form and hence

$$\omega^\alpha_{\ \beta} = \omega^\alpha_{\ \beta\mu}dx^\mu \tag{10}$$

is also a 1-form, known as the connection 1-form: it connects nearby frames. Given e, (9) enables us to determine ω. For the 2-sphere, $de^1 = 0$ and $de^2 = \cos\theta d\theta d\varphi$, and so the connection has only one nonvanishing component: $\omega^{12} = -\omega^{21} = -\cos\theta d\varphi$.

But you and I are free to choose whatever vielbein we like, as was already mentioned in connection with (4). Suppose that at a given point, my vielbein e^α is related to yours by $e^\alpha = \Lambda^\alpha_{\ \beta}\tilde{e}^\beta$. (It is worth emphasizing that this is merely a local Lorentz transformation, or rotation if we are in space rather than spacetime, of our orthonormal frame, not a coordinate transformation. Indeed, you can see from (1) that the metric is not changed.) Our connection 1-forms ω and $\tilde{\omega}$ better be related in such a way that (9) holds for both of us.

Suppressing indices, we plug $e = \Lambda\tilde{e}$ into (9), $de + \omega e = 0$, and plow ahead. Try doing it yourself. The first term becomes $d(\Lambda\tilde{e}) = \Lambda d\tilde{e} + (d\Lambda)\tilde{e} = -\Lambda\tilde{\omega}\tilde{e} + (d\Lambda)\Lambda^{-1}e$. Thus, requiring $d\tilde{e} + \tilde{\omega}\tilde{e} = 0$, we relate the two connection 1-forms by

$$\omega = \Lambda\tilde{\omega}\Lambda^{-1} - (d\Lambda)\Lambda^{-1} \tag{11}$$

Notice that ω does not transform as a Lorentz tensor due to the second term in (11). You should be reminded of the similar behavior of the Christoffel symbol under a coordinate transformation, as discussed in chapter V.6.

Cartan's formulation of Riemannian manifolds

The local curvature of the manifold is a measure of how the connection varies from point to point. Thus, we expect curvature to be given by something like $R^\alpha{}_\beta \sim d\omega^\alpha{}_\beta$, a 2-form. But we would like the curvature to transform nicely, as a Lorentz tensor (since $R^\alpha{}_\beta$ carries two Lorentz indices) under the local transformation Λ. But you can also see, just by looking at (11), that $d\omega^\alpha{}_\beta$ is not going to transform nicely (indeed, see (12) below).

Just about the only possibility is to add another 2-form to $d\omega$. We are severely limited in our choices, since we have available only the 1-forms e and ω. Looking at how the different possibilities transform, we easily arrive at the correct choice, namely ω multiplied by itself, that is, ω^2.

Confusio exclaims, "Wait! I thought that we showed that the square of a 1-form vanishes."

Ah, the point is that ω is a matrix 1-form: $\omega^\alpha{}_\beta = \omega^\alpha{}_{\beta\mu} dx^\mu$. Thus,

$$\omega^\alpha{}_\beta \omega^\beta{}_\gamma = \omega^\alpha{}_{\beta\mu} \omega^\beta{}_{\gamma\nu} dx^\mu dx^\nu = \frac{1}{2} \left(\omega^\alpha{}_{\beta\mu} \omega^\beta{}_{\gamma\nu} - (\mu \leftrightarrow \nu) \right) dx^\mu dx^\nu$$

which has no particular reason to vanish. Another way of saying this is to think of $\omega^\alpha{}_{\beta\mu}$ as d matrices $\hat\omega_\mu$ with matrix indices $\alpha\beta$. Then, suppressing the α and β indices, we can write what we just wrote as $\omega^2 = \frac{1}{2}[\hat\omega_\mu, \hat\omega_\nu] dx^\mu dx^\nu$. There is no reason for the matrix commutator to vanish. Note that this is consistent with what Confusio said. Were $\hat\omega_\mu$ not a matrix, ω would indeed vanish.

The upshot of all this is that the desired curvature 2-form can only be the sum of $d\omega$ and ω^2 with some relative coefficient, which turns out to be 1 (see below). Thus, we obtain the curvature 2-form $R = d\omega + \omega^2$. Restoring indices, we have $R^\alpha{}_\beta = d\omega^\alpha{}_\beta + \omega^\alpha{}_\gamma \omega^\gamma{}_\beta$.

We now check that $R = \Lambda \tilde{R} \Lambda^{-1}$ does transform nicely. This is one of the most famous calculations in physics history; I will do it, but you should try to do it before reading on.

Well, just plug (11) into $R = d\omega + \omega^2$ and plow ahead. First note that $0 = d(\Lambda\Lambda^{-1}) = (d\Lambda)\Lambda^{-1} + \Lambda d\Lambda^{-1}$, and so $d\Lambda^{-1} = -\Lambda^{-1}(d\Lambda)\Lambda^{-1}$. (You might recall that we have used this identity on more than one occasion, in (V.6.7), for example.) We have, using (11),

$$\begin{aligned}
d\omega &= d(\Lambda\tilde\omega\Lambda^{-1} - (d\Lambda)\Lambda^{-1}) \\
&= (d\Lambda)\tilde\omega\Lambda^{-1} + \Lambda(d\tilde\omega)\Lambda^{-1} - (-)\Lambda\tilde\omega\Lambda^{-1}(d\Lambda)\Lambda^{-1} - (dd\Lambda)\Lambda^{-1} + (d\Lambda)\Lambda^{-1}(d\Lambda)\Lambda^{-1}
\end{aligned} \quad (12)$$

and

$$\begin{aligned}
\omega^2 &= (\Lambda\tilde\omega\Lambda^{-1} - (d\Lambda)\Lambda^{-1})^2 \\
&= \Lambda\tilde\omega^2\Lambda^{-1} - \Lambda\tilde\omega\Lambda^{-1}(d\Lambda)\Lambda^{-1} - (d\Lambda)\tilde\omega\Lambda^{-1} + (d\Lambda)\Lambda^{-1}(d\Lambda)\Lambda^{-1}
\end{aligned} \quad (13)$$

The one tricky part of the calculation is the sign of the third and the fourth terms on the right hand side of (12). There are two extra minus signs, because we have to sail d past the 1-form $\tilde\omega$ to act on Λ^{-1} in the third term, and to sail d past the 1-form $d\Lambda$ to act on Λ^{-1} in the fourth term, as explained in (7). Did you miss these two minus signs? Good for you if you didn't. Also, note that the fourth term in (12) vanishes due to (8). Now add (12) and (13). After much cancellation, we find $R = d\omega + \omega^2 = \Lambda \tilde{R} \Lambda^{-1}$. Indeed, the 2-form $R^\alpha_{\ \beta}$ transforms like a Lorentz tensor.

Written out in components, $R^{\alpha\beta} = \frac{1}{2} R^{\alpha\beta}_{\ \ \mu\nu} dx^\mu dx^\nu$. (Don't forget the factor of $\frac{1}{2}$!) As explained earlier, we can trade Lorentz indices for world indices and vice versa. I leave it to you to verify that $R^{\alpha\beta}_{\ \ \mu\nu} e^\lambda_\alpha e^\sigma_\beta$ is our beloved Riemann curvature tensor $R^{\lambda\sigma}_{\ \ \mu\nu}$. In particular, $R^{\alpha\beta}_{\ \ \mu\nu} e^\mu_\alpha e^\nu_\beta$ is the scalar curvature.

For the sphere, we have

$$R^{12} = d\omega^{12} + \omega^{1\gamma} \omega^{\gamma 2} = d\omega^{12} = \sin\theta d\theta d\varphi$$

$$= e^1 e^2 = \frac{1}{2}\left(R^{12}_{\ \ 12} e^1 e^2 + R^{12}_{\ \ 21} e^2 e^1\right) = R^{12}_{\ \ 12} e^1 e^2 \tag{14}$$

(The second equality holds because $\omega^{1\gamma}\omega^{\gamma 2} = \omega^{11}\omega^{12} + \omega^{12}\omega^{22}$ vanishes due to the anti-symmetry of $\omega^{\alpha\beta}$. The fifth equality comes from expanding the 2-form R^{12}.) Comparing the final expression with $e^1 e^2$ gives $R^{12}_{\ \ 12} = 1$ and thus the scalar curvature $R = R^{\alpha\beta}_{\ \ \alpha\beta} = R^{12}_{\ \ 12} + R^{21}_{\ \ 21} = 2$. Alternatively, we can trade Lorentz indices for world indices and write

$$R^{12} = \sin\theta d\theta d\varphi = \frac{1}{2}\left(R^{12}_{\ \ \theta\varphi} d\theta d\varphi + R^{12}_{\ \ \varphi\theta} d\varphi d\theta\right) = R^{12}_{\ \ \theta\varphi} d\theta d\varphi$$

and so $R^{12}_{\ \ \theta\varphi} = \sin\theta$. This is of course consistent with $R^{12}_{\ \ 12} = R^{12}_{\ \ \theta\varphi} e^\theta_1 e^\varphi_2 = 1$.

Thus, in Cartan's formalism, Riemannian geometry can be elegantly summarized by the two statements* (suppressing Lorentz indices)

$$de + \omega e = 0 \tag{15}$$

and

$$R = d\omega + \omega^2 \tag{16}$$

Putting the Cartan formalism to work

In the following chapter, I will show you how to use the Cartan formalism to calculate curvature. In most cases, there is considerable reduction of labor. Also, I will postpone discussing the so-called Hodge star operation on differential forms until chapter X.5 for reasons that will become clear.

Here I give an example of how easily we can derive some of the identities we already know. Apply d to (15) and obtain $dde + (d\omega)e - \omega de = (d\omega)e - \omega de = 0$, remembering that $dd = 0$ and that moving d past a p-form produces (as shown in (7)) a sign $(-1)^p$. Use (15) to eliminate de, so that $(d\omega)e + \omega\omega e = 0 = Re$, where we recall Cartan's second equation (16) in the last step. Putting back the Lorentz indices, we learn that

$$R^\alpha_{\ \beta} e^\beta = 0 \tag{17}$$

---

* Incidentally, one of the most appealing features of this discussion is that it brings out the profound connection between curvature as given in (16) and the field strength in nonabelian gauge theories $F = dA + A^2$, with A a matrix 1-form. Because of this correspondence between ω and A, the latter is sometimes called the gauge connection. See *QFT Nut*.

or, more explicitly, $R^\alpha_{\ \beta\mu\nu}e^\beta_\lambda dx^\mu dx^\nu dx^\lambda = R^\alpha_{\ \lambda\mu\nu}dx^\mu dx^\nu dx^\lambda = e^\alpha_\sigma R^\sigma_{\ \lambda\mu\nu}dx^\mu dx^\nu dx^\lambda = 0$, that is, $R^\sigma_{\ \lambda\mu\nu}dx^\mu dx^\nu dx^\lambda = 0$. Since $dx^\mu dx^\nu dx^\lambda = dx^\nu dx^\lambda dx^\mu = dx^\lambda dx^\mu dx^\nu$ (since at each step we are "moving a dx past two dxs"), we can write the preceding as $R^\sigma_{\ \lambda\mu\nu}(dx^\mu dx^\nu dx^\lambda + dx^\nu dx^\lambda dx^\mu + dx^\lambda dx^\mu dx^\nu) = 0$. But by renaming the dummy indices we are summing over, we can also write this as $(R^\sigma_{\ \lambda\mu\nu} + R^\sigma_{\ \mu\nu\lambda} + R^\sigma_{\ \nu\lambda\mu})dx^\mu dx^\nu dx^\lambda = 0$. We recover the cyclic identity $R^\sigma_{\ \lambda\mu\nu} + R^\sigma_{\ \mu\nu\lambda} + R^\sigma_{\ \nu\lambda\mu} = 0$ found in chapter VI.1.

Appendix 1: Connecting the two connections

The vielbein e^α_μ and its inverse e^μ_α allow us to freely convert Lorentz indices to world indices and vice versa. Given an arbitrary vector V^μ, $V^\alpha = e^\alpha_\mu V^\mu$ is evidently a Lorentz vector and world scalar. In general, given an arbitrary tensor, say $T^{\mu\nu\zeta}_{\ \ \ \sigma\tau}$, we can construct objects with various mixed tensor structures by contracting with the vielbein and its inverse in appropriate combinations, such as $T^{\alpha\beta\zeta}_{\ \ \ \sigma\tau} = e^\alpha_\mu e^\beta_\nu e^\sigma_\gamma T^{\mu\nu\zeta}_{\ \ \ \sigma\tau}$.

The covariant derivative $D_\lambda V^\mu = \partial_\lambda V^\mu + \Gamma^\mu_{\lambda\nu}V^\nu$ of an arbitrary vector V^μ transforms, as you know well by now, as a world tensor and a Lorentz scalar. The Christoffel connection adjusts for the fact that the coordinate transformation varies from place to place. In contrast, the covariant derivative[5] $D_\lambda V^\alpha$ should transform like a world vector with a lower index and a Lorentz vector with an upper index. The ordinary derivative $\partial_\lambda V^\alpha$ transforms like a world vector, as it should, but not as a Lorentz vector under the location dependent Lorentz transformation $\Lambda(x)$. Instead, it turns out that we should write

$$D_\lambda V^\alpha = \partial_\lambda V^\alpha + \omega^\alpha_{\ \beta\lambda}V^\beta \tag{18}$$

As usual, when we transform, we have to compensate for the effect of ∂_λ acting on $\Lambda(x)$ in the first term by introducing the spin connection $\omega^\alpha_{\ \beta\lambda}$. Suppressing Lorentz indices and transforming $\tilde{V} = \Lambda V$, we have $D_\lambda \tilde{V} = \partial_\lambda \tilde{V} + \tilde{\omega}_\lambda \tilde{V} = \partial_\lambda(\Lambda V) + \tilde{\omega}_\lambda \Lambda V = \Lambda(\partial_\lambda V) + \tilde{\omega}_\lambda \Lambda V + (\partial_\lambda \Lambda)V$ under the transformation. Requiring that this be equal to $\Lambda D_\lambda V = \Lambda(\partial_\lambda V + \omega_\lambda V)$ gives us the transformation $\tilde{\omega} = \Lambda^{-1}\omega\Lambda + \Lambda^{-1}(d\Lambda)$, familiar to you by now from (11).

Consistency relates the Christoffel connection Γ and the spin connection ω. Since $D_\lambda V^\mu$ is a world tensor, we must have $D_\lambda V^\alpha = e^\alpha_\mu D_\lambda V^\mu$. The left hand side is equal to $D_\lambda \left(e^\alpha_\mu V^\mu\right) = \partial_\lambda\left(e^\alpha_\mu V^\mu\right) + \omega^\alpha_{\ \beta\lambda}e^\beta_\mu V^\mu$. Equating this to the right hand side $e^\alpha_\mu\left(\partial_\lambda V^\mu + \Gamma^\mu_{\lambda\nu}V^\nu\right)$, we obtain, after collecting terms and renaming indices, the rather satisfying and perhaps expected result

$$\partial_\mu e^\alpha_\nu + \omega^\alpha_{\ \beta\mu}e^\beta_\nu - \Gamma^\lambda_{\mu\nu}e^\alpha_\lambda = 0 \tag{19}$$

This relation tells us that as we move from x to $x + dx$, the d vectors e^α_ν "rotate" a bit in the Lorentz index α and a bit in the world index ν, each effected by the corresponding connection, ω and Γ, respectively. It may also be rewritten as

$$D_\mu e^\alpha_\nu = -\omega^\alpha_{\ \beta\mu}e^\beta_\nu \tag{20}$$

telling us that covariant differentiation on the vielbein generates an infinitesimal rotation of the local frame.

Recall that the metric is just the Lorentz dot product of two vielbein $g_{\nu\lambda} = e^\alpha_\nu \eta_{\alpha\gamma}e^\gamma_\lambda$, and so we can immediately conclude that the covariant derivative of the metric vanishes and thus recover (V.6.15). Applying D_μ on the metric and using the antisymmetry of the connection $\omega_{\gamma\beta\mu} = -\omega_{\beta\gamma\mu}$, we obtain, as expected, $D_\mu g_{\nu\lambda} = 0$, or in other words, $\partial_\mu g_{\nu\lambda} = \Gamma_{\lambda\cdot\mu\nu} + \Gamma_{\nu\cdot\mu\lambda}$ (where $\Gamma_{\lambda\cdot\mu\nu} \equiv g_{\lambda\rho}\Gamma^\rho_{\mu\nu}$ and the dot separates the two groups of indices in keeping with the notation of previous chapters). Recall from chapter II.2 that this equality amounts to the definition of the Christoffel symbol.

The relation (19) also leads immediately to Cartan's first equation (15): $de + \omega e = 0$. We simply have to compute: $de + \omega e = \left(\partial_\mu e^\alpha_\nu + \omega^\alpha_{\ \beta\mu}e^\beta_\nu\right)dx^\mu dx^\nu = \Gamma^\lambda_{\mu\nu}e^\alpha_\lambda dx^\mu dx^\nu = 0$, since the Christoffel symbol is symmetric in its two lower indices.

We can also use (19) to determine one connection in terms of the other:

$$\Gamma^\lambda_{\mu\nu} = e^\lambda_\alpha\left(\partial_\mu e^\alpha_\nu + \omega^\alpha_{\ \beta\mu}e^\beta_\nu\right) \tag{21}$$

and

$$\omega^{\alpha}{}_{\beta\mu} = -e^{\nu}_{\beta}\left(\partial_{\mu}e^{\alpha}_{\nu} - \Gamma^{\lambda}_{\mu\nu}e^{\alpha}_{\lambda}\right) \tag{22}$$

We can write (22) more compactly as $\omega^{\alpha}{}_{\beta} = \Gamma^{\alpha}{}_{\beta} + H^{\alpha}{}_{\beta}$ by defining the Christoffel 1-form $\Gamma^{\alpha}{}_{\beta} = e^{\alpha}_{\lambda}e^{\nu}_{\beta}\Gamma^{\lambda}_{\mu\nu}dx^{\mu}$ and the 1-form $H^{\alpha}{}_{\beta} = -e^{\nu}_{\beta}\partial_{\mu}e^{\alpha}_{\nu}dx^{\mu}$.

I leave to you to check that Cartan's second equation (16) also follows immediately.

For any curved spacetime, the symmetry group of the tangent space is always the Lorentz group. In other words, as explained earlier, the index α, by definition, responds to Lorentz transformation. The isometry group is of course another story entirely, and in fact may well be null. The confusion some students have stems from the fact that, for flat spacetime, the isometry group and the tangent space group are the same, namely the Lorentz group.

Appendix 2: Exact is closed, but closed is not necessarily globally exact

Here I mention an important concept somewhat outside the narrative of this book, involving differential forms. It is convenient to introduce some jargon. A p-form α is said to be closed if $d\alpha = 0$. It is said to be exact if there exists a $(p-1)$-form β such that $\alpha = d\beta$.

Talking the talk, we say that (8) tells us that exact forms are closed.

Is the converse of (8) true? Kind of. The Poincaré lemma states that a closed form is locally exact.[6] In other words, if $dH = 0$ with H some p-form, then locally (that is, within some coordinate patch)

$$H = dK \tag{23}$$

for some $(p-1)$-form K. However, it may or may not be the case that $H = dK$ globally, that is, everywhere. Actually, whether you knew it or not, you are probably already familiar with the Poincaré lemma. For example, surely you learned somewhere that if the curl of a vector field vanishes, the vector field is locally the gradient of some scalar field.

Forms are ready made to be integrated over. For example, given the 2-form $F = \frac{1}{2!}F_{\mu\nu}dx^{\mu}dx^{\nu}$, we can write $\int_M F$ for any 2-manifold M. Note the measure is already included and there is no need to specify a coordinate choice. Again, whether you knew it or not, you are already familiar with the important theorem*

$$\int_M dH = \int_{\partial M} H \tag{24}$$

with H a p-form and ∂M the boundary of a $(p+1)$-dimensional manifold M.

Back in appendix 4 to chapter III.6, I mentioned that, just as a current J^{μ} is associated with a point particle, a current $J^{\mu\nu} = -J^{\nu\mu}$ is associated with a string. It then follows that the analog of the electromagnetic potential A_{μ} coupling to J^{μ} is an antisymmetric tensor field $B_{\mu\nu}$ coupling to $J^{\mu\nu}$. Thus, string theory[7] contains a 2-form potential $B = \frac{1}{2}B_{\mu\nu}dx^{\mu}dx^{\nu}$ and the corresponding 3-form field $H = dB$. For some readers, this remark may clarify further the geometric character of differential forms.

Appendix 3: Spinors in curved spacetime

This appendix is strictly for readers familiar with the Dirac spinor and should be skipped by others. In relativistic field theory, spin $\frac{1}{2}$ particles, such as the electron, are described by Dirac spinors. As I explained starting in part V, Einstein's equivalence principle renders life easy for us. Given an action in Minkowskian spacetime (for example Maxwell's action, in which various Lorentz indices are contracted with $\eta_{\mu\nu}$ and its inverse), we simply replace $\eta_{\mu\nu}$ by $g_{\mu\nu}$ and, lo and behold, we obtain the action in curved spacetime. But this procedure works only with fields carrying Lorentz indices, such as A_{μ}. The Dirac spinor, as the name indicates, does not transform as a vector or a tensor under the Lorentz group, but instead carries a spinorial index, which we denote[†] by s. Let us write

---

* Namely, Stokes theorem in a more sophisticated form.

[†] It is commonly denoted by $\alpha, \beta, \cdots$, but those guys have already been pressed into service in this chapter.

the Dirac spinor as ψ_s. In the Dirac action in Minkowskian spacetime, the spinor ψ_s is acted upon by matrices known as Dirac gamma matrices $(\gamma^\alpha)_{rs}$, which, as indicated by the notation, are labeled by a Lorentz index α but are matrices in spinorial space, thus carrying two spinorial indices r and s. (In other words, combinations such as $(\gamma^\alpha)_{rs}\psi_s$ occur in the action.)

The issue is how to promote the Dirac action in Minkowskian spacetime to curved spacetime. Of course, spacetime derivatives ∂_μ also occur in the action, but we know how to promote them to curved spacetime, namely to covariant derivatives D_μ. To make a long story short, and to keep the discussion at the most pedestrian level, we can state the problem facing us as follows: in constructing an action for the Dirac spinor in curved spacetime, how do we connect these three types of indices, spinorial $(r, s, \cdots)$, Lorentz $(\alpha, \beta, \cdots)$, and world $(\mu, \nu, \cdots)$? I already told you that the Dirac gamma matrices connect spinorial and Lorentz indices, so half of the problem is solved.

Well, by now, you see how the other half is to be solved. The vielbein! Indeed, the vielbein e^α_μ connects Lorentz and world indices, and so the vielbein formalism is absolutely essential for the physics of the Dirac spinor in curved spacetime.

If, in spite of my warning, some intrepid reader, though unfamiliar with spinors, has insisted on going through this rather cryptic appendix, I hope that he or she will go on plumbing the mystery of the spinor.[8]

Appendix 4: Curvature and covariant derivative in the Cartan formalism

The discussion in the text suggests defining the covariant derivative $D = d + \omega$. For definiteness, consider a 0-form ϕ^β, that is, an object with a Lorentz index but without a world index. Write

$$D^\alpha_{\ \beta}\phi^\beta = d\phi^\alpha + \omega^\alpha_{\ \beta}\phi^\beta$$

We now show how curvature emerges in the Cartan formalism. Calculate

$$D^\gamma_{\ \alpha}D^\alpha_{\ \beta}\phi^\beta = d\left(d\phi^\gamma + \omega^\gamma_{\ \beta}\phi^\beta\right) + \omega^\gamma_{\ \alpha}\left(d\phi^\alpha + \omega^\alpha_{\ \beta}\phi^\beta\right)$$

The first term gives

$$dd\phi^\gamma + \left(d\omega^\gamma_{\ \beta}\right)\phi^\beta - \omega^\gamma_{\ \beta}d\phi^\beta$$

Since $dd = 0$, we have

$$D^\gamma_{\ \alpha}D^\alpha_{\ \beta}\phi^\beta = \left(d\omega^\gamma_{\ \beta} + \omega^\gamma_{\ \alpha}\omega^\alpha_{\ \beta}\right)\phi^\beta \equiv R^\gamma_{\ \beta}\phi^\beta$$

We see the curvature 2-form $R^\gamma_{\ \beta}$ emerging in front of our very eyes.

Exercises

1   For $ds^2 = f(y)^2 dx^2 + g(x)^2 dy^2$, calculate the curvature using differential forms.

2   By writing out the components explicitly, show that $dF = 0$ states something you are familiar with but is disguised in a compact notation.

3   Consider $F = \frac{g}{4\pi} d\cos\theta\, d\varphi$. By transforming to Cartesian coordinates, show that this describes a magnetic field pointing outward along the radial direction.

4   Calculate the curvature of the conformally flat 2-dimensional space $ds^2 = \Omega^2(x, y)(dx^2 + dy^2)$ using differential forms. Check your result using the 2-sphere.

5   Extend the calculation of exercise 4 to d-dimensional space, that is, calculate the curvature of the conformally flat space $ds^2 = \Omega^2(x^1, \cdots, x^d)((dx^1)^2 + \cdots + (dx^d)^2)$ using differential forms. Check your result using

the sphere. Also, show that for $\Omega(x^1, \cdots, x^d) = 1/x^1$, the scalar curvature is constant. We will discuss the corresponding spacetime, known as anti de Sitter spacetime, in chapter IX.11.

6 Calculate the curvature of the 2-dimensional space with $ds^2 = dr^2 + (f(r, \theta))^2 d\theta^2$ using differential forms. Recall that we showed, back in appendix 5 to chapter II.2, that we can construct a metric of this form in general for 2-dimensional spaces.

Notes

1. Some authors write "dreibeins," "vierbeins," and "vielbeins." Because the plural of "Bein" in German is "Beine," these people are using the English plural of a German word.

2. Some authors prefer the Greek words dyad, triad, and tetrad. As my friend Cecile DeWitt once said, "Why should German words be used for something discovered by a Frenchman?" It is odd indeed. One problem is the absence of a good substitute for vielbein: polytrad sounds a bit odd. Another term sometimes used is "frame field."

3. If we think of the vielbein as vectors $\vec{e}_\mu$, namely d (d-dimensional) vectors labeled by the index μ, with the components of the vector $\vec{e}_\mu$ given by e^α_μ, then $\vec{e}_\mu$ are just the d tangent vectors we encountered before in chapter I.7 in the context of surfaces.

4. What mathematicians would call Grassmann variables.

5. Our notation, using the same D_λ in $D_\lambda V^\mu$ and $D_\lambda V^\alpha$, is somewhat sloppy. But at the level of this book, it is a small price to pay to avoid going into fiber bundles and other fancy mathematical topics.

6. For the reader who wants to work through more examples of this statement, see B. Zumino, Y. S. Wu, and A. Zee, *Nucl. Phys.* B 239 (1984), p. 477, in particular, appendix A.

7. In fact, string theory typically contains numerous p-forms. See J. Polchinski, *String Theory*.

8. For further details, see more specialized texts. For an easy introduction, see *QFT Nut*, p. 445.

IX.8 | Differential Forms Applied

Calculating curvature with differential forms

In this chapter, I show you that differential forms provide a more efficient method to calculate curvature than the more traditional method of first working out the Christoffel symbols. The human mind appears to be ill evolved to handle 3-indexed objects like the Christoffel symbols. While the connection 1-form ω also nominally carries 3 indices,* the indices are actually of two types, and so in reality, you only have to handle either a 1-indexed object or a 2-indexed object, depending on how you look at it. Besides, the 1-form $\omega^{\alpha\beta} = \omega_\mu^{\alpha\beta} dx^\mu$ is antisymmetric in $\alpha\beta$, while a Christoffel symbol is symmetric in its two lower indices. In general, an antisymmetric object (lots of vanishing components!) is much easier to deal with than a symmetric one.

People used to argue about the relative merits of the two methods, but with the advent of symbolic manipulation software, the issue is now moot. It is easy to write a simple program to calculate the Riemann curvature tensor by brute force using the traditional method. Still, Cartan's method involving differential forms is well worth learning.

First, let's recall from the preceding chapter Cartan's first and second structural equations:

$$de + \omega e = 0 \tag{1}$$

and

$$R = d\omega + \omega^2 \tag{2}$$

As the emphasis in this chapter is learning to compute with differential forms, I will write it with a minimum of prose connecting the equations. You might call this chapter engineering with differential forms.

---

* If you expand it out into its components $\omega_\mu^{\alpha\beta}$.

A reminder about the indices: the world index is $\mu = 0, i$; the Lorentz index is $\alpha = 0, a$. For i, I often use its colloquial name, such as r or x. The symbol 0 does double duty. When I need to emphasize that a particular 0 is a world, rather than a Lorentz, index, I call it t, as for example in (22) below. Keep in mind that, with the $(-+++)$ signature used in this book, the raising and lowering of an a index entails a $+$ sign, while the raising and lowering of a 0 index entails a $-$ sign. If we are dealing with space (for example, the Poincaré half plane discussed below), rather than spacetime, then of course this $-$ sign does not come in.

Here are some useful relations based on antisymmetry:

$$\omega^0_{\ b} = +\omega^{0b} = -\omega^{b0} = \omega^b_{\ 0} \tag{3}$$

$$\omega^b_{\ c} = +\omega^{bc} = -\omega^{cb} = -\omega^c_{\ b} \tag{4}$$

$$\omega^0_{\ \gamma}\omega^\gamma_{\ b} = +\omega^b_{\ \gamma}\omega^\gamma_{\ 0}, \qquad \omega^a_{\ \gamma}\omega^\gamma_{\ b} = -\omega^b_{\ \gamma}\omega^\gamma_{\ a} \tag{5}$$

$$R^0_{\ a0a} = -R_{0a0a} = -R_{a0a0} = -R^a_{\ 0a0} \tag{6}$$

$$R^a_{\ bab} = +R_{abab} = +R_{baba} = R^b_{\ aba} \tag{7}$$

I hardly need say that if you want to learn to compute with Cartan's approach, you will just have to work through everything here and do the exercises.

In 2 dimensions, the curvature 2-form is just $R = d\omega$, since $\omega^2 = 0$ (note that $\omega^1_{\ a}\omega^a_{\ 2} = 0$).

Poincaré half plane

For $ds^2 = (dr^2 + dx^2)/r^2$, we have $e^1 = dr/r$, $e^2 = dx/r$, or in components, $e^1_r = \frac{1}{r}$, $e^2_x = \frac{1}{r}$. Thus, we have $de^1 = 0$ and $de^2 = -dr\,dx/r^2 = e^2 e^1$. Hence, using $de^2 = -\omega^2_{\ 1}e^1$, we obtain $\omega^1_{\ 2} = e^2$. We also obtain

$$R^1_{\ 2} = d\omega^1_{\ 2} = de^2 = -e^1 e^2 \tag{8}$$

so that $R^1_{\ 212} = -1$, from which we have

$$R_{22} = -1, \qquad R_{11} = -1 \tag{9}$$

that is, $R_{ab} = -\delta_{ab}$. Thus, the Poincaré half plane is a maximally symmetric space with constant negative curvature. Converting to $R_{\mu\nu} = e^a_\mu e^b_\nu R_{ab}$, we obtain

$$R_{rr} = -\frac{1}{r^2}, \qquad R_{xx} = -\frac{1}{r^2} \tag{10}$$

that is, $R_{\mu\nu} = -g_{\mu\nu}$, giving $R = -2$.

Expanding universe

From

$$ds^2 = -dt^2 + a(t)^2(dx^2 + dy^2 + dz^2) \tag{11}$$

we have

$$e^0 = dt, \quad e^1 = a(t)dx, \quad e^2 = a(t)dy, \quad e^3 = a(t)dz \tag{12}$$

and hence

$$de^0 = 0 = \omega^0{}_b e^b \tag{13}$$

which implies $\omega^0{}_b \propto e^b$, and

$$de^b = \dot{a}(t)dt dx^b = \dot{a}(t)e^0 dx^b = -\omega^b{}_0 e^0 - \omega^b{}_c e^c \tag{14}$$

We could have $\omega^b{}_c \propto e^c$, but then we cannot satisfy the antisymmetry of ω^{bc}, and so we conclude that $\omega^{bc} = 0$. Thus, we have

$$\omega^0{}_b = \omega^b{}_0 = \dot{a}(t)dx^b \tag{15}$$

$$R^0{}_b = d\omega^0{}_b + \omega^0{}_c \omega^c{}_b = \ddot{a}dt dx^b + 0 = \frac{\ddot{a}}{a}e^0 e^b \tag{16}$$

To guide you, I have indicated here (and in the following) that in this formalism, quite a few terms are equal to 0:

$$R^b{}_c = d\omega^b{}_c + \omega^b{}_d \omega^d{}_c + \omega^b{}_0 \omega^0{}_c = 0 + 0 + \dot{a}^2 dx^b dx^c = \frac{\dot{a}^2}{a^2}e^b e^c \tag{17}$$

Note that in this problem, there is a residual $SO(3)$ symmetry in the spatial indices. Thus, $R^0{}_b$ and $R^b{}_c$ must be proportional to $e^0 e^b$ and $e^b e^c$, respectively:

$$R^0{}_{b0b} = \frac{\ddot{a}}{a} = R_{0b0b} = R_{b0b0} = -R^b{}_{0b0} \quad \text{(no sum)} \tag{18}$$

$$R^b{}_{cbc} = \frac{\dot{a}^2}{a^2} \quad \text{(no sum)} \tag{19}$$

As indicated, the repeated indices are not summed. We temporarily suspend the Einstein summation convention. Then we have

$$R_{00} = \sum_b R^b{}_{0b0} = -3\frac{\ddot{a}}{a} \tag{20}$$

$$R_{bb} = R^0{}_{b0b} + \sum_c R^c{}_{bcb} = \frac{\ddot{a}}{a} + 2\frac{\dot{a}^2}{a^2} \tag{21}$$

You should understand where the 3 and 2 come from.

Now we can work out the nonvanishing components of the Ricci tensor:

$$R_{tt} = e^0{}_t e^0{}_t R_{00} = -3\frac{\ddot{a}}{a} \tag{22}$$

$$R_{ij} = \sum_b e^b{}_i e^b{}_j R_{bb} = a^2 \left(\frac{\ddot{a}}{a} + 2\frac{\dot{a}^2}{a^2} \right) \delta_{ij} = (a\ddot{a} + 2\dot{a}^2)\delta_{ij} \tag{23}$$

Finally, the scalar curvature, which we can work out more easily from (20) and (21) than from (22) and (23), is given by

$$R = -R_{00} + \sum_b R_{bb} = 3\frac{\ddot{a}}{a} + 3\left(\frac{\ddot{a}}{a} + 2\frac{\dot{a}^2}{a^2} \right) = 6\left(\frac{\ddot{a}}{a} + \frac{\dot{a}^2}{a^2} \right) \tag{24}$$

Maximally symmetric 3-spaces

Let's foliate 3-space with a series of spheres separated by the distance $F(r)dr$:

$$ds^2 = F(r)^2 dr^2 + r^2 d\theta^2 + r^2 \sin^2 \theta d\phi^2 \tag{25}$$

so that

$$e^1 = F(r)dr, \quad e^2 = rd\theta, \quad e^3 = r \sin \theta d\varphi \tag{26}$$

We obtain easily $de^1 = 0$, $de^2 = drd\theta$, $de^3 = \sin \theta drd\varphi + r \cos \theta d\theta d\varphi$.

Solving Cartan's first equation for the connection 1-form gets more involved as the dimension of space (or spacetime) increases. In general, write $\omega^a{}_b = \omega^a{}_{b\mu} dx^\mu = \omega^a{}_{bc} e^c$. Plug this into Cartan's first structural equation (1) and match terms. We obtain

$$\omega^1{}_2 = -\frac{1}{rF} e^2 = -\frac{1}{F} d\theta, \quad \omega^1{}_3 = -\frac{1}{rF} e^3 = -\frac{1}{F} \sin \theta d\varphi, \quad \omega^2{}_3 = -\frac{\cos \theta}{r \sin \theta} e^3 = -\cos \theta d\varphi \tag{27}$$

Cartan's second equation gives us

$$R^1{}_2 = \frac{F'}{rF^3} e^1 e^2, \quad R^1{}_3 = \frac{F'}{rF^3} e^1 e^3, \quad R^2{}_3 = \left(1 - \frac{1}{F^2}\right) \frac{1}{r^2} e^2 e^3 \tag{28}$$

Using differential forms, you have to exercise some judgment. For instance, the connection 1-forms are written in two equivalent versions in (27). To calculate R, it is somewhat easier to use the second version.

From (28), we simply read off $R^1{}_{212} = \frac{F'}{rF^3}$, $R^1{}_{313} = \frac{F'}{rF^3}$, and $R^2{}_{323} = \frac{1}{r^2}\left(1 - \frac{1}{F^2}\right)$. (Confused about the factors of 2, or absence thereof? See the preceding chapter.) Contracting Riemann, we arrive at the nicely symmetric Ricci tensor

$$R_{ab} = \frac{2F'}{rF^3} \delta_{ab} \tag{29}$$

If we require that the space be maximally symmetric and positively curved with some length scale L, so that $R_{ab} = \frac{2}{L^2} \delta_{ab}$, we obtain the differential equation $\frac{F'}{rF^3} = \frac{1}{L^2}$, with the solution $F^2 = \frac{1}{1 - \frac{r^2}{L^2}}$. As a check, plugging this back into (28), we find $R^1{}_2 = \frac{1}{L^2} e^1 e^2$, $R^1{}_3 = \frac{1}{L^2} e^1 e^3$, and $R^2{}_3 = \frac{1}{L^2} e^2 e^3$. With the change of variable $r = L \sin \chi$, the metric becomes

$$ds^2 = L^2 \left(d\chi^2 + \sin^2 \chi d\theta^2 + \sin^2 \chi \sin^2 \theta d\phi^2\right) \tag{30}$$

Here we have nothing other than S^3, of course, as you might recall from exercise I.5.9.

We can formally go to the maximally symmetric negatively curved space by setting $L^2 \to -L^2$, so that $F^2 = \frac{1}{1 + \frac{r^2}{L^2}}$. With $r = L \sinh \chi$, the metric is given by

$$ds^2 = L^2 \left(d\chi^2 + \sinh^2 \chi d\theta^2 + \sinh^2 \chi \sin^2 \theta d\phi^2\right) \tag{31}$$

Spherically symmetric static spacetimes

Start with the metric

$$ds^2 = -E(r)^2 dt^2 + \left(F(r)^2 dr^2 + r^2 d\theta^2 + r^2 \sin^2 \theta d\phi^2 \right) \tag{32}$$

or $e^0 = E dt$, $e^1 = F dr$, $e^2 = r d\theta$, $e^3 = r \sin \theta d\phi$. Note that the notation used in chapter VI.3 is related to that used here by $A = E^2$, $B = F^2$.

After some slightly tedious matching of terms, we solve Cartan's first structural equation to obtain

$$\omega^2_{\ 3} = -\frac{\cos \theta}{r \sin \theta} e^3, \quad \omega^1_{\ 3} = -\frac{1}{r F} e^3, \quad \omega^2_{\ 0} = 0$$

$$\omega^1_{\ 2} = -\frac{1}{r F} e^2, \quad \omega^1_{\ 0} = \frac{E'}{E F} e^0, \quad \omega^3_{\ 0} = 0 \tag{33}$$

Cartan's second equations now give us

$$R^2_{\ 3} = \frac{1}{r^2} \left(1 - \frac{1}{F^2} \right) e^2 e^3, \quad R^1_{\ 2} = \frac{F'}{r F^3} e^1 e^2, \quad R^1_{\ 3} = \frac{F'}{r F^3} e^1 e^3 \tag{34}$$

and

$$R^1_{\ 0} = \frac{1}{E F} \left(\frac{E'}{F} \right)' e^1 e^0, \quad R^2_{\ 0} = \frac{E'}{r E F^2} e^2 e^0, \quad R^3_{\ 0} = \frac{E'}{r E F^2} e^3 e^0 \tag{35}$$

Note that spherical symmetry relates $R^\alpha_{\ 2}$ and $R^\alpha_{\ 3}$, giving us a useful check on our arithmetic.

One nice feature of the differential form approach is that once we have the curvature 2-form, we can simply read off the Riemann curvature tensor:

$$R^2_{\ 323} = \frac{1}{r^2} \left(1 - \frac{1}{F^2} \right), \quad R^1_{\ 212} = \frac{F'}{r F^3}, \quad R^1_{\ 313} = \frac{F'}{r F^3} \tag{36}$$

and

$$R^1_{\ 010} = \frac{1}{E F} \left(\frac{E'}{F} \right)', \quad R^2_{\ 020} = \frac{E'}{r E F^2}, \quad R^3_{\ 030} = \frac{E'}{r E F^2} \tag{37}$$

The Ricci tensor follows immediately:

$$R_{00} = \frac{1}{E F} \left(\frac{E'}{F} \right)' + \frac{2 E'}{r E F^2}, \quad R_{11} = \frac{2 F'}{r F^3} - \frac{1}{E F} \left(\frac{E'}{F} \right)' \tag{38}$$

and

$$R_{22} = R_{33} = \frac{1}{r^2} \left(1 - \frac{1}{F^2} \right) + \frac{F'}{r F^3} - \frac{E'}{r E F^2} \tag{39}$$

Similar to the discussion in chapter VI.3, we see that second derivatives appear in R_{00} and R_{11} but not in the combination $R_{00} + R_{11} = \frac{2E'}{rEF^2} + \frac{2F'}{rF^3} = \frac{2}{rF^2} \left(\frac{E'}{E} + \frac{F'}{F} \right)$. Comparing the two discussions, we see that the appearance of this cancellation is packaged somewhat more clearly here. The content of the two formalisms is of course the same, as we see immediately by the substitution $A = E^2$, $B = F^2$ in (VI.3.5–7).

Anti de Sitter spacetime

We will discuss de Sitter and anti de Sitter spacetimes in detail in chapters IX.10 and IX.11. Here it suffices to note that the metric

$$ds^2 = \frac{-dt^2 + dx^2 + dy^2 + dr^2}{r^2} \tag{40}$$

generalizes the Poincaré half plane to $(3 + 1)$-dimensional spacetime. Here we have

$$e^0 = \frac{dt}{r}, \quad e^1 = \frac{dx}{r}, \quad e^2 = \frac{dy}{r}, \quad e^3 = \frac{dr}{r} \tag{41}$$

It is useful to decompose the index set $\alpha = 0, 1, 2, 3$ into $\alpha = \tilde{\alpha}, 3$, with $\tilde{\alpha} = 0, 1, 2$, and to recognize that there exists a residual $SO(2, 1)$ symmetry transforming the indices $\tilde{\alpha} = 0, 1, 2$ among themselves. The index 3 is special in this problem. It is convenient also to restrict the early letters a, b, and so on to denote 1, 2.

The residual $SO(2, 1)$ tells us that $\omega^{\tilde{\alpha}}_3 = ke^{\tilde{\alpha}}$ must have the form indicated. For the purpose of dimensional analysis here, we take t, x, y, and r to have dimensions of length, so that e^α and hence ω^α_β are dimensionless. Thus, k is dimensionless and by $SO(2, 1)$ can only be a constant. Furthermore, $SO(2, 1)$ and antisymmetry imply that $\omega^{0a} = 0$ and $\omega^{12} = 0$. These considerations render the arithmetic a snap.

Thus, for example, $de^0 = -\frac{drdt}{r^2} = -e^3 e^0 = -\omega^0_\alpha e^\alpha = -\omega^0_3 e^3$, giving us

$$\omega^0_3 = -e^0 \quad \text{and hence} \quad \omega^a_3 = -e^a \tag{42}$$

We can easily check the symmetry conclusion here by calculating de^1, for example.

The same kind of symmetry considerations tell us that $R^{\tilde{\alpha}\tilde{\beta}} = Ce^{\tilde{\alpha}}e^{\tilde{\beta}}$ and $R^{\tilde{\alpha}}_3 = De^{\tilde{\alpha}}e^3$, with numerical constants C and D. Direct evaluation of Cartan's second structural equation gives $C = D = -1$. For example, $R^0_3 = d\omega^0_3 + \omega^0_\alpha \omega^\alpha_3 = -de^0 = -e^0 e^3$. Collect these results:

$$R^0_a = -e^0 e^a, \quad R^1_2 = -e^1 e^2, \quad R^0_3 = -e^0 e^3, \quad R^a_3 = -e^a e^3 \tag{43}$$

We now read off immediately (it's that easy!):

$$R^0_{303} = -1 = -R^3_{030}, \quad R^a_{3a3} = -1 = R^3_{a3a}, \quad R^0_{a0a} = -1 = -R^a_{0a0}, \quad R^a_{bab} = -1 = R^b_{aba} \tag{44}$$

(Here we have used (6) and (7).) Contracting Riemann, we obtain Ricci:

$$R_{33} = R^0_{303} + \sum_a R^a_{3a3} = -1 - 1 - 1 = -3 \tag{45}$$

$$R_{00} = R^3_{030} + \sum_a R^a_{0a0} = +1 + 2 = +3 \tag{46}$$

$$R_{aa} = R^0_{a0a} + R^b_{aba} + R^3_{a3a} = -1 - 1 - 1 = -3 \tag{47}$$

These results are summarized by

$$R_{\alpha\beta} = -3\eta_{\alpha\beta} \tag{48}$$

The spacetime is maximally symmetric.

Exercises

1   Warped polar coordinates: using forms, calculate the curvature for $ds^2 = dr^2 + f(r)^2 d\theta^2$, with $\theta = \theta + 2\pi$ an angular coordinate. Determine the $f(r)$ that gives constant curvature. Verify that measuring the circumference of a small circle around the origin gives the same result.

2   Consider the class of spaces described by $ds^2 = y^{2p} dx^2 + x^{2p} dy^2$. Use differential forms to find the curvature as a function of p and determine the two values of p for which the space is flat. Hint: Euclidean space is a member of this class. Recall exercise VI.1.17.

3   Using forms, calculate the curvature of the torus. Recall that the metric was worked out in exercise I.5.16.

4   For $ds^2 = -dt^2 + A^2(t) dx^2 + B^2(t) dy^2 + C^2(t) dz^2$, calculate the curvature using differential forms. Solve for the Kasner universe in exercise VI.2.1. Extend your work to higher dimensions.

IX.9 | Conformal Algebra

Conformal transformation

The reader learning Einstein gravity for the first time can safely skip over this chapter. I will use conformal algebra only in the next chapter and in the chapter on twistors, and then only peripherally.

Recall that in chapter IX.6, an isometry is defined as a transformation $x \to x'(x)$ under which $g'_{\rho\sigma}(x') = g_{\rho\sigma}(x')$. Suppose we feel more relaxed, and, instead, impose the more forgiving condition that $g'_{\rho\sigma}(x') = \Omega^2(x')g_{\rho\sigma}(x')$ for some unknown function Ω. To use a terminology first introduced in chapter I.6, we do not demand that the two metrics $g'_{\mu\nu}$ and $g_{\mu\nu}$ are equal, but merely that they are conformally related.

In other words, we ask whether $g_{\mu\nu}(x)\frac{\partial x^\mu}{\partial x'^\rho}\frac{\partial x^\nu}{\partial x'^\sigma} = \Omega^2(x')g_{\rho\sigma}(x')$, an easier-going version of (IX.6.1), has any solutions. As before, we content ourselves with an infinitesimal transformation $x'^\mu = x^\mu + \varepsilon\xi^\mu(x)$. In the small ε limit, we expand $\Omega^2(x') \simeq 1 + \varepsilon\kappa(x') = 1 + \varepsilon\kappa(x) + O(\varepsilon^2)$. Collecting terms of order ε, we find that this condition amounts to what is known as the conformal Killing condition

$$g_{\mu\sigma}\partial_\rho\xi^\mu + g_{\rho\nu}\partial_\sigma\xi^\nu + \xi^\lambda\partial_\lambda g_{\rho\sigma} + \kappa g_{\rho\sigma} = 0 \tag{1}$$

We can eliminate the unknown function $\kappa(x)$ by contracting with $g^{\rho\sigma}$, so that this leads to a condition on the metric $g_{\mu\nu}$ and the vector field ξ, known as a conformal Killing vector field. For $\kappa = 0$, the conformal Killing condition reduces to the Killing or isometry condition of chapter IX.6.

As in chapter IX.6, we can write (1) more compactly as $\xi_{\sigma;\rho} + \xi_{\rho;\sigma} + \kappa g_{\rho\sigma} = 0$ using the covariant derivative, or as (recall (IX.6.4))

$$\mathcal{L}_\xi g_{\mu\nu} = -\kappa g_{\mu\nu} \tag{2}$$

using the Lie derivative[1] along the conformal Killing vector field ξ. (Also, as in chapter IX.6, we will often drop the word "field.")

Retreat to flat spacetime

If someone hands us a metric, we could in principle find its conformal Killing vectors by solving (1).

The simplest metric to deal with is the Minkowski metric, of course. In this introductory text, we are content to study this easy case, for which (1) simplifies to (with $\xi_\sigma \equiv \eta_{\sigma\mu}\xi^\mu$, as usual) $\partial_\rho\xi_\sigma + \partial_\sigma\xi_\rho + \kappa\eta_{\rho\sigma} = 0$. Contracting this with $\eta^{\rho\sigma}$, we obtain $\kappa = -2\partial \cdot \xi/d$ in d-dimensional spacetime. Hence the condition (1) becomes

$$\partial_\rho\xi_\sigma + \partial_\sigma\xi_\rho = \frac{2}{d}\eta_{\rho\sigma}\partial \cdot \xi \tag{3}$$

Infinitesimal transformations $x'^\mu = x^\mu + \varepsilon\xi^\mu(x)$ that satisfy (3) are said to generate the conformal algebra for Minkowski spacetime. (Clearly, with the substitution of $\delta_{\rho\sigma}$ for $\eta_{\rho\sigma}$, this entire discussion applies to flat space as well as flat spacetime.)

Compare this condition with the Killing equation $\partial_\rho\xi_\sigma + \partial_\sigma\xi_\rho = 0$ for Minkowski spacetime, with the most general solution $\xi^\mu = a^\mu + b^\mu_{\ \nu}x^\nu$ describing translations and the Lorentz transformations, namely rotations and boosts. Here $b^{\mu\nu} = b^\mu_{\ \lambda}\eta^{\lambda\nu} = -b^{\nu\mu}$ is required to be antisymmetric. (This goes way back to (I.3.7) and (III.3.13).)

In search of conformal generators

At this point, you can solve (3) for ξ^μ. Go ahead. Alternatively, we could wing it like a poor man. Stare at Minkowski spacetime: $ds^2 = \eta_{\mu\nu}dx^\mu dx^\nu$. What transformations on x would change the metric conformally?

By eyeball, we see that the scale transformation, or more academically, dilation,* $x^\mu \to \lambda x^\mu$ for λ a real number, would do the job, since ds^2 becomes $ds^2 = \lambda^2\eta_{\mu\nu}dx^\mu dx^\nu$. We have stretched spacetime by a constant factor. The conformal Killing condition is satisfied with $\Omega(x)$ constant. To identify the corresponding ξ^μ, consider an infinitesimal transformation with $\lambda = 1 + \varepsilon c$; then $\xi^\mu = cx^\mu$ with some (irrelevant) constant c. Sure enough, this satisfies (3), of course. Now, can you find another transformation? Think for a minute before reading on.

The clever poor man notices that inversion, $x^\mu = e^2y^\mu/y^2$, would work.† Plug in $dx^\mu = e^2(\delta^\mu_\lambda y^2 - 2y_\lambda y^\mu)dy^\lambda/(y^2)^2$. We obtain $ds^2 = \eta_{\mu\nu}dx^\mu dx^\nu = (e^4/(y^2)^2)\eta_{\mu\nu}dy^\mu dy^\nu$, which indeed is conformally flat. I introduced e to avoid confusing you, but now that its job is

---

* Or dilatation, if dilation is not academic enough for you.

† Another (irrelevant) constant e, with dimension of length, is introduced here to ensure that x and y both have dimensions of length.

done, we will set it to 1 and define inversion as the transformation (for $x^2 \neq 0$)

$$x^\mu \to \frac{x^\mu}{x^2} \tag{4}$$

You object, saying that the entire discussion has been couched in terms of infinitesimal transformations. The inversion is a discrete transformation and is in no way no how infinitesimal. How then can we identify the corresponding ξ^μ?

Now the poor man makes another clever move: invert, translate by some vector a^μ, then invert back. For $a^\mu = 0$, the two inversions knock each other out, and we end up with the identity transformation. Thus, the net result of these three transformations would indeed be an infinitesimal transformation as $a^\mu \to 0$. Let's work out what I just said:

$$x^\mu \to \frac{x^\mu}{x^2} \to \frac{x^\mu}{x^2} + a^\mu \to \left(\frac{x^\mu}{x^2} + a^\mu\right) \Big/ \eta_{\rho\sigma}\left(\frac{x^\rho}{x^2} + a^\rho\right)\left(\frac{x^\sigma}{x^2} + a^\sigma\right)$$

$$= \left(\frac{x^\mu}{x^2} + a^\mu\right) \Big/ \left(\frac{1}{x^2} + \frac{2a \cdot x}{x^2} + a^2\right)$$

$$= (x^\mu + a^\mu x^2)/(1 + 2a \cdot x + a^2 x^2) \simeq (x^\mu + a^\mu x^2)(1 - 2a \cdot x) + O(a^2)$$

$$= x^\mu + a_\lambda(\eta^{\mu\lambda}x^2 - 2x^\mu x^\lambda) + O(a^2) \tag{5}$$

The transformation $x^\mu \to x^\mu + a_\lambda(\eta^{\mu\lambda}x^2 - 2x^\mu x^\lambda)$ is known as a conformal transformation. You can verify that $\xi^\mu = a_\lambda(\eta^{\mu\lambda}x^2 - 2x^\mu x^\lambda)$ satisfies (3), of course, since inversion and translation both satisfy the condition we started out with.

As I said, you could have also simply solved (3) by brute force, and I am counting on you to have already done so. It is also instructive to act with $\partial^\rho \equiv \eta^{\rho\sigma}\partial_\sigma$ on (3); we obtain

$$d\partial^2\xi_\sigma = (2 - d)\partial_\sigma(\partial \cdot \xi) \tag{6}$$

Applying ∂^σ, we obtain further $\partial^2(\partial \cdot \xi) = 0$ (all for $d \neq 1$).

We can now draw two important conclusions.

1. The case $d = 2$ is special. We learn from (6) that any solutions of the generalized Laplace equation $\partial^2\xi_\nu = 0$ yield a conformal transformation. Indeed, for $d = 2$, either go to light cone coordinates for Minkowski spacetime or to complex coordinates for Euclidean space. With complex coordinates $z = x + iy$, we have $(\partial_x^2 + \partial_y^2)\xi_\sigma = (\partial_x + i\partial_y)(\partial_x - i\partial_y)\xi_\sigma = \frac{\partial}{\partial z^*}\frac{\partial}{\partial z}\xi_\sigma = 0$, and hence we can exploit the full power of complex analysis. This observation turns out to be of central importance in string theory.[2] For $d = 2$, there exists an infinite number of solutions of (3) for ξ^μ.

2. For $d \neq 2$, these equations tell us that ξ^μ can depend on x at most quadratically. Thus, we have in fact found all solutions of (3) for $d \neq 2$, namely

$$\xi^\mu = a^\mu + b^\mu_{\;\nu} x^\nu + c x^\mu + d_\nu (\eta^{\mu\nu} x^2 - 2 x^\mu x^\nu) \tag{7}$$

with $b^{\mu\nu}$ antisymmetric. We had noted all these terms already. Pleasingly, in (7), the constant term corresponds to translation, the linear terms to Lorentz transformation and to dilation, and the quadratic term to conformal transformation.

Generators of conformal algebra

Associated with each of these terms in (7), we have a generator of the Minkowskian conformal algebra. As in chapter I.3, it is convenient to use a differential operator representation. Recall that back in chapter III.3, by adding the generators* of translation P_μ to those of Lorentz transformation $J_{\mu\nu}$, we extended the Lorentz algebra to the Poincaré algebra, defined by commuting

$$P_\mu = \partial_\mu \quad \text{and} \quad J_{\mu\nu} = \left(x_\mu \partial_\nu - x_\nu \partial_\mu \right) \tag{8}$$

By adding the dilation generator D and the conformal generator K^μ,

$$D = x^\mu \partial_\mu \quad \text{and} \quad K^\mu = \left(\eta^{\mu\nu} x^2 - 2 x^\mu x^\nu \right) \partial_\nu \tag{9}$$

we can now, in turn, extend the Poincaré algebra to the conformal algebra, defined by commuting P_μ, $J_{\mu\nu}$, D, and K_μ.

In other words, the commutators between P, J, D, and K generate an algebra that contains the Poincaré algebra.

The commutators involving D are easy to compute: $[D, x^\nu] = [x^\mu \partial_\mu, x^\nu] = x^\mu [\partial_\mu, x^\nu] = x^\nu$ and $[D, \partial_\nu] = [x^\mu \partial_\mu, \partial_\nu] = [x^\mu, \partial_\nu] \partial_\mu = -\partial_\mu$. (To work out various commutators, keep in mind the identity $[A, BC] = [A, B]C + B[A, C]$.) Evidently, D, as is sensible for a dilation generator, simply counts the length dimension, $+1$ for x^ν and -1 for ∂_ν. Thus, $[D, J_{\mu\nu}] = 0$, since $J \sim x\partial$ has zero length dimension. Interestingly, another way of reading this is to write it as $[J_{\mu\nu}, D] = 0$, which says that D is a Lorentz scalar. Next, we can read off $[D, P^\mu] = -P^\mu$ and $[D, K^\mu] = +K^\mu$ just by counting powers of length dimension ($P \sim \partial$, $K \sim xx\partial$).

The commutators involving K^μ are not much harder to work out. First, $[J^{\mu\nu}, K^\lambda] = -\eta^{\mu\lambda} K^\nu + \eta^{\nu\lambda} K^\mu$ just tells us that K^μ transforms like a vector, as expected. The nontrivial commutator is $[K^\mu, P^\lambda] = -[\partial^\lambda, (\eta^{\mu\nu} x^2 - 2 x^\mu x^\nu) \partial_\nu] = -2(\eta^{\mu\nu} x^\lambda - \eta^{\lambda\mu} x^\nu - x^\mu \eta^{\lambda\nu}) \partial_\nu = 2(J^{\mu\lambda} + \eta^{\lambda\mu} D)$. Finally, verify that $[K^\mu, K^\nu] = 0$. Can you see why? (Recall that we constructed the conformal transformation as an inversion followed by a translation and then followed by another inversion.)

---

* Here I omit the overall factors of i commonly included in quantum mechanics. As explained in chapter III.3, you and I live in free countries and, according to what is convenient in a given context, could include or omit overall factors at will.

Collecting our results, we have the conformal algebra

$$[P^\mu, P^\nu] = 0, \qquad [K^\mu, K^\nu] = 0$$

$$[D, P^\mu] = -P^\mu, \qquad [D, J_{\mu\nu}] = 0, \qquad [D, K^\mu] = +K^\mu$$

$$[J^{\mu\nu}, P^\lambda] = -\eta^{\mu\lambda} P^\nu + \eta^{\nu\lambda} P^\mu, \qquad [J^{\mu\nu}, K^\lambda] = -\eta^{\mu\lambda} K^\nu + \eta^{\nu\lambda} K^\mu$$

$$[J^{\mu\nu}, J^{\lambda\rho}] = -\eta^{\mu\lambda} J^{\nu\rho} - \eta^{\nu\rho} J^{\mu\lambda} + \eta^{\nu\lambda} J^{\mu\rho} + \eta^{\mu\rho} J^{\nu\lambda}$$

$$[K^\mu, P^\nu] = 2(J^{\mu\nu} + \eta^{\mu\nu} D) \tag{10}$$

We see that, in some sense, K acts like the dual of P.

Identifying the conformal algebra

Now that we have used our eyeballs and brains, let's use our fingers. Count the number of generators (P, K, D, J): $d + d + 1 + \frac{1}{2}d(d-1) = \frac{1}{2}(d+2)(d+1)$. Do you know a group with this many generators?

Yes, $SO(d, 2)$. Good guess!

Remarkably, the conformal algebra of d-dimensional Minkowski spacetime with the Lorentz group $SO(d-1, 1)$ is the Lie algebra of $SO(d, 2)$. The two algebras are isomorphic. The rule is that given $SO(d-1, 1)$, the conformal algebra is $SO(d-1+1, 1+1) = SO(d, 2)$: we "go up" by $(1, 1)$. We can prove this assertion by the "what else could it be" argument. (The only uncertainty is the signature. Counting only tells us that it could be the algebra for the group $SO(p, q)$ with $p + q = d + 2$ and containing $SO(d-1, 1)$.) We can of course verify the assertion by direct computation and thus also ascertain the signature.

Denote the generators of $SO(d, 2)$ by J^{MN}, with $M, N = 0, 1, 2, \cdots, d-1, d, d+1$ (and $\mu, \nu = 0, 1, 2, \cdots, d-1$) satisfying

$$[J^{MN}, J^{PQ}] = -\eta^{MP} J^{NQ} - \eta^{NQ} J^{MP} + \eta^{NP} J^{MQ} + \eta^{MQ} J^{NP} \tag{11}$$

with $\eta^{dd} = -1$ and $\eta^{d+1,d+1} = +1$. The isomorphism between the two algebras is almost fixed by symmetry considerations. We already have the generators of the Lorentz group $SO(d, 1)$, namely $J^{\mu\nu}$. Now we want to identify the additional generators D, P^μ, and K^μ. By eyeball, we see that D is a scalar under $SO(d, 1)$, and so it can only be $J^{d,d+1}$. We identify $J^{d,d+1} = D$. Similarly, by eyeball, we see that P^μ and K^μ carry an index μ, and hence are vectors under $SO(d, 1)$. They can only be linear combinations of $J^{\mu,d}$ and $J^{\mu,d+1}$. So, let us make the educated guess $J^{\mu,d} = (K^\mu + P^\mu)/2$ and $J^{\mu,d+1} = (K^\mu - P^\mu)/2$. We will check only a few commutators to show that this assignment is correct. For example, (11) gives

$$[J^{\mu,d}, J^{\nu,d}] = -\eta^{dd} J^{\mu\nu} = J^{\mu\nu} = \frac{1}{4}[K^{\mu} + P^{\mu}, K^{\nu} + P^{\nu}]$$

$$= \frac{1}{4}([K^{\mu}, P^{\nu}] - [K^{\nu}, P^{\mu}]) = \frac{1}{4}(4J^{\mu\nu})$$

where in the last step, we used (10). Similarly,

$$[J^{\mu,d+1}, J^{\nu,d+1}] = -\eta^{d+1,d+1} J^{\mu\nu} = -J^{\mu\nu} = \frac{1}{4}\left[K^{\mu} - P^{\mu}, K^{\nu} - P^{\nu}\right] = -\frac{1}{4}\left(4J^{\mu\nu}\right)$$

As another example,

$$\left[D, \frac{1}{2}\left(K^{\mu} + P^{\mu}\right)\right] = \left[J^{d,d+1}, J^{\mu,d}\right] = \eta^{dd} J^{d+1,\mu} = \frac{1}{2}\left(K^{\mu} - P^{\mu}\right)$$

The poor man now speaks up, "It is easier to see through all this if we pick a definite d, say 6, and forget about signature, let it take care of itself. Just think about $SO(6)$." Evidently, $J^{\mu\nu}$, for μ, $\nu = 1, 2, 3, 4$, generates the rotation algebra for $SO(4)$. In addition, we have $J^{\mu,5}$ and $J^{\nu,6}$, clearly vectors labeled by 5 and 6. For $\mu \neq \nu$, they commute with each other, while for $\mu = \nu$, they commute to produce J^{56}. Recall, as we learned way back in chapter I.3, that (11) merely says that J^{MN} and J^{PQ} commute with each other, unless a pair of indices, one from each of the Js, are equal, in which case the commutator is a J carrying the remaining two indices. Thus, J^{56} commuted with $J^{\mu,5}$, and $J^{\mu,6}$ just turns one into the other.

(1+1)-dimensional Minkowski spacetime in light cone coordinates

It is instructive to work out the conformal algebra for a familiar spacetime written in not-so-familiar coordinates, namely the $(1 + 1)$-dimensional Minkowski spacetime written in light cone coordinates (as was described[3] in appendix 5 to chapter VII.2). Define $x^{\pm} = t \pm x$. Then $ds^2 = -dt^2 + dx^2 = -dx^+ dx^- = \eta_{\mu\nu} dx^{\mu} dx^{\nu}$, which tells us that $\eta_{+-} = \eta_{-+} = -\frac{1}{2}$ and $\eta^{+-} = \eta^{-+} = -2$. (For this discussion, we adopt the convention that the components we do not display, such as η^{++}, all vanish.) Also, define $\partial_{\pm} = \frac{1}{2}(\frac{\partial}{\partial t} \pm \frac{\partial}{\partial x})$, so that $\partial_+ x^+ = 1$ and $\partial_- x^- = 1$.

Then $P^{\pm} = \partial^{\pm} = 2\partial_{\mp}$, $D = x^+ \partial_+ + x^- \partial_-$, $J \equiv \frac{1}{2} J^{+-} = \frac{1}{2}(x^+ \partial^- - x^- \partial^+) = x^+ \partial_+ - x^- \partial_-$. Note that $D \pm J = 2x^{\pm} \partial_{\pm}$ works out nicely. (It is understood that the $\pm$ signs are correlated unless otherwise noted.) Can you guess what the conformal generators are? Let's find out; simply evaluate (9): $K^+ = x^2 \eta^{+-} \partial_- - 2x^+(x^+ \partial_+ + x^- \partial_-) = -2(x^+)^2 \partial_+$, and similarly, $K^- = -2(x^-)^2 \partial_-$.

Rather elegantly, the 6 generators of $SO(2, 2)$ can be taken to be

$$\partial^{\pm}, \quad x^{\pm} \partial_{\pm}, \quad \text{and} \quad -\left(x^{\pm}\right)^2 \partial_{\pm} \tag{12}$$

Recall that in chapter VII.2, we constructed the Penrose diagram for $M^{1,1}$, introducing the compact variables $X^{\pm}$ by $x^{\pm} = \tan X^{\pm}$. Note that $\frac{\partial}{\partial X^{\pm}} = (1 + (x^{\pm})^2)\partial_{\pm}$. In the conformally equivalent spacetime described by the cylinder $R \times S^1$, the time coordinate is given by $T = X^+ + X^-$. Time translation along the cylinder is then generated by

$$\frac{\partial}{\partial T} = \frac{1}{2} \left(\frac{\partial}{\partial X^+} + \frac{\partial}{\partial X^-} \right) = \frac{1}{2} \left(\partial_+ + \partial_- + (x^+)^2 \partial_+ + (x^-)^2 \partial_- \right) \tag{13}$$

You can now check the algebra in (10). For example, (10) gives $[K^+, P^-] = 2(J^{+-} + \eta^{+-}D) = 4(J + D) = 8x^+\partial_+$, and indeed, we compute $[K^+, P^-] = [-2(x^+)^2\partial_+, 2\partial_+] = 8x^+\partial_+$.

To the lost, angles are more important than distances

Some readers are no doubt already aware of the many motivations—historical, mathematical, and physical—for studying conformal transformations. Here I mention but a few. The key property is of course that conformal transformations preserve angles between line segments.

When you are lost, it matters more to you to know that you are going in the right direction than to know how far you are from your destination. To the lost, angles are more important than distances. Gerardus Mercator (1512–1594) (or "Jerry the merchant") fully appreciated this. As you worked out in exercise I.5.3, the Mercator map of the world is obtained by a conformal transformation of the spherical coordinates (θ, φ) on the globe. Mathematically, I already mentioned the connection to complex analysis and the consequent implications for string theory. A humbler, but no less beautiful, physical motivation for studying the conformal map is the method of images in electrostatics.

I cannot resist digressing a bit to remind you how it works. Consider the following problem. A charge q is located at $R\vec{e}_q$ in the presence of a conducting sphere of radius a grounded and centered at the origin. Calculate the potential $\phi(\vec{r})$ at an arbitrary point $\vec{r} = r\vec{e}$. (The unit vectors $\vec{e}$ and $\vec{e}_q$ point toward the observer and the charge, respectively.)

We take the potential due to the charge q and subtract from it the potential due to an image charge $\tilde{q}$ located at $\tilde{R}\vec{e}_q$. (Note that we invoke implicitly a symmetry argument to locate the image charge along the vector $\vec{e}_q$.) Then we have

$$\phi(\vec{r}) = \frac{q}{|r\vec{e} - R\vec{e}_q|} - \frac{\tilde{q}}{|r\vec{e} - \tilde{R}\vec{e}_q|} = \frac{q}{r|\vec{e} - \frac{R}{r}\vec{e}_q|} - \frac{\tilde{q}}{\tilde{R}|\vec{e}_q - \frac{r}{\tilde{R}}\vec{e}|} \tag{14}$$

Using the key observation that $|\vec{e} - K\vec{e}_q| = |\vec{e}_q - K\vec{e}|$, we see that we can satisfy the boundary condition $\phi(r = a, \theta, \varphi) = 0$ if we choose $\frac{\tilde{q}}{\tilde{R}} = \frac{q}{R}$ and $\frac{a}{\tilde{R}} = \frac{R}{a}$. The locations of the charge and of its image are related by an inversion $\tilde{R} = \frac{a^2}{R}$.

That this clever method works is due to the scale, or dilation, invariance of the Coulomb potential. It would not work, for example, with the short-ranged Yukawa potential $\propto e^{-mr}/r$. In other words, in electrostatics, we are solving Laplace's equation with appropriate boundary conditions, and Laplace's equation does not contain any intrinsic length scale. (Under inversion, we have to adjust the boundary condition accordingly; this is why we have to adjust the strength of the image charge.)

Scale and conformal invariances in particle and condensed matter physics

It is one thing to discuss the conformal algebra, but it is another to ascertain whether a given physical situation actually respects conformal invariance. In particle physics, one longstanding hope has been that at high energies, particle masses can be neglected, so that the physics would become scale invariant. It turns out that in a local field theory, it is true, more or less in general, that scale invariance typically leads to conformal invariance.* For example, we will check in appendix 1 that Maxwell's action, $\sim \int d^4x\, F_{\mu\nu} F^{\mu\nu}$, discussed in chapter IV.2, is both scale and conformal invariant. In condensed matter physics, intrinsic length scales are typically washed out at the critical point between two phases, so that scale and conformal invariances come in full blast into the theory of critical phenomena.[†]

I included this material on conformal algebra not so much because I will refer to it in the next chapter on de Sitter spacetime, but because the concepts involved are important. Indeed, as I am completing this book, the $\mathcal{N} = 4$ supersymmetric Yang-Mills theory is all the rage in the theoretical community. Not only does this theory have a conformal algebra, it also has a superconformal algebra.

Appendix 1: Maxwell's action is both scale and conformal invariant

Since $A'_\mu(x') = A_\rho(x) \frac{\partial x^\rho}{\partial x'^\mu}$, we have, under an infinitesimal transformation $x' = x - \xi$ with ξ arbitrary,

$$\delta A_\mu(x) \equiv A'_\mu(x) - A_\mu(x) = (A'_\mu(x) - A'_\mu(x')) + (A'_\mu(x') - A_\mu(x))$$
$$= \xi^\rho \partial_\rho A_\mu + A_\rho \partial_\mu \xi^\rho \tag{15}$$

(we are also suppressing ε and introducing a minus sign for convenience). The attentive reader will recall from chapter V.6 that this is just the Lie derivative $\mathcal{L}_\xi A_\mu(x)$ defined in (V.6.26). Similarly, $\delta F_{\mu\nu} = \mathcal{L}_\xi F_{\mu\nu} = \xi^\rho \partial_\rho F_{\mu\nu} + F_{\rho\nu} \partial_\mu \xi^\rho + F_{\mu\rho} \partial_\nu \xi^\rho$.

We can now evaluate the variation of the Maxwell Lagrangian $\mathcal{L} = -\frac{1}{4} F^{\mu\nu} F_{\mu\nu}$ for an arbitrary ξ. (A trivial heads-up: the symbols $\mathcal{L}$ and $\mathcal{L}_\xi$ denote entirely different beasts.) We have

$$\delta\left(F^{\mu\nu} F_{\mu\nu}\right) = 2F^{\mu\nu}\left(\xi^\rho \partial_\rho F_{\mu\nu} + F_{\rho\nu}\partial_\mu\xi^\rho + F_{\mu\rho}\partial_\nu\xi^\rho\right)$$

$$= \partial_\rho\left(\xi^\rho F^{\mu\nu} F_{\mu\nu}\right) - \partial\cdot\xi\left(F^{\mu\nu} F_{\mu\nu}\right) + 4F^{\mu\rho} F^\nu_{\ \rho}\partial_\mu\xi_\nu$$

$$= \partial_\rho\left(\xi^\rho F^{\mu\nu} F_{\mu\nu}\right) + 2F^{\mu\rho} F^\nu_{\ \rho}\left(\partial_\mu\xi_\nu + \partial_\nu\xi_\mu - \tfrac{1}{2}\eta_{\mu\nu}\partial\cdot\xi\right) \tag{16}$$

---

* This is because the violation of scale invariance and conformal invariance are both determined[4] by the trace $T^\mu_{\ \mu}$ of the energy momentum tensor. Indeed, we saw a hint of this in our discussion of the relativistic gas and of the electromagnetic field in chapters III.6 and VI.4.

[†] In both these types of physical applications, subtleties due to quantum and thermal fluctuations must be taken into account.

Thus far, ξ is arbitrary, but if ξ is a conformal Killing vector, that is, if it satisfies (3), then we obtain

$$\delta \mathcal{L} = \partial_\rho \left(\mathcal{L} \xi^\rho \right) + \left(\frac{4}{d} - 1 \right) (\partial \cdot \xi) \mathcal{L} \tag{17}$$

For $d = 4$, and only for $d = 4$, the variation $\delta\mathcal{L}$ of the Maxwell Lagrangian under a conformal Killing transformation is a total divergence, so that the variation of the action $\delta S = \int d^4x \, \delta\mathcal{L}$ vanishes (with the usual suitable boundary conditions at spacetime infinity). Thus, in the spacetime we live in, the Maxwell action* is both scale and conformal invariant. (In particular, it is also inversion invariant, so you can go on happily using the method of images.) As a bonus, we also reconfirm what we have known for a long time, ever since part IV, that it is translation and Lorentz invariant. And of course, the last property is what got us started on this amazing epic journey toward the heart of spacetime.

Appendix 2: Conformally related spacetimes

For the sake of pedagogical clarity, we hastily retreated from the conformal Killing condition (1) in all its glory to its humbler flat version (3). Nevertheless, it is often useful to study what happens in an arbitrary curved spacetime. To be specific, let us look at Maxwell's action $S_{\text{Maxwell}}(g_{\mu\nu}, A_\mu) = -\frac{1}{4e^2} \int d^d x \sqrt{-g} \, g^{\mu\nu} g^{\sigma\rho} F_{\mu\sigma} F_{\nu\rho}$ from this alternative point of view. We have also indicated explicitly that S_{Maxwell} is a functional of $g_{\mu\nu}$ and A_μ.

Now suppose that somebody hands you another metric $\tilde{g}_{\mu\nu}(x) = \Omega^2(x) g_{\mu\nu}(x)$ conformally related to the metric we have. (Perhaps it is still worthwhile to emphasize that the two metrics are not related by a coordinate transformation.) Since $\tilde{g}^{\mu\nu} = \Omega^{-2} g^{\mu\nu}$ and $\tilde{g} = \Omega^{2d} g$, we have $S_{\text{Maxwell}}(\tilde{g}_{\mu\nu}, A_\mu) = -\frac{1}{4e^2} \int d^d x \sqrt{-g} \, g^{\mu\nu} g^{\sigma\rho} \Omega^{d-4} F_{\mu\sigma} F_{\nu\rho}$. Thus, for $d = 4$, and only for $d = 4$, we have $S_{\text{Maxwell}}(\tilde{g}_{\mu\nu}, A_\mu) = S_{\text{Maxwell}}(g_{\mu\nu}, A_\mu)$.

In general, if we are given an action in curved spacetime such that $S(\Omega^2 g_{\mu\nu}, \cdots) = S(g_{\mu\nu}, \cdots)$, where the ellipses indicate various fields we are not touching at all (such as A_μ in the specific example just given), we can immediately take the infinitesimal limit $\Omega^2(x) \simeq 1 + \varepsilon(x)$, so that $\delta g_{\mu\nu}(x) \equiv \tilde{g}_{\mu\nu}(x) - g_{\mu\nu}(x) = \varepsilon(x) g_{\mu\nu}(x)$, and deduce

$$\delta S = 0 = \int d^4 x \frac{\delta S}{\delta g_{\mu\nu}(x)} \varepsilon(x) g_{\mu\nu}(x) = -\frac{1}{2} \int d^4 x \sqrt{-g} \, \varepsilon(x) g_{\mu\nu}(x) T^{\mu\nu}(x) \tag{18}$$

Since $\varepsilon(x)$ is arbitrary and local, we conclude that the trace $T(x) \equiv g_{\mu\nu}(x) T^{\mu\nu}(x) = 0$ vanishes, a result we have already derived in chapters III.6 and VI.4.

Here it is important to let $\Omega(x)$ depend on x, so that we can deduce from (18) that the trace $T(x)$ vanishes locally, that is, at any x. But to demonstrate that $S_{\text{Maxwell}}(g_{\mu\nu}, A_\mu)$ is invariant for $d = 4$, as we did a bit earlier, we could have taken a constant Ω independent of x. In our demonstration, we merely counted powers of Ω, and no derivative ever acted on Ω. In other words, we could consider simply multiplying the metric by a constant:[5] $g_{\mu\nu} \to \Omega^2 g_{\mu\nu}$, $g^{\mu\nu} \to \Omega^{-2} g^{\mu\nu}$, and $g \to \Omega^8 g$. At the risk of being repetitious, $S_{\text{Maxwell}} = -\frac{1}{4e^2} \int d^d x \sqrt{-g} \, g^{\mu\nu} g^{\sigma\rho} F_{\mu\sigma} F_{\nu\rho}$ is invariant because $\sqrt{-g} \, g^{\mu\nu} g^{\sigma\rho} \to \Omega^4 \Omega^{-2} \Omega^{-2} \sqrt{-g} \, g^{\mu\nu} g^{\sigma\rho} = \sqrt{-g} \, g^{\mu\nu} g^{\sigma\rho}$.

But now we can make a connection with high school dimensional analysis. Scale $x \to \omega x$, with ω an arbitrary real number. Then we have $\partial \to \omega^{-1} \partial$. Gauge invariance requires that A_μ scales the same way as ∂_μ, so that $A \to \omega^{-1} A$ and $F \to \omega^{-2} F$. In fact, you see that the powers of ω just correspond to the length dimensions of various quantities.[6] For example, x has length dimension $+1$ (by definition), and F has length dimension -2. So, the invariance of S_{Maxwell} for $d = 4$ is just the statement that $\int d^4 x \sqrt{-g} \, g^{\mu\nu} g^{\sigma\rho} F_{\mu\sigma} F_{\nu\rho} \to \omega^4 \omega^{-2} \omega^{-2} \int d^4 x \sqrt{-g} \, g^{\mu\nu} g^{\sigma\rho} F_{\mu\sigma} F_{\nu\rho}$ is dimensionless in length. By now, you also see the connection with the scaling in the preceding paragraph: the scaling $\sqrt{-g} \to \Omega^4 \sqrt{-g}$ corresponds to the scaling $d^4 x \to \omega^4 d^4 x$ for the proverbial high school student, and the scaling $g^{\mu\nu} g^{\sigma\rho} \to \Omega^{-4} g^{\mu\nu} g^{\sigma\rho}$ corresponds to the scaling $F_{\mu\sigma} F_{\nu\rho} \to \omega^{-4} F_{\mu\sigma} F_{\nu\rho}$. Note that, perhaps amusingly, in the high school approach, we do not touch the metric, while in the conformal transformation, we touch only the metric.

We now see that what we did in chapter VI.1 amounts to saying that $\int d^4 x \sqrt{-g} R$ in Einstein gravity has length dimension $+2$: four powers of length from $d^4 x$ and two negative powers of length from the two ∂s contained in R.

---

* You should convince yourself that for $d \neq 4$, the Maxwell action is manifestly not scale invariant. In contrast, Laplace's equation, in any spatial dimension, contains no scale.

Exercises

1 Show that under inversion, $(x_1 - x_2)^2 \to (x_1 - x_2)^2/(x_1^2 x_2^2)$ and thus the separation between spacetime points is not preserved. However, if the two points are null separated, they remain null separated under inversion. Null separation is a conformally invariant concept.

2 In the text, the poor man realizes that inversion $x^\mu = e^2 y^\mu/y^2$ of the Minkowski metric gives a conformally invariant metric. How about the transformation $x^\mu = f^4 y^\mu/(y^2)^2$, with f a constant with dimensions of length?

3 Sometimes the best way to learn a formalism is to apply it to a trivial problem to which we know the answer. Determine the conformal Killing vector fields of the Euclidean plane.

Notes

1. If ξ^a denotes a set of conformal Killing vectors for $a = 1, \cdots, n$ for some n, you can show that the commutators $[\mathcal{L}_{\xi^a}, \mathcal{L}_{\xi^b}]$ generate an algebra known as the conformal algebra.
2. See J. Polchinski, *String Theory*, chapter 2.
3. The coordinates (p, q) and (P, Q) there correspond to $x^\pm$ and $X^\pm$ here, respectively.
4. For a concise statement of when scale invariance implies conformal invariance, see Y. Nakayama, "Gravity Dual for a Model of Perception," http://arxiv.org/pdf/1003.5729.
5. You may recall that we did this kind of scaling to check our computations back in chapter VI.2.
6. Looking ahead, we will be using this sort of reasoning in chapter X.3.

Which curved spacetime is the most lovable?

Of all the curved spaces, we love the sphere most. This is of course due to the high degree of symmetry enjoyed by the sphere in all its manifestations, including the circle. In particular, every point on the sphere is identical to any other point: the sphere is a homogeneous space. Indeed, as explained in chapter IX.6 on isometry, it is maximally symmetric.

Among the curved spacetimes, which one should we love the most? Which spacetimes are closest to the spheres? Kepler talked about the music of the spheres; we've become somewhat more sophisticated 400 years later.

De Sitter spacetime

The d-dimensional sphere S^d of radius L is defined as the set of all points $(X^1, X^2, \cdots, X^{d+1})$ in $(d+1)$-dimensional Euclidean space E^{d+1} (that is, a space with $ds^2 = (dX^1)^2 + (dX^2)^2 + \cdots + (dX^d)^2 + (dX^{d+1})^2$) satisfying

$$(X^1)^2 + (X^2)^2 + \cdots + (X^d)^2 + (X^{d+1})^2 = L^2 \quad \text{(sphere } S^d\text{)} \tag{1}$$

In analogy, let us define the d-dimensional de Sitter spacetime[1,2] dS^d with length scale L as the set of all points $(X^0, X^1, X^2, \cdots, X^d)$ in $(d+1)$-dimensional Minkowskian space-time $M^{d,1}$ (that is, a spacetime with $ds^2 = -(dX^0)^2 + (dX^1)^2 + (dX^2)^2 + \cdots + (dX^d)^2$) satisfying

$$-(X^0)^2 + (X^1)^2 + (X^2)^2 + \cdots + (X^d)^2 = L^2 \quad \text{(de Sitter spacetime } dS^d\text{)} \tag{2}$$

We have renamed X^{d+1} as X^0 and by a feat of "imaginary magic" turned it into a time coordinate. Thus, de Sitter spacetime is sort of a Minkowskian version of the sphere living in Minkowski spacetime.

This flip of sign makes all the difference in the world: at a given X^0, the spatial coordinates $(X^1, X^2, \cdots, X^d)$ form a $(d-1)$-dimensional sphere S^{d-1} defined by $(X^1)^2 +$

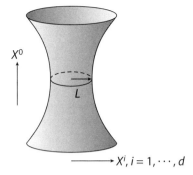

Figure 1 The d-dimensional de Sitter space-time dS^d embedded in $(d + 1)$-dimensional Minkowskian spacetime $M^{d,1}$.

$(X^2)^2 + \cdots + (X^d)^2 = L^2 + (X^0)^2$. Topologically, de Sitter spacetime is then $R \times S^{d-1}$: as the time coordinate X^0 goes from $-\infty$ to ∞, the radius $\sqrt{L^2 + (X^0)^2}$ of S^{d-1} starts at infinity, contracts to a minimum value L, and then expands again to infinity. See figure 1. Contrast the circles of constant latitude on the globe, expanding from the south pole to the equator and then contracting again toward the north pole. In the $(X^0\text{-}X^d)$ plane, we have a hyperbola, and so dS^d can also be regarded as a hyperboloid of rotation.

A word of caution about figures of this type: We naturally tend to look at it as if it were drawn in Euclidean space, while in fact, dS^d is constructed in Minkowskian spacetime $M^{d,1}$.

Maximal symmetry and coset manifold

The isometry group of S^d is clearly $SO(d + 1)$, the rotation group of the embedding space E^{d+1}, with the Killing generators $\left(X^M \frac{\partial}{\partial X^N} - X^N \frac{\partial}{\partial X^M}\right)$, $M, N = 1, 2, \cdots, X^{d+1}$. The sphere S^d can thus be regarded as the coset manifold $SO(d + 1)/SO(d)$, where the quotient group $SO(d)$ is the subgroup of $SO(d + 1)$ that leaves a point on S^d invariant, as we discussed in chapter IX.6. (Think about this for $d = 2$.)

Evidently, the isometry group[3] of de Sitter spacetime dS^d is $SO(d, 1)$, the Lorentz group of the embedding space $M^{d,1}$. The Killing generators fall into two sets, d-dimensional rotations and boosts:

$$X^M \frac{\partial}{\partial X^N} - X^N \frac{\partial}{\partial X^M} \quad \text{and} \quad X^M \frac{\partial}{\partial X^0} + X^0 \frac{\partial}{\partial X^M}, \quad \text{for } M, N = 1, 2, \cdots, d \tag{3}$$

obtained by letting $X^{d+1} \to iX^0$ formally. Note the sign flip between the two sets, just as in the familiar Lorentz algebra of special relativity.

Hence, just like the sphere S^d, de Sitter spacetime is also a coset manifold: $dS^d = SO(d, 1)/SO(d - 1, 1)$. Consider, for example, the point $X_* = (X^0, X^1, X^2, \cdots, X^d)_* = (0, 0, \cdots, 1)$ on dS^d: it is left invariant by the subgroup $SO(d - 1, 1)$, namely the Lorentz group acting on the d coordinates $(X^0, X^1, X^2, \cdots, X^{d-1})$. In particular, $dS^4 = SO(4, 1)/SO(3, 1)$. As a small check, note that $SO(4, 1)$ has $\frac{5 \cdot 4}{2} = 10$ generators, while

$SO(3, 1)$ has $\frac{4 \cdot 3}{2} = 6$ generators, so that $SO(4, 1)/SO(3, 1)$ is indeed $10 - 6 = 4$ dimensional.

The group $SO(d, 1)$ moves points on dS^d around. We thus conclude that, just like the sphere, de Sitter spacetime is maximally symmetric. So, according to the general theory of maximally symmetric spaces explained in chapter IX.6, the Riemann curvature tensor $R_{\mu\nu\lambda\sigma}$ must be equal to $(g_{\mu\lambda}g_{\nu\sigma} - g_{\mu\sigma}g_{\nu\lambda})$ up to an overall constant. (Here the Greek indices* range over $\mu = 0, 1, \cdots, d - 1$.) Now notice that we have constructed de Sitter spacetime but have yet to specify a set of coordinates on it.

Confusio: "Isn't $(X^0, X^1, X^2, \cdots, X^d)$ a set of coordinates?"

No, that is a set of coordinates for the ambient Minkowski space $M^{d,1}$.

Suppose we choose our coordinates on dS^d to have dimensions of length so that $g_{\mu\nu}$ is normalized to be dimensionless. Then by dimensional analysis, we must have

$$R_{\mu\nu\lambda\sigma} = \frac{1}{L^2}(g_{\mu\lambda}g_{\nu\sigma} - g_{\mu\sigma}g_{\nu\lambda}) \tag{4}$$

(We will show presently one way of determining the overall numerical coefficient.) The Ricci tensor, the scalar curvature, and the Einstein tensor are fixed, upon contraction of the indices in (4), to be

$$R_{\mu\nu} = \frac{(d-1)}{L^2}g_{\mu\nu}, \qquad R = \frac{d(d-1)}{L^2}, \qquad E_{\mu\nu} \equiv R_{\mu\nu} - \tfrac{1}{2}g_{\mu\nu}R = -\frac{(d-1)(d-2)}{2L^2}g_{\mu\nu} \tag{5}$$

respectively. Since (4) and (5) are equalities between tensors, they hold in every coordinate system.

Calculating the Riemann curvature tensor for de Sitter spacetime

One simple way to coordinatize de Sitter spacetime is to eliminate $W = X^d$ (precisely as we did in appendix 2 to chapter I.6) and use X^μ with $\mu = 0, 1, \cdots, d - 1$ as coordinates. Start with $W^2 = L^2 - X \cdot X$, where for convenience, I introduce the notation $A \cdot B = \eta_{\mu\nu}A^\mu B^\nu$. Then $W dW = -X \cdot dX$, so that $dW^2 = (X \cdot dX)^2/(L^2 - X \cdot X)$, leading to

$$ds^2 \equiv \eta_{\mu\nu}dX^\mu dX^\nu + dW^2 = \eta_{\mu\nu}dX^\mu dX^\nu - (X \cdot dX)^2/(X \cdot X - L^2)$$

$$= \left(\eta_{\mu\nu} - \frac{\eta_{\mu\lambda}\eta_{\nu\rho}X^\lambda X^\rho}{X \cdot X - L^2}\right)dX^\mu dX^\nu \tag{6}$$

We can now calculate the Riemann curvature tensor for de Sitter spacetime. As I just mentioned, we merely have to determine the overall coefficient in (4). Here is a rather nifty[4] approach. Let $X \to 0$ in (6), so that $g_{\mu\nu} \simeq (\eta_{\mu\nu} + \frac{1}{L^2}\eta_{\mu\lambda}\eta_{\nu\rho}X^\lambda X^\rho)$; in other words, the metric is locally flat at $X^\mu = 0$. But in chapter VI.1, we learned how to determine the curvature tensor in locally flat coordinates. Looking at (VI.1.11), we immediately

---

* You should realize that the index sets $(0, 1, \cdots, d - 1)$ and $(0, 1, \cdots, d - 1, d)$ are conceptually distinct and are labeled by different letters. (For example, for S^2, we have coordinates (θ, φ) on the sphere and (X, Y, Z) for the ambient embedding space.) But there are only so many letters and thus, a given set of letters often does double duty.

read off $B_{\mu\nu,\lambda\rho} = \frac{1}{2L^2}(\eta_{\mu\lambda}\eta_{\nu\rho} + \eta_{\mu\rho}\eta_{\nu\lambda})$. We then plug into (VI.1.14) to obtain $R_{\tau\rho\mu\nu} = \frac{1}{L^2}(\eta_{\tau\mu}\eta_{\rho\nu} - \eta_{\tau\nu}\eta_{\rho\mu})$. As usual, we simply promote $\eta_{\mu\nu}$ to $g_{\mu\nu}$ to obtain $R_{\tau\rho\mu\nu}$ at an arbitrary point. We have fixed the overall coefficient in (4). Nifty, eh? To summarize, we eliminate W and discover that the resulting coordinate system is locally flat. We look up chapter VI.1 and fix the only feature of (4) that does not follow from general principles.

The expanding universe once again

We see from (5) that de Sitter spacetime is a solution of Einstein's field equation $R_{\mu\nu} = 8\pi G\Lambda g_{\mu\nu}$ with a positive cosmological constant (see VI.5.14) given by

$$8\pi G\Lambda = \frac{3}{L^2} \tag{7}$$

To the extent that* $\sim 74\% \simeq 100\%$, we can say that our universe is observed to be almost maximally symmetric and de Sitter. (As was explained in chapter VIII.1, this approximate statement only applies to the future, not to the past.)

Topologically, de Sitter spacetime is $R \times S^3$, with a spatial section given by S^3, as just explained. In contrast, we know from chapters VI.2 and VI.5 that Einstein's field equation with a positive cosmological constant leads to $ds^2 = -dt^2 + e^{2Ht}(dx^2 + dy^2 + dz^2)$ with the Hubble constant given by $H = \left(\frac{8\pi G}{3}\Lambda\right)^{\frac{1}{2}}$, or in terms of the de Sitter length, $H = 1/L$. Thus, an interesting question poses itself: How is the 3-dimensional flat space with coordinates (x, y, z) hidden in the $(3 + 1)$-dimensional "Minkowskian sphere" defined in (2)? It must correspond to a rather nontrivial slice. Seems like quite a surprise that this "Minkowskian sphere" contains an exponentially expanding universe! Indeed, can you figure it out before reading on?

Angular coordinates on de Sitter spacetime and hyperbolic spaces

For ease of writing and for definiteness, I now specialize to $d = 4$. Our universe might very well be described by dS^4 to a good approximation, as discussed in chapter VIII.2 and in the preceding section. (Wait! Aren't you supposed to figure out something before reading on?) As we go along, you should, and could easily, work out dS^d for any integer d. (When confused, the beginner should also work out what happens for $d = 1, 2, 3$. From (2), we see that a spatial slice of dS^d at fixed X^0 is just the familiar sphere for $d = 3$, the circle for $d = 2$, and two points for $d = 1$.)

For $d = 4$, it is convenient to relinquish indices and name the coordinates: $X^0 = T$, $X^1 = X$, $X^2 = Y$, $X^3 = Z$, and $X^4 = W$, so that the defining equation (2) reads

$$-T^2 + X^2 + Y^2 + Z^2 + W^2 = L^2 \tag{8}$$

---

* Dark energy amounts to $\sim 74\%$ of the universe; see chapter VI.2. It better not be 100%, in which case there would be no matter to form physicists out of.

For X, Y, and Z, we can go over to the usual spherical coordinates $X = r \sin \theta \cos \varphi$, $Y = r \sin \theta \sin \varphi$, and $Z = r \cos \theta$, so the defining equation becomes

$$-T^2 + r^2 + W^2 = L^2 \tag{9}$$

(As always, $dX^2 + dY^2 + dZ^2 = dr^2 + r^2 d\Omega_2^2$, with $d\Omega_2^2 = d\theta^2 + \sin^2 \theta d\varphi^2$.)

Before attacking dS^4, let's warm up with S^3, a somewhat less familiar sphere than S^2. As explained in chapter I.6, the metric on S^3 with radius L is induced from the Euclidean metric $ds^2 = dX^2 + dY^2 + dZ^2 + dW^2$ of the embedding space E^4. We eliminate dW by differentiating the defining equation $X^2 + Y^2 + Z^2 + W^2 = L^2$, so that $-WdW = XdX + YdY + ZdZ = rdr$, using the usual spherical coordinates for X, Y, and Z in the last step. Then we have $dW^2 = \frac{(rdr)^2}{W^2} = \frac{r^2}{L^2 - r^2} dr^2$. Thus, we obtain $ds^2 = dr^2 + r^2 d\Omega_2^2 + \frac{r^2}{L^2 - r^2} dr^2 = \frac{L^2}{L^2 - r^2} dr^2 + r^2 d\Omega_2^2$ (as in I.6.11). This expression literally invites us to introduce spherical coordinates for S^3 by setting $r = L \sin \psi$ (note that ψ is a latitude like θ and so ranges from 0 to π, while φ is a longitude ranging from 0 to 2π), so that

$$ds^2 = L^2 \left(d\psi^2 + \sin^2 \psi d\Omega_2^2 \right) \equiv L^2 d\Omega_3^2 \tag{10}$$

We just constructed S^3 out of S^2, thus rediscovering what we have known since chapter I.6: the metric for S^d can be constructed iteratively. Indeed, a bright school child would have realized that the sphere S^2 can be built out of circles S^1.

After this exercise with S^3, we are now ready to induce, in exactly the same way, the metric on the de Sitter spacetime dS^4 from the metric $ds^2 = -dT^2 + (dX^2 + dY^2 + dZ^2 + dW^2) = -dT^2 + (dr^2 + r^2 d\Omega_2^2 + dW^2)$ of the embedding Minkowski space $M^{4,1}$. Differentiating the defining equation (9), we have $WdW = TdT - rdr$, so that $dW^2 = (TdT - rdr)^2/(L^2 + T^2 - r^2)$. Clearly, we are invited to introduce a hyperbolic angle ψ by $T = t \cosh \psi$ and $r = t \sinh \psi$, so that $T^2 - r^2 = t^2$, $TdT - rdr = tdt$, $dT^2 - dr^2 = dt^2 - t^2 d\psi^2$, and $dW^2 = t^2 dt^2/(L^2 + t^2)$. We obtain

$$ds^2 = -\frac{L^2}{L^2 + t^2} dt^2 + t^2 \left(d\psi^2 + \sinh^2 \psi d\Omega_2^2 \right) \equiv -\frac{L^2}{L^2 + t^2} dt^2 + t^2 dH_3^2 \tag{11}$$

where

$$dH_3^2 = d\psi^2 + \sinh^2 \psi d\Omega_2^2 \tag{12}$$

is the line element on the 3-dimensional hyperbolic space H^3 discussed in chapter I.6. Compare and contrast (10) with (12)! The $(3 + 1)$-dimensional spacetime dS^4 is here coordinatized by $(t, \psi, \theta, \varphi)$. (Clearly, the angular coordinates are just going along for the ride, so that, for example, for dS^3, we have $ds^2 = -\frac{L^2}{L^2 + t^2} dt^2 + t^2 (d\psi^2 + \sinh^2 \psi d\theta^2)$, and for dS^2, the surface pictured in figure 1, $ds^2 = -\frac{L^2}{L^2 + t^2} dt^2 + t^2 d\psi^2$.)

Just as clearly, we can obtain various related forms by changing variables in (11). For example, set $t = L \tan \theta$ to obtain

$$ds^2 = \frac{L^2}{\cos^2 \theta} \left(-d\theta^2 + \sin^2 \theta \, dH_3^2 \right) \tag{13}$$

These coordinates are defined by

$$T = L \tan \theta \cosh \psi, \ r = L \tan \theta \sinh \psi, \ W = \frac{L}{\cos \theta} \tag{14}$$

Another form,

$$ds^2 = L^2 \left(-d\rho^2 + \sinh^2 \rho \, dH_3^2\right) \tag{15}$$

is obtained by setting $t = L \sinh \rho$, so that

$$T = L \sinh \rho \cosh \psi, \ r = L \sinh \rho \sinh \psi, \ W = L \cosh \rho \tag{16}$$

Note that for all these coordinate choices, the slice at constant t (that is, constant θ in (13) and constant ρ in (15)) is now a hyperbolic surface (recall I.6.14) defined in (12). How about this spacetime? Do you recognize it?

Yes, it is the open universe described in chapter V.3 for a particular cosmic expansion factor $a(t)$.

De Sitter spacetime wears many disguises

As we will now see, we can write de Sitter spacetime in remarkably many[5] different forms, according to various choice of coordinates. Some, such as (13) and (15), are obtained by more or less obvious changes of variables. Others, such as the exponentially expanding universe, are far from obvious. For convenience, I list the many faces of de Sitter spacetime in the table near the end of this chapter.

Before we start going through these different choices, we should forewarn Confusio that there are only so many suitable letters in the alphabet, T and t for time, R and r for the radial coordinate. Inevitably, we are bound to use the same letter for conceptually different entities. We will have to trust Confusio to distinguish them by context.

For our first alternative coordinate choice, instead of S^2, we can go to S^3, setting $r = R \sin \psi$ and $W = R \cos \psi$ in the defining equation (9), which then becomes

$$-T^2 + R^2 = L^2 \tag{17}$$

In other words, we set $R^2 = X^2 + Y^2 + Z^2 + W^2$ in (8). Space is now described* by a series of spheres S^3 with radius $R \geq L$, since $R^2 = L^2 + T^2$, as shown in figure 1. Then, as per the above discussion, $ds^2 = -dT^2 + (dR^2 + R^2 d\Omega_3^2)$. Solving $R^2 = L^2 + T^2$ by writing $T = L \sinh t$, $R = L \cosh t$, so that again, as is familiar from Minkowskian geometry, we have $dT^2 - dR^2 = L^2 dt^2$. We thus obtain the metric

$$ds^2 = L^2 \left(-dt^2 + \cosh^2 t \, d\Omega_3^2\right) \tag{18}$$

---

* The rich man would say that space is foliated by spheres with $R \geq L$.

As was just described, a fixed t slice corresponds to a fixed X^0 slice of the hyperboloid in figure 1, giving for the spatial sections a series of spheres S^3 with radius varying as $\cosh t$ as t ranges from $-\infty$ to ∞. These coordinates, defined by

$$T = L \sinh t, \qquad r = L \cosh t \sin \psi, \qquad W = L \cosh t \cos \psi \tag{19}$$

cover the entire hyperboloid and are thus known as global coordinates.

Do you recognize this spacetime? Yes, it is the closed universe described in chapter V.3 for a particular cosmic expansion factor $a(t)$.

You could now go on to explore some other coordinate choices.

As another example, start with (9) and write $ds^2 = -dT^2 + (dr^2 + r^2 d\Omega_2^2 + dW^2)$. Define $T = \rho \sinh \chi$ and $r = \rho \cosh \chi$, so that (9) becomes $\rho^2 + W^2 = L^2$ and $dT^2 - dr^2 = \rho^2 d\chi^2 - d\rho^2$. Note that χ is a time coordinate and ρ a space coordinate. As T ranges from $-\infty$ to $+\infty$, the time variable χ ranges from $-\infty$ to $+\infty$. Then $dW^2 = \frac{(\rho d\rho)^2}{W^2} = \frac{\rho^2}{L^2 - \rho^2} d\rho^2$ (analogous to what we had above for the sphere). Putting it together, we obtain

$$ds^2 = -\rho^2 d\chi^2 + \left(\frac{1}{1 - \frac{\rho^2}{L^2}} d\rho^2 + \rho^2 \cosh^2 \chi \, d\Omega_2^2 \right)$$

$$= L^2 \left(-\sin^2 \psi d\chi^2 + d\psi^2 + \sin^2 \psi \cosh^2 \chi \, d\Omega_2^2 \right) \tag{20}$$

where we have written $\rho = L \sin \psi$, so that

$$T = L \sin \psi \sinh \chi, \qquad r = L \sin \psi \cosh \chi, \qquad W = L \cos \psi \tag{21}$$

Expanding flat universe as a de Sitter spacetime

I now finally come to the question raised earlier. Perhaps you have solved it already? Recall that by maximal symmetry, we know that de Sitter spacetime describes an Einstein universe driven by a positive cosmological constant. But the metrics we have shown thus far do not appear to look anything like the usual exponentially expanding Friedmann-Lemaître form we first met in chapter VI.2.

It turns out that the planar coordinates (t, r, θ, φ) (note: different t and r from before!) of the exponentially expanding universe are defined by

$$X^0 = L \left(\sinh t + \tfrac{1}{2} r^2 e^t \right), \qquad X^i = L r e^t \omega^i, \qquad X^4 = L \left(\cosh t - \tfrac{1}{2} r^2 e^t \right) \tag{22}$$

with $i = 1, 2, 3$ and $\omega^1 \equiv \sin \theta \cos \varphi$, $\omega^2 \equiv \sin \theta \sin \varphi$, $\omega^3 \equiv \cos \theta$ (recall the appendix to chapter I.7 from way back). Or, in Cartesian coordinates $x^i = (x, y, z)$, write $X^i = L e^t x^i$.

So, did you figure it out? You might not have readily guessed this rather bizarre and seemingly totally asymmetric coordinatization. In appendix 1, we will provide a group theoretic interpretation. (Of course, you could have found it by brute force. See appendix 2.)

At this stage, you are merely invited to plug and chug. First, check that the embedding equation (2) is satisfied. Second, show that, inserting (22) into the flat metric $ds^2 = \eta_{MN} dX^M dX^N$ in the embedding spacetime, the metric in our $(3 + 1)$-dimensional world has, lo and behold, the nice form

$$ds^2 = L^2 \left[-dt^2 + e^{2t} \left(dr^2 + r^2 d\Omega_2^2 \right) \right] = L^2 \left[-dt^2 + e^{2t} \left(dx^2 + dy^2 + dz^2 \right) \right] \tag{23}$$

Remarkably, this is precisely the expanding universe that we discussed in chapters VI.2 and VIII.1, and in which we may be living. Our universe may well be a Minkowskian sphere endowed with a sense of time! With these coordinates, a constant t slice gives flat Euclidean 3-space (as already noted in chapter V.3), hence the name "planar" coordinates.

For the familiar sphere, the embedding Cartesian coordinates (X, Y, Z) show the isometries but are not convenient to compute with, while the metric in spherical coordinates $ds^2 = d\theta^2 + \sin^2 \theta d\varphi^2$ hides the isometries, but these coordinates are better for many purposes. Similarly, while the embedding coordinates (2) display the isometries transparently, for various purposes other coordinates may be more convenient. For example, for cosmology, the planar coordinates in (23) are clearly appropriate, but they hide the underlying isometries. Note that we do not need to know the rather complicated transformation (22) at all to study cosmology. Indeed, we could have discovered, and did discover, (23) without talking about isometries and maximally symmetric spaces and spacetimes. It is, of course, illuminating to understand that the exponentially expanding universe we may be living in is nothing other than a glorified sphere.

Isometries and light cone coordinates

We started out in the embedding space $M^{d,1}$ with a manifold endowed with plenty of isometries. When we descend to a specific description tied to a particular coordinate choice, these isometries are still there, though harder to see. The metric in (23) provides a good example. It depends explicitly on time and so is definitely not invariant under time translation; however, it is invariant under $t \to t + \varepsilon$ accompanied by $x^i \to x^i - \varepsilon x^i$: $e^{2t} d\vec{x}^2 \to e^{2\varepsilon} (1-\varepsilon)^2 e^{2t} d\vec{x}^2 \simeq e^{2t} d\vec{x}^2 + O(\varepsilon^2)$. In other words, we have a Killing vector $\xi^\mu = (1, -x^i)$. This amounts to a fancy mathematical way of saying that the universe expands exponentially! The Killing vector tells us that the passage of time can be compensated for by shrinking the spatial coordinates.

Note that $g_{\mu\nu} \xi^\mu \xi^\nu = -1 + e^{2t} r^2$. The Killing vector stays timelike only in the region $e^t r < 1$ but becomes spacelike outside. Recalling our discussion of the Schwarzschild black hole, we recognize this switch of the Killing vector from timelike to spacelike as a hallmark of a horizon.

The reader with a good memory will recall that $e^t r < 1$ describes the region enclosed by the de Sitter horizon we deduced in chapter V.3 by physical reasoning! In that chapter, sitting at $r = 0$, we sent a message at time t to a friend located at r. The condition that she will receive our message was precisely $e^t r < 1$. For us, sitting at the origin, our de Sitter horizon is defined by

$$e^t r = 1 \tag{24}$$

Now, stare at (22). What does it suggest to you?

Noting the form of X^0 and X^4, we realize that it would be wise to go to light cone coordinates $X^\pm = X^0 \pm X^4$. In particular, $X^+ = X^0 + X^4 = Le^t$, an equation that describes a plane labeled by t in $M^{d,1}$. Thus, at a given t, our universe is the intersection of the Minkowskian sphere (2) with this plane. As t varies from $-\infty$ to ∞, the plane marches upward, and the universe expands. (For instance, for $\theta = 0$, the universe at the instant $t = 0$ consists of the surface $X = L(\frac{1}{2}r^2, r \cos \varphi, r \sin \varphi, 0, 1 - \frac{1}{2}r^2)$.)

So, rewrite (22) as (setting $L = 1$ for convenience)

$$X^+ = e^t, \qquad X^i = r\omega^i e^t = x^i e^t, \qquad X^- = e^{-t}(r^2 e^{2t} - 1) \tag{25}$$

This light cone construction suggests, at least in hindsight, one way for the poor man to proceed. Suppose that the poor man does not know about the expanding universe but is merely possessed by an unspeakable desire to slice the hyperboloid in figure 1 with "lightlike" planes $X^+ = \lambda(t)$, for λ some unknown function of some time coordinate. By rotational invariance, he writes $X^i = \sigma(t)\lambda(t)x^i$, with σ another unknown function of t. Then, he uses the defining equation $-X^+X^- + \vec{X}^2 = 1$ to determine $X^- = \sigma^2\lambda\vec{x}^2 - \lambda^{-1}$. Plugging all this into $ds^2 = -dX^+dX^- + d\vec{X}^2$, he finds that he can get rid of the cross term $\vec{x} \cdot d\vec{x}dt$ by setting $\sigma = 1$ (and absorbing an irrelevant constant into $\vec{x}$). Choosing t to be such that $g_{00} = -1$ fixes $\lambda(t)$ and gives the flat expanding universe in (23).

Since in this embedding, X^+ is always positive, the coordinates (22) cover only part of de Sitter spacetime. In particular, referring back to (22), we see that (again setting L to 1 for convenience) for $t \to \infty$, $X^4 = \cosh t - \frac{1}{2}r^2 e^t \to \frac{1}{2}(1 - r^2)e^t$, while for $t \to -\infty$, $X^4 \to \frac{1}{2}e^{-t} \sim +\infty$. Thus, as t ranges from $-\infty$ to ∞, for $r^2 > 1$, the coordinate X^4 ranges from ∞ to $-\infty$, but for $r^2 < 1$, it only ranges between ∞ and $\sqrt{1 - r^2}$, reaching its minimum value when $e^{2t} = 1/(1 - r^2)$.

Poincaré half plane and temporal boundary

To make contact with observational cosmology, let $t \to t/L$, $x \to x/L$, and so forth in (23), and write $ds^2 = -dt^2 + e^{2Ht}(dx^2 + dy^2 + dz^2)$. As expected, the Hubble constant $H = 1/L$ is just the inverse of the de Sitter length.

Introducing the conformal time u by $u = -e^{-Ht}/H$, we obtain another useful form:

$$ds^2 = \frac{1}{(Hu)^2}[-du^2 + (dx^2 + dy^2 + dz^2)] \tag{26}$$

Notice that as the cosmic time t runs from $-\infty$ to $+\infty$, the conformal time u runs from $-\infty$ to 0. (Often it is more convenient to work with $v = -u = e^{-Ht}/H$, even though as t runs from $-\infty$ to $+\infty$, the time v runs backward from ∞ to 0.)

Remembering chapter I.5, you realize that de Sitter spacetime is a Minkowskian version of the Poincaré half plane! Or, the Poincaré half plane is de Sitter spacetime Euclideanized. Just as the Poincaré half plane has a boundary, de Sitter spacetime has a temporal boundary* at $u = 0$, corresponding to $t = \infty$. At a fixed $u = 0^-$, the boundary consists of Euclidean 3-space with no time.

---

* This feature of de Sitter spacetime has attracted a great deal of attention.

The Poincaré coordinates, just like the closely related planar coordinates, cover only part of the Minkowski sphere in the embedding spacetime $M^{4,1}$.

Different slices give closed, flat, and open universes

We just saw that de Sitter spacetime describes an exponentially expanding flat universe driven by a positive cosmological constant.

Earlier, I asked you whether you recognized the spacetimes described by (15) and (18). Well, you might have if you did exercise VIII.1.2. They describe, respectively, an open and a closed universe driven by a positive cosmological constant. To see this, we need to see through various disguises.

For convenience, I write the two spacetimes here again (with a trivial change of notation):

$$ds^2 = L^2 \left(-dt^2 + \sinh^2 t \, dH_3^2 \right) \quad \text{(open)} \tag{27}$$

and

$$ds^2 = L^2 \left(-dt^2 + \cosh^2 t \, d\Omega_3^2 \right) \quad \text{(closed)} \tag{28}$$

Note, once again, that space is spherical for the closed universe and hyperbolic for the open universe.

First, for the closed case, go back to the definition $L^2 d\Omega_3^2 \equiv \frac{L^2}{L^2-r^2} dr^2 + r^2 d\Omega_2^2$ in the discussion leading to (10). Putting this into (28), we obtain

$$ds^2 = -dt^2 + \left(\cosh \frac{t}{L} \right)^2 \left(\frac{1}{1 - \frac{r^2}{L^2}} dr^2 + r^2 d\Omega_2^2 \right) \quad \text{(closed)} \tag{29}$$

after scaling $t \to t/L$.

For the open case, go back to (12) and set $\sinh \psi = r$ (note to Confusio: not the same r as in (16)) and so $d\psi^2 = dr^2/(1 + r^2)$. We then obtain (again after suitable scaling)

$$ds^2 = -dt^2 + \left(\sinh \frac{t}{L} \right)^2 \left(\frac{1}{1 + \frac{r^2}{L^2}} dr^2 + r^2 d\Omega_2^2 \right) \quad \text{(open)} \tag{30}$$

Since I am writing a textbook, I felt obliged to drag the de Sitter length around, at least for a while, but now I am finally fed up. Henceforth, let us use L as the length unit and set $L = 1$.

It is instructive to make contact with the cosmological equation determining the scale factor $a(t)$ (defined in chapter VIII.1, as you may recall). The functions $a(t) = \cosh t$, $a(t) = e^t$, and $a(t) = \sinh t$, in (29), (23), and (30), respectively, satisfy the elementary identities $\sinh^2 t + 1 = \cosh^2 t$, $e^{2t} = e^{2t}$, and $\cosh^2 t - 1 = \sinh^2 t$, respectively. But these represent just the three versions of the cosmological equation (written in appropriate units) for universes driven by a cosmological constant, namely $\dot{a}^2 + k = a^2$ for $k = +1, 0$, and -1, respectively. The universe boiled down to elementary identities!

As was explained in chapter VIII.1, a universe consisting purely of a cosmological constant could evade the argument given there that a Big Bang was inevitable in the past.

For $k = +1$, the equation just cited indicates that a cannot drop below 1. For $k = -1$, there was a Big Bang: the right hand side becomes negligible for small t and a grows linearly from the Bang. Interestingly, the closed, flat, and open universes driven by a positive cosmological constant correspond to different coordinatizations of de Sitter spacetime: different spatial curvature but the same Λ. As was also emphasized in chapter VIII.1, these are mathematical, rather than physical, universes, as any amount of radiation or matter would dominate over Λ in the past near the Big Bang.

Let us now track down the coordinate transformation that led to (15) and (18) and compare it with the corresponding transformation (22) in the flat case (which I rewrite here for convenience):

$$X^0 = \left(\sinh t + \tfrac{1}{2}r^2 e^t\right), \quad X^4 = \left(\cosh t - \tfrac{1}{2}r^2 e^t\right)$$

$$X^i = r e^t \, \omega^i, \quad i = 1, 2, 3 \quad \text{(flat)} \tag{31}$$

For the closed cosmological constant–dominated universe, we have (see (19))

$$X^0 = \sinh t, \quad X^4 = \sqrt{1 - r^2} \cosh t$$

$$X^i = r \cosh t \, \omega^i, \quad i = 1, 2, 3 \quad \text{(closed)} \tag{32}$$

For the open cosmological constant–dominated universe, we have (see (16), with a trivial change of notation)

$$X^0 = \sqrt{1 + r^2} \sinh t, \quad X^4 = \cosh t$$

$$X^i = r \sinh t \, \omega^i, \quad i = 1, 2, 3 \quad \text{(open)} \tag{33}$$

You might ask how the 3-dimensional space we live in, at a given instant in t, is embedded in the hyperboloid I refer to as the Minkowskian sphere. To ease visualization, you may wish to specialize to dS^2.

For the flat case, we already mentioned that space consists of the intersection of lightlike planes with the hyperboloid.

For the closed case, a given instant in t corresponds to a slice of the hyperboloid at a fixed X^0. Space is just a circle around the hyperboloid. In particular, for $t = 0$, $X = (0, \sqrt{1 - r^2}, r)$, understandably just the circle around the "waist" of the hyperboloid.

For the open case, a given instant in t corresponds to a slice of the hyperboloid at a fixed X^4. Space at $t = 0$ degenerates into the point $X = (0, 0, 1)$. It's the Big Bang!

Static coordinates

For another interesting coordinate system, solve the defining equation (9) $-T^2 + r^2 + W^2 = 1$ by writing

$$T = \sqrt{1 - r^2} \sinh t, \quad W = \sqrt{1 - r^2} \cosh t \tag{34}$$

and thus obtain

$$ds^2 = -\left(1 - r^2\right) dt^2 + \frac{dr^2}{1 - r^2} + r^2 d\Omega_2^2 \tag{35}$$

Remarkably, the metric has the same form as the Schwarzschild metric and furthermore is static. The time dependence has disappeared. In other words, $t \to t + \text{constant}$ is an isometry, or in somewhat fancier language, $\frac{\partial}{\partial t}$ is a Killing vector. Note that this Killing vector is timelike only for $r < 1$, that is, inside the de Sitter length. We see that $r = 1$ defines the de Sitter horizon. Indeed, the embedding (34) holds only for $r < 1$. Note also that $T + W = \sqrt{1 - r^2}e^t \geq 0$ and $-T + W = \sqrt{1 - r^2}e^{-t} \geq 0$, so that these coordinates cover only one quarter of the spacetime, namely the region $W \geq |T|$. Another quarter is covered by flipping the sign of W in (34), leading to the same metric.

As in the case of the Schwarzschild metric, as we formally cross the horizon into the region $r > 1$, t becomes a spatial coordinate and r a temporal coordinate.

Interestingly, we can put a spherical mass or a black hole in de Sitter spacetime. Indeed, recall from exercise VI.3.6 that

$$ds^2 = -\left(1 - \frac{2M}{r} - r^2\right) dt^2 + \frac{dr^2}{1 - \frac{2M}{r} - r^2} + r^2 d\Omega_2^2 \tag{36}$$

satisfies the Einstein field equation $R_{\mu\nu} = +3\Lambda g_{\mu\nu}$ outside the black hole, but not $R_{\mu\nu\lambda\sigma} = \frac{1}{L^2}(g_{\mu\lambda}g_{\nu\sigma} - g_{\mu\sigma}g_{\nu\lambda})$ of course, since this spacetime, known as the Schwarzschild–de Sitter spacetime (SdS^4 for short) is not maximally symmetric.

Kruskal-Szekeres–like coordinates for de Sitter spacetime

Starting with the de Sitter spacetime (35), we can go through steps similar to those we took in chapter VII.2 to obtain the Kruskal-Szekeres coordinates for the Schwarzschild black hole. Introduce $x^\pm = t \pm \frac{1}{2} \log(\frac{1+r}{1-r})$, for $0 \leq r < 1$. We have $dx^\pm = dt \pm \frac{dr}{1-r^2}$ and hence

$$ds^2 = -(1 - r^2)dx^+ dx^- + r^2 d\Omega^2 \tag{37}$$

where r is understood to be $r(x^+, x^-) = \frac{e^{(x^+ - x^-)} - 1}{e^{(x^+ - x^-)} + 1}$. Also, $2t = x^+ + x^-$.

Next, introduce $U = e^{x^-}$ and $V = -e^{-x^+}$, so that $UV = -e^{(x^- - x^+)}$ and hence

$$r = \frac{1 + UV}{1 - UV} \tag{38}$$

Also, we have

$$e^{2t} = -\frac{U}{V} \tag{39}$$

Plugging $dU = U dx^-$, $dV = -V dx^+$, and (38) into (37), we obtain

$$ds^2 = \frac{1}{(1 - UV)^2} \left(-4 dU dV + (1 + UV)^2 d\Omega^2\right) \tag{40}$$

Let us focus on the spacetime described by (40). In other words, now that we are done with $x^\pm$, we forget about them. In figure 2, we show the salient features of this spacetime. The U and V axes are drawn at $45°$ from the vertical and divide spacetime into four regions labeled I, II, III, and IV, as in the Schwarzschild case (discussed in chapter VII.2). According to (38), lines of constant r correspond to hyperbolas in the $(U$-$V)$ plane. In particular, the north and south poles, both with $r = 0$, correspond to the

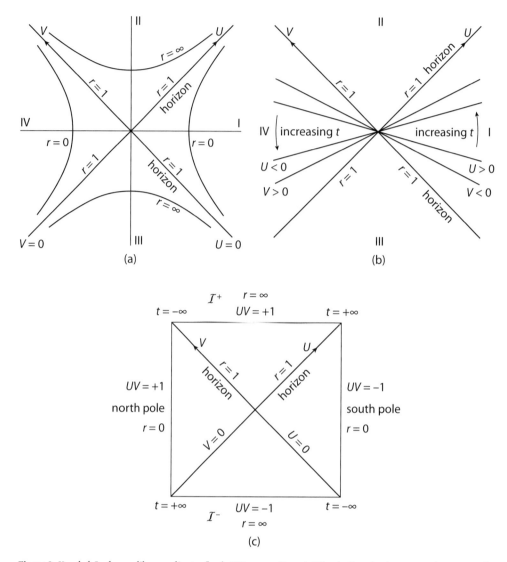

Figure 2 Kruskal-Szekeres–like coordinates for de Sitter spacetime. (a) The de Sitter horizon at $r = 1$ corresponds to $UV = 0$, namely the U and V axes, drawn at 45° from the vertical. Spatial infinity $r = \infty$ corresponds to $UV = 1$. (b) The lines of constant t are shown. In region I, time points upward, that is, the Killing vector is future directed, but in region IV, the Killing vector is past directed. (c) The Penrose diagram for de Sitter spacetime.

hyperbolas $UV = -1$ located in region IV and region I, respectively. The de Sitter horizon at $r = 1$ corresponds to $UV = 0$, namely the U- and V-axes. Finally, spatial infinity $r = \infty$ corresponds to $UV = 1$.

Notice that the metric in these (U, V) coordinates depends only on the product UV and not on U and V separately. Referring back to (35) and (39), we see that this is of course why the (t, r) coordinates are called static coordinates in the first place: the metric in (35) does not depend on t. Speaking in a fancier tongue, we would say that the spacetime admits a Killing vector

$$\frac{\partial}{\partial t} = U \frac{\partial}{\partial U} - V \frac{\partial}{\partial V} \tag{41}$$

Explicitly, scaling $U \to e^{\varepsilon} U$ and $V \to e^{-\varepsilon} V$ in (39) oppositely generates time translation $t \to t + \varepsilon$, leaving r unchanged. In figure 2b, we show the lines of constant t, namely $U = -e^{2t} V$. In region I, with $U > 0$ and $V < 0$, we see that with increasing t, the lines tilt upward, eventually ending up on the U-axis (namely the line $V = 0$). In other words, in region I, time points upward, or more accurately, the Killing vector is future directed.

Note also from (39) that as $t \to \infty$, $V \to 0$. So indeed, the U-axis corresponds to $t = \infty$, in agreement with what we just said. Similarly, the V-axis corresponds to $t = -\infty$.

Now we see an interesting phenomenon: the lines of constant t continue into region IV, with $U < 0$ and $V > 0$. As t increases, the lines tilt downward. In region IV, the Killing vector is past directed. Of course, this just means that $-t$ corresponds to time, and an observer in region IV would still move upward with the passage of proper time.

In more mundane language, the Killing vector (41) is given in component form by $\xi^{\mu} = (U, -V, 0, 0)$. We find that $\xi^2 = g_{\mu\nu}\xi^{\mu}\xi^{\nu} = 2g_{UV}\xi^U\xi^V = 8UV/(1 - UV)^2$. As expected, in regions I and IV, $\xi^2 < 0$ and ξ is timelike. But in regions II (with $U > 0$, $V > 0$) and III (with $U < 0$, $V < 0$), ξ is spacelike. Actually, we already knew that the time coordinate t and the space coordinate r exchange roles as we cross the horizon, which in this diagram corresponds to the U- and V-axes.

Finally, we can compactify and knead figure 2a into a square, that is, a Penrose diagram, as shown in figure 2c.

Thermal radiation from the de Sitter horizon

In 1976, Gibbons and Hawking showed that thermal radiation emanates from the de Sitter horizon, similar to the radiation emanating from the Schwarzschild horizon and to the radiation seen by an accelerated observer, discussed in chapter VII.3. The physics underlying each of these three cases is quite similar: quantum fluctuation produces a particle-antiparticle pair near the horizon (the Schwarzschild horizon, the Rindler horizon, and the de Sitter horizon, as the case may be), with one of them disappearing over the horizon, never to be seen by the observer. The other member of the pair is observed as thermal radiation. Indeed, we have already mentioned the striking similarity in form between de Sitter spacetime in static coordinates in (35) and the Schwarzschild metric, with the coordinate t changing from a temporal coordinate to a spatial coordinate.

The temperature of the Gibbons-Hawking radiation can again be estimated by dimensional analysis:

$$T_{\text{de Sitter}} \sim \frac{1}{L} \sim \frac{\hbar c}{L} \tag{42}$$

A detailed quantum field theoretic analysis, well beyond[6] the scope of this book, is needed only to determine the overall numerical coefficient (which happens to be $(2\pi)^{-1}$).

A great deal of mystery lurks behind the Gibbons-Hawking radiation, however, even beyond the mysteries behind Hawking radiation. To start with, the de Sitter horizon is observer dependent. For the black hole, we could invoke the possible microstates in its formation to account for the entropy. It is far from evident[7] what the corresponding counting of microstates would be for de Sitter spacetime.

Causal structure of de Sitter spacetime

Faced with this almost bewildering variety of coordinates, we evidently should choose wisely, using coordinates appropriate for the physics at hand. For visualizing and calculating geodesics, the embedding coordinates in figure 1 may actually be best (see appendix 3). But to understand the causal structure of de Sitter spacetime, clearly some sort of conformal coordinates (such as (26)) are best. Here we derive, by inserting $\cosh t = \frac{1}{\cos \tau}$ into (18), the metric in conformal coordinates (recall (10)):

$$ds^2 = \frac{1}{\cos^2 \tau} \left(-d\tau^2 + d\Omega_3^2 \right) = \frac{1}{\cos^2 \tau} \left(-d\tau^2 + d\psi^2 + \sin^2 \psi \, d\Omega_2^2 \right) \tag{43}$$

Here τ ranges from $-\frac{\pi}{2}$ to $\frac{\pi}{2}$, causing t to range from $-\infty$ to ∞, while the latitude ψ, as remarked earlier in connection with (10), ranges from 0 to π. Note that space, namely a constant τ slice of spacetime, is not flat. In terms of the coordinates of the embedding spacetime $M^{4,1}$, we have

$$T = \tan \tau, \qquad r = \frac{\sin \psi}{\cos \tau}, \qquad W = \frac{\cos \psi}{\cos \tau} \tag{44}$$

Consider the Penrose diagram in figure 3a. As depicted, the τ axis is vertical, the ψ axis horizontal. Each point in this 2-dimensional (τ, ψ) plot on a piece of paper represents a 2-sphere. The left and right hand sides correspond to the north ($\psi = 0$) and south ($\psi = \pi$) poles, respectively. Note that $r = 0$ for both the north and south poles, with W taking on opposite signs. The surfaces labeled $\mathcal{I}^-$ and $\mathcal{I}^+$ sit in the infinite past and future, respectively.

The attractive feature of a Penrose diagram is of course that light rays travel along lines at 45°. Indeed, in the $d = 2$ case, the trajectory a photon takes is given simply by $d\theta = \pm d\tau$.

Imagine yourself sitting at the south pole (which is of course equivalent to any other point on the sphere). Where in spacetime can you send a message to?

Clearly, you can send a message to any spacetime point in the shaded region labeled as $\mathcal{O}^+$ (figure 3b). For example, you can send a light pulse to B from the point A on your worldline, as shown in figure 3b. Of course, before you reach A, you can also send a message traveling at less than the speed of light to B, but once you live past A, you can no longer send anything to B.

Can the other observer, upon receipt of your message at B, send you a response? She cannot. Any response she sends will end up at $\mathcal{I}^+$.

Finally, you cannot send a message to the spacetime point C even if you had thought of it in your infinite past (namely, the lower right corner). A light pulse sent from the south pole in the infinite past would reach the north pole in the infinite future. However, if an observer stationed at C hurries, she could intercept your message upon crossing the diagonal line.

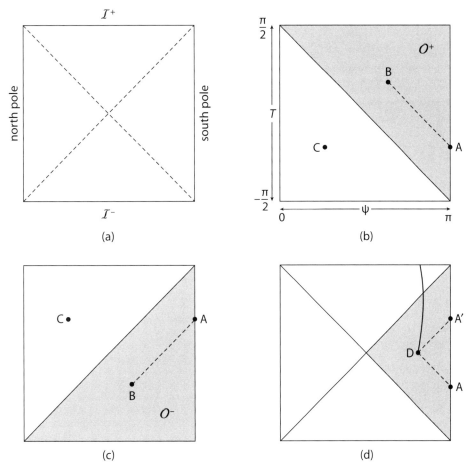

Figure 3 Penrose diagrams showing the causal structure of de Sitter spacetime. (a) Each point in this 2-dimensional (τ, ψ) plot represents a 2-sphere. The left and right hand sides correspond to the north $(\psi = 0)$ and south $(\psi = \pi)$ poles, respectively. (b) Suppose you are sitting at the south pole (which is in fact anywhere). You can send a light pulse to B from the point A on your worldline, but once you live past A, you can no longer send anything to B. An observer stationed at C can intercept your message if she hurries across the diagonal line into $\mathcal{O}^+$. (c) From any point in $\mathcal{O}^-$, a message can be sent to you, but not from outside $\mathcal{O}^-$. (d) The region in spacetime you can communicate with is known as the (southern) causal diamond, shown as the shaded region. You can send a message to D and actually get a response back.

Similarly, we could ask "From where in spacetime can a message be sent to you?" That region is shaded and labeled as $\mathcal{O}^-$ in figure 3c. For example, a message sent from the point C will end up in $\mathcal{I}^+$.

As many self-help books have assured us, communication is a two-way street. By this definition, the region in spacetime you can communicate with is given by the intersection of $\mathcal{O}^+$ and $\mathcal{O}^-$, known to the cognoscenti as the (southern) causal diamond and shown as the shaded region in figure 3d (and referred to as region I in figure 2). For example, you can send a message to D and actually get a response back.

In fact, we have already seen this causal structure of de Sitter spacetime by an explicit calculation in chapter V.3. In an exponentially expanding universe, everybody is moving away from us and will eventually move past our horizon forever, just as distant ships move over our everyday horizon. For example, unless the observer at D fires up her rocketship and hurries, she will eventually cross the 45° line in figure 3d and leave the southern causal diamond, headed toward the infinite future $\mathcal{I}^+$, as indicated by the solid curved line. If she really hurries, she could still meet you, but not until after you live past the point A'. If all goes well, she could then merge her worldline with yours, and you could travel to $\mathcal{I}^+$ together.

Iterative relationship between de Sitter spacetimes

All the way back in chapter I.5 you showed in exercise I.5.10 that the metrics on the spheres S^d enjoy an iterative relation between them: $ds_d^2 = d\theta^2 + \sin^2\theta ds_{d-1}^2$. (See also (10).) Not surprisingly, the metrics on de Sitter spacetimes dS^d also enjoy an iterative relation between them, since they are Minkowskian spheres.

Recall the angular coordinates you worked out (exercise I.5.9) on the sphere S^d:

$$X^1 = \cos\theta_1, \qquad X^2 = \sin\theta_1\cos\theta_2, \qquad \cdots$$
$$X^d = \sin\theta_1\cdots\sin\theta_{d-1}\cos\theta_d, \qquad X^{d+1} = \sin\theta_1\cdots\sin\theta_{d-1}\sin\theta_d \tag{45}$$

As already alluded to earlier in this chapter, and as explained in chapter III.3, we can jump immediately from the sphere to de Sitter spacetime by making Minkowski's "mystical" substitution $X^{d+1} = iX^0$, so that (1) becomes (2). Calling θ_d for brevity θ, we write* $\varphi = i\theta$ formally (so that $\cos\theta = \cosh\varphi$ and $\sin\theta = i\sinh\varphi$) and obtain

$$X^1 = \cos\theta_1, \qquad X^2 = \sin\theta_1\cos\theta_2, \qquad \cdots$$
$$X^d = \sin\theta_1\cdots\sin\theta_{d-1}\cosh\varphi, \qquad X^0 = \sin\theta_1\cdots\sin\theta_{d-1}\sinh\varphi \tag{46}$$

Replacing $d\theta_d^2$ in the spherical metric

$$ds_d^2 = d\theta_1^2 + \sin^2\theta_1 d\theta_2^2 + \cdots + \sin^2\theta_1\cdots\sin^2\theta_{d-1}d\theta_d^2$$

by $-d\varphi^2$, we obtain, for the de Sitter metric,

$$ds_d^2 = d\theta_1^2 + \sin^2\theta_1 d\theta_2^2 + \cdots - \sin^2\theta_1\cdots\sin^2\theta_{d-1}d\varphi^2$$

Thus, we arrive at the iterative relationship for dS^d:

$$ds_d^2 = d\theta^2 + \sin^2\theta ds_{d-1}^2 \tag{47}$$

which is formally precisely the same as the iterative relationship for S^d.

As remarked earlier in chapter I.5, this iterative relation just expresses the fact, known to many school children, that lines of constant latitude on the globe form circles. Similarly, you can see from figure 1 that if you slice the hyperboloid representing dS^d, you get a hyperboloid of one lower dimension.

---

* More precisely, we analytically continue. In quantum field theory, this is known as a Wick rotation.

I might mention here that, while we went, at the start of this chapter, from a sphere to a de Sitter spacetime by a feat of imaginary magic, we can certainly go from a de Sitter spacetime back to a sphere using the same trick. Letting $X^0 = i X^5$ while keeping X^i, X^4 fixed, we can turn a set of coordinates on dS^4 into a set of coordinates on S^4. For example, let $t \to it$ (so that $\sinh t \to i \sin t$ and $\cosh t \to \cos t$), then the coordinates in (32) for the closed cosmological constant–dominated universe become coordinates for S^4. In doing this, we have to make sure that X^i and X^4 remain the same, of course. Thus, for example, for the coordinates in (33) for the open cosmological constant–dominated universe, we have to let $r \to ir$ as well as $t \to it$. As another example, letting $t \to it$ and $\psi \to i\psi$ in the metric $ds^2 = -\frac{L^2}{L^2+t^2}dt^2 + t^2(d\psi^2 + \sinh^2 \psi d\theta^2)$ for dS^3 mentioned after (11), we recover the metric $ds^2 = \frac{L^2}{L^2-r^2}dr^2 + r^2d\Omega_2^2$ for S^3 mentioned just before (10).

Stereographic projection for de Sitter spacetime

Just as we can stereographically project the sphere (surely you did exercise I.5.13), we can stereographically project de Sitter spacetime by mapping $(X^0, X^1, X^2, X^3, X^4)$ into (x^0, x^1, x^2, x^3) as follows (here we reinstate L):

$$X^M = \frac{1}{1 + \frac{x^2}{4L^2}} \delta^M_\mu x^\mu, \qquad M = 0, 1, 2, 3 \tag{48}$$

and

$$X^4 = L \left(\frac{1 - \frac{x^2}{4L^2}}{1 + \frac{x^2}{4L^2}} \right) \tag{49}$$

where $x^2 \equiv -(x^0)^2 + (x^1)^2 + (x^2)^2 + (x^3)^2$. The Kronecker delta in (48) emphasizes that while the indices M and μ might be numerically identical, they are not conceptually the same. In other words, (48) says that, for example, X^3 and x^3 differ by the overall factor $\frac{1}{1+\frac{x^2}{4L^2}}$, so that

$$-\left(X^0\right)^2 + \left(X^1\right)^2 + \left(X^2\right)^2 + \left(X^3\right)^2 = \frac{-\left(x^0\right)^2 + \left(x^1\right)^2 + \left(x^2\right)^2 + \left(x^3\right)^2}{\left(1 + \frac{x^2}{4L^2}\right)^2} = \frac{x^2}{\left(1 + \frac{x^2}{4L^2}\right)^2}$$

You can now analytically continue the result from chapter I.5 on stereographic projection of the sphere to verify that the defining relation (2) is satisfied and that

$$ds^2 = \left(\frac{1}{1 + \frac{x^2}{4L^2}} \right)^2 \eta_{\mu\nu}dx^\mu dx^\nu \tag{50}$$

(Of course, you can also check this by brute force, plugging (48) and (49) into $ds^2 = \eta_{MN}dX^M dX^N$.) This shows explicitly that de Sitter spacetime, aka the Minkowskian sphere, is conformally flat in any dimension. Upon recalling that the garden variety sphere is also conformally flat,* we are perhaps not surprised.

The rise of de Sitter spacetime

De Sitter and anti de Sitter spacetimes have become the darlings of theoretical physicists for entirely different reasons. We will talk about anti de Sitter spacetime in the following chapter. As already mentioned on several occasions, observational cosmologists tell us that our universe is partly filled by a dark energy, presumably the cosmological constant. Thus, as we discussed in chapter VIII.2, our universe will eventually expand into a de Sitter spacetime. To me, it is strangely appealing that, having discovered that our world is Euclidean round, we now realize that our universe will become Minkowskian round.

Allow me to go into a bit more detail about the history of the de Sitter metric, first mentioned in chapter V.3 (see table for de Sitter spacetime). In 1917 de Sitter found that[8]

$$ds^2 = -\cos^2 \chi \, dt^2 + \left(d\chi^2 + \sin^2 \chi \, d\Omega_2^2 \right) \tag{51}$$

satisfies (4) and hence solves Einstein's field equation with a positive cosmological constant. The attentive reader recognizes that this metric is just the static metric (35) with the simple transformation $r = \sin \chi$. In 1922, Lanczos and Weyl independently and correctly wrote down $ds^2 = -dt^2 + \cosh^2 t (d\varphi^2 + \cos^2 \varphi d\psi^2 + \cos^2 \varphi \cos^2 \psi d\psi^2)$, which again the attentive reader recognizes as the global metric (18). Then, in 1925, Lemaître, while still a student, noted that de Sitter's coordinates were not comoving, namely that lines of constant χ, θ, and φ were not geodesics unless $\chi = 0$ (hence the corresponding "defect" plagues (35) also), and so de Sitter's coordinate choice was not homogeneous. Lemaître then discovered the metric (23) for the flat expanding universe that we have known and loved since chapter V.3. Later, in 1927, he extended his work to include closed and open universes. Perhaps with some justification, we should refer to the spacetime studied in this chapter as the de Sitter-Lanczos-Weyl-Lemaître spacetime.

Appendix 1: The group theory behind the exponentially expanding universe

As promised, I now give a more satisfying motivation[9] for the coordinate transformation given in (22). For this discussion, we will be a bit more general and discuss $dS^d = SO(d, 1)/SO(d-1, 1)$. With $L = 1$, (2) reads

$$\eta_{MN} X^M X^N = -\left(X^0 \right)^2 + \sum_{i=1}^{d-1} \left(X^i \right)^2 + \left(X^d \right)^2 = 1 \tag{52}$$

As always, we have the Lie algebra of $SO(d, 1)$ (see (III.3.21); M, N, P, and Q range over $0, 1, 2, \cdots, d$):

$$[J_{MN}, J_{PQ}] = i \left(\eta_{MP} J_{NQ} + \eta_{NQ} J_{MP} - \eta_{NP} J_{MQ} - \eta_{MQ} J_{NP} \right) \tag{53}$$

---

* At this stage, I need hardly remind you that conformally flat is not flat!

Table for de Sitter spacetime

Coordinates	Metric	Embedding relations
t,ψ,θ,φ	$-\dfrac{L^2}{L^2+t^2}dt^2 + t^2\,dH_3^2$	$T = t\cosh\psi, \quad r = t\sinh\psi$
$\theta,\psi,\theta,\varphi$	$\dfrac{L^2}{\cos^2\theta}\left(-d\theta^2 + \sin^2\theta\,dH_3^2\right)$	$T = L\tan\theta\cosh\psi, \quad r = L\tan\theta\sinh\psi, \quad W = \dfrac{L}{\cos\theta}$
ρ,ψ,θ,φ	$L^2(-d\rho^2 + \sinh^2\rho\,dH_3^2)$	$T = L\sinh\rho\cosh\psi, \quad r = L\sinh\rho\sinh\psi, \quad W = L\cosh\rho$
t,Ω	$L^2(-dt^2 + \cosh^2 t\,d\Omega_3^2)$	$T = L\sinh t, \quad r = L\cosh t\sin\psi, \quad W = L\cosh t\cos\psi$
χ,ψ,Ω	$L^2(-\sin^2\psi\,d\chi^2 + d\psi^2 + \sin^2\psi\cosh^2\chi\,d\Omega_2^2)$	$T = L\sin\psi\sinh\chi, \quad r = L\sin\psi\cosh\chi, \quad W = L\cos\psi$
χ,ρ,Ω	$-\rho^2 d\chi^2 + \left(\dfrac{1}{1-\frac{\rho^2}{L^2}}d\rho^2 + \rho^2\cosh^2\chi\,d\Omega_2^2\right)$	$T = \rho\sinh\chi, \quad r = \rho\cosh\chi, \quad W^2 = L^2 - \rho^2$
t,x,y,z	$L^2[-dt^2 + e^{2t}(dx^2 + dy^2 + dz^2)]$	$T = L\left(\sinh t + \tfrac{1}{2}r^2 e^t\right), \quad Z = Lr e^t\sin\theta, \quad W = L\left(\cosh t - \tfrac{1}{2}r^2 e^t\right)$
u,x,y,z	$\dfrac{L^2}{u^2}[-du^2 + (dx^2 + dy^2 + dz^2)]$	$u = e^{-t}$
t,r,Ω	$-dt^2 + \left(\cosh\tfrac{t}{L}\right)^2\left(\dfrac{1}{1-\frac{r^2}{L^2}}dr^2 + r^2 d\Omega_2^2\right)$	$T = Lr\cosh t, \quad Z = Lr\cosh t\cos\theta, \quad W = L\sqrt{1-r^2}\cosh t$
t,r,Ω	$-dt^2 + \left(\sinh\tfrac{t}{L}\right)^2\left(\dfrac{1}{1+\frac{r^2}{L^2}}dr^2 + r^2 d\Omega_2^2\right)$	$T = L\sqrt{1+r^2}\sinh t, \quad Z = Lr\sinh t\cos\theta, \quad W = L\cosh t$
t,r,Ω	$-(1-\tfrac{r^2}{L^2})dt^2 + \left(\dfrac{dr^2}{1-\frac{r^2}{L^2}} + r^2 d\Omega_2^2\right)$	$T = \sqrt{1-\tfrac{r^2}{L^2}}\sinh t, \quad W = \sqrt{1-\tfrac{r^2}{L^2}}\cosh t$
τ,Ω	$\dfrac{L^2}{\cos^2\tau}(-d\tau^2 + d\Omega_3^2)$	$T = L\sinh t, \quad \cosh t = \dfrac{1}{\cos\tau}$
x	$\left(\dfrac{1}{1+\frac{r^2}{4L^2}}\right)^2 \eta_{\mu\nu}dx^\mu dx^\nu$	$T = \dfrac{1}{1+\frac{r^2}{4L^2}}x^0, \quad W = L\left(\dfrac{1-\frac{r^2}{4L^2}}{1+\frac{r^2}{4L^2}}\right)$

Note: In the table, $d\Omega_3^2 = d\psi^2 + \sin^2\psi\,d\Omega_2^2$, and $dH_3^2 = d\psi^2 + \sinh^2\psi\,d\Omega_2^2$.

We now identify the $\frac{1}{2}d(d+1)$ generators $SO(d, 1)$ of acting on dS^d. In addition to the generators of rotation J_{ij} (here i and j range over $1, 2, \cdots, d-1$), we have the combinations

$$P_i \equiv J_{i0} + J_{d,i}, \qquad D \equiv J_{d,0}, \qquad K_i \equiv J_{i0} - J_{d,i} \tag{54}$$

which we identify as the generators of translation, dilation, and conformal transformation, respectively, as discussed in the preceding chapter. As the context is slightly different (and because of our inclusion of factors of i here), we again display the algebra, which you can deduce from (53):

$$[P_i, P_j] = 0, \qquad [K_i, K_j] = 0,$$

$$[D, P_i] = -i P_i, \qquad [D, J_{ij}] = 0, \qquad [D, K_i] = i K_i,$$

$$[J_{ij}, P_k] = i(\delta_{ik} P_j - \delta_{jk} P_i), \qquad [J_{ij}, K_k] = i(\delta_{ik} K_j - \delta_{jk} K_i),$$

$$[P_i, K_j] = 2i\delta_{ij} D - 2i J_{ij} \tag{55}$$

For example, $[P_i, P_j] = [J_{i0} + J_{d,i}, J_{j0} + J_{d,j}] = i(-J_{ij} + J_{ij} - \delta_{ij} D + \delta_{ij} D) = 0$. Note that to obtain this familiar result, we have to define translation P_i as a linear combination of a boost in the ith direction and a rotation in the $(d\text{-}i)$ plane. As another example, $[D, P_i] = [J_{d,0}, J_{i0} + J_{d,i}] = -i(J_{d,i} - J_{0i}) = -i P_i$.

These generators act linearly on the embedding coordinates X^M. As in chapter III.3, their action can be represented by $J_{MN} = i(X_M \partial_N - X_N \partial_M)$. Thus, each of these generators is represented by a $(d+1)$-by-$(d+1)$ matrix. We arrange the indices in the "natural" order $(0, \{i\}, d) = (0, 1, 2, \cdots, d-1, d)$ where, as indicated above, i ranges over $1, 2, \cdots, d-1$. For example, the boost $D \equiv J_{d,0}$ in the dth direction is

$$D = i \left[\begin{array}{c|c|c} 0 & 0 & -1 \\ \hline 0 & 0 & 0 \\ \hline -1 & 0 & 0 \end{array} \right] \tag{56}$$

The notation is such that along the diagonal, in the upper left, the 0 represents a 1-by-1 matrix with its entry equal to 0; in the center, the 0 represents a $(d-1)$-by-$(d-1)$ matrix with all its entries equal to 0; and finally, in the lower right, the 0 once again represents a 1-by-1 matrix with its entry equal to 0. Exponentiating the generator D to obtain the group element, we obtain

$$e^{iDt} = \left[\begin{array}{c|c|c} \cosh t & 0 & \sinh t \\ \hline 0 & I & 0 \\ \hline \sinh t & 0 & \cosh t \end{array} \right] \tag{57}$$

In other words, this is a $(d+1)$-by-$(d+1)$ matrix with a $(d-1)$-by-$(d-1)$ identity matrix in its center.

Similarly, we have

$$\vec{P} \cdot \vec{x} = i \left[\begin{array}{c|c|c} 0 & \vec{x}^T & 0 \\ \hline \vec{x} & 0 & \vec{x} \\ \hline 0 & -\vec{x}^T & 0 \end{array} \right] \tag{58}$$

In the matrix, $\vec{x}$ is to be interpreted as a $(d-1)$-dimensional column vector (so that $\vec{x}^T$ is an $(d-1)$-dimensional row vector). (Notice that as a linear combination of a boost and a rotation, $\vec{P} \cdot \vec{x}$ is symmetric in its upper left corner and antisymmetric in its lower right corner, so to speak.) Exponentiating, you will find

$$e^{i\vec{P}\cdot\vec{x}} = \left[\begin{array}{c|c|c} 1 + \frac{1}{2}\vec{x}^2 & \vec{x}^T & \frac{1}{2}\vec{x}^2 \\ \hline \vec{x} & I & \vec{x} \\ \hline -\frac{1}{2}\vec{x}^2 & -\vec{x}^T & 1 - \frac{1}{2}\vec{x}^2 \end{array} \right] \tag{59}$$

Notice that $(\vec{P} \cdot \vec{x})^3 = 0$, so that the exponential series terminates. You are invited to verify that $e^{i\vec{P}\cdot\vec{x}} e^{i\vec{P}\cdot\vec{y}} = e^{i\vec{P}\cdot(\vec{x}+\vec{y})}$.

Just as in chapter IX.6, we map the coset manifold $dS^d = SO(d, 1)/SO(d-1, 1)$ by acting with $g(t, \vec{x}) = \exp(i\vec{P} \cdot \vec{x}) \exp(iDt)$ on a reference point, which we choose to be $X_* = (0, \vec{0}, 1)$ (in analogy to the south pole for the familiar case of the sphere). (The logic here is that $X_* = (0, \vec{0}, 1)$ is left invariant by the $SO(d-1, 1)$ generated by $J_{ij}, J_{i0} = P_i + K_i$.) We obtain

$$X = (g X_*) = e^{i\vec{P}\cdot\vec{x}} e^{iDt}(0, \vec{0}, 1) = \left(\sinh t + \frac{1}{2}e^t \vec{x}^2, e^t \vec{x}, \cosh t - \frac{1}{2}e^t \vec{x}^2 \right) \tag{60}$$

We recognize that this is precisely what appears as the rather peculiar coordinatization (22) we encountered before, but now derived group theoretically. From (60), we obtain

$$ds^2 = \eta_{MN} dX^M dX^N = -dt^2 + e^{2t} d\vec{x}^2 \tag{61}$$

namely the metric for the expanding universe (23). In other words, to describe the expanding universe in the form (61), we coordinatize an event at $(t, \vec{x})$ by the group element $g(t, \vec{x}) = \exp(i\vec{P} \cdot \vec{x}) \exp(iDt)$ needed to bring the reference point X_* to our event.

Appendix 2: Discovering the expanding universe without knowing about Einstein's field equation

While we are all enamored of the beauty of group theory, the truly impoverished man ignorant of this wonderful subject can still obtain the coordinate transformation in (22) by brute force (of course). Starting with $ds^2 = -dT^2 + dX^2 + dY^2 + dZ^2 + dW^2$, we transform $T = f(t, r)$, $X = xh(t, r)$, $Y = yh(t, r)$, $Z = zh(t, r)$, $W = g(t, r)$ with $r^2 = x^2 + y^2 + z^2$ and try to get to $ds^2 = -dt^2 + a^2(t) d\vec{x}^2$, with $a(t)$ some unknown function.

The embedding equation $-T^2 + X^2 + Y^2 + Z^2 + W^2 = 1$ (set $L = 1$ for convenience) gives $-f^2 + g^2 + r^2 h^2 = 1$. Plugging (as usual, $\dot{f} = \frac{\partial f}{\partial t}$, $f' = \frac{\partial f}{\partial r}$, and so on) $dT = \dot{f} dt + f' dr$, $dX = hdx + x\dot{h} dt + xh' dr$, and so forth, into ds^2 and matching to the desired form gives us (a) $-\dot{f}^2 + \dot{g}^2 + r^2 \dot{h}^2 = -1$, (b) $-\dot{f} f' + \dot{g} g' + r^2 \dot{h} h' + r\dot{h} h = 0$, (c) $h^2 = a^2$, and (d) $f'^2 = g'^2$. This is straightforward to solve. For instance, (d) gives (with no loss of generality by flipping either T or W) $f(t, r) = g(t, r) + k(t)$, with $k(t)$ some unknown function. Eventually, we arrive at, as a bonus, $a(t) = e^t$ after absorbing an integration constant.

The point of this little exercise is to show that mathematical types thinking about the analogs of spheres for Minkowski spacetime could have, in principle, discovered the exponentially expanding universe long ago without knowing about Einstein's field equation. This suggests another extragalactic fable. We can imagine a smart guy in a civilization infatuated with the sphere arriving at the de Sitter universe, and then, by calculating the Ricci tensor (or knowing that the analog of the sphere is maximally symmetric), uncovering the cosmological constant and dark energy.

Appendix 3: Geodesics in the embedding space

Let us determine the geodesics in de Sitter spacetime using the embedding coordinates X^M satisfying $X^2 = \eta_{MN} X^M X^N = 1$. Instead of extremizing the integral $\int \sqrt{-\eta_{MN} dX^M dX^N}$, we extremize $\int d\zeta (\frac{1}{2} \dot{X}^2 + \frac{1}{2} \lambda (X^2 - 1))$, as explained in exercise II.2.6, imposing the constraint with the Lagrange multiplier λ. Here ζ is an appropriate parameter and $\dot{X}^M = dX^M/d\zeta$.

The reader may or may not recognize this as essentially the same problem of a particle on a sphere that we did back in appendix 4 to chapter II.3. We can lift many of the equations, suitably reinterpreted with a Minkowski metric, from that simple problem! The equation of motion reads $\ddot{X}^M = \lambda X^M$, to be solved with the constraint $X^2 = 1$, which, when differentiated, gives $X \cdot \dot{X} = 0$ (where we indicate the dot in the dot product to remind us that we are dealing with vectors). Using the equation of motion, we also have $\ddot{X} \cdot \dot{X} = 0$, thus concluding that $\dot{X}^2$ is a constant. As in chapter II.3, we verify by direct differentiation that $J^{MN} \equiv X^M \dot{X}^N - X^N \dot{X}^M$ is conserved. Define $2J^2 \equiv J^{MN} J_{MN} = 2(X^M \dot{X}^N - X^N \dot{X}^M) X_M \dot{X}_N = 2\dot{X}^2$. (Note that both signs are possible for J^2.) Hence we have $\dot{X}^2 = J^2$.

For definiteness, let us consider a timelike geodesic. Since $\dot{X}^2 = J^2$ is negative, write $K^2 = -J^2$, with K real. Then this last equation has the obvious solution $X^M = a^M e^{K\zeta} + b^M e^{-K\zeta}$, with $\vec{a}$ and $\vec{b}$ two constant real vectors. The constraint $1 = X^2 = a^2 e^{2K\zeta} + b^2 e^{-2K\zeta} + 2a \cdot b$ implies that $a^2 = b^2 = 0$, $2a \cdot b = 1$. The motion of the particle is completely solved:

$$X = ae^{K\zeta} + be^{-K\zeta}, \quad \text{with } a^2 = 0 = b^2, \quad \text{and } 2a \cdot b = 1 \tag{62}$$

Geometrically, the particle travels along the hyperbolic version of great circles on the surface defined by (2). You can verify this by following steps analogous to those given in chapter II.3. For a spacelike geodesic, J^2 is positive. Evidently, we replace $e^{\pm K\zeta}$ by $\cos J\zeta$ and $\sin J\zeta$.

For a lightlike geodesic, $J = 0$, and the equation of motion collapses to $\ddot{X} = 0$ with the solution $X = a + b\zeta$, with a and b two Minkowski vectors. In the embedding space, light travels along a straight line. The lightlike condition $(dX)^2 = 0$ implies that $b^2 = 0$. The condition that the photon stays on the de Sitter hyperboloid (2) implies $X^2 = a^2 + 2a \cdot b\,\zeta = 1$. Thus, lightlike geodesics are determined by

$$X = a + b\zeta, \quad \text{with } a^2 = 1,\, b^2 = 0, \quad \text{and } a \cdot b = 0 \tag{63}$$

Confusio looks puzzled for a moment, muttering "How can $(dX)^2 = 0$ and $X^2 = 1$ both be satisfied?" Yes, they can.

It is fun to verify that these geodesics are indeed followed in any specific coordinate system we use to map out de Sitter spacetime. Let us pick, for example, the expanding universe coordinates in (23), since the cosmologists like them.

Consider a cosmologist at rest at $r = 0$ (which is in fact anywhere in the universe) tracing a perfectly respectable timelike geodesic. Inspecting (22), we have (setting $L = 1$) $X = (\sinh t, 0, 0, 0, -\cosh t) = \frac{1}{2}e^t(1, 0, 0, 0, -1) - \frac{1}{2}e^{-t}(1, 0, 0, 0, 1)$, in agreement with (62) with $\zeta = t$, $K = 1$. Thus, $a = \frac{1}{2}(1, 0, 0, 0, -1)$, $b = -\frac{1}{2}(1, 0, 0, 0, 1)$. Indeed, $a^2 = 0 = b^2$ and $2a \cdot b = 1$.

Next, consider a photon moving along the x-axis, starting at $x = 0$, $t = t_S$. (Note that the expanding universe is not invariant in t, and recall that we studied this problem in chapter V.3.) Then $dx = e^{-t}dt$, so that $x = e^{-t_S} - e^{-t}$, with the photon reaching $x(t = \infty) = e^{-t_S}$ at $t = \infty$. We recover the de Sitter horizon. Now (22) gives $T = X^0 = \sinh t + \frac{1}{2}(e^{-t_S} - e^{-t})^2 e^t = \frac{1}{2}(1 + e^{-2t_S})e^t - e^{-t_S}$, $W = X^4 = \frac{1}{2}(1 - e^{-2t_S})e^t + e^{-t_S}$, and $X = X^1 = (e^{-t_S} - e^{-t})e^t = e^{t-t_S} - 1$. Indeed, (63) is satisfied with $a = (-e^{-t_S}, -1, 0, 0, e^{-t_S})$, $b = (\frac{1}{2}(1 + e^{-2t_S}), e^{-t_S}, 0, 0, \frac{1}{2}(1 - e^{-2t_S}))$, and $\zeta = e^t$.

Confusio is amazed, but you know that we are merely checking that the defining equation (2) is satisfied and that, if $ds^2 = 0$ in one set of coordinates, $ds^2 = 0$ in any other set of coordinates.

As yet another example, follow a photon starting at $r = 0$ and $t = 0$ in static coordinates (35). Integrating with $dt = dr/(1 - r^2)$ with these initial conditions, we obtain $r = \tanh t$. As expected, it reaches the horizon $r = 1$ at $t = \infty$. We check readily that (63) holds with $a = (0, 0, 1)$ and $b = (1, 1, 0)$ (with a minor abuse of notation).

Appendix 4: Space of spheres and de Sitter spacetime

In this appendix, we present an amusing tidbit regarding the space of spheres and de Sitter spacetime.[10]

Consider the space of spheres living in ordinary 3-dimensional Euclidean space. We need 3 numbers $\vec{x} = (x, y, z)$ to specify the location of the sphere and a number R to specify its radius. Thus, the space of spheres is 4-dimensional. Any guesses on what this 4-dimensional space is? Of course, the way I have set you up, and the mere fact that this appendix is in a chapter on de Sitter spacetime, you might suspect that it is dS^4. But as you will see, it is quite remarkable how the connection works.

Picture two spheres, one with radius R located at $\vec{x}$, the other with radius R' located at $\vec{x}'$. See figure 4. Suppose the two spheres intersect. The intersection is a circle perpendicular to $\vec{x} - \vec{x}'$. Pick any point V on this circle. (By rotational invariance, it will be clear that for our purposes, it does not matter which point we pick.) Consider the triangle formed by the centers C and C' of the two spheres and V. Denote the angle CVC' by Ω. (If you prefer, you can talk about the 3-dimensional space of circles, which is slightly easier to visualize. The intersection of two circles consists of two points, and we could pick either as V.) Then $(\vec{x} - \vec{x}')^2 = R^2 + R'^2 - 2RR' \cos \Omega$.

Let us associate the two spheres with two points X and X' on the hyperboloid that defines dS^4. We will work out the association as we go along. In the embedding 5-dimensional Minkowski spacetime, the distance between the two points is given by $(X - X')^2 = 2(1 - X \cdot X')$, where $X \cdot X' = -\frac{1}{2}(X^+X'^- + X^-X'^+) + \vec{X} \cdot \vec{X}'$ in light cone coordinates, with the dot product defined by the 5-dimensional Minkowski metric, of course. We can evaluate $X \cdot X'$ in any of the coordinate systems listed in this chapter, but with malice of hindsight, let us use the flat expanding universe coordinates given in (25):

$$2X \cdot X' = -\left(e^{t'}e^{-t}(r^2 e^{2t} - 1)\right) - (\ \leftrightarrow\ ') + 2e^t e^{t'}\vec{x} \cdot \vec{x}'$$
$$= -e^{t+t'}\left(\vec{x} - \vec{x}'\right)^2 + e^{t-t'} + e^{t'-t} \tag{64}$$

But in the space of spheres, we have $2\cos\Omega = \frac{R}{R'} + \frac{R'}{R} - \frac{(\vec{x} - \vec{x}')^2}{RR'}$, as mentioned above. Thus, if we make the association $R = e^{-t}$ and $R' = e^{-t'}$, we see that $\cos\Omega = X \cdot X'$.

For example, the condition that two spheres barely touch, namely $\cos\Omega = -1$, translates into the antipodal condition $X \cdot X' = -1$. We can now translate between the two mathematical constructs. As another example,

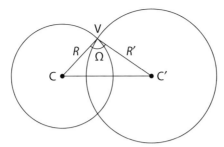

Figure 4 Two spheres intersecting, with the relation between various lengths and the angle Ω fixed by elementary trigonometry.

the invariant volume of spacetime $d^4x\sqrt{-g} = d^3x\,dt\,e^{3t} = d^3x\,dR/R^4$ gets translated into a measure that severely suppresses large spheres.

I find it quite surprising that the 19th century space of spheres somehow "knows" about the 20th century flat expanding universe, not to mention the Minkowski metric. Sitting in a flat expanding universe, we are each associated with a sphere whose radius shrinks with the inexorable passage of time.

Exercises

1 Using the "slicing" in (22), derive the standard form (23) of the exponentially expanding universe.

2 Verify for the various metrics of de Sitter spacetime the maximal symmetry relations (for $d = 4$) $R_{\mu\nu\lambda\sigma} = +\frac{1}{L^2}(g_{\mu\lambda}g_{\nu\sigma} - g_{\mu\sigma}g_{\nu\lambda})$, $R_{\mu\nu} = +\frac{3}{L^2}g_{\mu\nu}$, and $R = +\frac{12}{L^2}$.

3 Starting from the static coordinates (35), you can obtain the de Sitter analog of the coordinates we used for spherical black holes in chapter VII.2 by defining $dp = dt + \frac{dr}{1-r^2}$ and $dq = -dt + \frac{dr}{1-r^2}$. Show that

$$ds^2 = \left(1 - r^2\right)dp\,dq + r^2 d\Omega_2^2 \tag{65}$$

4 Show that the metric (36), with $\frac{2M}{r}$ replaced by $\frac{2M}{r^{d-3}}$, also solves the vacuum Einstein equation in d dimensions.

5 Show that in conformal coordinates, the event horizons of an observer at the south pole and of an observer at the north pole are given in terms of the embedding coordinates by $T \pm W = 0$, respectively, and $r = 1$.

6 Complete the brute force calculation in appendix 2.

7 Show that lines of constant χ, θ, and φ in de Sitter's original metric (51) are not geodesics unless $\chi = 0$.

Notes

1. A useful reference on de Sitter and anti de Sitter spacetimes is M. Spradlin, A. Strominger, and A. Volovich, *Proceedings of the LXXVI Les Houches School*, 2001.
2. Our treatment is resolutely from the physicist's point of view. For an entry to the mathematical literature, a starting point might be "A Geometrical Background for De Sitter's World" by H. S. M. Coxeter *Am. Math. Monthly* 50 (1943), pp. 217–228.

3. The isometry groups should be, strictly speaking, $O(d+1)$ and $O(d,1)$, respectively, but we won't be talking much about reflections.

4. I like this partly because I learned it from A. Einstein's *The Meaning of Relativity*—directly from the master, so to speak.

5. I was motivated to display this many forms of the de Sitter metric when I participated in various workshops and schools and realized that many in the audience were unaware of the variety of ways in which de Sitter and anti de Sitter spacetimes could be written.

6. Actually, it requires only a minimal amount of quantum field theory; I am almost tempted to devote an appendix to it. The clearest derivation I know of is in section III.2 of the article by M. Spradlin et al. Let me mention one key feature: for the observer sitting at $r = 0$, the passage of proper time between two events is measured by $X \cdot X' = -\sinh t \sinh t' + \cosh t \cosh t' = \cosh(t - t')$, but the function $\cosh(\Delta t) = \cosh(\Delta t + 2\pi i)$ is periodic in imaginary time. Thus, we can again invoke the "mystical argument" of time as an angle mentioned in appendix 1 of chapter VII.3.

7. People have argued that our lack of knowledge of what happens beyond the horizon amounts to a kind of entropy.

8. W. de Sitter, *Proc. Royal Acad. Amsterdam*, XIX (1917), p. 1217. De Sitter's motivation seems somewhat muddled to modern eyes. He began by saying that Einstein had proposed the boundary condition ($g_{00} = -1$, $g_{ij} = 0$) at infinity (which is clearly not invariant under coordinate transformation) and proposed instead that $g_{\mu\nu} = 0$ at infinity. Interestingly, he stated in a footnote, "The idea to make the 4-dimensional world spherical in order to avoid the necessity of assigning boundary-conditions, was suggested several months ago by Prof. Ehrenfest, in a conversation with the writer. It was, however, at that time not further developed." Also, in a postscript, de Sitter said that he communicated his result to Einstein, who wrote back objecting to a universe without matter. I am grateful to Gary Gibbons for showing me this paper.

9. I learned this from a paper by S. Deser and A. Waldron.

10. I am grateful to Gary Gibbons for telling me about this interesting connection between the space of spheres, which was developed in 19th century mathematics, and de Sitter spacetime.

IX.11 | Anti de Sitter Spacetime

A container for gravity

Now that you have mastered de Sitter spacetime, you are ready to tackle anti de Sitter spacetime. Incidentally, this antiterminology appears to be modern, as all these spacetimes were referred to as de Sitter in the older literature.[1]

As you well know, in theoretical physics, it is often useful to enclose in a box the system we want to study, be it the electromagnetic field or a quantum particle. Unfortunately, there is no known material out of which we can construct a box to contain the gravitational field. However, as you will learn in this chapter, anti de Sitter spacetime possesses a spatial boundary consisting of a Minkowskian spacetime of one lower dimension. For example, the 5-dimensional anti de Sitter spacetime has as boundary the 4-dimensional $M^{3,1}$ spacetime.[2] This striking feature prompts us to use anti de Sitter spacetime as a container[3] (the term tin can is sometimes used) for quantum gravity, the only way we know to confine the gravitational field and study its properties.

Interest in anti de Sitter spacetime has exploded[4] in recent years due to the amazing discovery by string theorists, notably Maldacena and others, that the physics of various theories of gravity in AdS^5 can be mapped onto the physics of certain gauge theories on the $M^{3,1}$ spacetime that forms the boundary of the AdS^5 spacetime. That this is even conceivable is intimately connected to the holographic principle mentioned in chapter VII.3. This surprising correspondence, known as* AdS/CFT correspondence, or more accurately, as the gauge/gravity duality, promises to shed light[5] on both quantum gravity and strongly coupled gauge theories. More recently, a great deal of excitement has also been generated by the possible relevance of this correspondence to condensed matter physics.

---

* AdS stands for anti de Sitter and CFT for conformal field theories: some special gauge theories are conformal invariant, hence the term.

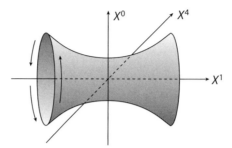

Figure 1 The d-dimensional anti de Sitter space-time AdS^d embedded in $(d + 1)$-dimensional Minkowski-type spacetime $M^{d-1,2}$.

Anti de Sitter spacetime

The d-dimensional anti de Sitter spacetime AdS^d with length scale L is defined as the set of all points $(X^0, X^1, X^2, \cdots, X^d)$ in $(d + 1)$-dimensional Minkowski-type spacetime $M^{d-1,2}$ (that is, a spacetime with $ds^2 = -(dX^0)^2 + (dX^1)^2 + (dX^2)^2 + \cdots + (dX^{d-1})^2 - (dX^d)^2$ satisfying $-(X^0)^2 + (X^1)^2 + (X^2)^2 + \cdots + (X^{d-1})^2 - (X^d)^2 = -L^2$, which we write as

$$\left(X^0\right)^2 - \sum_{i=1}^{d-1} \left(X^i\right)^2 + \left(X^d\right)^2 = L^2 \quad \text{(anti de Sitter spacetime)} \tag{1}$$

as shown in figure 1.

Compare and contrast this embedding equation with the one for de Sitter spacetime given in (IX.10.2), which I display here again for convenience:

$$-\left(X^0\right)^2 + \sum_{i=1}^{d-1} \left(X^i\right)^2 + \left(X^d\right)^2 = L^2 \quad \text{(de Sitter spacetime } dS^d) \tag{2}$$

In parallel with the discussion for de Sitter spacetime, we know that anti de Sitter spacetime is also maximally symmetric.

Note that the isometry group for anti de Sitter spacetime is $SO(d - 1, 2)$ rather than $SO(d, 1)$. Aside from this, much of the discussion for de Sitter spacetime could now be repeated. In particular, like de Sitter spacetime, anti de Sitter spacetime is also maximally symmetric. A point on anti de Sitter spacetime, $(0, \cdots, 0, 1)$ for example, is left invariant by $SO(d - 2, 2)$. In other words, AdS^d is the coset manifold $SO(d - 1, 2)/SO(d - 2, 2)$. For example, $AdS^5 = SO(4, 2)/SO(4, 1)$.

The isometry groups of de Sitter and anti de Sitter spacetimes are contrasted in this table:

AdS^d	$SO(d - 1, 2)$
dS^d	$SO(d, 1)$

The signs in (1), in contrast to those in (2), make all the difference in the world! Some authors treat de Sitter and anti de Sitter together by introducing the sign $\sigma = +1$ for dS and $\sigma = -1$ for AdS, so that the embedding equations (1) and (IX.10.2) are unified into $\eta_{\mu\nu}X^\mu X^\nu + \sigma(X^d)^2 = \sigma L^2$, with $\mu, \nu = 0, 1, \cdots, d - 1$. In general, I won't. I find the practice confusing, not worth saving some space in the exposition, but occasionally, it is instructive to compare the two spacetimes side by side, as I will do presently.

In the preceding chapter, we coordinatized de Sitter spacetime by eliminating $W = X^d$. We can do the same for anti de Sitter spacetime; indeed, it is illuminating to do them side by side, as I just outlined. With $W^2 = L^2 - \sigma X \cdot X$ (the notation is self-evident: $X \cdot X = \eta_{\mu\nu}X^\mu X^\nu$), we have $W dW = -\sigma X \cdot dX$ and $dW^2 = (X \cdot dX)^2/(L^2 - \sigma X \cdot X)$. Hence, we obtain

$$ds^2 = \eta_{\mu\nu}dX^\mu dX^\nu + \sigma dW^2 = \left(\eta_{\mu\nu} - \frac{\eta_{\mu\lambda}\eta_{\nu\rho}X^\lambda X^\rho}{X \cdot X - \sigma L^2}\right)dX^\mu dX^\nu \tag{3}$$

as in the preceding chapter, but now with $\sigma = \pm$. Thus, with the metric of AdS^d written in terms of d-dimensional coordinates, we can go back and forth between de Sitter and anti de Sitter spacetimes by formally letting $L^2 \to -L^2$.

In coordinates for which $g_{\mu\nu}$ is dimensionless, we thus have immediately

$$R_{\mu\nu\lambda\sigma} = -\frac{1}{L^2}\left(g_{\mu\lambda}g_{\nu\sigma} - g_{\mu\sigma}g_{\nu\lambda}\right) \tag{4}$$

and

$$R_{\mu\nu} = -\frac{(d-1)}{L^2}g_{\mu\nu}, \qquad R = -\frac{d(d-1)}{L^2}, \qquad E_{\mu\nu} = \frac{(d-1)(d-2)}{2L^2}g_{\mu\nu} \tag{5}$$

These expressions differ from the corresponding expressions (IX.10.4 and IX.10.5) for de Sitter spacetime by an overall sign, and so anti de Sitter spacetime solves Einstein's field equation $R_{\mu\nu} = 8\pi G\Lambda g_{\mu\nu}$ with a negative cosmological constant given by $8\pi G\Lambda = -\frac{3}{L^2}$. I remind you that, as explained in the preceding chapter, the observed cosmological constant is positive, leading to a de Sitter spacetime.

Again, we can calculate the Riemann curvature tensor, as in the preceding chapter, by letting $X \to 0$ in (3), so that $g_{\mu\nu} \simeq (\eta_{\mu\nu} + \frac{\sigma}{L^2}\eta_{\mu\lambda}\eta_{\nu\rho}X^\lambda X^\rho)$. The metric is locally flat at $X^\mu = 0$. Referring to (VI.1.14), we obtain $R_{\tau\rho\mu\nu} = \frac{\sigma}{L^2}(\eta_{\tau\mu}\eta_{\rho\nu} - \eta_{\tau\nu}\eta_{\rho\mu})$. Promoting $\eta_{\mu\nu}$ to $g_{\mu\nu}$, we obtain $R_{\tau\rho\mu\nu}$ at an arbitrary point. Setting $\sigma = +1$ for dS and $\sigma = -1$ for AdS, we obtain (IX.10.1) and (4), respectively.

For pedagogical clarity, instead of writing everything in terms of an arbitrary d, I will often specialize to whatever value of d suits my purpose best. I often call $X^0 = T$, and $X^d = W$. Here is a table comparing and contrasting AdS^4 and dS^4:

AdS^4	$-T^2 + X^2 + Y^2 + Z^2 - W^2 = -L^2$	$ds^2 = -dT^2 + dX^2 + dY^2 + dZ^2 - dW^2$
dS^4	$-T^2 + X^2 + Y^2 + Z^2 + W^2 = L^2$	$ds^2 = -dT^2 + dX^2 + dY^2 + dZ^2 + dW^2$

That the isometry group is $SO(3, 2)$ for AdS^4 but is $SO(4, 1)$ for dS^4 should now be as clear as day.

Two "times"?

Note that in (1), the embedding is not into the familiar Minkowski spacetime $M^{d,1}$, but into $M^{d-1,2}$ with two time coordinates. What do we do with two "times"? Very strange[6] indeed!

Physics as we know it does not admit two times.[7] Set $d = 3$ for definiteness. Let us understand how AdS^3 ends up having only one time coordinate, even though it started with two. According to (1), AdS^3 is defined by

$$\left(T^2 + W^2\right) - \left(X^2 + Y^2\right) = L^2 \tag{6}$$

with a metric induced from the two-time Minkowski metric $\eta = \operatorname{diag}(-1, +1, +1, -1)$ of the embedding space $M^{2,2}$, namely

$$ds^2 = -\left(dT^2 + dW^2\right) + \left(dX^2 + dY^2\right) \tag{7}$$

As shown in figure 1, we may picture AdS^3 as embedded in $M^{2,2}$, somewhat crudely, as a cylindrical tube with a radius that increases with increasing X and Y. In the (T, X) plane with $W = 0 = Y$, the defining equation traces out two hyperbolas $T = \pm\sqrt{L^2 + X^2}$. Compared to the tube shown in figure IX.10.1, the tube here is lying on its side, so to speak.

I have grouped the coordinates in (6) and (7) to render the isometry group $SO(2, 2)$ manifest. Replace the two time coordinates (T, W) by polar coordinates (R, t) by setting $T = R \cos t$ and $W = R \sin t$. Similarly, replace the two space coordinates (X, Y) by polar coordinates (r, θ) by setting $X = r \cos \theta$ and $Y = r \sin \theta$. Then $ds^2 = -(dR^2 + R^2 dt^2) + (dr^2 + r^2 d\theta^2)$. The apparent difficulty is that we have two time coordinates (R, t).

I have dragged L around long enough. Just as in the preceding chapter, for ease of writing, I am now going to unceremoniously set $L = 1$. The resolution of the puzzle of two times is that the apparent temporal coordinate R is not independent of the spatial coordinate r, since the defining equation $R^2 - r^2 = 1$ constrains them. Differentiating, we obtain $R dR = r dr$ and hence $dR^2 - dr^2 = (\frac{r^2}{R^2} - 1)dr^2 = -\frac{1}{1+r^2}dr^2$. Thus, we obtain

$$ds^2 = -\left(1 + r^2\right) dt^2 + \frac{dr^2}{1 + r^2} + r^2 d\theta^2 \tag{8}$$

We end up with only one time coordinate!

The metric (8) describes a manifestly respectable spacetime. Note also that the metric does not depend on time and hence these coordinates are known as static. They are defined by

$$T = \sqrt{1 + r^2} \cos t, \quad W = \sqrt{1 + r^2} \sin t, \quad X = r \cos \theta, \quad Y = r \sin \theta \tag{9}$$

Did you watch the magician carefully? How did one of the two time coordinates disappear, leaving us with only one time coordinate t? Even better, you should have done the calculation. Okay, the secret, so to speak, behind the two-timing $M^{2,2}$ becoming a respectable spacetime is that the spatial coordinate r is more spatial than the temporal coordinate R is temporal, in the sense that $R^2 > r^2$, so that $dR^2 < dr^2$.

Incidentally, something similar occurred back in chapter I.6. in the construction of hyperbolic spaces. They were embedded into what at that stage of the book we called "pseudo-Euclidean" spaces, but in fact, the hyperbolic spaces turned out to be perfectly Euclidean.

We can of course immediately jump to AdS^d by replacing in the preceding discussion $X^2 + Y^2$ by $\sum_{i=1}^{d-1}(X^i)^2$, and so on and so forth, to obtain $ds^2 = -(1+r^2)dt^2 + \frac{dr^2}{1+r^2} + r^2 d\Omega_{d-2}^2$. Just as in the de Sitter case (see (IX.10.35)), the metric has the same form $ds^2 = -f(r)dt^2 + f(r)^{-1}dr^2 + r^2 d\Omega_{d-2}^2$ as the Schwarzschild metric. But instead of $f(r) = 1 - r^2$ in the de Sitter case, we now have $f(r) = 1 + r^2 > 0$, and thus anti de Sitter spacetime does not have a horizon.

Time to unwind time!

I offered a word of caution about figure IX.10.1; so two words of double caution about figure 1. It is hard enough to have spatial intuition about $M^{3,1}$, let alone $M^{2,2}$.

Indeed, figure 1 indicates that there are closed timelike curves, which would also threaten physics as we know it.[8] We changed variables by $T = R\cos t$ and $W = R\sin t$, and so t started out as a periodic variable. However, in the spacetime defined by the resulting metric (8), the time coordinate t flows majestically from $-\infty$ to $+\infty$. For us physicists, then, we simply define anti de Sitter spacetimes by the metric in (8). Mathematicians would say that physicists have gone to the universal cover. Colloquially, just between you and me, we could say that we have unwrapped the circular time coordinates. Picture a roll of paper towels being unrolled into a long rectangular strip. Since the roll is not cylindrical, as shown in figure 1, the resulting strip of paper cannot be laid down flat, which is precisely what those factors of $(1 + r^2)$ in (8) are telling us.

The isometry group for AdS^d, $SO(d-1,2)$, contains $SO(d-1) \times SO(2)$ as its maximal compact subgroup. Referring to (1), we see that, evidently, $SO(d-1)$ rotates $(X^1, X^2, \cdots, X^{d-1})$, while $SO(2)$ rotates (X^0, X^d). In other words, $SO(2)$ rotates $T = R\cos t$ and $W = R\sin t$ into each other, thus translating $t \to t + \text{constant}$.

A hyperbolic radial coordinate

Those readers adept at doing integrals will recognize that the appearance of $1 + r^2$ in (8) is begging us to change variable to $r = \sinh \rho$. Doing so gives

$$ds^2 = -\cosh^2 \rho \, dt^2 + d\rho^2 + \sinh^2 \rho \, d\Omega_{d-2}^2 \tag{10}$$

Recall that $dH_{d-1}^2 = d\rho^2 + \sinh^2 \rho \, d\Omega_{d-2}^2$. With these coordinates, space is hyperbolic.

The original embedding coordinates are given by

$$T = \cosh \rho \cos t, \quad W = \cosh \rho \sin t, \quad X = \sinh \rho \cos \theta, \quad Y = \sinh \rho \sin \theta \tag{11}$$

(for $d = 3$). Note that since r ranges from 0 to $+\infty$, ρ also ranges from 0 to $+\infty$. As usual, the angular coordinates just go along for the ride.

Conformal coordinates for anti de Sitter spacetime

Like de Sitter spacetime, anti de Sitter spacetime wears many disguises, as already indicated by (8) and (10).

In parallel with the preceding chapter, I list the many faces of anti de Sitter spacetime in a table near the end of this chapter. Let us now find the conformal coordinates for anti de Sitter spacetime, namely the analog of (IX.10.43). Start with (8), set $r = \tan \psi$, and behold:

$$ds^2 = \frac{1}{\cos^2 \psi}\left(-dt^2 + d\psi^2 + \sin^2 \psi \, d\Omega_{d-2}^2\right) = \frac{1}{\cos^2 \psi}\left(-dt^2 + d\Omega_{d-1}^2\right) \tag{12}$$

Of course, we could also have started with (10) and set $\sinh \rho = \tan \psi$ and $\cosh \rho = (\cos \psi)^{-1}$. Compare and contrast with the de Sitter spacetime in (IX.10.43).

The time coordinate t runs from $-\infty$ to $+\infty$ (as it did in (10)). Thus, the conformal diagram, as shown in figure 2, is a strip extending to infinity in the time direction. As usual, light travels at $\pm 45°$ to the vertical. Note that each point in this 2-dimensional plot describes a sphere S^{d-2} of radius $\sin \psi$, which varies from 0 to 1 (also see below).

We see that AdS^d is conformally equivalent to a spacetime with $ds^2 = -dt^2 + d\Omega_{d-1}^2$. Recall from chapter VII.2 that the Minkowski spacetime $M^{d-1,1}$ is conformally equivalent to the same spacetime, topologically a cylinder for $d > 2$. We will see in appendix 3 that AdS^2 is special.

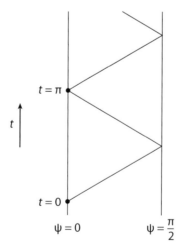

Figure 2 Anti de Sitter spacetime AdS^d represented by a conformal diagram consisting of a strip extending from $-\infty$ to $+\infty$ in the time coordinate t. Light travels at $\pm 45°$ to the vertical. Each point in this 2-dimensional plot describes a sphere S^{d-2} of radius $\sin \psi$, which varies from 0 to 1.

Note that ψ, which plays the role of a latitude, is a spatial coordinate, in contrast to its analog τ in (IX.10.43). As r (or $\sinh \rho$) goes from 0 to ∞, ψ goes from 0 to $\pi/2$.

Pay attention! What, $\pi/2$, not π?

Anti de Sitter spacetime has a boundary

Perhaps the existence of a boundary is the most striking feature of anti de Sitter spacetime, a feature that has been much exploited to contain gravity, as was mentioned at the start of this chapter. We will first show the boundary mathematically, and then more physically.

In getting to the form $ds^2 = \frac{1}{\cos^2 \psi}(-dt^2 + d\psi^2 + \sin^2 \psi \, d\Omega_{d-2}^2)$ in (12), we changed variables by setting $r = \tan \psi$. The radial coordinate r started out nice and easy, ranging from 0 to $+\infty$. But now comes something interesting, as just noted: with a seemingly innocuous change of variable, we have a latitude ψ that starts from 0 but gets up to only $\pi/2$, not π! Starting at the north pole, it reaches only the equator, not the south pole. What kind of a weakling latitude is that!

Thus, while it is locally correct to write $d\Omega_{d-1}^2$ for $d\psi^2 + \sin^2 \psi \, d\Omega_{d-2}^2$ in (12), it is misleading. Space covers only the northern hemisphere of S^{d-1}, with a boundary at the equator. In other words, we don't have the full sphere S^{d-1}. Rather, we have only a hemisphere, which is topologically the same as a $(d-1)$-dimensional generalized disk or ball B^{d-1}. It may be helpful to think of a familiar example: the northern hemisphere of the ordinary sphere S^2 is topologically the 2-dimensional disk, otherwise known as the ball B^2, with the equator as a boundary. (The layperson, in his or her infinite wisdom, understands by the word "ball" the 3-dimensional object B^3, which has S^2 as its boundary. Similarly, the 2-dimensional ball B^2 is the disk, with the circle S^1 as its boundary.)

Thus, the spatial sections of AdS^d are bounded by S^{d-2}, which may be thought of as Euclidean space E^{d-2} with spatial infinity identified as a single point. Adding back the time coordinate, we extend E^{d-2} to $M^{d-2,1}$. We finally conclude that the anti de Sitter spacetime AdS^d is bounded by $M^{d-2,1}$. (In appendix 1, I give you a slightly more intuitive demonstration of this statement. In appendix 3, you will see that AdS^2 has 2 boundaries. Can you figure it out now?)

Hence the slang expression "tin can" for anti de Sitter spacetime and the explosive development of the AdS/CFT correspondence. In particular, in this picture, we are to regard the good old $M^{3,1}$ spacetime we live and play in as the boundary of a $(4+1)$-dimensional AdS^5 spacetime. Wouldn't that be quite a laugh, if all this time we were actually living on the boundary of AdS^5 without knowing it? Sort of like the time when we realized that we were not living on E^2, but actually on the boundary of B^3.

For a more physical argument that a spatial boundary exists, picture the strip of paper we got from unrolling a roll of paper towels a short while ago, as shown in figure 1. The strip of paper, meant to represent AdS^2 and described by $ds^2 = -(1+r^2)dt^2 + \frac{dr^2}{1+r^2}$ (in other words, the line element in (8) with the angular component dropped), appears to be infinitely wide. It is important to note that instead of the embedding for AdS^3 in (9), we now have $T = \sqrt{1+r^2} \cos t$, $W = \sqrt{1+r^2} \sin t$, and $X = r$, and so the coordinate r, no longer a radial variable, actually ranges from $-\infty$ to ∞. (This is why AdS^2 is special, as we have alluded to; see appendix 3.)

But who is to say that the strip of paper is infinitely wide? As usual in relativity, we are to bounce light rays around to measure distances. Sitting at some fixed r_*, let us send a light beam to $r = \infty$ and wait for it to bounce back. Light follows the path $dt = \pm dr/(1 + r^2)$ and so comes back after the amount of proper time $2\sqrt{1 + r_*^2} \int_{r_*}^{\infty} dr/(1 + r^2)$ for us. The important point is that the integral converges at the upper limit, and so the paper is actually finite in width, with two edges or boundaries. For AdS^d with $d > 2$, the angular coordinates connect the "different" boundaries, and so AdS^d has only one boundary. In other words, in the discussion above, we concluded that the spatial sections of AdS^d are bounded by S^{d-2}, but S^0 consists of two points.

The conformal group of the bulk equals the isometry group of the boundary

With what little we know, we can already see a key group-theoretic feature that underlies the AdS/CFT correspondence. In chapter IX.9, we learned that the conformal group for $M^{3,1}$ is $SO(4, 2)$. But earlier in this chapter, we learned that the isometry group for AdS^5 is also $SO(4, 2)$. The conformal group is the manifestation of the isometry group on the boundary.

Poincaré coordinates

As we have already seen, anti de Sitter spacetime can be described using a variety of coordinates. We now turn to the Poincaré coordinates, which in recent years have enjoyed a resurgence of popularity, particularly in the context of AdS/CFT.

Slightly rewrite the defining equation (6) for AdS^3 as $(T^2 - X^2) + (W^2 - Y^2) = 1$, which we solve by writing[9] $T^2 - X^2 = \frac{t^2 - x^2}{w^2}$ and $W^2 - Y^2 = 1 + \frac{x^2 - t^2}{w^2}$, and

$$T = \frac{t}{w}, \quad X = \frac{x}{w}$$
$$Y = \frac{1}{2}\left(\frac{x^2 - t^2}{w} + w - \frac{1}{w}\right) = \frac{1}{2w}\left(x^2 - t^2 + w^2 - 1\right)$$
$$W = \frac{1}{2}\left(\frac{x^2 - t^2}{w} + w + \frac{1}{w}\right) = \frac{1}{2w}\left(x^2 - t^2 + w^2 + 1\right) \tag{13}$$

Note also that we start with four coordinates (T, X, Y, W) and end with three (t, x, w), since we are embedding a 3-dimensional spacetime into the 4-dimensional $M^{2,2}$.

Direct substitution of the seemingly awkward coordinate transformation (13) into $ds^2 = -dT^2 + dX^2 - dW^2 + dY^2$ leads to the amazingly simple form (do check it!)

$$ds^2 = \frac{1}{w^2}\left(-dt^2 + dx^2 + dw^2\right) \tag{14}$$

You may recognize that this is just the Minkowskian version of the Poincaré half plane (introduced back in chapter I.5) in one higher dimension.

Incidentally, we could have avoided some labor by noting that (13) is invariant under the scaling $(t, x, w) \to \lambda(t, x, w)$ and under a Lorentz transformation on (t, x), and thus (14) must be invariant under these same transformations.

Again, as in the corresponding discussion for de Sitter spacetime, it is a good idea to go to light cone coordinates for the pseudo-time coordinate W and the last spatial embedding coordinate Y:

$$W^+ \equiv W + Y = \frac{1}{w}\left(x^2 - t^2\right) + w, \qquad W^- \equiv W - Y = \frac{1}{w} \tag{15}$$

You can now see relatively simply that the defining equation $T^2 - X^2 + W^+ W^- = 1$ and (14) are satisfied.

It is now easy to generalize, going up in dimension. For AdS^4, write $T = \frac{t}{w}$, $X = \frac{x}{w}$, $Y = \frac{y}{w}$, and $W^+ \equiv W + Z = \frac{1}{w}(x^2 + y^2 - t^2) + w$, $W^- \equiv W - Z = \frac{1}{w}$. By now, it is almost immediate that AdS^5 is described by*

$$ds^2 = \frac{L^2}{w^2}\left(-dt^2 + dx^2 + dy^2 + dz^2 + dw^2\right) \tag{16}$$

Basically, given (14), the preceding follows almost trivially: y and z are just going along for the ride. We have restored L by dimensional analysis. Just like the Poincaré half plane, AdS^5 has a spatial boundary at $w = 0$. (In contrast, dS^4 as coordinatized in (IX.10.26) has a temporal boundary at $u = 0$.)

From the very simple form of the metric, you can see that the Christoffel symbol $\Gamma^{\cdot}_{\cdot\cdot}$ and the Riemann curvature tensor $R^{\cdot}_{\cdot\cdot\cdot} \sim \partial\Gamma^{\cdot}_{\cdot\cdot} + \Gamma^{\cdot}_{\cdot\cdot}\Gamma^{\cdot}_{\cdot\cdot}$ go like $\sim 1/w$ and $\sim 1/w^2$, respectively, and thus vanish[†] as $w \to \infty$.

A slice of this 5-dimensional spacetime at some specific value of w, say w_*, with the metric $ds^2 = \frac{1}{w_*^2}(-dt^2 + dx^2 + dy^2 + dz^2)$, is just the familiar 4-dimensional Minkowskian spacetime! See figure 3. (Contrast and compare with the exponentially expanding universe in (IX.10.23).)

Consider an object, for example a human, of physical size Δl measured from head to toes, lined up along the x-axis, say. Then his head is separated from his toes by $\Delta x = w_* \Delta l$. As w_* decreases toward the boundary at $w = 0$, the coordinate size Δx of the object shrinks.

Transforming coordinates $w = L^2/r$, we obtain the alternative form

$$ds^2 = \frac{r^2}{L^2}\left(-dt^2 + dx^2 + dy^2 + dz^2\right) + \frac{L^2}{r^2}dr^2$$

$$= \left(-r^2 dt^2 + \frac{1}{r^2}dr^2\right) + r^2 d\vec{x}^2 \quad \text{(setting } L = 1) \tag{17}$$

The boundary is at $r = \infty$, as indicated in figure 4. Note also that the metric is invariant under the scaling[‡] $t \to \lambda t$, $\vec{x} \to \lambda \vec{x}$, and $r \to \lambda^{-1}r$.

---

* In the literature, the notation $ds^2 = \frac{L^2}{z^2}(-dt^2 + d\vec{x}^2 + dz^2)$ is often used, with z, a letter we already used for something else, denoting the coordinate "perpendicular" to the boundary.

† The scalar curvature is of course constant.

‡ In recent applications to condensed matter physics (in an area of research known as AdS/CMP), which is in general not Lorentz invariant, the metric $ds^2 = (-r^{2z}dt^2 + \frac{1}{r^2}dr^2) + r^2 d\vec{x}^2$ is used. The real number z is known as a dynamical exponent. The time coordinate now scales as $t \to \lambda^z t$.

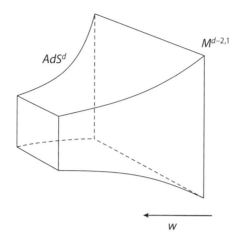

Figure 3 A slice of AdS^d at some specific value of w is the $(d-1)$-dimensional Minkowskian spacetime $M^{d-2,1}$.

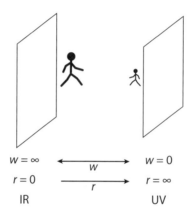

Figure 4 The head and toes of a human of physical size Δl located at w_* and lined up along the x-axis are separated by $\Delta x = w_* \Delta l$. IR = infrared; UV = ultraviolet.

You might have noticed that combinations such as $w + \frac{1}{w}$ in (13) are practically begging us to write w as an exponential and introduce cosh and sinh. Indeed, set $w = L e^{u/L}$ in (16) and write

$$ds^2 = e^{-\frac{2u}{L}} \left(-dt^2 + dx^2 + dy^2 + dz^2 \right) + du^2 \tag{18}$$

Note that the coordinate u ranges from $-\infty$ to $+\infty$. We will use this form in chapter X.2 in our discussion of brane worlds.

By now it is clear how the Poincaré coordinates are to be defined for AdS^d. (Compare and contrast with what we had for dS^d.) Let

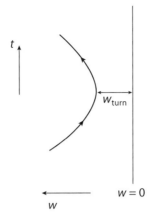

Figure 5 A massive particle moving in the $(t\text{-}w)$ plane toward the boundary cannot reach the boundary, but turns back at some w_{turn} determined by its initial position and speed.

$$X^\mu = e^{-u} x^\mu$$
$$X^+ \equiv X^d + X^{d-1} = e^{-u} \eta_{\rho\sigma} x^\rho x^\sigma + e^u$$
$$X^- \equiv X^d + X^{d-1} = e^{-u} \tag{19}$$

Here the indices μ, ρ, and σ run over 0, 1, $\cdots$, $d-2$. The defining equation $\eta_{\mu\nu} X^\mu X^\nu - X^+ X^- = -1$ is satisfied. Indeed, we do not even have to compute $ds^2 = \eta_{\mu\nu} dX^\mu dX^\nu - dX^+ dX^-$. Various symmetries essentially fix it to be $ds^2 = e^{-2u} \eta_{\mu\nu} dx^\mu dx^\nu + du^2$, as was already mentioned. For example, as noted, the metric must be invariant under scaling x^μ and translating u, and under Lorentz transformations of x^μ. The splitting of the embedding coordinates into the two sets X^μ and (X^d, X^{d-1}) reflects the two subgroups $SO(d-2, 1)$ and $SO(1, 1)$ contained in the isometry group $SO(d-1, 2)$.

Motion of light and massive particle in anti de Sitter spacetime

The Poincaré coordinates in (16) are particularly suitable for studying the motion of light and particles in anti de Sitter spacetime. That it is conformally equivalent to the Minkowski metric $d\tilde{s}^2 = (-dt^2 + dx^2 + dy^2 + dz^2 + dw^2)$ means that light follows a path determined by $ds^2 = 0 = d\tilde{s}^2$. A light beam sent by an observer located at $w = w_0$ toward the boundary at $w = 0$ will come back, if a mirror were placed appropriately, to her after coordinate time $t_{\text{return}} = 2w_0$. In contrast, consider a massive particle moving* in the $(t\text{-}w)$ plane toward the boundary, as shown in figure 5. The definition of proper time $g_{\mu\nu} \frac{dx^\mu}{d\tau} \frac{dx^\nu}{d\tau} = -1$ gives us $(\frac{dt}{d\tau})^2 - (\frac{dw}{d\tau})^2 = w^2$. The isometry under $t \to t+$ constant gives as usual the conservation law $\frac{d}{d\tau}(w^{-2} \frac{dt}{d\tau}) = 0$, and hence $\frac{dt}{d\tau} = \frac{w^2}{b}$, for some constant b. We thus obtain $(\frac{dw}{dt})^2 + \frac{b^2}{w^2} = 1$. The potential in the analog Newtonian problem is thus $V(w) = +\frac{b^2}{w^2}$, and

---

* You might realize that this is basically the same problem as one we worked out in chapter II.2.

we see that the particle cannot reach the boundary but turns back at $w_{\text{turn}} = b$, with b determined by its initial position and speed.

This simple exercise in classical general relativity already hints at a key feature that makes possible AdS/CFT. At first sight, it seems impossible that the physics of a $(3 + 1)$-dimensional theory could be mapped onto the physics of a $(4 + 1)$-dimensional theory. It turns out that particles carrying different energies in the boundary theory correspond to particles located at different positions along the w-axis in the bulk. The boundary theory is able to "grow" a spatial coordinate orthogonal to it.

By this point, you might be wondering if there is an analogous story for de Sitter spacetime. The answer is that some theoretical physicists are working intensively on establishing a dS/CFT correspondence. Intriguingly, de Sitter spacetime has a temporal boundary, as you may recall from the preceding chapter, rather than a spatial boundary, and so, if some kind of dS/CFT correspondence does in fact hold, the boundary theory would have to grow a temporal coordinate. Might this shed some light on the origin of time? Note that the boundary theory does not contain time, and thus represents some kind of statistical mechanics rather than a dynamical field theory.

Other forms of anti de Sitter spacetime

Just like de Sitter spacetime, anti de Sitter spacetime can be written in a bewildering variety of forms, as we have already mentioned.

We obtain an interesting form by changing the variable $r = 2\zeta/(1 - \zeta^2)$ in (8). Then $1 + r^2 = [(1 + \zeta^2)/(1 - \zeta^2)]^2$ and $dr = 2((1 + \zeta^2)/(1 - \zeta^2)^2 d\zeta$. We thus obtain the alternative form

$$ds^2 = \frac{-\left(1 + \zeta^2\right)^2 dt^2 + 4\left(d\zeta^2 + \zeta^2 d\Omega_{d-2}^2\right)}{\left(1 - \zeta^2\right)^2} = \frac{-\left(1 + \zeta^2\right)^2 dt^2 + 4\delta_{ij} dx^i dx^j}{\left(1 - \zeta^2\right)^2} \tag{20}$$

where $\zeta^2 = \delta_{ij} x^i x^j$. We have a ball defined by $\zeta < 1$, with its boundary at $\zeta = 1$.

To derive the next form, set $d = 3$ for definiteness and write $r^2 = X^2 + Y^2$. The defining equation $T^2 - r^2 + W^2 = 1$ invites us to define $T = \rho \sinh \chi$ and $r = \rho \cosh \chi$, so that $dT^2 - dr^2 = \rho^2 d\chi^2 - d\rho^2$. (You may recognize this as the Rindler coordinates that you worked on way back in exercise III.3.2 and that we discussed in chapter VII.3.) The defining equation then becomes $W^2 = 1 + \rho^2$. Differentiating, we have $W dW = \rho d\rho$, so that $dW^2 = \frac{\rho^2}{1 + \rho^2} d\rho^2$. We end up with

$$ds^2 = -\rho^2 d\chi^2 + \left(\frac{1}{1 + \rho^2} d\rho^2 + \rho^2 \cosh^2 \chi \, d\theta^2\right)$$
$$= -\sinh^2 \psi d\chi^2 + d\psi^2 + \sinh^2 \psi \cosh^2 \chi \, d\theta^2 \tag{21}$$

where $\rho = \sinh \psi$, so that we now have

$$T = \sinh \psi \sinh \chi, \quad r = \sinh \psi \cosh \chi, \quad W = \cosh \psi \tag{22}$$

For another form, let

$$T = \sin t \cosh \chi, \quad X = r \cos \theta, \quad Y = r \sin \theta, \quad W = \cos t \tag{23}$$

with $r = \sin t \sinh \chi$. Then we obtain

$$ds^2 = -dt^2 + \sin^2 t \left(d\chi^2 + \sinh \chi^2 d\theta^2\right) = -dt^2 + \sin^2 t \, dH_2^2 \tag{24}$$

More generally, $ds^2 = -dt^2 + \sin^2 t \, dH_{d-1}^2$ for AdS^d.

Anti de Sitter spacetime in hyperbolic coordinates

Again for definiteness, let us consider AdS^3. The trick is to rewrite the defining equation (6) and the metric (7) as $(T^2 - X^2) + (W^2 - Y^2) = 1$ and $ds^2 = (-dT^2 + dX^2) + (-dW^2 + dY^2)$, respectively. Let $T = R \cosh t$, $X = R \sinh t$, $W = r \cosh \psi$, and $Y = r \sinh \psi$. We have $ds^2 = dR^2 - R^2 dt^2 + dr^2 - r^2 d\psi^2$. Also, $R^2 = 1 - r^2$, $R dR = -r dr$, $dR^2 = \frac{r^2}{1-r^2} dr^2$, and $dR^2 + dr^2 = \frac{1}{1-r^2} dr^2$. Hence we obtain

$$ds^2 = -\left(r^2 - 1\right) dt^2 + \frac{dr^2}{r^2 - 1} + r^2 d\psi^2 \tag{25}$$

Note that this requires an analytic continuation: the metric (25) only makes sense for $r > 1$, which requires R^2 to be negative. This construction is readily generalized to AdS^d: $ds^2 = -(r^2 - 1)dt^2 + \frac{dr^2}{r^2-1} + r^2 \, dH_{d-2}^2$. (As in the preceding chapter, dH^2 denotes the hyperbolic line element defined in chapter I.6 that has already appeared in de Sitter spacetime.) Compare with (8) and (IX.10.35). In particular, for AdS^4, we have

$$ds^2 = -\left(r^2 - 1\right) dt^2 + \frac{dr^2}{r^2 - 1} + r^2 \left(d\psi^2 + \sinh^2 \psi d\varphi^2\right) \tag{26}$$

Stereographic projection for anti de Sitter spacetime

As for de Sitter spacetime, we can stereographically project anti de Sitter spacetime by mapping $(X^0, X^1, X^2, X^3, X^4)$ into (x^0, x^1, x^2, x^3) as follows (reinstating L):

$$X^M = \frac{1}{1 - \frac{x^2}{4L^2}} \delta^M_\mu x^\mu, \quad M = 0, 1, 2, 3 \tag{27}$$

and

$$X^4 = L \left(\frac{1 + \frac{x^2}{4L^2}}{1 - \frac{x^2}{4L^2}}\right) \tag{28}$$

where as before, $x^2 \equiv -(x^0)^2 + (x^1)^2 + (x^2)^2 + (x^3)^2$. Now (27) says that $-(X^0)^2 + (X^1)^2 + (X^2)^2 + (X^3)^2 = \frac{x^2}{1 - \frac{x^2}{4L^2}}$. Verify that the defining relation (1) is satisfied and that

$$ds^2 = \left(\frac{1}{1 - \frac{x^2}{4L^2}}\right)^2 \eta_{\mu\nu} dx^\mu dx^\nu \tag{29}$$

Anti de Sitter spacetime is conformally flat, just like de Sitter spacetime (see the table). No surprise there.

One of my students speaks up at this point. "You shouldn't say that," he says. "Everybody knows that the sphere is not flat; that it is conformally flat is kind of a surprise." But then, since de Sitter spacetime is sort of a Minkowskian sphere, it is arguably less surprising that it is also conformally flat. But that anti de Sitter spacetime is also conformally flat, now that, to him, is surprising. I suppose that it is fair to conclude that everybody has different thresholds for being surprised.[10]

Appendix 1: Euclidean anti de Sitter space and its boundary

At first sight, the appearance of a boundary in anti de Sitter spacetime is rather puzzling (but perhaps not so much to readers of this text, since we have already discussed the Poincaré half plane in part I). As far as I know, it is easiest to visualize this boundary if, using the stereographic coordinates in (29), we go to Euclidean anti de Sitter space.

Replace $\eta_{\mu\nu}$ in (29) by $\delta_{\mu\nu}$, thus going from Minkowski back to Pythagoras, which we can do by formally writing $X^0 = i X^T$ and $x^0 = i x^T$. Set

$$X^M = \frac{1}{1 - \frac{x^2}{4L^2}} \delta^M_\mu x^\mu, \qquad M = T, 1, 2, \cdots, d - 1 \quad \text{and} \quad X^d = \frac{L \left(1 + \frac{x^2}{4L^2}\right)}{\left(1 - \frac{x^2}{4L^2}\right)}$$

Here $x^2 \equiv (x^T)^2 + (x^1)^2 + (x^2)^2 + \cdots + (x^{d-1})^2$ denotes the Euclidean square of the d-dimensional vector $(x^T, x^1, x^2, \cdots, x^{d-1})$. (A word about notation: Perhaps it would have been more natural to denote x^T by x^d, but this would have led to a potential confusion, since X^d and x^d are not directly related.)

Then $(X^T)^2 + (X^1)^2 + (X^2)^2 + \cdots + (X^{d-1})^2 = x^2/(1 - \frac{x^2}{4L^2})^2$, so that the defining relation

$$\left(X^T\right)^2 + \sum_{i=1}^{d-1} \left(X^i\right)^2 - \left(X^d\right)^2 = -L^2$$

for Euclidean anti de Sitter space is satisfied. We obtain

$$ds^2 = \left(\frac{1}{1 - \frac{x^2}{4L^2}}\right)^2 ((dx^T)^2 + (dx^1)^2 + (dx^2)^2 + \cdots + (dx^{d-1})^2) \tag{30}$$

which we can also see by analytically continuing (29). Again, it is not a surprise that the Euclidean anti de Sitter space AdS^d_E is conformally related to Euclidean space. More importantly in this context, we note that it is topologically the Euclidean ball B^d defined by

$$x^2 \leq 4L^2 \tag{31}$$

which, as every child knows, has a boundary described by S^{d-1}. (The sphere S^{d-1} is just $(d - 1)$-dimensional Euclidean space E^{d-1} with infinity identified as a single point.) This is of course the Euclidean version of the statement that AdS^d has $M^{d-2,1}$ for its boundary. The metric tensor, and hence, by the basic theorem about maximally symmetric space, the curvature and the Ricci tensor (but not the scalar curvature) all diverge as we approach the boundary.

Appendix 2: Isomorphism between AdS^3 and $SL(2, R)$

The most general 2-by-2 matrix with real entries may be written as

$$\mathcal{U} = \begin{pmatrix} T + X & Y + W \\ Y - W & T - X \end{pmatrix}$$

The condition $\det \mathcal{U} = 1$ implies $T^2 - X^2 - Y^2 + W^2 = 1$. Under multiplication, the set of all 2-by-2 matrices with real entries and unit determinant clearly generates a group, known as $SL(2, R)$.

Table for anti de Sitter spacetime

t, r, Ω	$-\left(1+\frac{r^2}{L^2}\right)dt^2 + \frac{dt^2}{1+\frac{r^2}{L^2}} + r^2 d\Omega_{d-2}^2$	$T = \sqrt{1+r^2}\cos t$, $\quad W = \sqrt{1+r^2}\sin t$, $\quad X = r\cos\theta$, $\quad Y = r\sin\theta$
t, ρ, Ω	$-L^2\cosh^2\rho\, dt^2 + d\rho^2 + \sinh^2\rho\, d\Omega_{d-2}^2$	$T = \cosh\rho\cos t$, $\quad W = \cosh\rho\sin t$, $\quad X = \sinh\rho\cos\theta$, $\quad Y = \sinh\rho\sin\theta$
t, x, y, z, w	$\frac{L^2}{w^2}(-dt^2 + dx^2 + dy^2 + dz^2 + dw^2)$	$T = \frac{t}{w}$, $\quad X = \frac{x}{w}$, $\quad Y = \frac{1}{2}\left(\frac{x^2-t^2}{w} + w - \frac{1}{w}\right) = \frac{1}{2w}(x^2 - t^2 + w^2 - 1)$, $\quad W = \frac{1}{2}\left(\frac{x^2-t^2}{w} + w + \frac{1}{w}\right) = \frac{1}{2w}(x^2 - t^2 + w^2 + 1)$
t, x, y, z, v	$\frac{v^2}{L^2}(-dt^2 + dx^2 + dy^2 + dz^2) + \frac{L^2}{v^2}dv^2$	$w = L^2/v$
t, x, y, z, u	$e^{-\frac{2u}{L}}(-dt^2 + dx^2 + dy^2 + dz^2) + du^2$	$w = Le^{u/L}$
t, ζ, Ω	$\frac{-\left(1+\frac{\zeta^2}{L^2}\right)^2 dt^2 + 4(d\zeta^2 + \zeta^2 d\Omega^2)}{\left(1-\frac{\zeta^2}{L^2}\right)^2}$	$r = 2\zeta/(1-\zeta^2)$
χ, ρ, θ	$-\rho^2 d\chi^2 + \frac{1}{1+\rho^2}d\rho^2 + \rho^2\cosh^2\chi\, d\theta^2$	$T = \rho\sinh\chi$, $\quad r = \rho\cosh\chi$, $\quad w^2 = 1 + \rho^2$
χ, ψ, θ	$L^2(-\sinh^2\psi\, d\chi^2 + d\psi^2 + \sinh^2\psi\cosh^2\chi\, d\theta^2)$	$T = \sinh\psi\sinh\chi$, $\quad r = \sinh\psi\cosh\chi$, $\quad W = \cosh\psi$
t, r, ψ, φ	$-(r^2-1)dt^2 + \frac{dr^2}{r^2-1} + r^2(d\psi^2 + \sinh^2\psi\, d\varphi^2)$	$T = R\cosh t$, $\quad W = r\cosh\psi$, $\quad R^2 = 1 - r^2$
t, ψ, Ω	$L^2(-dt^2 + \sin^2 t(d\psi^2 + \sinh^2\psi\, d\Omega_2^2))$	$T = \sin t\cosh\chi$, $\quad X = r\cos\theta$, $\quad Y = r\sin\theta$, $\quad W = \cos t$, $\quad r = \sin t\sinh X$
u, ψ, Ω	$-\frac{L^2}{L^2-u^2}du^2 + u^2(d\psi^2 + \sinh^2\psi\, d\Omega_2^2)$	$u = L\sin t$
t, ψ, Ω	$\frac{1}{\cosh^2\frac{\ell}{L}}\left(-d\rho^2 + L^2\sinh^2\frac{\ell}{L}dH_3^2\right)$	$u = L\tanh\frac{\ell}{L}$
x	$\left(\frac{1}{1-\frac{x^2}{4L^2}}\right)^2 \eta_{\mu\nu}dx^\mu dx^\nu$	$X^M = \frac{1}{1-\frac{x^2}{4L^2}}\delta^M_{\ \mu}x^\mu$, $\quad M = 0, 1, 2, 3$, $\quad X^4 = L\left(\frac{1+\frac{x^2}{4L^2}}{1-\frac{x^2}{4L^2}}\right)$
t, ψ, Ω	$\frac{L^2}{\cos^2\psi}(dt^2 + d\psi^2 + \sin^2\psi\, d\Omega_{d-2}^2)$	$T = \sqrt{1+r^2}\cos t$, $\quad W = \sqrt{1+r^2}\sin t$, $\quad X = r\cos\theta$, $\quad Y = r\sin\theta$, $\quad r = \tan\psi$

Note: In the table, $d\Omega_d^2 = d\psi^2 + \sin^2\psi\, d\Omega_{d-1}^2$ and $dH_d^2 = d\psi^2 + \sinh^2\psi\, d\Omega_{d-1}^2$.

What we have just discovered is that AdS^3 is isomorphic to the universal cover of $SL(2, R)$: there is 1-to-1 correspondence between points on AdS^3 and elements of $SL(2, R)$.

Evidently, for $\mathcal{V}$ and $\mathcal{Z}$ any two elements of $SL(2, R)$, then $\mathcal{U}' = \mathcal{V}\mathcal{U}\mathcal{Z}$ also has unit determinant, and thus, as an element of $SL(2, R)$, corresponds to another point on AdS^3. In other words, the isometry group of AdS^3 is $SL(2, R) \times SL(2, R)$. But we know from the text that the isometry group* of AdS^3 is $SO(2, 2)$, yet this is consistent, since $SO(2, 2)$ is in fact (see below) isomorphic to $SL(2, R) \times SL(2, R)$. Incidentally, this beautiful piece of group theory is relevant to the recent use of twistors[11] to calculate scattering amplitudes in quantum field theory.

We now exhibit the isomorphism $SO(2, 2) = SL(2, R) \times SL(2, R)$ explicitly. In chapter IX.9, we saw that the 6 generators of $SO(2, 2)$, $\partial^\pm$, $x^\pm \partial_\pm$, $-(x^\pm)^2 \partial_\pm$, break into two mutually commuting sets, evidently corresponding to 2 copies of $SL(2, R)$. Consider an element $I + A$ of $SL(2, R)$ close to the identity. Using an identity we have encountered repeatedly, we evaluate its determinant to be $\det(I + A) = e^{Tr \log(I+A)} \simeq 1 + Tr A$. Thus, the generators of $SL(2, R)$ consist of 2-by-2 traceless matrices, among which we choose the linearly independent set

$$T_3 = \begin{pmatrix} 1 & 0 \\ 0 & -1 \end{pmatrix}, \qquad T_+ = \begin{pmatrix} 0 & 1 \\ 0 & 0 \end{pmatrix}, \qquad T_- = \begin{pmatrix} 0 & 0 \\ 1 & 0 \end{pmatrix} \tag{32}$$

You can verify that the desired identification is

$$\partial \sim T_-, \qquad x\partial \sim \tfrac{1}{2} T_3, \qquad -x^2 \partial \sim T_+ \tag{33}$$

Note the minus sign, as determined in the preceding chapter.

Appendix 3: AdS^2 and its two boundaries

Readers into group theory know that "smaller" groups often exhibit special features that the "larger" groups do not have. Similarly, AdS^2 differs from its higher dimensional counterparts by having two boundaries. This apparently puzzling assertion actually follows from an extremely elementary geometric fact. Consider the unit Euclidean ball B^D defined by $\sum_{j=1}^{D}(X^j)^2 \leq 1$. The boundary of the disk B^2 is the circle S^1, of the everyday ball B^3 is the sphere S^2, and so on. But the boundary of B^1, namely S^0, consists of two disconnected points. (Indeed, we already encountered this phenomenon in chapter VII.2.)

The boundary of AdS^d is more visible with some coordinate choices than with others. In particular, with the coordinates used in (12), we have for AdS^2

$$ds^2 = \frac{1}{\cos^2 \psi} \left(-dt^2 + d\psi^2\right) \tag{34}$$

The crucial difference with (12) is that now ψ runs from $-\pi/2$ to $\pi/2$. The rectangular strip that describes AdS^2 in the $(t$-$\psi)$ plane obviously has two boundaries at $\psi = \pm\pi/2$. (For the coordinates used in (10), with $\sinh \rho = \tan \psi$, the two boundaries are at $\rho = \pm\infty$. In terms of the original embedding coordinates $T = \cosh \rho \cos t$, $W = \cosh \rho \sin t$, and $X = \sinh \rho$, the boundaries correspond to the two end caps of the tube in figure 1 at $X = \pm\infty$.)

The presence of the two boundaries is less transparent in Poincaré coordinates:

$$ds^2 = \frac{1}{w^2} \left(-dt^2 + dw^2\right) \tag{35}$$

However, inspection of the coordinate transformation $T = \frac{t}{w}$, $W = \frac{1}{2w}(-t^2 + w^2 + 1)$, and $X = \frac{1}{2w}(-t^2 + w^2 - 1)$ reveals that it is actually at $w = 0^\pm$. Or, with $r = w^{-1}$, as in the text, we have $ds^2 = -r^2 dt^2 + r^{-2}dr^2$. Note that $-\infty < r < \infty$.

It is also instructive to relate the Poincaré coordinates to those used in (34) (but before we do that, we have to rename the time coordinate in the latter as τ). We find

$$w = \frac{\sin \psi + \sin \tau}{\cos^2 \psi - \cos^2 \tau} \cos \psi, \qquad t = \frac{\sin \psi + \sin \tau}{\cos^2 \psi - \cos^2 \tau} \cos \tau \tag{36}$$

From (36), we also see that $w = \pm\infty$ corresponds to the two lines $\psi = \tau$ and $\psi = \pi - \tau$.

---

* In comparison, the isometry group of dS^3 is $SO(3, 1) = SL(2, C)/Z_2$.[12]

Referring to the table contrasting isometry groups for de Sitter and anti de Sitter spacetimes given earlier in this chapter, we also observe the amusing fact that the isometry group $SO(1, 2)$ for AdS^2 and the isometry group $SO(2, 1)$ for dS^2 are actually the same, a fact that we can see geometrically from figures IX.10.1 and 1.

Appendix 4: Continuing from de Sitter to anti de Sitter spacetimes

From the defining equations (1) and (2), we see that formally we can go from de Sitter to anti de Sitter spacetime by analytically continuing $L \to iL$ and $X^d \to iX^d$. (By now, X^d is perhaps better known to us as W.) Alternatively, instead of thinking about the embedding spacetimes, we can consider a specific set of coordinates x^μ and solve Einstein's field equation. Evidently, a solution of $R_{\mu\nu} = -\frac{(d-1)}{L^2} g_{\mu\nu}$, which after all is a bunch of coupled partial (or ordinary) differential equations, becomes a solution of $R_{\mu\nu} = +\frac{(d-1)}{L^2} g_{\mu\nu}$ when we formally set $L \to iL$. However, this procedure may or may not result in a spacetime (see below), and further continuations in the coordinates x^μ will in general be needed.

Let us see how this works in a few cases. For example, taking the first entry $ds^2 = -\frac{L^2}{L^2+t^2} dt^2 + t^2 dH_3^2$ in the table for de Sitter spacetime in the preceding chapter, we plug in $L \to iL$ and encounter no trouble, thus reproducing an entry in the table for anti de Sitter spacetime given in this chapter. In contrast, taking the second entry $ds^2 = \frac{L^2}{\cos^2\theta}(-d\theta^2 + \sin^2\theta \, dH_3^2)$ and flipping the sign of L^2, we would encounter something with 3 time coordinates and 1 space coordinate. Thus, we obviously have to analytically continue θ also, $\theta \to i\rho$, thus obtaining $ds^2 = \frac{L^2}{\cosh^2\rho}(-d\rho^2 + \sinh^2\rho \, dH_3^2)$. (Useful identities in this context: $\cosh ix = \cos x$, $\sinh ix = i\sin x$.) Another approach is to use coordinates with dimension of length. So, first write $\theta = \frac{\rho}{L}$ and the de Sitter metric as $ds^2 = \frac{1}{\cos^2\frac{\rho}{L}}(-d\rho^2 + L^2 \sin^2\frac{\rho}{L} dH_3^2)$. Then continue $L \to iL$.

Another way of saying this is that if we use dimensionless coordinates, such as angles cyclic and hyperbolic, then by dimensional analysis, ds^2 has to be proportional to L^2, and so are the metric components $g_{\mu\nu}$. If we flip the sign of the metric, the Christoffel symbol $\Gamma^{\cdot}_{\cdot\cdot} \sim g^{\cdot\cdot}\partial g_{\cdot\cdot}$, the Riemann curvature tensor (with one upper and three lower indices) $R^{\cdot}_{\cdot\cdot\cdot}$, and the Ricci tensor $R_{\cdot\cdot}$ do not flip, but the scalar curvature R does. Einstein's field equation $R_{\cdot\cdot} \sim L^{-2}g_{\cdot\cdot}$ flips between de Sitter and anti de Sitter spacetime, as it must.

Appendix 5: Geodesics in the embedding space

As in appendix 3 in chapter IX.10, we can discuss geodesics as visualized with the embedding coordinates X^M satisfying $X^2 = \eta_{MN}X^M X^N = -1$. Go through the same steps as in that appendix, introducing a Lagrange multiplier and so forth but keeping in mind that in the present case, $\eta = (-++\cdots+-)$. In particular, in the embedding space* a photon zips merrily along a straight line in the sense that it follows $\ddot{X} = 0$. As in the preceding chapter, we can verify that the geodesics in the embedding space do describe geodesics in whatever coordinates are used to map anti de Sitter spacetime.

Confusio mumbles that, since he now understands how this went in the preceding chapter, there is no sense in checking it again. We respond that since we used a rather unnatural looking transformation (13) to define the Poincaré coordinates, we think that it is still fun to see how the laws of arithmetic work. A photon, for instance, traces out

$$X = a + b\zeta \tag{37}$$

with $a^2 = -1$, $b^2 = 0$, $a \cdot b = 0$ (to ensure that $X^2 = -1$).

Consider a photon moving along the x-axis. From $ds^2 = \frac{1}{w^2}(-dt^2 + dx^2 + dw^2)$, we have $dx = dt$, $x = t$, and $w = w_*$ (with w_* arbitrary). Note that anti de Sitter spacetime is translation invariant in t and x but not in w. Referring to (13), we translate $x = t$ into $X = (2t, 2t, w_*^2 - 1, w_*^2 + 1)/(2w_*)$, so that $a = (0, 0, w_*^2 - 1, w_*^2 + 1)/(2w_*)$ and $b = (1, 1, 0, 0)$. Note that the normalization of b can be absorbed into the definition of ζ. We see that indeed, $a^2 = -1$, $b^2 = 0$, and $a \cdot b = 0$.

Confusio: "You seem to use 'translate' in two senses."

Yes, literally and metaphorically.

---

* You might have noticed that I avoid using the term "embedding spacetime" in this chapter, since I wouldn't know what to call a space with two time coordinates.

Work out another case for fun. Let the photon move along the w-axis, so that $w = t$, which translates into $X = (1, 0, (2t)^{-1}, -(2t)^{-1})$. Remarkably, $T = X^0$ stays constant. But the other "time" W is not standing still! Thus, $a = (1, 0, 0, 0)$ and $b = (0, 0, 1, -1)$. Indeed, $a^2 = -1$, $b^2 = 0$, and $a \cdot b = 0$. No surprise that the laws of arithmetic work, even in general relativity!

Appendix 6: Scalar field in AdS/CFT

While the AdS/CFT correspondence is beyond the scope of this text, we can mention that one important step involves solving the equation of motion for a scalar field (of mass m) in AdS^{d+1}, namely $(\Box - m^2)\phi = \frac{1}{\sqrt{-g}}\partial_\mu(\sqrt{-g}g^{\mu\nu}\partial_\nu\phi) - m^2\phi = 0$. And this the devoted reader who has gotten this far is able to do.

Referring to the metric $ds^2 = (-r^2 dt^2 + \frac{1}{r^2}dr^2) + r^2 d\vec{x}^2$ given in (17), we have $-g = r^{2(d-1)}$ and $g^{tt} = 1/r^2$, $g^{xx} = 1/r^2, \cdots$, but $g^{rr} = r^2$. Thus, $\Box$ contains terms like $(-\partial_t^2 + \partial_x^2 + \cdots)/r^2$, terms which may be neglected near the boundary at $r = \infty$. In contrast, g^{rr} grows near the boundary. Hence, the equation of motion near the boundary reduces to

$$r^{-d+1}\partial_r\left(r^{d+1}\partial_r\phi\right) - m^2\phi = 0 \tag{38}$$

This equation, homogeneous in powers of r, can be solved by plugging in $\phi \sim r^K$. We obtain a quadratic equation with the roots $\Delta - d$ and $-\Delta$, where

$$\Delta \equiv \frac{d}{2} + \sqrt{\left(\frac{d}{2}\right)^2 + m^2} \tag{39}$$

Thus, we obtain

$$\phi(r \sim \infty, t, \vec{x}) = \alpha(t, \vec{x})r^{\Delta-d} + \beta(t, \vec{x})r^{-\Delta} \tag{40}$$

Since $\Delta - d > 0$, we have to impose the boundary condition $\alpha(t, \vec{x}) = 0$. The AdS/CFT correspondence states that the expectation value of a certain quantum field theoretic operator living on the boundary of AdS^{d+1} is then given by $\beta(t, \vec{x})$. As for the question why oh why, the answer is not contained in this textbook.

Appendix 7: Coset manifolds and the classification of space and spacetime

By the end of the 19th century, it was understood[13] that space could be Euclidean, spherical, or hyperbolic. In the language of coset manifolds, we can start with two empirical observations and arrive at these three possibilities. The isotropy of space implies that space is of the form $G/SO(3)$. The 3-dimensionality of space implies that G must have $3 + 3$ generators. There are 3 groups with 6 generators, namely $G = E(3)$ (the Euclidean group consisting of rotations and translations), $G = SO(4)$, and $G = SO(3, 1)$, corresponding to Euclidean, spherical, and hyperbolic, respectively. Note the appearance in this context of the Lorentz group long before special relativity!

Now let us generalize this discussion to spacetime. We know that spacetime is Lorentz invariant and 4-dimensional. Thus, if spacetime is homogeneous, it should be of the form $G/SO(3, 1)$, with G having $6 + 4 = 10$ generators. Again, there are 3 possibilities, namely $G = E(3, 1)$ (the Poincaré group consisting of Lorentz transformations and translations), $G = SO(4, 1)$, and $G = SO(3, 2)$, corresponding to Minkowski, de Sitter, and anti de Sitter spacetime, respectively.

Notes

1. For example, P. A. M. Dirac, "The Electron Wave Equation in de-Sitter Space," *Annals Math.* (1935), p. 657, and "A Remarkable Representation of the 3+2 de Sitter Group," *J. Math. Phys.* (1963), p. 901.
2. Strictly speaking, $M^{3,1}$ constitutes a patch of the boundary, which globally has the topology of $R \times S^3$. Recall the discussion in appendix 5 of chapter VII.2.
3. This aspect of anti de Sitter spacetime was emphasized by S. Hawking and D. Page, *Comm. Math. Phys.* 87 (1983), p. 577.

4. The reader is referred to many excellent reviews, in particular, O. Aharony, S. Gubser, J. Maldacena, H. Ooguri, and Y. Oz, *Phys. Reports* 323 (2000), p. 183. For applications to condensed matter physics, see various reviews by S. Hartnoll, by J. McGreavy, and by C. Herzog.

5. For general overviews, see J. Maldacena, "The Illusion of Gravity," *Scientific American*, November 2005, p. 57; and C. V. Johnson and P. Steinberg, *Physics Today*, May 2010, p. 29.

6. Have you ever wondered what a world with two time coordinates would be like? Science fiction writers have long played with time travel. You could be the first to toy with two times!

7. There have been speculations, of course, about more than one time. See papers by I. Bars, and by G. R. Dvali, G. Gabadadze, and G. Senjanovic.

8. Again, nothing prevents people from speculating about time travel and the like. Look up on the web the discussions surrounding the chronology protection conjecture.

9. See appendix 1 in the preceding chapter.

10. This reminds me of a story about a famous dictionary maker.

11. For an elementary introduction to this fascinating subject, see, for example, *QFT Nutshell,* chapter N.3 and appendix B.

12. *QFT Nutshell,* p. 532.

13. H. Helmholtz, "The Origin and Meaning of Geometrical Axioms," *Mind* 1 (1876), pp. 301–321, http://www.jstor.org/stable/2246591.

Recap to Part IX

Part IX consists of a collection of topics that hopefully amuses and amazes.

Transporting a vector by keeping it parallel to itself and eventually returning to its starting point tells us about curvature. A precessing gyroscope realizes parallel transport. How parallel straight lines approach or move away from each other also tells us about curvature.

Linearizing Einstein's field equation, we find that ripples in spacetime propagate as gravitational waves. Since a gravitational wave carries energy momentum, it inevitably acts upon itself. Starting with the linear theory, we soon discover that this inherent nonlinearity of gravity leads us to Einstein's theory.

The language for describing the symmetries of spacetime, often obscured by the coordinate choice, is developed through the notions of Killing vectors and isometry.

Just as the Lorentz algebra can be extended to the Poincaré algebra, the Poincaré algebra can be extended to the conformal algebra.

Differential forms provide us with a powerful method of calculating curvature.

For the same reasons that theoretical physicists like the circle and the sphere, we like de Sitter and anti de Sitter spacetimes.

Part X | Gravity Past, Present, and Future

X.1 | Kałuza, Klein, and the Flowering of Higher Dimensions

More than a new continent

> Yet I exist in the hope that these memoirs, in some manner, I
> know not how, may find their way to the minds of humanity in
> Some Dimension, and may stir up a race of rebels who shall
> refuse to be confined to limited Dimensionality.
>
> —narrator in E. A. Abbott's *Flatland*

Minkowski's innovative geometric view of special relativity as a 4-dimensional spacetime bringing together space and time was so obviously true that it was quickly accepted. In 1919, the German-Polish physicist Theodor Kałuza wrote to Einstein to say that he had added another dimension and moved up into a more spacious spacetime. Einstein was quite taken by this brilliant idea, but it did not prevent him from sitting on Kałuza's paper for a year[1] before sending it to the Prussian Academy for publication in 1921. Einstein confessed, perhaps somewhat ruefully, that dimensions higher than $4 = 3 + 1$ had never occurred to him.

Kałuza certainly recognized the far-reaching implication of his suggestion: the paper was titled "On the Problem of Unity in Physics." He managed to unify the two established interactions known at the time: gravity and electromagnetism. The Swedish physicist Oskar Klein, known for his many contributions[2] to physics, rediscovered[3] the Kałuza theory[4] in 1926 and later developed the theory further. Since then, the subject has been vastly extended and generalized. My strategy is to give, for the sake of pedagogical clarity, an overview of the essential concepts involved. To keep things as simple as possible, I focus mostly on the 5-dimensional case and relegate assorted technical details to appendices.

The profound idea of transformation

Of the several ways to motivate Kałuza's idea, I choose to start with the profound idea of transformation (instead of a brute force approach I will describe later). Theoretical physicists stand in awe of the fact that, insisting on invariance under gauge transformation, we are led to electromagnetism and, insisting on invariance under general coordinate transformation, we are led to Einstein gravity. Furthermore, generalizing* gauge transformation, particle physicists were led to the strong interaction and unification of the electromagnetic and weak interactions into a single electroweak interaction, which in turn opens the door to unifying all three nongravitational interactions—strong, weak, and electromagnetic—into a grand unified theory, as already mentioned in chapter VIII.3. Amazingly, insisting on invariance under these transformations has almost magically led to an understanding of the physical world.

For now, we go back to Kałuza's attempt to unify gravity and electromagnetism. Under an electromagnetic gauge transformation, the electromagnetic potential transforms by

$$A_\mu \to A'_\mu = A_\mu - \partial_\mu \Lambda \tag{1}$$

In contrast, under a coordinate transformation $x^\mu \to x'^\mu$, the metric transforms by $g'_{\rho\sigma}(x') = g_{\mu\nu}(x)\frac{\partial x^\mu}{\partial x'^\rho}\frac{\partial x^\nu}{\partial x'^\sigma}$. At first sight, these two transformations look totally different. But, as we saw in chapter IX.4, in the weak field limit, under the infinitesimal coordinate transformation $x^\mu \to x'^\mu = x^\mu + \varepsilon^\mu(x)$, the field $h_{\mu\nu} \equiv g_{\mu\nu} - \eta_{\mu\nu}$ transforms by

$$h_{\mu\nu} \to h'_{\mu\nu} = h_{\mu\nu} - \partial_\mu \varepsilon_\nu - \partial_\nu \varepsilon_\mu \tag{2}$$

The intriguing resemblance between (1) and (2) is striking to say the least, and almost begs for some kind of unification. But how to do it if $h_{\mu\nu}$ and A_μ don't even carry the same number of indices?

An invisible dimension

You rack your brain for a while, and if you're as smart as Kałuza, you might suddenly realize how to do it. Make one of the indices invisible! The desired relation (1) could have originated from an equation like (2) involving objects carrying two indices, if somehow one index is inert or invisible.

Kałuza's idea is to add to the existing coordinates x^0, x^1, x^2, and x^3 an extra coordinate† x^5. Denote the coordinates by $X^M = (x^\mu, x^5)$, with the index M running over 0, 1, 2, 3, and 5.

---

* Maxwell theory is thus generalized into Yang-Mills theory. While this fascinating development is largely beyond the scope of this text (see, for example, *QFT Nut,* part VII), I touch upon some aspects of this important subject, particularly in the appendices.

† You can see that the peculiar notation makes historical sense, since the time coordinate was once known as $x^4 = ict$ before getting renamed as $x^0 = ct$.

In 5-dimensional spacetime, Einstein gravity is invariant under $X^M \to X'^M = X^M + \varepsilon^M(X)$. Denote the metric by $G_{MN} = \eta_{MN} + h_{MN}$, where η_{MN} denotes the extension of the Minkowski metric to 5-dimensional spacetime, with $\eta_{55} = +1$ and $\eta_{\mu 5} = \eta_{5\mu} = 0$. (Note that the sign of η_{55} indicates that the extra dimension added is spatial, not temporal.) For now, we consider the weak field limit, in which (with ε the same order as h)

$$h_{MN} \to h'_{MN} = h_{MN} - \partial_M \varepsilon_N - \partial_N \varepsilon_M \tag{3}$$

under a coordinate transformation, where $\partial_M \equiv \frac{\partial}{\partial X^M}$.

With M, N restricted to μ, ν, (3) reduces to (2). But with M restricted to μ and N set to 5, (3) becomes

$$h_{\mu 5} \to h'_{\mu 5} = h_{\mu 5} - \partial_\mu \varepsilon_5 - \partial_5 \varepsilon_\mu \tag{4}$$

Compare this with (1).

First, under the usual Lorentz transformation in 4-dimensional spacetime, $h_{\mu 5}$ transforms as a vector, just like A_μ. So, let us identify $h_{\mu 5}$ as $l A_\mu$, with l some length. Since G_{MN} is dimensionless and A_μ has dimensions of an inverse length, dimensional analysis mandates the introduction of l, which sets the normalization of A_μ.

Now look at (4). Do you see what Kałuza saw?

Identify ε_5 as $l\Lambda$. Then, up to an overall factor of l, (4) would become (1) were it not for the last term $\partial_5 \varepsilon_\mu$ in (4). But we can get rid of this unwanted guy by simply supposing that ε_μ does not depend on x^5. In that case, we recover (1)!

In particular, we can take $\varepsilon_\mu = 0$. In other words, the familiar electromagnetic gauge transformation we know and love is just the 5-dimensional coordinate transformation

$$x'^\mu = x^\mu, \quad x'^5 = x^5 + l\Lambda(x^\mu) \tag{5}$$

Kałuza has managed to subsume electromagnetic gauge transformation into a general coordinate transformation! This elegant idea constitutes the essence of Kałuza-Klein theory.

The visibility problem and the escape problem

We are immediately confronted by two closely related problems, the visibility problem and the escape problem. How come we don't see the extra dimensions, in contrast to the glaringly obvious three that we deal with all the time? How come we can't escape, and we don't see any particle escaping, into the extra dimensions?

A crucial feature of Einstein's theory, that space can be curled up, almost makes it natural to contemplate higher dimensions. Let the fifth coordinate be curled up as a circle of radius a, so that x^5 ranges between 0 and $2\pi a$. Each spacetime point around us is actually* a tiny circle!

---

* At one time, the idea that solid rock may consist of largely empty space with tiny point particles whizzing around would have seemed equally fantastic.

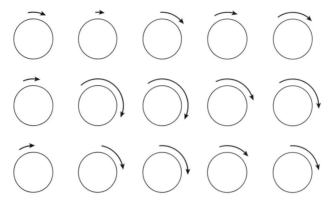

Figure 1 Ever since our ancestors first thought of space, we have been mistaking tiny circles for points. An electromagnetic gauge transformation amounts to twisting the circle at each spacetime point through a different angle.

If a is much smaller than any length scales our experimental friends have explored, then we could all have been fooled ever since our ancestors first thought of space. We have been mistaking tiny circles for points!

This reveals what an electromagnetic gauge transformation actually amounts to: we go around twisting the circle at each spacetime point through a different angle (see (5) and figure 1).

This solves the visibility problem, and now Heisenberg with his uncertainty principle solves the escape problem for us. For a photon to escape into the fifth dimension and be confined there, its momentum would have to be of order $p \sim 1/a$, which would be huge for a small enough. Classically, the frequency of the corresponding electromagnetic wave would be enormous. To squeeze* into such a tight space, the photon (or any of our favorite particles) would need to be terribly energetic.

The visibility problem and the escape problem are thus solved in one fell swoop. Indeed, if we assume that matter fields, such as the electromagnetic field, do not depend on the extra coordinates of the higher dimensions, then in some sense these fields do not "see" these hidden dimensions, and physics as we know it could go on happily as before.

Unifying gravity and electromagnetism

You might be wondering about h_{55}, which we have yet to talk about. In the weak field limit in which we have been working thus far, we obtain from (3) that $h_{55} \to h'_{55} = h_{55} - 2\partial_5\varepsilon_5$. But to get the electromagnetic gauge transformation to come out, we had taken ε_5 in (5) to be independent of x^5, and so $h_{55} = 0$ implies $h'_{55} = 0$. Thus, it is consistent to set $h_{55} = 0$.

---

* Note that in 1921, Heisenberg's uncertainty principle and Schrödinger's equation were both in the future, but this argument can rest only on de Broglie waves.

Kałuza-Klein theory can be implemented in any spacetime dimension $d > 4$, but for pedagogical clarity, it is best to stick with Kałuza's original $d = 5$. It is convenient to give the internal coordinate x^5 another name, say $y \equiv x^5$.

One remarkable feature of Einstein gravity is that the Einstein-Hilbert action $S_{\text{EH}} = + \int d^4x \sqrt{-g}\, M_{\text{P}}^2 R(g)$ can be written, without further ado, in any spacetime dimension. So, go ahead, write it for $d = 5$:

$$S_{\text{KK}} = + \int d^4x dy \sqrt{-G}\, M_5^3 R(G) \tag{6}$$

We denote the scalar curvature constructed out of the 5-dimensional metric G_{MN} by $R(G)$,* to distinguish it from the scalar curvature $R(g)$ constructed out of the 4-dimensional metric $g_{\mu\nu}$. Here the mass M_5 is the analog of the Planck mass M_{P}. To fix its power in (6), we invoke dimensional analysis as follows.

The scalar curvature R (in any spacetime dimension) contains two derivatives acting on the metric and hence has dimensions of an inverse length squared. Thus, as explained at length in chapter VI.1, we have to multiply $\int d^4x \sqrt{-g} R(g)$ by something with dimension of inverse length squared to make the Einstein-Hilbert action dimensionless, and we define that something as the square of the Planck mass M_{P}. (In natural units with $\hbar = 1$ and $c = 1$, mass has dimensions of an inverse length, as was explained way back in the introduction.) In 5-dimensional spacetime, we have to multiply $\int d^4x dy \sqrt{-G} R(G)$ by something with dimensions of inverse length cubed, that is, by some mass cubed. Hence M_5^3 in (6). To summarize, the mass M_5 sets the scale of 5-dimensional gravity, just as M_{P} sets the scale of 4-dimensional gravity.

Let us see how we recover the familiar 4-dimensional gravity and electromagnetism by evaluating the action S_{KK} for various choices of G_{MN}.

First, for $G_{MN}(x, y)$ of the form $G_{\mu\nu}(x, y) = g_{\mu\nu}(x)$, $G_{\mu 5}(x, y) = 0$, and $G_{55}(x, y) = 1$, we have $R(G) = R(g)$ and so the action S_{KK} reduces to $2\pi a M_5^3 \int d^4x \sqrt{-g}\, R(g)$, with the factor of $2\pi a$ coming from the integration over y. Identifying

$$M_{\text{P}}^2 = 2\pi a M_5^3 \tag{7}$$

we obtain the good old Einstein-Hilbert action. That S_{KK} must reduce correctly follows from general invariance considerations: the form of G_{MN} just given is maintained under the 4-dimensional general coordinate transformation $x^\mu = x^\mu(x'^\nu)$, $x^5 = x'^5$.

Next, consider the weak field configuration $G_{\mu\nu}(x, y) = \eta_{\mu\nu}$, $G_{\mu 5}(x, y) = l A_\mu(x)$ (with l the length introduced earlier), and $G_{55}(x, y) = 1$. The 4-dimensional action that results from plugging this into S_{KK} contains two derivatives acting on A_μ and does not change under the gauge transformation (1). This can only be Maxwell's action (recall (V.6.18)). Without doing any computation, we are guaranteed by invariance considerations that Maxwell's action must pop out!

---

* If you confused this G with Newton's constant, go back to square one.

In other words, with $F_{\mu\nu} \equiv \partial_\mu A_\nu - \partial_\nu A_\mu$ as always, S_{KK} must reduce to

$$2\pi a M_5^3 l^2 \zeta \int d^4x\, F_{\mu\nu} F^{\mu\nu} = (M_P l)^2 \zeta \int d^4x\, F_{\mu\nu} F^{\mu\nu}$$

with some unknown numerical coefficient ζ. A detailed calculation (we will do it in appendix 1) is required only if we want to determine ζ, which turns out to be $-\frac{1}{4}$. Thus, to obtain Maxwell's action $S = \int d^4x(-\frac{1}{4}) F_{\mu\nu} F^{\mu\nu}$ as commonly normalized, we set $M_P l = 1$, that is, $l = l_P$. (In other words, from the way A_μ was introduced, we are free to let $A \to \lambda A$ and $l \to l/\lambda$. We have now picked a particular normalization for A, thus fixing l.)

The Kałuza-Klein metric

Thus far, our discussion has been in the weak field expansion discussed in chapter IX.4, with G_{MN} given in two special cases. To construct the Kałuza-Klein metric G_{MN} in its full glory, we go back to (1) and note that under a gauge transformation, $A_\mu dx^\mu \to (A_\mu - \partial_\mu \Lambda)dx^\mu = A_\mu dx^\mu - d\Lambda$. (In a sense, we are going back to chapter IV.1, where you allegedly discovered electromagnetism.) Thus, the combination $(dy + lA_\mu dx^\mu)$ is gauge invariant if we also transform $y \to y + l\Lambda$ (as in (5)). You of course realize that we are just rewriting the $M = 5$ component of the coordinate transformation $X^M \to X'^M = X^M + \varepsilon^M(x)$ and rediscovering (4).

Having made the acquaintance of this invariant combination $(dy + lA_\mu dx^\mu)$, we are now ready to write down the 5-dimensional line element instantly. By invariance, it can only be

$$ds^2 = g_{\mu\nu}dx^\mu dx^\nu + (dy + lA_\mu dx^\mu)^2 \tag{8}$$

Comparing this with $ds^2 = G_{MN}dX^M dX^N = G_{\mu\nu}dx^\mu dx^\nu + 2G_{\mu 5}dx^\mu dy + G_{55}dy^2$, we can read off

$$G_{MN} = \begin{pmatrix} g_{\mu\nu} + l^2 A_\mu A_\nu & lA_\mu \\ lA_\nu & 1 \end{pmatrix} \tag{9}$$

In other words, we have constructed the Kałuza-Klein metric. Notice that $G_{\mu\nu}$ is equal to $g_{\mu\nu}$ only in the weak field expansion implicit in (3). In appendix 7, we give a more geometrical derivation of (8) for arbitrary dimensions. See (66).

Motion in the fifth dimension

We have yet to fix the radius a of the tiny circles all around us. Here is a clue: our fixing the normalization of A_μ by setting $l = l_P$ is an empty gesture unless A_μ actually couples to some charged particle or field. Thus far, this is absent from Kałuza-Klein theory. It behooves us now to work out the motion of a point particle in this theory. Intuitively, since

$G_{\mu 5} = A_\mu$, we expect that a particle moving along the fifth coordinate y would sense the electromagnetic potential. This turns out to be the case.

To keep life simple, let the $(3 + 1)$-dimensional spacetime be flat; after all, we now know how point particles move under gravity. The focus here is on the coupling to the electromagnetic field. So, set $g_{\mu\nu}$ in (8) to $\eta_{\mu\nu}$ and write down the action

$$S_{\text{particle}} = -m \int \left[-\eta_{\mu\nu}dx^\mu dx^\nu + \left(dy + lA_\mu dx^\mu \right)^2 \right]^{\frac{1}{2}} \tag{10}$$

Apply what you learned in part II of this book: vary S_{particle} with respect to y and x^μ to obtain the equations of motion for the particle. First, since the metric does not depend on y explicitly, we have the conservation law $\frac{d}{d\tau}\left(\frac{dy}{d\tau} + lA_\mu\frac{dx^\mu}{d\tau}\right) = 0$, so that

$$p \equiv m \left(\frac{dy}{d\tau} + lA_\mu \frac{dx^\mu}{d\tau} \right) \tag{11}$$

is a constant. Indeed, you recognize p as the conserved momentum in the y direction. Next, varying with respect to x^ν, we obtain, after using (11),

$$\frac{d}{d\tau} \left(-\eta_{\mu\nu}m\frac{dx^\mu}{d\tau} + (pl)A_\nu \right) = (pl)\partial_\nu A_\lambda \frac{dx^\lambda}{d\tau} \tag{12}$$

As in chapter IV.1, this equation of motion reduces to $m\frac{d^2x^\mu}{d\tau^2} = (pl)F^\mu{}_\nu\frac{dx^\nu}{d\tau}$. Comparing with (IV.1.23), we see that $pl \equiv q$ is the charge of the particle. Our intuition is vindicated: the momentum p along the y direction determines how strongly the particle senses A_μ.

But what is p? Classically, the momentum p can take on any value, and hence the charge q also. In quantum mechanics, however, the momentum of a particle moving around a circle is quantized. Earlier, we already noted that the uncertainty principle implies that p is of order $1/a$. But we can be more precise. Here I have to assume that the reader is sufficiently familiar with quantum mechanics to know that the wave function of a particle around a circle of radius a is given by $e^{ipy/\hbar}$. The reader who does not know this can simply skip this and the next section. (Note that we have momentarily restored $\hbar$.) Since y and $y + 2\pi a$ represent the same point, $e^{i2\pi ap/\hbar} = 1$, and so p must take on the quantized value $p = n\hbar/a$, with n any integer between $-\infty$ and $+\infty$. Since $q = pl$, we find that the charge of the particle is also quantized to have the allowed values $q_n = n\hbar l/a = ne$, with the fundamental unit of charge given by

$$e = \frac{l}{a} \tag{13}$$

(we set $\hbar = 1$ once again). In natural units, $e \simeq \sqrt{4\pi/137} \sim 0.3$. Since $l = l_{\text{p}}$, we find that $a = l_{\text{p}}/e \sim 3l_{\text{p}}$ is also of the order of the Planck length. Nicely, this accords with the undeniable fact that experimentalists have not seen the Kałuza-Klein circles to this very day.

With the internal space a circle, we are invited to introduce an angular variable θ defined by $y \equiv a\theta$. You are of course free to think of θ, rather than y, as the internal coordinate and rewrite (8) as* $ds^2 = g_{\mu\nu}dx^\mu dx^\nu + a^2(d\theta + eA_\mu dx^\mu)^2$.

Charge conjugation and direction of motion

That charge quantization pops out of the Kałuza-Klein framework is quite striking. Another nice feature of the theory is that the concept of charge conjugation, and hence of antimatter, also emerges naturally. This remark is directed to those readers who know about the three fundamental discrete symmetries of physics, namely parity (that is, reflection in space), time reversal (that is, reflection in time), and charge conjugation (that is, turning particles into antiparticles, thus flipping the sign of various conserved charges). Of these three discrete symmetries, charge conjugation stands apart from parity and time reversal in that it does not appear to have anything to do with spacetime. But in the Kałuza-Klein framework, it does. Charge conjugation just corresponds to reversing the motion of the particle along the y direction. The existence of antimatter follows from the possibility of going the other way around the circle!

Unfortunately, there is also a serious difficulty with this basic version of Kałuza-Klein theory: the charged particles found here all have y momentum of the order of the Planck mass M_P. This implies that these particles are all very massive (as we will make precise in the next section) and do not correspond to the observed charged particles. Incidentally, Kałuza already noted this difficulty in his paper, with a note thanking Einstein for pointing this out to him.

Extragalactic fable revisited

Time to revisit the extragalactic fable in chapter IV.1! Recall that the extragalactic version of you tried to include a potential in the Lorentz action for a free particle. You managed to come up with two, and only two, options: $S_E = -\int \{m\sqrt{-\eta_{\mu\nu}dx^\mu dx^\nu} + V(x)dt\}$ and $S_G = -m\int \sqrt{(1 + \frac{2V(x)}{m})dt^2 - d\vec{x}^2}$. You can put the potential V either outside or inside the square root. How could there possibly be another option?

After staring at this for years, you might, if you are as clever as Kałuza, realize one day that there is a third option unifying these two options. You extend the indices μ, $\nu = 0, 1, 2, 3$ to M, $N = 0, 1, 2, 3, 5$; write $S = -m\int \sqrt{-G_{MN}dx^M dx^N}$; and set $G_{\mu\nu} = \eta_{\mu\nu}$ and $G_{55} = 1$. Then

$$S = -m\int \sqrt{-\eta_{\mu\nu}dx^\mu dx^\nu - 2G_{5\mu}dx^5 dx^\mu - (dx^5)^2} = -m\int \sqrt{d\tau^2 - 2G_{5\mu}dx^5 dx^\mu - (dx^5)^2}$$

$$\simeq -m\int d\tau \left(1 - G_{5\mu}\frac{dx^5}{d\tau}\frac{dx^\mu}{d\tau} + \cdots\right) = \int \{-md\tau + eA_\mu dx^\mu + \cdots\}$$

---

* Readers familiar with quantum mechanics will be pleased to see the phase angle of the wave function for a charged particle emerging. We have $e^{ipy/\hbar} = e^{ipa\theta/\hbar} = e^{in\theta}$ (see, for example, QFT Nut, p. 79).

where in the last step you set $\frac{dx^5}{d\tau} = e/m$ and $G_{5\mu} = A_\mu$. This is the Kałuza idea in essence.

Kałuza-Klein towers

A striking prediction of Kałuza-Klein theory is the existence of entire towers of particles. In this section, we set not only $g_{\mu\nu} = \eta_{\mu\nu}$ but also $A_\mu = 0$. Consider the wave equation $\partial^2\Phi = \left(-\frac{\partial^2}{\partial t^2} + \vec{\nabla}^2 + \frac{\partial^2}{\partial y^2}\right)\Phi = 0$ satisfied by some field Φ whose identity we need not specify for this schematic discussion. The solution is $\Phi(t, \vec{x}, y) = e^{-i\omega t + i\vec{k}\cdot\vec{x} + i\kappa y}$, with $\omega^2 = (\vec{k}^2 + \kappa^2)$.

Here we essentially repeat the argument given in the previous section. Since y is curled up in a circle with radius a, the periodic boundary condition[5] specifies that $\kappa = n/a$, with n an integer. In quantum mechanics, de Broglie tells us that frequency and wave number become energy $E = \hbar\omega$ and momentum $\vec{p} = \hbar\vec{k}$, respectively. (Here $\vec{p}$ is the 3-momentum observed in the $(3 + 1)$-dimensional Minkowskian spacetime, while the momentum along the y direction that appeared in the preceding section is $p = \hbar\kappa$.) Thus, in Kałuza-Klein theory, each one of the particles we know—the electron, the photon, and so on—is associated with a tower of particles with masses given by $m_n^2 = E^2 - \vec{p}^2 = (n\hbar/a)^2$. (In appendix 4, we repeat the calculations in these two sections more carefully.)

For $n \neq 0$, these masses are enormous, of order M_P, since as we saw in the preceding section, a is of order l_P. The Kałuza-Klein towers are conveniently hidden away from the prying eyes of our experimental friends.

On the Planck scale, the known fundamental particles, quarks, leptons, and such, are essentially massless; in particle theory, their observed masses* are accounted for by the so-called Higgs mechanism. In this interpretation of Kałuza-Klein theory, the known particles correspond to $n = 0$, in which case the corresponding field $\Phi(\vec{x}, y)$ does not depend on y and thus in some sense does not know about the extra dimension.

We will not address further this difficulty of the theory not containing charged particles that are also massless (that is, with masses much less than M_P).

Note also that we have not inquired about the "mechanism" that breaks the 5-dimensional spacetime into 4 cosmologically large coordinates and one small coordinate.

Breathing circles

We have shown that it is consistent to set $h_{55} = 0$ and hence $G_{55} = 1$, but actually, we have the option of giving life to G_{55} and turning it into a field. Promote the fixed radius a to a field $\phi(x)$ (which evidently has dimension of a length) so that

$$ds^2 = g_{\mu\nu}dx^\mu dx^\nu + \phi(x)^2 \left(d\theta + eA_\mu dx^\mu\right)^2 \tag{14}$$

---

* As was emphasized in chapter III.5, one triumph of special relativity is that it allows us to talk about massless particles.

(Henceforth, we will absorb e into A_μ.) The Kałuza-Klein metric is correspondingly modified to

$$G_{MN} = \begin{pmatrix} g_{\mu\nu} + \phi^2 A_\mu A_\nu & \phi^2 A_\mu \\ \phi^2 A_\nu & \phi^2 \end{pmatrix} \tag{15}$$

(In comparing (15) with (9), note that one is written in the basis $(dx^\mu, d\theta)$, the other in (dx^μ, dy).) Throughout, $g_{\mu\nu}$, A_μ, and ϕ are all functions of x; we often suppress the x dependence to avoid clutter. For the record, the inverse metric is given by

$$G^{NP} = \begin{pmatrix} g^{\nu\rho} & -A^\nu \\ -A^\rho & \phi^{-2} + A^2 \end{pmatrix} \tag{16}$$

where $A^2 = A_\mu A^\mu$.

The scalar field $\phi(x)$ is sometimes called the dilaton or radion.[6] We suppose that in the ground state, $\phi(x) = a$.

Note that G, the determinant of G_{MN}, is given by $G = \phi^2 g$ (with g the determinant of $g_{\mu\nu}$, as usual), so that ϕ controls the volume of 5-dimensional spacetime. We could perhaps visualize this collection of Kałuza-Klein circles as an immense colony of minute marine organisms breathing, pulsating, and changing in size.

Note that we have now accounted for all $(5 \cdot 6)/2 = 15 = 10 + 4 + 1$ components of G_{MN}: 10 in $g_{\mu\nu}$, 4 in A_μ, and 1 in ϕ.

At the time of Kałuza and Klein, experimentalists had never heard of a scalar field, and $\phi(x)$ was seen as a fly in the ointment gluing gravity and electromagnetism together. In modern times, however, string theory contains a multitude of scalar fields, in particular the dilaton field, and the natural appearance of scalar fields was celebrated with much exuberance and joy. Nevertheless, experimentalists have not yet seen the Kałuza-Klein scalar field.

For some reason, while it is easy to excite $g_{\mu\nu}$ and A_μ, which we do endlessly each day, ϕ is very hard to excite. The Kałuza-Klein radius is extraordinarily rigid! Why? Nobody knows for sure.

I alluded earlier to a brute force plug-in approach. You can now see how this alternative presentation starts by plugging the G_{MN} displayed in (15), given without any motivation, into the action (6). After some tedious calculations, the action would be found to reduce to Einstein-Hilbert plus Maxwell plus an action for ϕ. I personally find this sort of approach not particularly illuminating. As to the inevitable question of why we should contemplate the form given in (15), the answer could be that we can always start with a general G_{MN}, demand that the action reduce to a nice form, and by trial and error arrive at (15). In appendix 7, I give a geometrical picture that leads to (15) and its higher dimensional generalization, to which we now turn.

Higher dimensional Kałuza-Klein and Yang-Mills theories

Back in 1921, only the gravitational and the electromagnetic interactions were known. It took decades before the strong and weak interactions were clearly recognized as such. Still later, in the 1970s, the strong, electromagnetic, and weak interactions were all discovered to be described by the nonabelian gauge theory* written down by Yang and Mills in 1954, as I have already mentioned in chapter VIII.3 and at the start of this chapter. I also mentioned in chapter VIII.3 that, furthermore, these three nongravitational interactions can be unified into a single gauge theory. Although experimentalists have yet to verify this grand unified theory,[7] many theorists have professed faith in its general structure.

Some readers may not be familiar with Yang-Mills theory.[8] For you to follow the rest of this chapter, all I ask of you is to know that the familiar electromagnetic gauge potential A_μ is generalized to a bunch of potentials A_μ^a, where the index[†] a is a group index associated with the nonabelian gauge group and labels the generators of the group. For example, for the group $SO(3)$, the index a ranges over 1, 2, and 3. Maxwell's theory corresponds to the simplest possible case, in which the gauge group is $U(1)$ and the index a can take on only one possible value (namely 1) and hence can be suppressed.

As already remarked, Kałuza's idea can be extended to any dimension. We simply start with a higher dimensional Einstein-Hilbert action

$$S_{\mathrm{KK}} = + \int d^d x \sqrt{-G} \, M_{\mathrm{TG}}^{d-2} R(G) = \int d^4 x \, d^{d-4} y \sqrt{-G} \, M_{\mathrm{TG}}^{d-2} R(G) \tag{17}$$

with M_{TG} the "true" mass scale of gravity. As explained earlier, the power of M_{TG} in (17) is fixed by dimensional analysis. The internal space (with coordinates $x^5, x^6, \cdots, x^d$, which we also refer to as ys) is compactified into a sphere rather than a circle, or more generally, a closed curved space characterized by some length a. As before, evaluating S_{KK} with a $G_{MN}(x, y)$ of the form $G_{\mu\nu}(x, y) = g_{\mu\nu}(x)$, $G_{\mu j}(x, y) = 0$, and $G_{ij}(x, y) = 1$ (with $i, j = 5, 6, \cdots, d$), we recover the familiar Planck mass

$$M_{\mathrm{P}}^2 = M_{\mathrm{TG}}^{d-2} V_{d-4} = M_{\mathrm{TG}}^{n+2} V_n \tag{18}$$

with V_n the volume of the n-dimensional internal space, thus generalizing the earlier relation $M_{\mathrm{P}}^2 = 2\pi a M_5^3$.

Perhaps not surprisingly, just as the Maxwell action pops out of Kałuza-Klein theory, the Yang-Mills action also pops out.[9] (Thus, in this context, you don't even have to know what the Yang-Mills action is. You could derive it from the Kałuza-Klein action.) To me, that is the truly beautiful feature of Kałuza-Klein[10] theory. In appendix 8, I show how the Yang-Mills field strength emerges algebraically and geometrically.

Our experience with the 5-dimensional example, in which $G_{\mu 5}$ is identified with A_μ, suggests that $4(d-4)$ components $G_{\mu j}(x, y)$, $j = 5, 6, \cdots, d$, must end up being born in $(3+1)$-dimensional spacetime as the Yang-Mills gauge potential A_μ^a.

The issue here is how to relate $G_{\mu j}(x, y)$ to $A_\mu^a(x)$. Both objects carry the index μ. But the indices j and a, carried by one but not the other, are manifestly different beasts. We have to find a way to connect them, using some mathematical entity that carries both j and a.

---

* In this terminology, Maxwell's theory is known as an abelian gauge theory.
[†] Not to be confused with the radius of the Kałuza-Klein circles of course!

Now your hard work learning about isometry and Killing vectors in chapter IX.6 pays off. Consider, for example, the sphere S^2. The isometry group, the group that leaves S^2 invariant, is $SO(3)$, the group of rotations in 3-dimensional space. On S^2, we have 3 Killing vectors $\vec{\xi}_a$, with $a = 1, 2, 3$, associated with the 3 generators. Indeed, they were explicitly displayed in chapter IX.6. Write the Killing vectors out in components $\xi_{aj} \equiv g_{jk}\xi_a^k$ (with g_{jk} the metric on the sphere). Since the vectors $\vec{\xi}_a$ live on S^2, the index j takes on two values $j = 1, 2$, corresponding to the two coordinates on S^2.

We see that, in general, the Killing vectors ξ_{aj} are precisely what we need: they carry both the j index and the a index. Given these 3 entities, the off-diagonal components of the higher dimensional metric $G_{\mu j}(x, y)$, the Yang-Mills gauge potential $A_\mu^a(x)$, and the Killing vectors $\xi_{aj}(y)$, the only relation we can write down is

$$G_{\mu j}(x, y) = \xi_{aj}(y)A_\mu^a(x) = g_{jk}(y)\xi_a^k(y)A_\mu^a(x) \tag{19}$$

The indices have to hang together right, and this places a powerful constraint on what is possible. Another example of symmetry considerations saving us a lot of work! We will see in appendix 8 that this relation is indeed correct.

Road signs to higher dimensions

Two major concepts lie at the foundation of modern physics: local coordinate invariance and local internal or gauge symmetry. The former leads to the theory of gravity; the latter leads to the gauge theory of strong, weak, and electromagnetic interactions.

The remarkable discovery* of Kałuza is that if we suppose that spacetime is embedded in a space with dimension higher than four, these two concepts may not be independent—the latter may be derived from the former. The physics is so astonishing and the mathematics so elegant that many theoretical physicists might be disappointed if Nature does not use the Kałuza mechanism at some level. Nature, please do not disappoint us!

Furthermore, string theory, the leading candidate theory for unifying all four fundamental interactions, can be consistently formulated only in higher dimensional spacetime and thus requires the Kałuza mechanism. A priori, this need not be; if somebody had written down a consistent theory of quantum gravity incorporating the known interactions in 4-dimensional spacetime, Kałuza-Klein theory would have disappeared into the dustbin of physics history. Also, if Kałuza-Klein theory could not incorporate Yang-Mills theory naturally, it also would have been kissed bye-bye. (See below, however.)

I mention in passing a historical curiosity. In chapter IX.5, we saw that we could follow either Nordström or Einstein to a relativistic theory of gravity. Interestingly, both roads lead to higher dimensions.

---

* In textbooks, theoretical physics is laid out, all polished and beautiful, as if it were almost logically inevitable. To counter this, I might mention that in his paper, Kałuza mentioned the possibility of his theory explaining terrestrial magnetism.

Idea for cartoon: Theoretical physicist driving down a highway and seeing an exit sign for "Higher Dimensions."

Recall that Nordström described the gravitational field by a scalar field Φ. In a brilliant insight, he noticed[11] in 1914 that gravity and electromagnetism can be unified in the $(4+1)$-dimensional Maxwell action $S = \int d^5x(-\frac{1}{4}F_{MN}F^{MN})$, M, $N = 0, 1, 2, 3, 5$. Let $y = x^5$ describe a circle with radius a, and let $A_\mu(x, y) = A_\mu(x)$ and $A_5(x, y) = \Phi(x)$ depend only on the $(3+1)$-dimensional coordinates x, so that $F_{\mu\nu} = \partial_\mu A_\nu - \partial_\nu A_\mu$ is the usual electromagnetic field and $F_{\mu 5} = \partial_\mu A_5 = \partial_\mu \Phi$. The action becomes $S = a \int d^4x \frac{1}{4}(-F_{MN}F^{MN} + \frac{1}{2}\partial_\mu\Phi\partial^\mu\Phi)$. Gravity and electromagnetism are unified under false pretense.

The moral of the story is that in theoretical physics, simplicity may not be all that it's cracked up to be.

Some negative notes

Considering how fundamental invariances[12] under (1) and (2) are to theoretical physics, it would be disappointing indeed if Nature does not make use of this beautifully simple way of unifying these two equations. Unhappily, higher dimensional theories currently being worked on typically come with their own sets of Yang-Mills gauge fields. For example, string theory already contains gauge fields among the vibrational modes of the string. Thus, the gauge fields produced by the Kałuza-Klein mechanism are not needed. If this should turn out to be correct, it would appear that Nature is "unreasonably" wasteful.

On this somewhat negative note, let me mention that we have avoided talking about the mechanism for compactifying the extra dimensions. What would have been the simplest (and cleanest) possibility, namely that gravity supplemented by a cosmological constant could do the job, turns out not to work. Indeed, look at the field equation $R_{MN} = \frac{1}{2}g_{MN}(R - \Lambda)$ and demand a flat spacetime, that is, $R_{\mu\nu} = 0$. But this implies $R = \Lambda$ and by the field equation, $R_{ij} = 0$. The internal space is Ricci flat and cannot curl up into a sphere. Numerous mechanisms[13] have been proposed, none compelling, involving the introduction of additional fields and using the energy momentum tensor of these fields to curl up the internal space. Some of these additional fields are in fact gauge fields or their generalizations.

I might as well mention another difficulty that diminishes interest in Kałuza-Klein theory, at least as traditionally formulated. (The following remarks are well beyond the scope of this book and are intended only for readers with at least a nodding acquaintance with particle physics.) In the so-called standard model of particle physics, fundamental fermions (namely the quarks and leptons) appear as left handed and right handed fields in unequal numbers. This fundamental lack of symmetry between left and right goes back to parity violation in the weak interaction. As should be intuitively[14] clear, however, if the internal manifolds are simple spheres, then the resulting Kałuza-Klein theory can hardly distinguish between left and right.[15]

This chapter contains a large number of appendices, many of which can be skipped upon first reading. Here is a list of the topics addressed: (1) a calculation of the 5-dimensional action using differential forms, verifying that it contains the Einstein-Hilbert and Maxwell actions; (2) symmetry arguments and the role of the dilaton ϕ in the action; (3) the Jordan frame versus the Einstein frame; (4) the charged scalar field; (5) the natural emergence of the Yang-Mills gauge potential; (6) the Kałuza-Klein metric viewed as foliation; (7) the Kałuza-Klein metric in the vielbein formalism; (8) a more geometrical view of Kałuza-Klein theory; (9) a glimpse of the Arnowitt-Deser-Misner formalism; and (10) some historical tidbits.

Appendix 1: Einstein-Hilbert contains Maxwell

As promised, we now calculate the 5-dimensional scalar curvature $R(G)$ in (17) for the metric

$$ds^2 = g_{\mu\nu}dx^\mu dx^\nu + \left(dy + lA_\mu dx^\mu\right)^2 \tag{20}$$

In this appendix, we absorb l into A_μ for ease of writing.

From general considerations, we have already argued in the text that $R(G) = R(g) + \zeta F^{\mu\nu}F_{\mu\nu}$ for some numerical constant ζ. You are invited to proceed by brute force, evaluating the Christoffel symbols, the Ricci tensor, and the scalar curvature, so as to determine ζ. In fact, as already mentioned in the text, to determine this numerical coefficient, you can ease your labor by calculating $R(G)$ for the special case of $g_{\mu\nu} = \eta_{\mu\nu}$.

Here, differential forms are presented with an opportunity to rise and shine. So instead of Christoffel symbols and the rest, here I will use differential forms. For the sake of pedagogy, and in a departure from my usual philosophy, I will actually calculate $R(G)$ for the metric (20) in its full glory, with a general 4-dimensional metric, even though we could simply set it to the Minkowskian metric. As we will soon see, though, the hard part is not getting $R(g)$, but getting $F^{\mu\nu}F_{\mu\nu}$.

For convenience, write $\tilde{R} \equiv R(g)$ and indicate quantities associated with the 4-dimensional metric with a tilde. Start with the 5-dimensional 1-forms $e^A = (e^\alpha, e^5)$ given by $e^\alpha = e^\alpha_\mu dx^\mu = \tilde{e}^\alpha$ and $e^5 = dy + A_\mu dx^\mu = dy + A_\alpha e^\alpha$, where the last step defines A_α. For convenience we also define ∂_α by $dx^\mu \partial_\mu = e^\alpha \partial_\alpha$. The 5-dimensional metric is then given as usual by

$$G_{MN} = \eta_{AB}e^A_M e^B_N = \eta_{\alpha\beta}e^\alpha_M e^\beta_N + e^5_M e^5_N \tag{21}$$

(Note in passing that we have $e^\alpha_\mu = \tilde{e}^\alpha_\mu$, $e^\alpha_5 = 0$, $e^5_\mu = A_\mu = A_\alpha e^\alpha_\mu$, and $e^5_5 = 1$.)

Now that we have set things up, let's use Cartan's first structural equation

$$de^A + \omega^A_{\ B}e^B = 0 \tag{22}$$

to determine the connection 1-forms $\omega^A_{\ B}$.

First, we have

$$de^5 = d\left(dy + A_\beta e^\beta\right) = 0 + \partial_\alpha A_\beta e^\alpha e^\beta = \tfrac{1}{2}\left(\partial_\alpha A_\beta - \partial_\beta A_\alpha\right)e^\alpha e^\beta = \tfrac{1}{2}F_{\alpha\beta}e^\alpha e^\beta \tag{23}$$

From $-de^5 = \omega^5_{\ \alpha}e^\alpha$, we obtain $\omega^5_{\ \alpha} = \tfrac{1}{2}F_{\alpha\beta}e^\beta$.

Next, $\tilde{e}^\alpha = e^\alpha$ and so we obtain

$$-d\tilde{e}^\alpha = \tilde{\omega}^\alpha_{\ \beta}\tilde{e}^\beta = \tilde{\omega}^\alpha_{\ \beta}e^\beta = -de^\alpha = \omega^\alpha_{\ A}e^A = \omega^\alpha_{\ \beta}e^\beta + \omega^\alpha_{\ 5}e^5 = \left(\omega^\alpha_{\ \beta} + \tfrac{1}{2}F^\alpha_{\ \beta}e^5\right)e^\beta \tag{24}$$

Hence $\omega^\alpha_{\ \beta} = \tilde{\omega}^\alpha_{\ \beta} - \tfrac{1}{2}F^\alpha_{\ \beta}e^5$. Note that because we absorbed l, A is dimensionless and F has dimensions of $1/L$, consistent with ω being dimensionless.

Now plug in Cartan's second structural equation

$$R^A_{\ B} = d\omega^A_{\ B} + \omega^A_{\ C}\omega^C_{\ B} \tag{25}$$

So we have

$$
\begin{aligned}
R^{\alpha}{}_{\beta} &= d\omega^{\alpha}{}_{\beta} + \omega^{\alpha}{}_{\gamma}\omega^{\gamma}{}_{\beta} + \omega^{\alpha}{}_{5}\omega^{5}{}_{\beta} \\
&= d\tilde{\omega}^{\alpha}{}_{\beta} - \tfrac{1}{2}\left(\partial_{\gamma}F^{\alpha}{}_{\beta}\right)e^{\gamma}e^{5} - \tfrac{1}{2}F^{\alpha}{}_{\beta}de^{5} \\
&\quad + \left(\tilde{\omega}^{\alpha}{}_{\gamma} - \tfrac{1}{2}F^{\alpha}{}_{\gamma}e^{5}\right)\left(\tilde{\omega}^{\gamma}{}_{\beta} - \tfrac{1}{2}F^{\gamma}{}_{\beta}e^{5}\right) + \left(-\tfrac{1}{2}F^{\alpha}{}_{\gamma}e^{\gamma}\right)\left(\tfrac{1}{2}F_{\beta\delta}e^{\delta}\right)
\end{aligned}
\tag{26}
$$

The tilde terms collect themselves nicely into $\tilde{R}^{\alpha}{}_{\beta}$. Also rather nicely, we note that to calculate the scalar curvature, we can ignore terms involving $e^{\gamma}e^{5}$. After cleaning up a bit, we obtain

$$
R^{\alpha}{}_{\beta} = \tilde{R}^{\alpha}{}_{\beta} - \tfrac{1}{4}F^{\alpha}{}_{\beta}F_{\gamma\delta}e^{\gamma}e^{\delta} - \tfrac{1}{8}\left(F^{\alpha}{}_{\gamma}F_{\beta\delta} - F^{\alpha}{}_{\delta}F_{\beta\gamma}\right)e^{\gamma}e^{\delta} + (e^{\gamma}e^{5}\ \text{terms})
\tag{27}
$$

Equating this to

$$
R^{\alpha}{}_{\beta} = \tfrac{1}{2}R^{\alpha}{}_{\beta\gamma\delta}e^{\gamma}e^{\delta} + \tfrac{1}{2}R^{\alpha}{}_{\beta\gamma5}e^{\gamma}e^{5} + \tfrac{1}{2}R^{\alpha}{}_{\beta5\gamma}e^{5}e^{\gamma}
\tag{28}
$$

we obtain

$$
R^{\alpha}{}_{\beta\gamma\delta} = \tilde{R}^{\alpha}{}_{\beta\gamma\delta} - \tfrac{1}{2}F^{\alpha}{}_{\beta}F_{\gamma\delta} - \tfrac{1}{4}\left(F^{\alpha}{}_{\gamma}F_{\beta\delta} - F^{\alpha}{}_{\delta}F_{\beta\gamma}\right)
\tag{29}
$$

Next, using (25) again, we evaluate

$$
\begin{aligned}
R^{5}{}_{\alpha} &= d\omega^{5}{}_{\alpha} + \omega^{5}{}_{\beta}\omega^{\beta}{}_{\alpha} \\
&= d\left(\tfrac{1}{2}F_{\alpha\beta}e^{\beta}\right) + \left(\tfrac{1}{2}F_{\beta\gamma}e^{\gamma}\right)\left(\tilde{\omega}^{\beta}{}_{\alpha} - \tfrac{1}{2}F^{\beta}{}_{\alpha}e^{5}\right) \\
&= \tfrac{1}{2}\left(\partial_{\gamma}F_{\alpha\beta}\right)e^{\gamma}e^{\beta} - \tfrac{1}{2}F_{\alpha\beta}\tilde{\omega}^{\alpha}{}_{\beta}e^{\gamma} - \tfrac{1}{2}F_{\beta\gamma}\tilde{\omega}^{\beta}{}_{\alpha}e^{\gamma} - \tfrac{1}{4}F_{\beta\gamma}F^{\beta}{}_{\alpha}e^{\gamma}e^{5}
\end{aligned}
\tag{30}
$$

Equating this to

$$
R^{5}{}_{\alpha} = \tfrac{1}{2}R^{5}{}_{\alpha CD}e^{C}e^{D} = \tfrac{1}{2}R^{5}{}_{\alpha\gamma\delta}e^{\gamma}e^{\delta} + \tfrac{1}{2}\cdot 2R^{5}{}_{\alpha5\gamma}e^{5}e^{\gamma}
\tag{31}
$$

we can determine $R^{5}{}_{\alpha\gamma\delta}$ and $R^{5}{}_{\alpha5\gamma}$. Note that to calculate the scalar curvature, we don't give two hoots about $R^{5}{}_{\alpha\gamma\delta}$. Rather, we need to extract $R^{5}{}_{\alpha5\gamma}$ from (30). But since $\tilde{\omega}$ does not involve e^{5}, we deduce that only the last term in (30) contributes. We thus obtain

$$
R^{5}{}_{\alpha5\gamma} = \tfrac{1}{4}F_{\beta\gamma}F^{\beta}{}_{\alpha}
\tag{32}
$$

We are now ready to calculate the Ricci tensors needed to obtain the scalar curvature. Using (29) and (32), we find

$$
R_{\beta\delta} = R^{\alpha}{}_{\beta\alpha\delta} + R^{5}{}_{\beta5\delta} = \tilde{R}_{\beta\delta} - \tfrac{1}{2}F^{\alpha}{}_{\beta}F_{\alpha\delta}
\tag{33}
$$

Next, using (32), we find

$$
R_{55} = R^{\alpha}{}_{5\alpha5} = \eta^{\alpha\beta}R_{\beta5\alpha5} = \eta^{\alpha\beta}R_{5\beta5\alpha} = \eta^{\alpha\beta}R^{5}{}_{\beta5\alpha} = \tfrac{1}{4}\eta^{\alpha\beta}F_{\gamma\alpha}F^{\gamma}{}_{\beta} = \tfrac{1}{4}F_{\gamma\alpha}F^{\gamma\alpha}
\tag{34}
$$

Finally, putting (33) and (34) together, we obtain the scalar curvature

$$
R = \eta^{\beta\delta}R_{\beta\delta} + R_{55} = \tilde{R} - \tfrac{1}{2}F^{\alpha\beta}F_{\alpha\beta} + \tfrac{1}{4}F^{\alpha\beta}F_{\alpha\beta} = \tilde{R} - \tfrac{1}{4}F^{\alpha\beta}F_{\alpha\beta} = \tilde{R} - \tfrac{1}{4}F^{\mu\nu}F_{\mu\nu}
\tag{35}
$$

Or, using various other notations, we have $R(G) = R(g) - \tfrac{1}{4}F^{\mu\nu}F_{\mu\nu}$ or $R^{(5)} = R^{(4)} - \tfrac{1}{4}F^{\mu\nu}F_{\mu\nu}$, as promised. All this for the $-\tfrac{1}{4}$!

Note that, in spite of what we said in the beginning of this appendix, we would have gained only marginally in the computation if we had set $\tilde{\omega} = 0$ to start with.

Appendix 2: The dilaton or radion in the action

We have left out the scalar field $\phi(x)$ thus far, but now let us consider $ds^2 = g_{\mu\nu}dx^\mu dx^\nu + \phi(x)^2(dy + A_\mu dx^\mu)^2$. (We absorb a into ϕ, just as we had absorbed l into A.) As I have mentioned on more than one occasion, we could always proceed by brute force, simply calculating the scalar curvature $R(G)$ with this metric. But since we prefer not to sweat, let us see how far we can get with symmetry considerations and the knowledge that two powers of spacetime derivatives must appear.

The scalar curvature is invariant under

$$G_{MN}(X) = G'_{PQ}(X')\frac{\partial X'^P}{\partial X^M}\frac{\partial X'^Q}{\partial X^N} \tag{36}$$

Consider the coordinate transformation $x'^\mu = x^\mu$ and $x'^5 = f(x^5)$, that is, we do not touch the usual spacetime coordinates. Then $G_{55}(X) = \phi(x)^2 = G'_{55}(X')\frac{\partial y'}{\partial y}\frac{\partial y'}{\partial y} = \phi'(x')^2\frac{\partial y'}{\partial y}\frac{\partial y'}{\partial y}$. Since we require $\phi(x)$ and $\phi'(x')$ to be independent of y, for the transformation to be allowed, $\frac{\partial y'}{\partial y}$ can only be equal to a constant K. Hence $\phi(x) = K\phi'(x') = K\phi'(x)$. Next, setting $(M, N) = (\mu, 5)$ in (36) and recalling that $G_{\mu 5} = \phi^2 A_\mu$, we see that $A_\mu(x) = K^{-1}A'_\mu(x)$. This coordinate transformation corresponds to multiplying ϕ and dividing A_μ by a constant K. Furthermore, $G_{\mu\nu}$ and hence $g_{\mu\nu}$ remain unchanged. Thus, coordinate invariance requires that the Maxwell term must now appear in the combination $\phi^2 F_{\mu\nu}F^{\mu\nu}$.

Let us now find those terms in the scalar curvature $R(G)$ containing ϕ and two spacetime derivatives. For now, set $g_{\mu\nu} = \eta_{\mu\nu}$ and $A_\mu = 0$. Invariance under $\phi(x) = K\phi'(x)$ indicates that there are only two possible terms: $\eta^{\mu\nu}\partial_\mu\partial_\nu\phi/\phi$ and $\eta^{\mu\nu}\partial_\mu\phi\partial_\nu\phi/\phi^2$. To determine the numerical coefficients of these two terms, we simply have to calculate the scalar curvature for $ds^2 = \eta_{\mu\nu}dx^\mu dx^\nu + \phi^2 dy^2$. What an easy calculation! You could have done it way back when you first saw the Riemann curvature tensor. At this point, the diligent reader with a good memory jumps up and exclaims, "I did do it as an exercise back in chapter VI.1! The answer was $R = -2\eta^{\mu\nu}\partial_\mu\partial_\nu\phi/\phi$."

For curved spacetime, we simply promote $\eta^{\mu\nu}\partial_\mu\partial_\nu\phi$ (the flat space d'Alembertian of ϕ) to

$$\Box\phi = \frac{1}{\sqrt{-g}}\partial_\mu(\sqrt{-g}\,g^{\mu\nu}\partial_\nu\phi)$$

(the curved space d'Alembertian of ϕ). We thus obtain

$$R(G) = R(g) - \tfrac{1}{4}\phi^2 F_{\mu\nu}F^{\mu\nu} - 2\frac{\Box\phi}{\phi} \tag{37}$$

where all contraction on the right hand side is done with $g_{\mu\nu}$.

Appendix 3: Who frames the action?

Recalling that $\sqrt{-G} = \sqrt{-g}\,\phi$, we can now use (37) to write the Kaluza-Klein action $S_{KK} = \int d^4x\,dy\sqrt{-G}\,M_5^3 R(G)$ as

$$S_{\text{Jordan}} = M_P^2 \int d^4x \sqrt{-g}\,\phi\left(R - \tfrac{1}{4}\phi^2 F_{\mu\nu}F^{\mu\nu} - 2\frac{\Box\phi}{\phi}\right) \tag{38}$$

This is known as the action in the Jordan frame after Pascual Jordan, one of the founders of quantum mechanics and quantum field theory. We could perfectly well work with this action, expanding the field ϕ around 1 and treating the deviation from 1 as a small fluctuation.

But we can also elect to get rid of the ϕ in front of the scalar curvature if we like. How can we do that? That question is actually a memory test for you. Do you remember?

Way way back in exercise I.6.10, you learned that two spaces described by the metric $\tilde{g}_{\mu\nu}$ and $g_{\mu\nu}$ are said to be conformally related if $\tilde{g}_{\mu\nu}(x) = \Omega^2(x)g_{\mu\nu}(x)$. Furthermore, you worked out (please don't tell me that you didn't!) in exercise VI.1.13 that the scalar curvatures in the two spaces (with dimension $d = 4$) are related by $\tilde{R} = \Omega^{-2}R - 6\Omega^{-3}\Box\Omega$, with $\Box$ the curved space d'Alembertian constructed using $g_{\mu\nu}$. Surely you remember that now? This shows that with a judicious choice of Ω, we can recover the Einstein-Hilbert action without a ϕ in front.

So, for notational convenience, let us now put a tilde on the metric in the Jordan frame. In other words, Kałuza-Klein theory with the dilaton field ϕ has given us the action $S_{\text{Jordan}} = M_P^2 \int d^4x \sqrt{-\tilde{g}} \phi (\tilde{R} + \phi^2 F_{\mu\nu} F_{\rho\sigma} \tilde{g}^{\mu\rho} \tilde{g}^{\nu\sigma} - 2\tilde{\Box} \phi / \phi)$, with $\tilde{\Box}$ evidently the d'Alembertian constructed with $\tilde{g}_{\mu\nu}$.

Now set $\tilde{g}_{\mu\nu} = \Omega^2 g_{\mu\nu}$, so that $\sqrt{-\tilde{g}} = \sqrt{-g}\Omega^4$ and $\tilde{g}^{\mu\nu} = \Omega^{-2} g^{\mu\nu}$. Thus, we have

$$\sqrt{-\tilde{g}}\phi\tilde{R} = \sqrt{-g}\Omega^4\phi\left(\Omega^{-2}R - 6g^{\mu\nu}\Omega^{-3}\partial_\mu\partial_\nu\Omega\right)$$

If we set $\Omega^2 = 1/\phi$, we recover the Einstein-Hilbert action at the cost of generating some more scalar terms. In fact, we obtain

$$\sqrt{-\tilde{g}}\phi\tilde{R} = \sqrt{-g}\left(R + \phi^{-1}\Box\phi - \tfrac{5}{2}\phi^{-2}g^{\mu\nu}\partial_\mu\phi\partial_\nu\phi\right) \tag{39}$$

For future reference, we now have $\sqrt{-\tilde{g}} = \sqrt{-g}/\phi^2$ and $\tilde{g}^{\mu\nu} = \phi g^{\mu\nu}$.

We turn next to the Maxwell term:

$$\sqrt{-\tilde{g}}\phi\left(\phi^2 F_{\mu\nu} F_{\rho\sigma} \tilde{g}^{\mu\rho} \tilde{g}^{\nu\sigma}\right) = \sqrt{-g}\phi^{-2}\phi\left(\phi^2 F_{\mu\nu} F_{\rho\sigma} g^{\mu\rho} g^{\nu\sigma}\phi^2\right) = \sqrt{-g}\phi^3 F_{\mu\nu} F^{\mu\nu}$$

(Note that the raising of indices in $F^{\mu\nu}$ is now done with the metric $g^{\mu\nu}$.)

Recall that in the two preceding appendices, we absorbed the two Kałuza-Klein lengths a and l. Now put them back by scaling $\phi \to \phi/a$ and $A_\mu \to l A_\mu$. Putting all this together, we obtain (using $M_P l = 1$) the action in the so-called Einstein frame:

$$S_{\text{Einstein}} = \int d^4x \sqrt{-g}\left(M_P^2 R - \tfrac{1}{4}a^{-3}\phi^3 F_{\mu\nu} F^{\mu\nu} + M_P^2\left(\phi^{-1}\Box\phi - \tfrac{5}{2}\phi^{-2}g^{\mu\nu}\partial_\mu\phi\partial_\nu\phi\right)\right) \tag{40}$$

The multiplicative invariance $\phi \to K\phi$ in the preceding appendix clearly suggests a change of variable $\phi = \zeta e^{\lambda\varphi}$, with ζ and λ some real numbers. We now have an additive invariance under $\varphi \to \varphi + \text{constant}$. By invariance (or simple differentiation) we deduce that the two purely scalar terms in (40) can only become a linear combination of $\Box\varphi$ and $g^{\mu\nu}\partial_\mu\varphi\partial_\nu\varphi$. The first term goes away upon integration by parts. Thus, we finally obtain

$$S_{\text{Einstein}} = \int d^4x \sqrt{-g}\left(M_P^2 R - \tfrac{1}{4}e^{\sqrt{3}l_P\varphi} F_{\mu\nu} F^{\mu\nu} - \tfrac{1}{2}g^{\mu\nu}\partial_\mu\varphi\partial_\nu\varphi\right) \tag{41}$$

(We have chosen $\zeta = a$ and $\lambda = l_P/\sqrt{3}$, so that the kinetic energy term for φ and the Maxwell term for $\varphi = 0$ have their standard forms.)

Appendix 4: Charged scalar field

As promised in the text, here we study a charged scalar field Φ with mass m in $(4+1)$ dimensions more carefully. Its action is given by

$$S = \int d^4x \int dy \sqrt{-G}\left(-\partial_M \Phi^\dagger G^{MN} \partial_N \Phi - m^2 \Phi^\dagger \Phi\right) \tag{42}$$

For simplicity, we will let the $(3+1)$-dimensional spacetime be flat and set $g_{\mu\nu} = \eta_{\mu\nu}$. To evaluate this action, we first have to invert the matrix (9) to obtain

$$G^{MN} = \begin{pmatrix} \eta^{\mu\nu} & -l A^\mu \\ -l A^\nu & 1 + l^2 A^2 \end{pmatrix} \tag{43}$$

where $A^2 = A_\mu A^\mu = \eta^{\mu\nu} A_\mu A_\nu$. (Compare with (16).) Let us now specialize to the particular mode $\Phi(x, y) = \varphi_n(x)e^{iny/a}$ and insert this into (42). Integrating trivially over y, using the relation $l = ea$ (and suppressing the subscript n on the complex scalar field φ, which evidently should not be confused with the real scalar field φ in the preceding appendix), we obtain

$$S = 2\pi a \int d^4x \left(-\left(\partial_\mu + ine A_\mu\right)\varphi^\dagger \eta^{\mu\nu}\left(\partial_\nu - ine A_\nu\right)\varphi - m_n^2 \varphi^\dagger \varphi\right) \tag{44}$$

Those readers familiar with the action of a complex scalar field in the presence of an electromagnetic field will see that φ describes a particle with charge ne and mass m_n given by $m_n^2 = m^2 + (n/a)^2$. This makes precise the discussion in the text.

Appendix 5: Emergence of Yang-Mills structure

After Kaluza proposed the hidden internal space to be a circle, it seems glaringly obvious, at least in hindsight, that we should consider the sphere next, and that was exactly what Klein did. As mentioned in the text, Yang-Mills structure naturally emerges from Kaluza-Klein theory. Indeed, we could already have guessed from the index structure alone how the construction would go.

The discussion here can be couched in fairly general terms, but for pedagogical clarity and definiteness, the reader seeing this for the first time should imagine the internal space as the sphere S^2. Other readers might want to think of an arbitrary coset manifold G/H, as explained in appendix 1 to chapter IX.6. As in the text, divide the coordinates: $X^M = (x^\mu, y^i)$. The letter x when unspecified will refer to x^μ only.

Generalize $ds^2 = g_{\mu\nu}dx^\mu dx^\nu + (dy + lA_\mu dx^\mu)^2$ in (8) to

$$ds^2 = g_{\mu\nu}(x)dx^\mu dx^\nu + g_{ij}(y)\left(dy^i + A_\mu^a(x)\xi_a^i(y)dx^\mu\right)\left(dy^j + A_\nu^b(x)\xi_b^j(y)dx^\nu\right) \tag{45}$$

(with $l = 1$ for maximal clarity). The internal space is coordinatized by y^k and has Killing vectors $\xi_a^k(y)$ indexed by a.

Start by remembering how isometry works in the absence of the gauge potentials. So set A_μ^a to 0. The transformation $y^i \to y^i + \Lambda^a \xi_a^i(y)$ (with Λ^a infinitesimal constants) is supposed to leave $g_{ij}dy^i dy^j$ invariant. Let us now verify this. We have $dy^i \to dy^i + \Lambda^a \partial_k \xi_a^i dy^k$, or in other words, $\delta(dy^i) = \Lambda^a \partial_k \xi_a^i dy^k$. Then we obtain

$$\delta\left(g_{ij}dy^i dy^j\right) = \Lambda^a \left(\xi_a^k \partial_k g_{ij}dy^i dy^j + g_{ij}\partial_k \xi_a^i dy^k dy^j + g_{ij}\partial_k \xi_a^j dy^i dy^k\right) \tag{46}$$

This vanishes if the Killing equation (IX.6.2) $\xi_a^k \partial_k g_{ij} + g_{kj}\partial_i \xi_a^k + g_{ki}\partial_j \xi_a^k = 0$ holds. Indeed, this provides an alternative (but essentially the same, of course) derivation of the Killing equation.

Now turn on the gauge potentials, so that dy^i is replaced by $Dy^i = dy^i + A_\mu^a \xi_a^i dx^\mu$. Consider the transformation $y^i \to y^i + \Lambda^a(x)\xi_a^i(y)$, with the infinitesimals $\Lambda^a(x)$ allowed to depend on x. We want to know how the gauge potentials A_μ^a should transform for $g_{ij}Dy^i Dy^j$ to remain unchanged.

Let us split the calculation of the change $\delta(Dy^i)$ into two pieces:

$$\delta\left(dy^i\right) = d\left(\Lambda^a \xi_a^i\right) = \Lambda^a \partial_k \xi_a^i dy^k + \xi_a^i \partial_\mu \Lambda^a dx^\mu \tag{47}$$

and

$$\delta\left(A_\mu^a \xi_a^i dx^\mu\right) = \left(\delta A_\mu^a\right)\xi_a^i dx^\mu + A_\mu^a \Lambda^b \xi_b^k \partial_k \xi_a^i dx^\mu \tag{48}$$

Evidently, it is important to keep in mind that Λ^a and A_μ^a depend only on x, while ξ_a^i depends only on y. Our job is to determine δA_μ^a by requiring $\delta(g_{ij}Dy^i Dy^j) = 0$.

One clue is that we must use our knowledge that the isometries form a group, namely Lie's equation, as discussed in chapter IX.6. There we worked out the Killing vectors for the sphere $S^2 = SO(3)/SO(2)$ and showed that they satisfy $[\xi_a, \xi_b] = \varepsilon_{ab}{}^c \xi_c$. When written out in components, this means, in the notation used here,

$$\xi_a^k \frac{\partial \xi_b^i}{\partial y^k} - \xi_b^k \frac{\partial \xi_a^i}{\partial y^k} = \varepsilon_{ab}{}^c \xi_c^i \tag{49}$$

The preceding exercise gives us another clue: thanks to the Killing equation, the requirement $\delta(g_{ij}Dy^i Dy^j) = 0$ will be satisfied if $\delta(Dy^i) = \Lambda^b \partial_k \xi_b^i Dy^k$. Staring at (47) and (48), we see that we want δA_μ^a to contain two pieces. First, we need a piece $-\partial_\mu \Lambda^a$ to cancel off the second term in (47). This is more or less expected, since the electromagnetic gauge potential transforms similarly: $\delta A_\mu = -\partial_\mu \Lambda$. Second, we need another piece to convert the combination $\xi_b^k \partial_k \xi_a^i$ in the second term in (48) to $\xi_a^k \partial_k \xi_b^i$, that is, to interchange the indices a and b. How can you do that? Think for a moment before reading on. You are about to discover Yang-Mills theory.

Of course, I have set it all up for you. Come on, use (49)! Write it in the form $\xi_b^k \partial_k \xi_a^i + \varepsilon_{ab}{}^c \xi_c^i = \xi_a^k \partial_k \xi_b^i$. We see that we need

$$\delta A_\mu^c = \varepsilon_{ab}{}^c A_\mu^a \Lambda^b - \partial_\mu \Lambda^c \tag{50}$$

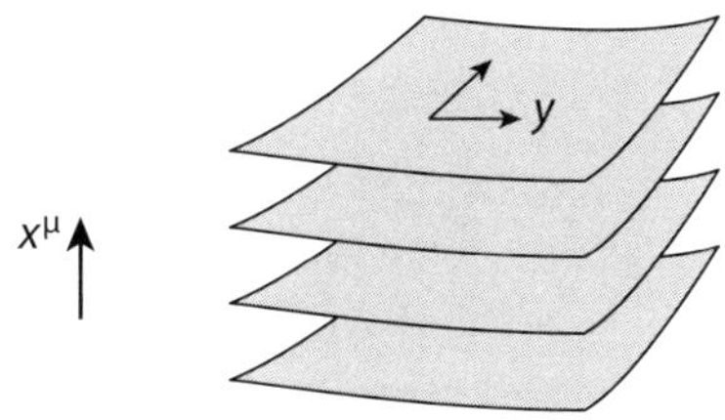

Figure 2 Construct our spacetime by piling sheets on top of one another, with each sheet labeled by x^μ. Within each sheet, points are located by the internal coordinates y^i.

Plug this in and watch the right hand side of (48) become $-\xi_c^i \partial_\mu \Lambda^c dx^\mu + A_\mu^a \Lambda^b \xi_a^k \partial_k \xi_b^i dx^\mu$. Hence $\delta(Dy^i) = \Lambda^b \partial_k \xi_b^i Dy^k$, as desired.

Just as the transformation (1) fixes the electromagnetic field strength $F_{\mu\nu}$, as was discussed back in chapter IV.2, we expect the transformation (50) to fix the Yang-Mills field strength $F_{\mu\nu}^a$ (which in fact carries, as you could have guessed, an extra group index a compared to the electromagnetic field strength). Given $F_{\mu\nu}^a$ (see appendix 8), the action follows almost immediately.[16]

For the record, from (45), we can read off

$$G_{ij} = g_{ij}, \qquad G_{\mu j} = A_\mu^a g_{ij} \xi_a^i, \qquad G_{\mu\nu} = g_{\mu\nu} + \left(g_{ij} \xi_a^i \xi_b^j \right) A_\mu^a A_\nu^b \tag{51}$$

Appendix 6: Kałuza-Klein as foliation

I now give[17] a geometrical and pictorial derivation of the Kałuza-Klein metric. In general, we have

$$G_{MN}(x, y) = \begin{pmatrix} G_{\mu\nu} & G_{\mu j} \\ G_{i\nu} & G_{ij} \end{pmatrix} \tag{52}$$

Our goal here is to identify the components of G_{MN} in terms of the metric $g_{\mu\nu}$ of the space we live in and the metric g_{ij} of the internal space.

Think of a patch of the internal space as a sheet. Picture our spacetime constructed by piling sheets on top of one another (see figure 2). The sheets are labeled and distinguished by x^μ. Within each sheet, points are located by the coordinates y^i.

Infinitesimal displacements characterized by δy^i lie completely within a given sheet (that is, $\delta x^\mu = 0$: no translating in the space we live in). The distance squared is then $ds^2 = G_{MN} \delta X^M \delta X^N = G_{ij} \delta y^i \delta y^j$. This means, somewhat trivially, that the metric for the internal space is given by

$$g_{ij} = G_{ij} \tag{53}$$

Less trivially, suppose we want to translate purely in spacetime. By "purely," we mean that the translation is to be perpendicular to the internal space. But we need to be careful about what we mean by "perpendicular." In fact, as we will see presently, a displacement perpendicular to the internal space will involve displacing in y as well. Roughly speaking, the y coordinate markings in the sheet labeled by $x + \delta x$ will not in general be lined up with the y coordinate markings in the sheet labeled by x. In other words, the desired displacement $\delta X^M = (\delta x^\mu, \delta y^i)$ must be perpendicular to any infinitesimal internal displacement* $\delta' X^M = (0, \delta' y^i)$. With the invariant definition of "perpendicular," this means that $\delta X^M G_{MN}(0, \delta' y)^N = \delta X^M G_{Mj} \delta' y^j = 0$. Since $\delta' y$ is arbitrary, it follows that

$$0 = \delta x^\mu G_{\mu j} + \delta y^i G_{ij} \tag{54}$$

---

* Here δ' is to be thought of as another Greek letter. The prime is not an operation on δ. In other words, $\delta' y$ is just some arbitrary infinitesimal change in y different from δy (like $\delta_1 x$ and $\delta_2 x$ in chapter VI.1). We want to find the restriction perpendicularity imposes on δX^M.

We can solve for

$$\delta y^i = -g^{ij} G_{j\mu} \delta x^\mu \tag{55}$$

where, importantly, g^{ij} is the inverse of $g_{ij} = G_{ij}$, not the ij component of G^{MN}, the inverse of G_{MN}. (Got that?)

Now that we have determined a displacement in spacetime $\delta X^M = (\delta x^\mu, -g^{ij} G_{j\mu} \delta x^\mu)$, we can calculate its length squared:

$$G_{MN} \delta X^M \delta X^N = G_{M\nu} \delta X^M \delta x^\nu = \left(G_{\mu\nu} \delta x^\mu + G_{i\nu} \delta y^i \right) \delta x^\nu = \left(G_{\mu\nu} - G_{i\mu} g^{ij} G_{j\nu} \right) \delta x^\mu \delta x^\nu \tag{56}$$

(The first equality holds since by construction, $G_{Mj} \delta X^M \delta y^j = 0$ for any δy. In the third expression, δy^i is the specific infinitesimal change determined in (55).) In other words, we can identify the spacetime metric as $g_{\mu\nu} \equiv G_{\mu\nu} - G_{i\mu} g^{ij} G_{j\nu}$. Note that it is not just $G_{\mu\nu}$ (as poor Confusio might have thought).

Collecting, we can finally write

$$G_{MN} = \begin{pmatrix} g_{\mu\nu} + G_{i\mu} g^{ij} G_{j\nu} & G_{\mu j} \\ G_{i\nu} & g_{ij} \end{pmatrix} \tag{57}$$

We now understand the geometrical origin of the Kałuza-Klein form in (9) and (15)! Needless to say, this is also entirely consistent with (51).

At this point, our friend the Jargon Guy interjects and tells us that we have been describing foliation. Thanks, Jargon Guy!

Appendix 7: The Kałuza-Klein metric in the vielbein formalism

Here we derive the Kałuza-Klein metric (9) and (57) for arbitrary dimension using the vielbein formalism of chapter IX.7. Denote the vielbein for the extended spacetime by e_M^A. Here $M = (\mu, i)$, and $A = (\alpha, a)$. The indices μ and α run over 0, 1, 2, 3, and the indices i and a run over 5, $\cdots$, d. The metric is given by $G_{MN} = \eta_{AB} e_M^A e_N^B$. The world index M is contracted with G_{MN}, and the Lorentz index A with the extended Minkowski metric η_{AB}, containing the usual Minkowski metric $\eta_{\alpha\beta}$ and $\eta_{ab} = \delta_{ab}$. (This means that the indices $A = (\alpha, a)$ can be raised and lowered at will (up to a sign).) For example, $e_A^M = \eta_{AB} e^{BM} = \eta_{AB} G^{MN} e_N^B$, with G^{MN} the inverse of G_{MN}. The discussion in this appendix complements that in the preceding appendix to some extent.

For some arbitrary $\delta\lambda_A$, we displace X^M in the Ath direction by

$$\delta X^M = \delta\lambda_A e^{AM} \tag{58}$$

We say that $\delta\lambda_a$ generates a displacement in the internal space, with $\delta x^\mu = 0$ by definition, thus implying

$$e^{a\mu} = 0 \Rightarrow e_a^\mu = 0 \tag{59}$$

We then have

$$e_A^M = \begin{pmatrix} e_\alpha^\mu & e_\alpha^i \\ 0 & e_a^i \end{pmatrix} \tag{60}$$

Note that this introduces an asymmetry between external and internal spaces.

The orthonormality of the vielbein $e^{AM} e_M^B = \eta^{AB}$ implies $0 = \eta^{a\beta} = e^{a\mu} e_\mu^\beta + e^{ai} e_i^\beta = e^{ai} e_i^\beta = 0$, where we have used (59). Thus we have

$$e_i^\beta = 0 \Rightarrow e_{\beta i} = 0 \tag{61}$$

This gives

$$e_M^A = \begin{pmatrix} e_\mu^\alpha & 0 \\ e_\mu^a & e_i^a \end{pmatrix} \tag{62}$$

In what follows, we evaluate the metric $G_{MN} = \eta_{AB} e_M^A e_N^B$, using (59) and (61) repeatedly.

First, we obtain

$$G_{ij} = \eta_{\alpha\beta} e_i^\alpha e_j^\beta + \eta_{ab} e_i^a e_j^b = \eta_{ab} e_i^a e_j^b \equiv g_{ij} \tag{63}$$

The last step amounts to the definition of the internal metric $g_{ij} \equiv \eta_{ab} e_i^a e_j^b$.

Next, introducing the notation $N_{iv} \equiv G_{iv}$ and $N_{vi} \equiv G_{vi}$, we have

$$N_{vi} = N_{iv} = G_{iv} = \eta_{\alpha\beta} e_i^\alpha e_v^\beta + \eta_{ab} e_i^a e_v^b = \eta_{ab} e_i^a e_v^b = e_i^a e_{av} \tag{64}$$

Since (59) implies $e_b^i e_i^a = e_b^M e_M^a = \delta_b^a$, we can write $e_{av} = e_a^i N_{iv}$.

Finally, we have

$$G_{\mu v} = \eta_{\alpha\beta} e_\mu^\alpha e_v^\beta + \eta_{ab} e_\mu^a e_v^b = g_{\mu v} + \eta^{ab} e_{a\mu} e_{bv}$$
$$= g_{\mu v} + \eta^{ab} e_a^i e_b^j N_{i\mu} N_{jv} = g_{\mu v} + N_{i\mu} g^{ij} N_{jv} \tag{65}$$

In the second equality, we defined $g_{\mu v} \equiv \eta_{\alpha\beta} e_\mu^\alpha e_v^\beta$; in the third equality, we used the result of the previous step; and in the fourth equality, we defined $g^{ij} \equiv \eta^{ab} e_a^i e_b^j$. It is worth emphasizing that g^{ij} is the inverse of g_{ij}, not the ij component of G^{MN}, namely G^{ij}. (If this does not sound familiar, read the preceding appendix again.) Let's check this: $g^{ij} g_{jk} = \eta^{ab} e_a^i e_b^j \eta_{cd} e_j^c e_k^d = \eta^{ab} e_a^i \eta_{cd} e_k^d \delta_b^c = \delta_k^i$.

So in summary, we have

$$G_{MN} = G_{\mu i, vj} = \begin{pmatrix} g_{\mu v} + N_{i\mu} g^{ih} N_{hv} & N_{\mu j} \\ N_{iv} & g_{ij} \end{pmatrix} \tag{66}$$

We have obtained once again the Kałuza-Klein metric (57) (and (9)), with $g_{\mu v} \equiv \eta_{\alpha\beta} e_\mu^\alpha e_v^\beta$.

The form of the metric in (66) generalizes an expression[18] found in the celebrated textbook by Misner, Thorne, and Wheeler, described there as "pushing forward the many fingers of Time." I don't know about you, but as an undergraduate studying with Wheeler, I had considerable difficulty in picturing the many fingers of Time. Here we derived (66) by following our nose. At this point, our friend the Jargon Guy kindly informs us that various quantities in (66) have names like "lapse" and "shift."

Again, for the record, the inverse of G_{MN} is given by

$$G^{NP} = G^{vj, \rho k} = \begin{pmatrix} g^{v\rho} & -g^{v\sigma} N_{l\sigma} g^{lk} \\ -g^{jl} N_{l\sigma} g^{\sigma\rho} & g^{jk} + g^{jm} N_{mv} g^{v\sigma} N_{l\sigma} g^{lk} \end{pmatrix} \tag{67}$$

where $g^{v\rho}$ is the inverse of $g_{\mu v}$.

Appendix 8: A more geometrical view of Kałuza-Klein theory and emergence of Yang-Mills structure

As is made clear by the discussion in the text, we could simply insert $G_{\mu j}(x, y) = g_{jk}(y) \xi_a^k(y) A_\mu^a(x)$ into the higher dimensional Einstein-Hilbert action and watch the Yang-Mills action emerge. Just plug and chug. (The calculation is similar to, but more involved than, the calculation we trudged through in appendix 1. At some point, you clearly have to use the properties of the Killing vectors.) Since this is readily worked out (and also available in a number of places), I elect to follow a rather different, and more geometrical, approach.

Let us contemplate the following question. Given G_{MN} in (66), what must it satisfy for us to be able to bring it to the block diagonal form

$$G'_{MN} = \begin{pmatrix} \times & 0 \\ 0 & \times \end{pmatrix} \tag{68}$$

by a coordinate transformation? In other words, under what condition do the internal and external geometries decouple?

Here we are inspired by the electromagnetic case. Back in (15), the internal and external geometries decouple if $A_\mu = 0$. But this is not a gauge invariant statement; the correct condition* is that $F_{\mu v} = 0$. Thus, we expect that the condition for the internal and external geometries to decouple is the vanishing of the analog of $F_{\mu v}$, and we can identify that object as the Yang-Mills field strength.

---

* If you are ever asked the question, "What is the electromagnetic field?" you can answer that within Kałuza-Klein theory, it is that which links the internal and external geometries.

We want to preserve (61) stating that $e_i^\alpha = 0$ under the desired coordinate transformation. As before, we use the notation $X^M = (x^\mu, y^j)$, that is, with the internal coordinates denoted by y. Recall from chapter IX.7 how the vielbein transforms:

$$e_N'^A(X') = \frac{\partial X^M}{\partial X'^N} e_M^A(X) \tag{69}$$

Hence we require $0 = e_i'^\alpha(X') = \frac{\partial X^N}{\partial X'^i} e_N^\alpha(X) = \frac{\partial x^\mu}{\partial y'^i} e_\mu^\alpha(X)$. From (59), we have $e_\mu^A e_A^\nu = e_\mu^\alpha e_\alpha^\nu = \delta_\mu^\nu$. Thus, multiplying by e_α^ν, we see that the preceding requirement implies that we should restrict ourselves to those coordinate transformations in which x^μ does not depend on y'^i.

See if you can work out the condition for decoupling before reading further.

From $G'_{MN}(X') = \frac{\partial X^P}{\partial X'^M} \frac{\partial X^Q}{\partial X'^N} G_{PQ}(X)$, we have

$$G'_{\mu j}(X') = \frac{\partial X^P}{\partial x'^\mu} \frac{\partial X^Q}{\partial y'^j} G_{PQ}(X) = \frac{\partial y^k}{\partial y'^j} \left(\frac{\partial x^\nu}{\partial x'^\mu} G_{\nu k} + \frac{\partial y^l}{\partial x'^\mu} G_{lk} \right) \tag{70}$$

since x^μ does not depend on y'^j.

In what follows, we must keep an eagle eye out for what variables are held fixed in the various partial derivatives (as when doing thermodynamics). Clearly, the quantity $\frac{\partial y^l}{\partial x'^\mu}$ in (70) is evaluated with y'^k held fixed. Instead of thinking of it as a function of x'^μ and y'^k, it is more convenient for our purposes to think of it as a function of x^ν and y'^k: $x^l = x^l(x^\nu, y'^k)$. So in (70), write $\frac{\partial y^l}{\partial x'^\mu}$ as $\frac{\partial y^l}{\partial x^\nu}|_{y'^k} \frac{\partial x^\nu}{\partial x'^\mu}$. Then we obtain

$$G'_{\mu j}(x', y') = \frac{\partial y^k}{\partial y'^j} \frac{\partial x^\nu}{\partial x'^\mu} \left(G_{\nu k} + \frac{\partial y^l}{\partial x^\nu}\bigg|_{y'^k} G_{lk} \right) \tag{71}$$

Now we impose the stated form (68) and demand that $G'_{\mu j}(x', y') = 0$. Assuming that the matrices $\frac{\partial y^k}{\partial y'^j}$ and $\frac{\partial x^\nu}{\partial x'^\mu}$ are nonsingular, we peel them off in (70) to obtain $G_{\nu k} + \frac{\partial y^l}{\partial x^\nu}|_{y'^k} G_{lk} = 0$. Using this, and recognizing that $G_{\nu k} = N_{\nu k}$ and $G_{lk} = g_{lk}$, we obtain

$$-\frac{\partial y^l}{\partial x^\nu}\bigg|_{y'^k} = N_{\nu k} g^{kl} \equiv N_\nu^l \tag{72}$$

(where, as before, g^{kl} is the inverse of g_{ln}). Differentiating (72), we have

$$-\frac{\partial^2 y^l}{\partial x^\mu \partial x^\nu}\bigg|_{y'^k} = \frac{\partial}{\partial x^\mu}\bigg|_{y'^k} N_\nu^l = \frac{\partial N_\nu^l}{\partial x^\mu}\bigg|_{y^j} + \frac{\partial y^j}{\partial x^\mu}\bigg|_{y'^k} \frac{\partial N_\nu^l}{\partial y^j}\bigg|_{x^\mu} = \frac{\partial N_\nu^l}{\partial x^\mu} - N_\mu^j \frac{\partial N_\nu^l}{\partial y^j} \tag{73}$$

In this final form, we regard N_ν^l as a function of x^μ and y^j, as indeed it is. So let's give this expression a name:

$$F_{\mu\nu}^l \equiv \frac{\partial N_\nu^l}{\partial x^\mu} - N_\mu^j \frac{\partial N_\nu^l}{\partial y^j} - (\mu \leftrightarrow \nu) \tag{74}$$

The condition for the geometries to decouple is then simply $F_{\mu\nu}^l = 0$.

As I explained, we should identify the expression for $F_{\mu\nu}^l$ in (74) as the analog of the electromagnetic field strength. The reader who has studied nonabelian gauge theory will recognize that this almost looks like the Yang-Mills field strength. It is certainly pleasing to see something like that emerging from considerations of geometries decoupling. Nevertheless, it is clear that we should not yet identify $F_{\mu\nu}^l$ as the desired field strength. After all, $F_{\mu\nu}^l$ carries only geometrical indices. We have to connect geometry to algebra.

As explained in the text, the Killing vectors do precisely that. So write[19]

$$N_\mu^j = A_\mu^a \xi_a^j \tag{75}$$

and then (74) becomes

$$F_{\mu\nu}^l = \frac{\partial A_\nu^a}{\partial x^\mu} \xi_a^l - A_\mu^a \xi_a^j A_\nu^b \frac{\partial \xi_b^l}{\partial y^j} - (\mu \leftrightarrow \nu) \tag{76}$$

Using Lie's equation (49), we obtain

$$F_{\mu\nu}^l = \left(\partial_\mu A_\nu^c - \varepsilon_{ab}^c A_\mu^a A_\nu^b - (\mu \leftrightarrow \nu) \right) \xi_c^l \equiv F_{\mu\nu}^c \xi_c^l \tag{77}$$

The Yang-Mills field strength $F_{\mu\nu}^c$ emerges naturally in the Kaluza-Klein framework.

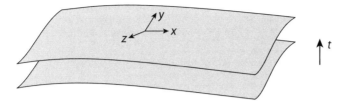

Figure 3 The dynamics of geometry: an instant in time is represented by a spacelike 3-dimensional hypersurface in curved spacetime.

Appendix 9: The dynamics of geometry and the ADM formulation

This may be a good place to mention an important subject, namely the Arnowitt-Deser-Misner (known as ADM) formulation of gravitational dynamics, even though it is not directly related to Kałuza-Klein theory. In physics, you are used to specifying a dynamical system at some initial time and then asking how it evolves in time. In general relativity, time t is one of the coordinates, and an instant in time is represented by a spacelike 3-dimensional hypersurface in spacetime, in general curved. Thus, we are to specify the 3-dimensional geometry on an initial spacelike hypersurface specified by t equal to some constant and then to follow the dynamics of the geometry—what Wheeler called geometrodynamics*—as we move from one hypersurface to another.

Well, you might have noticed that the sheets in figure 2 can represent these hypersurfaces. We flip a switch in our brains and rename (see figure 3) the "internal coordinates" y^i as the spatial coordinates x^i and the spacetime coordinates x^μ as the single time coordinate t. We can immediately take over the metric in (66) and write

$$g_{\mu\nu} = \begin{pmatrix} -N^2 + N_i g^{ij} N_j & N_i \\ N_j & g_{ij} \end{pmatrix} \tag{78}$$

where $N^2 = -g_{00}$ and $N_i = g_{0i}$, with N known as lapse and N_i as shift. Pictorially, the shift measures how one hypersurface is not lined up with the next. (Now the phrase "pushing forward the many fingers of Time" probably makes sense to you.) We now plug $g_{\mu\nu}$ into the Einstein-Hilbert action, identify the conjugate momentum variables, and then pass from the action to a Hamiltonian formulation[†] of Einstein gravity

This initial value formulation is important in the burgeoning field of numerical relativity. Another reason for its importance is that once we have the Hamiltonian, we can apply Heisenberg's formalism to quantize gravity. The ADM Hamiltonian also gives us one way of defining energy. Thus, the positive energy theorem states that the ADM energy[20] of an asymptotically flat (nonsingular) spacetime satisfying the field equation with a $T_{\mu\nu}$ obeying the dominant energy condition is not negative (recall chapter IX.3). This all important but vast subject of the ADM formulation lies way beyond the scope of an introductory text; I refer[21] the reader to more specialized monographs.

Appendix 10: Letters from Einstein to Kałuza 1919–1925

Theodor Kałuza's son made public[22] the letters his father received from Einstein. I find it quite interesting to see how Einstein's thinking evolved, particularly in comparing Kałuza's idea with Weyl's idea. I give here a few excerpts with the corresponding dates.

April 21, 1919: "The idea [of unifying electromagnetism with gravity] has also frequently and persistently haunted me. The idea, however, that this can be achieved through a five dimensional cylinder-world has never occurred to me and would seem to be altogether new. I like your idea at first sight very much. From a physical point of view it appears to me more promising than the <u>mathematically</u> so penetrating Ansatz of Weyl, because

---

* Which, in my humble opinion, is a better name than Einstein gravity, and certainly far superior to the historical general relativity.

† Conceptually, this procedure is the same as, but far more sophisticated than, the passage from the Lagrangian $L(\dot{q}, q)$ to the Hamiltonian $H(p, q)$ you learned in classical mechanics.

it concerns itself with the electric <u>field</u> and not with the, in my opinion, physically meaningless four-potential." (The underlining is Einstein's.) Einstein turned out to be wrong in that last sentence! He offered to present Kałuza's paper to the Berlin academy if it could be shortened to less than 8 printed pages (the limit imposed on nonmembers).

April 28, 1919: Einstein started with "I have read through your paper and find it really interesting" but went on to state "the arguments . . . do not appear convincing enough." He then suggested a further calculation to clear up a "question of the geodesic lines," saying that "You must not be offended by this because if I present your work [to the academy] I am backing it up with my name." He also mentioned that he knew the editor of the "newly founded *Mathematische Zeitschrift*" quite well and could get Kałuza's paper published there instead.

May 29, 1919: "It is true that I made a blunder with [some remark Einstein made in a previous letter]. . . . I see that you thought the matter over quite carefully. I have great respect for the beauty and boldness of your idea. But you will understand that I cannot take side with it <u>as originally planned</u> given the present factual doubts." Again, he offered to "put in a word" for Kałuza with the editor of the *Mathematische Zeitschrift*.

October 14, 1921: Notice that this letter was written more than 2 years after the first one. Interestingly, this letter carries the salutation "Most revered Dr. Kałuza" instead of the "Dear colleague" used in the previous letters. The letter was direct and to the point. "I am having second thoughts about having restrained you from publishing your idea on a unification of gravitation and electricity two years ago. Your approach seems in any case to have more to it than the one by H. Weyl. If you wish I shall present your paper to the academy after all, provided you sent it to me."

February 27, 1925: More than 3 years later, Einstein wrote: "I am still of the opinion that your idea . . . is of great originality and merits the serious interest of academic colleagues. . . . I myself have so far struggled with this problem in vain. It often appears to me that the magnetic field of the earth is based upon an as yet unknown connection between gravitation and electromagnetism, but I cannot come out of the inconsistencies." To a present-day theoretical physicist such as myself, that last sentence sounds rather astonishing, to say the least.

Notes

1. Perhaps Einstein's reluctance was explained in a 1922 letter he wrote to Hermann Weyl, saying that "Kałuza seems to me to have come closest to reality, even though he too fails to provide the singularity free electron" (quoted in J. van Dungen, *Einstein's Unification,* Cambridge University Press, 2010, p. 134). Einstein's dream of seeing the electron emerge as a solution of his field equation has not been realized and seems more remote than ever. The lesson here is not to demand too much of a promising theory.

2. The scalar field ϕ introduced in chapter II.3 and more recently in chapter IX.5 satisfies the Klein-Gordon equation. Klein also anticipated something like the Yang-Mills field strength (which we will discuss briefly in appendix 5) in 1938. O. Klein, *New Theories in Physics,* International Institute of Intellectual Cooperation, League of Nations, 1938, pp. 77–93.

3. Independently, the Russian physicist H. Mandel did the same in 1926. For this and some other historical tidbits, see the introduction in T. Appelquist, A. Chodos, and P. G. O. Freund, *Modern Kaluza-Klein Theory,* Addison-Wesley, 1987.

4. Klein later said that it was Pauli who told him that Kałuza had anticipated his work.

5. The physics here is essentially the same as that in a wave guide.

6. P. Jordan was the first to introduce this scalar field, but before he could return the proofs of his paper, the building housing the journal, *Physikalische Zeitschrift,* was bombed. Fortunately, a copy of the proofs was sent to Pauli, who showed them to Einstein and Bergmann.

7. See *QFT Nut,* chapter VII.5–7.

8. See, for example, *QFT Nut,* chapter IV.5.

9. As far as I know, this was first published in 1968 by R. Kerner from the University of Warsaw (*Ann. Inst. H. Poincaré* 9 (1968), p. 143). A complete and general derivation was first given in 1975 by Y. M. Cho and P. G. O. Freund, *Phys. Rev.* D 12 (1975), p. 1711. Earlier, it was given as homework problem number 77 in B. DeWitt's 1963 Les Houches lectures "Dynamical Theory of Groups and Fields." A personal note: The volume of the proceedings of the 1963 Les Houches school (*Relativity, Groups, and Topology,* ed. C. DeWitt and B. DeWitt, Gordon and Breach, 1964) is in fact one of the first physics books I owned as a sophomore in college. (J. A. Wheeler probably gave me a copy; I couldn't possibly have had the wits or the means to buy this huge book with almost 1,000 pages.) I remember poring over Wheeler's lectures "Geometrodynamics and the Issue of the Final State," trying to make sense of the whole thing.

10. One reason that Klein did not obtain the Yang-Mills structure earlier than he did was that he considered a 5, rather than a 6, dimensional theory.

11. G. Nordström, *Physikalische Zeitschrift* 15 (1914), pp. 504–506.

12. As you have probably also heard, string theory can only be formulated in 10 or 11 dimensions, thus sparking a tidal wave of contemporary interest in higher dimensions.

13. See, for example, the reprint volume by T. Appelquist et al., *Modern Kaluza-Klein Theory,* cited in note 3.

14. E. Witten has rendered this conclusion mathematically precise with index theorems.

15. This is one of the reasons string theorists abandoned the spheres for the mathematically more sophisticated but physically less friendly Calabi-Yau manifolds.

16. See the first footnote in this chapter.

17. Adapted from A. Zee, "Grand Unification and Gravity," in *Grand Unified Theories and Related Topics,* ed. M. Konuma and T. Maskawa, World Scientific, 1981, p. 143.

18. For example, C. W. Misner et al., *Gravitation,* p. 506.

19. It is important to note that, as was shown clearly in the step-by-step calculation, it is not $N_{\mu j}$ but N_μ^j that appears here. Recall especially (72). Some authors even mistakenly write G_μ^j, which is of course identically 0. The original Kałuza $5 = 4 + 1$ example emphasizes this point: $N_{\mu 5} = G_{\mu 5} = \phi^2 A_\mu$, but $N_\mu^5 = G_{\mu 5} g^{55} = A_\mu$, since $G_{55} = \phi^2 = g_{55}$ and hence $g^{55} = 1/\phi^2$.

20. In particular, this defines the ADM mass of an object. For a discussion of the different definitions of mass in general relativity, see N. O. Murchadha et al., arXiv: 0912.4001.

21. See particularly *Deserfest: A Celebration of the life and works of Stanley Deser,* ed. J. T. Liu, M. J. Duff, and K. S. Stelle.

22. Facsimiles and translations of the letters may be found in *Unified Field Theories of More Than 4 Dimensions,* ed. V. De Sabbata and E. Schmutzer, World Scientific, 1983.

X.2 | Brane Worlds and Large Extra Dimensions

Escape and visibility

The escape and visibility problems confront all those who want to have more than $(3 + 1)$ dimensions. In Kałuza-Klein theory, they are solved by supposing that the size of the higher dimensions is characterized by the Planck length. How else can you solve these problems?

Another possibility is that all the particles we know and love are somehow nailed down to the $(3 + 1)$-dimensional spacetime we call home, which amounts merely to a slice[1] in a higher dimensional spacetime. The graviton, in contrast, roams over all of spacetime, since it has to do with fluctuations of the entire metric, not just the metric on the particular slice humans live on.

You might think that somebody would have raised this possibility at some point after, say, 1920, but as far as I know, nobody did until recent times, at least not in print. This solution to the escape and visibility problems probably would have struck most theoretical physicists, until now, as rather contrived and ad hoc, with all but one of the particles forbidden to roam over all of spacetime. But, within string theory, this scenario occurs rather naturally, as was pointed out by Polchinski. In one realization, quarks, leptons, gauge bosons (such as the photon and the gluons), and so on are all described by open strings, while the graviton is described by a closed string. The $(3 + 1)$-dimensional spacetime we live in is known[2] as a 3-brane, to which the ends of the open string must be attached. A closed string, in contrast, has no ends to attach to anything. Consequently, the graviton is free to roam. Needless to say, this sketch hardly does justice to the glory and splendor of string theory.

This scenario is known variously as large extra dimensions[3] or brane worlds. Let me say right off that the term "large extra dimensions" merely means that the extra dimensions are large compared to the Planck length. The extra dimensions are still tiny on the scale of everyday life (see below). So, no, you could not wander off into the extra dimensions as in some science fiction story. I might also mention that you do not need to know string theory to read this chapter; indeed, one could readily make up purely field theoretic mechanisms for confining all particles except for the graviton to the $(3 + 1)$-dimensional slice of the larger spacetime. The brane world scenario is inspired by string theory but does not depend on it.

We will start with some general considerations of gravity in a higher dimensional world rather than a specific brane world model.

Newton's inverse square law

Way back in chapter II.1, we discussed the answer to the question physics students often ask (or fail to ask): why an inverse square law, and not inverse cube, say? The deep answer is that the power 2 follows from rotational invariance and the dimensions of space.

The physical origin of the inverse square law goes back to Faraday and his flux picture. He was talking about electric flux, but it could just as well be gravitational flux. Consider a sphere of radius r surrounding a charge. There is a fixed amount of electric flux coming out of the charge. Since the area of the sphere is given by $4\pi r^2$, the flux going through the sphere per unit area varies like $(4\pi r^2)^{-1} \propto 1/r^2$. That's it, the inverse square law! We see that it comes from a geometrical fact about area and clearly depends on the dimensions of space.

More formally, recall the discussion given in chapter II.1. Newton's gravitational potential V around a point mass M satisfies

$$\nabla^2 V(x, y, z) = 4\pi G M \delta^{(3)}(x, y, z) \tag{1}$$

The poor man solves this dimensionally (as we did in chapter II.1) by setting $\nabla^2 \sim 1/r^2$ and $\delta^{(3)}(x, y, z) \sim 1/r^3$, so that the equation becomes $V/r^2 \propto 1/r^3$, thus giving $V \sim 1/r$ and hence the inverse square force. A richer man, but not necessarily a rich man, would solve (1) by Fourier analysis, using the integral representation[4] of the delta function $\delta^{(3)}(x, y, z) = \int \frac{d^3k}{(2\pi)^3} e^{i\vec{k}\cdot\vec{x}}$ to obtain

$$V \propto \int d^3k \, e^{i\vec{k}\cdot\vec{x}} \frac{1}{\vec{k}^2} \propto \frac{1}{r} \tag{2}$$

By dimensional analysis, the integral has dimension of $k^3/k^2 \sim k \sim 1/r$. The inverse square law then follows.

Note, once again, rotational invariance and the 3 dimensions of space, as reflected in the factor d^3k in (2). This is of course just a more sophisticated formulation of Faraday's flux picture.

A bit of digression for the benefit of some readers. I might mention that in quantum field theory, the potential V is given by[5] the Fourier transform of the Feynman amplitude for the exchange of a graviton between 2 external masses. Again, rotational invariance and the 3 dimensions of space imply the inverse square law. The attractive,[6] rather than repulsive, character of gravity is due to the 2 units of spin carried by the graviton. Whether a force is repulsive or attractive can be traced back, in some sense, to the difference between space and time.[7]

Brane world

Given this discussion, you might worry that the brane world is untenable. Won't the inverse square law get unacceptably modified? As we just saw, the inverse square depends on the dimensions of space. Indeed, the reader might remember working this out in exercise II.1.2.

So suppose there are n extra dimensions, with coordinates $x^4, x^5, \cdots, x^{3+n}$. Since (1) follows from rotational invariance, it continues to hold, except that the delta function $\delta^{(3)}$ on the right hand side is to be replaced by $\delta^{(3+n)}$, appropriate for the $(3+n)$-dimensional space we are now in.

No sweat, the poor man says, I can solve this equation in an instant too: now $\delta^{(3+n)} \sim 1/r^{3+n}$, so that $V/r^2 \propto 1/r^{3+n}$, thus giving $V \sim 1/r^{n+1}$. The richer man, who has learned Fourier analysis, also obtains this result with scarcely more labor. He writes

$$V(r) \propto \int d^{3+n}k \, e^{i\vec{k}\cdot\vec{x}} \frac{1}{\vec{k}^2} \propto \frac{1}{r^{n+1}} \tag{3}$$

and obtains a force decreasing like $\frac{1}{r^{n+2}}$. This sure is not your grandparents' gravitational force law!

Doesn't (3) contradict observation immediately? Well, no, because Newton's law continues to hold for $r \gg R$, where R denotes the characteristic length scales associated with the extra coordinates. In this regime, the extra dimensions are so small, compared with the length scale r of the phenomenon we are interested in, that the gravitational flux cannot spread far in the direction of the n extra coordinates. Think of the flux being forced to spread in only the 3 spatial directions we know, just like the electromagnetic field in a wave guide is forced to propagate down the tube. Effectively, we are back in $(3+1)$-dimensional spacetime, and $V(r)$ reverts back to a $1/r$ dependence. Another way of seeing this is that in the limit $R \to 0$, effectively there aren't any extra dimensions.

You can put it somewhat paradoxically by saying that, in all actual situations involving gravity, not only is the large extra dimension not large, but it is effectively zero.

The new law of gravity (3) holds only in the opposite regime $r \ll R$, when the two masses are very close to each other. Heuristically, when R is much larger than the separation between the two particles, the flux does not know that the extra coordinates are finite in extent and thinks that it lives in a $(3+n+1)$-dimensional universe. Thus, we should look for deviation from the inverse square law at short distances. At present, the force law has only been checked[8] down to $r \sim 1$ millimeter or so, which is huge compared to the Planck length. Because of the weakness of gravity, Newton's force law has not been tested to much accuracy at laboratory distance scales, and so there is plenty of room for theorists to speculate.

The true scale of gravity

I already mentioned, way back in the introduction to this book, that the immensity of the Planck mass M_P, numerically $\sim 10^{19}$ times larger than the proton mass M_p, is responsible for the Mother of All Headaches plaguing fundamental physics today. Why is the intrinsic mass scale of gravity so large compared to anything else* we know?

An attractive feature of the large extra dimension scenario is that the true mass scale of gravity M_{TG} may be lowered considerably past M_P and thus alleviate this so-called hierarchy problem.

The preceding discussion and (3) tell us that the gravitational potential between two objects of masses m_1 and m_2 separated by a distance $r \ll R$ is given by

$$V(r) = -\frac{m_1 m_2}{(M_{TG})^{2+n}} \frac{1}{r^{1+n}} \qquad \text{for } r \ll R \tag{4}$$

Note that the dependence on M_{TG} follows from dimensional analysis: two powers to cancel $m_1 m_2$ and n powers to match the n extra powers of $1/r$.

In contrast, for $r \gg R$, as we have argued, the geometric spread of the gravitational flux is cut off by R, and the potential reverts to the familiar $1/r$ dependence. Thus motivated, we replace n powers of r in (4) by n powers of R to obtain[†]

$$V(r) = -\frac{m_1 m_2}{(M_{TG})^{2+n}} \frac{1}{R^n} \frac{1}{r} \qquad \text{for } r \gg R \tag{5}$$

Comparing this with the observed law $V(r) = -\frac{m_1 m_2}{M_P^2} \frac{1}{r}$, we manage to determine the true scale of gravity: $M_P^2 = (M_{TG})^{2+n} R^n = M_P^{2+n} l_P^n$. In the last step, we rewrote the left hand side using $l_P = 1/M_P$. In other words,

$$M_{TG} = (l_P/R)^{\frac{n}{2+n}} M_P \tag{6}$$

so that, if R/l_P could be made large enough, we would have the intriguing possibility that the fundamental scale of gravity M_{TG} might be much lower than what our grandparents thought. Equivalently, the size of the large extra dimensions is given in terms of the Planck length by

$$R = \left(\frac{M_P}{M_{TG}}\right)^{\frac{2+n}{n}} l_P \tag{7}$$

Suppose the true scale of gravity is as low as $M_{TG} \sim 10 \text{ TeV} = 10^4 \text{ GeV}$, this being the energy regime that the Large Hadron Collider can explore[‡] then $R \sim (10^{15})^{\frac{2+n}{n}} 10^{-33}$ centimeter. We see that the $n = 1$ case is already ruled out, but $n \geq 2$ is still allowed. Evidently, as $n \to \infty$, $R^{-1} \to M_{TG}$. Thus, R is bounded on one side by our desire to lower the fundamental scale of gravity and on the other by experiments.

You might have noticed that R large implies the appearance of a small mass $\mu = (\frac{M_{TG}}{M_P})^{\frac{2+n}{n}} M_P = R^{-1}$. In particular, for $M_{TG} \sim 10 \text{ TeV}$ and $n = 2$, we have

$$\mu = (10^{-15})^2 10^{19} \text{ GeV} = 10^{-2} \text{ ev}$$

---

* In particular, the electroweak scale at $\sim 10^3 M_P$.

[†] We are not interested in numerical factors of $0(1)$ here.

[‡] Outside the theoretical physics community, this is known as wishful thinking.

much smaller* than the typical scale of particle physics. In this sense, this scenario is not an entirely satisfactory solution of the hierarchy problem.

The beginning student might wish to skip the rest of this chapter. Unlike the material in the bulk of this book, the ultimate value of any specific brane world model to physics is far from certain.

A 2-brane model

As an interesting alternative to the picture outlined in the introduction, Randall and Sundrum proposed a 2-brane model.[9] Consider a 5-dimensional spacetime with the fifth coordinate $y \equiv x^5$ restricted to $\pi R \geq y \geq -\pi R$, with the points y and $-y$ identified. A circle with pairs of points thus identified, namely S^1/Z_2, is known as an orbifold.[10] The action is taken to be

$$S = \int d^4x \, dy \sqrt{-G} \left(\tfrac{1}{2} M_5^3 R(G) + \Lambda_5 \right) + S_{\text{branes}}$$

(8)

As in chapter X.1, we denote the scalar curvature constructed out of the $(4+1)$-dimensional metric G_{MN} by[†] $R(G)$ to distinguish it from the scalar curvature $R(g)$ constructed out of the $(3+1)$-dimensional metric $g_{\mu\nu}$. (We denote the 5-dimensional coordinates by $x^M = (x^\mu, y)$, with $M = 0, 1, 2, 3, 5$ and $\mu = 0, 1, 2, 3$.) Here the mass M_5 is the analog of the Planck mass M_P for 5-dimensional spacetime. (As in chapter X.1, dimensional analysis dictates that this mass, which sets the scale of 5-dimensional gravity, be cubed, since the scalar curvature R (in any spacetime dimension) contains two derivatives acting on the metric and hence has mass dimension 2.) Also, we denote the determinant of $g_{\mu\nu}$ and G_{MN} by g and G, respectively. Note that Λ_5, the 5-dimensional analog of the cosmological constant, has dimensions of mass to the fifth power.

Our brane, namely the brane we live on, is placed at one end of this spacetime at $y = \pi R$, and another brane, known as the Planck brane, is located at the other end at $y = 0$. In other words, we write

$$P_{\text{branes}} = \int d^4x \, dy \sqrt{-G} \left(\delta(y) \left(\Lambda_P - \mathcal{L}_P \right) + \delta(y - \pi R) \left(\Lambda_O - \mathcal{L}_O \right) \right)$$

(9)

Here Λ_P, $\mathcal{L}_P$, Λ_O, and $\mathcal{L}_O$ denote the cosmological constant and the Lagrangian of the matter fields on the Planck brane and on our brane, respectively. In what follows, we mostly set all the matter fields on the two branes to 0, that is, we simply ignore $\mathcal{L}_P$ and $\mathcal{L}_O$.

The 5-dimensional Einstein field equations are obtained, as usual, by varying S. Following Randall and Sundrum, let us look for a solution of the form

$$ds^2 = -e^{-2w(y)} \eta_{\mu\nu} dx^\mu dx^\nu + dy^2$$

(10)

---

* Notice in passing that this happens to be roughly of order of the cosmological constant mass scale and also the "typical" neutrino mass. Is there something here?

[†] Not to be confused with the radius R of the orbifold!

with a function $w(y)$ known picturesquely as the warp function. In other words, $G_{\mu\nu} = -e^{-2w(y)}\eta_{\mu\nu}$, $G_{\mu5} = 0$, $G_{55} = 1$. In contrast to the simple picture presented earlier, $G_{\mu\nu}$ depends on y.

A straightforward computation shows that the only nonvanishing Christoffel symbols are $\Gamma^5_{\mu\nu} = \eta_{\mu\nu}e^{-2w}w'$ and $\Gamma^\mu_{\nu5} = -w'$, with $w' \equiv dw/dy$. Since the Riemann curvature tensor involves the sum of $\partial\Gamma$ and $\Gamma\Gamma$, we are not surprised that the nonvanishing components of the Ricci tensor work out to be

$$R_{\mu\nu} = -\left(4w'^2 - w''\right)e^{-2w}\eta_{\mu\nu}$$

$$R_{55} = 4\left(w'^2 - w''\right) \tag{11}$$

Away from the branes, that is, for $y \neq 0$ and $y \neq \pi R$, S_{branes} do not contribute, and the field equations read simply

$$R_{MN} = G_{MN}\Lambda_5/M_5^3 \tag{12}$$

To solve this, we first note that we can rewrite the result in (11) as $R_{\mu\nu} = (4w'^2 - w'')G_{\mu\nu}$ and $R_{55} = 4(w'^2 - w'')G_{55}$. For $R_{\mu\nu}$ and R_{55} to be proportional to $G_{\mu\nu}$ and G_{55}, respectively, with the same proportionality constant as required by (12), we see that w'' must vanish, so that $w(y)$ is a linear function of y. Furthermore, $4w'^2 = -\Lambda_5/M_5^3$, so that $w(y) = \pm y/L$, with L a length scale defined by

$$L = \left(-4M_5^3/\Lambda_5\right)^{\frac{1}{2}} \tag{13}$$

thus indicating that this whole setup works only if $\Lambda_5 < 0$. Imposing the orbifold symmetry $w(y) = w(-y)$, we obtain $w(y) \propto |y|$, for $y \neq 0$ and $y \neq \pi R$.

Next, the 5-dimensional spacetime has to notice the presence of the branes at $y = 0$ and $y = \pi R$. In other words, we have to solve Einstein's field equation (12) amended by $\Lambda_5 \to \Lambda_5 + \delta(y)\Lambda_P + \delta(y - \pi R)\Lambda_O$. Observe that with the solution $w(y) \propto |y|$, the slope $w'(y)$ flips sign as we cross $y = 0$, so that indeed, $w''(y)$ behaves like a delta function at $y = 0$ and similarly at $y = \pi R$. (Some readers might be reminded of solving the delta function potential problem in quantum mechanics.) Integrating the $MN = 55$ equation in (12)

$$4M_5^3(w'^2 - w'') = -\left(\Lambda_5 + \delta(y)\Lambda_P + \delta(y - \pi R)\Lambda_O\right) \tag{14}$$

across $y = 0$ from $y = 0^-$ to $y = 0^+$, we obtain

$$\int_{0^-}^{0^+} dy\, w''(y) = w'|_{0^-}^{0^+} = \Lambda_P/\left(4M_5^3\right)$$

Similarly, integrating across $y = \pi R$ from $y = \pi R^-$ to $y = \pi R^+$, we obtain

$$\int_{\pi R^-}^{\pi R^+} dy\, w''(y) = w'|_{\pi R^-}^{\pi R^+} = \Lambda_O/\left(4M_5^3\right)$$

Recalling the condition $w(y) = w(-y)$ and sketching the function $w(y)$, you see that the jumps $w'|_{0^-}^{0^+}$ and $w'|_{\pi R^-}^{\pi R^+}$, and hence the cosmological constants Λ_P and Λ_O on the two branes, must have opposite signs. There is also a sign choice at this point; for the considerations below to work, we choose $w(y) = +|y|/L$, and thus obtain[11]

$$\Lambda_P = -\Lambda_O = 8M_5^3/L = 4\left(-M_5^3\Lambda_5\right)^{\frac{1}{2}} \tag{15}$$

In this brane world scenario, a fine tuning between the parameters Λ_P, Λ_O, and M_5 is required.

Some readers may have already noticed that the metric (10) with $w(y)$ linear is just the anti de Sitter spacetime studied in chapter IX.11; see in particular (IX.11.18). This also explains why the Ansatz (10) leads to a solution of Einstein's field equation so readily.

We are now ready to extract some physics from this model. First, allow the metric on our $(3 + 1)$-dimensional world to fluctuate,* replacing (10) by $ds^2 = -e^{-2|y|/L}g_{\mu\nu}(x) + dy^2$. How much action do we have to "pay" for this fluctuation?

Simply plug $G_{\mu\nu}(x, y) = e^{-2|y|/L}g_{\mu\nu}(x)$, $G_{\mu 5}(x, y) = 0$, and $G_{55}(x, y) = 1$ into (16) and evaluate the gravitational action

$$S \sim M_5^3 \int d^4x \left(\int_{-\pi R}^{\pi R} dy\, e^{-2|y|/L}\right) \sqrt{-g}\, R(g) \tag{16}$$

We thus identify the Planck mass by

$$M_P^2 = 2M_5^3 \int_0^{\pi R} dy\, e^{-2y/L} = M_5^3 L\left(1 - e^{-2\pi R/L}\right) \tag{17}$$

In the large R/L limit, $M_P^2 \simeq M_5^3 L$ and, once again, the true scale of gravity M_5 can be made to be much lower than the Planck scale. Note that, in contrast to the discussion earlier, due to the presence of Λ_5, we have 2 length scales available, R and L. In the large R/L limit, L plays the role of R in (6).

Next, consider a scalar field with mass m on our brane, that is, set $\mathcal{L}_O = -\{(\partial\varphi)^2 + m^2\varphi^2\}$ in (9). Since $G_{\mu\nu}(x, y = \pi R) = e^{-2\pi R/L}g_{\mu\nu}(x)$, the effective action we see in our world is given by

$$S \sim \int d^4x \sqrt{-g}\, e^{-4\pi R/L}\left(e^{+2\pi R/L}g^{\mu\nu}\partial_\mu\varphi\partial_\nu\varphi + m^2\varphi^2\right)$$

$$= \int d^4x \sqrt{-g}\left(g^{\mu\nu}\partial_\mu\tilde{\varphi}\partial_\nu\tilde{\varphi} + m^2 e^{-2\pi R/L}\tilde{\varphi}^2\right) \tag{18}$$

where $\tilde{\varphi} = e^{-\pi R/L}\varphi$ is normalized to have the correct kinetic energy term. Supposing the "true mass" m of the scalar field to be of order M_P, we can lower the physical mass $m_p = e^{-\pi R/L}m$ to the electroweak scale by choosing $R/L \sim 10$.

Speculations on brane worlds

Subsequently, Randall and Sundrum realized that the separation between the 2 branes could be taken to infinity, thus effectively leaving us with a 1-brane model,[12] which

---

* Note that we are keeping frozen other degrees of freedom in G_{MN}.

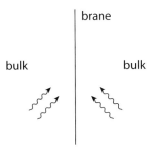

Figure 1 You are sitting on the brane minding your own business, and some wave could come in from the bulk and hit you.

has become more "fashionable" than their 2-brane model. This work stimulated a vast literature, amounting to literally thousands of papers. I obviously cannot provide a survey of the literature here. I merely mention that one interesting application is to cosmology. At this point, it ought to be clear to you that any specific brane world model should be taken as merely suggestive. You, the astute reader, probably realize that, with the vast expertise on classical general relativity you now have, you too could join in the fun. So, a word of encouragement: The full story of higher dimensional spacetime is yet to be written, and it could well be written by you!

Apparent violation of causality

The brane world model discussion given here is entirely static. Obviously, it would be interesting to introduce time dependence. In the appendix, I discuss one early attempt. Here I point out what I regard as a serious difficulty with all such attempts.

Dynamical discussions of our universe as a brane generically suffer from the awkward feature that evolution requires not just initial data on the brane, but also initial data for the bulk fields as well. In other words, if you are sitting on the brane minding your own business, some wave could come in from the bulk and hit you (figure 1). I like to call this the "finger of God" problem. Observers living on the brane would see apparent violation of causality, not to mention violation of energy conservation. Various happenings that have never been seen! In other words, there are a large number of degrees of freedom living in the bulk that we do not have direct access to. To me, this is one of the least attractive features* of the brane world.

---

* Of course, this has not prevented any number of persons from authoring any number of papers discussing the apparent violation of any number of "sacred" concepts. When and if experimentalists see these violations, you and I could always revisit the discussion here.

Appendix: Outgoing brane wave model

As promised, from the vast literature on brane worlds, here I briefly describe the outgoing brane wave model[13] to give you a flavor of the sort of calculations one can indulge in. Also, the following discussion involves solving Einstein's equations in light cone coordinates, something that might be of methodological interest. Let me first set up the model and then describe its physical properties.

Consider a $(4 + 1)$-dimensional world (with coordinates t, x^i, and y) containing a $(3 + 1)$-dimensional brane at $y = 0$ and described by the low energy effective action*

$$S = \int d^5x \sqrt{-G} \left[R(G) - \tfrac{4}{3}(\partial\varphi)^2 \right] + \int d^4x \sqrt{-g}\, e^{b\varphi} \mathcal{L}_{3+1} \tag{19}$$

The metric $g_{\mu\nu} = \delta_\mu^M \delta_\nu^N G_{MN}$ is the $(3 + 1)$-dimensional metric induced on the brane, and $\mathcal{L}_{3+1}$ denotes the Lagrangian of the $(3 + 1)$-dimensional world. (Here we set M_5 to 1.) Note that we do not include a bulk cosmological constant. In contrast, a dilaton field φ, as suggested by string theory, is included, with b taken here as a free phenomenological parameter that ultimately may be determined by string theory.

Einstein's equations read

$$R_{MN} - \tfrac{1}{2} G_{MN} R = \tfrac{4}{3} \left[\partial_M \varphi \partial_N \varphi - \tfrac{1}{2} G_{MN} (\partial\varphi)^2 \right] + \tfrac{1}{2} \sqrt{\tfrac{g}{G}}\, T_{\mu\nu} \delta_M^\mu \delta_N^\nu\, \delta(y) \tag{20}$$

(Evidently, $\partial_M = \frac{\partial}{\partial x^M}$ and so on.)

Assume the geometry on the brane to be homogeneous and isotropic, so that the stress energy tensor is required to take the form

$$T^\mu_{\ \nu} = -e^{b\varphi} \Lambda \delta^\mu_{\ \nu} + e^{b\varphi} \mathrm{diag}(-\rho, P, P, P) \tag{21}$$

describing a $(3 + 1)$-dimensional world containing a cosmological constant or vacuum energy Λ and a perfect fluid with energy density ρ and pressure P, with Λ, ρ, and P in general functions of t.

Upon varying the action with respect to the dilaton field φ, we obtain its equation of motion

$$\tfrac{8}{3} \Box\, \varphi = -\sqrt{\tfrac{g}{G}}\, b\, e^{b\varphi} \mathcal{L}_{3+1} \delta(y) \tag{22}$$

The original paper on this model includes a perfect fluid. But if I include the perfect fluid, at this point, I would have to digress and explain to you what $\mathcal{L}_{3+1}$ is for a perfect fluid.[14] Without the perfect fluid, we have, in (22), simply $\mathcal{L}_{3+1} = -\Lambda$. But since this only affects the dilaton equation of motion, much of the following discussion goes through even if the perfect fluid included. Thus, I will proceed to discuss the more general situation, with a perfect fluid included. For those readers who like to see everything derived rather than simply stated (and I am a member of this class), I will simply alert you when the explicit form of $\mathcal{L}_{3+1}$ for a perfect fluid is actually needed.

Consider solutions depending on both t and y that are homogeneous and isotropic in the three transverse directions, and write the metric in the form

$$ds^2 = e^{2A(t,y)} \left(-dt^2 + dy^2 \right) + e^{2B(t,y)} \left(dx_1^2 + dx_2^2 + dx_3^2 \right)$$
$$= -e^{2A(u,v)}\, du\, dv + e^{2B(u,v)} \left(dx_1^2 + dx_2^2 + dx_3^2 \right) \tag{23}$$

with the light cone coordinates $u \equiv t - y$ and $v \equiv t + y$. The metric for the $(3 + 1)$-dimensional universe on the brane can be written in the standard form of a Friedmann-Lemaître-Robertson-Walker universe $ds^2 = -d\tau^2 + a(\tau)^2(dx_1^2 + dx_2^2 + dx_3^2)$, with $\tau = \int dt\, e^{A(t,0)}$ and $a(\tau) = e^{B(t(\tau),0)}$.

Now solve Einstein's equation (20) and the dilaton equation (22) in the bulk, that is, away from the brane, to obtain $A(t, y)$, $B(t, y)$, and $\varphi(t, y)$ for $y > 0$ and $y < 0$ separately, and then match the solutions across the brane. It is convenient to use the (u, v) light cone coordinates to solve the equations in the bulk, and the (t, y) coordinates to do the matching across the brane. Before reading on, you are cordially invited to solve these equations.

---

* The $\tfrac{4}{3}$ is conventional and can be absorbed in b.

The bulk equations of motion are (with the notation $B_{,u} = \frac{\partial B}{\partial u}$, $B_{,uv} = \frac{\partial^2 B}{\partial u \partial v}$, and so forth)

$$B_{,uv} + 3B_{,u} B_{,v} = 0 \tag{24}$$

$$2\varphi_{,uv} + 3\left(B_{,u}\varphi_{,v} + B_{,v}\varphi_{,u}\right) = 0 \tag{25}$$

$$A_{,u}B_{,u} - \tfrac{1}{2}\left(B_{,u}^2 + B_{,uu}\right) - \tfrac{2}{9}\varphi_{,u}^2 = 0 \tag{26}$$

$$A_{,v}B_{,v} - \tfrac{1}{2}\left(B_{,v}^2 + B_{,vv}\right) - \tfrac{2}{9}\varphi_{,v}^2 = 0 \tag{27}$$

$$2\varphi_{,u}\varphi_{,v} + 3\left(A_{,uv} + 2B_{,uv} + 3B_{,u}B_{,v}\right) = 0 \tag{28}$$

These equations are, respectively, the uv component of Einstein's equation; the dilaton equation of motion; and the uu component, the vv component, and the ii component of the Einstein equation. (The last equation is not independent of the first four by virtue of the Bianchi identity.)

As mentioned above, we have to match the solutions for $y < 0$ and $y > 0$. The matching conditions for the metric and for the dilaton at the brane can be obtained by writing out (20) and (22) in the (t, y) coordinates and integrating across $y = 0$ at fixed t:

$$6\left.\frac{\partial A}{\partial y}\right] = -e^{A+b\varphi}\Big|\,(\Lambda - 2\rho - 3P) \tag{29}$$

$$6\left.\frac{\partial B}{\partial y}\right] = -e^{A+b\varphi}\Big|\,(\Lambda + \rho) \tag{30}$$

$$\tfrac{8}{3}\left.\frac{\partial \varphi}{\partial y}\right] = be^{A+b\varphi}\Big|\,(\Lambda + \rho) \tag{31}$$

Alert! To write down the right hand side of (31), you integrate (22), and hence you have to know what $\mathcal{L}_{3+1}$ for a perfect fluid is. (Note that, in contrast, to write down (29) and (30), you merely have to know what the stress energy tensor of a perfect fluid is, which you have known since part III.) So, if you insist on not taking anybody's word for what this is, you could simply set $\rho = 0$ and $P = 0$ in (29)–(31).

Note that these matching conditions have to be satisfied at any instant in t, with the two sides in (29)–(31) functions of t. Thus, we have the rather strong constraint that

$$\varphi_{,y}] = -\tfrac{9}{4}\,bB_{,y}] \tag{32}$$

In many solutions, this condition forces φ to be proportional to B.

Without the perfect fluid, we see from (29) and (30) that A equals B up to a possible additive constant.

We now have to deal with the "finger of God" problem mentioned in the text. The simplest and most natural assumption is that the bulk degrees of freedom simply respond to motion on the brane, but they do not act on the brane. In this spirit, the authors of the outgoing brane wave model simply decree by fiat that there are no incoming bulk waves, only outgoing waves.

A simple class of solutions to the bulk equations (24)–(28) consists of taking A, B, and φ to depend only on u. Of the five bulk equations, only one, namely (26), survives:

$$A_{,u}B_{,u} - \tfrac{1}{2}\left(B_{,u}^2 + B_{,uu}\right) - \tfrac{2}{9}\varphi_{,u}^2 = 0 \tag{33}$$

This corresponds to a plane wave propagating to the right. Similarly, solutions depending only on v exist, representing plane waves moving to the left.

But if A depends only on u, we see by inspecting (23) that it can be absorbed by reparametrizing u. Then (33) determines B in terms of φ, or vice versa. Physically, it makes sense that there are no independent gravitational degrees of freedom, since we have assumed isotropy in the transverse x^i space. (Gravitational plane waves, as discussed in chapter IX.4, expand some directions while contracting others.)

We now construct a solution of (20) and (22) by matching a solution depending only on u (for $y > 0$) to a solution depending only on v (for $y < 0$). The result is a solution in which the bulk spacetime consists of plane waves moving away from the brane on both sides.

Write $B(u, v) = \log h(u)$ on the $y > 0$ side. The continuity of B implies that $B(u, v) = \log h(v)$ on the $y < 0$ side. Since, according to (32), the jump in $\varphi_{,y}$ must be proportional to the jump in $B_{,y}$ for all t, φ itself must be proportional to B:

$$\varphi(u, v) = -\tfrac{9}{4}\,b\,\log h(u) \tag{34}$$

for $y > 0$, and $\varphi(u, v) = -\frac{9}{4} b \log h(v)$ for $y < 0$. Note that an additive constant in $\varphi(u, v)$ can be absorbed by scaling h, and the consequent additive constant in B can be absorbed by scaling x^i. We can now immediately integrate (33) to obtain*

$$A(u, v) = \frac{1}{2} \left(1 + \frac{9}{4} b^2 \right) B(u, v) + \frac{1}{2} \log B_{,u}(u, v)$$

$$= \frac{9}{8} b^2 \log h(u) + \frac{1}{2} \log h'(u) \tag{35}$$

for $y > 0$, and $A(u, v) = \frac{9}{8} b^2 \log h(v) + \frac{1}{2} \log h'(v)$ for $y < 0$. An additive constant can be absorbed by scaling u and v.

The function h is determined once we pick an equation of state for the matter on the brane. In other words, the equation of state fixes both the amplitude of the bulk waves and the dynamics of the brane geometry.

A simple example is to let the total pressure and total energy density be linearly related: $P - \Lambda = \gamma(\rho + \Lambda)$, with γ a constant. Then the matching conditions (30) and (29) imply that the jump in $A_{,y}$ is proportional to the jump in $B_{,y}$. Since this must hold for all time, we obtain

$$A = -(2 + 3\gamma)B + k \tag{36}$$

for some constant k. This yields a first order equation for h that can be integrated explicitly.

Without a perfect fluid, P and ρ vanish and so $\gamma = -1$, leading to $A = B + k$ as anticipated.

To illustrate some of the features of this model, we start with a particularly simple special case. Inspection of (33) shows that we can choose φ, A, and B to be linear functions of u on the $y > 0$ side and linear functions of v on the $y < 0$ side, respectively. This corresponds to setting

$$h(t) = e^{\lambda t} \tag{37}$$

where we assume that the constant λ is positive. Thus, on the $y > 0$ side, $B = \lambda u$, $\varphi = -\frac{9}{4} b \lambda u$, and

$$A = \frac{1}{2} \left(1 + \frac{9}{4} b^2 \right) \lambda u + \frac{1}{2} \log \lambda \tag{38}$$

and similarly on the $y < 0$ side. We now set $b = \pm \frac{2}{3}$. (The case of general b will be considered below.) Then $A = B + \text{const}$, and it follows from (36) that $\gamma = -1$, so the stress energy on the brane is a pure cosmological constant. From (30), the vacuum energy is

$$\Lambda = 12 \lambda^{\frac{1}{2}} \tag{39}$$

The bulk metric is

$$ds^2 = e^{2\lambda(t-y)} \left[-\lambda dt^2 + \lambda dy^2 + dx_i dx^i \right] \tag{40}$$

for $y > 0$ and

$$ds^2 = e^{2\lambda(t+y)} \left[-\lambda dt^2 + \lambda dy^2 + dx_i dx^i \right] \tag{41}$$

for $y < 0$. Changing to cosmological time $\tau = e^{\lambda t}/\sqrt{\lambda}$, we see that the metric on the brane at $y = 0$ has the Friedmann-Lemaître-Robertson-Walker form $ds^2 = -d\tau^2 + \lambda \tau^2 dx_i dx^i$, which after scaling x^i, gives

$$ds^2 = -d\tau^2 + \tau^2 dx_i dx^i \tag{42}$$

Surprise! Even with a cosmological constant Λ, we have a universe with the scale factor growing linearly $a(\tau) = \tau$ rather than the usual exponential growth. Remarkably, the expansion rate is independent of the value of the vacuum energy Λ (as long as it is nonzero). In the literature, this is known as self-tuning.

The solution here does not have naked timelike singularities in the bulk, which appear to be a generic feature of the static solutions discussed in the literature. Rather, the factor $e^{2\lambda(t-y)} = e^{2\lambda u}$ in (40) shows that the spacetime has a null singularity at $u = -\infty$ on the right of the brane. Similarly, it also has a null singularity at $v = -\infty$ on the left. Null geodesics from these singularities reach the brane by finite affine parameters, so they are not really at infinity. This is most easily seen by introducing a new coordinate $U = e^{2\lambda u}/2$, so that the metric for $y > 0$ becomes

$$ds^2 = -dU dv + 2U dx_i dx^i \tag{43}$$

---

* Throughout, it is understood that all functions inside the logarithm have absolute value signs.

The singularity is now at $U = 0$. The brane is located at $u = v$ or $U = e^{2\lambda v}/2$, so the brane never actually hits the singularity, but instead becomes asymptotically null as $v \to -\infty$. The geometry on the left of the brane is similar, with the roles of u and v interchanged.

Next, let us determine the Newton's constant G seen by observers on the $(3 + 1)$-dimensional brane. We go through the same discussion as in the text, considering a fluctuation of the metric $g_{\mu\nu}$ on the brane and evaluating the corresponding action. As in (17), we end up with an integral over y. Since the null singularity cuts off space in the fifth direction, we obtain a nonzero effective G. But since the distance of the singularity to the brane changes with time, G will be time dependent, posing a serious problem for this model.

Remarkably, we can find solutions of the bulk equations (24)–(28) that are much more general than the plane waves discussed so far. The key is to observe that (24) involves only B and can be solved to give

$$B(u, v) = \tfrac{1}{3} \log(f(u) + g(v)) \tag{44}$$

where f and g are two arbitrary functions. (A possible additive integration constant can always be absorbed by an overall scaling of f and g.) Given $B(u, v)$, we can solve the dilaton equation of motion (25). Using separation of variables, the general solution can be written as

$$\varphi(u, v) = \int dk \frac{c(k)}{\sqrt{(f(u) - k)(g(v) + k)}} \tag{45}$$

with some smooth function $c(k)$. These solutions can be used to study a phase transition in which the vacuum energy changes on an initially static, Poincaré invariant brane. In one solution, the brane becomes time dependent after the transition.

I refrain from giving more details here and simply refer the interested reader to the original paper on the outgoing brane wave model. It is at least mathematically interesting that the coupled Einstein and dilaton equations appear to have many solutions. As I said earlier, you now know enough to contribute, if you wish, to the brane world literature.

Notes

1. At the level of a popular book, I offer the following analogy. A biologist puts a few drops of water from a pond between 2 thin glass plates. To the life forms in the 2-dimensional world between the plates, the world outside the glass plates is beyond comprehension. Yet streams of photons can pass back and forth between the 2-dimensional world and the world beyond. See *Toy/Universe*, p. 250.

2. See the original papers by J. Polchinski. For a textbook on brane physics, see C. Johnson, *D-Branes*, Cambridge University Press, 2006.

3. The notion of large extra dimensions goes back to V. Rubakov and M. Shaposnikov in 1983 and to M. Visser. More recently, the subject was developed in 1990 by I. Antoniadis, and then in 1998 by N. Arkani-Hamed, S. Dimopoulos, and Dvali, and by G. Shiu and S.-H. Tye.

4. Let me remind you how this works. The 1-dimensional integral $f(x) \equiv \frac{1}{2\pi} \int_{-K}^{K} dk\, e^{ikx} = \frac{1}{2\pi} \int_{-K}^{K} dk\, \cos(kx) = \sin(Kx)/(\pi x)$. In the limit $K \to \infty$, this function $f(x)$ approaches the delta function $\delta(x)$, since $f(0) = K/\pi \to \infty$, and for $x \neq 0$, it oscillates rapidly with an amplitude that quickly tends to 0. The identity used in the text is the trivial 3-dimensional generalization of this.

5. *QFT Nut,* chapter I.4.

6. *QFT Nut,* chapter I.5.

7. More precisely, the repulsion between like electric charges is due to this sign flip between η_{00} and η_{ii}.

8. At this distance scale, the gravitational force is easily overwhelmed by the electromagnetic force, and on the nanoscale, even by Casimir forces.

9. L. Randall and R. Sundrum, *Phys. Rev. Lett.* 83 (1999), p. 3370.

10. D. Tymoczko, *A Geometry of Music*, Oxford University Press, 2011. See p. 410. Note that Pythagoras, whose influence pervades this entire textbook, was also into geometry and music.

11. This relation has emerged from some papers by P. Horava and E. Witten.

12. L. Randall and R. Sundrum, *Phys. Rev. Lett.* 83 (1999), p. 4690.

13. G. Horowitz, I. Low, and A. Zee, *Phys. Rev.* D62 (2000), p086005.

14. It is $\mathcal{L} = -\rho$. See, for example, S. Endlich, A. Nicolis, R. Rattazzi, and J. Wang, *JHEP* 4 (2011), p. 102.

X.3 | Effective Field Theory Approach to Einstein Gravity

Powers of derivatives and the long distance expansion

Back when I airlifted you to the Einstein-Hilbert action, you might have asked, "The scalar curvature R is not the only coordinate scalar we could have formed out of the metric. What about other possibilities?" When I teach Einstein gravity, someone usually asks this question. I would respond that R is the only coordinate scalar involving two powers of derivatives ∂. True, we also have the scalars R^2, $R_{\mu\nu}R^{\mu\nu}$, and $R_{\mu\nu\rho\sigma}R^{\mu\nu\rho\sigma}$, but they all involve four powers of derivatives. To understand better what to do with these possible terms in the action, let us step back and examine the much simpler case of Newtonian gravity.

Way way back, in chapters II.1 and II.3, I reminded you that in Newtonian gravity, the gravitational potential Φ satisfies Poisson's equation $\vec{\nabla}^2\Phi(\vec{x}) = 4\pi G\rho(\vec{x})$. Let's see how a really poor man, an impoverished man who doesn't know how to solve partial differential equations, would determine Φ around an object of mass M and radius R. The density is easy, he says, $\rho \sim M/R^3$. Mired in poverty but nevertheless smart, he next approximates the derivative $\nabla\Phi$ by Φ divided by the relevant distance scale $\sim R$, so that $\nabla\Phi \sim \Phi/R$ and $\nabla^2\Phi \sim \Phi/R^2$. Then he writes*

$$\nabla^2\Phi \sim \Phi/R^2 \sim G\rho \sim GM/R^3 \tag{1}$$

which requires only algebra to solve, giving $\Phi \sim GM/R$, the right answer. No need to take a fancy† course in partial differential equations!

Now the same guy who wants to know why we didn't add terms involving four powers of derivatives, such as R^2, to the Einstein-Hilbert action might also ask why Newton (or

---

* You will recall that we used a similar argument in the preceding chapter.

† Beginning students often snicker at this sort of getting an answer by "winging it," compared to solving a partial differential equation in all its glory, complete with factors of 2π and what not. But in fact, in cutting edge research, the ability to do the former is often much more prized than the ability to do the latter. On the cutting edge, the analog of the partial differential equation is typically not known, but the truly great theorists are often able to grope for what they want in the dark "by the seat of their pants."

Poisson) did not add terms involving four powers of derivatives to the equation determining Φ. Indeed, we could invite ourselves to write

$$\vec{\nabla}^2\Phi(\vec{x}) + l^2\vec{\nabla}^2\vec{\nabla}^2\Phi = 4\pi G\rho(\vec{x}) \tag{2}$$

By dimensional analysis, we have to introduce an unknown length scale l characterizing the deviation from Newtonian gravity. The impoverished man kindly solves this equation for us:

$$\nabla^2\Phi + l^2\nabla^2\nabla^2\Phi \sim \frac{\Phi}{R^2} + l^2\frac{\Phi}{R^4} \sim \frac{GM}{R^3} \tag{3}$$

Hence

$$\Phi \sim \frac{GM}{R\left(1 + \frac{l^2}{R^2}\right)} \sim \frac{GM}{R}\left(1 - \frac{l^2}{R^2} + \cdots\right)$$

and we see that the effect of the added term is negligible for $l \ll R$. Indeed, we can reach the same conclusion by looking directly at the postulated equation (2).

We can also turn the argument around. The fact that deviation from Newtonian gravity has not been observed down to a certain length scale allows us to set an upper bound* on the unknown length l.

Einstein-Hilbert action as merely effective

This simple but elegant argument forms the basis of the so-called effective field theory approach[1] emphasized by Ken Wilson and others and is much used in contemporary theoretical physics.

In the context of gravity, yes, we are certainly more than welcome to add higher derivative terms to the Einstein-Hilbert action, so that

$$S = M_P^2 \int d^4x \sqrt{-g}\, R$$

$$\rightarrow M_P^2 \int d^4x \sqrt{-g}\left(R + l^2\left(\alpha R^2 + \beta R_{\mu\nu}R^{\mu\nu} + \gamma R_{\mu\nu\rho\sigma}R^{\mu\nu\rho\sigma}\right) + \cdots\right) \tag{4}$$

Again, high school dimensional analysis forces us to introduce a length l, and 3 numbers[2] α, β, and γ of order unity. The ellipsis indicates terms involving cubes and ever higher powers of the curvature tensor, objects such as $R_{\mu\nu\rho\sigma}R^{\mu\rho}R^{\nu\sigma}$. Since the only length scale we know associated with gravity is the Planck length l_P, we naturally assume that $l = l_P$. However, in theoretical physics, we should of course always keep an open mind. The verity of this commonly made assumption has to be checked by experimentalists. Indeed, this is one of the issues concerning gravity, an issue we will come back to in chapters X.7 and X.8.

---

* The length l characterizes the distance scale at which deviations from our present knowledge of gravity might show up. As of this writing, $l \lesssim 1$ mm, as I have already mentioned in chapter X.2.

Let us go ahead and assume $l = l_P$. We now use the same argument as before. In working out the gravitational field in some given physical situation, we effectively convert the derivative ∂ acting on the metric in the action, and hence in the equations of motion, into $\sim 1/L$, with L some characteristic length scale over which the metric varies. Thus, we expect the effects of the higher order terms R^2, $R_{\mu\nu}R^{\mu\nu}$, and $R_{\mu\nu\rho\sigma}R^{\mu\nu\rho\sigma}$ (known as the Weyl-Eddington terms[3]) to be suppressed by $\sim (l/L)^2$, which normally is almost infinitesimal if $l = l_P$. The effects of the terms represented by the ellipsis in (4) are suppressed even more severely. This explains why it sufficed to keep only the Einstein-Hilbert term in the action back in chapter VI.1.

Effective field theory

The example of gravity suffices to show how the effective field theory approach, which pervades contemporary particle and condensed matter physics, works. We classify all possible terms in an action by powers of derivatives. The relative coefficients of these terms are then fixed by dimensional analysis to be some inherent length l (possibly l_P in the case of gravity) raised to the appropriate powers. The effects of various terms are then controlled by various powers of (l/L), with L some characteristic length scale of the physical phenomenon* we are studying.

Condensed matter physicists like to think in terms of distance, but particle physicists tend to think about an energy or mass scale.[4] Thanks to Planck's $\hbar$, classification in terms of an energy scale is equivalent to classification in terms of a distance scale, but conceptually they should be kept distinct. For example, the scalar curvature R has mass dimension 2, while R^2, $R_{\mu\nu}R^{\mu\nu}$, and $R_{\mu\nu\rho\sigma}R^{\mu\nu\rho\sigma}$ have mass dimension 4, and so on.

Our discussion above indicates that a term of mass dimension[5] p in the action must have a coefficient that, according to high school dimensional analysis, goes like $1/M^{p-4}$. Here M denotes some (usually unknown) mass scale at which the physics associated with that term kicks in. Thus, in a process characterized by energy E, the effects of that term would be of order $(E/M)^{p-4}$. This is one reason particle physicists are always clamoring for higher energy accelerators. We will come back to this point in chapter X.8.

Our friend the Smart Experimentalist speaks up, "Indeed, it would be the height of hubris, almost inimical to the spirit of physics, for you theorists to suppose that your action[6] du jour is actually the ultimate. The established actions in physics describe Nature only at the length or energy scales we have explored experimentally."

We totally agree. In quantum field theory, all possible terms not explicitly forbidden by the symmetries of the theory are mandated, as was already mentioned in chapter VI.2. The eternal hope of theoretical physics is that, for a given set of phenomena, keeping only a few dominant terms in the action suffices. All the actions studied in this book should be regarded in this light.

---

* We will return to this point when we talk about the quantum Hall fluid in chapter X.5.

Appendix 1: The cosmological constant paradox once again

In the text, we added terms with higher mass dimensions than the scalar curvature R. What about terms with lower mass dimensions? In fact, the term 1, with mass dimension 0, is also allowed. We are free to write

$$S = M_P^2 \int d^4x \sqrt{-g} \left(\frac{1}{l_C^2} + R + l^2 (\alpha R^2 + \beta R_{\mu\nu} R^{\mu\nu} + \gamma R_{\mu\nu\rho\sigma} R^{\mu\nu\rho\sigma}) + \cdots \right) \tag{5}$$

where, again, by high school dimensional analysis we were forced to introduce a length scale l_C which a priori may or may not be the same as l. Compare with (4).

Recall from chapter VI.2 that the "1" term is, once again, the dreaded cosmological constant, with the identification of the energy density Λ as

$$\Lambda \sim \frac{M_P^2}{l_C^2} \tag{6}$$

Without bothering to plug in numbers, we can see that l_C is enormous. From chapters VI.2 and VIII.1, we know that to a first approximation, our universe is dominated by the cosmological constant, aka dark energy, with the scale factor determined by Einstein's field equation $(\dot{a}/a)^2 \sim G\rho$, so that* $H^2 \sim G\Lambda \sim \Lambda/M_P^2$. Comparing this with (6), we see that l_C is the Hubble size of the universe, and so is almost inconceivably larger than l, even if, in a departure from conventional wisdom, we take l to be much larger than the Planck length l_P. This humongous discrepancy amounts to another statement of the cosmological constant paradox.[7] Nature flagrantly violates the theorist's cherished naturalness doctrine.

Since Λ is an energy density with dimensions $M/L^3 \sim M^4 \sim 1/L^4$, we can define a length scale l_Λ associated with the cosmological energy density by $\Lambda \equiv l_\Lambda^{-4}$. A possibly illuminating way of writing (6) is

$$l_\Lambda \sim \sqrt{l_P l_U} \tag{7}$$

where we renamed l_C as l_U, the length scale or size of the universe. Einstein tells us that the length scale associated with the dark energy or the cosmological constant is the geometric mean[8] of the smallest and the largest lengths known in physics. Rather mysterious!

The cosmological constant paradox unmasks theoretical physicists as double-talking snake oil salesmen:[†] in the effective action for gravity, they want l to be tiny on the one hand and l_C to be enormous on the other.

We will come back to the cosmological constant paradox in chapter X.7.

Appendix 2: Reversal of fortune

This appendix is for those readers with some knowledge of quantum field theory. Other readers should skip this upon first reading of the book.

The view of quantum field theory as a low energy effective theory sketched here represents a remarkable shift in attitude toward quantum field theory over the past 30 years. Traditionally, a term in the action for a quantum field theory is classified according to whether its mass dimension is < 4, $= 4$, or > 4, known respectively as superrenormalizable, renormalizable, and nonrenormalizable. Textbooks taught that superrenormalizable interactions were nice, renormalizable interactions were what we want, while nonrenormalizable interactions should fill us with fear and loathing.

The reason is simple. As already explained in the text, terms with mass dimension p lead to contributions going like $(E/M)^{p-4}$, and so nonrenormalizable terms with $p > 4$ diverge badly at high energies.

In an astonishing reversal of fortune, the nonrenormalizable terms are now welcomed and well liked as terms that are inevitably here with us. They are regarded as innocuous, since they are suppressed by powers of some

---

* Which of course in this context is just the statement that the first two terms in (5) battle each other to a standstill.

† Well, not quite. I exaggerate totally. Forget that I said that.

higher mass scale $\frac{1}{M^{p-4}}$. In contrast, our former friends the superrenormalizable terms are now regarded as nasty guys.

Since these nasty guys have nominal mass dimension < 4, there are fortunately only a finite number of them. They represent the challenges confronting fundamental physics today, and are in turn known as the Higgs mass term, the Einstein-Hilbert term, and the cosmological constant term. The Higgs mass term has dimension 2. The Einstein-Hilbert term has nominal dimension 2, which after rescaling[9] by the Planck mass, becomes dimension $4 + 5 + 6 + \cdots$. The cosmological constant term has nominal dimension 0, which after rescaling, becomes dimension $0 + 1 + 2 + \cdots$. Perhaps there is something seriously wrong with this picture.

Our present understanding of physics is based on this notion of effective field theory, to which all we know can be reduced. Yet there are many questions, many doubts, but no clear answers. Field theory itself, and Einstein gravity as an effective field theory, could fail at truly long distances. More in chapter X.7.

Appendix 3: Nonlocal cosmology

With a universe dominated by a cosmological constant Λ or dark energy, adding local terms to Einstein gravity as in (4) will not significantly change large scale Lemaître–de Sitter cosmology. The field equation is modified by additional local terms of the form $R_{\mu\lambda\nu\rho}R^{\lambda\rho}$, for example. But with the maximally symmetric de Sitter form of the Riemann curvature tensor, all these terms reduce to some combinations of the Hubble parameter H times $g_{\mu\nu}$, and so only the relation between H and Λ is modified.

One way around this situation is to introduce nonlocal terms, for example replacing the Einstein-Hilbert term R by $Rf(\frac{1}{D^2}R)$, with $D^2 = (1/\sqrt{-g})\partial_\mu(\sqrt{-g}g^{\mu\nu}\partial_\nu)$ the covariant version of ∂^2. With a suitable choice[10] of the function $f(x)$, but without having to introduce a cosmological constant Λ or dark energy, this type of nonlocal action can reproduce current observations, including the accelerating expansion. To me, an attractive feature of this approach is that quantum field theory with known physics naturally generates this type of nonlocal term via loop corrections involving the massless graviton. One may regard these nonlocal terms as due to the cumulative effect of the fluctuating graviton, an effect that manifests itself only on cosmological distance scales. A drawback is of course that this comes with the freedom of adjusting an entire function[11] to fit data.

Appendix 4: More on the scalar field

Starting in part V, I have extolled the power of Einstein's equivalence principle: given a Lagrangian in Minkowskian spacetime, in which various Lorentz indices are contracted with $\eta_{\mu\nu}$ and its inverse, we simply replace $\eta_{\mu\nu}$ by $g_{\mu\nu}$ and immediately obtain the corresponding Lagrangian in curved spacetime. For a scalar field $\varphi(x)$, we go immediately from $\mathcal{L} = -\frac{1}{2}\eta^{\mu\nu}\partial_\mu\varphi\partial_\nu\varphi - V(\varphi)$ to $\mathcal{L} = -\frac{1}{2}g^{\mu\nu}\partial_\mu\varphi\partial_\nu\varphi - V(\varphi)$.

But in the spirit of effective field theory, we are also free to add to $\mathcal{L}$ the term $\xi R\varphi^2$, with ξ some numerical constant. Note that this term also has mass dimension 4, just like the term $g^{\mu\nu}\partial_\mu\varphi\partial_\nu\varphi$. Let the characteristic length scale over which φ and the scalar curvature R vary be l_φ and L_R, respectively. Then the relative importance of this additional nonminimal term versus the standard kinetic term $g^{\mu\nu}\partial_\mu\varphi\partial_\nu\varphi$ is given by $\sim \xi(l_\varphi/L_R)^2$. This is another example underlining why the equivalence principle is always formulated with the caveat "in a small enough region of spacetime," as first discussed in the prologue to book 2. Here, the region of spacetime over which φ varies has to be small compared with the region over which the curvature varies for the equivalence principle to hold.

Note that the energy momentum tensor $T^{\mu\nu}$ of the scalar field is corrected by this additional term, resulting in what became known as the "new and improved"[12] energy momentum tensor, much discussed in the field theory literature in the 1970s.

Exercise

1   Describe how higher derivative terms can be added to Maxwell's theory of electromagnetism and discuss their physical manifestations.

Notes

1. *QFT Nut*, chapter VIII.3, pp. 452 ff.
2. Evidently, one of them can be absorbed into *l*.
3. As you can imagine, there has been a vast literature regarding these terms going back to our forebears in theoretical physics. For an example I know particularly well, see A. Zee, "A Theory of Gravity Based on the Weyl-Eddington Action," *Phys. Lett.* B 109 (1982), p. 183.
4. Phil Anderson once remarked to me that particle physicists renaming themselves high energy physicists was a stroke of genius in terms of getting more funding (in the United States, that is). The name "long distance physicists" hardly sounds thrilling, and there may be people so ignorant in Congress as to think that condensed matter physics has something to do with condensed milk or the gunk one finds underneath kitchen sinks.
5. The scaling dimensions of various possible terms in the action play a central role in quantum field theory. See, for example, *QFT Nut*, chapters III.2 and VI.8.
6. Indeed, instead of the venerable R, some authors have proposed $f(R)$, for some arbitrary function f that had been revealed to them in the middle of the night.
7. See *QFT Nut*, chapter VIII.2.
8. This fact has been noted by a number of authors. See, for example, S. Hsu and A. Zee, arXiv:hep-th/0406142 (*Mod. Phys. Lett.* A 20 (2005), pp. 2699–2704).
9. As explained in chapter IX.5, $h \equiv \hat{h}/M_P$.
10. See S. Deser and R. P. Woodard, arXiv:0706.215v2, and related literature for details.
11. It is important to distinguish this proposal from proposals to replace the Einstein-Hilbert term R by $f(R)$.
12. It turns out that for some special choice of ξ, the resulting $T^{\mu\nu}$ possesses properties much desired by particle theorists.

X.4 | Finite Sized Objects and Tidal Forces in Einstein Gravity

Motion of extended objects

Most texts on Einstein gravity treat the motion of point particles exclusively. So, good, watch the particles move happily along geodesics in curved spacetime.

But in some physical situations, we may have to take into account the finite size of the "particles." One example is the emission of gravitational waves from binary systems. As we saw in chapter IX.4, one astounding prediction of Einstein gravity is the existence* of gravitational waves. Various sources of gravitational waves have been studied intensively. One possible source consists of a black hole of size r_S (its Schwarzschild radius) moving with velocity v a distance r_O from another object, possibly another black hole of similar size. As the black holes spiral into each other, they emit gravitational waves with a characteristic wavelength λ determined by the orbital period according to $\lambda = 2\pi r_O/v$. Thus, the physics contains three distance scales: r_S, r_O, and λ. We will stay within the simple "post-Newtonian" regime $r_S \ll r_O \ll \lambda$. To leading approximation, the black hole may be regarded as pointlike, but we might want to include corrections governed by the small parameter r_S/r_O.

Blue sky and finite sized objects in electromagnetism

For pedagogical clarity, let us start by retreating to the corresponding problem in electromagnetism. Also, take spacetime to be flat. For a point particle with charge e moving in an electromagnetic field, the relevant term in the action, as we discussed in chapter IV.1, is given by $S_{pp} = \int d\tau e A_\mu \dot{X}^\mu$, with $\dot{X}^\mu \equiv \frac{dX^\mu}{d\tau}$.

Instead, consider an extended object, such as an atom or a molecule, that is, an electrically neutral assembly of charged particles. Let us construct an action for the motion

---

* Recall also Newton's snide remark about "competent faculty of thinking" in chapter IX.5.

of this object, say an atom for definiteness, in an electromagnetic field. Since the overall charge vanishes, the point particle term S_{pp} is absent from the action. The individual charged particles in the assembly are of course sensitive to A_μ, but the atom as an overall neutral collection of charged particles cannot be. Rather, as the worldlines of the individual charged particles in the atom traverse different locations in spacetime, the atom can only be sensitive to the spacetime variations of A_μ, not to A_μ itself. By gauge invariance, these spacetime variations must package themselves into $F_{\mu\nu}$. In fact, let us define $E_\mu \equiv F_{\mu\nu}\dot{X}^\nu$ and $B_\mu \equiv \tilde{F}_{\mu\nu}\dot{X}^\nu$, where $\tilde{F}_{\mu\nu}(x) \equiv \frac{1}{2}\epsilon_{\mu\nu\sigma\eta}F^{\sigma\eta}$. Going to the rest frame of the particle, where $\dot{X}^0 = 1$ and $\dot{X}^i = 0$, we see that $E_0 = 0$, $E_i = F_{i0}$, $B_0 = 0$, and $B_i = \tilde{F}_{i0} = -\frac{1}{2}\epsilon_{ijk}F^{jk}$. Thus, as the notation suggests, this is just the familiar decomposition of the electromagnetic field into electric and magnetic fields.

Given these considerations, we see that, since Lorentz indices must be contracted, the action can only be

$$S = \int d\tau \left(-m + c_E E_\mu E^\mu + c_B B_\mu B^\mu + \cdots \right) \tag{1}$$

given as an expansion in terms of the size of the atom. We will not be concerned with the higher order corrections (as indicated by the dots in (1)) due to the size of the atom, only with the leading order. (Note that a possible term like $\int d\tau\, F_{\mu\nu}F^{\mu\nu}$ can be absorbed into the two terms quadratic in $F_{\mu\nu}$ already displayed.) The fields E and B are to be evaluated on the worldline $X^\lambda(\tau)$ of the particle, of course. Physically, from the discussion above, we know that, in the limit where we can neglect the size of the atom, the coefficients c_E and c_B must vanish. Thus, you will be hardly surprised to learn that they are related to elementary concepts, such as the electric dipole moment and the magnetic moment of the atom.

Using this action, we can derive a result familiar to everybody, including the proverbial guy and gal in the street, namely that the sky is blue. Consider an electromagnetic wave of frequency ω scattering on the atom. The quantities $E_\mu E^\mu$ and $B_\mu B^\mu$ each contain two powers of derivatives, which, acting on the electromagnetic wave, translate into two powers of ω in the scattering amplitude. Upon squaring the scattering amplitude to obtain the cross section, we conclude that the cross section for an electromagnetic wave or a photon of frequency ω to scatter on an atom or a molecule goes like ω^4. Thus, as the light from the sun traverses the atmosphere, blue light (higher frequency) scatters more than red light (lower frequency). As is well known, this explains why the sky is blue.

The "electric" and "magnetic" components of a gravitational field

Now that we have derived the action governing the motion of a finite sized object moving in an electromagnetic field to leading order in the object's size, we are ready to move on to the gravitational case, keeping in mind the essential differences between electromagnetism and gravitation. As mentioned at the beginning of this chapter, one application would be to study the finite sized corrections to the motion of a black hole.

One key difference is of course that there is no analog of positive and negative charges for gravity: masses are acted upon equally by the gravitational field without regard to race or creed. Thus, the action necessarily starts out with the point particle action, which is $S_{\mathrm{pp}} = -m \int d\tau$ in this context. We don't have objects that are neutral under gravity!

As in the electromagnetic case, a finite sized object would also be sensitive to the variations of the metric in spacetime, and by general coordinate invariance these variations, to leading order, must get packaged into the Riemann curvature tensor. Would the Riemann curvature appear already contracted into the Ricci tensor and the scalar curvature? Time for you to pause and think!

Well, these two quantities both vanish, according to Einstein's field equation, in the empty spacetime the black hole is moving through. In other words, the black hole can only sample the Riemann curvature tensor $R_{\mu\lambda\nu\rho}$ itself, not the Ricci tensor and the scalar curvature. Thus, any terms we add to S_{pp} must involve the Riemann curvature tensor $R_{\mu\lambda\nu\rho}$, with the indices not allowed to be contracted with each other. What can they be contracted with, then?

The only thing around is the 4-velocity of the object $\dot{X}^\mu$. Due to the antisymmetry of $R_{\mu\lambda\nu\rho}$, we are not able to contract all 4 indices of $R_{\mu\lambda\nu\rho}$ with $\dot{X}^\mu$. We can contract at most 2 indices with $\dot{X}^\mu$ to form the two objects

$$E_{\mu\nu}(X) \equiv R_{\mu\lambda\nu\rho}(X)\dot{X}^\lambda\dot{X}^\rho \quad \text{and} \quad B_{\mu\nu}(X) \equiv \tilde{R}_{\mu\lambda\nu\rho}(X)\dot{X}^\lambda\dot{X}^\rho \tag{2}$$

where $\tilde{R}_{\mu\lambda\nu\rho}(x) \equiv \epsilon_{\mu\lambda\sigma\eta}R^{\sigma\eta}{}_{\nu\rho}(x)/(2\sqrt{-g})$ denotes the dual* of the curvature tensor. Indeed, these correspond to $E_\mu \equiv F_{\mu\nu}\dot{X}^\nu$ and $B_\mu \equiv \tilde{F}_{\mu\nu}\dot{X}^\nu$, respectively, in the electromagnetic case. By analogy, the fields $E_{\mu\nu}$ and $B_{\mu\nu}$ represent the decomposition of curvature into its "electric" and "magnetic" components, as suggested by the notation. Perhaps the reader is not surprised that these fields now carry two indices instead of one. We now need to square them to form scalars to put into the action.

Finite sized objects and tidal forces

Hence, to leading order in the size of the object, the action governing its motion in a gravitational field is given by

$$S_{\mathrm{p}} = \int d\tau \left(-m + c_E E_{\mu\nu}E^{\mu\nu} + c_B B_{\mu\nu}B^{\mu\nu} + \cdots \right) \tag{3}$$

with two unknown constants c_E and c_B. Compare with (1). Note that, since $E_{..}$ or $B_{..} \sim R_{....}\dot{X}^2$ has dimensions $L^{-2} = M^2$ in natural units, c_E and c_B must have dimensions of $M^{-3} = L^3$ to match the dimensions of the first term in (3). As expected, they vanish as the size of the object goes to zero.

---

* The appearance of $\sqrt{-g}$ will be explained in chapter X.5.

In chapter V.3, we varied the first term in (3) to obtain the standard geodesic equation that is at the heart of Einstein's theory. Here we obtain

$$\frac{d^2 X^\rho}{d\tau^2} + \Gamma^\rho_{\mu\nu}(X(\tau))\frac{dX^\mu}{d\tau}\frac{dX^\nu}{d\tau} = f^\mu(X(\tau)) \tag{4}$$

where $f^\mu(X(\tau))$ comes from varying the E and B terms in (3). A finite sized body experiences a tidal force f^μ due to the varying gravitational force acting on it. It no longer follows a geodesic. Everything makes sense.

The blue sky effect gets squared in gravity

The fact that we had to square the "electric" and "magnetic" components of the curvature to form the effective action (3) means that the effects of these correction terms are highly suppressed. Since the Riemann curvature contains 2 derivatives, the correction terms involve four derivatives. The blue sky effect gets squared in gravity: for the scattering of a gravitational wave or a graviton of frequency ω on a finite sized object, that part of the amplitude due* to the finite size goes like ω^4! You are not surprised, are you?

Appendix

To estimate the magnitude of c_E and c_B for a black hole, we exploit a rather cute argument[1] as follows (cute in the sense that it does not involve any tedious computation at all).

Consider the scattering of a graviton with frequency ω off a finite sized object (which, remember, is a black hole in the problem we are studying). The interaction between the particle and the gravitational wave indicated in (3) contributes a term to the scattering amplitude $\mathcal{M}$ as indicated by

$$\mathcal{M} \sim \cdots + c_{E,B}\omega^4/M_P^2 + \cdots \tag{5}$$

The ellipses in $\mathcal{M}$ represent effects of the interactions we have not included explicitly, for example, the one originating from the first term in (3) (namely the term responsible for keeping us down to earth!). A nice feature of the argument I am about to give is that we don't even need to know what the $(\cdots)$ are. Here $c_{E,B}$ denotes the two unknown couplings $c_E \sim c_B$ generically. We have derived the ω^4 dependence just a moment ago. So the only unexplained feature here is the power of M_P, which I will derive presently.

Imagine calculating the total scattering cross section for a graviton on a finite sized object. Squaring the amplitude $\mathcal{M}$ and so forth, we end up with $\sigma(\omega) \sim \cdots + c_{E,B}^2\omega^8/M_P^4 + \cdots$. We can use dimensional analysis to determine the power of M_P here (and hence the power of M_P in $\mathcal{M}$) as follows. The cross section σ has dimension of an area and hence the dimension M^{-2}. We had determined earlier that $c_{E,B}$ has dimension M^{-3}, so that $c_{E,B}^2\omega^8$ has dimension $M^{-6}M^8 = M^2$. Thus, to get the dimension to match, we need to divide by M_P^4. Note that the mass m of the object is not available to make up the dimension.

We are now able to estimate $c_{E,B}$ for a black hole. The preceding treatment of the black hole as almost a point particle is only valid for $\omega r_S \ll 1$ of course, that is, with the Schwarzschild radius much less than the wavelength of the gravitational wave. But we argue that by dimensional analysis, the cross section must have the form $\sigma(\omega) = r_S^2 f(\omega r_S)$, since the only length scale in the Schwarzschild metric is r_S. Expanding the unknown function $f(\omega r_S)$ in powers of its argument, we have $\sigma(\omega) = \cdots + \gamma\omega^8 r_S^{10} + \cdots$, with γ some numerical constant.

---

* This sentence is more awkward to write than the corresponding sentence in the electromagnetic case for a very physical reason: Even if the size of the object goes to zero, it still cannot hide from the gravitational field. There is no escape from gravity. In contrast, a zero sized atom is invisible to the electromagnetic field.

(A technical aside: the massless graviton could produce infrared factors like $\log \omega r_S$, which we ignore for our purposes.)

Requiring that the two expressions agree, we obtain $c_{E,B} \sim M_P^2 r_S^5$. Indeed, as expected, the couplings $c_{E,B}$ are highly suppressed as $r_S \to 0$.

Exercise

1   Work out the "electric" and "magnetic" components of the gravitational field we feel every day.

Note

1. This treatment is based on P. Goldberger and I. Rothstein, arXiv: 0409156. I neglect various technicalities, such as field redefinition. See *QFT Nut* for more details.

X.5 | Topological Field Theory

Not having clocks and rulers means that you are topological

To do physics, we need clocks and rulers.

By specifying the separation between events in spacetime, the metric in effect provides us with clocks and rulers. Indeed, in the action, we have to use the metric to contract spacetime indices, and even if there were no indices* to contract, the spacetime volume $d^d x \sqrt{-g}$ knows about the metric. It would appear that the metric is indispensable for writing down the action.

But is that necessarily so? Dear reader, please pause and think.

Recall the antisymmetric or Levi-Civita symbol $\epsilon^{\mu\nu\lambda\cdots\zeta}$ first introduced in chapter I.4, which we have since met repeatedly, for example in chapter IV.2. Recall also that in d-dimensional spacetime, $\epsilon^{\mu\nu\lambda\cdots\zeta}$ carries d indices with $\epsilon^{012\cdots,d-1} = 1$, and the rest determined by antisymmetry. For example, for $d = 4$, $\epsilon^{2031} = -\epsilon^{2013} = +\epsilon^{0213} = -\epsilon^{0123} = -1$. So, besides the metric, we can also use the antisymmetric symbol to contract indices.

Offered the antisymmetric symbol, we could contract it with a bunch of vectors or tensors to form an object with no free uncontracted indices, for instance $\mathcal{T} \equiv \epsilon^{\mu\nu\lambda\cdots\zeta} A_\mu B_\nu C_\lambda \cdots Z_\zeta$. To see clearly what is going on, we specialize to $d = 2$ and study $\mathcal{T} = \epsilon^{\mu\nu} A_\mu B_\nu = A_0 B_1 - A_1 B_0$.

How does $\mathcal{T}$ transform? Our friend Confusio might have naively thought that since this object does not carry any indices, it transforms like a scalar. But that's not so: while it looks like a duck, it does not quack like a duck. Here we need the definition of the determinant: for a matrix M, $\epsilon^{\rho\sigma} M^\mu{}_\rho M^\nu{}_\sigma = (\det M) \epsilon^{\mu\nu}$. You can easily verify that this definition coincides with the high school definition. Set $\mu = 0$, $\nu = 1$, for example: $\epsilon^{\rho\sigma} M^0{}_\rho M^1{}_\sigma = M^0{}_0 M^1{}_1 - M^0{}_1 M^1{}_0 = \det M$, so that $\det M$ is indeed the determinant of the matrix $M^\sigma{}_\rho$.

---

* Such as in the cosmological constant term in chapter VI.2.

Now we can work out how $\mathcal{T}$ transforms. Regarding $\frac{\partial x^\mu}{\partial x'^\rho}$ as the matrix $M^\mu_{\ \rho}$, we obtain

$$\mathcal{T}'(x') = \epsilon^{\rho\sigma} A_\rho(x') B_\sigma(x') = \epsilon^{\rho\sigma} \frac{\partial x^\mu}{\partial x'^\rho} \frac{\partial x^\nu}{\partial x'^\sigma} A_\mu(x) B_\nu(x) = \det\left(\frac{\partial x}{\partial x'}\right) \epsilon^{\mu\nu} A_\mu(x) B_\nu(x)$$

$$= \det\left(\frac{\partial x}{\partial x'}\right) \mathcal{T}(x) \tag{1}$$

We learned that, in spite of $\mathcal{T}$ carrying no indices, it does not transform as a scalar.

How to deal with that pesky determinant in (1)? As we had already noted when we discussed area and volume in chapter I.5, taking the determinant of both sides of the equation $g'_{\rho\sigma}(x') = g_{\mu\nu}(x) \frac{\partial x^\mu}{\partial x'^\rho} \frac{\partial x^\nu}{\partial x'^\sigma}$ telling us how the metric transforms, we obtain $g'(x') = g(x)\left(\det\left(\frac{\partial x}{\partial x'}\right)\right)^2$, where, as always, $g(x) \equiv \det g_{\mu\nu}(x)$ is also somebody who fails to be a scalar.

In a manner reminiscent of our discussion in chapter V.6, we can now form the combination* $\mathcal{T}(x)/\sqrt{g(x)}$, which does transform like a scalar, since $\mathcal{T}'(x')/\sqrt{g'(x')} = \mathcal{T}(x)/\sqrt{g(x)}$.

Convince yourself that while this discussion was carried out for $d = 2$ for the sake of pedagogical clarity, our conclusion holds for any d. (You can write down more Greek letters, can't you?) Specifically, if we have available in a d-dimensional theory an antisymmetric tensor $T_{\mu\nu\lambda\cdots\zeta}(x)$, then we can form a scalar† $\epsilon^{\mu\nu\lambda\cdots\zeta} T_{\mu\nu\lambda\cdots\zeta}(x)/\sqrt{g(x)}$. We are thus free to add to our action the term

$$S_{\text{topological}} \equiv \int d^d x \sqrt{g(x)} \epsilon^{\mu\nu\lambda\cdots\zeta} T_{\mu\nu\lambda\cdots\zeta}(x)/\sqrt{g(x)} = \int d^d x \epsilon^{\mu\nu\lambda\cdots\zeta} T_{\mu\nu\lambda\cdots\zeta}(x) \tag{2}$$

which is invariant under general coordinate transformations.

The point of this discussion is that, remarkably, the volume factor of $\sqrt{g}$ associated with $d^d x$ has disappeared. Indeed, $S_{\text{topological}}$ does not know anything about the metric $g_{\mu\nu}$, and for that matter, it does not even know about the flat Minkowski metric $\eta_{\mu\nu}$. In other words, it does not know about clocks and rulers!

We can stretch and deform spacetime without $S_{\text{topological}}$ noticing anything different: that guy is topological! The physics it describes is sensitive only to the topology of spacetime, not to the metric and the curvature.

Topological terms in gauge theories

Let us illustrate how this works with a theory we know and love, namely 4-dimensional electromagnetism. Indeed, when the professor in a course on electromagnetism showed you the Maxwell action (V.6.18) $S_{\text{Maxwell}} = -\frac{1}{4} \int d^4 x \sqrt{-g} g^{\lambda\mu} g^{\rho\nu} F_{\lambda\rho} F_{\mu\nu}$ (perhaps written

---

* In this chapter, I find it convenient to write, on occasion, $\sqrt{g}$ instead of $\sqrt{-g}$ to minimize clutter.

† Another way of stating this result is that while $\epsilon^{\mu\nu\lambda\cdots\zeta}$ does not transform as a tensor, $\hat{\epsilon}^{\mu\nu\lambda\cdots\zeta}(x) \equiv \epsilon^{\mu\nu\lambda\cdots\zeta}/\sqrt{g(x)}$ does.

only for the Minkowskian case $g_{\mu\nu} = \eta_{\mu\nu}$), you could have raised your hand and asked about adding the term

$$\frac{1}{4} \int d^4x \, \epsilon^{\lambda\rho\mu\nu} F_{\lambda\rho} F_{\mu\nu} = \int d^4x \, \epsilon^{\lambda\rho\mu\nu} \left(\partial_\lambda A_\rho\right) \left(\partial_\mu A_\nu\right) = \int d^4x \, \partial_\lambda \left(\epsilon^{\lambda\rho\mu\nu} A_\rho \partial_\mu A_\nu\right)$$

a term mentioned only in some of the better texts on electromagnetism. Over the past few decades, particle and condensed matter theorists have come to appreciate the role played by this term (an example of the term described in general in (2)) and its various generalizations.

Well, a better professor of electromagnetism might have pointed out that the integrand in this extra term is equal to $\partial_\lambda(\epsilon^{\lambda\rho\mu\nu} A_\rho \partial_\mu A_\nu)$ and is therefore a total divergence. The extra term you clamored for only depends on A_μ at spacetime infinity. Since the action principle involves local variations, this extra term does not contribute to Maxwell's equations of motion, and for this reason is normally not mentioned in standard texts on electromagnetism. (This action has a number of other interesting features, but I do not wish to pursue them here. I might mention only that it is not invariant under time reversal* $t \to -t$ and space reflection† $\vec{x} \to -\vec{x}$.)

The Chern-Simons term in (2+1)-dimensional spacetime

As I emphasized, the discussion so far applies for any d. Suppose we are in $(2+1)$-dimensional, rather than $(3+1)$-dimensional, spacetime, and suppose that physics is governed by the analog of the Maxwell action $S_{\text{Maxwell}} = -\frac{1}{4} \int d^3x \, \sqrt{-g} g^{\lambda\mu} g^{\rho\nu} f_{\lambda\rho} f_{\mu\nu}$, where $f_{\mu\nu} = \partial_\mu a_\nu - \partial_\nu a_\mu$. I write a_μ rather than A_μ here to emphasize that I am not talking about the electromagnetic potential but rather some gauge potential that describes the degree of freedom in some $(2+1)$-dimensional physical situation. In appendix 1, I will tell you that there are $(2+1)$-dimensional condensed matter systems that can be described by a gauge potential a_μ, but for the moment, our discussion is purely theoretical.

As explained, we can now add the topological term (known as the Chern-Simons term[1])

$$S_{\text{CS}} \equiv \frac{k}{4\pi} \int d^3x \, \epsilon^{\lambda\mu\nu} a_\lambda f_{\mu\nu} = \frac{k}{2\pi} \int d^3x \, \epsilon^{\lambda\mu\nu} a_\lambda \partial_\mu a_\nu \tag{3}$$

to the Maxwell action. By the way, using the differential forms introduced in chapter IX.7, we can write this compactly as

$$S_{\text{CS}} = \frac{k}{2\pi} \int a\,da \tag{4}$$

using the identity $dx^\mu dx^\nu dx^\lambda = \epsilon^{\mu\nu\lambda} d^3x$.

---

* I have already mentioned time reversal on several occasions, including chapters III.1 and VIII.1.

† In odd-dimensional space, space reflection (also known as parity) is equivalent to reflection along a particular axis (say, $x^1 \to -x^1$, $x^i \to +x^i$, $i = 2, \cdots, D-1$, D odd) followed by a rotation. To see this, note that the determinants of the transformations involved are variously ± 1.

Long distance dominance

Before we investigate this truly amazing state of affairs further, I need to bring up another important point, namely how various terms behave at long distances. As explained in chapter X.3, at long distances, terms with higher length dimensions (that is, terms with lower mass dimensions, to use the language favored in particle physics) dominate terms with lower length dimensions (that is, terms with higher mass dimensions). Indeed, let's review the argument placed in the present context. Consider a system whose effective field theoretic description[2] of the system is given by

$$S = \frac{k}{2\pi} \int d^3x \left(\epsilon^{\lambda\mu\nu} a_\lambda \partial_\mu a_\nu - l\sqrt{-g} g^{\lambda\mu} g^{\rho\nu} f_{\lambda\rho} f_{\mu\nu} + \cdots \right) \tag{5}$$

namely the sum of a Chern-Simons term, a Maxwell term, and so on.

Since the Maxwell term has two powers of derivatives while the Chern-Simons term has only one (and they both have two powers of the gauge potential a_μ), high school dimensional analysis demands the introduction of a length l characteristic of the system we are studying. When we study physical phenomena on a distance scale of L, the effect of the Maxwell term is thus suppressed by a factor of l/L relative to the Chern-Simons term for $L \gg l$. The ellipsis in (5) indicates terms of even lower length dimensions. They are thus multiplied by even higher powers[3] of l and are even more strongly suppressed at long distances.

Suppose we have some 2-dimensional solid state structure with complicated microscopic physics but such that its long distance degree of freedom is described by a gauge potential $a_\mu(x)$ whose dynamics is gauge invariant (that is, invariant under the transformation $a_\mu \to a_\mu + \partial_\mu \Lambda$). Actual solid state structures are of course not Lorentz invariant. Thus, the Maxwell term in (5) should be replaced by something like $(f_{0i})^2 - \beta(f_{ij})^2$, with β a coefficient determined by the microscopic dynamics. You might expect that the Chern-Simons term would similarly break up into $\epsilon^{ij} a_i \partial_0 a_j + \gamma a_0 \partial_i a_j$). But remarkably, as you can readily verify, gauge invariance fixes the coefficient γ to have precisely the value that allows the 2 terms to combine into $\epsilon^{\tau\mu\nu} a_\tau \partial_\mu a_\nu$. For the Chern-Simons term—but not for the Maxwell term—gauge invariance implies Lorentz invariance.

We conclude that, amazingly, the long distance physics of such a system, if it exists, is topological and does not depend on the nasty microscopic physics (such as band structure and the effect of impurities) that our solid state colleagues revel in. The physics is universal and determined completely by the parameter k. More on this in appendix 1.

Note that in $(3 + 1)$-dimensional spacetime, the added topological term, as described in the preceding section, has the same mass dimensions as the Maxwell term f^2, and hence it does not dominate at long distances. In $(4 + 1)$-dimensional space, the topological term $\epsilon^{\rho\mu\nu\lambda\sigma} a_\rho f_{\mu\nu} f_{\lambda\sigma}$ is less important at long distances than the Maxwell term $\sim f^2$.

It is also worth remarking that with differential forms, we can write the topological term compactly as $(da)^n$ in $d = 2n$-dimensional spacetime, and as $a(da)^n$ in $d = (2n + 1)$-dimensional spacetime.

Appendix 1: Quantum Hall fluid and ground state degeneracy

Is topological field theory merely a theoretical possibility, a curiosity for theorists, or can it be realized physically? In fact, the long distance effective theory of the quantum Hall fluid is topological. Unquestionably, a detailed discussion of the theory of the quantum Hall fluid lies beyond the scope[4] of a textbook on Einstein gravity. Here I limit myself to saying that the long distance physics for a system of electrons confined to 2-dimensional structures in the presence of a strong magnetic field turns out to be given by the Chern-Simons action S_{CS} (3).

A topological field theory must feel peculiarly out of place in a book on gravity and curved spacetime! It doesn't know about the metric, a concept central to Riemannian geometry and Einstein gravity. We learned early on that the energy momentum tensor is defined by varying the action with respect to $g_{\mu\nu}$. What if the action does not depend on $g_{\mu\nu}$? Inescapably, in a topological field theory, the energy momentum tensor and hence the Hamiltonian is identically zero! As I already said, to determine the Hamiltonian we need clocks and rulers.

What does it mean for a quantum system to have a Hamiltonian $H = 0$? Well, when we took a course on quantum mechanics, if the professor assigned an exam problem to find the spectrum of the Hamiltonian 0, we could do it easily! All states have energy $E = 0$. We are ready to hand in the solution.

But the nontrivial problem is to determine how many states there are. This number, known as the ground state degeneracy, depends only on the topology of the manifold, not on whatever metric we might put on the manifold. It is beyond the scope of a book on gravity to calculate this quantum degeneracy, but it turns out[5] to be equal to k^g, where g here denotes the genus of the manifold. (Recall that $g = 0$ for the sphere, $g = 1$ for the torus,* and so on.)

Note that this result implies that k has to be an integer. Otherwise, it would be senseless to say that there are k^g states with $E = 0$. This fascinating phenomenon is known as topological quantization.[6]

Appendix 2: The Hodge star operation on differential forms

Here I discuss the Hodge star operation $*$ on differential forms. While this topic properly belongs in chapter IX.7, I had to postpone it until we have discussed the antisymmetric symbol in curved spacetime. Again, for ease of writing and clarity of presentation, I will specialize temporarily to $d = 2$. The diligent reader can readily generalize the discussion to arbitrary d as we go along.

As in the text, we adopt the convention that $\epsilon^{01} = 1$. You might have noticed that we did not ever have to introduce the totally antisymmetric symbol $\epsilon_{\mu\nu}$ with lower indices, but here it comes finally. We define it by specifying $\epsilon_{01} = -1$. (That ϵ^{01} and ϵ_{01} have opposite signs is to avoid an overall minus sign in the definition of the star operation given below.) With this convention, we have

$$\epsilon^{\mu\nu}\epsilon_{\lambda\rho} = -\delta^\mu_\lambda \delta^\nu_\rho + \delta^\mu_\rho \delta^\nu_\lambda \quad \text{and} \quad \epsilon^{\mu\nu}\epsilon_{\nu\lambda} = +\delta^\mu_\lambda \tag{6}$$

Another identity comes from the definition of the determinant

$$g_{\mu\lambda}g_{\nu\rho}\epsilon^{\lambda\rho} = -g\epsilon_{\mu\nu} \quad \text{and} \quad g^{\mu\lambda}g^{\nu\rho}\epsilon_{\mu\nu} = -\frac{1}{g}\epsilon^{\lambda\rho} \tag{7}$$

as was already mentioned in the text. You can verify (6) and (7) by evaluating them for various values of μ and ν. In particular, in flat spacetime, $\epsilon_{\mu\nu} = \eta_{\mu\lambda}\eta_{\nu\rho}\epsilon^{\lambda\rho}$. Multiplying the first identity in (7) by $g^{\nu\sigma}$, we obtain $g_{\mu\lambda}\epsilon^{\lambda\sigma} = -g\epsilon_{\mu\nu}g^{\nu\sigma}$. Multiplying this by $\epsilon_{\sigma\rho}$ then yields

$$\epsilon_{\mu\nu}g^{\nu\sigma}\epsilon_{\sigma\rho} = -\frac{1}{g}g_{\mu\rho} \tag{8}$$

(Some readers may recognize this as Cramer's rule for finding the inverse of a matrix in this context.)

I use the convention in which $\epsilon^{\mu\nu}$ and $\epsilon_{\mu\nu}$ are numerical, that is, with components given variously by $\pm 1, 0$. The price we pay is that when we raise and lower indices, factors of g will appear as in (6) and (7). We can

---

* Incidentally, it is not as far-fetched as it might seem that theoretical physicists would consider systems living on a torus. If you study a quantum system in a rectangular domain and impose periodic boundary conditions $\psi(x, y) = \psi(x + L, y) = \psi(x, y + L)$ on the wave function, you are effectively putting the system on a torus, namely a square with opposite sides identified.

define, alternatively, $\hat{\epsilon}^{\mu\nu} \equiv \epsilon^{\mu\nu}/\sqrt{-g}$ and $\hat{\epsilon}_{\mu\nu} \equiv \epsilon_{\mu\nu}\sqrt{-g}$, which behave like tensors, as was already mentioned in a footnote. As you can easily imagine, there are advantages and disadvantages to both conventions.

Given a 1-form $V = V_\mu dx^\mu$, define

$$*V = (*V)_\mu dx^\mu \equiv \left(\sqrt{-g}\epsilon_{\mu\lambda}V^\lambda\right)dx^\mu \tag{9}$$

with $V^\lambda = g^{\lambda\rho}V_\rho$ as usual.

Let us compute

$$d*V = \partial_\mu\left(\sqrt{-g}\epsilon_{\nu\lambda}V^\lambda\right)dx^\mu dx^\nu = \partial_\mu\left(\sqrt{-g}V^\mu\right)d^2x = \left(\frac{1}{\sqrt{-g}}\partial_\mu\left(\sqrt{-g}g^{\mu\lambda}V_\lambda\right)\right)\sqrt{-g}d^2x \tag{10}$$

where in the second equality, we used

$$dx^\mu dx^\mu = \epsilon^{\mu\nu}d^2x \tag{11}$$

and (6). We see that the operation $d*$ acting on a 1-form gives us a 0-form proportional to the covariant divergence $D_\mu V^\mu = \frac{1}{\sqrt{-g}}\partial_\mu(\sqrt{-g}g^{\mu\lambda}V_\lambda)$, and so it is clearly going to be useful for physics.

Proceed now to d-dimensional spacetime. Given a p-form V (with $p \leq d$), define the $(d-p)$-form $*V$ by generalizing (9)

$$*V \equiv \left(\sqrt{-g}\epsilon_{\mu_1\cdots\mu_{d-p}\mu_{d-p+1}\cdots\mu_d}V^{\mu_{d-p+1}\cdots\mu_d}\right)dx^{\mu_1}\cdots dx^{\mu_{d-p}} \tag{12}$$

Here we use $\epsilon_{\mu_1\cdots\mu_d}$ instead of $\epsilon_{\mu\lambda}$ (of course!). (Some authors define the $*$ operation with an overall factor $(p!(d-p)!)^{-1}$ in (12). We will not get all uptight about these factors. You can fill them in if you so desire.)

For V a d-form, we obtain the 0-form $*V = (\sqrt{-g}\epsilon_{\mu_1\cdots\mu_d}V^{\mu_1\cdots\mu_d})$. Consider the 0-form denoted by 1. Note the d-form $*1 = \sqrt{-g}\epsilon_{\mu_1\cdots\mu_d}dx^{\mu_1}\cdots dx^{\mu_d}$. We readily check that, up to a factor, $**$ takes 1 back to itself: $**1 = (\sqrt{-g})^2\epsilon_{\mu_1\cdots\mu_d}g^{\mu_1\nu_1}\cdots g^{\mu_d\nu_d}\epsilon_{\nu_1\cdots\nu_d} = ((-g)/(-g))\epsilon_{\mu_1\cdots\mu_d}\epsilon^{\mu_1\cdots\mu_d} = d!$. Indeed, you can check that $**$ takes any p-form back to itself. I will verify a simple case: act with $*$ on (9) to obtain $**V = \sqrt{-g}\epsilon_{\nu\rho}g^{\rho\mu}(\sqrt{-g}\epsilon_{\mu\lambda}V^\lambda)dx^\nu = V_\nu dx^\nu = V$, as claimed (we used (8) in the next to last equality).

Back in chapter IX.7, you learned that the Bianchi identity $dF = ddA = 0$ corresponds to half of Maxwell's equations. You might wonder, in the language of forms, where the other half is, the half sourced by the current J_μ in contrast to the half corresponding to the Bianchi identity. Here is the answer:

$$d*F = *J \tag{13}$$

where $F = dA$ denotes the electromagnetic field strength 2-form, and $J = J_\mu dx^\mu$ the current 1-form. First, check that the two sides match. The right hand side is a $(d-1)$-form. As for the left hand side, $*F$ is a $(d-2)$-form, and so $d*F$ is also a $(d-2)+1 = (d-1)$-form.

The left hand side of our proposed equation contains one power of derivative ∂_ν and one power of $F_{\mu\nu}$, so it has got to be the divergence of the field strength, with possibly some factors of $\sqrt{-g}$ (coming from the definition of $*$) thrown in. Between us friends, it is hardly necessary to verify this claim, but let's do it anyway, for arbitrary d. However, do be a mature adult and not worry about the factorials and signs that are irrelevant for our purposes here. We have

$$d*F = d\left(\sqrt{-g}\epsilon_{\mu_1\cdots\mu_{d-2}\lambda\rho}F^{\lambda\rho}dx^{\mu_1}\cdots dx^{\mu_{d-2}}\right)$$
$$= \partial_\sigma\left(\sqrt{-g}\epsilon_{\mu_1\cdots\mu_{d-2}\lambda\rho}F^{\lambda\rho}\right)dx^\sigma dx^{\mu_1}\cdots dx^{\mu_{d-2}} \tag{14}$$

while

$$*J = \left(\sqrt{-g}\epsilon_{\mu_1\cdots\mu_{d-2}\mu_{d-1}\kappa}J^\kappa\right)dx^{\mu_1}\cdots dx^{\mu_{d-2}}dx^{\mu_{d-1}} \tag{15}$$

Multiplying (14) and (15) by dx^ν and using the generalization of (11) and (6) to d dimensions, we obtain (without worrying about signs and such) $\partial_\mu(\sqrt{-g}F^{\mu\nu}) = \sqrt{-g}J^\nu$, as expected. Note that Maxwell's equations work in any spacetime dimension.

At this point, you might wonder how to write Maxwell's action using forms. The answer is $S = \int F*F$. (Check this!) Note that since $*F$ is a $(d-2)$-form, $F*F$ is a d-form, just ripe for integrating over d-dimensional spacetime. It is easy to add a current: $S = \int F*F + A*J$. The equation of motion we laboriously derived in (13) follows by writing $F = dA$ and varying S with respect to A formally. Again, without worrying about irrelevant factors, we obtain immediately $\delta S = \int(d*F + *J)\delta A = 0$ and hence Maxwell's equations.

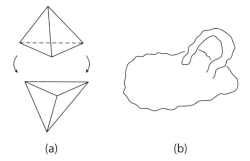

Figure 1 (a) Gluing 2 tetrahedra together. (b) A spherical blob grows a trunk.

What about the Einstein-Hilbert action? Let's try $S = \int R_{\alpha\beta} * (e^\alpha e^\beta)$. Again, the answer is easy to guess. We need a d-form to integrate over, involving 1 power of the curvature 2-form and no power of derivative. To obtain a d-form, we multiply the curvature 2-form by a $(d-2)$-form: the only possibility in this context is the star of the 2-form $e^\alpha e^\beta$. Again, between us friends, do you doubt this? Okay, let's check (but, to save writing and to spare you the stream of indices, only for 4-dimensional spacetime):

$$
\begin{aligned}
R_{\alpha\beta} * \left(e^\alpha e^\beta \right) &= \left(R_{\alpha\beta\omega\psi} dx^\omega dx^\psi \right) \left(\epsilon_{\rho\sigma}^{\ \ \mu\nu} \sqrt{-g} e_\mu^\alpha e_\nu^\beta dx^\rho dx^\sigma \right) \\
&= R_{\mu\nu\omega\psi} \epsilon_{\rho\sigma}^{\ \ \mu\nu} \sqrt{-g} dx^\omega dx^\psi dx^\rho dx^\sigma = R_{\mu\nu\omega\psi} \epsilon_{\rho\sigma}^{\ \ \mu\nu} \sqrt{-g} \epsilon^{\omega\psi\rho\sigma} d^4x \\
&= R_{\mu\nu\omega\psi} g^{\mu\omega} g^{\nu\psi} \sqrt{-g} d^4x = \sqrt{-g} d^4x R
\end{aligned}
\tag{16}
$$

(using the generalization of (6) and again ignoring overall numerical factors), as expected.

Appendix 3: Topological invariants: Euler characteristic, Gauss-Bonnet theorem, and all that

We are now ready to look at some celebrated invariants in topology. Our discussion will be heuristic rather than rigorous, hitting some highlights rather than being exhaustive. I keep the discussion as elementary as possible.

I suspect that many readers probably first encountered the Euler characteristic, like me, in a popular book of mathematics. For me, it was a real eye opener. First, let us look at the empirical data. The cube has 8 vertices or corners, 12 edges (4 on the top, 4 on the bottom, and 4 on the side), and 6 faces. Hence, $V = 8$, $E = 12$, $F = 6$. Next, the tetrahedron has 4 vertices, 6 edges (3 on the bottom and 3 on the side), and 4 faces, that is, $V = 4$, $E = 6$, $F = 4$. We see that the combination, $\chi = V - E + F$, known as the Euler characteristic, is equal to 2 in both cases. (Note that in the sum, we add the number of geometrical entities, vertices, edges, and faces, with an alternating sign according to whether the dimension of the entity is even or odd.) The joke is that theoretical physicists would proclaim this to be a theorem at this point, but in fact it is easy to prove. Let me give a proof that would satisfy most physicists (but not mathematicians) and is suitable for elementary school children.

Glue another tetrahedron to the tetrahedron we have, producing a 6 faced "diamond-shaped" object. See figure 1a. We gain 4 faces (the four of the second tetrahedron) but lose 2 faces (the two that are glued together). Thus, $\Delta F = +4 - 2 = 2$. Similarly, we can see that $\Delta V = +4 - 3 = 1$ and $\Delta E = +6 - 3 = 3$. Hence, we have $\Delta\chi = \Delta V - \Delta E + \Delta F = 1 - 3 + 2 = 0$. Now we can glue zillions of these tetrahedra together to approximate any object we like. At every gluing, the Euler characteristic does not change. Hence $\chi = 2$ for any spherical-looking blob. The discussion makes clear that we are counting the V, E, and F on the surface. (The interior of the object, consisting of the faces we have glued together and their edges and vertices, has been "lost forever," so to speak.) Throughout, we will be studying a surface or a 2-manifold, not the 3-dimensional blob enclosed by the surface.

The Euler characteristic is evidently a topological quantity. We can lengthen and shorten the edges of the tetrahedra we are gluing together without changing V, E, and F.

Suppose we now go beyond deforming the spherical blob to changing its topology. We show presently that χ "measures" the topology of the object.

By gluing more tetradehedra on, we can make the spherical blob grow a trunk like that of an elephant. Let us slowly extend the trunk back toward another part of the blob, sort of like an elephant about to scratch its back with its trunk. The Euler characteristic χ remains equal to 2, until we glue the tetrahedron at the tip of the trunk to a tetrahedron on the back of the elephant. See figure 1b. Now we lose 2 faces, 3 vertices, and 3 edges, and thus $\Delta\chi = \Delta V - \Delta E + \Delta F = -3 + 3 - 2 = -2$. The resulting surface has the topology of a torus and Euler characteristic $\chi = 2 - 2 = 0$.

Growing another trunk and attaching it somewhere else causes the Euler characteristic χ to decrease further, with $\Delta\chi = -2$ every time we do it. We have thus derived the general result $\chi = V - E + F = 2(1 - g)$, where g denotes the genus ($g = 0$ for the sphere, $g = 1$ for the torus, and so on). Some people call the genus the number of "holes." (The torus is said to have 1 hole, but as we will soon see, in this context, the word "genus" or "handle" is preferable to the word "hole.") The Euler characteristic χ is manifestly a topological invariant, independent of the size or "shape" of the surface, and only dependent on its genus.

A trivial generalization is to punctured surfaces. (In everyday parlance, a punctured sphere is a sphere with a hole in its surface, like a rapidly shrinking balloon with a hole in it. This is one reason why the use of the word "hole" for genus or handle, as indulged in by some, is ill advised.) Since we can puncture a surface by removing a triangular face from the surface (so that $\Delta V = 0$, $\Delta E = 0$, and $\Delta F = -1$ and hence $\Delta\chi = -1$), we have, more generally,

$$\chi = 2 - 2g - h \tag{17}$$

with h the number of holes or punctures in the surface.[7]

Another way of proving (17) is to start with $\chi = 2$ for a spherical surface, and to punch any number of holes in it, thus obtaining $\chi = 2 - h$. Deforming the surface to bring 2 holes near each other and then to glue them together, we decrease h by 2 and increase the genus by 1. Hence we have (17). This derivation also makes clear the relative factor of 2 in the coefficients of g and h in χ.

We obtain what is known as a triangulation of the surface. (Indeed, that is what surveyors do: they triangulate the surface of the earth.) At the level of rigor of physics, any surface can be approximated to arbitrary accuracy by making the triangles small enough. In other words, we physicists would take the continuum limit without further ado.

At the same level of rigor, we can also approximate spacetime by a large number of discrete elements. This represents the first step in a program to discretize Einstein gravity and to put it on the computer for numerical analysis.*

That we used tetrahedra is not essential. Take the tetrahedron we started with. Call the vertices on the triangle on its "base" A, B, and C. Pick a point X on the edge joining A and B, and draw a line connecting X to the other vertex C. Then $\Delta V = 1$, $\Delta E = 2$, and $\Delta F = 1$, and so $\Delta\chi = 0$. By "pulling" on the point X, we can deform the tetrahedron, if we feel like it, to a pyramid with a square base. As another example, we can pick a triangle on the surface we are studying and draw a line from one side of the triangle to another side, so that we divide the triangle into a smaller triangle and a quadrilateral. In the process, $\Delta V = 2$, $\Delta E = 3$, and $\Delta F = 1$, and so again $\Delta\chi = 0$. You can make up your own "moves" and show that instead of triangles, the surface could be composed of polygons with any number of sides you like.

Indeed, Descartes had already published, in his progymnasmata to the study of solids, a theorem on angular deficits that foreshadowed the Euler characteristic. Here is what Descartes said. At each vertex of a cube, 3 squares meet. The 3 angles at the vertex add up to $3(\frac{1}{2}\pi) = \frac{3}{2}\pi$. The amount by which this is less than 2π is known as the angular deficit, in this case equal to $(2 - \frac{3}{2})\pi = \frac{1}{2}\pi$. The total angular deficit, namely the sum of the angular deficits at each of the 8 vertices, is then equal to $8(\frac{1}{2}\pi) = 4\pi$. Descartes stated that the total angular deficit of any polyhedron topologically equivalent to the sphere is equal to 4π.

Let's try it for a tetrahedron. At each vertex, 3 equilateral triangles meet, with angles adding up to $3(\frac{\pi}{3}) = \pi$, so that the deficit equals $(2 - 1)\pi = \pi$. There are 4 vertices, and so the total deficit is indeed 4π. We have verified 2 cases, so it is surely a theorem. We are proud physicists, but still, perhaps a third example would be good.

So consider the dodecahedron with 12 pentagonal faces. Since there are $E = (12 \times 5)/2 = 30$ edges, according to Euler's theorem, the number of vertices equals $V = 2 + E - F = 2 + 30 - 12 = 20$. The 12 pentagons have $12 \times 5 = 60$ vertices, and hence $60/20 = 3$ pentagons meet at each vertex. For a regular polygon with n sides, the angle α at each vertex is given by $n\alpha + 2\pi = n\pi$ (to see this, divide the polygon into n triangles), that

---

* Lattice gravity is a thriving subject of research. Historically, this first step is known as the Regge calculus.

is, $\alpha = (n-2)\pi/n$. In particular, for a pentagon, we have $\alpha = \frac{3}{5}\pi$. The angular deficit at each vertex is then $\left(2 - 3\left(\frac{3}{5}\right)\right)\pi = \frac{1}{5}\pi$. Thus, the total angular deficit equals to $20 \times \frac{1}{5}\pi = 4\pi$. Descartes was right!

Let's now show that the total angular deficit is topologically invariant. Consider some triangulated surface as described earlier, and visualize it as a wire framework. (For ease of presentation, let the faces be triangles. Thus, for the cube mentioned above, simply divide the squares by their diagonals.) Pick an edge AB, shared by two triangles ABC_1 and ABC_2. Call the 6 angles contained in these 2 triangles $\alpha_j, \beta_j, \gamma_j$, with $j = 1, 2$, using an almost self-evident notation (for example, α_1 is the angle of ABC_1 at the vertex A, γ_1 is the angle of ABC_1 at the vertex C_1, and so on). Now imagine lengthening the edge AB slightly, thus increasing γ_j and decreasing α_j and β_j. Thus, the change in the angular deficit at the vertex A is $-(\delta\alpha_1 + \delta\alpha_2)$. But since the angles in a triangle have to add up to π, $\delta(\alpha_j + \beta_j + \gamma_j) = 0$. The angular deficit at the vertices A, B, C_1, and C_2 all vary, but the total angular deficit stays the same: it is a topological invariant, as Descartes taught us.

Here is the previous proof dressed up to make it look more sophisticated. Label the vertices by p. At vertex p, a set $s(p)$ of triangles meet. The sum of the angles meeting at vertex p is then $\sum_{i \in s(p)} \alpha_i$, with α_i the angle extended by the ith triangle at that vertex. The total angular deficit is then $\sum_p \left(2\pi - \sum_{i \in s(p)} \alpha_i\right)$. Let us now deform the surface infinitesimally by lengthening or shortening each of the zillions of edges. The variation in the total angular deficit is equal to $-\sum_p \sum_{i \in s(p)} \delta\alpha_i$. Rearranging to sum over 1 triangle at a time, we see that this vanishes, since the sum of the angles in each triangle is constrained to add up to π.

With this background explanation of what the Euler characteristic χ is, we now return to the subject of this appendix and of this chapter. Can we write χ as an integral? In other words, how do we calculate χ, originally defined to be $V - E + F$, in the continuum limit where V, E, and F are not defined?

We are given a closed surface, that is, a 2-manifold M without boundary (in other words, we are setting $h = 0$ for simplicity). What is a 2-form that we could integrate over M? The curvature 2-form $R^{\alpha\beta}$ comes to mind; so let's try $\int_M \varepsilon_{\alpha\beta} R^{\alpha\beta}$. (It is also instructive to try the other possibility: $e^\alpha e^\beta$. It turns out that $\int_M \varepsilon_{\alpha\beta} e^\alpha e^\beta$ is the area of the surface, as you might have guessed from the fact that it does not contain any derivative.) We first work out the integrand in terms of a more elementary notation:

$$\varepsilon_{\alpha\beta} R^{\alpha\beta} = \varepsilon_{\alpha\beta} R^{\alpha\beta}_{\mu\nu} dx^\mu dx^\nu = \varepsilon_{\alpha\beta} e^\alpha_\rho e^\beta_\sigma R^{\rho\sigma}_{\mu\nu} dx^\mu dx^\nu$$

$$= (\det e)\varepsilon_{\rho\sigma} R^{\rho\sigma}_{\mu\nu} \varepsilon^{\mu\nu} d^2x = 2d^2x\sqrt{g}\delta^\mu_\rho \delta^\nu_\sigma R^{\rho\sigma}_{\mu\nu} = d^2x\sqrt{g}R \tag{18}$$

(In the next to last step, we used (6).)

We have known, for quite a while now, that for a sphere of radius a, the scalar curvature $R = 2/a^2$. The area is $\int_M d^2x\sqrt{g} = 4\pi a^2$. Thus, the radius cancels out in the integral $\int_M \varepsilon_{\alpha\beta} R^{\alpha\beta} = 2\int_M d^2x\sqrt{g}R$, and this integral is indeed, as we might suspect, topological in character, equal to some constant like 16π.

The scalar curvature of a torus was calculated back in exercise I.6.2 and again in exercise I.5.16; it assumes both positive and negative values. I will let you check that the integral gives 0 for a torus. We claim that, up to some overall factor, this integral gives the Euler characteristic χ.

In fact, Descartes' theorem is just the statement, up to some overall constant, that the Euler characteristic χ equals 2 for a surface with the topology of the sphere. Go back to the spherical-looking blob. Under the microscope, we see that the surface is formed out of zillions of triangles. Inside any triangle, we have a flat surface; the surface curvature is concentrated on the edges and at the vertices. Indeed, the angular deficit measures our intuitive understanding of curvature: that which we cannot iron flat is curvature. The smaller the angular deficit is at a vertex, the less that vertex sticks out. When the angular deficit vanishes, the surface around that vertex is flat. The angular deficit "measures" the curvature.

The generalization to a 4-manifold M almost immediately suggests itself. Let's parallel the discussion embodied in (18) and integrate the 4-form $\varepsilon_{\alpha\beta\gamma\delta} R^{\alpha\beta} R^{\gamma\delta}$ over M. As in the discussion above, the characteristic length a of M will cancel out in the integral $\int_M \varepsilon_{\alpha\beta\gamma\delta} R^{\alpha\beta} R^{\gamma\delta}$. I will let you have the fun of working this out in an exercise. That this integral is a topological invariant[8] is known as the Gauss-Bonnet theorem in the physics literature.

In the interest of keeping this appendix to a manageable size, I have not proven why the 2 integrals mentioned here are topological invariant,[9] but instead, have "merely" shown you why they must be so (in the spirit of what the American Mathematical Society said about my field theory textbook, as quoted in a footnote in chapter V.6).

Exercises

1   Check Descartes' theorem for the icosahedron, constructed out of 20 equilateral triangular faces. Hint: Use Euler's theorem.

2 Show that the integral $\int_M d^2x\,\sqrt{g}\,R$ vanishes for a torus.

3 Just as in (18), evaluate the 4-form $\varepsilon_{\alpha\beta\gamma\delta}R^{\alpha\beta}R^{\gamma\delta}$ in a more elementary notation.

Notes

1. The first appearance of this term in the physics literature, as far as I know, is in S. Deser, R. Jackiw, and S. Templeton, *Phys. Rev. Lett.* 48 (1982), pp. 975–978.
2. It is implicitly assumed here that such a description is in fact appropriate.
3. Here we implicitly assume that the microscopic physics produces only one length scale l.
4. For a glimpse of this fascinating subject, the interested reader is referred to *QFT Nut*, chapter VI.2. He or she could then move on to more specialized monographs.
5. See, for example, A. Zee, "Quantum Hall Fluids," in *Field Theory, Topology and Condensed Matter Physics*, ed. H. B. Geyer, Springer, 1994, with references to the original literature.
6. For further discussion of topological effects in quantum Hall fluids, see the reference in endnote 5. In particular, a topological quantity called the shift was introduced in X. G. Wen and A. Zee, *Phys. Rev. Lett.* 69 (1992), p. 953, 3600(E).
7. It turns out that this theorem about punctured surfaces is useful in studying RNA folding. See M. Bon, G. Vernizzi, H. Orland, and A. Zee, *J. Mol. Biol.* 379 (2008), p. 900.
8. Thus, this integral can be added to the Hilbert-Einstein action and the resulting action studied. See, for example, I. Low and A. Zee, *Nucl. Phys.* B585 (2000), p. 395; P. Binétruy, C. Charmousis, S. Davis, J. F. Dufaux, *Phys. Lett.* B 544 (2002), p. 185.
9. For the reader eager for a proof, I give a few hints on how to produce a proof, sketched in the briefest possible way. The essential physics (and mathematics) goes back to Faraday's entirely intuitive picture of magnetic flux lines and their conservation. Consider the magnetic flux (or electric flux for that matter) going through a surface $\mathcal{A}$ with boundary $\mathcal{C}$, namely $\int_{\mathcal{A}} d\vec{a} \cdot \vec{B}$, with $d\vec{a}$ an infinitesimal area element. Now distort the surface $\mathcal{A}$ to a surface $\mathcal{A}'$ with the same boundary $\mathcal{C}$. Then Faraday tells us that $\int_{\mathcal{A}} d\vec{a} \cdot \vec{B} = \int_{\mathcal{A}'} d\vec{a} \cdot \vec{B}$. Equivalently, write $0 = (\int_{\mathcal{A}} - \int_{\mathcal{A}'})d\vec{a} \cdot \vec{B} = \int_{S} d\vec{a} \cdot \vec{B}$, where in the last expression $S = \mathcal{A} - \mathcal{A}'$ denotes the closed surface enclosed by $\mathcal{A}$ and $-\mathcal{A}'$. For example, S could be the 2-sphere S^2, with $\mathcal{A}$ and $-\mathcal{A}'$ its northern or southern hemisphere, respectively, and $\mathcal{C}$ the equator. What we just said is simply the elementary fact that the magnetic flux enclosed by S^2 vanishes. Now imagine a magnetic monopole sitting inside S^2. Then the magnetic flux enclosed by S^2 would be equal to 1 in some suitable units. But the conservation of magnetic flux lines now tells us that we could distort S^2 to any surface S with the topology of the sphere, and as long as S encloses the magnetic monopole, the total flux $\int_S d\vec{a} \cdot \vec{B}$ will continue to be equal to 1 regardless of the shape of S. In a sense, this is the first hint that topology is relevant to physics. We can deform the surface S, up to some limit, without changing the total flux S encloses, be it equal to 0, 1, or some other value.

   In the language of forms, the preceding discussion is intimately related to what we touched upon in appendix 2 to chapter IX.7. The electromagnetic 2-form F is closed, that is, $dF = 0$ (corresponding to magnetic flux conservation), but F is only locally—but not globally—exact, that is, $F = dA$ only locally, not globally. Otherwise, according to (IX.7.24), $\int_S F$ would vanish for any surface without a boundary, such as the sphere S^2. The integral $\int_S F$ can be nonzero precisely because, under some circumstances, we cannot define an electromagnetic 1-form A over the entire S, but have to divide S into overlapping "patches." These considerations led Dirac to conclude that $\int_S F$ must be quantized to take on only integer values, just like the Euler characteristic. For details of this argument, which I do not have room to go into here, see, for example, *QFT Nut*, p. 248. To show that the 2 integrals discussed in this appendix are also quantized to take on integer values, we follow essentially the same steps (with d replaced by the covariant D).

A Brief Introduction to Twistors

Twistors

Here I introduce you to twistors. After lying dormant for decades, twistors have recently returned to fundamental physics amid tremendous excitement.[1] Introductory texts on Einstein gravity do not normally cover twistors, but I cannot resist giving readers who have gotten this far at least a flavor of what the recent excitement is about. Actually, you are well equipped, as you will see, by the discussion in, for example, chapters III.3 and VII.2, to embark on a journey into twistor space. We won't get very far, but my hope is that this brief introduction will inspire you to venture deeper into this beautiful subject.

In the following, we will need a few concepts that some readers may be unfamiliar with. For the benefit of these readers, I will collect these topics in appendix 1, which you might want to read first before going on. And of course, if you find these concepts too alien, you could simply skip this chapter.

Also, while I find the mathematical foundation of twistors fascinating and beautiful, here I adopt a down-to-earth and pedestrian approach, dealing with twistors entirely at the "arithmetical level" that most theoretical physicists favor, without inessential mathematical embellishments. I provide the motivation for studying twistors as we move along. (I prefer to avoid the common practice of some writers telling the reader what something is good for before the reader has any idea what that something is.)

The discussion here is restricted to flat spacetime.

Covering the Lorentz group

Given four real numbers $p^\mu = (p^0, p^1, p^2, p^3)$ (which we can regard as the momentum of a particle), consider the matrix

$$
p_{\alpha\dot\alpha} = \begin{pmatrix} p^0 - p^3 & -p^1 + ip^2 \\ -p^1 - ip^2 & p^0 + p^3 \end{pmatrix}_{\alpha\dot\alpha} \tag{1}
$$

Here the two indices α and $\dot{\alpha}$ run over 1, 2. Notice that the matrix p is hermitean and thus can be written as a linear combination of the matrices σ^μ: $p_{\alpha\dot{\alpha}} = -p_\mu(\sigma^\mu)_{\alpha\dot{\alpha}} = (p^0 I - p^i \sigma^i)_{\alpha\dot{\alpha}}$ (see appendix 1). Clearly, there is an unavoidable notational overload: the single letter p denotes both the vector and the matrix, but you should be able to tell from the context which object is being referred to.

By inspection, we see that the determinant of (1), $\det p = (p^0)^2 - (p^1)^2 - (p^2)^2 - (p^3)^2 = -\eta_{\mu\nu}p^\mu p^\nu = -p \cdot p$, is just the Minkowskian square of the 4-momentum p.

Let us now indulge in a few steps of elementary linear algebra. Let L denote an arbitrary 2-by-2 complex matrix with determinant equal to 1. Assuming that L_1 and L_2 are two such matrices, the product $L_1 L_2$ is also a 2-by-2 complex matrix with $\det(L_1 L_2) = (\det L_1)(\det L_2) = 1$. Thus, the set of all such matrices form a group known[2] as $SL(2, C)$ to the cognoscenti, with the letters indicating that this is the special linear group of 2-by-2 matrices over the complex numbers.

Given the matrix p, let us consider

$$p' = L^\dagger p L \tag{2}$$

for some element L of $SL(2, C)$. Manifestly, p' is also hermitean, since $(p')^\dagger = (L^\dagger p L)^\dagger = L^\dagger p^\dagger L = L^\dagger p L = p'$ and thus can be written as $p' = (p'^0 I - p'^i \sigma^i)$. In this way, an element L defines a transformation on 4-vectors, taking p into p'. Now observe that $\det p' = (\det L^\dagger)(\det p)(\det L) = \det p$, or in other words, the transformation preserves the Minkowskian square of the 4-momentum: $p'^2 = p^2$. As you might have expected, it is a Lorentz transformation.

This shows that an element L of $SL(2, C)$ corresponds to an element $\Lambda(L)$ of the Lorentz group $SO(3, 1)$. However, it is not a 1-to-1 map, since L and $-L$ give the same transformation $p \to p'$. Mathematicians say that $SL(2, C)$ double covers $SO(3, 1)$. Furthermore, if L is also unitary, that is, $L^\dagger = L^{-1}$, then from (2) we have $p'^0 = p^0$, and the transformation is a rotation. In other words, the $SU(2)$ subgroup of $SL(2, C)$ double covers the rotation subgroup $SO(3)$ of the Lorentz group $SO(3, 1)$. (Readers familiar with quantum mechanics will recognize that here we are extending and generalizing the standard discussion of how spin $\frac{1}{2}$ particles transform under rotation.) If L is not unitary, we have $p'^0 \neq p^0$, and the transformation involves a Lorentz boost. (For more details, see appendix 1.)

Penrose and the twistor

Penrose, in inventing twistors, was motivated by the thought that, in a general spacetime, lightlike or null lines traced by light might be more fundamental than points. Given your familiarity with Penrose diagrams by now, you will not be surprised that this is one and the same Roger Penrose, who incidentally has also authored a number of well-known popular books. After all, Penrose diagrams emphasize the causal web constructed out of null lines between various events in spacetime. In this chapter, we restrict ourselves to Minkowski spacetime (as I've already mentioned).

So let's consider a lightlike vector p^μ. The preceding discussion simplifies enormously! We have det $p = -p^2 = 0$, and hence the matrix p generically has one zero eigenvalue. (In fancy talk, the matrix has rank 1 rather than 2.) From elementary linear algebra, we recall that a 2-by-2 matrix m of rank 1 can always be written as $m_{ij} = v_i w_j$, with v and w two 2-component vectors (since the vector orthogonal to w provides the zero eigenvector). Thus, for a lightlike vector p^μ, we can write

$$p_{\alpha\dot\alpha} = \lambda_\alpha \tilde\lambda_{\dot\alpha} \tag{3}$$

The two 2-component objects λ and $\tilde\lambda$ are sometimes called helicity spinors. (Our friend the Jargon Guy is beside himself with joy in this chapter.)

Upon first exposure, the formalism appears quite opaque, but actually, like a lot of formalisms, it is fairly simple or perhaps even trivial. If you are confused at any point in the following exposition, just work things out explicitly. For example, consider a physical momentum with $p^0 = E > 0$. With no loss of generality, call the direction of $\vec{p}$ the third axis, so that (with a trivial abuse of notation) $p = \begin{pmatrix} E-p & 0 \\ 0 & E+p \end{pmatrix}$, which for p lightlike collapses to the rank 1 matrix $p = 2E \begin{pmatrix} 0 & 0 \\ 0 & 1 \end{pmatrix} = 2E \begin{pmatrix} 0 \\ 1 \end{pmatrix} (0\ 1)$. Thus, in this case, λ and $\tilde\lambda$ are both equal to $\sqrt{2E} \begin{pmatrix} 0 \\ 1 \end{pmatrix}$ numerically. (To make sure you get it, work this out for $\vec{p}$ pointing in some other direction.)

You can think of the Pauli spinors λ and $\tilde\lambda$ as the* "square root" of the Lorentz vector p^μ.

Another motivation for studying twistors (note that I haven't told you what they are yet!) comes from particle physics, specifically, quantum field theory. Fear not, the only knowledge of quantum field theory I ask of you is minimal. First, just as in quantum mechanics, one of the tasks of quantum field theory is to calculate the scattering amplitude $M(p_1, p_2, p_3, \cdots, p_n)$ involving particles with momentum p_a. (Here we have written the amplitude for a process of the form $p_1 + p_2 \to p_3 + \cdots + p_n$.) The second thing I need you to know is that when we quantize the electromagnetic field, we obtain photons,[†] and when we quantize the gravitational field, we obtain gravitons, something I already mentioned back in chapter IX.4.

In our titanic struggle to tame quantum gravity (see chapter X.8), one (fairly down-to-earth) approach is to study the scattering of gravitons off each other and see what happens. Gravitons are of course, just like photons, massless, and they carry null momentum, so that in the scattering amplitude $M(p_1, p_2, p_3, \cdots, p_n)$, the momenta p_a, for $a = 1, \cdots, n$, are all null. Henceforth, we simply write the amplitude as $M(p_a)$, and in fact, even as $M(p)$. This standard abuse of notation is not as distasteful as you might think, as we will be focusing on one specific momentum at a time. The preceding discussion indicates that we can also write the amplitude as $M(\lambda_a, \tilde\lambda_a)$, or more compactly, $M(\lambda, \tilde\lambda)$.

---

* Some sophisticated readers might realize that this rather nontrivial possibility of taking a square root of a vector is foreordained by the structure of the Lorentz group. In a sense, this represents Dirac's great discovery. See *QFT Nut*, chapter II.3, for example.

† Einstein's Nobel Prize for the photoelectric effect!

Complexification, or two times

Since momentum is characterized by 4 real numbers p^μ, the matrix $p_{\alpha\dot\alpha} = -p_\mu(\sigma^\mu)_{\alpha\dot\alpha}$ is hermitean (indeed, that was how we started this chapter), which implies that $\tilde\lambda = \lambda^*$ is the complex conjugate of λ. The spinor $\tilde\lambda$ is not independent of λ. As students of physics, we know that constraints are, generically, bad news. Theorists (and mathematicians) need more freedom! We prefer to keep the variables λ and $\tilde\lambda$ in $M(\lambda_a, \tilde\lambda_a)$ independent of each other.

Our esteemed experimentalist friends insist that the momentum components p^μ must be real, and they are absolutely right, of course. Theorists, on the other hand, are free* to analytically continue the variables in the scattering amplitude $M(p_1, p_2, p_3, \cdots, p_n)$ to complex values. As you learned in a course on complex analysis, to evaluate an integral over a real variable, it is often useful to analytically continue the integrand into the complex plane and use Cauchy's theorem. We are proceeding in the same spirit here. It is known in quantum field theory (and in quantum mechanics) that scattering amplitudes are analytic functions of their kinematic variables.[3] Thus, quantum field theorists often continue analytically without a moment's thought. Theoretical physicists love that guy Cauchy! Of course, all momenta in the scattering amplitude are to be set back to reality at the end of the calculation.

I invite you to verify that the discussion in this chapter thus far goes through even if p^μ are complex, so that λ and $\tilde\lambda$ are no longer yoked to each other.

An alternative approach is to change the signature of spacetime, from $(-+++)$ to $(--++)$, so that, instead of the Lorentz group $SO(3, 1)$, we consider the group $SO(2, 2)$. (I mentioned the groups $SO(m, n)$ as far back as chapter III.3.) As Minkowski already noted in his famous paper referred to in chapter III.3, it is a simple matter of removing (or adding) an i here and there. Let us strip the Pauli matrix σ^2 (kind of a troublemaker or at least an odd man out) of its i and define (for our purposes here) $\sigma^2 \equiv \begin{pmatrix} 0 & -1 \\ 1 & 0 \end{pmatrix}$. Any real 2-by-2 matrix p (this matrix is of course to be distinguished from the matrix p in (1)) can be decomposed as

$$p = p^4 I + \vec{p} \cdot \vec{\sigma} = \begin{pmatrix} p^4 + p^3 & p^1 - p^2 \\ p^1 + p^2 & p^4 - p^3 \end{pmatrix}$$

with (p^1, p^2, p^3, p^4) four real numbers. Now we have $\det p = (p^4)^2 + (p^2)^2 - (p^3)^2 - (p^1)^2$. Instead of $SL(2, C)$, consider $SL(2, R)$, consisting of all 2-by-2 real matrices with unit determinant. For any two elements L_l and L_r of this group, transform $p \to p' = L_l p (L_r)^T$. Evidently, $\det p' = (\det L_l)(\det p)(\det L_r) = \det p$. Thus, the transformation preserves the quadratic invariant $(p^4)^2 + (p^2)^2 - (p^3)^2 - (p^1)^2$. This shows explicitly that

---

* What we are doing in this chapter is following the three ways of the warrior theorist; see *QFT Nut*, p. 522.

the group $SO(2, 2)$ is locally isomorphic to $SL(2, R) \otimes SL(2, R)$, where the two factors of $SL(2, R)$ reflect the fact that L_l and L_r can be chosen independently of each other.

For a null $SO(2, 2)$ vector p, that is, a real 4-vector such that $(p^4)^2 + (p^2)^2 - (p^3)^2 - (p^1)^2 = 0$, we can write $p_{\alpha\dot\alpha} = \lambda_\alpha \tilde\lambda_{\dot\alpha}$, with λ and $\tilde\lambda$ two independent real spinors. Indeed, λ and $\tilde\lambda$ transform independently, according to

$$\lambda_\alpha \to \left(L_l\right)_\alpha^{\ \beta} \lambda_\beta \quad \text{and} \quad \tilde\lambda_{\dot\alpha} \to \left(L_r\right)_{\dot\alpha}^{\ \dot\beta} \tilde\lambda_{\dot\beta} \tag{4}$$

Incidentally, we are changing the signature of spacetime in the same spirit as complexifying a manifestly real variable. At the end of the calculation, the signature is to be switched back to the physical signature. Nobody is suggesting that we live in a spacetime with two time dimensions and two space dimensions.

Both approaches, complexifying momentum and changing signature, are used in the literature. We will jump back and forth between the two approaches.

Freedom to rescale

You learned in school that the ordinary square root has a sign ambiguity. Analogously, in (3), p does not determine λ and $\tilde\lambda$ uniquely. We can always rescale

$$\lambda \to t^{-1}\lambda \quad \text{and} \quad \tilde\lambda \to t\tilde\lambda \tag{5}$$

for any complex number t. (You might have wondered what fixed the overall constant in λ and $\tilde\lambda$ in the simple explicit example above; I made an arbitrary choice.) This freedom to rescale will play an important role.

By the way, for real momentum, $\lambda = \tilde\lambda^*$, and so the rescaling parameter t is restricted to be a phase factor $e^{i\gamma}$. In this case, the condition that p has rank 1 allows for two solutions: $p_{\alpha\dot\alpha} = \pm\lambda_\alpha\tilde\lambda_{\dot\alpha}$, with the two possible signs corresponding to whether p^0 is positive or negative. With the $SO(2, 2)$ signature, the rescaling parameter t is restricted to be a real number.

It is instructive to count the number of real degrees of freedom for these two different approaches.

A complex lightlike momentum depends on $4 \times 2 - 2 = 6$ real numbers, since the condition $p^2 = 0$ now amounts to two real conditions, while λ and $\tilde\lambda$ each contain 2 complex numbers. But with rescaling, we are left with $2 \times 2 - 1 = 3$ complex numbers, that is, 6 real numbers.

A real lightlike momentum depends on $4 - 1 = 3$ real numbers, while $\tilde\lambda$ and λ each contain 2 complex numbers. But now they are tied to each other, so altogether, they contain 2 complex numbers, which get reduced to 3 real numbers after rescaling by a phase factor.

On the other hand, for a (real) lightlike vector transforming under $SO(2, 2)$, we have 2 real spinors, which after rescaling contain $2 \times 2 - 1 = 3$ real numbers. So it all works out, of course.

Lorentz invariance

In terms of helicity spinors, Lorentz invariance takes on a particularly simple form. Under a Lorentz transformation, $\lambda \to L\lambda$, with L an arbitrary 2-by-2 complex matrix with determinant equal to 1, that is, an element of $SL(2, C)$. Back in chapter III.3, given 2 vectors p and q, we constructed the Lorentz invariant quantity $p \cdot q = \eta_{\mu\nu}p^\mu q^\nu$. Given 2 helicity spinors λ and μ, what is the Lorentz invariant quantity we can construct out of them?

Once we realize that the only property of L that we have to work with is its unit determinant, the answer becomes clear. Define

$$\langle \lambda, \mu \rangle \equiv \varepsilon^{\alpha\beta} \lambda_\alpha \mu_\beta = -\langle \mu, \lambda \rangle \tag{6}$$

with the antisymmetric symbol $\varepsilon^{12} = -\varepsilon^{21} = -1$, $\varepsilon^{11} = \varepsilon^{22} = 0$. Under a Lorentz transformation, we have $\varepsilon^{\alpha\beta} \lambda_\alpha \mu_\beta \to \varepsilon^{\alpha\beta} L_\alpha^{\ \alpha'} L_\beta^{\ \beta'} \lambda_{\alpha'} \mu_{\beta'} = (\det L)\varepsilon^{\alpha'\beta'} \lambda_{\alpha'} \mu_{\beta'} = \varepsilon^{\alpha'\beta'} \lambda_{\alpha'} \mu_{\beta'}$, where we have used the definition of the determinant (as mentioned in chapter X.5). The quantity $\langle \lambda, \mu \rangle$ is manifestly Lorentz invariant.

Similarly, given $\tilde{\lambda}$ and $\tilde{\mu}$, we have the Lorentz invariant

$$[\lambda, \mu] \equiv \varepsilon^{\dot\alpha\dot\beta} \tilde{\lambda}_{\dot\alpha} \tilde{\mu}_{\dot\beta} = -[\mu, \lambda] \tag{7}$$

(A trivial notational remark: You might be inclined to write $[\tilde{\lambda}, \tilde{\mu}]$, but then the twiddles are redundant. The square bracket is defined only for twiddled spinors.)

In contracting helicity spinors, the antisymmetric symbol plays the role as the metric $\eta_{\mu\nu}$ in contracting vectors and tensors. In parallel with the discussion in chapter III.3, we are clearly invited to define helicity spinors with an upper index according to $\mu^\alpha = \varepsilon^{\alpha\beta} \mu_\beta$. Then the invariant $\langle \lambda, \mu \rangle$ can be written as $\lambda_\alpha \mu^\alpha$.

Polarization and helicity

You know that an electromagnetic wave has two polarizations. After quantization, the resulting photon has two helicity states labeled by $+1$ or -1: it can spin either clockwise or counterclockwise around the direction of its 3-momentum. As discussed in chapter IX.4, the situation in gravity is entirely analogous. A gravitational wave has two polarizations, and the graviton has two helicity states labeled by $+2$ or -2. (The photon has spin 1, while the graviton has spin 2, a fact that can be traced back to A_μ and $g_{\mu\nu}$ carrying one and two indices, respectively.) Thus, in talking about the graviton scattering amplitude, I have to specify the helicity of each graviton and write $M(p_1, h_1, p_2, h_2, p_3, h_3, \cdots, p_n, h_n)$ with $h_a = \pm 2$ for $a = 1, \cdots, n$.

Now I have to tell you something about the scattering amplitude $M(\lambda_a, \tilde{\lambda}_a, h_a)$ when expressed in terms of helicity spinors. At this point, I appeal to your knowledge of quantum mechanics. Take a quantum state with angular momentum h around some axis. Rotate it around that axis through an angle ξ. Then the quantum state acquires a phase $e^{ih\xi} = t^{2h}$, where $t \equiv e^{\frac{1}{2}i\xi}$.

Let's focus on a specific particle and omit the subscript a, writing simply $M(\lambda, \tilde{\lambda}, h)$. Imagine rotating the quantum state of that specific particle around the direction of its momentum $\vec{p}$ through an angle ξ. Referring to appendix 1, we see that under this rotation, $\lambda \to t^{-1}\lambda$ and $\tilde{\lambda} \to t\tilde{\lambda}$. (This is a particular case of the rescaling expressed in (5).) Note that the momentum $p = \lambda\tilde{\lambda}$ is left unchanged, as it better be, since we are rotating using $\vec{p}$ as the rotation axis. What I just told you about quantum mechanics states that the scattering amplitude must satisfy

$$M\left(t^{-1}\lambda, t\tilde{\lambda}, h\right) = t^{2h}M\left(\lambda, \tilde{\lambda}, h\right) \tag{8}$$

Keep in mind the suppressed subscript a. By analytic continuation, we argue that this scaling property should hold for arbitrary t and will serve to severely restrict the scattering amplitude. Under rotation, λ and $\tilde{\lambda}$ transform oppositely, and so what we are doing is simply counting the powers of λ minus the powers of $\tilde{\lambda}$ in the scattering amplitude.*

Power of helicity spinors

After talking about scattering amplitudes for all this time, I regret to inform you that we can't actually calculate one. That's kind of a bummer, but to calculate a scattering amplitude, you would have to learn field theoretic methods, such as Feynman diagrams (just as, in nonrelativistic quantum mechanics, to calculate a scattering amplitude you would have to master stuff like perturbation theory), which are way beyond the scope of this book. However, I can impress upon you the power of the helicity spinor formalism.

Instead of graviton scattering, let's talk about the far simpler case of gluon scattering. I already mentioned, in chapter X.1 for example, that the strong interaction is described by a Yang-Mills theory, with the gluon playing the role of the photon. Consider two gluons scattering, ending up with 3 gluons (in the notation used earlier, this is described by $p_1 + p_2 \to p_3 + p_4 + p_5$). We refer to this as 5-gluon scattering. Suppose you want to calculate this to lowest order in perturbation, in the simplest possible case (for example, without any quarks around). If you used the traditional Feynman diagram method, the result contains something on the order of 7,000 terms.[†] When the result is expressed in terms of helicity spinors, the scattering amplitude, for a particular choice of helicities, simplifies dramatically to

$$M(1^-, 2^-, 3^+, 4^+, 5^+) = \frac{\langle 12 \rangle^4}{\langle 12 \rangle \langle 23 \rangle \langle 34 \rangle \langle 45 \rangle \langle 51 \rangle} \delta^{(4)}\left(\sum_{a=1}^{5} \lambda_a \tilde{\lambda}_a\right) \tag{9}$$

---

* This sounds more mysterious than it actually is. The reason for that is because, for fear of confusing some readers, I have not digressed into a discussion of how to write polarization vectors in terms of helicity spinors. See *QFT Nut*, pp. 489 and 493.

[†] Part of the result is shown as a black smudge on p. 484 of *QFT Nut*.

Here I use the compact notation favored in the research literature of writing $M(\cdots, p_a, h_a, \cdots)$ as $M(\cdots, a^{h_a}, \cdots)$ and $\langle \lambda_a, \lambda_b \rangle$ as $\langle ab \rangle$. The delta function specifies that $\sum_{a=1}^{5} p_a = 0$ and hence momentum conservation.*

Behold, several thousand terms have collapsed into one single term! I trust that you are impressed. The point here is not how this amplitude is derived, but how it simplifies drastically when expressed in terms of "correct" variables.

While I cannot derive[4] (9) for you, I can point out that this remarkable expression satisfies all our invariance requirements. Lorentz invariance is satisfied, since $\langle ab \rangle$, as defined in (6), is a Lorentz scalar. Let's check the scaling requirement (8). Letting $\lambda_3 \rightarrow t^{-1}\lambda_3$, we have $M(1^-, 2^-, 3^+, 4^+, 5^+) \rightarrow t^2 M(1^-, 2^-, 3^+, 4^+, 5^+)$. In contrast, letting $\lambda_1 \rightarrow t^{-1}\lambda_1$, we have $M \rightarrow t^{-4} t^2 M = t^{-2} M$. You could see that scaling severely restricts M. Scaling and Lorentz invariance almost fix M uniquely.

The ambitwistor representation

After these many pages, I still haven't told you what a twistor is. I needed to set up helicity spinors first. Finally, we are ready to build twistors out of helicity spinors.

Consider a scattering amplitude M and again focus on the particle a. Write $M(\lambda_a, \tilde{\lambda}_a)$, suppressing the dependence on the other particles. Let us Fourier transform M in two possible ways (and overuse the letter M somewhat):

$$M(W_a) = \int d^2\lambda_a \exp\left(i \tilde{\mu}_a^\alpha \lambda_{a\alpha}\right) M\left(\lambda_a, \tilde{\lambda}_a\right) \tag{10}$$

and

$$M(Z_a) = \int d^2\tilde{\lambda}_a \exp\left(i \mu_a^{\dot\alpha} \tilde{\lambda}_{a\dot\alpha}\right) M\left(\lambda_a, \tilde{\lambda}_a\right) \tag{11}$$

We have defined two 4-component objects (suppressing the subscript a):

$$W \equiv \begin{pmatrix} \tilde{\mu}^\alpha \\ \tilde{\lambda}_{\dot\alpha} \end{pmatrix} \quad \text{and} \quad Z \equiv \begin{pmatrix} \lambda_\alpha \\ \mu^{\dot\alpha} \end{pmatrix} \tag{12}$$

The intent here is to transform M sequentially for $a = 1, 2, \cdots, n$, using either (10) or (11). Consider $SO(2, 2)$ here instead of $SO(3, 1)$, so that the spinors λ and $\tilde{\lambda}$ are real, and hence we can take μ and $\tilde{\mu}$ to be real as well. Thus, these integral transforms are no more and no less than the Fourier transforms you have long been familiar with, and the variable μ is conjugate to the variable $\tilde{\lambda}$ in the same sense that p is conjugate to q in quantum mechanics. The objects W and Z are known as a dual twistor and a twistor, respectively.[5]

What is the point of Fourier transforming and packaging 2-component objects into 4-component objects? One advantage is that the scaling requirement (8) comes out nicer. Instead of λ and $\tilde{\lambda}$ scaling oppositely, we now have, thanks to Mr. Fourier, λ and μ scaling the same way. Similarly for the pair $(\tilde{\mu}, \tilde{\lambda})$.

---

* To write momentum conservation in this form, I have reversed the signs of p_1 and p_2. I have also omitted mentioning various quantum field theoretic notions and technicalities, such as crossing and color stripping.

We find easily that

$$M(tZ, h) = \int d^2\tilde{\lambda} \exp\left(it\mu\tilde{\lambda}\right) M\left(t\lambda, \tilde{\lambda}, h\right) = t^{-2} \int d^2\tilde{\lambda}' \exp\left(i\mu\tilde{\lambda}'\right) M\left(t\lambda, t^{-1}\tilde{\lambda}', h\right)$$

$$= t^{-2(h+1)} M(Z, h) \tag{13}$$

(displaying the helicity h of particle i while suppressing the index a). We used elementary calculus in the second equality and (8) in the third equality. Similarly,

$$M(tW, h) = \int d^2\lambda \exp\left(it\tilde{\mu}\lambda\right) M\left(\lambda, t\tilde{\lambda}, h\right) = t^{-2} \int d^2\lambda' \exp\left(i\tilde{\mu}\lambda'\right) M\left(t^{-1}\lambda', t\tilde{\lambda}, h\right)$$

$$= t^{2(h-1)} M(W, h) \tag{14}$$

This scaling result indicates that we should favor a mixed or ambitwistor representation for the scattering amplitude, using W when the particle carries $+$ helicity and Z when the particle carries $-$ helicity. (In particular, for gluons, $h = \pm 1$, and so we have $M(tW, +) = M(W, +)$ and $M(tZ, -) = M(Z, -)$. We return to this remarkable result in a minute.)

$SL(4, R)$ suddenly appears

Another advantage of the twistor formalism is that these objects (12) with 4 real components (we are still sticking to $SO(2, 2)$ for the moment) naturally invite us to consider transformation under the special linear group $SL(4, R)$ over real numbers. In other words, transform $Z \to \mathcal{L}Z$ with $\mathcal{L}$ a real 4-by-4 matrix with real entries and det $\mathcal{L} = 1$.

But wait! The physics we started out with is supposed to be invariant under $SO(2, 2) = SL(2, R) \otimes SL(2, R)$, evidently a subgroup of $SL(4, R)$. Indeed, from (4), we have

$$Z = \begin{pmatrix} \lambda_\alpha \\ \mu^{\dot{\alpha}} \end{pmatrix} \to \begin{pmatrix} (L_l)_\alpha^{\ \beta} & 0 \\ 0 & (\varepsilon L_r \varepsilon^{-1})^{\dot{\alpha}}_{\ \dot{\beta}} \end{pmatrix} \begin{pmatrix} \lambda_\beta \\ \mu^{\dot{\beta}} \end{pmatrix} \tag{15}$$

where $(\varepsilon L_r \varepsilon^{-1})^{\dot{\alpha}}_{\ \dot{\beta}} = ((L_r^{-1})^T)^{\dot{\alpha}}_{\ \dot{\beta}} = (L_r^{-1})^{\ \dot{\alpha}}_{\dot{\beta}}$ (see appendix 1). (The reader struggling with this material should not be overly concerned with the ε and ε^{-1}; we merely have to raise and lower some spinor indices.) What is important here is that the 4-by-4 matrices in the subgroup $SO(2, 2)$ are constructed by placing 2-by-2 blocks along the diagonal.

Similarly, W transforms under $SL(4, R)$. Indeed, as indicated by the indices matching up, the product $W \cdot Z = \tilde{\mu}^\alpha \lambda_\alpha + \tilde{\lambda}_{\dot{\alpha}} \mu^{\dot{\alpha}}$ is invariant under $SL(4, R)$.

Given more than one W and Z, we also have the Lorentz invariants $Z_1 I Z_2 \equiv <\lambda_1, \lambda_2>$ and $W_1 I W_2 \equiv [\lambda_1, \lambda_2]$. (Here I, in a slightly abused notation used in the literature, evidently denotes either the 4-by-4 matrix $\begin{pmatrix} I & 0 \\ 0 & 0 \end{pmatrix}$ or $\begin{pmatrix} 0 & 0 \\ 0 & I \end{pmatrix}$, depending on whether it acts on W or Z.) Note that these two quantities $Z_1 I Z_2$ and $W_1 I W_2$ are not $SL(4, R)$ invariant.

What is this mysterious group $SL(4, R)$ that contains the "Lorentz" group $SL(2, R) \otimes SL(2, R)$? I will let you figure it out. Here is a hint: Count the number of generators. The group $SL(4, R)$ has $4^2 - 1 = 15$ generators, while $SL(2, R) \otimes SL(2, R)$ has $2(2^2 - 1) = 6$ generators. What could the remaining $15 - 6 = 9$ generators possibly be? Do think for a while.

Power of the ambitwistor

I now show you the power of the ambitwistor. Consider the 4-gluon scattering amplitude $M(1^+, 2^-, 3^+, 4^-)$ to lowest order, the calculation of which using the traditional Feynman diagram method can be done by hand but still involves about 100 terms. As explained above, we should use the variable W for particles 1 and 3, and Z for 2 and 4.

Now apply the remarkable result $M(tW, +) = M(W, +)$ and $M(tZ, -) = M(Z, -)$ we derived following (13) and (14). We have

$$M\left(tW_1^+, Z_2^-, W_3^+, Z_4^-\right) = M\left(W_1^+, tZ_2^-, W_3^+, Z_4^-\right) = M\left(W_1^+, Z_2^-, tW_3^+, Z_4^-\right)$$
$$= M\left(W_1^+, Z_2^-, W_3^+, tZ_4^-\right) = M\left(W_1^+, Z_2^-, W_3^+, Z_4^-\right)$$

Naively, it would appear that $M(W_1^+, Z_2^-, W_3^+, Z_4^-)$ does not depend on W_1, Z_2, W_3, and Z_4 at all. We are tempted to conclude that, in the ambitwistor representation, this scattering amplitude is, up to an irrelevant overall constant, just 1! Not so fast, though. It could also be -1. The sign depends on which kinematic regime we are in. More carefully, we conclude that

$$M\left(W_1^+, Z_2^-, W_3^+, Z_4^-\right) = \text{sign}\left(W_1 \cdot Z_2\right) \text{sign}\left(Z_2 \cdot W_3\right) \text{sign}\left(W_3 \cdot Z_4\right) \text{sign}\left(Z_4 \cdot W_1\right) \qquad (16)$$

As an exercise, you can Fourier transform back to the λ and $\tilde{\lambda}$ representation.

The result (16) is truly amazing: it tells us that the 4-gluon scattering amplitude, when written in appropriate variables, is just equal to $+1$ or -1, depending on the kinematic regime. The 100 or so terms in the Feynman approach, alluded to above, are struggling to tell us that they will sum up to ± 1 when we translate everything into the language of twistors.

Interaction among gravitons

I hope that by these examples, I have convinced you plenty that the traditional Feynman diagram approach is almost hopeless when it comes to gluon scattering. The situation with gravity is far worse.

Back in chapter IX.5, I mentioned that if we plug $g_{\mu\nu} = \eta_{\mu\nu} + h_{\mu\nu}$ into the Einstein-Hilbert action and expand to $O(h^3)$, we obtain cubic terms of the form $h\partial h\partial h$, with indices suppressed. Since there are 8 indices contracted every which way, the schematic form $h\partial h\partial h$ actually contains many terms. I also explained that the infinite number of terms of the form $h \cdots h\partial h\partial h$ describe the complicated interaction of many gravitons with one another. In the traditional Feynman approach in quantum field theory, one Fourier transforms $h\partial h\partial h$ to momentum space to obtain the interaction amplitude $M(p_1, h_1, p_2, h_2, p_3, h_3)$ with (p_a, h_a) the momentum and helicity of the 3 interacting gravitons. Take my word for it, the whole thing is a horrible mess.

What does the basic cubic vertex for gravity come out to be in the ambitwistor representation? Well, the scaling relations (13) and (14) tell us that, for $h = \pm 2$, we have

$M(tW, ++) = t^2 M(W, ++)$ and $M(tZ, --) = t^2 M(Z, --)$. Thus $M(Z_1^{--}, Z_2^{--}, W_3^{++})$ must be quadratic in Z_1, in Z_2, and in W_3. The only possibility[6] is

$$M\left(Z_1^{--}, Z_2^{--}, W_3^{++}\right) = \left| (Z_1 \cdot W_3)(Z_2 \cdot W_3)(Z_1 I Z_2) \right| \tag{17}$$

This amplitude describes the basic interaction of gravitons with one another. (If you are not that impressed, it is because you have never dealt with the mess referred to in the preceding paragraph.) In other words, this cubic vertex, as expressed in the language of twistors in (17), embodies the Einstein-Hilbert action, and thus, in some sense, provides a compact summary of this entire book.

Another extragalactic fable suggests itself. In some other civilization, after the discovery of special relativity, some mathematically inclined physicist could have written Lorentz vectors in terms of helicity spinors and then constructed twistors out of them. Another theorist showed that a massless spin 2 particle generates the inverse square law of gravity. The cubic vertex for 3 interacting gravitons (17) could then be written down, and then Fourier transformed back to an expression involving helicity spinors. Expressing this in terms of momentum and then Fourier transforming to spacetime, some bright young guy could have discovered Einstein gravity (and then Riemannian geometry while he or she was at it) via this route!

By the way, did you figure out what the group $SL(4, R)$ is? If you didn't, you should have remembered chapter IX.9. Its 9 extra generators not in the Lorentz algebra of $SL(2, R) \otimes SL(2, R)$ describe 4 translations, 4 conformal transformations, and 1 dilation, corresponding to

$$\begin{pmatrix} 0 & 0 \\ X & 0 \end{pmatrix}, \quad \begin{pmatrix} 0 & X \\ 0 & 0 \end{pmatrix}, \quad \text{and} \quad \begin{pmatrix} I & 0 \\ 0 & -I \end{pmatrix} \tag{18}$$

respectively. (Here X denotes the 4 linearly 2-by-2 matrices, including the identity.)

Where is spacetime?

In our discussion, we approached twistors by a purely utilitarian approach. We express the physics in terms of ever shinier and better variables, from p^μ to $p_{\alpha\dot\alpha}$, to λ and $\tilde\lambda$, and then to W and Z.* In this pedestrian approach, the beautiful geometric essence† of twistors is completely obscured.

We have been acting mostly like particle physicists, talking about scattering amplitudes and living happily in momentum space. But where is the spacetime we know and love hiding in these scattering amplitudes?

---

* In the literature, people have gone one step further to supertwistors $\mathcal{W}$ and $\mathcal{Z}$ by adjoining Grassmannian variables. This subject naturally invites the inclusion of supersymmetry, upon which it becomes, perhaps not surprisingly, even more elegant and compact.

† The geometric origin of twistors has been illuminated by R. Penrose, A. Hodges, and others.

Well, Emmy Noether gives us a hint. All scattering amplitudes contain the momentum conservation delta function, but we learned way way back in chapter I.2 that momentum conservation, according to Noether's theorem, encodes the translation invariance of spacetime. Hence, one starting point might be to plug the scattering amplitude (9) into the Fourier transform (11) and watch what happens. Replace the 4-dimensional delta function $\delta^{(4)}(p)$ in (9) by its integral representation[7] $\int d^4 X e^{-ip \cdot X} = (2\pi)^4 \delta^{(4)}(p)$ to obtain

$$
\begin{aligned}
M(Z_a) &= f(\lambda) \int \prod_a d^2\tilde{\lambda}_a \, e^{i\mu_a^{\dot{\alpha}}\tilde{\lambda}_{a\dot{\alpha}}} \delta^{(4)} \left(\sum_a \lambda_{a\alpha}\tilde{\lambda}_{a\dot{\alpha}} \right) \\
&= f(\lambda)(2\pi)^{-4} \int \prod_a d^2\tilde{\lambda}_a \, e^{i\mu_a^{\dot{\alpha}}\tilde{\lambda}_{a\dot{\alpha}}} \int d^4 X \, e^{-iX^{\alpha\dot{\alpha}}\left(\sum_a \lambda_{a\alpha}\tilde{\lambda}_{a\dot{\alpha}}\right)} \\
&= f(\lambda) \int d^4 X \prod_a \delta^{(2)} \left(\mu_a^{\dot{\alpha}} - X^{\alpha\dot{\alpha}}\lambda_{a\alpha} \right)
\end{aligned}
\tag{19}
$$

In this context, we could care less[8] about the factor $f(\lambda) = \langle 12 \rangle^4/(\langle 12 \rangle \langle 23 \rangle \langle 34 \rangle \langle 45 \rangle \langle 51 \rangle)$ from (9), which describes gluon scattering in detail and which the ~7,000 terms in the Feynman approach were desperately trying to sum up to. If we want to compare with experimental data on gluon scattering, we need $f(\lambda)$, but that's not what we want to do. Instead, we want to find spacetime!

What we have learned from (19) is that the two spinors λ and μ, contained in each of the twistors $Z_a = (\lambda_a, \mu_a)$, are constrained by the equality $\mu_a^{\dot{\alpha}} = X^{\alpha\dot{\alpha}}\lambda_{a\alpha}$. The variable X appears as the Fourier dual of the momentum p and so quite plausibly should be interpreted as a spacetime coordinate.

We have found our beloved spacetime: X is the thing that connects λ and μ.

Let's give a simpler example to bolster our case. Consider the wave equation $\partial^2 \phi = 0$. The solution is given by the integral representation $\phi(X) = \int d^4 p \delta(p^2) f(p) e^{ipX}$, where $f(p)$ is some smooth function we don't particularly care about in this context. That $\phi(X)$ satisfies the wave equation is because of the delta function $\delta(p^2)$: namely $\partial^2 \phi(X) = -\int d^4 p p^2 \delta(p^2) f(p) e^{ipX} = 0$.

The presence of the delta function allows us to express p in terms of λ and $\tilde{\lambda}$, as in (3), and to write

$$
\begin{aligned}
\phi(X) &= \int_P d^2\lambda d^2\tilde{\lambda} f(\lambda, \tilde{\lambda}) e^{iX\lambda\tilde{\lambda}} = \int_P d^2\lambda d^2\tilde{\lambda} e^{iX\lambda\tilde{\lambda}} \int d^2\mu \hat{f}(\lambda, \mu) e^{-i\mu\tilde{\lambda}} \\
&= \int_P d^2\lambda \int d^2\mu \hat{f}(\lambda, \mu) \int d^2\tilde{\lambda} e^{i(X\lambda-\mu)\tilde{\lambda}} = (2\pi)^2 \int_P d^2\lambda \int d^2\mu \hat{f}(\lambda, \mu) \delta^2(\mu - X\lambda) \\
&= (2\pi)^2 \int_P d^2\lambda \hat{f}(\lambda, X\lambda)
\end{aligned}
\tag{20}
$$

(We have to mention a technical detail that doesn't much matter for the main point we are trying to make: the subscript P on the integral sign indicates that we are really integrating over projective space due to the rescaling freedom in (5). Concretely, this simply means that we can set one of the components in λ to 1 by scaling. This makes sense, since the integral over p we started with was over 3 real variables due to $\delta(p^2)$.)

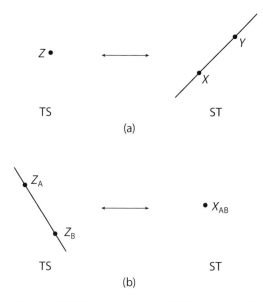

Figure 1 (a) Points in twistor space (TS) represent null lines in spacetime (ST). (b) A line in twistor space corresponds to a point in spacetime.

These two examples indicate that, given a twistor $Z = (\lambda, \mu)$, we can define a point or an event in spacetime by

$$\mu^{\dot{\alpha}} = X^{\alpha\dot{\alpha}}\lambda_{\alpha} \tag{21}$$

The geometry of twistor space

Spacetime has appeared, but is the solution to (21) unique? Physicists often neglect to ask such refined questions, but here it is crucial to bow to the mathematicians. Suppose that there also exists a Y satisfying $\mu^{\dot{\alpha}} = Y^{\alpha\dot{\alpha}}\lambda_{\alpha}$. Subtracting, we obtain $(X - Y)^{\alpha\dot{\alpha}}\lambda_{\alpha} = 0$, which tells us that the 2-by-2 matrix $(X - Y)$ has a zero eigenvalue, and hence $\det(X - Y) = 0$. This in turn tells us that the vector $(X^{\mu} - Y^{\mu})$ is lightlike or null. In other words, given a solution X, any point Y in spacetime null separated from X is also a solution. A point in twistor space (TS in figure 1) corresponds to a null line in spacetime (ST in figure 1) going through points X defined by (21). See figure 1a.

Points in twistor space thus represent null lines in spacetime. This fact realizes Penrose's vision of a representation in which light rays are somehow more fundamental than spacetime events. Notice that if we scale the twistor $Z = (\lambda, \mu)$ by any complex number t, that is, let $Z \rightarrow tZ$, then the solution X of (21) remains unchanged.

At this point, it is also convenient to follow the mathematicians and complexify Z, that is, think of $Z = (\lambda, \mu)$ as 4 complex, rather than 4 real, numbers. Then the matrix X will in general not be hermitean, and the corresponding X^{μ} are complex, describing a complexified Minkowski spacetime. (Because of the scaling freedom, Z does not actually live in the 8-dimensional space C^4, but in the 6-dimensional complex projective space denoted by CP^3.) Note that since a line is characterized by two points that lie on it (call them X and Y), we can think of the point-to-line map from twistor space to spacetime as a 1-to-2 map $Z \to (X, Y)$.

If a point in twistor space corresponds to a null line in spacetime, what does a line in twistor space correspond to in spacetime? You might want to think about this before reading on.

Consider the straight line in twistor space going through two points Z_A and Z_B. The two points describe two null lines in spacetime. Do they intersect? In other words, do the two equations $\mu_A = X\lambda_A$ and $\mu_B = X\lambda_B$ share an X as a common solution?

They do, and the common solution is given by

$$X_{AB}^{\alpha\dot\alpha} = \frac{\lambda_A^\alpha \mu_B^{\dot\alpha} - \lambda_B^\alpha \mu_A^{\dot\alpha}}{\langle \lambda_A \lambda_B \rangle} \tag{22}$$

which you can verify by direct substitution. (Recall that the "metric" for spinor indices is antisymmetric.) Actually, this solution is essentially fixed by symmetry and scaling considerations. For example, from $\mu_A = X\lambda_A$ and $\mu_B = X\lambda_B$, we see that if we scale $\mu_A \to t\mu_A$ and $\mu_B \to t\mu_B$, then clearly we have $X \to tX$. On the other hand, under $\lambda_A \to t\lambda_A$ and $\lambda_B \to t\lambda_B$, we should have $X \to t^{-1}X$. Also, X should be symmetric under $A \leftrightarrow B$.

Thus, we have a map $(Z_A, Z_B) \to X_{AB}$. Indeed, take any point $Z_C = uZ_A + (1 - u)Z_B$, for u an arbitrary complex number, and you can easily show that $X_{AC}^{\alpha\dot\alpha} = X_{AB}^{\alpha\dot\alpha}$. Thus, rather pleasingly, a line in twistor space corresponds to a point in spacetime. See figure 1b.

To summarize, a point in twistor space corresponds to a null line in spacetime, and a line in twistor space corresponds to a point in spacetime. Cool, eh?

Incidentally, (22) indicates that two complex null lines in complex Minkowski spacetime generically intersect. Note that this is not true of two arbitrary null lines in real Minkowski spacetime.

Now that we have defined points and straight lines joining two points in twistor space, we can go on to study planes, triangles, polygons, tetrahedrons, polyhedrons, and more generally, polytopes, in direct analogy to the familiar objects in Euclidean space. In a truly amazing discovery, Hodges[9] realized that the scattering amplitudes we have been talking about can be interpreted as the volumes of polytopes in momentum-twistor space.*

Appendix 1: A quick review of matrix algebra

As promised, here I go over some concepts that you may be unfamiliar with. I hate to lose anybody who has gotten this far.

The hermitean conjugate of a complex matrix M, written as $M^\dagger$, is defined to be the complex conjugate of its transpose, thus $M^\dagger = (M^T)^*$. The matrix M is said to be hermitean if it is equal to its hermitean conjugate:

---

* Explaining what momentum-twistor space is would take us too far beyond the scope of an introductory textbook on gravity.

$M = M^\dagger$. Apply this to an arbitrary 2-by-2 complex matrix:

$$\begin{pmatrix} u & v \\ w & z \end{pmatrix} = \begin{pmatrix} u & w \\ v & z \end{pmatrix}^* = \begin{pmatrix} u^* & w^* \\ v^* & z^* \end{pmatrix} \tag{23}$$

The matrix is hermitean if u and z are real, and $v = w^*$. Thus, given four real numbers $p^\mu = (p^0, p^1, p^2, p^3)$, the matrix in (1) is indeed the most general 2-by-2 hermitean matrix. Define the three Pauli matrices

$$\sigma^1 = \begin{pmatrix} 0 & 1 \\ 1 & 0 \end{pmatrix}, \quad \sigma^2 = \begin{pmatrix} 0 & -i \\ i & 0 \end{pmatrix}, \quad \sigma^3 = \begin{pmatrix} 1 & 0 \\ 0 & -1 \end{pmatrix} \tag{24}$$

These three matrices, together with the 2-by-2 identity matrix I (for convenience, define $\sigma^0 \equiv I$), form a complete basis in the sense that any 2-by-2 hermitean matrix can be written as $(p^0 I - p^i \sigma^i) = -p_\mu \sigma^\mu$, with 4 real numbers $p_\mu = \eta_{\mu\nu} p^\nu$. Our convention is such that the index on a Pauli matrix can be freely raised and lowered, for example $\sigma_2 = \sigma^2$. The product of two Pauli matrices is given by

$$\sigma^i \sigma^j = \delta^{ij} + i\varepsilon^{ijk}\sigma^k \tag{25}$$

which you can verify by direct computation. Here $\varepsilon^{123} = +1$ denotes the antisymmetric symbol.

Write $\vec{\sigma} = (\sigma^1, \sigma^2, \sigma^3)$, and let $\vec{z} = (z^1, z^2, z^3)$ be 3 complex numbers. Verify that $(\vec{z} \cdot \vec{\sigma})^2 = \sum_{i,j=0}^3 z^i z^j \sigma^i \sigma^j = \sum_{i,j=0}^3 z^i z^j (\delta^{ij} + i\varepsilon^{ijk}\sigma^k) = \vec{z}^2$. Define $|\vec{z}| = (\vec{z}^2)^{\frac{1}{2}}$ and $\hat{z} = \vec{z}/|\vec{z}|$. (Note that $|\vec{z}|$ defined here is in general a complex number.) Expand in the usual Taylor series the exponential $L = e^{i\vec{z}\cdot\vec{\sigma}} = \sum_{n=0}^\infty (i\vec{z} \cdot \vec{\sigma})^n/n! = \cos|\vec{z}| + i\hat{z} \cdot \vec{\sigma} \sin|\vec{z}|$, where we arrived at the last step by separating the sum into two sums, one over even n, the other over odd n. Using this result, you can check that $\det L = 1$. I suspect that many readers have seen this for $\vec{z}$ a real vector. If you have never seen Pauli matrices before, you might wish to skip this chapter entirely at a first reading and come back to it later.

Fine. Now let us go back to the transformation (2): $p' = L^\dagger p L$. As explained in the text, this produces a Lorentz transformation of p^μ into p'^μ, provided that $\det L = 1$. Using the representation $L = e^{i\vec{z}\cdot\vec{\sigma}}$, we see that if $\vec{z}$ is real, the transformation is a rotation, while if $\vec{z}$ is imaginary, the transformation is a boost. Work this out!

We can also count. The statement $\det L = 1$ imposes 2 real conditions on a matrix with 4 complex numbers, so that L is characterized by $8 - 2 = 6$ real numbers. In other words, the Lie algebra of $SL(2, C)$ has 6 generators. On the other hand, we know that the Lorentz transformations consist of 3 rotations and 3 boosts. Indeed, the discussion just given already indicates what the precise correspondence is.

In the text, we raise spinor indices with the antisymmetric symbol* $\varepsilon^{\alpha\beta}$ according to $\mu^\alpha = \varepsilon^{\alpha\beta}\mu_\beta$. We wish to lower spinor indices with $\varepsilon_{\alpha\beta}$ according to $\mu_\gamma = \varepsilon_{\gamma\alpha}\mu^\alpha$. This requires $\varepsilon_{\gamma\alpha}\varepsilon^{\alpha\beta} = \delta_\gamma^\beta$ and thus $\varepsilon_{12}\varepsilon^{21} = 1 = -\varepsilon_{12}\varepsilon^{12}$. Thus, we have to define ε_{12} and ε^{12} with opposite signs, which leads to all kinds of pesky signs when dealing with spinors. I would advise you not to worry too much about signs when reading this chapter.[10] Typically, in this subject in particular, and in theoretical physics in general, the relative signs matter, but not the overall signs (unless you are building a bridge or something like that).

A useful identity is $\sigma_2 \sigma_i^T \sigma_2 = -\sigma_i$, which you can verify by evaluating this expression for the three different values of i. In parallel with $\sigma^\mu = (I, \vec{\sigma})$, define $\bar{\sigma}^\mu = (I, -\vec{\sigma})$, then $\text{tr }\sigma^\mu\bar{\sigma}^\nu = -2\eta^{\mu\nu}$. Using this identity, we can show that the scalar product of two vectors p and q is given by

$$-2p \cdot q = \varepsilon^{\alpha\beta}\varepsilon^{\dot{\alpha}\dot{\beta}}p_{\alpha\dot{\alpha}}q_{\beta\dot{\beta}} = -p_{\alpha\dot{\alpha}}\varepsilon^{\dot{\alpha}\dot{\beta}}q_{\dot{\beta}\beta}^T\varepsilon^{\beta\alpha} = \text{tr}(p\sigma_2 q^T\sigma_2) = p_\mu q_\nu \text{ tr}(\sigma^\mu\bar{\sigma}^\nu) \tag{26}$$

For $q = p$, we see, upon recognizing the definition of the determinant, that this reduces to $p \cdot p = \varepsilon^{\alpha\beta}\varepsilon^{\dot{\alpha}\dot{\beta}}p_{\alpha\dot{\alpha}}p_{\beta\dot{\beta}} = \det p$.

Appendix 2: Inversion

In chapter IX.9, we discussed the inversion of spacetime. Perhaps it would not surprise you that inversion comes out quite elegantly in the language of twistors. From (19), we learned that the spacetime coordinates X are

---

* Do not confuse the two different antisymmetric symbols ε^{ijk} and $\varepsilon^{\alpha\beta}$! The former carries vector indices $i, j, k = 1, 2, 3$, while the latter carries spinor indices $\alpha, \beta = 1, 2$.

determined by the relation (with the irrelevant index a suppressed)

$$\mu_{\dot{\alpha}} = X_{\alpha\dot{\alpha}}\lambda^{\alpha} \qquad (27)$$

It follows* that $X^{\beta\dot{\alpha}}\mu_{\dot{\alpha}} = X^{\beta\dot{\alpha}}X_{\alpha\dot{\alpha}}\lambda^{\alpha} = X^2\lambda^{\beta}$. We obtain

$$\lambda^{\beta} = \frac{X^{\beta\dot{\alpha}}}{X^2}\mu_{\dot{\alpha}} \qquad (28)$$

Thus, inversion corresponds to the interchange $\lambda \leftrightarrow \mu$.

Exercises

1 For two lightlike vectors p and q, write $p_{\alpha\dot{\alpha}} = \lambda_{\alpha}\tilde{\lambda}_{\dot{\alpha}}$ and $q_{\alpha\dot{\alpha}} = \mu_{\alpha}\tilde{\mu}_{\dot{\alpha}}$. Calculate the Lorentz scalar product $p \cdot q$ in terms of λ, μ, $\tilde{\lambda}$, and $\tilde{\mu}$.

2 Fourier transforming (16), show that the 4-gluon scattering is

$$M(1^+, 2^-, 3^+, 4^-) = \frac{\langle 24 \rangle^4}{\langle 12 \rangle \langle 23 \rangle \langle 34 \rangle \langle 41 \rangle}$$

to lowest order. Comparing with (9), do you see a pattern?

3 Fourier transform the cubic vertex for gravity (17) to show that

$$M\left(1^{--}, 2^{--}, 3^{++}\right) = \left(\frac{\langle 12 \rangle^4}{\langle 12 \rangle \langle 23 \rangle \langle 31 \rangle}\right)^2 \qquad (29)$$

Notes

1. See, for example, http://online.kitp.ucsb.edu/online/qcdscat11/. For a pedagogical introduction, see chapters N.2–4 in *QFT Nut*.
2. Names are not so important; I give them just so that you can chat at a cocktail party.
3. See, for example, *QFT Nut*, chapter III.8.
4. It is derived using a recursion technique explained in, for example, *QFT Nut*, chapters N.2 and N.3.
5. Or vice versa, as you like.
6. The need for the absolute value involves an argument that we cannot go into here. Here is a cryptic explanation almost designed to add to your puzzlement: starting with the representation $\delta(x) = (2\pi)^{-1}\int dp e^{ipx}$, we can write formally $\text{sign}(x) = 2(2\pi)^{-1}\int dp e^{ipx} p^{-1}$ and $|x| = 2(2\pi)^{-1}\int dp e^{ipx} p^{-2}$, facts that you can verify by differentiating these two integrals with respect to x. The statement is that the second integral in this sequence appears in Yang-Mills theory, while the third appears in Einstein gravity. I refer you to N. Arkani-Hamed, F. Cachazo, C. Cheung, and J. Kaplan, arXiv:0903.2110v2, for more details.
7. As explained in endnote 4 in chapter X.2, we have $\delta(x) = \frac{1}{2\pi}\int_{-\infty}^{\infty} dk e^{-ikx} = \lim_{K\to\infty}\frac{1}{2\pi}\int_{-K}^{K} dk e^{-ikx} = \lim_{K\to\infty}\frac{1}{2\pi}\int_{-K}^{K} dk \cos(kx) = \lim_{K\to\infty}\sin(Kx)/(\pi x)$. The representation used in the text is the 4-dimensional generalization of this representation: $\delta^{(4)}(x) = \delta(t)\delta(x)\delta(y)\delta(z) = \int \frac{d^4k}{(2\pi)^4} e^{-ikx}$ with $kx = \eta_{\mu\nu}k^{\mu}x^{\nu}$ and a slight abuse of notation (making x represent more than one thing).
8. And even less about the $(2\pi)^4$.
9. A. Hodges, arXiv:0905.1473; N. Arkani-Hamed, J. Bourjaily, F. Cachazo, A. Hodges, and J. Trnka, arXiv: 1012.6030.
10. Readers worried about signs are referred to appendix E in *QFT Nut*, second edition, in which the signs are allegedly correct. It contains additional material related to the discussion in this appendix.

---

* Remember to raise and lower spinor indices with the antisymmetric symbol! (Here it is $X^2 = \eta_{\mu\nu}X^{\mu}X^{\nu}$ as usual.)

X.7 | The Cosmological Constant Paradox

The graviton knows about everything

Gravity knows about everything, whatever its origin, luminous or dark, even the energy contained in fluctuating quantum fields.

This omniscience of gravity lies at the root of the gravest, or if you prefer, one of the gravest, puzzles of theoretical physics, namely the cosmological constant paradox.[1] According to quantum field theory, spacetime is a boiling sea of quantum fluctuations, and according to Einstein, gravity should know all about this.

Allegedly, quantum field theorists can reliably estimate the energy density of this boiling sea, but somehow the theoretical value they are led to disagrees enormously with observation, and by enormously, we are not talking about a mere few orders of magnitude. In this chapter, we discuss how this dismal and embarrassing situation for theoretical physics comes about.

The vacuum as a boiling sea of quantum fluctuations

In chapter VI.2, I already mentioned quantum fluctuations contributing to the cosmological constant. Another way of expressing this is that in quantum field theory, good old Minkowskian spacetime is unstable. It gets driven to de Sitter spacetime.

A full understanding of quantum fluctuations requires some acquaintance with quantum field theory, but you can readily grasp the origin of the cosmological constant paradox with a rudimentary knowledge of quantum mechanics. Consider the harmonic oscillator. Classically, a mass attached to a spring attains its lowest energy, namely 0 by definition, when it sits quietly at the bottom of the potential well. In quantum mechanics, however, due to the Heisenberg uncertainty principle, there is constant and unavoidable fluctuation in the particle's position, and the lowest energy the particle can attain is not 0 but

$\frac{1}{2}\hbar\omega$, where ω denotes the (circular) frequency of the oscillator.[2] This irreducible amount of energy is known as the zero point energy.

You probably know that the electromagnetic field can be treated as a superposition of waves known as modes, each with some wave vector $\vec{k}$ (defined by the inverse of the wavelength λ times 2π) and vibrating with frequency $\omega(\vec{k}) = c|\vec{k}|$. (Recall also chapters II.3 and IX.4.) Each of these modes corresponds to a harmonic oscillator.* When the electromagnetic field is quantized, each mode contributes $\frac{1}{2}\hbar\omega(\vec{k})$ to the zero point energy. This represents the minimum amount of energy in any given mode, present even when the electromagnetic field is not excited, in the same way that $\frac{1}{2}\hbar\omega$ represents the energy of the harmonic oscillator even when it is not excited. In other words, in quantum electrodynamics, the electromagnetic field contributes an energy to spacetime even when there is no electromagnetic field present! This energy verily deserves the name vacuum energy.

To determine the total vacuum energy, we simply sum over all modes, one for each value of $\vec{k}$: thus, $E_{vacuum} \sim \sum_{\vec{k}} \hbar\omega(\vec{k}) \sim \sum_{\vec{k}} |\vec{k}|$ in natural units. (To do the sum, we follow a standard procedure in quantum mechanics: put the system in a box of volume V and impose periodic boundary conditions on the electromagnetic wave. Then $\vec{k}$ becomes a discrete, rather than continuous, variable, so that the sum makes sense. In the limit $V \to \infty$, the sum $\sum_{\vec{k}}$ tends to the integral† $V \int d^3k$.)

The end result[3] is that quantum fluctuations of a field contribute to the vacuum energy per unit volume by an amount $\Lambda \sim (V \int d^3k \, \omega(\vec{k}))/V \sim \int^{M_c} dk \, k^2 k \sim M_c^4$. Here M_c (traditionally known as a cutoff[4] in quantum field theory) expresses our threshold of ignorance. We are saying that we understand[5] the electromagnetic field up to an energy or momentum scale M_c, corresponding to some maximum value of k, beyond which we dare not go, so that we integrate only up to M_c. Thus, M_c^4 represents a conservative estimate.

In summary, each quantum field‡ contributes M_c^4 to the vacuum energy density, possibly with different values of M_c for different fields.

A humongous discrepancy between expectation and observation

We do not know precisely the mass scale M_c at which our current understanding of quantum field theory starts to break down. Traditionally, people take for M_c the Planck mass $M_P \sim 10^{19}$ GeV, at which quantum gravity kicks in (see the following chapter). But this gives a vacuum energy density of $M_P^4 = M_P/l_P^3$, and we don't have to bother to put in any numbers to see that this is way way off.

Indeed, go ahead, take your best guess of what M_c might be. If you are inclined to be conservative, you might think that M_P, all the way up in the clouds, is way too high. OK, how about ~ 1 GeV, about equal to the proton mass? Or perhaps $\sim \frac{1}{2}$ MeV, close to the

---

* As was mentioned in appendix 2 of chapter VII.3. In chapter VI.4, we also alluded to the fact that the electromagnetic field may be regarded as an infinite number of harmonic oscillators.

† Note that this is dimensionally correct, since $\vec{k}$ has dimensions of inverse length.

‡ Another way of expressing the difficulty is to say that a quantum field, such as the electromagnetic field, contains a very large number of oscillators, one for each $\vec{k}$, or equivalently, one at each point in space.

electron mass m_e? Surely, the basic principles that go into quantum field theory have been verified experimentally up to that kind of energy scale. You then predict a vacuum energy density of $m_e^4 = m_e/(1/m_e)^3$.

You don't even have to bother looking up the observational data. Just look around you. Even with M_c as low as $\sim m_e$, you are still way way off. Do you see the vacuum filled with one $m_e c^2$ worth of energy in a volume the size of the Compton wavelength of the electron?

By any measure, this is the Mother of all Discrepancies between what the theorists think and what the experimentalists observe. Our theoretical expectation is not the result of some crummy calculations based on somebody's pennyworth model. In fact, forget field theory, all we need is good old dimensional analysis. In natural units, energy density has dimension of mass to the fourth power. The only natural mass associated with gravity is the Planck mass, but whatever smaller mass we put in, even m_e, we still get an unacceptably large energy density. This nasty discrepancy is known as the cosmological constant paradox.

Rightly or wrongly, I presumed in chapter VI.2 that the observed dark energy is the fabled cosmological constant. The evidence seems increasingly to favor this simplest of hypotheses. Even if this were not the case, the paradox still remains. Why is the contribution of quantum fields to the vacuum energy so small?

Instead of giving you the observational value of the dark energy density Λ in some units such as pounds per parsec cubed, I find it more convenient to define the mass scale M_Λ according to $\Lambda \equiv M_\Lambda^4$. Observationally, the mass scale associated with the dark energy density comes out to be $M_\Lambda \sim 10^{-3}$ eV. Expressing the observational data in this way shows clearly how humongous the discrepancy is. Even if we take M_c to be as small as the electron mass, the ratio between theoretical expectations and experimental reality would be $\sim (\frac{1}{2} 10^6 / 10^{-3})^4 \sim 10^{35}$.

Another way of expressing the cosmological constant paradox is that M_Λ is much smaller than anything that is considered reasonable in particle physics. The observation of dark energy appears to suggest that there is a hitherto unknown mass scale of $\sim 10^{-3}$ eV in physics. Here is a curious fact. In the late 1990s (strangely, around the same time dark energy was discovered), neutrinos, which up until that time were thought to be massless, were experimentally found to oscillate, which implies, according to standard particle theory, that they are massive. Since there are 3 kinds of neutrinos, their masses, which have not yet been completely nailed down experimentally, can span quite a range. But they appear to have generic values, very roughly, of order 10^{-3} eV. Is this pure coincidence?[6] In any case, there might be some physics we have yet to understand at a mass scale of $\sim 10^{-3}$ eV.

The largest and the smallest masses

I also find it convenient to express M_Λ by repackaging a remark from appendix 1 of chapter X.3. Define $M_U \equiv 1/L_{universe}$, with $L_{universe}$ the size of the universe, say the Hubble radius, as some sort of Compton mass of the universe. Then (X.3.7) becomes $M_\Lambda \sim \sqrt{M_P M_U}$. With $M_P \sim 10^{19}$ GeV and $M_U \sim 2 \times 10^{-33}$ eV, we find $M_\Lambda \sim 4 \times 10^{-3}$ eV, which is of course just the statement that the dark energy almost singlehandedly closes the universe. Well, M_P is the largest mass considered in fundamental physics, and surely, M_U is the smallest, and so interestingly, M_Λ is the geometric mean between the largest and the smallest, as already remarked on in chapter X.3 but in terms of length scales. (By the way, between us friends, when I say "largest" and "smallest," you know what I am talking about. To the nitpickers: yes, I know about ∞ and 0.)

I define Λ as an energy density by writing the Einstein-Hilbert action as $\int d^4x \sqrt{-g}(\Lambda + \frac{1}{G}R)$. Trivially, we can also regard it as a sort of curvature by writing the action as $\int d^4x \sqrt{-g}\frac{1}{G}(\lambda + R)$. Then λ is given by the inverse square of some length, call it L_λ. Again, observationally, we know that the two terms in the action have comparable weight, and hence the length scale associated with the cosmological constant is on the order of the size of the universe. In other words, the radius of curvature associated with the cosmological constant is given by $L_\lambda = M_P/M_\Lambda^2 \sim 1/M_U \sim L_{universe}$.

For the record, let us also restore the 3 fundamental constants. Then we have $\Lambda \sim c^5 M_\Lambda^4/\hbar = (M_\Lambda c^2)/(\hbar/M_\Lambda c)^3$ and the curvature of the universe $L_{universe}^{-2} \sim Gc^5 M_\Lambda^4/\hbar$. We note in passing that although all three fundamental constants appear here, it is not clear, in spite of the cube of physics mentioned in the introduction to this book, whether quantum gravity is essential in unraveling the cosmological constant paradox. At least naively, gravity appears to be merely acting as a probe. (Recall that analogous remarks were made in connection with Hawking radiation in chapter VII.3.) We would of course prefer to think that the cosmological constant paradox and Hawking radiation will eventually prove to be indispensable keys for unlocking the mystery of quantum gravity.

Dead as a door nail

The cosmological constant paradox has been with us for a long time. To the best of my knowledge, Pauli was the first to worry about the gravitational effect of the zero point energy filling space. He used for M_c the inverse of the classical radius of the electron and concluded that the resulting universe could not even reach to the moon![*] Many of the greats of quantum physics were also skeptical of the zero point energy. At the 1913 Solvay Congress, Einstein declared that he did not believe in the zero point energy, writing to Ehrenfest that the concept was dead as a door nail. However, the experiment $\gamma + H_2 \rightarrow H + H$ convinced Pauli and others. For energy to be conserved, $\frac{1}{2}\hbar\omega$ has to be included in the energy of the H_2 nucleus.

At present, one could hardly doubt the reality of the zero point energy. Theoretically, it comes directly from the Heisenberg uncertainty principle. Experimentally, the liquidity of helium at zero temperature provides direct evidence, according to standard textbooks. People also often cite the Casimir effect,[7] namely the force between two conducting plates generated by quantum fluctuations, as showing that the vacuum energy is perfectly real.

---

[*] Surely, for Pauli, the zero point energy $\frac{1}{2}\hbar\omega$ was in the category of beautifully and intriguingly wrong, way beyond the infamous category of "not even wrong."

One word of caution, however. Experiments on the Casimir effect measure the force, that is, the variation of the vacuum energy contained between the two plates as we vary the separation between them.[8]

With the passage of time, people found better things to worry about, and the issue was forgotten until Y. B. Zel'dovich raised it again in the late 1960s. I would say that general awareness that a paradox was indeed lurking did not occur till the 1970s, particularly in the West. (One reason was that particle theorists in the United States by and large did not worry[9] about gravity and cosmology until the publication of Weinberg's influential books.) Until the observation of dark energy in the late 1990s, there was only an upper bound to the vacuum energy density. Since in natural units, this upper bound is on the order[*] of 10^{-123} in natural units, particle theorists generally declared that, for some unknown reason, the cosmological constant is mathematically zero. (An ultimate example of proof by authority!) For decades, many pinned their hopes first on supersymmetry, then on supergravity, and finally on superstrings. Unfortunately, nobody was able to produce a compelling argument for $\Lambda = 0$.

The cosmological constant paradox may thus be summarized as follows. In some suitable units, the cosmological constant was expected to have the value $\sim 10^{123}$. This is so huge that it was decreed to be zero identically, while the measured value (here the presumption that the dark energy is the cosmological constant comes in) turned out to be ~ 1.

Incidentally, while Λ was decreed to be identically zero by particle theorists, it was never banished by observational cosmologists, who needed it to reconcile various discrepancies in the data (for example, a universe younger than the earth due to an erroneous value of the Hubble constant in the 1930s and the clustering of the redshift data of quasars in the 1960s). This contrarian, but data based, point of view was particularly championed by P. J. E. Peebles.

Naturalness

In discussing the cosmological constant paradox, I should mention briefly the naturalness dogma or doctrine in high energy theory, as was alluded to in passing in chapter X.3. It is sometimes said jokingly that there are only two dimensionless numbers in fundamental physics: 1 and 0 (∞ being of course the inverse of 0). Again, between friends, the symbol 1 is understood to encompass numbers like 2π. In other words, if you choose units appropriately, physical quantities should have the magnitude you would reasonably expect. If a dimensionless number is exceptionally small, you should have an explanation for it. (One of my favorite examples is the ratio of the speed of sound in metals to the speed of light c_s/c. Solid state theory explains why this number is small: it is composed of the electromagnetic coupling $\alpha \sim 1/137$ and the ratio of the electron mass to the proton mass $m_e/m_p \sim 10^{-3}$. See also appendix 3.) Indeed, this naturalness doctrine is what makes the

---

[*] Since the discrepancy is so large, it hardly matters what nominal number I put here.

art of dimensional analysis possible. Stated thus, the naturalness doctrine sounds rather plausible, and, duh, perhaps even natural.

In high energy theory, the naturalness doctrine was sharpened and forcefully articulated by 't Hooft. The statement is that if a dimensionless number ε is unexpectedly small, then a new symmetry ought to emerge when $\varepsilon = 0$. (An example is the electron mass: if $m_e = 0$, then a continuous chiral symmetry appears.) In the case of the cosmological constant, the natural candidate symmetry is scale invariance. Unfortunately, scale invariance excludes not only the cosmological constant, but also the Einstein-Hilbert action by the very fact that M_P sets a mass scale. Furthermore, quarks and leptons are not massless (but acquire mass through the Higgs field). Incidentally, this is intimately connected with the remark in chapter X.3 about terms with mass dimensions less than 4. Over the decades, theorists have searched in vain, as I said, for a symmetry principle that would guarantee $\Lambda = 0$. The discovery that Λ is small but not zero complicates the situation further, as mentioned earlier.

The extreme ultra infrared

Particle physicists, who also call themselves high energy physicists, readily profess ignorance about physics at high energies and short distances, namely the ultraviolet regime, and so ask for ever more energetic accelerators. But they generally claim that they understand physics at low energies and long distances, namely the infrared regime, at least in principle and in broad outline. The cosmological constant paradox indicates that there may be a serious flaw in this view. Truth be told, we know almost nothing about physics in what we may call the extreme ultra infrared, namely physics on cosmological distance scales. One plausible approach to the cosmological constant paradox is that somehow in the extreme ultra infrared, which we may define as corresponding to distances beyond the galactic scale, gravity responds to vacuum energy differently.

The most naive approach is to soften the contribution of the vacuum energy to the right hand side of Einstein's equation $R^{\mu\nu} - \frac{1}{2}g^{\mu\nu}R = 8\pi G T^{\mu\nu}$ by acting with some differential operator $f(L^2 D^2)$ on $T^{\mu\nu}$, where D denotes the covariant derivative and L some cosmological length scale. The right hand side is effectively multiplied by $f(L^2/L^2_{\text{phenomenon}})$, where $L_{\text{phenomenon}}$ denotes the length scale of the phenomenon under study. The strategy is then to require f to have the properties $f(\sim\infty) = 1$ (to retain the success of the solar system tests and so forth, for $L_{\text{phenomenon}} \ll L$) and $f(\sim 0) = 0$ (to switch off gravity's awareness of the vacuum energy, for $L_{\text{phenomenon}} \gg L$).

Needless to say, the various proposals that have been discussed in the literature are all rather ad hoc, arbitrary, and unattractive to varying degrees, particularly given the elegant structure of Einstein gravity. A differential operator of the form $f(L^2 D^2)$ would almost invariably imply that the resulting equation is highly nonlocal. Furthermore, equations of this type tend not to be derivable from any reasonable action principle and are to be regarded as phenomenological rather than fundamental.

Another approach is to add nonlocal terms directly to the action. We already briefly discussed one such proposal for nonlocal cosmology in chapter X.3. If the goal is to merely fit observation, then we can certainly craft an action that would do the job.

The discussion of effective field theory in chapter X.3 also makes clear the need for nonlocal terms. Local terms with mass dimensions higher than the Einstein-Hilbert term are important only at short spacetime distances. As for local terms with mass dimensions lower than the Einstein-Hilbert term, there is only the cosmological constant term. If we insist on locality, then to have any cosmological impact, we are squeezed, so to speak, between the Einstein-Hilbert term and the cosmological constant term.

Another possibility is to violate some cherished principles, such as Lorentz invariance. We should keep an open mind, as we are dealing with almost unfathomably large distances in space and time here. In the appendices, we mention this and other possibilities.

The coincidence problem and inflation

The cosmological constant paradox is made even more mysterious by the cosmic coincidence problem. As explained in chapters VIII.1 and VIII.2, the energy density ρ in matter varies with the scale factor a of the expanding universe like $1/a^3$, while the energy density in the cosmological constant varies like $1/a^0$. It is remarkable that they are comparable now. Why now?

Inflation adds to the mystery. As explained in chapter VIII.4, inflation is essentially driven by a vacuum energy, which amounts to an effective cosmological constant. How is it that after the universe exits from inflation, the vacuum energy manages, not to turn itself off, but to shrink to an infinitesimal shadow of its former self? Theorists speaking of both inflation and of the cosmological constant may be exhibiting a severe case of the "wanting the cake and eating it too" syndrome.

The only plausible "explanation" is the anthropic principle, or if you prefer, "anthropic lack of principle," as some physicists call it. The anthropic principle states that physics must be consistent with the existence of physicists (which you can define in whatever way you like).

A strong version states that there are certain physical phenomena that physicists will not be able to explain using (what most physicists would agree as) the traditional approach of physics. The bold claim is that the cosmological constant paradox is one of them. Of course, one could legalistically take apart every word in the statement of the principle just given. For instance, what do you mean by "will not"? Is the implied time scale forever and forever, or is it merely until the advocate of the strong anthropic principle ceases to exist? A priori, how do we know which phenomena fall into the category of inexplicable by physics as we know it?

A weak version of the anthropic principle states that the goal of physics is to correlate observed phenomena (such as cannonballs falling from the Tower of Pisa and the precession of the perihelion of Mercury) and that to the list of observed phenomena, we should add the existence of humans. Certainly, most people would not object to this version. For instance, we could use it to calculate the distance of our planet from its sun, given various inputs about the properties of the sun, the temperature range in which biochemical processes can operate, and so on. The calculation yields only an upper and a lower bound on the distance, but it is a calculation nonetheless.

I will discuss the anthropic principle further in appendix 9, but for now, I mention that, in some sense, the smallness of Λ was predicted by Weinberg using a very weak version of the anthropic principle. This very weak version of the anthropic principle should be acceptable to most theoretical physicists (certainly to me, for what it's worth!): it merely correlates two observations, namely that galaxies formed and Λ is very small. If Λ were larger than a certain critical value, which turns out to be not much larger than the observed value, galaxies would not have formed. You are then free to extend Weinberg's reasoning to say that had galaxies not formed, then humans would not exist.

Linkage between the infrared and the ultraviolet

Quantum field theorists speak of ultraviolet (that is, high energy) physics versus infrared (that is, low energy) physics. Typically, in calculating a Feynman diagram, one encounters an integral of the form $\int d^4 k f(k)$, and if the dominant contribution comes from the large k (that is, high momentum or high frequency) region, we say that the relevant physics is ultraviolet, or UV for short. Contrariwise, if the dominant contribution comes from the small k (low momentum or low frequency) region, we say that the relevant physics is infrared, or IR for short. The quantum field theoretic prediction we had for the cosmological constant, $\Lambda \sim \int d^4 k \sim \int^{M_c} dk k^3 \sim M_c^4$, is manifestly a UV effect: the dominant region comes from the region $k \sim M_c$, from quantum fluctuations with momentum comparable to the cutoff. (Although we obtain our prediction using a handwaving argument about oscillators, a calculation using Feynman diagrams gives essentially the same result, which, after all, is basically fixed by dimensional analysis.) If we follow tradition and take M_c to be M_P, the physics underlying the cosmological constant is about as UV as it can be.

One fundamental feature of quantum field theory is that the physics at different energy scales naturally segregate themselves,[10] speaking very roughly. The general belief is that if we are studying the UV, we don't have to worry about the IR. And vice versa: if we are studying the IR, we don't have to worry about the UV.

The cosmological constant paradox appears to be the first exception to this general picture. Although the cosmological constant is generated by UV physics, it controls the expansion of the universe, which is definitely an IR phenomenon, indeed, what we called the extreme ultra infrared regime.

What is vacuum energy?

I prefer to banish speculations to the appendices. (My list of speculations on the cosmological constant paradox is far from complete and heavily biased toward what I know. The appendices are meant to give you a flavor of the sort of things that have been considered.) Instead, I end this chapter with a few general remarks. We think that we know how to calculate the vacuum energy using quantum field theory, following established rules. The cosmological constant paradox, however, indicates that we will ultimately have to face up to the question, "What is vacuum energy?"

The question reminds me of an earlier question: "What is heat?" (Or perhaps also, "What is the ether?") Everybody but physicists knew what heat was, but it proved to be extremely elusive to define. It was not a substance or a fluid (the caloric), as was once thought. The true answer had to wait until the nature of matter was understood. As you and I know, physics progresses through asking first the what, then the how, and finally the why questions, for example, "What is matter made of?" "How do these atoms behave?" "Why are there atoms?" Now we know the answers to all three questions, but the why question had to wait until the advent of baryogenesis and leptogenesis (as was discussed in chapter VIII.3), and skeptics certainly still abound. At the least, we don't know the detailed answer.

It would be a bit disappointing if dark energy or the cosmological constant proves to be merely due to some mundane mechanism, such as the presence of dime-a-dozen scalar fields (see chapter VIII.4), which ultimately have to be fine-tuned. I think that most theoretical physicists would hope that the cosmological constant paradox, like the great paradoxes of the late 19th century, will lead us to a deeper understanding of physics.

The universe says to the quantum field theorist, "I am doing just fine, thank you, but something is wrong with your understanding of the vacuum energy, or your understanding of how the gravitational field responds to the vacuum energy."

A distinguished colleague said to me recently, "The cosmological constant paradox is more than a paradox; it's a profound public humiliation of theoretical physicists."

Appendix 1: Scaling at cosmological distances

The history of physics is full of examples of reasoning by analogy that turn out to be fruitful. As explained in the text, the cosmological constant paradox can be summarized as follows:

expected value	enormous
decreed value	mathematically 0
observed value	tiny but not 0

(1)

Have we ever encountered something similar? I proposed long ago that the story of proton decay may provide such an analogy.[11]

I will not go into the particle physics behind proton decay here. Suffice it to say that, at one time, the expected value of the proton decay rate was enormous, then it was decreed to be mathematically 0, while the observed value* turns out to be extremely tiny but nonzero. The important question is how theorists managed to reduce the enormous expected value down to the extremely tiny observed value.

---

* I am fudging slightly here: at the moment, we only have an upper bound for the observed value. Experimentalists have yet to observe proton decay, but that unfortunate fact might merely be due to the fact that the detectors constructed thus far are too small. As I explained in chapter VIII.3, theorists have compelling reason to believe that the proton does decay, so we can easily imagine that experimentalists in some other civilization were not as unlucky and had observed proton decay soon after grand unified theory was proposed. In any case, the particular details of how particle physics evolved in our civilization do not concern us here.

The secret is scaling (as embodied in the renormalization group ideas in quantum field theory developed by M. Gell-Mann and F. Low, K. Wilson, and others). As was explained in chapter VIII.3, the physics responsible for proton decay originates at some grand unified distance scale, while the actual decay occurs on a distance scale on the order of the proton size. In going from one distance scale to the other, we traverse more than 16 orders of magnitude, which suffices to reduce the expected value enormously to the appropriate level.

Might we try the same trick[12] and scale[13] the cosmological constant term to make it less relevant at large distances compared with the Einstein-Hilbert curvature term?

In one particular scheme, we have to require Einstein gravity to start to deviate from Lorentz invariance beyond a length scale $L_{IR} \sim 1$–10^3 kpc, on the order of the galactic or cluster scale. It is then possible to scale the cosmological constant by a factor $\sim(\frac{L_{IR}}{L_{EuIR}})^{z-1} \sim (10^4$–$10^7)^{z-1}$, where we take the extreme ultra infrared length scale $L_{EuIR} \sim 10^4$ Mpc to be the size of the visible universe. Here z measures the deviation from Lorentz invariance and corresponds to what is called the dynamical exponent in condensed matter physics.[14] To screen the cosmological constant to the desired value, we need $z \sim 20$–30, which is at least not outrageously large.

There are many serious difficulties with this picture; the interested reader is referred to the literature for details.[15] For one thing, the resulting action is nonlocal in time at cosmological distances. Perhaps an optimist would think that this could provide a hint about the nature of time. For another, while we may be able to scale the vacuum energy away at cosmological distances, the vacuum energy can still make its effects felt over smaller regions. As one possible speculation, we can imagine each local region of the universe trying to expand and pressing against other regions in "rebellious symphony," perhaps something like a cluster of soap bubbles.

Einstein curved spacetime. Here we are suggesting that the logical next step might be to endow spacetime with some "substance," such as would be the case in some kind of foamy picture of emergent spacetime.

Appendix 2: The universe is secretly acausal, but only the universe knows about it

Arkani-Hamed et al.[16] have proposed modifying Einstein's equation to

$$M_P^2 \left(R_{\mu\nu} - \tfrac{1}{2}g_{\mu\nu}R\right) - \tfrac{1}{4}\bar{M}^2 g_{\mu\nu}\bar{R} = T_{\mu\nu} \tag{2}$$

where $\bar{R}$ denotes the spacetime averaged scalar curvature $\bar{R} \equiv \int d^4x \sqrt{-g}R / \int d^4x \sqrt{-g}$. This equation is manifestly nonlocal and acausal: physics now depends not only on what happened in the past but also on what will happen in the far future. But by construction, the modification to Einstein's equation takes effect only if the future is de Sitter with constant scalar curvature determined by the cosmological constant $\bar{R} = -4\Lambda/(M_P^2 + \bar{M}^2)$. To account for observation, the new mass scale $\bar{M}$ has to be huge, taking values ranging from $\sim 10^{48}$ GeV to $\sim 10^{80}$ GeV, depending on the assumed value of the cosmological constant one wishes to "neutralize." Unhappily, another enormous mass scale has to be introduced into physics.

In this approach, the modification is clearly designed not to matter for any situation other than cosmological. For the solar system, for example, $\bar{R}$ would come out to be practically zero. The universe is secretly acausal but only the universe knows about it! I must say that in recent years, theoretical physicists have become increasingly adept at hiding new physics from experimentalists.

Arkani-Hamed et al. argue that any mechanism to neutralize the cosmological constant must be acausal: when a vacuum energy density turns on, the alleged mechanism must wait for a cosmological time period to find out whether the energy density is indeed a cosmological constant. I am very much troubled by the thought that physics may be ultimately nonlocal, even if it is only on the cosmological scale.

Appendix 3: Possibility of an algebraic solution

Another possibly relevant historical analogy involves the inverse light speed $\zeta \equiv c^{-1}$. Consider the expected value of ζ, before it was measured, say, in some civilization in a galaxy far far away. The expected value is enormous in natural units, if propagation in the ether is assumed to be similar to sound waves in ordinary materials, let alone ocean waves. By the naturalness dogma, we might have expected ζ to be comparable to ζ_{sound}. Just as in the cosmological constant paradox, we can see that this is way off merely by looking around us. Evidently, $\zeta \ll \zeta_{sound}$. Given this, physicists would have been tempted to decree (proof by authority) that ζ is mathematically 0. But eventually, it was observed by the extragalactic version of Ole Rømer[17] that the observed value turned out to be tiny but not 0 (as both Galileo and Newton had thought). In this case, the naturalness dogma would have been off by a measly 6 orders of magnitude or so.

How was this ζ paradox resolved? It was resolved by making c part of the kinematics. We went from the Galilean to the Lorentz group, and c became a conversion factor between space and time. The unification of space and time into spacetime allows us to chose units in which $c = 1$, a value protected by Lorentz invariance. In other words, it does not get renormalized! (In contrast, in nonrelativistic theories, c would get renormalized.) Quantum fluctuations do not affect $\zeta \equiv c^{-1}$, thanks to its being part of an algebra.

Does this analogy tell us anything? To solve the ζ paradox, we had to go from the Galilean group to the Lorentz group. Perhaps we need to go one step further and extend the Lorentz group to the de Sitter group! The cosmological constant Λ, like c before it, would then become a fundamental constant of nature. Just as c is a fixed constant in the Lorentz algebra, Λ then becomes a fixed constant in the de Sitter algebra. In this sense, the question of why the cosmological constant is so small compared to what the naturalness dogma would lead us to expect might eventually turn out to be the wrong question to ask, or at least the wrong way of phrasing the question.

Another analogy might be illuminating. Imagine a civilization on a very large planet, much larger than our own. Physicists in this civilization could have developed physics to a high level of sophistication without realizing that their world was actually round. The symmetry group of physics was found to be the Euclidean group, consisting of two translations and a rotation about the vertical axis, generated by $P_x = \partial_x$, $P_y = \partial_y$, and $J = x\partial_y - y\partial_x$. But technology kept advancing, and with the development of powerful binoculars, a new phenomenon was discovered: ships going out to sea did not simply become smaller and smaller, but vanished over the horizon. The rate was eventually measured to be tiny but definitely not zero, as leading theorists had decreed. But all the efforts theorists put in trying to calculate this rate from known physics was in vain.

Later (who knows how much later), it was realized that the invariance group of physics was not the Euclidean group, but the rotation group $SO(3)$, generated by $J_x = y\partial_z - z\partial_y$, $J_y = z\partial_x - x\partial_z$, and $J_z = x\partial_y - y\partial_x$. The Euclidean group, previously held to be "sacred," turned out to be generated by $P_x \simeq (z\partial_x - x\partial_z)/R$, $P_y \simeq -(y\partial_z - z\partial_y)/R$, and $J = J_z = x\partial_y - y\partial_x$ in the limit $z \simeq R \simeq \infty$, where the very large length R was revealed to be the radius of the planet. Furthermore, R was not renormalized by quantum fluctuations.

If this analogy contains some elements of truth, then it also suggests, like the previous analogy, that the cosmological constant should be built into the invariance group of physics. It is then perfectly understandable that our continuing struggle to calculate the cosmological constant would fail. I have in mind a formulation of gravity based on the de Sitter group, not the study of Einstein gravity in a de Sitter spacetime, much as Einstein gravity is a formulation of gravity based on the Lorentz group, not the study of Newtonian gravity in some Minkowskian setting.

Appendix 4: Unimodular gravity

In chapter VI.2, I mentioned that we can always sneak an additive constant into the Lagrangian, but only when gravity is not around. Since gravity knows about this additive constant through $\sqrt{-g}$, perhaps we can solve the cosmological constant paradox by nailing g down, not allowing it to vary. The result is known as unimodular gravity.[18] The ugly part of the proposal is that we are then no longer allowed to make any coordinate transformation $x \to x'(x)$ that we please, but only those that preserve g.

Fixing g to be equal to -1 would seem to render the cosmological constant term impotent and hence irrelevant. But in fact, it comes back!

To obtain Einstein's field equation, we varied the action in chapter VI.5 and used the identity $\delta \int d^4x \sqrt{-g} R = -\int d^4x \sqrt{-g} (R^{\mu\nu} - \frac{1}{2} g^{\mu\nu} R) \delta g_{\mu\nu}$. Now we are told that we cannot vary arbitrarily, but consider only those variations $\delta g_{\mu\nu}$ that do not change the determinant and hence satisfy the constraint $\delta g = 0$, which (with the use of an identity derived back in chapter V.6) works out to be $g^{\mu\nu} \delta g_{\mu\nu} = 0$; in other words, $\delta g_{\mu\nu}$ is traceless. Let us split the Einstein tensor $R^{\mu\nu} - \frac{1}{2} g^{\mu\nu} R$ into a traceless part and a traceful part: $(R^{\mu\nu} - \frac{1}{4} g^{\mu\nu} R) - \frac{1}{4} g^{\mu\nu} R$. When multiplied by $\delta g_{\mu\nu}$ in the variation of the action, the traceful part drops out. Thus, we only get the traceless part of Einstein's equation

$$R_{\mu\nu} - \tfrac{1}{4} g_{\mu\nu} R = T_{\mu\nu} - \tfrac{1}{4} g_{\mu\nu} T \tag{3}$$

To see that the cosmological constant comes back, write the $-\frac{1}{4}$ on the left hand side of (3) as $-\frac{1}{2} + \frac{1}{4}$ and covariantly differentiate. Using the fact that the covariant derivative of the Einstein tensor and of the energy momentum tensor both vanish, we obtain $\partial_\mu R = -\partial_\mu T$, which can be solved to give $R = -T + C$. Insert this back into (3) and watch the integration constant C reappear as the cosmological constant.

Thus, unimodular gravity does not solve the problem but makes some people "feel more comfortable," because in theoretical physics, we have the license, supposedly, to set integration constants to whatever we want.

Instead of nailing g down, we might consider a softer constraint by including in the action a term like $\int d^4x\, V(g)$, where the function $V(g)$ has a deep minimum at $g = -1$ (for example, $V(g) = A(g+1)^2$, with A large). This would serve to encourage g to stay close to -1. Again, we would no longer be allowed to make just any coordinate transformation we feel like.

Appendix 5: Decaying cosmological constant

Over the years, many physicists have had many ("crazy") thoughts about gravity. One possibility, once considered highly speculative, was to entertain a decaying cosmological constant[19] $\frac{d\Lambda}{dt} \neq 0$. But these days, with a multitude of scalar fields[20] around, rolling down this hill or that, this possibility would be considered commonplace rather than outrageous.

A dynamical realization of this might be to have the vacuum energy in de Sitter spacetime dissipate by producing particle antiparticle pairs. The mechanism would be similar to that involved in Hawking radiation. However, an order of magnitude estimate would seem to suggest that the effect is far too small.

Appendix 6: Breaking free of local field theory

The cosmological constant paradox suggests to some people that we might have to break free of local field theory entirely. One possibility is to add terms not of the form $\int d^4x(\cdots)$ to the action, in a vaguely Landau-Ginzburg sort of approach to the action.[21] Interestingly, without too much contortion, we could obtain the relation $M_\Lambda \sim \sqrt{M_P M_U}$ already mentioned in the text.

Appendix 7: Lagrange multiplier for the volume of spacetime

Here is a remark I find intriguing. What is the cosmological constant Λ? It is the Lagrange multiplier for the volume $\int d^4x \sqrt{-g}$ of spacetime, whatever that means: $S_{\text{cosmological}} = -\int d^4x \sqrt{-g}\,\Lambda = -\Lambda \int d^4x \sqrt{-g}$.

In statistical or thermal physics, the Lagrange multiplier for the volume $\int d^3x$ of the system—picture a balloon filled with gas—has a name: the pressure P. We certainly understand the concept of pressure well. It is also under the experimentalist's control. But here the universe is not some container pushing into some external space, at least not in the standard view.[22]

Appendix 8: Deleting Feynman diagrams and the equivalence principle

This appendix is for those readers with some familiarity with Feynman diagrams. Let us consider the diagram responsible for the cosmological constant. Start with a matter field (for example, an electron field) loop. This diagram describes the vacuum fluctuation of the electron field. An electron and a positron pop out of the vacuum, propagate, and then the two annihilate each other. As explained in chapter VII.3, this goes on all the time. Now a graviton wanders by and couples to the electron line: the graviton is sampling the energy generated by this particular vacuum fluctuation. Ultimately, it is this diagram that causes all our hand-wringing over the cosmological constant. Suppose you were to work long and hard and come up with a rule or theory that cleverly deletes this diagram, thus solving the cosmological constant paradox. As emphasized by J. Polchinski, any such rule or theory would always be doomed to fail because of the equivalence principle.

The argument is as follows. Connect the diagram by, say, two photon lines to the propagator of some atomic nucleus, say, aluminum or iron. This diagram thus contributes to the gravitational mass of the nucleus. On the other hand, consider the same diagram with the atomic nucleus but with the graviton removed, a diagram that presumably has nothing to do with gravity. But this diagram contributes to the inertial mass of the nucleus.

Thus, with the enormous accuracy to which the equivalence principle has been tested, we already know that the diagram with the graviton attached cannot be deleted. But we are claiming that, to resolve the cosmological constant paradox, we have some rule to delete this diagram. Basically, Einstein gravity is so tightly constructed that we cannot easily bend the rules without upsetting something else.

The trouble is once again that physics, as we understand it, should be local: at the point where the graviton couples to the electron, how can the graviton "know" what the electron loop is going to do? It cannot know whether the electron is just going to loop back upon itself, or that before looping back, the electron is "planning" to emit two photons, which subsequently will be absorbed by a nucleus.

The local nature of Feynman diagrams, plus the constraint from the experimental verification of the equivalence principle, make it difficult to imagine how any rule could be invented to delete one Feynman diagram and not another. Perhaps one loophole is offered by the phrase "nothing to do with gravity"; perhaps even a diagram without the graviton is subject to the requirements of some ultimate theory of gravity.

Appendix 9: Argument using the anthropic principle

Here I repeat, and elaborate on, some of the remarks made in the text regarding the anthropic principle. The anthropic principle states that the laws of physics must be consistent with the fact that there are physicists around to discuss them. Opinions on this principle differ enormously, and I do not wish to go into this raging controversy here. Suffice it to say that many find it vaguely distasteful, perhaps even unprincipled. At one level, the statement is almost trivially true, something of a tautology.

The goal of physics is to relate apparently disjoint phenomena, for example the moon orbiting the earth and the apple falling. One of the great triumphs of physics is to relate these two particular observations. In the text, I mentioned that Weinberg showed that the very fact that galaxies formed allowed him to put an upper bound on Λ: if Λ were too large, the universe would have expanded too fast for galaxies to have formed. (See chapter VIII.3.) In fact, the observed value is not too far from this upper bound.

Put this way, I don't see how the statement can be objectionable: physics relates two different phenomena. (But notice that a theoretical framework is needed, namely the expansion of the universe and a scenario for how galaxies came to be.) Similarly, if we take the earlier statement and replace in it the phrase "galaxies formed" by "humans exist," the resulting statement, namely that the very fact that humans exist allowed an anthropic theorist to put an upper bound on Λ, is hardly more objectionable. This is no different from using the fact that humans exist on this particular planet to set bounds on how far we live from our sun.

One important aspect of the anthropic principle is that to even entertain this principle, one has to be able to conceive of different universes with different laws of physics. This is why, although the principle originated during the mid-20th century in the study of nucleosynthesis in stars, it did not come into bloom until the advent of gauge theories of the strong, weak, and electromagnetic interactions. With gauge theories, one can conceive of changing the gauge group and the parameters contained in the theories. In this sense, string theory appears to support, or at least to permit, the anthropic principle.

In the early days, string theory faithfuls hoped that they would be led to a unique ground state, so that all fundamental constants of physics, including the gauge coupling strengths, the quark and lepton masses, the cosmological constant, and so on and so forth (of course what I mean here are the dimensionless ratios formed out of these quantities) would be uniquely fixed. This hope has not been realized, to say the least. In fact, almost the exact opposite has occurred. At last count, string theory is said to have 10^{500} (the precise number hardly matters to the innocent bystander) possible ground states, each corresponding to a universe with a different set of laws of physics. The only way we know to choose between this plethora of ground states is, allegedly, the anthropic principle. Indeed, string theorists have turned this apparent defect of the theory, its inability to predict a unique ground state, into a virtue: it is only with this vast wealth of ground states that we can "understand" anthropically why the cosmological constant is so tiny.

My distaste of the anthropic principle—or, at least, discomfort with it—is that it provides a disincentive to theoretical physicists to search for explanations in the traditional sense of the word. At one time, some people invoked the anthropic principle to explain why the neutron is more massive than the proton, even though naive reasoning[23] involving the energy contained in the electromagnetic field surrounding the proton would have predicted the precise opposite. The argument was that, if the neutron were less massive, the proton would have decayed into the neutron, rather than the other way around (through the process $n \to p + e^- + \bar{\nu}_e$ mentioned in chapter VIII.3). The hydrogen atom would disintegrate, and there would be no physicists around. (This argument is hardly watertight, since other nuclei could still be stable, and who is to say that we could not build physicists without hydrogen.) While we still do not understand exactly why the neutron is more massive than the proton, we have at least pushed physics to a deeper level and reduced the question to why the down quark is more massive than the up quark. Similarly, the use of the anthropic principle in the context of stellar nucleosynthesis (roughly, if an excited level in some nucleus did not exist, nucleosynthesis could not have proceeded at the rate that it in fact does) might have discouraged the development of nuclear theory. I trust that nuclear theory can be developed to the point that the existence of this level could be understood, at least in principle. It is fair to say that most physicists, if presented with an anthropic and a traditional explanation for a given phenomenon, would probably choose the latter. In an ideal world, in a galaxy far far away, perhaps universities could afford to have separate departments of physics and of anthropic studies.

Notes

1. This chapter is adapted from A. Zee, "Gravity and Its Mysteries: Some Thoughts & Speculations," in *Proceedings of the Conference in Honor of C. N. Yang's 85th Birthday,* ed. M. L. Ge, C. H. Oh, and K. K. Phua, World Scientific, 2008. The gist of the story outlined here was told over three birthday parties: Dirac's 80th (see endnote 11), Yang's 85th, and Gell-Mann's 80th (see endnote 15).

2. For the reader who knows some quantum mechanics, this result can be derived immediately from the uncertainty principle $\delta p \sim 1/\delta x$. The energy of the oscillator is then $E = \frac{p^2}{2m} + \frac{1}{2}m\omega^2 x^2 \sim \frac{1}{mx^2} + m\omega^2 x^2$. Minimizing this as a function of x gives the desired result.

3. The actual result is $\Lambda \sim M_c^4 + M_c^2 m_e^2 \log(\frac{M_c}{m_e}) + m_e^4 \log(\frac{M_c}{m_e}) + \cdots$.

4. See, for example, *QFT Nut*, p. 163.

5. For this argument, we need to invoke merely free field theory: we are just adding up zero point energies of harmonic oscillators. We don't even have to learn how the electromagnetic field is coupled to the electron field and all the other charged fields. Or do we?

6. There is another possible coincidence if one is willing to play fast and loose with numbers: $M_P/M_{EW} \sim M_{EW}/M_\Lambda$ where $M_{EW} \sim 10^3$ GeV is the scale of the electroweak interaction. The left hand side is $\sim 10^{19}/10^3 = 10^{16}$, while the right hand side is $\sim 10^3$ GeV$/10^{-3}$ eV $= 10^{15}$. Of course, we would be off by 3 orders of magnitude if we took $M_{EW} \sim 10^2$ GeV. The actual scale, which is a somewhat loosely defined concept anyway, is perhaps about 300 GeV.

7. For a simple explanation, see *QFT Nut*, p. 70. The discussion also provides a beautiful realization of the concept of cutoff in quantum field theory.

8. In quantum mechanics, experiments typically measure only energy differences ΔE and not the energies themselves. The Casimir effect is a case in point: it does not measure the vacuum energy itself (as is sometimes erroneously stated).

9. One well-known saying by a leading theorist of the time went, "I am uninterested in gravity and I am superuninterested in supergravity."

10. This is the basic notion behind the renormalization group. See, for example, *QFT Nut*.

11. A. Zee, "Remarks on the Cosmological Constant Paradox," in *High Energy Physics in Honor of P. A. M. Dirac in His Eightieth Year,* ed. S. L. Mintz and A. Perlmutter, Plenum Press, 1983.

12. In the proton decay story, the recognition that hadrons and leptons are distinct provided the first step. One difficulty in the cosmological constant story is that the two terms involved, $\int d^4x \sqrt{-g} R$ and $\int d^4x \sqrt{-g}$, are made of the same kind of stuff. The only difference (within our present understanding of gravity) is that curvature involves derivatives, while volume doesn't. This suggests that perhaps a foamlike structure could distinguish between the two.

13. See A. Zee, *Int. J. Mod. Phys.* A 23 (2008), p. 1295. Also see endnote 14.

14. Condensed matter physicists are used to systems scaling differently in space and time: $t \to b^z t$ and $x \to bx$, with a dynamical critical exponent z, or in Fourier space $\omega \to b^{-z}\omega$ and $k \to b^{-1}k$. Lorentz invariance would tell us that $z = 1$. The speculation is that in the extreme ultra infrared, gravity breaks Lorentz invariance and scales with z not equal to 1. In the discussion of linearized gravity in chapter IX.4, the response of the gravitational field is governed by $\omega^2 - \vec{k}^2$ in Fourier space; the proposal is to replace this by $\omega^2 + \cdots + \omega^{2/z} - \vec{k}^2$. Note that the usual nonrelativistic physics corresponds to the choice $z = 2$.

15. R. Porto and A. Zee, *Class. Quant. Grav.* 27 (2010) 065006 arXiv:0910.3716 [hep-th]. See also the talk given by A. Zee at Murray Gell-Mann's 80th birthday celebration held in Singapore, February 2010. See *Proceedings of the Conference in Honor of Murray Gell-Mann's 80th Birthday: Quantum Mechanics, Elementary Particles, Quantum Cosmology and Complexity,* ed. H. Fritzsch and K. K. Phua, World Scientific, 2011.

16. N. Arkani-Hamed, S. Dimopoulos, G. Dvali, and G. Gabadadze, arXiv:hep-th/0209227.

17. Once, when I was lecturing in Copenhagen and talking about this analogy, I looked up and saw a picture of Rømer looking down at me.

18. The notion goes back to Einstein in some sense, and was developed later by H. van der Bij, H. van Dam, Y. J. Ng, F. Wilczek, A. Zee, A. Dolgov, S. Weinberg, and many others.

19. I was inspired by Dirac's large number hypothesis. See endnote 11.

20. I have always been bothered by the liberal and indiscriminate use of scalar fields in particle theory and cosmology. Quantum field theory textbooks start with scalar fields precisely because they are "without quality." If Nature wanted to show us an elementary scalar field, wouldn't she have shown us one long ago? We have encountered elementary spin 1 fields, an elementary spin 2 field, and in a mysterious twist, even elementary spin 1/2 fields. We know about meson fields, but they are clearly composite. An interesting question might be whether the Higgs field can be regarded as composite. I have speculated elsewhere that perhaps quantum field theory somehow forbids elementary scalar fields. In an improved formulation of quantum field theory, might elementary scalar fields not be allowed?

21. S. Hsu and A. Zee, *Mod. Phys. Lett.* A20 (2005), pp. 2699–2703, arXiv:hep-th/0406142.

22. Invoking an analogy between quantum hydrodynamics and quantum gravity, G. Volovik has argued that the cosmological constant paradox could be resolved. Some would maintain, however, that the paradox does not depend directly on the quantum nature of gravity, and that gravity merely provides a probe of the fluctuating energy in the vacuum. Exploring problems analogous to the cosmological constant in condensed matter systems may nevertheless provide a fruitful avenue for further understanding. See G. E. Volovik, arXiv:gr-qc/0505104; F. R. Klinkhamer and G. E. Volovik, arXiv:0711.3170.

23. See for example A. Zee, *Phys. Reports* 3C (1972), p. 127.

X.8   Heuristic Thoughts about Quantum Gravity

In search of quantum gravity

Almost the entirety of this book is devoted to the classical theory of gravity. The quantum appeared on only a few occasions, namely in our discussion of Planck's natural units way back in the introduction to the book, of Hawking radiation in chapter VII.3, (somewhat peripherally) of Kałuza-Klein theory, and of the cosmological constant paradox in chapter X.7. I have kept the knowledge required of quantum mechanics and quantum field theory to an absolute minimum. However, if I am going to talk about quantum gravity, obviously[1] I will have to mention quantum mechanics and quantum field theory. Given that at various points in this chapter I am assuming that you know quantum field theory, many readers will have to take my word for it in connection with various statements, but I try to minimize the number of these assertions. The reader who has had no exposure to quantum mechanics should skip this chapter.

Of Einstein's two offsprings, special relativity has been joined with the quantum since the late 1940s, leading to relativistic quantum field theory. Meanwhile, general relativity has stubbornly resisted being quantized. Even with the intensive development in recent decades of two candidate theories, string theory and loop gravity, a theory of quantum gravity[2] remains elusive. Certainly, I do not have room[3] here to discuss these candidate theories. I also do not discuss various other approaches[4] to quantum gravity, notably lattice gravity[5] and the notion of asymptotic[6] safety.[7] Instead, we will chat heuristically about the root cause (or causes) of the difficulty in constructing a theory of quantum gravity.

The appearance of fundamental scales

The seed of discord between gravity and the quantum had already been sown in the introduction to this book. In a world without gravity, that is, a world with $\hbar$ and c, but without gravity (a world in which Newton's constant G vanishes), we could happily do physics using relativistic quantum field theory (and its many limits thereof, such as nonrelativistic quantum mechanics or classical mechanics).

But as soon as gravity enters into the discussion, the Planck mass $M_P = \sqrt{\frac{\hbar c}{G}}$, the Planck length $l_P = \frac{\hbar}{M_P c} = \sqrt{\frac{\hbar G}{c^3}}$, and the Planck time $t_P = \frac{l_P}{c} = \sqrt{\frac{\hbar G}{c^5}}$ burst upon the scene. By the way, since by now even mass circulation magazines can mention lightyears without explanation,[8] we can set $c = 1$ without risk of conceptual confusion. Thus, we henceforth work with only the Planck mass $M_P = \sqrt{\frac{\hbar}{G}}$ and the Planck length $l_P = \frac{\hbar}{M_P} = \sqrt{\hbar G}$. As $G \to 0$, we note that $M_P \to \infty$ and $l_P \to 0$, and we lose our units. Note also that $M_P l_P = \hbar$. (As $\hbar \to 0$, $M_P \to 0$ and $l_P \to 0$, as we would expect in the classical world.)

The Planck mass spells trouble

To see that the appearance of the Planck mass spells trouble, consider the following traditional, and fairly well-known, gedanken experiment. Scatter two gravitons elastically off each other. Let us now use high school dimensional analysis to determine the scattering amplitude $\mathcal{M}$ (a notion you encountered in the preceding chapter), but to do this, you would have to know that $\mathcal{M}$ is dimensionless,[9] something I dare say the typical high school student would not know. You have to take my word for it—$\mathcal{M}$ is dimensionless. If you don't want to, I will give in appendix 2 a more elaborate version of this argument not dependent on this particular piece of knowledge.

To leading order, $\mathcal{M}$ is proportional to G. After all, G measures the strength of the gravitational interaction. In the center-of-mass frame, $\mathcal{M}$ depends on the center-of-mass energy E and the scattering angles. Since the graviton is massless, G and $\hbar$ are the only other quantities $\mathcal{M}$ can depend on. So go ahead, see if you can write down a dimensionless function of E and of the scattering angles that is linear in G.

You are forced to

$$\mathcal{M} = \alpha(\theta) G E^2 / \hbar + O(G^2) = \alpha(\theta) \left(E/M_P \right)^2 + O(G^2) \tag{1}$$

with $\alpha(\theta)$ a dimensionless function of the scattering angles θ.

The important point is that out of G and $\hbar$, we can form the combination $G/\hbar$ with dimensions of inverse mass or energy squared, which in turn requires the amplitude $\mathcal{M}$ to go like E^2. As we crank up the energy E past the Planck energy $M_P = \sqrt{\frac{\hbar}{G}}$, the amplitude increases past unity. But in quantum mechanics, the absolute square of the amplitude gives the probability for the process to occur, and probability cannot exceed unity. Hence, the leading order expression (1) for $\mathcal{M}$ eventually violates the unitarity bound basic to quantum physics.

Perhaps the most concise way of describing this problem is as follows: in a quantum world with gravity, the Planck mass sets an intrinsic energy scale, at which something we don't understand is bound to happen. As I have remarked elsewhere,[10] I find it sobering that theories in physics have the ability to announce their own eventual failure and hence their domains of validity, in contrast to theories in some other areas of human thought. So Einstein's theory is literally crying out, telling us that it will fail around the Planck energy.

Minimum length

Ever since Louis de Broglie astonished the physics world (and won a Nobel Prize in the process[11]) with his assertion that a particle with momentum p behaves like a wave with wavelength of the order $\hbar/p$, particle physicists have been pestering heads of governments (and taxpayers) that they need to build larger and larger accelerators to probe shorter and shorter distances. Given a beam of particles with energy E, they can probe distances of the order $l_{dB} \sim \hbar/p \sim \hbar/E$. Allowed enough resources in a world without gravity, they could keep on increasing the energy and happily probe smaller and smaller distances.

But in a world with gravity, we have black holes!

A concentration of mass or energy E in a region smaller than the corresponding Schwarzschild radius $r_S \sim GE$ is expected to collapse into a black hole. Thus, the colliding beams we use would collapse when $GE \gtrsim l_{dB} \sim \hbar/E$, precisely when $E \gtrsim M_P$. This suggests that the Planck length l_P represents a minimum length below which we cannot probe.[12]

In the quantum world, physical quantities are constantly fluctuating. The appearance of a fundamental length as soon as gravity is turned on suggests that in quantum gravity, spacetime itself is fluctuating on the scale of l_P, thus leading to the picturesque notion of spacetime foam. Does this mean, as was just suggested, that l_P represents the minimum length that we can probe? It seems plausible, but let's try to make this assertion somewhat more precise.

Historically, the de Broglie relation grew into the uncertainty principle, which I have already mentioned on a couple of occasions (particularly in chapters VII.3 and X.7). Starting from the fundamental commutation relation $[\hat{x}, \hat{p}] = i\hbar$ between the position operator $\hat{x}$ and the momentum operator $\hat{p}$, Heisenberg showed that the uncertainty in the position Δx and the uncertainty in the momentum Δp satisfy the inequality

$$\Delta x \Delta p \gtrsim \hbar \qquad (2)$$

Since the uncertainty principle is so fundamental to quantum physics, it would be good to recall how it is derived[13] and to have a precise statement of it. Given a quantum (hermitean) operator A, we define $\Delta A \equiv A - \langle A \rangle$, a quantity specific to the state in which we take the expectation value $\langle A \rangle$. Then the mean square deviation is given by $\langle (\Delta A)^2 \rangle = \langle A^2 \rangle - \langle A \rangle^2$. The relevant mathematical theorem, which you could look up or challenge yourself to prove, states that

$$\langle (\Delta A)^2 \rangle \langle (\Delta B)^2 \rangle \geq \tfrac{1}{4} |\langle [A, B] \rangle|^2 \qquad (3)$$

Measuring device collapsing into a black hole

When we are faced with a notion that is typically mumbled fast and loose, a good move is to call upon our friend the Smart Experimentalist.[14]

We ask SE, "How would you determine if there is a minimum length?"

SE: "I would take two distance measurements, and try to push the difference between them down to arbitrarily small values."

So, consider[15] a measuring device of size L and mass M. To determine the position of something, you have to know the position of the measuring device. We can also think of the object whose position we want to measure as part of the device. SE proceeds to measure the position of the device at time 0 and at time t, take the difference $s \equiv x(t) - x(0)$, and see if that can be made arbitrarily small.

For simplicity of analysis, assume that the device can move freely, so that the relevant Heisenberg operators are related by

$$\hat{x}(t) - \hat{x}(0) = \frac{\hat{p}}{M}t \tag{4}$$

(If the device is not free, but tied to another mass by a heavy spring, we can always regard the mass and the spring as part of the device.) Commuting (4) with $\hat{x}(0)$, we obtain $[\hat{x}(0), \hat{x}(t)] = i\hbar\frac{t}{M}$. (Note that by assumption, $\hat{p}$ does not depend on time.) It follows from (3) that

$$\langle(\Delta x(0))^2\rangle\langle(\Delta x(t))^2\rangle \geq \left(\frac{\hbar t}{2M}\right)^2 \tag{5}$$

In other words, $x(0)$ and $x(t)$ form a complementary pair and obey the uncertainty relation $\Delta x(0)\Delta x(t) \gtrsim \frac{\hbar t}{2M}$.

If SE tries to get the uncertainty in her measurement of $x(0)$ down, the uncertainty in her measurement of $x(t)$ necessarily goes up. Thus, try as she may, the uncertainty in her measurement of $s \equiv x(t) - x(0)$ is limited by the larger of the two uncertainties $\Delta x(0)$ and $\Delta x(t)$. The best she can do is bring the uncertainty Δs down to $\sqrt{\frac{\hbar t}{2M}}$, that is,

$$\Delta s \gtrsim \sqrt{\frac{\hbar t}{M}} \tag{6}$$

Now comes the key point. In a world without gravity, SE could make Δs as small as we like. Just take the two measurements quickly in succession and build the most massive measuring device (so it does not quantum jiggle too much) the funding agency would allow. In other words, make t as small as possible and M as large as possible.

But now we feel the wrath of Einstein's two intellectual offsprings!

Special relativity tells us that t cannot be smaller than the time it takes light to traverse the device (otherwise only a part of the device can be regarded as "the device"); so $t > L$, the size of the device.

General relativity tells us that if we crank up M too much, the device will collapse into a black hole, and we will not be able to receive the result of the measurement. For the device not to be a black hole, we require $L > GM$.

Thus, we conclude that

$$\Delta s \gtrsim \sqrt{\frac{\hbar t}{M}} > \sqrt{\frac{\hbar L}{M}} > \sqrt{\hbar G} = l_{\mathrm{P}} \tag{7}$$

The Planck length is indeed the smallest distance experimentalists can measure. Note that the first inequality comes from special relativity, the second from general relativity. In a world with gravity, we cannot measure distances less than the Planck length. Note the power of this argument: it does not depend on details of how the device was constructed.

In appendix 1, I give an alternative argument.

Black holes are strange

> He who does not believe it owes one dollar.
> —M. Bronstein

The reader might recognize that all these arguments, including those to be given in appendix 1, amount to essentially different versions of the same argument. Ultimately, they all come around to the fundamental fact that gravity introduces a natural energy or mass scale and a corresponding length scale. This type of argument goes way back, to a little-known paper[16] published in 1935 by the brilliant Russian physicist Matvei Bronstein, who was purged and executed at the age of 31 in 1938.

Historically, Heisenberg and Pauli quantized the electromagnetic field in 1929, concluding rather optimistically that "the quantization of the gravitational field . . . may be carried out without any new difficulties by means of a formalism fully analogous to that applied here."[17] Ha! Even quantum electrodynamics was not so easy, let alone quantum gravity. As you probably know, this early attempt at quantum electrodynamics was afflicted by infinities and various inconsistencies, difficulties that were not cleared up until the late 1940s by the generation consisting of Schwinger, Feynman, Tomonaga, and others.

But the general belief[18] throughout the 1930s was that, once quantum electrodynamics came under control, quantum gravity would follow readily, with perhaps some trivial modifications. With deep insight, Bronstein pointed out emphatically[19] that the electromagnetic and the gravitational fields are intrinsically different, because of what was then known as the "gravitational radius" of massive objects.

Black holes are strange, in more ways than one. The founders of quantum physics taught us that the quantum size of a particle of mass m is of order $\hbar/m$: the more massive the particle, the smaller it is in the quantum world. But a black hole of mass M has size $GM = (M/M_{\mathrm{P}})l_{\mathrm{P}}$. The more massive the black hole, the larger it is, a behavior that is precisely the opposite that of all other particles, including the graviton. This peculiar fact underlies the arguments given here.

The presence of the Planck length l_P indicates that the theory of quantum gravity, whatever it turns out to be, cannot possibly be a quantum field theory. For one thing, quantum field theory is based on the notion of local observables, described by fields defined at points in spacetime. But with spacetime itself fluctuating wildly according to the "dance of the quantum," we cannot even locate precisely where we are. In other words, to formulate quantum field theory, we need slices of spacelike surfaces to succeed one another in an orderly progression along a timelike coordinate axis. Bronstein in his 1935 paper advocated "a radical reconstruction of the theory . . . and the rejection of [a] Riemannian geometry, . . . and perhaps also the rejection of our ordinary concepts of space and time, replacing them by some much deeper and nonevident concepts."[20] In the early 21st century, string theorists are saying pretty much the same thing. Indeed, you can readily understand that with a fluctuating metric, fundamental concepts that we take for granted in doing physics, such as the arrow of time, the signature of the metric, and the topology of spacetime all become problematic.

Unitarization and ultraviolet completion

Interestingly, Fermi's theory of the weak interaction is also characterized by a coupling strength G_F, which has dimensions, in natural units, of inverse squared mass, just like Newton's constant G. The same dimensional reasoning that led to (1) can be applied to the scattering of two neutrinos, say. Again, we expect something dramatic to happen at the energy scale $\sqrt{\frac{\hbar}{G_F}} \sim 10^2$ GeV. (Compare this with the Planck mass $\sqrt{\frac{\hbar}{G}} \sim 10^{19}$ GeV.)

But in contrast to the case of quantum gravity, we have known what that something is since the 1970s, not only theoretically but also experimentally. At that energy scale, the weak interaction becomes unified with the electromagnetic interaction into the electroweak interaction,[21] and unitarity is restored. In this deeper and more complete theory, Fermi's coupling turns out to be $G_F \sim e^2/M_W^2$, where e denotes the electromagnetic coupling constant and M_W the mass of the intermediate vector boson responsible for the weak interaction. A fashionable terminology is that the electroweak interaction "ultraviolet completes" the weak interaction.

Unfortunately, the lesson we learned in dealing with the weak interaction does not appear to carry over to quantum gravity. At the moment, we do not know what the ultraviolet completion of quantum gravity might be. If it turns out to be string theory, the mechanism is to replace the graviton by a closed loop of vibrating string.[22]

An interesting possibility is that the scattering of gravitons at high energies is unitarized by the formation of a black hole. Our understanding of black holes certainly leads us to expect that when an amount of energy E is dumped into a region of size $GE \sim (E/M_P)l_P$ (namely the Schwarzschild radius corresponding to E), a black hole will form. Explicit calculation[23] shows that this indeed occurs. Indeed, we have already used this "fact" a number of times in this chapter.

Historically, when physics ventured into atomic distance scales, classical physics gave way to quantum physics. Various classical theories were quantized. Intriguingly, it may be that when quantum gravity enters into Planck distance scales, quantum physics will in turn be replaced by classical physics somehow. G. Dvali and his collaborators have referred to this possibility as the classicalization of gravity.[24] Quantum gravity may be classicalized. All these are just words, of course, at this point.

Effective field theory of gravity

As mentioned, historically, people were upset by the divergent behavior of quantum gravity treated perturbatively in G. One practical attitude is: "Who cares if the scattering amplitude goes bad at energy of the order of M_P? As long as we deal only with low energies, the theory works perfectly well." More formally, this attitude is embodied in the more modern outlook of effective field theory, as described in chapter X.3 (and also to be mentioned in appendix 2). Recall that in this view, the Einstein-Hilbert term is the first in an infinite series of terms $R + M_P^{-2}(\alpha R^2 + \beta R_{\mu\nu} R^{\mu\nu} + \gamma R_{\mu\nu\rho\sigma} R^{\mu\nu\rho\sigma}) + \cdots$ in the action. As the energy E in the scattering process approaches M_P, the higher derivative terms kick in. But since they appear with coefficients of order M_P^{-2}, their effects are suppressed by $(E/M_P)^2$ and so are entirely negligible until $E \sim M_P$. This is of course just another way of saying that we can pretty much forget about quantum gravity in our low energy world.

I think that a rough analogy might be the following. Suppose that some other civilization had a rudimentary understanding of quantum physics, say at the level attained circa 1910 in our civilization, not long after Maxwell wrote down his theory of electromagnetism. The perturbative correction to the electromagnetic scattering between point charges (call them electrons) also grows with energy, although only logarithmically (like $\alpha \log(E/m_e)$, where $\alpha \simeq 1/137$ measures the electromagnetic interaction strength). So at some point, we also lose control over our scattering amplitude when the correction becomes of order unity, namely when the energy approaches $E \sim e^{137} m_e$. People might wring their hands over the "inconsistency" of quantum electrodynamics until their hands got all swollen, but as we know in our civilization, this difficulty was eventually resolved and revealed to be totally harmless.

Thus, more recently, the focus has been less on the divergent behavior of quantum gravity, but rather on the strange behavior of black holes. Yes, black holes are strange, as we have seen again and again. As a reminder, consider once again the bizarre behavior of black hole entropy S. We already noted way back in the introduction to this volume that, by dimensional analysis, $S \sim GM^2 = (GM)^2/G \sim A/l_P^2$. Again, the puzzling fact that it goes like the area follows almost immediately[25] from the fact that G defines a natural length scale. As discussed in chapter VII.3, there has been some progress on the question of where black hole entropy, which certainly does not leap out at you from the Schwarzschild solution, comes from.

Quantum gravitational corrections to the Newtonian potential

It is important not to get the impression that the bad high energy behavior of gravity prevents us from calculating anything. In fact, for measurable physical processes, it should be possible to segregate the high energy contribution from the low energy contribution, using an effective field theory type of approach. For instance, in quantum field theory, the Newtonian potential $V = -Gm_1m_2/r$ between two particles of mass m_1 and m_2 results from the exchange of a single graviton between the two particles.[26] If you know what Feynman diagrams are, you can easily draw diagrams in which two gravitons are exchanged. These have been calculated to give the result[27]

$$V(r) = -\frac{Gm_1m_2}{r}\left(1 + 3\frac{G(m_1+m_2)}{c^2r} + \frac{41}{10\pi}\frac{G\hbar}{c^3r^2} + \cdots\right) \tag{8}$$

I have restored c and $\hbar$, so as to indicate which corrections go away in the limits $\hbar \to 0$ and $c \to \infty$. Note that the quantum correction is of the form $(l_P/r)^2$. If experimentalists could measure these miniscule corrections, this would represent an eminently falsifiable prediction of our understanding of the low energy effects of quantum gravity.

Unification with the other three interactions

I already mentioned in passing the so-called fine structure constant $\alpha = \frac{e^2}{\hbar c} \simeq 1/137$ introduced by Sommerfeld[28] in 1916. This quantity characterizes the coupling strength of the electromagnetic interaction and is known as the electromagnetic coupling constant, except that we have now understood for a long time that it is not constant.* Our friend SE would measure electromagnetic coupling strength by scattering, say, two electrons off each other. So clearly $\alpha(E)$ is a function of the energy involved.[29] (This fact was not apparent before high energy accelerators were built, because physicists had explored only a tiny range of energies over which $\alpha(E)$ was approximately equal to $\alpha(E = 0) \simeq 1/137$.)

The coupling strengths of the strong and the weak interactions are characterized similarly. Hence, we have three coupling functions† $\alpha_1(E)$, $\alpha_2(E)$, and $\alpha_3(E)$ all varying logarithmically[30] with energy E. I had mentioned (in chapter VIII.3, chapter X.1, and elsewhere) that the three nongravitational interactions have been unified. One indication of the unification is that these three coupling functions meet at grand unified theory energy scale $E_{\mathrm{GUT}} \sim 10^{16}$ GeV (as already alluded to in chapter VIII.3). In other words, although $\alpha_1(E)$, $\alpha_2(E)$, and $\alpha_3(E)$ are quite different in our low energy world, they become equal in the grand unified world. (I am necessarily painting the picture with broad brush strokes here, omitting all the ifs and buts.)

---

* Hence the term "coupling constant" belongs with "recombination" (see chapter VIII.3) on the list of top ten worst physics terms.
† They are not called $\alpha_{\mathrm{strong}}(E)$, $\alpha_{\mathrm{weak}}(E)$, and $\alpha_{\mathrm{electromagnetic}}(E)$ for reasons I do not go into here.

So, how does gravity fit into this picture? The standard answer is that it does not. For one thing, gravity is exceedingly feeble compared to the other three interactions.

But as you can see, in the context of this discussion, the bad behavior in (1) tells us that the coupling strength of gravity should be, in some sense, $\alpha_G(E) \equiv GE^2$ rather than just G. If so, then, as we go up the energy scale, the gravitational coupling will shoot up compared to the three couplings ambling along logarithmically. Thus, all four coupling functions could become equal, so that the four interactions we know, love, and fear could become unified into one happy theory at the Planck energy M_P. Indeed, people have also speculated about the effect of the opening up of higher dimensions on $\alpha_G(E)$.

People often confound the taming of the bad high energy behavior of gravity and its possible unification with the other three interactions. The two issues are logically distinct. While it would be nice indeed to achieve both of these objectives within one elegant theory, we should keep in mind that it is possible to have the first without the second.

Discord between Einstein gravity and the quantum

In the discord between Einstein gravity and quantum physics, somehow it is gravity that gets blamed. Most attempts to reconcile the two have involved modifying or extending Einstein gravity. For example, string theory is formulated by assuming that quantum physics as we know it will continue to hold all the way up to the Planck energy. It is of course entirely possible that it is quantum physics that has to be changed. People have suggested the breakdown of quantum mechanics at high energies, but it is entirely possible that it fails in some hitherto unexplored regime.

One thought that appeals to me is that quantum mechanics as we know it breaks down when the splitting between energy levels ΔE is less than the inverse of some cosmological time scale, such as the age of the universe.

Certainly people have tried for a long time to change the rules of quantum physics as laid down around 1926, but it turns out to be extraordinarily difficult to produce a consistent and compelling extension that does not run into some kind of contradictions or difficulties. You are of course free to speculate.

A dissenting attitude, perhaps articulated most forcefully by Freeman Dyson, is that gravity should not be quantized at all. I will let Dyson speak for himself.

> The essence of any theory of quantum gravity is that there exists a particle called the graviton which is a quantum of gravity, just like the photon which is a quantum of light. It is easy to detect individual photons, as Einstein showed, by observing the behavior of electrons kicked out of metal surfaces by light incident on the metal. The difference between photons and gravitons is that gravitational interactions are enormously weaker than electromagnetic interactions. If you try to detect individual gravitons by observing electrons kicked out of a metal surface by incident gravitational waves, you find that you have to wait longer than the age of the universe before you are likely to see a graviton. If individual gravitons cannot be observed in any conceivable

experiment, then they have no physical reality and we might as well consider them non-existent. They are like the ether, the elastic solid medium which nineteenth-century physicists imagined filling space. Einstein built his theory of relativity without the ether, and showed that the ether would be unobservable if it existed. He was happy to get rid of the ether, and I feel the same way about gravitons. According to my hypothesis, the gravitational field described by Einstein's theory of general relativity is a purely classical field without any quantum behavior.[31]

Note that Dyson is dismissing quantum gravity because of its weakness, but gravity is weak precisely because M_P is so huge, as we have seen since the introduction to this volume. So it is basically the same attitude mentioned above: "Thanks but no thanks, we are already quite happy with our low energy calculation of, say, the perihelion shift of Mercury."

You, I, and everybody else—we are all free to form our own opinions. I would take issue with the statement "If individual gravitons cannot be observed . . . we might as well consider them nonexistent." Imagine uttering this sentence in the 19th century with the word "atoms" substituted for "gravitons." As it turned out, the concept of atoms eventually did lead to a much deeper understanding of nature. Perhaps Dyson is advocating a utilitarian philosophy here. What does it buy us? Would quantizing gravity lead to a deeper understanding of nature? We will have to see, evidently.

I discuss a bit more in appendix 3 whether gravity must be quantized.

Appendix 1: More handwaving arguments

Here I recount briefly an argument given by Mead.[32] According to the textbook argument leading to the uncertainty principle, to localize a particle to within Δx, we need to use a short wavelength, high frequency photon with energy E satisfying $\Delta x \sim \hbar/E$. This is all fine in a world without gravity, but in a world with gravity, the photon will exert a gravitational force on the particle, causing it to accelerate with acceleration $a \sim GE/r^2$. Here r denotes a vaguely defined characteristic distance describing the interaction between the photon and the particle. (Fortunately, r will drop out.) The photon traverses this interaction region in time r, during which the particle acquires a velocity $v \sim ar$ and travels a distance $d \sim vr \sim ar^2 \sim GE$. Combining this with Heisenberg's uncertainty principle, we conclude that our knowledge of the position of the particle is limited by what might be called the generalized uncertainty principle

$$\Delta x \sim \frac{\hbar}{E} + GE \tag{9}$$

In other words, in addition to the uncertainty imposed by the wavelength of the photon, the particle we are trying to observe has also moved due to its gravitational interaction with the photon. Minimizing this, we see that the best we can do is $\Delta x \sim \sqrt{\hbar G} = l_P$. Again, notice that we have implicitly used special relativity here, equating the gravitational mass with energy.

To me, this kind of handwaving argument is rather fast and loose, and should (and could) be refined. Indeed, Mead did refine his argument, first taking into account momentum conservation and then replacing Newtonian gravity by Einsteinian gravity.

Another argument given by Mead (and already mentioned in the text) involves an attempt to confine a particle to a small region of size s. By the uncertainty principle, the energy of particle $E \gtrsim p \sim \hbar/s$. For this region not to become a black hole, we require $s > GE \sim G\hbar/s$, giving $s \gtrsim \sqrt{\hbar G} = l_P$.

Interestingly, string theory also naturally leads to (9). Imagine using a graviton, allegedly a closed loop of string. As we pump energy into it, it expands to have size GE, thus accounting for the term in (9) that is added to Heisenberg's term.

Appendix 2: Failure of the perturbative expansion and effective field theory

Suppose you refuse to let me simply assert, as in the text, that the amplitude $\mathcal{M}$ for the elastic scattering of two gravitons is dimensionless. In that case, I will start with the weaker statement that $\mathcal{M} \propto G$, but now let us calculate the order G^2 correction to this amplitude:

$$\mathcal{M} \propto G \left\{ 1 + \beta G E^2 / \hbar + O\left(G^2\right) \right\} = G \left\{ 1 + \beta \left(E/M_P\right)^2 + O\left(G^2\right) \right\} \tag{10}$$

with β some dimensionless function of the scattering angles. By definition, the correction to the 1 in the curly brackets is linear in G, and so the correction term has to go like E^2.

As we crank up the energy E past the Planck energy $M_P = \sqrt{\frac{\hbar}{G}}$, the second order term becomes larger than the first order term. We lose control over the perturbative expansion. Again, you recognize this as essentially the same argument given in the text: gravity introduces an energy scale M_P.

Here are a few comments on this argument:

1. Historically, this argument is confusingly phrased in terms of infinities. In more modern treatments of quantum field theory, there are no infinities in physics, only cutoffs.[33] The more sensible way to regard the difficulty we face is the violation of unitarity, as was explained in the text.

2. We can readily extend this argument to cover the higher order terms. Thus, the $O(G^2)$ terms in (10) must have, again by dimensional analysis, the form $\gamma (E/M_P)^4$, with γ some other dimensionless function of the scattering angles.

3. One possible reaction to this unitarity argument could be "So what? The perturbation expansion fails." It is certainly possible that one day, but a day that theoretical physicists can only dream of at this point, we will know how to treat quantum gravity nonperturbatively. The series in the curly brackets in (10) might turn out to be the expansion of a function $f((E/M_P)^2)$, which behaves with decency even for $E \gtrsim M_P$. It is also possible that the function is nonanalytic and does not admit a perturbative expansion. But these are merely words.

4. As mentioned in the text, the modern view is to regard the series in (10) as due to some effective theory of the type described in chapter X.3. In quantum field theory, powers of derivatives in the action get converted into powers of momentum or energy in the scattering amplitude.

Appendix 3: Induced gravity

The perennial question of whether gravity must be quantized has a long history. Here we give an extremely schematic overview. First, you may know that there are three equivalent formulations for quantum physics, known as the Heisenberg, the Schrödinger, and the Dirac-Feynman pictures.

In the Dirac-Feynman or path integral formulation, one integrates over $e^{iS(q)/\hbar}$, where S denotes the classical action as defined in part II of this book, over all possible paths or histories that the dynamical variable $q(t)$ can follow. In other words, one has to evaluate an integral of the form $\int Dq\, e^{iS(q)/\hbar}$. In the limit $\hbar \to 0$, classical physics is recovered by evaluating the integral in the stationary phase approximation.[34]

The problem of quantum gravity can thus be stated as follows. Let the world be described by a set of "matter fields" ψ and the metric $g_{\mu\nu}$. We envisage doing a giant integral of the form (we have set $\hbar$ to 1) $\int Dg \int D\psi\, e^{i(S_{EH}(g)+S(g,\psi))}$, where $S_{EH}(g)$ denotes the Einstein-Hilbert action and $S(g,\psi)$ the action for the fields in a spacetime described by the metric g. The general strategy is to do the integral in two stages, that is, to write it as $\int Dg\, e^{iS_{EH}(g)} \int D\psi\, e^{iS(g,\psi)}$.

At this point, quantum field theorists and people of that ilk claim that they more or less know how to do the integral over ψ, including the electron field, the quark fields, the gauge fields, and so forth. The difficulty of quantum gravity amounts to doing, or even defining, the integral over g.

The proposal of induced gravity is simply this: what if we don't integrate over g?

Since $S(g,\psi)$ is invariant under general coordinate transformation, we are guaranteed that $\int D\psi\, e^{iS(g,\psi)}$ is also invariant under general coordinate transformation, and hence must have the form

$$\int D\psi\, e^{iS(g,\psi)} = e^{i\int d^4x \sqrt{-g}(\Lambda + M_P^2 R + \cdots)} \tag{11}$$

with some mass scale M_P as mandated by dimensional analysis. In other words, the integration over ψ produces the effective field theory of gravity as described in chapter X.3.

There is no question that integration over the matter fields ψ will generate the Einstein-Hilbert term, that is, the scalar curvature R: this is merely a consequence of general symmetry considerations. The difficulty is that it is accompanied by Λ, which comes out naturally large, of order M_P^4, but this is of course just the cosmological constant problem biting us again. Another difficulty is that the classical equation of motion of the gravitational field no longer emerges automatically as Planck's constant approaches zero, but has to be introduced by hand. Perhaps there is nothing wrong with this, but it sure is unattractive.

A more extreme version of induced gravity is that we don't even have to include S_{EH}: Einstein gravity is induced by the dynamics of the matter field. An analogy would be an elastic medium, such as the vibrating string or membrane we talked about in part II. We can certainly write down an action for an elastic medium, but we know perfectly well that this action is not fundamental. It is produced or induced by microscopic physics. Similarly (almost blasphemous to say), perhaps gravity is not a fundamental interaction but is induced by the other three interactions. The fact that gravity stands apart from the other three can be regarded as supportive of this view.

At one time, induced gravity appeared to offer a way out of our problems with gravity and thus enjoyed a following. But not quantizing gravity leads to other problems, as we will see in the next appendix.

Appendix 4: Gravity as a classical probe

In the Heisenberg picture, classical observables are replaced by quantum operators. In particular, the quantities appearing in Einstein's field equation are to be treated as quantum operators. Thus, not quantizing gravity means that we continue to regard $g_{\mu\nu}$ as classical, but we treat $T_{\mu\nu}$ (which is constructed out of other fields, such as the electromagnetic field) as a quantum operator. C. Møller in 1962 and A. Rosenfeld in 1963 proposed the equation

$$R_{\mu\nu} - \tfrac{1}{2}g_{\mu\nu}R = 8\pi \langle \text{state}|T_{\mu\nu}|\text{state}\rangle \tag{12}$$

In other words, instead of a quantum operator, the right hand side is to be replaced by the expectation value of the quantum operator in some state.

Once again, if we invoke the naturalness dogma, this produces a huge cosmological constant on the right hand side: $\langle \text{state}|T_{\mu\nu}|\text{state}\rangle = \Lambda g_{\mu\nu} + \cdots$. But let us leave that aside. The objection to this equation is that it violates the uncertainty principle.

If gravity is not quantized, then it acts as a classical probe, and we could use a massive ball attached to a torsion balance to measure the position and momentum of a passing electron. In other words, the uncertainty principle, if strictly interpreted, does not allow the world to be part quantum and part classical. Conceptually, there may be nothing wrong with this. Let the uncertainty principle hold only in the quantum world. In his reasoning, Heisenberg used only quantum probes.

However, in 1981 Page and Geilker[35] experimentally demonstrated the difficulty one runs into. Consider a Cavendish experiment in which the heavy ball is moved from one position "here" to another position "there," as determined by some radioactive decay. This amounts to a Schrödinger's cat experiment with the quantum state in (12) given by $|\text{state}\rangle = \frac{1}{\sqrt{2}}(|\text{here}\rangle + |\text{there}\rangle)$. The torsion pendulum would then point to a phantom ball situated halfway between here and there.[36]

Appendix 5: Quantum particles in a classical gravitational field

This may not be the best place, but it will have to do, to mention that a series of elegant and fascinating experiments[37] were performed, starting in the 1960s, to study the behavior of quantum particles in a classical gravitational field, such as that of the earth. Typically, a beam of neutrons is split into two, with one sub-beam made to travel at a higher altitude than the other. The two sub-beams are then allowed to come together and interfere. Thus, theoretically, we are to study the Schrödinger equation $\left(-\frac{\hbar^2}{2m}\nabla^2 + mgz\right)\psi = E\psi$, with z the vertical axis. In another type of experiment, neutrons are literally bounced off the floor like basketballs.

While these experiments confirm quantum mechanics resoundingly, they do not shed any light on quantum gravity.

Appendix 6: Absence of local observables in quantum gravity

Another difficulty with quantum gravity is the absence of local observables. Heuristically, due to the possibility of making general coordinate transformations, we can move spacetime points around. People sometimes say, rather sloppily, that in an observable $O(x)$, the x cannot be specified. I will try to be more specific.

To read this appendix, you need to know that in quantum physics, observables are represented by operators $\hat{O}(x)$ in the Heisenberg picture mentioned in appendix 3. Our task is to calculate the functions[38] $G(x_1, \cdots, x_n) \equiv \langle \hat{O}(x_1) \cdots \hat{O}(x_n) \rangle$ defined as the expectation value of a product of operators in the vacuum state. In the Dirac-Feynman or path integral formalism, we do not speak of operators, but instead of functional integrals, as was also mentioned in appendix 3. The expectation value $\langle \hat{O}(x_1) \cdots \hat{O}(x_n) \rangle$ is then represented by an integral, so that $G(x_1, \cdots, x_n) = \int Dg D\phi \cdots e^{iS(g,\phi,\cdots)} O(x_1) \cdots O(x_n)$. Physically measurable quantities, such as scattering cross sections, are determined by the functions $G(x_1, \cdots, x_n)$. Dear reader, if you have had no exposure to quantum physics, and if this paragraph is complete gibberish to you, then you should certainly skip this appendix.

After this brief preliminary, let $O(x)$ be a scalar observable; in other words, if we make a general coordinate transformation, then we have $O'(x') = O(x)$. Specifically, if $O(x)$ is constructed out of $g_{\mu\nu}(x)$, a scalar field $\phi(x)$, and so forth, then $O'(x')$ is constructed out of $g'_{\mu\nu}(x')$, $\phi'(x') = \phi(x)$, and the like. Then we have

$$
\begin{aligned}
G\left(x_1, \cdots, x_n\right) &\equiv \langle \hat{O}\left(x_1\right) \cdots \hat{O}\left(x_n\right) \rangle \\
&\equiv \int Dg D\phi \cdots e^{iS(g,\phi,\cdots)} O\left(x_1\right) \cdots O\left(x_n\right) \\
&= \int Dg D\phi \cdots e^{iS(g,\phi,\cdots)} O'\left(x_1'\right) \cdots O'\left(x_n'\right) \\
&= \int Dg' D\phi' \cdots e^{iS(g',\phi',\cdots)} O'\left(x_1'\right) \cdots O'\left(x_n'\right) \\
&= \int Dg D\phi \cdots e^{iS(g,\phi,\cdots)} O\left(x_1'\right) \cdots O\left(x_n'\right) \\
&\equiv \langle \hat{O}\left(x_1'\right) \cdots \hat{O}\left(x_n'\right) \rangle \\
&\equiv G\left(x_1', \cdots, x_n'\right)
\end{aligned}
\tag{13}
$$

Note that the first equality (not counting the definitions) is just $O'(x') = O(x)$. The second equality follows from the invariance of the action $S(g', \phi', \cdots) = S(g, \phi, \cdots)$ and of the integration measure $Dg' D\phi' = Dg D\phi$ under general coordinate transformation. The third, and crucial, equality is due to the elementary calculus theorem stating that under integrals, we can rename the dummy integration variables at will. Note, however, that we do not erase* the primes on $x_1', \cdots, x_n'$. The equality $G(x_1, \cdots, x_n) = G(x_1', \cdots, x_n')$ implies[39] that $G(x_1, \cdots, x_n)$ does not depend on its arguments and thus can only be a constant. Thus, quantum gravity cannot be based on local observables, but instead has to be built out of nonlocal observables. This statement provides one of the starting points of the approach to quantum gravity known as loop quantum gravity.[40]

The argument depends essentially on the fact that quantum physics involves an integration over all configurations.

Notes

1. I try hard to avoid the use of words like "obviously" in my textbooks, but surely the reader would agree that my usage of the dreaded word here is justified.

2. For those readers who want to get into the subject, a recommended starting point is A. Strominger, "Five Problems in Quantum Gravity," arXiv:0906.1313 (2009). See also *Approaches to Quantum Gravity: Toward a*

---

* Compare and contrast with the discussion in appendix 1 to chapter VI.4.

New Understanding of Space, Time and Matter, ed. D. Oriti (Cambridge University Press, 2009) for a wide variety of viewpoints.

3. Not to mention my assumption that the typical reader of this book has only a limited knowledge of quantum physics.

4. For a list, see, for example, p. 5 in C. Rovelli, *Quantum Gravity,* Cambridge University Press, 2007.

5. In one sentence: spacetime is discretized and the distances between lattice points are dynamical. To get into the literature, look at review articles by R. Loll and others.

6. The word is appropriate for a book on gravity: from a-sym-ptotos = falling together. See the lament of Qfwfq in chapter IX.3. The reference to falling persists in the medical term ptosis, a drooping of the upper eyelid.

7. In one sentence: as advocated by S. Weinberg (http://arXiv.org/abs/arXiv:0908.1964), quantum gravity may be governed by an attractive ultraviolet fixed point at some finite value of the coupling. To me, it is an attractive idea, but unfortunately, to explain it properly, I would have to assume a great deal of knowledge about quantum field theory, particularly the notion of the renormalization group. To get into the literature, look at review articles by M. Reuter and others.

8. But Lightfoot, as in Gordon Lightfoot, is not a unit of time, since foot is not necessarily a unit of length.

9. See any textbook on quantum mechanics and quantum field theory, for example, *QFT Nut,* pp. 139ff.

10. *QFT Nut,* p. 172.

11. In 1929, a year after the death of his mother, who thought that her youngest son would never amount to anything. Born in 1892 and having died in 1987, he was one of the longest-lived theoretical physicists.

12. As is appropriate for a textbook, I am presenting the standard mainstream view here. The statement that we cannot go smaller than l_P is far from settled and has a controversial literature. For a small sampling, see H. Salecker and E. P. Wigner, *Phys. Rev.* 109 (1958), pp. 571–577; R. Gambini and R. Porto, http://arXiv.org/pdf/gr-qc/0603090.pdf, sec. II; Y. J. Ng, *Ann. N.Y. Acad. Sci.* 755 (1995), pp. 579–584; R. Gambini, J. Pullin, and R. Porto, http://arXiv.org/abs/hep-th/0406260; and G. Amelino-Camelia and L. Doplicher, *Class. Quant. Grav.* 21 (2004), pp. 4927–4940, hep-th/0312313.

13. For example, J. J. Sakurai and J. Napolitano, *Modern Quantum Mechanics,* p. 34.

14. She helped us crucially in understanding renormalization in *QFT Nut,* chapter III, and has already appeared in this book on several occasions.

15. I adapted this argument from X. Calmet, M. Graesser, and S. D. H. Hsu, *Phys. Rev. Lett.* 93 (2004), to which I refer the reader for caveats and further details.

16. See G. Gorelik, *Physics Uspekhi* 48 (2005), p. 1039.

17. Translated from W. Heisenberg and W. Pauli, "Zur Quantendynamic der Wellenfelder," *Zeit. Physik* 56 (1929), p. 3 [in German].

18. Keep in mind the enormous confusion at the time, such as Bohr's proposal that energy is not conserved, and the issue of whether the uncertainty principle could be applied to fields. L. Rosenfeld was apparently the first to show that quantum field theory and classical relativity are not consistent together: http://www.sciencedirect.com/science/article/pii/0029558263902797.

19. He ended his paper with the statement "Wer's nicht glaubt, bezahlt einen Thaler." [He who does not believe it owes one dollar.] Compare the stories of J. Grimm and W. Grimm.

20. Translated from M. Bronstein, "Quantization of Gravitational Waves," *J. Expt. Theor. Phys.* 6 (1936), p. 195 [in Russian].

21. See any number of textbooks on particle physics for an explanation. For a concise summary, see *QFT Nut,* chapter VII.2.

22. See, for example, J. Polchinski, *String Theory.*

23. By D. Eardley and S. Giddings, and by S. Hsu. For one speculation on what high energy scattering of gravitons may look like, see S. Giddings and R. Porto, http://arXiv.org/abs/0908.0004.

24. You may wish to look at some of G. Dvali's papers in the physics archive.

25. We need to argue from the expectation that $S \to 0$ as $G \to 0$ that S is linear in G. See the introduction to this book.

26. See, for example, *QFT Nut,* chapter I.5.

27. N. E. Bjerrum-Bohr, J. F. Donoghue, and B. R. Holstein, arXiv:hep-th/0211072. The earlier literature can be traced from this paper. In particular, the philosophy of treating general relativity as an effective field theory was outlined by J. F. Donoghue, arXiv:gr-qc/9405057.

28. A story about $\hbar$: I once stayed at a physics institute in Munich, where a commemorative metal plaque inscribed with Sommerfeld's formula was set in the lobby. Some friends who were not physicists came to visit, and I asked them what the funny symbol $\hbar$ meant. The craftsman carved the plaque in such a way

that the bar in "h bar" was a short horizontal line, which crossed the long vertical line that forms the spine of the letter h. A German woman, evidently an antinuclear and peace activist, immediately responded that physicists were contrite about inventing nuclear reactors: the "rounded arch" in h represented a nuclear reactor, right next to which was erected a Christian cross memorializing all the people physicists had killed indirectly. Very creative deconstruction of $\hbar$!

29. See, for example, *QFT Nut*, p. 164.

30. This is related to the assertion in the preceding section that the perturbative correction to electromagnetic scattering grows logarithmically with energy.

31. F. Dyson, "The World on a String," *New York Review of Books*, May 13, 2004. Copyright © 2004 by Freeman Dyson.

32. C. Alden Mead, *Phys. Rev.* 135 (1964), p. B849.

33. This point is explained here. See, for example, *QFT Nut*, chapters III.1 and III.2.

34. This description is of course way too schematic for anyone not already familiar with the subject. For a brief introduction, see *QFT Nut*, chapter I.2.

35. D. N. Page and C. D. Geilker, "Experimental Evidence for Quantum Gravity," *Phys. Rev. Lett.* 47 (1981), p. 979.

36. I don't know enough about quantum measurement theory to decide whether I should worry about this.

37. For a detailed review, see D. M. Greenberger and A. W. Overhauser, *Rev. Mod. Phys.* 51 (1979), p. 43. See also the textbook on quantum mechanics by J. J. Sakurai and J. J. Napolitano.

38. Known as some form of Green's function. But I don't want to confuse some readers even further with unnecessary names.

39. The crucial point is that general coordinate transformations form a very large set of transformations indeed. If we only have invariance under, say, translation, namely $x'_j = x_j + a$, then we can only conclude that $G(x_1, \cdots, x_n)$ is a function of $x_1 - x_n, x_2 - x_n, \cdots, x_{n-1} - x_n$, as usual. Similarly for invariance under Lorentz transformation.

40. R. Gambini and J. Pullin, *A First Course in Loop Quantum Gravity*, Oxford University Press, 2011.

Recap to Part X

As I warned you in the preface, part X contains more speculative topics, including some that may not be of lasting value.

I like the Kałuza-Klein idea so much that I will be disappointed if Nature does not make use of it somehow. I feel the same way about twistors, but less strongly. The treatment of finite sized objects, the effective field theory approach to physics, and topological field theory are all topics that in all likelihood will last.

In contrast, the chapters on the cosmological constant paradox and on quantum gravity are wildly speculative and, some might say, do not belong in a textbook. But I disagree: textbooks should not consist exclusively of material that has been carved in stone, or even worse, embalmed.

Closing Words

I admire Einstein's theory of gravity as a work of art.

—Max Born

In his last years, as I knew him, Einstein was a twentieth century Ecclesiastes, saying with unrelenting and indomitable cheerfulness, "Vanity of vanities, all is vanity."

—Freeman Dyson[1]

Here I collect some closing words,[2] a few random thoughts that constitute neither a conclusion nor a summary, just a bit of purple prose.

Theoretical physicists have been bowled over, not only by the aesthetic appeal and observational successes of Einstein gravity, but also by its profound impact. As we have seen, Einstein's theory is characterized by four fabulous features:

1. its mathematics is strikingly elegant;

2. its input consists of one single long-established fact that would otherwise be deeply puzzling;

3. its predictions were immediately verified; and

4. it has profound things to say about our understanding of the world, the very nature of spacetime.

As mentioned in chapter V.2, my enthusiasm is based not merely on the impact of Einstein gravity on physics, but also on the impact of Einstein gravity on how we do theoretical physics. With its great success, Einstein gravity has in time become a model for theoretical physics. It remains to be seen how fruitful this approach will prove to be, but there is no denying its appeal for theoretical physicists. Latch onto a well-established but not understood physical fact, start with an attractive mathematical framework, get the whole enchilada in one fell swoop, and enjoy a dramatic, almost immediate, confirmation. When this approach works, as it did for Einstein, it's fabulous, no question.

We may call this the Einstein mode of theoretical physics, exemplified later by Dirac, for example, albeit at a somewhat lower level. Theoretical physicists have strived for this ideal[3] ever since, but thus far, always stumbling on one or the other of these four features.

———

The most beautiful experience we can have is the mysterious. It is the fundamental emotion that stands at the cradle of true art and true science. Whoever does not know it and can no longer wonder, no longer marvel, is as good as dead.

—A. Einstein

Einstein revealed to us two mysteries, the mystery of gravity and the mystery of the cosmos. The two are logically separate, but we feel vaguely that they are somehow intimately linked. At our present level of understanding, while the universe certainly needs gravity, gravity appears to be indifferent to the universe: gravity operates within the universe, growing structures and making it expand. All essential tasks as far as the universe is concerned, but even if the universe consists of only the sun and the earth and nothing else, the Einstein-Hilbert action could still work its magic. We don't even know what a linkage between the two mysteries might look like, if there is one. The Dirac large number hypothesis—that fundamental constants could conceivably age with the universe—may offer a primitive example of this. To me, a decaying cosmological constant might be an attractive resolution of the cosmological constant paradox.

The cosmological constant paradox may or may not be the key to a deeper understanding of gravity, but let us hope that the dark energy is not due to a random bunch of scalar fields.[4] Einstein said, "Physics should be as simple as possible, but not any simpler." To this, we say, "The solution to the cosmological constant paradox should be as crazy as possible, but not any crazier."

Is our present understanding of cosmology too simple? With a first order ordinary differential equation and the liberal use of the Gamow principle, we have conquered the universe! It certainly cannot be denied that we have achieved an astonishingly quantitative understanding of cosmological data.[5] In a way, it is cause for celebration. Physics triumphs. But if there is no more cosmic mystery, might that not cause a sense of bitter disappointment, even among the most rationalist physicists? Is that all there is to it?

Perhaps our present cosmology has already been made too simple for what Einstein would have liked: is it simpler than "as simple as possible"?

So I am glad, and I suppose many others are also, to feel that, in spite of the fabulous success of the standard model of cosmology,[6] a sense of unease remains. I suspect that many, deep down, are elated by the coincidence problem.[7] In the unfathomable and unceasing parade of the eons, why[8] now?

There is something seriously incomplete in our understanding. Both dark matter and dark energy were largely unexpected and are an embarrassment for particle physics, but it is by and large the particle physicists who think that almost everything is understood, not the astronomers and the cosmologists.

Einstein spoke of the asylum more than once. We have grown to be far more self-confident and conceited than he.

Wigner once wrote about the unreasonable effectiveness of mathematics in physics (see also chapter VII.3). Here we could speak about the unreasonable effectiveness of physics in understanding the universe. A first order ordinary differential equation suffices! Of course, we understand this as a consequence of the perfect cosmological principle,[9] but still.

Is the universe we understand not the whole universe, but the universe filtered through the human mind? The distinguished cosmologist E. R. Harrison, in his final book *Masks of the Universe,* suggests that our current cosmology, with its dark and light sectors, was yet another mask obscuring the true universe. My impression is that most in the physics community are inclined to dismiss Harrison, but I sympathize with his mystical views to some extent.

The universe may have its own mysteries that gravity knows not.

––––––––

Wheeler once argued that the universe could only be closed. Open and flat universes troubled him, Wheeler said, because the infinite space implied that everything that could have happened would have happened. Another version of you would have read not only this book, but also this book in all its many drafts. The same argument could have been invoked to rule out the flat earth with its edge infinitely far away. As we go on an endless quest to the edge, we would have encountered every imaginable goblin and monstrosity. Just as some films are not suitable for young minds, infinity is not suitable for the human mind. In theoretical physics, we must have cutoffs.

With an accelerating expansion, however, the universe separates into isolated regions that cannot communicate with each other. So, Wheeler's concern might be mollified.

We are often awed by the power of aesthetic or "philosophic" arguments, but on the other hand, we should not succumb to selective bias and fail to remember the ones that failed. But has Wheeler's argument really failed? No. The universe has not been proven to be flat.[10]

The underlying algebra of physics was extended from the Galilean algebra to the Lorentz algebra. As the flat earth once gave way to the round world, will the Lorentz algebra be replaced by the de Sitter algebra?[11]

––––––––

As for the mystery of gravity, let us not forget that we actually have a pretty good understanding of gravity; in particular, the connection with spacetime curvature is nothing short of astonishing. We have merely gotten used to it. But by its very nature as a quest, theoretical physics always wants more. Mainly, theorists would like gravity (1) to be quantized, and (2) to be unified with the other interactions.

The internal consistency of physics appears to demand the quantization of gravity. Some simply cannot tolerate to see the world split up into two pieces, a quantum world and a classical world. The best argument I have heard is that the uncertainty principle would be violated if gravity is not quantized: gravity could then be used as a classical probe.[12] But dissenters abound.[13] One line of thought is that gravity may not be fundamental; another, similar, line of thought is that gravity may be induced. Then there is the Dysonian view that the quantization of gravity is inconsequential[14] for physics.

We talked about quantum gravity in chapter X.8—a mishmash of thoughts about quantum gravity. An intriguing possibility is that when we get into the Planckian domain, we will have to classicalize physics, in contrast to that previous occasion when we got into the atomic domain and had to quantize physics.

———

While the quantization of gravity may be required, the unification with the other three interactions does not appear to be.[15] Gravity stands apart from the other three interactions. As Einstein said, the gravitational field is first among equals. While the other three interactions operate within spacetime, Einstein gravity is spacetime. For me (and of course also for many other theoretical physicists), the most puzzling aspect of Einstein gravity is its ability to alter the causal structure of spacetime completely. When we focus on gravity as a small perturbation on Minkowski spacetime, as in chapter IX.4, then it behaves just like the other three interactions. When quantized, gravitational waves give rise to the graviton, which conceptually does not differ vastly from the photon. But when gravity is allowed to curve spacetime globally, all manner of strange goings-on torment theoretical physicists.

Is unification a prerequisite for understanding gravity? We don't know. Historically, neither electricity nor magnetism could be fully understood until they were unified.

———

Perhaps a greater mystery than either the mystery of gravity or the mystery of the universe is the mystery of the quantum. We know how to calculate but not how to interpret. We teach and learn the Copenhagen interpretation, but many prefer the many worlds interpretation. Yet, to many, the concept of the many many, very many, worlds somehow exudes a sleaziness that cannot be expressed in words.[16]

Quantum field theory in curved spacetime is a well-developed subject and leads to Hawking radiation, for example, but again I have lingering doubts. In calculating a loop diagram for some quantity, say the electron's magnetic moment at the horizon, are there subtleties involving virtual particles propagating inside the horizon and then out again? Presumably it is okay over a distance scale on the order of the Compton wavelength of the particles involved.

In the path integral formalism, to calculate the propagator of a particle from point A to point B, we are instructed to sum over all classical paths going from A to B. Suppose A and B are both near but outside the horizon of a black hole. Do we sum over the paths in which the particle goes inside the horizon and then out again? These paths are forbidden by classical physics.

Here is a toy problem. Consider the quantum mechanics problem of a particle near a sharp cliff, that is, a potential of the form $V(x) = 0$ if $x > 0$ and $V(x) = -w$ if $x \leq 0$. We all know how to do this using the Schrödinger formalism, but in the path integral formalism, do we sum over the classical paths in which the particle falls off and can't come back (or, depending on the initial and final condition, paths in which the particle stays down and doesn't know about life in the upper crust)? One approach would be to regularize, that is, to round off the sharp edge of the cliff. Admittedly, this problem does not have the causal richness of black hole physics.

More generally, in doing a quantum gravity path integral sum over all gravitational field configurations, are we to include configurations containing black holes or not? The glib response is of course, in the same way that when we do the path integral sum for a quantum field theory, we are to include the solitons, if any, in the spectrum. But the internal world of a black hole is outside "knowable" physics.[17]

Many frontier questions involve quantum field theory. Is quantum field theory solid? We should think so, at least in the infrared. But yet, as explained in chapter X.8, quantum field theory as we understand it does not appear to accord with observation, even if we trust it up to merely the electron mass scale, let alone the Planck or string scale. Students often think that quantum field theory is a closed subject, on which (too?) many textbooks have been written. But if history is a guide, there ought to be a wealth of phenomena in quantum field theory yet to be dreamed of, just as there is a wealth of phenomena in quantum field theory we now know of that were quite unknown in the 1950s and 1960s. There is the additional mystery in the general belief that quantum gravity cannot be a local quantum field theory, since quantum gravity does not have any local observables.[18]

Is it possible for Planck's constant $\hbar(M)$ to depend on the relevant mass or energy scale M? Does it make sense to raise this possibility? Why not? After all, $\hbar$ is the parameter that controls the proximity of classical and quantum physics. At one time, anybody suggesting that the fine structure "constant" $\alpha(M)$ was not constant might also have been accused of being crazy (but not crazier).

It is tempting to blame the woes of quantum gravity on quantum mechanics, as already said in chapter X.8. The blame game is certainly inappropriate in many human situations; it may also be inappropriate in theoretical physics. Consider the ultraviolet catastrophe as an example. Around the start of the 20th century, would you have blamed electromagnetism or statistical mechanics?[19] Place your bet! As it turned out, neither was to be blamed. A novel kind of physics had to arrive on the scene. Similarly, perhaps the cosmological constant paradox and the difficulties of quantum gravity are to be blamed neither on quantum mechanics nor on gravity, but are crying out for a novel kind of physics.

The existence of spin $\frac{1}{2}$ fermions is no doubt another deep mystery of physics.[20] The existence of the representation $(\frac{1}{2}, 0) + (0, \frac{1}{2})$ of the Lorentz group allows the existence of something that becomes the negative of itself upon a 2π rotation, but only in a quantum world! Suppose we didn't know about the electron. Could we have imagined that Nature would fill this representation? This suggests another one of our extragalactic fables: in some alternative history of physics, the existence of a spin $\frac{1}{2}$ particle could have been another famous prediction of special relativity, like $E = mc^2$.

If we didn't know about the electron, would we have known that there is another formulation of Einstein gravity using the vielbein instead of the metric? (The answer is yes, because Cartan did invent the vielbein without invoking the electron.)

The fermion puzzles me. Sometimes I feel that the world ought to contain only bose fields. Perhaps half integral spins are emergent.[21]

Traditionally, particle physicists have focused on the bad high energy behavior of gravity, but that may not be the real issue, as mentioned in chapter X.8. Furthermore, work using the twistor formalism sketched in chapter X.6 indicates that superficially more complicated theories, like Einstein gravity and Yang-Mills theory, may have better ultraviolet behavior than a simple scalar field theory. People have discovered amazing cancellations and tantalizing simplifications[22] in calculating amplitudes for the scattering of gravitons. One particularly intriguing hint is that amplitudes in Einstein gravity can be regarded as the square, or sum of squares, of amplitudes[23] in Yang-Mills theory. Not only is Einstein gravity deeply geometrical, but also the gauge theories that underlie high energy physics may be geometrical, at least much more so than we have appreciated thus far.

This book adores the action. But, as was made quite stark in the twistor chapter, dramatic simplification occurs when, and only when, we restrict the 4-momenta in the scattering amplitude to be lightlike, that is, to be on-shell. The action carries a lot of off-shell information; in other words, using the action, we can calculate quantum amplitudes $\mathcal{A}(p_1, p_2, \cdots, p_n)$, with p_a^2 taking on arbitrary values that are not necessarily 0. The Einstein-Hilbert action certainly does not look like the square of anything. A lot of relevant physics might be hidden inside the action.[24]

As a low energy effective theory, Einstein gravity is rather rigid, which is both good and bad. In the effective field theory approach, there is little room on either side of the Einstein-Hilbert action: the higher dimensional terms are relevant only at short distances, while the only lower dimensional term is the mysterious cosmological constant. No room to fool around in. To deal with the cosmological constant paradox, we may be compelled to add nonlocal terms, and they can readily be designed to account for observations. Alternatively, we could abandon Lorentz invariance, thus opening up the gap[25] between the Einstein-Hilbert and the cosmological constant terms.

Theoretical physics as we now know it rests on many pillars: the quantum principle, the action principle, locality, causality, Lorentz invariance, general coordinate invariance, the gauge principle, and perhaps a few other odds and ends. When people make a list like this, it is hard to see why certain things are included and others not. These concepts or principles intertwine and are mutually dependent, to some extent. For example, the action as usually formulated involves a local Lagrangian and has locality and causality built in, and if by the quantum principle, we mean the path integral formalism, the quantum principle is dependent on the action. And of course general coordinate invariance and the gauge principle are not principles at all but merely a statement that $g_{\mu\nu}$ and A_μ contain "nonphysical" degrees of freedom. So, in the future, if one of these pillars were to crack and fall, which one would it be? Perhaps there are already some tantalizing hints that locality might fail. (Some might even argue that string theory is not a local theory but predicts locality at low energies.) Indeed, it is causality that we want to preserve, not locality.

I am particularly respectful of, perhaps even awed by, the action principle. It is truly amazing that, while many phenomenological theories cannot be derived from an action, all the fundamental interactions we know—gravity, strong, weak, and electromagnetic—can be. A priori, there is no transparent reason why all the foundational equations we know can be written as the extrema of something.

Our friends the observational cosmologists have given us much comfort regarding whether physics operates the same way throughout the universe. Nevertheless, many theoretical physicists might not be bent too much out of shape if some of their cherished concepts fail on the cosmological scale, as mentioned in chapters X.7 and X.8. The universe may be secretly acausal, but only the universe knows about it.

———

In special relativity, young Einstein was able to accomplish what Lorentz and Poincaré were not able to do, even though the two established giants had most of it worked out, at least mathematically. After all, Lorentz had the Lorentz transformation in all its glory. The two older physicists were not able to abandon the perfectly sensible notion that if there is a wave, something must be waving. (Incidentally, Maxwell believed in the ether, even though his equations did not need it.[26]) So they had the ether as a dynamical variable. Einstein simply trashed the ether and asserted that nothing could also wave.[27]

Nowadays, any student is able to accept, without blinking twice, that an electromagnetic wave consists of A_μ waving—yes, just a mathematical symbol A_μ known as a field waving. Of course, there are energy and momentum densities associated with the wave, and so it is real in that sense.

But what is a field? After spending years writing a textbook on quantum field theory, I understand a field as something that does what a field does. No more, no less. A recent textbook[28] on electromagnetism asserts that the electromagnetic field is as real as a rhino. My response is that a quantum field is as real as a quantum rhino.

To move forward, physics had to abandon an apparently ironclad piece of common sense: where there is a wave, something must be waving. I would not be at all surprised if it turns out that to move forward, we have to abandon an equally ironclad piece of common sense.

Another reason that the old guards such as Lorentz and Poincaré failed while Einstein succeeded was that they tried to derive special relativity from some dynamical theory of the electron now mercifully forgotten. The establishment's first reaction to Einstein's work was that the young fellow merely imposed the answer, which he had not derived in any way.

Einstein curved spacetime. Perhaps the next step is to endow spacetime with substance, so that spacetime in neighboring regions can push against one another.[29]

———

Many have made careers out of worrying about quantum gravity. But classical gravity is already plenty puzzling. When we first studied physics, we were told that physics should be local, that something happening here can only affect something happening nearby, and for a physical effect to propagate across spacetime, a field is needed.[30] But the horizon around a black hole is a strikingly nonlocal concept. Nothing happens locally. Observers falling in do not notice anything. The puzzle is that the Riemann curvature is nice and smooth at the horizon and can be made arbitrarily small for massive black holes. But somehow the other fields in the world know about the metric $g_{\mu\nu}$ directly, not about Riemann curvature.

The horizon is an inherently nonlocal concept.[31] By drawing a Penrose diagram, we can see that we could be sitting peacefully while an incoming shell of matter far away threatens to form a black hole soon, and we could be inside the horizon even before the black hole forms.

Can we possibly modify general relativity so as to avoid having a horizon? Once again, apparently not, because a black hole is a low energy phenomenon. Naively, we might also think that the addition of local terms would not remove a nonlocal phenomenon like a horizon. But perhaps one should still try—it is certainly conceivable, at least to me, that the naive view is wrong.

———

The founders of quantum field theory wrote profound equations such as $A_\mu = 0 + A_\mu$ and $\varphi = 0 + \varphi$. Fields execute quantum fluctuation around vanishing classical values. But then physicists became more sophisticated in the 1960s and wrote fancier equations like $\varphi = v + h$, with $v = \langle \varphi \rangle$. The basic equation for the graviton field has the same form: $g_{\mu\nu} = \eta_{\mu\nu} + h_{\mu\nu}$. This naturally suggests that $\eta_{\mu\nu} = \langle g_{\mu\nu} \rangle$ and perhaps some sort of spontaneous symmetry breaking. But gravity exhibits a fundamentally new feature: $g_{\mu\nu}$ is a matrix and hence has a signature. Large fluctuations of $h_{\mu\nu}$ can change the signature of $g_{\mu\nu}$, and there could be regions with two times. An obvious thing to write down would be a potential for $g_{\mu\nu}$ (which breaks general coordinate invariance) of the form $V(g) = \lambda(g_{\mu\nu} - \eta_{\mu\nu})^2$, or more generally, a potential with a deep well pinning $g_{\mu\nu}$ to values close to $\eta_{\mu\nu}$. This induces a graviton mass of order $m_g^2 \sim \lambda/M_P^2$, so that $\lambda^{\frac{1}{2}}$ is given by the product of the largest mass and possibly the smallest mass known to physics.

This line of thought raises the possibility that the potential $V(g)$ might have minima elsewhere. Perhaps there is a phase with $g_{\mu\nu} = 0$. That could be the ultimate terrorist plot: to unleash a $g_{\mu\nu} = 0$ bomb that would annihilate spacetime in the victimized country. Compared to this catastrophic transition, heading back into the Big Bang merely causes g_{ij} to vanish; the Bang created space but not time. Could the universe have begun at a singularity "where" $g_{00} = 0$ as well as $g_{ij} = 0$? Not only no space, but also no time. This is not as wild as it may sound; indeed, as mentioned in an endnote in chapter IX.10, both Einstein and de Sitter contemplated different versions of this for use as boundary conditions.

———

In the beginning, the strong interaction was written in terms of nucleons and pions. Decades passed before the correct dynamical variables were discovered. Writing the action in terms of quarks and gluons rather than nucleons and pions turned out to be the crucial step in understanding the strong interaction. It is conceivable that a similar step has to be taken for gravity. At the simplest level, we have already seen that the action can be written in terms of either the metric or the vielbein. But a more drastic step may be needed, and the discovery that the scattering amplitudes for gravitons are equal to the square of the scattering amplitudes for gluons may offer a hint. Perhaps the correct dynamical variables[32] have yet to be found.

———

Could Einstein gravity be replaced by something more fundamental, which could lead to $\frac{1}{G}\sqrt{-g}\,R$ effectively at low energy, much as quantum chromodynamics leads to the Yukawa pion-nucleon theory? Suppose particle physics experiments had stopped in the mid-1950s. Could we have leapt from Yukawa theory to quantum chromodynamics? It is conceivable that, by thinking about the proton decay paradox, we could have. This may turn out to shed some light on the cosmological constant paradox.[33]

The question, stated in the format of an IQ test question, is then "What is to gravity as quantum chromodynamics is to pion-nucleon theory?"

I am not necessarily suggesting here that the graviton is composite. Indeed, a theorem by Weinberg and Witten states, with rather general assumptions, that the graviton cannot be a bound state. By the way, the AdS/CFT correspondence exemplifies a way around this theorem. The gauge theory at the boundary of anti de Sitter spacetime could produce a graviton, but only by growing a spatial direction at the same time.[34]

Even if this theorem could be somehow evaded, the possible compositeness of the graviton appears to be irrelevant to the cosmological constant paradox. Let the graviton's compositeness be characterized by a^*. In quantum field theory, this would be revealed in the graviton's propagator at a characteristic momentum of $q^* \sim 1/a^*$. But in the cosmological constant paradox, the relevant momentum carried by the graviton is in what I called the extreme ultra infrared, with $q_{cosmological} \sim 1/L_{cosmological} \sim 0$, where $L_{cosmological}$ is a cosmological distance scale. Presumably, we have $L_{cosmological} \ggg a^*$. In other words, the universe could care less if the graviton is composite at an energy scale of, say, 1 Tev.

It is not compositeness that we are after. For example, in the proposal mentioned in appendix 1 of chapter X.7, the graviton propagator is modified at $q_{cosmological}$ by abandoning Lorentz invariance rather than by appealing to compositeness.

————

Could gravity be part of a larger structure?

Note that this is a different question from the one asked in the preceding section. We now understand the electromagnetic field as part of a larger structure.[35] Gravity could be part of a larger structure in a mathematical sense, as electromagnetism, based on the gauge group $U(1)$, is in fact part of a larger structure based on the gauge group $SU(5)$ or $SO(10)$. The larger structure reveals itself only at higher energies. But even if the structure is not seen at low energies, it imposes physical consequences. Thus, electric charge is quantized if the larger structure is a grand unified theory based on a simple group, and we understand why $Q_{electron} = -Q_{proton}$ exactly, a fact of cosmological significance. There is no way of understanding this fact within electromagnetism itself.

This is an example of an unintended consequence in theoretical physics, in this case a consequence of unifying electromagnetism with the strong and the weak interactions. Perhaps the answers to some of the questions we are asking also have unintended consequences. It is conceivable, for example, that unifying gravity with the other three interactions is possible only if there are three families of quarks and leptons, the existence of which poses one of the most puzzling questions in particle physics.[36]

The question, again stated in the format of an IQ test question, is then "What is to gravity as grand unified theory is to electromagnetism?"

In chapter X.7, I mentioned two analogies to the cosmological constant paradox. Here is yet another. The history of physics contains a number of logical impasses. One of my favorite examples is radiant heat. The leading theory of heat at one time held that matter contained a mysterious fluid known as the caloric, but the boring of cannons (see endnote 1 in chapter VI.4) showed that the amount of caloric in iron appeared to be unlimited. The alternative theory held that heat was due to the motion of molecules, which we now know is the correct explanation, but this theory suffered from a fatal logical impasse: the phenomenon of radiant heat. How could molecular motion be transmitted across empty[37] space? The logical contradiction seemed insupportable.

The paradox was solved by the discovery that radiant heat was a form of electromagnetic energy. Perhaps the cosmological constant paradox will be solved in an analogous way: a piece of physics is missing from our present understanding.

———

Then there is the mystery of time. In our current description, space is created at the Big Bang, but not time. In spacetime, space and time are unified, but time, to paraphrase what Einstein said about gravity, appears to be first among equals. Various authors[38] have suggested that time may be discrete, but discrete time appears to be conceptually more difficult than discrete space, although numerical workers use it routinely.[39]

In the AdS/CFT correspondence mentioned in chapter IX.11, from the point of view of the physicists living on the boundary, a spatial coordinate appears to emerge (as mentioned earlier). A subject of current research is a possible dS/CFT correspondence. If this were to be realized, then we would have the intriguing scenario of time emerging.

———

In Einstein gravity, the origin of spacetime is intimately linked to the origin of gravity. Emergent spacetime has been discussed in a variety of contexts. It has long been known in condensed matter physics that various lattice Hamiltonians lead to emergent gauge fields[40] in the low energy effective theory. And it was speculated that the gauge fields responsible for the three nongravitational interactions could all be emergent from an underlying lattice system containing only quantum spins.[41] Given this background, it is natural to speculate that the graviton also emerges from some underlying lattice system.[42] But while it is surprisingly easy[43] for a gauge field to emerge from a condensed matter system, it is very difficult, because of the Weinberg-Witten theorem, for a gravitational field to emerge.

———

A year before Einstein's death, John Wheeler asked the old man to speak to a select group of students. Besides repeating his opposition to quantum mechanics, Einstein also made a cryptic comment: "There is much reason to be attracted to a theory with no space, no time. But nobody has any idea how to build it up."[44]

Perhaps we have to go beyond space and time. But these are just words. As the old man said, nobody knows how.[45]

Over time, many speculative thoughts about gravity have been thought, some from the great, some from the not-so-great, and more than plenty from the cracked. To paraphrase the grand old man, future speculations should be as wild as possible—in the way that quantum physics would have seemed utterly wild to prequantum physicists (such as Maxwell) and that curved spacetime would have seemed utterly wild to prerelativistic physicists (such as Newton)—but not any wilder. That is the difference between theoretical physics and cracked pottery: the "not any wilder" part.

This has been a long trek, in spite of what I said at the beginning of chapter VI.1. And now I end, with an exhortation to the reader, quoting Henry David Thoreau: "What old people say you cannot do, you try and find that you can. Old deeds for old people, and new deeds for new."[46]

Notes

1. F. Dyson, UNESCO lecture, 1965.
2. In the same sense as the closing words in my textbook on quantum field theory.
3. String theory satisfies feature 2, inputting the existence of gravity, but fails at feature 3.
4. Scalar fields are fields without qualities, colorless individuals with no character or personality who could blend in anywhere. Quantum field theory textbooks start with scalar fields for pedagogical clarity, precisely because they are without qualities.

    We could use scalar fields to do practically anything we want. They fit in anywhere. In cosmology, scalar fields are used all the time and all over the place. They could drive inflation. They could account for dark energy and perhaps even dark matter. Almost anything could be explained with scalar fields. Attempted long distance modifications of gravity essentially all amount to adding scalar fields. Scalar fields are way too cheap and so so painless. Just throw them in. People get them for free.

    Perhaps we should feel a bit uneasy? I have no objection to composite scalar fields, of course, but then a deeper dynamical understanding is called for.
5. Such as the location of the acoustic peaks in the microwave background discussed in chapter VIII.3.
6. Sometimes compared to and contrasted with the standard model of particle physics.
7. I touched upon the coincidence problem in chapter VIII.1.
8. But recall from chapter VIII.3 that there is another coincidence problem that we apparently don't need to worry about: photon decoupling and matter dominance also occurred at roughly the same time.
9. Sounds a bit like the perfect celestial dome that our predecessors talked about. See chapter VIII.1.
10. As explained in chapter V.3, observations can only set a lower bound on the universe's characteristic length scale.
11. As mentioned in appendix 3 of chapter X.7.
12. See appendix 4 of chapter X.8.
13. As mentioned in chapter VII.3, for example, and in chapter X.8.
14. I don't subscribe to this idea. Even if the measurable effect of quantization is far too small for actual measurement, a consistent quantization might require something else, for example, that the dimension of spacetime be 4. See endnote 15.
15. As mentioned in chapter X.8, the two are logically distinct issues.
16. The Dao that can be expressed in words is not the true Dao.
17. And, indeed, this assertion is supposed to account for the entropy of black holes. See the discussion of entanglement entropy in the literature.
18. As explained in chapter X.8.

19. Apparently, Planck did not believe in the equipartition theorem, so that for physicists like him, the ultraviolet catastrophe was not a catastrophe at all.

20. Jordan's anticommutation manuscript languished in Born's pocket for a whole year. As if noncommutation were not shocking enough!

21. See, for example, A. Zee in *Quantum Coherence: 30 Years of Aharonov-Bohm Effect,* ed. J. Anandan, Consider the effect discovered by 't Hooft et al. that binding a boson to a magnetic monopole produces a fermion. The group theory of $SO(10)$ is also highly suggestive. See *QFT Nut,* p. 428.

22. As alluded to in chapter X.6.

23. Appropriately color stripped.

24. Fermat's least time principle does not know that light is a wave. But here the Einstein-Hilbert action certainly knows about the graviton and its interaction with other gravitons. Could it be that, analogous to Fermat's principle, there are things that the Einstein-Hilbert action does not know about?

25. As described in appendix 1 of chapter X.7.

26. So what is the lesson? If your equations do not need it, does it exist?

27. This reminds me of an old puzzle. What is greater than God, and if you eat it, you will die? (Sometimes I use this to puzzle young children.)

28. A. Garg, *Electromagnetism in a Nutshell,* Princeton University Press, 2012.

29. As in the rebellious symphony alluded to in appendix 1 of chapter X.7.

30. However, the mysteries of quantum mechanics have also led to entanglement and the Aharonov-Bohm effect.

31. But confusingly, while we cannot directly perform local measurements to detect the presence of a horizon, we can do so indirectly. By measuring whether light rays tend to converge or diverge, we can detect the presence of a trapped surface (or apparent horizon). A sequence of highly plausible theorems (each of which nevertheless involves some technical assumptions) by Penrose, Ellis, and others, combined with the unproven cosmic censorship conjecture, states that the presence of a trapped surface implies the presence of a horizon.

32. Imagine a civilization in some other galaxy that developed a theory of light based on the intensity and polarization of light beams. Color could be expressed as a 2-dimensional vector based on something like our color wheel. The theory could account for most observations but would eventually be found to be lacking when confronted with wave phenomena. Is our theory of gravity an analog of this kind of theory?

33. As explained in appendix 1 to chapter X.7.

34. See, for example, G. Horowitz and J. Polchinski, in D. Oriti (as cited in chapter X.8, note 2).

35. Gerard 't Hooft has given an elegant expression for the Maxwell field $\mathcal{F}_{\mu\nu}$ in terms of the Yang-Mills field $F^a{}_{\mu\nu}$. Is there an analog for gravity? Can $g_{\mu\nu}$ be written in terms of some more elaborate object $\mathcal{G}^a_{\mu\nu}$?

36. In the dream of the ultimate theory, everything will be fixed, not just the fact that there are three families. That would be the final response to the anthropic alternative to physics.

37. I am not enough of a historian to know whether attempts were made to measure the transfer of radiant heat across a chamber with its air pumped out.

38. Including T. D. Lee and G. 't Hooft.

39. As explained in our discussion of the initial value problem in chapter VI.6.

40. There is an extensive literature starting from the late 1980s with the discovery of fractional quantum Hall fluids and of high temperature superconductivity.

41. For one particular example, see A. Zee, "Emergence of Spinor from Flux and Lattice Hopping," in *M. A. B. Bég Memorial Volume,* ed. A. Ali and P. Hoodbhoy, World Scientific, 1990.

42. See especially the work of X. G. Wen and his collaborators.

43. For example, write the spin field (a unit vector $\vec{n}$) as $\vec{n}(x) = z^\dagger(x)\vec{\sigma}z(x)$ in terms of a spinor field $z(x)$ and the Pauli matrices $\vec{\sigma}$ (as introduced in chapter X.6). The local symmetry $z(x) \to e^{i\theta(x)}z(x)$ leads naturally to a gauge potential. See, for example, X. G. Wen and A. Zee, "Possible T and P Breaking Vacua of O(3) Non-Linear σ-Model and Spin Charge Separation," *Phys. Rev. Lett.* 63 (1989), p. 461.

44. T. Damour, O. Darrigol, B. Duplantier, and V. Rivasseau, eds., *Einstein, 1905–2005, Poincaré Seminar 2005,* Birkhäuser, 2006, p. 174.

45. The proposal mentioned in appendix 1 of chapter X.7 might conceivably be a first tentative step in this direction.

46. H. D. Thoreau, *Walden.*

Timeline of Some of the People Mentioned

Galileo Galilei (1564–1642)

René Descartes (1596–1650)

Pierre Fermat (1601 or 07/08?–1665)

Isaac Newton (1643–1727 [1642–1726])

Gottfried Leibniz (1646–1716)

Johann Bernoulli (1677–1748)

Leonhard Euler (1707–1783)

Jean-Baptiste le Rond d'Alembert (1717–1783)

John Michell (1724–1793)

Joseph-Louis Lagrange (1736–1813)

D'Amondans Charles de Tinseau (1748–1822)

Pierre-Simon, Marquis de Laplace (1749–1827)

Carl Friedrich Gauss (1777–1855)

Friedrich Wilhelm Bessel (1784–1846)

Urbain Jean Joseph Le Verrier (1811–1877)

Jean Frenet (1816–1900)

Joseph Serret (1819–1885)

Georg Friedrich Bernhard Riemann (1826–1866)

Elwin Bruno Christoffel (1829–1900)

Julius Weingarten (1836–1910)

Marius Sophus Lie (1842–1899)

Wilhelm Killing (1847–1923)

Woldemar Voigt (1850–1919)

Gregorio Ricci-Curbastro (1853–1925)

Hendrik Antoon Lorentz (1853–1928)

Jules Henri Poincaré (1854–1912)

Luigi Bianchi (1856–1928)

Max Planck (1858–1947)

David Hilbert (1862–1943)

Hermann Minkowski (1864–1909)

Élie Cartan (1869–1951)

Gustav Mie (1869–1957)

Willem de Sitter (1872–1934)

Karl Schwarzschild (1873–1916)

Max Abraham (1875–1922)

Albert Einstein (1879–1955)

Gunnar Nordström (1881–1923)

Amalie Emmy Noether (1882–1935)

Arthur Eddington (1882–1944)

Hermann Weyl (1885–1955)

Theodor Kałuza (1885–1954)

Johannes Droste (1886–1963)

Alexander Alexandrovich Friedman (1888–1925)

Attilio Palatini (1889–1949)

Kornel Lanczos (1893–1974)

Georges Lemaître (1894–1966)

Oskar Klein (1894–1977)

George Szekeres (1911–2005)

Richard Feynman (1918–1988)

Solutions to Selected Exercises

In the book of life, the answers aren't in the back.
—Charles M. Schulz, speaking through
Charlie Brown

Prologue

1 With x and L as labeled in figure 1 (with sand replaced by air), the time getting from F to G is given by $T = c_a^{-1}\sqrt{x^2 + A^2} + c_w^{-1}\sqrt{(L-x)^2 + B^2}$. (Since the math involved is high school level, I won't even bother to define A and B.) Setting the derivative of T with respect to x to 0, we obtain $c_w x/\sqrt{x^2 + A^2} = c_a(L-x)/\sqrt{(L-x)^2 + B^2}$, which we recognize as $c_w \sin\theta_a = c_a \sin\theta_w$.

I.1 Newton's Laws

1 The first part is obvious since $\delta(x)$ is sharply spiked at $x = 0$, so that in the integrand we can replace $f(x)$ by $f(0)$ and then do the integral. In the second part, change variable to $y = ax$ and note that the limits of integration depend on the sign of a.

2 Write $r' \equiv \frac{dr}{d\theta} = -\frac{1}{u^2}u'$, so that the equation $(\frac{dr}{d\theta})^2 = \frac{2r^4}{l^2}(\epsilon - v(r))$ becomes

$$u'^2 + u^2 - \frac{2\kappa}{l^2}u = \frac{2\epsilon}{l^2}$$

You recognize this as just the shifted harmonic oscillator, which you solve instantly as

$$\frac{1}{r} = u = \frac{\kappa}{l^2}(1 + e\cos\theta)$$

with the eccentricity e given by $e^2 = 1 + \frac{2\epsilon l^2}{\kappa^2}$. That the orbit closes is now obvious.

3 We could still use (19) except that the root r_{min} is now negative, which is not physical since $u = 1/r > 0$. A moment's thought indicates that the lower limit for the u integral in (19) should be set to 0. Changing integration variable as before, we obtain $\Delta\theta = 4\int_{\zeta_{min}}^{\frac{\pi}{2}} d\zeta$, with ζ_{min} determined by $\sin^2\zeta_{min} = -\frac{u_{min}}{u_{max} - u_{min}} = \frac{1}{2} - \frac{\kappa}{2\sqrt{2\epsilon l^2 + \kappa^2}}$. First, let's check that we are on the right track by turning off gravity: set $\kappa = 0$, then $\zeta_{min} = \pi/4$ and $\Delta\theta = \pi$, the correct answer for light moving in a straight line. Next, expanding to leading order in κ, we obtain $\Delta\theta = \pi + \frac{2\kappa}{l\sqrt{2\epsilon}}$. We now express the deflection angle (as usually understood, that is, straight line corresponds to no deflection) $\Delta\theta = \frac{2\kappa}{l\sqrt{2\epsilon}}$ in terms of the impact parameter b defined by saying that as $x \to \infty$, the light ray moves along a path specified by $y = b$. Translating into polar coordinates, we have, as $r \to \infty$, from (9) $b \simeq r\theta$, from (18) $\dot{r}^2 \simeq 2\epsilon$ and $\dot{r} \simeq -\sqrt{2\epsilon}$. Using (15) and $\frac{dr}{d\theta} = \frac{\dot{r}}{\dot{\theta}}$, we determine $l = b\sqrt{2\epsilon}$. Thus, $\Delta\theta = \frac{\kappa}{\epsilon b}$.

Newton of course did not know what the speed of light was, but if we set $\dot{r}^2 = c^2$ so that $\epsilon = c^2/2$, we find the Newtonian result for the deflection of starlight by the sun

$$\Delta\theta = \frac{2GM}{c^2 b}$$

4-5 Any spherical mass distribution can be built up by stacking up spherical shells. The potential at $(x = 0, y = 0, z = R)$ due to a single Newtonian shell of radius a, thickness δ, and mass density ρ is then given by

$$V(R) = \int \frac{G\rho a^2 da d\theta \sin\theta d\varphi}{\sqrt{R^2 + a^2 - 2Ra\cos\theta}} = 2\pi G\rho a^2 \delta \int_{-1}^{1} \frac{du}{\sqrt{R^2 + a^2 - 2Rau}}$$

$$= 2\pi(G\rho a/R)\delta \left[(R^2 + a^2 + 2Ra)^{1/2} - (R^2 + a^2 - 2Ra)^{1/2}\right]$$

Outside the shell, $R > a$ and the square bracket evaluates to $(R + a) - (R - a) = 2a$ and so $V(R)_{\text{outside}} = G\rho(4\pi a^2\delta)/R = GM_{\text{shell}}/R$, the first superb theorem. Inside the shell, $R < a$ and the square bracket evaluates to $(R + a) - (a - R) = 2R$ and so $V(R)_{\text{inside}} = G\rho(4\pi a\delta)$, which means the shell exerts no force on an observer inside—he is tugged in all directions—the second superb theorem.

I.3   Rotation: Invariance and Infinitesimal Transformation

2 Intuitively, it should be obvious, since $\int dx dy \delta(x)\delta(y) f(x, y) = f(0, 0)$ just picks out the value of the function f at the origin. More formally, we have

$$\delta(x')\delta(y') = \delta(\cos\theta\, x + \sin\theta\, y)\delta(-\sin\theta\, x + \cos\theta\, y)$$

$$= \delta(\cos\theta\, x + \sin\theta\, y)\delta\left(-\sin\theta\, x - \frac{\cos^2\theta}{\sin\theta}\, x\right)$$

$$= \delta(\cos\theta\, x + \sin\theta\, y)\delta\left(\frac{1}{\sin\theta}\, x\right) = \delta(\sin\theta\, y)\delta\left(\frac{1}{\sin\theta}\, x\right) = \delta(x)\delta(y)$$

where the second equality follows since the first delta function forces $y = -\frac{\cos\theta}{\sin\theta}\, x$. This result can be generalized to any dimension. For example, in 3-dimensional space, $\delta(x')\delta(y')\delta(z') = \delta(x)\delta(y)\delta(z)$, a result we will use in chapter II.1.

5 We could either perform the integral after writing down the components of $\vec{p}$ explicitly, or argue by rotational invariance that the integral must be proportional to δ^{ij}. The proportionality constant could be then determined by contracting with δ^{ij} (using the repeated index summation convention).

I.4   Who Is Afraid of Tensors?

4 Write $\vec{\mathcal{L}} \equiv \vec{l} \times \dot{\vec{r}} + C(r)\vec{r}$ with $C(r) \equiv \frac{\kappa}{r}$ and differentiate: $\dot{\vec{\mathcal{L}}} = \vec{l} \times \ddot{\vec{r}} + C(r)\dot{\vec{r}} + C'(r)\dot{r}\vec{r}$. Use $\ddot{\vec{r}} = -\frac{\kappa}{r^3}\vec{r}$, $r^2 = \vec{r}^2$, so that $r\dot{r} = \vec{r} \cdot \dot{\vec{r}}$, and the identity derived in the text, so that $\vec{l} \times \ddot{\vec{r}} = r^2\dot{\vec{r}} - (\vec{r} \cdot \dot{\vec{r}})\vec{r}$. From here a few lines of arithmetic lead to $\dot{\vec{\mathcal{L}}} = 0$. I am dealing with the Laplace-Runge-Lenz vector per unit mass here. Of course, if you like, you can multiply everything by m and write $\vec{p} = m\dot{\vec{r}}$.

5 $S^{ij}A^{ij} = -S^{ji}A^{ji} = -S^{ij}A^{ij} = 0$, since something equal to its own negative has to vanish. Note that the second equality follows from relabeling the dummy summation indices.

8   Cyclically permute the definition of H:

$$H^{k \cdot ij} = G^{ki \cdot j} + G^{kj \cdot i}$$
$$H^{i \cdot jk} = G^{ij \cdot k} + G^{ik \cdot j}$$
$$H^{j \cdot ki} = G^{jk \cdot i} + G^{ji \cdot k}$$

Add the first two lines and subtract the third. Then $G^{ki \cdot j} = \frac{1}{2}(H^{k \cdot ij} + H^{i \cdot jk} - H^{j \cdot ki})$.

9   For example, for $D = 2$, let us evaluate $\varepsilon^{ij} R^{ip} R^{jq}$ for $p = 1, q = 2$. We have $\varepsilon^{ij} R^{i1} R^{j2} = \varepsilon^{12} R^{11} R^{22} + \varepsilon^{21} R^{21} R^{12} = \varepsilon^{12}(R^{11} R^{22} - R^{21} R^{12}) = \varepsilon^{12} \det R$.

I.5   From Change of Coordinates to Curved Spaces

1   We have $ds^2 = g_{\mu\nu}(x + dx)(-dx^\mu)(-dx^\nu) = g_{\mu\nu}(x)dx^\mu dx^\nu + \cdots$, with the dots indicating higher order terms.

3 $\Omega = \dfrac{2\pi}{W \cosh \frac{2\pi y}{W}}$

4   As explained in appendix 2, this space is just E^3. Transform coordinates by $x = \sqrt{r^2 + a^2} \sin\theta \cos\varphi$, $y = \sqrt{r^2 + a^2} \sin\theta \sin\varphi$, $z = r \cos\theta$. Note that $r = 0$ represents a disk of radius a in the $(x$-$y)$ plane. The surfaces of constant r are ellipsoids, and the lines of fixed θ and φ are hyperbolas.

5 $+1$ and -1, respectively.

10   This follows immediately from the result of exercise 9 with θ_1 renamed θ. We can also see this more geometrically by noting that the defining equation for S^d, namely $(X^1)^2 + (X^2)^2 + \cdots + (X^{d+1})^2 = 1$, may be written as $(X^1)^2 + (X^2)^2 + \cdots + (X^d)^2 = 1 - (X^{d+1})^2$, that is, as made of a collection of S^{d-1} with radius $\sqrt{1 - (X^{d+1})^2}$ as X^{d+1} ranges from -1 to $+1$.

16   Let the torus be formed out of a flexible cylindrical tube of radius a and length $2\pi L$. It is embedded in E^3 according to $X = (L + a \sin\theta) \cos\varphi$, $Y = (L + a \sin\theta) \sin\varphi$, $Z = a \cos\theta$. Note that the two coordinates θ and φ on the torus run from 0 to 2π, with θ winding around the tube and φ running around the "hole" of the torus. Then $ds^2 = dX^2 + dY^2 + dZ^2 = a^2 d\theta^2 + (L + a \sin\theta)^2 d\varphi^2$.

17   Given $ds^2 = Adu^2 + Bdv^2 + 2Cdudv$, with A, B, C functions of u, v. Let $u = f(x, y)$, $v = g(x, y)$, with two unknown functions f and g, so that

$$du = f_x dx + f_y dy, \, dv = g_x dx + g_y dy$$

Plugging in, we have

$$ds^2 = A(f_x dx + f_y dy)^2 + B(g_x dx + g_y dy)^2 + 2C(f_x dx + f_y dy)(g_x dx + g_y dy)$$

Collecting terms and setting the coefficient of $dxdy$ to 0 and the coefficients of dx^2 and of dy^2 equal to each other, we obtain two equations that we can solve for f_y and g_y. In other words, we have two equations giving $\partial_y f$ and $\partial_y f$ in terms of f_x, g_x and A, B, C. Now think of this as an initial value problem with y playing the role of time. Let us specify the two functions $f(x, y_0)$ and $g(x, y_0)$ at some initial time y_0. Our two equations tell us what $\partial_y f$ and $\partial_y f$ are, which allows us to determine the two functions $f(x, y_0 + \delta y)$ and $g(x, y_0 + \delta y)$ at some infinitesimally later time $y_0 + \delta y$. In other words, we can integrate to obtain the unknown functions $f(x, y)$ and $g(x, y)$, at least within some local region. (Of course, in the integration, the functions A, B, C are to be treated as functions of x, y, that is, $A = A(f(x, y), g(x, y))$ and so forth.) Thus, within some coordinate patch, the metric can be written in the conformally flat form $ds^2 = \Omega^2(x, y)(dx^2 + dy^2)$.

18   Change coordinates by $u = \log r$, $v = \theta$, and we obtain $ds^2 = (dr^2 + r^2 d\theta^2)$. The space is just the plane in disguise.

I.6   Curved Spaces: Gauss and Riemann

4   With no loss of generality, we can pick the point P to be $(x, y) = (0, y_*)$. A particular set of locally flat coordinates (u, v) is given by $x = y_*(u + uv + \cdots)$, $y = y_*(v + \frac{1}{2}(v^2 - u^2) + \cdots)$, where the dots represent terms cubic and higher in u, v. We could of course trivially rotate (u, v) (and also translate) to obtain another set of locally flat coordinates.

5   From $g_{11} = (1 + \mu^2 u^2)$, we have $B_{11,22} = 0$. Similarly, $B_{22,11} = 0$. From $B_{12,12} = \frac{1}{2}\mu v$. So $2B_{12,12} - B_{11,22} - B_{22,11} = \mu v$, which is indeed the intrinsic curvature.

8   We have $ds^2 = dx^2 + dz^2 / \left(pz^{\frac{p-1}{p}} \right)^2$. For example, for $z = y^2$, $g_{zz} = 1/(4z)$ blows up at $z = 0$.

10   Let one line segment go from the point x to $x + (\Delta x)_1$, and the other to $x + (\Delta x)_2$. The angle between the two line segments is given by

$$\cos \theta = (\Delta x)_1 \cdot (\Delta x)_2 / \sqrt{((\Delta x)_1 \cdot (\Delta x)_1)((\Delta x)_2 \cdot (\Delta x)_2)}$$

(defining $(\Delta x)_1 \cdot (\Delta x)_2 \equiv g_{\mu\nu}(\Delta x)_1^\mu (\Delta x)_2^\nu$ and generalizing the standard high school formula for the scalar dot product). Suppose we now calculate the angle with the metric $\tilde{g}$: the factors of Ω evidently cancel between the numerator and the denominator.

11   Solution by counting. We can choose D functions, but we have to satisfy $\frac{1}{2}D(D+1) - 1$ conditions. We have enough freedom for $D = 2$, but not for $D > 2$.

13 $R = \frac{x+y}{2x^2y^2}$

I.7   Differential Geometry Made Easy, But Not Any Easier!

2   We calculate the components of $\vec{V}$ along the two basis vectors, namely $\vec{V} \cdot \vec{e}_\mu(y)$, and then use these two components to form a linear combination of the two basis vectors. These words translate into an expression for $\vec{V}$ projected into the tangent plane: $\vec{V}_P(y) \equiv (\vec{V} \cdot \vec{e}_\mu(y))g^{\mu\nu}(y)\vec{e}_\nu(y)$, where $g^{\mu\nu}(y)$ is the inverse of the 2-by-2 matrix $g_{\mu\nu}(y)$ defined by $g^{\mu\nu}(y)g_{\nu\lambda}(y) = \delta^\mu_\lambda$. (For the sphere, $g_{\mu\nu} = \begin{pmatrix} 1 & 0 \\ 0 & \sin^2\theta \end{pmatrix}$ and $g^{\mu\nu} = \begin{pmatrix} 1 & 0 \\ 0 & 1/\sin^2\theta \end{pmatrix}$.) To see that the inverse is needed here, take the dot product of $\vec{V}_P(y)$ and a basic vector $\vec{e}_\lambda(y)$:

$$\vec{V}_P(y) \cdot \vec{e}_\lambda(y) = (\vec{V} \cdot \vec{e}_\mu(y))g^{\mu\nu}(y)\vec{e}_\nu(y) \cdot \vec{e}_\lambda(y) = (\vec{V} \cdot \vec{e}_\mu(y))g^{\mu\nu}(y)g_{\nu\lambda}(y) = (\vec{V} \cdot \vec{e}_\mu(y))\delta^\mu_\lambda = (\vec{V} \cdot \vec{e}_\lambda(y))$$

In other words, $(\vec{V}_P(y) - \vec{V}) \cdot \vec{e}_\lambda(y) = 0$, which is just another way of saying that $\vec{V}_P(y)$ and $\vec{V}$ differ by a vector normal to the surface. In other words, we subtracted out the component of $\vec{V}$ normal to the surface from $\vec{V}$ to obtain $\vec{V}_P(y)$.

3   Multiply the two eigenvalue equations (for $i = 1, 2$)

$$(K_{\mu\nu} - k_i g_{\mu\nu})t_i^\nu = 0$$

by t_j^μ (with $j \neq i$) and subtract one from the other. We obtain $(k_1 - k_2)g_{\mu\nu}t_1^\mu t_2^\nu = 0$. For those readers who have studied quantum mechanics, does this remind you of the proof of wave function orthogonality?

6   Take the dot product: $\vec{e}_\rho \cdot \vec{e}_{\mu,\nu} = g_{\rho\lambda}\Gamma^\lambda_{\mu\nu} = \Gamma_{\rho\cdot\mu\nu}$. Interchange $\rho \leftrightarrow \mu$: $\vec{e}_\mu \cdot \vec{e}_{\rho,\nu} = \Gamma_{\mu\cdot\rho\nu}$. Add to obtain $\partial_\nu(\vec{e}_\rho \cdot \vec{e}_\mu) = \partial_\nu g_{\rho\mu} = \Gamma_{\rho\cdot\mu\nu} + \Gamma_{\mu\cdot\rho\nu}$, which we can solve for Γ using I.4.15.

II.1   The Hanging String and Variational Calculus

2   Once again, we can solve $\nabla^2\Phi = GM\delta^{(3)}(\vec{x})$ by dimensional analysis. It is also easy to verify the solution explicitly. By rotational invariance, Φ can only depend on $r^2 \equiv \sum_{i=1}^D x^i x^i$. Differentiate this to obtain $r\,dr = \sum_{i=1}^D x^i dx^i$, so that $\frac{\partial r}{\partial x^i} = \frac{x^i}{r}$. Then we have

$$\sum_{i=1}^D \frac{\partial}{\partial x^i}\frac{\partial}{\partial x^i}\frac{1}{r^p} = \sum_{i=1}^D \frac{\partial}{\partial x^i}\left(-\frac{px^i}{r^{p+2}}\right) = \frac{p(p+2-D)}{r^{p+2}}$$

We see that $\Phi \propto 1/r^p$ with $p = D - 2$ solves the equation for $\vec{x}$ away from the origin. The potential goes like $1/r^{D-2}$, and so the force law is an inverse $(D-1)$ law.

3   Varying S with respect to b gives $a' = 0$, and varying it with respect to a gives $(r(1-b))' = 0$. Fitting to the boundary conditions at spatial infinity gives $a = 1$ and $b = 1 - \frac{2M}{r}$.

5   Describe the desired curve by $y(x)$, with y the vertical axis. Released at rest at $y = 0$, the bead attains a speed of $v(y) = \sqrt{2gy}$ after falling a distance of y (with the coordinate y chosen to point downward). We see that the transit time (in suitable units) is given by

$$T = \int \sqrt{\frac{dx^2 + dy^2}{y}} = \int dx \left[\frac{1}{y}\left(1 + \left(\frac{dy}{dx}\right)^2\right)\right]^{\frac{1}{2}}$$

You could proceed from here, derive the Euler-Lagrange equation, solve for $y(x)$, and "rival" Newton more than 300 years later. Note that in spite of my remark in the text, in this context, the notation $y(x)$ seems quite natural.

Here is the instructive part of the problem. We could just as well have chosen y as the variable and solved for $x(y)$. Then

$$T = \int dy \left[\frac{1}{y}\left(1 + \left(\frac{dx}{dy}\right)^2\right)\right]^{\frac{1}{2}}$$

You can verify that, in contrast to the case with the previous choice, the second order differential equation can now be integrated trivially to yield the first order differential equation

$$\left(\frac{dx}{dy}\right)^2 = \frac{y}{y^* - y}$$

with y^* an integration constant. You could solve this easily with the change of variable $y = y^* \sin^2\theta$.

Moral of the story: in solving variational problems, it pays to choose the independent variable wisely.

II.2   The Shortest Distance between Two Points

6   Varying the stated quantity, we obtain $\frac{d}{d\lambda}\left(2g_{\sigma\nu}\frac{dX^\nu}{d\lambda}\right) = (\partial_\sigma g_{\mu\nu})\frac{dX^\mu}{d\lambda}\frac{dX^\nu}{d\lambda}$ almost instantly, but this is just (16).

8   Simply plug $g_{\mu\sigma} = e^{2\Omega}\delta_{\mu\sigma}$ into (24) to obtain $\frac{d}{dl}\left(e^{2\Omega}\delta_{\mu\sigma}V^\mu\right) = \partial_\sigma\Omega$ (since $g_{\mu\nu}V^\mu V^\nu = 1$). Noting that $\frac{d}{dl}e^{2\Omega} = 2V^\nu\partial_\nu\Omega e^{2\Omega}$, we find, after cleaning up a bit,

$$\frac{dV^\lambda}{dl} + 2\left(V^\nu \partial_\nu \Omega\right) V^\lambda = \partial^\lambda \Omega$$

where $\partial^\lambda = g^{\lambda\sigma} \partial_\sigma \Omega$.

II.3   Physics Is Where the Action Is

4 $\delta S = 2 \int dx^+ dx^- \left(\frac{\partial \delta\phi}{\partial x^+} \frac{\partial \phi}{\partial x^-} + \frac{\partial \phi}{\partial x^+} \frac{\partial \delta\phi}{\partial x^-} \right) = 4 \int dx^+ dx^- \frac{\partial^2 \phi}{\partial x^+ \partial x^-} \delta\phi$

III.1   Galileo versus Maxwell

1   For convenience, write the velocity of the incoming ball as $2\vec{v}$. In the center of mass frame, one ball has velocity $\vec{v}$, the other $-\vec{v}$. After the collision, one ball has velocity $\vec{u}$, the other $-\vec{u}$. Energy conservation gives $\vec{u}^2 = \vec{v}^2$. In the lab frame, after the collision, one ball has velocity $\vec{v} + \vec{u}$, the other $\vec{v} - \vec{u}$. Since $(\vec{v} + \vec{u}) \cdot (\vec{v} - \vec{u}) = \vec{v}^2 - \vec{u}^2 = 0$, the angle between the velocities of the two balls is 90°.

Of course, the problem is so elementary that we can also easily do it in the lab frame. Momentum and energy conservation give $\vec{v} = \vec{u}_1 + \vec{u}_2$, $\vec{v}^2 = \vec{u}_1^2 + \vec{u}_2^2$. Squaring the first equation and comparing with the second yields $\vec{u}_1 \cdot \vec{u}_2 = 0$ immediately.

III.3   Minkowski and the Geometry of Spacetime

1   This is given in the preceding chapter.

3   Differentiate $\eta_{\mu\nu} V^\mu V^\nu = -1$: $\frac{d}{d\tau} \eta_{\mu\nu} V^\mu V^\nu = 0 = 2\eta_{\mu\nu} a^\mu V^\nu = a_\nu V^\nu$. In the rest frame of the particle $V^\mu = (1, \vec{0})$ and hence $a^0 = 0$.

4   We can choose $\vec{V}$ to point in the 1-direction and $x^2 = x^3 = 0$. Solving the three equations $V_\mu V^\mu = -1$, $a_\mu V^\mu = 0$, and $a_\mu a^\mu = g^2$, we obtain $a^0 = \frac{dV^0}{d\tau} = gV^1$ and $a^1 = \frac{dV^1}{d\tau} = gV^0$, giving the solution $t(\tau) = x^0(\tau) = g^{-1} \sinh g\tau$, $x(\tau) = x^1(\tau) = g^{-1} \cosh g\tau$. Thus, $x^\mu(\tau)$ traces out a hyperbola $x(\tau)^2 = t(\tau)^2 + g^{-2}$. We have $V^\mu = (\cosh g\tau, \sinh g\tau, 0, 0)$ and $a^\mu = g(\sinh g\tau, \cosh g\tau, 0, 0)$, which satisfy all the stated equations. Note that with some suitable adjustments, this shows that for fixed ρ, this coordinate transformation amounts to a transformation to the frame of an accelerating observer, with $T = g\tau$ and $\rho = g^{-1}$.

6   As explained in the text, we have $F^{\mu\nu} \to F'^{\mu\nu} = \Lambda^\mu{}_\sigma \Lambda^\nu{}_\omega F^{\sigma\omega}$. Hence, $F'^{0i} = \Lambda^0{}_\sigma \Lambda^i{}_\omega F^{\sigma\omega}$ and $F'^{ij} = \Lambda^i{}_\sigma \Lambda^j{}_\omega F^{\sigma\omega}$. For example, for a boost along the 1-axis, Λ is given explicitly in the text, and you merely have to write out the repeated index sums.

8   Denote by $\Lambda(x, \varphi)$ a boost in the x direction by the rapidity parameter φ, and so on and so forth. Then, with the abbreviation $c = \cosh \varphi$, $s = \sinh \varphi$, $c' = \cosh \varphi'$, $s' = \sinh \varphi'$, we have

$$\Lambda(x, \varphi)\Lambda(y, \varphi') = \begin{pmatrix} c & s & 0 \\ s & c & 0 \\ 0 & 0 & 1 \end{pmatrix} \begin{pmatrix} c' & 0 & s' \\ 0 & 1 & 0 \\ s' & 0 & c' \end{pmatrix} = \begin{pmatrix} cc' & s & cs' \\ sc' & c & ss' \\ s' & 0 & c' \end{pmatrix} \tag{29}$$

Next we follow what we did in appendix 2 to chapter I.3. Compare $\Lambda(y, \varphi')\Lambda(x, \varphi)$ with $\Lambda(x, \varphi)\Lambda(y, \varphi')$ by calculating

$$(\Lambda(y, \varphi')\Lambda(x, \varphi))^{-1}\Lambda(x, \varphi)\Lambda(y, \varphi') \simeq I + \varphi\varphi' \begin{pmatrix} 0 & 0 & 0 \\ 0 & 0 & 1 \\ 0 & -1 & 0 \end{pmatrix}$$

$$= I + \varphi\varphi'[K_x, K_y] + \cdots$$

$$= I - i\varphi\varphi' J_z \tag{30}$$

Here we used the symmetry of Λ, so that $\Lambda(y, \varphi')\Lambda(x, \varphi) = (\Lambda(x, \varphi)\Lambda(y, \varphi'))^T$, and thus in computing the left hand side of (30) to the order indicated, we need only keep the ss' term in (29). We have thus verified (24).

III.5 The Worldline Action and the Unification of Material Particles with Light

1 As in (15), we vary with respect to the auxiliary variable $\gamma_{\alpha\beta}$, which we then eliminate. We would like to vary S with respect to $\gamma_{\alpha\beta}$. Using a matrix identity we have used again and again, we have $\delta\gamma^{\alpha\beta} = -\gamma^{\alpha\varepsilon}\delta\gamma_{\varepsilon\eta}\gamma^{\eta\beta}$ and $\delta\gamma = \gamma\gamma^{\beta\alpha}\delta\gamma_{\alpha\beta}$. For ease of writing, define $h_{\alpha\beta} \equiv \partial_\alpha X^\mu \partial_\beta X_\mu$. The variation of the integrand in (21) thus gives $\delta[\gamma^{\frac{1}{2}}\gamma^{\alpha\beta}h_{\alpha\beta}] = \gamma^{\frac{1}{2}}[\frac{1}{2}\gamma^{\eta\varepsilon}\delta\gamma_{\varepsilon\eta}(\gamma^{\alpha\beta}h_{\alpha\beta}) - \gamma^{\alpha\varepsilon}\delta\gamma_{\varepsilon\eta}\gamma^{\eta\beta}h_{\alpha\beta}]$. Setting the coefficient of $\delta\gamma_{\varepsilon\eta}$ to 0, we obtain

$$h_{\varepsilon\eta} = \frac{1}{2}\gamma_{\varepsilon\eta}(\gamma^{\alpha\beta}h_{\alpha\beta})$$

where the indices on h are raised and lowered by the metric γ. Multiplying this equation by $h^{\eta\varepsilon}$ (and summing over repeated indices), we find $\gamma^{\alpha\beta}h_{\alpha\beta} = 2$ and thus $\gamma_{\varepsilon\eta} = h_{\varepsilon\eta}$. Plugging this into (21), we find that $S = \frac{1}{2}T\int d\tau d\sigma (\det h)^{\frac{1}{2}}2$. Thus, S and $S_{\text{Nambu-Goto}}$ are indeed equivalent in the sense that they lead to the same equation of motion.

III.6 Completion, Promotion, and the Nature of the Gravitational Field

1 Consider a head-on collision $p + k \to p' + k'$ with $p = (E, 0, 0, p)$ (note the trivial abuse of notation here), $k = \omega(1, 0, 0, -1)$, and $k' = \omega'(1, 0, \sin\theta, \cos\theta)$. Minkowski squaring $p' = p + k - k'$ gives us, for $\omega \ll p$, $\omega' = \frac{\omega(E+p)}{E+\omega-(p-\omega)\cos\theta}$, which is maximized when $\cos\theta = 1$. For a highly relativistic particle, $p = \sqrt{E^2 - m^2} \simeq E - \frac{m^2}{2E}$, and we obtain the stated result.

2 The identity is $s + t + u = \sum_a m_a^2$.

3 In calculating $\partial_\mu T^{\mu\nu}$, we see that the first few steps are the same as in calculating $\partial_\mu n^\mu$ as given in the text. The reason is that, during these steps, the quantity $\frac{dq_a^\nu}{d\tau_a}$ in (7) is just going along for the ride. We arrive at

$$\partial_\mu T^{\mu\nu} = -\sum_a \int d\tau_a m_a \frac{dq_a^\nu}{d\tau_a}\frac{d}{d\tau_a}\delta^{(4)}(x - q_a(\tau_a)) = \sum_a \int d\tau_a m_a \frac{d^2 q_a^\nu}{d\tau_a^2}\delta^{(4)}(x - q_a(\tau_a)) = 0$$

upon using the equation of motion $\frac{d^2 q_a^\nu}{d\tau_a^2} = 0$.

4 $T^{0i}(x) = (\rho + P)\frac{v^i}{1-\vec{v}^2}$, $T^{ij}(x) = (\rho + P)\frac{v^i v^j}{1-\vec{v}^2} + P\delta^{ij}$

5 The number of components is given by $\frac{1}{2}d(d-1)$, which for $d = 4$ is equal to 6. First, F^{0i} is a 3-vector. We are then left with the 3 components $F^{ij} = -F^{ji}$. Form the 3-vector $\varepsilon^{ijk}F^{jk}$.

IV.2 Electromagnetism Goes Live

4   We have

$$\partial_0 T^{00} = \frac{1}{2}\frac{\partial}{\partial t}(\vec{E}^2 + \vec{B}^2) = \vec{E}\cdot\frac{\partial\vec{E}}{\partial t} + \vec{B}\cdot\frac{\partial\vec{B}}{\partial t}$$

$$= -\vec{E}\cdot\vec{J} + \vec{E}\cdot\vec{\nabla}\times\vec{B} - \vec{B}\cdot\vec{\nabla}\times\vec{E}$$

$$= -\vec{E}\cdot\vec{J} - \vec{\nabla}\cdot(\vec{E}\times\vec{B})$$

As you may know from a course on electromagnetism, the vector $\vec{E}\times\vec{B}$ is known as the Poynting vector and measures the momentum flow in an electromagnetic field. Note that the equation we just derived is the $\nu = 0$ component of the equation $\partial_\mu T^{\mu\nu}_{\text{electromagnetic}} = -J_\lambda F^{\nu\lambda}$ we derived in exercise 2b. The relativistic notation is far more compact!

5   Using the antisymmetry of the ϵ symbol and $\epsilon_{0123} = -1$, we have $\tilde{F}_{01} = -\epsilon_{0123}F^{23} = B^1$, after referring to (IV.1.17). Next, $\tilde{F}_{12} = -\epsilon_{1203}F^{03} = E^3$. Comparing with (IV.1.17), we see that going from $F_{\mu\nu}$ to the dual tensor $\tilde{F}_{\mu\nu}$, we exchange the roles of the electric and the magnetic fields up to a sign: $\vec{E}\to -\vec{B}$ and $\vec{B}\to\vec{E}$.

7   Simply plug the identity in the preceding exercise into the expression for the energy momentum tensor of the electromagnetic field.

8   The two invariants under Lorentz transformation with the stated property are $F_{\mu\nu}F^{\mu\nu}$ and $F_{\mu\nu}\tilde{F}^{\mu\nu}$. We have already encountered the first one in exercise 1. Up to an overall constant, the second invariant is equal to $\vec{E}\cdot\vec{B}$.

10   As in the derivation of the virial theorem in classical mechanics, we want to take the time average of various quantities. Define $\langle A\rangle \equiv \frac{1}{T}\int_0^T dt\, A$ for T large. Note that, provided that a time dependent quantity $B(T)$ remains bounded, $\langle\frac{dB}{dt}\rangle = \frac{1}{T}(B(T) - B(0))\to 0$ for T large. This is where the assumption that the motion of the particles is confined to a finite region comes in.

Let $T^{\mu\nu} = T^{\mu\nu}_{\text{particles}} + T^{\mu\nu}_{\text{electromagnetic}}$, as in exercise 2b. Time averaging the conservation law $\partial_\mu T^{\mu j} = \partial_0 T^{0j} + \partial_i T^{ij} = 0$, we obtain $\partial_i\langle T^{ij}\rangle = 0$. Therefore, $\int d^3x\, x^j\partial_i\langle T^{ij}\rangle = 0 = -\int d^3x\langle T^{ii}\rangle$. In the last step, we integrated by parts. Here repeated indices are summed.

Next, using the result of exercise 9 and the expression for $T^{\mu\nu}_{\text{particles}}$ in (III.6.7), we have

$$T \equiv \eta_{\mu\nu}T^{\mu\nu} = \eta_{\mu\nu}T^{\mu\nu}_{\text{particles}} = -\sum_a\int d\tau_a m_a\delta^{(4)}(x - q_a(\tau_a))$$

$$= -\sum_a m_a\sqrt{1 - \vec{v}^2_{N,a}}\,\delta^{(3)}(\vec{x} - \vec{q}_a(\tau_a))$$

where τ_a is understood as the solution of $q_a^0(\tau_a) = x^0 = t$. For the last equality, we used (III.6.11) and the discussion that follows it to integrate over τ_a.

Finally, putting things together, we obtain

$$-\int d^3x\langle T\rangle = \sum_a m_a\left\langle\sqrt{1 - \vec{v}^2_{N,a}}\right\rangle = \int d^3x\langle(T^{00} - T^{ii})\rangle = \int d^3x\langle T^{00}\rangle = E$$

where E denotes the total energy of the system. Thus, we obtain the relativistic virial theorem

$$E = \sum_a m_a\left\langle\sqrt{1 - \vec{v}^2_{N,a}}\right\rangle$$

In the nonrelativistic limit, we recover the usual virial theorem $E - \sum_a m_a = -\frac{1}{2}\sum_a m_a\langle\vec{v}^2_{N,a}\rangle$. While the right hand side is equal to minus the time averaged kinetic energy $\langle K\rangle$, the left hand side is the total nonrelativistic energy, which we can write as the time averaged kinetic plus potential energy $\langle K + V\rangle$, since $K + V$ is conserved. We thus recognize the more familiar nonrelativistic form of the virial theorem

$$\langle V \rangle = -2\langle K \rangle$$

As the total energy decreases, the kinetic energy, that is, the temperature, increases.

Prologue to Book Two: The Happiest Thought

1   First of all, we have to understand how a burning candle works normally. The hot gas produced by the burning candle, being less dense than air, rushes upward. The upward rush of the glowing gas is what we see as the flame. The candle is thus assured of a steady supply of oxygen from the ambient air as the gas rushes out of the way. The second point is that the upward rush of the gas can be better interpreted as due to gravity pulling the denser air down. By moving downward, the ambient air is actually displacing the gas upward. The falling candle feels no gravity, and neither does the air around it. The hot gas expands outward rather than rushing upward out of the way. For a moment, the candle is deprived of air supply and goes out. Watch this on the web! (http://www.youtube.com/watch?v=NlBp21fqguU)

V.1   Spacetime Becomes Curved

1   The helium in the balloon does not know, momentarily, that the car has stopped and tries to continue its forward motion. But the air in the car is trying to do the same. When the air reaches the front part of the interior of the car, it flows back. Since the density of air is higher than the density of helium, it pushes the helium balloon back. Unlike other massive objects in the car, such as the driver and the passengers, the helium balloon jerks backward rather than forward.

V.2   The Power of the Equivalence Principle

1   Let $R =$ earth radius, $\omega =$ earth's angular velocity, and $h =$ altitude of plane. With $dr = 0$ and $d\theta = 0$, the proper time interval is given by

$$d\tau^2 = \left(1 - \frac{2GM}{r}\right) dt^2 - r^2 d\phi^2 = \left(1 - \frac{2GM}{r} - r^2 \left(\frac{d\phi}{dt}\right)^2\right) dt^2$$

Since $\frac{d\phi}{dt} \simeq \frac{(R+h)\omega+v}{R+h}$, we have

$$d\tau = \left(1 - \frac{GM}{R+h} - \frac{1}{2}(R\omega + v + h\omega)^2\right) dt$$

with the proper time interval $d\tau_g$ of the clock on the ground given by this expression with v and h set to 0. The fractional shift is thus equal to

$$\Delta = \frac{d\tau - d\tau_g}{d\tau_g} \simeq \frac{GMh}{R^2} - R\omega - \frac{1}{2}v^2$$

We find that the fractional shift between the eastward flying clock and the westward clock is $\Delta_{ew} = \frac{d\tau_e - d\tau_w}{d\tau_g} \simeq -v^2$.

V.3   The Universe as a Curved Spacetime

5   With $r = L \sin \psi$, for example, we have $L d\psi = \frac{dr}{\sqrt{1 - \frac{r^2}{L^2}}}$.

6   Do I have to teach you how to integrate?

7   The trajectory of the light ray is determined by $\frac{dt}{a(t)} = \frac{dr}{\sqrt{1-k\frac{r^2}{L^2}}}$, and thus the proper time interval Δt_R between the two pulses is now given by

$$\int_0^R \frac{dr}{\sqrt{1-k\frac{r^2}{L^2}}} = \int_{t_S}^T \frac{dt}{a(t)} = \int_{t_S+\Delta t_S}^{T+\Delta t_R} \frac{dt}{a(t)}$$

The derivation then proceeds as in the text. We don't care about the r integral, only the equality between the two t integrals.

V.4   Motion in Curved Spacetime

2   We obtain $\Gamma^\nu_{\ r\nu} = \frac{A'}{2A} + \frac{B'}{2B} + \frac{2}{r}$ and $\Gamma^\nu_{\ \theta\nu}\cot\theta$, and we verify that $\Gamma^\nu_{\ \mu\nu} = \frac{1}{\sqrt{g}}\partial_\mu\sqrt{g}$ is satisfied, since $g = r^4 AB\sin^2\theta$.

4   With $A = 1$ and θ, φ constant, $\epsilon = 1$, r constant solves the radial equations of motion.

5   A plot of $v(r)$ shows that for $r > 2GM$, particles "fall" toward larger r.

6   Plugging $\tilde{g}_{\mu\nu}(x) = \Omega^2(x)g_{\mu\nu}(x)$ into the expression for the Christoffel symbol, we obtain

$$\tilde{\Gamma}^\mu_{\ \nu\lambda} = \Gamma^\mu_{\ \nu\lambda} + \left(\delta^\mu_\nu\partial_\lambda + \delta^\mu_\lambda\partial_\nu - g_{\nu\lambda}g^{\mu\rho}\partial_\rho\right)\log\Omega$$

Thus, in the "twiddle" spacetime, the geodesic equation $\frac{d^2X^\mu}{d\tau^2} + \tilde{\Gamma}^\mu_{\ \nu\lambda}\frac{dX^\nu}{d\tau}\frac{dX^\lambda}{d\tau} = 0$ is manifestly not the same as the geodesic equation $\frac{d^2X^\mu}{d\tau^2} + \Gamma^\mu_{\ \nu\lambda}\frac{dX^\nu}{d\tau}\frac{dX^\lambda}{d\tau} = 0$ in the "nontwiddle" spacetime.

However, for a massless particle, if X^μ describes a trajectory so that $g_{\mu\nu}(X)dX^\mu dX^\nu = 0$ according to (19), then clearly $\tilde{g}_{\mu\nu}(X)dX^\mu dX^\nu = \Omega^2(X)g_{\mu\nu}(X)dX^\mu dX^\nu = 0$.

It is also instructive to show that the geodesic equation (20) holds in both spacetimes. Suppose that it holds in the "twiddle" spacetime. Then we have

$$\frac{d^2X^\mu}{d\zeta^2} + \tilde{\Gamma}^\mu_{\ \nu\lambda}\frac{dX^\nu}{d\zeta}\frac{dX^\lambda}{d\zeta} = \frac{d^2X^\mu}{d\zeta^2} + \Gamma^\mu_{\ \nu\lambda}\frac{dX^\nu}{d\zeta}\frac{dX^\lambda}{d\zeta} + 2\frac{dX^\mu}{d\zeta}\frac{dX^\lambda}{d\zeta}\partial_\lambda\log\Omega = 0$$

where we used $dX \cdot dX = 0$ for a massless particle. This does not look like the geodesic equation in the "nontwiddle" spacetime.

But suppose we write $\zeta(\eta)$. Then we have

$$\frac{dX^\mu}{d\zeta} = \frac{d\eta}{d\zeta}\frac{dX^\mu}{d\eta} \quad \text{and} \quad \frac{d^2X^\mu}{d\zeta^2} = \frac{d\eta}{d\zeta}\frac{d}{d\eta}\left(\frac{d\eta}{d\zeta}\frac{dX^\mu}{d\eta}\right) = \left(\frac{d\eta}{d\zeta}\right)^2\frac{d^2X^\mu}{d\eta^2} + \frac{d^2\eta}{d\zeta^2}\frac{dX^\mu}{d\eta}$$

Now note that the last term on the left hand side is equal to

$$2\frac{dX^\mu}{d\zeta}\frac{d}{d\zeta}\log\Omega = 2\left(\frac{d\eta}{d\zeta}\right)^2\frac{dX^\mu}{d\eta}\left(\frac{d\Omega}{d\eta}/\Omega\right)$$

(Here Ω is evaluated on the trajectory of the particle and hence may be regarded as either a function of ζ or η.) Thus, by choosing $\eta(\zeta)$ to satisfy $\frac{d\eta}{d\zeta} = \frac{1}{\Omega^2}$, we can knock off this unwanted last term and obtain $\frac{d^2X^\mu}{d\eta^2} + \Gamma^\mu_{\ \nu\lambda}\frac{dX^\nu}{d\eta}\frac{dX^\lambda}{d\eta} = 0$.

V.6   Covariant Differentiation

4   Varying the action

$$S = \int d^4x \sqrt{-g} \left(-\frac{1}{2} g^{\lambda\mu} g^{\rho\nu} F_{\lambda\rho} \partial_\mu A_\nu + J^\nu A_\nu \right)$$

we obtain

$$\partial_\mu \left(\sqrt{-g} g^{\lambda\mu} g^{\rho\nu} F_{\lambda\rho} \right) = -\sqrt{-g} J^\nu$$

that is,

$$D_\mu F^{\mu\nu} = \frac{1}{\sqrt{-g}} \partial_\mu \left(\sqrt{-g} F^{\mu\nu} \right) = -J^\nu$$

Compare with (IV.2.13).

5 Note that in curved spacetime, F_{01} and F_{12}, are no longer equal to $-F^{01}$ and F^{12}, respectively. Our convention is to define $\vec{E} = (E^1, E^2, E^3)$ and $\vec{B} = (B^1, B^2, B^3)$ by $E^1 = -F_{01}$ and $B^3 = F_{12}$ and their cyclic analogs. For $ds^2 = dt^2 - a(t)^2 d\vec{x}^2$, $-g = a^6$, $g^{00} = 1$, $g^{11} = 1/a^2$, and so $S = -\frac{1}{2} \int dt d^3x (a(t)\vec{E}^2 - \frac{1}{a(t)}\vec{B}^2)$.

VI.1 To Einstein's Field Equation, as Quickly as Possible

1 Simply plug in the Christoffel symbols for the sphere $\Gamma^\theta_{\varphi\varphi} = -\sin\theta\cos\theta$ and $\Gamma^\varphi_{\theta\varphi} = \frac{\cos\theta}{\sin\theta}$ into the expressions $D_\lambda W_\mu = \partial_\lambda W_\mu - \Gamma^\sigma_{\lambda\mu} W_\sigma$ and $D_\lambda U^\mu \equiv \partial_\lambda U^\mu + \Gamma^\mu_{\lambda\nu} U^\nu$ to obtain

$$D_\theta W_\theta = \partial_\theta W_\theta, \qquad D_\theta W_\varphi = \partial_\theta W_\varphi - \frac{\cos\theta}{\sin\theta} W_\varphi,$$

$$D_\varphi W_\theta = \partial_\varphi W_\theta - \frac{\cos\theta}{\sin\theta} W_\varphi, \qquad D_\varphi W_\varphi = \partial_\varphi W_\varphi + \sin\theta\cos\theta W_\theta$$

and

$$D_\theta U^\theta = \partial_\theta U^\theta, \qquad D_\theta U^\varphi = \partial_\theta U^\varphi + \frac{\cos\theta}{\sin\theta} U^\varphi,$$

$$D_\varphi U^\theta = \partial_\varphi U^\theta - \sin\theta\cos\theta U^\varphi, \qquad D_\varphi U^\varphi = \partial_\varphi U^\varphi + \frac{\cos\theta}{\sin\theta} U^\theta$$

The suggested check is then easily done.

2 In calculating the left hand side of (5), $[D_\mu, D_\nu]S_\rho$, we could choose ρ to be either θ or φ. I will do one case and let you do the other. We have

$$D_\varphi D_\theta S_\theta = \partial_\varphi (D_\theta S_\theta) - \frac{\cos\theta}{\sin\theta} D_\varphi S_\theta - \frac{\cos\theta}{\sin\theta} D_\theta S_\varphi$$

$$= \partial_\varphi \partial_\theta S_\theta - \frac{\cos\theta}{\sin\theta} \partial_\varphi S_\theta + \left(\frac{\cos\theta}{\sin\theta}\right)^2 S_\varphi - \frac{\cos\theta}{\sin\theta} \partial_\theta S_\varphi + \left(\frac{\cos\theta}{\sin\theta}\right)^2 S_\varphi$$

and

$$D_\theta D_\varphi S_\theta = \partial_\theta (D_\varphi S_\theta) - \frac{\cos\theta}{\sin\theta} D_\varphi S_\theta$$

$$= \partial_\theta (\partial_\varphi S_\theta) - \partial_\theta \left(\frac{\cos\theta}{\sin\theta} S_\varphi\right) - \frac{\cos\theta}{\sin\theta} \partial_\varphi S_\theta + \left(\frac{\cos\theta}{\sin\theta}\right)^2 S_\varphi$$

Subtracting, we find $[D_\theta, D_\varphi]S_\theta = -S_\varphi$. Equating this to $-R^\sigma_{\theta\varphi\theta} S_\sigma$, and invoking the symmetry property of the curvature tensor and the diagonality of the metric, we obtain $R^\varphi_{\theta\varphi\theta} = 1$. From this $R_{\theta\theta} = 1$ follows. We also have $R^\varphi_{\theta\varphi\theta} = g^{\varphi\varphi} R_{\varphi\theta\varphi\theta} = g^{\varphi\varphi} g_{\theta\theta} R^\theta_{\varphi\theta\varphi}$, and so $R^\theta_{\varphi\theta\varphi} = \sin^2\theta$. Hence we have $R_{\varphi\varphi} = \sin^2\theta$.

Finally, we obtain $R = g^{\theta\theta} R_{\theta\theta} + g^{\varphi\varphi} R_{\varphi\varphi} = 2$, as expected.

6–7 In 2 dimensions, (14) simplifies to the expression stated. If we are so fortunate that we are already in locally flat coordinates, we can read off the single component of the Riemann curvature tensor. Let us check the example in (I.6.2): we read off $B_{11,22} = c^2$, $B_{12,12} = \frac{1}{2}(ab + c^2)$, $B_{22,11} = c^2$, and so $R_{1212} = ab - c^2$, in complete agreement with (I.6.3).

8 The first one is flat (the transformation from Cartesian coordinates is $x = u + v^2$, $y = v - u^2/2$). The second has a scalar curvature given by $-4v/(1 + 4uv - 2v^2 + 2u^2v^2)^2$.

10 The antisymmetry in (10) and (15) implies that the indices A and B in the Petrov notation can each take on $\frac{1}{2}d(d-1)$ values. Next, (16) implies that the matrix R_{AB} is symmetric, and thus contains $\frac{1}{2}\{\frac{1}{2}d(d-1)\}\{\frac{1}{2}d(d-1) + 1\} = d(d-1)(d^2 - d + 2)/8$ independent components. Finally, after imposing the constraints from exercise 3, we are left with $d(d-1)(d^2 - d + 2)/8 - d(d-1)(d-2)(d-3)/24 = d^2(d^2 - 1)/12$, in agreement with chapter I.6.

11 Equating the number of independent components in the Riemann curvature tensor $d^2(d^2 - 1)/12$ to the number of independent components in the Ricci curvature tensor $d(d + 1)/2$, we obtain the cubic $d^3 - 7d - 6 = (d - 3)(d + 1)(d + 2) = 0$. Thus, for $d = 3$, the two tensors have the same number of components.

13 Plugging $\tilde{g}_{\mu\nu} = \Omega^2 g_{\mu\nu}$ into the definition of the Christoffel symbol, we obtain the first equality in (26) immediately as a result of the product rule of differentiation. Writing it schematically as $\tilde{\Gamma} \sim \Gamma + \Omega^{-1}\partial\Omega$ and plugging it into the definition $\tilde{R}^{\cdot}{}_{...} \sim \partial\tilde{\Gamma} + \tilde{\Gamma}\tilde{\Gamma}$, we have $\tilde{R}^{\cdot}{}_{...} \sim \partial\Gamma + \Gamma\Gamma + \Omega^{-1}\partial\partial\Omega + \Gamma\Omega^{-1}\partial\Omega + \Omega^{-1}\partial\Omega\Omega^{-1}\partial\Omega$. Convincing ourselves that the third and fourth terms combine into $\Omega^{-1}D\partial\Omega$ (as they must and as we can readily verify by keeping track of the indices for one specific combination), we obtain schematically $\tilde{R}^{\cdot}{}_{...} \sim R^{\cdot}{}_{...} + \Omega^{-1}D\partial\Omega + \Omega^{-2}\partial\Omega\partial\Omega$. Once we realize this, we can simplify the rest of the calculation drastically by letting $g_{\mu\nu}$ be the flat metric (that is, $\eta_{\mu\nu}$ for spacetime or $\delta_{\mu\nu}$ for space), so that the problem reduces to that of calculating the Riemann curvature tensor for the metric $g_{\mu\nu} = \Omega^2\eta_{\mu\nu}$, collecting the two sets of terms of the forms $\partial\partial\Omega$ and $\partial\Omega\partial\Omega$. Once we have that result, we can then promote $\eta_{\mu\nu}$ to $g_{\mu\nu}$, and so on, to obtain $\tilde{R}^{\mu}{}_{\nu\lambda\sigma}$.

VI.2 To Cosmology as Quickly as Possible

1 Calculate the Christoffel symbol, then the Ricci tensor. For example,

$$\Gamma^x_{tx} = p/t, \quad R_{tt} = (-p^2 + p - q^2 + q - r^2 + r)/t^2 \quad \text{and} \quad R_{xx} = p(p + q + r - 1)t^{2(p-1)}$$

(By the way, we will calculate the curvature for the Kasner universe using differential forms in chapter IX.8.) Einstein's equations $R_{\mu\nu} = 0$ are solved for $p + q + r = p^2 + q^2 + r^2 = 1$. These 2 equations could of course be immediately solved by eliminating q and r in terms of p, giving a 1-parameter family of universes.

There exists, however, a more elegant and symmetrical solution based on the identity $e^{i\pi/3} + e^{-i\pi/3} + e^{i\pi} = 0$. Write $p = a + 2b\cos(\theta + (\pi/3)) = a + b(e^{i\theta}e^{i\pi/3} + e^{-i\theta}e^{-i\pi/3})$, $q = a + 2b\cos(\theta - (\pi/3))$, $r = a + 2b\cos(\theta - \pi)$. Then $p + q + r = 1$ if $a = 1/3$. Next, using the identity $e^{2i\pi/3} + e^{-2i\pi/3} + e^{2i\pi} = 0$, we have $p^2 + q^2 + r^2 = 3a^2 + 2ab(0) + 3(2b^2) = 1$ if $b = 1/3$.

Interestingly, the solution has the following geometrical interpretation. Draw a circle of radius $2/3$ centered at $(x, y) = (1/3, 0)$. Inscribe an equilateral triangle inside the circle and oriented at some suitable angle. The projections of the 3 vertices on the x-axis give p, q, r.

A more obvious geometrical construction would be to go to 3-dimensional Euclidean space and label the axes as p, q, r. Then $p + q + r = p^2 + q^2 + r^2 = 1$ describes the circle formed by intersecting a unit sphere centered by the plane passing through $(1, 0, 0)$, $(0, 1, 0)$, and $(0, 0, 1)$.

VI.3   The Schwarzschild-Droste Metric and Solar System Tests of Einstein Gravity

2   For example, $R^t_{\ rtr} = -\frac{A''(r)}{2A(r)} + \frac{A'(r)B'(r)}{4A(r)B(r)} + \frac{A'(r)^2}{4A(r)^2}$.

3   Use the transformation $r = \rho(1 + \frac{GM}{2\rho})^2$. The horizon occurs at $\rho = \frac{1}{2}GM$, where g_{00} vanishes, and which translates into $r = 2GM$ as expected.

4   Use the transformation $R = r - GM$ with $\vec{x}$ related to (R, θ, φ) by the usual Cartesian to spherical coordinate transformation.

5   Repeat the calculation in the text using the metric used in the post-Newtonian parametrization. All the steps are conceptually the same, but arithmetically, such parameters as β and γ appear here and there.

VI.4   Energy Momentum Distribution Tells Spacetime How to Curve

4   Using (2), we obtain

$$\delta S_{\text{scalar}} = \frac{1}{2}\int d^4x \sqrt{-g}\left\{g^{\lambda\mu}\delta g_{\mu\nu}g^{\nu\rho}\partial_\lambda\varphi\partial_\rho\varphi - g^{\mu\nu}\delta g_{\mu\nu}\left(\tfrac{1}{2}(\partial\varphi)^2 + V(\varphi)\right)\right\}$$

and hence the stated result. We have in flat spacetime $T^{00} = \partial^0\varphi\partial^0\varphi + \eta^{00}(\frac{1}{2}(\partial\varphi)^2 + V(\varphi)) = \frac{1}{2}\{(\partial^0\varphi)^2 + (\vec{\nabla}\varphi)^2\} + V(\varphi)$.

5   We have

$$\int d^4x \sqrt{-g(x)}g^{\sigma\zeta}(x)g^{\lambda\rho}(x) = \int d^4x' \sqrt{-g'(x')}g'^{\sigma\zeta}(x')g'^{\lambda\rho}(x')$$

$$= \int d^4x \sqrt{-g'(x)}g'^{\sigma\zeta}(x)g'^{\lambda\rho}(x)$$

where the first equality follows from invariance under coordinate transformation and the second from renaming the dummy integration variable. Using the leading order expression $g'_{\rho\sigma}(x) - g_{\rho\sigma}(x) = -(g_{\mu\sigma}(x)\partial_\rho\varepsilon^\mu(x) + g_{\rho\nu}(x)\partial_\sigma\varepsilon^\nu(x) + \varepsilon^\lambda\partial_\lambda g_{\rho\sigma}(x)) + O(\varepsilon^2)$ in (18), we arrive at (20), with $T^{\rho\sigma}$ the energy momentum tensor associated with the cosmological constant term.

7   Recall that

$$D_\mu T^{\mu\nu} = \partial_\mu T^{\mu\nu} + \Gamma^\mu_{\mu\lambda}T^{\lambda\nu} + \Gamma^\nu_{\mu\lambda}T^{\mu\lambda} = \frac{1}{\sqrt{-g}}\partial_\mu\left(\sqrt{-g}T^{\mu\nu}\right) + \Gamma^\nu_{\mu\lambda}T^{\mu\lambda}$$

Plugging in the given form of $T^{\mu\nu}$ and suppressing an overall factor of $\frac{m}{\sqrt{-g(x)}}$ for the moment, we obtain for the first term on the right hand side

$$\int d\tau \frac{dX^\mu}{d\tau}\frac{dX^\nu}{d\tau}\partial_\mu\delta^4(x - X(\tau)) = -\int d\tau \frac{dX^\nu}{d\tau}\frac{dX^\mu}{d\tau}\frac{\partial}{\partial X^\mu}\delta^4(x - X(\tau))$$

$$= -\int d\tau \frac{dX^\nu}{d\tau}\frac{d}{d\tau}\delta^4(x - X(\tau)) = \int d\tau \frac{d^2X^\nu}{d\tau^2}\delta^4(x - X(\tau))$$

where we integrated by parts in the last step. Putting it together, we find that $D_\mu T^{\mu\nu} = 0$ gives $\frac{d^2X^\nu}{d\tau^2} + \Gamma^\nu_{\mu\lambda}\frac{dX^\mu}{d\tau}\frac{dX^\lambda}{d\tau} = 0$, which is just the geodesic equation of motion. The result is hardly surprising: the energy momentum tensor we were given did not drop from the sky but was derived from the action for the particle, while the equation of motion follows from the very same action. Physically, we expect the energy momentum tensor to be conserved only if the particle does what it is supposed to do, rather than moving around capriciously.

8 Start with $0 = D_\mu T^{\mu\nu} = D_\mu(\rho U^\mu) U^\nu + \rho U^\mu D_\mu U^\nu$. Contract with U_ν and use $U_\nu D_\mu U^\nu = 0$ (since $U_\nu U^\nu = -1$). We obtain $D_\mu(\rho U^\mu) = 0$. Plugging this back into the above, we find the stated result $U^\mu D_\mu U^\nu = 0$, which tells us that the dust particles follow geodesics, as might be expected. See also exercise 7.

VI.5   Gravity Goes Live

2 According to exercise VI.1.6, in 2-dimensional spacetime $R_{\tau\rho\mu\nu} = \frac{1}{2} R(g_{\tau\mu}g_{\rho\nu} - g_{\tau\nu}g_{\rho\mu})$ (this follows immediately since the Riemann curvature tensor has only one component). Contracting, we find that $R_{\tau\mu} = g_{\tau\mu} R/2$ and thus $E_{\mu\nu} = 0$.

VI.6   Initial Value Problems and Numerical Relativity

1 To obtain $E^{0\mu}$, we have to calculate the Riemann curvature tensor. From $R^{\cdot}_{\cdot\cdot\cdot} \sim \partial\Gamma + \Gamma\Gamma$, we obtain $R_{\cdots} \sim g\partial\Gamma + g\Gamma\Gamma$. Since $\Gamma \sim g^{\cdot\cdot}\partial g_{\cdot\cdot}$, we encounter in $\partial\Gamma$ terms involving $\partial\partial g_{\cdot\cdot}$ and terms involving $\partial g^{\cdot\cdot} \sim g^{\cdot\cdot}\partial g_{\cdot\cdot}g^{\cdot\cdot}$ (which, if you want, you could express in terms of Γ and g using the definition of Γ, but you don't even have to bother for the purposes at hand). Thus, $R_{\cdots} \sim \partial\partial g_{\cdot\cdot} + \Gamma\Gamma$.

Since we are hunting for ∂_0^2, we could care less about the $\Gamma\Gamma$ terms. So far so good, but it would appear that we still have to slave away to obtain the $\partial\partial g_{\cdot\cdot}$ terms. Now we appeal to Professor Flat for help. Go to a locally flat coordinate system. Back in chapter VI.1, we obtained $R_{\cdots}$ locally in terms of the $\partial\partial g_{\cdots}$. The general expression for $R_{\cdots}$ must reduce to the expression in chapter VI.1; hence we conclude that

$$R_{\tau\rho\mu\nu} = \tfrac{1}{2}(g_{\tau\nu,\mu\rho} - g_{\rho\nu,\mu\tau} - g_{\tau\mu,\nu\rho} + g_{\rho\mu,\nu\tau}) + \Gamma\Gamma \text{ terms}$$

where we have switched to the comma notation for partial derivatives: $g_{\tau\nu,\mu\rho} \equiv \partial_\mu\partial_\rho g_{\tau\nu}$. Keeping in mind the antisymmetric properties of $R_{\tau\rho\mu\nu}$, we see that we won't encounter $g_{00,00}$ and $g_{i0,00}$. We are down to $R_{i0j0} = \tfrac{1}{2}(g_{i0,j0} - g_{00,ji} - g_{ij,00} + g_{0j,0i}) + \Gamma\Gamma$ terms, leading to R_{i0j0} "$=$" $-\tfrac{1}{2}g_{ij,00}$. To save writing, we introduce the symbol "$=$" to mean equal up to terms not containing ∂_0^2.

Contracting indices, we find

$$R_{00} \text{ "}=\text{" } -\tfrac{1}{2}g^{ij}g_{ij,00} \qquad R_{0i} \text{ "}=\text{" } +\tfrac{1}{2}g^{0j}g_{ij,00} \qquad R_{ij} \text{ "}=\text{" } -\tfrac{1}{2}g^{00}g_{ij,00}$$

and R "$=$" $(g^{0i}g^{0j} - g^{00}g^{ij})g_{ij,00}$.

Confusio is busily calculating in the corner. That guy does every exercise in the book to set a good example for the students. Now he cries out, "But I get $E_{00} = R_{00} - \tfrac{1}{2}g_{00}R$ "$=$" $-\tfrac{1}{2}(g^{ij} + g_{00}g^{0i}g^{0j} - g_{00}g^{00}g^{ij})g_{ij,00}$ and this contains ∂_0^2!"

Perhaps you could help Confusio out. You chide him, "By now, you should know the importance of distinguishing upper and lower indices! In the text, the statement is that $E^{0\mu}$ does not contain ∂_0^2."

We will raise the indices in two steps. First,

$$R_0^0 = g^{00}R_{00} + g^{0i}R_{0i} \text{ "}=\text{" } \frac{1}{2}(g^{0i}g^{0j} - g^{00}g^{ij})g_{ij,00}$$

Indeed,

$$E_0^0 = R_0^0 - \frac{1}{2}R \text{ "}=\text{" } 0$$

Similarly,

$$E_i^0 = R_i^0 = g^{00}R_{0i} + g^{0j}R_{ji} \text{ "}=\text{" } 0$$

Now we are almost done:

$$E^{00} = g^{00}E_0^0 + g^{0i}E_i^0 \text{ "}=\text{" } 0 \qquad \text{and} \qquad E^{0i} = g^{i0}E_0^0 + g^{ij}E_j^0 \text{ "}=\text{" } 0$$

Phew! It's nice to have the Bianchi identity proof!

2   Using the contracted Bianchi identity as in the text, we have $\partial_0(E^{0\nu} - T^{0\nu}) = -\partial_i E^{i\nu} +$ terms involving the Christoffel symbols, Γ times (various Es minus $\partial_0 T^{0\nu}$). Next, use the field equation $E^{\mu\nu} = T^{\mu\nu}$ to write this as $-\partial_i T^{i\nu} + \Gamma T - \partial_0 T^{0\nu}$. If there is any justice in the world, the Γs should convert the ∂s into covariant derivatives and turn this into $-D_\mu T^{\mu\nu} = 0$.

VII.1   Particles and Light around a Black Hole

3   Following the by now familiar steps, we obtain for the radially plunging observer the equations of motion

$$\left(\frac{dT}{d\tau}\right)^2 - \left(\frac{dr}{d\tau} + v\frac{dT}{d\tau}\right)^2 = 1 \quad \text{and} \quad \frac{dT}{d\tau} - v\left(\frac{dr}{d\tau} + v\frac{dT}{d\tau}\right) = 1$$

Solving, we obtain $\frac{dr}{d\tau} = -v = -\sqrt{r_S/r}$ and $\frac{dT}{d\tau} = 1$. We also see that if we set $dr = -vdT$ in (13), we obtain $d\tau = dT$.

4   The path of a radially infalling photon is determined by $ds = 0$, which implies $dT = \pm(dr + vdT)$, with $v = \sqrt{r_S/r}$. We thus obtain $\frac{dr}{dT} = -(1 + v)$ for an infalling photon (and $\frac{dr}{dT} = (1 - v)$ for an outgoing photon). This proves that a photon falls faster than the plunging observer. Note that in these coordinates, we have, for an outgoing photon, $\frac{dr}{dT} = 0$ at the horizon, as expected.

VII.3   Hawking Radiation

1   From chapters V.4 and VI.3, we have

$$\left(\frac{dr}{d\tau}\right)^2 = -\left(1 - \frac{r_S}{r}\right) + \epsilon^2, \quad \text{with} \quad \epsilon = 1 - \frac{r_S}{r_S + a}$$

Integrating, we obtain

$$\Delta\tau = \int_{r_S}^{r_S+a} dr(\frac{r_S}{r} - \frac{r_S}{r_S + a})^{-\frac{1}{2}} = 2(r_S a)^{\frac{1}{2}}$$

The time it takes to reach the horizon scales like $a^{\frac{1}{2}}$. So Heisenberg tells us that the characteristic energy is $\sim \hbar/2(r_S a)^{\frac{1}{2}}$. Multiplying by the gravitational redshift factor

$$\left(g_{00}(r_S + a)\right)^{\frac{1}{2}} = \left(1 - \frac{r_S}{r_S + a}\right)^{\frac{1}{2}} \simeq \sqrt{a/r_S}$$

derived in chapter V.4, we obtain that the characteristic energy measured at spatial infinity is given by $T_H \sim \hbar/r_S \sim \hbar/GM$. Nicely, the dependence on a cancels out.

VII.5   Rotating Black Holes

1   We have $dr = f_{\tilde{r}}d\tilde{r} + f_\theta d\theta$, where $f_{\tilde{r}} = \partial f/\partial\tilde{r}$, and so forth. Getting rid of the cross term $d\tilde{r}d\theta$ requires $f_\theta = -c/a$, which fixes f up to an arbitrary additive function of $\tilde{r}$. We then drop the tilde sign.

2   We obtain

$$g_{tt} \to -1 + \frac{r_S}{r} - \frac{a^2 r_S \cos^2\theta}{r^3} + O\left(\frac{1}{r^5}\right)$$

$$g_{t\varphi} \to -\frac{a r_S \sin^2\theta}{r} + \frac{a^3 r_S \sin^2\theta \cos^2\theta}{r^3} + O\left(\frac{1}{r^5}\right)$$

$$g_{\varphi\varphi} \to r^2 \sin^2\theta \left(1 + \frac{a^2}{r^2} + \frac{a^2 r_S \sin^2\theta}{r^3} + O\left(\frac{1}{r^5}\right)\right)$$

$$g_{rr} \to \left(1 + \frac{r_S}{r} + \frac{a^2 \cos^2\theta - a^2 + r_S^2}{r^2} + \frac{a^2 r_S \cos^2\theta - 2a^2 r_S + r_S^3}{r^3} + O\left(\frac{1}{r^4}\right)\right)$$

$$g_{\theta\theta} \to r^2 \left(1 + \frac{a^2 \cos^2\theta}{r^2} + O\left(\frac{1}{r^4}\right)\right)$$

VIII.1 The Dynamic Universe

1   Use the fact that $\rho = \rho_0 (R_0/R)^4$ to write $\dot{R}^2 = \frac{T^2}{4R^2} - 1$, where we define the time scale T by $T^2 = (4/3)(8\pi G \rho_0 R_0^4)$. Then the solution is $R(t) = \sqrt{t(T-t)}$. For small t, $R(t) \propto t^{\frac{1}{2}}$, in agreement with what we had in the text. The universe ends at time T.

3   In this case, $\rho = \rho_0 (R_0/R)^3$. Define $T = 4\pi G \rho_0 R_0^3/3$, so that the cosmological equation becomes $\dot{R}^2 = \frac{2T}{R} - 1$. The solution is given parametrically by $R(\eta) = T(1 - \cos\eta)$ and $t(\eta) = T(\eta - \sin\eta)$. For small η, $R \simeq T\eta^2/2$ and $t(\eta) \simeq T\eta^3/6$, so that in the early universe, $R(t) \propto t^{\frac{2}{3}}$.

   Amusingly, the resulting curve $R(t)$ is a cycloid, namely the curve traced by a point on the rim of a rolling wheel, with $R(\eta)$ corresponding to the height of the point and $t(\eta)$ to the distance traveled by the wheel.

VIII.2 Cosmic Struggle between Dark Matter and Dark Energy

1   Set $\Omega_{m,0} = 1$ and $\Omega_{r,0} = \Omega_{\Lambda,0} = 0$ in (32) to find $t_{age} = \frac{1}{H_0} \int_0^1 da\, a^{1/2}$. From $a \propto 1/t^{\frac{2}{3}}$, we have $H = -\frac{2}{3t}$. Plugging into (7), we have $\rho = (\frac{3}{8\pi G})(\frac{4}{9t^2})$ and thus the stated result.

2   Set $\Omega_{r,0} = 1$ and $\Omega_{m,0} = \Omega_{\Lambda,0} = 0$ in (32) to find $t_{age} = \frac{1}{H_0} \int_0^1 da\, a$. From $a \propto 1/t^{\frac{1}{2}}$ we have $H = -\frac{1}{2t}$. Plugging into (7), we have $\rho = (\frac{3}{8\pi G})(\frac{1}{4t^2})$ and thus the stated result.

3   Setting $\ddot{R} = 0$ in (VIII.1.16), we obtain the condition $\rho + 3P = 0 = \rho_m - 2\rho_\Lambda$ and thus $\rho = \rho_m + \rho_\Lambda = 3\rho_\Lambda = 3\Lambda$. Setting $\dot{R} = 0$ in (VIII.1.18), we obtain $k = 1$ necessarily and the radius $R = (\frac{3}{8\pi G\rho})^{\frac{1}{3}} = (\frac{1}{8\pi G\Lambda})^{\frac{1}{3}}$.

4   From $\Omega_k = -\frac{k}{H^2 R^2}$, we have $\Omega_k(z) = \Omega_{k,0}(1+z)^2(H_0/H(z))^2$. Using (15) and $a = (1+z)^{-1}$, we obtain the stated result.

5   According to (32), we have $\int dt = (H_0)^{-1} \int da (\Omega_{m,0} a^{-1} - |\Omega_{\Lambda,0}|a^2)^{-\frac{1}{2}}$, and thus the expansion of the universe stops when the denominator in the integrand, which is in fact proportional to H, vanishes. For a larger than $a_{max} = (\Omega_{m,0}/(-\Omega_{\Lambda,0}))^{1/3}$, the denominator goes negative. The time necessary to go from a_{max} to the Big Crunch is thus given by

$$(H_0)^{-1} \int_{a_{max}}^0 (-da) \left(\Omega_{m,0} a^{-1} - |\Omega_{\Lambda,0}|a^2\right)^{-\frac{1}{2}} = \frac{2}{3H_0 |\Omega_{\Lambda,0}|^{\frac{1}{2}}} \int_0^{a_{max}^{\frac{3}{2}}} du \left(a_{max} - u^2\right)^{-\frac{1}{2}} = \frac{\pi}{3H_0 |\Omega_{\Lambda,0}|^{\frac{1}{2}}}$$

The total life of the universe, as measured from the Big Bang to the Big Crunch, is thus $T_{universe} = \frac{2\pi}{3H_0 |\Omega_{\Lambda,0}|^{\frac{1}{2}}}$.

IX.1   Parallel Transport

1   Let the rectangle be of width a and b with its "lower left" corner at (x^*, y^*) and its "upper right" corner at $(x^* + a, y^* + b)$. We arbitrarily pick the counterclockwise direction to integrate in. Then $a^{xy} = \int_{x^*}^{x^*+a} dx(y^* - (y^* + b)) = -ab$. As a check, $a^{yx} = \int_{y^*}^{y^*+b} dy(x^* + a - x^*) = ab$.

3   Denote the vertices of the triangle by (A, B, C) and the corresponding interior angles by (a, b, c). The sides of the triangle are straight lines, of course, namely geodesics. Along a geodesic, the tangent vector is parallel transported, and so, along each side of the triangle, the angle between the vector we are parallel transporting, call it $\vec{S}$, and the tangent vector remains constant. Let's say we are moving along the side CA. When the tangent vector reaches the vertex A, it turns through an angle of $(\pi - a)$ to point along the side AB. Thus, when we get back to where we started, what we call the tangent vector has turned through an angle of $(\pi - a) + (\pi - b) + (\pi - c) = 3\pi - (a + b + c) = \pi - (a + b + c)$. Since the angle between $\vec{S}$ remains the same, $\vec{S}$ has also turned through $\pi - (a + b + c)$. Thus, the angular excess $(a + b + c) - \pi$ measures the curvature.

IX.3   Geodesic Deviation

2   Plug $y^\mu(\tau) = x^\mu(\tau) + \epsilon^\mu(\tau)$ into (IX.3.2), expand in ϵ, and subtract (IX.3.1) to obtain

$$\frac{d^2\epsilon^\mu}{d\tau^2} + \epsilon^\rho \partial_\rho \Gamma^\mu_{\nu\lambda} \frac{dx^\nu}{d\tau} \frac{dx^\lambda}{d\tau} + 2\Gamma^\mu_{\nu\lambda} \frac{dx^\nu}{d\tau} \frac{d\epsilon^\lambda}{d\tau} = 0 \tag{33}$$

Note that $\frac{d^2\epsilon^\mu}{d\tau^2}$ is not a vector; indeed, none of the terms in (33) is a vector.

We would like to rewrite this as an equation between vectors, and so let's evaluate

$$\frac{D^2\epsilon^\mu}{D\tau^2} = \frac{d}{d\tau}\left(\frac{d\epsilon^\mu}{d\tau} + \Gamma^\mu_{\nu\lambda} \frac{dx^\nu}{d\tau} \epsilon^\lambda\right) + \Gamma^\mu_{\nu\lambda} \frac{dx^\nu}{d\tau}\left(\frac{d\epsilon^\lambda}{d\tau} + \Gamma^\lambda_{\omega\kappa} \frac{dx^\omega}{d\tau} \epsilon^\kappa\right) \tag{34}$$

After the differentiation is carried out, the first two terms in (34) become

$$\frac{d^2\epsilon^\mu}{d\tau^2} + \left(\partial_\rho \Gamma^\mu_{\nu\lambda}\right)\frac{dx^\rho}{d\tau} \frac{dx^\nu}{d\tau} \epsilon^\lambda + \Gamma^\mu_{\nu\lambda} \frac{d^2x^\nu}{d\tau^2} \epsilon^\lambda + \Gamma^\mu_{\nu\lambda} \frac{dx^\nu}{d\tau} \frac{d\epsilon^\lambda}{d\tau} \tag{35}$$

Using (33), we can write the first term in (35) as

$$-\left(\epsilon^\rho \partial_\rho \Gamma^\mu_{\nu\lambda} \frac{dx^\nu}{d\tau} \frac{dx^\lambda}{d\tau} + 2\Gamma^\mu_{\nu\lambda} \frac{dx^\nu}{d\tau} \frac{d\epsilon^\lambda}{d\tau}\right)$$

The second term in this expression knocks off the third term in (34) and the fourth term in (35). Next, using the geodesic equation in (IX.3.1), we can write the third term in (35) as

$$-\Gamma^\mu_{\nu\lambda} \Gamma^\nu_{\sigma\rho} \frac{dx^\sigma}{d\tau} \frac{dx^\rho}{d\tau} \epsilon^\lambda$$

Collecting terms, we find that (34) becomes

$$\frac{D^2\epsilon^\mu}{D\tau^2} = -\epsilon^\rho \partial_\rho \Gamma^\mu_{\nu\lambda} \frac{dx^\nu}{d\tau} \frac{dx^\lambda}{d\tau} + \left(\partial_\rho \Gamma^\mu_{\nu\lambda}\right)\frac{dx^\rho}{d\tau} \frac{dx^\nu}{d\tau} \epsilon^\lambda - \Gamma^\mu_{\nu\lambda} \Gamma^\nu_{\sigma\rho} \frac{dx^\sigma}{d\tau} \frac{dx^\rho}{d\tau} \epsilon^\lambda + \Gamma^\mu_{\nu\lambda} \frac{dx^\nu}{d\tau} \Gamma^\lambda_{\rho\sigma} \frac{dx^\rho}{d\tau} \epsilon^\sigma \tag{36}$$

Renaming indices and recalling the definition of the Riemann curvature tensor, we watch with satisfaction the terms on the right hand side gathering themselves into a particularly nice form: indeed, what else but (IX.3.6)?

This calculation, while slightly tedious, shows quite clearly where the $\partial\Gamma$ and the $\Gamma\Gamma$ terms in the definition of the Riemann curvature tensor $R^{\cdot}_{\cdot\cdot\cdot} \sim \partial_{\cdot}\Gamma^{\cdot}_{\cdot\cdot} + \Gamma^{\cdot}_{\cdot\cdot}\Gamma^{\cdot}_{\cdot\cdot}$ come from.

4 The stated energy conditions are invariant under $V^\mu \to aV^\mu$, and so we could normalize the arbitrary timelike vector V^μ by $V^\mu V_\mu = -1$. Then, with

$$T_{\mu\nu} = (\rho + P)U_\mu U_\nu + Pg_{\mu\nu} \quad \text{we have} \quad T_{\mu\nu}V^\mu V^\nu = (\rho + P)(U \cdot V)^2 - P$$

Go to the frame in which

$$U^\mu = (1, 0, 0, 0)/\sqrt{-g_{00}} \quad \text{so that} \quad (U \cdot V)^2 = -g_{00}(V^0)^2$$

which ranges from 1 to ∞ for an arbitrary timelike vector V^μ. Thus, for the weak energy condition, we obtain $\rho \geq 0$ and $(\rho + P) \geq 0$ as claimed.

Next,

$$T_{\mu\nu}V^\nu = (\rho + P)U_\mu(U \cdot V) + PV_\mu$$

Thus, for the dominant energy condition,

$$-g^{\mu\rho}(T_{\mu\nu}V^\mu)(T_{\rho\sigma}V^\sigma) = (\rho + P)^2(U \cdot V)^2 - 2P(\rho + P)(U \cdot V)^2 + P^2 \geq 0$$

For $(U \cdot V)^2$ equal to ∞ and 1, we obtain $\rho^2 \geq P^2$ and $\rho \geq 0$, respectively.

Since $T = -(\rho - 3P)$, the strong energy condition says that $(\rho + P)(U \cdot V)^2 - P \geq \frac{1}{2}(\rho - 3P)$. For $(U \cdot V)^2$ equal to ∞ and 1, we obtain $(\rho + P) \geq 0$ and $(\rho + 3P) \geq 0$, respectively.

IX.4 Linearized Gravity, Gravitational Waves, and the Angular Momentum of Rotating Bodies

1 Start with the equation of geodesic deviation

$$\frac{D^2 s^\mu}{D\tau^2} = R^\mu{}_{\sigma\rho\lambda}\frac{dx^\sigma}{d\tau}\frac{dx^\rho}{d\tau}s^\lambda$$

(where I have changed the separation between the two particles from ϵ to s to avoid possible confusion with the polarization vector of the gravitational wave). To leading order, $\frac{dx^\sigma}{d\tau} = (1, \vec{0})$ and so

$$\frac{D^2 s^\mu}{D\tau^2} \simeq R^\mu{}_{00\lambda}s^\lambda$$

Now $\Gamma \sim O(h)$ and so we have

$$R^\mu{}_{\sigma\rho\lambda} = \partial_\rho\Gamma^\mu{}_{\sigma\lambda} - \partial_\lambda\Gamma^\mu{}_{\sigma\rho} + O(h^2) \quad \text{giving} \quad R^\mu{}_{00\lambda} \simeq \partial_0\Gamma^\mu{}_{0\lambda}$$

since $\Gamma^\mu{}_{00}$ vanishes to this order in the TT gauge. Furthermore, since in TT gauge, $h_{0\kappa}$ vanishes, we have

$$\Gamma^\mu{}_{0\lambda} \simeq \frac{1}{2}\eta^{\mu\kappa}\partial_0 h_{\lambda\kappa} = \frac{1}{2}\partial_0 h^\mu{}_\lambda$$

Thus,

$$R^\mu{}_{00\lambda} \simeq \frac{1}{2}\frac{\partial^2}{\partial t^2}h^\mu{}_\lambda$$

and we obtain to leading order

$$\frac{d^2 s^\mu}{dt^2} \simeq \frac{1}{2}\left(\frac{\partial^2}{\partial t^2}h^\mu{}_\lambda\right)s^\lambda$$

For a plane wave $h_{\mu\nu} = \varepsilon_{\mu\nu}(k)\sin(\omega t - kx)$ moving along the x-axis,

$$\frac{d^2 s^\mu}{dt^2} \simeq -\frac{1}{2}\omega^2\varepsilon^\mu{}_\lambda s^\lambda \sin(\omega t - kx)$$

For example, for the plus polarization,

$$\frac{d^2 s^x}{dt^2} \simeq -\frac{1}{2}\omega^2 \varepsilon_+ s^x \sin(\omega t - kx) \quad \text{and} \quad \frac{d^2 s^y}{dt^2} \simeq +\frac{1}{2}\omega^2 \varepsilon_+ s^y \sin(\omega t - kx)$$

with the relative minus sign between the two equations characteristic of a tidal force.

3   Under the stated conditions, (10) reduces to $\nabla^2 \tilde{h}_{\mu\nu} = 0$, which has the immediate solution $\tilde{h}_{\mu\nu} = k_{\mu\nu}/r$, with $k_{\mu\nu}$ some constant tensor, as is familiar to students of elementary physics. Now impose the harmonic gauge condition $\partial^\mu \tilde{h}_{\mu\nu} = 0$.

If you started to write $k_{r\nu} \partial^r (1/r) = 0$, stop! You are making an error. This is the subtlety I alluded to in a footnote in connection with (4). Think about how (1) was derived in chapter VI.5: we used $\partial_\lambda \eta_{\mu\nu} = 0$ repeatedly. Thus, we must take for $\eta_{\mu\nu}$ the diagonal matrix with diagonal elements $(-1, 1, 1, 1)$ rather than $(-1, 1, r^2, r^2 \sin^2 \theta)$. But since the problem has spherical symmetry and since r has already appeared, it is easy to fall in the trap of deducing that $k_{r\nu} = 0$.

Instead, we have $k_{\mu\nu} \partial^\mu (1/r) = 0 = -k_{i\nu} x^i/r^3$ and hence $k_{ij} = 0$ and $k_{i0} = 0$. (In other words, ∂^μ in this context refers to Cartesian coordinates, not spherical coordinates.) Since $k_{\mu\nu}$ is symmetric, only k_{00} is nonvanishing, and hence the only nonvanishing component of $\tilde{h}_{\mu\nu}$ is $\tilde{h}_{00} = 2r_S/r$, where, with the malice of afterthought, we have renamed $k_{00} = 2r_S$. Note that $\tilde{h} = -2r_S/r$. We next have to go from $\tilde{h}_{\mu\nu}$ to $h_{\mu\nu} = \tilde{h}_{\mu\nu} - \frac{1}{2}\eta_{\mu\nu}\tilde{h}$. We obtain

$$h_{00} = \tilde{h}_{00} + \frac{1}{2}\tilde{h} = r_S/r, \qquad h_{ij} = -\eta_{ij} r_S/r$$

Thus,

$$ds^2 = -\left(1 - \frac{r_S}{r}\right) dt^2 + \left(1 + \frac{r_S}{r}\right)\left(dx^2 + dy^2 + dz^2\right) = -\left(1 - \frac{r_S}{r}\right) dt^2 + \left(1 + \frac{r_S}{r}\right)\left(dr^2 + r^2 d\Omega^2\right)$$

What? You exclaim that this is not the Schwarzschild metric—the coefficient of dr^2 and $r^2 d\Omega^2$ are the same. After all this work?

But this agrees with (19) with $h_{0i} = 0$, so our result is in fact correct. The resolution is that we can perform a coordinate transformation. Call the coefficient of $d\Omega^2$ in the ds^2 given above R^2. So, $R^2 = (1 + \frac{r_S}{r})r^2$; that is, $R \simeq r + \frac{1}{2}r_S$. Then $dR \simeq dr$ to this order and we have

$$\frac{r_S}{R} \simeq \frac{r_S}{r + \frac{1}{2}r_S} \simeq \frac{r_S}{r} + O\left(\left(\frac{r_S}{r}\right)^2\right)$$

We obtain the Schwarzschild metric

$$ds^2 \simeq -\left(1 - \frac{r_S}{R}\right) dt^2 + \left(1 + \frac{r_S}{R}\right) dR^2 + R^2 d\Omega^2$$

to leading order, as expected.

IX.6   Isometry, Killing Vector Fields, and Maximally Symmetric Spaces

3a   For a scalar, (19) collapses to $\xi^\lambda \partial_\lambda S = 0$. Since this holds for all Killing vectors, we have $\partial_\lambda S = 0$.

3b–c   At some arbitrary point X, consider the $D(D-1)/2$ rotational Killing vectors, for which $\xi^\mu(X) = 0$. All subsequent statements are meant to hold at the arbitrary point X (and thus hold everywhere). The covariant derivative of these Killing vectors simplifies to $\xi^\mu_{;\rho} = \partial_\rho \xi^\mu$ (since $\xi^\mu = 0$ at that point). Write this as $\partial_\rho \xi^\mu = g^{\mu\omega}\xi_{\omega;\rho}$. Thus, we can write the first term in (19) as

$$T_{\mu\sigma\cdots\tau} g^{\mu\omega}\xi_{\omega;\rho} = T^\omega_{\ \sigma\cdots\tau}\xi_{\omega;\rho} = T^\omega_{\ \sigma\cdots\tau}\delta^\zeta_\rho \xi_{\omega;\zeta}$$

(The insertion of the Kronecker delta in the last step is merely for later convenience.) Similarly, we can write the second term as $T_\rho^{\ \omega}_{\cdots\tau}\delta^\zeta_\sigma \xi_{\omega;\zeta}$, and so on, until we get to the last term, which vanishes, since

$\xi^{\mu}(X) = 0$. Thus, at the point X, we have

$$\left(T_{\sigma \cdots \tau}{}^{\omega} \delta_{\rho}^{\zeta} + T_{\rho \cdots \tau}{}^{\omega} \delta_{\sigma}^{\zeta} + \cdots \right) \xi_{\omega;\zeta} = 0$$

Since $\xi_{\omega;\zeta}$ span the basis of all D-by-D antisymmetric matrices, we conclude that the parenthetical expression in this equation must be symmetric under $\omega \leftrightarrow \zeta$. We now apply this result to various specific cases.

For a maximally form invariant vector, we have $V^{\omega} \delta_{\rho}^{\zeta} = V^{\zeta} \delta_{\rho}^{\omega}$. Contracting ζ and ρ, we have $(D - 1) V^{\omega} = 0$, thus proving the stated assertion.

For a maximally form invariant 2-indexed tensor $T_{\mu\nu}$, the expression in parentheses above simplifies to

$$\left(T_{\sigma}{}^{\omega} \delta_{\rho}^{\zeta} + T_{\rho}{}^{\omega} \delta_{\sigma}^{\zeta} \right) = \left(T_{\sigma}{}^{\zeta} \delta_{\rho}^{\omega} + T_{\rho}{}^{\zeta} \delta_{\sigma}^{\omega} \right)$$

Contracting ζ and ρ and lowering ω, we obtain $(D - 1) T_{\omega\sigma} + T_{\sigma\omega} = g_{\omega\sigma} T_{\rho}^{\rho}$. Decompose $T_{\omega\sigma} = S_{\omega\sigma} + A_{\omega\sigma}$ into its symmetric and antisymmetric parts. The preceding equation then becomes

$$D S_{\omega\sigma} + (D - 2) A_{\omega\sigma} = g_{\omega\sigma} S_{\rho}^{\rho}$$

giving us

$$(D - 2) A_{\omega\sigma} = 0 \quad \text{and} \quad D S_{\omega\sigma} = s g_{\omega\sigma}$$

with $s \equiv S_{\rho}^{\rho}$. For $D \neq 2$, $A_{\omega\sigma} = 0$.

It remains to show that s is a constant. Plug the result $T_{\omega\sigma} = S_{\omega\sigma} = s g_{\omega\sigma}$ (absorbing a trivial factor) back into (19). The last term begets two terms:

$$\xi^{\lambda} \partial_{\lambda} T_{\rho\sigma} = \xi^{\lambda} \partial_{\lambda} (s g_{\rho\sigma}) = \left(\xi^{\lambda} \partial_{\lambda} g_{\rho\sigma} \right) s + g_{\rho\sigma} \left(\xi^{\lambda} \partial_{\lambda} s \right)$$

The term $(\xi^{\lambda} \partial_{\lambda} g_{\omega\sigma}) s$ combines with the other two terms in (19) to yield zero, thanks to the Killing condition (2). We are thus left with $\xi^{\lambda} \partial_{\lambda} s = 0$, which implies that s is constant.

Finally, we have to deal with the special case of $D = 2$, for which $A_{\omega\sigma}$ needs not vanish. Indeed, besides the metric tensor, we have the form invariant 2-indexed tensor $\varepsilon_{\mu\nu} / \sqrt{g}$, namely the Levi-Civita tensor. See chapter X.5 for further discussion.

IX.7 Differential Forms and Vielbein

1 We have $e^{1} = f(y) dx$ and $e^{2} = g(x) dy$, so that

$$\omega^{1}{}_{2} = \frac{f'(y)}{g(x)} dx - \frac{g'(x)}{f(y)} dy \quad \text{and so} \quad R^{1}{}_{2} = - \left(\frac{g''(x)}{f(y)} + \frac{f''(y)}{g(x)} \right) dx dy$$

Converting to world indices, we find $R_{xyxy} = -(g(x) g''(x) + f(y) f''(y))$.

2 We have $dF = \frac{1}{2} \partial_{\lambda} F_{\mu\nu} dx^{\lambda} dx^{\mu} dx^{\nu} = 0$. Since dx^{λ}, dx^{μ}, and dx^{ν} anticommute, this is equivalent to $\epsilon^{\sigma\lambda\mu\nu} \partial_{\lambda} F_{\mu\nu} = 0$, which you should recognize as an identity (since $F_{\mu\nu} = \partial_{\mu} A_{\nu} - \partial_{\nu} A_{\mu}$) and as the "other half" of Maxwell's equations.

4 We have $e^{1} = \Omega dx$ and $e^{2} = \Omega dy$. Since we are dealing with a space rather than a spacetime, we have Euclidean indices $a, b, \cdots$ rather than Minkowski indices, and we can freely move indices up and down. Thus we write, for example,

$$de^{1} = -\omega^{12} e^{2} = \Omega_{y} dy dx = - \left(\Omega_{y} / \Omega^{2} \right) e^{1} e^{2}$$

where we use the notation $\mathcal{O}_{y} = \partial_{y} \mathcal{O}$ (and similarly $\mathcal{O}_{x}$). Thus we have $\omega^{12} = (\Omega_{y} / \Omega) e^{1} + (e^{2} \text{ term})$. By symmetry, $\omega^{21} = -\omega^{12} = (\Omega_{x} / \Omega^{2}) e^{2} + (e^{2} \text{ term})$, and hence we obtain

$$\omega^{12} = (\Omega_{y} e^{1} - \Omega_{x} e^{2}) / \Omega^{2} = (\Omega_{y} dx - \Omega_{x} dy) / \Omega$$

Thus we have

$$R^{12} = -((\Omega_x/\Omega)_x + (\Omega_y/\Omega)_y)dxdy = -(\nabla^2 \log \Omega)e^1 e^2/\Omega^2$$

The curvature is therefore given by $\frac{1}{2}R = R_{11} = R_{22} = R^1_{212} = -(\nabla^2 \log \Omega)/\Omega^2$.

As a check, we learned from exercise I.5.13 that the metric for the sphere can be written as $ds^2 = (d\rho^2 + \rho^2 d\theta^2)/(1 + \frac{\rho^2}{4})^2$. Plugging in $\Omega = (1 + \frac{x^2+y^2}{4})^{-1}$, we obtain $R = 2$, as expected.

5   To be concise, we will abuse notation as noted below. Also, as noted in exercise 3, since we are dealing with a space, we have Euclidean indices $a, b, \cdots$, which we will freely move up and down. World indices, of course, have to be moved using the metric. From $e^a = \Omega dx^a$ (this is already notational abuse: strictly speaking, we should write $e^a = \Omega \delta^a_i dx^i$ to distinguish between the world index i and the Euclidean index a), we have $de^a = \partial_i \Omega dx^i dx^a = -\omega^{ab} e^b = -\omega^{ab} \Omega dx^b$. We will write $\Omega_i \equiv \partial_i \Omega$. Using the fact that ω^{ab} is antisymmetric, we obtain

$$\omega^{ab} = \Omega^{-1} \left(\Omega_b dx^a - \Omega_a dx^b \right)$$

and hence we have

$$d\omega^{ab} = \Omega^{-1} \left(\Omega_{bc} dx^c dx^a - \Omega_{ac} dx^c dx^b \right) - \Omega^{-2} \left(\Omega_b \Omega_c dx^c dx^a - \Omega_a \Omega_c dx^c dx^b \right)$$

Also, we have

$$\omega^{ac} \omega^{cb} = \Omega^{-2} \left(\Omega_c dx^a - \Omega_a dx^c \right) \left(\Omega_b dx^c - \Omega_c dx^b \right)$$

$$= \Omega^{-2} \left(\Omega_a \Omega_c dx^c dx^b - \Omega_b \Omega_c dx^c dx^a - \Omega_c \Omega_c dx^a dx^b \right)$$

Putting this together, we obtain the curvature 2-form

$$R^{ab} = d\omega^{ab} + \omega^{ac} \omega^{cb}$$

$$= \Omega^{-1} \left(\Omega_{bc} dx^c dx^a - \Omega_{ac} dx^c dx^b \right)$$

$$- 2\Omega^{-2} \left(\Omega_b \Omega_c dx^c dx^a - \Omega_a \Omega_c dx^c dx^b \right) - \Omega^{-2} \Omega_c \Omega_c dx^a dx^b$$

It is instructive to compare our work here with that in exercise 3. Note in particular that the nonabelian term $\omega^{ac} \omega^{cb}$ is absent there.

Next, we have to extract the Riemann curvature tensor R^{ab}_{cd} defined by $R^{ab} = \frac{1}{2} R^{ab}_{cd} e^c e^d$. So, replace dx^c by $\Omega^{-1} e^c$ in the expression for R^{ab} above and read off

$$R_{abcd} = \left(\Omega^{-3} \left(\Omega_{bc} \delta_{ad} - \Omega_{ac} \delta_{bd} \right) - 2\Omega^{-4} \left(\Omega_b \Omega_c \delta_{ad} - \Omega_a \Omega_c \delta_{bd} \right) - \Omega^{-4} \Omega_f \Omega_f \delta_{ac} \delta_{bd} \right)$$

$$- (c \leftrightarrow d)$$

From this we obtain $R_{ac} = R_{abcd} \delta_{bd}$ and $R = R_{ac} \delta^{ac}$. To obtain the "usual" Riemann and Ricci tensors, remember to convert Euclidean indices to world indices with the vielbein, thus for example, $R_{\mu\nu} = e^a_\mu e^b_\nu R_{ab} = \Omega^2 R_{ab} \delta^a_\mu \delta^b_\nu$. Keep in mind that with our abuse of notation, we have $\Omega_{12} = \frac{\partial}{\partial x^1} \frac{\partial}{\partial x^2} \Omega$; that is, the indices on Ω_{ac} in the expression above are world indices to begin with.

Following this procedure, we obtain the Riemann tensor (which we won't display, since it can be read off from what is given above), the Ricci tensor

$$R_{\mu\nu} = 2(d-2)\Omega^{-2} \partial_\mu \Omega \partial_\nu \Omega - \left(\delta_{\mu\nu} \left(\Omega^{-1} \partial^2 \Omega + (d-3)\Omega^{-2} \partial_\lambda \partial^\lambda \Omega \right) + (d-2)\Omega^{-1} \partial_\mu \partial_\nu \Omega \right)$$

and the scalar curvature

$$R = -2(d-1)\Omega^{-3} \partial^2 \Omega - (d-1)(d-4)\Omega^{-4} \partial_\mu \Omega \partial^\mu \Omega$$

(Note that the world indices are raised and lowered with the Euclidean metric $\delta_{\mu\nu}$.)

This exercise suggests a relatively easy way to obtain the results of exercise VI.1.13. We simply promote, in the expressions given here, $\delta_{\mu\nu}$ to $g_{\mu\nu}$ and the partial derivatives to covariant derivatives. Finally, the leading term in the expressions given in exercise VI.1.13 can be determined by setting Ω to a constant.

For example, there we had $\bar{R} = \Omega^{-2}R + \cdots$. The factor Ω^{-2} is determined by noting that for Ω constant, we are simply scaling the coordinates by Ω.

Finally, we can check the results given above. For the d-dimensional sphere, $\Omega = (1 + \frac{r^2}{4})^{-1}$, and we obtain $R = d(d-1)$, as we have long known. For anti de Sitter spacetime, write x^1 for convenience as x, then $\Omega = 1/x$. The only derivatives we have to calculate are then $\partial_x \Omega = -1/x^2$ and $\partial^2 \Omega = \partial_x \partial_x \Omega = 2/x^3$, and the messy expressions above collapse nicely to $R_{\mu\nu} = -(d-1)\delta_{\mu\nu}$ and $R = -d(d-1)$, in agreement with what we will learn in chapter IX.11.

6   With $e^1 = dr$ and $e^2 = f(r,\theta)d\theta$, we obtain $de^1 = 0 = -\omega^{12}e^2$, which implies $\omega^{12} \propto e^2$, and

$$de^2 = (\partial_r f)drd\theta = (\partial_r f/f)e^1 e^2 = -\omega^{21}e^1$$

We obtain

$$\omega^{12} = -(\partial_r f/f)e^2 = -(\partial_r f)d\theta$$

Thus we have

$$R^{12} = d\omega^{12} = -\left(\partial_r^2 f\right)drd\theta = -\left(\partial_r^2 f/f\right)e^1 e^2$$

from which we find the scalar curvature $R = 2R^1_{212} = -\partial_r^2 f/f$.

Let us check this against the two examples given in chapter II.2. For polar coordinates on the plane, we find $f(r,\theta) = r$ and indeed $R = 0$. For spherical coordinates on the sphere, with suitable renaming of the coordinates, we have $f(r,\theta) = \sin r$, and indeed $R = 1$.

IX.8   Differential Forms Applied

1   We have $e^1 = dr$ and $e^2 = f(r)d\theta$, so that $de^1 = 0 = -\omega^1_{\ 2}e^2$, which tells us that $\omega^1_{\ 2}$ is proportional to e^2, and $de^2 = f'(r)drd\theta = -\omega^2_{\ 1}e^1$, which tells that $\omega^2_{\ 1} = f'(r)d\theta$. Use the antisymmetry to see that $\omega^2_{\ 1} = -\omega^1_{\ 2}$, and so $\omega^2_{\ 1}$ cannot contain a piece proportional to e^1. We then obtain

$$R^2_{\ 1} = d\omega^2_{\ 1} = f''(r)drd\theta = \frac{f''(r)}{f(r)}e^1 e^2$$

$$R^\alpha_{\ \beta\gamma\delta} : R^2_{\ 121} = -\frac{f''(r)}{f(r)}$$

$$R_{\alpha\beta} : R_{11} = -\frac{f''(r)}{f(r)}, \quad R_{22} = -\frac{f''(r)}{f(r)},$$

$$R = -\frac{2f''(r)}{f(r)}$$

Setting $R = 2C$ gives us the differential equation $f'' = -Cf$, whose solutions are given by either trigonometric or hyperbolic sine and cosine, depending on whether C is positive or negative.

But now the global condition $\theta = \theta + 2\pi$ tells us that for the space to be locally flat as discussed in chapter I.6, we must have $f(r) \to r$ as $r \to 0$. This not only fixes $f(r)$ but also requires C to be either 1 or -1. (Around the tip of a cone, we could have $f(r) \to kr$ as $r \to 0$ for $k < 1$, but then the coordinates and the curvature would be singular at the tip and our formalism breaks down.) For positive constant curvature, $f(r) = \sin r$, and so the space here, as you have already seen in chapter I.5, is just the sphere in disguise (with the usual coordinates $\theta \to r$ and $\varphi \to \theta$). For negative constant curvature, $f(r) = \sinh r$. Satisfyingly, this agrees with the result you got for exercise I.5.5.

It is also instructive, noting that a circle of radius a centered at the origin has circumference $2\pi f(a)$, to use the mites' formula introduced in the prologue:

$$R = \lim_{radius \to 0} \frac{6}{(radius)^2}\left(1 - \frac{circumference}{2\pi \, radius}\right) = \lim_{r \to 0} \frac{6}{r^2}\left(1 - \frac{\sin r}{r}\right) = \frac{6}{r^2}\left(\frac{r^3}{3!r}\right) = 1$$

(We now see that the mite professor of geometry included the overall factor of 6 so that the unit sphere would have unit curvature.)

2   We have $e^1 = y^p dx$ and $e^2 = x^p dx$. Then

$$de^1 = py^{p-1}dydx = -pe^1e^2/(yx^p) = -\omega^1{}_2 e^2$$

and thus

$$\omega^1{}_2 = pe^1/(yx^p) + (e^2 \text{ term})$$

By symmetry,

$$\omega^2{}_1 = pe^2/(xy^p) + (e^1 \text{ term})$$

and so

$$\omega^1{}_2 = -\omega^2{}_1 = p\left(e^1/\left(yx^p\right) - e^2/\left(xy^p\right)\right) = p\left(\left(y^{p-1}dx/x^p\right) - \left(x^{p-1}dy/y^p\right)\right)$$

the last form being easier to differentiate. Differentiating, we have

$$R^1{}_2 = d\omega^1{}_2 = -\omega^2{}_1 = -p(p-1)\left(\left(x^{p-2}/y^p\right) + \left(y^{p-2}/x^p\right)\right)dxdy$$

$$= -p(p-1)\left(\left(1/\left(x^2y^{2p}\right)\right) + \left(1/\left(y^2x^{2p}\right)\right)\right)e^1e^2$$

Since $R^a{}_b = \frac{1}{2}R^a{}_{bcd}e^c e^d$, we find $R^1{}_{212} = -p(p-1)((1/(x^2y^{2p})) + (1/(y^2x^{2p}))) = R_{22}$. Thus, we finally obtain

$$R = -\frac{2p(p-1)}{x^2y^2}\left(\frac{1}{x^{2(p-1)}} + \frac{1}{y^{2(p-1)}}\right)$$

The space is flat for $p = 0$ (Pythagoras) and for $p = 1$ (see exercise VI.1.17). For $p = 1/2$, $R = \frac{1}{2xy}(\frac{1}{x} + \frac{1}{y})$.

3   We have $e^1 = ad\theta$, $e^2 = (L + a\sin\theta)d\varphi$, and so $de^1 = 0$, telling us that $\omega^{12} \propto d\varphi$, $de^2 = a\cos\theta d\varphi = -\omega^{21}e^1 = -\omega^{21}ad\theta$, so that $\omega^{21} = \cos\theta d\varphi$. Thus we have $R^{21} = d\omega^{21} = -\sin\theta d\varphi = \frac{\sin\theta}{a(L+a\sin\theta)}e^2e^1$, giving $R^{21}_{21} = \frac{\sin\theta}{a(L+a\sin\theta)} = R_{11} = R_{22}$. Hence we obtain $R = \delta^{ab}R_{ab} = \frac{2\sin\theta}{a(L+a\sin\theta)}$. Note that, as might be expected, at $\theta = 0$ or π, $R = 0$; at $\theta = \pi/2$, $R = \frac{2}{a(L+a)}$; and at $\theta = 3\pi/2$, $R = -\frac{2}{a(L-a)}$. The "outer half" of the torus has positive curvature, while the "inner half" has negative curvature.

4   With $e^0 = dt$, $e^1 = A(t)dx$, $e^2 = B(t)dy$, $e^3 = C(t)dz$, we have $de^0 = 0 = -\omega^0{}_a e^a$, so that $\omega^0{}_a = (e^a$ term, no e^0 term). Next we have $de^1 = \dot{A}dtdx = (\dot{A}/A)e^0e^1 = -\omega^1{}_0e^0 - \omega^1{}_b e^b$, which implies $\omega^1{}_0 = (\dot{A}/A)e^1 = \omega^0{}_1$. Here a possible e^0 term in $\omega^1{}_0$ is disallowed by our earlier conclusion. Note that while it is possible here for $\omega^1{}_2$ to be proportional to e^2, this would imply that $\omega^2{}_1$ is proportional to e^1, but this is ruled out by the antisymmetry of ω^{12}. (Note that we have used repeatedly the fact that the metric is diagonal, so that we can raise and lower indices easily.) We thus conclude that the only nonvanishing components of $\omega^\alpha{}_\beta$ are $\omega^0{}_1 = \omega^1{}_0 = (\dot{A}/A)e^1 = \dot{A}dx$ and the components obtained from it by permuting 1, 2, 3 and A, B, C.

We next obtain

$$R^0{}_1 = d\omega^0{}_1 + \omega^0{}_a\omega^a{}_1 = \ddot{A}dtdx + 0 = (\ddot{A}/A)e^0e^1 = \frac{1}{2}R^0{}_{1\alpha\beta}e^\alpha e^\beta$$

and hence we have $R^0{}_{101} = \ddot{A}/A = -R^1{}_{010}$. Since $\omega^1{}_2 = 0$, we might be tempted to conclude that $R^1{}_2 = 0$ also, but we would be wrong. In fact,

$$R^1{}_2 = d\omega^1{}_2 + \omega^1{}_0\omega^0{}_2 + \omega^1{}_a\omega^a{}_2 = 0 + (\dot{A}/A)(\dot{B}/B)e^1e^2 + 0$$

and hence we have

$$R^1{}_{212} = (\dot{A}/A)(\dot{B}/B) = R^2{}_{121}$$

All other nonvanishing components of $R^\alpha{}_\beta$ are obtained from $R^0{}_1$ and $R^1{}_2$ by permuting 1, 2, 3 and A, B, C.

Finally, $R_{\alpha\beta}$ (note that this is not a form and is not to be confused with the curvature 2-forms we had before) is given by

$$-R_{00} = -R^a{}_{0a0} = (\ddot{A}/A) + (\ddot{B}/B) + (\ddot{C}/C)$$

and

$$R_{11} = R^0{}_{101} + R^2{}_{121} + R^3{}_{131} = (\ddot{A}/A) + (\dot{A}/A)((\dot{B}/B) + (\dot{C}/C))$$

(Again, the other nonvanishing components of $R_{\alpha\beta}$ are obtained from R^0_1 and R^1_2 by permuting 1, 2, 3 and A, B, C.)

So the Einstein field equation $R_{\alpha\beta} = 0$ consists of 4 equations for 3 unknown functions A, B, C, but we know that 1 linear combination of the 4 equations corresponds to the Bianchi identity. By inspection, we see that power laws $A = t^p$, $B = t^q$, $C = t^r$ solve these equations: $R_{00} = 0$ gives $p + q + r = p^2 + q^2 + r^2$, and $R_{11} = 0$ gives $p^2 + p(-1 + q + r) = 0$ implying that either $p = 0$ or $p + q + r = 1$.

We can easily extend this to higher dimensions. For $ds^2 = -dt^2 + f_1(t)dx^2 + \cdots$, we obtain

$$R_{00} = -\sum_a \ddot{f}_a/f_a \quad \text{and} \quad R_{aa} = (\ddot{f}_a/f_a) + (\dot{f}_a/f_a)\sum_{b \neq a}(\dot{f}_b/f_b)$$

Einstein's field equation is solved by $f_a = t^{p_a}$, with p_a satisfying $\sum_a p_a^2 = \sum_a p_a = 1$.

Let us also ask what happens when $d = 2$. Then R^α_β contains only 1 component, namely R^0_1. Hence $\ddot{A} = 0$ and after shifting the origin of t and so forth, we have $ds^2 = -dt^2 + t^2dx^2$. Analytically continuing and giving the coordinates more familiar names, we see that we have found the plane in polar coordinates.

IX.9   Conformal Algebra

1   Simply calculate $\eta_{\mu\nu}\left(x_1^\mu/x_1^2 - x_2^\mu/x_2^2\right)\left(x_1^\nu/x_1^2 - x_2^\nu/x_2^2\right)$. If the two points are null separated, $(x_1 - x_2)^2 \to 0$ and hence remains 0 under inversion.

3   For the Euclidean plane, (1) simplifies to $\xi_{\mu,\nu} + \xi_{\nu,\mu} = 2c\delta_{\mu\nu}$, with c some constant. Then $\xi_{1,1} = c$, with the solution $\xi_1 = cx + f_1(y)$. Similarly, $\xi_{2,2} = c$ gives $\xi_2 = cy + f_2(x)$. Plugging into $\xi_{1,2} + \xi_{2,1} = 0$ gives $f_1'(y) + f_2'(x) = 0$, and hence $f_1 = a_1y + b_1$ and $f_2 = a_2x + b_2$, with $a_1 + a_2 = 0$. Therefore, $\xi_1 = cx + ay + b_1$ and $\xi_2 = cy - ax + b_2$. The three conformal Killing vectors are $\xi = (b_1, b_2)$ (translation), $\xi = (y, -x)$ (rotation), and $\xi = (x, y)$ (dilation).

IX.10   De Sitter Spacetime

5   The event horizon of the observer at the south pole is given by the diagonal $\tau = \psi - \frac{\pi}{2}$. Plugging this into (44), we obtain $T = \tan\tau$, $r = 1$, and $W = -\tan\tau$, and thus the intersection of $T + W = 0$ and $r = 1$. The event horizon of the observer at the north pole is given by the other diagonal $\tau = \frac{\pi}{2} - \psi$.

X.3   Effective Field Theory Approach to Einstein Gravity

1   Lorentz and gauge invariance allow us to construct, in addition to the mass dimension 4 Maxwell scalar $F_{\mu\nu}F^{\mu\nu}$, also the mass dimension 6 scalar $\partial_\lambda F_{\mu\nu}\partial^\lambda F^{\mu\nu}$. Moving further up in mass scale, we also have the mass dimension 8 scalars $(F_{\mu\nu}F^{\mu\nu})^2$ and $(\tilde{F}_{\mu\nu}F^{\mu\nu})^2$ (where $\tilde{F}_{\mu\nu} \equiv -\frac{1}{2}\epsilon_{\mu\nu\rho\sigma}F^{\rho\sigma}$, as you might recall from chapter IV.2). If we add them to the action to form

$$S = \int d^4x \left(-\tfrac{1}{4}F_{\mu\nu}F^{\mu\nu} + \alpha l^2 \partial_\lambda F_{\mu\nu}\partial^\lambda F^{\mu\nu} + \cdots + l'^4\left(\beta\left(F_{\mu\nu}F^{\mu\nu}\right)^2 + \gamma\left(\tilde{F}_{\mu\nu}F^{\mu\nu}\right)^2\right) + \cdots + eA_\mu J^\mu\right)$$

we have to introduce, by high school dimensional analysis, two lengths l and l', which a priori may not be the same. Here α, β, and γ are dimensionless numbers. Since we understand quantum electrodynamics (but not quantum gravity), we can in fact determine all these unknown quantities (see *QFT Nut*, chapters III.7 and VIII.3 and p. 460). The lengths l and l' are set by the electron mass.

X.5   Topological Field Theory

3   Let us evaluate

$$\varepsilon_{\alpha\beta\gamma\delta}R^{\alpha\beta}R^{\gamma\delta} = \varepsilon_{\alpha\beta\gamma\delta}R^{\alpha\beta}_{\mu\nu}R^{\gamma\delta}_{\psi\omega}dx^{\mu}dx^{\nu}dx^{\psi}dx^{\omega}$$

$$= \varepsilon_{\alpha\beta\gamma\delta}e^{\alpha}_{\rho}e^{\beta}_{\sigma}e^{\gamma}_{\tau}e^{\delta}_{\zeta}R^{\rho\sigma}_{\mu\nu}R^{\tau\zeta}_{\psi\omega}\varepsilon^{\mu\nu\psi\omega}d^{4}x$$

$$= (\det e)\varepsilon_{\rho\sigma\tau\zeta}R^{\rho\sigma}_{\mu\nu}R^{\tau\zeta}_{\psi\omega}\varepsilon^{\mu\nu\psi\omega}d^{4}x$$

$$\propto d^{4}x\sqrt{g}\left(\delta^{\mu}_{\rho}\delta^{\nu}_{\sigma}\delta^{\psi}_{\tau}\delta^{\omega}_{\zeta} \pm \text{permutations}\right)R^{\rho\sigma}_{\mu\nu}R^{\tau\zeta}_{\psi\omega}$$

$$= \left(d^{4}x\sqrt{g}\right)4\left(R^{2} - 4R^{\mu\nu}R_{\mu\nu} + R^{\mu\nu\rho\sigma}R_{\mu\nu\rho\sigma}\right)$$

In the next-to-last step, we used the fact that $\varepsilon_{\rho\sigma\tau\zeta}\varepsilon^{\mu\nu\psi\omega}$ is equal to ± 1 or 0 according to whether the two sets of indices $(\rho\sigma\tau\zeta)$ and $(\mu\nu\psi\omega)$ match or not, up to some permutation. In the last step, we add up the various possibilities. The combination $(R^{2} - 4R^{\mu\nu}R_{\mu\nu} + R^{\mu\nu\rho\sigma}R_{\mu\nu\rho\sigma})$ is known as the Gauss-Bonnet term. Incidentally, this computation shows eloquently the advantage of using differential forms.

X.6   A Brief Introduction to Twistors

2   This is worked out on p. 493 in *QFT Nut*, 2nd edition.

3   This is worked out on p. 509 in *QFT Nut*, 2nd edition.

Bibliography

Books on gravity

S. Carroll. *Spacetime and Geometry: An Introduction to General Relativity*. Addison-Wesley, 2003.

T.-P. Cheng. *Relativity, Gravitation and Cosmology: A Basic Introduction*. Oxford University Press, 2010.

R. d'Inverno. *Introducing Einstein's Relativity*. Oxford University Press, 1992.

P.A.M. Dirac. *General Theory of Relativity*. Princeton University Press, 1996.

S. Dodelson. *Modern Cosmology*. Academic Press, 2003.

A. Einstein. *The Meaning of Relativity*, 5th edition. Princeton University Press, 2004.

A. Einstein, H. A. Lorentz, H. Weyl, and H. Minkowski. *The Principle of Relativity*. www.bnpublishing .com, 2008.

R. Feynman, F. Morinigo, W. Wagner, B. Hatfield, and D. Pines. *Feynman Lectures on Gravitation (Frontiers in Physics)*. Westview Press, 2002.

J. Foster and J. D. Nightingale. *A Short Course in General Relativity*. Springer, 2001.

R. Freedman and W. Kauffmann. *Universe*. W. H. Freeman, 2007.

Ø. Grøn and S. Hervik. *Einstein's General Theory of Relativity: With Modern Applications in Cosmology*. Springer, 2007.

J. B. Hartle. *Gravity: An Introduction to Einstein's General Relativity*. Benjamin Cummings, 2003.

S. Hawking and R. Penrose. *The Nature of Space and Time*. Princeton University Press, 2010.

M. P. Hobson, G. P. Efstathiou, and A. N. Lasenby. *General Relativity: An Introduction for Physicists*. Cambridge University Press, 2006.

I. R. Kenyon. *General Relativity*. Oxford University Press, 1990.

C. W. Misner, K. S. Thorne, and J. A. Wheeler. *Gravitation*. W. H. Freeman, 1973.

V. Mukhanov. *Physical Foundations of Cosmology*. Cambridge University Press, 2005.

H. C. Ohanian and R. Ruffini. *Gravitation and Spacetime*, 2nd edition. W. W. Norton and Company, 1994.

T. Padmanabhan. *Gravitation: Foundations and Frontiers*. Cambridge University Press, 2010.

J. Plebanski and A. Krasinski. *An Introduction to General Relativity and Cosmology*. Cambridge University Press, 2006.

E. Poisson. *A Relativist's Toolkit: The Mathematics of Black-Hole Mechanics*. Cambridge University Press, 2004.

W. Rindler. *Relativity: Special, General, and Cosmological*. Oxford University Press, 2006.

L. Ryder. *Introduction to General Relativity*. Cambridge University Press, 2009.

B. F. Schutz. *A First Course in General Relativity*. Cambridge University Press, 1985.

H. Stephani. *General Relativity: An Introduction to the Theory of Gravitational Fields*. Cambridge University Press, 1990.

N. Straumann. *General Relativity: With Applications to Astrophysics*. Springer, 2004.

S. Weinberg. *Gravitation and Cosmology: Principles and Applications of the General Theory of Relativity*. Wiley, 1972.

———. *Cosmology*. Oxford University Press, 2008.

D. L. Wiltshire, M. Visser, and S. M. Scott, eds. *The Kerr Spacetime: Rotating Black Holes in General Relativity*. Cambridge University Press, 2009.

Popular books and historical accounts

R. Baierlein. *Newton to Einstein: The Trail of Light: An Excursion to the Wave-Particle Duality and the Special Theory of Relativity*. Cambridge University Press, 2001.

P. Coles. "Einstein, Eddington and the 1919 Eclipse," in *APS Conference Proceedings,* vol. 252, 2001.

F. M. Cornford, S. E. Eddington, and W. Dampier. *Background to Modern Science*. Nabu Press, 2011.

T. Damour. *Once upon Einstein*. A. K. Peters, 2006.

A. Einstein. *Essays in Science*. Philosophical Library, 1934.

———. *Einstein's Miraculous Year: Five Papers That Changed the Face of Physics,* John Stachel, ed. Princeton University Press, 2005.

P. Galison. *Einstein's Clocks, Poincaré's Maps: Empires of Time*. W. W. Norton and Company, 2004.

G. Gamow. *My World Line: An Informal Autobiography*. Viking Adult, 1970.

D. Goldsmith and M. Bartusiak. *E = Einstein: His Life, His Thought, and His Influence on Our Culture*. Sterling, 2008.

J. T. Liu, M. J. Duff, and K. S. Stelle, eds. *Deserfest: A Celebration of the Life and Works of Stanley Deser*. World Scientific, 2006.

E. Maor. *The Pythagorean Theorem: A 4,000-Year History*. Princeton University Press, 2007.

H. Nussbaumer and L. Bieri. *Discovering the Expanding Universe*. Cambridge University Press, 2009.

K. S. Thorne. *Black Holes and Time Warps: Einstein's Outrageous Legacy*. W. W. Norton and Company, 1995.

J. Waller. *Einstein's Luck: The Truth behind Some of the Greatest Scientific Discoveries*. Oxford University Press, 2004.

J. A Wheeler. *A Journey into Gravity and Spacetime*. Scientific American Library. W. H. Freeman, 1990.

A. Zee. *An Old Man's Toy: Gravity at Work and Play in Einstein's Universe*. Macmillan, 1990.

———. *Fearful Symmetry*. Princeton University Press, 1999.

Other physics books

R. P. Feynman and A. R. Hibbs. *Quantum Mechanics and Path Integrals*. McGraw-Hill, 1965.

J. Polchinski. *String Theory*. Cambridge University Press, 1998.

J. J. Sakurai. *Invariance Principles and Elementary Particles*. Princeton University Press, 1964.

J. J. Sakurai and J. J. Napolitano. *Modern Quantum Mechanics,* 2nd edition. Addison-Wesley, 2010.

J. H. Schwartz and P. M. Schwartz. *Special Relativity: From Einstein to Strings*. Cambridge University Press, 2004.

S. Weinberg. *The Quantum Theory of Fields, Volume 1: Foundations*. Cambridge University Press, 2005.

A. Zee. *Quantum Field Theory in a Nutshell,* 2nd edition. Princeton University Press, 2010. (First edition published 2003.)

B. Zwiebach. *A First Course in String Theory,* 2nd edition. Cambridge University Press, 2009.

Index

Page numbers followed by letters e, f, and n refer to exercises, figures, and notes, respectively.

Collection of Formulas and Conventions

The following is a loosely organized list of formulas used in this text.

$$\phi'(x') = \phi(x) \tag{1}$$

$$dx'^{\mu} = \left(\frac{\partial x'^{\mu}}{\partial x^{\nu}}\right) dx^{\nu} \equiv S^{\mu}_{\ \nu}(x) dx^{\nu} \tag{2}$$

$$\partial'_{\mu} = \frac{\partial x^{\nu}}{\partial x'^{\mu}} \partial_{\nu} \equiv (S^{-1})^{\nu}_{\ \mu} \partial_{\nu} \tag{3}$$

$$S^{\mu}_{\ \nu} \equiv \frac{\partial x'^{\mu}}{\partial x^{\nu}} \tag{4}$$

$$(S^{-1})^{\mu}_{\ \rho} \equiv \frac{\partial x^{\mu}}{\partial x'^{\rho}} \tag{5}$$

$$W'^{\mu}(x') = S^{\mu}_{\ \nu}(x) W^{\nu}(x) = \frac{\partial x'^{\mu}}{\partial x^{\nu}} W^{\nu}(x) \tag{6}$$

$$W'_{\rho}(x') = W_{\mu}(x)(S^{-1})^{\mu}_{\ \rho}(x) = W_{\mu}(x) \frac{\partial x^{\mu}}{\partial x'^{\rho}} \tag{7}$$

$$ds^2 = g'_{\rho\sigma}(x') dx'^{\rho} dx'^{\sigma} = g_{\mu\nu}(x) dx^{\mu} dx^{\nu} = g_{\mu\nu}(x) \frac{\partial x^{\mu}}{\partial x'^{\rho}} \frac{\partial x^{\nu}}{\partial x'^{\sigma}} dx'^{\rho} dx'^{\sigma} \tag{8}$$

$$g'_{\rho\sigma}(x') = g_{\mu\nu}(x) \frac{\partial x^{\mu}}{\partial x'^{\rho}} \frac{\partial x^{\nu}}{\partial x'^{\sigma}} = g_{\mu\nu}(x)(S^{-1})^{\mu}_{\ \rho}(S^{-1})^{\nu}_{\ \sigma} \tag{9}$$

$$g'^{\mu\nu}(x') = S^{\mu}_{\ \rho} S^{\nu}_{\ \sigma} g^{\rho\sigma}(x) \tag{10}$$

Infinitesimal transformation $x'^{\mu} = x^{\mu} + \xi^{\mu}(x)$

$$g'_{\rho\sigma}(x') = g_{\rho\sigma}(x) - g_{\mu\sigma}\partial_{\rho}\xi^{\mu} - g_{\rho\nu}\partial_{\sigma}\xi^{\nu} \tag{11}$$

Example of tensor transformation

$$T'^{\mu}_{\ \lambda\sigma}(x') = S^{\mu}_{\ \rho} T^{\rho}_{\ \eta\nu}(x)(S^{-1})^{\eta}_{\ \lambda}(S^{-1})^{\nu}_{\ \sigma} \tag{12}$$

Geodesic

$$\frac{d^2 X^{\rho}}{dl^2} + \Gamma^{\rho}_{\mu\nu}(X(l))\frac{dX^{\mu}}{dl}\frac{dX^{\nu}}{dl} = 0 \tag{13}$$

Christoffel symbol

$$\Gamma^{\rho}_{\mu\nu} \equiv \tfrac{1}{2}g^{\rho\lambda}(\partial_{\mu}g_{\nu\lambda} + \partial_{\nu}g_{\mu\lambda} - \partial_{\lambda}g_{\mu\nu}) \tag{14}$$

$$\Gamma_{\lambda\cdot\mu\nu} \equiv \tfrac{1}{2}(\partial_{\mu}g_{\nu\lambda} + \partial_{\nu}g_{\mu\lambda} - \partial_{\lambda}g_{\mu\nu}) \tag{15}$$

$$\Gamma^{\rho}_{\mu\rho} = \frac{1}{\sqrt{g}}\partial_{\mu}\sqrt{g} \tag{16}$$

Christoffel symbols for polar coordinates

$$\Gamma^{r}_{\theta\theta} = -r, \qquad \Gamma^{\theta}_{r\theta} = \frac{1}{r} \tag{17}$$

Christoffel symbols for the sphere

$$\Gamma^{\theta}_{\varphi\varphi} = -\sin\theta\cos\theta, \qquad \Gamma^{\varphi}_{\theta\varphi} = \frac{\cos\theta}{\sin\theta} \tag{18}$$

$$\Gamma'^{\mu}_{\ \lambda\kappa} = S^{\mu}_{\ \eta}(S^{-1})^{\omega}_{\ \lambda}(S^{-1})^{\sigma}_{\ \kappa}\Gamma^{\eta}_{\ \omega\sigma} + S^{\mu}_{\ \eta}(S^{-1})^{\rho}_{\ \lambda}\partial_{\rho}(S^{-1})^{\eta}_{\ \kappa} \tag{19}$$

Covariant derivatives

$$D_{\lambda}W^{\mu} \equiv \partial_{\lambda}W^{\mu} + \Gamma^{\mu}_{\lambda\nu}W^{\nu} \tag{20}$$

$$D_{\lambda}W_{\mu} = \partial_{\lambda}W_{\mu} - \Gamma^{\sigma}_{\lambda\mu}W_{\sigma} \tag{21}$$

Covariant divergence of a tensor

$$D_{\mu}T^{\mu\nu} = \partial_{\mu}T^{\mu\nu} + \Gamma^{\mu}_{\mu\lambda}T^{\lambda\nu} + \Gamma^{\nu}_{\mu\lambda}T^{\mu\lambda} = \frac{1}{\sqrt{-g}}\partial_{\mu}(\sqrt{-g}T^{\mu\nu}) + \Gamma^{\nu}_{\mu\lambda}T^{\mu\lambda} \tag{22}$$

Sphere S^2

$$ds^2 = d\theta^2 + \sin^2\theta\, d\varphi^2 = \frac{dr^2}{1-r^2} + r^2 d\varphi^2 \tag{23}$$

Stereographic projection for S^2

$$r = \frac{\rho}{1 + \frac{\rho^2}{4}} \tag{24}$$

$$ds^2 = \frac{1}{\left(1 + \frac{\rho^2}{4}\right)^2}(d\rho^2 + \rho^2 d\varphi^2) \tag{25}$$

Iterative relation for sphere

$$ds_d^2 = d\theta^2 + \sin^2\theta\, ds_{d-1}^2 \tag{26}$$

Locally flat coordinate system

$$g_{\tau\mu}(x) = \eta_{\tau\mu} + B_{\tau\mu,\lambda\sigma}x^\lambda x^\sigma + \cdots, \qquad B_{\tau\mu,\lambda\sigma} = \frac{1}{2}\partial_\lambda\partial_\sigma g_{\tau\mu} \tag{27}$$

$$R_{\tau\rho\mu\nu} = B_{\tau\nu,\mu\rho} - B_{\rho\nu,\mu\tau} - B_{\tau\mu,\nu\rho} + B_{\rho\mu,\nu\tau} \tag{28}$$

Riemann curvature tensor

$$R^\sigma{}_{\rho\mu\nu} = (\partial_\mu\Gamma^\sigma{}_{\nu\rho} + \Gamma^\sigma{}_{\mu\kappa}\Gamma^\kappa{}_{\nu\rho}) - (\partial_\nu\Gamma^\sigma{}_{\mu\rho} + \Gamma^\sigma{}_{\nu\kappa}\Gamma^\kappa{}_{\mu\rho}) \tag{29}$$

Ricci tensor

$$R_{\mu\nu} = R^\sigma{}_{\mu\sigma\nu} = (\partial_\sigma\Gamma^\sigma{}_{\mu\nu} + \Gamma^\sigma{}_{\kappa\sigma}\Gamma^\kappa{}_{\mu\nu}) - (\partial_\nu\Gamma^\sigma{}_{\mu\sigma} + \Gamma^\sigma{}_{\kappa\nu}\Gamma^\kappa{}_{\mu\sigma}) \tag{30}$$

Bianchi identity

$$D_\nu R_{\rho\mu\sigma\lambda} + D_\sigma R_{\rho\mu\lambda\nu} + D_\lambda R_{\rho\mu\nu\sigma} = 0 \tag{31}$$

$$D^\mu(R_{\mu\nu} - \tfrac{1}{2}g_{\mu\nu}R) = D^\mu E_{\mu\nu} = 0 \tag{32}$$

Einstein's field equation and action

$$S_{\text{EH}} = +\frac{1}{16\pi G}\int d^4x\sqrt{-g}\,R \tag{33}$$

$$\delta g^{\sigma\rho} = -g^{\sigma\mu}\delta g_{\mu\nu}g^{\nu\rho} \tag{34}$$

$$\delta\sqrt{-g} = \tfrac{1}{2}\sqrt{-g}\,g^{\mu\nu}\delta g_{\mu\nu} \tag{35}$$

$$R^{\mu\nu} - \tfrac{1}{2}g^{\mu\nu}R = +8\pi G T^{\mu\nu} \tag{36}$$

$$R^{\mu\nu} = +8\pi G(T^{\mu\nu} - \tfrac{1}{2}g^{\mu\nu}T) \tag{37}$$

$$\delta S_{\text{matter}} = +\tfrac{1}{2}\int d^4x\sqrt{-g}\,T^{\mu\nu}\delta g_{\mu\nu} \tag{38}$$

Newton's field equation

$$\nabla^2\Phi = 4\pi G\rho \tag{39}$$

$$g_{00} = -(1 + 2\Phi), \qquad \Phi = -\frac{GM}{r} \tag{40}$$

Static symmetric spacetime

$$ds^2 = -A(r)dt^2 + B(r)dr^2 + r^2 d\theta^2 + r^2\sin^2\theta\,d\varphi^2 \tag{41}$$

$$\Gamma^t{}_{tr} = \frac{A'}{2A}, \qquad \Gamma^r{}_{tt} = \frac{A'}{2B}, \qquad \Gamma^r{}_{rr} = \frac{B'}{2B}, \qquad \Gamma^r{}_{\theta\theta} = -\frac{r}{B}, \qquad \Gamma^r{}_{\varphi\varphi} = -\frac{r\sin^2\theta}{B},$$

$$\Gamma^\theta{}_{r\theta} = \frac{1}{r}, \qquad \Gamma^\varphi{}_{r\varphi} = \frac{1}{r},$$

$$\Gamma^\theta{}_{\varphi\varphi} = -\sin\theta\cos\theta, \qquad \Gamma^\varphi{}_{\theta\varphi} = \cot\theta \tag{42}$$

$$R_{tt} = \frac{A''}{2B} + \frac{A'}{rB} - \frac{A'}{4B}\left(\frac{A'}{A} + \frac{B'}{B}\right)$$

$$R_{rr} = -\frac{A''}{2A} + \frac{B'}{rB} + \frac{A'}{4A}\left(\frac{A'}{A} + \frac{B'}{B}\right)$$

$$R_{\theta\theta} = 1 - \frac{1}{B} - \frac{r}{2B}\left(\frac{A'}{A} - \frac{B'}{B}\right) \tag{43}$$

Schwarzschild solution ($r_S = 2GM$)

$$ds^2 = -\left(1 - \frac{r_S}{r}\right)dt^2 + \frac{1}{1 - \frac{r_S}{r}}dr^2 + r^2 d\theta^2 + r^2 \sin^2\theta d\varphi^2$$

$$= -\left(\frac{r - r_S}{r}\right)(d\bar{t} + dr)\left(d\bar{t} - \frac{r + r_S}{r - r_S}dr\right) + r^2 d\Omega^2 \tag{44}$$

$$\Gamma^t_{tr} = \frac{r_S}{2r(r - r_S)}, \qquad \Gamma^r_{tt} = \frac{r_S}{2r^3}(r - r_S), \qquad \Gamma^r_{rr} = -\frac{r_S}{2r(r - r_S)},$$

$$\Gamma^r_{\theta\theta} = -(r - r_S), \qquad \Gamma^r_{\varphi\varphi} = -(r - r_S)\sin^2\theta,$$

$$\Gamma^\theta_{r\theta} = \frac{1}{r}, \qquad \Gamma^\varphi_{r\varphi} = \frac{1}{r}, \qquad \Gamma^\theta_{\varphi\varphi} = -\sin\theta\cos\theta, \qquad \Gamma^\varphi_{\theta\varphi} = \cot\theta \tag{45}$$

Kruskal-Szekeres coordinates

$$ds^2 = -\frac{4r_S^3}{r}e^{-r/r_S}(dV^2 - dU^2) + r^2 d\Omega^2 \tag{46}$$

For $r > r_S$,

$$V = \left(\frac{r}{r_S} - 1\right)^{1/2} e^{r/2r_S}\sinh\left(\frac{t}{2r_S}\right), \qquad U = \left(\frac{r}{r_S} - 1\right)^{1/2} e^{r/2r_S}\cosh\left(\frac{t}{2r_S}\right) \tag{47}$$

For $r < r_S$,

$$V = \left(1 - \frac{r}{r_S}\right)^{1/2} e^{r/2r_S}\cosh\left(\frac{t}{2r_S}\right), \qquad U = \left(1 - \frac{r}{r_S}\right)^{1/2} e^{r/2r_S}\sinh\left(\frac{t}{2r_S}\right) \tag{48}$$

$$V^2 - U^2 = \left(1 - \frac{r}{r_S}\right)e^{r/r_S} \tag{49}$$

Tolman-Oppenheimer-Volkoff equation of relativistic stellar structure

$$\frac{dP}{dr} = -\frac{G\mathcal{M}(r)\rho(r)}{r^2}\left(1 + \frac{P(r)}{\rho(r)}\right)\left(1 + \frac{4\pi r^3 P(r)}{\mathcal{M}(r)}\right)\left(1 - \frac{2G\mathcal{M}(r)}{r}\right)^{-1} \tag{50}$$

$$\frac{d\mathcal{M}(r)}{dr} = 4\pi r^2 \rho(r) \tag{51}$$

Kerr black hole

$$ds^2 = -\left(1 - \frac{rr_S}{\rho^2}\right)dt^2 - \frac{2r_S ar\sin^2\theta}{\rho^2}dt d\varphi + \frac{\rho^2}{\Delta}dr^2 + \rho^2 d\theta^2$$

$$+ \left(r^2 + a^2 + \frac{r_S r a^2 \sin^2\theta}{\rho^2}\right)\sin^2\theta d\varphi^2 \tag{52}$$

where

$$r_S = 2M, \qquad a = \frac{J}{M} = \frac{2J}{r_S}, \qquad \rho^2 = r^2 + a^2 \cos^2 \theta, \qquad \Delta = r^2 + a^2 - r r_S \tag{53}$$

Reissner-Nordström black hole

$$ds^2 = -\left(1 - \frac{2M}{r} + \frac{Q^2}{r^2}\right) dt^2 + \left(\frac{1}{1 - \frac{2M}{r} + \frac{Q^2}{r^2}}\right) dr^2 + r^2 d\Omega^2 \tag{54}$$

Perfect fluid

$$T^{\mu\nu} = (\rho + P) U^\mu U^\nu + P g^{\mu\nu} \tag{55}$$

Cosmology

$$ds^2 = -dt^2 + a(t)^2 \left(\frac{1}{1 - k\frac{r^2}{L^2}} dr^2 + r^2 d\Omega^2\right) \tag{56}$$

$$\dot{R}^2 + k = \frac{8\pi G}{3} \rho R^2 \tag{57}$$

$$\frac{\ddot{R}}{R} = -\frac{4\pi G}{3}(\rho + 3P) \tag{58}$$

$$\dot{\rho} + 3(\rho + P)\frac{\dot{R}}{R} = 0 \tag{59}$$

$$\rho \propto \frac{1}{a^{3(1+w)}} \tag{60}$$

$$H^2 = H_0^2 \sum_n \frac{\Omega_{n,0}}{a^{\gamma_n}} = H_0^2 \left(\frac{\Omega_{m,0}}{a^3} + \frac{\Omega_{r,0}}{a^4} + \Omega_{\Lambda,0} + \frac{\Omega_{k,0}}{a^2}\right) \tag{61}$$

$$q = -\frac{\ddot{R}/R}{\dot{R}^2/R^2} = +\frac{1}{2}\sum_j (1 + 3w_j)\Omega_j = \frac{1}{2}(2\Omega_r + \Omega_m - 2\Omega_\Lambda) \tag{62}$$

$$\dot{\Omega}_j = +H\Omega_j \left(-3w_j - 1 + \sum_i (1 + 3w_i)\Omega_i\right) \tag{63}$$

For $\Omega_r \simeq 0$,

$$\dot{\Omega}_m = H\Omega_m(\Omega_m - 2\Omega_\Lambda - 1), \qquad \dot{\Omega}_\Lambda = H\Omega_\Lambda(\Omega_m - 2\Omega_\Lambda + 2) \tag{64}$$

Weak field

$$g_{\mu\nu} = \eta_{\mu\nu} + h_{\mu\nu}, \qquad R_{\mu\nu} = -\frac{1}{2}\partial^2 h_{\mu\nu} \quad \text{harmonic gauge} \tag{65}$$

Killing condition

$$\xi_{\sigma;\rho} + \xi_{\rho;\sigma} = 0 \tag{66}$$

$$g_{\mu\sigma}\partial_\rho \xi^\mu + g_{\rho\nu}\partial_\sigma \xi^\nu + \xi^\lambda \partial_\lambda g_{\rho\sigma} = 0 \tag{67}$$

$$\mathcal{L}_\xi g_{\rho\sigma} = 0 \tag{68}$$

Maximal symmetry

$$R_{\tau\rho\mu\nu} = K(g_{\tau\mu}g_{\rho\nu} - g_{\tau\nu}g_{\rho\mu})$$

$$R_{\rho\nu} = (D-1)Kg_{\rho\nu}$$

$$R = D(D-1)K \tag{69}$$

Differential forms and vielbein

$$g_{\mu\nu}(x) = \eta_{\alpha\beta}e^{\alpha}_{\mu}(x)e^{\beta}_{\nu}(x) \tag{70}$$

$$e^{\alpha} = e^{\alpha}_{\mu}dx_{\mu} \tag{71}$$

Cartan's first and second structural relations

$$de + \omega e = 0 \tag{72}$$

$$R = d\omega + \omega^2 \tag{73}$$

Antisymmetry

$$\omega^0_{\ b} = +\omega^{0b} = -\omega^{b0} = \omega^b_{\ 0} \tag{74}$$

$$\omega^b_{\ c} = +\omega^{bc} = -\omega^{cb} = -\omega^c_{\ b} \tag{75}$$

Conformally related metrics

$$\tilde{g}_{\mu\nu}(x) = \Omega^2(x)g_{\mu\nu} \tag{76}$$

$$\tilde{\Gamma}^{\mu}_{\nu\lambda} = \Gamma^{\mu}_{\nu\lambda} + \frac{1}{\Omega}(\delta^{\mu}_{\nu}\partial_{\lambda}\Omega + \delta^{\mu}_{\lambda}\partial_{\nu}\Omega - g_{\nu\lambda}g^{\mu\rho}\partial_{\rho}\Omega) \tag{77}$$

$$\tilde{R}^{\mu}_{\ \nu\lambda\sigma} = R^{\mu}_{\ \nu\lambda\sigma} - \left(\delta^{\mu}_{\lambda}\delta^{\rho}_{\sigma}\delta^{\omega}_{\nu} - \delta^{\mu}_{\sigma}\delta^{\rho}_{\lambda}\delta^{\omega}_{\nu} + g_{\nu\sigma}g^{\mu\omega}\delta^{\rho}_{\lambda} - g_{\nu\lambda}g^{\mu\omega}\delta^{\rho}_{\sigma}\right)\frac{D_{\rho}\partial_{\omega}\Omega}{\Omega}$$

$$+ \left(2\delta^{\mu}_{\lambda}\delta^{\rho}_{\sigma}\delta^{\omega}_{\nu} - 2g_{\nu\lambda}g^{\mu\omega}\delta^{\rho}_{\sigma} + 2g_{\nu\sigma}g^{\mu\omega}\delta^{\rho}_{\lambda} - 2\delta^{\mu}_{\sigma}\delta^{\omega}_{\nu}\delta^{\rho}_{\lambda} + g_{\nu\lambda}g^{\rho\omega}\delta^{\mu}_{\sigma} - g_{\nu\sigma}g^{\rho\omega}\delta^{\mu}_{\lambda}\right)\frac{(\partial_{\rho}\Omega)(\partial_{\omega}\Omega)}{\Omega^2} \tag{78}$$

$$\tilde{R}_{\nu\lambda} = R_{\nu\lambda} - \left[(d-2)\delta^{\rho}_{\nu}\delta^{\omega}_{\lambda} + g_{\nu\lambda}g^{\rho\omega}\right]\frac{D_{\rho}\partial_{\omega}\Omega}{\Omega} + \left[2(d-2)\delta^{\rho}_{\nu}\delta^{\omega}_{\lambda} - (d-3)g_{\nu\lambda}g^{\rho\omega}\right]\frac{(\partial_{\rho}\Omega)(\partial_{\omega}\Omega)}{\Omega^2} \tag{79}$$

$$\tilde{R} = \frac{R}{\Omega^2} - 2(d-1)g^{\rho\omega}\frac{D_{\rho}\partial_{\omega}\Omega}{\Omega^3} - (d-1)(d-4)g^{\rho\omega}\frac{(\partial_{\rho}\Omega)(\partial_{\omega}\Omega)}{\Omega^4} \tag{80}$$

with $D_{\rho}\partial_{\omega}\Omega = D_{\rho}D_{\omega}\Omega$

Weyl tensor

$$C_{\mu\nu\rho\sigma} \equiv R_{\mu\nu\rho\sigma} + (d-2)^{-1}(g_{\mu\sigma}R_{\rho\nu} + g_{\nu\rho}R_{\sigma\mu} - g_{\mu\rho}R_{\sigma\nu} - g_{\nu\sigma}R_{\rho\mu})$$

$$+ ((d-1)(d-2))^{-1}(g_{\mu\rho}g_{\sigma\nu} - g_{\mu\sigma}g_{\rho\nu})R \tag{81}$$

Lie derivative

$$\mathcal{L}_V W^{\mu} = [V, W]^{\mu} \equiv V^{\nu}\partial_{\nu}W^{\mu} - W^{\nu}\partial_{\nu}V^{\mu} = V^{\nu}D_{\nu}W^{\mu} - W^{\nu}D_{\nu}V^{\mu} \tag{82}$$

$$\mathcal{L}_V W_{\mu\lambda} = V^{\nu}\partial_{\nu}W_{\mu\lambda} + W_{\nu\lambda}\partial_{\mu}V^{\nu} + W_{\mu\nu}\partial_{\lambda}V^{\nu} \tag{83}$$